MCKNIGHT'S
Physical Geography

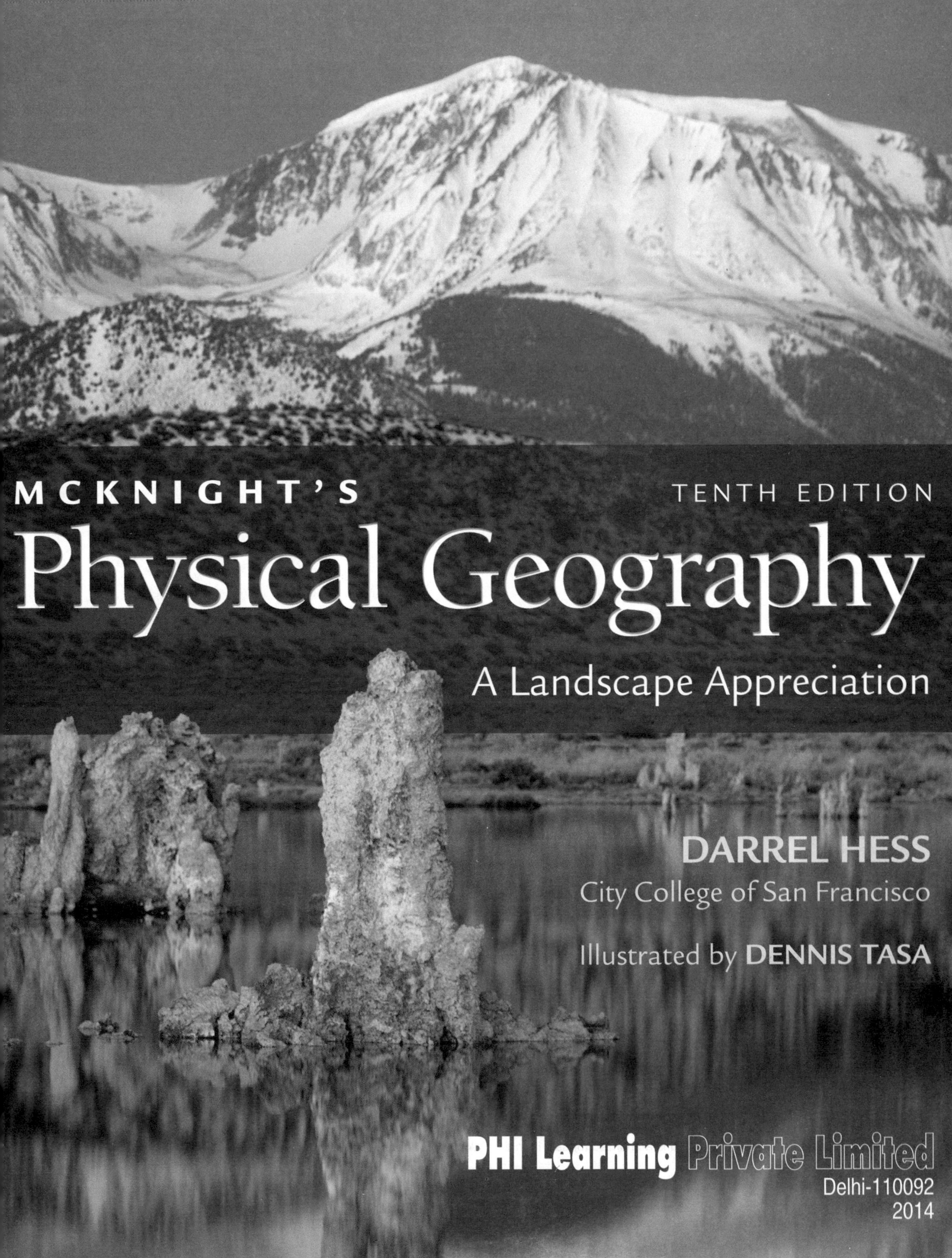
MCKNIGHT'S
TENTH EDITION
Physical Geography
A Landscape Appreciation
DARREL HESS
City College of San Francisco
Illustrated by DENNIS TASA
PHI Learning Private Limited
Delhi-110092
2014

This Indian Reprint—₹ 895.00
(Original U.S. Edition—₹ 8710.00)

MCKNIGHT'S PHYSICAL GEOGRAPHY—A Landscape Appreciation, 10th ed.
by Darrel Hess
Illustrated by Dennis Tasa

Original edition, entitled *Mcknight's Physical Geography—A Landscape Appreciation, 10th ed.* by Darrel Hess Illustrated by Dennis Tasa, published by Pearson Education, Inc., publishing as Pearson Prentice Hall.

ISBN-978-81-203-4412-9

Cover photo credit: Mono County, California, USA. © David Muench/Corbis

Indian edition published by PHI Learning Private Limited.

This edition is manufactured in India, and is authorized for sale in India, Pakistan, Sri Lanka, Bhutan, Bangladesh, Nepal and the Maldives only.

Published by Asoke K. Ghosh, PHI Learning Private Limited, Rimjhim House, 111, Patparganj Industrial Estate, Delhi-110092 and Printed by Rajkamal Electric Press, Plot No. 2, Phase IV, HSIDC, Kundli-131028, Sonepat, Haryana.

Brief Contents

1 Introduction to Earth
Solar System Formation
Earth-Sun Relations

2 Portraying Earth
Map Projections

3 Introduction to the Atmosphere
Ozone Depletion
Coriolis Effect

4 Insolation and Temperature
Atmospheric Energy Balance
Ocean Circulation Patterns
Global Warming

5 Atmospheric Pressure and Wind
Development of Wind Patterns
Cyclones and Anticyclones
Global Atmospheric Circulation
The Jet Stream and Rossby Waves
Seasonal Pressure and Precipitation Patterns
El Niño

6 Atmospheric Moisture
Hydrologic Cycle
Phase Changes of Water
Adiabatic Processes and Atmospheric Stability
Seasonal Pressure and Precipitation Patterns

7 Atmospheric Disturbances
Cold Fronts and Warm Fronts
Midlatitude Cyclones
Hurricanes
Tornadoes

8 Climate and Climate Change
Seasonal Pressure and Precipitation Patterns
End of the Last Ice Age
Orbital Variations and Climate Change

9 The Hydrosphere
Hydrologic Cycle
Tides
Ocean Circulation Patterns
The Water Table
Groundwater Cone of Depression

10 Cycles and Patterns in the Biosphere
Net Primary Productivity
Biological Productivity in Midlatitude Oceans

13 Introduction to Landform Study
Metamorphic Rock Foliation
Isostasy

14 The Internal Processes
Plate Boundaries
Seafloor Spreading
Paleomagnetism
Convection and Plate Tectonics
Divergent Boundaries
Subduction Zones
Collision of India with Eurasia
Transform Faults and Boundaries
Assembly and Breakup of Pangaea
Mantle Plumes
Terrane Formation
Eruption of Mount St. Helens
Volcanoes
Formation of Crater Lake
Igneous Features
Folding
Faulting
Earthquake Waves
Seismographs

15 Preliminaries to Erosion: Weathering and Mass Wasting
Mechanical Weathering
Mass Wasting
Eruption of Mount St. Helens

16 The Fluvial Processes
Stream Sediment Movement
Meandering Streams
Floods and Natural Levee Formation
Stream Rejuvenation

18 The Topography of Arid Lands
Wind Transportation of Sediment
Desert Sand Dunes

19 Glacial Modification of Terrain
End of the Last Ice Age
Isostasy
Flow of Ice Within a Glacier
Glacial Processes

20 Coastal Processes and Terrain
Wave Motion and Wave Refraction
Tsunami
Tides
Coastal Sediment Transport
Coastal Stabilization Structures

Contents

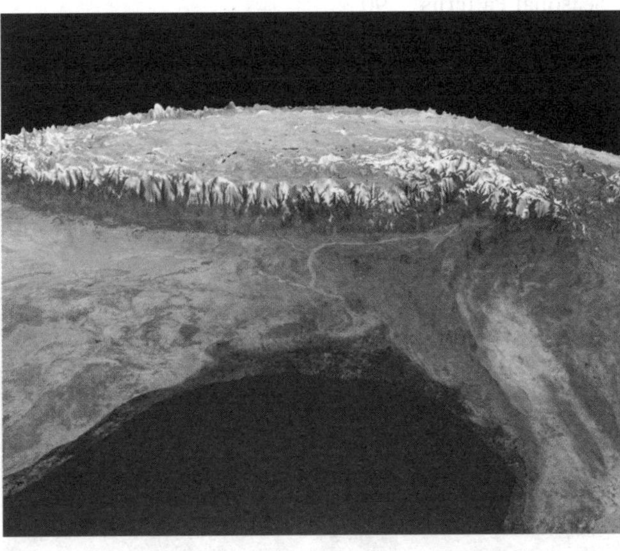

3 Introduction to the Atmosphere 48

4 Insolation and Temperature 66

5 Atmospheric Pressure and Wind 96

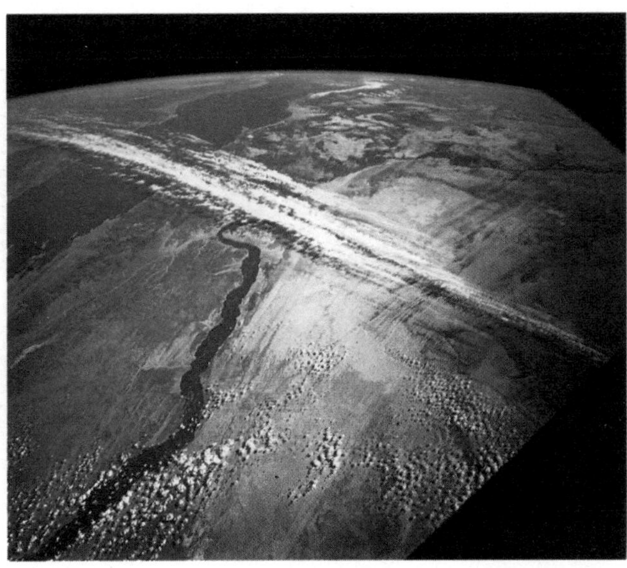

6 Atmospheric Moisture 126

7 Atmospheric Disturbances 160

8 Climate and Climate Change 188

9 The Hydrosphere 234

14 The Internal Processes 374

15 Preliminaries to Erosion: Weathering and Mass Wasting 416

16 Fluvial Processes 436

17 Karst and Hydrothermal Processes 466

18 The Topography of Arid Lands 480

19 Glacial Modification of Terrain 502

Preface

To the Student

Welcome to *Physical Geography: A Landscape Appreciation*! If you opened this book expecting that the study of geography was going to be memorizing names and places on maps, you'll be surprised to find that geography is much more than that. Geographers study the location and distribution of things—tangible things such as rainfall, mountains, and trees, as well as less tangible things such as language, migration, and voting patterns. In short, geographers look for and explain patterns in the physical and human landscape.

In this book you'll learn about fundamental processes and patterns in the natural world—the kinds of things you can see whenever you walk outside: clouds in the sky, mountains, streams and valleys, and the plants and animals that inhabit the landscape. You'll also learn about human interactions with the natural environment—how events such as hurricanes, earthquakes, and floods affect our lives and the world around us, as well as how human activities are increasingly altering our environment. By the time you finish this book you'll understand and "appreciate" the landscape in new ways.

If you take a minute to skim through the book, you'll see some of the features that will help you learn this material:

- The book includes many diagrams, maps, and photographs. Physical geography is a visual discipline, so the figures and captions will allow you to get the most out of each chapter.
- Many photographs have "locator maps" to help you learn the locations of the many places we mention in the book.
- Each chapter includes a Learning Review to test your understanding of major concepts and key terms that are printed in **bold type** throughout the text.
- The alphabetical glossary at the end of the book provides definitions for all of the key terms.
- Most of the chapters include icons that direct you to online interactive animations that accompany the book. These animations explain important concepts in physical geography. Each animation includes a written and an audio narration, as well as a short quiz to test your understanding.
- A reference map of physical features of the world is found inside the front cover of the book, and a reference map of the countries of the world is found inside the back cover.

To the Instructor

Physical Geography: A Landscape Appreciation presents the concepts of physical geography in a clear, readable way to help students comprehend the landscape around them. The 10th edition of the book has undergone a thorough revision, while maintaining the time-proven approach to physical geography first presented by Tom McKnight more than 25 years ago.

New to the 10th Edition

Users of earlier editions will see that the overall sequence of chapters and topics remains the same, with material added or updated in several key areas. Changes to the new edition include:

- The entire art program has been thoroughly revised and updated by illustrator Dennis Tasa. Dozens of new diagrams, maps, and photographs are found throughout.
- A new series of boxed essays on *Renewable Energy* has been added. Contributed by professors from across the country, the essay topics include: Solar Energy, Geothermal Energy, Wind Energy, Biofuels, Hydroelectric Power, and Tidal Power.
- In Chapter 2 the material on GPS and GIS has been updated and expanded.
- Chapter 8, *Climate and Climate Change*, has been substantially revised, and includes expanded coverage of climate change.
- In Chapter 14 the material on faulting has been updated, as has the material on floods in Chapter 16, desert geomorphology in Chapter 18, and glacial processes in Chapter 19.
- Some key material previously found in Focus Boxes has been integrated directly into the text. Updated and revised Focus Boxes include: The UV Index; Monitoring Earth's Radiation Budget; Measuring Earth's Surface Temperature by Satellite; Global Dimming; Hurricanes and Global Warming; Thawing Permafrost in Alaska; Signs of Climate Change in the Arctic; Paleoclimatology at Dome C, Antarctica; Subsidence from Groundwater Extraction; Changing Climate Affects Bird Populations; Earthquake Forecasting; The Changing Mississippi River Delta; Disintegration of Antarctic Ice Shelves; Shrinking Glaciers; The Sumatra-Andaman Earthquake and Tsunami of 2004; and Imperiled Coral Reefs.

- New Focus Boxes include: Portraying the Three-Dimensional Landscape; Geographic Research Using *Google Earth™* and *The National Map;* Strategies for Reducing Greenhouse Gas Emissions; The Great Pacific Garbage Patch; Bark Beetles Killing Forests in Western North America; Using Soil Profiles to Decipher Past Environments; and The La Conchita Landslides.
- The end of chapter Learning Review now integrates all key terms into study questions to help students learn key concepts.
- Finally, online **Multimedia and Assessment** for *Physical Geography: A Landscape Appreciation*, found at www.mygeoscienceplace.com, now includes a fully-integrated eBook, Interactive Geoscience Animations and Videos with assessment, Flashcards and Glossary, annotated resources for further exploration, RSS Feeds, Web links, student self-study Exercises and Problems, and Class Manager and GradeTracker Gradebook functionality for instructors.

The Teaching and Learning Package

For the Instructor

- *Instructor Manual* (0-321-67246-1): Available for download, this resource for both new and experienced instructors includes a Chapter Overview, list of highlighted lecture topics, Key Terms, a detailed Chapter Outline, suggested Study Questions for Key Concepts, and answers to end of chapter review questions.
- *TestGen/Test Bank*® (0-321-66774-3): TestGen is a computerized test generator that lets instructors view and edit *Test Bank* questions, transfer questions to tests, and print the test in a variety of customized formats. This *Test Bank* includes approximately 1000 multiple choice, true/false, and short answer/essay questions. Questions are correlated against the U.S. National Geography Standards and Bloom's Taxonomy to help instructors to better map the assessments against both broad and specific teaching and learning objectives. The *Test Bank* is also available in Microsoft Word©, and is importable into Blackboard and WebCT.
- *Instructor Resource Center on DVD* (0-321-67247-X): Everything instructors need, where they want it. The *Instructor Resource Center on DVD* helps make instructors more effective by saving them time and effort. All digital resources can be found in one well-organized, easy-to-access place, and include:

  Figures—All textbook images as JPGs, PDFs, and PowerPoint™ Slides

  Preauthored Lecture Outline PowerPoint Presentations, which outline the concepts of each chapter with embedded art and can be customized to fit instructors' lecture requirements

  CRS "Clicker" Questions in PowerPoint format correlated against U.S. National Geography Standards and Bloom's Taxonomy

  TestGen—The TestGen software, questions, and answers for both MACs and PCs

  Electronic Files of the *Instructor Manual* and *Test Bank*

  This Instructor Resource content is also available completely online via the Instructor Resources section of www.mygeoscienceplace.com and www.pearson highered.com/irc.

- Answer Key to *Laboratory Manual* (0-321-67248-8): Available for download, the answer key provides answers to problem sets presented in the Laboratory Manual.
- *Television for the Environment Earth Report Geography Videos on DVD* (0-321-66298-9): This three-DVD set is designed to help students visualize how human decisions and behavior have affected the environment, and how individuals are taking the steps toward recovery. With topics ranging from poor land management promoting the devastation of river systems in South America, to the struggles for electricity in China and Africa, these 13 videos from Television for the Environment's global *Earth Report* series recognize the efforts of individuals around the world to unite and protect the planet.
- *Aspiring Academics: A Resource Book for Graduate Students and Early Career Faculty* (0-13-604891-9): Drawing on several years of research, this set of essays is designed to help graduate students and early career faculty start their careers in geography and related social and environmental sciences. This teaching aid stresses the interdependence of teaching, research, and service in faculty work—and the importance of achieving a healthy balance in professional and personal life—and does not view it as a collection of unrelated tasks. Each chapter provides accessible, forward-looking advice on topics that often cause the most stress in the first years of a college or university appointment.
- *Teaching College Geography: A Practical Guide for Graduate Students and Early Career Faculty* (0-13-605547-1): Provides a starting point for becoming an effective geography teacher from the very first day of class. Divided in two parts, the first set of chapters addresses "nuts-and-bolts" teaching issues in the context of the new technologies, student demographics, and institutional expectations that are the hallmarks of higher education in the twenty-first century. The second part explores other important issues: effective teaching in the field; supporting critical thinking with GIS and mapping technologies; engaging learners in large geography classes; and promoting awareness of international perspectives and geographic issues.

- *AAG Community Portal for Aspiring Academics and Teaching College Geography:* This Web site is intended to support community-based professional development in geography and related disciplines. Here you will find activities providing extended treatment of the topics covered in both books. The activities can be used in workshops, graduate seminars, brown bags, and mentoring programs offered on campus or within an academic department. You can also use the discussion boards and contributions tool to share advice and materials with others: www.pearsonhighered.com/aag/.
- *Course Management:* Pearson Prentice Hall is proud to partner with many of the leading course management system providers on the market today. These partnerships enable us to provide our testing materials already formatted for easy importation into the powerful course management tools Blackboard and CourseCompass. Please contact your local Pearson Prentice Hall representative for details:
www.pearsonhighered.com/elearning/.

For the Student

- *Mygeoscienceplace Premium Website:* A dedicated Premium Website (www.mygeoscienceplace.com) offers a variety of resources for students and professors, including Interactive Geoscience Animations, Videos, Flashcards and Glossary, annotated resources for further exploration, RSS Feeds, Web links, student self-study exercises and problems, and Class Manager and GradeTracker Gradebook functionality for instructors. The Student Home page is organized by a four-step process: Review Chapter Objectives, Read correlating Pearson eText chapters, Visualize key concepts with Animations and Videos, and Test Yourself with Review Exercises and Critical Thinking Problems.
- *Laboratory Manual* (0-321-67836-2): Written by Darrel Hess, this manual offers a comprehensive set of lab exercises to accompany any physical geography class. The manual is organized to meet the personal needs of instructors, as they can pick and choose a series of short lab exercises each week.
- *Study Guide* (0-321-67834-6): Incorporating text and art from the text, Darrel Hess has created the essential study tool for students. Included are chapter notes, with key terms and page references, study questions, self-check exercises, and a vocabulary review.
- ***Encounter Physical Geography* Workbook and Premium Website (0-321-68175-4):** *Encounter Physical Geography* provides rich, interactive explorations of physical geography concepts through Google Earth™ explorations. All chapter explorations are available in print format as well as online quizzes, accommodating different classroom needs. All worksheets are accompanied with corresponding Google Earth™ media files, available for download from www.mygeoscienceplace.com.
- *Goode's World Atlas* (0-321-65200-2): *Goode's World Atlas* has been the world's premiere educational atlas since 1923, and for good reason. It features over 250 pages of maps, from definitive physical and political maps to important thematic maps that illustrate the spatial aspects of many important topics. The 22nd edition includes 160 pages of new, digitally produced reference maps, as well as new thematic maps on global climate change, sea level rise, CO_2 emissions, polar ice fluctuations, deforestation, extreme weather events, infectious diseases, water resources, and energy production.
- *Dire Predictions* (978-0-13-604435-2): Periodic reports from the Intergovernmental Panel on Climate Change (IPCC) evaluate the risk of climate change brought on by humans. But the sheer volume of scientific data remains inscrutable to the general public, particularly to those who may still question the validity of climate change. In just over 200 pages, this practical text presents and expands upon the essential findings in a visually stunning and undeniably powerful way to the lay reader. Scientific findings that provide validity to the implications of climate change are presented in clear-cut graphic elements, striking images, and understandable analogies.

Acknowledgments

First of all, I want to express my great appreciation to illustrator Dennis Tasa—I've admired his work in other textbooks for many years, and it has been a great pleasure to work with him on the 10th edition of this book.

Over the years, scores of colleagues, students, and friends have helped me and the founding author of this book, Tom McKnight, update and improve this textbook. Their assistance has been gratefully acknowledged previously. Here we acknowledge those who have provided assistance in recent years by acting as reviewers of the text and animations that accompany it, or by providing helpful critiques and suggestions:

Casey Allen, *Weber State University*

Sergei Andronikov, *Austin Peay State University*

Greg Bierly, *Indiana State University*

Mark Binkley, *Mississippi State University*

Peter Blanken, *University of Colorado*

Margaret Boorstein, *Long Island University*

James Brey, *University of Wisconsin Fox Valley*

Sean Cannon, *Brigham Young University–Idaho*

Wing Cheung, *Palomar College*

Jongnam Choi, *Western Illinois University*

Glen Conner, *Western Kentucky University*

Carlos E. Cordova, *Oklahoma State University*

Richard A. Crooker, *Kutztown University of Pennsylvania*

Mike DeVivo, *Grand Rapids Community College*

Bryan Dorsey, *Weber State University*

Don W. Duckson, Jr., *Frostburg State University*

Tracy Edwards, *Frostburg State University*

Steve Emerick, *Glendale Community College*

Basil Gomez, *Indiana State University*

Michael Grossman, *Southern Illinois University–Edwardsville*

Perry J. Hardin, *Brigham Young University*

Barbara Holzman, *San Francisco State University*

Paul Hudson, *University of Texas*

Catherine Jain, *Palomar College*

Steven Jennings, *University of Colorado at Colorado Springs*

Dorleen B. Jenson, *Salt Lake Community College*

Kris Jones, *Saddleback College*

Ryan Kelly, *Lexington Community College*

Rob Kremer, *Metropolitan State College of Denver*

Steve LaDochy, *California State University*

Michael Madsen, *Brigham Young University–Idaho*

Kenneth Martis, *West Virginia University*

Mandy Munro-Stasiuk, *Kent State University*

Paul O'Farrell, *Middle Tennessee State University*

Thomas Orf, *Las Positas College*

Michael C. Pease, *University of New Mexico*

Stephen Podewell, *Western Michigan University*

Nick Polizzi, *Cypress College*

Robert Rohli, *Louisiana State University*

Randall Schaetzl, *Michigan State University*

Jeffrey Schaffer, *Napa Valley College*

John H. Scheufler, *Mesa College*

Robert A. Sirk, *Austin Peay State University*

Stephen Stadler, *Oklahoma State University*

Herschel Stern, *Mira Costa College*

Timothy Warner, *West Virginia University*

Shawn Willsey, *College of Southern Idaho*

My thanks go out to contributors of new short boxed essays included in this edition: Richard Greene of Northern Illinois University, Jennifer Rahn of Samford University, Birmingham, Valerie Sloan of the University of Colorado, and Nancy Wilkinson of San Francisco State University. Randall Schaetzl of Michigan State University contributed a new boxed essay, as well as a detailed review of the material on soils and geomorphology, and Barbara Holzman of San Francisco State University provided a valuable critique of the material on biogeography. Shawn Willsey of College of Southern Idaho offered a detailed critique of the entire manuscript.

Special thanks go to Karl Byrand of the University of Wisconsin, Sheboygan, and Stephen Stadler of Oklahoma State University. In addition to contributing essays to this edition, both have long shared their expertise by providing student- and instructor-support materials for this textbook series.

Many of my colleagues at City College of San Francisco offered valuable suggestions on sections of the previous and current edition of the book: Carla Grandy, Dack Lee, Joyce Lucas-Clark, Robert Manlove, Kathryn Pinna, Todd Rigg-Carriero, Carole Toebe, and Katryn Wiese. I also extend my appreciation to my many students over the years—their curiosity, thoughtful questions, and cheerful acceptance of my enthusiasm for geography have helped me as a teacher and as a textbook author.

Textbooks of this scope cannot be created without a production team that is as dedicated to quality as the authors. I had the great fortune of working with a group of outstanding professionals at Pearson Prentice Hall. First of all, my thanks and appreciation go to Geography Acquisitions Editor Christian Botting who provided skillful leadership—with both good insight and good humor—throughout this project. My thanks also go to Production Editors Cindy Miller and Chitra Ganesan, Photo Researchers Clare Maxwell and Maureen Raymond, Art Project Manager Connie Long, Assistant Editor Jennifer Aranda, Editorial Assistant Christina Ferraro, Marketing Manager Maureen McLaughlin, Proofreader Linda Berger, and Media Producer Ziki Dekel. The legacy of many individuals who worked on earlier editions of this book remains a part of this edition, especially that of Dan Kaveney, Chris Rapp, Andrew Sobel, Amanda Brown, and all at Cadre Design.

I also want to extend my special gratitude to three individuals who worked with me on this and previous editions of the book—their professionalism is fully ingrained into this work. The oversight of Production Project Manager Shari Toron ensured that the elements needed to create this book came together with the highest of possible standards. Copy Editor Marcia Youngman has worked with me on so many books that she now knows how to figure out what I *really* mean to say. And Tim Flem, Geography Project Manager, has provided me unfailing support, sound advice, and unwavering encouragement. Thanks to you all.

Finally, I wish to express my appreciation for my wife, Nora. Her help, understanding, and support have once again seen me through the long hours and many months of work that went into this book.

Darrel Hess
Earth Sciences Department
City College of San Francisco
50 Phelan Avenue
San Francisco, CA 94112
dhess@ccsf.edu

For my parents, Tharon and Weaver

D.H.

About the Authors

Darrel Hess began teaching geography at City College of San Francisco in 1990 and served as chair of the Earth Sciences Department from 1995 to 2009. After earning his bachelor's degree in geography at the University of California, Berkeley, in 1978, he served for two years as a teacher in the Peace Corps on the Korean island of Cheju-do. Upon returning to the United States, he worked as a writer, photographer, and audiovisual producer. His association with Tom McKnight began as a graduate student at UCLA, where he served as one of Tom's teaching assistants. Their professional collaboration developed after Darrel graduated from UCLA with a master's degree in geography in 1990. He first wrote the *Study Guide* that accompanied the fourth edition of *Physical Geography: A Landscape Appreciation*, and then the *Laboratory Manual* that accompanied the fifth edition. Darrel has been authoring both works ever since. In 1999 Tom asked Darrel to join him as coauthor of the textbook. As did Tom, Darrel greatly enjoys the outdoor world. Darrel and his wife Nora, are avid hikers, campers, and scuba divers.

Tom L. McKnight taught geography at UCLA from 1956 to 1993. He received his bachelor's degree in geology from Southern Methodist University in 1949, his master's degree in geography from the University of Colorado in 1951, and his Ph.D. in geography and meteorology from the University of Wisconsin in 1955. During his long academic career, Tom served as chair of the UCLA Department of Geography from 1978 to 1983, and was director of the University of California, Education Abroad Program in Australia from 1984 to 1985. Passionate about furthering the discipline of geography, he helped establish the UCLA/Community College Geography Alliance and generously funded awards for both undergraduate and graduate geography students. His many honors include the California Geographical Society's Outstanding Educator Award in 1988, and the honorary rank of Professor Emeritus upon his retirement from UCLA. In addition to *Physical Geography: A Landscape Appreciation*, his other college textbooks include *The Regional Geography of the United States and Canada*; *Oceania: The Geography of Australia, New Zealand, and the Pacific Islands*; and *Introduction to Geography*, with Edward F. Bergman. Tom passed away in 2004—the geographic community misses him enormously.

1 Introduction to Earth

People live on the surface of Earth in a physical environment that is extraordinarily complex, extremely diverse, infinitely renewing, and yet ultimately fragile. This habitable environment exists over almost the entire land surface of Earth, which means that its horizontal dimensions are vast. Its vertical extent, however, is very limited (Figure 1-1). The great majority of all life inhabits a zone less than 5 kilometers (3 miles) thick, and the total vertical extent of the life zone is about 30 kilometers (20 miles).

Geography as a Field of Learning

It is within this shallow life zone that geographers focus their interest and do their work. In fact, the word *geography* comes from the Greek words for "earth description." Geography has always been (and remains) a generalized—as opposed to a specialized—discipline. Its viewpoint is one of broad understanding.

Several thousand years ago many scholars were more truly "Earth describers" than anything else. However, over the centuries there has been a trend away from generalized Earth description toward more specific scholarly specializations. This narrowing of focus led to the growth of a variety of more specialized disciplines—such as geology, meteorology, economics, and biology—along with an associated eclipsing of geography. If "mother geography gave birth to many offshoot sciences," as some geographers have said, these newer disciplines soon became better known than their predecessor. It was not until the 1600s that a rekindling of interest in geography in the European world began, and it was another two centuries before there was a strong impetus given to the discipline by geographers around the world.

The prominent geographic theme that has persisted through the centuries is that geographers study how things differ from place to place—in other words, geography is the spatial variation of Earth's surface. There is a multiplicity of "things" to be found on Earth, and nearly all of them are distributed unevenly, thus providing a spatial, or geographic, aspect to the planet.

Table 1-1 lists the kinds of things to which we are referring. These things are divided into two columns signifying the two principal branches of geography. The elements of **physical geography** are natural in origin, and for this

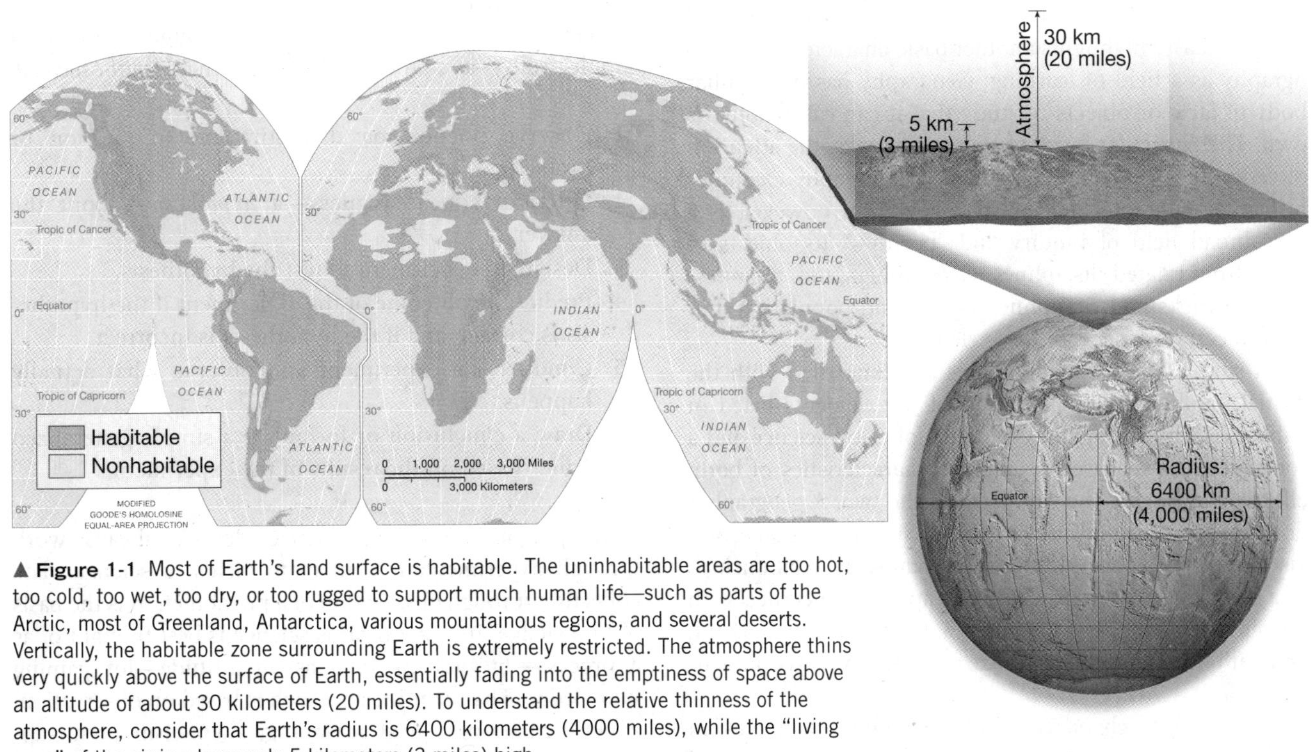

▲ **Figure 1-1** Most of Earth's land surface is habitable. The uninhabitable areas are too hot, too cold, too wet, too dry, or too rugged to support much human life—such as parts of the Arctic, most of Greenland, Antarctica, various mountainous regions, and several deserts. Vertically, the habitable zone surrounding Earth is extremely restricted. The atmosphere thins very quickly above the surface of Earth, essentially fading into the emptiness of space above an altitude of about 30 kilometers (20 miles). To understand the relative thinness of the atmosphere, consider that Earth's radius is 6400 kilometers (4000 miles), while the "living zone" of the air is a layer only 5 kilometers (3 miles) high.

◀ Earth as it appears from space in a composite satellite image.

TABLE 1-1 The Elements of Geography

Physical Elements	Cultural Elements
Rocks and Minerals	Population
Landforms	Settlements
Water	Economic activities
Weather and Climate	Transportation
Flora	Recreation activities
Fauna	Languages
Soil	Religion
	Political systems
	Traditions
	And many others

reason physical geography is sometimes called *environmental geography*. The elements of **cultural geography** are those of human endeavor, so this branch is sometimes referred to as *human geography*. The list of physical geography elements is essentially complete as a broad tabulation. The list of cultural geography elements, on the other hand, is merely suggestive of a much longer inventory of both material and nonmaterial features.

All the items listed in Table 1-1 are familiar to us, and this familiarity highlights another basic characteristic of geography as a field of learning: Geography has no peculiar body of facts or objects of study that it can call wholly its own. The particular focus of geology is rocks, the attention of economics is economic systems, demography examines human population, and so on. Geography, however, is a very broad field of inquiry and "borrows" its objects of study from related disciplines. Geographers, too, are interested in rocks and economic systems and population, but only in certain aspects of these elements.

In simplest terms, geographers are concerned with the spatial, or distributional, aspects of the elements listed in Table 1-1. Thus, geography is both a physical science and a social science because it combines characteristics of both, effectively bridging the gap between the two. A geographer does not specialize in studying the environment as an ecologist does, or in concentrating on human social relationships as a sociologist does. Instead, geography is concerned with the environment as it provides a home for humankind, and with the way humans utilize and alter this environmental home.

Another basic characteristic of geography is its interest in interrelationships. One cannot understand the distribution of soils, for example, without knowing something of the rocks from which the soils were derived, the slopes on which the soils developed, and the climate and vegetation under which they developed. Similarly, it is impossible to comprehend the distribution of agriculture without an understanding of climate, topography, soil, drainage, population, economic conditions, technology, historical development, and a host of other factors, both physical and cultural. Thus, the elements in Table 1-1 are enmeshed in an intricate web of interrelationships, all of which are encompassed in the discipline we call *geography*. In short, the fundamental questions of geographic inquiry are, "Why what is where and so what?"

Science and Geography

The subject matter of this book is physical geography. We shall examine the various components of the natural environment, the nature and characteristics of the physical elements of the landscape, the processes involved in their development, their distribution over Earth, and their basic interrelationships. We shall largely ignore interrelationships among the various elements of cultural geography except where they help to explain the development or contemporary distribution patterns of the physical elements, and the ways in which humans influence or alter the physical environment.

The Process of Science

Because physical geography is concerned with processes and patterns in the natural world, knowledge in physical geography is advanced through the study of science, and so it may be useful for us to begin with a few words about science in general.

Science is often described—although somewhat simplistically—as a process that follows the *scientific method*:

1. Observe phenomena that stimulate a question or problem.
2. Offer an educated guess—a *hypothesis*—about the answer.
3. Design an experiment to test the hypothesis.
4. Predict the outcome of the experiment if the hypothesis is correct, and if the hypothesis is incorrect.
5. Conduct the experiment and observe what actually happens.
6. Draw a conclusion or formulate a simple generalized "rule" based on the results of the experiment.

In practice, however, science doesn't always work through experimentation; in many fields of science, data collection through observation of a phenomenon is the basis of knowledge. In some regards science is best thought of as a process—or perhaps even as an attitude—for gaining knowledge. The scientific approach is based on observation, experimentation, logical reasoning, skepticism of unsupported conclusions, and the willingness to modify or even reject long-held ideas when new evidence contradicts them.

Although the term "scientific proof" is sometimes used by the general public, strictly speaking, science does not "prove" ideas. Instead, science works by eliminating alternative explanations—eliminating explanations that aren't supported by evidence. In fact, in order for a hypothesis to be "scientific," there must be some test or possible observation that could *disprove* it—if there is no way to disprove an idea, then that idea simply cannot be supported by science.

The word "theory" is often used in everyday conversation to mean a "hunch" or conjecture. However, in science a *theory* represents the highest order of understanding for a body of information—a logical, well-tested explanation that encompasses a wide variety of facts and observations. Thus, as we will see in Chapter 14, the "theory of plate tectonics" represents an empirically supported, broadly accepted, overarching framework for understanding processes operating within Earth.

The acceptance of scientific ideas and theories is based on a preponderance of evidence, not on "belief" and not on the pronouncements of "authorities." New observations and new evidence often cause scientists to revise their conclusions and theories or those of others. Much of this self-correcting process for refining scientific knowledge takes place through peer-reviewed journal articles. Peers—that is, fellow scientists—scrutinize a scientific report for sound reasoning, appropriate data collection, and solid evidence before it is published; reviewers need not agree with the author's conclusions, but they attempt to ensure that published articles meet rigorous standards of scholarship.

Since new evidence may prompt scientists to change their ideas, good science tends to be somewhat cautious in the conclusions that are drawn. For this reason, the findings of many scientific studies are prefaced by phrases such as "the evidence suggests," or "the results most likely show." In some cases, different scientists interpret the same data quite differently and so disagree in their conclusions. Frequently, studies find that "more research is needed." The kind of uncertainty sometimes inherent in science may lead the general public to question the conclusions of scientific studies—especially when presented with a simple, and perhaps comforting nonscientific alternative. It is, however, this very uncertainty that often compels scientists to push forward in the quest for knowledge and understanding!

In this book we present the fundamentals of physical geography as it is supported by scientific research and evidence. In some cases, we will describe how our current understanding of a phenomenon developed over time; in other cases we will point out where uncertainty remains, where there is still disagreement among scientists, or where intriguing questions still remain.

Numbers and Measurement Systems

Because so much of science is based on observation and measurable data, any thorough study of physical geography entails the use of mathematics. Although this book introduces physical geography primarily in a conceptual way without the extensive use of mathematical formulas, numbers and measurement systems are nonetheless important for us. Throughout the book, we use numbers and simple formulas to help illustrate concepts—the most obvious of which are numbers used to describe distance, size, weight, and temperature.

There are two quite different systems of measurement used around the world today. In the United States much of the general public is most familiar with the so-called *English System* of measurement—using measurements such as miles, pounds, and degrees Fahrenheit. However, most of the rest of the world—and the entire scientific community—uses the **International System** of measurement (abbreviated **S.I.** from the French *Système International*; also sometimes called the "metric system")—using measurements such as kilometers, kilograms, and degrees Celsius. You have probably noticed that this book gives measurements in both S.I. and English units. If you are not yet familiar with both systems, with just a little effort you can learn the basic equivalents in each. Tables 1-2 and 1-3 provide some simple conversion formulas; a detailed table of equivalents between English and S.I. units appears in Appendix I.

TABLE 1-2 Conversions: S.I. to English

	Quick Approximation	Conversion Formulas
Distance:	1 centimeter = a little less than 1/2 inch	cm × 0.394 = inches
	1 meter = a little more than 3 feet	m × 3.281 = feet
	1 kilometer = about 2/3 mile	km × 0.621 = miles
	1 cm (centimeter) = 10 mm (millimeters)	
	1 m (meter) = 100 cm	
	1 km (kilometer) = 1000 m	
Volume:	1 liter = about 1 quart	liters × 1.057 = quarts
Mass (Weight):	1 gram = about 1/30 ounce	g × 0.035 = ounces
	1 kilogram = about 2 pounds	kg × 0.205 = pounds
	1 kg (kilogram) = 1000 g (grams)	
Temperature:	1°C change = 1.8°F change	(°C × 1.8) + 32 = °F

TABLE 1-3 Conversions: English to S.I.

	Quick Approximation	Conversion Formulas
Distance:	1 inch = about $2\frac{1}{2}$ centimeters	inches × 2.540 = centimeters
	1 foot = about $\frac{1}{3}$ meters	feet × 0.305 = meters
	1 yard = about 1 meter	yards × 0.914 = meters
	1 mile = about $1\frac{1}{2}$ kilometers	miles × 1.609 = kilometers
	1′ (foot) = 12″ (inches)	
	1 yard = 3′	
	1 mile = 5280′	
Volume:	1 quart = about 1 liter	quarts × 0.946 = liters
	1 gallon = about 4 liters	gallons × 3.785 = liters
	1 gallon = 4 quarts	
Mass (Weight):	1 ounce = about 30 grams	ounces × 28.350 = grams
	1 pounds = about $\frac{1}{2}$ kilogram	pounds × 0.454 = kilograms
	1 lb (pound) = 16 oz (ounces)	
Temperature:	1°F change = about 0.6°C change	(°F − 32) ÷ 1.8 = °C

Global Environmental Change

Some characteristics of the global environment are in a state of perpetual change and apparently have been throughout its existence—this is very obvious when we consider Earth's atmosphere. As with most other natural phenomena, the rate of change of the atmosphere has generally been very slow, although the effects have been profound. One has only to look at the occurrence of past ice ages to recognize that the habitability of various parts of Earth has varied remarkably through time as temperature and precipitation conditions have changed.

Throughout this book, the topic of global environmental change—both human-induced and natural—will be addressed. We pay special attention to specific aspects of the accelerating impact that human activities have on the environment. For example, in the chapters on the atmosphere we discuss such issues as global warming, ozone depletion, and acid rain, while in later chapters we look at issues such as rainforest removal and coastal erosion. Rather than treat global environmental change as a separate topic, we integrate this theme throughout the book.

We also supplement the main text material with short boxed essays, such as those entitled, "People and the Environment," which focus on specific cases of human interaction with the natural environment, as well as boxes entitled, "Renewable Energy," which focus on the growing role of renewable or "sustainable" sources of energy. These essays serve to illustrate the connections between various aspects of the environment, such as the relationships between changing global temperatures, changing sea level, changing quantities of polar ice, and the changing distribution of plant and animal species, as well as the relationships between the environment, the economy, and human society.

The Environmental Spheres

Although there are billions of celestial bodies in the universe, insofar as we know, the environment of our planet is different from that of all the others. This unique environment has produced a unique landscape: a combination of solids, liquids, and gases; organic and inorganic matter; a bewildering variety of life forms; and an extraordinary diversity of physical features that is not even approximated on any other known planet. In this book, our study of geography is built around an attempt to understand (in other words, "appreciate") this unique landscape. As we proceed from chapter to chapter, the notion of landscape development and landscape modification will serve as a central theme.

For the remainder of this chapter, our attention is focused on a few broad concepts: the place of Earth in the solar system, the basic physical characteristics of this planet, and the functional relationship between Earth and its prime source of energy, the Sun. Understanding these concepts will help us comprehend the remarkable landscapes we inhabit.

From the standpoint of physical geography, the surface of Earth is a complex interface where the four principal components of the environment meet and to some degree overlap and interact (Figure 1-2). The solid, inorganic portion of Earth

▲ **Figure 1-2** The physical landscape of Earth's surface is composed of four overlapping and interacting systems called "spheres." The atmosphere is the air we breathe. The hydrosphere is the water of rivers, lakes, and oceans, the moisture in soil and air, as well as the snow and ice of the cryosphere. The biosphere is the habitat of all earthly life, as well as the life forms themselves. The lithosphere is the soil and bedrock that cover Earth's surface. This scene shows Wonder Lake and Mt. McKinley (Denali) in Denali National Park, Alaska.

is sometimes referred to as the **lithosphere** [1] (*litho* is Greek for "stone"), comprising the rocks of Earth's crust as well as the broken and unconsolidated particles of mineral matter that overlie the solid bedrock. The lithosphere's surface is shaped into an almost infinite variety of landforms, both on the seafloors and on the surfaces of the continents and islands.

The gaseous envelope of air that surrounds Earth is the **atmosphere** (*atmo* is Greek for "air"). It contains the complex mixture of gases needed to sustain life. Most of the atmosphere adheres closely to Earth's surface, being densest at sea level and rapidly thinning with increased altitude. It is a very dynamic sphere, kept in almost constant motion by solar energy and Earth's rotation.

The **hydrosphere** (*hydro* is Greek for "water") comprises water in all its forms. The oceans contain the vast majority of the water found on Earth and are the moisture source for most precipitation. A subcomponent of the hydrosphere is known as the **cryosphere** (*cry* comes from the Greek word for "cold")—water frozen as snow and ice.

The **biosphere** (*bio* is Greek for "life") encompasses all the parts of Earth where living organisms can exist; in its broadest and loosest sense, the term also includes the vast variety of earthly life forms (properly referred to as *biota*).

These "spheres" are not discrete and separated entities but rather are considerably interconnected. This intermingling is readily apparent when considering an ocean—a body that is clearly a major component of the hydrosphere and yet may contain a vast quantity of fish and other organic life that are part of the biosphere. An even better example is soil, which is composed largely of bits of mineral matter (lithosphere) but also contains life forms (biosphere), soil moisture (hydrosphere), and air (atmosphere) in pore spaces. These spheres can serve to broadly organize concepts for the systematic study of Earth's physical geography and will be used that way in this book.

Before focusing our attention on the spheres, however, we set the stage by considering the planetary characteristics of Earth and noting the most important relationships between our planet and its basic source of energy, the Sun.

The Solar System

Solar System Formation

We live on an extensive rotating mass of mostly solid material that orbits the enormous ball of superheated gases we call the Sun. The geographer's concern with spatial relationships properly begins with the relative location of this "spaceship Earth" in the universe.

Earth is one of eight plants of our solar system, which also contains more than 150 moons revolving around the planets, an uncertain number of smaller *plutoids* such as Pluto, scores of comets ("dirty snowballs" composed of frozen liquid and gases together with small pieces of rock and metallic minerals), more than 50,000 asteroids (small, rocky objects, mostly less than a few kilometers in diameter), and millions of meteoroids (most of them the size of sand grains).

The medium-sized star we call our Sun is the central body of the solar system and makes up more than 99 percent of its total mass. The solar system is part of the Milky Way Galaxy, which consists of at least 200,000,000,000 stars arranged in a disk-shaped spiral that is 100,000 light-years in diameter (1 light-year equals about 9 trillion kilometers—the distance a beam of light travels over a period of one year) and 10,000 light-years thick at the center (Figure 1-3). The Milky Way Galaxy is only one of many billions of galaxies in the universe.

To begin to develop an understanding for astronomical distances, we might consider a reduced-scale model of the universe: if the distance between Earth and the Sun, which is about 150,000,000 kilometers (93,000,000 miles), is taken to be 2.5 centimeters (1 inch), then the distance

[1]As we will see in Chapter 13, in the context of *plate tectonics* and our study of landforms, the term "lithosphere" is used specifically to refer to large "plates" consisting of Earth's crustal and upper mantle rock.

▲ **Figure 1-3** The structure of the Milky Way Galaxy showing the approximate location of our Sun on one of the spiral arms.

from Earth to the nearest star would be 7.2 kilometers (4.5 miles), and the distance from Earth to the next galaxy beyond the Milky Way would be about 240,000 kilometers (150,000 miles)!

The origin of Earth, and indeed of the universe, is incompletely understood. It is generally accepted that the universe began with a cosmic "explosion" called the *big bang*. The most widely held view is that the big bang took place some 13.7 billion years ago—similar to the age of the oldest known stars. The big bang began in a fraction of a second after an infinitely dense fireball started to expand away in all directions at extraordinary speeds, pushing out the fabric of space and filling the universe with the energy and matter we see today.

Our solar system apparently originated between 4.5 and 5 billion years ago when a *nebula*—a huge, cold, diffuse cloud of gas and dust—began to contract inward, owing to its own gravitational collapse, forming a hot, dense *protostar* (Figure 1-4). This hot center—our Sun—was surrounded by a cold, revolving disk of gas and dust that eventually condensed and coalesced to form the planets.

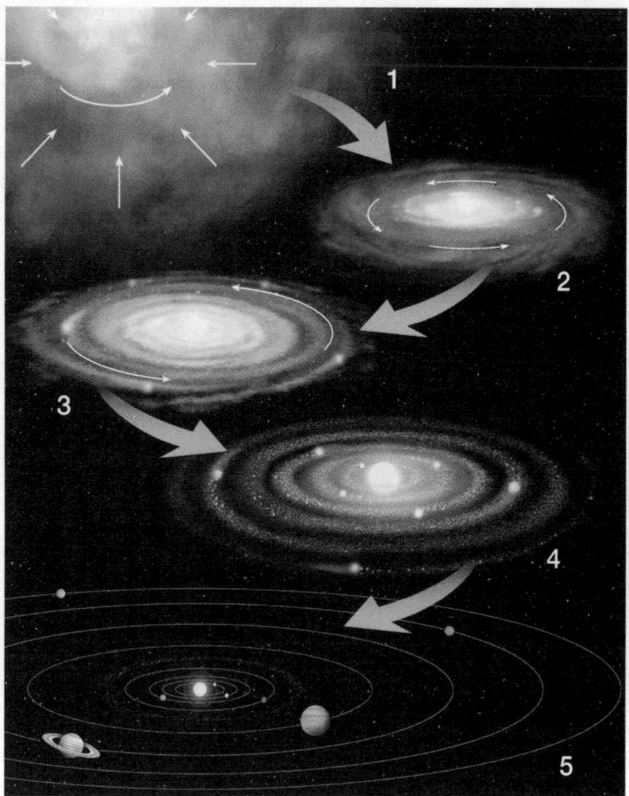

▲ **Figure 1-4** The birth of the solar system. (1) Diffuse gas cloud, or nebula, begins to contract inward. (2) Cloud flattens into nebular disk as it spins around a central axis. (3) Particles in the outer parts of the disk collide with each other to form protoplanets. (4) Protoplanets coalesce into planets and settle into orbits around the hot center. (5) The final product: a central Sun surrounded by eight orbiting planets. (Solar system not shown in correct scale.)

All of the planets revolve around the Sun in elliptical orbits, with the Sun located at one focus (looking "down" on the solar system from a vantage point high above the North Pole of Earth, the planets orbit in a counterclockwise direction around the Sun). All the planetary orbits are in nearly the same plane (Figure 1-5), perhaps revealing their relationship to the original nebular disk. The Sun

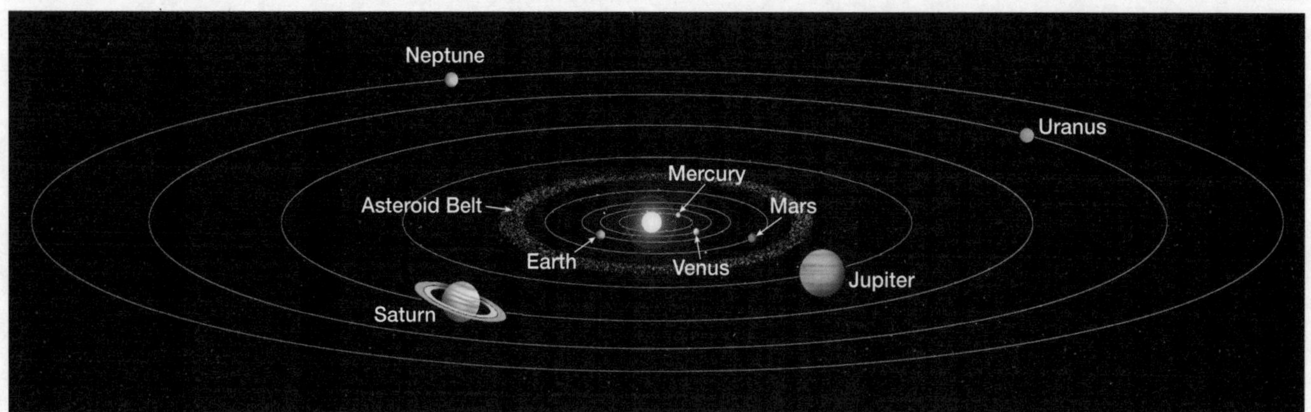

▲ **Figure 1-5** The solar system (not drawn to correct scale). The Sun is not exactly at the center of the solar system—the planets revolve around the Sun in elliptical orbits.

rotates on its axis from west to east. Moreover, most of the planets rotate from west to east on their own axes (Uranus rotates "sideways" with its rotational axis almost parallel to its orbital plane; Venus rotates from east to west). Generally, planets revolve more slowly and have a lower temperature as their distance from the Sun increases.

The four inner *terrestrial planets*—Mercury, Venus, Earth, and Mars—are generally smaller, denser, less oblate (more nearly spherical), and rotate more slowly on their axes than the four outer *Jovian planets*—Jupiter, Saturn, Uranus, and Neptune. Also, the inner planets are composed principally of mineral matter and, except for airless Mercury, have diverse but relatively shallow atmospheres.

By contrast, the four Jovian planets tend to be much larger, more massive (although they are less dense), much more oblate, and they rotate more rapidly. The Jovian planets are mostly composed of gases such as hydrogen and helium—perhaps liquid near the surface, but frozen toward the interior—as well as ices of compounds such as methane and ammonia. The Jovian planets generally have atmospheres that are dense, turbulent, and relatively deep.

It was long thought that tiny Pluto was the ninth and outermost planet in the solar system. In recent years, however, astronomers have discovered other icy bodies, such as distant Eris, Makemake, and Haumea that are similar to Pluto and orbiting the Sun beyond Neptune in what is referred to as the *Kuiper Belt*. In June 2008 the International Astronomical Union reclassified Pluto and other similar-size objects within the solar system as "plutoids" (after briefly classifying these objects as "dwarf planets"). Some astronomers speculate that there may be several dozen yet to be discovered plutoids in the outer reaches of the solar system.

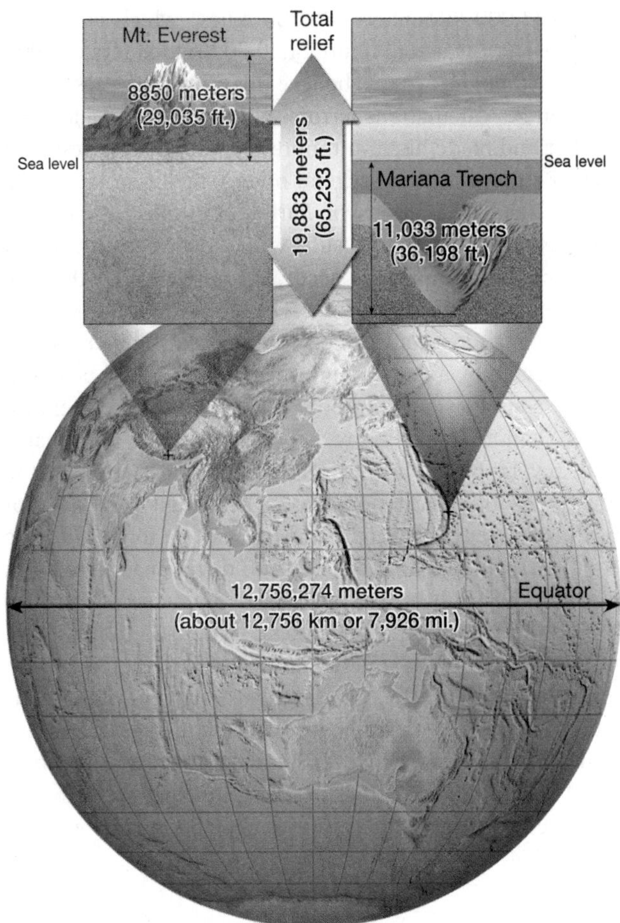

▲ **Figure 1-6** Earth's maximum relief (the difference in elevation between the highest and lowest points) is 19,883 meters (65,233 feet) or about 20 kilometers (12 miles) from the top of Mount Everest to the bottom of the Mariana Trench in the Pacific Ocean.

The Size and Shape of Earth

Is Earth large or small? The answer to this question depends on one's frame of reference. If the frame of reference is the universe, Earth is almost infinitely small. The diameter of our planet is only about 13,000 kilometers (7900 miles), a tiny distance at the scale of the universe—for instance, the Moon is 385,000 kilometers (239,000 miles) from Earth, the Sun is 150,000,000 kilometers (93,000,000 miles) away, and the nearest star is 40,000,000,000,000 kilometers (25,000,000,000,000 miles) distant.

In a human frame of reference, however, Earth is a very impressive mass. Its surface varies in elevation from the highest mountain peak, Mount Everest, at 8850 meters (29,035 feet) above sea level, to the deepest ocean trench, the Mariana Trench of the Pacific Ocean, at 11,033 meters (36,198 feet) below sea level, a total difference in elevation of 19,883 meters (65,233 feet).

Although prominent on a human scale of perception, this difference is minor on a planetary scale, as Figure 1-6 illustrates. If Earth were the size of a basketball, Mount Everest would be an imperceptible pimple no greater than 0.17 millimeter (about 7 thousandths of an inch), high. Similarly, the Mariana Trench would be a tiny crease only 0.21 millimeter (about 8 thousandths of an inch) deep—this represents a depression smaller than the thickness of a sheet of paper. Our perception of the relative size of topographic irregularities on Earth is often distorted by three-dimensional wall maps and globes that emphasize such landforms. To portray any noticeable appearance of topographic variation, the vertical distances on such maps are usually exaggerated 8 to 20 times their actual proportional dimensions—as are many diagrams used in this book. Further, many diagrams illustrating features of the atmosphere also exaggerate relative sizes to convey important concepts.

The Shape of Earth More than 2600 years ago Greek scholars correctly reasoned Earth to have a spherical shape. About 2200 years ago, Eratosthenes, the director of the Greek library at Alexandria, calculated the circumference of Earth trigonometrically. He determined the angle of the noon Sun rays at Alexandria and at another city, Syene, measured the distance between the two localities (960 kilometers [600 miles]), and from these angular and linear distances he was able to estimate an Earth circumference of almost 43,000 kilometers (26,700 miles) which is reasonably close to the actual figure of 40,000 kilometers (24,900 miles). Some historians believe that Eratosthenes made errors in his calculations, errors that partially canceled one another and made his result more accurate than it deserved to be. Be that as it may, several other Greek scholars made independent calculations of Earth's circumference during this period, and all their results were close to reality, which indicates the antiquity of accurate knowledge of the size and shape of Earth.

Earth is almost, but not quite, spherical. The cross section revealed by a cut through the equator would be circular, but a similar cut from pole to pole would be an ellipse rather than a circle (Figure 1-7). Any rotating body has a tendency to bulge around its equator and flatten at the polar ends of its rotational axis. Although the rocks of Earth may seem quite rigid and immovable to us, they are sufficiently pliable to allow Earth to develop a bulge in its midriff. The slightly flattened polar diameter of Earth is 12,714 kilometers (7900 miles), whereas the slightly bulging equatorial diameter is 12,756 kilometers (7926 miles), a difference of only about 0.3 percent. Thus, our planet is properly described as an *oblate spheroid* rather than a true sphere. However, because the variation from true sphericity is exceedingly small, in most cases in this book we can treat Earth as if it were a perfect sphere.

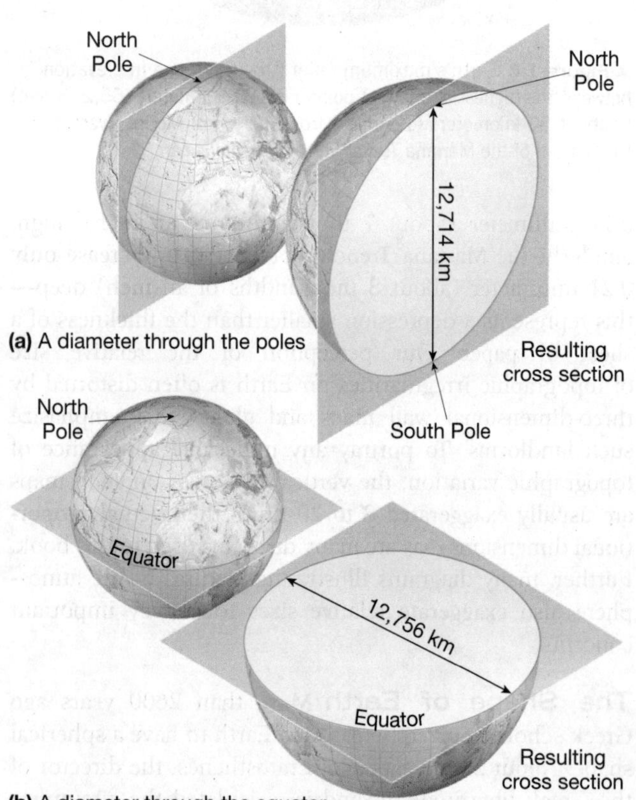

▲ **Figure 1-7** Earth is not quite a perfect sphere because its surface flattens a bit at the North Pole and the South Pole and bulges out slightly around the equator. Thus, a cross section through the poles, shown in (a), has a diameter slightly less than the diameter of a cross section through the equator, shown in (b).

The Geographic Grid—Latitude and Longitude

Any understanding of the distribution of geographic features over Earth's surface requires some system of accurate location. The simplest technique for achieving this precision is to design a grid system consisting of two sets of lines that intersect at right angles, thus permitting the location of any point on the surface to be described by the appropriate intersection, as shown in Figure 1-8. Such a rectangular grid system has been reconfigured for Earth's spherical surface, forming a geographic grid consisting of east–west and north–south lines.

If our planet were a nonrotating body, the problem of describing precise surface locations would be more difficult than it is: imagine trying to describe the location of a particular point on a perfectly round, perfectly clean Ping-Pong ball. Fortunately, Earth does rotate, and we can use its rotation axis as a starting point to describe locations on Earth.

Earth's rotation axis is an imaginary line that connects the points on Earth's surface called the **North Pole** and the **South Pole** (Figure 1-9). Further, if we visualize an imaginary plane passing through Earth halfway between the poles and perpendicular to the axis of rotation, we have another valuable reference feature: the *plane of the equator*. Where this plane intersects Earth's surface is the imaginary midline of Earth, called simply the **equator**. We use the North Pole, South Pole, rotational axis, and equatorial plane as natural reference features for measuring and describing locations on Earth's surface.

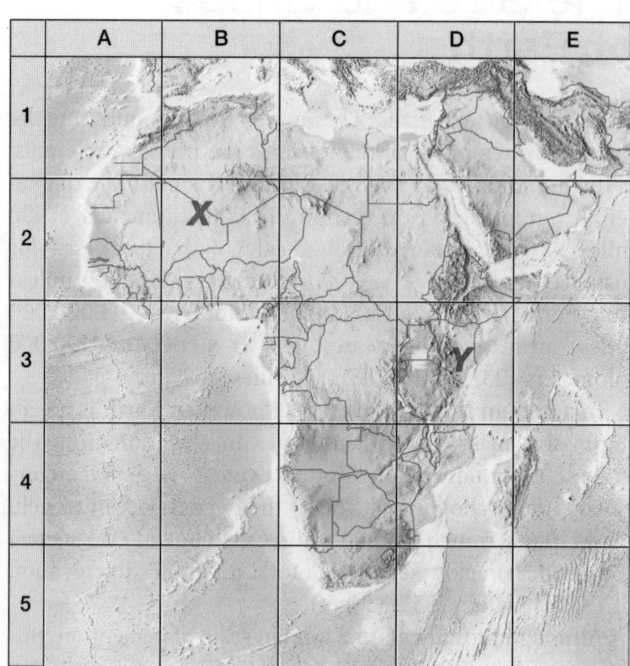

▲ **Figure 1-8** An example of a grid system. The location of point X can be described as 2B or as B2; the location of Y is 3D or D3.

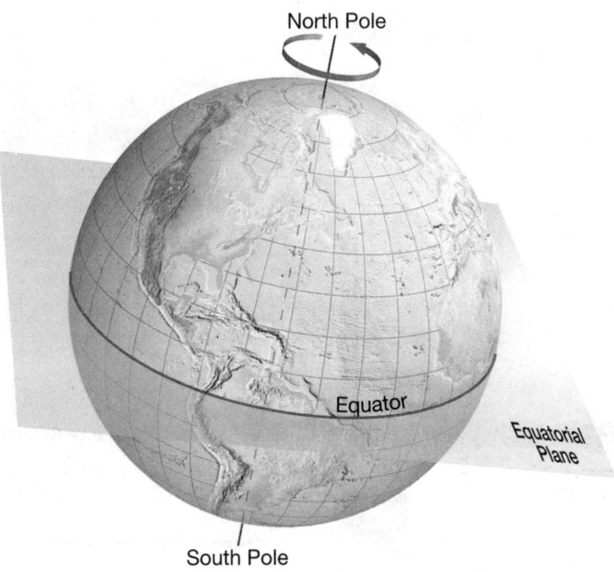

▲ **Figure 1-9** Earth spins around its rotation axis, an imaginary line that passes through the North Pole and the South Pole. An imaginary plane bisecting Earth midway between the two poles defines the equator.

Great Circles Any plane that is passed through the center of a sphere bisects that sphere (divides it into two equal halves) and creates what is called a **great circle** where it intersects the surface of the sphere (Figure 1-10a). Planes passing through any other part of the sphere produce what are called *small circles* where they intersect the surface (Figure 1-10b). Great circles have two principal properties of special interest.

1. A great circle is the largest circle that can be drawn on a sphere; it represents the circumference of that sphere and divides its surface into two equal halves or hemispheres. For example, the Sun illuminates one-half of Earth at any given moment. The edge of the sunlit hemisphere, called the *circle of illumination*, is a great circle that divides Earth between a light half and a dark half.
2. Only one great circle can be drawn on a sphere to include any two given points (not diametrically opposite each other). As a consequence, the segment, or arc, of the great circle connecting those two points is always the shortest route between the points. Such routes on Earth are known as *great circle routes* (great circle routes will be discussed in more detail in Chapter 2).

The geographic grid used as the locational system for Earth is based on the principles just discussed. Furthermore, the system is closely linked with the various positions assumed by Earth in its orbit around the Sun. The grid system of Earth is referred to as a *graticule* and consists of lines of latitude and longitude.

Latitude

Latitude is a description of location expressed as an angle north or south of the equator. As shown in Figure 1-11, we can project a line from any location on Earth's surface to the center of Earth. The angle between this line and the equatorial plane is the latitude of that location.

Latitude is expressed in degrees, minutes, and seconds. There are 360 degrees (°) in a circle, 60 minutes (′) in one

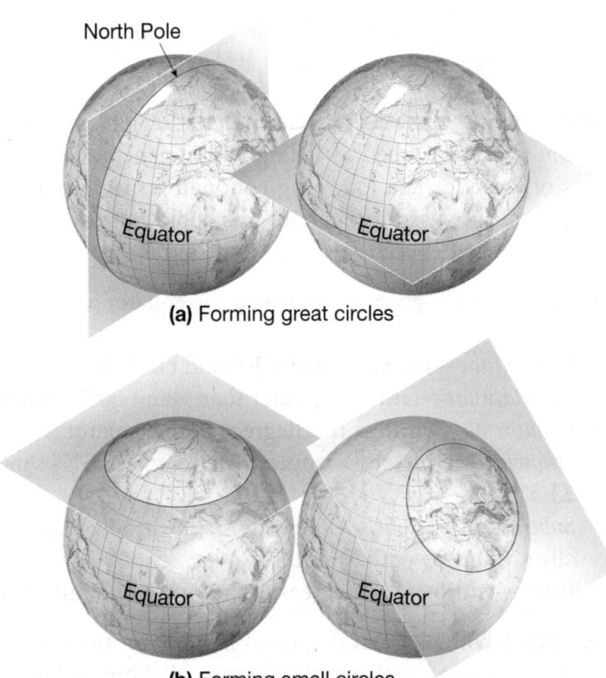

▲ **Figure 1-10** Comparison of great and small circles. (a) A great circle results from the intersection of Earth's surface with any plane that passes through Earth's center. (b) A small circle results from the intersection of Earth's surface with any plane that does not pass through Earth's center.

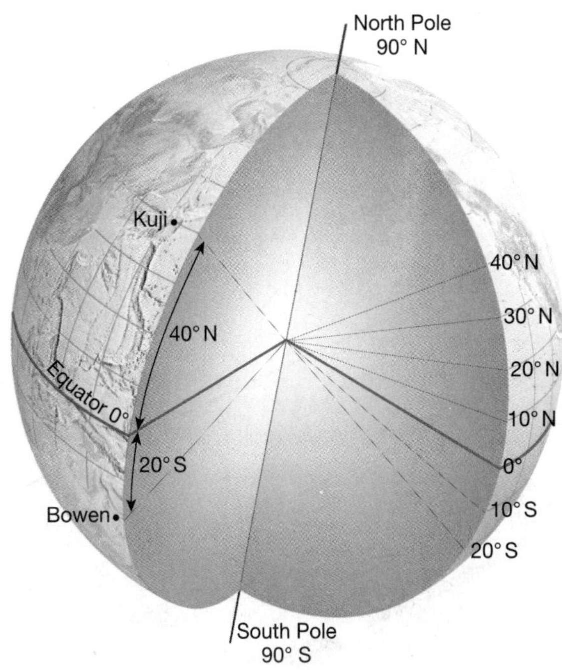

▲ **Figure 1-11** Measuring latitude. An imaginary line from Kuji, Japan, to Earth's center makes an angle of 40° with the equator. Therefore, Kuji's latitude is 40° N. An imaginary line from Bowen, Australia, to Earth's center makes an angle of 20°, giving this city a latitude of 20° S.

degree, and 60 seconds (″) in one minute. With the advent of GPS navigation (discussed in Chapter 2), it is increasingly common to see latitude and longitude designated using decimal notation, for example, 38°22′47″ N can be written 38°22.78′ N or even 38.3797° N.

Latitude varies from 0° at the equator to 90° north at the North Pole and 90° south at the South Pole. Any position north of the equator is north latitude, and any position south of the equator is south latitude (the equator itself is simply referred to as having a latitude of 0°).

A line connecting all points of the same latitude is called a **parallel** (because it is parallel to all other lines of latitude). The equator is the parallel of 0° latitude, and it, alone of all parallels, constitutes a great circle. All other parallels are small circles—all aligned in true east–west directions on Earth's surface. Since latitude is expressed as an angle, it can be infinitely subdivided—parallels can be constructed for every degree of latitude, or even for fractions of a degree of latitude.

Although it is possible to either construct or visualize an unlimited number of parallels, seven latitudes are of particular significance in a general study of Earth (Figure 1-12):

1. Equator, 0°
2. Tropic of Cancer, 23.5° N
3. Tropic of Capricorn, 23.5° S (Figure 1-13)
4. Arctic Circle, 66.5° N
5. Antarctic Circle, 66.5° S
6. North Pole, 90° N
7. South Pole, 90° S

The North Pole and South Pole are of course points rather than lines, but can be thought of as infinitely small parallels. The importance of these seven parallels will be explained later in this chapter.

▲ **Figure 1-13** The Tropic of Capricorn; like all other parallels of latitude, is an imaginary line. As a significant parallel, however, its location is often commemorated by a sign. This scene is near Alice Springs in the center of Australia.

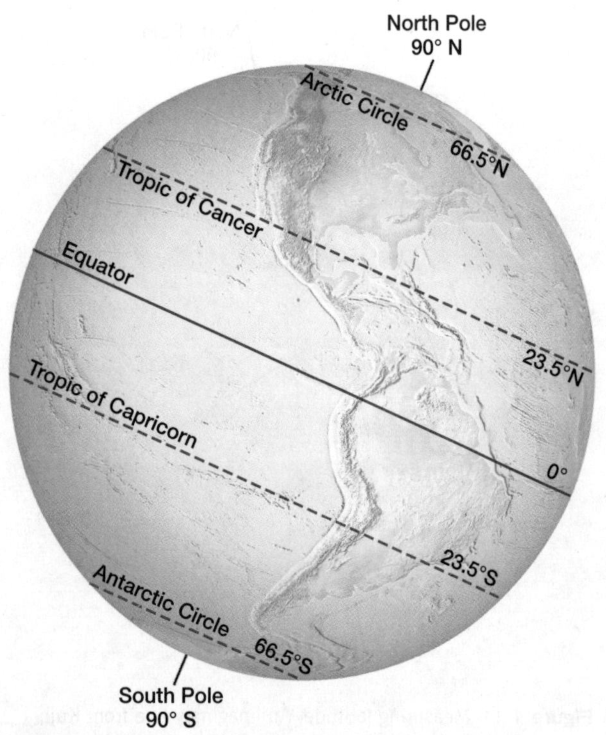

▲ **Figure 1-12** Seven important parallels. As discussed later, these latitudes represent special points where rays from the Sun strike Earth's surface on certain days of the year.

Descriptive Zones of Latitude Regions on Earth are sometimes described as falling within general bands or zones of latitude. The following common terms associated with latitude are used throughout this book (note that there is some overlap between several of these terms):

- *Low latitude*—generally between the equator and 30° N and S
- *Midlatitude*—between about 30° and 60° N and S
- *High latitude*—latitudes greater than about 60° N and S
- *Equatorial*—within a few degrees of the equator
- *Tropical*—within the tropics (between 23.5° N and 23.5° S)
- *Subtropical*—slightly poleward of the tropics, generally around 25– 30° N and S
- *Polar*—within a few degrees of the North or South Pole

Nautical Miles Each degree of latitude on the surface of Earth covers a north–south distance of about 111 kilometers (69 miles; Table 1-4). The distance varies slightly with latitude because of the flattening of Earth at the poles. The distance measurement of a *nautical mile*—and the description of speed known as a *knot* (one nautical mile per hour)—is defined by the distance covered by one minute

TABLE 1-4 Length of Degrees of Latitude and Longitude

Latitude (degrees)	Length of 1° of Latitude Measured along a Meridian: Kilometers	Miles	Length of 1° of Longitude Measured along a Parallel: Kilometers	Miles
0	110.567	68.703	111.321	69.172
10	110.605	68.726	109.641	68.129
20	110.705	68.789	104.649	65.026
30	110.857	68.883	96.488	59.956
40	111.042	68.998	85.396	53.063
50	111.239	69.121	71.698	44.552
60	111.423	69.235	55.802	34.674
70	111.572	69.328	38.188	23.729
80	111.668	69.387	19.394	12.051
90	111.699	69.407	0	0

of latitude (1′), the equivalent of about 1.15 statute ("ordinary") miles or about 1.85 kilometers.

Longitude

Latitude comprises one-half of Earth's grid system, the north–south component. The other half is **longitude**—an angular description of east–west location, also measured in degrees, minutes, and seconds.

Longitude is represented by imaginary lines extending from pole to pole and crossing all parallels at right angles. These lines, called **meridians**, are not parallel to one another except where they cross the equator. Any pair of meridians is farthest apart at the equator, becoming increasingly close together northward and southward and finally converging at the poles.

The distance between any two meridians varies predictably. At the equator, the surface length of one degree of longitude is about the same as that of one degree of latitude. However, because meridians converge at the poles, the distance covered by one degree of longitude decreases poleward (Table 1-4), diminishing to zero at the poles where all meridians meet at a point.

The equator is a natural baseline from which to measure latitude, but no such natural reference line exists for longitude. Consequently, for most of recorded history, there was no accepted longitudinal baseline; each country would arbitrarily select its own "prime meridian" as the reference line for east–west measurement. Thus, the French measured from the meridian of Paris, the Italians from the meridian of Rome, and so forth. There were at least 13 prime meridians in use in the 1880s. Not until the late 1800s was standardization finally achieved.

United States and Canadian railway executives adopted a standard time system for all North American railroads in 1883, and the following year an international conference was convened in Washington, D.C., to achieve the same goal on a global scale and to agree upon a single prime meridian. After weeks of debate, the delegates chose the meridian passing through the Royal Observatory at Greenwich, England, just east of London, as the **prime meridian** for all longitudinal measurement (Figure 1-14). The principal argument for adopting the Greenwich meridian as the prime meridian was a practical one: more than two-thirds

▲ **Figure 1-14** The prime meridian of the world, longitude 0°0′0″ at Greenwich, England, which is about 32 km (20 miles) down the river Thames from the heart of London.

of the world's shipping lines already used the Greenwich meridian as a navigational base.

Thus, an imaginary north–south plane passing through Greenwich and through Earth's axis of rotation represents the plane of the prime meridian. The angle between this plane and a plane passed through any other point and the axis of Earth is a measure of longitude. For example, the angle between the Greenwich plane and a plane passing through the center of the city of Freetown (in the western African country of Sierra Leone) is 13 degrees, 15 minutes, and 12 seconds. Since the angle is formed west of the prime meridian, the longitude of Freetown is written 13°15′12″ W (Figure 1-15).

Longitude is measured both east and west of the prime meridian to a maximum of 180° in each direction. Exactly halfway around the globe from the prime meridian, in the middle of the Pacific Ocean, is the 180° meridian (Figure 1-16). All places on Earth, then, have a location that is either east longitude or west longitude, except for points exactly on the prime meridian (described simply as 0° longitude) or exactly on the 180th meridian (described as 180° longitude).

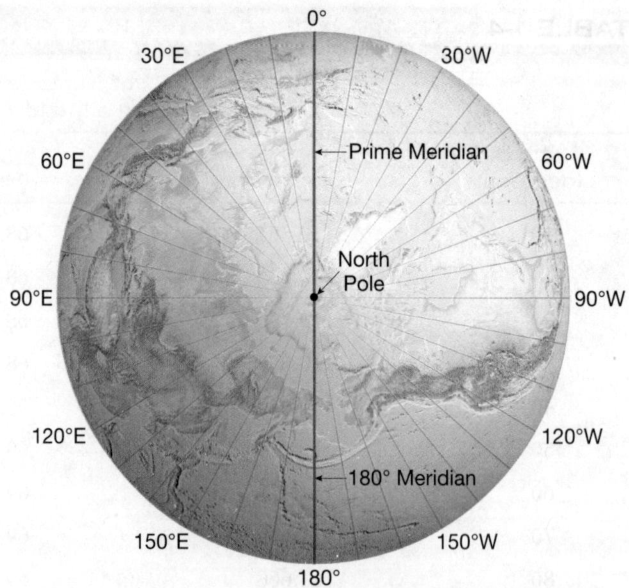

▲ **Figure 1-16** A polar view of meridians radiating from the North Pole. Think of each line as the top edge of an imaginary plane passing through both poles. All the planes are perpendicular to the plane of the page.

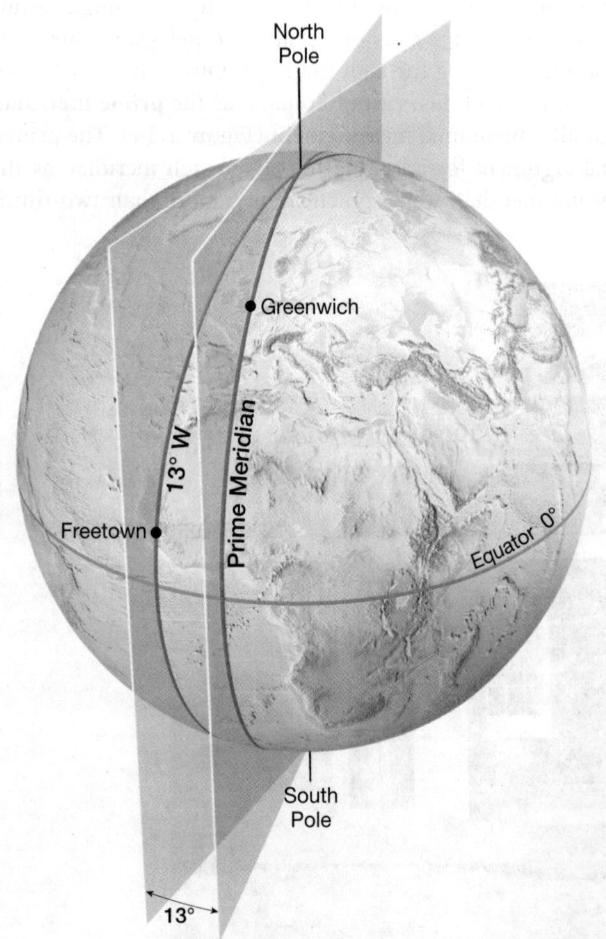

▲ **Figure 1-15** The meridians that mark longitude are defined by intersecting imaginary planes passing through the poles. Here are shown the planes for the prime meridian through Greenwich, England, and the meridian through Freetown, Sierra Leone, at 13° west longitude.

The network of intersecting parallels and meridians creates a geographic grid over the entire surface of Earth (Figure 1-17). The location of any place on Earth's surface can be described with great precision by reference to detailed latitude and longitude data. For example, at the 1964 World's Fair in New York City, a time capsule (a container filled with records and memorabilia of contemporary life) was buried. For reference purposes, the U.S. Coast and Geodetic Survey determined that the capsule was located at 40°28′34.089″ north latitude and 73°43′16.412″ west longitude. At some time in the future, if a hole were to be dug at the spot indicated by those coordinates, it would be within 15 cm (6 inches) of the capsule.

Earth–Sun Relations

Earth–Sun Relations

Life on Earth is dependent on solar energy; therefore, the functional relationship between Earth and the Sun is of vital importance. Because of the perpetual motions of Earth, this relationship does not remain the same throughout the year. Two basic Earth movements—its daily rotation on its axis and its annual revolution around the Sun—along with the inclination and "polarity" of Earth's rotation axis, combine to change Earth's orientation to the Sun during the year and thus are of special interest to geographers.

Earth's Rotation on Its Axis

Earth rotates from west to east on its axis (Figure 1-18), a complete **rotation** requiring 24 hours (from the vantage point of looking down at the North Pole from space, Earth

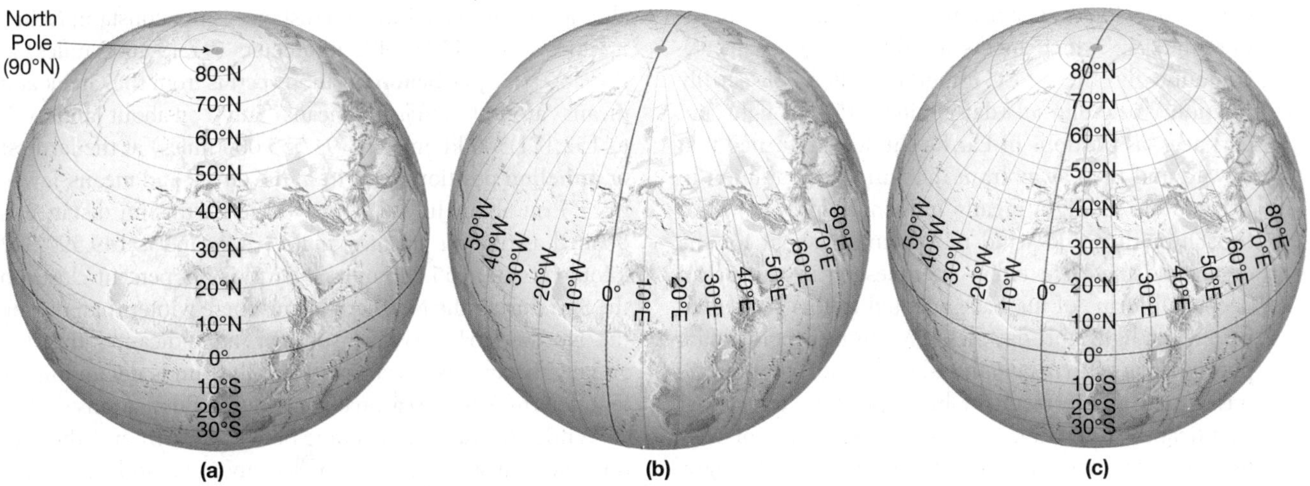

▲ **Figure 1-17** Development of the geographic grid, or graticule. (a) Parallels of latitude. (b) Meridians of longitude. (c) The two combined to form a complete grid system.

is rotating in a counterclockwise direction). The Sun, the Moon, and the stars appear to rise in the east and set in the west—this is, of course, an illusion created by the steady eastward spin of Earth.

Rotation causes all parts of Earth's surface except the poles to move in a circle around Earth's axis. Table 1-5 shows the speed of this motion at various latitudes. Although the speed of rotation varies by latitude, it is constant at any given place on Earth and so we experience no sense of motion. This is the same reason that we have little sense of motion on a smooth jet airplane flight at cruising speed—only when speed changes, such as during takeoff and landing, does motion become apparent.

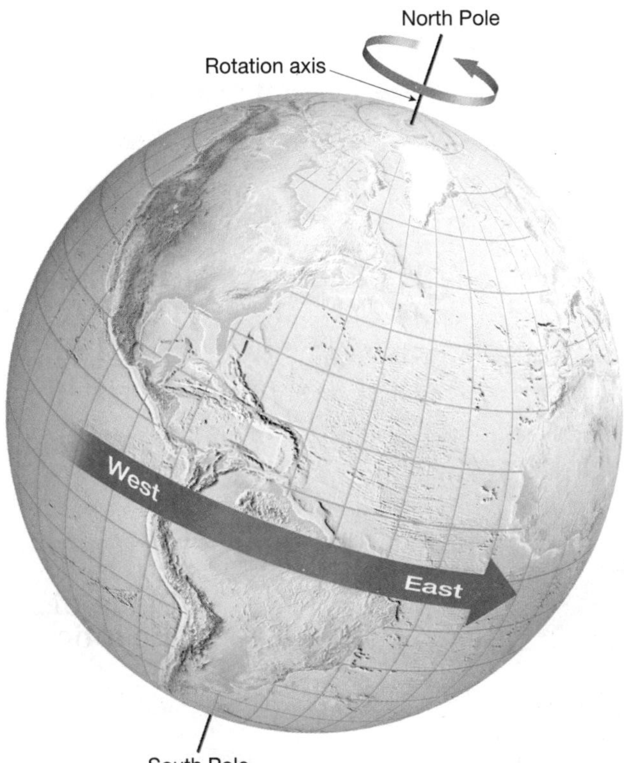

▲ **Figure 1-18** Earth rotates from west to east on a rotation axis that is tilted from the vertical. Looking down at the North Pole from above, Earth would appear to rotate in a counterclockwise direction.

Rotation has several striking effects on the physical characteristics of Earth's surface. Most important are the following:

1. The constancy of Earth's rotation in the same direction causes an apparent deflection in the flow paths of both wind and ocean currents. The deflection is to the right in the Northern Hemisphere and to the left in the Southern Hemisphere. This phenomenon is called the *Coriolis effect* and is discussed in detail in Chapter 3.
2. The rotation of Earth brings any point on the surface through the increasing and then decreasing gravitational pull of the Moon and the Sun. Although the land areas of Earth are too rigid to be significantly moved by these oscillating gravitational attractions, oceanic waters move onshore and then recede in a rhythmic pattern as a result of the interplay of earthly rotation with these

TABLE 1-5 Speed of Rotation of Earth's Surface at Selected Latitudes

Latitude	Kilometers per Hour	Miles per Hour
0	1669.9	1037.6
10	1642.0	1021.9
20	1569.7	975.4
30	1447.3	899.3
40	1280.9	795.9
50	1075.5	668.3
60	837.0	520.1
70	572.8	355.9
80	291.0	180.8
90	0	0

gravitational forces. The rise and fall of water level constitutes the *tides*, which are discussed further in Chapter 9.

3. Undoubtedly the most important effect of earthly rotation is the *diurnal* (daily) alternation of light and darkness, as portions of Earth's surface are turned first toward and then away from the Sun. This variation in exposure to sunlight greatly influences local temperature, humidity, and wind movements. Except for the organisms that live either in caves or in the ocean deeps, all forms of life have adapted to this sequential pattern of light and darkness. We human beings fare poorly when our *circadian* (24-hour cycle) rhythms are misaligned as the result of high-speed air travel that significantly interrupts the normal sequence of daylight and darkness. We are left with a sense of fatigue known as "jet lag," which can include unpleasant changes in our usual patterns of appetite and sleep.

Earth's Revolution around the Sun

Another significant Earth motion is its **revolution** around the Sun. Each revolution takes 365 days, 5 hours, 48 minutes, and 46 seconds, or 365.242199 days. This is known officially as the *tropical year* and for practical purposes is usually simplified to 365.25 days. (Astronomers define the year in other ways as well, but the duration is very close to that of the tropical year and need not concern us here.)

The path followed by Earth in its journey around the Sun is not a true circle but an ellipse (Figure 1-19). Because of this elliptical orbit, the Earth–Sun distance is not constant; rather, it varies from 147,166,480 kilometers (91,455,000 miles) at the closest or **perihelion** position (*peri* is from the Greek and means "around" and *helios* means "Sun") on about January 3, to 152,171,500 kilometers (94,555,000 miles) at the farthest or **aphelion** position (*ap* is from the Greek and means "away from") on about July 4. The average Earth–Sun distance is defined as one *astronomical unit* (1 AU) and is 149,597,892 kilometers (92,955,806 miles). Earth is 3.3 percent closer to the Sun during the Northern Hemisphere winter than during the Northern Hemisphere summer, an indication that the varying distance between Earth and the Sun is not an important determinant of seasonal temperature fluctuations.

While the varying distance between Earth and the Sun does not cause the change of seasons, two additional factors in the relationship of Earth to the Sun—inclination and polarity—work together with rotation and revolution to produce the changes of seasons.

Inclination of Earth's Axis

The imaginary plane defined by the orbital path of Earth around the Sun is called the **plane of the ecliptic** (see Figure 1-19). However, the plane of the ecliptic does not match the plane of the equator because Earth's rotation axis is not perpendicular to the ecliptic plane. Rather, the axis is tilted about 23.5° away from the perpendicular (Figure 1-20) and maintains this tilt throughout the year. This tilt is referred to as the **inclination of Earth's axis**.

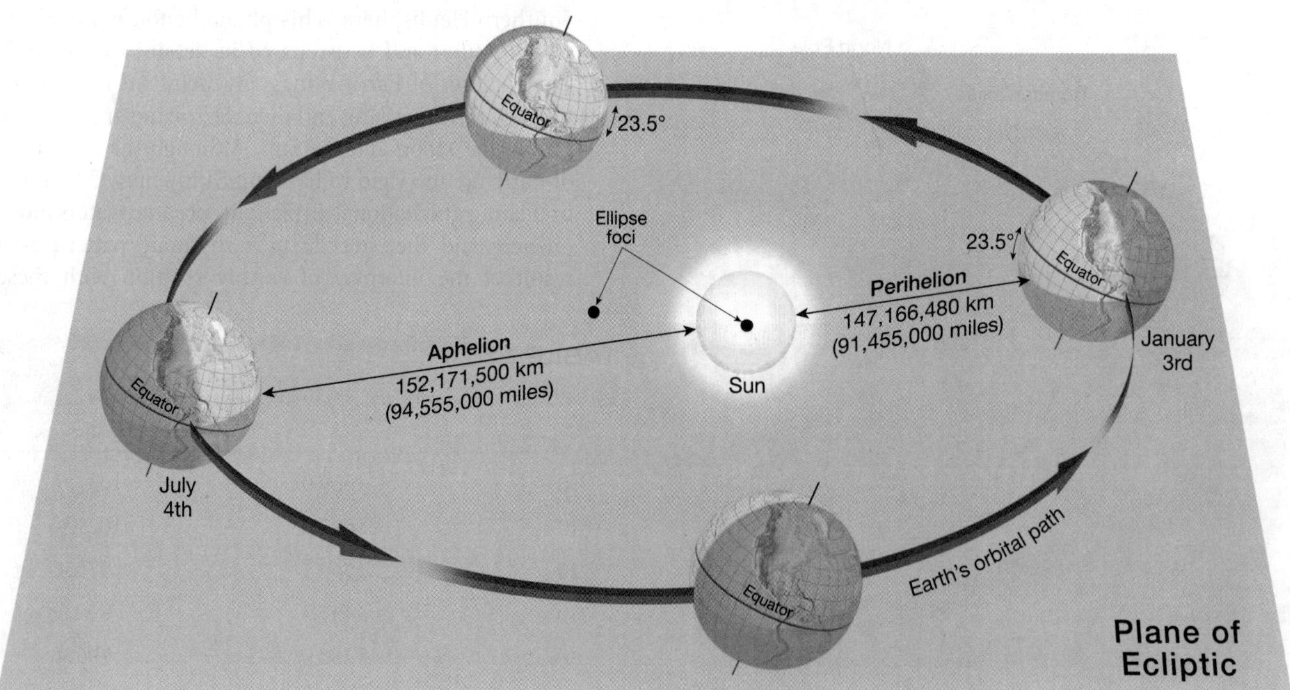

▲ **Figure 1-19** The plane of the ecliptic is the orbital plane of Earth. Because Earth's rotation axis is tilted, the plane of the ecliptic and the equatorial plane do not coincide. The path Earth follows in its revolution around the Sun is an ellipse having the Sun at one focus. Earth reaches perihelion (its closest point to the Sun) on about January 3rd at a distance of 147,166,480 kilometers (91,455,000 miles). Earth reaches aphelion (its farthest point from the Sun) on about July 4th at a distance of 152,171,500 kilometers (94,555,000 miles). (In this diagram the elliptical shape of Earth's orbit is greatly exaggerated.)

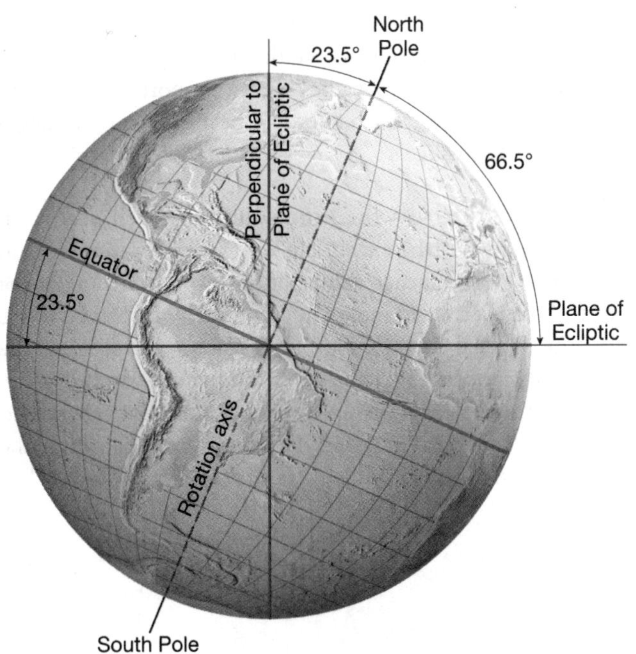

▲ **Figure 1-20** Earth's rotation axis forms an angle of 66.5° with the plane of the ecliptic. However, the inclination of Earth's axis is most commonly described as being 23.5° from a line perpendicular to the plane of the ecliptic.

Polarity of Earth's Axis

Not only is Earth's rotation axis inclined relative to its orbital path, no matter where Earth is in its orbit around the Sun the axis always points in the same direction relative to the stars—toward the North Star, Polaris. In other words, at any time during the year, Earth's rotation axis is parallel to its orientation at all other times. This characteristic is called the **polarity of Earth's axis** (or **parallelism**) (Figure 1-21).

The combined effects of rotation, revolution, inclination, and polarity result in the seasonal patterns experienced on Earth. Notice in Figure 1-21 that at one point in Earth's orbit, on about June 21, the North Pole is oriented most directly toward the Sun, while six months later, on about December 21, the North Pole is oriented most directly away from the Sun—this is the most fundamental feature of the annual march of the seasons.

The Annual March of the Seasons

During a year, the changing relationship of Earth to the Sun results in variations in day length and in the angle at which the Sun's rays strike the surface of Earth. These changes are

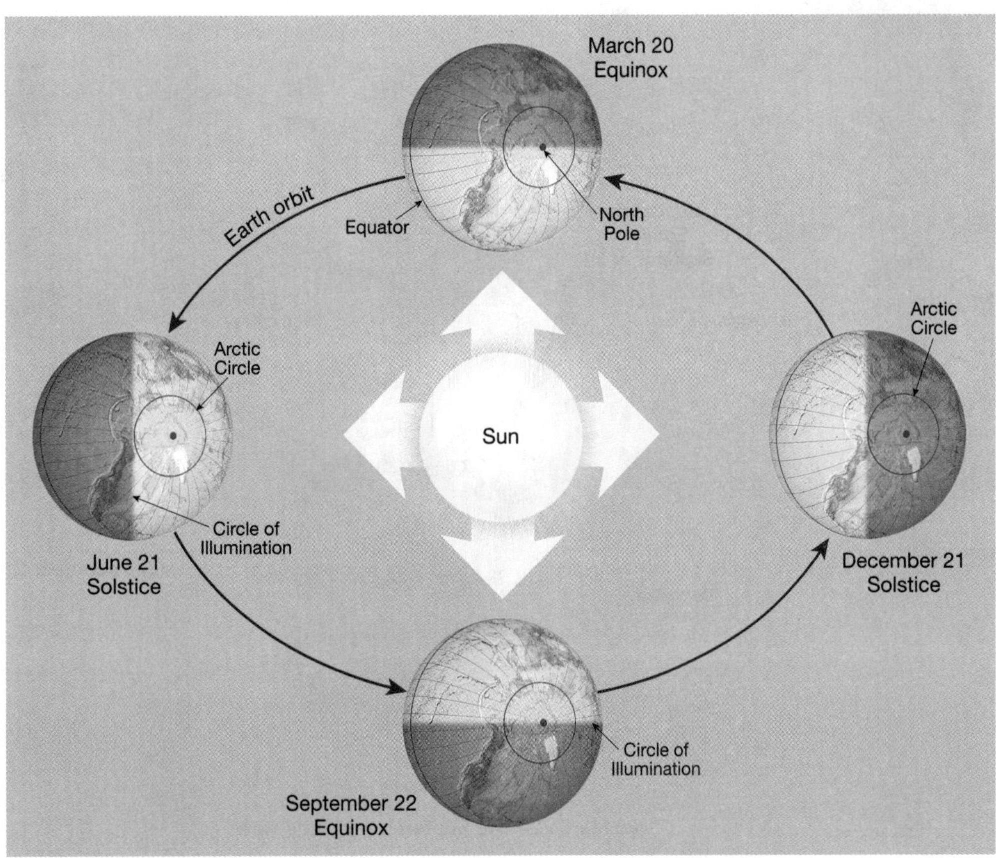

▲ **Figure 1-21** A "top view" of the march of the seasons. Earth's rotational axis maintains polarity (points in the same direction) throughout the year, so on the June solstice the North Pole leans most directly toward the Sun, while on the December solstice the North Pole leans most directly away from the Sun. One-half of Earth is illuminated at all times during the year. The line between the two halves is called the circle of illumination.

most obvious in the mid- and high latitudes, but important variations take place within the tropics as well.

As we discuss the annual march of the seasons, we will pay special attention to three conditions:

1. The latitude receiving the vertical rays of the Sun (also referred to as the *subsolar point* or the *declination of the Sun*).
2. The **solar altitude** (the height of the Sun above the horizon) at different latitudes.
3. The length of day (number of daylight hours) at different latitudes.

Initially, we emphasize the conditions on four special days of the year: The March equinox, the June solstice, the September equinox, and the December solstice (Figure 1-22), although the transitions between these special days are important for us as well. As we describe the change of seasons, the significance of the "seven important parallels" discussed earlier in this chapter will become clear. We begin with the June solstice.

June Solstice On the **June solstice**, which occurs on or about June 21 (the exact date varies slightly from year to year), the Earth reaches the position in its orbit where the North Pole is oriented most directly toward the Sun. On this day, the vertical rays of the Sun at noon are striking the **Tropic of Cancer**, 23.5° north of the equator (Figure 1-22b). Were you at the Tropic of Cancer on this day, the Sun would be directly overhead in the sky at noon (in other words, the solar altitude would be 90°). However, were you a few

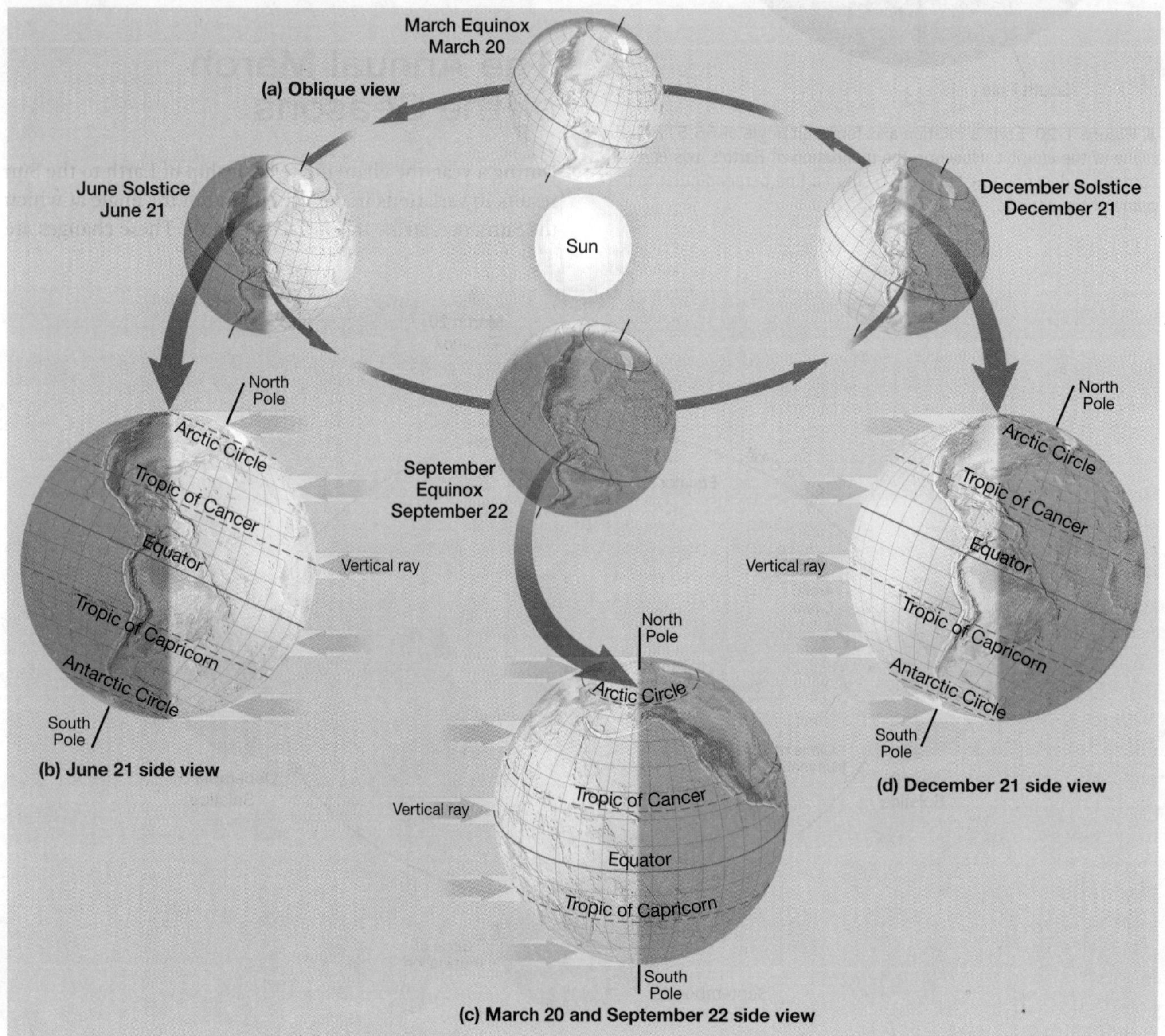

▲ **Figure 1-22** (a) The annual march of the seasons showing Earth–Sun relations on the June solstice, September equinox, December solstice, and March equinox. (b) On the June solstice the vertical rays of the noon Sun strike 23.5° N latitude. (c) On the March equinox and September equinox, the vertical rays of the noon Sun strike the equator. (d) On the December solstice, the vertical rays of the noon Sun strike 23.5° S latitude.

degrees of latitude north or south of the Tropic of Cancer, the noon Sun would be a few degrees lower in the sky—the farther north or south of the Tropic of Cancer you were, the lower in the sky the noon Sun would be. The Tropic of Cancer marks the northernmost location reached by the vertical rays of the Sun during the year.

You will recall that the dividing line between the daylight half of Earth and nighttime half of Earth is known as the **circle of illumination**. The relationship of the circle of illumination to the parallels on any given day of the year determines the day length experienced at different latitudes. On the June solstice, the circle of illumination bisects the equator (Figure 1-22b), so on this day the equator receives equal day and night—12 hours of daylight and 12 hours of darkness. However, as we move north of the equator, the portion of each parallel in daylight increases—in other words, as we move north of the equator, day length increases. Conversely, day length decreases as we move south of the equator.

Notice in Figure 1-22b that on the June solstice, the circle of illumination reaches 23.5° *beyond* the North Pole but stops 23.5° short on the near side of the South Pole. As Earth rotates, all points north of 66.5° N (in other words, within 23.5° of the North Pole) remain continuously within the circle of illumination, thus experiencing 24 continuous hours of daylight. By contrast, all points south of 66.5° S are always outside the circle of illumination and thus have 24 continuous hours of darkness. These special parallels defining the 24 hours of light and dark on the solstice dates are called the *polar circles*. The northern polar circle, at 66.5° N, is the **Arctic Circle**; the southern polar circle, at 66.5° S, is the **Antarctic Circle**.

The June solstice is called the *summer solstice* in the Northern Hemisphere and the *winter solstice* in the Southern Hemisphere (what are commonly called the "first day of summer" and the "first day of winter" in their respective hemispheres).

September Equinox Three months after the June solstice, on approximately September 22 (as with solstice dates, this date also varies slightly from year to year), Earth experiences the **September equinox**. Notice in Figure 1-22c that the vertical rays of the Sun are striking the equator. Notice also that the circle of illumination just touches both poles, bisecting all other parallels—on this day all locations on Earth experience 12 hours of daylight and 12 hours of darkness (the word "equinox" comes from the Latin, meaning "the time of equal days and equal nights"). At the equator—and only at the equator—every day of the year has 12 hours of daylight and 12 hours of darkness; all other locations have equal day and night only on an equinox.

The September equinox is called the *autumnal equinox* in the Northern Hemisphere and the *vernal equinox* in the Southern Hemisphere (and what are commonly called the "first day of fall" and the "first day of spring" in their respective hemispheres).

December Solstice On the **December solstice**, which occurs on or about December 21, the Earth reaches the position in its orbit where the North Pole is oriented most directly away from the Sun; the vertical rays of the Sun now strike 23.5° S, the **Tropic of Capricorn** (Figure 1-22d). Once again, the circle of illumination reaches to the far side of one pole and falls short on the near side of the other pole, but this time the illuminated hemispheres are reversed. Areas north of the Arctic Circle are in continuous darkness, whereas areas south of the Antarctic Circle are in daylight for 24 hours.

Although the latitude receiving the vertical rays of the Sun has shifted 47° from June 21 to December 21, the relationships between Earth and the Sun on the June solstice and the December solstice are very similar—the conditions in each hemisphere are simply reversed. The December solstice is called the *winter solstice* in the Northern Hemisphere and the *summer solstice* in the Southern Hemisphere (what are commonly called the "first day of winter" and the "first day of summer," respectively).

March Equinox Three months after the December solstice, on approximately March 20, Earth experiences the **March equinox**. The relationships of Earth and the Sun are virtually identical on the March equinox and the September equinox (Figure 1-22c). The March equinox is called the *vernal equinox* in the Northern Hemisphere and the autumnal equinox in the Southern Hemisphere (what are commonly called the "first day of spring" and the "first day of fall," respectively).

Table 1-6 summarizes the conditions present during the solstices and equinoxes. The enormous variations in length of daylight and noon Sun angle at different latitudes at the time of a solstice are shown in Table 1-7.

Seasonal Transitions

In the preceding discussion of the solstices and equinoxes, we mainly emphasized the conditions on just four special days of the year. It is important to understand the transitions in day length and Sun angle that take place between those days as well.

Latitude Receiving the Vertical Rays of the Sun The vertical rays of the Sun can only strike Earth between the Tropic of Cancer and the Tropic of Capricorn. Following the March equinox, the vertical rays of the Sun migrate from the equator northward and strike the Tropic of Cancer on the June solstice (although latitudes north of the Tropic of Cancer never experience the vertical rays of the Sun, the June solstice marks the day of the year when the Sun is highest in the sky in those latitudes). After the June solstice, the vertical rays migrate south, striking the equator again on the September equinox and finally to their southernmost latitude on the December solstice (the December solstice marks the day of the year when the Sun is lowest in the sky in the Northern Hemisphere). Following the December solstice, the vertical rays migrate northward, reaching the equator once again on the March equinox.

Day Length Only at the equator is day length constant throughout the year—virtually 12 hours of daylight every day of the year.

TABLE 1-6 Conditions on Equinoxes and Solstices

	March Equinox	June Solstice	September Equinox	December Solstice
Latitude of Vertical Rays of Sun	0°	23.5° N	0°	23.5° S
Day length at Equator	12 hours	12 hours	12 hours	12 hours
Day length in midlatitudes of Northern Hemisphere	12 hours	Day length becomes longer with increasing latitude north of equator.	12 hours	Day length becomes shorter with increasing latitude north of equator.
Day length in midlatitudes of Southern Hemisphere	12 hours	Day length becomes shorter with increasing latitude south of equator.	12 hours	Day length becomes longer with increasing latitude south of equator.
24 hours of daylight	Nowhere	From Arctic Circle to North Pole	Nowhere	From Antarctic Circle to South Pole
24 hours of darkness	Nowhere	From Antarctic Circle to South Pole	Nowhere	From Arctic Circle to North Pole
Season in Northern Hemisphere	Spring	Summer	Autumn	Winter
Season in Southern Hemisphere	Autumn	Winter	Spring	Summer

For all regions in the Northern Hemisphere up to the latitude of the Arctic Circle, following the shortest day of the year on the December solstice, the period of daylight gradually increases, reaching 12 hours of daylight on the March equinox. After the equinox, day length continues to increase until the longest day of the year on the June solstice. (During this period, day length is diminishing in the Southern Hemisphere.)

Following the longest day of the year in the Northern Hemisphere on the June solstice, the pattern is reversed, with the days getting shorter in the Northern Hemisphere—reaching 12 hours on the September equinox, and then diminishing until the shortest day of the year on the December solstice. (During this period, day length is increasing in the Southern Hemisphere.)

Overall, the annual variation in day length is the least in the tropics and greatest in the high latitudes.

Day Length in the Arctic and Antarctic The patterns of day and night in the Arctic and Antarctic deserve special mention. For an observer at the North Pole, the Sun rises on the March equinox and is above the horizon continuously for the next six months—circling the horizon higher and higher each day until the June solstice, after which it circles lower and lower until setting on the September equinox.

Week by week after the March equinox, the region experiencing 24 hours of daylight extends south from the North Pole until the June solstice—when the entire region from the North Pole to the Arctic Circle experiences 24 hours of daylight. Following the June solstice, the region in the Arctic experiencing 24 hours of daylight diminishes week by week until the September equinox—when the Sun sets at the North Pole and remains below the horizon continuously for the next six months.

Week by week following the September equinox, the region experiencing 24 hours of darkness extends south from the North Pole until the December solstice—when the entire region from the North Pole to the Arctic Circle experiences 24 hours of darkness. Following the December solstice, the region experiencing 24 hours of darkness diminishes week by week until the March equinox—when the Sun again rises at the North Pole.

In the Antarctic region of the Southern Hemisphere, these seasonal patterns are simply reversed.

Significance of Seasonal Patterns

Both day length and the angle at which the Sun's rays strike Earth are principal determinants of the amount of solar energy received at any particular latitude. As a generalization, the higher the angle at which the sunlight strikes Earth, the more effective is the resultant heating. Where the vertical rays of the Sun strike Earth, solar energy is concentrated onto the smallest possible surface area. Where the Sun's rays strike Earth at smaller angles, the same amount of energy is spread over a larger surface area—thus, the amount of energy reaching a particular surface area is significantly smaller. Day length influences patterns of solar energy receipt on Earth as well. For example, short periods of daylight in winter and long periods of daylight in summer contribute to seasonal differences in temperature in the mid- and high latitude regions of Earth.

Thus, the tropical latitudes are generally always warm because they always have high Sun angles and consistent day lengths that are close to 12 hours long. Conversely, the

TABLE 1-7 Day Length at Time of June Solstice

Latitude	Day Length	Noon Sun Angle (degrees above horizon)
90° N	24 h	23.5
80° N	24 h	33.5
70° N	24 h	43.5
60° N	18 h 53 min	53.5
50° N	16 h 23 min	63.5
40° N	15 h 01 min	73.5
30° N	14 h 05 min	83.5
20° N	13 h 21 min	86.5
10° N	12 h 43 min	76.5
0°	12 h 07 min	66.5
10° S	11 h 32 min	56.5
20° S	10 h 55 min	46.5
30° S	10 h 12 min	36.5
40° S	09 h 20 min	26.5
50° S	08 h 04 min	16.5
60° S	05 h 52 min	6.5
70° S	0	0
80° S	0	0
90° S	0	0

Source: After Robert J. List, *Smithsonian Meteorological Tables*, 6th rev. ed. Washington, D.C.: Smithsonian Institution, 1963, Table 171.

polar regions are consistently cold because they always have low Sun angles—even the 24-hour days in summer do not compensate for the low angle of incidence of sunlight. Seasonal temperature differences are large in the midlatitudes because of sizable seasonal variations in Sun angles and length of day. This topic will be explored further in Chapter 4.

Telling Time

Malcolm Thomson, a Canadian authority on the physics of time, has noted that there are only three natural units of time: the *tropical year*, marked by the return of the seasons; the *lunar month*, marked by the return of the new moon; and the *day*, marked by passage of the Sun. All other units, such as the hour, minute, and second, are human-made to meet the needs of society.

In prehistoric times, the rising and setting of the Sun were probably the principal means of telling time. As civilizations developed, however, more precise timekeeping was required. Early agricultural civilizations in Egypt, Mesopotamia, India, China, and even England, as well as the Aztec and Mayan civilizations in the New World, observed the Sun and the stars to tell time and keep accurate calendars.

Local *solar noon* can be determined by watching for the moment when objects cast their shortest shadows. The Romans used sundials to tell time (Figure 1-23) and gave great importance to the noon position, which they called the *meridian*—the Sun's highest (*meri*) point of the day (*diem*). Our use of A.M. (*ante meridian*: "before noon") and P.M. (*post meridian*: "after noon") was derived from the Roman world.

When nearly all transportation was by foot, horse, or sailing vessel, it was difficult to compare time at different localities. In those days, each community set its own time by correcting its clocks to high noon at the moment of the shortest shadow. A central public building, such as a temple in India or a county courthouse in Kansas, usually had a large clock or loud bells to toll the hour. Periodically, this time was checked against the shortest shadow.

Standard Time

As the telegraph and railroad began to speed words and passengers between cities, the use of local solar time created increasing problems. A cross-country rail traveler in the United States in the 1870s might have experienced as many as 24 different local time standards between the Atlantic and Pacific coasts. Eventually, the railroads stimulated the development of a standardized time system. At the

▲ **Figure 1-23** A typical sundial. A sundial is the oldest known device for measuring time, having been used in Babylon at least as early as 4000 years ago. The two main parts of any sundial are a horizontal face and a vertical piece of metal—called a *gnomon*—in its center. The upper edge of the gnomon must slant upward from the dial face at an angle equal to the latitude at the place where the sundial is installed. The gnomon points toward the North Pole in the Northern Hemisphere and toward the South Pole in the Southern Hemisphere. When the Sun shines on it, the gnomon casts a shadow that tells the time. As the Sun appears to move across the sky during the course of a day, the position of the shadow changes. The time shown in this photograph of a sundial in Cornwall, United Kingdom, is about 11:00 A.M.

1884 International Prime Meridian Conference in Washington, D.C., countries agreed to divide the world into 24 standard **time zones**, each extending over 15° of longitude. The mean local solar time of the Greenwich (prime) meridian was chosen as the standard for the entire system. The prime meridian became the center of a time zone that extends 7.5° of longitude to the west and 7.5° to the east of the prime meridian. Similarly, the meridians that are multiples of 15°, both east and west of the prime meridian, were set as the central meridians for the 23 other time zones (Figure 1-24).

Although **Greenwich Mean Time (GMT)** is now referred to as **Universal Time Coordinated (UTC)**, the prime meridian is still the reference for standard time. Since it is always the same number of minutes after the hour in all standard time zones (keeping in mind that a few countries, such as India, do not adhere to standard one-hour-interval time zones), we usually need to know only the correct time at Greenwich and the number of hours that our local time zone is later or earlier than the Greenwich meridian to know exact local time. Figure 1-24 shows the number of hours later or earlier than UTC it is in each time zone of the world.

Most of the countries of the world are sufficiently small in their east–west direction so as to lie totally within a single time zone. However, large countries may encompass several zones: Russia occupies nine time zones; including Alaska and Hawai'i, the United States spreads over six (Figure 1-25); Canada, six; and Australia, three. In international waters, time zones are defined to be exactly 7°30′ to the east and 7°30′ to the west of the central meridians. Over land areas, however, zone boundaries vary to coincide with appropriate political and economic boundaries. For example, continental Europe from Portugal to Poland shares one time zone, although longitudinally covering about 30°. At the extreme, China extends across four 15° zones, but the entire nation, at least officially, observes the time of the 120th east meridian, which is the one closest to Beijing.

In each time zone, there is a *controlling meridian* along which clock time is the same as mean Sun time (i.e., the Sun reaches its highest point in the sky at 12:00 noon). On either side of that meridian, of course, clock time does not coincide with Sun time. The deviation between the two is shown for one U.S. zone in Figure 1-26.

From the map of time zones of the United States (Figure 1-25), we can recognize a great deal of manipulation of the time zone boundaries for economic and political convenience. For example, the Central Standard Time Zone, centered on 90°W, extends all the way to 105°W

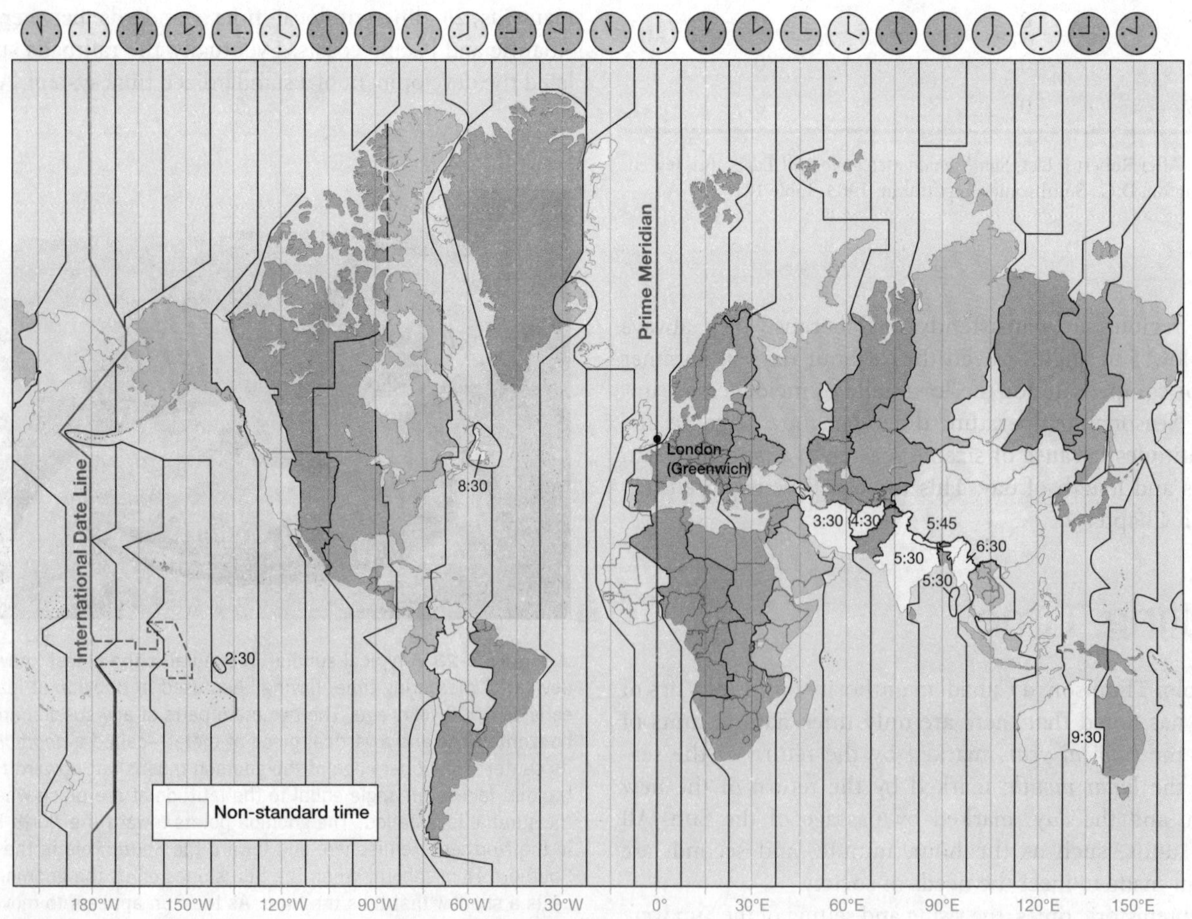

▲ **Figure 1-24** The 24 time zones of the world, each based on meridians spaced 15° apart. Especially over land areas, these boundaries have been significantly adjusted.

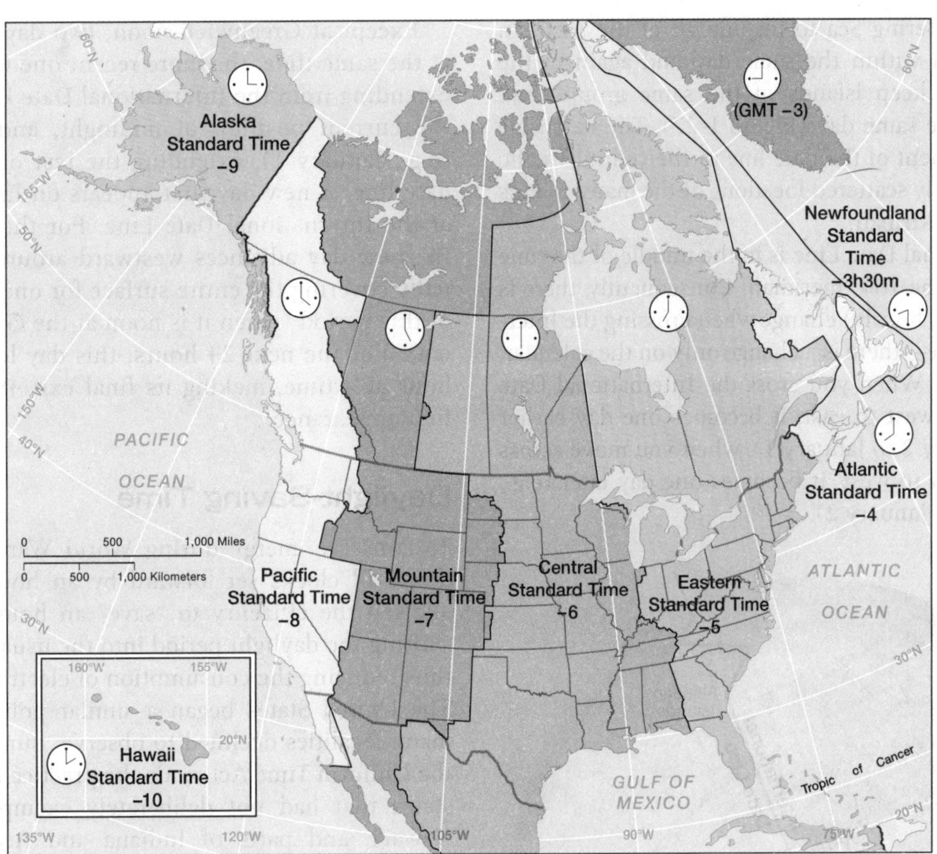

▲ **Figure 1-25** Times zones for Canada, the United States, and northern Mexico. The number in each time zone refers to the number of hours earlier than UTC (GMT).

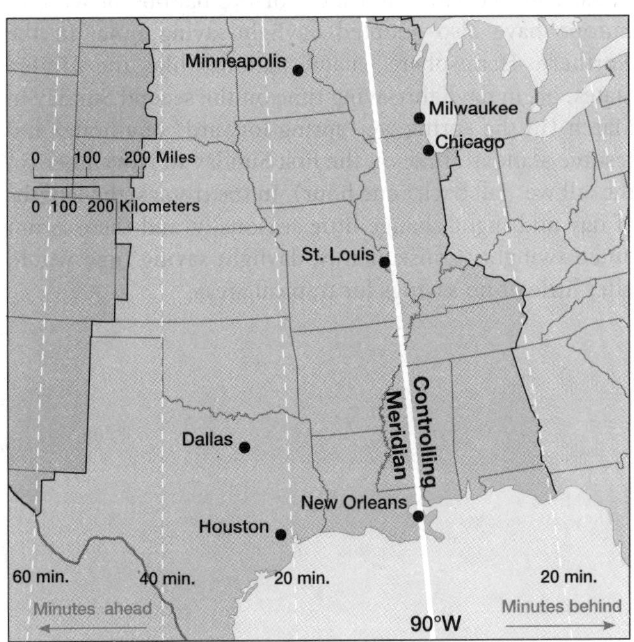

▲ **Figure 1-26** Standard clock time versus Sun time. The Sun reaches its highest point in the sky at 12:00 noon in St. Louis and New Orleans because these two cities lie on the controlling meridian. For places not on the controlling meridian, the Sun is highest in the sky at some time other than standard time noon. In Chicago, for instance, the Sun is highest in the sky at 11:50 A.M. and in Dallas it is highest in the sky at 12:28 P.M.

(which is the central meridian of the Mountain Standard Time Zone) in Texas to keep most of that state within the same zone. By contrast, El Paso, Texas, is officially within the Mountain Standard Time Zone in accord with its role as a major market center for southern New Mexico, which observes Mountain Standard Time. In the same vein, northwestern Indiana is in the Central Standard Time Zone with Chicago.

International Date Line

In 1519, Ferdinand Magellan set out westward from Spain, sailing for East Asia with 241 men in five ships. Three years later, the remnants of his crew (18 men in one ship) successfully completed the first circumnavigation of the globe. Although a careful log had been kept, the crew found that their calendar was one day short of the correct date. This was the first human experience with time change on a global scale, the realization of which eventually led to the establishment of the **International Date Line**.

One advantage of establishing the Greenwich meridian as the prime meridian is that its opposite arc is in the Pacific Ocean. The 180th meridian, transiting the sparsely populated mid-Pacific, was chosen as the meridian at which new days begin and old days exit from the surface of Earth. The International Date Line deviates from the 180th meridian in the Bering Sea to include all of the Aleutian Islands of Alaska within the same day and again in the South Pacific to keep islands of the same group (Fiji, Tonga) within the same day (Figure 1-27). The extensive eastern displacement of the date line in the central Pacific is due to the widely scattered locations of the many islands of the country of Kiribati.

The International Date Line is in the middle of the time zone defined by the 180° meridian. Consequently, there is no time zone (i.e., hourly) change when crossing the International Date Line—there is a change only on the calendar, not on the clock. When you cross the International Date Line going from west to east, it becomes one day earlier (e.g., from January 2 to January 1); when you move across the line from east to west, it becomes one day later (e.g., from January 1 to January 2).

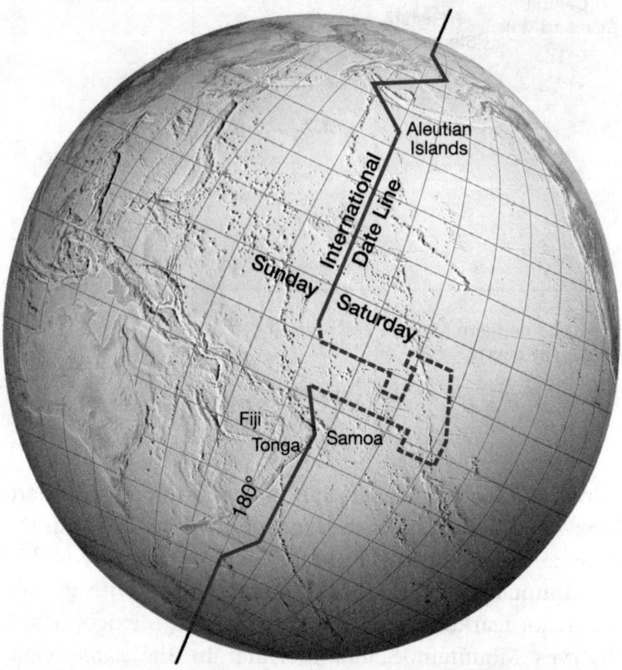

▲ **Figure 1-27** The International Date Line generally follows the 180th meridian, but it deviates around various island groups, most notably Kiribati.

Except at Greenwich noon, two days exist on Earth at the same time: the more recent one (e.g., January 2), extending from the International Date Line westward to the current position of midnight, and the older one (e.g., January 1), extending the rest of the way to the date line. A new day first appears on Earth at midnight at the International Date Line. For the next 24 hours, the new day advances westward around the world, finally covering the entire surface for one hour at the end of this period (when it is noon at the Greenwich meridian). For the next 24 hours, this day leaves Earth, one hour at a time, making its final exit 48 hours after its first appearance.

Daylight-Saving Time

To conserve energy during World War I, Germany ordered all clocks set forward by an hour. This practice allowed the citizenry to "save" an hour of daylight by shifting the daylight period into the usual evening hours, thus reducing the consumption of electricity for lighting. The United States began a similar policy in 1918, but many localities declined to observe "summer time" until the Uniform Time Act made the practice mandatory in all states that had not deliberately exempted themselves. Hawai'i, and parts of Indiana and Arizona, have exempted themselves from observance of **daylight-saving time** under this act.

Russia has adopted permanent daylight-saving time (and double daylight-saving time—two hours ahead of Sun time—in the summer). In recent years, Canada, Australia, New Zealand, and most of the nations of western Europe have also adopted daylight-saving time. In the Northern Hemisphere, many nations, like the United States, begin daylight-saving time on the second Sunday in March (in the spring we "spring forward" one hour) and resume standard time on the first Sunday in November (in the fall we "fall back" one hour). In the tropics, the lengths of day and night change little seasonally, and there is not much twilight. Consequently, daylight-saving time would offer little or no savings for tropical areas.

CHAPTER 1 LEARNING REVIEW

After studying this chapter, you should be able to answer the following questions. Key terms from each text section are shown in bold type. Definitions for key terms are also found in the glossary at the back of the book

Geography as a Field of Learning (p. 3)

1. Contrast **physical geography** and **cultural geography**.

Science and Geography (p. 4)

2. If an idea cannot be disproven by some possible observation or test, can such an idea be supported by science? Explain.
3. Using formulas found in Table 1-2, convert the following from **International System (S.I.)** measurement units to English units:
   - 12 centimeters = _______ inches
   - 140 kilometers = _______ miles

The Environmental Spheres (p. 6)

4. Briefly describe the environmental "spheres": **atmosphere**, **hydrosphere**, **cryosphere**, **biosphere**, and **lithosphere**.

The Solar System (p. 7)

5. In what ways do the inner and outer planets (the terrestrial and Jovian planets) of our solar system differ from each other?

The Size and Shape of Earth (p. 9)

6. Compare the size of Earth to that of its surface features and atmosphere.
7. Is Earth perfectly spherical? Explain.

The Geographic Grid—Latitude and Longitude (p. 10)

8. Define the following terms: **latitude**, **longitude**, **parallel**, **equator**, **North Pole**, **South Pole**, **Tropic of Cancer**, **Tropic of Capricorn**, **Arctic Circle**, **Antarctic Circle**, **meridian**, **prime meridian**.
9. What is a **great circle**? A small circle? Provide examples of both.
10. Latitude ranges from _______° to _______° north and south, while longitude ranges from _______° to _______° east and west.
11. Using a world map or globe, estimate the latitude and longitude of both New York City and Sydney, Australia. Be sure to specify if these locations are north or south latitude, and east or west longitude.

Earth–Sun Relations (p. 14)

12. Describe and explain the four factors in Earth–Sun relations associated with the change of seasons: **rotation**, **revolution** around the Sun, **inclination of Earth's axis**, and **polarity (parallelism) of Earth's axis**.
13. Does the **plane of the ecliptic** coincide with the plane of the equator? Explain.
14. On which day of the year is Earth closest to the Sun **(perihelion)**? Farthest from the Sun **(aphelion)**?

The Annual March of the Seasons (p. 17)

15. Provide the approximate date for the following special days of the year: **March equinox**, **June solstice**, **September equinox**, and **December solstice**.
16. What is the **circle of illumination**?
17. Beginning with the March equinox, describe the changing latitude of the vertical rays of the Sun throughout the year.
18. For the equator, describe the relative angle of the noon Sun above the horizon (the **solar altitude**) and the approximate number of daylight hours on the following days of the year: March equinox, June solstice, September equinox, and December solstice.
19. For the midlatitudes of the Northern Hemisphere, describe the relative angle of the noon Sun above the horizon and the approximate number of daylight hours on the following days of the year: March equinox, June solstice, September equinox, and December solstice.
20. For the North Pole, describe the relative angle of the noon Sun above the horizon and the approximate number of daylight hours on the following days of the year: March equinox, June solstice, September equinox, and December solstice.
21. In terms of the change of seasons, explain the significance of the Tropic of Cancer, the Tropic of Capricorn, the Arctic Circle, and the Antarctic Circle.
22. What is the longest day of the year (the day with the greatest number of daylight hours) in the midlatitudes of the Northern Hemisphere? What is the longest day of the year in the midlatitudes of the Southern Hemisphere?
23. What would be the effect on the annual march of the seasons if Earth's axis was not inclined relative to the plane of the ecliptic?

Telling Time (p. 21)

24. What happens to the hour when crossing a **time zone** boundary moving from west to east?
25. What happens to the day when crossing the **International Date Line** moving from east to west?
26. Using the map of North American time zones (Figure 1-25) for reference, if it is 5:00 P.M. standard time on Thursday in New York City (41°N, 74°W), what is the day and time in Los Angeles (34°N, 118°W)?
27. Using the map of world time zones (Figure 1-24) for reference, if it is 11:00 A.M. **UTC (Universal Time Coordinated or Greenwich Mean Time)**, what is the standard time in Seattle (48°N, 122°W)?
28. When **daylight-saving time** begins in the spring, you would adjust your clock from 2:00 A.M. to _______.

Log in to **www.mygeoscienceplace.com** for animations, videos, RSS feeds, maps, and self-study quizzes to enhance your study of **Introduction to Earth.**

2 Portraying Earth

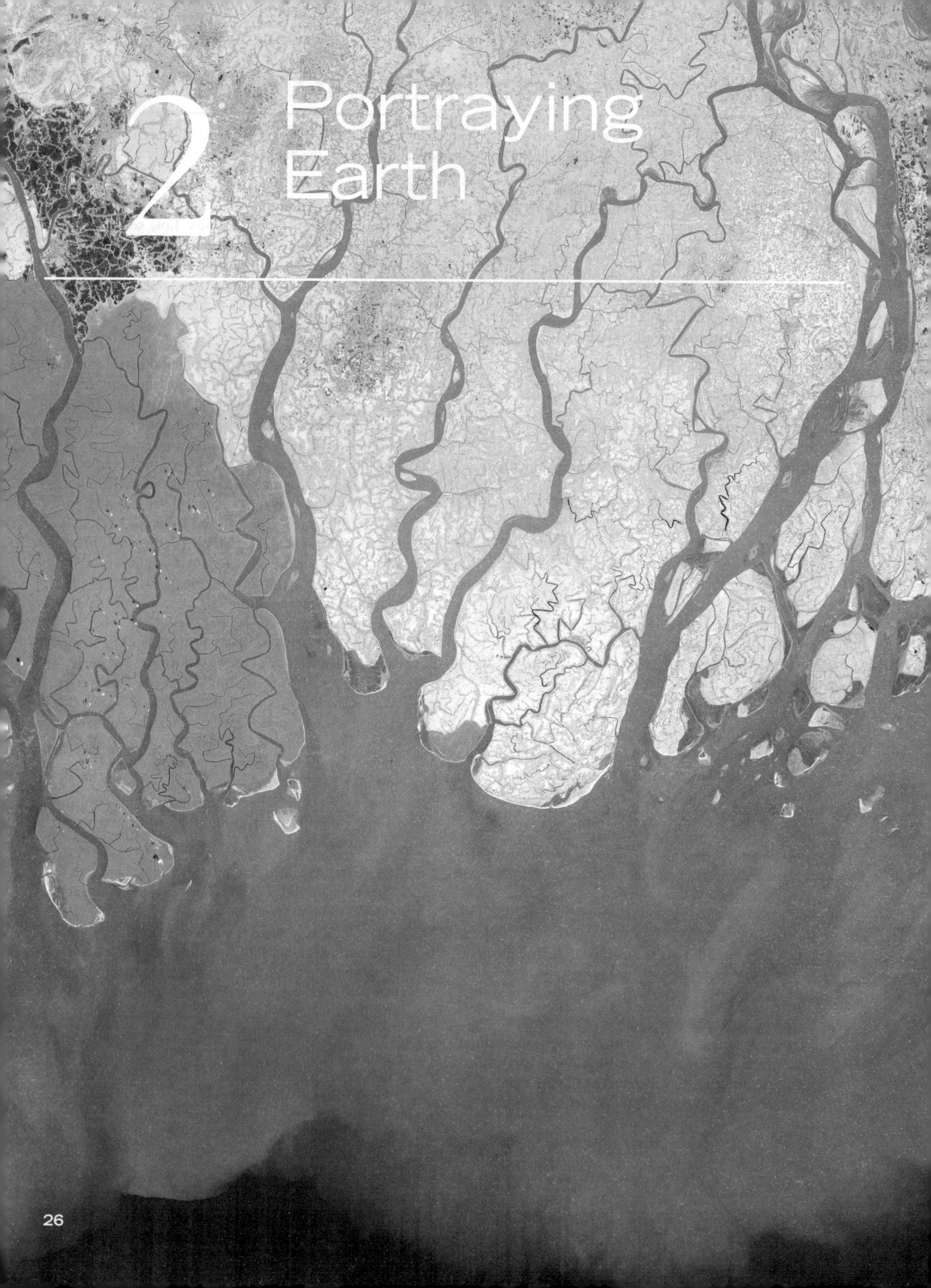

The surface of Earth is the focus of the geographer's interest. The enormity and complexity of this surface would be difficult to comprehend and analyze without tools and equipment to aid in systematizing and organizing the varied data. Although many kinds of tools are used in geographic studies, the most important and universal are maps because the mapping of any geographic feature is normally essential to understanding the spatial distributions and relationships of that feature.

In some cases, geographers deal with maps as an end in themselves, but more often than not maps serve geographers as a means to some end. This book is a case in point. It contains numerous maps of various kinds, each inserted in the book to further your understanding of some fact, concept, or relationship. In each case, the map serves to enhance learning.

The purpose of this chapter is twofold: (1) To describe the basic characteristics of maps, including their capabilities and limitations; (2) To describe the various ways a landscape can be portrayed—through map projections, globes, photographs, and remotely sensed imagery (Figure 2-1).

▲ **Figure 2-1** Earth is indeed a sphere. This composite satellite image was created using data from the MODIS (Moderate Resolution Imaging Spectroradiometer) instrument on NASA's Terra satellite, along with data from other satellites.

The Nature of Maps

A **map** is a two-dimensional representation of Earth and the spatial distribution of selected phenomena. In essence, a map is a scaled drawing of a portion of a landscape, representing the area at a reduced scale and showing only selected data. A map serves as a surrogate (a substitute) for any surface we wish to portray or study. Although any surface can be mapped—the lunar surface, for instance, or that of Mars—all the maps we are concerned with in this book portray portions of Earth's surface.

The basic attribute of maps is their ability to show distance, direction, size, and shape in their horizontal (that is to say, two-dimensional) spatial relationships. In addition to these fundamental graphic data, most maps show other kinds of information as well. Maps nearly always have a special purpose, and that purpose is usually to show the distribution of one or more phenomena (Figure 2-2). Thus, a map may be designed to show street patterns, the distribution of Tasmanians, the ratio of sunshine to cloud, the number of earthworms per cubic meter of soil, or any of an infinite number of other facts or combinations of facts. Because they depict graphically "what is where" and because they are often helpful in providing clues as to "why" such a distribution occurs, maps are indispensable tools for geographers.

Even so, it is important to realize that maps have their limitations. First of all, no map is perfectly accurate. Although most people understand that not everything we might read in a book, in a newspaper, or on the Internet is necessarily valid (thus the somewhat cynical adage, "Don't believe everything you read"), these same people may uncritically accept all information portrayed on a map as being true. Most people have insufficient experience with maps to view them with the same critical eye as they have for the written word. Nevertheless, the inaccuracy of maps is ubiquitous simply because it is impossible to portray the curved surface of Earth on a flat piece of paper without distortion. The extent to which this geometric impossibility becomes a problem when interpreting a map depends especially on two related variables: how much of Earth is being shown on a map (for example, these distortions are always significant on a world map, but less significant in a map showing a very limited region of Earth), and the *scale* of the map—the topic we turn to next.

◀ Landsat 7 satellite image of the Ganges River Delta in the Bay of Bengal.

Map Scale

A map is always smaller than the portion of Earth's surface it represents, and so any user who is to understand the areal relationships (distances or relative sizes, for example) depicted on that map must know how to use a **map scale**. The scale of a map gives the relationship between length measured on the map and the corresponding actual distance on the ground. Knowing the scale of a map makes it possible to measure distance, determine area, and compare sizes.

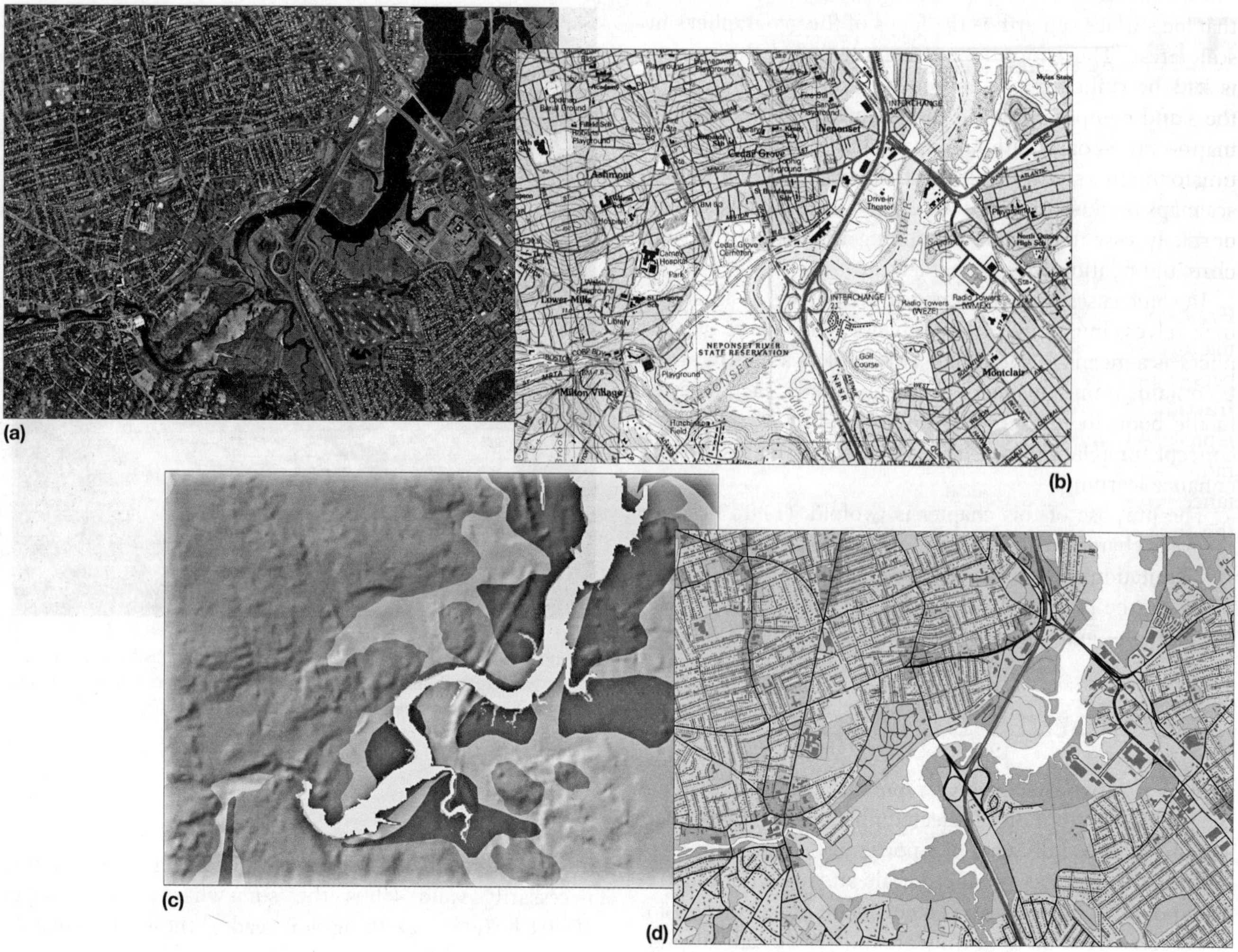

▲ **Figure 2-2** A sampler of maps of the Boston-Quincy-Milton area in Massachusetts at an original scale of 1:25,000. (a) Orthophoto map derived from aerial photographs. (b) Topographic map with elevation contour lines. (c) Geologic map showing rock types. (d) Road map.

Scale can never be represented with perfect accuracy on a map because of the impossibility of rendering the curve of a sphere on the flatness of a sheet of paper without distortion. Therefore, a map scale cannot be constant (in other words, the same) over the entire map. If the map is of a small area, the scale is so nearly perfect that it can be accepted uncritically throughout the map. If the map represents either a large portion or Earth's entire surface, however, there may be enormous scale variations because of the significant distortions involved (such a map, for example, might need to list different scales for different latitudes). Thus, it is important for us to understand the capabilities and recognize the limitations of different kinds of maps and maps at different scales.

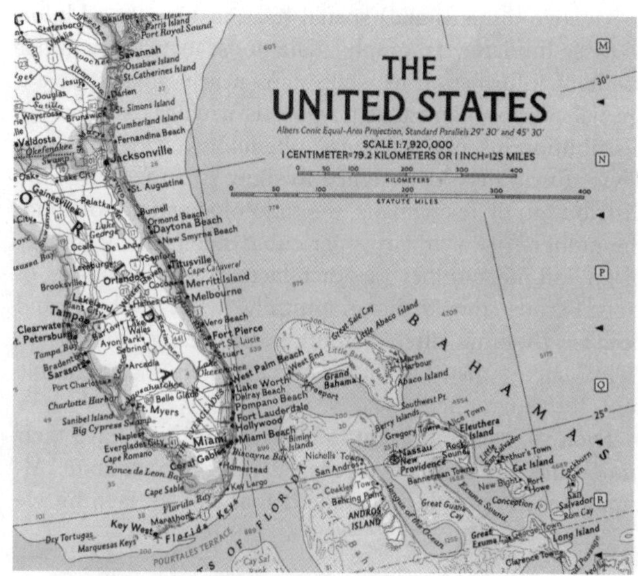

▲ **Figure 2-3** All three expressions of scale are shown on this map. Included are a fractional scale (1:7,920,000), a verbal scale (1 centimeter = 79.2 kilometers or 1 inch = 125 miles), and a graphic scale (shown in both miles and kilometers).

Scale Types

There are several ways to portray scale on a map, but only three are widely used: the graphic scale, the fractional scale, and the verbal scale (Figure 2-3).

Graphic Map Scales A **graphic scale** uses a line marked off in graduated distances. To determine the distance between two points on Earth's surface, one measures the distance between the points on the map and compares that length with a measurement along the line of the graphic scale belonging to the map. The advantage of a graphic scale is its simplicity: you determine approximate distances on the surface of Earth by measuring them directly on the map—graphic scales are commonly used by motorists to estimate travel distances on a road map. Moreover, a graphic scale remains correct when a map is reproduced in a larger or smaller size because the length of the graphic scale line is changed precisely as the map size is changed.

Fractional Map Scales A **fractional scale** compares map distance with ground distance by proportional numbers expressed as a fraction or ratio called a **representative fraction**. For example, a common fractional scale uses the representative fraction 1/63,360, usually expressed as the ratio 1:63,360. This notation means that one unit of distance on the map represents an actual distance of 63,360 of the same units on the ground. Thus, 1 millimeter on the map represents an actual distance of 63,360 millimeters on the ground, while 1 inch on the map represents an actual distance of 63,360 inches on the ground, and so forth.

Verbal Map Scales A **verbal scale** (or *word scale*) states in words the ratio of the map scale length to the actual distance on Earth's surface, such as "five centimeters to ten kilometers" or "one inch to five miles." A verbal scale is simply a mathematical manipulation of the fractional scale. For example: There are 63,360 inches in 1 mile, so on a map with a fractional scale of 1:63,360, we can say that "1 inch represents 1 mile."

Large and Small Map Scales

The adjectives *large* and *small* are comparative rather than absolute. In other words, scales are "large" or "small" only in comparison with other scales (Figure 2-4). A **large-scale**

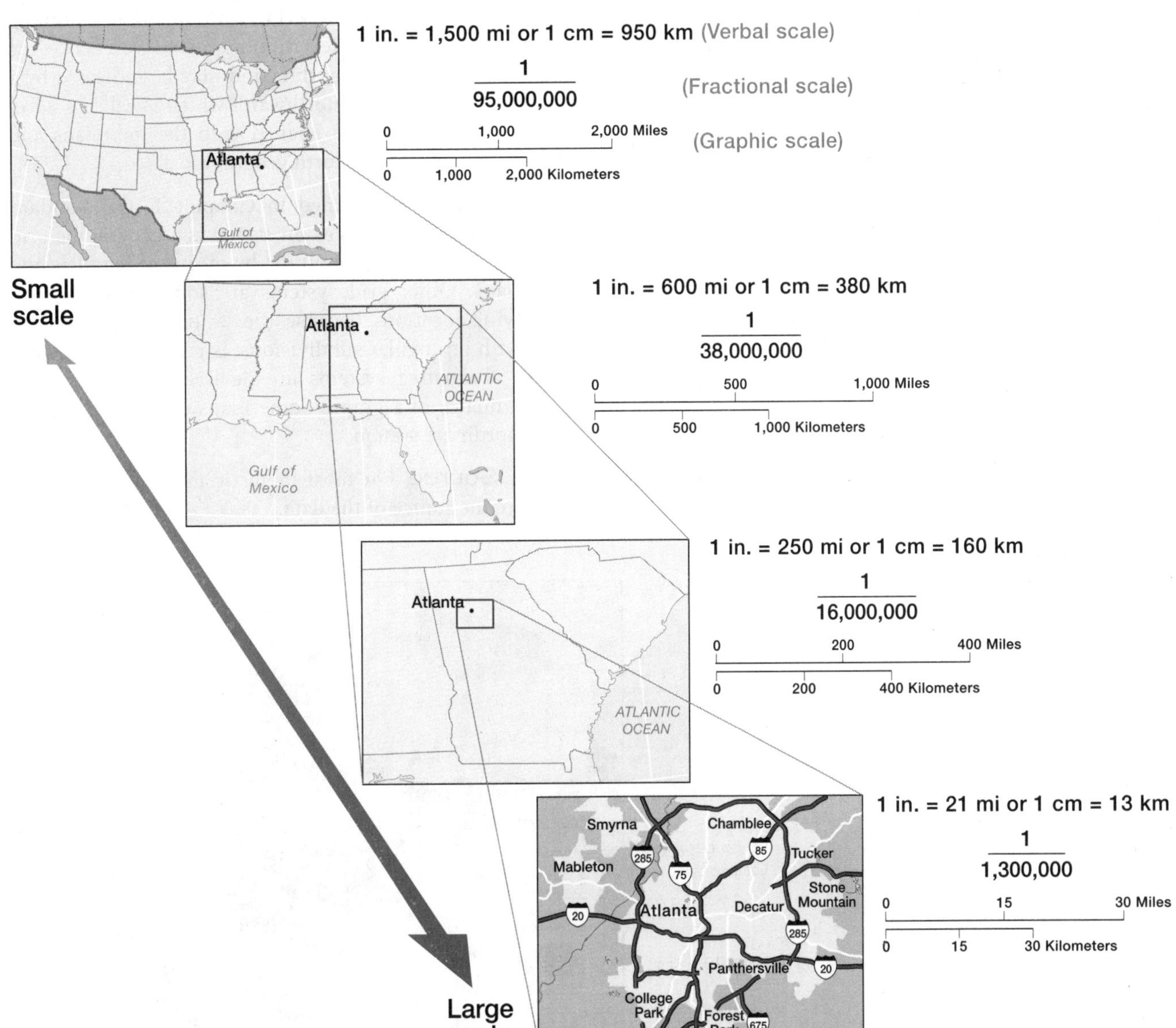

▲ **Figure 2-4** Comparisons of distance and area at various map scales. A small-scale map portrays a large part of Earth's surface but depicts only the most important features, whereas a large-scale map shows only a small part of the surface but in considerably more detail.

map is one that has a relatively large representative fraction, which means that the denominator is small. Thus, 1/10,000 is a larger value than, say, 1/1,000,000, and so a scale of 1:10,000 is large in comparison with one of 1:1,000,000; consequently, a map at a scale of 1:10,000 is called a large-scale map. Such a map portrays only a small portion of Earth's surface but portrays it in considerable detail. For example, if this page were covered with a map having a scale of 1:10,000, it would be able to show just a small part of a single city, but that part would be rendered in great detail.

A **small-scale map** has a small representative fraction—in other words, one having a large denominator. A map having a scale of 1:10,000,000 is classified as a small-scale map. If it were covered with a map of that scale, this page would be able to portray about one-third of the United States, but only in limited detail.

Map Essentials

Maps come in an infinite variety of sizes and styles and serve a limitless diversity of purposes. Some are general reference maps—a map of the world, say, or one of the western coast of Africa. Others are *thematic maps*, which means that they show the location or distribution of particular phenomena, as shown in Figure 2-5. Regardless of type, however, every map must contain a few basic components to facilitate their use. Omission of any of these essential components decreases the clarity of the map and makes it more difficult to interpret.

Title This should be a brief summary of the map's content or purpose. It should identify the area covered and provide some indication of content, such as "Road Map of Kenya," "River Discharge in Northern Europe," or "Seattle: Shopping Centers and Transit Lines."

Date This should indicate the time span over which the information was collected. In addition, some maps also give the date of publication of the map. Most maps depict conditions or patterns that are temporary or even momentary. For a map to be meaningful, therefore, the reader must be informed when the data were gathered, as this information indicates how timely or out of date the map is.

Legend Most maps use symbols, colors, shadings, or other devices to represent features or the amount, degree, or proportion of some quantity. Some symbols are self-explanatory, but it is usually necessary to include a legend box in a corner of the map to explain the symbolization.

Scale Any map that serves as more than a pictogram must be drawn to scale, at least approximately. A graphic, verbal, or fractional scale is therefore necessary.

Direction Direction is normally shown on a map by means of the geographic grid, with meridians running north to south and parallels running east to west across the map. If no grid is shown, direction may be indicated by a straight arrow pointing northward, which is called a *north arrow*. A north arrow is aligned with the meridians and thus points toward the north geographic pole.

Location As we learned in Chapter 1, the standard system for locating places on a map is a geographic grid showing latitude and longitude by means of parallels and meridians. Other grid systems are sometimes used for specifying locations because the latitude/longitude system, with its angular subdivisions, is cumbersome to use. These alternative systems are devised like the *x*- and *y*-coordinates of a graph. Some maps display more than one coordinate system.

Data Source For most thematic maps, it is useful to indicate the source of the data.

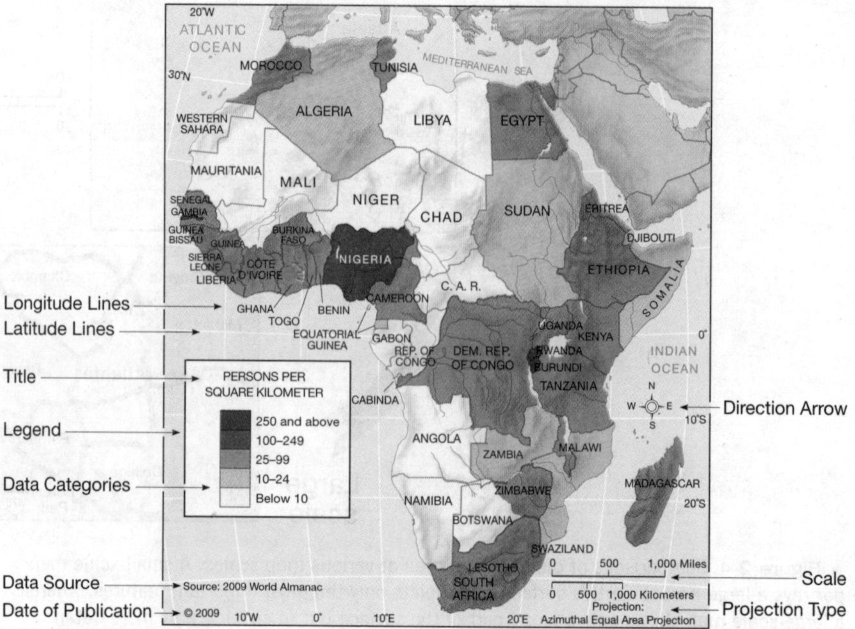

▶ **Figure 2-5** A typical thematic map containing all the essentials.

Projection Type On many maps, particularly small-scale ones, the type of map projection is indicated (map projections are discussed later in this chapter).

The Role of Globes

For properly portraying the sphericity of Earth, there is no substitute for a model globe (Figure 2-6). If manufactured carefully, a globe can be an accurate representation of the shape of our planet. The only thing changed in the transition from the immensity of Earth to the manageable proportions of a model globe is size. A globe is capable of maintaining the correct geometric relationships of meridian to parallel, of equator to pole, of continents to oceans. It can show comparative distances, comparative sizes, and accurate directions. It can represent, essentially without distortion, the spatial relationships of the features of Earth's surface.

A globe is not without disadvantages, however. For one thing, only half of it can be viewed at one time, and the periphery of the visible half is not easy to see. Moreover, almost any globe must be constructed at a very small scale, which means that it is incapable of portraying much detail. The principal problem with a globe, then, is that it is too cumbersome for almost any use other than classroom study or quiet contemplation. Because maps are much more portable and versatile than globes, there are literally billions of maps in use over the world, whereas globes are extremely limited both in number and variety.

▲ **Figure 2-6** A model globe provides a splendid broad representation of Earth at a very small scale, but no details can be portrayed.

Map Projections

The challenge to the *cartographer* (mapmaker) is to try to combine the geometric exactness of a globe with the convenience of a flat map. This melding has been attempted for many centuries, and further refinements continue to be made. The fundamental problem is always the same: to transfer data from a spherical surface to a flat piece of paper with a minimum of distortion.

A **map projection** is a system whereby the spherical surface of Earth is transformed to display it on a flat surface. The basic principle of map projection is direct and simple. Imagine a transparent globe on which are drawn meridians, parallels, and continental boundaries; also imagine a light bulb in the center of this globe. A piece of paper, either held flat or rolled into some shape such as a cylinder or cone, is placed over the globe as in Figure 2-7, 2-8, and 2-9. When the bulb is lighted, all the lines on the

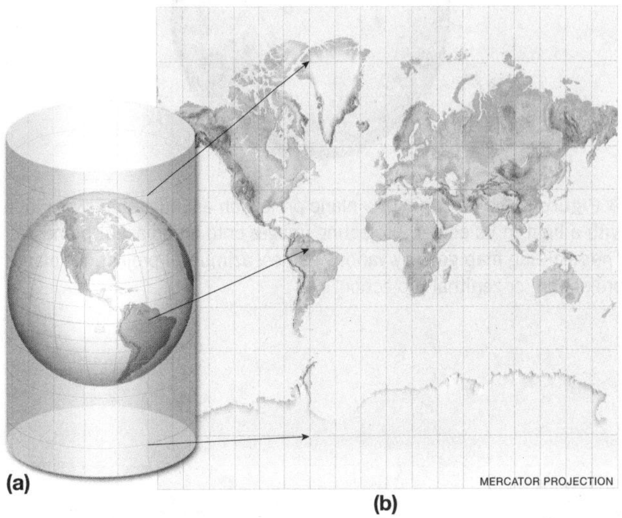

▲ **Figure 2-7** The origin of a cylindrical projection, as illustrated by a globe with a light in its center (a), projecting images onto an adjacent cylinder. (b) The resulting map is called a cylindrical projection.

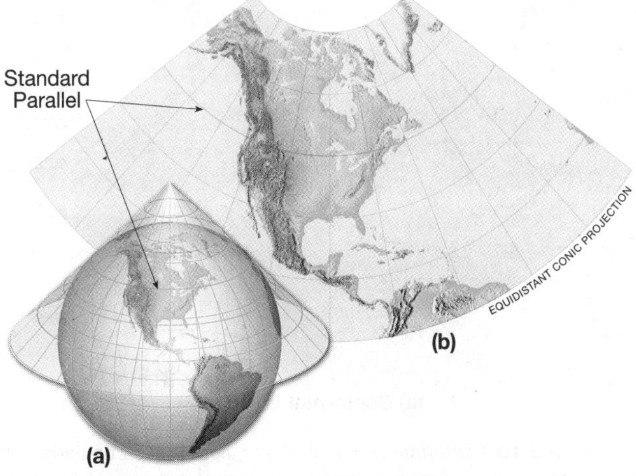

▲ **Figure 2-8** The origin of a conic projection, as illustrated by a globe with a light in its center (a), projecting images onto a cone. (b) The resulting map is called a conic projection.

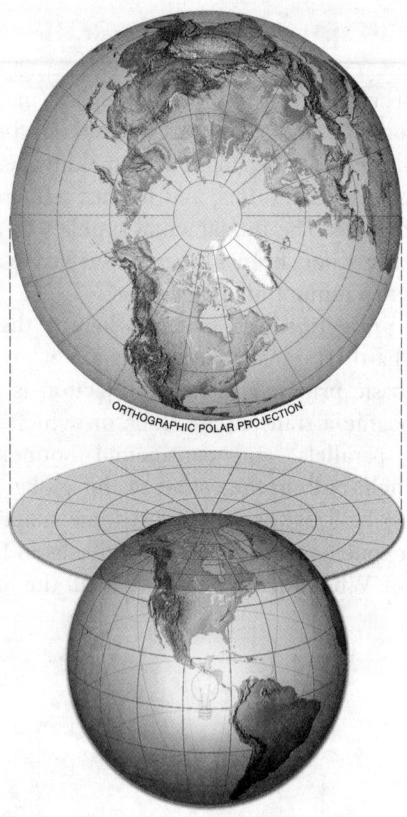

▲ **Figure 2-9** The origin of a plane projection as illustrated by a globe with a light in its center, projecting images onto an adjacent plane. The resulting map goes by various names: azimuthal projection, plane projection, or zenithal projection.

globe are projected outward onto the paper. These lines are then sketched on the paper. When the paper is laid out flat, a map projection has been produced. (Very few map projections have ever been constructed by such an actual optical projection from a globe onto a piece of paper—nearly all are derived by mathematical computation.)

Because a piece of paper cannot be closely fitted to a sphere without wrinkling or tearing, no matter how the transformation is done, data from a globe (parallels, meridians, continental boundaries, and so forth) cannot be transferred to a map without distortion of shape, relative area, distance, and/or direction. However, the cartographer can choose to control or reduce one or more of these distortions—although not all distortions can be controlled on a single map. Cartographers often strive especially to maintain accuracy of either size or shape.

The Major Dilemma: Equivalence versus Conformality

The cartographer wants any map to portray distances and directions accurately so that the sizes and shapes shown on the map are correct. However, such perfection is impossible, and so a compromise must be struck. Which to emphasize: size or shape? Which to sacrifice: shape or size? This is one of the central problems both in constructing and in choosing a map projection. The projection properties involved are called *equivalence* and *conformality* (Figure 2-10).

Equivalence In an **equivalent projection** (also called an **equal area projection**), the size ratio of any area on the map to the corresponding area on the ground is the same all over the map. To illustrate, suppose you have a world

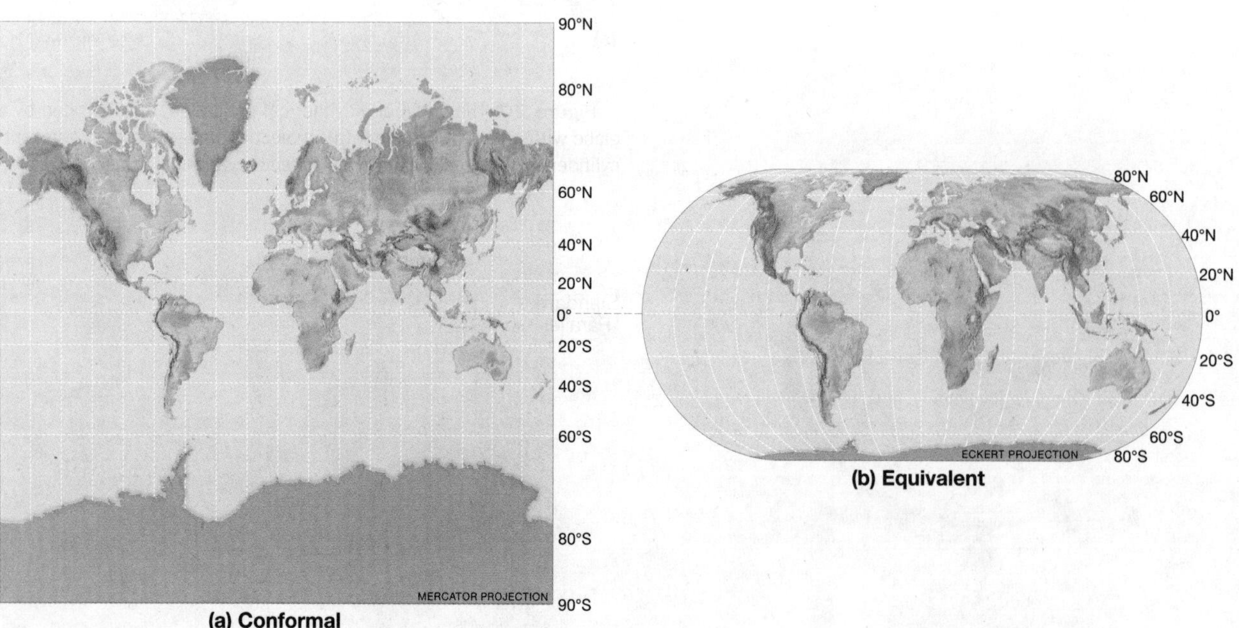

▲ **Figure 2-10** Conformal and equivalent maps. It is particularly difficult to portray the whole world accurately on a map. Compare the sizes and shapes of Antarctica, Alaska, and Greenland in these examples. (a) A conformal projection (the *Mercator*) depicts accurate shapes, but the sizes are severely exaggerated in high latitudes. (b) An equivalent projection (the *Eckert*) is accurate with regard to size, but shapes are badly distorted in high latitudes.

map before you and you place four dimes at different places (perhaps one on Brazil, one on Australia, one on Siberia, and one on South Africa). Calculate the area covered by each coin. If it is the same in all four cases, there is a good chance that the map is an equivalent projection—in other words, that there are equal areal relationships all over it.

Equivalent projections are very desirable because, with them, misleading impressions of size are avoided. (The world maps in this book are mostly equivalent projections because they are so useful in portraying distributions of the various geographic features we will be studying.) They are by no means perfect, however. Equivalence is difficult to achieve on small-scale maps because shapes must be sacrificed to maintain proper areal relationships. Most equivalent world maps, which are small-scale maps, therefore display disfigured shapes. For example, as Figure 2-10b shows, Greenland and Alaska are usually shown as more "squatty" than they actually are.

Conformality A **conformal map projection** is one in which proper angular relationships are maintained so that the shape of something on the map is the same as its shape on Earth. It is impossible to depict true shapes for large areas such as a continent, but they can be approximated, and for small areas the true shape can be shown on a conformal map. All conformal projections have meridians and parallels crossing each other at right angles, just as they do on a globe.

The outstanding problem with conformal projections is that the size of an area must often be considerably distorted to depict the proper shape. Thus, the scale necessarily changes from one region to another. For example, a conformal map of the world normally greatly enlarges sizes in the higher latitudes. Figure 2-10a shows the conformal projection known as a *Mercator projection* (discussed in greater detail later in this chapter).

Except for maps of very small areas (in other words, large-scale maps), where both can be closely approximated, conformality and equivalence cannot be maintained on the same projection, and thus the art of mapmaking, like politics, is an art of compromise. For example, Figure 2-11 shows a *Robinson projection*, which is a compromise between equivalence and conformality that shows reasonably accurate shapes and has a reasonably constant scale, especially in the middle and lower latitudes. The Robinson projection is a popular choice as a general purpose classroom map.

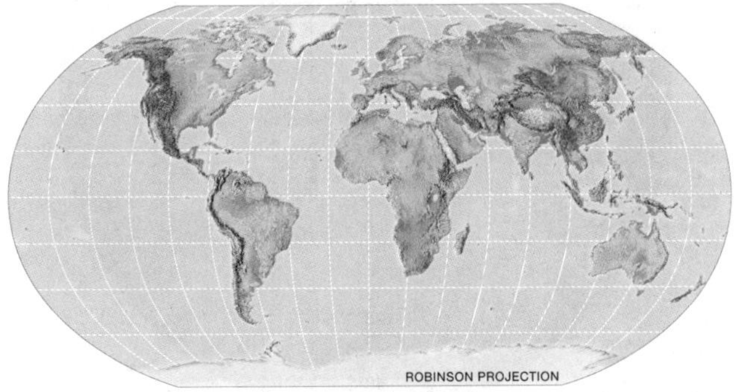

▲ **Figure 2-11** Many world maps are neither purely conformal nor purely equivalent, but a compromise between the two. One of the most popular compromises is the Robinson projection shown here.

As a rule of thumb, it can be stated that some projections are purely conformal, some are purely equivalent, none are both conformal and equivalent, and many are neither purely conformal nor purely equivalent but rather a compromise between the two.

Families of Map Projections

Because there is no possible way to avoid distortion completely, no map projection is perfect. More than a thousand types of map projections have been devised for one purpose or another. Most of them can be grouped into just a few families. Projections in the same family generally have similar properties and related distortion characteristics.

Cylindrical Projections

As Figure 2-7 shows, a **cylindrical projection** is obtained by mathematically "wrapping" the globe with a cylinder of paper in such a way that the paper touches the globe only at the globe's equator. We say that paper positioned this way is *tangent* to the globe at the equator, and the equator is called the *circle of tangency* (some cylindrical projections choose a circle of tangency other than the equator). The curved parallels and meridians of the globe then form a perfectly rectangular grid on the map. Having the equator as the tangency line produces a right-angled grid (meridians and parallels meet at right angles) on a rectangular map. There is no size distortion at the circle of tangency, but size distortion increases progressively with increasing distance from this circle, a characteristic clearly exemplified by the Mercator projection.

Mercator: The Most Famous Projection

Although some map projections were devised centuries ago, there has been a continuing refining of projection techniques right up to the present day. Thus, it is remarkable that the most famous of all map projections, the **Mercator projection**, originated in 1569 by a Flemish geographer and cartographer, is still in common usage today without significant modification (see Figure 2-10a).

Gerhardus Mercator produced some of the best maps and globes of his time. His place in history, however, is based largely on the fact that he developed a special-purpose projection that became inordinately popular for general-purpose use. The Mercator projection is essentially a navigational chart of the world designed to facilitate oceanic navigation.

The prime advantage of a Mercator map is that it shows *loxodromes* as straight lines. A **loxodrome**, also called a **rhumb line**, is a curve on the surface of a sphere that crosses all meridians at the same angle and represents a line of constant compass direction. A navigator first plots the shortest distance between origin and destination on a map projection in which great circles are shown as straight lines, as shown in Figure 2-12 (great circle routes are discussed in Chapter 1), and then transfers that route to a Mercator projection with straight-line loxodromes. This procedure allows the navigator to generally follow a great circle route by simply making periodic changes in the compass course of the airplane or ship. Today, of course, these calculations are all done by computer.

A Mercator map is relatively undistorted in the low latitudes. However, because this is a conformal projection, size distortion increases rapidly in the middle and high latitudes. Because the projection method causes the meridians to appear as parallel lines rather than lines that converge at the poles, there is extreme east–west distortion in the higher latitudes. And, to maintain conformality, Mercator compensated for the east–west stretching by spacing the parallels of latitude increasingly farther apart so that north–south stretching occurs at the same rate. This procedure allowed shapes to be approximated with reasonable accuracy, but at great expense to proper size relationships. Area is distorted by 4 times at the 60th parallel of latitude and by 36 times at the 80th parallel. If the North Pole were shown on a Mercator projection, it would be a line as long as the equator rather than a single point.

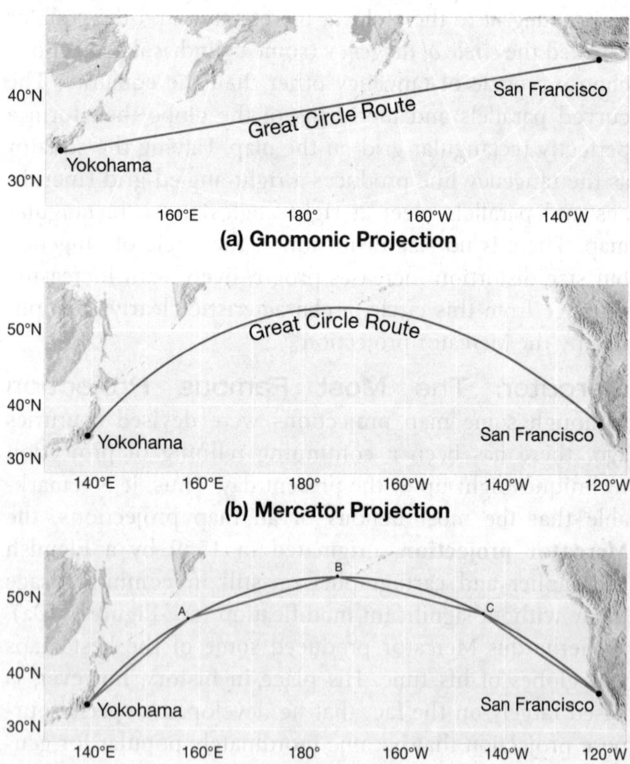

▲ **Figure 2-12** The prime virtue of the Mercator projection lies in is usefulness for straight-line navigation. (a) The shortest distance between two locations—here San Francisco and Yokohama—can be plotted on a *gnomonic* projection (on which great circles are shown as straight lines). (b) The great circle route can be transferred to a Mercator projection. (c) On the Mercator projection, straight-line loxodromes can then be substituted for the curved great circle. The loxodromes allow the navigator to maintain constant compass headings over small distances while still approximating the curve of the great circle.

It is clear then that the Mercator projection is excellent for straight-line navigation but not for most other uses. Despite the obvious flaws associated with areal distortion in the high latitudes, Mercator projections have been widely used in American classrooms and atlases. Indeed, several generations of American students have passed through school with their principal view of the world provided by a Mercator map. This has created many misconceptions, not the least of which is confusion about the relative sizes of high-latitude landmasses. For example, on a Mercator projection, the island of Greenland appears to be as large as or larger than Africa, Australia, and South America. In actuality, however, Africa is 14 times larger than Greenland, South America is 9 times larger, and Australia is 3.5 times larger.

The Mercator projection was devised several centuries ago for a specific purpose, and it still serves that purpose well. Its fame, however, is significantly due to its misuse.

Plane Projections

A **plane projection** (also called with an *azimuthal projection* or a *zenithal projection*) is obtained by projecting the markings of a center-lit globe onto a flat piece of paper that is tangent to the globe at some point. Such a projection is shown in Figure 2-9. The point of tangency can be any spot on the globe, but it is usually either the North or South Pole or some point on the equator. There is no distortion immediately around the point of tangency, but distortion increases progressively away from this point.

No more than one hemisphere can be displayed with any success on a plane projection. These projections may often show a perspective similar to the view one gets when looking at a globe in the classroom and the same view an astronaut sees from distant space (Figure 2-13). This half-view-only characteristic can be a drawback, of course, just as it is with a globe, although plane projections can be useful for focusing attention on a specific region.

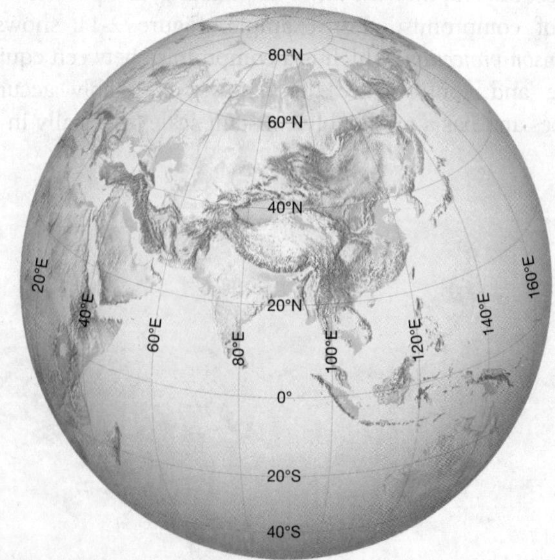

▲ **Figure 2-13** An orthographic plane projection showing Earth as it would appear from space.

Conic Projections

As Figure 2-8 shows, a **conic projection** is obtained by projecting the markings of a center-lit globe onto a cone wrapped tangent to, or intersecting, a portion of the globe. Normally the apex of the cone is positioned above a pole, which means that the circle of tangency coincides with a parallel. This parallel then becomes the *principal parallel* of the projection; distortion is least in its vicinity and increases progressively as one moves away from it. Consequently, conic projections are best suited for regions of east–west orientation in the middle latitudes, being particularly useful for maps of the United States, Europe, or China.

It is impractical to use conic projections for more than one-fourth of Earth's surface (a semihemisphere), and they are particularly well adapted for mapping relatively small areas, such as a state or county.

Pseudocylindrical Projections

A **pseudocylindrical projection** (also called an *elliptical* or *oval projection*) is a roughly football-shaped map, usually of the entire world (see the Eckert in Figure 2-10b and the Robinson in Figure 2-11), although sometimes only the central section of a pseudocylindrical projection is used for maps of lesser areas.

In most pseudocylindrical projections, a central parallel (usually the equator) and a central meridian (often the prime meridian) cross at right angles in the middle of the map, which is a point of no distortion; distortion in size and/or shape normally increases progressively as one moves away from this point in any direction. All of the other parallels are then drawn parallel to the central one. All the other meridians are drawn so that they begin at a common point at the top of the oval and end at a common point at the bottom, mathematically "wrapping around" the sphere of Earth; the result, of course, is that all meridians except the central one must be drawn as curved lines.

Interrupted Projections One technique used with pseudocylindrical projections to minimize distortion of the continents is to "interrupt" oceanic regions—*Goode's interrupted homolosine projection* (Figure 2-14) is a popular example of this. Goode's projection is equivalent, and, although it is impossible for this map to be conformal, the shapes of continental coastlines are very well maintained, even in high latitudes. When global distributions are mapped, the continents are often more important than the oceans, and yet the oceans would occupy most of the map space in a normal projection. Hence, the projection can be interrupted in the Pacific, Atlantic, and Indian oceans and can be based on central meridians that pass through each major landmass. With no land area far from a central meridian, shape and size distortion is greatly decreased. The interruption of the projection in the oceans creates a void in the map, one that is simply filled with information not part of the map. The result is that some of the oceanic portions of the map are torn apart and otherwise distorted, but the major landmasses are shown with relatively little distortion, and the overall accuracy of the distribution pattern is enhanced. (For world maps that emphasize ocean areas, continents can be interrupted instead of ocean basins.) You'll see that many of the maps used in this book employ variations of Goode's interrupted projection.

Isolines

Geographers employ a variety of cartographic devices to display data on maps. One of the most widespread devices for portraying the spatial distribution of some phenomenon is the **isoline** (from the Greek *isos*, "equal"), which is also called by a variety of related terms, such as *isarithm, isogram, isopleth*, and *isometric line*, all of which can be considered as synonymous for our purposes. The word isoline is a generic term that refers to any line that joins points of equal value of something. More than 100 kinds of isolines have been identified by name, ranging from *isoamplitude* (used to describe radio waves) to *isovapor* (water vapor content of the air).

Some isolines represent tangible surfaces, such as the **elevation contour lines** on a topographic map (Figure 2-15). Most, however, signify such intangible features as temperature and precipitation, and some express relative values such as ratios or proportions (Figure 2-16). Only a few types of isolines are important in an introductory physical geography course:

- *Elevation contour line*—a line joining points of equal elevation (see Appendix II for a description of U.S. Geological Survey topographic maps).

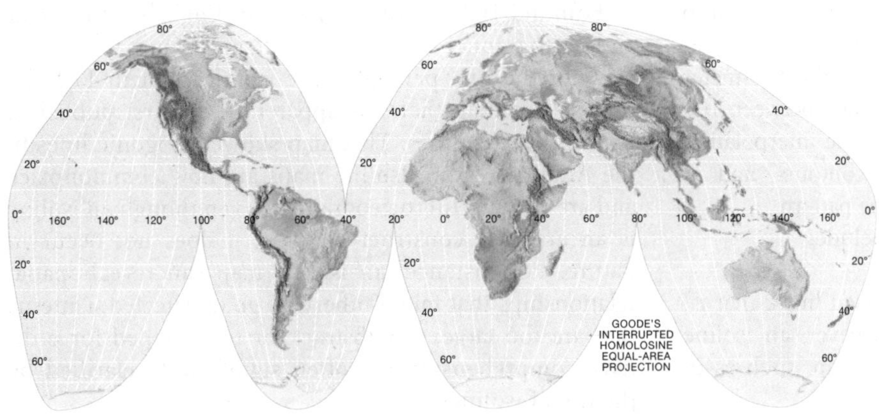

◀ **Figure 2-14** An interrupted projection of the world. The purpose of the interruptions is to portray certain areas (usually continents) more accurately, at the expense of portions of the map (usually oceans) that are not important to the map's theme. The map shown here is a Goode's interrupted homolosine equal-area projection. A variation of this projection is used for many maps in this book.

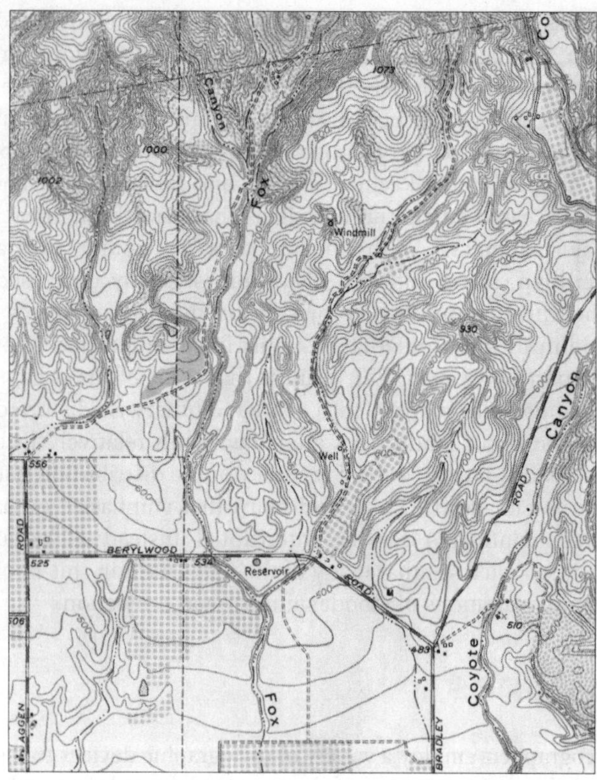

▲ **Figure 2-15** This portion of a typical United States Geological Survey topographic map quadrangle illustrates the use of contour lines. This is a section of the Santa Paula, California, quadrangle. (The original map scale was 1:62,500; the contour interval is 20 feet [7 meters].)

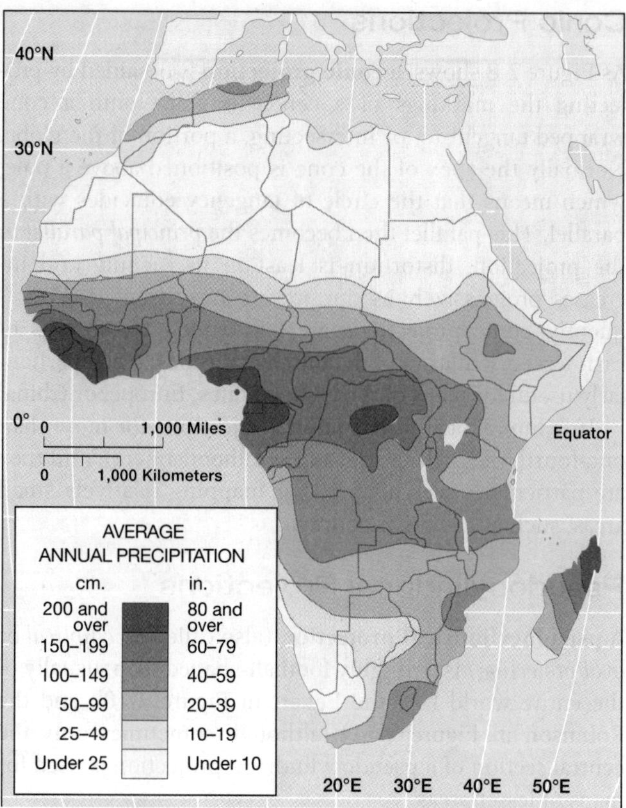

▲ **Figure 2-16** Isolines can be used to show the spatial variation of even intangible features, such as in this map that shows average annual precipitation for the continent of Africa (on this map the areas between isolines have been shaded to clarify the pattern).

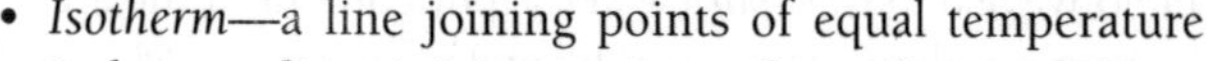

- *Isotherm*—a line joining points of equal temperature
- *Isobar*—a line joining points of equal atmospheric pressure
- *Isohyet*—a line joining points of equal quantities of precipitation (*hyeto* is from the Greek, meaning "rain")
- *Isogonic line*—a line joining points of equal magnetic declination

To construct an isoline, it is necessary to estimate values that are not available. As a simple example, Figure 2-17 illustrates the basic steps in constructing an isoline map—in this case, an elevation contour map. Each dot in Figure 2-17a represents a data collection location, and the number next to each dot is the elevation above sea level in meters. We begin by drawing the 110-meter elevation contour: the 110-meter contour line passes between 109 and 113, and between 106 and 111 (Figure 2-17b). In Figure 2-17c, the interpolation process is repeated for other elevation contours, and in Figure 2-17d shading is added to clarify the pattern.

The basic characteristics of isolines include:

1. Conceptually, isolines are always closed lines; that is, they have no ends. In practice, however, an isoline often extends beyond the edge of a map, such as in Figure 2-17.
2. Because they represent gradations in quantity, isolines can never touch or cross one another except under special circumstances.
3. The numerical difference between one isoline and the next is called the *interval*. Although intervals can be varied according to the wishes of the mapmaker, it is normally more useful to maintain a constant interval all over a given map.
4. Isolines close together indicate a steep gradient (in other words, a rapid change); isolines far apart indicate a gentle gradient.

Edmund Halley (1656–1742), an English astronomer and cartographer (for whom Halley's Comet is named), was not the first person to use isolines, but in 1700 he produced a map that was apparently the first published map to have isolines. This map showed isogonic lines in the Atlantic Ocean. Isoline maps are now commonplace and are very useful to geographers even though an isoline is an artificial construct—that is, it does not occur in nature. For instance, an isoline map can reveal spatial relationships that might otherwise go undetected. Patterns that are too large, too abstract, or too detailed for ordinary comprehension are often significantly clarified by the use of isolines.

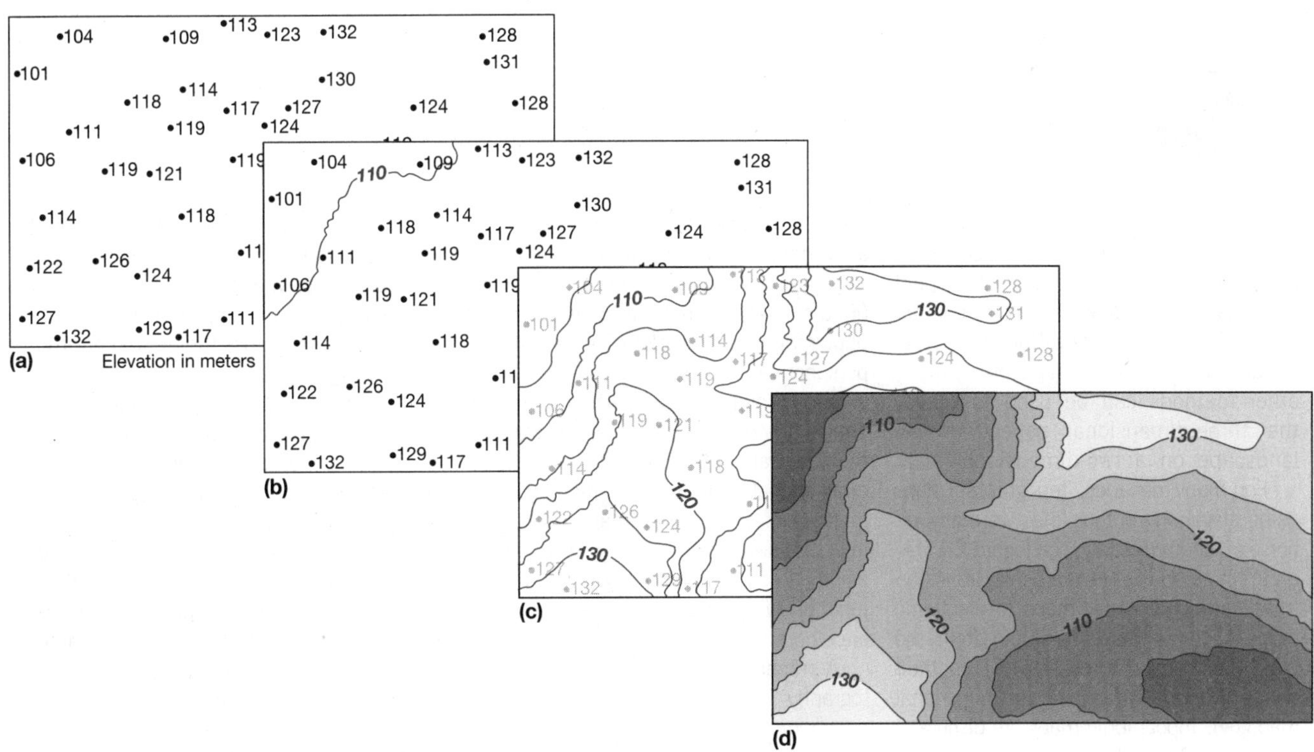

▲ **Figure 2-17** Drawing isolines. (a) Each dot represents a data location point, in this case, elevation above sea level in meters. (b) The location of the 110-meter elevation contour is interpolated and drawn. (c) The other contour lines at 5-meter intervals are interpolated and drawn. (d) Shading is added for clarity.

GPS—The Global Positioning System

The **global positioning system**, or simply **GPS**, is a global navigation satellite system for determining accurate positions on or near Earth's surface. It was developed in the 1970s and 1980s by the U.S. Department of Defense to aid in navigating aircraft, guiding missiles, and controlling ground troops. The first receivers were the size of a file cabinet, but continued technological improvement has reduced them to the size of a paperback book, and they are still shrinking (Figure 2-18). The cost is also diminishing rapidly: at the time of this writing, a good receiver sells for less than $150.

The system (formally called NAVSTAR GPS [Navigation Signal Timing and Ranging Global Positioning System]) is based on a constellation of at least 24 high-altitude satellites configured so that a minimum of four—and preferably six—are in view of any position on Earth (currently there are 30 active satellites, with several older satellites still in orbit as backups). Each satellite continuously transmits both identification and positioning information that can be picked up by receivers on Earth (Figure 2-19). The distance between a given receiver and each member in a group of four or more satellites is calculated by comparing clocks stored in both units, and then the three-dimensional coordinates of the receiver's position are calculated through triangulation. The greater the number of channels in a GPS unit (even inexpensive units now have 12) the greater the number of satellites that can be tracked, and so the better the accuracy. The system already has accuracy greater than that of the best base maps. Even the simplest GPS units determine position to within 15 meters (49 feet).

Wide Area Augmentation System (WAAS)

Increased GPS accuracy is gained when the *Wide Area Augmentation System* (*WAAS*) is employed. Originally developed in cooperation with the Federal Aviation Administration (FAA) and the U.S. Department of Transportation, WAAS was implemented to increase the accuracy of instrument-based flight approaches for airplanes. Several dozen ground-based stations across North America monitor GPS signals from the satellites and then generate a correction message that is transmitted to GPS units. With WAAS, GPS units achieve a position accuracy of 3 meters (about 10 feet) about 95 percent of the time. WAAS capability is built in to virtually all new GPS receivers today. WAAS service is not yet available around the world, although similar systems are being implemented in Asia (Japan's *Multi-Functional Satellite Augmentation System*) and Europe (the *Euro Geostationary Navigation Overlay Service*).

Continuously Operating GPS Reference Stations (CORS) The National Oceanic and Atmospheric Administration (NOAA) manages a system of permanently-installed GPS receiving stations known as

FOCUS

Portraying the Three-Dimensional Landscape

Most maps are simply flat representations of Earth, but especially in physical geography, the vertical aspect of the landscape is an important component of study. In addition to actual raised-relief models of landforms, many other methods can be used to convey the three-dimensional aspect of the landscape on a two-dimensional map.

For many decades, topographic maps using elevation contour lines were a workhorse of landform study (see Figure 2-15)—and remain so today even as we transition from traditional paper maps to electronic maps such as those available from the U.S. Geological Survey (USGS) on their online *National Map* site (http://nationalmap.gov). Topographic maps are discussed in detail in Appendix II.

Stereo aerial photographs have also been used for decades to study the landscape. To create a stereopair (or *stereogram*), vertical aerial photographs are taken at regular intervals, close enough to allow considerable overlap in neighboring photographs—typically about 60 percent (Figure 2-A). When a pair of such photographs—each taken from a different perspective—is properly aligned and viewed simultaneously with a binocular optical instrument called a *stereoscope*, the mental impression of the three-dimensional landscape results. The majority of topographic maps made during the twentieth century were made using stereo aerial photographs in conjunction with ground-based survey reference points.

A remarkable recent advance in cartography has been the use of *digital elevation models* (DEM) to convey topography. The starting point for creating a DEM image is a detailed data base of precise elevations. For example, the USGS maintains such a database for the United States at several different spatial resolutions—a 30-meter grid being one of the most commonly used (meaning that elevation data are available at distance intervals of 30 meters, both north–south and east–west, across the entire country). Similar digital elevation data are increasingly available for the entire world.

From digital elevation data, a computer can generate a shaded-relief image of the landscape by portraying the landscape as if it were illuminated from the northwest by the Sun (Figure 2-B). Although shaded relief maps have been drafted by hand in the past, one of the great virtues of a DEM is that the parameters of the image—such as its orientation, scale, and vertical exaggeration of the topography—can be readily manipulated. Further, various kinds information or images can be overlain on the topography to create maps that were once impossible to conceive (for example, see Figure 2-30).

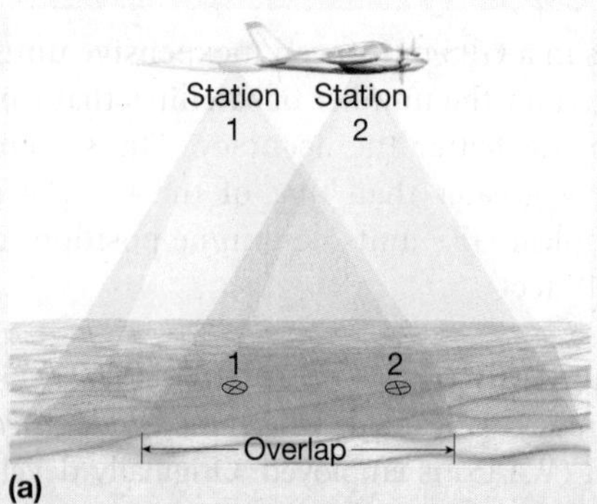

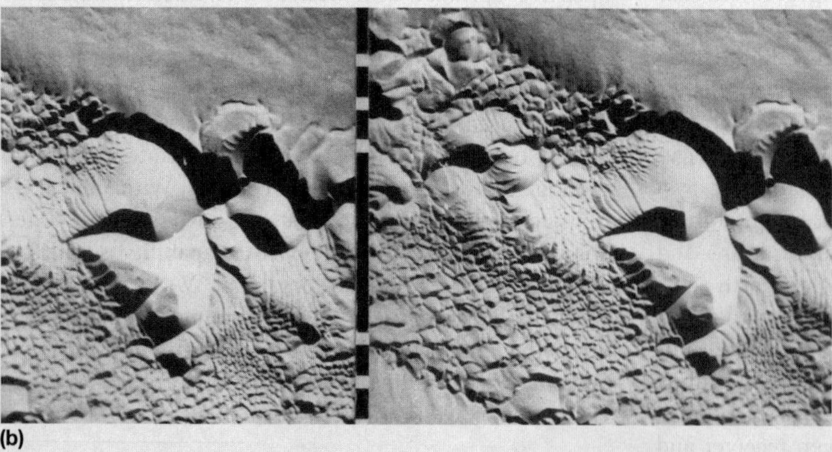

▲ **Figure 2-A** (a) Stereo aerial photographs are made from pairs of photographs showing the same part of the landscape, but taken from different angles. (b) A stereogram of star sand dunes in the Erg Er Raoui, Algeria.

▲ **Figure 2-B** An oblique shaded-relief digital elevation model of post-1980 eruption Mount St. Helens.

▲ **Figure 2-18** A handheld GPS receiver. It receives signals sent by the network of global positioning system satellites, calculating its position anywhere in the world to within 10 meters (33 feet).

▲ **Figure 2-19** Global positioning system (GPS) satellites circling 17,700 kilometers (11,000 miles) above Earth broadcast signals that are picked up by the receiver in an ambulance and used to pinpoint the location of the ambulance at any moment. A transmitter in the ambulance then sends this location information to a dispatch center. Knowing the location of all ambulances at any given moment, the dispatcher is able to route the closest available vehicle to each emergency and then direct that vehicle to the nearest appropriate health facility.

Continuously Operating GPS Reference Stations (*CORS*). These highly accurate units are capable of detecting location differences of less than 1 centimeter of latitude, longitude and elevation. They are used, for example, for the long-term monitoring of slight changes in the ground surface caused by lithospheric plate movement or the bulging of magma below a volcano.

GPS Applications In 1983, President Ronald Reagan made access to the system free to the public, and astounding commercial growth has resulted. It is anticipated that eventually practically everything that moves in our society—airplane, truck, train, car, bus, ship, cell phone—will be equipped with a GPS receiver. Meanwhile, GPS has been employed in earthquake forecasting, ocean floor mapping, volcano monitoring, and a variety of mapping projects. For example, recognizing that GPS is a relatively inexpensive way of collecting data, the Federal Emergency Management Agency (FEMA) has used the system for damage assessment following such natural disasters as floods and hurricanes. GPS was used by workers to catalog items found in the enormous heaps of rubble at Ground Zero following the World Trade Center disaster of September 11, 2001.

Commercial applications now far outnumber military uses of the system. The sale of GPS services is now a multibillion dollar a year industry in the United States. What was born as a military system has become a national economic resource.

Because of the growing importance of GPS applications, other global navigation satellite systems are being implemented around the world. Russia's GLONASS system is partially operational at this writing, and Europe and China are planning systems of their own.

In part because of the great accuracy of even inexpensive GPS units, latitude and longitude are increasingly being reported in decimal form, such as 94°45.5′ W or even 94.7583° W rather than in its traditional form of 94°45′30″ W. Even the simplest handheld GPS units can provide location coordinates with a resolution of 0.01′ (1/100th minute) or even 0.001′ (1/1000th minute) of latitude and longitude (for reference, a difference in latitude of 0.001′ represents a distance of less than 2 meters [about 6 feet]).

Remote Sensing

Throughout most of history, maps have been the only tools available to depict anything more than a tiny portion of Earth's surface with any degree of accuracy. However, sophisticated technology developed in recent years permits precision recording instruments to operate from high-altitude vantage points, providing a remarkable new set of tools for the study of Earth. **Remote sensing**, broadly considered, is any measurement or acquisition of information by a recording device that is not in physical contact with the object under study—in this case, Earth's surface.

Originally utilizing only airplanes, the use of satellites is revolutionizing the remote sensing industry. Since the Soviets launched the *Sputnik* in 1957, satellites have become increasingly sophisticated and versatile. We now have hundreds of satellites from dozens of countries perched high in the atmosphere where they either are circling Earth in a "low" orbit (an altitude of 20,000 kilometers [12,400 miles] or less) or in a lofty *geosynchronous orbit* (usually about 36,000 kilometers [22,400 miles] high) that allows a satellite to remain over the same spot on Earth at all times. These satellites gather data and produce images that provide communications, global positioning, weather data, and a variety of other information for a wide range of commercial and military applications.

Aerial Photographs

Aerial photography was almost the only form of remote sensing used for geographic purposes until the 1960s. An **aerial photograph** is one taken from an elevated platform, such as a balloon, airplane, or rocket. The earliest aerial photographs were taken from balloons in France in 1858 and in the United States in 1860. By World War I (1914–1918), systematic aerial photographic coverage from airplanes was possible.

Depending on the camera angle, aerial photographs are classified as either *oblique* or *vertical* (Figure 2-20). The advantage of oblique photographs, where the camera angle is less than 90°, is that features are seen from a more or less familiar point of view. The disadvantage is that, because of perspective, measurement is more difficult than on vertical photographs, which are taken with the optical axis of the camera approximately perpendicular to the surface of Earth.

Precise measurement is possible on vertical aerial photographs. **Photogrammetry** is the science of obtaining reliable measurements from photographs and, by extension, the science of mapping from aerial photographs. Color photogrammetry was developed in the 1940s and 1950s, with many of the improvements coming as a result of the importance of color aerial photographs during World War II (1939–1945).

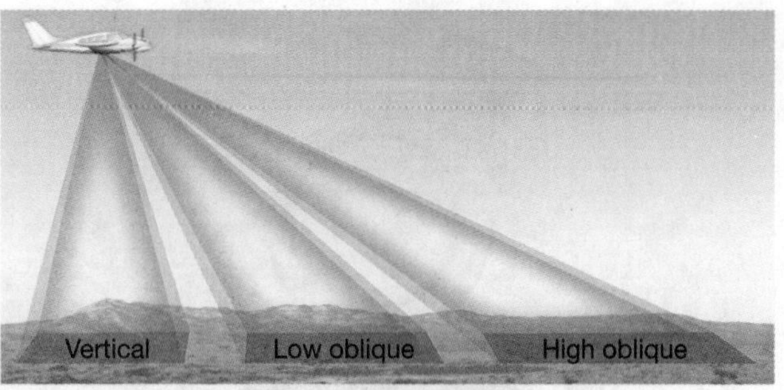

▲ **Figure 2-20** On the basis of the angle between the camera and Earth's surface, aerial photographs are classified as vertical, low oblique, or high oblique.

Orthophoto Maps

Orthophoto maps are multicolored, distortion-free photographic maps prepared from aerial photographs or digital images. Displacements caused by camera tilt or differences in terrain elevations have been removed, which gives the orthophoto the geometric characteristics of a map (Figure 2-21). Thus, an orthophoto can show the landscape in much greater detail than a conventional map, but retains the map characteristic of a common scale that allows precise measurement of distances. Orthophoto maps are particularly useful in flat-lying coastal areas because they can show subtle topographic detail in areas of very low relief, such as marshlands.

Visible Light and Infrared Sensing

One of the most important advancements in remote sensing came when wavelengths of radiation other than visible light were first utilized. As we will see in Chapter 4, *electromagnetic radiation* includes a wide range of wavelengths of energy emitted by the Sun and other objects (Figure 2-22). The human eye (and conventional photographic film) is only sensitive to the narrow portion of the electromagnetic spectrum

▲ **Figure 2-21** Orthophoto map of Wilmington, North Carolina; original scale: 1:24,000.

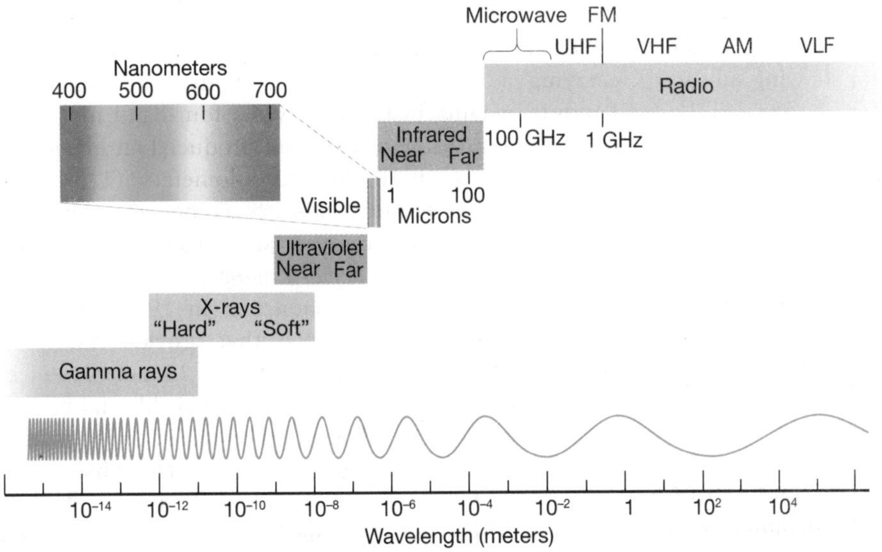

◀ **Figure 2-22** The electromagnetic spectrum. The human eye can only sense radiation from the visible-light region. Conventional photography also can use only a small portion of the total spectrum. Various specialized remote-sensing scanners are capable of "seeing" radiation from other parts of the spectrum.

known as *visible light*—the colors seen in a rainbow. However, a wide range of other wavelengths of energy—such as X-rays, ultraviolet radiation, infrared radiation, and radio waves—are emitted, reflected, or absorbed by surfaces and can be detected by special films or instruments, yielding a wealth of information about the environment.

Color infrared (color IR) imagery uses electronic sensors or photographic film sensitive to radiation in the *near infrared* portion of the electromagnetic spectrum—wavelengths of radiation just longer than the human eye can sense. With color IR imagery, sensitivity to visible blue light is replaced by sensitivity to near infrared wavelengths. The images produced in this way, even though they are "false-color" images (e.g., living vegetation appears red instead of green), are still extremely valuable. Color IR film was first widely used in World War II when it was often called "camouflage-detection" film because of its ability to discriminate living vegetation from the withering vegetation used to hide objects during the war. Today, one of the major uses of color IR imagery remains the identification and evaluation of vegetation (Figure 2-23).

Thermal Infrared Sensing

None of the middle or far infrared part of the electromagnetic spectrum, called the *thermal infrared* (thermal IR), can be sensed with traditional photographic film; as a result, special supercooled scanners are needed. Thermal scanning senses the radiant temperature of objects and may be carried out either day or night. The photograph-like images produced in this process are particularly useful for showing diurnal temperature differences between land and water, and between bedrock and alluvium, for studying thermal water pollution, and for detecting forest fires.

By far the greatest use of thermal IR scanning systems has been on meteorological satellites (for example, see "Focus: GOES Weather Satellites," in Chapter 6). Although the spatial resolution (the size of the smallest feature that can be identified) is not as high as some other kinds of sensing systems, it is more than sufficient to provide details that allow weather forecasting that is far more accurate and complete than ever before.

Microwave Sensing

Systems that sense wavelengths even longer than infrared are used widely in Earth sciences, including microwave

▲ **Figure 2-23** Advanced Spaceborne Thermal Emission and Reflection Radiometer (ASTER) image of the cities of Palm Springs, Cathedral City, and Palm Desert, California. In this false-color infrared image, healthy vegetation is shown in red; bare ground is shown in gray-blue.

radiometry, which senses radiation in the 100-micrometer to 1-meter range. Although such systems have low spatial resolution, they are particularly useful for showing subsurface characteristics such as moisture.

Multispectral Remote Sensing

Today, most sophisticated remote sensing satellites are **multispectral** or *multiband* (the latter name because the various regions of the electromagnetic spectrum are sometimes called *bands*). These instruments image more than one region of the electromagnetic spectrum simultaneously from precisely the same location. Thus, while traditional photographic film is sensitive to only a narrow band of visible radiation, a satellite equipped with a multiband instrument images the surface of Earth in several spectrum regions at once, each designated for a different application.

A multispectral satellite image is digital, conveyed through a matrix of numbers, with each number representing a single value for a specific pixel (picture element) and band. These data are stored in the satellite, eventually transmitted to an Earth receiving station, numerically manipulated by a computer, and produced as a set of gray values and/or colors on a screen or hard-copy printout (Figure 2-24).

Landsat The early NASA space missions (Mercury, Gemini, and Apollo) used multiband photography obtained through multicamera arrays. These imaging experiments were so successful that NASA then developed what was initially called the *Earth Resources Technology Satellite* series (ERTS) and later renamed *Landsat*. The 1970s and 1980s saw the launch of five Landsat satellites carrying a variety of sensor systems. The basic early Landsat imaging instrument was a four-band multispectral scanning system (MSS) that produced an image covering an area of 183-by-170 kilometers (115-by-106 miles). Landsat 4 and Landsat 5, launched in the early 1980s, carried an improved seven-band multispectral scanner known as the *thematic mapper*.

Landsat 7, which was launched in 1999, carries an *enhanced thematic mapper plus* that provides images in eight spectral bands with a resolution of 15 meters (49 feet) in the panchromatic band (sensitive to visible and near infrared wavelengths), 30 meters (98 feet) in the six narrow bands of visible and short infrared wavelengths, and 60 meters (197 feet) in thermal infrared (Figure 2-25). A description of the primary applications for the various bands is provided in Table 2-1. Although the satellites were originally designed for a life of less than 10 years, as of this writing both Landsat 5 and Landsat 7 remain in active operation, however, equipment malfunctions now limit their capabilities somewhat. The next generation Landsat satellite, known as the *Landsat Data Continuity Mission*, is scheduled for launch in late 2012.

Earth Observing System Satellites In 1999 NASA launched the first of its *Earth Observing System* (EOS) satellites known as *Terra*. The key instrument of these satellites is the Moderate Resolution Imaging

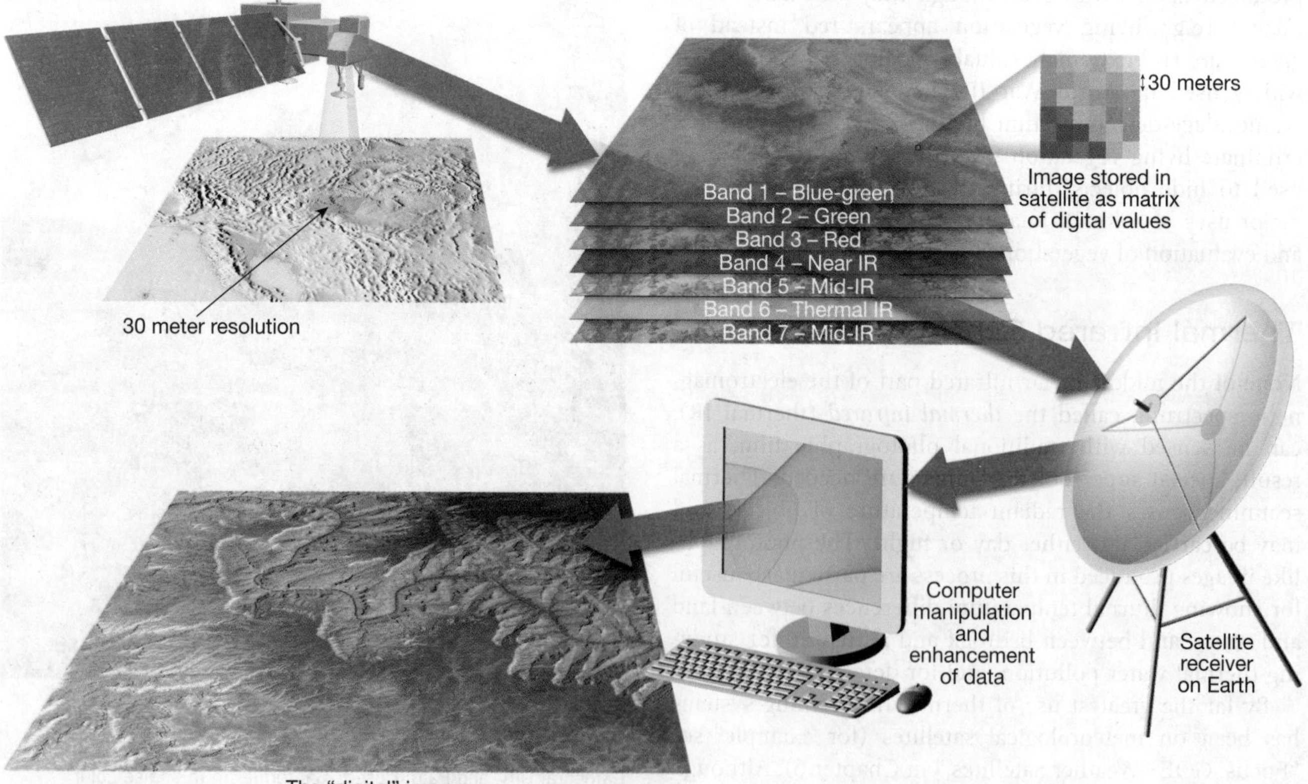

▲ **Figure 2-24** The sequence of events that takes place as a multispectral satellite scan is converted to a digital image.

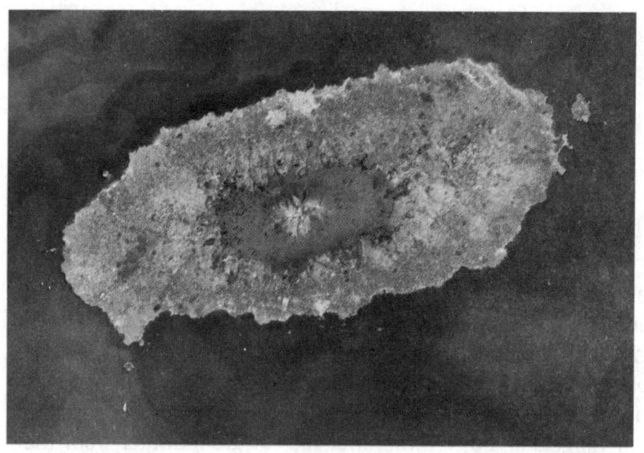

▲ **Figure 2-25** Landsat 7 satellite image of the island of Cheju-do, South Korea, taken with the Enhanced Thematic Mapper Plus in April 2000. The central shield volcano of Mount Halla rises to an elevation of 1950 meters (6398 ft.).

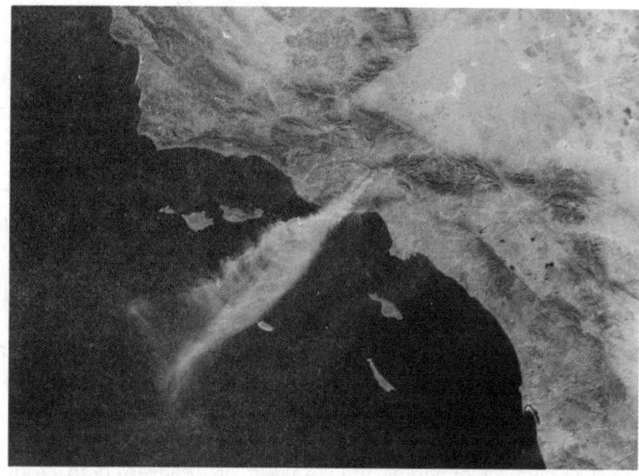

▲ **Figure 2-26** Fires fanned by Santa Ana winds blow a plume of smoke off the coast of southern California near Los Angeles. This image was taken on October 13, 2008, with the MODIS (Moderate Resolution Imaging Spectroradiometer) instrument aboard NASA's Aqua satellite.

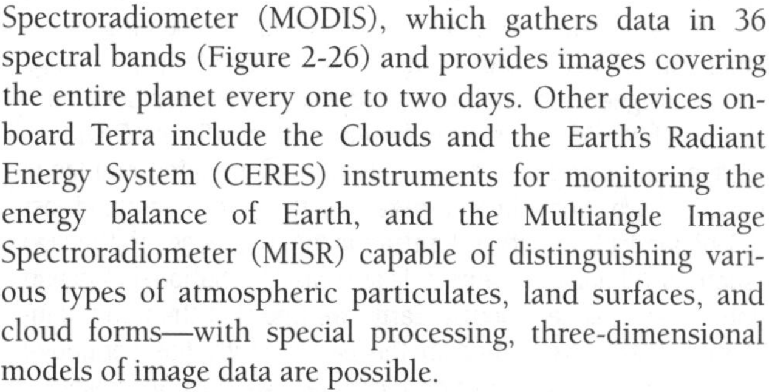

Spectroradiometer (MODIS), which gathers data in 36 spectral bands (Figure 2-26) and provides images covering the entire planet every one to two days. Other devices onboard Terra include the Clouds and the Earth's Radiant Energy System (CERES) instruments for monitoring the energy balance of Earth, and the Multiangle Image Spectroradiometer (MISR) capable of distinguishing various types of atmospheric particulates, land surfaces, and cloud forms—with special processing, three-dimensional models of image data are possible.

The more recently launched EOS satellite, *Aqua*, is designed to enhance our understanding of Earth's water cycle by monitoring water vapor, clouds, precipitation, glaciers, and soil wetness. In addition to instruments such as MODIS, Aqua includes the Atmospheric Infrared Sounder (AIRS), designed to permit very accurate temperature measurements throughout the atmosphere.

Many satellite images are now easily available for viewing and downloading via the Internet from NASA and NOAA. For example, you can visit http://earthobservatory.nasa.gov/ and http://www.goes.noaa.gov/.

Commercial High Resolution Satellites In addition to imagery from government-operated satellites that is often available either free of charge or for a nominal fee (such as the GOES satellites, Landsat, and the EOS satellites), a number of satellites now offer very high resolution imagery (up to 50 to 60 centimeter [20 to 24 in.] resolution) for commercial applications, including *SPOT* (Satellite Pour l'Observation de la Terre), *GeoEye-1*, *QuickBird*, and *WorldView*. The market for these images seems to be growing remarkably.

Radar and Sonar Sensing All the systems mentioned so far work by sensing the natural radiation emitted by or reflected from an object and are therefore characterized as *passive systems*. Another type of system, called an *active system*, has its own source of electromagnetic radiation. The most important active sensing system used in the Earth sciences is **radar**, the acronym for *radio detection and ranging*. Radar senses wavelengths longer than 1 millimeter, using the principle that the time it takes for an emitted signal to reach a target and then return to the sender can be converted to distance information.

TABLE 2-1 Bands of the Landsat 7 Enhanced Thematic Mapper Plus

Band Number	Bandwidth (micrometers)	Spectral Region	Resolution (meters)	Applications
1	0.45–0.52	Blue	30	Water penetration and vegetation analysis
2	0.52–0.60	Green	30	Vegetation analysis
3	0.63–0.69	Red	30	Vegetation analysis
4	0.77–0.90	Near IR	30	Biomass and soil analysis
5	1.55–1.75	Middle IR	30	Soil moisture and hydrologic analysis
6	10.4–12.5	Thermal	60	Geothermal resources and vegetation stress
7	2.08–2.35	Middle IR	30	Geologic features
8	0.52–0.90	Panchromatic	15	High resolution images

FOCUS

Geographic Research Using *Google Earth* and *The National Map*

Richard Greene, Northern Illinois University

Google Earth and *The National Map* have transformed how geographers approach research using maps. Google Earth™ (http://earth.google.com) is a three-dimensional geographic data viewer produced by Google, Inc. Google Earth is best known for providing lifelike views of urban landscapes and mountainous regions through the use of terrain models and aerial photography. The National Map (http://nationalmap.gov) is a popular mapping program developed by the United States Geological Survey (USGS) that gives users access to a wide range of federal geographic information including aerial photographs, elevation, transportation, and land cover.

The National Map differs from Google Earth in that it is a Web site viewed through a Web browser, whereas Google Earth works as a program that must be installed on your computer before you can explore its map content stored online. What is common to both platforms is the way in which maps are displayed and viewed: they both allow map customization by toggling map layers on and off, and by zooming in or out to change map scale.

There are strengths and weaknesses of each program. For example, Google Earth offers more flexibility and speed when customizing a map, whereas The National Map allows users to download a map layer that is preformatted for use in a geographic information system (GIS). Combining the strengths of both Google Earth and The National Map provides a powerful tool for scientific investigations of Earth processes. For example, Google Earth allows users to display layers downloaded from The National Map through KML files (a file format that allows geographic data to be displayed in Google Earth) created in a GIS program.

In the following example, The National Map and Google Earth were used to update an earlier study of the sustainability of America's agricultural land resources*. The earlier study found that approximately 11 million acres of U.S. cropland were converted to urban land between 1982 and 1997, and that in turn, an equal amount of arid rangeland was converted to cropland that relies heavily on irrigation. Unfortunately, the sample size of the database used in the earlier study was reduced, making it difficult to monitor the problem beyond 1997.

Google Earth and The National Map provided the tools needed to update this study. We start with a map of 2001 land cover types in "The National Map Viewer" in which a different color is assigned to each type of land cover—in this case, urban land is red, grassland is yellow, and cropland is brown (Figure 2-C). We then repeat these steps for 1992 land cover. Zooming in on an area such as Alliance, Nebraska, can reveal land cover change as the 2001 and 1992 map layers are toggled on and off. Land use change in western Nebraska appears circular. The circular fields are an indicator that this cropland uses central pivot irrigation with groundwater pumped from the Ogallala Aquifer (see Chapter 9), and more importantly, that some of it (shown in blue in the Google Earth inset) was nonirrigated rangeland only nine years earlier. This trend is alarming because the withdrawal rate of groundwater from the Ogallala aquifer for irrigated agriculture often exceeds its natural recharge rate.

If we download a land cover change map from "The National Map Seamless Server" to display in Google Earth, we can investigate a county on the edge of Chicago that shows land converted from cropland to urban land. Zooming in on the county on the edge of Chicago reveals that new subdivisions have replaced nonirrigated fertile cropland (Figure 2-D).

By examining new irrigated croplands in Nebraska and the prime farmland lost to urbanization on the edge of Chicago using tools such as The National Map and Google Earth, we find that the trend identified in the earlier study has continued beyond 1997. As we see in many of Earth's physical systems, human activities may limit the sustainability of the economic systems we are trying to propagate, in this case agriculture.

*Greene, R. P. and J. Stager, 2001, "Rangeland to cropland conversions as replacement land for prime farmland lost to urban development," *The Social Science Journal* 38, 543–555.

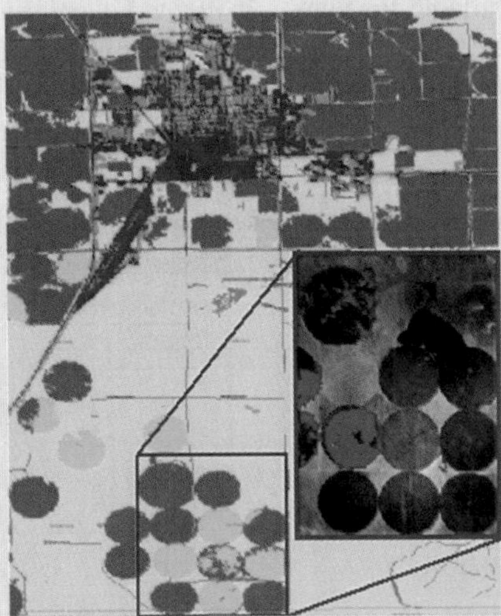

▲ **Figure 2-C** Land cover in 2001 for Alliance, Nebraska, derived from the National Map with Google Earth inset showing cropland on former rangeland in blue.

▲ **Figure 2-D** Google Earth display of crop to urban land use change shown in red at the edge of Chicago, 1992–2001.

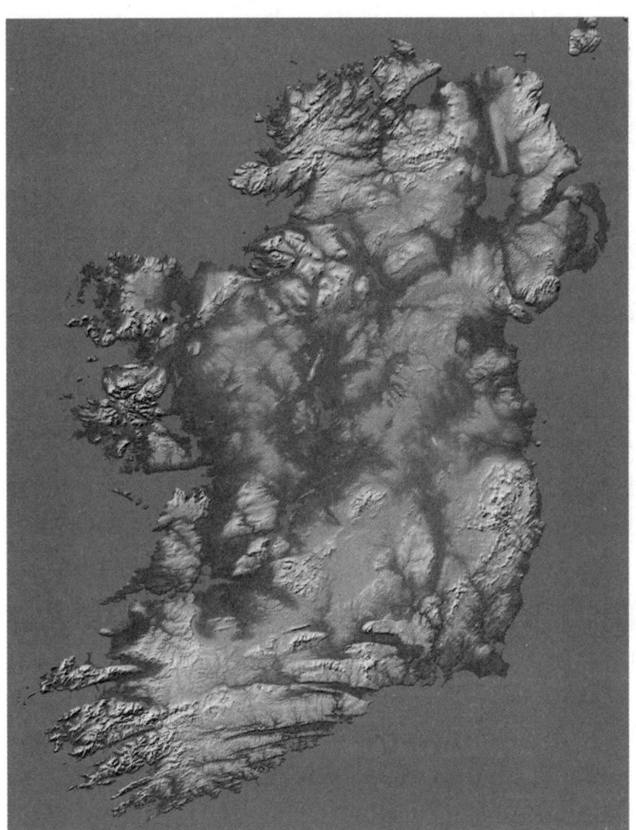

▲ **Figure 2-27** Radar image showing the topography of the island of Ireland. The data were gathered from the Shuttle Radar Topography Mission using synthetic aperture radar aboard Space Shuttle Endeavour in 2000. The data were processed with elevations represented by different colors, ranging from green for lowlands to white for high mountain tops. Shaded relief was added to highlight the topography.

Initially, radar images were viewed only on a screen, but they are now available in photograph-like form (Figure 2-27). In common with some other sensors, radar is capable of operating by day or night, but it is unique in its ability to penetrate atmospheric moisture. Thus, some wet tropical areas that could never be sensed by other systems have now been imaged by radar. Radar imagery is particularly useful for terrain analysis and meteorology. Another active remote sensing system, **sonar** (*sound navigation and ranging*), permits underwater imaging so that scientists can determine the form of that part of Earth's crust hidden by the world ocean.

Geographic Information Systems (GIS)

Cartographers have been at work since the days of the early Egyptians, but it is only in the last half century, with the introduction of computers in the 1950s that their technology has advanced beyond manual drawing on a piece of paper. Computers have provided incredible improvements in speed and image handling ability—as one example, all of the maps in this textbook were made with desktop computers. Of all the technological advances in cartography over the last few decades, however, one of the most revolutionary has been *geographic information systems*.

Geographic information systems (GIS) are computer systems designed to analyze and display spatial data. GIS involves specialized hardware and software that allows users to collect, store, retrieve, reorganize, analyze, and map geographic data from the real world (Figure 2-28).

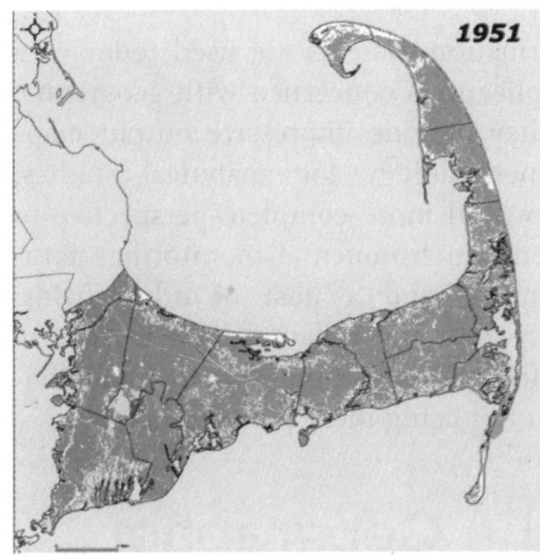

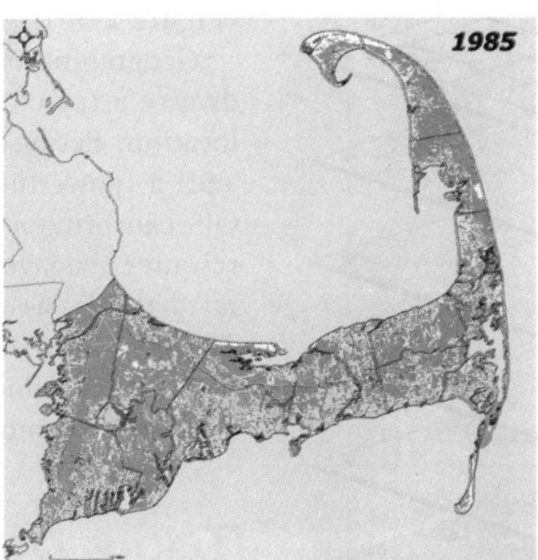

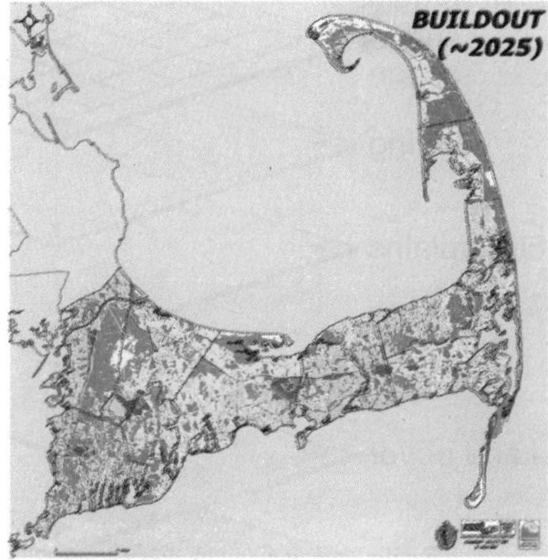

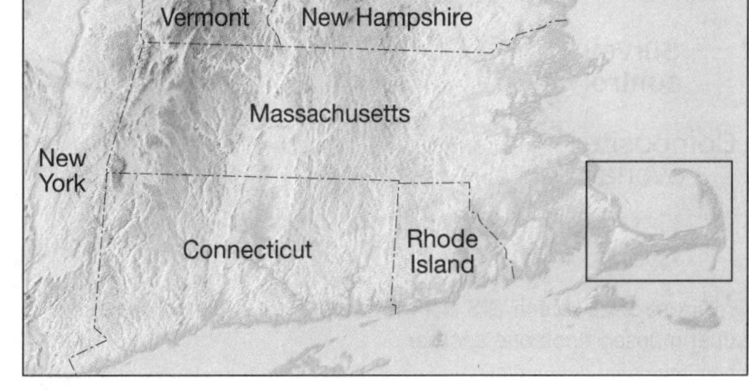

▲ **Figure 2-28** Land uses changes on Cape Cod from 1951 to 1985, with the projected build out by the year 2025. Dark green shows areas of undisturbed vegetation, yellow shows areas of residential housing, and purple shows areas of industrial development.

Geographic information systems originally developed out of computer science, geography, and cartography, and found their greatest early uses in surveying, photogrammetry, spatial statistics, and remote sensing. So commonly are they now used in geographical analysis that GIS has become a science of spatial analysis by itself, known as *geographic information science*, and the software has spun off a multibillion-dollar industry in spatial data and spatial information.

Geographic information systems are libraries of information that use maps to organize, store, view, and analyze information in an intuitive, visual manner. Just as an ordinary computer database management system can manipulate rows and columns of data in tabular form, a GIS allows data management using the link between data and a map. This means that the map and data are encoded, usually as numbers representing coordinates of locations at points on a grid covering the mapped area. Once the data and the map are inside the GIS, the user can organize or search the data using the map, or the map using the data. An important attribute is the capability of GIS data from different maps and sources, such as field data, map data, and remotely sensed images, to be registered together at the correct geographic location within a common database, with a common map scale and map projection. In this way, one map layer, such as the locations of rivers, can then be cross-referenced to another, such as geology, soils, or slope.

GIS is frequently used in overlay analysis, where two or more layers of data are superimposed or integrated.

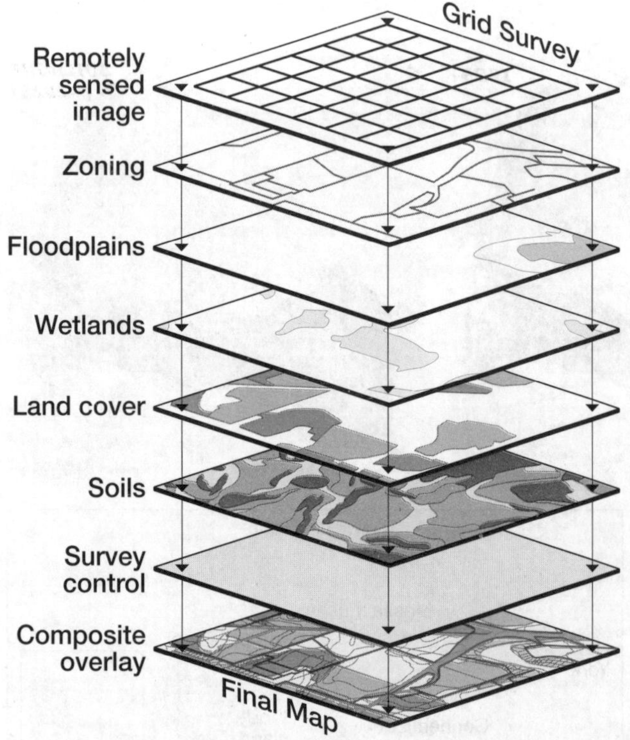

▲ **Figure 2-29** Much GIS work involves layers of spatial data superimposed upon one another.

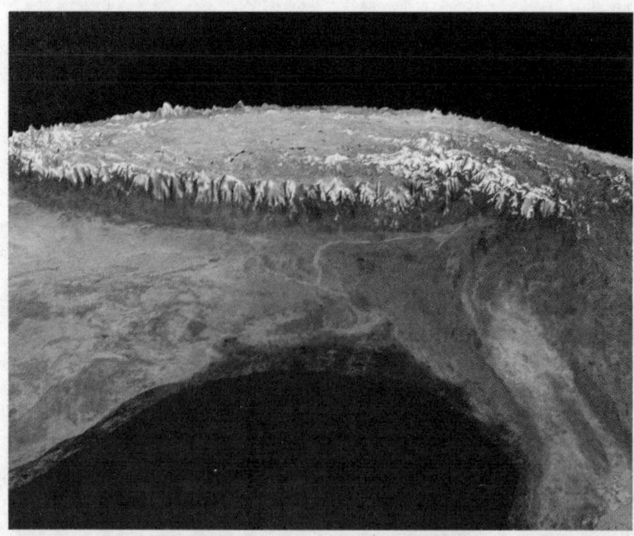

▲ **Figure 2-30** This oblique view of Bangladesh and the Himalayas was created by configuring MODIS images from the Terra satellite over a 50-times vertically exaggerated digital elevation model of the topography.

GIS treats each spatially distributed variable as a particular layer in a sequence of overlays. As shown in Figure 2-29, input layers bring together such diverse elements as topography, vegetation, land use, land ownership, and land survey. Details of these various components are converted to digital data and are synthesized onto a reference map or data set. Particularly useful images for the study of physical geography can be developed with GIS when data or satellite images are overlain on topography generated with a *digital elevation model*, offering oblique views of the landscape previously impossible to obtain (Figure 2-30).

Geographic information systems are used today in a diverse array of applications concerned with geographic location. Because they provide impressive output maps and a powerful methodology for analytical studies, GIS can bring a new and more complete perspective to resource management, environmental monitoring, natural hazards assessment—and a host of other fields. The growth of GIS is so rapid today, that there are few fields of academic study, sectors of the economy, and divisions of government not using these powerful tools.

Tools of the Geographer

As we have just seen, a vast array of maps, remotely sensed imagery, satellite data, and computer applications are now available, making the tools of the geographer more widely used than ever before. The effective use of these tools, however, still entails thoughtful consideration. Although it is easy to download a satellite image or quickly print a handsome-looking map from the Internet, those images and maps may or may not be useful for analysis—and might actually be deceptive—unless care is taken to choose an appropriate map projection, an appropriate scale, and an appropriate selection of data to depict.

Certain types of imagery are useful for particular purposes. For example, when studying major features of the lithosphere, high-altitude space imagery is especially valuable, although this type of imagery might have limited use in detailed local terrain studies where large-scale oblique aerial photographs or topographic maps might be more appropriate. For studying the hydrosphere, multiband satellite images of an entire hemisphere can tell us much about the water content in clouds, air masses, glaciers, and snowfields at a given time, while detailed conventional color images might be better for discriminating complicated shoreline features. Vegetation patterns in the biosphere are often best appreciated with color IR imagery—overall vegetation patterns on small-scale satellite images and detailed aerial photographs for crop and forest inventory studies. Features of human creation are generally not evident on very high-altitude imagery, but they become increasingly clear as one approaches Earth, and so survey patterns, transportation lines, rural settlements, and cities are best interpreted on imagery of intermediate or large scale. And in all cases, GIS may be used to uncover or highlight geographic relationships that may not be obvious when employing any single source of data or kind of imagery.

In using these tools, the geographer should never lose sight of our major objective: to better understand Earth. Such understanding does not come simply through the application of technology, however. Understanding comes from a carefully designed investigation, often using technology, but frequently supported by such traditional sources of information as field study and observation.

CHAPTER 2 LEARNING REVIEW

After studying this chapter, you should be able to answer the following questions. Key terms from each text section are shown in bold type. Definitions for key terms are also found in the glossary at the back of the book.

The Nature of Maps (p. 27)

1. Why is it impossible for a **map** of the world to portray Earth as accurately as can be done with a globe?

Map Scale (p. 27)

2. Describe and explain the concept of **map scale**.
3. Contrast **graphic scales**, **fractional scales**, and **verbal scales**.
4. What is meant by a map scale with a **representative fraction** of 1/10,000 (also written 1:10,000)?
5. Explain the difference between **large-scale maps** and **small-scale maps**.

Map Projections (p. 31)

6. What is meant by a **map projection**?
7. Explain the differences between an **equivalent map projection** and a **conformal map projection**.
8. Is it possible for a map to be both conformal and equivalent? Neither conformal nor equivalent?

Families of Map Projections (p. 33)

9. Briefly describe the four major families of map projections: **cylindrical projections**, **plane projections**, **conic projections**, and **pseudocylindrical projections**.
10. Why is a **Mercator projection** useful as a navigation map? Why is it not ideal for use as a general purpose classroom map?
11. What is a **loxodrome (rhumb line)**?
12. Why are there so many types of map projections?

Isolines (p. 35)

13. Explain the concept of an **isoline**.
14. What characteristics on maps are shown by isotherms, isobars, and **elevation contour lines**?

GPS—The Global Positioning System (p. 37)

15. Briefly explain how the **Global Positioning System (GPS)** works.

Remote Sensing (p. 40)

16. What is **remote sensing**?
17. Briefly define the following terms: **aerial photograph**, **photogrammetry**, **orthophoto map**.
18. What are some of the applications of color infrared imagery?
19. What are some of the applications of thermal infrared imagery?
20. Describe **multispectral** remote sensing.
21. Compare and contrast **radar** and **sonar**?

Geographic Information Systems (p. 45)

22. Distinguish between GPS and **GIS (geographic information systems)**.

Log in to **www.mygeoscienceplace.com** for animations, videos, RSS feeds, maps, and self-study quizzes to enhance your study of **Portraying Earth**.

3 Introduction to the Atmosphere

Earth is different from all other known planets in a variety of ways. One of the most notable differences is the presence around our planet of an atmosphere distinctive from other planetary atmospheres. It is our atmosphere that makes life possible on Earth. The atmosphere supplies most of the oxygen that animals must have to survive, as well as the carbon dioxide needed by plants. It helps maintain a water supply, which is essential to all living things. It insulates Earth's surface against temperature extremes and thus provides a livable environment over most of the planet. It also shields Earth from much of the Sun's ultraviolet radiation, which otherwise would be fatal to most life forms.

Air, generally used as a synonym for atmosphere, is not a specific gas, but rather a mixture of gases—mainly nitrogen and oxygen. In addition, most air contains minor but varying quantities of solid and liquid particles that can be thought of as impurities. The individual particles are mostly microscopic and therefore held in suspension in the air. Most air also contains some gaseous impurities.

Pure air is odorless and tasteless, and is invisible because the gases in it are colorless. Gaseous impurities, on the other hand, can often be smelled, and the air may even become visible if enough microscopic solid and liquid impurities coalesce (stick together) to form particles large enough to either reflect or scatter sunlight. Clouds, by far the most conspicuous visible features of the atmosphere, represent the coalescing of water droplets or ice crystals around microscopic particulates that act as condensation nuclei.

▲ **Figure 3-1** The atmosphere completely surrounds Earth in this composite satellite image; beyond the narrow band of the atmosphere is the blackness of outer space.

Size of Earth's Atmosphere

The atmosphere completely surrounds Earth and can be thought of as a vast ocean of air with Earth at its bottom (Figure 3-1). It is held to Earth by gravitational attraction and therefore accompanies our planet in all its celestial motions. The attachment of Earth and atmosphere is a loose one, however, and the atmosphere can therefore move on its own, doing things that the solid Earth cannot do.

Although the atmosphere extends outward at least 10,000 kilometers (6000 miles), most of its mass is concentrated at very low altitudes. More than half of the mass of the atmosphere lies below the summit of North America's highest peak, Mount McKinley (Denali) in Alaska, which reaches an elevation of 6.2 kilometers (3.8 miles), and more than 98 percent of it lies within 26 kilometers (16 miles) of sea level (Figure 3-2). Therefore, relative to Earth's diameter of about 13,000 kilometers (8000 miles), the "ocean of air" we live in is a very shallow one.

◀ The Moon seen through the top of Earth's atmosphere photographed from the International Space Station.

In addition to reaching upward above Earth's surface, the atmosphere also extends slightly downward. Because air expands to fill empty spaces, it penetrates into caves and crevices in rocks and soil. Moreover, it is

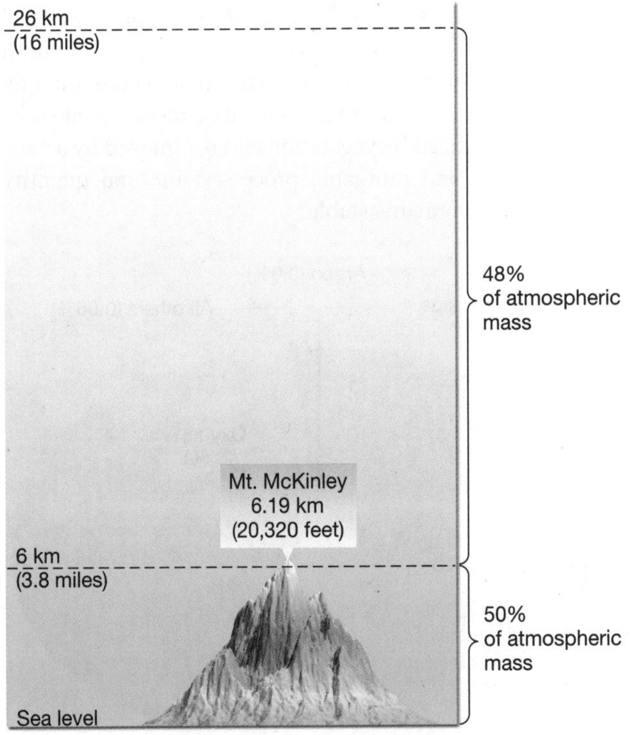

▲ **Figure 3-2** Most of the atmospheric mass is close to Earth's surface. More than half of the mass is below the elevation of Mount McKinley, North America's highest peak.

dissolved in the waters of Earth and in the bloodstreams of organisms.

The atmosphere interacts with other components of the earthly environment, and it is instrumental in providing a hospitable milieu for life. Whereas we often speak of human beings as creatures of Earth, it is perhaps more accurate to consider ourselves creatures of the atmosphere. As surely as a crab crawling on the sea bottom is a resident of the ocean, a person living at the bottom of the ocean of air is a resident of the atmosphere.

Composition of the Atmosphere

The chemical composition of pure, dry air at lower altitudes (altitudes lower than about 80 kilometers or 50 miles) is simple and uniform, and the concentrations of the major components are basically unvarying over time. Certain minor gases and nongaseous particles vary markedly from place to place and from time to time, however, as does the amount of moisture in the air.

Permanent Gases

Nitrogen and Oxygen The two most abundant gases in the atmosphere are nitrogen and oxygen (Figure 3-3 and Table 3-1). Nitrogen makes up more than 78 percent of the total, and oxygen makes up nearly 21 percent. Nitrogen is added to the air by the decay and burning of organic matter, volcanic eruptions, and the chemical breakdown of certain rocks, and it is removed by certain biological processes and by being washed away in rain or snow. Overall, the addition and removal of nitrogen gas are balanced, and consequently the quantity present in the air remains constant over time. Oxygen is produced by vegetation and is removed by a variety of organic and inorganic processes; its total quantity also apparently remains stable.

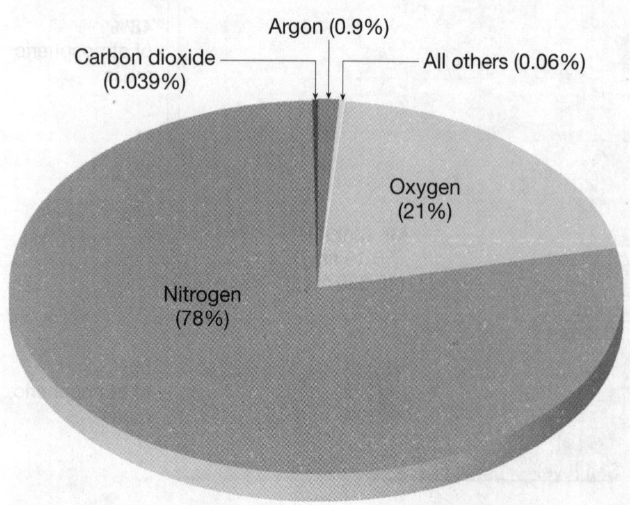

▲ **Figure 3-3** Proportional volume of the gaseous components of the atmosphere. Nitrogen and oxygen are the dominant components.

TABLE 3-1 Principal Gases of Earth's Atmosphere

	Percent of Volume of Dry Air	Concentration in Parts per Million Parts of Air
Permanent gases		
Nitrogen (N_2)	78.084	
Oxygen (O_2)	20.946	
Argon (Ar)	0.934	
Neon (Ne)	0.00182	18.2
Helium (He)	0.00052	5.2
Krypton (Kr)	0.00011	1.1
Hydrogen (H_2)	0.00005	0.5
Variable gases		
Water vapor (H_2O)	0–4	
Carbon dioxide (CO_2)	0.039	390
Carbon Monoxide (CO)		<100
Methane (CH_4)	0.000178	1.78
Ozone (O_3)		<2
Sulfur dioxide (SO_2)		<1
Nitrogen dioxide (NO_2)		<0.2

The remaining 1 percent of the atmosphere's volume consists mostly of the inert gas argon. These three principal atmospheric gases—nitrogen, oxygen, argon—have a minimal effect on weather and climate and therefore need no further consideration here. The trace gases neon, helium, krypton, and hydrogen also have little effect on weather and climate.

Variable Gases

Several other gases occur in sparse but highly variable quantities in the atmosphere, and their influence on weather and climate is significant.

Water Vapor Water in the form of a gas is known as **water vapor**. Water vapor is invisible—the visible forms of water in the atmosphere, such as clouds and precipitation, consist of water in its liquid or solid form. Water vapor is most abundant in air overlying warm, moist surface areas such as tropical oceans, where water vapor may amount to as much as 4 percent of total volume. Over deserts and in polar regions, the amount of water vapor is but a tiny fraction of 1 percent.

In the atmosphere as a whole, the total amount of water vapor remains nearly constant. Thus, its listing as a "variable gas" in Table 3-1 means variable in location, not variable in time. Water vapor has a significant effect on weather and climate in that it is the source of all clouds and precipitation; water vapor also plays important roles in a number of heating and cooling processes in the atmosphere.

Carbon Dioxide Another important atmospheric component is **carbon dioxide** (CO_2). Like water vapor, carbon dioxide also has a significant influence on climate, primarily because of its ability to absorb infrared radiation and thereby help warm the lower atmosphere. Carbon dioxide is distributed fairly uniformly in the lower layers of the atmosphere, although its concentration has been increasing steadily for the last century or so because of the increased burning of fossil fuels—the proportion of carbon dioxide in the atmosphere has been increasing at a rate of about 0.0002 percent (2 parts per million) per year and at present is about 390 parts per million. Most atmospheric scientists conclude that the increased levels of CO_2 are causing the lower atmosphere to warm up enough to produce significant, although still somewhat unpredictable, global climatic changes (the topic of "global warming" will be presented in greater detail in Chapter 4).

Ozone Another minor but vital gas in the atmosphere is **ozone**, which is a molecule made up of three oxygen atoms (O_3) instead of the more common two oxygen atoms (O_2). For the most part, ozone is concentrated in a layer of the atmosphere called the *ozone layer*, which lies between 15 and 48 kilometers (9 and 30 miles) above Earth's surface. Ozone is an excellent absorber of ultraviolet solar radiation; it filters out enough of these rays to protect life forms from potentially deadly effects (a discussion of the recent thinning of the ozone layer follows later in this chapter).

The other variable gases listed in Table 3-1—carbon monoxide, sulfur dioxide, nitrogen oxides, and various hydrocarbons—are increasingly being introduced into the atmosphere by emission from factories and automobiles. All of them are hazardous to life and may possibly have some effect on climate. Methane, introduced into the atmosphere both naturally and through human activity, absorbs certain wavelengths of radiation and so plays a role in regulating the temperature of the atmosphere.

Particulates (Aerosols)

The larger nongaseous particles in the atmosphere are mainly liquid water and ice, which form clouds, rain, snow, sleet, and hail. There are also dust particles large enough to be visible, which are sometimes kept aloft in the turbulent atmosphere in sufficient quantity to cloud the sky (Figure 3-4), but they are too heavy to remain long in the air. Smaller particles, invisible to the naked eye, may remain suspended in the atmosphere for months or even years.

The solid and liquid particles found in the atmosphere are collectively called **particulates** or **aerosols**. They have innumerable sources, some natural and some the result of human activities. Volcanic ash, windblown soil and pollen grains, meteor debris, smoke from wildfires, and salt spray from breaking waves are examples of particulates from natural sources. Particulates coming from human sources consist mostly of industrial and automotive emissions and smoke and soot from fires of human origin.

These tiny particles are most numerous near their places of origin—above cities, seacoasts, active volcanoes, and some desert regions. They may be carried great distances, however, both horizontally and vertically by the restless atmosphere. They affect weather and climate in two major ways:

1. Many are *hygroscopic* (which means they absorb water), and water vapor condenses around such *condensation nuclei*. This accumulation of water molecules

◀ **Figure 3-4** Dust particles sometimes cloud the sky for a short time over a limited part of Earth's surface. On some occasions, as in this scene from the central part of New South Wales, Australia, the term "dust storm" is very appropriate, and the visual effect is imposing, if not menacing.

is a critical step in cloud formation, as we shall see in Chapter 6.
2. Some either absorb or reflect sunlight, thus decreasing the amount of solar energy that reaches Earth's surface.

Vertical Structure of the Atmosphere

The next five chapters of this book deal with atmospheric processes and their influence on climatic patterns. Our attention in these chapters will be devoted primarily to the lower portion of the atmosphere, which is the zone in which most weather phenomena occur. Even though the upper layers of the atmosphere usually affect the environment of Earth's surface only minimally, it is still useful to have some understanding of the total atmosphere. Therefore, we now note some general characteristics of the atmosphere in its total vertical extent, emphasizing vertical patterns of temperature, pressure, and composition.

A given layer (altitudinal zone) of the atmosphere has different names depending on the characteristic or feature under discussion. In this section, we are discussing thermal layers; later we introduce names for other overlapping layers.

Thermal Layers

Most of us have had some personal experience with temperature variables associated with altitude. As we climb a mountain, for instance, we sense a decrease in temperature. Until about a century ago, it was generally assumed that temperature decreased with increasing altitude throughout the whole atmosphere, but now we know that such is not the case.

As shown in Figure 3-5, the vertical pattern of temperature is complex, consisting of a series of layers in which temperature alternately decreases and increases. From the surface of Earth up, these thermal layers are called the troposphere, stratosphere, mesosphere, thermosphere, and exosphere. In addition to these five principal names, we also have special names for the top of the first three layers: tropopause, stratopause, and mesopause. We use the *-sphere* name when talking about an entire layer and the *-pause* name when our interest is either in the upper portion of a layer or in the boundary between two layers.

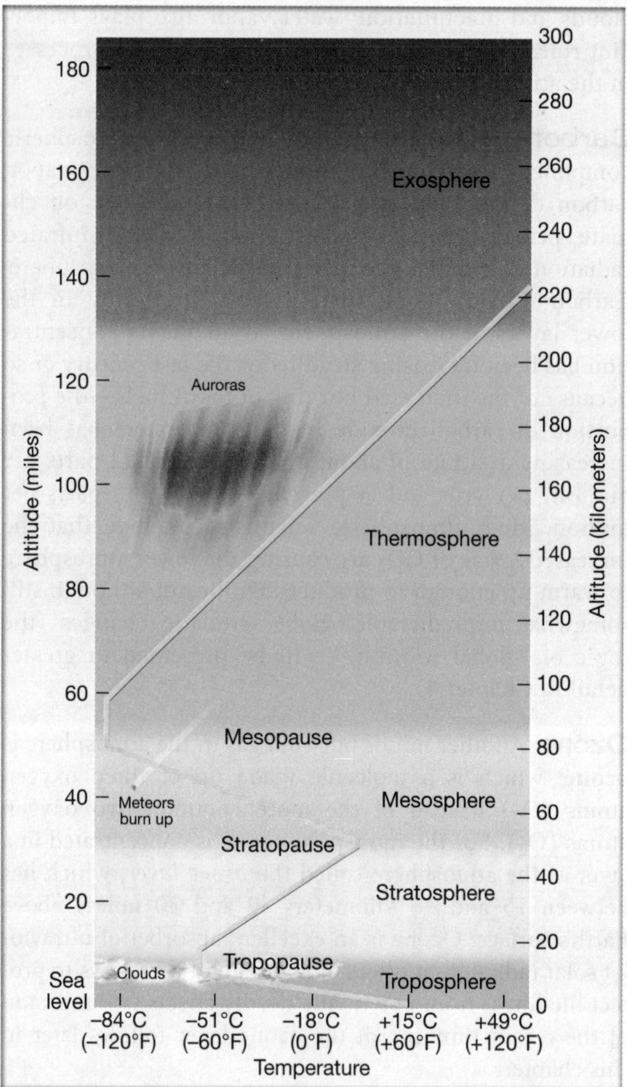

▲ **Figure 3-5** Thermal structure of the atmosphere. Air temperature (shown with the yellow line) decreases with increasing altitude in the troposphere and mesosphere and increases with increasing altitude in the stratosphere and thermosphere.

Troposphere The lowest layer of the atmosphere—and the one in contact with Earth's surface—is known as the **troposphere**. The names troposphere and *tropopause* (the top of the troposphere) are derived from the Greek word *tropos* ("turn") and imply an overturning of the air in this zone as a result of vertical mixing and turbulence. The depth of the troposphere varies in both time and place (Figure 3-6). It is deepest over tropical regions and shallowest over the poles, deeper in summer than in winter, and varies with the passage of warm and cold air masses. On the average, the top of the troposphere (including the tropopause) is about 18 kilometers (11 miles) above sea level at the equator and about 8 kilometers (5 miles) above sea level over the poles.

Stratosphere The names **stratosphere** and *stratopause* come from the Latin *stratum* ("a cover"), implying a layered or stratified condition without vertical mixing—if we describe the air in the troposphere as being "turbulent," we can describe the air in the stratosphere as being "stagnant." The stratosphere extends from an altitude of 18 kilometers (11 miles) above sea level to about 48 kilometers (30 miles).

Upper Thermal Layers The names *mesosphere* and *mesopause* are from the Greek *meso* ("middle"). The mesosphere begins at 48 kilometers (30 miles) and extends to

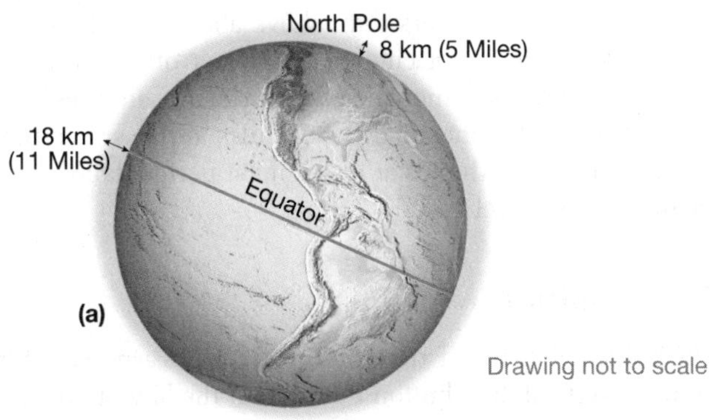

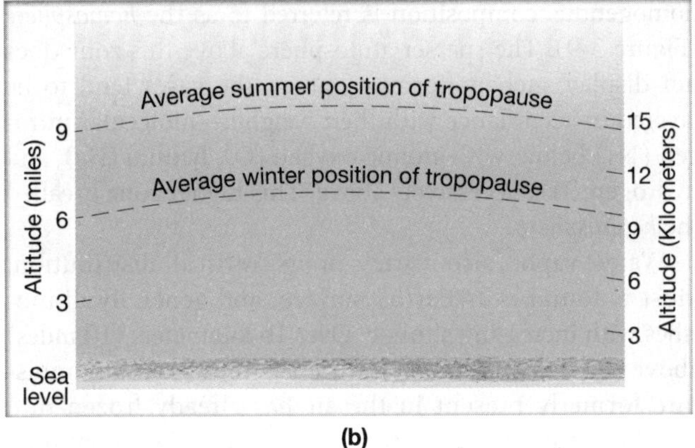

▲ **Figure 3-6** The depth of the troposphere is variable. (a) This thermal layer is deepest over the equator where surface temperatures are warm and thermal mixing is greatest, and shallowest over the poles. (b) It is deeper in summer than in winter. (The thickness of the atmosphere is greatly exaggerated in this drawing.)

about 80 kilometers (50 miles) above sea level. Above the mesopause is the *thermosphere* (from the Greek *therm*, meaning "heat"), which begins at an altitude of 80 kilometers (50 miles) above sea level and has no definite top. Instead it merges gradually into the region called the *exosphere*, which in turn blends into interplanetary space. Traces of atmosphere extend for literally thousands of kilometers higher. Therefore, "top of the atmosphere" is a theoretical concept rather than a reality, with no true boundary between atmosphere and outer space.

Temperature Patterns in the Atmosphere

Air temperature changes with altitude (see Figure 3-5). Beginning at sea level, where the average global temperature is about 15°C (59°F), temperature first decreases steadily with increasing altitude through the troposphere, declining to an average temperature of about −57°C (−71°F) at the tropopause. Temperature remains constant through the tropopause and for some distance into the stratosphere. At an altitude of about 20 kilometers (12 miles), air temperature begins increasing with increasing altitude, reaching a maximum at 48 kilometers (30 miles) at the bottom of the mesosphere, where the temperature is about −2°C (28°F). Then the temperature decreases with increasing altitude all through the mesosphere, reaching a minimum at the top of that layer at an altitude of 80 kilometers (50 miles). Temperature remains constant for several kilometers into the thermosphere and then begins to increase until, at an altitude of 200 kilometers (125 miles), it is higher than the maximum temperature in the troposphere. In the exosphere, the normal concept of temperature no longer applies.

Each "warm zone" in this temperature gradient has a specific source of heat. In the lower troposphere, the heat source is the surface of Earth itself—solar energy warms the surface of Earth, and this energy is in turn transferred through a number of different processes to the troposphere immediately above. The warm zone at the stratopause is near the top of the ozone layer, where ozone is absorbing the ultraviolet portion of sunlight and thereby warming the atmosphere. In the thermosphere, various atoms and molecules also absorb ultraviolet rays from the Sun and are thus split and heated. The "cold zones" that separate these warm zones are cold simply because they lack such sources of heat.

Although there are many interesting physical relationships in the stratosphere, mesosphere, thermosphere, and exosphere, our attention in this book is directed almost entirely to the troposphere because storms and essentially all the other phenomena we call "weather" occur here. Occasionally, however, we must consider atmospheric conditions above the troposphere—especially the ozone layer.

Pressure

Atmospheric pressure can be thought of, for simplicity's sake, as the "weight" of the overlying air. (In Chapter 5, we explore the concept of pressure in much greater detail.) The taller the "column of air" above an object, the greater the air pressure exerted on that object, as Figure 3-7 shows. Because air is highly compressible, the lower layers of the atmosphere are compressed by the air above, and this compression increases both the pressure

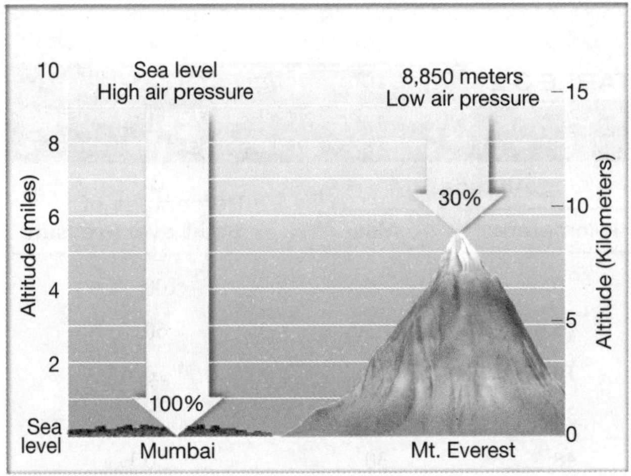

▲ **Figure 3-7** Atmospheric pressure is highest at sea level and diminishes rapidly with increasing altitude.

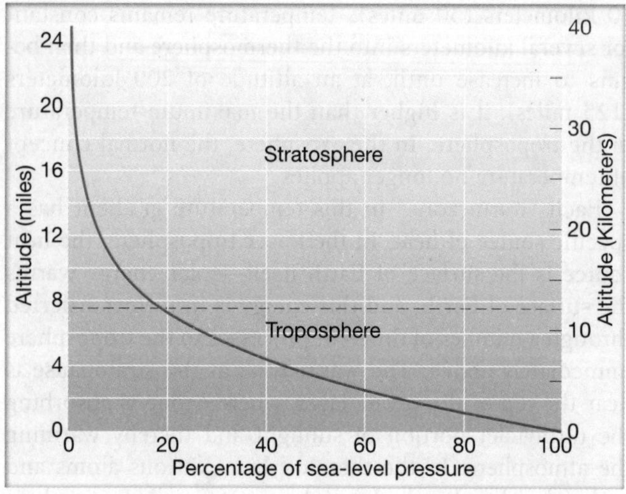

▲ **Figure 3-8** Air pressure decreases with increasing altitude but not at a constant rate.

exerted by the lower layers and the density of these layers. Air in the upper layers is subjected to less compression and therefore has a lower density and exerts a lower pressure.

Air pressure is normally highest at sea level and decreases rapidly with increasing altitude. The change of pressure with altitude is not constant, however. As a generalization, pressure decreases upward at a decreasing rate, as Figure 3-8 shows. Table 3-2 expresses the pressure at various altitudes as a percentage of sea-level pressure. From it we see, for instance, that at an altitude of 16 kilometers (10 miles)—a typical elevation of the tropopause—atmospheric pressure is only 10 percent of its sea-level value; this is just another way to say that most of the mass of the atmosphere is found relatively close to the ground.

One-half of all the gas molecules making up the atmosphere lie below 5.6 kilometers (3.5 miles), and 90 percent of them are concentrated in the first 16 kilometers (10 miles) above sea level (see Figure 3-2). Pressure becomes so slight in the upper layers that, above about 80 kilometers (50 miles), there is not enough to register on an ordinary barometer, the instrument used to measure air pressure. Above this level, atmospheric molecules are so scarce that air pressure is less than that in the most perfect laboratory vacuum at sea level.

TABLE 3-2 Atmospheric Pressure at Various Altitudes Expressed as Percentage of Sea-Level Pressure

Altitude		
Kilometers	Miles	Percentage of Sea-Level Pressure
0	0	100
5.6	3.5	50
16	10	10
32	20	1
48	30	0.1
80	50	0.001
96	60	0.00001

Composition

The principal gases of the atmosphere have a remarkably uniform vertical distribution throughout the lowest 80 kilometers (50 miles) or so of the atmosphere. This zone of homogenous composition is referred to as the *homosphere* (Figure 3-9). The sparser atmosphere above this zone does not display such uniformity; rather, the gases tend to be layered in accordance with their weights—molecular nitrogen (N_2) below, with atomic oxygen (O), helium (He), and hydrogen (H) successively above. This higher zone is called the *heterosphere*.

Water vapor also varies in its vertical distribution. Most is found near Earth's surface, and generally diminishes with increasing altitude. Over 16 kilometers (10 miles) above sea level, the temperature is so low that any moisture formerly present in the air has already frozen into ice. At these altitudes, therefore, there is rarely enough moisture to provide the raw material to make even a wisp of a cloud. If you have done any flying, you may recall the remarkable sight of a cloudless sky overhead once the plane breaks through the top of a solid cloud layer below.

Two other vertical compositional patterns are worthy of mention here.

1. The **ozone layer**, which, as stated above, lies between 15 and 48 kilometers (9 and 30 miles) up, is sometimes called the *ozonosphere*. Despite its name, the ozone layer is not composed primarily of ozone. It gets its name because that is where the concentration of ozone relative to other gases is at its maximum. Even in the section of the ozone layer where the ozone attains its greatest concentration, at about 25 kilometers (15 miles) above sea level, this gas only accounts for no more than about 15 parts per million of the atmosphere.
2. The *ionosphere* is a deep layer of electrically charged molecules and atoms (which are called *ions*) in the middle and upper mesosphere and the lower thermosphere, between about 60 and 400 kilometers (40 and 250 miles). The ionosphere is significant because it aids long-distance communication by reflecting radio waves back to Earth. It is also known for its auroral displays, such as the "northern lights" (Figure 3-10) that develop when charged atomic particles from the Sun are trapped by the magnetic field of Earth near the poles. In the ionosphere, these particles "excite" the nitrogen molecules and oxygen atoms, causing them to emit light, not unlike a neon light bulb.

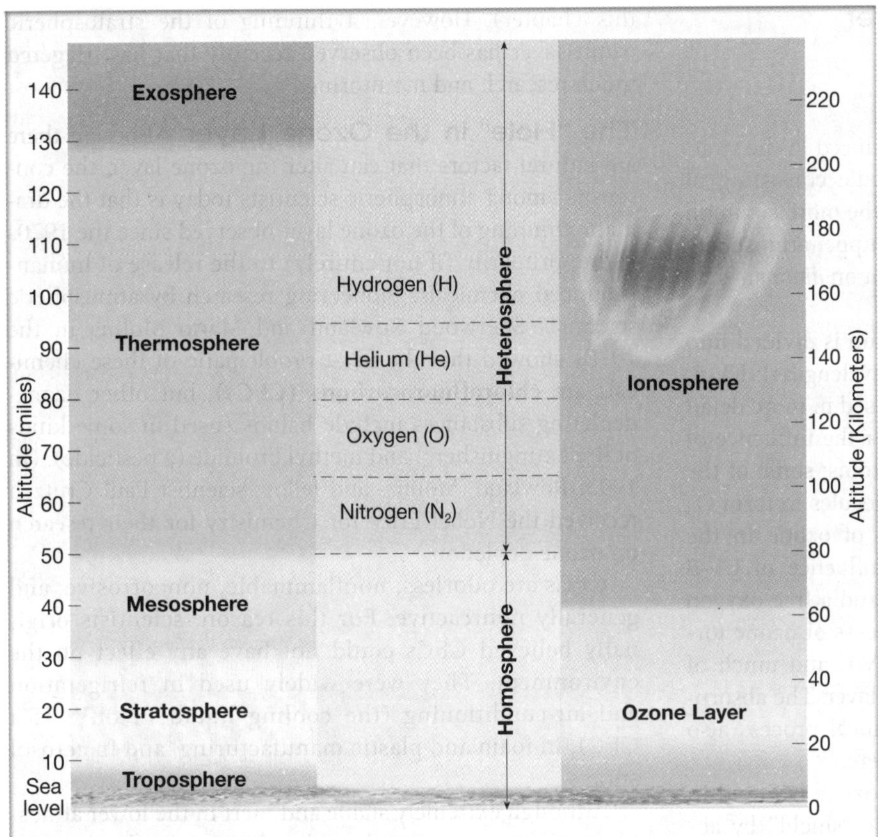

◀ **Figure 3-9** Relationships among the layers of the atmosphere. The homosphere is a zone of uniform vertical distribution of gases; in the heterosphere, however, the distribution lacks this uniformity. The ozone layer contains significant concentrations of ozone; the ionosphere is a deep layer of ions, which are electrically charged molecules and atoms.

Human-Induced Atmospheric Change

As world population grows and the use of industrial technology intensifies, human activity has increasingly had unintended and uncontrolled effects on the atmosphere—effects seen around the globe. This human impact, in simplest terms, consists of the introduction of impurities into the atmosphere at a pace previously unknown—impurities capable of altering global climate or harming forms of life. Although for many years a concern of atmospheric scientists, over the last decade especially, the consequences of human-produced changes in the atmosphere—especially global climate change—have received international attention from not only the scientific community but from the general public as well.

▲ **Figure 3-10** The Aurora Borealis (Northern Lights) in the ionosphere, as seen from Bear Lake, Alaska.

In June 2009, the United States Global Change Research Program, a joint scientific effort involving more than a dozen federal agencies and the White House, issued its latest report, entitled *Global Climate Change Impacts in the United States*. Building on the findings of the *Fourth Assessment Report* of the Intergovernmental Panel on Climate Change (discussed in subsequent chapters), the report offers a candid assessment of climate change in the United States:

> Climate change is apparent now across our nation. Trends observed in recent decades include rising temperatures, increasing heavy downpours, rising sea level, longer growing seasons, reductions in snow and ice, and changes in the amounts and timing of river flows. These trends are projected to continue, with larger changes resulting from higher amounts of heat-trapping gas emissions, and smaller changes from lower amounts of these emissions. The observed changes in climate are already causing a wide range of impacts, and these impacts are expected to grow.

In subsequent sections of the book, we will highlight a number of aspects of human-induced atmospheric change and some of the steps that can be—and are being—taken to ameliorate them. The first major topic of global environmental change we spotlight is the depletion of the ozone layer.

Depletion of the Ozone Layer

Ozone Depletion

As we saw earlier, ozone is naturally produced in the stratosphere. It is a form of oxygen molecule consisting of three atoms of oxygen (O_3) rather than the more common two atoms (O_2). Ozone is created in the upper atmosphere by the action of ultraviolet solar radiation on *diatomic oxygen* (O_2) molecules.

Ultraviolet (UV) radiation from the Sun is divided into three bands (from longest to shortest wavelengths): *UV-A*, *UV-B*, and *UV-C* (radiation will be discussed in more detail in Chapter 4). In the stratosphere, under the influence of UV-C, O_2 molecules split into oxygen atoms; some of the free oxygen atoms combine with O_2 molecules to form O_3 (Figure 3-11). The natural breakdown of ozone in the stratosphere occurs when, under the influence of UV-B and UV-C, ozone breaks down into O_2 and a free oxygen atom. Through this ongoing natural process of ozone formation and breakdown, nearly all of UV-C and much of UV-B radiation is absorbed by the ozone layer. The absorption of UV radiation in this photochemical process also serves to warm this layer of the atmosphere.

About 90 percent of all atmospheric ozone is found in the stratosphere where it forms a fragile "shield" by absorbing most of the potentially dangerous ultraviolet radiation from the Sun. Ultraviolet radiation can be biologically harmful in many ways. Prolonged exposure to UV radiation is linked to skin cancer—both the generally curable nonmelanoma varieties as well as much more serious melanoma; it is also linked to increased risk for cataracts; it can suppress the human immune system, diminish the yield of many crops, disrupt the aquatic food chain by killing microorganisms such as phytoplankton on the ocean surface, and may have other negative effects still undiscovered.

Ozone is also produced near Earth's surface in the troposphere through human activities, forming one of components of photochemical smog (discussed later in this chapter). However, a thinning of the stratospheric ozone layer has been observed recently that has triggered much research and monitoring.

The "Hole" in the Ozone Layer Although there are natural factors that can alter the ozone layer, the consensus among atmospheric scientists today is that the dramatic thinning of the ozone layer observed since the 1970s is due primarily (if not entirely) to the release of human-produced chemicals. Pioneering research by atmospheric scientists Sherwood Rowland and Mario Molina in the 1970s showed that the most problematic of these chemicals are **chlorofluorocarbons (CFCs)**, but other ozone-depleting substances include halons (used in some kinds of fire extinguishers) and methyl bromide (a pesticide). (In 1995, Rowland, Molina, and fellow scientist Paul Crutzen received the Nobel Prize for Chemistry for their research on ozone depletion.)

CFCs are odorless, nonflammable, noncorrosive, and generally nonreactive. For this reason, scientists originally believed CFCs could not have any effect on the environment. They were widely used in refrigeration and air-conditioning (the cooling liquid Freon™ is a CFC), in foam and plastic manufacturing, and in aerosol sprays.

Although extremely stable and inert in the lower atmosphere, CFCs are broken down by ultraviolet radiation once they reach the ozonosphere. Under the influence of UV radiation, a chlorine atom is released from a CFC molecule (Figure 3-12); the chlorine atom then reacts with ozone,

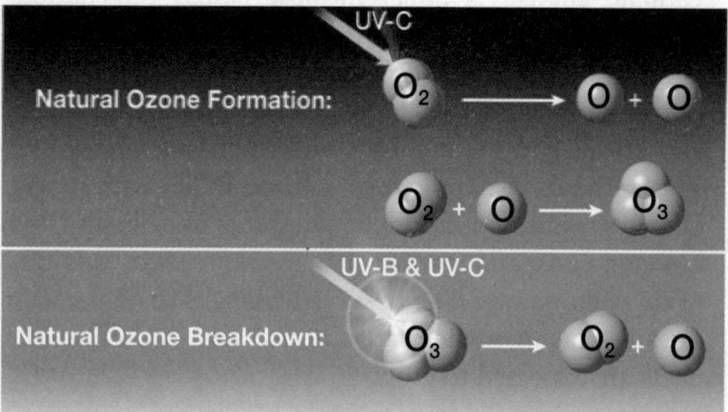

▲ **Figure 3-11** The natural formation and breakdown of ozone. Ultraviolet radiation splits oxygen molecules (O_2) into free oxygen atoms (O), some of which combine with other O_2 molecules to form ozone (O_3); also under the influence of UV light, ozone naturally breaks back down into O_2 and a free oxygen atom.

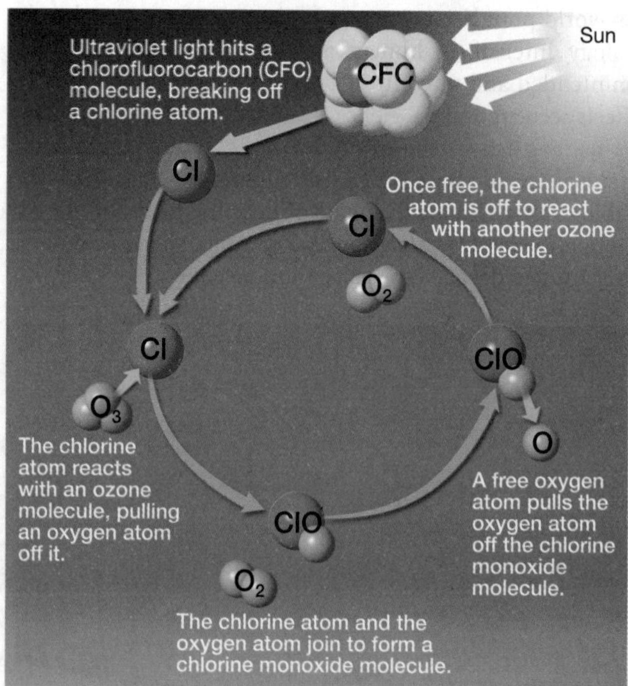

▲ **Figure 3-12** The destruction of ozone in the stratosphere by chlorine atoms derived from the breakdown of CFCs in the atmosphere. Chlorine atoms are unchanged by the reaction and can repeat the process. Thus a single chlorine atom can destroy tens of thousands of ozone molecules.

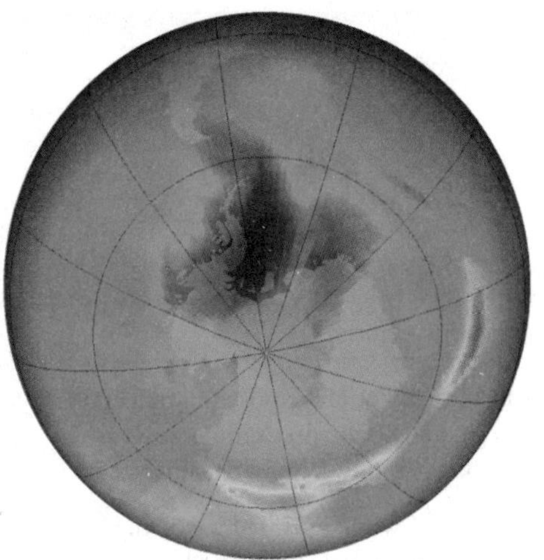
September 1979

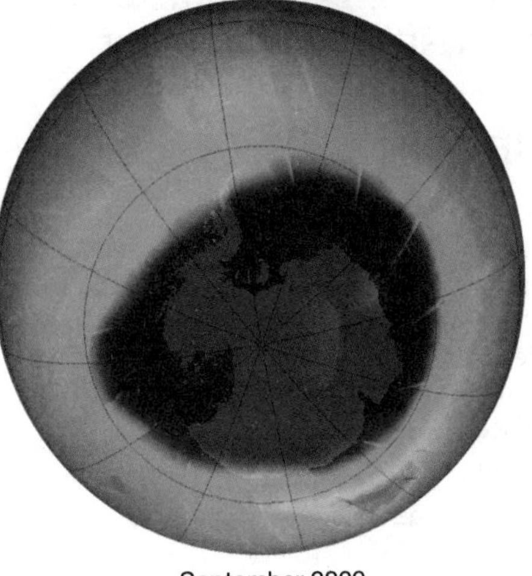
September 2009

◀ **Figure 3-13** The Antarctic ozone hole in 1979 and 2009. The area over Antarctica, shown in dark blue and purple, has the lowest concentration of ozone. As a result of international efforts to limit the production of ozone-depleting chemicals, after growing in size since the 1970s the Antarctic ozone hole seems to be stabilizing.

breaking it apart to form one chlorine monoxide (ClO) molecule and one O_2 molecule. The chlorine monoxide molecule can then react with a free atom of oxygen, forming a diatomic oxygen molecule while freeing the chlorine atom to react with another ozone molecule. As many as 100,000 ozone molecules can be destroyed for every chlorine atom released.

Not only is the ozone layer thinning, in some places it has temporarily disappeared almost entirely. Monitored by a satellite instrument called *TOMS* (Total Ozone Mapping Spectrometer), the annual ebb and flow of the ozone layer has been continuously mapped since 1979 with a "hole" developing and persisting over Antarctica longer and longer each year (Figure 3-13). By the late 1980s, an ozone "hole" was found over the Arctic as well.

Why is ozone depletion more severe over the polar regions, particularly Antarctica? In part because of the extreme cooling of Antarctica during the winter, a whirling wind pattern known as the *polar vortex* develops, effectively isolating the polar air from the atmosphere in lower latitudes. Within the stratosphere, ice crystals form thin *polar stratospheric clouds* (PSC)—the presence of these clouds can dramatically accelerate the process of ozone destruction. The ice crystals in PSCs provide surfaces on which a number of reactions can take place, including the accumulation of chlorine-based molecules. With the return of sunlight in the polar spring (September in the Southern Hemisphere), UV radiation triggers the catalytic reaction, and ozone depletion begins.

The thinning of ozone over the Arctic has been generally less severe than over the Antarctic because comparable atmospheric conditions to those in Antarctica are less well developed over the North Pole.

Stratospheric ozone depletion has been correlated with increased levels of ultraviolet radiation reaching ground level in Antarctica, Australia, mountainous regions of Europe, central Canada, and New Zealand. In part because of the increased health risks posed by higher levels of UV radiation reaching the surface, a *UV Index* has been established to provide the public with information about the intensity of UV radiation in an area (see the box, "People and the Environment: The UV Index").

The Montreal Protocol These discoveries were sufficiently alarming that a number of countries including the United States banned the use of CFCs in aerosol sprays in 1978. A major international treaty—the Montreal Protocol on Substances That Deplete the Ozone Layer—was negotiated in 1987 to set timetables for phasing out the production of the major ozone-depleting substances. More than 189 countries, including all major producers of ozone-depleting substances, have ratified the proposal. Following stipulations of the treaty and its more recent amendments, the industrialized countries of the world had banned CFC production by 1996. Moreover, the protocol signatories pledged a fund of more than $700 million to help developing countries implement alternatives to CFCs and end their production by 2010.

Even with the Montreal Protocol fully implemented, the ozone layer will not recover immediately because the reservoir of CFCs in the atmosphere may persist for 50 or 100 years. The largest measurable Antarctic ozone hole was observed in 2006, and since then it appears that ozone loss is stabilizing. However, some studies suggest that it may be 2050 before recovery is well under way.

Addressing the depletion of the ozone layer is considered by many scientists to be an example of an environmental success story: a human-produced problem was identified, and a global strategy was implemented to counteract it. In 2009, an international team of scientists led by NASA's Paul Newman used computer models to predict what would have happened if CFC use had *not* been curtailed by the Montreal Protocol. Their study suggested that without the ban on CFC use, by the year 2100 ozone levels over the tropics would have collapsed to levels found today only in the Arctic and Antarctic—resulting in summertime UV radiation levels more than three times greater than what is considered "very high" today.

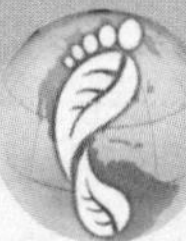

PEOPLE AND THE ENVIRONMENT

The UV Index

The UV Index, or *UVI*, was developed in the 1990s by the Environmental Protection Agency (EPA) and National Weather Service of the United States to inform the public about levels of harmful ultraviolet radiation reaching the surface. The index was revised in 2004 to conform to international reporting standards coordinated by the World Health Organization.

The UVI is a forecast of the next day's level of ultraviolet radiation, given on a scale of 1 to 11+, with 1 representing relatively low UV exposure risk, and levels of 8 or greater representing relatively high UV exposure risk. Each risk category of the UVI is accompanied by recommended precautions that should be taken by people exposed to outside sunlight (Table 3-A).

TABLE 3-A UV Index Scale

UVI Range	Exposure Risk	Recommendations
Less than 2	Low	If you burn easily or are exposed to reflections off snow or water, take precautions such as wearing sunglasses and applying sunscreen.
3–5	Moderate	Protect yourself while outside by wearing protective clothing, hats and sunglasses. Limit midday Sun exposure.
6–7	High	Apply sunscreen of SPF 15 or greater. Use sunglasses and wide-brim hat. Reduce midday Sun exposure.
8–10	Very High	Take extra precautions. Use SPF 15+ sunscreen liberally. Minimize Sun exposure between 10 A.M. and 4 P.M.
11 or greater	Extreme	Take all precautions. Avoid midday Sun exposure between 10 A.M. and 4 P.M. Reapply SPF 15+ sunscreen liberally every two hours.

Source: U.S. Environmental Protection Agency, SunWise Program.

Current research indicates that skin damage from exposure to ultraviolet radiation is cumulative. It is especially important that young children—of all skin types—be protected from excessive Sun exposure through the use of sunscreens of SPF (Sun Protection Factor) 15 or greater, and by taking other measures such as wearing hats and protective clothing.

The UV Index forecast for a city or region is based on measured levels of ozone in the atmosphere, as well as the amount of cloud cover and the elevation of the location (Figure 3-A). In general, the lower the concentration of stratospheric ozone and the clearer the skies, the greater the level of harmful UV radiation that will reach the surface (Figure 3-B).

The ozone concentration data used in the forecast are obtained with satellites using equipment such as the SBUV/2 (Solar Backscatter Ultraviolet/2) instruments on the National Oceanic and Atmospheric Administration's (NOAA) Polar Orbiting Environmental Satellites, and the OMI (Ozone Monitoring Instrument) onboard the NASA Aura satellite.

To find the UVI forecast for your location, go to http://www.epa.gov/sunwise/uvindex2.html.

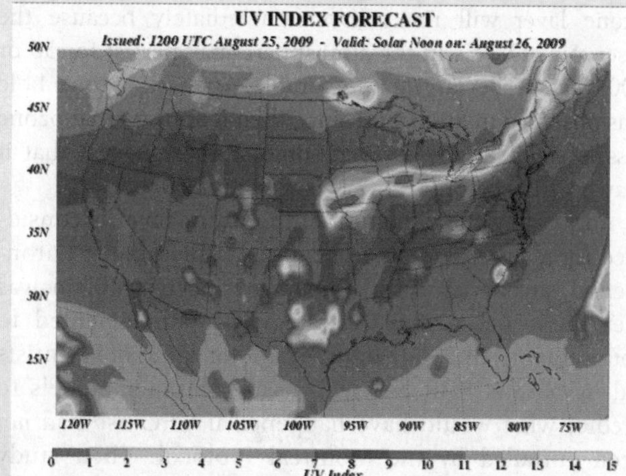

▲ **Figure 3-A** UV Index forecast map.

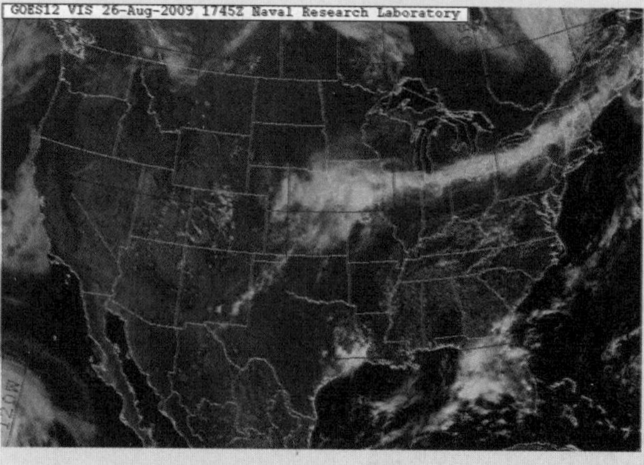

▲ **Figure 3-B** Visible light satellite image showing the location of cloud cover over the United States at the time of the UVI forecast in Figure 3-A. The areas of cloudiness were generally forecast to have lower amounts of UV radiation reaching the surface than areas with clearer skies.

Air Pollution

In addition to diminishing stratospheric ozone, humans have altered the composition of the atmosphere in other ways as well, most obviously in the form of photochemical smog and other forms of air pollution.

The atmosphere has never been without pollutants. There are many natural sources of contaminants in the atmosphere: smoke from wildfires, ash from volcanic eruptions, windblown dust, pollen from plants, and salt particles from breaking waves. Humans, however, have dramatically increased both the frequency and the magnitude of pollutants released into the atmosphere. By far the greatest problems are associated with cities because of the concentration of people and activities, and particularly from internal combustion engines and industrial establishments. The presence of pollutants in the air is most obviously manifest by impaired visibility due to fine particulate matter and associated water droplets. More critical, however, is the health hazard imposed by increasing concentrations of chemical impurities in the air.

With the advent of the Industrial Revolution, pollution became more widespread and intense, and by the twentieth century it was recognized as a major societal problem. Air-pollution "episodes" have often been death-dealing: 63 people killed over a five-day period in Belgium's Meuse Valley in 1930; 6000 illnesses and 20 deaths in Donora, Pennsylvania, during a week in 1948; 4000 deaths in London during a five-day ordeal in 1952. In addition to these isolated episodes, however, is the general increase in foul air over most large urban areas of the world.

Primary and Secondary Pollutants Two general categories of air pollutants can be identified. **Primary pollutants** are contaminants released directly into the air, such as particulates, sulfur compounds, nitrogen compounds, carbon oxides, and hydrocarbons. In the United States alone, nearly 130 million tons of primary pollutants were released in 2007 (Figure 3-14). **Secondary pollutants**, such as photochemical smog, are not released directly into the air but instead form as a consequence of chemical reactions or other processes in the atmosphere.

Particulates Particulates (or aerosols) are tiny solid particles or liquid droplets suspended in the air. Primary sources of particulates from human activities include smoke from combustion and dust emitted from industrial activities. The concentration of particulates may also increase through secondary processes, such as when small particles coalesce to form larger particles, or when liquid droplets develop around condensation nuclei. It appears that many of the health hazards associated with particulates are greatest when the particles are less than 2.5 micrometers (<2.5 millionths of a meter) in diameter. In 1997, the U.S. Environmental Protection Agency revised its regulations to take into account the harmful effects of such fine particulates, known as $PM_{2.5}$, although recent research indicates that still smaller particulates may also pose a significant hazard.

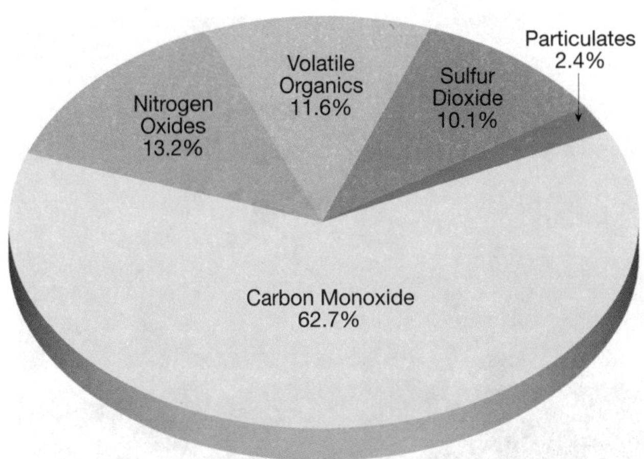

▲ **Figure 3-14** The emissions of primary atmospheric pollutants by the United States in 2007. Values given represent the percentage of the total weight of pollutants emitted, estimated at 129 million tons in 2007.

Carbon Monoxide Carbon monoxide (CO) is the most plentiful primary pollutant, formed by the incomplete combustion of carbon-based fuels, especially by motor vehicles. Because this gas is odorless and colorless, people exposed to carbon monoxide in confined spaces can be quickly overcome after CO enters the bloodstream and decreases the amount of oxygen available to their brain and other organs.

Sulfur Compounds A large portion of the sulfur compounds found in the atmosphere are of natural origin, released from volcanoes or hydrothermal vents such as those in Yellowstone National Park—hydrogen sulfide (H_2S) with its familiar "rotten egg" smell is one example of this. However, especially over the last century, human activity has increased the release of sulfur compounds into the atmosphere, primarily through the burning of fossil fuels such as coal and petroleum. Sulfur is one of the minor impurities in coal and oil; when those fuels are burned, sulfur compounds such as sulfur dioxide (SO_2) are released. Sulfur dioxide itself is a lung irritant and corrosive, but it also may react in the atmosphere to produce secondary pollutants such as sulfur trioxide (SO_3) and sulfuric acid (H_2SO_4)—one of the contributors to acid rain (discussed in Chapter 6).

Nitrogen Compounds Nitric oxide (NO) is a gas that can form in water and soil as a natural byproduct of biological processes; it generally breaks down quite quickly. However, NO may also form through combustion that takes place at high temperatures and pressures—such as in an automobile engine. Nitric oxide reacts in the atmosphere to form nitrogen dioxide (NO_2) a corrosive gas that gives some polluted air its yellow and reddish-brown color. Although NO_2 itself tends to break down quickly, it may in turn react under the influence of sunlight to form several different components of smog.

▲ **Figure 3-15** Photochemical smog over the city of Santiago, Chile.

Photochemical Smog A number of gases react to ultraviolet radiation in strong sunlight to produce secondary pollutants, making up what is known as **photochemical smog** (Figure 3-15). (The word "smog" was originally derived from the combination of the words *smoke* and *fog*; photochemical smog usually includes neither.) Nitrogen dioxide and hydrocarbons (also known as *volatile organic compounds* or VOC)—both of which can result from the incomplete burning of fuels such as gasoline—are major contributors to photochemical smog. Nitrogen dioxide breaks down under ultraviolet radiation to form nitric oxide, which may then react with VOC to form peroxyacetyl nitrate (PAN), which has become a significant cause of crop and forest damage in some areas.

The breakdown of NO_2 into NO also frees an oxygen atom that can react with O_2 molecules to form ozone—the main component of photochemical smog. Ozone has an acrid, biting odor that is distinctive of photochemical smog, causing damage to vegetation, corroding building materials (such as paint, rubber, plastics), and damaging sensitive tissues in humans (eyes, lungs, noses).

The state of the atmosphere is also an important determinant of the level of air pollution, especially photochemical smog. If there is considerable air movement, the pollutants can be quickly and widely dispersed. On the other hand, stagnant air allows for a rapid accumulation of pollutants. Thus, air pollution is particularly notable when high-pressure (anticyclonic) conditions prevail, with air subsiding from above and little surface wind flow. If the air is particularly stable, *temperature inversions* (cooler air below warmer) develop which function as "stability lids" that inhibit updrafts and general air movement. Almost all the cities that are persistently among the smoggiest in the world, such as Mexico City and Los Angeles, are characterized by a high frequency of temperature inversions and anticyclonic development.

Consequences of Anthropogenic Air Pollution Although the effects of atmospheric pollutants vary, in general all forms of air pollution adversely affect lung function. Carbon monoxide, sulfur dioxide, and particulates can contribute to cardiovascular disease, while prolonged exposure to some particulates may promote lung cancer. Nitrogen oxides and sulfur dioxide are the principal contributors to acid rain. Tropospheric ozone damages crops and trees and is now the most widespread air pollutant, and it continues on a rising trend. The Environmental Protection Agency (EPA) reports that perhaps one-fifth of all hospital cases involving respiratory illness in the summer are a consequence of ozone exposure.

In the last few decades in the United States, there has been a downward trend in the emission of all pollutants except nitrogen oxides and ozone, largely as a result of increasingly stringent emission standards imposed by the EPA. Although significant steps are now being taken to reverse the growth of atmospheric pollution, as global population expands, the pollutant load expands with it, and an ongoing effort on the part of all countries will be required to curtail the problem.

Weather and Climate

Now that we have described the composition and structure of the atmosphere, we turn more specifically to the broad set of processes operating within this ocean of air.

The vast and invisible atmospheric envelope is energized by solar radiation, stimulated by earthly motions, and affected by contact with Earth's surface. The atmosphere reacts by producing an infinite variety of conditions and phenomena known collectively as *weather*—the study of weather is known as *meteorology*. The term **weather** refers to short-run atmospheric conditions that exist for a given time in a specific area. It is the sum of temperature, humidity, cloudiness, precipitation, pressure, winds, storms, and other atmospheric variables for a short period of time. Thus, we speak of the weather of the moment, the week, the season, or perhaps even of the year or the decade.

Weather is in an almost constant state of change, sometimes in seemingly erratic fashion, yet in the long view, it is possible to generalize the variations into a composite pattern which is termed *climate*. **Climate** is the aggregate of day-to-day weather conditions over a long period of time. It encompasses not only the average characteristics, but also the variations and extremes of weather. To describe the climate of an area requires weather information over an extended period, normally at least three decades.

Weather and climate, then, are related but not synonymous terms. The distinction between them is the difference between immediate specifics and protracted generalities. As the country philosopher once said, "Climate is what you expect; weather is what you get." Stated more sarcastically, "It is the climate that attracts people to a location, and the weather that makes them leave."

Weather and climate have direct and obvious influences on agriculture, transportation, and human life in general. Moreover, climate is a significant factor in the development of all major aspects of the physical landscape—soils, vegetation, animal life, hydrography, and topography.

Because climate and weather are generated in the atmosphere, our ultimate goal in studying the atmosphere is to understand the distribution and characteristics of climatic types over Earth. To achieve this understanding, we must consider in detail many of the processes that take place in the atmosphere. While our ultimate goal is an understanding of long-run atmospheric conditions (climate), we will spend the next four chapters gaining an appreciation for the dynamics involved in the momentary state of the atmosphere (weather).

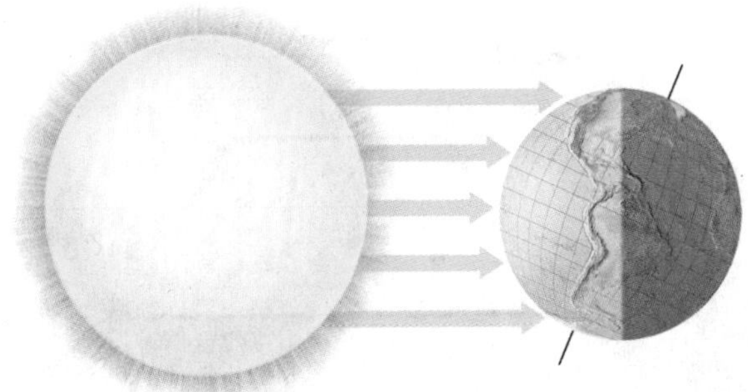

▲ **Figure 3-16** Solar energy coming to Earth. See Figure 1-22 and the accompanying text for a review of this topic.

The Elements of Weather and Climate

The atmosphere is a complex medium, and its mechanisms and processes are sometimes very complicated. Its nature, however, is generally expressed in terms of only a few variables, which are measurable. The data thus recorded provide the raw materials for understanding both temporary (weather) and long-term (climate) atmospheric conditions.

These variables can be thought of as the **elements of weather and climate**. The most important are (1) temperature, (2) moisture content, (3) pressure, and (4) wind (Table 3-3). These are the basic ingredients of weather and climate—the ones you hear about on the nightly weather report. Measuring how they vary in time and space makes it possible to decipher at least partly the complexities of weather dynamics and climatic patterns.

The Controls of Weather and Climate

Variations in the climatic elements are frequent, if not continuous, over Earth. Such variations are caused, or at least strongly influenced, by certain semipermanent attributes of our planet, which are often referred to as the **controls of weather and climate** (Table 3-3). The principal controls are briefly described in the paragraphs that follow and are explained in more detail in subsequent chapters. Although they are discussed individually here, it should be emphasized that there often is much overlap and interaction among them, with widely varying effects.

TABLE 3-3 The Elements and Controls of Weather and Climate

Elements of Weather and Climate	Controls of Weather and Climate
Temperature	Latitude
Pressure	Distribution of land and water
Wind	General circulation of the atmosphere
Moisture content	General circulation of the oceans
	Altitude
	Topographic barriers
	Storms

Latitude We noted in Chapter 1 that the continuously changing positional relationship between the Sun and Earth brings continuously changing amounts of sunlight, and therefore of radiant energy, to different parts of Earth's surface. Thus, the basic distribution of heat over Earth is first and foremost a function of latitude, as indicated in Figure 3-16. In terms of elements and controls, we say that the control latitude strongly influences the element temperature. Overall, latitude is the most fundamental control of climate.

Distribution of Land and Water Probably the most fundamental distinction concerning the geography of climate is the distinction between continental climates and maritime (oceanic) climates. Oceans heat and cool more slowly and to a lesser degree than do landmasses, which means that maritime areas experience milder temperatures than continental areas in both summer and winter. For example, Seattle, Washington, and Fargo, North Dakota, are at approximately the same latitude (47° N), with Seattle on the western coast of the United States and Fargo deep in the interior. Seattle has an average January temperature of 5°C (41°F), while the January average in Fargo is −13°C (7°F). In the opposite season, Seattle has a July average temperature of 19°C (66°F), whereas in Fargo the July average is 22°C (71°F).

Oceans are also a much more prolific source of atmospheric moisture. Thus, maritime climates are normally more humid than continental climates. The uneven distribution of continents and oceans over the world, then, is a prominent control of the elements moisture content and temperature.

General Circulation of the Atmosphere The atmosphere is in constant motion, with flows that range from transitory local breezes to vast regional wind regimes. At the planetary scale, a semipermanent pattern of major wind and pressure systems dominates the troposphere and greatly influences most elements of weather

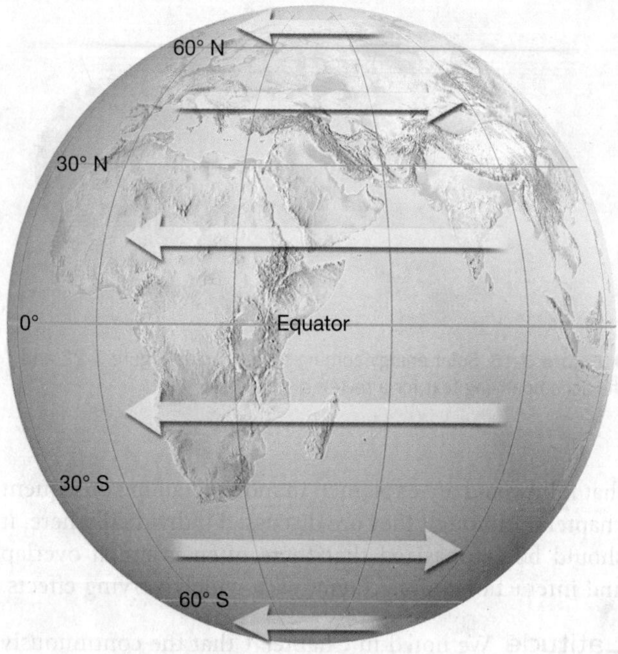

▲ **Figure 3-17** The general circulation of the atmosphere is an important climatic control. This diagram is a highly simplified version of the actual circulation, which is discussed in Chapter 5.

and climate. As a simple example, most surface winds in the tropics come from the east, whereas the middle latitudes are characterized by flows that are mostly from the west, as Figure 3-17 shows.

General Circulation of the Oceans Somewhat analogous to atmospheric movements are the motions of the oceans (Figure 3-18). Like the atmosphere, the oceans have many minor motions, but they also have a broad general pattern of currents. These currents assist in heat transfer by moving warm water poleward and cool water equatorward. Although the influence of currents on climate is much less than that of atmospheric circulation, ocean currents are not inconsequential. For example, warm currents are found off the eastern coasts of continents, and cool currents occur off western coasts—a distinction that has a profound effect on coastal climates.

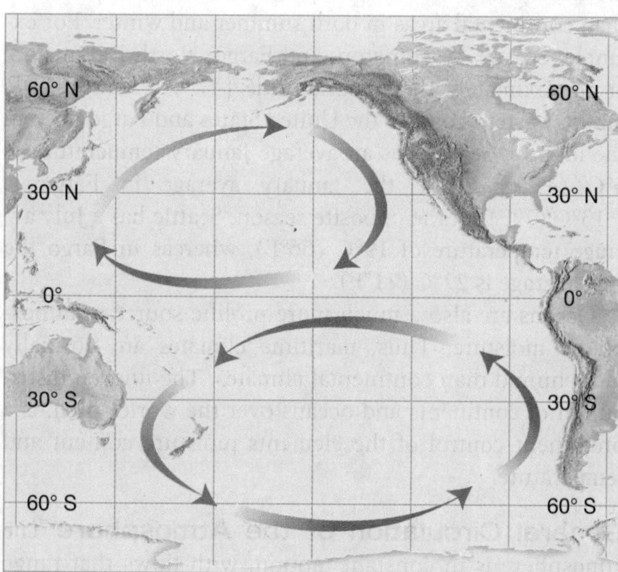

▲ **Figure 3-18** The general circulation of the oceans involves the movement of large amounts of warm water (red arrows) and cool water (blue arrows). These surface ocean currents have a significant climatic effect on neighboring landmasses.

▲ **Figure 3-19** Increasing altitude affects many components of the environment, as indicated by the variety of natural vegetation patterns on the slopes of Blanca Peak in south-central Colorado.

Altitude We have already noted that three of the four weather elements—temperature, pressure, and moisture content—generally decrease upward in the troposphere and are therefore under the influence of the control altitude. This simple relationship between the three elements and the control has significant ramifications for many climatic characteristics, particularly in mountainous regions (Figure 3-19).

Topographic Barriers Mountains and large hills sometimes have prominent effects on one or more elements of climate by diverting wind flow (Figure 3-20). The side of a mountain range facing the wind (the "windward" side), for example, is likely to have a climate vastly different from that of the sheltered ("leeward") side.

Storms Various kinds of storms occur over the world; some have very widespread distribution, whereas others are localized (Figure 3-21). Although they often result from interactions among other climate controls, all storms create specialized weather circumstances and so are considered to be a control. Indeed, some storms are prominent and frequent enough to affect not only weather but climate as well.

Before we move on to Chapter 4 to discuss temperature, the first of the elements of weather and climate, we need to consider one additional control of weather and climate—or perhaps more correctly, a control of some of the controls of weather and climate—the rotation of Earth.

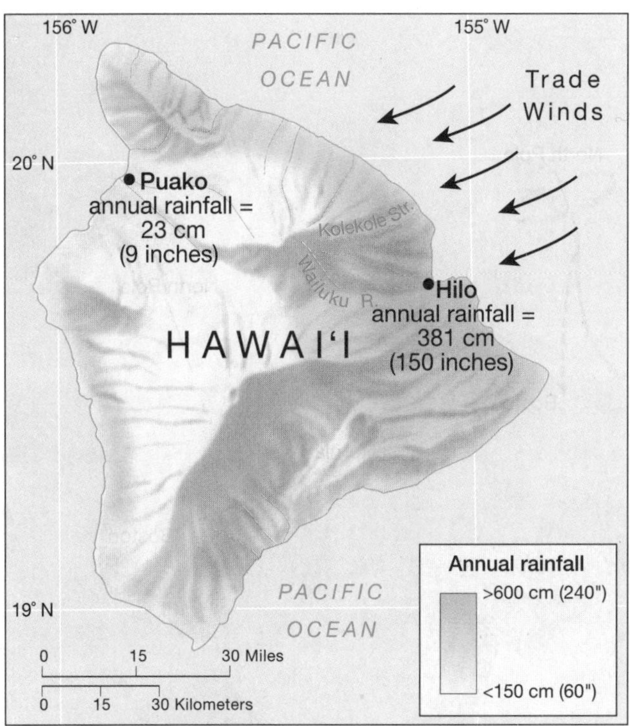

▲ **Figure 3-20** A topographic barrier as a control of climate. The difference in average annual rainfall in these two locations on the island of Hawai'i is caused by the mountain range separating them. Moisture-laden trade winds coming in from the northeast drop their moisture when forced to rise by the eastern face of the mountains. The result is a very wet eastern side of the island and a very dry western side.

▲ **Figure 3-21** A prominent midlatitude cyclone storm system over the British Isles is counterpointed by localized thunderstorms over North Africa, Sicily, Italy, and Greece.

The Coriolis Effect

Coriolis Effect

Everyone is familiar with the unremitting force of gravity. Its powerful pull toward the center of Earth influences all vertical motion that takes place near Earth's surface. Much less well known, however, because of its inconspicuous nature, is a pervasive horizontal influence on earthly motions. All things that move over the surface of Earth or in Earth's atmosphere appear to drift sideways as a result of Earth's rotation. George Hadley described this apparent deflection in the 1730s, but it was not explained quantitatively until Gaspard G. Coriolis (1792–1843), a French civil engineer and mathematician, did so a century later. The phenomenon is called the **Coriolis effect** in his honor (Figure 3-22a). The Coriolis effect is frequently, and correctly, referred to as the "Coriolis force," especially when mathematically calculating its consequences. In this book we retain the more general term Coriolis effect when referring to this phenomenon.

The nature of the deflection due to the Coriolis effect can be demonstrated by imagining a rocket fired toward Boston from the North Pole. During the few minutes that the rocket is in the air, earthly rotation will have moved the target some kilometers to the east because the planet rotates from west to east. If the Coriolis effect was not included in the ballistic computation, the rocket would pass some distance to the west of Boston (Figure 3-22b). To a person standing at the launch point and looking south, the uncorrected flight path appears to deflect to the right.

This rightward deflection (in the Northern Hemisphere) applies no matter in which direction the rocket moves. If a rocket were aimed at Boston from a location on the same parallel of latitude—say, northern California—there would also be a drift to the right (as viewed by a person at the launch site and looking east). During the time the rocket is in motion, both the California firing point and the Boston target rotate eastward. Because Earth rotates, the landing point of an uncorrected flight path is southwest of Boston (Figure 3-22c).

The Coriolis principle applies to any freely moving object—ball, bullet, airplane, automobile, even a person walking. For these and other short-range movements, however, the deflection is so minor and counterbalanced by other factors (such as friction, air resistance, and initial impetus) as to be insignificant. Long-range movements, on the other hand, can be significantly influenced by the Coriolis effect. The accurate firing of artillery shells or launching of rockets and spacecraft requires careful compensation for the Coriolis effect if the projectiles are to reach their targets.

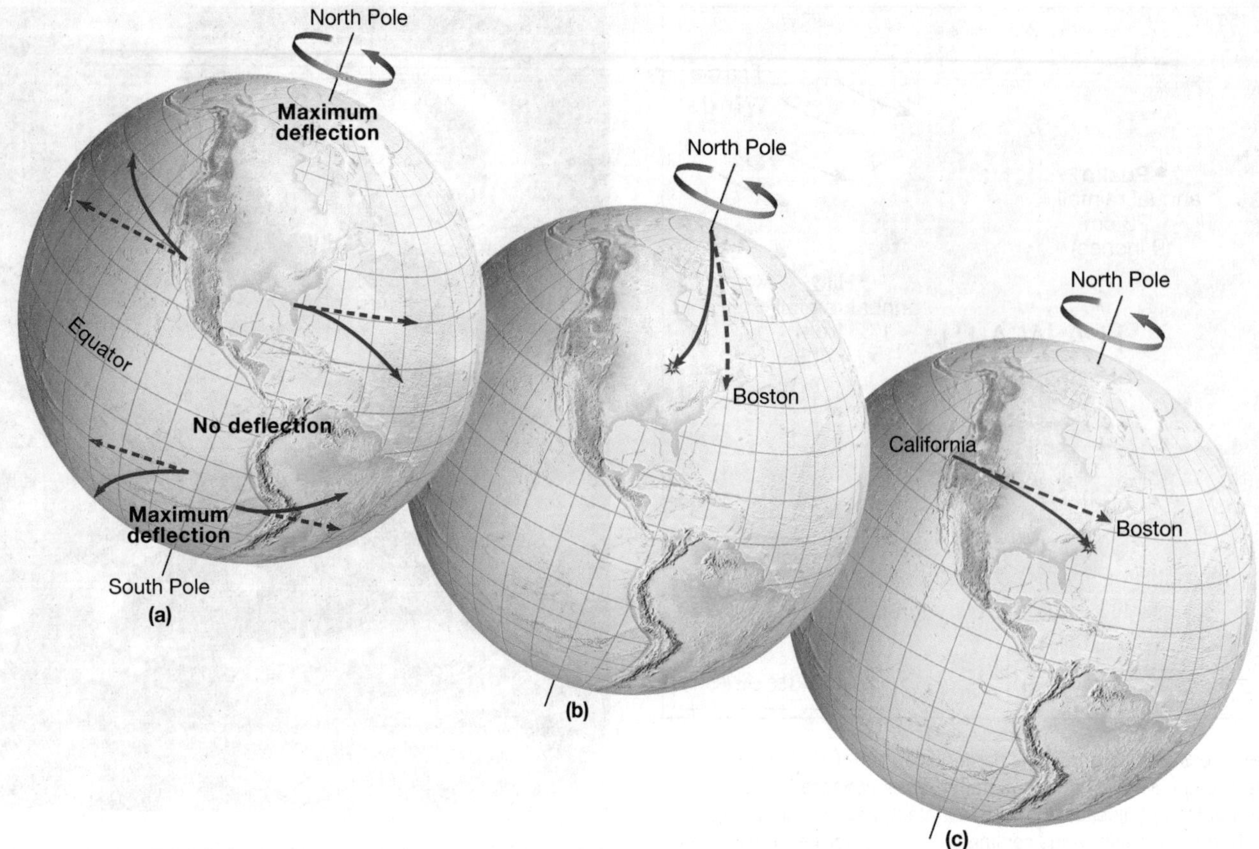

▲ **Figure 3-22** (a) The deflection caused by the Coriolis effect is to the right in the Northern Hemisphere and to the left in the Southern Hemisphere. The dashed lines represent the planned route, and the solid lines represent actual movement. (b) A rocket launched from the North Pole toward Boston lands to the west of the target if the Coriolis effect is not considered when the flight path is computed. (c) A rocket fired toward Boston from a point at the same latitude in California also appears to curve to the right because of the Coriolis effect deflection.

The following are four basic points to remember:

1. Regardless of the initial direction of motion, any freely moving object appears to deflect to the right in the Northern Hemisphere and to the left in the Southern Hemisphere.
2. The apparent deflection is strongest at the poles and decreases progressively toward the equator, where there is zero deflection.
3. The Coriolis effect is proportional to the speed of the object, and so a fast-moving object is deflected more than a slower one.
4. The Coriolis effect influences direction of movement only; it has no influence on speed.

The major importance of the Coriolis effect in our study of climate involves its influence on winds and ocean currents: All winds and ocean currents are influenced by the Coriolis effect. As we will see in Chapter 4, Northern Hemisphere ocean currents trend to the right and Southern Hemisphere currents to the left. The Coriolis effect deflection is a factor associated with the *upwelling* of cold water that takes place in subtropical latitudes where cool currents veer away from continental coastlines. The surface water that moves away from the shore is replaced by cold water rising from below.

One phenomenon that the Coriolis effect does not appear to influence is the circulation pattern of water that drains out of a sink or bathtub. There is a folk tale that Northern Hemisphere sinks drain clockwise and Southern Hemisphere sinks counterclockwise. The time involved is so short and the speed of the water so slow, however, that the Coriolis effect cannot be offered to explain these movements—the characteristics of the plumbing system, the shape of the washbowl, and pure chance are more likely to determine the flow patterns. A reader can test this hypothesis, of course, by filling and emptying several sinks and recording the results. Geographer Tom McKnight spent 5 months in Australia in 1992–1993 and tested 100 sinks. The results: 34 that emptied clockwise, 39 that emptied counterclockwise, and 27 that simply gushed down without a swirl in either pattern.

CHAPTER 3 LEARNING REVIEW

After studying this chapter, you should be able to answer the following questions. Key terms from each text section are shown in bold type. Definitions for key terms are also found in the glossary at the back of the book.

Size of Earth's Atmosphere (p. 49)

1. Why is the question "How deep is the atmosphere?" difficult to answer?

Composition of the Atmosphere (p. 50)

2. What are meant by the terms, "constant gases" and "variable gases" in the atmosphere?
3. Describe the most important constant gases of the atmosphere.
4. Briefly describe some of the roles that **water vapor**, **carbon dioxide**, **ozone**, and **particulates (aerosols)** play in atmospheric processes.
5. Describe both the vertical distribution of water vapor in the atmosphere and its horizontal (geographic) distribution near Earth's surface.

Vertical Structure of the Atmosphere (p. 52)

6. Discuss the size and general temperature characteristics of the **troposphere** and **stratosphere**.
7. Describe how atmospheric pressure changes with increasing altitude.
8. What is the **ozone layer**, and where is it located?
9. In our study of physical geography, why do we concentrate primarily on the troposphere rather than on other zones of the atmosphere?

Human-Induced Atmospheric Change (p. 55)

10. How is ozone formed and why is it important in the atmosphere?
11. What is meant by the "hole" in the ozone layer and what role have **chlorofluorocarbons (CFCs)** played in this?
12. Describe and contrast **primary pollutants** and **secondary pollutants** in the atmosphere.
13. Describe and explain the causes of **photochemical smog**.

Weather and Climate (p. 60)

14. What is the difference between **weather** and **climate**?
15. What are the four **elements of weather and climate**?
16. Briefly describe the seven dominant **controls of weather and climate**.
17. Describe the **Coriolis effect** and its cause.

Log in to **www.mygeoscienceplace.com** for animations, videos, RSS feeds, maps, and self-study quizzes to enhance your study of **Introduction to the Atmosphere**.

4 Insolation and Temperature

The temperature of the air at any time and at any place in the atmosphere is the result of the interaction among a variety of complex factors. In this chapter, our attention is focused on the energetics of the atmosphere—the important processes involved in bringing *insolation* (incoming solar radiation) to the atmosphere, in determining the extent of heating (and cooling) that takes place, and in transferring heat from one place to another. Understanding these processes will help us understand the distribution of temperature over Earth.

The Impact of Temperature on the Landscape

All organisms have certain temperature tolerances, and most are harmed by wide fluctuations in temperature. Thus, when the weather becomes particularly hot or cold, mobile organisms are likely to search for shelter, and their apparent presence in the landscape is diminished. Humans also seek haven from temperature extremes, although they have other options (such as specialized clothing) that allow them to brave the elements. In a broader view, temperature has a more profound effect on the landscape because both animals and plants often evolve in response to hot or cold climates. For this reason, the inventory of flora and fauna in any area of temperature extremes is influenced by the capability of the various species to withstand the long-term temperature conditions. An example of such an adaptation is shown in Figure 4-1a. The human-built landscape is also influenced by temperature considerations, as demonstrated by architectural styles and building materials (Figure 4-1b).

Most inorganic components of the landscape also are affected by long-run temperature conditions. For example, temperature is a basic factor in soil development, and repeated fluctuations of temperature are a prominent cause of the breakdown of exposed bedrock.

(a)

(b)

▲ **Figure 4-1** (a) A yellow-headed collared lizard in the desert of western Colorado. It strikes an erect pose to keep its belly off the hot rock. (b) The town museum in Igloolik in Canada's Northwest Territories is built in a style that is adjusted to the cold climate.

Energy, Heat, and Temperature

We begin by clarifying a few terms and concepts that we will use in our discussion of atmospheric heating. In the most basic sense, the universe is made up of just two kinds of "things": *matter* and *energy*. The concept of matter is fairly easy to comprehend. Matter is the "stuff" of the universe: the solids, liquids, gases—and atomic particles—from which all things are made. Matter has mass and volume; we can easily see and feel many kinds of matter. The concept of energy, on the other hand, may be more difficult to grasp.

◀ Sand dunes in Sossusvlei, Namibia.

Energy

Energy is commonly defined as the "ability to do work," but in practice, it may be better to describe what energy does rather than trying to define it. Energy refers to anything that changes the state or condition of matter. For example, it takes energy to cause something to move faster, or cause it to change direction, or cause it to break apart.

There are many forms of energy: *kinetic energy*, *chemical energy*, and *radiant energy*, among many others. Although energy can neither be created nor destroyed, it can, and frequently does, change from one form to another. In our discussion of the heating of the atmosphere, we will look at just a few forms of energy and the transformations between these forms.

To understand the heating of the atmosphere—or anything else—we first need to consider what is happening at the atomic or molecular level—a scale that we do not readily comprehend in our everyday lives. A great simplification of reality: all substances are composed of extraordinarily tiny, constantly "jiggling" *atoms*, commonly bonded together into combinations of atoms called *molecules*. The state of those substances—whether they are solid, liquid, gas, (or plasma, which is an ionized gas)—depends in part on how vigorously the molecules are jiggling back and forth in place. Because of this constant movement, the molecules in all substances possess energy. This kind of internal energy is a form of **kinetic energy**—the energy of movement.

The average amount of kinetic energy possessed by the molecules in a substance is closely associated with a physical property that we can often readily sense in our everyday lives: how hot or cold something is.[1] When a substance becomes warmer, it indicates that the average kinetic energy of the molecules in that substance has increased. In other words, energy has been added that has caused the molecules to jiggle back and forth more vigorously. So this brings us to several important definitions.

Temperature and Heat

Temperature is a description of the average kinetic energy of the molecules in a substance (strictly speaking, the average "back and forth" or *translational kinetic energy* of the molecules). The more vigorous the jiggling of the molecules is (and therefore the greater the internal kinetic energy), the higher the temperature of a substance.

Heat refers to energy that transfers from one object or substance to another because of a difference in temperature. Sometimes the term **thermal energy** is used interchangeably with the term heat. A substance doesn't "contain" heat—heat is simply the energy that is transferred from an object with a higher temperature to an object with a lower temperature, thereby decreasing the internal energy of the hotter object and increasing the internal energy of the cooler object.

Measuring Temperature

Instruments used to measure temperature are called *thermometers*. The temperature of an object is relative to one of three temperature scales that are in concurrent use (Figure 4-2). Each permits a precise measurement, but the existence of three scales creates an unfortunate degree of confusion.

Fahrenheit Scale The temperature scale most widely understood by the general public in the United States is the Fahrenheit scale (named after Gabriel Daniel Fahrenheit, the eighteenth-century German physicist who devised it). Public weather reports from the National Weather Service in the United States and the news media usually state temperatures in degrees Fahrenheit. Reference points on this scale include the sea-level freezing and boiling points of pure water, which are 32° and 212°, respectively. The United States is one of only a few countries that still use the Fahrenheit scale.

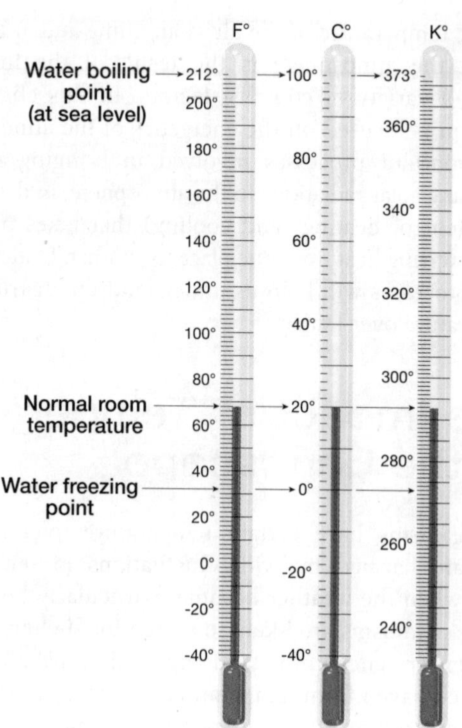

▲ **Figure 4-2** The Fahrenheit, Celsius, and Kelvin temperature scales.

Celsius Scale In most other countries, the Celsius scale (named for Anders Celsius, the eighteenth-century Swedish astronomer who devised it) is used either exclusively or predominantly. It is an accepted component of the International System of measurement (S.I.) because it is a decimal scale with 100 units (degrees) between the freezing and boiling points of water. The Celsius scale has long been used for scientific work in the United States and is now slowly being established to supersede the Fahrenheit scale in all usages in this country.

To convert from degrees Celsius to degrees Fahrenheit, the following formula is used:

$$\text{degrees Fahrenheit} = (\text{degrees Celsius} \times 1.8) + 32°$$

To convert from degrees Fahrenheit to degrees Celsius, the formula is:

$$\text{degrees Celsius} = (\text{degrees Fahrenheit} - 32°) \div 1.8$$

Kelvin Scale For many scientific purposes, the Kelvin scale (named for the nineteenth-century British physicist William Thomson, also known as Lord Kelvin) has long been used because it measures what are called *absolute temperatures*, which means that the scale begins at *absolute zero* (0 K or "zero Kelvin") the lowest possible temperature.[2] The scale maintains a 100-degree range between the boiling and freezing points of water. There are no negative values. Because this scale is not normally used by climatologists and

[1]In this book, the adjectives hot, cold, warm, and cool are used frequently; like other descriptive words that appear (high, low, wet, dry, fast, slow, to name a few), they are often used in a comparative sense; something is hot or cold in comparison with something else.

[2]In classical physics *absolute zero* is the temperature at which molecules have their *zero-point energy* (have no kinetic energy that can be given up). In quantum physics, however, the zero-point energy (lowest energy state for a system) need not be actually zero.

meteorologists, we ignore it in this book except to compare it to the Fahrenheit and Celsius scales. On the Celsius scale, absolute zero is at −273° and so the conversion is simple:

$$\text{degrees Celsius} = \text{degrees Kelvin} - 273^\circ$$
$$\text{degrees Kelvin} = \text{degrees Celsius} + 273^\circ$$

Solar Energy

The Sun is the only significant source of energy for Earth's atmosphere. Millions of other stars radiate energy, but they are too far away to affect Earth. Energy is also released from inside Earth, primarily as radioactive decay of minerals, and energy is released on the ocean floor through hydrothermal vents, although probably not enough to influence the atmosphere significantly. Thus, the Sun supplies essentially all of the energy that drives most of the atmospheric processes. Further, we will see that it is the unequal heating of Earth by the Sun that ultimately puts the atmosphere in motion and is responsible for the most fundamental patterns of weather and climate.

The Sun is a star of average size and average temperature, but its proximity to Earth gives it a far greater influence on our planet than that exerted by all other celestial bodies combined. The Sun is a prodigious generator of energy. In a single second, it produces more energy than the amount used by humankind since civilization began. The Sun functions as an enormous thermonuclear reactor, producing energy through *nuclear fusion*—under extremely high temperatures and pressures, nuclei of hydrogen are fused together to form helium—a process that utilizes only a very small portion of the Sun's mass but provides an immense and continuous flow of energy that is dispersed in all directions.

Electromagnetic Radiation

The Sun gives off energy in the form of **electromagnetic radiation**—sometimes referred to as **radiant energy**. We have experience almost every day with many different kinds of electromagnetic radiation: visible light, microwaves, X-rays, and radio waves are all forms of electromagnetic radiation.

Electromagnetic radiation entails the flow of energy in the form of waves. These waves of energy move through space by way of rapidly oscillating electromagnetic fields. These electromagnetic fields are oscillating in the same rhythm as the vibrations of the electrical charges that form them—for example, it is the oscillation of electrons within an atom that can generate visible light. Electromagnetic radiation does not require a medium (the presence of matter) to pass through. The electromagnetic waves traverse the great voids of space in unchanging form. The waves travel outward from the Sun in straight lines at the *speed of light*—300,000 kilometers (186,000 miles) per second.

Only a tiny fraction of the Sun's radiant output is intercepted by Earth. The waves travel through space without loss of energy, but since they are diverging from a spherical body their intensity continuously diminishes with increased distance from the Sun (Figure 4-3). As a result of this intensity decrease and the distance separating Earth from the Sun, less than one two-billionth of total solar output reaches the outer limit of Earth's atmosphere, having traveled 150,000,000 kilometers (93,000,000 miles) in just over 8 minutes. Although it consists of only a minuscule portion of total solar output, in absolute terms the amount of solar energy Earth receives is enormous: the amount received in 1 second is approximately equivalent to all the electric energy generated on Earth in a week.

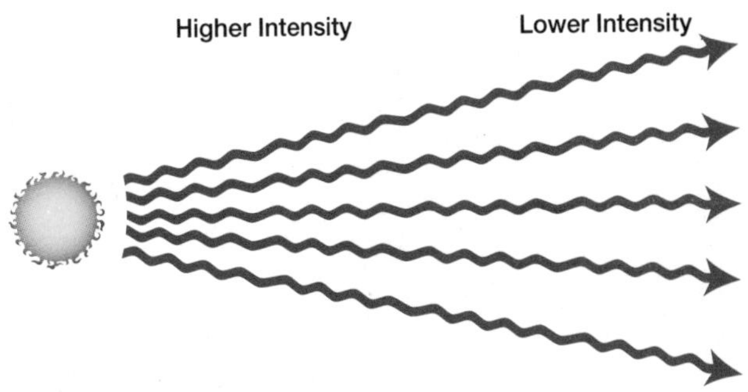

▲ **Figure 4-3** Electromagnetic waves from the Sun spread out as they move outward, thus their intensity decreases as they get farther and farther from the Sun.

Electromagnetic radiation can be classified on the basis of *wavelength*—the distance between the crest of one wave and the crest of the next (Figure 4-4). Collectively, electromagnetic radiation of all wavelengths comprises what is called the **electromagnetic spectrum** (Figure 4-5). Electromagnetic radiation varies enormously in wavelength—ranging from the exceedingly short wavelengths of gamma rays and X-rays (with some wavelengths less than one-billionth of a meter), to the exceedingly long wavelengths of television and radio waves (with some wavelengths measured in kilometers). For the physical geographer, however, only three areas of the spectrum are of importance:

1. **Visible light**: Wavelengths of radiation to which the human eye is sensitive make up a fairly narrow band of the electromagnetic spectrum known as **visible light**, and include wavelengths between about 0.4 and 0.7 micrometers (μm; 1 micrometer equals one-millionth

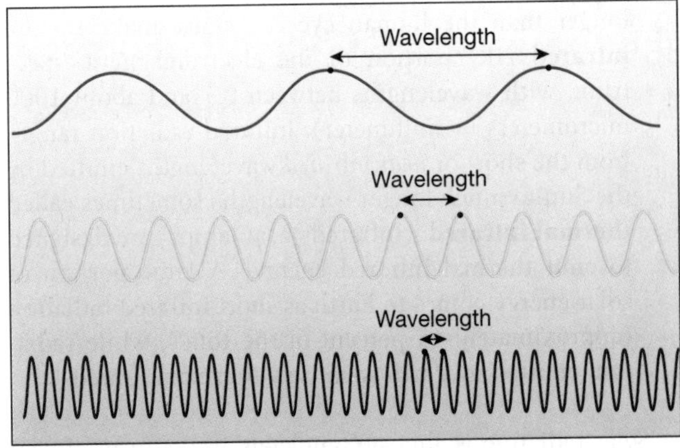

▲ **Figure 4-4** Electromagnetic waves can be of almost any length. The distance from one crest to the next is called the *wavelength*.

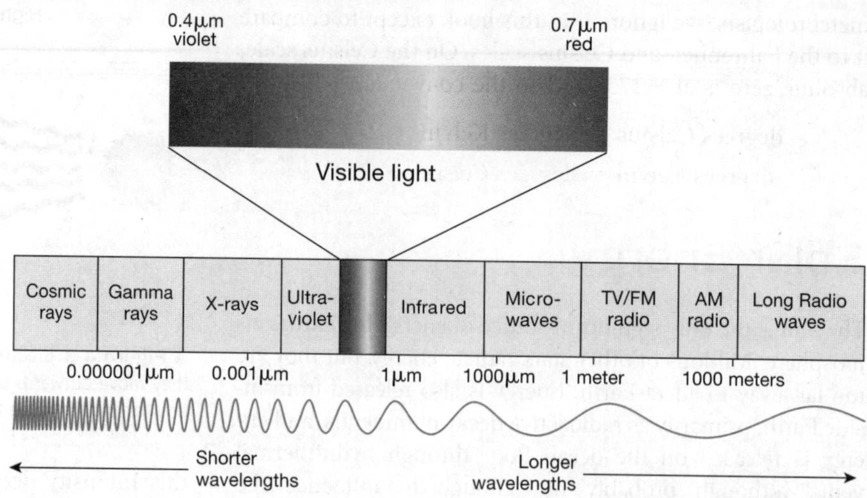

▶ **Figure 4-5** The electromagnetic spectrum.

of a meter). Visible light ranges from the shortest wavelength of radiation the human eye can sense—violet—through the progressively longer wavelengths of blue, green, yellow, orange, and finally red, the longest wavelength of radiation the human eye can see (you will notice that this sequence of color is the same as you see in a rainbow, from inner to outer; see page 126). While visible light makes up a narrow band on the electromagnetic spectrum, the peak intensity of electromagnetic radiation arriving from the Sun is in the visible portion of the spectrum, and approximately 47 percent of the total energy coming from the Sun arrives at Earth as visible light.

2. **Ultraviolet Radiation**: Wavelengths of radiation just shorter than the human eye can sense, with wavelengths from about 0.01 to 0.4 micrometers, make up the **ultraviolet (UV)** portion of the electromagnetic spectrum. The Sun is a prominent natural source of ultraviolet rays, and solar radiation reaching the top of our atmosphere contains a considerable amount (approximately 8 percent of the total energy coming from the Sun). However, as we saw in Chapter 3, much of the UV radiation from the Sun is absorbed by the ozone layer, and so the shortest ultraviolet wavelengths do not reach Earth's surface, where they could cause considerable damage to most living organisms.
3. **Infrared Radiation**: Wavelengths of radiation just longer than the human eye can sense make up the **infrared** (IR) portion of the electromagnetic spectrum, with wavelengths between 0.7 and about 1000 micrometers (1 millimeter). Infrared radiation ranges from the short or *near infrared* wavelengths emitted by the Sun to much longer wavelengths sometimes called **thermal infrared**. (Infrared "heat lamps" are designed to emit thermal infrared energy.) A large portion of solar energy comes to Earth as short infrared radiation (approximately 45 percent of the total), while radiation emitted by Earth is entirely thermal infrared.

Solar radiation is almost completely in the form of visible light, ultraviolet and short infrared radiation, which as a group is referred to as **shortwave radiation** (Figure 4-6). Radiation emitted by Earth—or **terrestrial radiation**—is entirely in the thermal infrared portion of the spectrum and is referred to as **longwave radiation**. A wavelength of about 4 micrometers is considered the boundary on the spectrum separating long waves from short ones. Thus, all terrestrial radiation is longwave radiation, while virtually all solar radiation is shortwave radiation.

Insolation

The total **insolation** (*in*coming *sol*ar radi*ation*) received at the top of the atmosphere is believed to be constant when averaged over a year, although it may vary slightly over long periods of time with fluctuations in the Sun's temperature. This constant amount of incoming energy—referred to as the *solar constant*—is about 1372 watts per square meter (W/m^2). One watt is equal to 1 joule per second (a *joule* is a unit of energy equivalent to about 0.239 calories; 1 *calorie* being the amount of heat required to raise the temperature of 1 gram of water [at 15°C] by 1°C).

The entrance of insolation into the upper atmosphere is just the beginning of a complex series of events in the atmosphere and at Earth's surface. Some of the insolation is reflected off the atmosphere back out into space, where it is lost. The remaining insolation may pass through the atmosphere, where it can be transformed either before or after reaching Earth's surface. This mixed reception of solar energy—and the resulting energy cascade that ultimately heats Earth's surface and atmosphere—are discussed after a brief digression to define our terms.

Basic Heating and Cooling Processes in the Atmosphere

Before looking at the events that occur after energy travels from the Sun to Earth, we first need to examine the physical processes involved in the movement of heat energy. Our goal is to provide practical explanations of the most

RENEWABLE ENERGY

Solar Power

Karl Byrand, University of Wisconsin, Sheboygan

During a 24-hour day, an average of 164 watts of solar energy strikes each square meter of Earth's surface, providing more than enough energy to meet the electrical generation needs of the entire planet. At present, humans are making little use of this clean and relatively cheap form of energy. Overall, about 0.1 percent of the global energy is provided via solar power, with the world's photovoltaic capacity reaching 15 gigawatts in 2008, which is about what is generated by 15 nuclear power plants.

Traditionally, photovoltaic (PV) cells are constructed of inorganic materials such as silicon. The silicon is processed to eliminate any contaminates. However, silicon is a poor electrical conductor, so a minute amount of other elements—such as arsenic, boron, and indium—is added to increase the silicon's conductivity and to determine polarity. Alternate layers of positive and negative polarity ("p-type" and "n-type" semiconductors) are stacked and contacts are added so the cell basically becomes a "silicon sandwich" that comprises a complete electrical circuit. When photons ("packets" of light energy) strike the silicon cell's surface, some of the electrons are displaced from the circuit's negative layer and are forced to flow into the cell's positive layer. The flow of these electrons creates an electrical current that can then be used as a power source (Figure 4-A).

Although the electrical output of an individual PV cell is rather small, many cells may be wired together to form a module, and many modules can be connected together to form an array. Depending on the array's size, the cumulative output can be quite large.

Using photovoltaics for electrical generation offers several advantages. Compared to other power plants, such as coal and nuclear, a photovoltaic power plant has a low capital cost, and once installed, requires little maintenance of the solar panels. Likewise, no fuel or other energy sources, except for the Sun, are required to create the electricity. Perhaps one of the largest advantages is photovoltaic's potential to create decentralized energy electrical generation, in that individual pieces of equipment, or structures overall, do not have to be wired into a grid but instead can generate their own energy on site (Figure 4-B).

Despite these advantages, PV cells must overcome several challenges if they are going to replace other forms of electrical generation. One is efficiency. Today's photovoltaic cells are only capable of capturing a limited portion of the solar spectrum, with a sizable amount of the captured photon energy being lost as heat. Presently, an individual cell can obtain between 15–25 percent efficiency. Furthermore, when the cells are wired together into modules, and the modules into arrays, that efficiency is further impeded. Photovoltaics' capability as an electrical generation medium is further compromised because they possess limited electrical generation capacity under cloudy and hazy conditions, and they generate no electricity when it is dark, so storage batteries must provide power during down hours.

Scientists specializing in solar energy are investigating new technologies to boost the potential of photovoltaics as a viable electrical generation medium. For instance, simply by using mirrors and lenses to better focus the Sun's rays onto photovoltaic arrays, researchers have been able to boost photovoltaic output. However, to guarantee maximum efficiency, these concentration devices must continually track the Sun and therefore are more costly and less maintenance-free than their static counterparts. In addition, research is underway to develop multispectral PV cells that can capture a wider array of the electromagnetic spectrum—including infrared energy—and to incorporate organic elements that will allow photovoltaic capability to be integrated into thin, flexible, and in some instances, even transparent materials. Light-sensitive materials possessing these characteristics have the potential to allow everyday objects, such as paint, windows, curtains, and even clothing to generate electricity in productive ways.

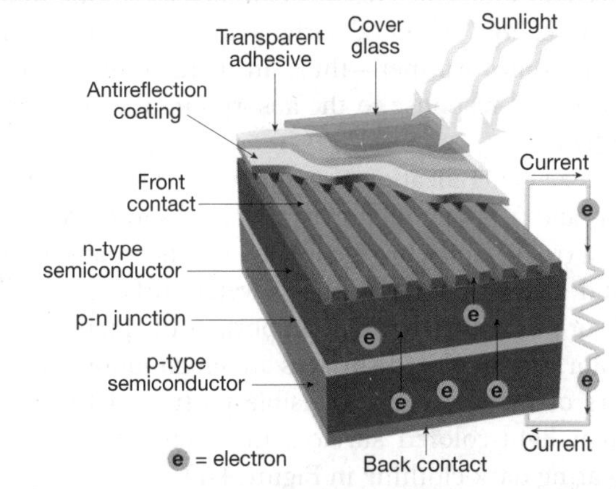

▲ **Figure 4-A** When sunlight strikes a photovoltaic cell, some electrons are displaced from the negative layer to the positive layer, generating an electrical current.

▲ **Figure 4-B** Photovoltaic panels installed on the roof of a house.

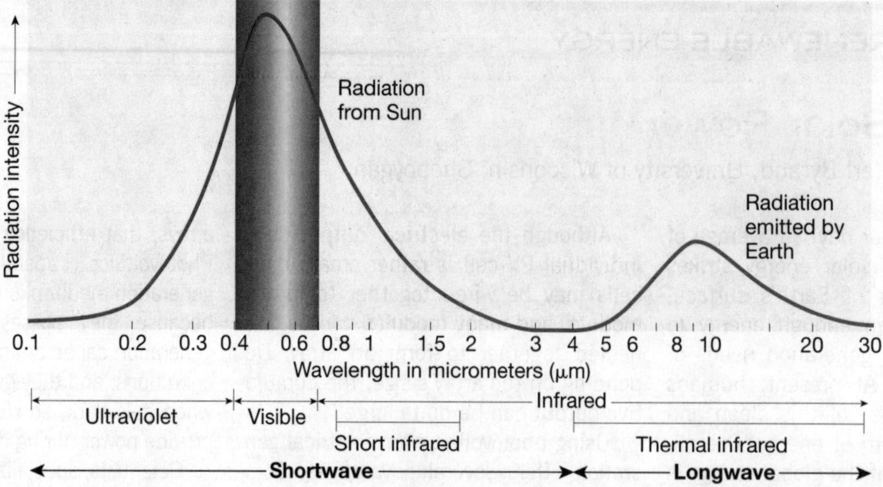

▲ **Figure 4-6** Comparison of solar and terrestrial radiation. Shortwave radiation from the Sun consists of ultraviolet wavelengths (about 8 percent of total solar energy), visible light (about 47 percent of total solar energy), and short infrared wavelengths (about 45 percent of total solar energy). Longwave radiation emitted by Earth consists entirely of thermal infrared wavelengths. Note: To show the relationship between solar and terrestrial radiation on the same chart (shown here in micrometers [μm]), the wavelength scale used is logarithmic, compressing the scale dramatically to the right and distorting the apparent proportions of energy from the various wavelengths of solar radiation—if a linear wavelength scale had been used, the peak of longwave radiation emitted by Earth would be off the page, about 50 cm (20 in.) to the right of visible light!

important processes associated with the heating and cooling of the atmosphere. As such, in some cases we limit our discussion of a process to the aspects of direct importance to meteorology.

Radiation

Radiation—or **emission**—is the process by which electromagnetic energy is emitted from an object. All objects radiate electromagnetic energy, but hotter bodies are more potent radiators than cooler ones. In general, the hotter the object, the more intense its radiation—because the Sun is much hotter than Earth, it emits about 2 billion times more energy than Earth. In addition, the hotter the object, the shorter the wavelengths of that radiation. Hot bodies radiate mostly short waves, while cooler bodies radiate mostly long waves. The Sun is the ultimate "hot" body of our solar system, and so nearly all of its radiation is in the shortwave portion of the electromagnetic spectrum. Earth, on the other hand, is cooler than the Sun, so Earth emits longer wavelengths of radiant energy (thermal infrared).

Temperature, however, is not the only control of radiation effectiveness. Objects at the same temperature may vary considerably in their radiating capability. A body that emits the maximum possible amount of radiation at all wavelengths is called a *blackbody radiator*. Both the Sun and Earth function very nearly as blackbodies, that is, as perfect radiators. They radiate with almost 100 percent efficiency for their respective temperatures. The atmosphere, on the other hand, is not as efficient a radiator as either the Sun or Earth's surface.

Absorption

Electromagnetic waves striking an object may be assimilated by that object—this process is called **absorption**. Different materials have different absorptive capabilities, with the variations depending in part on the wavelength of radiation involved. Although it is a great simplification, when electromagnetic waves strike a material, atoms or even whole molecules in that material may be forced into vibration by the frequency of the incoming electromagnetic waves. (Since all wavelengths of electromagnetic radiation travel at the same speed [the speed of light], short wavelengths of radiation arrive with a higher frequency than do longer wavelengths.) The increased vibrations of the atoms and molecules result in an increase in the internal kinetic energy of the absorbing material, and so the material becomes warmer—thus, an increase in temperature is a typical response to the absorption of electromagnetic radiation.

As a basic generalization, a good radiator is also a good absorber, and a poor radiator is a poor absorber. Mineral materials (rock, soil) are generally excellent absorbers; snow and ice are poor absorbers; water surfaces vary in their absorbing efficiency. One important distinction concerns color. Dark-colored surfaces are much more efficient absorbers of radiation in the visible portion of the spectrum than light-colored surfaces (as exemplified by the skier wearing dark clothing in Figure 4-7).

As we will see, both water vapor and carbon dioxide are efficient absorbers of certain wavelengths of radiation, while nitrogen, the most abundant gas in the atmosphere, is not.

▲ **Figure 4-7** Most solar radiation that reaches Earth's surface is either absorbed or reflected. In this example, the white clothes of one skier reflect much of the solar energy, keeping him cool, whereas the dark clothes of the other skier absorb energy, thereby raising his temperature.

Reflection

Reflection is the ability of an object to repel electromagnetic waves without altering either the object or the waves. When insolation is reflected by the atmosphere or the surface of Earth, it is deflected away, unchanged, in the general direction from which it came. A mirror, for example, is designed to be highly efficient in reflecting visible light.

In our context, reflection is the opposite of absorption. If the wave is reflected, it cannot be absorbed. Hence, an object that is a good absorber is a poor reflector, and vice versa (see Figure 4-7). A simple example of this principle is the existence of unmelted snow on a sunny day. Although the air temperature may be well above freezing, the snow does not melt rapidly because its white surface reflects rather than absorbs a large share of the solar energy that strikes it.

Scattering

Gas molecules and particulate matter in the air can deflect light waves and redirect them in a type of reflection known as **scattering** (Figure 4-8). This deflection involves a change in the direction of the light wave, but no change in wavelength. Some of the waves are backscattered into space and thus are lost to Earth, but many of them continue through the atmosphere in altered but random directions eventually striking the surface as *diffuse radiation*.

The amount of scattering that takes place depends on the wavelength of the light as well as on the size, shape, and composition of the molecule or particulate. In general, shorter wavelengths are more readily scattered than longer ones by the gases in the atmosphere. One prominent kind of scattering, known as *Rayleigh scattering*, takes place when the shortest wavelengths of visible light—violet and blue—are scattered more easily in all directions by the gas molecules in the atmosphere than are the longer wavelengths of visible light—orange and red. This is why on a clear day the sky is blue. (The sky is blue and not violet because of the greater prevalence of blue wavelengths in solar radiation and because our eyes are less sensitive to violet light.) Were there no scattering of blue light by the atmosphere, the sky would appear black. When the Sun is low in the sky, the light has passed through so much atmosphere that nearly all of the blue wavelengths have been scattered away, leaving only the longest wavelengths of visible light, orange and red—the dominant colors of light we see at sunrise and sunset (Figure 4-9).

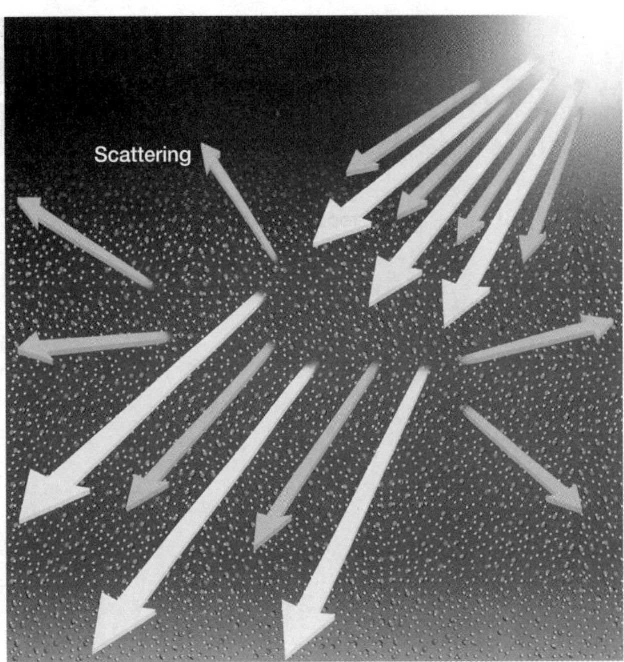

▲ **Figure 4-8** Gas molecules and impurities in the atmosphere scatter and redirect light waves. Some waves are scattered into space and therefore lost to Earth; others are scattered but continue through the atmosphere in altered directions. There is greater scattering of blue light than the longer wavelengths of visible light, resulting in a blue sky.

When the atmosphere contains large quantities of larger particles, such as suspended aerosols, all wavelengths of visible light are more equally scattered (a process known as *Mie scattering*), leaving the sky looking gray rather than blue.

In terms of atmospheric heating, one of the consequences of scattering—especially Rayleigh scattering—is that the intensity of solar radiation striking the surface of Earth is less than that of the solar radiation striking the upper atmosphere because scattering redirects a portion of the insolation back out to space (Figure 4-10).

Transmission

Transmission is the process whereby electromagnetic waves pass completely through a medium, as when light waves are transmitted through a pane of clear, colorless glass. There is obviously considerable variability among mediums in their capacity to transmit electromagnetic waves. Earth materials, for example, are very poor transmitters of insolation; sunlight is absorbed at the surface of rock or soil and does not

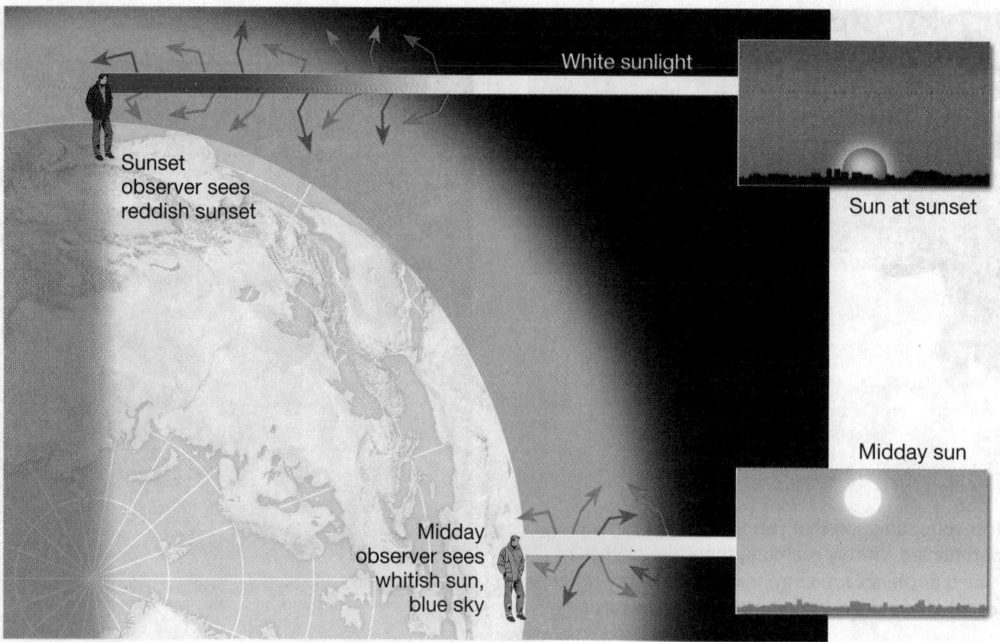

▲ **Figure 4-9** Short wavelengths (blue and violet) of visible light are scattered more easily than longer wavelengths. Consequently, when the Sun is overhead on a clear day, one can see blue sky in all directions because blue light was selectively scattered. At sunrise or sunset, however, the path that light must take through the atmosphere is much longer, with the result that most of the blue light is scattered out before the light waves reach Earth's surface. Thus, the Sun appears reddish in color.

penetrate at all. Water, on the other hand, transmits sunlight well: even in very murky water, light penetrates some distance below the surface, and in clear water, sunlight may illuminate to considerable depths.

The transmission ability of a medium generally depends on the wavelength of radiation. For example, glass has high transmissivity for shortwave radiation but not for longwave radiation. Heat builds up in a closed automobile left parked in the Sun because shortwave radiation is transmitted through the window glass and is absorbed by the upholstery. The longwave radiation then emitted by the interior of the car does not readily transmit out through the glass, causing the inside of the car to warm up (Figure 4-11). This is commonly called the **greenhouse effect**.[3]

▲ **Figure 4-10** The predominant colors of a sunset are orange and red, due to the scattering away of blue light, as in this scene of the Sun setting behind the Golden Gate Bridge in San Francisco.

The Greenhouse Effect The greenhouse effect is at work in the atmosphere (Figure 4-12). A number of gases in the atmosphere, known as **greenhouse gases**, readily transmit incoming shortwave radiation from the Sun but do not easily transmit outgoing longwave terrestrial radiation. The most important greenhouse gas is water vapor, followed by carbon dioxide. Many other trace gases such as methane also play a role, as do some kinds of clouds.

[3]The term greenhouse effect was given to this process because it was long thought that greenhouses maintained heat in the same manner—the glass roof transmitting shortwave solar energy inhibited the passage of longwave radiation out. Recently, it has been shown that this is not the full story. For example, greenhouses having windows made of rock salt, which permits equal transmission of longwaves and shortwaves, experience a heat buildup approximately as great as that of ordinary glass greenhouses. Further investigation showed that glass greenhouses maintain high temperatures largely because the warm air in the building is trapped and does not dissipate through mixing with the cooler air outside. Thus, the term greenhouse effect is based on a misconception, and the retention of heat in the lower troposphere because of differential transmissivity for the shortwaves and longwaves should probably be called something else. "Atmospheric effect" has been suggested, but greenhouse effect continues to be the customary term.

▲ **Figure 4-11** Transmission is illustrated by the sunlight being transmitted through the car windows and into the car interior. Because glass allows transmission of only shortwave radiation, the longwave radiation emitted by the heated upholstery is not able to easily transmit through the windows. This is known as the greenhouse effect.

In the simplest terms, incoming shortwave solar radiation transmits through the atmosphere to Earth's surface, where this energy is absorbed, raising the temperature of the surface. However, the longwave radiation emitted by Earth's surface is inhibited from transmitting back through the atmosphere by the greenhouse gases. Much of this outgoing terrestrial radiation is absorbed by greenhouse gases and clouds and then reradiated back toward the surface, hence delaying this energy loss to space.

The greenhouse effect is one of the most important heating processes in the troposphere. The greenhouse effect keeps Earth's surface and lower troposphere much warmer than would be the case if there were no atmosphere—without the greenhouse effect, the average temperature of Earth would be about −15°C (5°F) rather than the present average of 15°C (59°F).

While the ongoing, natural greenhouse effect in the atmosphere makes life as we know it possible, over the last century or so a significant increase in greenhouse gas concentration—especially carbon dioxide—has been measured. This increase in atmospheric carbon dioxide is closely associated with human activity, especially the burning of fossil fuels such as petroleum and coal (carbon dioxide is one of the by-products of combustion). The increase in greenhouse gas concentration has been accompanied by a slight, yet nonetheless measurable, increase in average global temperature, raising the likelihood that humans are altering the global energy balance of the atmosphere. This important issue, generally referred to as *global warming*, will be addressed in more detail at the end of this chapter after we have concluded our discussion of atmospheric heating processes and patterns.

Conduction

The movement of heat energy from one molecule to another without changes in their relative positions is called **conduction**. This process enables heat to be transferred from one part of a stationary body to another, or from one object to a second object when the two are in contact.

Conduction comes about through molecular collision, as the blown-up view in Figure 4-13 illustrates. A "hot" molecule becomes increasingly agitated as heat is added to it and collides against a cooler, calmer molecule, transferring kinetic energy to it. In this manner, the heat is passed from one place to another. When two molecules of unequal temperature are in contact with one another, heat energy transfers from the hotter to the cooler until they attain the same temperature.

The ability of different substances to conduct heat is quite variable. For example, most metals are excellent conductors, as can be demonstrated by pouring hot coffee into a metal cup and then touching your lips to the edge of the cup. The heat of the coffee is quickly conducted throughout the metal

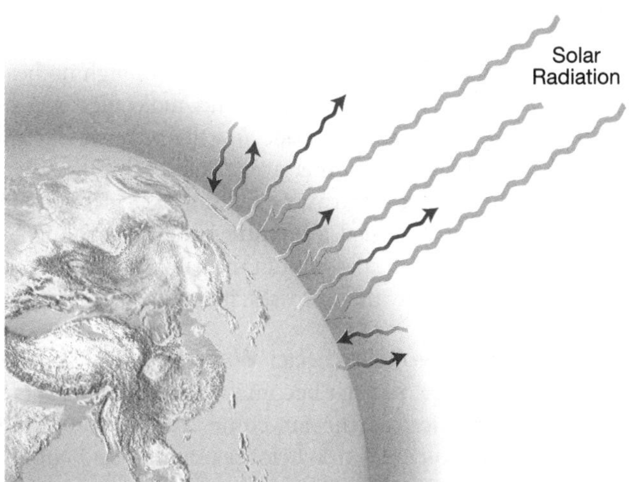

▲ **Figure 4-12** The atmosphere easily transmits shortwave radiation from the Sun but is a poor transmitter of longwave radiation from Earth's surface. This selective transmission causes the greenhouse effect in the atmosphere.

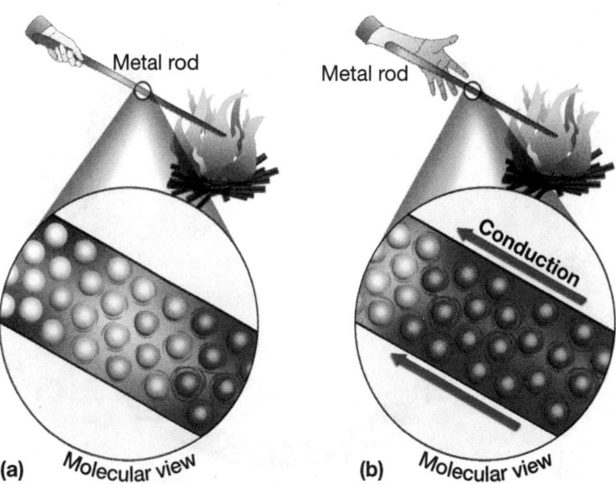

▲ **Figure 4-13** Energy is conducted from one place to another by molecular agitation. (a) One end of the metal rod is in the flame and so becomes hot. (b) This heat is then conducted the length of the rod and will burn the hand of the incautious rod-holder.

and burns the lips of the drinker. In contrast, hot coffee poured into a ceramic cup heats the cup very slowly because such earthy material is a poor conductor.

Earth's land surface warms up rapidly during the day because it is a good absorber of incoming shortwave radiation, and some of that warmth is transferred away from the surface by conduction. A small part is conducted deeper underground, but not much, because Earth materials are not good conductors. Another portion of this absorbed heat is transferred to the lowest portion of the atmosphere by conduction from the ground surface. Air, however, is a poor conductor, and so only the air layer touching the ground is heated very much (perhaps a layer just a few millimeters thick); physical movement of the air is required to spread the heat around. On the other hand, when the ground surface is very cold, heat can transfer from the air to the ground through conduction, chilling the air above.

Moist air is a slightly more efficient conductor than dry air. If you are outdoors on a winter day, you will stay warmer if there is little moisture in the air to conduct heat away from your body.

Convection

In the process of **convection**, heat is transferred from one point to another by the predominately vertical circulation of a fluid, such as air or water. Convection involves movement of the heated molecules from one place to another. (Do not confuse the movement of molecules from one place to another in the process of *convection* with the back-and-forth vibratory movement and molecular collisions associated with *conduction*—in convection, the molecules physically move away from the heat source; in conduction, they do not.)

If your room is heated by a "radiator" (Figure 4-14), you have taken advantage of convection. The heated air immediately above the radiator rises because the heating has caused it to expand and therefore become less dense than the nearby air. Surrounding air then moves in to fill the void left by the rising warm air. In turn, cooler air from above descends to replace that which has moved in from the side, and a cellular circulation is established—up, out, down, and in.

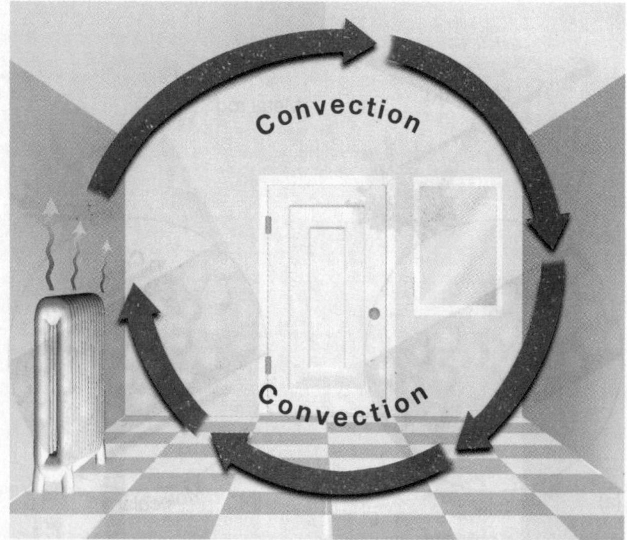

▲ **Figure 4-14** A "radiator" in a closed room sets up a convective circulation system. Air warmed by the radiator rises, and cooler air flows toward the radiator to replace the heated air that has risen.

A similar convective pattern frequently develops in the atmosphere. Unequal heating (for a variety of reasons) may cause a parcel of surface air to become warmer than the surrounding air, and thus the warm air will rise. The heated air expands and moves upward in the direction of lower pressure. The cooler surrounding air then moves in toward the heat source, and air from above sinks down to replace that which has moved in from the sides, thus establishing a convective circulation system, also called a **convection cell**. The prominent elements of the system are an updraft of warm air and a downdraft of air after it has cooled. Convection is common in each hemisphere during its summer and throughout the year in the tropics.

Advection

When the dominant direction of heat transfer in a moving fluid is horizontal, the term **advection** is applied. In the atmosphere, wind may transfer warm or cool air horizontally from one place to another through the process of advection. As we will see in the following chapter, some wind systems develop as part of large atmospheric convection cells: the horizontal component of air movement within such a convection cell is properly called advection.

Adiabatic Cooling and Warming

Whenever air ascends or descends, its temperature changes. This invariable result of vertical movement is due to the change in pressure. When air rises, it expands because there is less air above it, and so less pressure is exerted on it (Figure 4-15). When air descends, it is compressed because there is more air above it, and so more pressure is exerted on it.

Expansion—Adiabatic Cooling The expansion that occurs in rising air is a cooling process even though no heat is lost. As air rises and expands, the molecules spread through a greater volume of space and this requires energy. As the molecules spread farther away from each other their frequency of collision decreases, resulting in a decrease in temperature. This is called **adiabatic cooling**—cooling by expansion (*adiabatic* means without the gain or loss of heat). In the atmosphere, any time air rises, it cools adiabatically.

Compression—Adiabatic Warming Conversely, when air descends, it must become warmer. The descent causes compression as the air comes under increasing pressure. The molecules draw closer together and collide more frequently. The result is an increase in temperature even though no heat is added from external sources. This is called **adiabatic warming**—warming by compression. In the atmosphere, any time air descends it warms adiabatically.

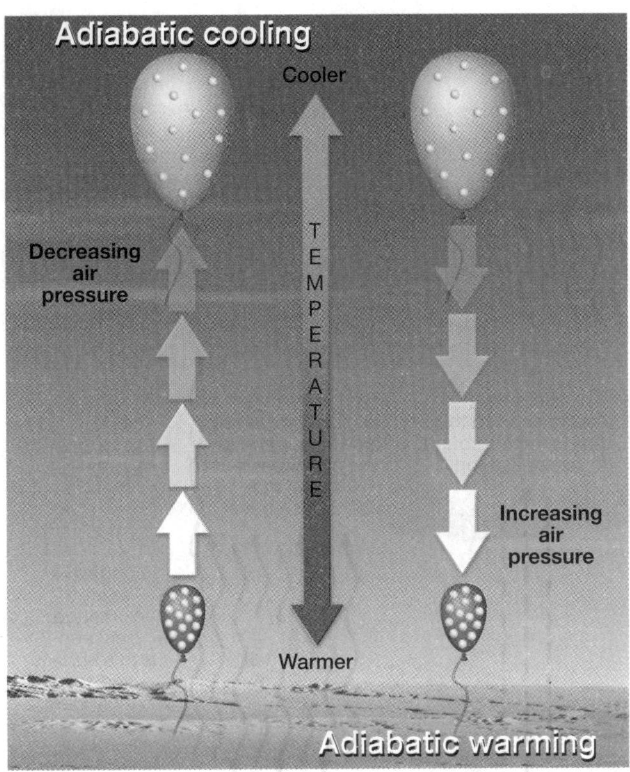

▲ **Figure 4-15** Rising air represented by a balloon cools adiabatically due to expansion and descending air warms adiabatically due to compression. No heat transfer is involved in either process.

As we will see in Chapter 6, adiabatic cooling of rising air is one of the most important processes involved in cloud development and precipitation, while the adiabatic warming of descending air has just the opposite effect.

Latent Heat

The physical state of water in the atmosphere frequently changes—ice changes to liquid water, liquid water changes to water vapor, and so forth. Depending on the process, these phase changes involve either the storage or release of energy known as **latent heat** (*latent* is from the Latin, "lying hidden"). The two most common phase changes are **evaporation**, in which liquid water is converted to gaseous water vapor, and **condensation**, in which gaseous water vapor is converted to liquid water. During the process of evaporation, latent heat energy is stored and so evaporation is, in effect, a cooling process. On the other hand, during condensation latent heat energy is released and so condensation is, in effect, a warming process.

As we will explore more fully in Chapter 6 and 7, a great deal of energy is transferred from one place to another in the atmosphere in the form of water vapor. Energy that is stored in one location through evaporation can be released as heat in another location far away. We will also see that many storms, such as hurricanes, are fueled by the release of latent heat.

The Heating of the Atmosphere

Atmospheric Energy Balance

We now turn to the specifics of atmospheric heating. What happens to solar radiation when it enters Earth's atmosphere? How is it received and distributed? What are the dynamics of converting electromagnetic radiation to atmospheric heat?

In the long run, there is a balance between the total amount of insolation received by Earth and its atmosphere on one hand, and the total amount of radiation returned to space on the other (Figure 4-16). (As we suggested earlier, humans are likely altering the energy balance of the atmosphere slightly through global warming, to be discussed at the end of this chapter. For the purposes of understanding atmospheric heating processes, we will ignore that possibility for the moment.)

The annual balance between incoming and outgoing radiation is the *global energy budget*, which can be illustrated by using 100 "units" of energy to represent total insolation (100% of insolation) received at the outer edge of the atmosphere and tracing its dispersal (Figure 4-17). Keep in mind that the values shown here are annual averages for the entire globe and do not apply to any specific location.

Most of the insolation that enters the atmosphere does not heat it directly. About 31 units of total insolation are reflected (or scattered) back into space by the atmosphere and the surface; the radiation that is bounced back into space is called Earth's **albedo**. (Albedo is the reflectivity of an object; the higher the albedo value, the more radiation the object reflects.) Only 24 units of incoming radiation heat the atmosphere directly, 3 of them heating the ozone layer and 21 heating the rest of the atmosphere. About 45 units of insolation—nearly half of the total—simply pass through the atmosphere to Earth's surface where they are absorbed, heating the surface. The heated surface of Earth in turn transfers energy to the atmosphere above in a number of ways.

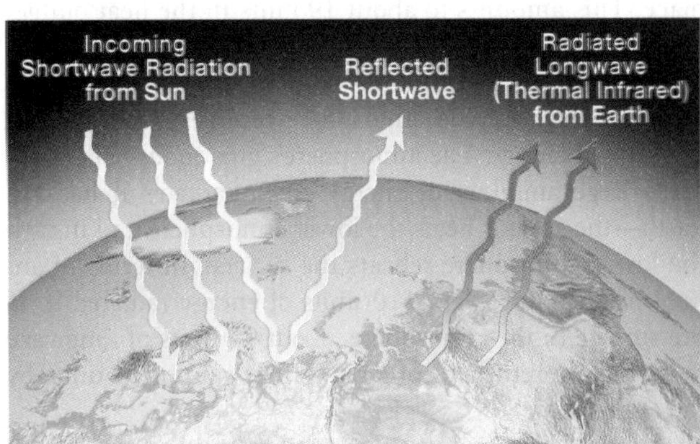

▲ **Figure 4-16** The big picture: Earth's energy budget in simplified form. Incoming shortwave radiation and outgoing longwave radiation are in a long-term balance.

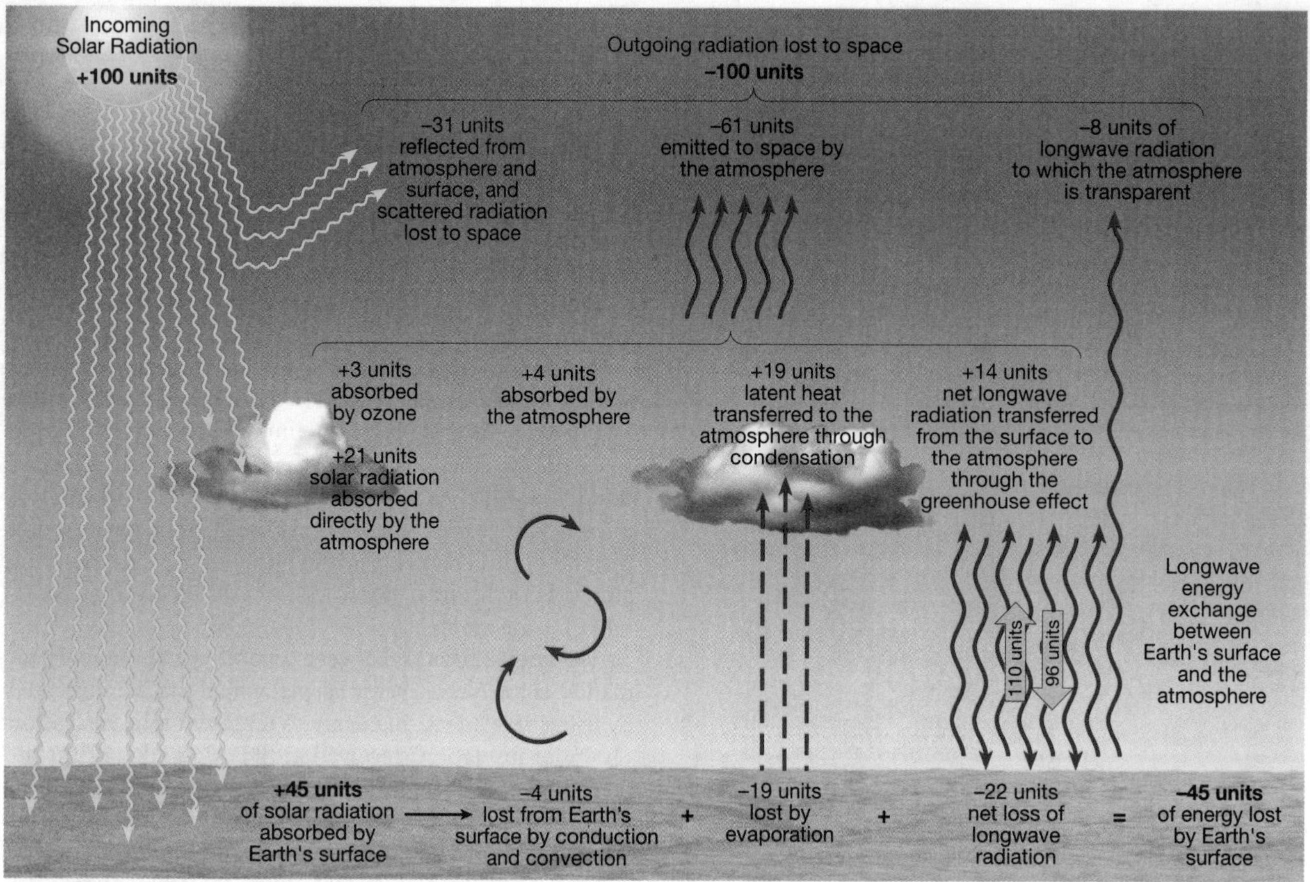

▲ **Figure 4-17** The generalized energy budget of Earth and its atmosphere.

About 4 units of energy are conducted from Earth's surface back into the atmosphere, where it is dispersed by convection. Energy is also transferred from the surface to the atmosphere through the transport of latent heat in water vapor. About three-fourths of all sunshine falls on a water surface when it reaches Earth. Much of this energy is utilized in evaporating moisture from oceans, lakes, and other bodies of water. The energy passed into the atmosphere as latent heat stored in water vapor is released subsequently when condensation takes place. This amounts to about 19 units in the heat budget equation.

Through the absorption of terrestrial radiation by greenhouse gases, the atmosphere receives a net gain of 14 units of energy. The atmosphere's greenhouse effect absorbs large amounts of energy and reradiates it back to Earth—the surface heats the lower atmosphere and in turn the lower atmosphere reheats the surface. In terms of annual balance, for every 110 units of energy radiated from the surface to the atmosphere, about 96 units of longwave radiation are returned to Earth—a net gain of 14 units by the atmosphere.

Some of the longwave radiation emitted by Earth's surface, however, is transmitted directly through the atmosphere without being absorbed by the greenhouse gases. Approximately 8 units of energy in the form of longwave radiation with wavelengths between about 8 and 12 micrometers transmit through what is called the *atmospheric window*—a range of wavelengths of infrared radiation that is not strongly absorbed by any atmospheric component.

For the most part, then, the atmosphere is heated by the surface of Earth, although the Sun is the original source of the energy. Thus, there is an intimate link between troposphere temperatures and Earth surface conditions. The air temperature at any given time represents the balance between insolation and terrestrial radiation.

This complicated sequence of atmospheric heating has many ramifications. One of the most striking is that the atmosphere is heated mostly from below rather than from above. The result is a troposphere in which cold air overlies warm air. This "unstable" situation (explored further in Chapter 6) creates an environment of almost constant convective activity and vertical mixing. If the atmosphere were heated directly from the Sun, producing warm air at the top and cold air near Earth's surface, the situation would be stable, essentially without vertical air movements. The result would be a troposphere that is largely motionless, apart from the effects of Earth's rotation.

FOCUS

Monitoring Earth's Radiation Budget

Earth's radiation budget is monitored on an ongoing basis by several permanently orbiting satellites. Sensors onboard the satellites detect both reflected shortwave radiation and emitted longwave radiation in several wavelength bands to gather information about Earth's surface and atmosphere.

One of the most important instruments used is the *Advanced Very High Resolution Radiometer* (AVHRR), first carried onboard TIROS-N launched in October of 1978. A much more advanced instrument, the AVHRR/3, was first carried onboard NOAA-15, launched in May of 1998. Currently, NOAA-18 is one of the satellites operated by the National Oceanic and Atmospheric Administration (NOAA). First operational in 2005, NOAA-18 is one of several Polar Orbiting Environmental Satellites (POES), each making about 14 orbits a day. Polar orbits allow these satellites to provide complete global coverage four times each day. Data gathered from the POES is used to monitor weather and ocean conditions, as well as to provide a database for climate change research. Earth's radiation budget is monitored day and night using the AVHRR/3 instrument.

Total incoming shortwave radiation—known as the *available solar energy*—is measured at the top of the atmosphere in watts per square meter (W/m^2; Figure 4-C). These values are influenced only by the angle of the Sun and the number of hours of daylight. Notice in this image from June that the highest average daily insolation occurs over the Arctic—the Sun's angle of incidence is quite low, but 24 hours of daylight result in a very high daily average.

The total amount of shortwave energy absorbed by the atmosphere and surface is indicated by *absorbed solar energy*—this is the difference between the total shortwave energy at the top of the atmosphere and the total shortwave radiation reflected back to space. In Figure 4-D notice the high absorption in subtropical latitudes where the Sun is highest in the sky.

Figure 4-E shows outgoing longwave radiation at night. Notice the very high emission of longwave radiation from the subtropical and midlatitude desert regions in the southwestern United States, northern Africa, and east-central Eurasia. The clear nighttime skies and the generally low water vapor content of the air permit significant transmission of longwave radiation away from the surface into space.

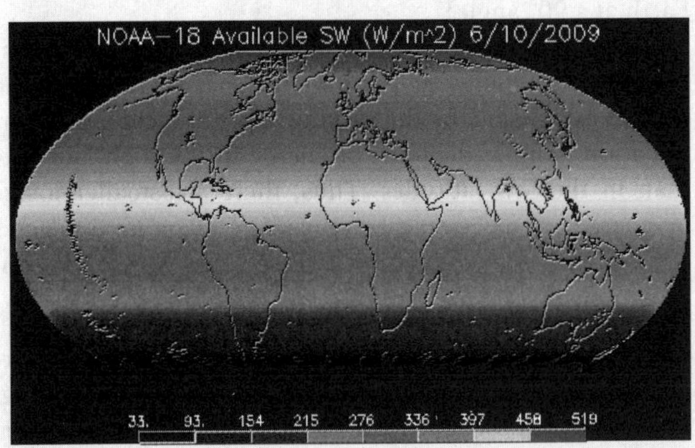

▲ **Figure 4-C** Available solar energy at the top of the atmosphere in W/m^2, taken by the NOAA-18 satellite.

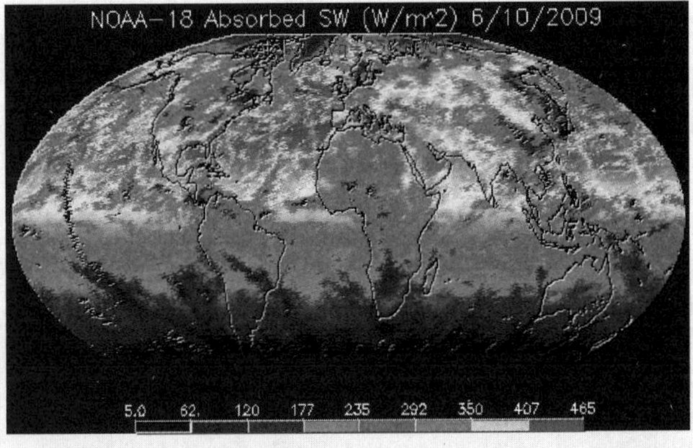

▲ **Figure 4-D** Absorbed solar radiation in W/m^2 on the same day as Figure 4-C, taken by the NOAA-18.

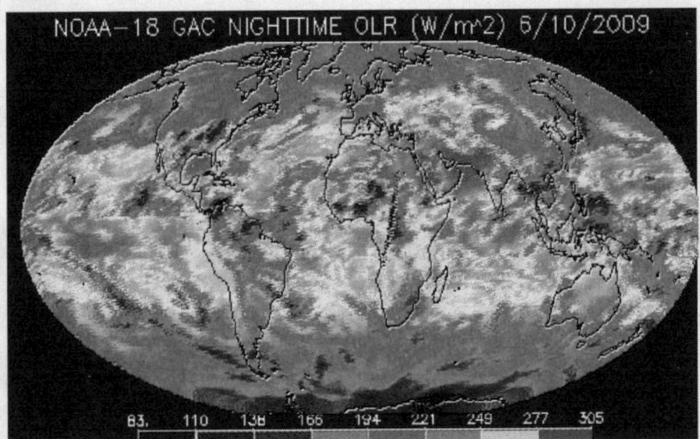

▲ **Figure 4-E** NOAA-18 satellite image showing nighttime emission of outgoing longwave radiation in W/m^2 on the same day as Figures 4-C and 4-D.

Variations in Heating by Latitude and Season

The radiation budget we have been discussing is broadly generalized. There are, however, many latitudinal and vertical imbalances in this budget, and these are among the most fundamental causes of weather and climate variations. In essence, we can trace a causal continuum wherein radiation differences lead to temperature differences that lead to air-density differences that lead to pressure differences that lead to wind differences that often lead to moisture differences. It has already been noted that world weather and climate differences are fundamentally caused by the unequal heating of Earth and its atmosphere. This unequal heating is the result of latitudinal and seasonal variations in how much energy is received by Earth.

Latitudinal and Seasonal Differences

There are only a few basic reasons for the unequal heating of different latitudinal zones.

Angle of Incidence The angle at which rays from the Sun strike Earth's surface is called the **angle of incidence**. This angle is measured from a line drawn tangent to the surface, as Figure 4-18 shows. By this definition, a ray striking Earth's surface vertically, when the Sun is directly overhead, has an angle of incidence of 90°, a ray striking the surface at a slant has an angle of incidence smaller than 90°, and for a ray striking Earth tangent to the surface (as at sunrise and sunset) the angle of incidence is zero.

Because Earth's surface is curved and because the positional relationship between Earth and the Sun is always changing, the angle of incidence for any given location on Earth also changes. The angle of incidence is the primary determinant of the intensity of solar radiation received at any spot on Earth. If a ray strikes Earth's surface vertically, the energy is concentrated in a small area; if the ray strikes Earth obliquely, the energy is spread out over a larger portion of the surface. The more nearly perpendicular the ray (in other words, the closer to 90° the angle of incidence), the smaller the surface area heated by a given amount of insolation and the more effective the heating. Averaged over the year as a whole, the insolation received by high-latitude regions is much less intense than that received by tropical areas (Figure 4-19).

Atmospheric Obstruction We have already noted that clouds, particulate matter, and gas molecules in the atmosphere may absorb, reflect, or scatter insolation. The result is a reduction in the intensity of the energy received at Earth's surface. On the average, sunlight received at Earth's surface is only about half as strong as it is at the top of Earth's atmosphere.

This weakening effect varies from time to time and from place to place depending on two factors: the amount of atmosphere the radiation has to pass through and the transparency. The distance a ray of sunlight travels through the atmosphere (commonly referred to as *path length*) is determined by the angle of incidence (Figure 4-20). A high-angle ray traverses a shorter course through the atmosphere than a low-angle one. A tangent ray (one having an incidence angle of zero) must pass through nearly 20 times as much atmosphere as a vertical ray (one striking Earth at a 90° angle).

The effect of atmospheric obstruction on the distribution of solar energy at Earth's surface is to reinforce the pattern established by the varying angle of incidence. Solar radiation is more depleted of energy in the high latitudes than in the low latitudes. Thus, there are smaller losses of energy in the tropical atmosphere than in the polar atmosphere—however, as we will see, this general pattern is complicated by patterns of cloud cover.

Day Length The duration of sunlight is another important factor in explaining latitudinal inequalities in heating. Longer days allow more insolation to be received and thus more heat to be absorbed. In tropical regions, this factor is

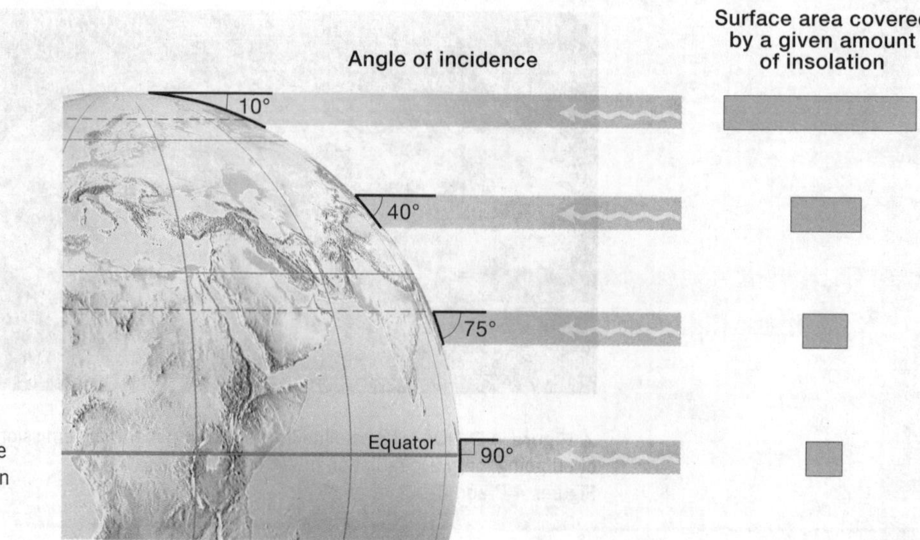

▶ **Figure 4-18** The angle at which solar rays hit Earth's surface varies with latitude. The higher the angle, the more concentrated the energy and therefore the more effective the heating. The day shown in this diagram is the March/September equinox.

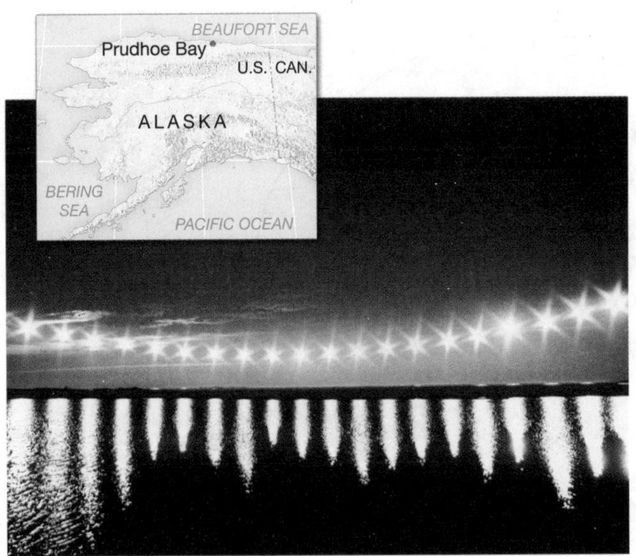

▲ **Figure 4-19** The Sun is low in the sky even during the summer in high latitudes. This time-lapse photograph shows the "midnight Sun" on the June solstice at Prudhoe Bay, Alaska (70° N).

relatively unimportant because the number of hours between sunrise and sunset does not vary significantly from one month to another; at the equator, of course, daylight and darkness are equal in length (12 hours each) every day of the year. In middle and high latitudes, however, there are pronounced seasonal variations in day length. The conspicuous buildup of heat in summer in these regions is largely a consequence of the long hours of daylight, and the winter cold is a manifestation of limited insolation being received because of the short days.

Latitudinal Radiation Balance

As the vertical rays of the Sun shift northward and southward across the equator during the course of the year, the

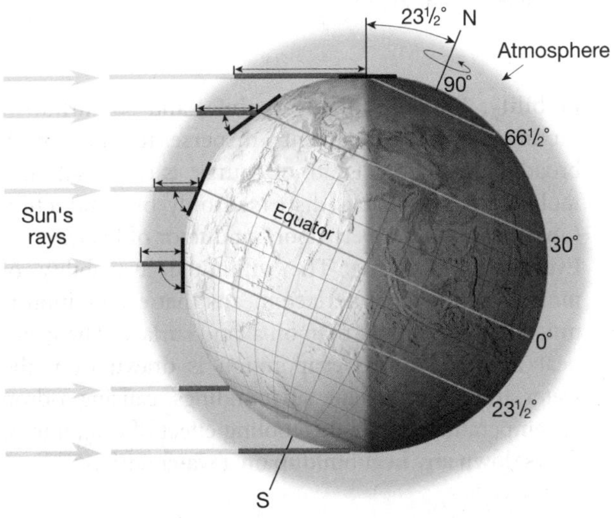

▲ **Figure 4-20** Atmospheric obstruction of sunlight. Low-angle rays (such as in the high latitudes) must pass through more atmosphere than high-angle rays; thus, the former are subject to more depletion through reflection, scattering, and absorption. The day shown in this diagram is the December solstice.

belt of maximum solar energy swings back and forth through the tropics. Thus, in the low latitudes, to about 28° N and 33° S, there is an energy surplus, with more incoming than outgoing radiation. In the latitudes north and south of these two parallels, there is an energy deficit, with more radiant loss than gain. The surplus of energy in low latitudes is directly related to the consistently high angle of incidence, and the energy deficit in high latitudes is associated with low angles.

Figure 4-21 shows the distribution of average daily insolation at the surface around the world for December and June, and Figure 4-22 shows the annual average annual daily insolation for the United States and parts of Canada and Mexico. The maps show the average daily insolation received in watts per square meter (W/m^2; 1 watt = 0.239 calories per second). The variations are largely latitudinal, as is to be expected. The principal interruptions to the simple latitudinal pattern are based on the presence or absence of frequent cloud cover, where insolation is reflected, diffused, and scattered. In Figure 4-22, for example, it can be seen that insolation is greatest in the southwestern United States, where clouds are consistently sparse, and is least in the northwestern and northeastern corners of the country, where cloud cover is frequent.

Despite the variable pattern shown on the maps, there is a long-term balance between incoming and outgoing radiation for the Earth–atmosphere complex as a whole; in other words, the net radiation balance for Earth is zero. The mechanisms for exchanging heat between the surplus and deficit regions involve the general circulation patterns of the atmosphere and oceans, which are discussed later in the book.

Land and Water Contrasts

As we've seen, the atmosphere is heated mainly by heat reradiated and transferred from Earth's surface rather than by heat directly received from the Sun; thus, the heating of Earth's surface is a primary control of the heating of the air above it. To comprehend variations in air temperatures, it is useful to understand how different kinds of surfaces react to solar energy. There is considerable variation in the absorbing and reflecting capabilities of the almost limitless kinds of surfaces found on Earth—soil, water, grass, trees, cement, rooftops, and so forth. Their varying receptivity to insolation in turn causes differences in the temperature of the overlying air.

By far the most significant contrasts are those between land and water surfaces. The generalization is that land heats and cools faster and to a greater degree than water (Figure 4-23).

Heating

A land surface heats up more rapidly and reaches a higher temperature than a comparable water surface subject to the same insolation. In essence, a thin layer of land is heated to relatively high temperatures, whereas a thick layer of water is heated more slowly to moderate

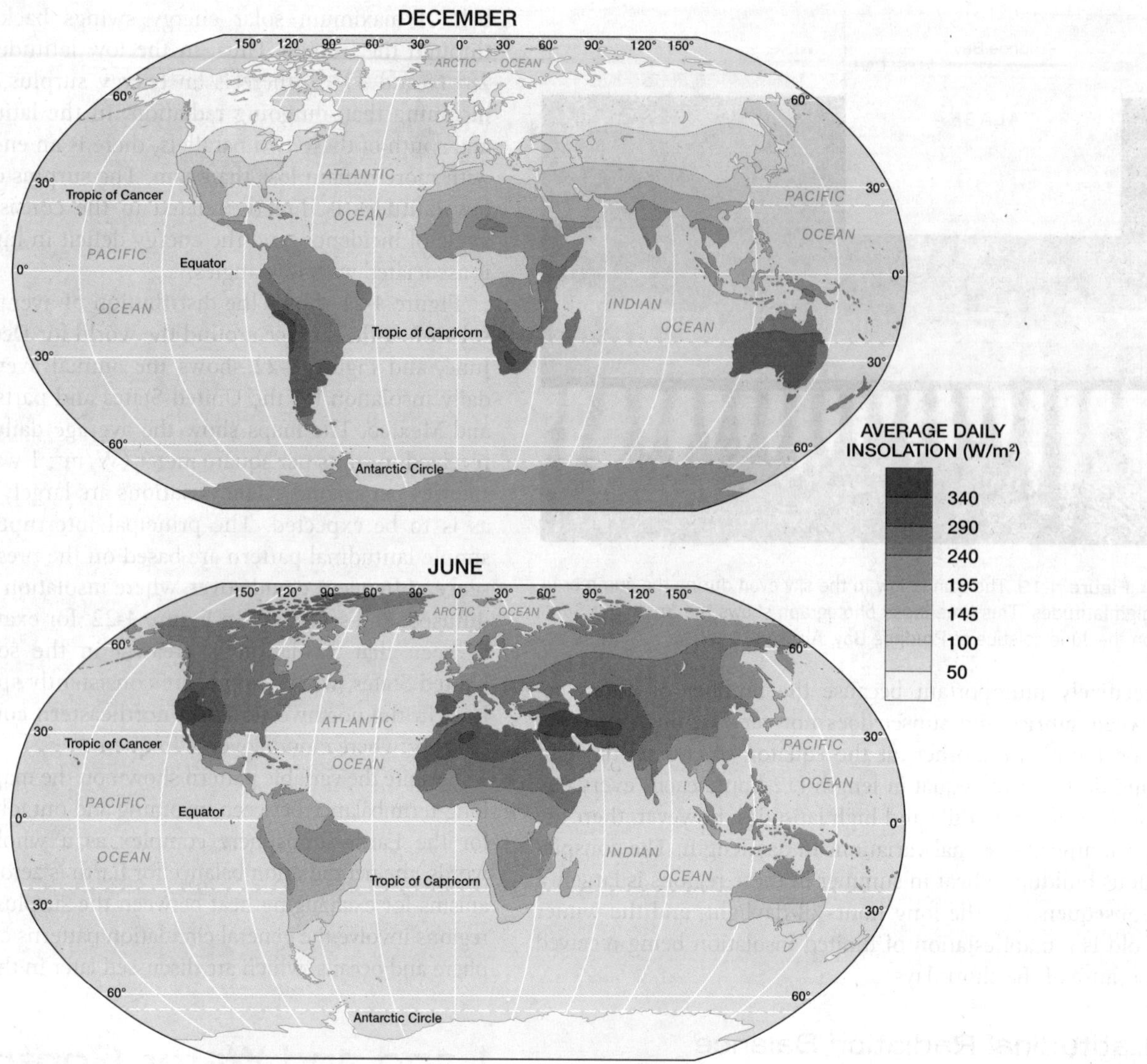

▲ **Figure 4-21** World distribution of average daily insolation for December and June in watts per square meter. The pattern is determined mainly by latitude (note the low December levels in northern areas) and amount of cloudiness (note the high June levels in such cloud-free desert areas as the southwestern United States and southern Eurasia).

temperatures. There are several significant reasons for this difference.

1. **Specific Heat**: Water has a higher **specific heat** than land. Specific heat is the amount of energy required to raise the temperature of 1 gram of a substance by 1°C. The specific heat of water is about five times as great as that of land, which means that water can absorb much more solar energy without its temperature increasing.
2. **Transmission**: Sun rays penetrate water more deeply than they do land; that is, water is a better transmitter of radiation than land. Thus, in water solar energy is absorbed through a much greater volume of matter, and maximum temperatures remain considerably lower than they do on land, where the heat is concentrated at the surface and maximum temperatures can be much higher.
3. **Mobility**: Water is highly mobile, and so turbulent mixing and ocean currents disperse the heat both broadly and deeply. Land, of course, is essentially immobile, and so heat is dispersed only by conduction (and land is a relatively poor conductor of heat).
4. **Evaporative Cooling**: The unlimited availability of moisture on a water surface means that evaporation is much more prevalent than on a land surface. The latent heat needed for this evaporation is drawn from the water and its immediate surroundings, causing a drop in temperature. Thus, the cooling effect of evaporation slows down any heat buildup on a water surface.

Cooling

When both are overlain by air at the same temperature, a land surface cools more rapidly and to a lower temperature than a water surface. During winter, the shallow, heated

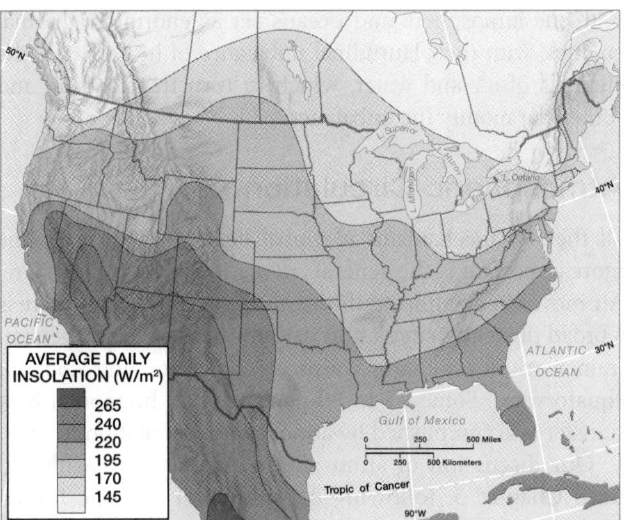

▲ **Figure 4-22** Average annual daily insolation in watts per square meter received in the 48 conterminous states and adjacent parts of Canada and Mexico. The sunny southwestern areas receive the most radiation; the cloudy northeastern and northwestern areas receive the least.

layer of land radiates its heat away quickly. Water loses its heat more gradually because the heat has been stored deeply and is brought only slowly to the surface for radiation. As the surface water cools, it sinks and is replaced by warmer water from below. The entire water body must be cooled before the surface temperatures decrease significantly.

Implications

The significance of these contrasts between land and water heating and cooling rates is that both the hottest and coldest areas of Earth are found in the interiors of continents, distant from the influence of oceans. In the study of the atmosphere, probably no single geographic relationship is more important than the distinction between continental and maritime climates. A continental climate experiences greater seasonal extremes of temperature—hotter in summer, colder in winter—than a maritime climate.

These differences are shown in Figure 4-24, which portrays average monthly temperatures for San Diego and Dallas. These two cities are at approximately the same latitude and experience almost identical lengths of day and angles of incidence. Although their annual average temperatures are almost the same, the monthly averages vary significantly. Dallas, in the interior of the continent, experiences notably warmer summers and cooler winters than San Diego, which enjoys the moderating influence of an adjacent ocean.

The oceans, in a sense, act as great reservoirs of heat. In summer they absorb heat and store it. In winter they give off heat and warm the air. Thus, they function as a sort of global thermostat, moderating temperature extremes.

The ameliorating influence of the oceans can also be demonstrated, on a totally different scale, by comparing latitudinal temperature variations in the Northern Hemisphere with those in the Southern Hemisphere. The Northern Hemisphere is often thought of as a "land hemisphere" because 39 percent of its area is land surface; the Southern Hemisphere is a "water hemisphere," with only 19 percent of its area as land. Table 4-1 shows the average *annual temperature range* (difference in average temperature of the coldest and warmest months) for comparable parallels in each hemisphere. It is obvious that the land hemisphere has greater extremes.

Mechanisms of Heat Transfer

If there were no mechanisms for moving heat poleward in both hemispheres, the tropics would become progressively warmer until the amount of heat energy absorbed equaled the amount radiated from Earth's surface, and the high latitudes would become progressively colder. Such temperature trends do not occur because there is a persistent shifting of

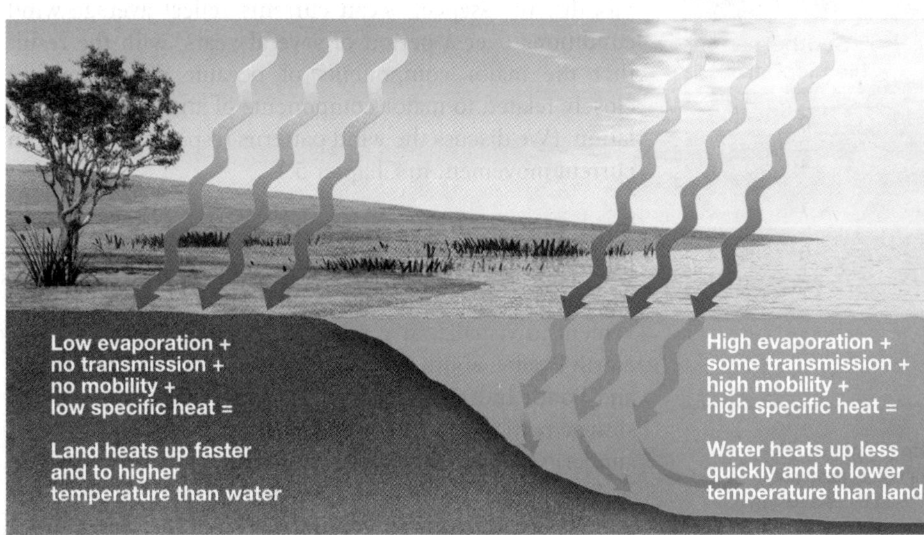

◀ **Figure 4-23** Some contrasting characteristics of the heating of land and water.

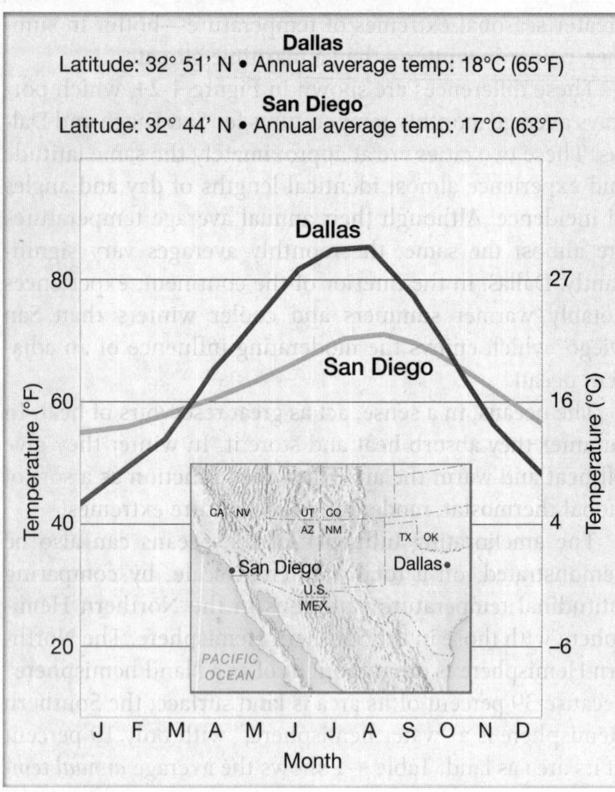

▲ **Figure 4-24** Annual temperature curves for San Diego, California, and Dallas, Texas. Although both have similar average annual temperatures, they have very different annual temperature regimes. In both summer and winter, San Diego, situated on the coast, experiences milder temperatures than inland Dallas.

warmth toward the high latitudes and the consequent cooling of the low latitudes. This shifting is accomplished by circulation patterns in the atmosphere and in the oceans. The broad-scale or planetary circulation of these two mediums moderates the buildup of heat in equatorial regions and the loss of heat in polar regions, thereby making both those latitudinal zones more habitable than they would otherwise be.

TABLE 4-1 Average Annual Temperature Range by Latitude, in Degrees Celsius

Latitude	Northern Hemisphere	Southern Hemisphere
0	0	0
15	3	4
30	13	7
45	23	6
60	30	11
75	32	26
90	40	31

Source: From Frederick K. Lutgens and Edward J. Tarbuck, *The Atmosphere: An Introduction to Meteorology*, 9th ed., Upper Saddle River, NJ: Pearson-Prentice Hall, 2004, p. 71. Used by permission of Prentice Hall.

Both the atmosphere and oceans act as enormous thermal engines, with their latitudinal imbalance of heat driving the currents of air and water, which in turn transfer heat and somewhat modify the imbalance.

Atmospheric Circulation

Of the two mechanisms of global heat transfer, by far the more important is the general circulation of the atmosphere. Air moves in an almost infinite number of ways, but there is a broad planetary circulation pattern that serves as a general framework for moving warm air poleward and cool air equatorward. Some 75 to 80 percent of all horizontal heat transfer is accomplished by atmospheric circulation.

Our discussion of atmospheric circulation is withheld until Chapter 5, following consideration of some fundamentals concerning pressure and wind.

Oceanic Circulation

Ocean Circulation Patterns

Winds disturb the surface of the sea with swells and waves, but winds also can propel large bodies of water. These sometimes become currents. Surface water can flow at about 1 to 2 percent of wind speed, meaning that water in a surface current might travel some tens or even hundreds of kilometers in a day.

There is a close relationship between the general circulation patterns of the atmosphere and oceans. Various kinds of oceanic water movements are categorized as ocean currents, and it is air blowing over the surface of the water that is the principal force driving the major surface ocean currents. In the other direction, the heat energy stored in the oceans has important effects on atmospheric circulation.

For our purposes in understanding heat transfer by the oceans, we are concerned primarily with the broad-scale surface currents that make up the general circulation of the oceans (Figure 4-25). These major currents respond to changes in wind direction, but they are so broad and ponderous that the response time normally amounts to many months. In essence, ocean currents reflect average wind conditions over a period of several years, with the result that the major components of oceanic circulation are closely related to major components of atmospheric circulation. (We discuss the wind patterns responsible for ocean current movement in Chapter 5.)

The Basic Pattern All the oceans of the world are interconnected. Because of the location of landmasses and the pattern of atmospheric circulation, however, it is convenient to visualize five relatively separate ocean basins—North Pacific, South Pacific, North Atlantic, South Atlantic, and South Indian. Within each of these basins, there is a similar pattern of surface current flow based on a general similarity of prevailing wind patterns.

Despite variations based on the size and shape of the various ocean basins and on the season of the year, a single

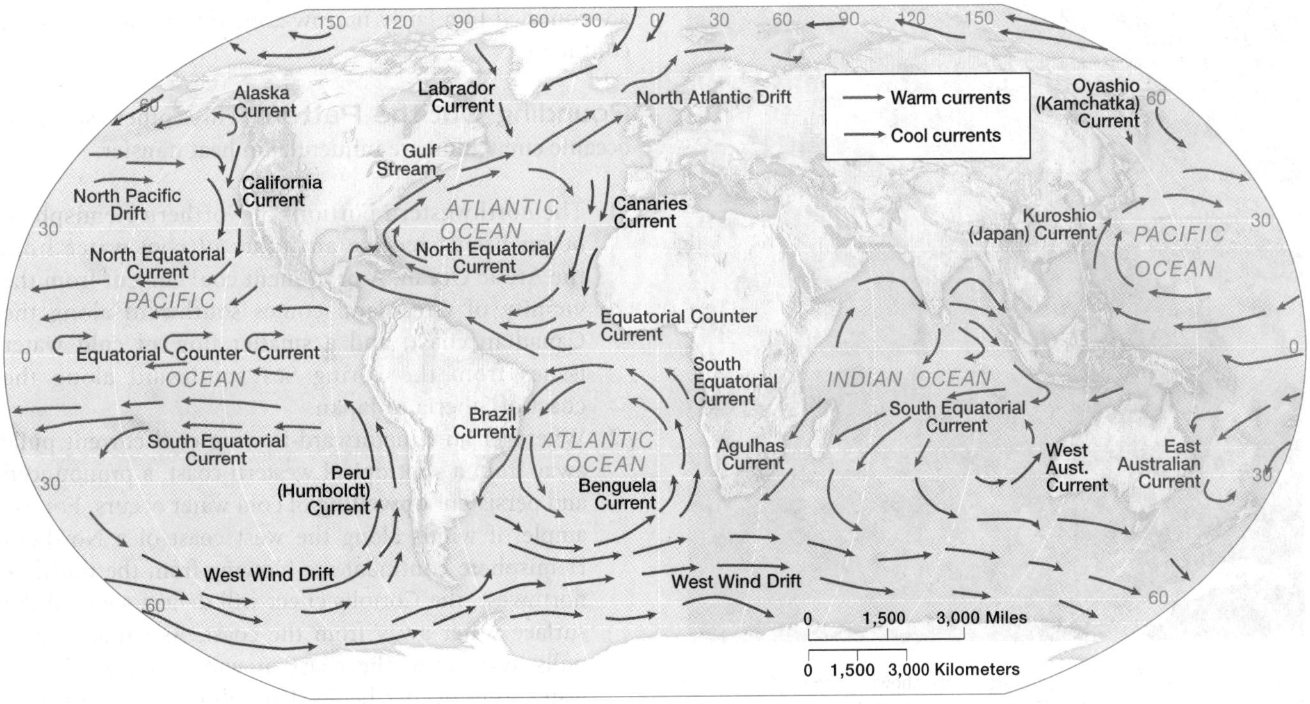

▲ **Figure 4-25** The major surface ocean currents. Warm currents are shown with red arrows and cool currents with blue arrows.

simple pattern of surface currents is characteristic of all the basins. It consists of a series of enormous elliptical loops elongated east-west and centered approximately at 30° of latitude (except in the Indian Ocean, where it is centered closer to the equator). These loops, called **subtropical gyres**, flow clockwise in the Northern Hemisphere and counterclockwise in the Southern Hemisphere (see Figure 4-25).

On the equatorward side of each subtropical gyre is the *Equatorial Current*, which moves steadily toward the west. The two equatorial currents have an average position 5° to 10° north or south of the equator and are separated by the *Equatorial Countercurrent*, which is an east-moving flow approximately along the equator in each ocean. The equatorial currents feed the Equatorial Countercurrent near its western margin in each basin. Water from the Equatorial Countercurrent in turn drifts poleward to feed the Equatorial Current near the eastern end of its path.

Near the western margin of each ocean basin, the general current curves poleward. As these currents approach the poleward margins of the ocean basins, they curve east; as they reach the eastern edges of the basins they curve back toward the equator, producing an incompletely closed loop in each basin.

The movement of these currents, although impelled by the wind, is also influenced by the deflective force of Earth's rotation—the Coriolis effect (discussed in Chapter 3). The Coriolis effect dictates that the ocean currents are deflected to the right in the Northern Hemisphere and to the left in the Southern Hemisphere. A glance at the basic pattern shows that the current movement around the gyres responds precisely to the Coriolis effect.

Northern and Southern Variations In the two Northern Hemisphere basins—North Pacific and North Atlantic—the bordering continents lie so close together at the northern basin margin that the bulk of the current flow is prevented from entering the Arctic Ocean. This effect is more pronounced in the Pacific than in the Atlantic. The North Pacific has very limited flow northward between Asia and North America, whereas in the North Atlantic a larger proportion of the flow escapes northward between Greenland and Europe.

In the Southern Hemisphere, the continents are far apart. Thus, the southern segments of the gyres in the South Pacific, South Atlantic, and South Indian oceans are connected as one continuous flow in the uninterrupted belt of ocean that extends around the world in the vicinity of latitude 60° S. This circumpolar flow is called the *West Wind Drift*.

Current Temperatures Of utmost importance to our understanding of latitudinal heat transfer are the temperatures of the various currents. Each major current can be characterized as *warm* or *cool* relative to the surrounding water at that latitude. The generalized temperature characteristics are as follows:

1. Low-latitude currents (Equatorial Current, Equatorial Countercurrent) have warm water.
2. Poleward-moving currents on the western sides of ocean basins (off the east coast of continents) carry warm water toward higher latitudes (Figure 4-26).
3. The high-latitude currents in the Northern Hemisphere gyres carry warm water to the east, while the high-latitude currents in the Southern Hemisphere

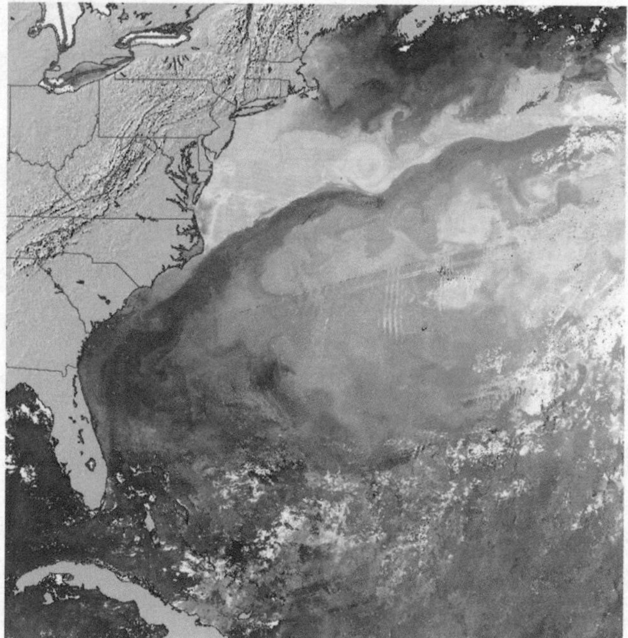

▲ **Figure 4-26** Multi-pass satellite image showing the Gulf Stream off the east coast of North America. Red represents relatively high water temperature while blue represents relatively low water temperatures. Several vertical and horizontal bands are not actual temperature features but are caused by incomplete satellite data.

gyres (generally combined into the West Wind Drift) carry cool water to the east.

4. Equatorward-moving currents on the eastern sides of ocean basins (off the west coasts of continents) carry cool water toward the equator.

In summary, from the perspective of the margins of the continents, the general circulation of the oceans is a poleward flow of warm tropical water along the east coasts of continents and an equatorward movement of cool, high-latitude water along the west coasts of continents.

Western Intensification In addition to differences in temperature, the poleward moving warm currents off the east coast of continents tend to be narrower, deeper, and faster than the equatorward moving cool currents flowing off the west coast of continents. This phenomenon is called *western intensification* because it occurs on the western side of the subtropical gyres (in other words, it occurs in the currents flowing poleward off the east coasts of continents in the midlatitudes). This intensification of poleward moving warm currents arises for a number of reasons, including the Coriolis effect. Because the Coriolis effect is greater in higher latitudes, the eastward moving high-latitude current flow is deflected back toward the equator more strongly than the westward moving equatorial current flow is deflected toward the poles. This means that cool water is slowly flowing back toward the equator across much of the eastward-moving high latitude currents, whereas the poleward-flowing warm currents are confined to a fairly narrow zone off the east coasts of continents.

Rounding Out the Pattern Three other aspects of oceanic circulation are influential in heat transfer:

1. The northwestern portions of Northern Hemisphere ocean basins receive an influx of cool water from the Arctic Ocean. A prominent cool current from the vicinity of Greenland comes southward along the Canadian coast, and a smaller flow of cold water issues from the Bering Sea southward along the coast of Siberia to Japan.
2. Wherever an equatorward-flowing cool current pulls away from a subtropical western coast, a pronounced and persistent **upwelling** of cold water occurs. For example, if winds along the west coast of a Northern Hemisphere continent are blowing from the north or northwest, the Coriolis effect will deflect some of the surface water away from the coast. As surface water pulls away from the coast, it will be replaced with water from deeper below. Upwelling brings nutrient-rich water to the surface, making west-coast marine ecosystems highly productive. The upwelling also brings colder water to the surface, generally decreasing the surface temperature of the already cool currents off the western coast of continents. Upwelling is most striking off South America but is also notable off North America, northwestern Africa, and southwestern Africa. It is much less developed off the coast of western Australia.
3. In addition to the surface ocean currents we have just described, there is a deep ocean circulation pattern—sometimes called the *global conveyor belt circulation*—that influences global climate in subtle, but nonetheless important ways. The deep ocean conveyor belt circulation is discussed in greater detail in Chapter 9.

Vertical Temperature Patterns

As we study the geography of weather and climate most of our attention is directed to the horizontal dimension; in other words, we are concerned with the spatial distribution of phenomena over the surface of Earth. From time to time, however, we must focus on vertical patterns or processes because of their influence on surface features. This statement is particularly pertinent with regard to the atmosphere.

Environmental Lapse Rate

Temperature change with increasing altitude within the troposphere is relatively predictable. As we learned in Chapter 3, throughout the troposphere, under typical conditions, a general decrease in temperature occurs with increasing altitude (Figure 4-27a). However, there are many exceptions to

FOCUS

Measuring Earth's Surface Temperature by Satellite

The temperature of Earth's surface is a direct and significant influence on the atmosphere above, and so monitoring surface temperatures is a key to both weather prediction and climate analysis. Before the advent of remote sensing, scientists relied on ships, buoys, and land-based instruments to gather this data, leaving large gaps in coverage, especially over ocean areas. In recent years, however, scientists have used the Moderate Resolution Imaging Spectroradiometer (MODIS) on NASA's Aqua and Terra satellites to gather surface temperature data around the globe.

Using the midinfrared and thermal infrared channels of the MODIS instruments, along with computer algorithms to compensate for factors such as absorption and scattering in the atmosphere, the "skin" temperature of ocean and land surfaces can be estimated by measuring emitted thermal infrared radiation.

Sea surface temperature (SST) is one of the most important influences of weather and climate (Figure 4-F). The SST influences not only the temperature of air masses that originate over ocean areas (and so the weather of continental areas over which these air masses may pass), but also the intensity of storms such as hurricanes. Regular monitoring of SST helps scientists anticipate the onset of El Niño and La Niña events (discussed in Chapter 5) and provides important information about long-term changes to Earth's environment.

As is the case with sea surface temperature, the day and night land temperatures (Figures 4-G and 4-H) measure the temperature of the surface itself, not the temperature of the air above. In these images, yellow represents the highest land temperatures while blue represents the lowest. Notice the high daytime surface temperatures in the Northern Hemisphere subtropical and midlatitude deserts—a consequence of high Sun and sparse cloud cover. These same areas cool off significantly at night, as do high mountain areas, such as the Himalayas, where the thinner atmosphere allows rapid heat loss at night.

▲ **Figure 4-F** Global sea surface temperature in May 2009 derived from the MODIS instruments on NASA's Aqua and Terra satellites.

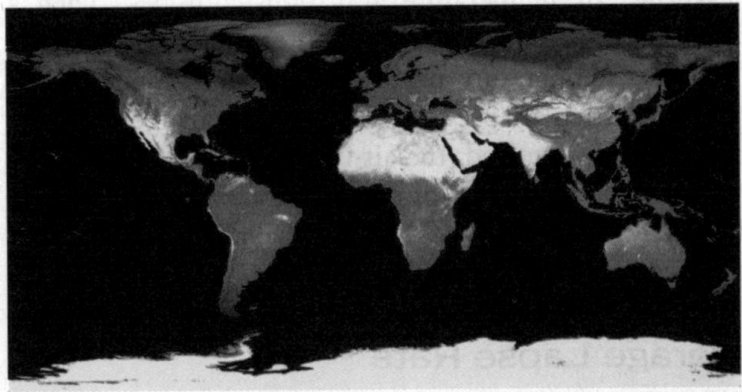

▲ **Figure 4-G** Daytime land surface temperature in May 2009 derived from the MODIS instruments on NASA's Aqua and Terra satellites.

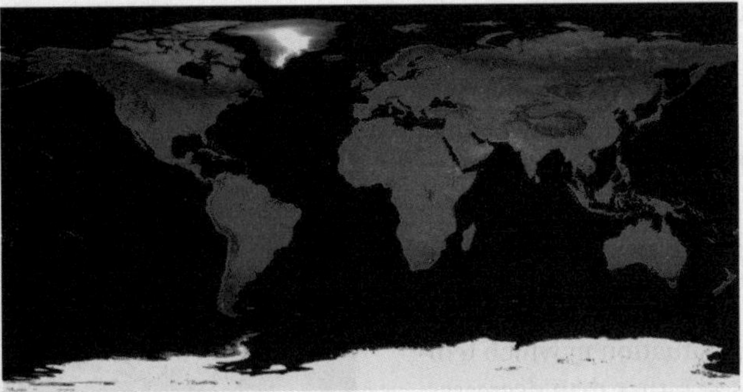

▲ **Figure 4-H** Nighttime land surface temperature in May 2009 derived from the MODIS instruments on NASA's Aqua and Terra satellites.

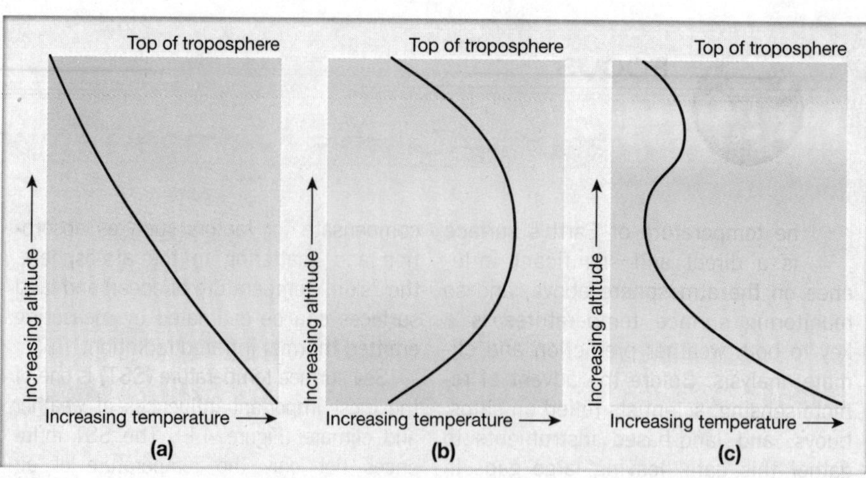

▶ **Figure 4-27** A comparison of normal and inverted lapse rates. (a) Tropospheric temperature normally decreases with increasing altitude. (b) In a surface inversion, temperature increases with increasing altitude from ground level to some distance above the ground. (c) In an upper-air inversion, temperature first decreases with increasing altitude as in a normal lapse rate, but then at some altitude well below the tropopause begins to increase with increasing altitude.

this general statement. Indeed, the rate of vertical temperature decline can vary according to season, time of day, amount of cloud cover, and a host of other factors. In some cases, there is even an opposite trend, with the temperature increasing upward for a limited distance.

The observed trend of vertical temperature change in the atmosphere is called the **environmental lapse rate**. Determining the lapse rate of a column of air involves measuring air temperature at various altitudes. Then a graph of temperature as a function of height is drawn to produce a temperature profile of that air column. When measuring such a lapse-rate temperature change, only the thermometer is moved; the air is at rest. If the air is moving vertically, expansion or compression will cause an adiabatic temperature change—such adiabatic lapse rates are explored more fully in Chapter 6.

Average Lapse Rate

Although the environmental lapse rate varies from place to place and from time to time, particularly in the lowest few hundred meters of the troposphere, the average rate of temperature change is about 6.5°C per 1000 meters (3.6°F per 1000 feet). This is called the **average lapse rate**, or average vertical temperature gradient within the troposphere. The average lapse rate tells us that if a thermometer measures the temperature 1000 meters above a previous measurement, the reading will be, on the average, 6.5°C cooler. Conversely, if a second measurement is made 1000 meters lower than the first, the temperature will be about 6.5°C warmer.

Temperature Inversions

The most prominent exception to an average lapse-rate condition is a **temperature inversion**, a situation in which temperature in the troposphere increases, rather than decreases, with increasing altitude. Inversions are relatively common in the troposphere but are usually of brief duration and restricted depth. They can occur near Earth's surface, as in Figure 4-27b, or at higher levels, as in Figure 4-27c.

Inversions influence weather and climate. As we shall see in Chapter 6, an inversion inhibits vertical air movements and greatly diminishes the possibility of precipitation. Inversions also contribute significantly to increased air pollution because they create stagnant air conditions that greatly limit the natural upward dispersal of urban-industrial pollutants (Figure 4-28).

Surface Inversions The most readily recognizable inversions are those found at ground level. These are often *radiation inversions* that can develop on a long, cold winter night when a land surface rapidly emits longwave radiation into a clear, calm sky. The cold ground then cools the air above by conduction. In a relatively short time, the lowest few hundred meters of the troposphere become colder than the air above and a temperature inversion is in effect. Radiational inversions are primarily winter phenomena because there is only a short daylight period for incoming solar heating and a long night for radiational cooling. They are therefore much more prevalent in high latitudes than elsewhere.

▲ **Figure 4-28** Downtown Los Angeles on a "mild" smog day. The top of the inversion is visible where the mountains rise above the smog layer.

Advectional inversions develop where there is a horizontal inflow of cold air into an area. This condition commonly is produced by cool maritime air blowing into a coastal locale. Advectional inversions are usually short-lived (typically overnight) and shallow. They may occur at any time of year, depending on the location of the relatively cold surface and on wind movement.

Another type of surface inversion results when cooler air slides down a slope into a valley, thereby displacing slightly warmer air. This fairly common occurrence during winter in some midlatitude regions is called a *cold-air-drainage inversion*.

Upper-Air Inversions Temperature inversions well above the ground surface nearly always are the result of air descending from above. These *subsidence inversions* are usually associated with high-pressure conditions, which are particularly characteristic of subtropical latitudes throughout the year and of Northern Hemisphere continents in winter. A subsidence inversion can be fairly deep (sometimes several thousand meters), and its base is usually a few hundred meters above the ground, as low-level turbulence prevents it from sinking lower.

Global Temperature Patterns

The goal of this and the four succeeding chapters is to examine the world distribution pattern of climate. With the preceding pages as background, we now turn our attention to the worldwide distribution of temperature, the first of the four climate elements.

Maps of global temperature patterns usually show seasonal extremes rather than annual averages. January and July are the months of lowest and highest temperatures for most places on Earth, and so maps portraying the average temperatures of these two months provide a simple but meaningful expression of thermal conditions in winter and summer (Figure 4-29 and Figure 4-30). Temperature distribution is shown by means of **isotherms**, lines joining points of equal temperature. Temperature maps are based on monthly averages, which are based on daily averages; the maps do not show the maximum daytime heating or the maximum nighttime cooling. Although the maps are on a very small scale, they permit a broad understanding of temperature patterns for the world.

Prominent Controls of Temperature

Gross patterns of temperature are controlled largely by four factors—altitude, latitude, land-water contrasts, and ocean currents.

Altitude Because temperature responds sharply to altitudinal changes, it would be misleading to plot actual temperatures on a temperature map, since high-altitude stations would almost always be colder than low-altitude stations. The complexity introduced by hills and mountains would make the map more complicated and difficult to comprehend. Consequently, the data for most maps displaying world temperature patterns are modified by adjusting the temperature to what it would be if the station were at sea level. This is done by using the average lapse rate, a method that produces artificial temperature values but eliminates the complication of terrain differences. Maps plotted in this way are useful in showing world patterns, but they are not satisfactory for indicating actual temperatures for locations that are not close to sea level.

Latitude Clearly the most conspicuous feature of any world temperature map is the general east-west trend of the isotherms, roughly following the parallels of latitude. If Earth had a uniform surface and did not rotate, the isotherms would probably coincide exactly with parallels, showing a progressive decrease of temperature poleward from the equator. However, Earth does rotate, and it has ocean waters that circulate and land that varies in elevation. Consequently, there is no precise temperature correlation with latitude. Nevertheless, the fundamental cause of temperature variation the world over is insolation, which is governed primarily by latitude, and the general temperature patterns reflect latitudinal control.

Land-Water Contrasts The different heating and cooling characteristics of land and water are also reflected conspicuously on a temperature map. Summer temperatures are higher over the continents than over the oceans, as shown by the poleward curvature of the isotherms over continents in the respective hemispheres (July in the Northern Hemisphere, January in the Southern Hemisphere). Winter temperatures are lower over the continents than over the oceans; the isotherms bend equatorward over continents in this season (January in the Northern Hemisphere, July in the Southern Hemisphere). Thus, in both seasons, isotherms make greater north-south shifts over land than over water.

Another manifestation of the land–water contrast is the regularity of the isothermal pattern in the midlatitudes of the Southern Hemisphere, in contrast to the situation in the Northern Hemisphere. There is very little land in these Southern Hemisphere latitudes, and so contrasting surface characteristics are absent.

Ocean Currents Some of the most obvious bends in the isotherms occur in near-coastal areas of the oceans, where prominent warm or cool currents reinforce the isothermal curves caused by land-water contrasts. Cool currents deflect isotherms equatorward, whereas warm currents deflect them poleward. Cool currents produce the greatest isothermal bends in the warm season: note the January situation off the western coast of South America and the southwestern coast of Africa, or the July conditions off the western coast of North America. Warm currents have

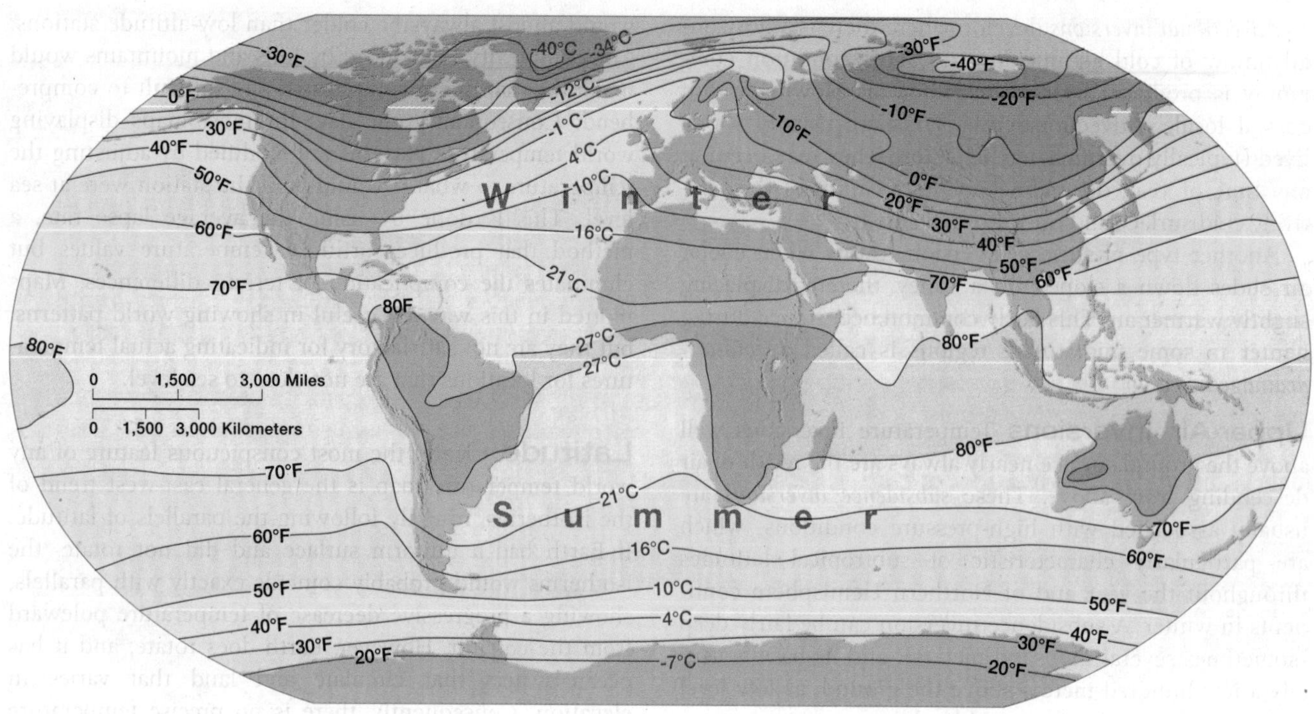

▲ **Figure 4-29** Average January sea-level temperatures.

their most prominent effects in the cool season: witness the isothermal pattern in the North Atlantic Ocean in January.

Seasonal Patterns

Apart from the general east-west trend of the isotherms, probably the most conspicuous feature of Figure 4-29 and Figure 4-30 is the latitudinal shift of the isotherms from one map to the other. The isotherms follow the changing balance of insolation during the course of the year, moving northward from January to July and returning southward from July to January. Note, for example, the 10°C (50°F) isotherm in southernmost South America: in January (midsummer), it is positioned at the southern tip of the continent, whereas in July (midwinter), it is shifted considerably to the north. This isothermal shift is much more pronounced in high latitudes

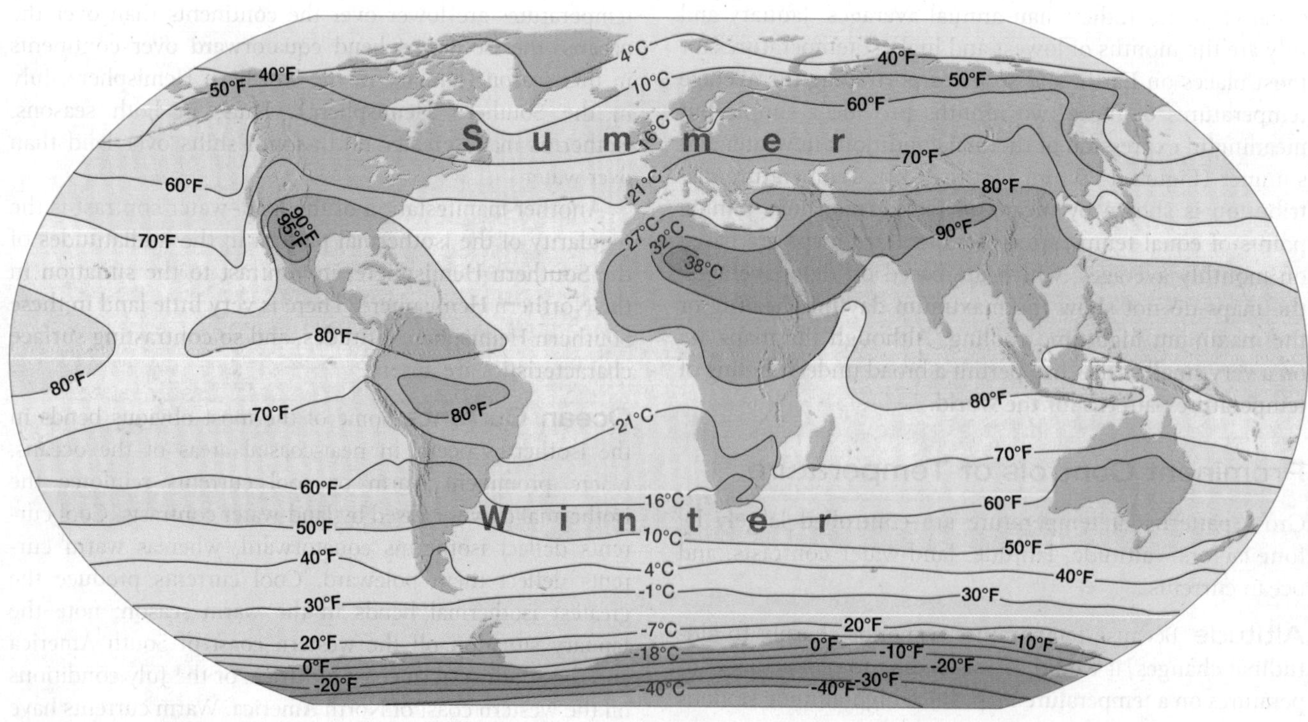

▲ **Figure 4-30** Average July sea-level temperatures.

than in low and also much more pronounced over the continents than over the oceans. Thus, tropical areas, particularly tropical oceans, show relatively small displacement of the isotherms from January to July, whereas over middle- and high-latitude landmasses, an isotherm may migrate northward or southward more than 4000 kilometers (2500 miles)—some 14° of latitude—as illustrated in Figure 4-31.

Isotherms are also more tightly packed in winter. This close line spacing indicates that the temperature gradient (rate of temperature change with horizontal distance) is steeper in winter than in summer, which in turn reflects the greater contrast in radiation balance in winter. The temperature gradient is also steeper over continents than over oceans.

The coldest places on Earth are over landmasses in the higher latitudes. During July, the polar region of Antarctica is the dominant area of coldness. In January, the coldest temperatures occur many hundreds of kilometers south of the North Pole, in subarctic portions of Siberia, Canada, and Greenland. The principle of greater cooling of land than water is clearly demonstrated.

The highest temperatures also are found over the continents. The locations of the warmest areas in summer, however, are not equatorial. Rather, they are in subtropical latitudes, where descending air maintains clear skies most of the time, allowing for almost uninterrupted insolation. Frequent cloudiness precludes such a condition in the equatorial zone. Thus, the highest July temperatures occur in northern Africa and in the southwestern portions of Asia and North America, whereas the principal areas of January heat are in subtropical parts of Australia, southern Africa, and South America.

Average annual temperatures are highest in equatorial regions, however, because these regions experience so little winter cooling. Subtropical locations cool substantially on winter nights, and so their annual average temperatures are lower. The ice-covered portions of the Earth—Antarctica and Greenland—remain quite cold throughout the year.

Annual Temperature Range

Another map useful in understanding the global pattern of air temperature is one that portrays the average annual range of temperatures (Figure 4-32). **Average annual temperature range** is the difference between the average temperatures of the warmest and coldest months (normally July and January). The data are portrayed on the map by isolines that resemble isotherms. Enormous seasonal variations in temperature occur in the interiors of high-latitude continents, and continental areas in general experience much greater ranges than do equivalent oceanic latitudes. At the other extreme, the average temperature fluctuates only slightly from season to season in the tropics, particularly over tropical oceans.

Global Warming and the Greenhouse Effect

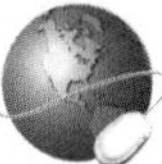

Global Warming

As we described earlier in this chapter, the "natural" greenhouse effect has been part of the basis of life on Earth since the early atmosphere formed. Without it, our planet would be a frozen mass, perhaps 30°C (54°F) colder than it is today. Over the last three decades, the greenhouse effect has received considerable attention from the media and the general public. Data gathered from surface weather stations, ships, buoys, balloons, satellites, ice cores, and other paleoclimatological sources indicate that the climate of Earth is becoming warmer. This warming trend has become known to the public as **global warming.**

Over the twentieth century, average global temperatures increased by more than 0.7°C (1.3°F)—in the last quarter of the twentieth century alone temperatures increased 0.2–0.3°C (0.4°F) Figure 4-33. This temperature increase

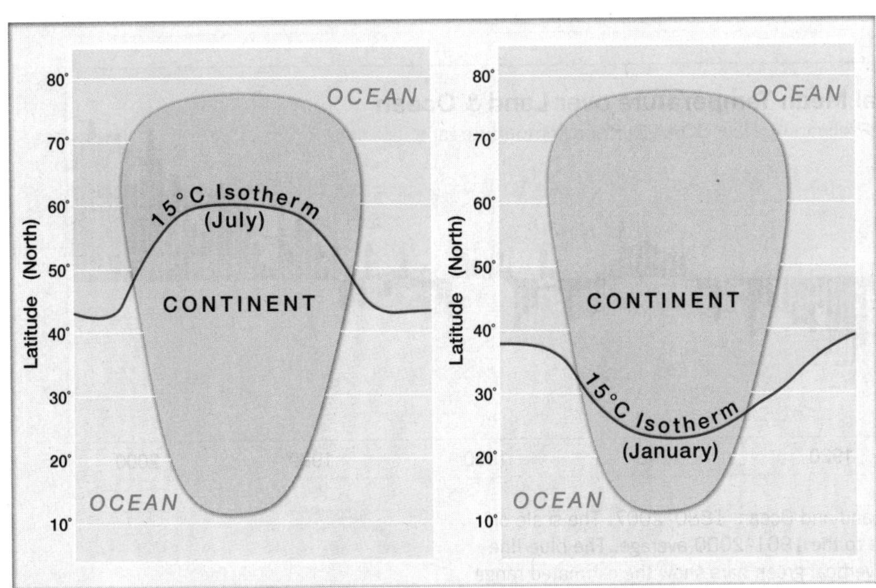

◀ **Figure 4-31** Idealized seasonal migration of the 15°C (59°F) isotherm over a hypothetical Northern Hemisphere continent. The latitudinal shift is greatest over the interior of the continent and least over the adjacent oceans. For example, over the western ocean, the isotherm moves only from latitude 38° N in January to 42° N in July, but over the continent the change is from 22° N in January all the way to 60° N in July.

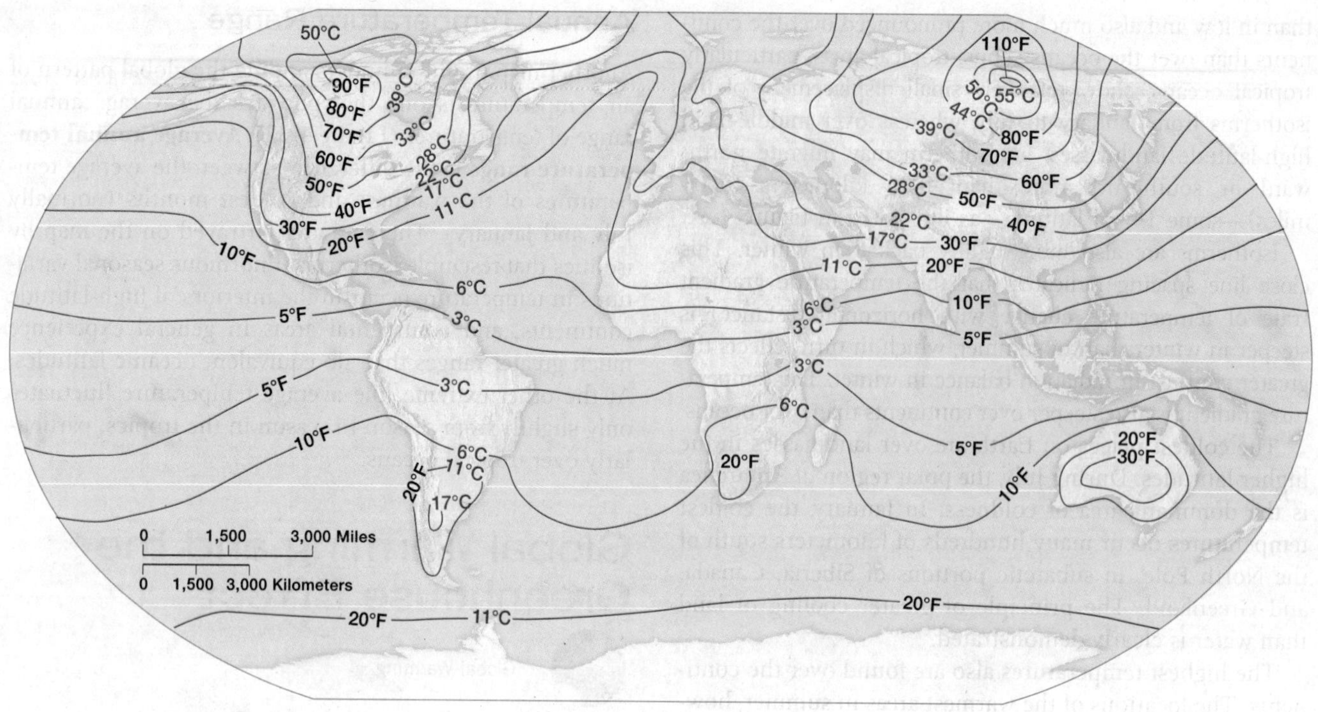

▲ **Figure 4-32** The world pattern of average annual temperature range. The largest ranges occur in the interior of high-latitude landmasses.

over the last 100 years is apparently greater than that of any other century in at least the last 1000 years, and the rate of temperature increase over the last three decades may be greater than at any time in the last 800,000 years. Overall, global temperatures are higher today than they have been in at least 100,000 years, and the last two decades have been the hottest since widespread instrument readings began about 140 years ago—8 of the 10 hottest years on record have occurred since the year 2000. Because direct instrument measurements of Earth's temperature only go back a few centuries, past temperature patterns are calculated using "proxy" measures—such as data deciphered from polar ice cores. Although there is a margin of error in these figures, evidence clearly points to warming.

The cause of global warming appears to be *human-enhanced greenhouse effect*. Since the industrial era began in the mid 1700s, human activities have increased the concentrations of greenhouse gases—such as carbon dioxide,

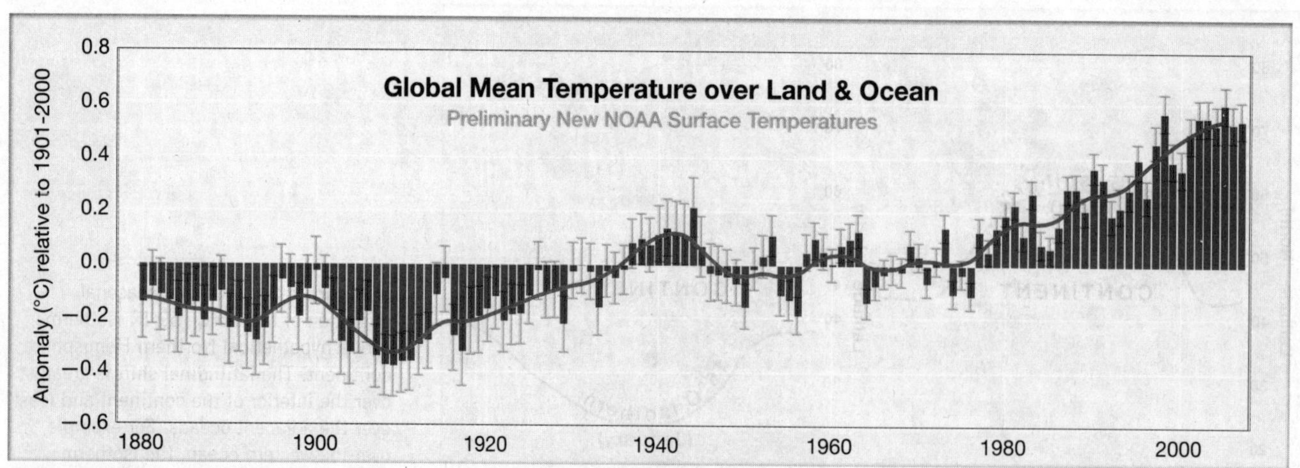

▲ **Figure 4-33** Global Mean Temperature over Land and Ocean, 1880–2007. The scale on the left shows the temperature difference relative to the 1901–2000 average. The blue line shows the smoothed temperature trend, and the vertical green bars show the estimated range of error.

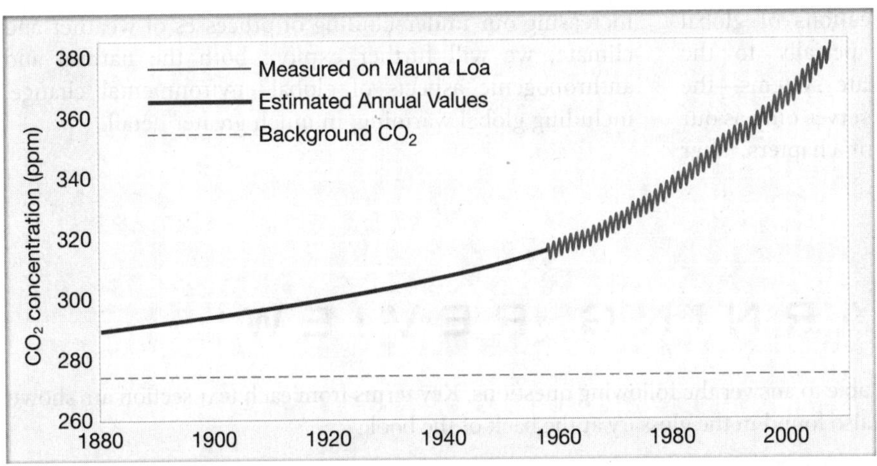

◀ **Figure 4-34** Change in atmospheric carbon dioxide concentration from 1880–2008. The fluctuations in the measured values from Mauna Loa, Hawai'i, result from seasonal variations in plant photosynthesis and respiration. The dashed blue line shows pre-industrial levels of carbon dioxide.

methane, tropospheric ozone, and chlorofluorocarbons in the atmosphere. As greenhouse gas concentrations in the atmosphere increase, more terrestrial radiation is retained in the lower atmosphere, thereby raising global temperatures.

Carbon dioxide (CO_2) is thought to be responsible for about 64 percent of the human-enhanced greenhouse effect. CO_2 concentrations have been rising steadily since the Industrial Revolution began in the mid-1700s (Figure 4-34). Carbon dioxide is a principal by-product of combustion of anything containing carbon, such as coal and petroleum. Since 1750, carbon dioxide levels in the atmosphere have increased by more than 30 percent. The latest paleoclimatological data indicates that the current concentration of CO_2 in the atmosphere of about 390 parts per million is greater than at any time in the last 800,000 years.

Many other greenhouse gases have been added to the atmosphere by human activity. Methane—produced by grazing livestock and rice paddies and a by-product of the combustion of wood, natural gas, coal, and oil—has more than doubled since 1750. Nitrous oxide—which comes from chemical fertilizers and automobile emissions—has increased by 17 percent since 1750. Chlorofluorocarbons (CFCs) are synthetic chemicals that were widely used as refrigerants and as propellants in spray cans until quite recently (see Chapter 3 for a discussion of another consequence of CFCs in the atmosphere—ozone depletion). Many of these gases, and others, are being released into the atmosphere at accelerating rates.

The increase in greenhouse gas concentrations, especially carbon dioxide, correlates well with the observed increase in global temperature: as CO_2 has increased, so have average global temperatures (Figure 4-35).

It is well known that climate undergoes frequent natural fluctuations, becoming warmer or colder regardless of human activities. There is, however, an increasing body of evidence that indicates that *anthropogenic* (human-induced) factors are largely, if not wholly, responsible for the recent temperature increase. The Intergovernmental Panel on Climate Change (IPCC) is the most important international organization of atmospheric scientists and policy analysts assessing global climate change. In 2007, in recognition of their many years of work on climate change, the IPCC was the corecipient of the Nobel Peace Prize. The *Fourth Assessment Report* of the IPCC released in 2007 concluded bluntly:

> Warming of the climate system is unequivocal, as is now evident from observations of increases in global average air and ocean temperatures, widespread melting of snow and ice, and rising global average sea level.

With regard to the causes of this climate change, the IPCC concluded:

> Most of the observed increase in globally averaged temperatures since the mid-20th century is very likely [greater than 90% probability] due to the observed increase in anthropogenic greenhouse gas emissions.

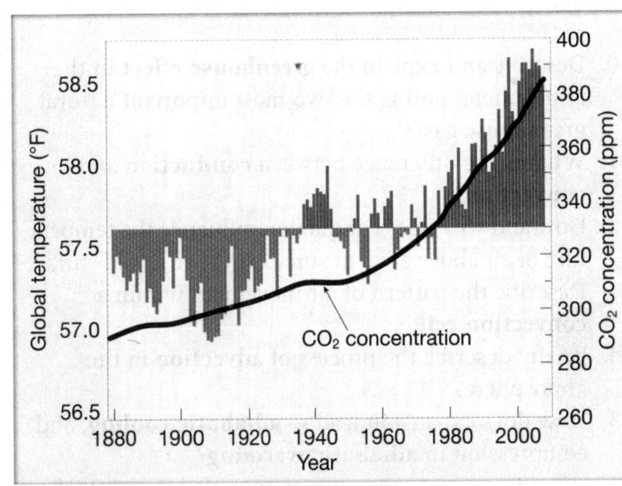

▲ **Figure 4-35** The relationship between atmospheric carbon dioxide concentration and mean global temperature.

Because both the causes and implications of global warming are so complicated—due especially to the many feedback loops involved in climate systems—the preceding description of global warming serves only as our introduction to the topic. In subsequent chapters, after increasing our understanding of processes of weather and climate, we will further explore both the natural and anthropogenic aspects of global environmental change, including global warming, in much greater detail.

CHAPTER 4 LEARNING REVIEW

After studying this chapter, you should be able to answer the following questions. Key terms from each text section are shown in bold type. Definitions for key terms are also found in the glossary at the back of the book.

Energy, Heat, and Temperature (p. 67)

1. What is the difference between **heat (thermal energy)** and **temperature**?
2. What is the relationship between the internal **kinetic energy** of a substance and its temperature?

Solar Energy (p. 69)

3. Briefly describe the following bands of **electromagnetic radiation (radiant energy): visible light**, **ultraviolet (UV)**, **infrared**.
4. Describe and contrast the portions of the **electromagnetic spectrum** referred to as **shortwave radiation** and **longwave radiation (terrestrial radiation)**.
5. What is **insolation**?

Basic Heating and Cooling Processes in the Atmosphere (p. 70)

6. Describe and contrast the following processes associated with electromagnetic energy: **radiation (emission)**, **absorption**, **reflection**, and **transmission**.
7. What generally happens to the temperature of an object as a response to the absorption of electromagnetic radiation?
8. How is **scattering** different from reflection?
9. Why is the sky blue? Why are sunsets orange and red?
10. Describe and explain the **greenhouse effect** in the atmosphere, noting the two most important natural greenhouse gases.
11. What is the difference between **conduction** and **convection**?
12. How and why does conduction influence the temperature of air above a warm surface? Above a cold surface?
13. Describe the pattern of air movement within a **convection cell**.
14. Briefly describe the process of **advection** in the atmosphere.
15. How does expansion lead to **adiabatic cooling**, and compression to **adiabatic warming**?
16. What happens to the temperature of rising air? Of descending air? Why?
17. What is **latent heat**?
18. Why do we say that **evaporation** is a cooling process and **condensation** a warming process?

The Heating of the Atmosphere (p. 77)

19. Briefly describe how the troposphere is heated by the Sun.
20. What is **albedo**?
21. Why does temperature generally decrease with increasing altitude in the troposphere?
22. "The atmosphere is mostly heated by Earth's surface rather than directly by the Sun." Comment on the validity of this statement.

Variations in Heating by Latitude and Season (p. 80)

23. What is meant by the **angle of incidence** of the Sun's rays?
24. Explain the reasons for the unequal heating (by latitude) of Earth by the Sun.
25. Why are there greater seasonal temperature differences in the high latitudes than in the tropics?

Land and Water Contrasts (p. 81)

26. How does the **specific heat** of a substance influence its rate of heating?
27. Explain why land heats and cools faster and more than water.

Mechanisms of Heat Transfer (p. 83)

28. What are the two dominant mechanisms of heat transfer around the world?
29. How would global temperature patterns be different without this heat transfer?
30. What is the relative temperature of the ocean current flowing along the west coast of a continent in the midlatitudes? Along the east coast of a continent?
31. Describe the basic pattern of ocean currents (the **subtropical gyres**) around the margins of a major ocean basin (including the relative temperature of each current—either "cool" or "warm"). You should be able to sketch the direction of movement and note the relative temperature of major ocean currents on a blank map of an ocean basin.
32. Describe **upwelling** and its cause.

Vertical Temperature Patterns (p. 86)

33. What is meant by the **environmental lapse rate**?
34. What is the **average lapse rate** in the troposphere?

35. What is a **temperature inversion**?
36. What is the difference between a radiational inversion and an advectional inversion?

Global Temperature Patterns (p. 89)

37. What is an **isotherm**?
38. Using the isotherm maps of average January and July sea level temperature (Figure 4-29 and Figure 4-30) describe the influence of latitude, season, land–water contrasts, and ocean currents on global temperature patterns. For example, explain why the following isotherms vary in latitude across the maps:
    - January −1°C (30°F) isotherm in the Northern Hemisphere
    - July 21°C (70°F) isotherm in the Northern Hemisphere
39. Where in the world do we find the greatest **average annual temperature ranges** and where do we find the smallest average annual temperature ranges? Why?

Global Warming and the Greenhouse Effect (p. 91)

40. What is meant by the term **global warming**?
41. How might humans be enhancing the natural greenhouse effect?

Log in to **www.mygeoscienceplace.com** for animations, videos, RSS feeds, maps, and self-study quizzes to enhance your study of **Insolation and Temperature**.

5 Atmospheric Pressure and Wind

To the layperson, atmospheric pressure is the most difficult climate element to comprehend. The other three—temperature, wind, and moisture—are more readily understood because our bodies are much more sensitive to them. We can "feel" heat, air movement, and humidity, and we are quick to recognize variations in these elements. Pressure, on the other hand, is a phenomenon of which we are usually unaware; its variations are considerably less noticeable to our senses. Pressure usually impinges on our sensitivity only when we experience rapid vertical movement, as in an elevator or an airplane because of the difference in pressure inside and outside our ears, which is sometimes relieved only by "popping" them.

Despite its inconspicuousness, pressure is an important feature of the atmosphere. It is tied closely to the other weather elements, acting on them and responding to them. Pressure has an intimate relationship with wind: spatial variations in pressure are responsible for air movements. Hence, pressure and wind are often discussed together, as is done in this chapter.

The Impact of Pressure and Wind on the Landscape

The influence of atmospheric pressure on the landscape is significant but indirect; this influence is manifested mostly by wind and temperature as these two elements respond to pressure changes. The impact of wind is more direct and explicit than the impact of temperature because wind has the energy to transport solid particles in the air and thus has a visible component to its activity. Vegetation may bend in the wind and loose material of whatever kind may be shifted from one place to another. The results are nearly always short-run and temporary, however, and usually have no lasting effect on the landscape except at the time of a severe storm. Nevertheless, pressure and wind are major elements of weather and climate, and their interaction with other atmospheric components and processes cannot be overestimated.

The Nature of Atmospheric Pressure

Gas molecules, unlike those of a solid or liquid, are not strongly bound to one another. Instead, they are in continuous motion, colliding frequently with one another and with any surfaces to which they are exposed. Consider a container in which a gas is confined, as in Figure 5-1. The molecules of the gas zoom around inside the container and collide again and again with the walls. The *pressure* of the gas is defined as the force the gas exerts on some specified area of the container walls.

◀ Wind driven storm wave strikes a lighthouse at Seaford in Sussex, England.

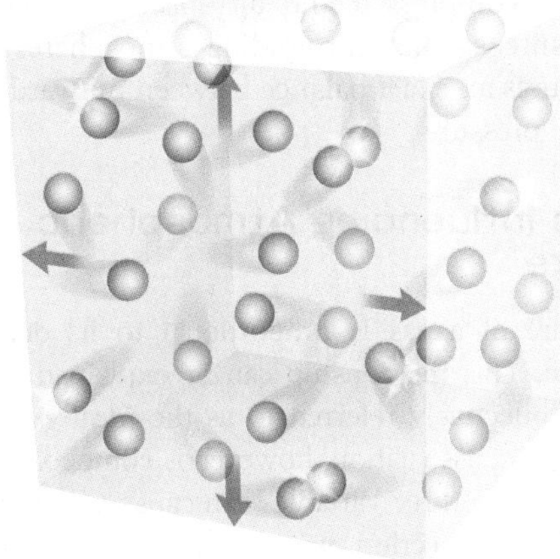

▲ **Figure 5-1** Gas molecules are always in motion. In this closed container, they bounce around, colliding with one another and with the walls of the container. These collisions give rise to the pressure exerted by the gas.

The atmosphere is made up of gases, of course, and so **atmospheric pressure** is the force exerted by the gas molecules on some area of Earth's surface or on any other body—including yours! At sea level, the pressure exerted by the atmosphere is slightly more than 1 kilogram per square centimeter (about 14.7 pounds per square inch). This value drops with increasing altitude because the farther away you get from Earth and its gravitational pull, the fewer gas molecules are present in the atmosphere.

The atmosphere exerts pressure on every solid or liquid surface it touches. The pressure is exerted equally in all directions—up, down, sideways, and obliquely. In other words, atmospheric pressure is not simply a "weight" from above. This means that every square centimeter of any exposed surface—animal, vegetable, or mineral—at sea level is subjected to that much pressure (Figure 5-2). We are not sensitive to this ever-present burden of pressure because our bodies contain solids and liquids that are not

▲ **Figure 5-2** The empty plastic bottle on the left was opened and then sealed tightly at an elevation of 3030 meters (9945 feet); when brought down to sea level the surrounding higher atmospheric pressure partially collapses the bottle. The bottle on the right contains air at sea level pressure.

significantly compressed and air spaces that are at the same pressure as the surrounding atmosphere; in other words, there is an exact balance between outward pressure and inward pressure.

Factors Influencing Atmospheric Pressure

The pressure of a gas is proportional to its density and temperature. This relationship can be explained by several equations collectively referred to as the *ideal gas law*. The cause-and-effect association, however, is complex. Variations in any one of the three variables can cause variations in the other two. In this section, we examine how pressure varies with density, temperature, and the ways in which vertical movement of air influences surface atmospheric pressure.

Density and Pressure *Density* is the mass of matter in a unit volume. For example, if you have a 10-kilogram cube of material with edge lengths of 1 meter, the density of that material is 10 kilograms per cubic meter (10 kg/m^3). The density of solid material is the same on Earth or the moon or in space, that of liquids varies very slightly from one place to another, but that of gases varies greatly with location. Gas density changes so easily because a gas expands as far as the environmental pressure will allow. If you have 10 kg of gas in a container that has a volume of 1 cubic meter, the gas density is 10 kg/m^3. If you then transfer all the gas to a container having a volume of 5 cubic meters, the gas expands to fill the larger volume. There are the same number of gas molecules that were in the smaller container (in other words, the same mass), but they are now spread out over a volume five times as large. Therefore, the gas density in the larger container is only 2 kg/m^3 (10 kg divided by 5 cubic meters).

The density of a gas is proportional to the pressure on it. The reverse of this statement is also true: the pressure a gas exerts is proportional to its density. The denser the gas, the greater the pressure it exerts.

The atmosphere is held to Earth by the force of gravity which prevents the gaseous molecules from escaping into space. At lower altitudes, the gas molecules of the atmosphere are packed more densely together (Figure 5-3). Because the density is higher, there are more molecular collisions and therefore higher pressure at lower altitudes. At higher altitudes, the air is less dense and there is a corresponding decrease in pressure. At any level in the atmosphere, then, the pressure is directly proportional to the air density at that altitude.

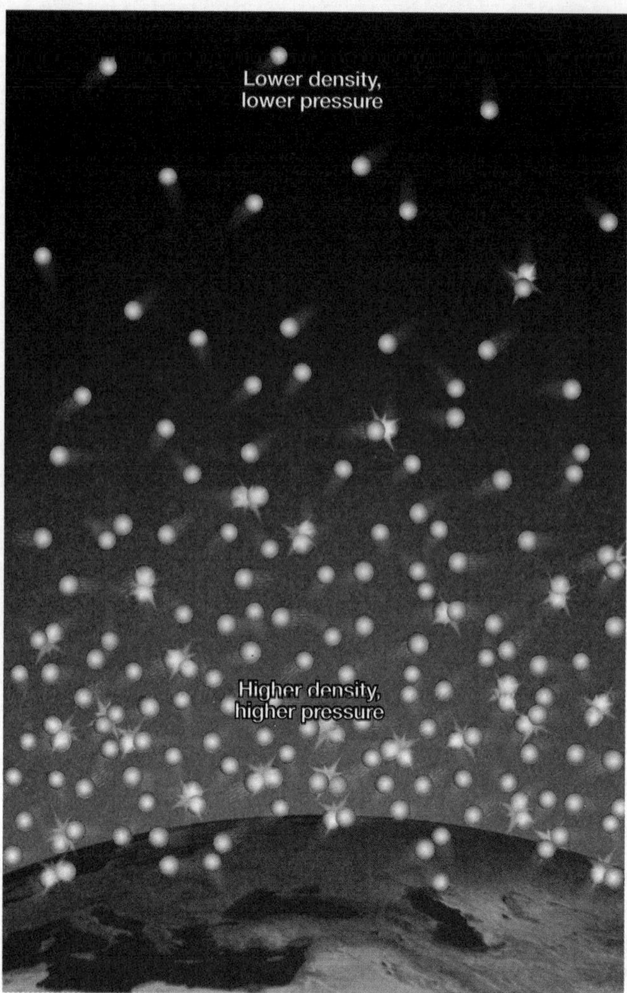

▲ **Figure 5-3** In the upper atmosphere, gas molecules are far apart and collide with each other infrequently, a condition that produces relatively low pressure. In the lower atmosphere, the molecules are closer together, and there are many more collisions, a condition that produces high pressure.

Temperature and Pressure If air is heated, as we noted in the preceding chapter, the molecules become more agitated and their speed increases. This increase in speed produces a greater force to their collisions and results in higher pressure. Therefore, if other conditions remain the same (in particular, if volume is held constant), an increase in the temperature of a gas produces an increase in pressure, and a decrease in temperature produces a decrease in pressure. Knowing this, you might conclude that the air pressure will be high on warm days and low on cold days. Such is not usually the case, however: warm air is generally associated with low atmospheric pressure and cool air with high atmospheric pressure. While this seems contradictory, recall that we made the qualifying statement "if other conditions remain the same" in describing how temperature and pressure are related. When air is heated and no control is kept on its volume, it expands, which decreases its density. Thus, the increase in temperature may be accompanied by a decrease in pressure caused by the decrease in density.

Dynamic Influences on Air Pressure Surface air pressure may also be influenced by "dynamic" factors. In other words, air pressure may be influenced by the movement of the air—especially the vertical movement of air associated with different rates of air convergence and divergence at the surface and in the upper troposphere. As a generalization, descending air tends to be associated with relatively high pressure at the surface, while rising air tends to be associated with relatively low pressure at the surface.

In short, atmospheric pressure is affected by differences in air density, air temperature, and air movement; and the relationship among these variables is complex. It is important for us to be alert to these linkages, but it is often difficult to predict how a change in one variable will influence the others in a specific instance. Nevertheless, some useful generalizations about the factors associated with areas of high pressure and low pressure near the surface can be made:

- Strongly descending air is usually associated with high pressure at the surface—a **dynamic high**.
- Very cold surface conditions are often associated with high pressure at the surface—a **thermal high**.
- Strongly rising air is usually associated with low pressure at the surface—a **dynamic low**.
- Very warm surface conditions are often associated with relatively low pressure at the surface—a **thermal low**.

Surface pressure conditions often can be traced to one of these factors being dominant.

Mapping Pressure with Isobars

Atmospheric pressure is measured with instruments called **barometers**. The first liquid-filled barometers date back to the 1600s, and measurement scales based on the height of a column of mercury are still in use (average sea-level pressure using a mercury barometer is 760 millimeters or 29.92 inches). Today, however, one of the most common units of measure for atmospheric pressure is the **millibar**. The millibar (mb) is an expression of force per surface area. One millibar is defined as 1000 dynes per square centimeter (one dyne is the force required to accelerate 1 gram of a mass 1 centimeter per second per second).[1] Average sea-level pressure is 1013.25 millibars. In many countries air pressure is described with the *kilopascal* (kPa; 1 kPa = 10 mb).

Once pressure in millibars is plotted on a weather map, it is then possible to draw isolines of equal pressure called **isobars**, as shown in Figure 5-4. The pattern of the isobars reveals the horizontal distribution of pressure in the region under consideration. Prominent on such maps are roughly circular or oval areas characterized as being either "high pressure" or "low pressure." These **highs** and **lows** represent relative conditions—pressure that is higher or lower than that of the surrounding areas. In a similar way, a **ridge** is an elongated area of relatively high pressure, while a **trough** is an elongated area of relatively low pressure. It is important to keep the relative nature of pressure centers in mind. For example, a pressure reading of 1005 millibars could be either "high" or "low" depending on the pressure of the surrounding areas.

[1]In the definition of a dyne the term "per second per second" may seem strange. However, in this case we are describing the force required to *accelerate* a mass, in other words, the force required to change its speed or direction. Thus, the dyne describes the rate of change, not the speed.

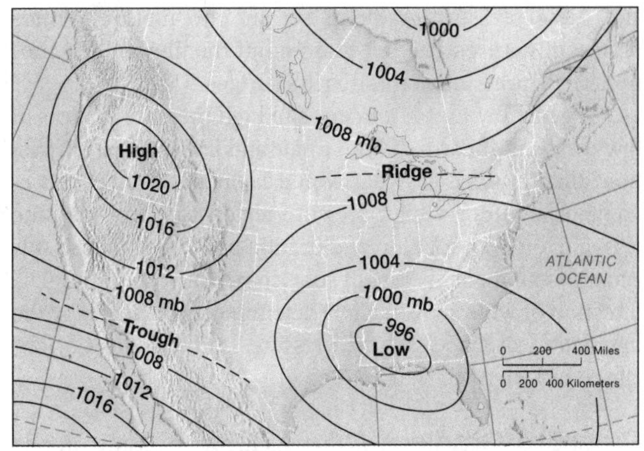

▲ **Figure 5-4** Isobars are lines connecting points of equal atmospheric pressure. When they have been drawn on a weather map, it is easy to determine the location of high-pressure and low-pressure centers. This simplified weather map shows pressure in millibars.

On most maps of air pressure, actual pressure readings are adjusted to represent pressures at a common elevation, usually sea level. This is done because pressure decreases rapidly with increasing altitude, and consequently significant variations in pressure readings are likely at different weather stations simply because of differences in elevation.

As with other types of isolines, the relative closeness of isobars indicates the horizontal rate of pressure change, or **pressure gradient**. The pressure gradient can be thought of as representing the "steepness" of the pressure "slope" (or more correctly, the abruptness of the pressure change over a distance), a characteristic that has a direct influence on wind, the topic to which we turn next.

The Nature of Wind

Development of Wind Patterns

The atmosphere is virtually always in motion. Air is free to move in any direction, its specific movements being shaped by a variety of factors. Some airflow is lackadaisical and brief; some is strong and persistent. Atmospheric motions often involve both horizontal and vertical displacement. **Wind** refers to horizontal air movement; it has been characterized as "air in a hurry." Instead of being called wind, small-scale vertical motions are normally referred to as *updrafts* and *downdrafts*; large-scale vertical motions are *ascents* and *subsidences*. The term wind is applied only to horizontal movements. Although both vertical and horizontal motions are important in the atmosphere, much more air is involved in horizontal movements than in vertical.

Direction of Movement

Insolation is the ultimate cause of wind because all winds originate from the same basic sequence of events: unequal heating of different parts of Earth's surface brings about temperature gradients that generate pressure gradients, and these pressure gradients set air into motion. Winds represent nature's attempt to even out the uneven distribution of air pressure across Earth's surface.

Air generally flows from areas of high pressure to areas of low pressure. If Earth did not rotate and if there were no friction, that is precisely what would happen—a direct movement of air from a high-pressure region to a low-pressure region. However, rotation and friction both exist, and so this general statement is usually not completely accurate. The direction of wind movement is determined principally by the interaction of three factors: pressure gradient, the Coriolis effect, and friction.

Pressure Gradient If there is higher pressure on one side of a parcel of air than on the other, the air will begin to move from the higher pressure toward the lower pressure, as shown in Figure 5-5. If you visualize a high-pressure area as a pressure "hill" and a low-pressure area as a pressure "valley," it is not difficult to imagine air flowing "down" the pressure gradient in the same manner that water flows down a hill (keep in mind that these terms are metaphorical—air is not necessarily actually flowing downhill).

The *pressure-gradient force* acts at right angles to the isobars in the direction of the lower pressure. If there were no other factors to consider, that is the way the air would move, crossing the isobars at 90° (Figure 5-6a). However, such a flow rarely occurs in the atmosphere.

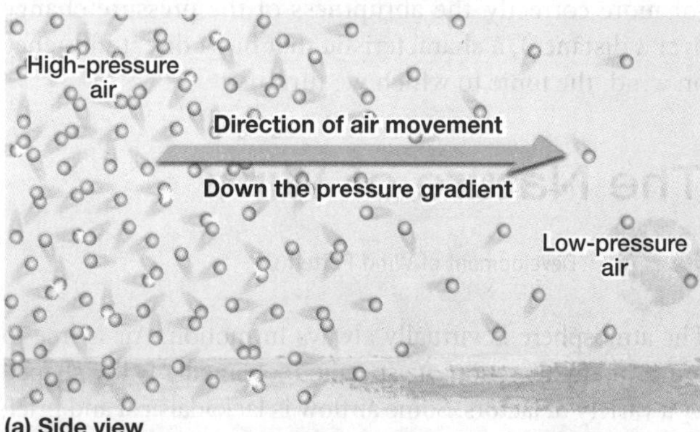

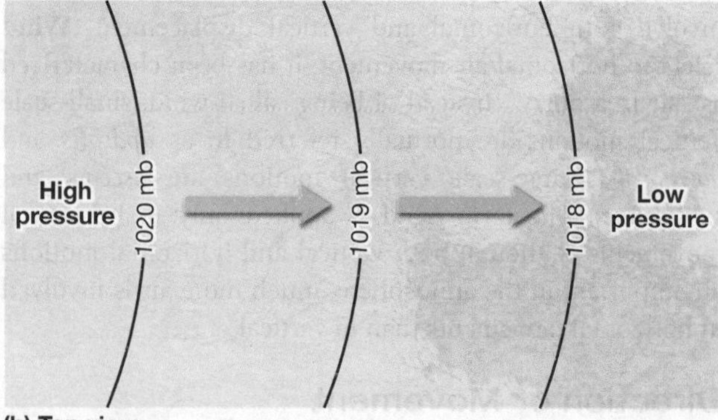

▲ **Figure 5-5** (a) Air tends to move from areas of higher pressure toward areas of lower pressure. We say this movement is "down the pressure gradient." (b) If pressure gradient were the only force involved, air would flow perpendicular to the isobars.

The Coriolis Effect Because Earth rotates, any object moving freely near Earth's surface appears to deflect to the right in the Northern Hemisphere and to the left in the Southern Hemisphere. The Coriolis effect has an important influence on the direction of wind flow (for a review of the Coriolis effect see Chapter 3). The Coriolis effect deflection acts at 90° from the direction of movement—to the right in the Northern Hemisphere and to the left in the Southern Hemisphere. There is an eternal battle, then, between the pressure-gradient force

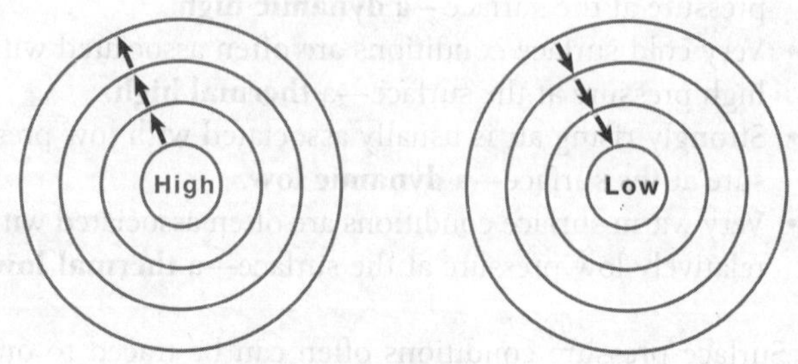

(a) Pressure Gradient Force only

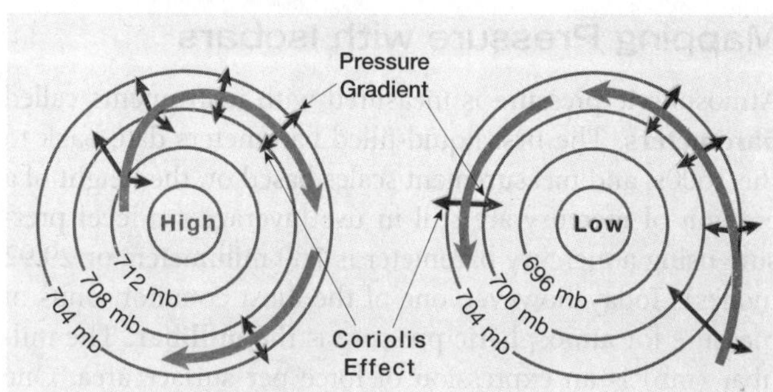

(b) Upper Atmosphere—Pressure Gradient Force and Coriolis Effect

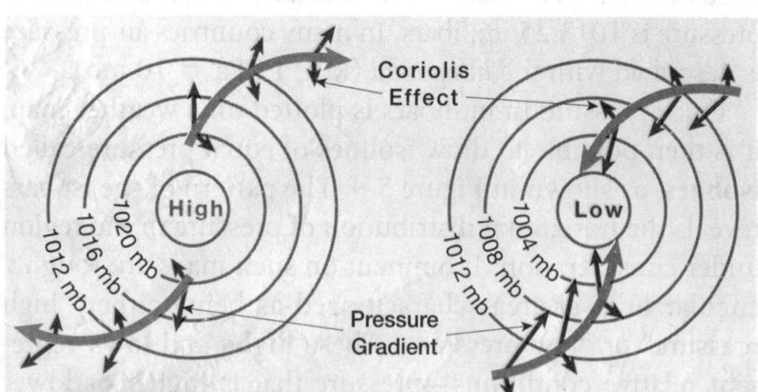

(c) Lower Atmosphere—Pressure Gradient Force, Coriolis Effect, and Friction

▲ **Figure 5-6** The direction of wind flow is determined by a combination of three factors. (a) If the pressure gradient force were the only factor influencing wind direction, air would flow down the pressure gradient away from high pressure and toward low pressure, crossing the isobars at an angle of 90°. (b) In the upper atmosphere (above about 1000 m/3300 ft.), an approximate balance develops between the pressure gradient force and the Coriolis effect deflection, resulting in geostrophic wind blowing parallel to the isobars; wind circulates clockwise around a high and counterclockwise around a low in the Northern Hemisphere. (c) In the lower atmosphere, friction slows the wind (which results in less Coriolis effect deflection) and so wind diverges clockwise out of a high and converges counterclockwise into a low in the Northern Hemisphere.

moving air from high toward low pressure and the deflection of the Coriolis effect 90° from its pressure-gradient path: the Coriolis effect keeps the wind from flowing directly down a pressure gradient, while the pressure gradient force prevents the Coriolis effect from turning the wind back up the pressure slope. Where these two factors are in balance—as is usually the case in the upper atmosphere—wind moves parallel to the isobars and is called a **geostrophic wind**[2] (Figure 5-6b).

Most winds in the atmosphere are geostrophic or nearly geostrophic in that they flow nearly parallel to the isobars. Only near the ground is there another significant factor—friction—to further complicate the situation.

Friction In the lowest portions of the troposphere, a third force influences wind direction—*friction*. The frictional drag of Earth's surface slows wind movement and so the influence of the Coriolis effect is reduced (you will recall that rapidly moving objects are deflected more by the Coriolis effect than are slowly moving objects). Instead of blowing perpendicular to the isobars (in response to the pressure gradient) or parallel to them (where pressure gradient force and the Coriolis effect are in balance), the wind takes an intermediate course between the two and crosses the isobars at angles between 0° and 90° (Figure 5-6c). In essence, friction reduces wind speed, which in turn reduces the Coriolis effect deflection—thus, although the Coriolis effect does introduce a deflection to the right (in the Northern Hemisphere) the pressure gradient "wins the battle" and air flows into an area of low pressure and away from an area of high pressure.

As a general rule, the frictional influence is greatest near Earth's surface and diminishes progressively upward (Figure 5-7). Thus, the angle of wind flow across the isobars is greatest (closest to 90°) at low altitudes and becomes smaller at increasing elevations. The **friction layer** of the atmosphere extends to only about 1000 meters (approximately 3300 feet) above the ground. Higher than that, most winds follow a geostrophic or near-geostrophic course.

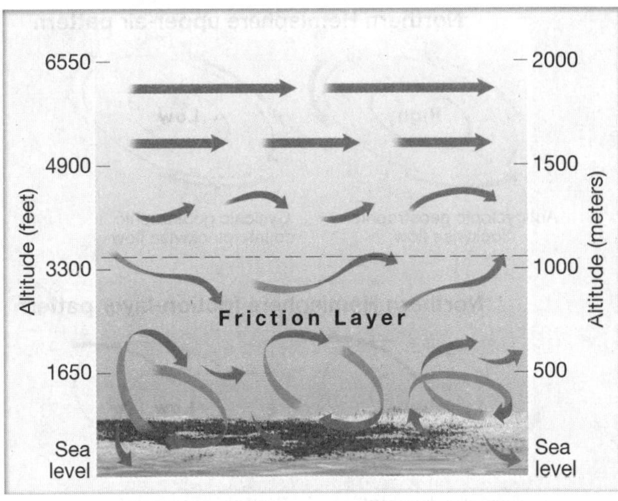

▲ **Figure 5-7** Near Earth's surface friction causes wind flow to be turbulent and irregular. At higher altitudes, above the friction layer, the lines of wind flow are much straighter.

Cyclones and Anticyclones

Cyclones and Anticyclones

Distinct and predictable wind-flow patterns develop around all high-pressure and low-pressure centers—patterns determined by the pressure gradient, Coriolis effect, and friction. A total of eight circulation patterns are possible: four in the Northern Hemisphere and four in the Southern Hemisphere. Within each hemisphere there are two patterns associated with high pressure centers and two patterns associated with low pressure centers, as shown in Figure 5-8.

High-Pressure Circulation Patterns A high-pressure center is known as an **anticyclone**, and the flow of air associated with it is described as being *anticyclonic*. The four patterns of anticyclonic circulation, shown in Figure 5-8, are as follows:

1. In the upper atmosphere of the Northern Hemisphere, the winds move clockwise in a geostrophic manner parallel to the isobars.
2. In the friction layer (lower altitudes) of the Northern Hemisphere, there is a divergent clockwise flow, with the air spiraling out away from the center of the anticyclone.
3. In the upper atmosphere of the Southern Hemisphere, there is a counterclockwise, geostrophic flow parallel to the isobars.
4. In the friction layer of the Southern Hemisphere, the pattern is a mirror image of the Northern Hemisphere case. The air diverges in a counterclockwise pattern.

Low-Pressure Circulation Patterns Low-pressure centers are called **cyclones**, and the associated wind movement is said to be *cyclonic*. As with anticyclones, Northern Hemisphere cyclonic circulations are mirror images of their Southern Hemisphere counterparts:

5. In the upper atmosphere of the Northern Hemisphere, air moves counterclockwise in a geostrophic pattern parallel to the isobars.
6. In the friction layer of the Northern Hemisphere, a converging counterclockwise flow exists.
7. In the upper atmosphere of the Southern Hemisphere, a clockwise, geostrophic flow occurs paralleling the isobars.
8. In the friction layer of the Southern Hemisphere, the winds converge in a clockwise spiral.

[2]Strictly speaking, geostrophic wind is only found in areas where the isobars are parallel and straight; the term *gradient wind* is a more general term used to describe wind flowing parallel to the isobars. In this book we use the term geostrophic to mean all wind blowing parallel to the isobars.

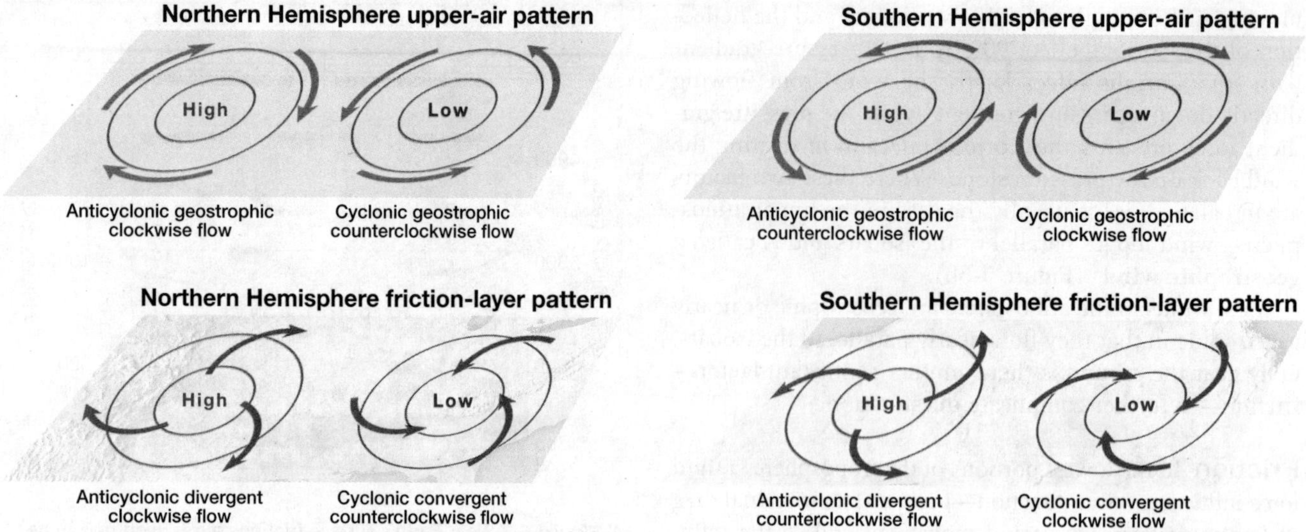

Northern Hemisphere upper-air pattern

High — Anticyclonic geostrophic clockwise flow

Low — Cyclonic geostrophic counterclockwise flow

Southern Hemisphere upper-air pattern

High — Anticyclonic geostrophic counterclockwise flow

Low — Cyclonic geostrophic clockwise flow

Northern Hemisphere friction-layer pattern

High — Anticyclonic divergent clockwise flow

Low — Cyclonic convergent counterclockwise flow

Southern Hemisphere friction-layer pattern

High — Anticyclonic divergent counterclockwise flow

Low — Cyclonic convergent clockwise flow

▲ **Figure 5-8** The eight basic patterns of air circulation around pressure cells in the upper and lower atmosphere of the Northern and Southern Hemispheres.

Cyclonic patterns in the lower troposphere may at first glance appear to be misleading because the arrows seem to defy the Coriolis effect—in the Northern Hemisphere, the arrows seem to bend to the left, whereas we know that the Coriolis deflection is to the right. Remember, however, that wind patterns are the balance of several different forces acting in different directions: as air begins to flow down the pressure gradient into a low, the Coriolis effect does indeed deflect wind to the right—and this introduces the counterclockwise flow in the Northern Hemisphere (see Figure 5-6c and the animation *Development of Wind Patterns*).

Thus, large-scale winds do not blow directly from high pressure to low. Indeed, whether the wind is geostrophic or not, there is always high pressure on one side and low pressure on the other.

Vertical Movement within Cyclones and Anticyclones A prominent vertical component of air movement is also associated with pressure centers. As Figure 5-9 shows, air descends in anticyclones and rises in cyclones. Such motions are particularly notable in the lower troposphere. The anticyclonic pattern can be visualized as upper air sinking down into the center of the high and then diverging near the ground surface. Opposite conditions prevail in a low-pressure center, with the air converging horizontally into the cyclone and then rising. Notice in Figure 5-9 that cyclones and rising air are associated with clouds, while anticyclones and descending air are associated with clear conditions—the reasons for this will be explained in Chapter 6.

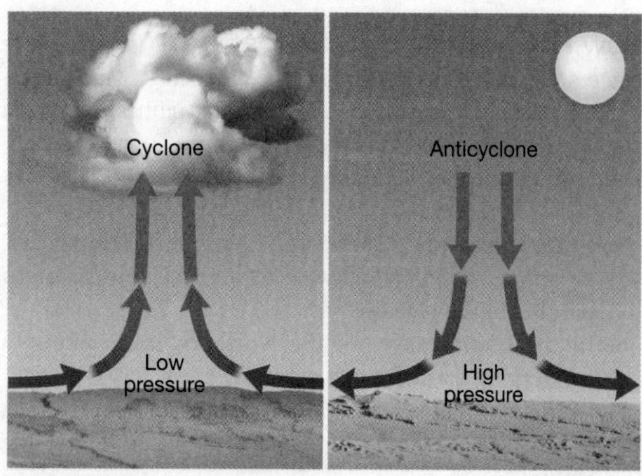

▲ **Figure 5-9** In a cyclone (low-pressure cell), air converges and rises. In an anticyclone (high-pressure cell), air descends and diverges.

Wind Speed

Thus far, we have been considering the direction of wind movement and paying little attention to speed. Although some complications are introduced by inertia and other factors, it is accurate to say that the speed of wind flow is determined primarily by the pressure gradient. If the gradient is steep, the air moves swiftly; if the gradient is gentle, movement is slow. This relationship can be portrayed in the simple diagram of Figure 5-10. The closeness of the isobars indicates the steepness of the pressure gradient.

Over most of the world most of the time, surface winds are relatively gentle. As Figure 5-11 shows, for instance, annual average wind speed in North America is generally

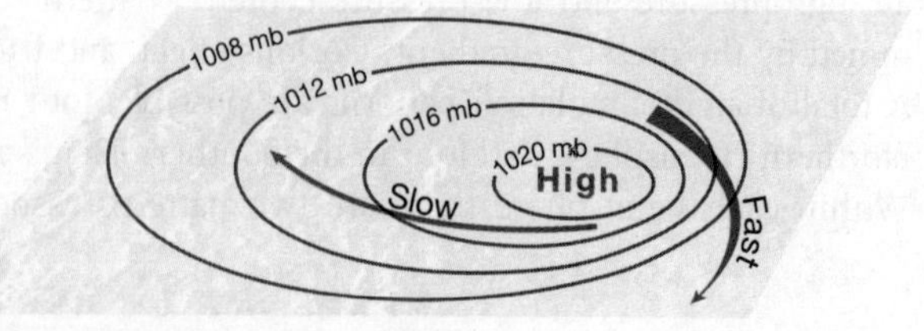

▲ **Figure 5-10** Wind speed is determined by the pressure gradient, which is indicated by the spacing of isobars. Where isobars are close together, the pressure gradient is steep and wind speed is high; where isobars are far apart, the pressure gradient is gentle and wind speed is low.

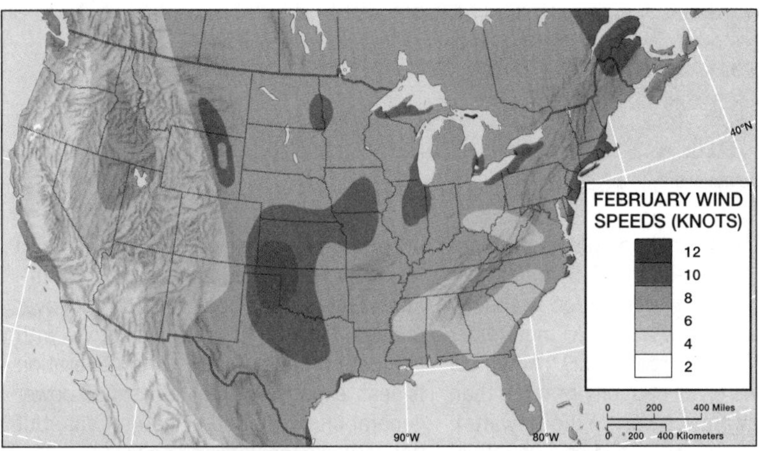

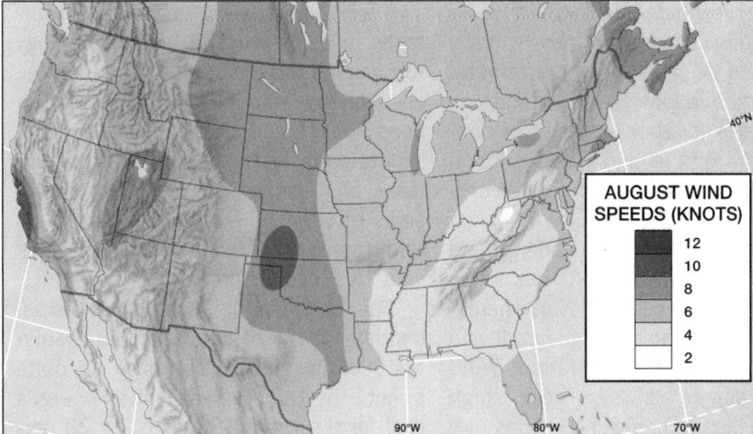

◀ **Figure 5-11** Average North American wind speeds in February and August. Wind speed tends to be higher in winter than in summer because pressure changes are more abrupt (due to greater temperature contrasts) and storms are more frequent in winter than in summer. The Great Plains tend to have the highest speed winds in all seasons. The strong coastal winds in California during summer largely result from heating in the interior Central Valley (such *sea breezes* are discussed later in this chapter).

between 6 and 12 knots.[3] Cape Dennison in Antarctica holds the dubious distinction of being the windiest place on Earth, with an annual average wind speed of 38 knots. The most persistent winds are usually in coastal areas or high mountains.

Vertical Variations in Pressure and Wind

Although the main topic of this chapter is the horizontal distribution of pressure and wind, it is worthwhile to note major features of the vertical pattern as well.

Atmospheric pressure, with only minor localized exceptions, decreases rapidly with height (Table 5-1). The pressure change is most rapid at lower altitudes, the rate of decrease diminishing significantly above about 3 kilometers (10,000 feet). Further, prominent surface pressure centers (anticyclones and cyclones) often "lean" with height, which is to say that they are not absolutely vertical in orientation.

Wind speed is quite variable from one altitude to another and from time to time, usually increasing with height. Winds tend to move faster above the friction layer. As we shall see in subsequent sections, the very strongest tropospheric winds are usually found at intermediate levels in what are called *jet streams*, or in violent storms near Earth's surface.

The General Circulation of the Atmosphere

Global Atmospheric Circulation

Earth's atmosphere is an extraordinarily dynamic medium. It is constantly in motion, responding to the various forces described previously as well as to a variety of more localized conditions. Some atmospheric motions are broadscale and sweeping; others are minute and momentary. Most important to an understanding of geography is the general pattern of circulation, which involves major semipermanent conditions of both wind and pressure. This circulation is the principal mechanism for both longitudinal and latitudinal heat transfer and is exceeded only by the global pattern of insolation as a determinant of world climates.

If Earth were a nonrotating sphere with a uniform surface, we could expect a very simple circulation pattern

[3]Recall that a *knot* is a unit of speed equivalent to 1 nautical mile per hour. A nautical mile is a bit longer than a "statute" mile. One nautical mile is equal to 1852 meters (6076 feet) and one statute mile is equal to 1609 meters (5280 feet). A knot is also equivalent to 1.85 km/hr (1.15 statute miles per hour).

RENEWABLE ENERGY

Wind Power

Stephen Stadler, Oklahoma State University

Wind is a universal feature of Earth's environment and represents the circulation of the atmosphere because of local and global energy imbalances. Humans have long employed the wind to ease our physical labors. Sailboats were in use by 6000 years ago. By the ninth century A.D. the Persians (modern Iran) ground grain and pumped water in enclosures into which wind channeled and moved blades made from reeds attached to a vertical shaft. Twelfth-century Europeans used cloth blades attached to a horizontal axis—this new technology allowed windmill blades to be lighter and much more aerodynamically efficient.

The first electrical generation wind turbine was invented in 1887 and could generate about 12 kilowatts (12,000 watts) of power. These first wind turbines were made of metal so that their power-to-weight ratio was very high. After World War II, the Danish invented the modern turbine as a response to the high costs of fuels in their country. Their work resulted in large three-bladed devices having low weight per blade and good aerodynamic design. The United States became interested in wind power during the oil shortages of the 1970s. Some wind farms were installed, but after federal tax credits ended the farms were economically untenable.

A number of engineering improvements has resulted in a tenfold decrease in the cost of wind power since the early 1980s. Blades have become much lighter per area because of the use of advanced composite materials. Turbines have become much larger, thus lessening the per kilowatt cost of construction (i.e., a single new turbine takes the place of several old turbines). Finally, computer control maximizes blade efficiency by turning the blades toward the wind and adjusting blade angles for maximum efficiency.

Modern turbines are of many sizes but commercial wind farms have very large machines on towers 50 meters (165 feet) or more high, with blade lengths in excess of 30 meters (100 feet). The machines face into the wind and, after start-up wind speeds of about 2 meters per second (4.5 miles per hour), reach full generation capacity with wind speeds of about 20 meters per second (45 miles per hour). A typical large turbine has a capacity of more than 1.5 megawatts (1.5 million watts), enough to supply the need of a couple of thousand households (Figure 5-A).

The blades of large turbines typically turn at 10 to 20 revolutions per minute. This high-torque/low speed rotation is imparted to a gearbox where it is converted to low-torque/high-speed rotation which directly drives an onboard electrical generator. The electricity is taken through the tower and away underground to be placed on transmission lines.

Although wind blows "everywhere," there is a known physical geography to the best locations for generation. Wind farm developers always seek high ground because higher elevations are associated with higher-speed winds. They also locate in areas where vegetation is minimal over long distances; trees, for instance, create significant friction and slow regional winds.

The wind is a viable source of electrical energy over much of Earth. Because there is no ongoing fuel cost, the question of economic viability becomes whether the large amount of funding needed for construction of a wind farm—typically a couple of hundred million dollars—can be spread out over enough time so it is competitive with competing fuels. Today, wind generation is less expensive than electrical power generation by fuel oil or natural gas, but not as inexpensive as by coal.

There are several inherent advantages in using wind turbine technology for generating electricity. There is no cost for the wind itself. The generation does not depend on the discovery and extraction of fossil fuels. There is no release of greenhouse gases. The downsides of wind power include wind's intermittency and the "not in my back yard" feelings of some citizens.

The world wind industry grew exponentially at the beginning of the twenty-first century and the outlook is quite bright. The U.S. Department of Energy's goal for the United States is for 20 percent of its power generated by wind by the year 2030. The wind resources of the United States are immense, but there is a geographic problem: the best U.S. wind resources are above the Great Plains, but the countryís population density (and electrical usage) is greatest on the coasts. Therefore, much larger and more interconnected power transmission lines must be built.

▲ **Figure 5-A** Electricity generating wind turbines in Weatherford Wind Farm, Oklahoma.

TABLE 5-1 Atmospheric Pressure Variation with Altitude

Altitude		
Kilometers	Miles	Pressure (millibars)
18	11	76
16	10	104
14	8.7	142
12	7.4	194
10	6.2	265
9.0	5.6	308
8.0	5.0	356
7.0	4.3	411
6.0	3.7	472
5.0	3.1	540
4.0	2.5	617
3.5	2.2	658
3.0	1.9	701
2.5	1.6	747
2.0	1.2	795
1.5	0.9	846
1.0	0.6	899
0.5	0.3	955
0	0	1013

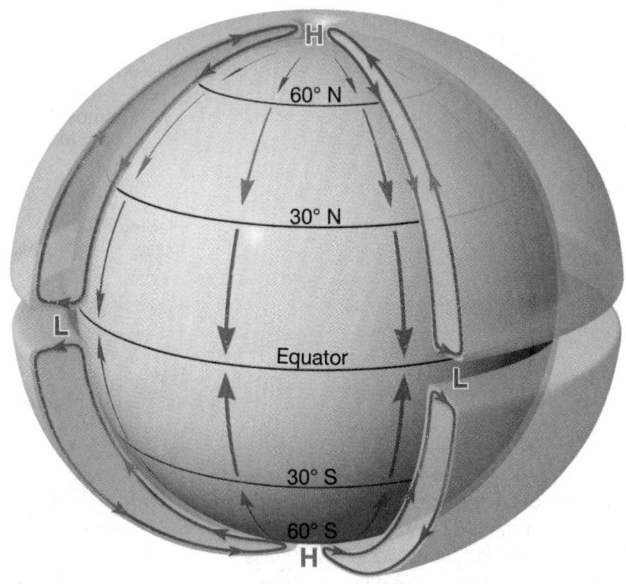

▲ **Figure 5-12** Wind circulation patterns would be simple if Earth's surface were uniform (no distinction between continents and oceans) and if the planet did not rotate. High pressure at the poles and low pressure at the equator would produce northerly surface winds in the Northern Hemisphere and southerly surface winds in the Southern Hemisphere.

(Figure 5-12). The greater amount of solar heating in the equatorial region would produce a girdle of low pressure around the world, and radiational cooling at the poles would develop a cap of high pressure in those areas. Surface winds in the Northern Hemisphere would flow directly down the pressure gradient from north to south, whereas those in the Southern Hemisphere would follow a similar gradient from south to north. Air would rise at the equator in a large convection cell and flow toward the poles (south to north in the Northern Hemisphere and north to south in the Southern Hemisphere), where it would subside into the polar highs.

Earth does rotate, however, and in addition has an extremely varied surface. Consequently, the broadscale circulation pattern of the atmosphere is much more complex than that shown in Figure 5-12. Apparently only the tropical regions have a complete vertical convective circulation cell. Similar cells have been postulated for the middle and high latitudes, but observations indicate that the midlatitude and high-latitude cells either do not exist or are weakly and sporadically developed. The low-latitude cells—one north and one south of the equator—can be thought of as gigantic convection systems (Figure 5-13). These two prominent tropical

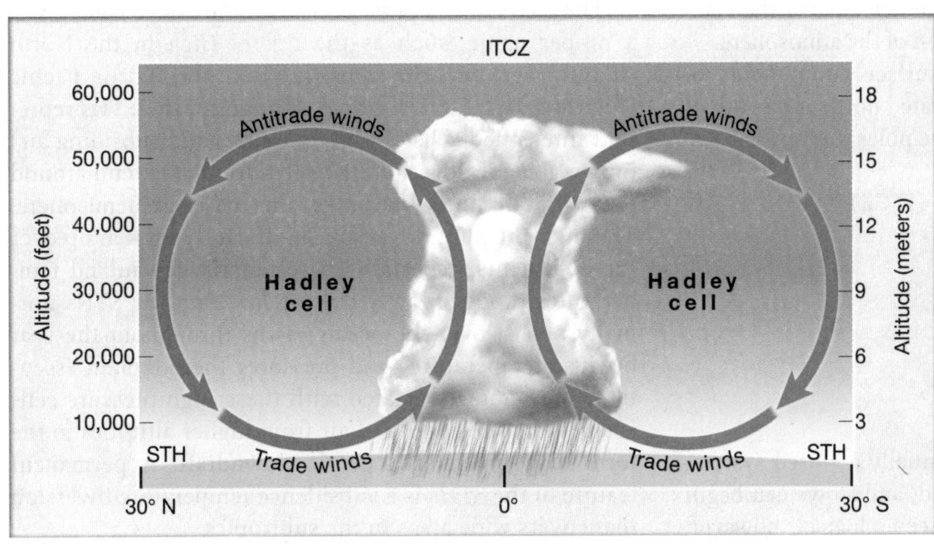

◀ **Figure 5-13** Distinct cells of vertical circulation occur in tropical latitudes; they are called Hadley cells. The equatorial air rises to some 12 to 15 kilometers (40,000 to 50,000 feet) in the intertropical convergence zone (ITCZ) before spreading poleward. This air descends at about 30° N and S into subtropical high-pressure cells (STH). The vertical dimension of the Hadley cells is considerably exaggerated in this idealized diagram.

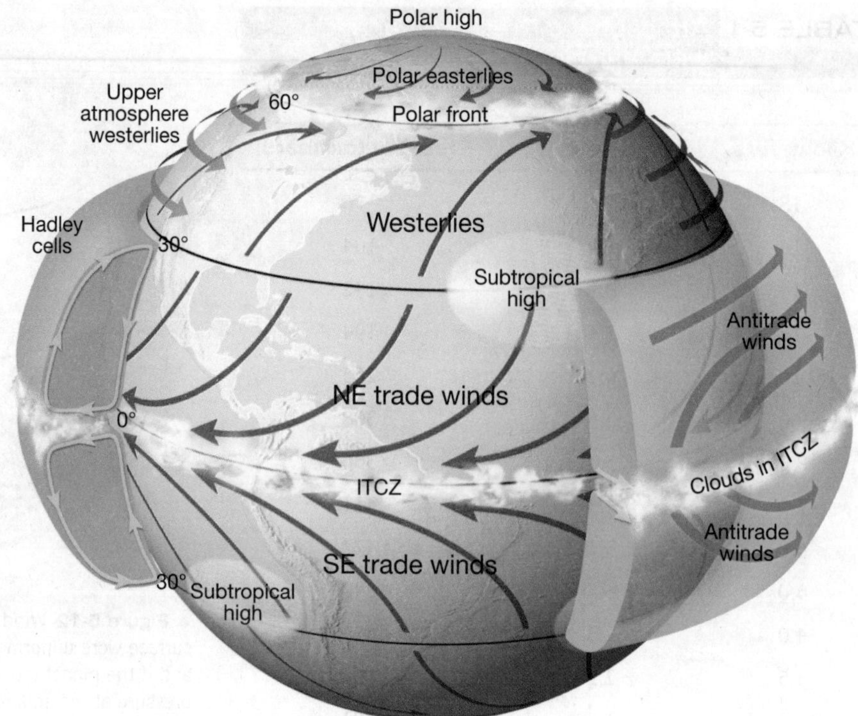

▶ **Figure 5-14** Idealized global circulation. Rising air of the ITCZ in the Hadley cell is deflected aloft and forms westerly *antitrade winds*, while surface winds diverging from the STH form easterly trade winds and the westerlies. The upper atmosphere flow of the westerlies is shown with faint blue arrows. (Vertical dimension exaggerated considerably.)

circulations are called **Hadley cells**, after George Hadley (1685–1768), an English meteorologist who first conceived the idea of enormous convective circulation cells in 1735.

Around the world in equatorial latitudes, warm air rises, producing a region of relative low pressure at the surface. This air ascends to great heights, mostly in thunderstorm updrafts. By the time this air reaches the upper troposphere at elevations of about 15 kilometers (50,000 feet), it has cooled. The air then spreads north and south and moves poleward, eventually descending at latitudes of about 30° N and S, where it forms bands of high pressure at the surface (Figure 5-14). One portion of the air diverging from these surface high-pressure zones flows toward the poles, while another portion flows back toward the equator—where the Northern and Southern Hemisphere components converge and the warm air rises again.

Although the Hadley cell model is a simplification of reality, it is a useful starting point for understanding the main components of the general circulation of the atmosphere.

The basic pattern has seven surface components of pressure and wind, which are replicated north and south of the equator. From the equator to the poles, they are:

1. Intertropical convergence zone (ITCZ)
2. Trade winds
3. Subtropical highs
4. Westerlies
5. Polar front (Subpolar lows)
6. Polar easterlies
7. Polar highs

Tropospheric circulation is essentially a closed system, with neither a beginning nor an end, and so we can begin describing it almost anywhere. It seems logical, however, to begin in the subtropical latitudes of the five major ocean basins because these areas serve as the "source" of the major surface winds of the planet.

Subtropical Highs

Each ocean basin has a large semipermanent high-pressure cell centered at about 30° of latitude called a **subtropical high (STH)** (Figure 5-15). These gigantic anticyclones, with an average diameter of perhaps 3200 kilometers (2000 miles), develop from the descending air of the Hadley cells. They are usually elongated east-west and tend to be centered in the eastern portions of an ocean basin (in other words, just off the west coasts of continents). Their latitudinal positions vary from time to time, shifting a few degrees poleward in summer and a few degrees equatorward in winter.

The STHs are so persistent that some have been given a proper name, such as the *Azores High* in the North Atlantic and the *Hawaiian High* in the North Pacific (Figure 5-16). From a global standpoint, the STHs represent intensified cells of high pressure (and subsiding air) in two general ridges of high pressure that extend around the world in these latitudes, one in each hemisphere. The high-pressure ridges are significantly broken up over the continents, especially in summer when inland temperatures produce lower air pressure, but the STHs normally persist over the ocean basins throughout the year because temperatures and pressures there remain essentially constant. Associated with these high-pressure cells is a general subsidence of air from higher altitudes in the form of a broadscale, gentle downdraft. A permanent feature of the STHs is a subsidence temperature inversion that covers wide areas in the subtropics.

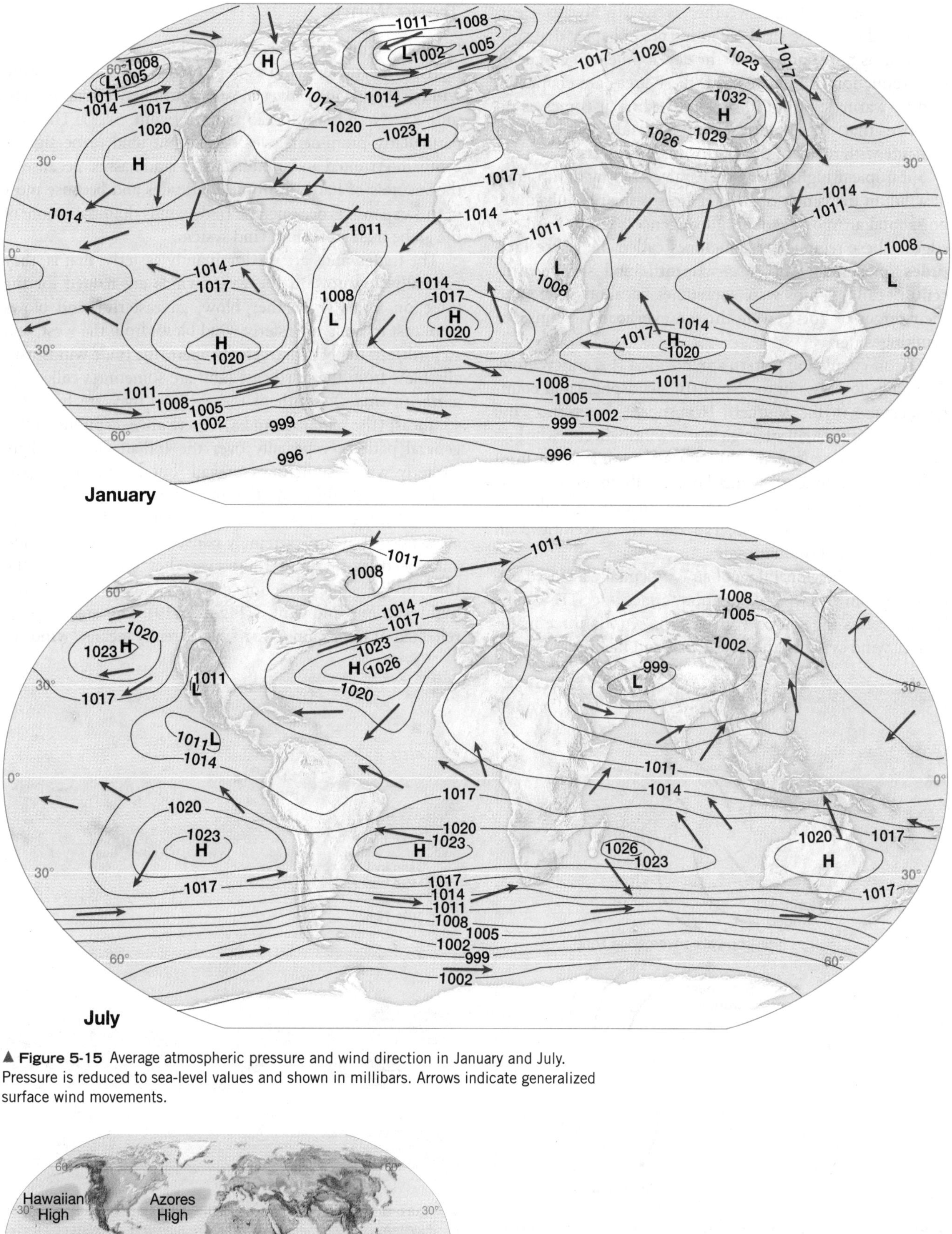

January

July

▲ **Figure 5-15** Average atmospheric pressure and wind direction in January and July. Pressure is reduced to sea-level values and shown in millibars. Arrows indicate generalized surface wind movements.

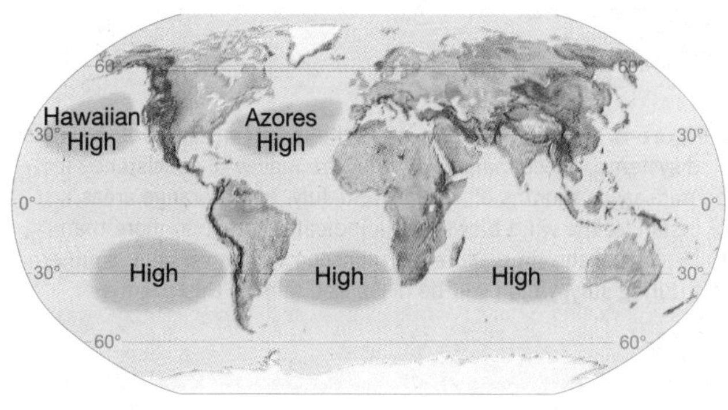

◀ **Figure 5-16** The subtropical highs are generally located over ocean basins at latitudes of about 30° N and S, also called the *horse latitudes*.

Within an STH, the weather is nearly always clear, warm, and calm. We shall see in the next chapter that subsiding air is not conducive to the development of clouds or the production of rain. Instead, these areas are characterized by warm, subtropical sunshine. Thus, it comes as no surprise that these anticyclonic, subsiding-air regions coincide with many of the world's major deserts.

Subtropical highs are also characterized by an absence of wind: in the center of an STH air is primarily subsiding; horizontal air movement and divergence begin toward the edges. These regions are sometimes called the **horse latitudes**, presumably because sixteenth- and seventeenth-century sailing ships were sometimes becalmed there and their cargos of horses were thrown overboard to conserve drinking water.

The air circulation pattern around an STH is anticyclonic: diverging clockwise in the Northern Hemisphere and counterclockwise in the Southern Hemisphere. In essence, the STHs can be thought of as gigantic "wind wheels" whirling in the lower troposphere, fed with air sinking down from above and spinning off winds horizontally in all directions (Figure 5-17). The winds are not dispersed uniformly around an STH, however; instead, they are concentrated on the northern and southern sides.

Although the global flow of air is essentially a closed circulation from a viewpoint at Earth's surface, the STHs can be thought of as the source of two of the world's three major surface wind systems: the *trade winds* and the *westerlies*.

Trade Winds

Diverging from the equatorward sides of the subtropical highs is the major wind system of the tropics—the **trade winds**. These winds cover most of Earth between about latitude 25° N and latitude 25° S (see Figure 5-17). They are particularly prominent over oceans but tend to be significantly interrupted and modified over landmasses. Because of the vastness of Earth in tropical latitudes and because most of this expanse is oceanic, the trade winds dominate more of the globe than any other wind system.

The trade winds are predominantly easterly; that is, they generally flow toward the west (winds are named for the direction from which they blow; an easterly wind blows from east to west, a westerly wind blows from the west, and so forth). In the Northern Hemisphere, the trade winds usually blow from the northeast (and are sometimes called the *northeast trades*); south of the equator, they are from the southeast (the *southeast trades*). There are exceptions to this general pattern, especially over the Indian Ocean, where westerly winds sometimes prevail, but for the most part there is easterly flow over the tropical oceans.

Indeed, the trade winds are by far the most "reliable" of all winds. They are extremely consistent in both direction and speed, as Figure 5-18 shows. They blow most of the time in the same direction at the same speed, day and night, summer and winter (Figure 5-19). This steadiness is reflected in their name: trade winds really means "winds of

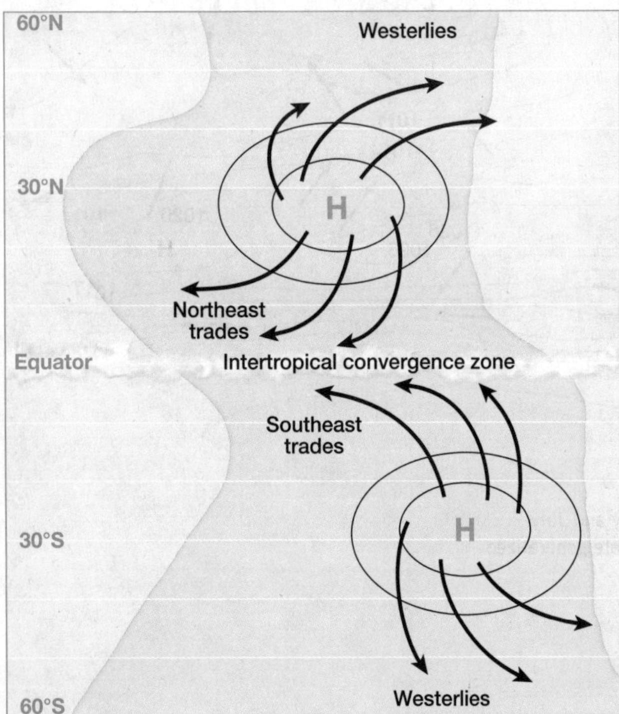

▲ **Figure 5-17** The air that descends and diverges out of the subtropical highs is the source of the surface trade winds and westerlies. This map shows the generalized location of the intertropical convergence zone, trade winds, subtropical highs, and westerlies in hypothetical Northern and Southern Hemisphere ocean basins.

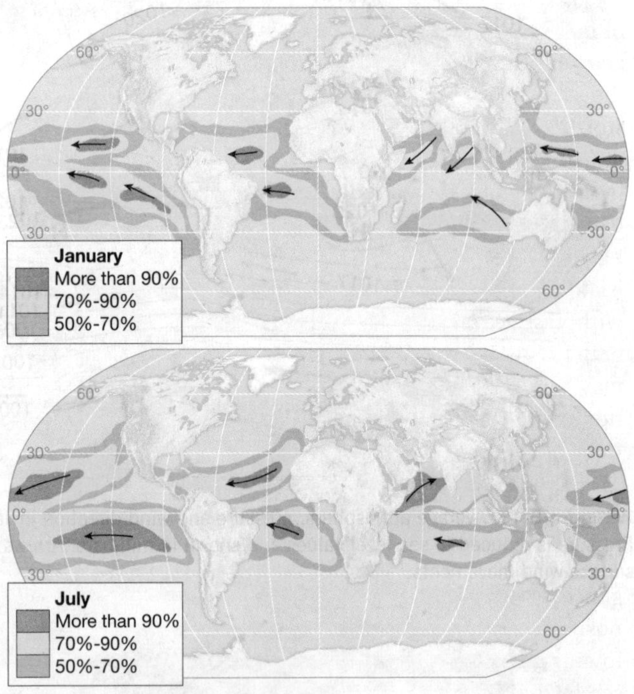

▲ **Figure 5-18** The trade winds are the most consistent of the major wind systems. These maps show their frequency of consistency for the midseason months of January and July. In the orange areas, for instance, the wind blows in the indicated direction more than 90 percent of the time. Note the monsoon wind reversal in southern Asia during July, which will be discussed later in this chapter.

◀ **Figure 5-19** Tropical coastal areas often experience the ceaseless movement of the trade winds. This breezy scene is at Cairns on the northeastern coast of Queensland, Australia.

commerce." Mariners of the sixteenth century recognized early that the quickest and most reliable route for their sailing vessels from Europe to the Americas lay in the belt of northeasterly winds of the southern part of the North Atlantic Ocean. Similarly, the trade winds were used by Spanish galleons in the Pacific Ocean, and the name became applied generally to these tropical easterly winds.

The trades originate as warming, drying winds capable of holding an enormous amount of moisture. As they blow across the tropical oceans, they evaporate vast quantities of moisture and therefore have a tremendous potential for storminess and precipitation. They do not release the moisture, however, unless forced to do so by being uplifted by a topographic barrier or some sort of pressure disturbance (the reasons for this will be explained in Chapter 6). Low-lying islands in the trade-wind zone are often desert islands because the moisture-laden winds pass over them without dropping any rain. If there is even a slight topographic irregularity, however, the air that is forced to rise may release abundant precipitation (Figure 5-20). Some of the wettest places in the world are windward slopes in the trade winds, such as in Hawai'i (see Figure 3-20).

Intertropical Convergence Zone (ITCZ)

The northeast and southeast trades come together in the general vicinity of the equator, although the latitudinal position shifts seasonally northward and southward following the Sun. This shift is greater over land than over sea because the land heats more. The zone where the air from the Northern Hemisphere and Southern Hemisphere meet is usually called the **intertropical convergence zone**, or simply the **ITCZ**, but it is also referred to as the **doldrums** (this last name is attributed to the fact that sailing ships were often becalmed in these latitudes).

The ITCZ is a zone of convergence and weak horizontal airflow characterized by feeble and erratic winds. It is a globe-girdling zone of warm surface conditions, low pressure associated with high rainfall, instability, and rising air in the Hadley cells (see Figure 5-14). It is not a region of continuously ascending air, however. Almost all the rising air of the tropics ascends in the updrafts that occur in thunderstorms in the ITCZ, and these updrafts pump an enormous amount of sensible heat and latent heat of condensation into the upper troposphere, where much of it spreads poleward.

The ITCZ often appears as a well-defined, relatively narrow cloud band over the oceans near the equator (Figure 5-21). Over continents, however, it is likely to be more diffused and indistinct, although thunderstorm activity is common.

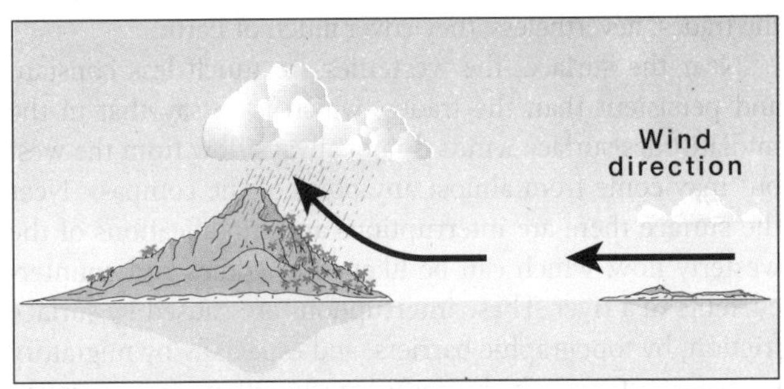

▲ **Figure 5-20** Trade winds are usually heavily laden with moisture, but usually do not produce clouds and rain unless forced to rise. Thus, they may blow across a low-lying island with little or no visible effect. An island of greater elevation, however, causes the air to rise up the side of the mountain, and the result is usually a heavy rain.

▲ **Figure 5-21** A well-defined band of clouds marks the ITCZ zone over equatorial Africa in this infrared satellite image (darker shades of gray indicate warmer surface temperatures). This image was taken during the Northern Hemisphere summer so the ITCZ has shifted slightly north of the equator. Note the generally clear skies off the west coasts of northern and southern Africa, corresponding to the areas of the subtropical highs, and the cloudiness and storms in the band of westerlies in the midlatitudes.

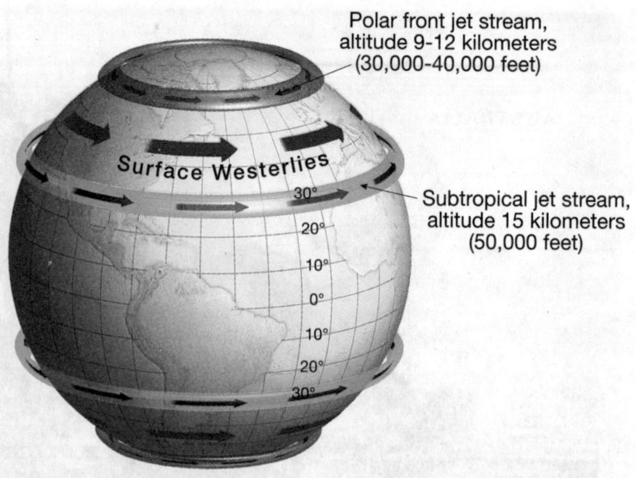

▲ **Figure 5-22** Neither jet stream is centered in the band of the westerlies. The polar front jet stream is closer to the poleward boundary, and the subtropical jet stream is closer to the equatorward boundary of this wind system. The two jet streams are not at the same altitude; the subtropical jet stream is at a higher altitude than the polar front jet stream.

The Westerlies

The Jet Stream and Rossby Waves

The fourth component of the general atmospheric circulation is represented by the arrows that issue from the poleward sides of the STHs in Figure 5-17. This is the great wind system of the midlatitudes, commonly called the **westerlies**. These winds flow basically from west to east around the world in the latitudinal zone between about 30° and 60° both north and south of the equator. Because the globe is smaller at these latitudes than in the tropics, the westerlies are less extensive than the trades; nevertheless, they cover much of Earth.

Near the surface, the westerlies are much less constant and persistent than the trades, which is to say that in the midlatitudes surface winds do not always flow from the west but may come from almost any point of the compass. Near the surface there are interruptions and modifications of the westerly flow, which can be likened to eddies and countercurrents in a river. These interruptions are caused by surface friction, by topographic barriers, and especially by migratory pressure systems, which produce airflow that is not westerly.

Jet Streams Although the surface westerlies are somewhat variable, the geostrophic winds aloft, however, blow very prominently from the west. Moreover, there are two remarkable "cores" of high-speed winds in each hemisphere called **jet streams**: one called the *polar front jet stream* (or simply the *polar jet stream*) and the other called the *subtropical jet stream*, at high altitudes in the westerlies (Figure 5-22). The belt of the westerlies can therefore be thought of as a meandering river of air moving generally from west to east around the world in the midlatitudes, with the jet streams as its fast-moving cores.

The polar front jet stream, which usually occupies a position 9 to 12 kilometers (30,000 to 40,000 feet) high, is not centered in the band of the westerlies; it is displaced poleward, as Figure 5-22 shows (the name comes from its location near the *polar front*). This jet stream is a feature of the upper troposphere located over the area of greatest horizontal temperature gradient—that is, cold just poleward and warm just equatorward.

A jet stream is not always the sharply defined narrow ribbon of wind as often portrayed on weather maps; rather it is a zone of strong winds within the upper troposphere westerly flow. Jet stream speed is variable. Sixty knots is generally considered as the minimum speed required for recognition as a jet stream, but speeds as much as five times that number have been recorded (Figure 5-23).

Commercial air travel can be significantly influenced by the high-speed flow of upper tropospheric winds. The cruising altitude of commercial jetliners is typically 9 to 12 kilometers (30,000 to 40,000 feet)—a typical elevation for the polar front jet stream. It generally takes longer to fly from east to west across North America than it does to fly from west to east. When traveling from the east, a "headwind" is likely to impede progress, while traveling from the west, a "tailwind" may reduce travel time.

Rossby Waves The polar front jet stream shifts its latitudinal position with some frequency, and this change has considerable influence on the path of the westerlies. Although the basic direction of movement is west to east, frequently sweeping undulations develop in the westerlies and produce a meandering jet stream path that wanders

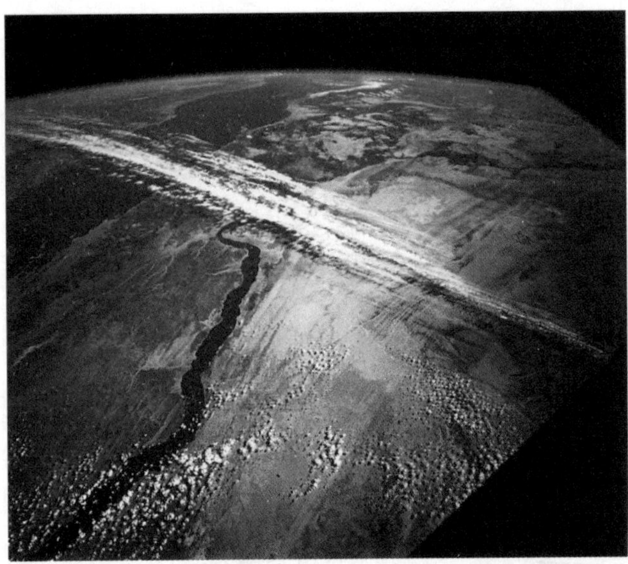

▲ **Figure 5-23** A jet stream sometimes generates a distinctive cloud pattern that is conspicuous evidence of its presence, as in this photograph taken over the Nile Valley and Red Sea. Equatorward of the axis of the jet there is a tendency for air to rise, a condition that can produce thin clouds. Poleward of the axis, the air is clear.

widely north and south (Figure 5-24). These curves are very large and are generally referred to as *long waves* or **Rossby waves** (after the Chicago meteorologist C. G. Rossby, who first explained their nature). At any given time, there are usually from three to six Rossby waves in the westerlies of each hemisphere. These waves can be thought of as separating cold polar air from warmer tropical air. When the polar front jet stream path is more directly west–east, there is a *zonal flow* pattern in the weather, with cold air poleward of warm air. However, when the jet stream begins to oscillate and the Rossby waves develop significant amplitude (which means a prominent north-south component of movement) there is a *meridional flow*: cold air is brought equatorward and warm air moves poleward, bringing frequent and severe weather changes to the midlatitudes.

The subtropical jet stream is usually located at high altitudes—just below the tropopause, as Figure 5-25a shows—over the poleward margin of the subsiding air of the STH. It has less influence on surface weather patterns because there is less temperature contrast in the associated air streams. Sometimes, however, the polar front jet and the subtropical jet merge as shown in Figure 5-25b to produce a broad belt of high-speed winds in the upper troposphere—a condition that can intensify the weather conditions associated with either zonal or meridional flow of the Rossby waves.

All things considered, no other portion of Earth experiences such short-run variability of weather as the midlatitudes.

Polar Highs

Situated over both polar regions are high-pressure cells called **polar highs** (see Figure 5-14). The Antarctic high, which forms over an extensive, high-elevation, very cold continent, is strong, persistent, and almost a permanent feature above the Antarctic continent. The Arctic high is much less pronounced and more transitory, particularly in winter. It tends to form over northern continental areas rather than over the Arctic Ocean. Air movement associated with these cells is typically anticyclonic. Air from above sinks down into the high and diverges horizontally near the surface, clockwise in the Northern Hemisphere and counterclockwise in the Southern Hemisphere, forming the third of the world's wind systems, the polar easterlies.

Polar Easterlies

The third broad-scale global wind system occupies most of the area between the polar highs and about 60° of latitude (Figure 5-26). The winds move generally from east to west

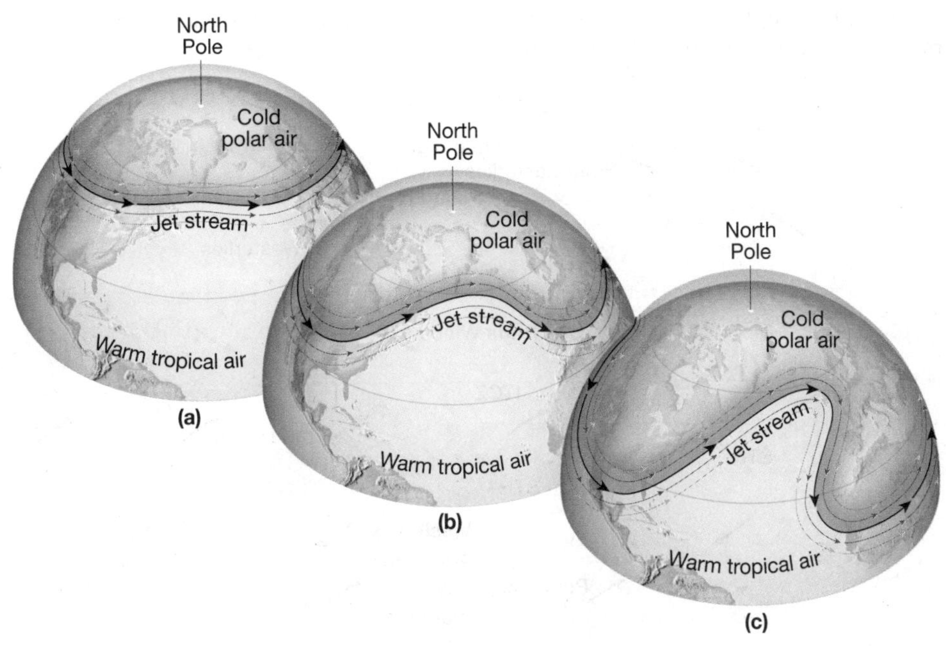

◀ **Figure 5-24** Rossby waves as part of the general flow (particularly the upper-air flow) of the westerlies. (a) When there are few waves and their amplitude (north-south component of movement) is small, cold air usually remains poleward of warm air. (b) This distribution pattern begins to change as the Rossby waves grow. (c) When the waves have great amplitude, cold air pushes equatorward and warm air moves poleward.

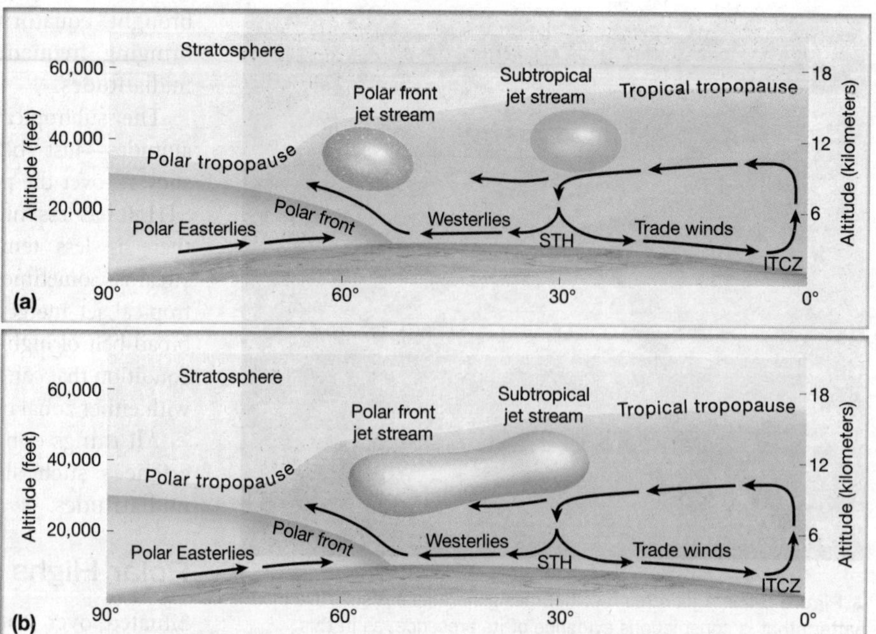

▶ **Figure 5-25** (a) A vertical cross section of the atmosphere from the equator to the poles showing the usual relative positions of the two jet streams. (b) The two jet streams sometimes merge, and the result is intensified weather conditions.

and are called the **polar easterlies**. They are typically cold and dry but quite variable.

Polar Front

The final surface component of the general pattern of atmospheric circulation is a zone of low pressure at about 50° to 60° of latitude in both Northern and Southern Hemispheres. The zone is commonly called the **polar front**, although it is sometimes most clearly visible by the presence of semipermanent zones of low pressure called the **subpolar lows**. The polar front is a meeting ground and zone of conflict between the cold winds of the polar easterlies and the relatively warmer westerlies. The subpolar low of the Southern Hemisphere is nearly continuous over the uniform ocean surface of the cold seas surrounding Antarctica. In the Northern Hemisphere, however, the low-pressure zone is discontinuous, being interrupted by the continents. It is much more prominent in winter than in summer and is best developed over the northernmost reaches of the Pacific and Atlantic oceans, forming the Aleutian Low and the Icelandic Low respectively.

The polar front area is characterized by rising air, widespread cloudiness, precipitation, and generally unsettled or stormy weather conditions (Figure 5-27). Many of the migratory storms that travel with the westerlies have their origin in the conflict zone of the polar front.

Vertical Patterns of the General Circulation

As we have seen, over tropical regions, between the equator and 20° to 25° of latitude, surface winds generally blow from the east. In the midlatitudes, the surface winds are generally westerly, while in the highest latitudes, surface winds are again easterly. In the upper elevations of the troposphere, however, the wind patterns are somewhat different from the surface winds (Figure 5-28).

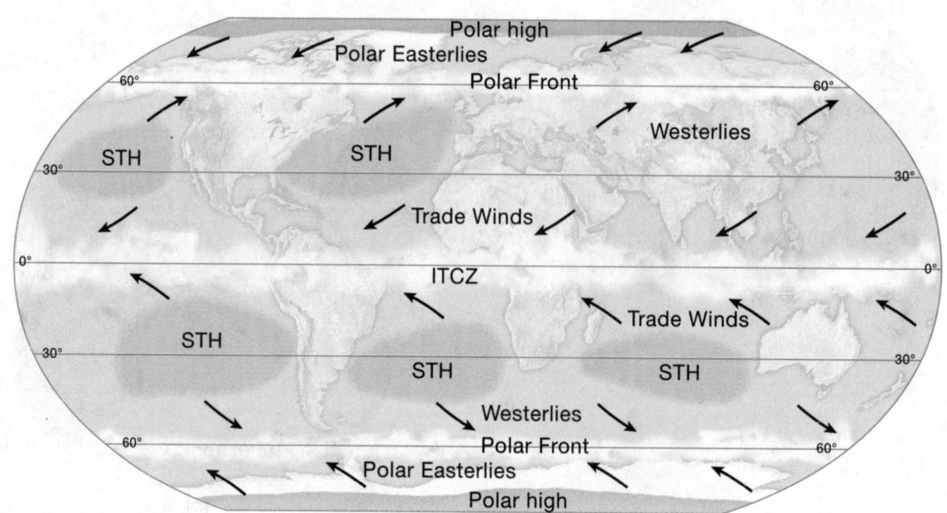

▶ **Figure 5-26** Map showing the generalized locations of the seven components of the general circulation patterns of the atmosphere. (In this map projection, the areal extent of the high-latitude components—polar high, polar easterlies, polar front—is considerably exaggerated.)

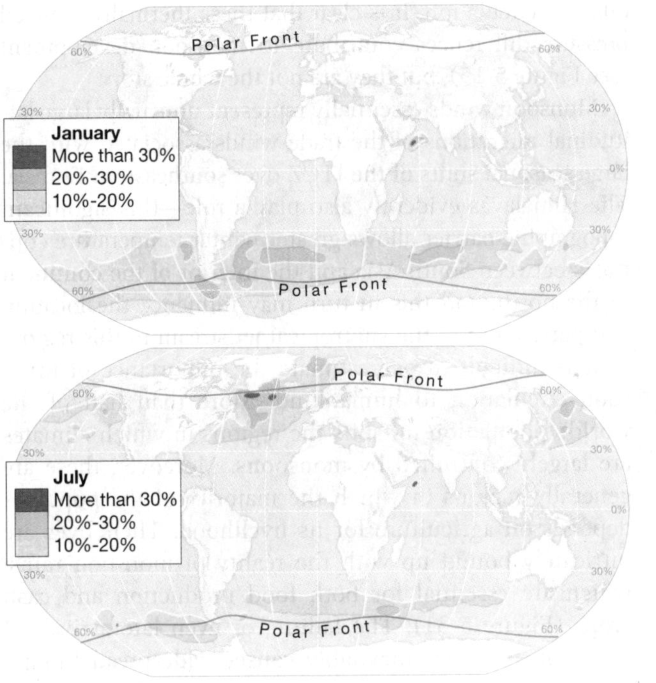

▲ **Figure 5-27** Frequency of gale-force (34 knot) winds over the oceans in January and July. It is clear from this map that the strongest oceanic winds are associated with activities along the polar front (subpolar lows).

The most dramatic difference is seen over the tropics. After equatorial air has risen in the ITCZ, the high-elevation poleward flow of air in the Hadley cell is deflected by the Coriolis effect (Figure 5-14). This results in upper-elevation winds blowing from the southwest in the Northern Hemisphere and from the northwest in the Southern Hemisphere in the **antitrade winds**. This flow eventually becomes more westerly and encompasses the subtropical jet stream. Thus, at the surface within the tropics, winds are generally from the east, while high above the antitrade winds are blowing from the west.

Modifications of the General Circulation

There are many variations to the pattern discussed on the preceding pages, and all features of the general circulation may appear in altered form, much different from the idealized description. Indeed, components sometimes disappear from sizable parts of the atmosphere where they are expected to exist. Even the troposphere sometimes "disappears" (for example, during a high-latitude winter with very cold surface temperatures, the atmospheric temperature may steadily increase with height into the stratosphere; in such cases the tropopause cannot be identified). Nevertheless, the generalized pattern of global wind and pressure systems comprises the seven components described above. To understand how real-world weather and climate differ from this general picture, it is necessary to discuss two important modifications of the generalized scheme.

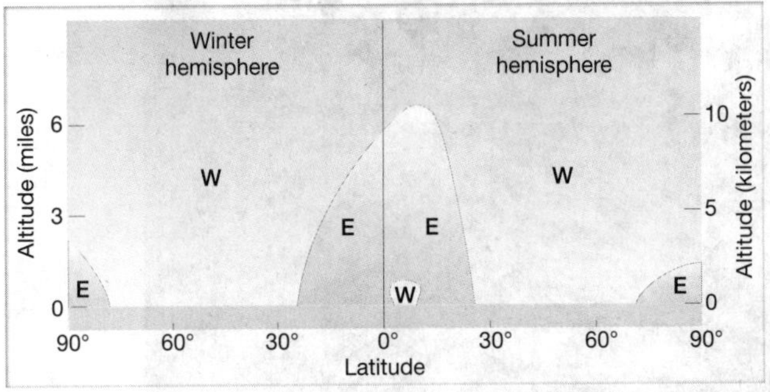

▲ **Figure 5-28** A generalized cross section through the troposphere, showing the horizontal component of air movement. Legend: E = easterly winds; W = westerly winds.

Seasonal Variations in Location

Seasonal Pressure and Precipitation Patterns

The seven surface components of the general circulation shift latitudinally with the changing seasons. When sunlight, and therefore surface heating, is concentrated in the Northern Hemisphere (Northern Hemisphere summer), all components are displaced northward; during the opposite season (Southern Hemisphere summer), everything is shifted southward. The displacement is greatest in the low latitudes and least in the polar regions. The ITCZ, for example, can be found as much as 25° north of the equator in July and 20° south of the equator in January (Figure 5-29), while the polar highs experience little or no latitudinal displacement from season to season.

Weather is affected by shifts in the general circulation components only minimally in equatorial and polar regions, but the effects can be quite significant in the midlatitudes and their fringes. For example, as we will see in Chapter 8, regions of *mediterranean climate*, which are centered at about 35° of latitude, have warm, rainless summers while under the influence of the STH; in winter, however, the belt of the westerlies shifts equatorward, bringing changeable and frequently stormy weather to these regions (see Figure 5-15). Also, as we will see, the shift of the ITCZ is closely tied to seasonal rainfall patterns in large areas within the tropics.

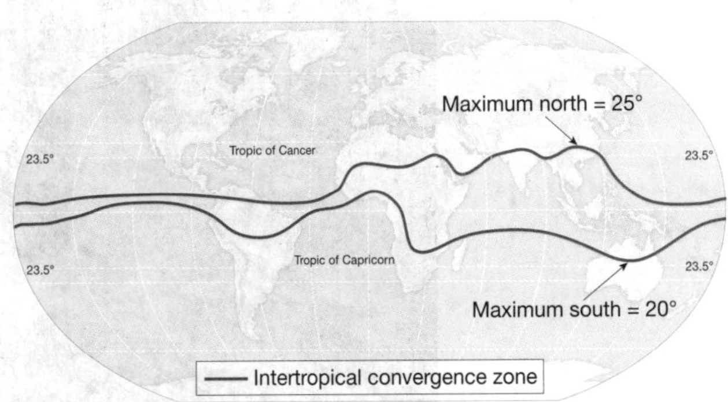

▲ **Figure 5-29** Typical maximum poleward positions of the intertropical convergence zone at its seasonal extremes. The greatest variation in location is associated with monsoon activity in Eurasia and Australia.

Monsoons

By far the most significant deviation from the pattern of general circulation is the development of **monsoons** in certain parts of the world, particularly southern and eastern Eurasia (Figure 5-30). The word monsoon is derived from the Arabic *mawsim* (meaning "season") and has come to mean a seasonal reversal of winds, a general sea-to-land movement—called *onshore flow*—in summer and a general land-to-sea movement—called *offshore flow*—in winter. Associated with the monsoon wind pattern is a distinctive seasonal precipitation regime—heavy summer rains derived from the moist maritime air of the onshore flow and a pronounced winter dry season when continental air moving seaward dominates the circulation.

It would be convenient to explain monsoon circulation simply on the basis of the unequal heating of continents and oceans. A strong thermal (in other words, heat-produced) low pressure cell generated over a continental landmass in summer attracts oceanic air onshore; similarly, a prominent thermal anticyclone in winter over a continent produces an offshore circulation. It is clear that these thermally induced pressure differences contribute to monsoon development (see Figure 5-15), but they are not the whole story.

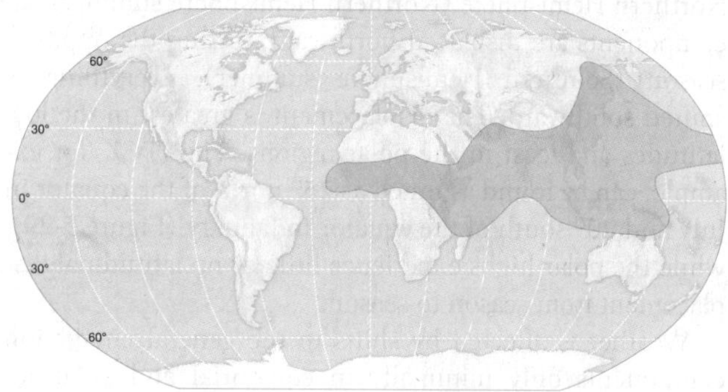

▲ **Figure 5-30** The principal monsoon areas of the world.

Monsoon winds essentially represent unusually large latitudinal migrations of the trade winds associated with the large seasonal shifts of the ITCZ over southeastern Eurasia. The Himalayas evidently also play a role—this significant topographic barrier allows greater winter temperature contrasts between South Asia and the interior of the continent to the north, and this in turn may influence the location and persistence of the subtropical jet stream in this region.

It is difficult to overestimate the importance of monsoon circulation to humankind. More than half of the world's population inhabits the regions in which climates are largely controlled by monsoons. Moreover, these are generally regions in which the majority of the populace depends on agriculture for its livelihood. Their lives are intricately bound up with the reality of monsoon rains, which are essential for both food production and cash crops (Figure 5-31). The failure, or even late arrival, of monsoon moisture inevitably causes widespread hunger and economic disaster.

Although the causes are complex, the characteristics of monsoons are well known, and it is possible to describe the monsoon patterns with some precision. There are two major monsoon systems (one in South Asia and the other in East Asia), two minor systems (in Australia and West Africa), and several other regions where monsoon tendencies develop (especially in Central America and the southeastern United States).

The most notable environmental event each year in South Asia is the annual burst of the summer monsoon, illustrated in Figure 5-32a. In this first of the two major monsoon systems, prominent onshore winds spiral in from the Indian

▲ **Figure 5-31** Flooding caused by summer monsoon rains in the central business district of Dhaka, Bangladesh.

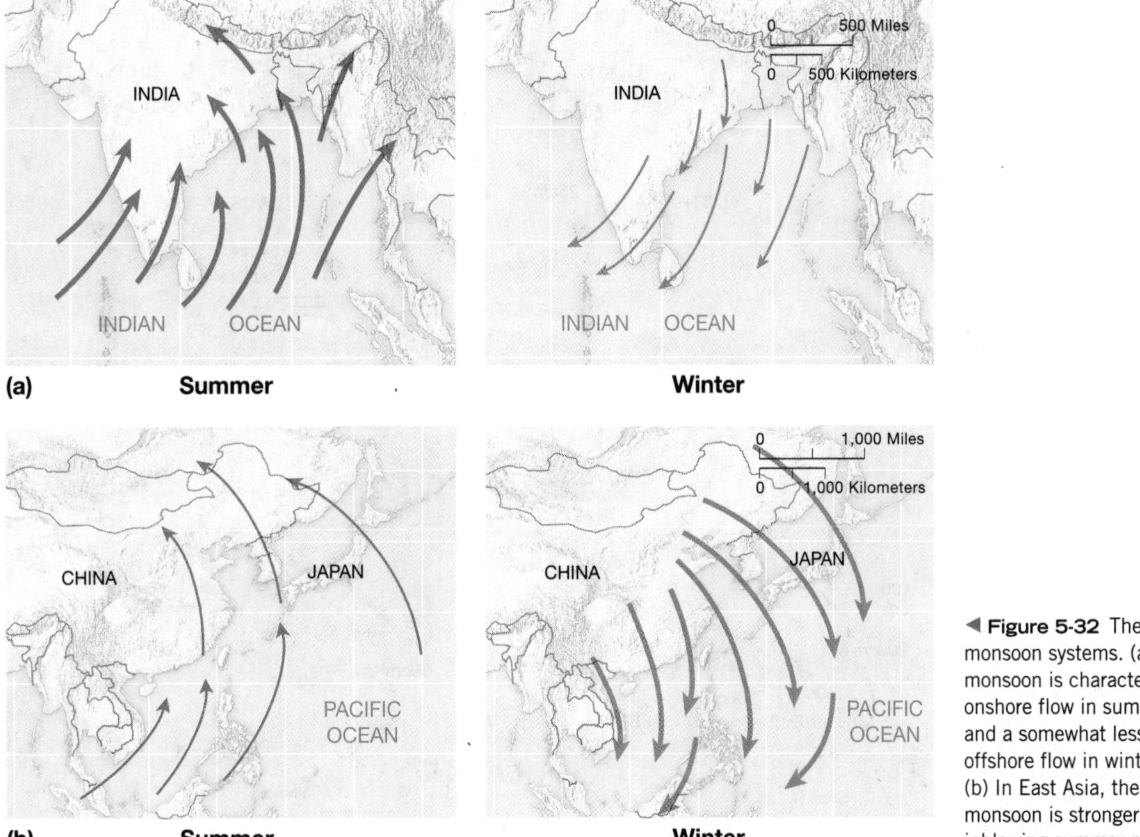

◀ **Figure 5-32** The two major monsoon systems. (a) The South Asian monsoon is characterized by a strong onshore flow in summer (rainy season) and a somewhat less pronounced offshore flow in winter (dry season). (b) In East Asia, the outblowing winter monsoon is stronger than the inblowing summer monsoon.

Ocean, bringing life-giving rains to the parched subcontinent. In winter, South Asia is dominated by outblowing dry air diverging generally from the northeast. This flow is not very different from normal northeast trades except for its low moisture content.

Turning to the second of the two major monsoon systems, we see that winter is the more prominent season in the East Asian monsoon system, which primarily affects China, Korea, and Japan and is illustrated in Figure 5-32b. A strong outflow of dry continental air, largely from the northwest, is associated with anticyclonic circulation around the massive thermal high-pressure cell over western Eurasia called the Siberian High. The onshore flow of maritime air in summer is not as notable as that in South Asia, but it does bring southerly and southeasterly winds, as well as considerable moisture, to the region.

In one of the two minor systems, the northern quarter of the Australian continent experiences a distinct monsoon circulation, with onshore flow from the north during the height of the Australian summer (December through March) and dry, southerly, offshore flow during most of the rest of the year. This system is illustrated in Figure 5-33a.

The south-facing coast of West Africa is dominated within about 650 kilometers (400 miles) of the coast by the second minor monsoonal circulation. This system is shown in Figure 5-33b. Moist oceanic air flows onshore from the south and southwest during summer, and dry, northerly, continental flow prevails in the opposite season.

Localized Wind Systems

The preceding sections have dealt with only the broadscale wind systems that make up the global circulation and influence the climatic pattern of the world. There are many kinds of lesser winds, however, that are of considerable significance to weather and climate at a more localized scale. Such winds are the result of local pressure gradients that develop in response to topographic configurations in the immediate area, sometimes in conjunction with broadscale circulation conditions.

Sea and Land Breezes

A common local wind system along tropical coastlines and to a lesser extent during the summer in midlatitude coastal areas is the cycle of **sea breezes** during the day and **land breezes** at night (Figure 5-34). (As is usual with winds, the name tells the direction from which the wind comes. A sea breeze blows from sea to land, and a land breeze blows from land to sea.) This is essentially a convectional circulation caused by the differential heating of land and water surfaces. The land warms up rapidly during the day, heating the air above by conduction and reradiation. This heating causes the air to expand and rise, creating low pressure that attracts surface breezes from over the adjacent water body. Because the onshore flow is relatively cool and moist, it holds down daytime temperatures in the coastal zone and

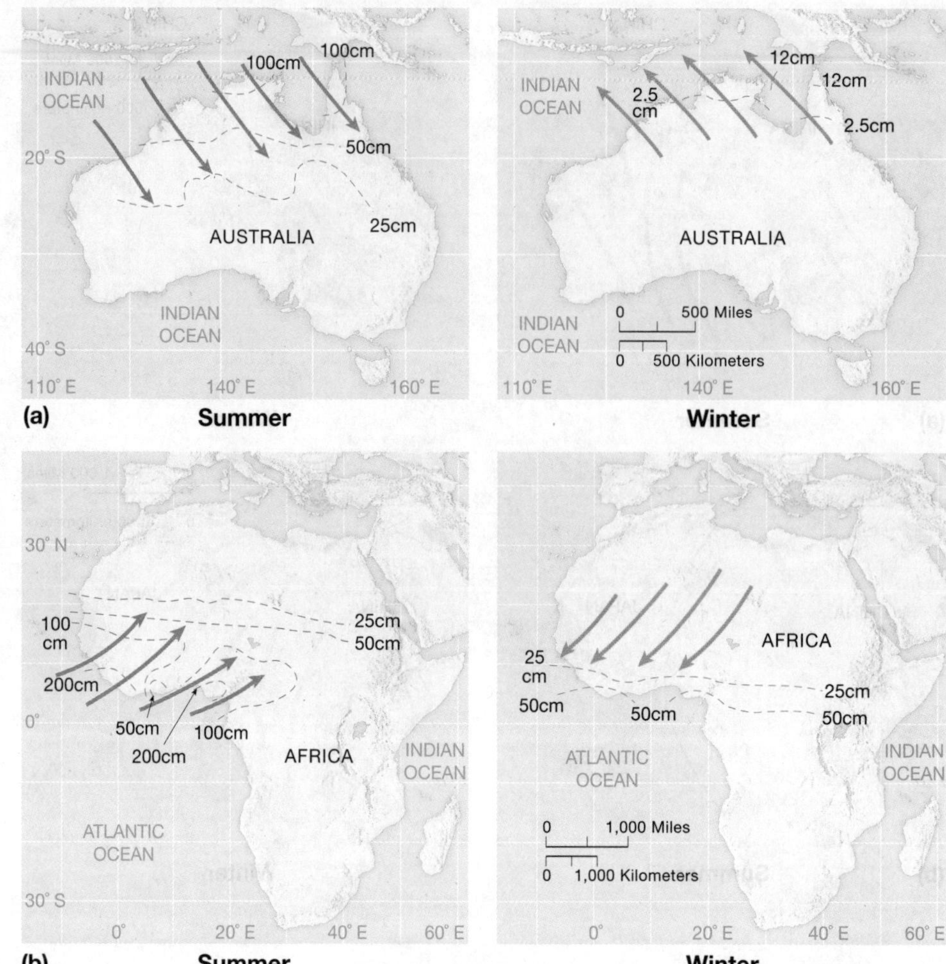

▶ **Figure 5-33** The two minor monsoon systems, showing 3-month seasonal rainfall isohyets (lines of equal rainfall). (a) In Australia, northwesterly summer winds bring the wet season to northern Australia; dry southeasterly flow dominates in winter. (b) In West Africa, summer winds are from the southwest and winter winds are from the northeast.

provides moisture for afternoon showers. Sea breezes are sometimes strong, but they rarely are influential for more than 15 to 30 kilometers (10 to 20 miles) inland.

The reverse flow at night is normally considerably weaker than the daytime wind. The land and the air above it cool more quickly than the adjacent water body, producing relatively higher pressure over land. Thus, air flows offshore in a land breeze.

Valley and Mountain Breezes

Another notable daily cycle of airflow is characteristic of many hill and mountain areas. During the day, conduction and reradiation from the land surface cause air near the mountain slopes to heat up more than air over the valley floor (Figure 5-35). The heated air rises, creating a low-pressure area, and then cooler air from the valley floor flows

▲ **Figure 5-34** In a typical sea-land breeze cycle, daytime heating over the land produces relatively low pressure there, and this low-pressure center attracts an onshore flow of air from the sea. Later, nighttime cooling over the land causes high pressure there, a condition that creates an offshore flow of air.

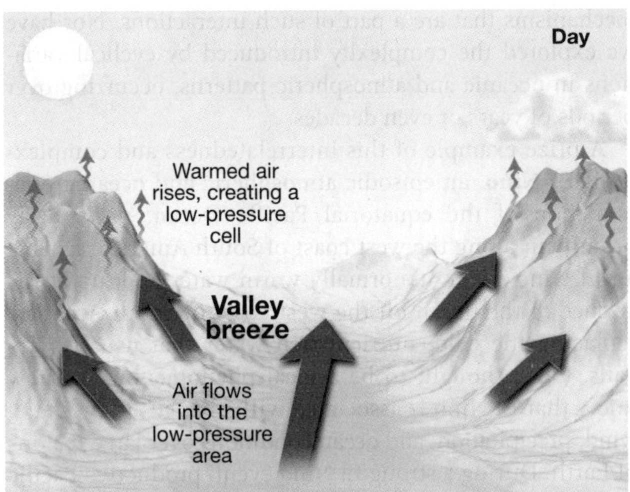

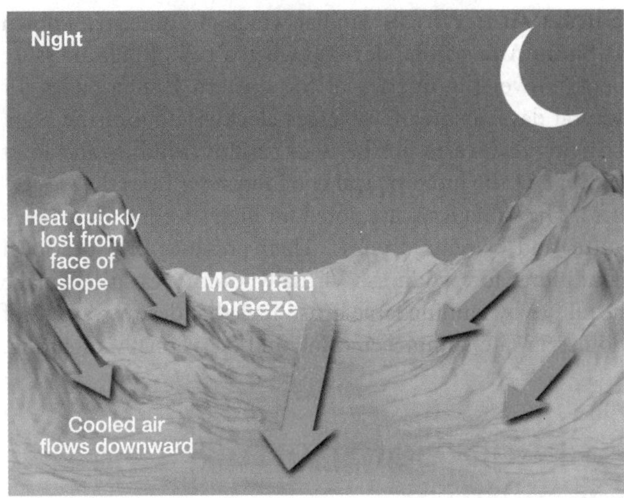

▲ **Figure 5-35** Daytime heating of the mountain slopes causes the air above the slopes to warm and rise, creating lower air pressure than over the valley. Cooler valley air then flows along the pressure gradient (in other words, up the mountain slope) in what is called a valley breeze. At night, the slopes radiate their heat away and as a result the air just above them cools. This cooler, denser air flows down into the valley in what is called a mountain breeze.

upslope from the high-pressure area to the low-pressure area. This upslope flow is called a **valley breeze**. The rising air often causes clouds to form around the peaks, and afternoon showers are common in the high country as a result. After dark, the pattern is reversed. The mountain slopes lose heat rapidly through radiation, which chills the adjacent air, causing it to slip downslope as a **mountain breeze**.

Valley breezes are particularly prominent in summer, when solar heating is most intense. Mountain breezes are often weakly developed in summer and are likely to be more prominent in winter. Indeed, a frequent winter phenomenon in areas of even gentle slope is *cold air drainage*, which is simply the nighttime sliding of cold air downslope to collect in the lowest spots; this is a modified form of mountain breeze.

Katabatic Winds

Related to simple air drainage is the more general and powerful spilling of air downslope in the form of **katabatic winds** (from the Greek *katabatik*, which means "descending"). These winds originate in cold upland areas and cascade toward lower elevations under the influence of gravity; they are sometimes referred to as *gravity-flow winds*. The air in them is dense and cold, and although warmed adiabatically as it descends, it is usually colder than the air it displaces in its downslope flow.

Katabatic winds are particularly common in Greenland and Antarctica, especially where they come whipping off the edge of the high, cold ice sheets. Sometimes a katabatic wind will become channeled through a narrow valley where it may develop high speed and considerable destructive power. An infamous example of this phenomenon is the *mistral*, which sometimes surges down France's Rhône Valley from the Alps to the Mediterranean Sea. Similar winds are called *bora* in the Adriatic region and *taku* in southeastern Alaska.

Foehn/Chinook Winds

Another downslope wind is called a **foehn** (pronounced as in fern but with a silent "r") in the Alps, and a **chinook** in the Rocky Mountains. It originates only when a steep pressure gradient develops with high pressure on the windward side of a mountain and a low-pressure trough on the leeward side. Air moves down the pressure gradient, which means from the windward side to the leeward side, as shown in Figure 5-36. The downflowing air on the leeward side is dry and relatively warm: it has lost its moisture through precipitation on the windward side, and it is warm relative to the air on the windward side because it contains all the latent heat of condensation given up by the condensing of the snow or rain that fell at the peak. As the wind blows down the leeward slope, it is further warmed adiabatically, and so it arrives at the base of the range as a warming, drying wind. It can produce a remarkable rise of temperature leeward of the mountains in just a few minutes. It is known along the Rocky Mountains front as a "snoweater" because it not only melts the snow rapidly but also quickly dries the resulting mud.

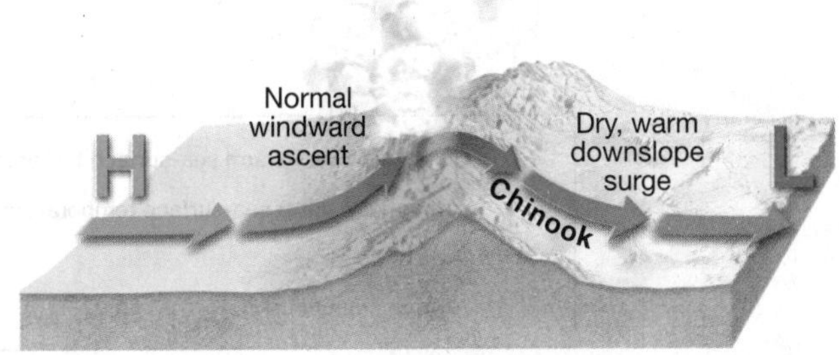

▲ **Figure 5-36** A chinook is a rapid downslope movement of relatively warm air. It is caused by a pressure gradient on the two faces of a mountain.

Santa Ana Winds Similar winds in California, known as **Santa Ana winds**, develop when a cell of high pressure persists over the interior of the western United States for several days. The wind diverges clockwise out of the high, bringing dry, warm northerly or easterly winds to the coast (instead of the more typical cool, moist air from the westerlies). The Santa Anas are noted for high speed, high temperature, and extreme dryness. Their presence provides ideal conditions for wildfires. Virtually every year they make headlines by fanning large brush fires that destroy dozens of homes in late summer and fall and occasionally in spring.

El Niño-Southern Oscillation

El Niño

One of the basic tenets of physical geography is the interrelatedness of the various elements of the environment. Nowhere is environmental interrelatedness shown better than at the air-ocean interface—that vast horizontal surface that marks the meeting of atmosphere and hydrosphere over 70 percent of Earth. These two great fluid systems interact in intricate and often perplexing fashion so that cause and effect become almost meaningless concepts. Thus far in our look at weather and climate, we have introduced patterns of temperature, ocean circulation, wind, and pressure without fully exploring important feedback mechanisms that are a part of such interactions. Nor have we explored the complexity introduced by cyclical variations in oceanic and atmospheric patterns, occurring over periods of years or even decades.

A prize example of this interrelatedness and complexity is **El Niño**, an episodic atmospheric and oceanic phenomenon of the equatorial Pacific Ocean, particularly prominent along the west coast of South America. During an El Niño event, abnormally warm water appears at the surface of the ocean off the west coast of South America, replacing the cold, nutrient-rich water that usually prevails. Once thought to be a local phenomenon, we now know that El Niño is associated with changes in pressure, wind, precipitation, and ocean conditions over large regions of Earth. During a strong El Niño event, productive Pacific fisheries off South America are disrupted and heavy rains come to some regions of the world while drought comes to others.

A slight periodic warming of Pacific coastal waters has been noticed by South American fishermen for many generations. It typically occurs around Christmastime—hence the name *El Niño* (Spanish for "the boy" in reference to the Christ child). Every three to seven years, however, the warming of the ocean is much greater, and fishing is likely to be much poorer. Historical records have documented the effects of El Niño for several hundred years, with archeological and paleoclimatological evidence pushing the record back several thousand years.

It was not until the major 1982–1983 event, however, that El Niño received worldwide attention (Figure 5-37).

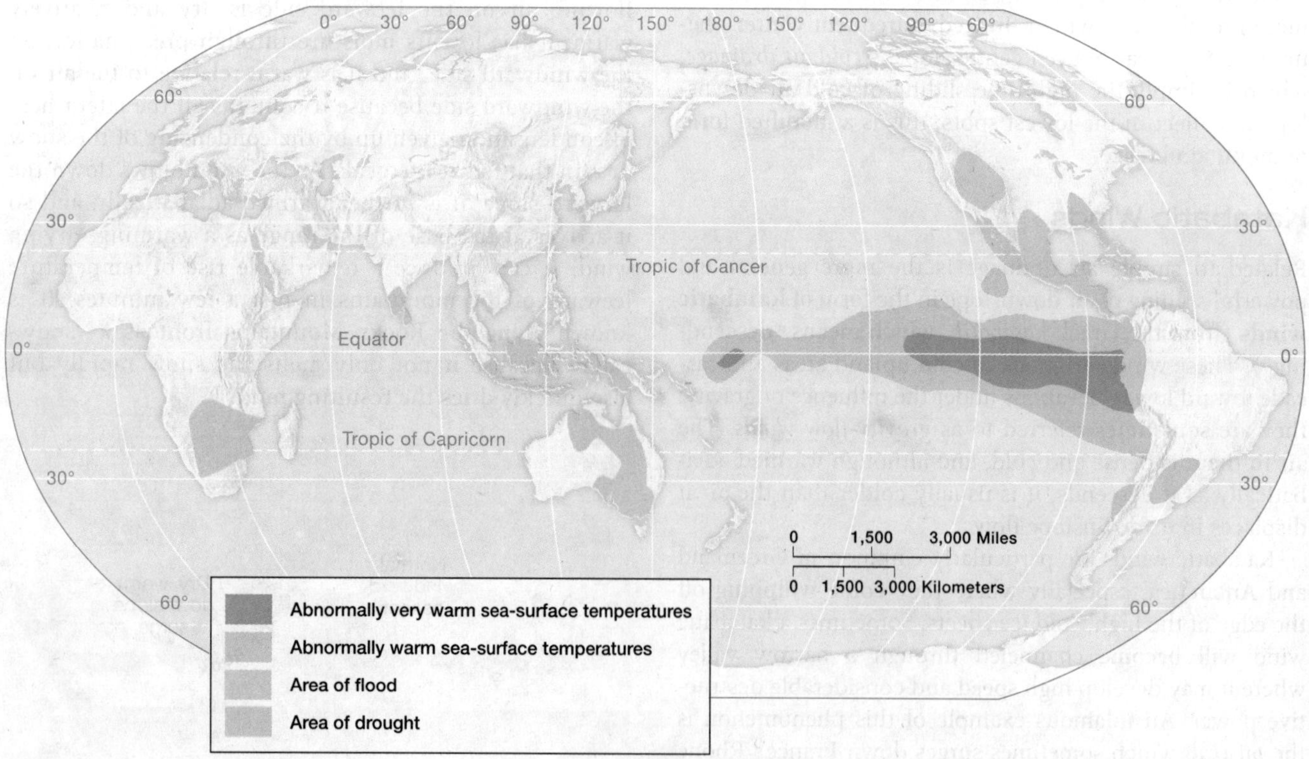

▲ **Figure 5-37** Some major weather events associated with El Niño during its 1982–1983 outbreak. Abnormally warm ocean water in the eastern equatorial Pacific is the most readily recognizable characteristic.

Over a period of several months, there were crippling droughts in Australia, India, Indonesia, the Philippines, Mexico, Central America, and southern Africa; devastating floods in the western and southeastern United States, Cuba, and northwestern South America; destructive tropical cyclones in parts of the Pacific (such as in Tahiti and Hawai'i) where they are normally rare; and a vast sweep of ocean water as much as 8°C (14°F) warmer than normal stretched over 13,000 kilometers (8000 miles) of the equatorial Pacific, causing massive die-offs of fish, seabirds, and coral. Directly attributable to these events were more than 1500 human deaths, damage estimated at nearly $9 billion, and vast ecological changes.

In 1997–1998, another strong El Niño cycle took place. This time worldwide property damage exceeded $30 billion, at least 2100 people died, and tens of thousands of people were displaced. There were severe blizzards in the Midwest, devastating tornadoes in the southeastern United States, and much higher than average rainfall in California.

So what happens during El Niño that causes such widespread changes in the weather?

Normal Pattern

To understand El Niño, we begin with a description of the normal conditions in the Pacific Ocean basin (Figure 5-38a). As we saw in Chapter 4, usually the waters off the west coast of South America are cool. The wind and pressure patterns in this region are dominated by the persistent subtropical high (STH) associated with the subsiding air of the Hadley cell circulation (see Figure 5-17). As the trade winds diverge from the STH, they flow from east to west across the Pacific—this tropical airflow drags surface ocean water westward across the Pacific basin in the warm Equatorial Current (introduced in Chapter 4; see Figure 4-25). As surface water pulls away from the coast of South America, an upwelling of cold, nutrient-rich ocean water rises into the already cool Peru current. This combination of cool water and high pressure result in relatively dry conditions along much of the west coast of South America.

In contrast to the cold water and high pressure near South America, in a normal year on the other side of the Pacific Ocean near Indonesia things are quite different. The trade winds and the equatorial current pile up warm water, raising sea level in the Indonesian region as much as 60 centimeters (about 2 feet) higher than near South America, turning the tropical western Pacific into an immense storehouse of energy and moisture. Warm water and persistent low pressure prevail around northern Australia and Indonesia; local convective thunderstorms develop in the intertropical convergence zone (ITCZ), producing high annual rainfall in this region of the world. After this air rises in the ITCZ, it begins to flow poleward but is deflected by the Coriolis effect into the upper-atmosphere westerly antitrade winds; some of this air flow aloft eventually subsides into the STH on the other side of

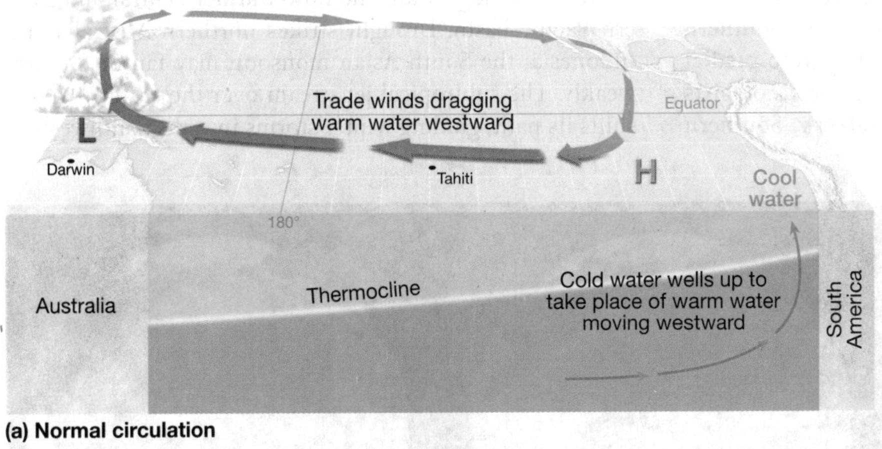

(a) Normal circulation

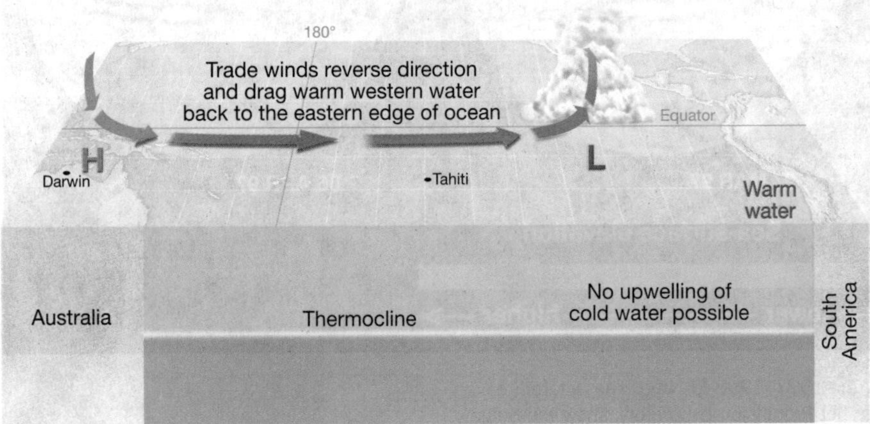

(b) Circulation during El Niño

◀ **Figure 5-38** (a) Normal conditions in the South Pacific. The trade winds carry warm equatorial water across the Pacific from west to east. (b) These conditions either weaken or reverse during an El Niño event. The upwelling of cold water off of South America diminishes, the thermocline boundary between near-surface and cold deep water lowers, and much warmer water than usual is present there.

the Pacific (see Figure 5-14). This general circuit of air flow is called the **Walker Circulation**, after the British meteorologist Gilbert Walker (1868–1958) who first described these circumstances. (Although Figure 5-38a shows the Walker Circulation as a closed convection cell, recent studies suggest that this is probably too simplistic—while the upper atmosphere is generally flowing from west to east, a closed "loop" of airflow probably does not exist.)

El Niño Pattern

Every few years, the normal pressure patterns in the Pacific change (Figure 5-38b). High pressure develops over northern Australia and low pressure develops to the east near Tahiti. This "seesaw" of pressure is known as the **Southern Oscillation**. It was first recognized by Gilbert Walker in the first decade of the twentieth century. Walker had become director of the Meteorological Service in colonial India in 1903, where a search was underway for a method to predict the monsoon—when the life-giving South Asian monsoon failed to develop, drought and famine ravaged India. In the global meteorological records, Walker thought that he saw a pattern: in most years pressure is low over northern Australia (specifically, Darwin, Australia) and high over Tahiti, and in these years, the monsoon usually comes as expected. However, in some years pressure is high in Darwin and low in Tahiti, and in these years the monsoon would often—but not always—fail. As it turned out, Walker's observed correlation between the monsoons in India and the Southern Oscillation of pressure was not reliable enough to predict the monsoons. However, by the 1960s meteorologists recognized a connection between Walker's Southern Oscillation and the occurrence of strong El Niño warming near South America. This overall coupled ocean-atmosphere pattern is now known as the **El Niño-Southern Oscillation** or simply **ENSO**.

While no two ENSO events are exactly alike, we can describe a typical El Niño cycle. For many months before the onset of an El Niño, the trade winds pile up warm water in the western Pacific near Indonesia. A bulge of warm equatorial water perhaps 25 centimeters (10 inches) high then begins to move to the east across the Pacific toward South America. Such slowly moving bulges of warm water are known as *Kelvin waves*. A Kelvin wave might take two or three months to arrive off the coast of South America (Figure 5-39). The bulge of warm water in a Kelvin wave spreads out little as it moves across the ocean since the Coriolis effect effectively funnels the eastward-moving water toward the equator in both hemispheres.

When the Kelvin wave arrives at South America, sea level rises as the warm water pools. The usual high pressure in the subtropics has weakened; upwelling no longer brings cold water to the surface, so ocean temperature increases still further—an El Niño is underway. By this time the trade winds have weakened or even reversed directions and started to flow from the west—blowing moist air into the deserts of coastal Peru. The *thermocline* boundary between near-surface and cold deep ocean waters lowers. Pressure increases over Indonesia and the most active portion of the ITCZ in the Pacific shifts from the now-cooler western Pacific, toward the now-warmer central and eastern Pacific basin. Drought strikes northern Australia and Indonesia; the South Asian monsoon may fail or develop weakly. The subtropical jet stream over the eastern Pacific shifts its path, guiding winter storms into the southwestern

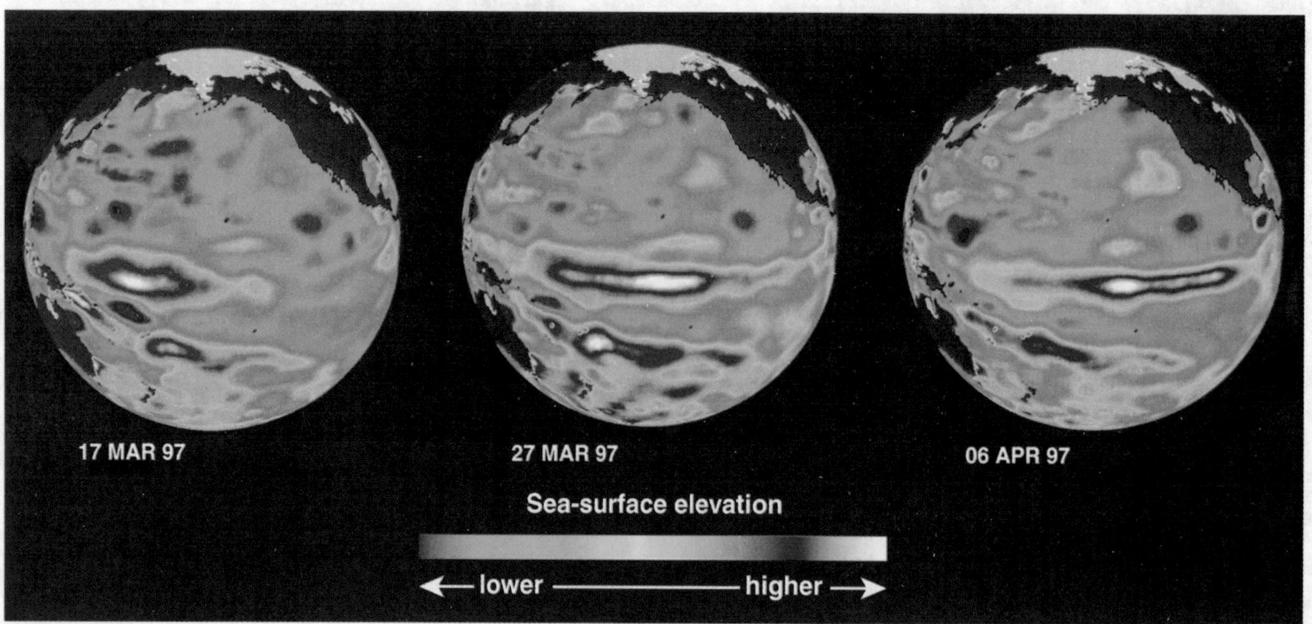

▲ **Figure 5-39** Progress of a Kelvin wave during the 1997–1998 El Niño. These satellite images from the California Institute of Technology, Jet Propulsion Laboratory, show the bulge of warm water in a Kelvin wave slowly moving across the equatorial Pacific Ocean.

United States—California and Arizona experience more powerful winter storms than usual, resulting in high precipitation and flooding.

La Niña

Adding to the complexity is a more-recently recognized component of the ENSO cycle, **La Niña**. In some ways, La Niña is simply the opposite of El Niño: the waters off South America become unusually cool; the trade winds are stronger than usual; the waters off Indonesia are unusually warm; the southwestern United States is drier than usual while Southeast Asia and northern Australia are wetter. However, La Niña conditions are not as prominent or as predictable as those of El Niño.

Causes of ENSO

So which comes first with the onset of an El Niño event—the change in ocean temperature or the change in pressure and wind? The "trigger" of an ENSO event is not clear since atmospheric pressure and wind patterns are tied together with the ocean in a complex feedback loop with no clear starting point: if atmospheric pressure changes, wind changes; when wind changes, ocean currents and ocean temperature may change; when ocean temperature changes, atmospheric pressure changes—and that in turn may change wind patterns still more. In short, the causes of ENSO are not fully understood.

Not only are the causes of ENSO elusive, even the effects are not completely predictable. For example, while a strong El Niño generally brings high precipitation to the southwestern United States, a mild or moderate El Niño could bring either drought or floods. Further, although we can generalize that during a strong El Niño high rainfall is more likely in California than in La Niña years, very wet winters can occur in any year no matter what ENSO is doing. In other words, El Niño might "open the storm door" every few years, but there is no guarantee that the storms will actually come.

Teleconnections

As more has been learned about ENSO over the last few decades, its connections with oceanic and atmospheric conditions inside and outside the Pacific basin are increasingly being recognized (Figure 5-40). Drought in Brazil; cold winters in the southeastern United States; high temperatures in the Sahel; a weak monsoon in India; tornadoes in Florida; fewer hurricanes in the North Atlantic—all seem to correlate quite well with a strong El Niño event. Such coupling of weather and oceanic events in one part of the world with those in another are termed **teleconnections**. Adding to the complexity of these teleconnections is growing evidence that long-term ENSO patterns may be influenced by other ocean-atmosphere cycles, such as the *Pacific Decadal Oscillation* (discussed below).

Over the last century, El Niño events have occurred on average once every two to seven years. It appears that El Niños have been becoming more frequent and progressively warmer in recent decades—the 1997–1998 event was probably the strongest El Niño of the last 200 years, and it developed more rapidly than any in the last 50 years—although the reasons for this are not yet clear. A much weaker El Niño took place in 2002–2003, and a mild El Niño appeared underway in mid-2009. There is speculation that global warming may be influencing the intensity of the El Niño cycle, but a clear connection has not yet been found.

Over the last 25 years, great strides have been taken toward forecasting El Niño events months in advance—largely because of better satellite monitoring of ocean conditions and the establishment of the TAO/TRITON array of oceanic buoys in the tropical Pacific (see the box, "People and the Environment: Forecasting El Niño"). While there is much left to be learned, a clearer understanding of cause and effect, of countercause and countereffect of ENSO, is gradually emerging.

Other Multiyear Atmospheric and Oceanic Cycles

El Niño is the best known multiyear atmospheric/oceanic cycle, although a number of other cyclical patterns have now been recognized.

Pacific Decadal Oscillation

The *Pacific Decadal Oscillation* (*PDO*) is a long-term pattern of sea surface temperature change between the northern/west tropical and eastern tropical Pacific Ocean. Approximately every 20 to 30 years the sea surface temperatures in these zones abruptly shift. From the late 1940s to late 1970s, the northern/west tropical Pacific was relatively warm while the eastern tropical Pacific was relatively cool (the PDO "negative" or "cool" phase); from the late 1970s through the mid-1990s, this pattern switched, with cooler sea surface temperatures in the northern/west tropical Pacific and warmer conditions in the east tropical Pacific (the PDO "positive" or "warm" phase; Figure 5-41). In the late 1990s the switch back to the negative phase was underway, but by the mid-2000s it was not clear if this negative phase was going to be short-lived.

Although the causes of the PDO are not well understood, the pattern of relatively warm and relatively cool ocean water seems to influence the location of the jet stream (and so storm tracks across North America) as well as the intensity of El Niño events. For example, when the PDO is in a positive phase with warmer water in the eastern tropical Pacific, the influence of El Niño on regional weather patterns appears to be more significant.

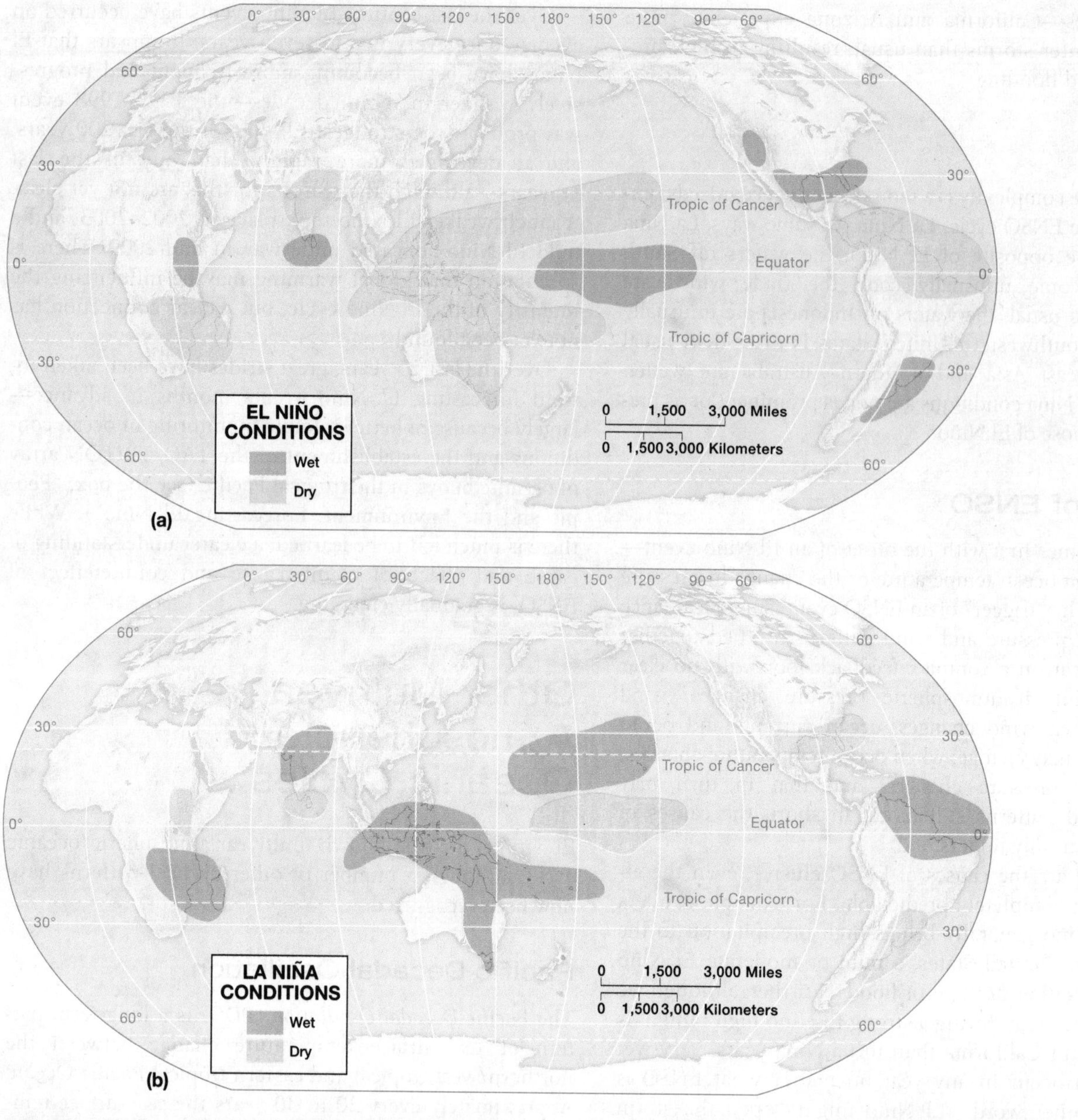

▲ **Figure 5-40** Generalized precipitation patterns during (a) El Niño conditions and (b) La Niña conditions.

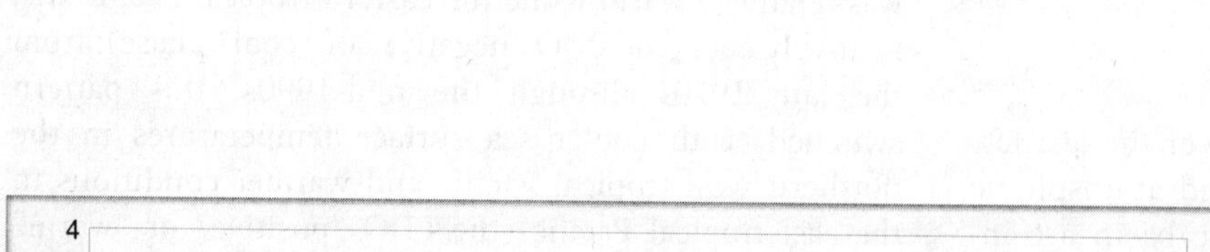

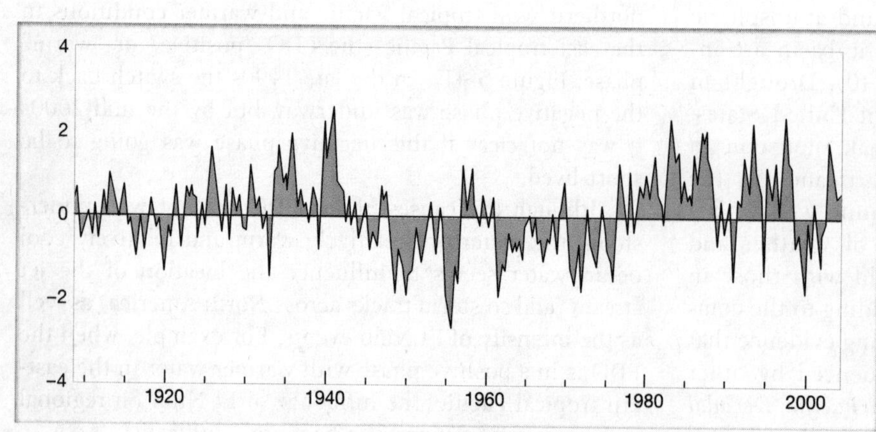

◀ **Figure 5-41** Pacific decadal oscillation index. Positive values indicate warm water in the eastern tropical Pacific, while negative values indicate relatively cool water in the same area.

PEOPLE AND THE ENVIRONMENT

Forecasting El Niño

Since the powerful El Niño of 1982–1983 caught large populations unprepared, a concerted multinational effort has been undertaken to understand El Niño and its teleconnections. Part of this effort included anchoring some 70 instrument buoys in the tropical Pacific Ocean beginning in 1985. The initial installations were part of the Tropical Atmosphere Ocean Array (TOA) administered by the National Oceanic and Atmospheric Administration (NOAA), but the array was soon expanded to include the Triangle Trans Ocean Buoy Network (TRITON) maintained by Japan. The combined TAO/TRITON array monitors ocean and atmospheric conditions—especially sea surface temperature and wind direction—across the tropical Pacific Ocean (Figure 5-B). By 1994, sufficient data had been gathered to begin to develop computer models to predict the onset of an El Niño event several months in advance.

These efforts were rewarded in 1997: by that spring, the TAO/TRITON buoys were recording a surge of warm water moving to the east across the equatorial Pacific, replacing the normally cool water off the west coast of South America (Figures 5-C and 5-D). Months in advance of its arrival, the onset of the 1997–1998 El Niño was announced by NOAA's Climate Prediction Center, and preparations were made for the series of powerful storms that eventually struck southwestern North America that winter. Even with this forewarning, the 1997–1998 El Niño resulted in billions of dollars of damage and is blamed for the deaths of thousands of people around the world.

The TAO/TRITON array has become an indispensable tool for the study of El Niño and its companion phenomenon, La Niña (Figure 5-E), as well as the atmospheric and oceanic conditions in the tropical Pacific in general. Because of its early success, the network has recently been expanded to include the Research Moored Array for African-Asian-Australian Monsoon Analysis (RAMA) in the Indian Ocean, and the Prediction and Research Moored Array in the Atlantic (PIRATA) forming a complete Global Tropical Moored Buoy Array (see Figure 5-B).

Scientists do not fully understand all aspects of El Niño, but the greatly improved forecasting tools should prevent the public from being taken by surprise again.

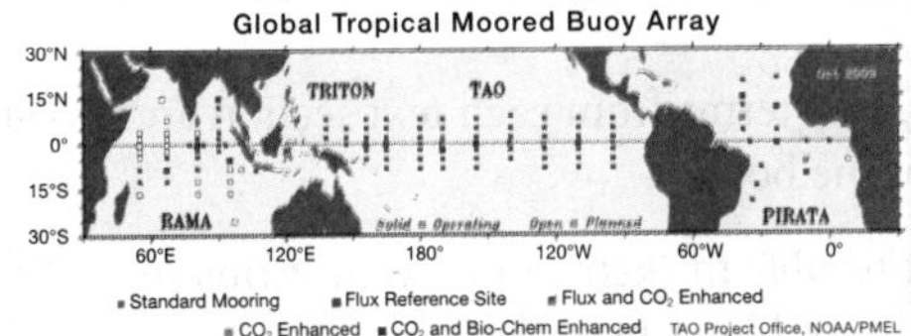

▲ **Figure 5-B** The TAO/TRITON Array of instrument buoys in the tropical Pacific Ocean is part of a Global Tropical Moored Buoy Array.

TAO Monthly Mean SST (°C) and Winds (m s⁻¹)

December 1993 Means

December 1993 Anomalies

Normal Conditions

TAO Project Office/PMEL/NOAA

▲ **Figure 5-C** Normal conditions in the tropical Pacific Ocean in December 1993; sea surface temperature and wind measured by the TAO/TRITON Array.

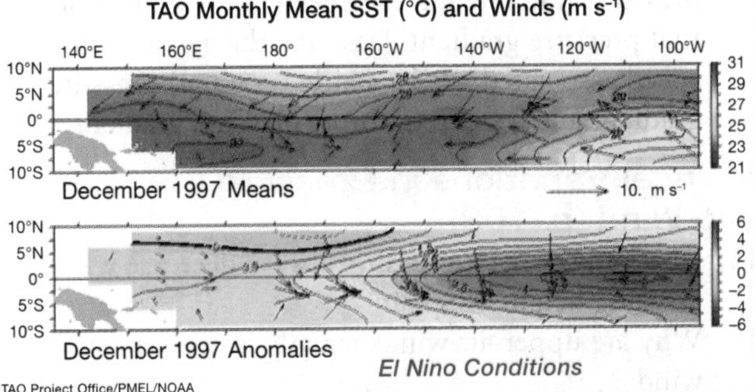

▲ **Figure 5-D** El Niño conditions in the tropical Pacific Ocean in December 1997; sea surface temperature and wind measured by the TAO/TRITON Array.

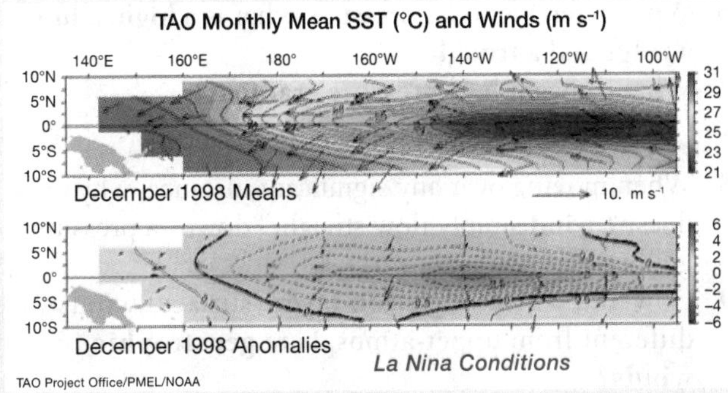

▲ **Figure 5-E** La Niña conditions in the tropical Pacific Ocean in December 1998; sea surface temperature and wind measured by the TAO/TRITON Array.

The North Atlantic Oscillation and the Arctic Oscillation

In the North Atlantic Ocean basin two related but somewhat irregular multiyear cycles of pressure, wind patterns, and temperature exist: the North Atlantic Oscillation and the Arctic Oscillation.

The *North Atlantic Oscillation* (*NAO*) is an irregular "seesaw" of pressure differences between two regional components of the general atmospheric circulation in the North Atlantic Ocean basin: the Icelandic Low and the subtropical high (the Azores High; see Figure 5-15 and 5-16). In the "positive" phase of the NAO, a greater pressure gradient exists between the Icelandic Low and the Azores High (in other words, the low exhibits lower than usual pressure, while the high exhibits higher than usual pressure). During such a positive phase, winter storms tend to take a more northerly track across the Atlantic, bringing mild, wet winters to Europe and the eastern United States but colder, drier conditions in Greenland.

In the "negative" phase of the NAO, both the Icelandic Low and the Azores High are weaker. During such a negative phase, winter storms tend to bring higher than average precipitation to the Mediterranean and colder winters in northern Europe and the eastern United States, while Greenland experiences milder conditions.

The *Arctic Oscillation* alternates between warm and cold phases that are closely associated with the NAO. During the Arctic Oscillation "warm" phase (associated with the NAO positive phase), the polar high is weaker, so cold air masses do not move as far south and sea surface temperatures tend to be warmer in Arctic waters—although Greenland tends to be colder than usual. During the Arctic Oscillation "cold" phase (associated with the NAO negative phase), the polar high is strengthened, bringing cold air masses farther south and leaving Arctic waters colder—although Greenland tends to be warmer than usual.

As with El Niño and the Pacific Decadal Oscillation, the causes of the North Atlantic Oscillation and the Arctic Oscillation are not well understood. Nor are all of the possible teleconnections between atmospheric and oceanic cycles in the North Atlantic and other parts of the world.

CHAPTER 5 LEARNING REVIEW

After studying this chapter, you should be able to answer the following questions. Key terms from each text section are shown in bold type. Definitions for key terms are also found in the glossary at the back of the book.

The Nature of Atmospheric Pressure (p. 97)

1. What generally happens to **atmospheric pressure** with increasing altitude?
2. Explain how atmospheric pressure is related to air density and air temperature.
3. What causes a **thermal high** near the surface? A **thermal low**?
4. What causes a **dynamic high** near the surface? A **dynamic low**?
5. Define the following terms: **barometer**, **millibar**, **isobar**.
6. When referring to air pressure, what is a **high**, a **low**, a **ridge** and a **trough**?
7. What is meant by a **pressure gradient**?

The Nature of Wind (p. 99)

8. When moving over any significant distance, why doesn't **wind** simply flow straight "down" a pressure gradient?
9. How and why are **friction layer** (surface) winds different from upper-atmosphere **geostrophic winds**?
10. Describe and explain the pattern of wind flow in the Northern Hemisphere around:
    - a surface high
    - a surface low
    - an upper atmosphere high
    - an upper atmosphere low

    (You should be able to sketch in wind direction on isobar maps of highs and lows near the surface and in the upper atmosphere for both the Northern and Southern Hemispheres.)
11. What is the reason for the difference in wind flow patterns in the Northern Hemisphere and the Southern Hemisphere?
12. What is a **cyclone**? An **anticyclone**?
13. Describe the pattern of vertical air movement within a cyclone and within an anticyclone.
14. Describe the relationship between the "steepness" of a pressure gradient and the speed of the wind along that pressure gradient. Describe the general wind speed associated with a gentle (gradual) pressure gradient and a steep (abrupt) pressure gradient.

Vertical Variations in Pressure and Wind (p. 103)

15. Does atmospheric pressure always decrease with increasing height? Explain.
16. Why are upper-air winds usually faster than surface winds?

The General Circulation of the Atmosphere (p. 103)

17. What are the **Hadley cells**, and what generally causes them?

18. Describe the general location and characteristics of the following atmospheric circulation components:
    - **intertropical convergence zone (ITCZ)**
    - **trade winds**
    - **subtropical highs**
    - **westerlies**

    (You should be able to sketch in the location of these four components on a blank map of an ocean basin.)
19. Discuss the characteristic weather associated with the ITCZ and the characteristic weather associated with subtropical highs.
20. Why do trade winds cover such a large part of the globe?
21. What are meant by the **horse latitudes** and the **doldrums**?
22. Describe the general location and characteristics of the **jet streams** of the westerlies.
23. What are **Rossby waves**?
24. Briefly describe the location and general characteristics of the high-latitude components of the general circulation patterns of the atmosphere:
    - **polar front (subpolar lows)**
    - **polar easterlies**
    - **polar highs**
25. Differentiate between trade winds and **antitrade winds**.

Modifications of the General Circulation (p. 113)

26. Describe and explain the seasonal shifts of the general circulation patterns; especially note the significance of the seasonal shifts of the ITCZ and the subtropical highs.
27. Describe and explain the South Asian **monsoon**.

Localized Wind Systems (p. 115)

28. Explain the origin of **sea breezes** and **land breezes**.
29. In what ways are sea breezes and land breezes similar to **valley breezes** and **mountain breezes**?
30. What is a **katabatic wind** and where are they commonly found?
31. In what ways are **Santa Ana winds** similar to **foehn** and **chinook** winds?

El Niño-Southern Oscillation (p. 118)

32. What is the **Walker Circulation**?
33. Why is **El Niño** commonly referred to as **El Niño-Southern Oscillation (ENSO)**?
34. Contrast the oceanic and atmospheric conditions in the tropical Pacific Ocean basin during an El Niño event with those of a normal pattern.
35. Contrast the oceanic and atmospheric conditions during an El Niño event with those of a **La Niña** event.
36. What is meant by **teleconnections**?

Log in to **www.mygeoscienceplace.com** for animations, videos, RSS feeds, maps, and self-study quizzes to enhance your study of **Atmospheric Pressure and Wind**.

6 Atmospheric Moisture

The fourth element of weather and climate is moisture. Although it might seem that we are all familiar with water, most atmospheric moisture occurs not as liquid water but rather as water vapor, which is much less conspicuous and much less familiar. One of the most distinctive attributes of water is that it occurs in the atmosphere in three physical states: solid (snow, hail, sleet, ice), liquid (rain, water droplets in clouds), and gas (water vapor). Of the three states, the gaseous state is the most important insofar as the dynamics of the atmosphere are concerned.

The Impact of Atmospheric Moisture on the Landscape

When the atmosphere contains enough moisture, water vapor may condense to form haze, fog, clouds, rain, sleet, hail, or snow, producing a skyscape that is both visible and tangible. Precipitation produces dramatic short-run changes in the landscape whenever rain puddles form, streams and rivers flood, or snow and ice blanket the ground. The long-term effect of atmospheric moisture is even more fundamental. Water vapor stores energy that can incite the atmosphere into action. Rainfall and snowmelt in soil and rock are an integral part of weathering and erosion. In addition, the presence or absence of precipitation is critical to the survival of almost all forms of terrestrial vegetation.

The Hydrologic Cycle

Hydrologic Cycle

The widespread distribution of water vapor in the atmosphere reflects the ease with which moisture can change from one state to another at the pressures and temperatures found in the lower troposphere. Moisture can leave Earth's surface as a gas and return as a liquid or solid. Indeed, there is a continuous interchange of moisture between Earth and the atmosphere (Figure 6-1). This unending circulation of our planet's water supply is referred to as the **hydrologic cycle**, and its essential feature is that liquid water (primarily from the oceans) evaporates into the air, condenses to the

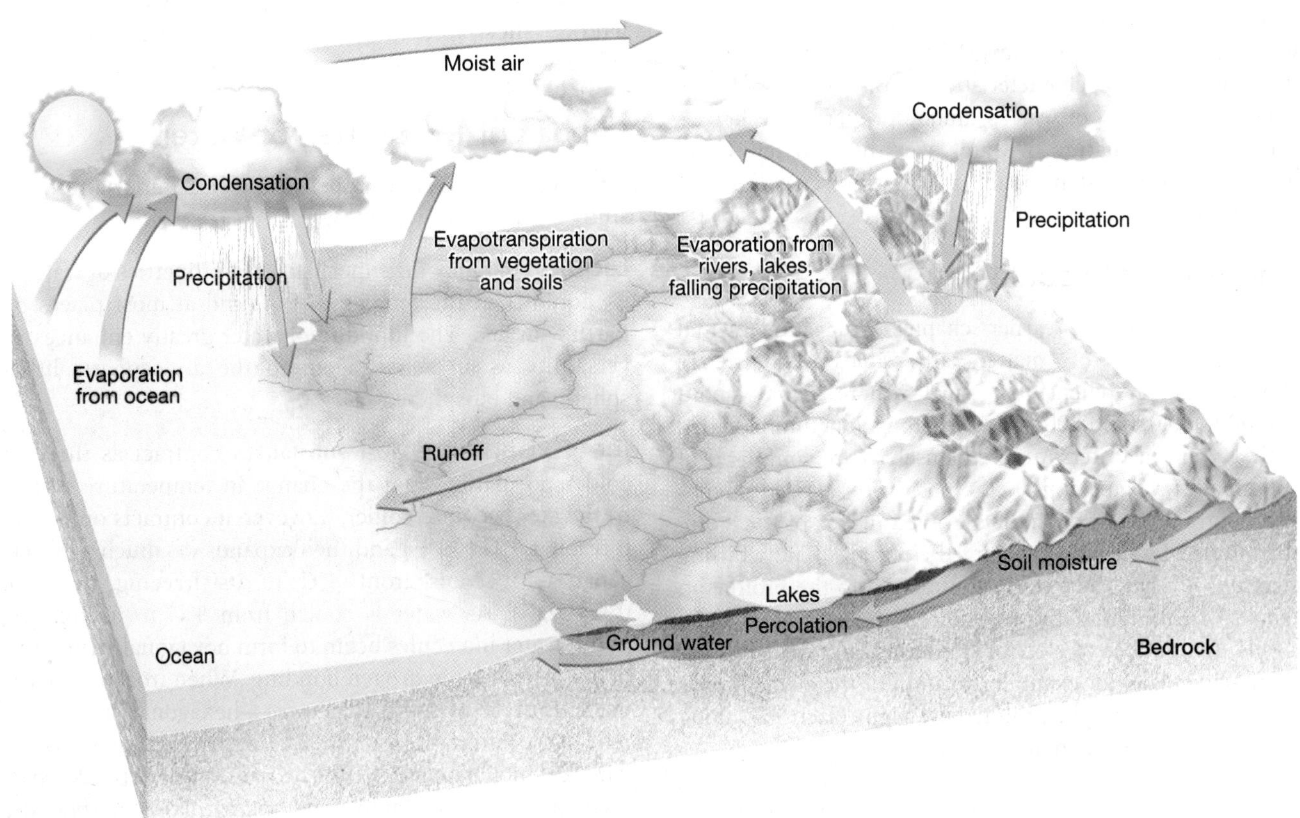

▲ **Figure 6-1** The hydrologic cycle is a continuous interchange of moisture between the atmosphere and Earth.

◀ Rainbow and clouds near the Alaska Range, Alaska.

liquid (or solid) state, and returns to Earth as some form of precipitation. The movement of moisture through the cycle is intricately related to many atmospheric phenomena and is an important determinant of climate because of its role in rainfall distribution and temperature modification. We will discuss the complete hydrologic cycle in greater detail in Chapter 9.

The Nature of Water: Commonplace but Unique

Water is the most widespread substance on the surface of Earth, occupying more than 70 percent of the surface area of the planet. Water is also perhaps the most distinctive substance found on Earth: pure water has no color, no taste, and no smell. It turns to a solid at 0°C (32°F) and boils at sea level at 100°C (212°F). The density of liquid water at 4°C is 1 gram per cubic centimeter (1 g/cm^3), while the density of ice is only 0.92 g/cm^3—meaning that ice floats in liquid water. It has a very high heat capacity and is an extremely good solvent. It has the ability to move upward into narrow openings. Water set the stage for the evolution of life and is still an essential ingredient of all life today.

Before we look specifically at the role of water in weather and climate, we must look at the water molecule itself, because the characteristics of the water molecule help explain many of water's unique properties. What we learn about the nature of water over the next few pages will help us understand many processes discussed in the remaining chapters of the book.

The Water Molecule

As we discussed in earlier chapters, the fundamental building blocks of matter are *atoms*. Atoms are almost incomprehensibly small—there are about 100,000,000,000,000,000,000,000 atoms in a thimble filled with 1 gram of water. Atoms themselves are composed of still smaller *subatomic particles*: positively charged *protons* and neutrally charged *neutrons* in the nucleus of the atom, surrounded by negatively charged *electrons*.[1] The whirling electrons are held by electrical attraction to the protons in "shells" surrounding the nucleus of the atom. In many atoms the number of electrons, protons, and neutrons are normally equal—for example, an atom of the element oxygen typically has eight protons and eight electrons, along with eight electrically neutral neutrons.[2]

[1]Although it was once thought that the smallest particles of matter were electrons, protons, and neutrons, we know now that some subatomic particles themselves are made of still smaller particles such as *quarks*.

[2]Atoms with a large number of protons will usually have a greater number of neutrons than protons. The number of neutrons can and do vary, without affecting the chemical behavior of the atom—atoms of an element with identical numbers of protons but different numbers of neutrons are called *isotopes*. If the number of electrons is different from the number of protons, the neutral atom becomes an electrically charged *ion*.

Two or more atoms can be held together to form a *molecule* by bonding. There are many different types of bonds in nature. Inside the water molecule two atoms of hydrogen and one atom of oxygen (H_2O) are held together by *covalent bonds*, in which the oxygen and hydrogen atoms share electrons (some of the electrons move between energy shells of both atoms; Figure 6-2a). Because of the shape of the electron shells around the oxygen, the resulting structure of a water molecule is such that the hydrogen atoms are not opposite one another, but rather are on the same side of the molecule (separated by an angle of 105°). As a consequence of this geometry, a water molecule has electrical *polarity*: the oxygen side of the molecule has a slight negative charge, while the hydrogen side has a slight positive charge (Figure 6-2b).

It is the weak electrical polarity of the molecules that gives water many of its interesting properties. For example, water molecules tend to orient themselves toward each other so that the negatively charged oxygen side of one molecule is next to the positively charged hydrogen side of another molecule. This attraction forms a **hydrogen bond** between adjacent water molecules (Figure 6-2b). Although hydrogen bonds are relatively weak (compared with the covalent bonds that hold together the atoms in an individual water molecule), it does mean that water molecules tend to "stick" to each other (when molecules of the same substance stick together it is called *cohesion*).

Important Properties of Water

Water has a number of properties that are important in our study of physical geography.

Liquidity One of the most striking properties of water is its liquidity at the temperatures found at most places on Earth's surface. The liquidity of water greatly enhances its versatility as an active agent in the atmosphere, lithosphere, and biosphere.

Ice Expansion Most substances contract as they get colder no matter what the change in temperature. When freshwater becomes colder, however, it contracts only until it reaches 4°C (39°F) and then expands (as much as 9 percent) as it cools from 4°C to its freezing point of 0°C (32°F). As water is cooled from 4°C to its freezing point, water molecules begin to form hexagonal structures, held together by hydrogen bonding. When frozen, water is made entirely of these structures—hexagonal snowflakes reflect the internal structure of these ice crystals. As we will see in Chapter 15, this expansion can break apart rocks and so is an important process of *weathering* (the disintegration of rock exposed to the atmosphere).

Because water expands as it approaches freezing, ice is less dense than liquid water. As a result, ice floats on and near the surface of water. If it were denser than water, ice would sink to the bottom of lakes and oceans, where melting would be virtually impossible, and eventually many water bodies would become ice choked. In fact, because freshwater becomes less dense as it approaches its freezing

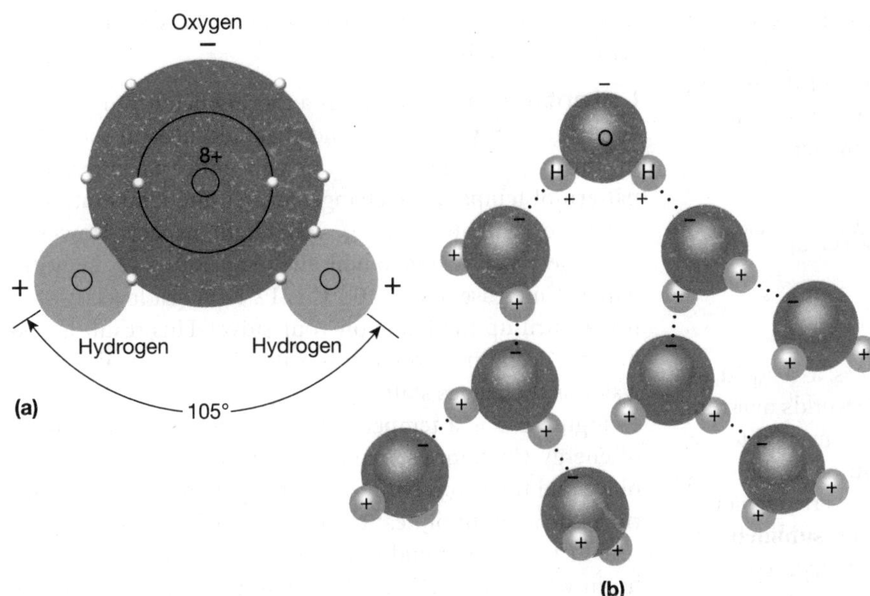

◀ **Figure 6-2** (a) A water molecule is made up of two hydrogen atoms and one oxygen atom held together with covalent bonds (the small dots represent electrons). The oxygen side of the molecule has a slight negative charge while the hydrogen side has a slight positive charge. (b) Hydrogen bonds form between water molecules because the negatively charged oxygen side of one molecule is attracted to the positively charged hydrogen side of another molecule.

point, water that is ready to freeze rises to the tops of lakes, and hence all lakes freeze from the top down. (Note: the expansion of liquid water is not a factor for ocean water because the high salinity prevents the hexagonal structures from forming until the water is completely frozen.)

Surface Tension Because of its electrical polarity, liquid water molecules tend to stick together, giving water extremely high **surface tension—**a thin "skin" of molecules forms on the surface of liquid water causing it to "bead." Some insects use the stickiness of water to stride atop the surface of a body of water—the weight of the insect spread over the surface area of the water is less than the force of the hydrogen bonds sticking the water together (Figure 6-3). Later in this chapter we'll see that surface tension influences the way that water droplets grow within a cloud.

Capillarity Water molecules also "stick" easily to many other substances—a characteristic known as *adhesion*. Surface tension combined with adhesion allows water to climb upward in narrow openings. Confined in this way, water can sometimes climb upward for many centimeters or even meters, in an action called **capillarity**. Capillarity enables water to circulate upward through rock cracks, soil, and the roots and stems of plants.

Solvent Ability Water can dissolve almost any substance and it is sometimes referred to as the "universal solvent." Because of the polarity of water molecules, they not only are attracted to each other but to other polar chemical compounds as well. Water molecules attach themselves quickly to the ions that constitute the outer layers of solid materials and in some cases can overcome the strength of those bonds, tearing the ions out of the solid and eventually dissolving the material. As a result, water in nature is nearly always impure, by which we mean that it contains various other chemicals in addition to its hydrogen and oxygen atoms. As water moves through the atmosphere, on the surface of Earth, and in soil, rocks, plants, and animals, it carries with it a remarkable diversity of dissolved minerals and nutrients as well as tiny solid particles in suspension.

Specific Heat Another environmentally important characteristic of water is its great heat capacity. As we saw in Chapter 4, *specific heat* (or *specific heat capacity*) is defined as the amount of energy required to raise the temperature of 1 gram of a substance (at 15°C) by 1 degree Celsius. When water is warmed, it can absorb an enormous amount of energy with only a small increase in temperature. Water's specific heat (1 calorie/gram) is exceeded by that of no other common substance except ammonia. The high heat capacity is a consequence of the relatively large amount of kinetic energy required to overcome the hydrogen bonds between water molecules. The practical result, as we saw in Chapter 4, is that bodies of water are very slow

▲ **Figure 6-3** Weight of this raft spider (*Dolomedes fimbriatus*) is supported by the surface tension of the water surface.

to warm up during the day or in summer, and very slow to cool off during the night or in winter. Thus, water bodies have a moderating effect on the temperature of the overlying atmosphere by serving as reservoirs of warmth during winter and having a cooling influence in summer.

Phase Changes of Water

Phase Changes of Water

Water of Earth is found naturally in three states: as a liquid, as a solid, and as a gas. The great majority of the world's moisture is in the form of liquid water, which can be converted to the gaseous form (water vapor) by **evaporation** or to the solid form (ice) by freezing. Water vapor can be converted to liquid water by **condensation** or directly to ice by **sublimation**. (Sublimation is the process whereby a substance converts either from the gaseous state directly to the solid state, or from the solid state directly to the gaseous state without ever passing through the liquid state.)[3] Ice can be converted to liquid water by melting or to water vapor by sublimation. In each of these phase changes, there is an exchange of *latent heat* energy, a concept we first introduced in Chapter 4 (Figure 6-4). An understanding of phase changes and latent heat are central to understanding several atmospheric processes we will describe later in this chapter.

Latent Heat If you insert a thermometer into a pan of cool water on your stove, and then add heat to the water by turning on the burner, you will notice an interesting pattern of temperature change: at first, the temperature of the water will increase, as you might expect. However, once the water begins to boil, the temperature of the water will not increase above 100°C (212°F; at sea level)—even if you turn up the flame on your stove! This result is just one of several important observations we can make about water as it changes states.

Figure 6-5 is a temperature chart showing the amount of energy (in calories) required to melt a one-gram block of ice and then convert it to water vapor (recall that 1 calorie is the amount of heat required to increase the temperature of 1 gram of liquid water by 1°C). In this example, we begin with our block of ice at a temperature of −40°C. As heat is added, the temperature of the ice increases quickly; only 20 calories of heat are needed to increase the temperature of the ice to its melting point of 0°C. Once the ice begins to melt, it absorbs 80 calories of heat per gram, but the temperature does not increase above 0°C until all of the ice has melted. We then add another 100 calories of heat to increase the temperature of the liquid water from 0°C to 100°C, its boiling point at sea level. Once the water is boiling, it absorbs 540 calories per gram, but once again, the temperature does not increase above 100°C until all of the liquid water is converted to water vapor.

[3]Sometimes the term *sublimation* is restricted to a description of the phase change from a solid state directly to a gaseous state, and *deposition* used to describe the phase change from a gaseous state directly to a solid state. In this book, as is common practice, we use the term sublimation for both.

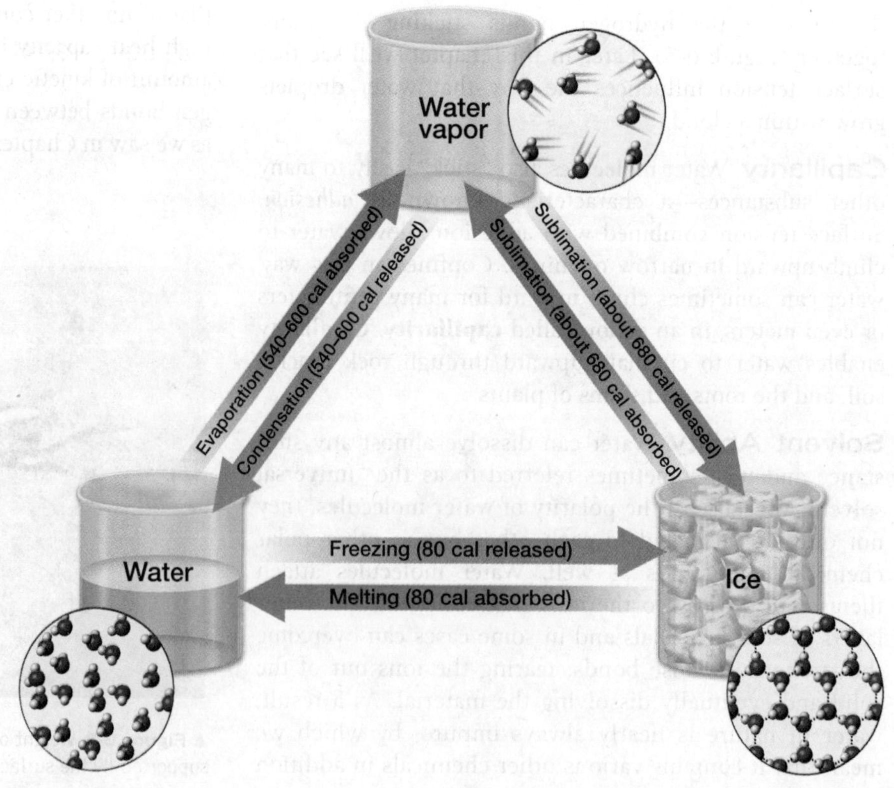

▶ **Figure 6-4** Phase changes of water are accompanied by the exchange of latent heat. The red arrows indicate that latent heat is absorbed; the blue arrows indicate that latent heat is released. The values given for heat absorbed or released during evaporation, condensation, and sublimation are for one gram of water.

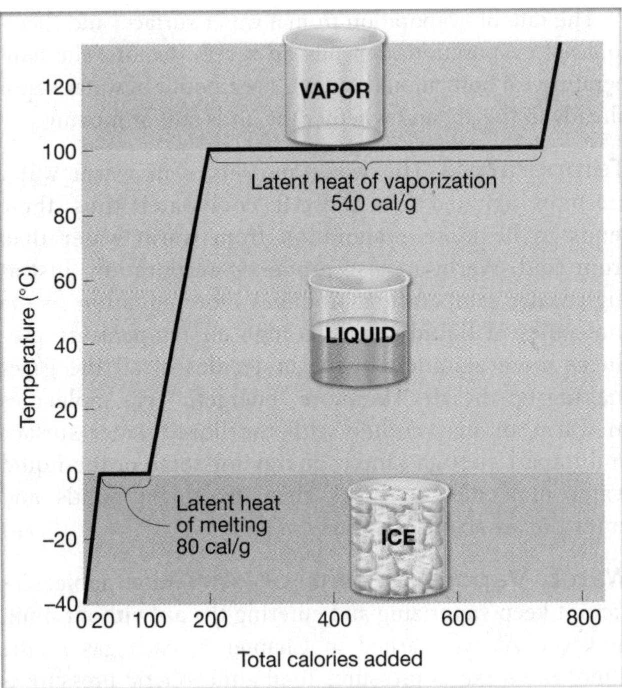

▲ **Figure 6-5** The energy input (in calories) and the associated temperature changes as one gram of ice starting at a temperature of −40°C is melted and then converted to water vapor. The latent heat of vaporization is much greater than the latent heat of melting.

Even as heat is added, the temperature of water does not increase while it is undergoing a *phase change*. Here's why: in order for ice to melt, energy must be added to "agitate" water molecules enough to break some of the hydrogen bonds that are holding these molecules together as ice crystals—the energy added does not increase the temperature of the ice but instead increases the internal structural energy of water molecules so that they can break free to become liquid.[4] As we saw in Chapter 4, the energy exchanged during a phase change is called **latent heat**. The energy required to melt ice is called the *latent heat of melting*. The opposite is also true: when water freezes, the liquid water molecules must give up some of their internal structural energy in order to revert to a less agitated condition in which ice can form. The energy released as water freezes is called the *latent heat of fusion*. For each gram of ice, 80 calories of heat are absorbed when ice melts, and 80 calories of heat are released when water freezes.

In a similar way, energy must be added to agitate liquid water molecules enough for them to escape into the surrounding air as water vapor—the energy added does not increase the temperature of the liquid water, but instead increases the internal structural energy of water molecules so that they can break free to become vapor. The energy required to vaporize liquid water is called the *latent heat of vaporization*. And, once again, the opposite is also true: when water vapor condenses back to liquid water, the highly agitated water vapor molecules must give up some of their internal structural energy in order to revert to a less agitated liquid state. The energy released during condensation is called the **latent heat of condensation**. For each gram of liquid water at a temperature of 100°C, 540 calories of heat are absorbed as water vaporizes, and 540 calories of heat are released when water vapor condenses.

The value for the latent heat of vaporization given above refers to circumstances in which water is boiling. *Boiling* occurs when vaporization takes place beneath the surface of the liquid water—not just at the surface. In nature, however, most water vapor is added to the atmosphere through simple evaporation off the surface of water bodies at temperatures below 100°C. In this case, the energy required for evaporation is greater than when water is boiling. The **latent heat of evaporation** ranges from 540 calories to about 600 calories depending on the temperature of the water (it is approximately 585 calories when the liquid water is at a temperature of 20°C [68°F]).

Notice in Figures 6-4 and 6-5 that about seven times more heat is needed to evaporate one gram of liquid water than is needed to melt one gram of ice. Notice also that when sublimation takes place, the latent heat exchange is simply the total of the solid-liquid and liquid-gas exchanges.

Importance of Latent Heat in the Atmosphere The significance of the exchange of latent heat during phase changes—especially between liquid water and water vapor—is straightforward. Whenever evaporation takes place, energy is removed from the liquid to vaporize some of the water, and so the temperature of the remaining liquid is reduced. Because latent heat energy is "stored" in water vapor during evaporation, evaporation is, in effect, a cooling process. The effect of such evaporative cooling is experienced when a swimmer leaves a swimming pool on a dry, warm day. The dripping wet body immediately loses moisture through evaporation to the surrounding air, and the skin feels the resulting drop in temperature.

Conversely, because latent heat energy must be released during condensation, condensation is, in effect, a warming process. Water vapor represents a "reservoir" of heat—whenever and wherever condensation takes place, this heat is added back to the atmosphere. As we will begin to see later in this chapter, the release of latent heat during condensation plays important roles in the stability of the atmosphere and in the power of many storms.

Water Vapor and Evaporation

Although the vast majority of water on Earth is in liquid form—and although the ultimate goal of this chapter is to understand the development of clouds and precipitation—for the moment we will direct our attention to water vapor, the source of moisture for clouds and precipitation.

[4] It may be helpful to think of *latent heat* as the energy used to break or form the bonds (the rearrangement of the structure between two phases), while *specific heat* is the energy used to raise the temperature (the "speed" of the molecules in one phase).

Water vapor is a colorless, odorless, tasteless, invisible gas that mixes freely with the other gases of the atmosphere. We are only likely to become aware of water vapor when humidity is high because the air feels sticky, clothes feel damp, and our skin feels clammy, or when humidity is low because our lips chap and our hair will not behave. Water vapor is a minor constituent of the atmosphere, with the amount present being quite variable from place to place and from time to time. It is virtually absent in some places but constitutes as much as 4 percent of the total atmospheric volume in others. Essentially, water vapor is restricted to the lower troposphere. More than half of all water vapor is found within 1.5 kilometers (about 1 mile) of Earth's surface, and only a tiny fraction exists above 6 kilometers (about 4 miles).

Now we turn specifically to the process that puts water vapor into the atmosphere: evaporation.

Evaporation and Rates of Evaporation

In our description of the phase changes of water above, we simplified things a bit. As it turns out, both evaporation and condensation may be taking place at the same time. Strictly speaking, water vapor is added to the air when the rate of evaporation exceeds the rate of condensation, in other words, when there is *net evaporation* (Figure 6-6).

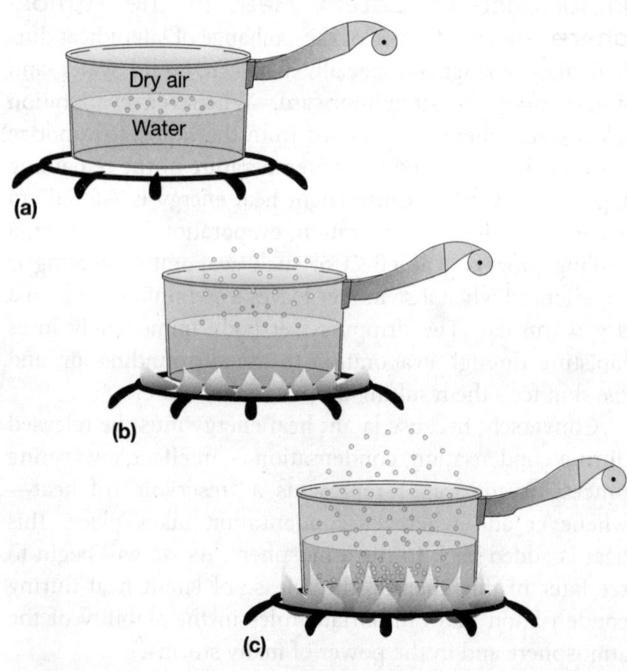

▲ **Figure 6-6** Evaporation involves the escape of water molecules from a liquid surface into the air as water vapor. It can take place at any temperature, but high temperatures increase the energy of the molecules in the liquid and therefore increase the rate of evaporation. (a) With the burner off, some molecules break free from the surface as water vapor, whereas some return from the air to the liquid surface. (b) As the water temperature increases, all of the molecules become more agitated, and so more of the water molecules break free from the surface as water vapor than return to the liquid surface from the air. (c) Continued heating increases molecular activity so much that the air may become saturated with water vapor.

The rate of evaporation from a water surface, and therefore net evaporation, depends on several factors: the temperature (of both air and water), the amount of water vapor already in the air, and whether the air is still or moving.

Temperature The water molecules in warm water are more agitated than those in cool water; thus, there tends to be more evaporation from warm water than from cold. Warm air also promotes evaporation. Just as high water temperature produces more agitation in the molecules of liquid water, so high air temperature produces more agitation in the molecules of all the gases making up the air. The more "energetic" gas molecules in warm air may collide with the liquid water surface and impart enough kinetic energy for some of the liquid water molecules to break their hydrogen bonds and enter the air above as vapor.

Water Vapor Content of Air Water molecules cannot keep vaporizing and entering the air without limit, however. As we learned in Chapter 5, each gas in the atmosphere exerts pressure. Total atmospheric pressure is simply the sum of the pressures exerted by all of the individual gases in the atmosphere. The pressure exerted by water vapor is called the **vapor pressure**. At any given temperature, there is a maximum vapor pressure that water molecules can exert. The higher the temperature, the higher the maximum vapor pressure (in other words, there can be more water vapor in warm air than in cold air). When water molecules in the air are exerting the maximum possible vapor pressure at a given temperature, the air is "saturated" with water vapor—at this point, the rate of evaporation and the rate of condensation are the same. If this maximum vapor pressure is exceeded, more water vapor molecules will leave the air through condensation than are added to the air through evaporation—*net condensation* will take place until the rates of evaporation and condensation are again matched and the air again has its maximum vapor pressure. In practice, this means that evaporation tends to take place more rapidly when there is relatively little water vapor in the air, and that the rate of evaporation drops off as the air gets closer to saturation.

Windiness If the air overlying a water surface is almost saturated with water vapor, the rate of evaporation is about the same as the rate of condensation, and so very little further evaporation can take place. If the air remains calm and the temperature does not change, there is no net evaporation. If the air is in motion, however, through windiness and/or turbulence, the water vapor molecules in it are dispersed more widely. This dispersing of water vapor molecules originally in the air at the air–water interface means that air is now further from saturation and so the rate of evaporation increases.

To summarize, the rate of evaporation from a water surface is influenced by the temperature of the water and air, the amount of water vapor already in the air, and the degree of windiness. Higher temperatures, drier air, and greater windiness lead to greater net evaporation.

FOCUS

GOES Weather Satellites

The satellite images commonly seen on television weather reports come from a pair of satellites known as GOES, or Geostationary Operational Environmental Satellites. The GOES satellites are operated by the National Oceanic and Atmospheric Administration (NOAA) and have become essential tools for weather forecasting. The GOES satellites are geostationary, orbiting at a distance of 35,800 kilometers (22,300 miles) in fixed locations relative to the surface of Earth below. For example, GOES-East orbits above the equator in South America (75° W), where it can see the conterminous United States, as well as much of the North and South Atlantic Ocean. GOES-West orbits above the equator in the Pacific (135° W), where it can see most of the Pacific Ocean from Alaska to New Zealand. Similar weather satellites are maintained by other countries.

The primary instruments onboard the GOES satellites are the Sounder and the Imager. The sounder provides information about vertical temperature and moisture variations within the atmosphere, as well as data about ozone distribution. The imager has sensors to detect both radiant and reflected electromagnetic energy in several different bands of wavelengths. The intensity of radiation in each wavelength band is measured, recorded, and then transmitted down to the surface, where the images are processed. The GOES satellites send back several images each hour, allowing time-lapse satellite "movie" loops to be produced. Up-to-date GOES satellite images are now readily available to the public through many Internet sites, such as http://www.goes.noaa.gov.

Satellite images showing visible light as well those showing various bands of infrared wavelengths are useful in weather forecasting since each provides different kinds of information about Earth's surface and the atmosphere. Visible light satellite images show sunlight that has been reflected off the surface of Earth or by clouds in the atmosphere (Figure 6-A). The brightness of a surface depends both on its albedo (reflectance) and the angle of the light striking it. The brightest (high albedo) surfaces in visible light are typically the tops of clouds and snow or ice-covered surfaces. The darkest (low albedo) surfaces are typically land areas (especially unvegetated land surfaces), and the oceans—which are usually the darkest surfaces seen on visible light satellite images.

The infrared images of Earth and its atmosphere are produced by the GOES satellites day and night and are among the most widely used in meteorology. Infrared images show the longwave ("thermal infrared") radiation that has been emitted by the surface of Earth or by clouds in the atmosphere (Figure 6-B). Warm objects emit more longwave radiation than cold objects, and so infrared images show us, in effect, differences in temperature. Differences in infrared intensity are reproduced on black-and-white satellite images so that cooler surfaces (those emitting relatively little longwave radiation) are shown in white, while warmer surfaces are shown in black. It is also common to assign colors to different intensities of infrared radiation. In infrared images, the tops of high clouds are easy to distinguish from low clouds and fog. The tops of high clouds, such as massive cumulonimbus clouds, are much colder than low clouds and fog, and so will appear brighter (white) on infrared images. Low clouds and fog tend to have similar temperatures to that of the surface, and so will appear as nearly the same shade of gray as the surface below.

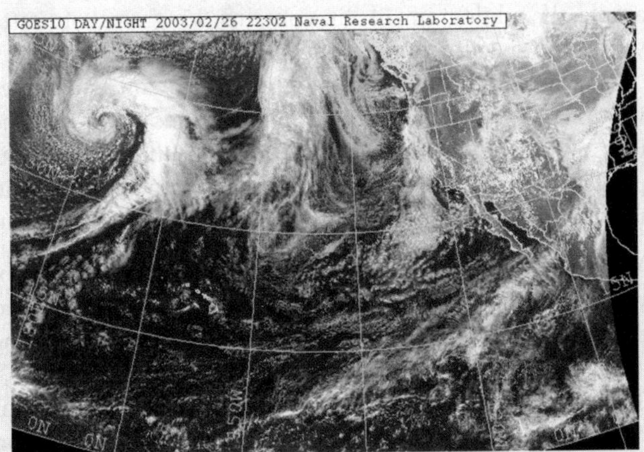

▲ **Figure 6-A** GOES-West visible light satellite image showing a midlatitude cyclone approaching the west coast of North America.

▲ **Figure 6-B** GOES-West infrared satellite image taken at the same time as Figure 6-A.

Evapotranspiration

Although most of the water that evaporates into the air comes from bodies of water, a relatively small amount comes from the land. This evaporation from land has two sources: (1) soil and other inanimate surfaces, and (2) plants. The amount of moisture that evaporates from soil is relatively minor, and thus most of the land-derived moisture present in the air comes from plants. The process whereby plants give up moisture through their leaves is called *transpiration*, and so the combined process of water vapor entering the air from land sources is called **evapotranspiration** (see Figure 6-1). Thus, the water vapor in the atmosphere was added through evaporation from bodies of water and evapotranspiration from land surfaces.

Whether a given land location is wet or dry depends on the rates of evapotranspiration and precipitation. To analyze these rates, we need to know about a concept called *potential evapotranspiration*. This is the amount of evapotranspiration that would occur if the ground at the location in question were sopping wet all the time. To determine a value for the potential evapotranspiration at any location, data on temperature, vegetation, and soil characteristics at that location are added to the actual evapotranspiration value in a formula that results in an estimate of the maximum evapotranspiration that could result under local environmental conditions if the moisture were available.

In locations where the precipitation rate exceeds the potential evapotranspiration rate, a water surplus accumulates in the ground. In many parts of the world, however, there is no groundwater surplus, except locally and/or temporarily, because the potential evapotranspiration rate is higher than the precipitation rate. Where potential evapotranspiration exceeds actual precipitation, there is no water available for storage in soil and in plants; dry soil and brown vegetation are the result.

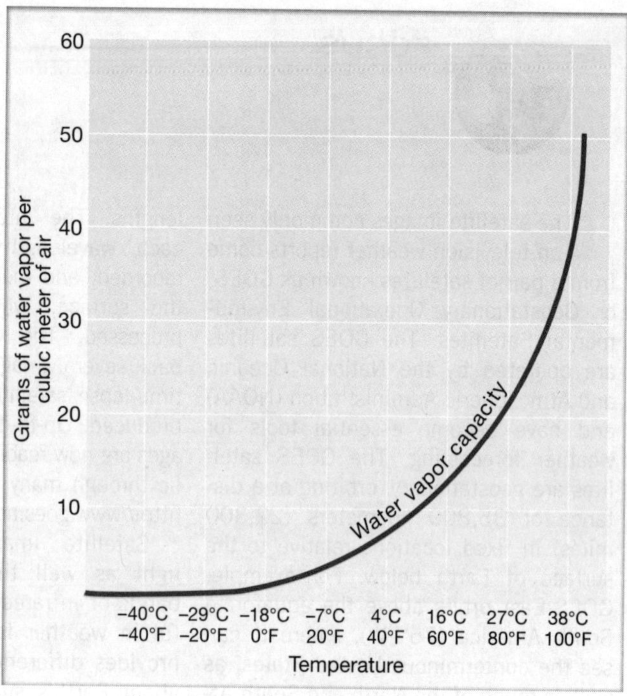

▲ **Figure 6-7** The maximum amount of water vapor that can be in the air (water vapor capacity) increases as temperature increases. This chart shows the saturation absolute humidity in g/m³.

Measures of Humidity

The amount of water vapor in the air is referred to as *humidity*. It can be measured and expressed in a number of ways, each useful for certain purposes.

Actual Water Vapor Content

The actual amount of water vapor in the air can be described in several different ways.

Absolute Humidity One direct measure of the water vapor content of air is **absolute humidity**—the mass of water vapor in a given volume of air. Absolute humidity is usually expressed in grams of water vapor per cubic meter of air (g/m^3; 1 gram is approximately 0.035 ounces, and 1 cubic meter is about 35 cubic feet). For example, if a cubic meter of air contains 12 grams of water vapor, the absolute humidity would be 12 g/m^3. The maximum possible absolute humidity (the water vapor "capacity") for a parcel of air is limited by the temperature: cold air has a small maximum absolute humidity while warm air has a great maximum absolute humidity (Figure 6-7).

If the volume of air changes (as happens when air expands or compresses as it moves vertically), the value of the absolute humidity also changes even though there is no change in the total amount of water vapor present. For this reason, absolute humidity is generally not used to describe moisture in air that is rising or descending.

Specific Humidity The mass of water vapor in a given mass of air is called the **specific humidity** and is usually expressed in grams of water vapor per kilogram of air (g/kg; for comparison one cubic meter of air at sea level has a mass of about 1.4 kg at room temperature). For example, if a kilogram of air contains 15 grams of water vapor, the specific humidity is 15 g/kg.

Specific humidity changes only as the quantity of water vapor varies; it is not affected by variations in air volume in the way that absolute humidity is. Specific humidity is particularly useful in studying the characteristics and movements of air masses (discussed in Chapter 7).

Vapor Pressure As we saw earlier, the contribution of water vapor to the total pressure of the atmosphere is called the vapor pressure. Vapor pressure can be expressed in the same way as total atmospheric pressure, in millibars (mb). The maximum possible vapor pressure (the water vapor "capacity") at a given temperature is called the **saturation vapor pressure**. Notice in Figure 6-8 that at a temperature of 10°C (50°F) the saturation vapor pressure is a little over 10 mb, while at 30°C (86°F) the saturation vapor pressure is about 40 mb—illustrating again that warm air has the potential to contain much more water vapor than cold air.

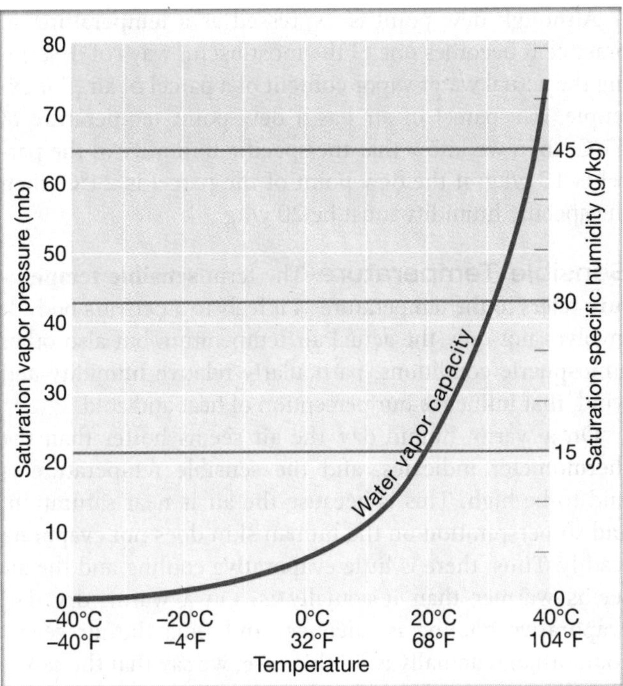

▲ **Figure 6-8** Saturation vapor pressure (in millibars) and saturation specific humidity (in g/kg). As temperature increases, the water vapor capacity of the air increases.

Absolute humidity, specific humidity, and vapor pressure are all ways of expressing the actual amount of water vapor in the air—and as such, are indications of the quantity of water that could be extracted by condensation and precipitation. However, before we discuss condensation and precipitation, we need to introduce the important concept of *relative humidity*.

Relative Humidity

The most familiar of humidity measures is **relative humidity**. Unlike absolute humidity, specific humidity, and vapor pressure, however, relative humidity is not a description of the actual water vapor content of the air. Rather, relative humidity describes how close the air is to saturation with water vapor.

Relative humidity is a ratio (expressed as a percentage) that compares the actual amount of water vapor in the air to the water vapor "capacity" of the air. **Capacity** is the maximum amount of water vapor that can be in the air at a given temperature.[5] As we have seen in Figure 6-7 and Figure 6-8, cold air has a low water vapor capacity, while warm air has a high water vapor capacity. (In popular terms, it is sometimes said that warm air can "hold" more water vapor than cold air, but this is somewhat misleading. The air does not actually hold water vapor as if it were in a sponge. Water vapor is simply one of the gaseous components of the atmosphere. The water vapor capacity of the air is determined by the temperature, which determines the rate of vaporization of water.)

Relative humidity is calculated with a simple formula:

$$\text{Relative humidity} = \frac{\text{Actual water vapor in air}}{\text{Capacity}} \times 100$$

For example, suppose that a kilogram of air contains 10 grams of water vapor (in other words, the specific humidity is 10 g/kg). If the temperature is 24°C (75.2°F), the capacity of the air is about 20 g/kg (see Figure 6-8; note that the g/kg scale is on the right side of chart), and so the relative humidity is

$$\frac{10 \text{ g}}{20 \text{ g}} \times 100 = 50\%$$

A relative humidity of 50 percent means that the air contains half of the maximum possible water vapor at that temperature. In other words, the air is 50 percent of the way to saturation.

Relative humidity changes if either the water vapor content or the water vapor capacity of the air changes. If, in our example above, 5 grams of water vapor are added to the air through evaporation while the temperature remains constant (so keeping the capacity unchanged), relative humidity will increase:

$$\frac{15 \text{ g}}{20 \text{ g}} \times 100 = 75\%$$

Conversely, if water vapor is removed from the air by condensation or dispersal, the relative humidity can decrease.

Relative humidity will also change when the temperature changes—even if the actual amount of water vapor in the air remains the same. Again, beginning with our initial example above, if the temperature increases from 24°C to 32°C (89.6°F), the water vapor capacity increases from 20 to 30 grams, and so relative humidity decreases:

$$\text{at } 24°\text{C } (75.2°\text{F}): \frac{10 \text{ g}}{20 \text{ g}} \times 100 = 50\%$$

$$\text{at } 32°\text{C } (89.6°\text{F}): \frac{10 \text{ g}}{30 \text{ g}} \times 100 = 33\%$$

On the other hand, if the temperature decreases from 24°C to 15°C (59.0°F), the capacity decreases from 20 to 10 grams, and so relative humidity increases:

$$\text{at } 24°\text{C } (75.2°\text{F}): \frac{10 \text{ g}}{20 \text{ g}} \times 100 = 50\%$$

$$\text{at } 20°\text{C } (68.0°\text{F}) \ \frac{10 \text{ g}}{15 \text{ g}} \times 100 = 67\%$$

$$\text{at } 15°\text{C } (59.0°\text{F}) \ \frac{10 \text{ g}}{10 \text{ g}} \times 100 = 100\%$$

Notice that the air can be brought to saturation (100 percent relative humidity) simply through a decrease in temperature—no water vapor has been added. As we will see, the most common way that air is brought to the point of saturation and condensation is through cooling.

[5]Depending on the measure of actual water vapor content being used, capacity is also called the *saturation absolute humidity*, *saturation specific humidity*, or *saturation vapor pressure*.

Temperature–Relative Humidity Relationship

The relationship between temperature and relative humidity is one of the most important in all of meteorology: as temperature increases, relative humidity decreases; as temperature decreases, relative humidity increases (at least until condensation begins). This inverse relationship is portrayed in Figure 6-9, which demonstrates the fluctuation in temperature and relative humidity during a typical day (assuming no variation in the amount of water vapor in the air). In the early morning, the temperature is low and the relative humidity is high—because the air's water vapor capacity is low. As the air warms up during the day, the relative humidity declines—because the warm air has a higher water vapor capacity than cool air. With the approach of evening, air temperature decreases, the air's water vapor capacity diminishes, and relative humidity increases.

See Appendix III for a description of how relative humidity can be determined using a simple instrument known as a psychrometer.

Related Humidity Concepts

Two other concepts related to relative humidity are useful in a study of physical geography: *dew point temperature* and *sensible temperature*.

Dew Point Temperature As we have seen, when air is cooled the water vapor capacity decreases and relative humidity increases. Cooling can bring formerly unsaturated air to the saturation point. The temperature at which saturation is reached is called the **dew point temperature**, or simply the **dew point**. The dew point temperature varies with the moisture content of the air. In our example above, we saw that air containing 10 grams of water vapor per kilogram of air reaches its dew point when chilled to about 15°C (59.0°F); air containing 20 grams of water vapor per kilogram of air reaches its dew point at about 24°C (75.2°F).

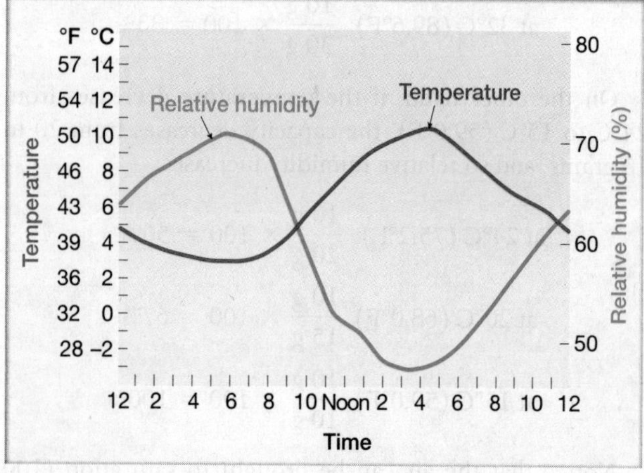

▲ **Figure 6-9** Typically there is an inverse relationship between temperature and relative humidity on any given day. As the temperature increases, the relative humidity decreases. Thus, relative humidity tends to be lowest in midafternoon and highest just before dawn.

Although dew point is expressed as a temperature, in practice it becomes one of the most useful ways of describing the actual water vapor content of a parcel of air. For example, if a parcel of air has a dew point temperature of 15°C, then we know that the specific humidity of the parcel is 10 g/kg; if the dew point of the parcel is 24°C, then the specific humidity must be 20 g/kg.

Sensible Temperature The term **sensible temperature** refers to the temperature as it feels to a person's body. It involves not only the actual air temperature but also other atmospheric conditions, particularly relative humidity and wind, that influence our perception of heat and cold.

On a warm, humid day, the air seems hotter than the thermometer indicates, and the sensible temperature is said to be high. This is because the air is near saturation, and so perspiration on the human skin does not evaporate readily. Thus, there is little evaporative cooling and the air seems warmer than it actually is. On a warm, dry day, evaporative cooling is effective, and thus the air seems cooler than it actually is; in this case, we say that the sensible temperature is low.

On a cold, humid day, the coldness seems more piercing because body heat is conducted away more rapidly in damp air; the sensible temperature is again described as low. On a cold, dry day, body heat is not conducted away as fast. The temperature seems warmer than it actually is, and we say that the sensible temperature is relatively high.

The amount of wind movement also affects sensible temperature, primarily by its influence on evaporation and the convecting away of body heat. This is especially true when the air temperature is below the freezing point of water—on windy cold days the wind may lower the apparent temperature significantly. See Appendix III for a description of the *heat index* and *wind chill*.

Condensation

Condensation is the opposite of evaporation. It is the process whereby water vapor is converted to liquid water. In other words, it is a change in state from gas to liquid. In order for condensation to take place, the air must be saturated. In theory, this saturated state can come about through the addition of water vapor to the air, but in practice it is usually the result of the air being cooled to a temperature below the dew point.

Saturation alone is not enough to cause condensation, however. Surface tension makes it virtually impossible to grow droplets of pure water. Because surface tension inhibits an increase in surface area, it makes it very difficult for additional water molecules to enter a droplet. (On the other hand, molecules can easily leave a small droplet by evaporation, thereby decreasing its area.) Thus, it is necessary to have a surface on which condensation can take place. If no such surface is available, no condensation occurs. In such a situation, the air becomes **supersaturated** (has a relative humidity greater than 100 percent) if cooling continues.

FOCUS

Water Vapor Satellite Images

Although water vapor in the atmosphere is invisible to the human eye, sensors onboard weather satellites can detect water vapor by measuring selected wavelengths of electromagnetic radiation. Satellite water vapor images show regions of dry air and moist air in the atmosphere—even if these areas are cloud free and so will not show clearly on conventional visible light or infrared weather satellite images (Figures 6-C, 6-D, and 6-E).

Recall from Chapter 4 that the "atmospheric window" is a portion of the longwave band of the electromagnetic spectrum (between about 8 and 12 micrometers [μm]) for which the atmosphere is nearly transparent. In other words, longwave radiation emitted by Earth's surface or from clouds with wavelengths between 8 and 12 μm is not readily absorbed by greenhouse gases such as water vapor. However, wavelengths of radiation outside the atmospheric window—specifically 6.7 μm and 7.3 μm— are prominently absorbed and then reradiated by water vapor.

By detecting the wavelengths of infrared radiation emitted by water vapor, scientists have a way to estimate the quantity of water vapor in the atmosphere. We infer that regions with high emission of infrared radiation at 6.7 μm and 7.3 μm contain relatively large amounts of water vapor, while regions with low emission of those wavelengths contain relatively small amounts of water vapor.

In gray-scale ("black and white") water vapor images, darker shades of gray show areas of relatively dry air, while lighter shades of gray show areas of relatively moist air. In color water vapor images, relatively dry air is often shown as dark blue, while relatively moist air is shown in shades of orange and red.

Because water vapor images detect infrared emission from water vapor in the middle and upper troposphere (above about 4500 meters [15,000 feet]), in addition to providing information about the moisture content of the atmosphere, these images also provide information about wind movement in the upper atmosphere, especially when time-sequence "movie loops" are used.

You can view satellite water vapor images on the Internet from many sites, including http://www.nrlmry.navy.mil/sat_products.html.

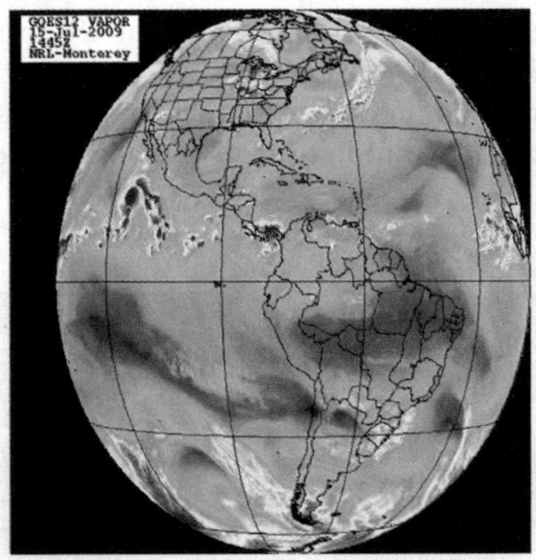

▲ **Figure 6-C** Water vapor satellite image of North and South America taken by NOAA's GOES-East satellite. Note the relatively dry areas (shown in dark blue) in the North Atlantic off of north Africa, over the Amazon Basin, and off the west coast of South America. The small patch of very moist air (shown in orange) just west of Mexico is Tropical Storm Dolores; a cold front in the Atlantic just east of the United States is visible as a narrow band of moist air.

▲ **Figure 6-D** Infrared satellite image taken at the same time as Figure 6-C.

▲ **Figure 6-E** Visible light satellite image taken at the same time as Figure 6-C.

Normally, plenty of surfaces are available for condensation. At ground level, availability of a surface is obviously no problem. In the air above the ground, there is also usually an abundance of "surfaces," as represented by tiny particles of dust, smoke, salt, pollen, bacteria, and other compounds. Most of these various particles are microscopic and therefore invisible to the naked eye (Figure 6-10). They are most concentrated over cities, seacoasts, and volcanoes, which are the source of much particulate matter but are present in lesser amounts throughout the troposphere. They are referred to as *hygroscopic particles* or **condensation nuclei**, and they serve as collection centers for water molecules during condensation.

As soon as the air temperature cools to the dew point, water vapor molecules begin to condense around condensation nuclei. The droplets grow rapidly as more and more water vapor molecules stick to them, and as they become larger, they bump into one another and coalesce as the colliding droplets stick together. Continued growth can make them large enough to be visible, forming haze or cloud particles. The diminutive size of these particles can be appreciated by realizing that a single raindrop may contain a million or more condensation nuclei plus all their associated moisture.

Clouds often may be composed of liquid water droplets even when their temperature is below freezing. Although water in large quantity freezes at 0°C (32°F), if it is dispersed as fine droplets, it can remain in liquid form at temperatures as cold as −40°C (−40°F). Water that persists in liquid form at temperatures below freezing is said to be "supercooled." **Supercooled water** droplets are important to condensation because they promote the growth of ice particles in cold clouds by freezing around them or by evaporating into vapor from which water molecules are readily added to the ice crystals.

Condensation nucleus
0.0002 millimeters
Large cloud droplet
0.05 millimeters
Typical cloud droplet
0.02 millimeters
Typical raindrop
2 millimeters

▲ **Figure 6-10** Comparative sizes of condensation nuclei and condensation particles. Condensation nuclei can be particles of dust, smoke, salt, pollen, bacteria, or any other microscopic matter found in the air.

Adiabatic Processes

Adiabatic Processes and Atmospheric Stability

One of the most significant facts in physical geography is that the only way in which large masses of air can be cooled to the dew point temperature is by expansion as the air masses rise. Thus, the only prominent mechanism for the development of clouds and the production of rain is *adiabatic cooling*. As we noted in Chapter 4, when air rises, its pressure decreases, and so it expands and cools adiabatically.

As a parcel of unsaturated air rises, it cools at the relatively steady rate of 10°C per 1000 meters (5.5°F per 1000 feet). This is known as the **dry adiabatic rate** (also called the *dry adiabatic lapse rate*). The term is a misnomer: the air is not necessarily "dry"; it is simply unsaturated. If the air mass rises high enough, it cools to the dew point temperature, condensation begins, and clouds form. The altitude at which this occurs is known as the **lifting condensation level** (LCL). Under normal circumstances, the LCL is clearly visible as the base of the clouds that form (Figure 6-11).

As soon as condensation begins, latent heat is released (this heat was stored originally as the latent heat of evaporation). If the air continues to rise, cooling due to expansion continues but release of the latent heat during condensation counteracts some of the adiabatic cooling and lessens the rate of cooling. This diminished rate of cooling is called the **saturated adiabatic rate** (also called the *saturated adiabatic lapse rate* or *moist adiabatic rate* (Figure 6-12) and depends on temperature and pressure but averages about 6°C per 1000 meters (3.3°F per 1000 feet).

Adiabatic warming occurs when air descends. Typically, descending air will warm at the dry adiabatic rate

▲ **Figure 6-11** The flat bottom of this cumulus cloud represents the lifting condensation level.

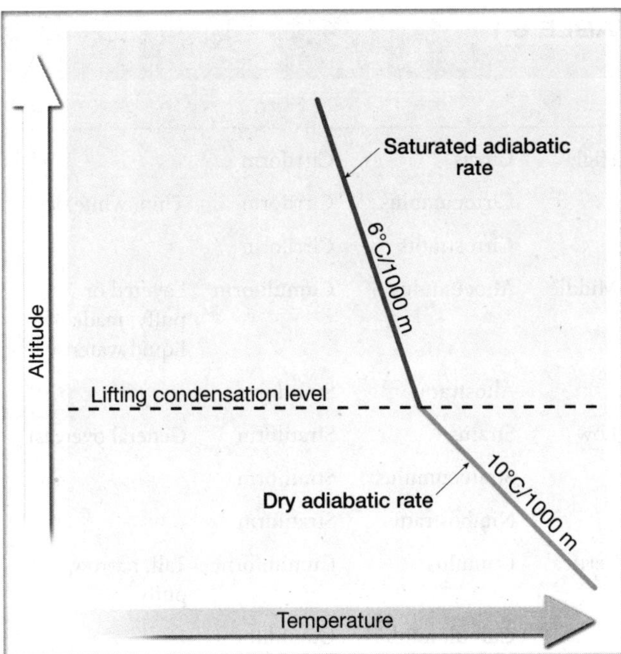

▲ **Figure 6-12** Unsaturated rising air cools at the dry adiabatic rate. Above the lifting condensation level, rising saturated air cools at the saturated adiabatic rate.

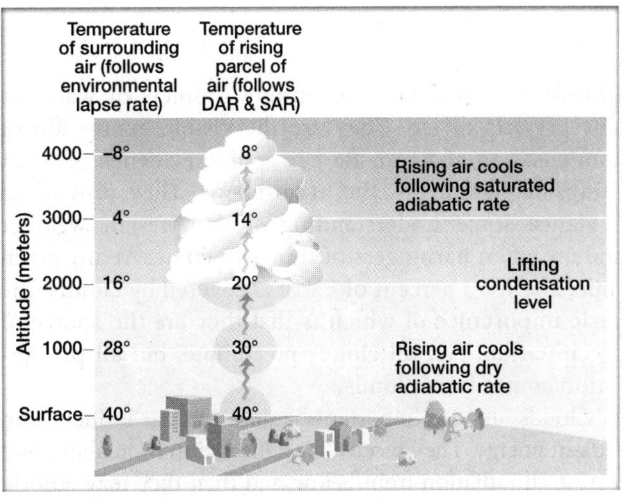

▲ **Figure 6-13** A hypothetical comparison of lapse rates. The column of temperatures on the left represents the vertical temperature gradient (the environmental lapse rate in °C) of the surrounding air through which a parcel of air is rising and cooling adiabatically. The rising parcel of air first cools at the dry adiabatic rate (DAR = 10°C per 1000 meters). As the parcel of air rises above the lifting condensation level, the release of latent heat lessens the rate of cooling and so the parcel cools at the saturated adiabatic rate (SAR = 6°C per 1000 meters).

of 10°C/1000 meters (5.5°C/1000 feet).[6] The increasing temperature of descending air increases the water vapor capacity of the air and thus causes saturated air to become unsaturated. In short, this is why descending air does not make clouds.

In any consideration of adiabatic temperature changes, remember that we are dealing with air that is rising or descending. The adiabatic rates are not to be confused with the *environmental lapse rate* or the *average lapse rate* discussed in Chapter 4, that describe the temperature of the still air at different altitudes in the atmosphere (Figure 6-13).

Some of the implications of adiabatic temperature changes can be seen in Figure 6-14, showing the temperature changes associated with a hypothetical parcel of air moving up and over a mountain range. The parcel is unsaturated when it begins to rise over the mountain, so it cools at the dry adiabatic rate. Once the lifting condensation level is reached, the air continues to rise and cool at the saturated adiabatic rate as condensation forms a cloud. Once the air reaches the summit, it begins to descend down the lee side of the mountain. The descending air warms at the dry adiabatic rate, so by the time the air has reached sea level again, it is significantly warmer and significantly drier—in both relative and absolute terms—than when it started out (in this simple example, we assume that moisture condensed out of the rising air is left as precipitation or clouds on the windward side of the mountain, and that no evaporation takes place as the air descends). We will see that this circumstance is one way in which deserts are formed.

[6]Although descending air typically warms following the dry adiabatic rate, there is a circumstance when this may not be the case. If air descends through a cloud, some water droplets may evaporate and the evaporative cooling will counteract some of the adiabatic warming. As a result, such descending air can warm at a rate close to the saturated adiabatic rate. As soon as evaporation of water droplets ceases, this descending air will warm following the dry adiabatic rate.

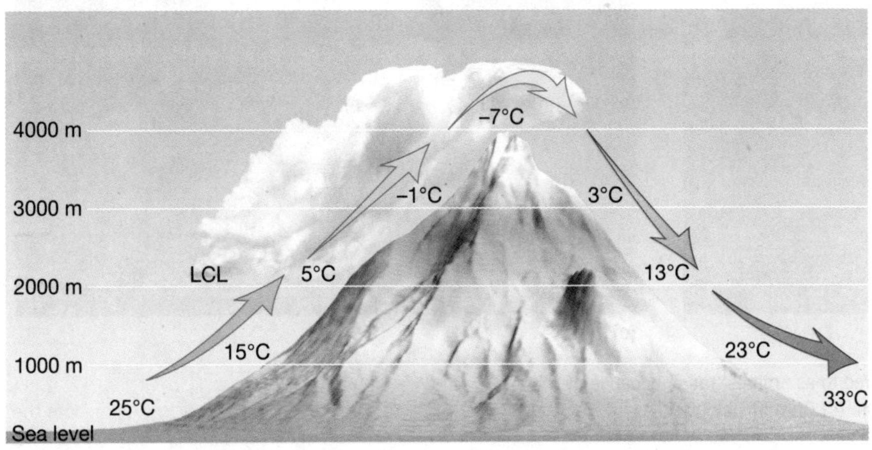

◀ **Figure 6-14** Temperature changes in a hypothetical parcel of air passing over a 4000 meter (13,100 foot) high mountain (assuming no evaporation as the air descends down the lee side of the mountain). The lifting condensation level of the parcel is 2000 meters, the dry adiabatic rate is 10°C/1000 m and the saturated adiabatic rate is 6°C/1000 m. Notice that because of the release of latent heat during condensation on the windward side of the mountain, by the time the air has descended back down to sea level on the leeward side, it is warmer than before it started up the windward side.

Clouds

Clouds are collections of minute droplets of water or tiny crystals of ice. They are the visible expression of condensation and provide perceptible evidence of other things happening in the atmosphere. They provide at a glance some understanding of the present weather and are often harbingers of things to come. At any given time, about 50 percent of Earth is covered by clouds, the basic importance of which is that they are the source of precipitation. Not all clouds precipitate, but all precipitation comes from clouds.

Clouds are also important because of their influence on radiant energy. They receive both insolation from above and terrestrial radiation from below, and then they may absorb, reflect, scatter, or reradiate this energy. Thus understanding the function of clouds in the global energy budget is important, and clouds must be taken into account when trying to anticipate the causes or consequences of climate change.

Classifying Clouds

Although clouds occur in an almost infinite variety of shapes and sizes, certain general forms recur commonly. Moreover, the various cloud forms are normally found only at certain generalized altitudes, and it is on the basis of these two factors—form and altitude—that clouds are classified (Table 6-1).

TABLE 6-1 The International Classification Scheme for Clouds

Family	Type	Form	Characteristics
High	Cirrus	Cirriform	
	Cirrocumulus	Cirriform	Thin, white, icy
	Cirrostratus	Cirriform	
Middle	Altocumulus	Cumuliform	Layered or puffy; made of liquid water
	Altostratus	Stratiform	
Low	Stratus	Stratiform	General overcast
	Stratocumulus	Stratiform	
	Nimbostratus	Stratiform	
Vertical	Cumulus	Cumuliform	Tall, narrow, puffy
	Cumulonimbus	Cumuliform	

Cloud Form The international classification scheme for clouds recognizes three forms (Figure 6-15).

1. *Cirriform* clouds (Latin *cirrus*, "a lock of hair") are thin and wispy and composed of ice crystals rather than water droplets.
2. *Stratiform* clouds (Latin *stratus*, "spread out") appear as grayish sheets that cover most or all of the sky, rarely being broken up into individual cloud units.

(a)

(b)

(c)

▲ **Figure 6-15** (a) The classic cirrus clouds referred to as "mares' tails" because of their shape. (b) Low stratus clouds over the Arctic Ocean. (c) Cumulus clouds.

3. *Cumuliform* clouds (Latin *cumulus*, "mass" or "pile") are massive and rounded, usually with a flat base and limited horizontal extent but often billowing upward to great heights.

These three cloud forms are subclassified into 10 types based on shape (Figure 6-16). The types overlap, and cloud development frequently is in a state of change, so that one type may evolve into another. Three of the 10 types are purely of one form, and these are called **cirrus clouds**, **stratus clouds**, and **cumulus clouds**. The other seven types may be combinations of these three. Cirrocumulus clouds, for example, have the wispiness of cirrus clouds and the puffiness of cumulus clouds.

Precipitation comes from clouds that have "nimb" in their name, specifically *nimbostratus* or *cumulonimbus*. Normally these types develop from other types; that is, cumulonimbus clouds develop from cumulus clouds, and nimbostratus clouds develop from stratus clouds.

Cloud Families As the final detail of the international classification scheme, the 10 cloud types are divided into four families on the basis of altitude.

1. *High clouds* are generally found above 6 kilometers (20,000 feet). Because of the small amount of water vapor and low temperature at such altitudes, these clouds are thin, white, and composed of ice crystals.

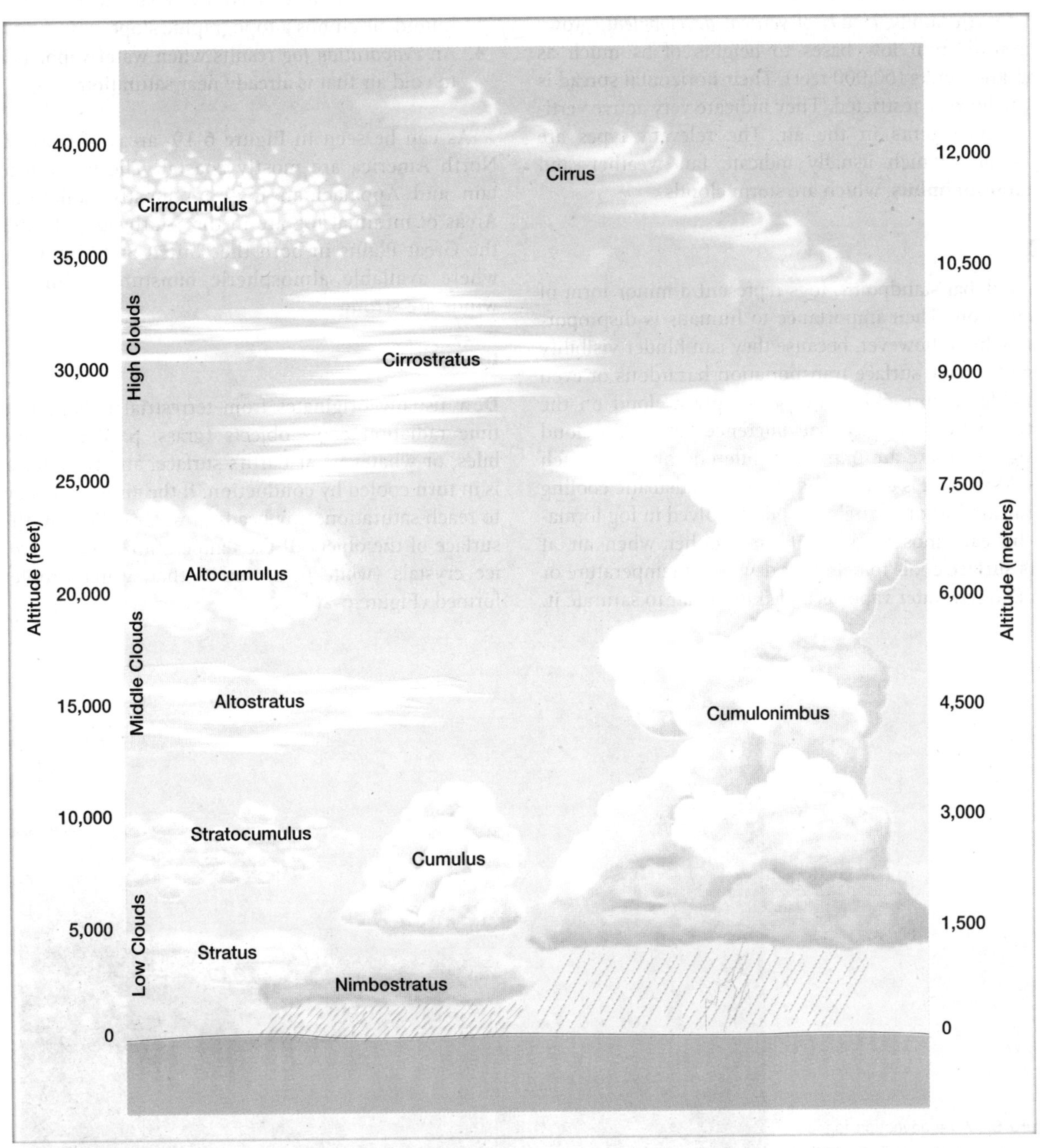

▲ **Figure 6-16** Typical shapes and altitudes of the 10 principal cloud types.

Included in this family are *cirrus*, *cirrocumulus*, and *cirrostratus*. These high clouds often are harbingers of an approaching weather system or storm.

2. *Middle clouds* normally occur between about 2 and 6 kilometers (6500 and 20,000 feet). They may be either stratiform or cumuliform and are composed of liquid water. Included types are altocumulus and altostratus. The puffy *altocumulus* clouds usually indicate settled weather conditions, whereas the lengthy *altostratus* are often associated with changing weather.
3. *Low clouds* usually are below 2 kilometers (6500 feet). They sometimes occur as individual clouds but more often appear as a general overcast. Low cloud types include *stratus*, *stratocumulus*, and *nimbostratus*. These low clouds often are widespread and are associated with somber skies and drizzly rain.
4. A fourth family, *clouds of vertical development*, grows upward from low bases to heights of as much as 15 kilometers (60,000 feet). Their horizontal spread is usually very restricted. They indicate very active vertical movements in the air. The relevant types are *cumulus*, which usually indicate fair weather, and **cumulonimbus**, which are storm clouds.

Fog

From a global standpoint, fogs represent a minor form of condensation. Their importance to humans is disproportionately high, however, because they can hinder visibility enough to make surface transportation hazardous or even impossible (Figure 6-17). **Fog** is simply a cloud on the ground. There is no physical difference between a cloud and fog, but there are important differences in how each forms. Most clouds develop as a result of adiabatic cooling in rising air, but only rarely is uplift involved in fog formation. Instead, most fogs are formed either when air at Earth's surface cools to below its dew point temperature or when enough water vapor is added to the air to saturate it.

There are four generally recognized types of fog (Figure 6-18):

1. A *radiation fog* results when the ground loses heat through radiation, usually at night. The heat radiated away from the ground passes through the lowest layer of air and into higher areas. The air closest to the ground cools as heat flows conductively from it to the relatively cool ground, and fog condenses in the cooled air at the dew point.
2. An *advection fog* develops when warm, moist air moves horizontally over a cold surface, such as snow-covered ground or a cold ocean current. Air moving from sea to land is the most common source of advection fogs.
3. An *upslope fog*, or *orographic fog* (from the Greek *oro*, "mountain"), is created by adiabatic cooling when humid air climbs a topographic slope.
4. An *evaporation fog* results when water vapor is added to cold air that is already near saturation.

As can be seen in Figure 6-19, areas of heavy fog in North America are mostly coastal. The western-mountain and Appalachian fogs are mostly radiation fogs. Areas of minimal fog are in the Southwest, Mexico, and the Great Plains in both the United States and Canada, where available atmospheric moisture is limited and winds are strong.

Dew

Dew usually originates from terrestrial radiation. Nighttime radiation cools objects (grass, pavement, automobiles, or whatever) at Earth's surface, and the adjacent air is in turn cooled by conduction. If the air is cooled enough to reach saturation, tiny beads of water collect on the cold surface of the object. If the temperature is below freezing, ice crystals (*white frost*) rather than water droplets are formed (Figure 6-20).

▶ **Figure 6-17** An advection fog enveloping the harbor at Ketchikan, Alaska.

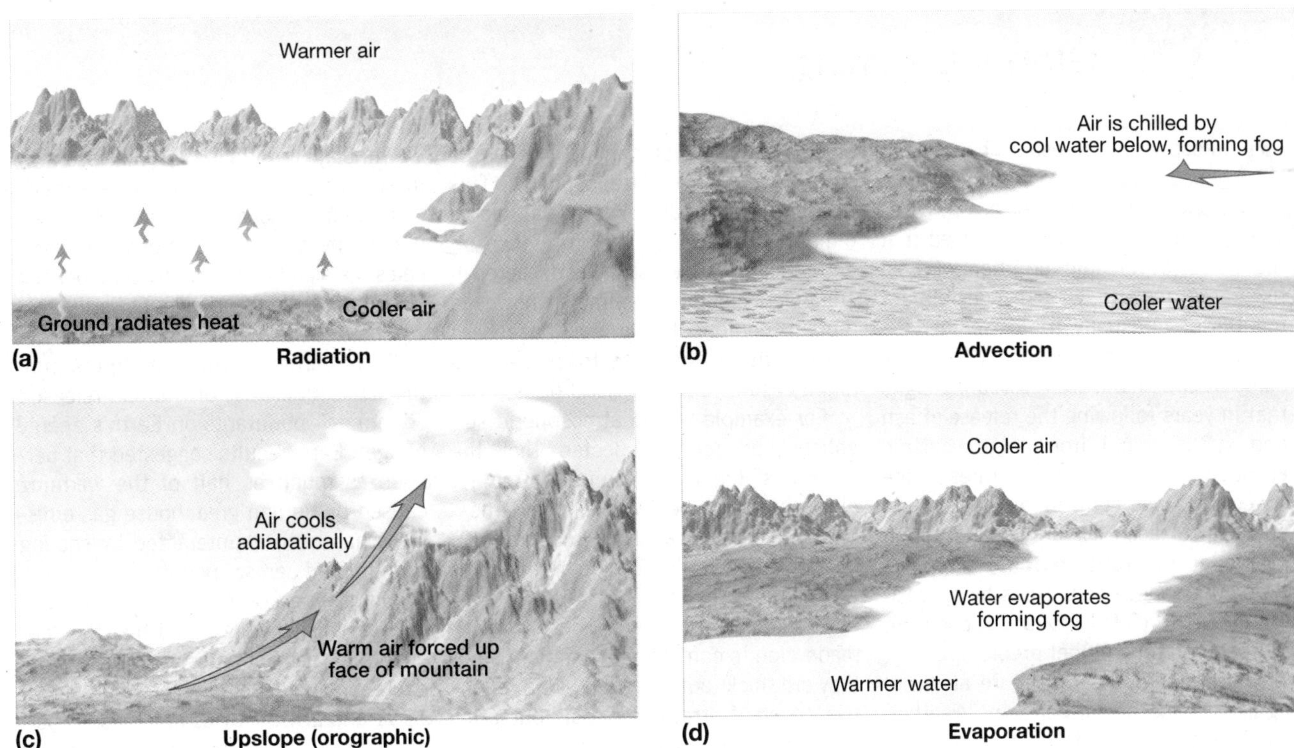

▲ **Figure 6-18** The four principal types of fog: (a) radiation, (b) advection, (c) upslope (orographic), (d) evaporation.

The Buoyancy of Air

Because most condensation and precipitation are the result of rising air, conditions that promote or hinder upward movements in the troposphere are obviously of great importance to weather and climate. Depending on how buoyant it is, air rises more freely and extensively under some circumstances than under others. Therefore, the concept of air buoyancy is one of the most significant in physical geography.

Atmospheric Stability

The tendency of any object to rise in a fluid is called the *buoyancy* of that object. For the present discussion, you should picture a given parcel of air (one having imaginary boundaries) as being an "object" and the surrounding air as being the fluid. As with other gases (and liquids, too), an air parcel tends to seek its equilibrium level. This means that a parcel of air moves vertically until it reaches an altitude at which the surrounding air is of equal density.

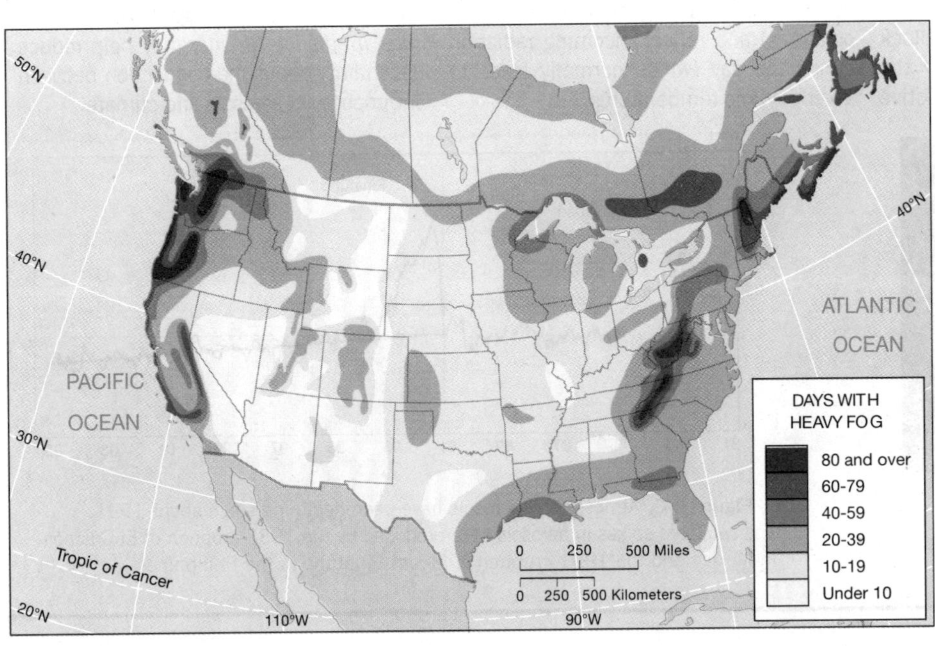

◀ **Figure 6-19** Distribution of fog in North America.

Global Dimming

Atmospheric aerosols have long been recognized to be an influence on weather and climate. As early as 1783, Benjamin Franklin proposed that the massive volcanic eruption of Laki fissure in Iceland was responsible for months of unusually frigid weather experienced in Europe. By the 1980s, ice cores in Greenland were revealing signs that in years following the release of ash and sulfuric acid from large volcanic eruptions, global temperatures were lower. This was confirmed when sulfate aerosols released by the 1991 eruption of Mount Pinatubo in the Philippines lowered global temperatures by approximately 0.5°C (0.9°F) for about one year, closely matching model predictions.

Anthropogenic aerosols are also recognized to be an influence on weather. By the 1950s, clouds were being artificially "seeded" with silver iodide smoke to increase condensation and precipitation. And in the late 1980s, satellite images were showing clouds persisting over oceanic shipping lanes: the sulfate aerosols from ship smokestacks were in effect "seeding" clouds.

Project INDOEX, a multinational study led by atmospheric scientist Veerabhadran Ramanathan in the 1990s, showed that the plume of sulfates, smoke particles, and other anthropogenic aerosols blowing over the Indian Ocean was blocking sunlight and promoting cloud formation, effectively reducing the amount of sunlight reaching the surface by 10 percent.

In short, it appears that human-released aerosols may directly block sunlight, absorb solar energy high in the atmosphere, and create more reflective clouds—all of which can act to cool the surface and lower troposphere. This phenomenon became known as "global dimming." Gauging the influence of anthropogenic aerosols and aerosol-enhanced clouds on climate, however, remains complicated, in part because it is difficult to eliminate these aerosols and measure "control" conditions.

For example, some atmospheric scientists had long suspected that the "contrails" (condensation trails) from jet airliners might locally alter the atmospheric energy budget slightly (Figure 6-F). Condensation trails develop when water vapor in jet exhaust cools in the cold air of the upper troposphere. These human-made clouds can reflect radiation when they are thick (acting much like a reflective cover of stratus clouds), but after some time they may turn into a wispier cover of cirrus-like clouds that absorb outgoing longwave radiation and so act as "warming" clouds.

Estimating the specific effects of contrails proved difficult since there are few opportunities to study the atmosphere when contrails are absent from the sky—until September 11, 2001. The grounding of all commercial airline traffic over the United States in the three days following 9/11 showed that in the absence of contrails, the daily temperature range across the country increased sharply by more than 1°C (2°F)—the days were hotter and the nights were colder without contrails to absorb longwave radiation at night (which normally leads to higher nighttime temperatures) and reflect incoming radiation during the day (which normally leads to lower daytime temperatures).

Although contrails are not seen by most atmospheric scientists to be a major contributor to global dimming, their measurable effect on local temperatures indicates the significance of human-produced clouds to the energy budget of Earth.

The Aqua satellite launched by NASA in 2002 carries instruments to measure the consequences of human-released "dimming" pollutants on Earth's energy budget. Early results suggested that perhaps as much as half of the warming caused by human greenhouse gas emissions has been counteracted by cooling from human aerosol pollution.

However, a 2007 study led by NASA's Michael Mishchenko found that the "optical thickness" of aerosols in the atmosphere declined steadily between about 1991 and 2005—by as much as 20 percent from previous decades—largely because of reduced emissions from industrialized countries (Figure 6-G). That decrease in aerosols may explain why, after decades of decline, the measured amount of sunlight striking Earth's surface began to slowly increase after about 1990. This raises a paradoxical concern for atmospheric scientists: if human-produced global dimming has been masking some of the effects of global warming, if we continue to reduce "dimming pollutants" we will face higher temperatures from global warming.

NASA plans to launch its Glory satellite in 2010. Designed to monitor the effects of atmospheric aerosols more precisely than before, the data gathered from its *Total Irradiance Monitor* may help reduce uncertainty about the connection between anthropogenic aerosols and climate.

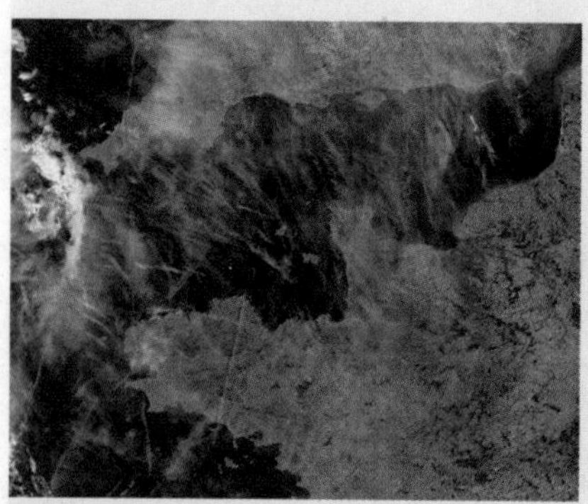

▲ **Figure 6-F** Airplane contrails over the English Channel are shown in this MODIS satellite image.

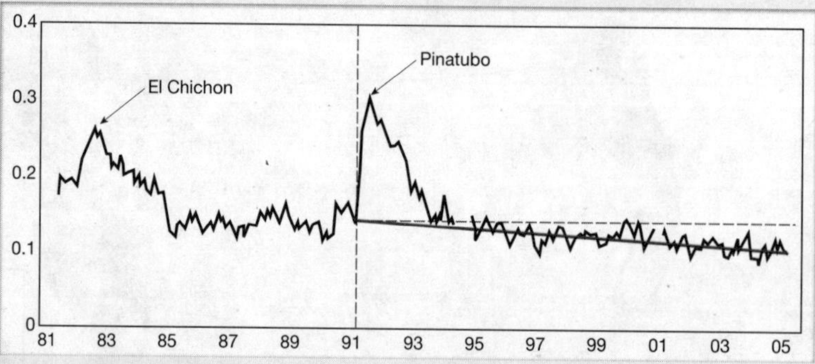

▲ **Figure 6-G** Atmospheric aerosols have been declining since about 1991. The two large spikes in aerosols were produced by the 1982 eruption of El Chichón in Mexico and the 1991 eruption of Mount Pinatubo in the Philippines.

(a)

(b)

▲ **Figure 6-20** (a) Dewdrops on a wild daisy. (b) White frost emphasizes the delicate lacework of a spider web in Yellowstone Park's Hayden Valley.

Said another way, if a parcel of air is warmer, and thus less dense, than the surrounding air, it tends to rise. If a parcel is cooler, and therefore denser, than the surrounding air, it tends either to sink or at least resist uplift. Thus, we say that warm air is more buoyant than cool air.[7]

If a parcel of air resists uplift it is said to be **stable** (Figure 6-21). If stable air is forced to rise, perhaps by coming up against a mountain slope, it does so only as long as the force is applied. Once the force is removed, the air sinks back to its former position. In other words, stable air is nonbuoyant. When unstable air comes up against the same mountain, it continues to rise once it has passed the peak.

In the atmosphere, high stability is promoted when cold air is beneath warm air, a condition most frequently observed during a temperature inversion. With colder, denser air below warmer, lighter air, upward movement is unlikely. A cold winter night is typically a highly stable situation, although high stability can also occur in the daytime.

[7]Water vapor content may also slightly influence air buoyancy. Because water vapor molecules have a lower molecular weight than nitrogen molecules (N_2) or oxygen molecules (O_2), when air has a high water vapor content, heavier N_2 and O_2 molecules have been displaced by lighter H_2O molecules and so such moist air is just slightly "lighter" than dry air. Overall, however, air temperature is by far the most important determinant of air buoyancy.

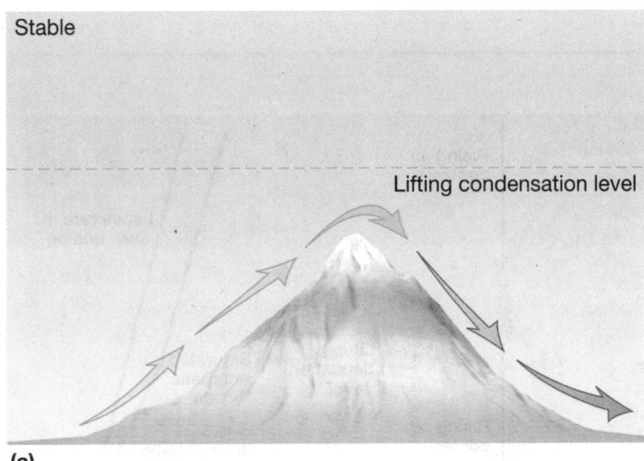

(a)

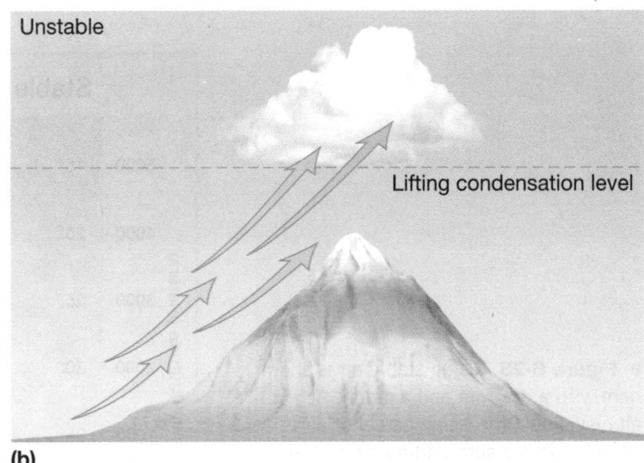

(b)

▲ **Figure 6-21** (a) As stable air blows over a mountain, it rises only as long as it is forced to do so by the mountain slope. On the leeward side, it moves downslope. (b) When unstable air is forced up a mountain slope, it is likely to continue rising of its own accord until it reaches surrounding air of similar temperature and density; if it rises to the lifting condensation level, clouds form.

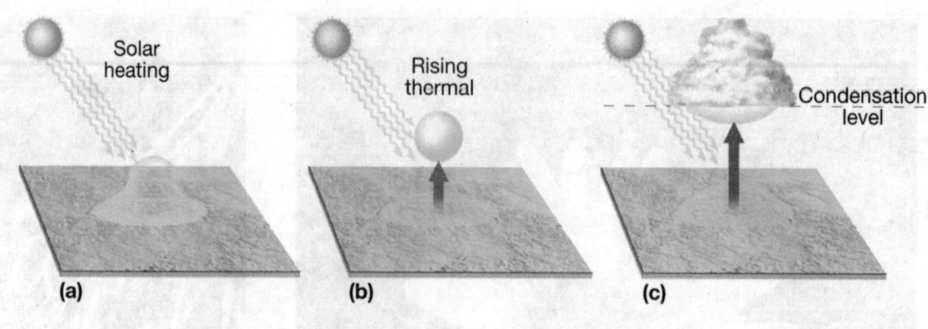

▶ **Figure 6-22** (a) Localized heating of Earth's surface warms a parcel of air. (b) Because this parcel of air is warmer than the surrounding air, it is unstable and so begins to rise. (c) If this "thermal" rises high enough it may reach the lifting condensation level, forming a cloud.

Because it does not rise, and therefore stays at an essentially constant pressure and volume, highly stable air obviously provides little opportunity for adiabatic cooling unless there is some sort of forced uplift. Highly stable air is normally not associated with cloud formation and precipitation.

Air is said to be **unstable** if it either rises without any external force other than the buoyant force or continues to rise after such an external force has ceased to function. In other words, unstable air is buoyant. When a mass of air is heated enough so that it is warmer than the surrounding air, it becomes unstable. This is a typical condition on a warm summer afternoon (Figure 6-22). The unstable air rises until it reaches an altitude where the surrounding air has similar temperature and density, which is referred to as the *equilibrium level*. While ascending, it will be cooled adiabatically. In this situation, clouds are likely to form.

There is an intermediate condition, called *conditional instability*, between absolute stability and absolute instability. Near the surface, such a parcel of air is the same temperature or cooler than the surrounding air and so acts like stable air. If forced to rise above the lifting condensation level, however, the release of latent heat during condensation may provide enough buoyancy to make the parcel unstable. It then behaves like unstable air, rising until it reaches an altitude where the surrounding air has density and temperature similar to its own.

Determination of Stability via Temperature and Lapse Rate An accurate determination of the stability of any mass of air depends on temperature measurements. The temperature of the rising air can be compared with the temperature of surrounding nonrising air by a series of thermometer readings at different elevations. The rising air cools (at least initially) at the dry adiabatic rate of 10°C per 1000 meters (5.5°F per 1000 feet). The environmental lapse rate of the surrounding (nonrising) air depends on many things and may be different from the dry adiabatic rate.

For example, if the environmental lapse rate of the surrounding air is less than the dry adiabatic rate of the rising air, as in Figure 6-23, at every elevation the rising air is cooler than the surrounding air and therefore stable. Under such conditions, the air rises only when forced to do so. Once the lifting force is removed, the air ceases to rise and will actually sink.

On the other hand, if the environmental lapse rate of the surrounding air is greater than the dry adiabatic rate of the rising air, at every elevation the rising air is

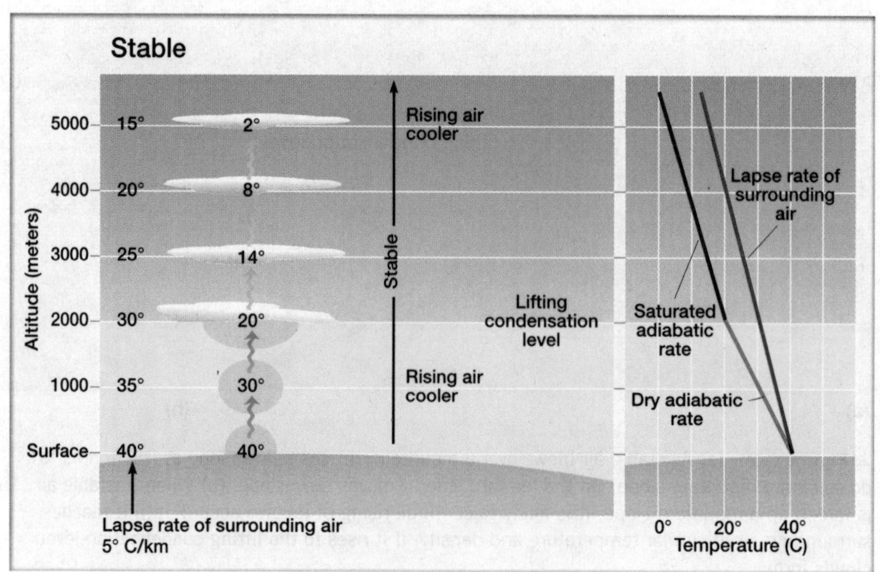

▶ **Figure 6-23** Rising stable air is shown here with a diagram and a graph. At all elevations, the rising parcel of air is cooler than the surrounding air, so the parcel is stable and will rise only if forced. (Dry adiabatic rate = 10°C/1000 m; saturated adiabatic rate = 6°C/1000 m; lifting condensation level = 2000 m; lapse rate of surrounding air [environmental lapse rate] = 5°C/1000 m.)

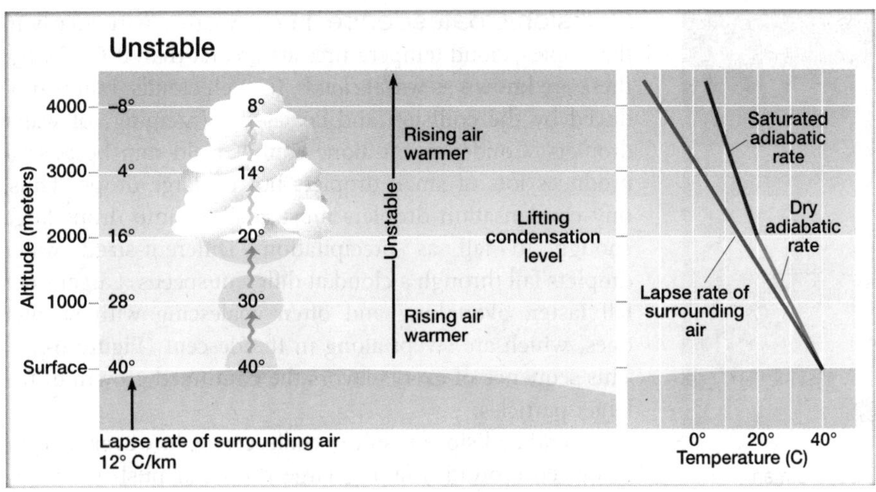

◀ **Figure 6-24** Rising unstable air. At all elevations, the rising parcel of air is warmer than the surrounding air, so the parcel is unstable and will rise because of its buoyancy. (Dry adiabatic rate = 10°C/1000 m; saturated adiabatic rate = 6°C/1000 m; lifting condensation level = 2000 m; lapse rate of surrounding air = 12°C/1000 m.)

warmer than the surrounding air and so it is unstable. The unstable air rises until it reaches an elevation where the surrounding air is of similar temperature and density (Figure 6-24).

The situation may become more complicated after rising stable air is cooled to its dew point temperature at the lifting condensation level. Once condensation begins, latent heat is released. This situation increases the tendency toward instability and reinforces the rising trend (Figure 6-25).

Visual Determination of Stability The cloud pattern in the sky is often indicative of air stability. Unstable air is associated with distinct updrafts, which are likely to produce vertical clouds (Figure 6-26). Thus, the presence of cumulus clouds suggests instability, and a towering cumulonimbus cloud is an indicator of pronounced instability. Horizontally developed clouds, most notably stratiform, are characteristic of stable air that has been forced to rise, and a cloudless sky may be an indicator of stable air that is immobile.

Regardless of stability conditions, no clouds form unless the air is cooled to the dew point temperature—unstable rising air will not produce any clouds if the dew point of the rising air is not reached. Thus the mere absence of clouds is not certain evidence of stability; it is only an indication. The general features of stable and unstable air are summarized in Table 6-2.

Precipitation

All **precipitation** originates in clouds, but most clouds do not yield precipitation. Exhaustive experiments have demonstrated that condensation alone is insufficient to produce raindrops. The tiny water droplets that make up clouds cannot fall to the ground as rain because their size makes them

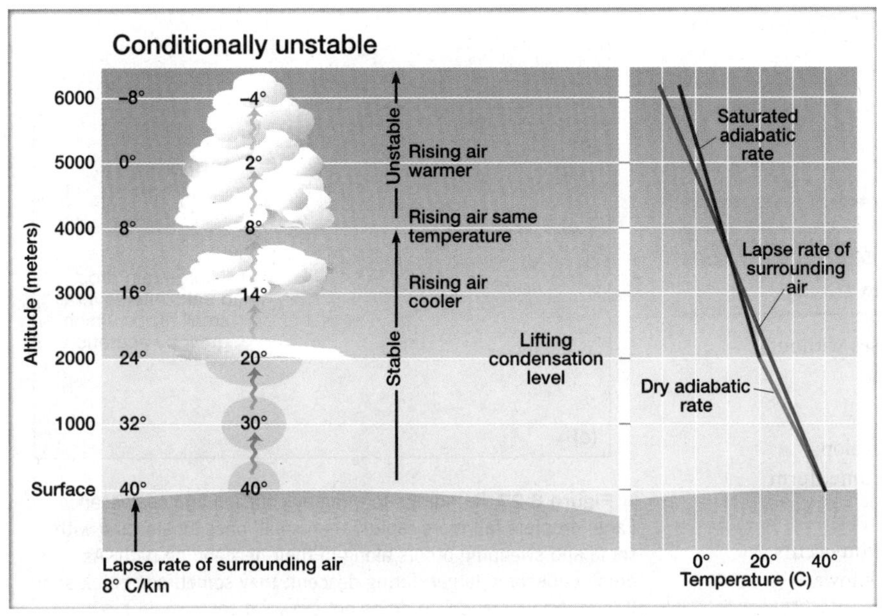

◀ **Figure 6-25** Conditionally unstable air. In this case, the rising parcel of air is cooler than the surrounding air and so is stable up to an elevation of 4000 meters. However, above that elevation, the release of latent heat during condensation warms the rising air enough to make it unstable—thus the rising air is conditionally unstable. (Dry adiabatic rate = 10°C/1000 m; saturated adiabatic rate = 6°C/1000 m; lifting condensation level = 2000 m; lapse rate of surrounding air = 8°C/1000 m.)

▲ **Figure 6-26** A large cumulonimbus cloud as seen from an airplane cockpit window. The anvil-shaped top is characteristic of cumulonimbus clouds.

very buoyant and the normal turbulence of the atmosphere keeps them aloft. Even in still air, their fall would be so slow that it would take many days for them to reach the ground from even a low cloud. Besides that, most droplets would evaporate in the drier air below the cloud before they made a good start downward.

Despite these difficulties, rain and other forms of precipitation are commonplace in the troposphere. What is it, then, that produces precipitation in its various forms?

The Processes

An average-sized raindrop contains several million times as much water as the average-sized water droplet found in any cloud. Consequently, great multitudes of droplets must join together to form a drop large enough to overcome both turbulence and evaporation and thus be able to fall to Earth under the influence of gravity.

Two mechanisms are believed to be principally responsible for producing precipitation particles: (1) collision and coalescence of water droplets, and (2) ice-crystal formation.

TABLE 6-2 Characteristics of Stable and Unstable Air

Stable Air	Unstable Air
Nonbuoyant; remains immobile unless forced to rise	Buoyant; rises without outside force
If clouds develop, tend to be stratiform or cirriform	If clouds develop, tend to be cumuliform
If precipitation occurs, tends to be drizzly	If precipitation occurs, tends to be showery

Collision/Coalescence In many cases, particularly in the tropics, cloud temperatures are greater than 0°C (32°F); these are known as *warm clouds*. In such clouds, rain is produced by the collision and coalescing (merging) of water droplets. Condensation alone cannot yield rain because it produces lots of small droplets but no large drops. Thus, tiny condensation droplets must coalesce into drops large enough to fall as precipitation. Different-sized water droplets fall through a cloud at different speeds. Larger ones fall faster, overtaking and often coalescing with smaller ones, which are swept along in the descent (Figure 6-27). This sequence of events favors the continued growth of the larger particles.

Not all collisions result in coalescence, however. The air displaced around a falling large drop can push very tiny

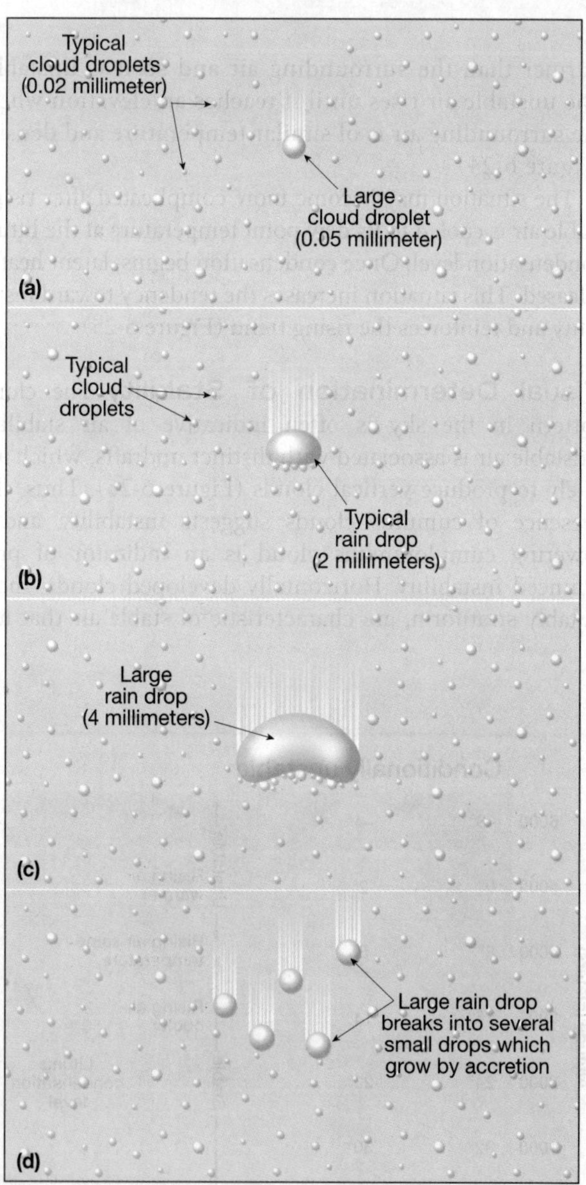

▲ **Figure 6-27** Raindrops forming by collision and coalescence. Large droplets fall more rapidly than small ones, coalescing with some and sweeping others along in their descending path. As droplets become larger during descent, they sometimes break apart. (Cloud droplets and rain drops not drawn at same scale.)

droplets out of its path. In addition, there is evidence to suggest that differences in the electrical charges of droplets may influence coalescence as well.

Collision/coalescence is the process most responsible for precipitation in the tropics, and it also produces some of the precipitation in the middle latitudes.

Ice-Crystal Formation Many clouds or portions of clouds extend high enough to have temperatures well below the freezing point of liquid water (these are known as *cold* or *cool clouds*). In this situation, ice crystals and supercooled water droplets often coexist in the cloud. These two types of particles are in direct "competition" for the water vapor that is not yet condensed. There is lower saturation vapor pressure around the ice crystals than around liquid water droplets—this means that if the air around a liquid water droplet is saturated (100 percent relative humidity), that same air is supersaturated around an ice crystal. Thus, the ice crystals attract most of the water vapor, and the liquid water droplets, in turn, evaporate to replenish the diminishing supply of vapor, as shown in Figure 6-28. Therefore, the ice crystals grow at the expense of the water droplets until the crystals are large enough to fall. As they descend through the lower, warmer portions of the cloud, they pick up more moisture and become still larger. They may then either precipitate from the cloud as snowflakes or melt and precipitate as raindrops.

Precipitation by ice-crystal formation was first proposed by the Swedish meteorologist Tor Bergeron more than half a century ago. It is now known as the *Bergeron process* and is believed to account for the majority of precipitation outside of tropical regions.

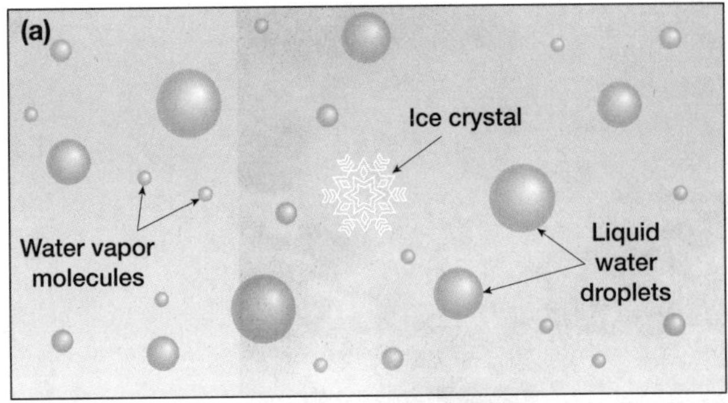

▲ **Figure 6-28** Precipitation by means of ice-crystal formation in clouds (the Bergeron process). Ice crystals grow by attracting water vapor to themselves, causing the liquid water droplets that make up the cloud to evaporate to replenish the water vapor supply. The process of growing ice crystals and shrinking cloud droplets may continue until the ice crystals are large and heavy enough to fall. (Particle sizes are greatly exaggerated.)

Forms of Precipitation

Several forms of precipitation can result from the processes just described depending on air temperature and turbulence.

Rain By far the most common and widespread form of precipitation is **rain**, which consists of drops of liquid water. Most rain is the result of condensation and precipitation in ascending air that has a temperature above freezing, but some results from the melting of ice crystals as they descend through warmer air (Figure 6-29).

Meteorologists often make a distinction among "rain," which goes on for a relatively long time; *showers*, which are relatively brief and involve large drops; and *drizzle*, which consists of very small drops and usually lasts for some time.

Snow The general name given to solid precipitation in the form of ice crystals, small pellets, or flakes is **snow**. It is formed when water vapor is converted directly to ice through sublimation without an intermediate liquid stage. However, the water vapor may have evaporated from supercooled liquid cloud droplets inside cold clouds.

Sleet In the United States, *sleet* refers to small raindrops that freeze during descent and reach the ground as small pellets of ice. In other countries, the term is often applied to a mixture of rain and snow.

Glaze *Glaze* (or *freezing rain*) is rain that turns to ice the instant it collides with a solid object. Raindrops fall through a shallow layer of subfreezing air near the ground. Although the drops do not freeze in the air (in other words, they do not turn to sleet), they become supercooled while in this cold layer and are instantly converted to an icy surface when they land. The result can be a thick coating of ice that makes both pedestrian and vehicular travel hazardous as well as breaks tree limbs and transmission lines.

Hail The precipitation form with the most complex origin is **hail**, which consists of either small pellets or larger lumps of ice (Figure 6-30). Hailstones are usually composed of roughly concentric layers of clear and cloudy ice. The cloudy portions contain numerous tiny air bubbles among small crystals of ice, whereas the clear parts are made up of large ice crystals.

▲ **Figure 6-29** A towering cumulonimbus cloud produces a small but intense thunderstorm over the desert of southern Arizona near Tombstone.

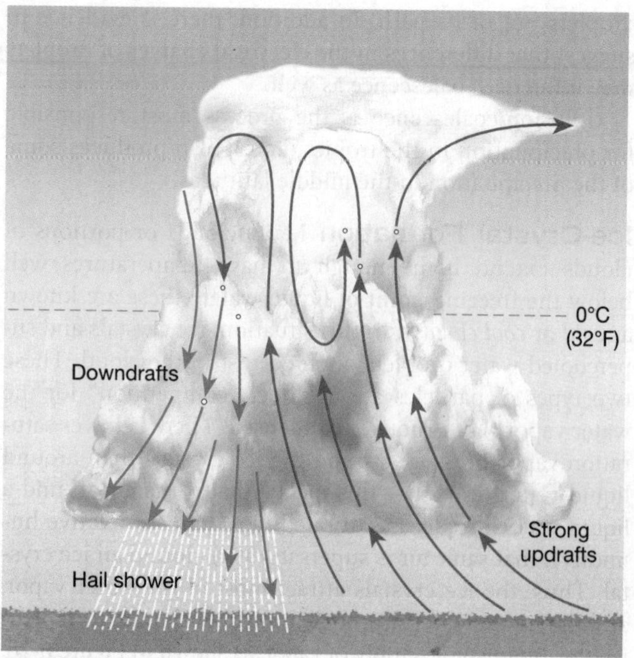

▲ **Figure 6-31** Hail is produced in cumulonimbus clouds with strong updrafts that are partly at a temperature above the freezing point of water and partly at a temperature below the freezing point of water. The curved and spiral arrows indicate paths a hailstone takes as it is forming.

Hail is produced in cumulonimbus clouds as a result of great instability and strong vertical air currents (updrafts and downdrafts) (Figure 6-31). Because highly unstable air is needed for it to form, hail tends to be more common in summer than in the middle of winter. For hail to form from a cloud, the cloud must have a lower part that is warmer than the freezing point of 0°C (32°F) and an upper part that is colder than this. Updrafts carry water from the above-freezing layer or small ice particles from the lowest part of the below-freezing layer upward, where they grow by collecting moisture from supercooled cloud droplets. When the particles become too large to be supported in the air, they fall, gathering more moisture on the way down. If they encounter a sufficiently strong updraft, they may be carried skyward again, only to fall another time. This sequence may be repeated several times, as indicated by the spiral path in Figure 6-31.

▲ **Figure 6-30** Enormous hailstones gathered by geographer Bob Solomon in his backyard in Sydney, Australia, after an intense hailstorm in September 1999. The average hailstone in this collection weighed 85 grams (3 ounces) and had a diameter of 8 cm (3 inches).

A hailstone normally continues to grow whether it is rising or falling in the below-freezing layer, providing it passes through portions of the cloud that contain supercooled droplets. If there is considerable supercooled moisture available, the hailstone, which is ice, becomes surrounded by a wet layer that freezes relatively slowly, producing large ice crystals and forcing the air out of the water. The result is clear-ice rings. If there is a more limited supply of supercooled droplets, the water may freeze almost instantly around the hailstone. This fast freezing produces small crystals with tiny air bubbles trapped among them, forming opaque rings of ice.

The eventual size of a hailstone depends on the amount of supercooled water in the cloud, the strength of the updrafts, and the total length (up, down, and sideways) of the path taken by the stone through the cloud. The largest documented hailstone fell in Nebraska in 2003, and was 17 centimeters (7 inches) in diameter; the heaviest documented hailstone, which fell in Kansas in 1970, was 14 centimeters (5.5 inches) in diameter and weighed 766 grams (1.67 pounds).

Atmospheric Lifting and Precipitation

The role of rising air and adiabatic cooling has been stressed in this chapter. Only through these events can any significant amount of precipitation originate. It remains for us to consider the causes of rising air. There are four principal types of atmospheric lifting. One type is spontaneous, and the other three require the presence of some external force. More often than not, however, the various types operate in conjunction.

Convective Lifting

Because of unequal heating of different surface areas, a parcel of air near the ground may be warmed by conduction more than the air around it. The density of the heated air is reduced as the air expands, and so the parcel rises toward a lower-density layer, in a typical **convective lifting** situation, as shown in Figure 6-32a. The pressure of the unstable air decreases as the air rises, and so it cools adiabatically to the dew point temperature. Condensation begins and a cumulus cloud forms. With the proper humidity, temperature, and stability conditions, the cloud is likely to grow into a towering cumulonimbus thunderhead, with a downpour of showery raindrops and/or hailstones accompanied sometimes by lightning and thunder (see Figure 6-29).

An individual convective cell is likely to cover only a small horizontal area, although sometimes multiple cells are formed very close to each other, close enough to form a much larger cell. *Convective precipitation* is typically showery, with large raindrops falling fast and furiously but only for a short duration. It is particularly associated with the warm parts of the world and warm seasons.

In addition to the spontaneous uplift just described, various kinds of forced uplift, such as air moving over a mountain range, can trigger formation of a convective cell if the air tends toward instability. Thus convective uplift often accompanies other kinds of uplift.

Orographic Lifting

Topographic barriers that block the path of horizontal air movements are likely to cause large masses of air to travel upslope, as Figure 6-32b shows. This kind of forced ascent from **orographic lifting** can produce *orographic precipitation* if the ascending air is cooled to the dew point.

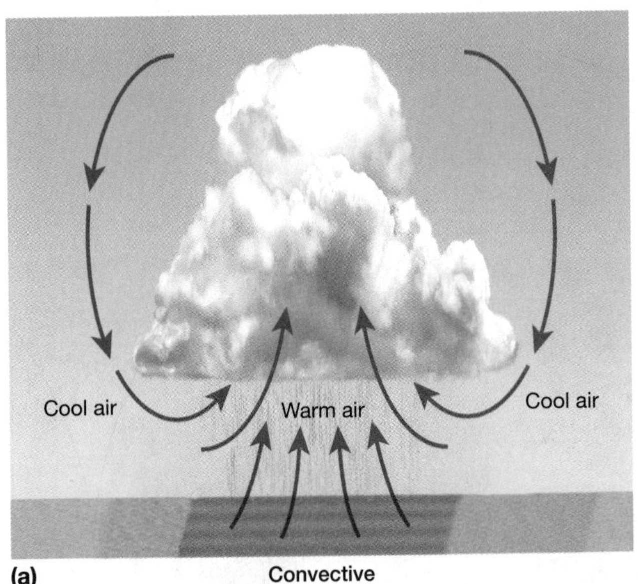

(a) Convective

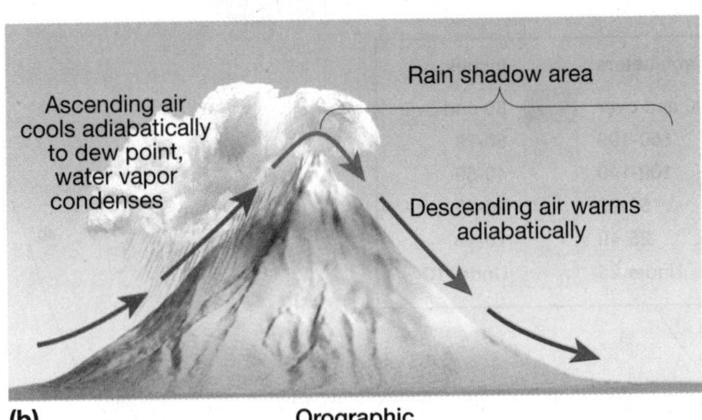

(b) Orographic

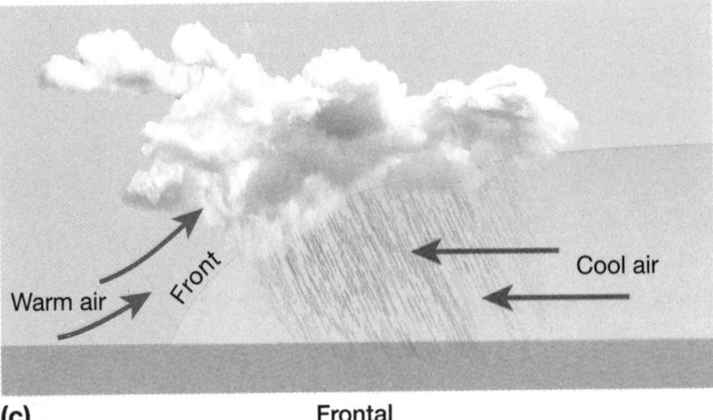

(c) Frontal

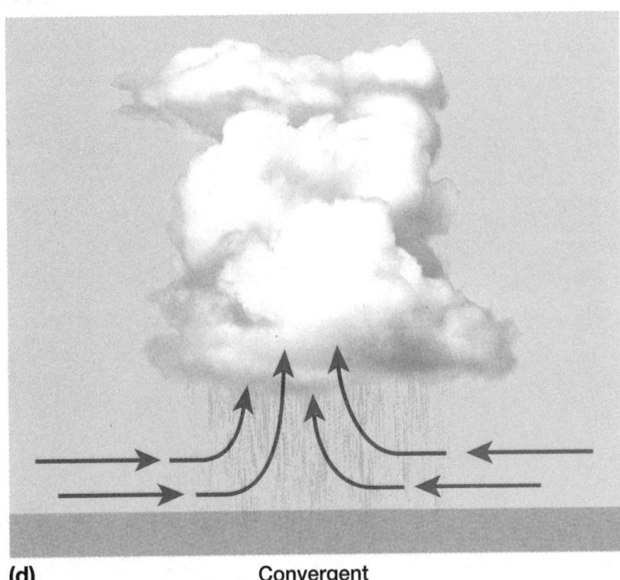
(d) Convergent

▲ **Figure 6-32** The four basic types of atmospheric lifting and precipitation: (a) convective, (b) orographic, (c) frontal, (d) convergent.

As we learned above, if significant instability has been triggered by the upslope motion, the air keeps rising when it reaches the top of the slope and the precipitation continues. More often, however, the air descends the leeward side of the barrier. As soon as it begins to move downslope, adiabatic cooling is replaced by adiabatic warming and condensation and precipitation ceases. Thus the windward slope of the barrier is the wet side, the leeward slope is the dry side, and the term **rain shadow** is applied to both the leeward slope and the area beyond as far as the drying influence extends (Figure 6-33).

Orographic precipitation can occur at any latitude, any season, any time of day. The only requisite conditions are a topographic barrier and moist air to move over it. Orographic precipitation is likely to be prolonged because there is a relatively steady upslope flow of air.

Frontal Lifting

When unlike air masses meet, they do not mix. Rather, a zone of discontinuity called a *front* is established between them, and the warmer air rises over the cooler air, as shown in Figure 6-32c. As the warmer air is forced to rise, it may be cooled to the dew point with resulting clouds and

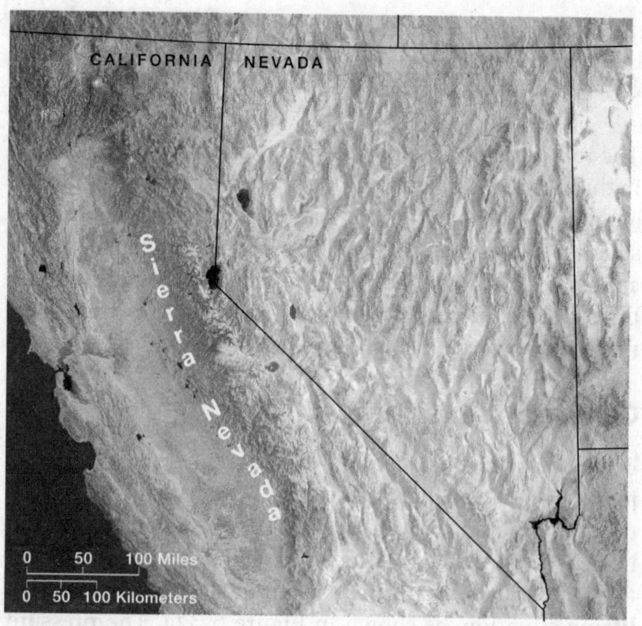

▲ **Figure 6-33** Nevada is in the rain shadow of the Sierra Nevada Mountains of California. The west slope of the Sierra has high rainfall because of the orographic lifting of the moist westerly winds coming from the Pacific Ocean. East of the Sierra, the climate is arid.

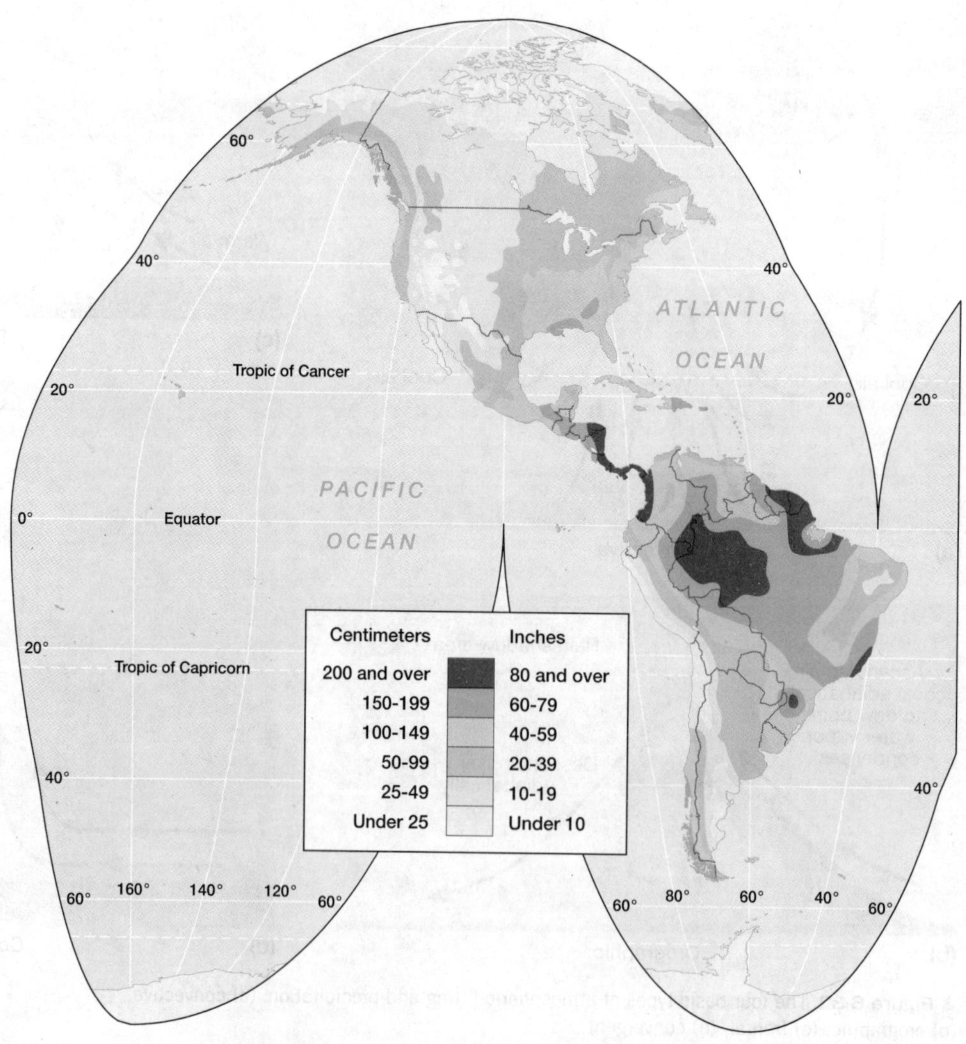

▶ **Figure 6-34** Average annual precipitation over the land areas of the world.

precipitation. Precipitation that results from such **frontal lifting** is referred to as *frontal precipitation*. We shall discuss frontal precipitation in greater detail in the next chapter. It tends to be widespread and protracted, but it is also frequently associated with convective showers.

Frontal activity is most characteristic of the midlatitudes, and so frontal precipitation is particularly notable in those regions, which are meeting grounds of cold polar air and warm tropical air. It is less significant in the high latitudes and rare in the tropics because those regions contain air masses that tend to be like one another.

Convergent Lifting

Less common than the other three types, but nevertheless significant in some situations, is **convergent lifting** and the accompanying *convergent precipitation*, illustrated in Figure 6-32d. Whenever air converges, the result is a general uplift because of the crowding. This forced uplift enhances instability and is likely to produce showery precipitation. It is frequently associated with cyclonic storm systems and is particularly characteristic of the low latitudes. It is common, for example, in the intertropical convergence zone (ITCZ; discussed in Chapter 5) and is notable in such tropical disturbances as hurricanes and easterly waves.

Global Distribution of Precipitation

Seasonal Pressure and Precipitation Patterns

The most important geographic aspect of atmospheric moisture is the spatial distribution of precipitation. The broad-scale zonal pattern is based on latitude, but many other factors are involved and the overall pattern is complex. This section of the chapter focuses on a series of maps that illustrate worldwide and U.S. precipitation distribution. A major cartographic device used on these maps is the **isohyet**, a line joining points of equal quantities of precipitation.

Average Annual Precipitation

The amount of precipitation on any part of Earth's surface is determined by the nature of the air mass involved and the degree to which that air is uplifted. The humidity, temperature, and stability of the air mass are mostly dependent on where the air originated (over land or water, in high or low latitudes) and on the trajectory it has followed. The amount of uplifting and whether that uplifting takes place are determined largely by zonal pressure patterns, topographic barriers, storms, and other atmospheric disturbances (Figure 6-34).

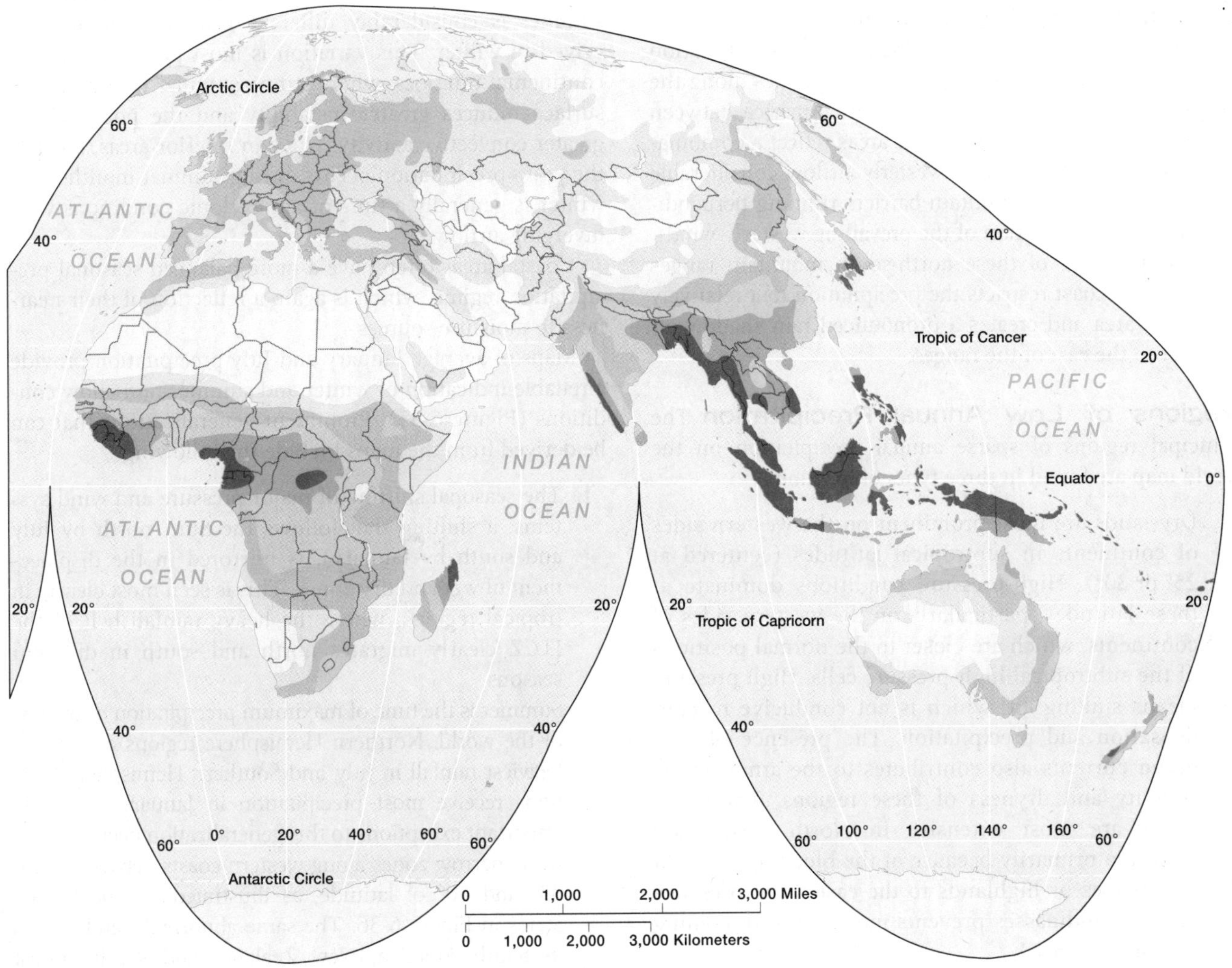

Regions of High Annual Precipitation High annual precipitation is generally found in two types of locations:

1. The most conspicuous feature of the worldwide annual precipitation pattern is that the tropical latitudes contain most of the wettest areas. The warm easterly trade winds are capable of carrying enormous amounts of moisture, and where they are forced to rise, very heavy rainfall is usually produced. Equatorial regions particularly reflect these conditions where warm ocean water easily vaporizes and warm, moist, unstable air is uplifted in the ITCZ. Considerable precipitation also results where trade winds are forced to rise by topographic obstacles. As the trades are easterly winds, it is the eastern coasts of tropical landmasses—for example, the east coast of Central America, northeastern South America, and Madagascar—where this orographic effect is most pronounced. Where the normal trade-wind pattern is modified by monsoons, the onshore trade-wind flow may occur on the opposite coast. Thus, the wet areas on the western coast of southeastern Asia, India, and what is called the Guinea Coast of West Africa are caused by the onshore flow of southwesterly winds that are nothing more than trade winds diverted from a "normal" pattern by the South Asian and West African monsoons.
2. The only other regions of high annual precipitation shown on the world map are narrow zones along the western coasts of North and South America between 40° and 60° of latitude. These areas reflect a combination of frequent onshore westerly airflow, considerable storminess, and mountain barriers running perpendicular to the direction of the prevailing westerly winds. The presence of these north-south mountain ranges near the coast restricts the precipitation to a relatively small area and creates a pronounced rain shadow effect to the east of the ranges.

Regions of Low Annual Precipitation The principal regions of sparse annual precipitation on the world map are found in three types of locations.

1. Dry lands are most prominent on the western sides of continents in subtropical latitudes (centered at 25° or 30°). High-pressure conditions dominate at these latitudes, particularly on the western sides of continents, which are closer to the normal positions of the subtropical high-pressure cells. High pressure means sinking air, which is not conducive to condensation and precipitation. The presence of cool ocean currents also contributes to the atmospheric stability and dryness of these regions. These dry zones are most extensive in North Africa and Australia primarily because of the blocking effect of landmasses or highlands to the east. (The presence of such landmasses prevents moisture from coming in from the east.)
2. Dry regions in the midlatitudes are most extensive in central and southwestern Asia, but they also occur in western North America and southeastern South America. In each case, the dryness is due to lack of access for moist air masses. In the Asian situation, this lack of access is essentially a function of distance from any ocean where onshore airflow might occur. In North and South America, there are rain shadow situations in regions of predominantly westerly airflow.
3. In the very high latitudes, there is not much precipitation anywhere. Water surfaces are scarce and cold, and so little opportunity exists for moisture to evaporate into the air. As a result, polar air masses have low absolute humidities and precipitation is slight. These regions are referred to accurately as cold deserts.

One further generalization on precipitation distribution is the contrast between continental margins and interiors. Because coastal regions are much closer to moisture sources, they usually receive more precipitation than interior regions.

Seasonal Precipitation Patterns

A geographic understanding of climate requires knowledge of seasonal as well as annual precipitation patterns. Over most of the globe, the amount of precipitation received in summer is considerably different from the amount received in winter. This variation is most pronounced over continental interiors, where strong summer heating at the surface induces greater instability and the potential for greater convective activity. Thus, in interior areas, most of the year's precipitation occurs during summer months, and winter is generally a time of anticyclonic conditions with diverging airflow.

Coastal areas often have a more balanced seasonal precipitation regime, which is again a reflection of their nearness to moisture sources.

Maps of average January and July precipitation provide a reliable indication of winter and summer rain-snow conditions (Figure 6-35). Prominent generalizations that can be derived from the maps include the following:

1. The seasonal shifting of major pressure and wind systems, a shifting that follows the Sun (north by July and south by January), is mirrored in the displacement of wet and dry zones. This is seen most clearly in tropical regions, where the heavy rainfall belt of the ITCZ clearly migrates north and south in different seasons.
2. Summer is the time of maximum precipitation over most of the world. Northern Hemisphere regions experience heaviest rainfall in July, and Southern Hemisphere locations receive most precipitation in January. The only important exceptions to this generalization occur in relatively narrow zones along western coasts between about 35° and 50° of latitude, as illustrated for the United States in Figure 6-36. The same abnormal trend occurs in South America, New Zealand, and southernmost

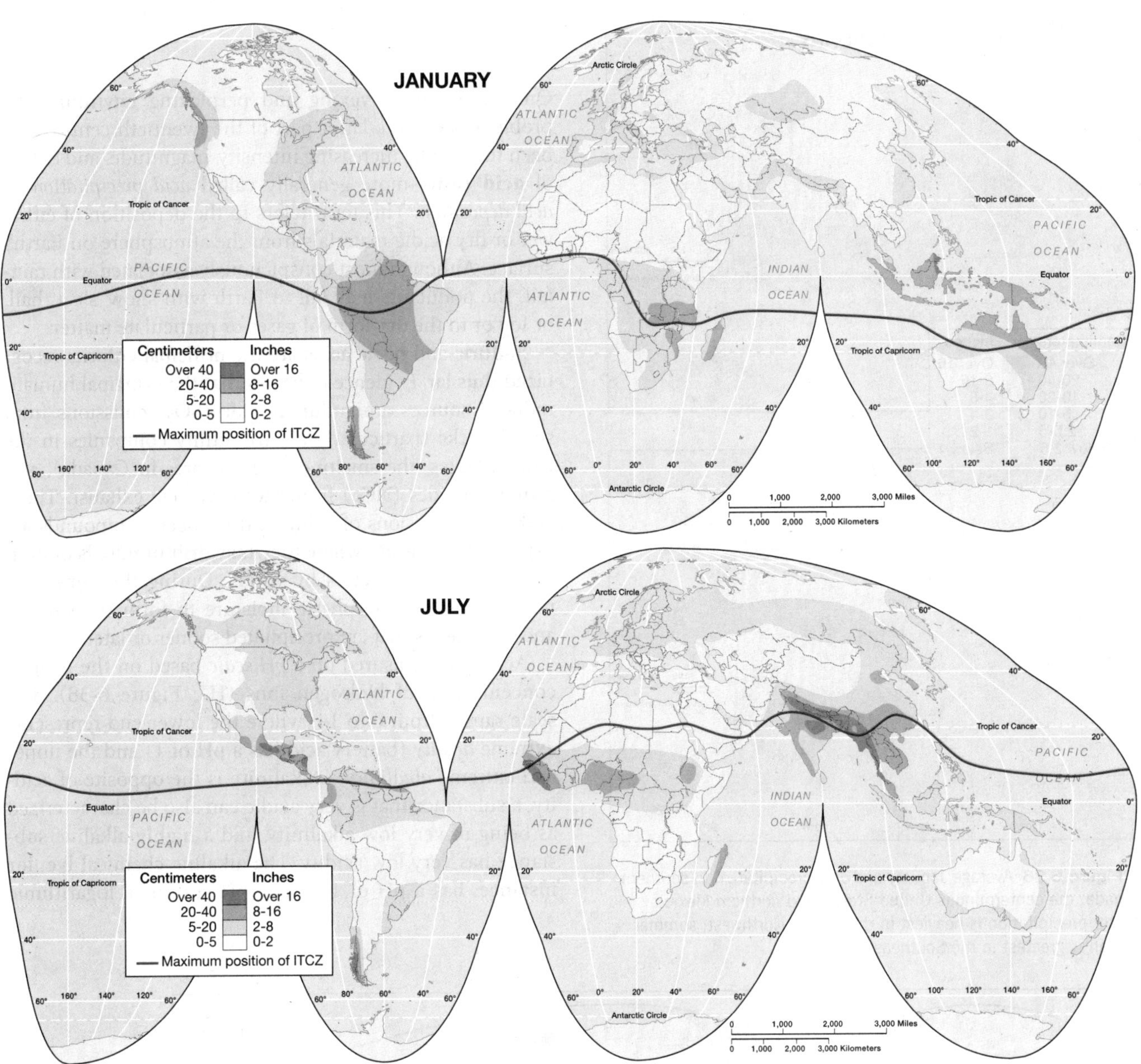

▲ **Figure 6-35** Average January and July precipitation over the land areas of the world. The red line marks the typical maximum poleward position of the intertropical convergence zone (ITCZ).

Australia. These regions experience the summer dryness associated with the seasonal shift of the STH.

3. The most conspicuous variation in seasonal precipitation is found, predictably, in monsoon regions (principally southern and eastern Asia, northern Australia, and West Africa), where summer tends to be very wet and winter is generally dry.

Precipitation Variability

The maps considered thus far all portray average conditions. The data on which they are based were gathered over decades, and thus the maps represent abstraction rather than reality. In any given year or any given season, the amount of precipitation may or may not be similar to the long-term average.

Precipitation variability is the expected departure from average precipitation in any given year, expressed as a percentage above or below average. For example, a precipitation variability of 20 percent means that a location expects to receive either 20 percent more or 20 percent less precipitation than average in any given year. If a location has a long-term average annual precipitation of 50 cm (20 in.) and has a precipitation variability of 20 percent, the "normal" rainfall for a year would be either 40 cm (16 in.) or 60 cm (24 in.)—in the long run, the average comes out to 50 cm.

Figure 6-37 reveals that regions of normally heavy precipitation experience the least variability and normally dry regions experience the most. Said another way, dry regions experience great fluctuations in precipitation from one year to the next.

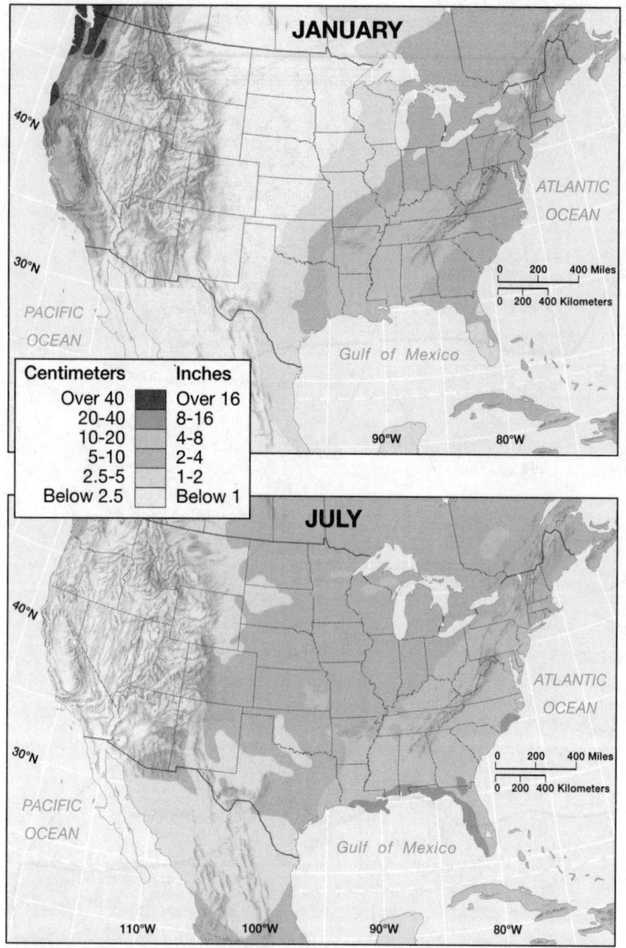

▲ **Figure 6-36** Average January and July precipitation in southern Canada, the conterminous United States, and northern Mexico. Winter precipitation is heaviest in the Pacific Northwest; summer rainfall is greatest in the Southeast.

Acid Rain

One of the most vexing and perplexing environmental problems since the latter part of the twentieth century has been the rapidly increasing intensity, magnitude, and extent of **acid rain**—more generally called *acid precipitation* or *acid deposition*. This term refers to the deposition of either wet or dry acidic materials from the atmosphere on Earth's surface. Although most conspicuously associated with rainfall, the pollutants may fall to Earth with snow, sleet, hail, or fog or in the dry form of gases or particulate matter.

Sulfuric and nitric acids are the principal culprits recognized thus far. Evidence indicates that the principal human-induced sources are sulfur dioxide (SO_2) emissions from smokestacks (particularly electric utility companies in the United States, the smelting of metal ores in Canada), and nitrogen oxides (NO_x) from motor vehicle exhaust. These and other emissions of sulfur and nitrogen compounds are expelled into the air, where they may drift hundreds or even thousands of kilometers by winds. During this time they may mix with atmospheric moisture to form the sulfuric and nitric acids that are precipitated sooner or later.

Acidity is measured on a *pH scale* based on the relative concentration of hydrogen ions (H^+; Figure 6-38). The scale ranges from 0 to 14, where the lower end represents extreme *acidity* (battery acid has a pH of 1) and the upper end extreme *alkalinity*. (Alkalinity is the opposite of acidity; a substance that is very acidic can also be characterized as being of very low alkalinity, and a highly alkaline substance has very low acidity. The alkaline chemical lye, for instance, has a pH of 13.) The pH scale is a logarithmic

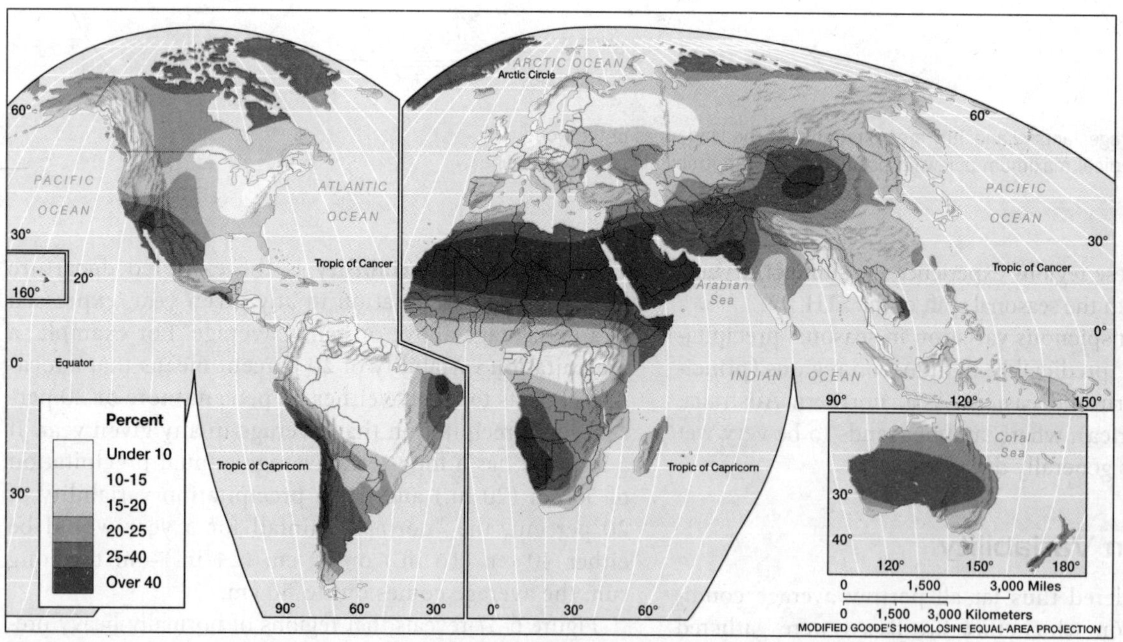

▲ **Figure 6-37** Precipitation variability is the expected departure from average precipitation in any given year, expressed as a percentage above or below average. Dry regions (such as northern Africa, the Arabian peninsula, southwestern Africa, central Asia, and much of Australia) experience greater variability than humid areas (such as the eastern United States, northern South America, central Africa, and western Europe).

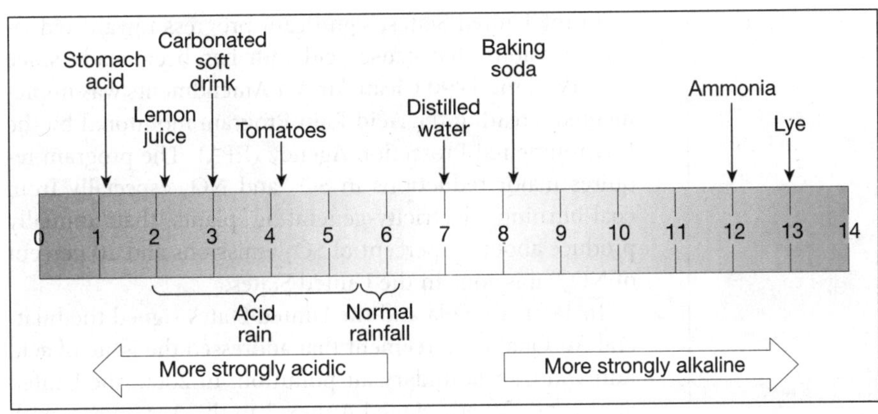

◄ **Figure 6-38** The pH scale. Rainfall in clean air has a pH of about 5.6 (slightly acidic). Acid rain can have a pH as low as 2.0.

scale, which means that a difference of one whole number on the scale reflects a 10-fold change in absolute values.

Rainfall in clean, dust-free air has a pH of about 5.6. Thus any precipitation that has a pH value of less than 5.6 is considered to be acid precipitation. Normal rain is slightly acidic (because slight amounts of carbon dioxide dissolve in raindrops to form *carbonic acid* [H_2CO_3], a mild acid), but acid rain can be as much as 100 times more acidic. Increasingly, precipitation with a pH of less than 4.5 (the level below which most fish perish) is being recorded. For example, around Washington, D.C., the average pH of rain is 4.2 to 4.4 (Figure 6-39).

Many parts of Earth's surface have naturally alkaline soil or bedrock that neutralizes acid precipitation. Soils developed from limestone, for example, contain calcium carbonate, which can neutralize acid. Granitic soils, on the other hand, have no neutralizing component (Figure 6-40).

Damage from Acid Precipitation Acid precipitation is a major hazard to the environment. The most conspicuous damage is being done to aquatic ecosystems. Thousands of lakes and streams are now acidic, and hundreds of lakes in the eastern United States and Canada became biological deserts over the last 30 years, primarily due to acid rain. Forest diebacks are currently taking place on every continent except Antarctica. In some parts of eastern and central Europe, 30 to 50 percent of the forests have been affected or killed by acid rain. Even buildings and monuments are being destroyed; acid deposition has caused more erosion on the marble Parthenon in Athens in the last 30 years than in the previous 30 centuries.

One of the great complexities of the situation is that much of the pollution is deposited at great distances from its source. Downwind locations receive unwanted acid deposition from upwind origins. Thus Scandinavians and Germans complain about British pollution; Canadians blame U.S. sources; New Englanders accuse the Midwest.

One of the thorniest issues in North American international relations during the 1980s was Canadian dissatisfaction with U.S. government efforts to mitigate acid rain. Acid rain was viewed by Canadians as a grave environmental concern, and at the time perhaps half the acid rain falling on Canada came from U.S. sources, particularly older coal-burning power plants in the Ohio and Tennessee River valleys.

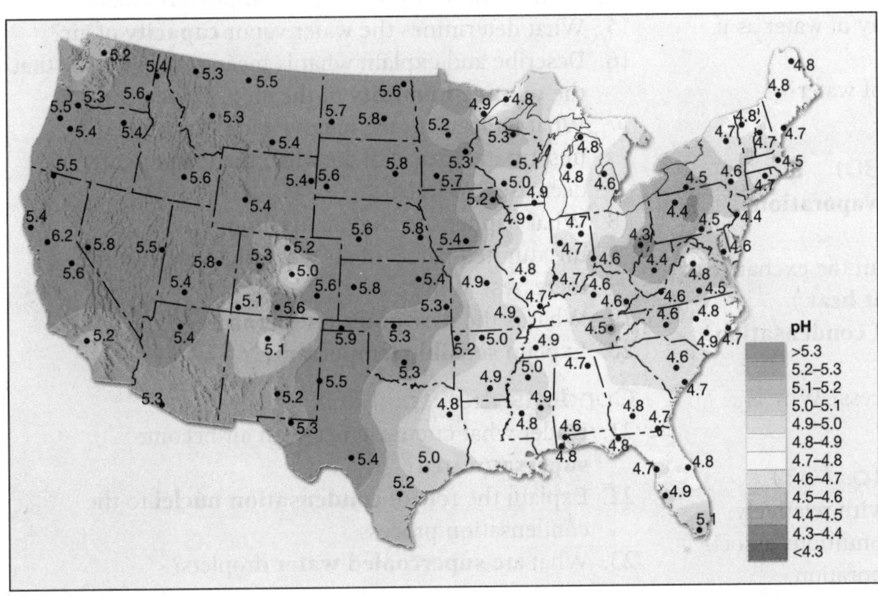

◄ **Figure 6-39** Map showing acidity of rain in the conterminous United States in 2006.

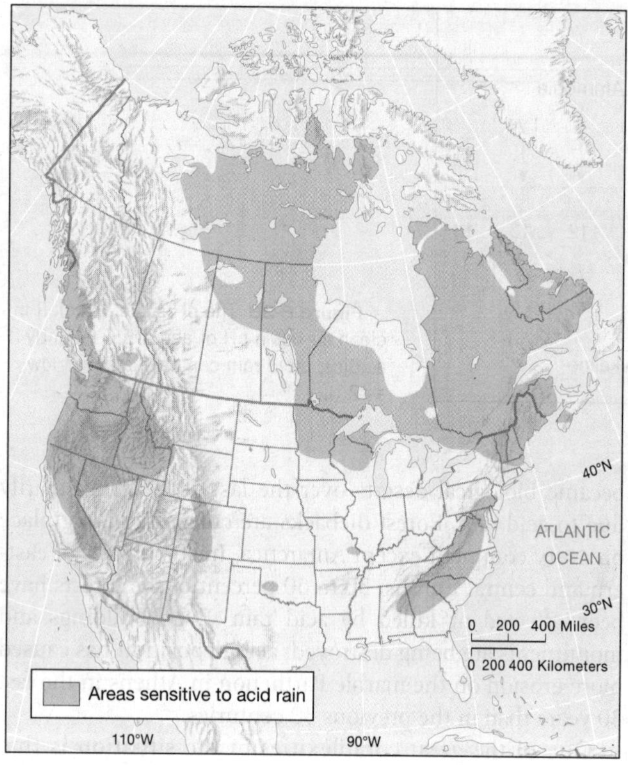

▲ **Figure 6-40** Areas in the United States and Canada particularly sensitive to acid rain because of a scarcity of natural buffers.

In the United States, significant progress toward reducing emissions that cause acid rain has been made since Title IV of the 1990 Clean Air Act Amendments was implemented, creating the Acid Rain Program monitored by the Environmental Protection Agency (EPA). The program requires major reductions in SO_2 and NO_x especially from coal-burning electricity-generating plants that annually produce about 70 percent of SO_2 emissions and 20 percent of NO_x emissions in the United States.

In 1991, Canada and the United States signed the bilateral Air Quality Agreement that addressed the issue of acid rain and transboundary air pollution. In 2000, the United States and Canada signed Annex 3 to the Agreement, with the goal of reducing emission of both NO_x and volatile organic compounds (VOC); in 2008 discussions were held regarding adding a particulate matter annex to the agreement as well.

These programs have been quite successful: The EPA reports that by 2007, SO_2 emissions had dropped to 50 percent of 1980 levels—achieving the emissions target three years ahead of the 2010 deadline; NO_x emissions have been reduced significantly as well. Such reductions in emissions are central to the sustained effort needed to reduce the effects of acid rain.

CHAPTER 6 LEARNING REVIEW

After studying this chapter, you should be able to answer the following questions. Key terms from each text section are shown in bold type. Definitions for key terms are also found in the glossary at the back of the book.

The Hydrologic Cycle (p. 127)

1. Briefly describe how water moves through the **hydrologic cycle**.

The Nature of Water: Commonplace but Unique (p. 128)

2. What is a **hydrogen bond** between water molecules?
3. Describe what happens to the density of water as it freezes.
4. What is meant by **surface tension** of water?
5. What is **capillarity**?

Phase Changes of Water (p. 130)

6. Briefly define the following terms: **evaporation**, **condensation**, **sublimation**.
7. How do phase changes of water entail the exchange of heat? (In other words, explain **latent heat**.)
8. What is meant by the **latent heat of condensation**? **Latent heat of evaporation**?
9. Why is evaporation a "cooling" process? Why is condensation a "warming" process?

Water Vapor and Evaporation (p. 131)

10. Describe the conditions associated with relatively high rates of evaporation, and the conditions associated with relatively low rates of evaporation.
11. What is meant by the **vapor pressure** of water in the atmosphere?
12. What is **evapotranspiration**?

Measures of Humidity (p. 134)

13. What is **absolute humidity**? **Specific humidity**?
14. What is meant by **saturation vapor pressure**?
15. What determines the water vapor **capacity** of air?
16. Describe and explain what is meant when we say that the **relative humidity** of the air is 50 percent.
17. What happens to the relative humidity of an unsaturated parcel of air when the temperature decreases? Why?
18. What happens to the relative humidity of an unsaturated parcel of air when the temperature increases? Why?
19. What is the **dew point temperature**?
20. Explain **sensible temperature**.

Condensation (p. 136)

21. Under what circumstances can air become **supersaturated**?
22. Explain the role of **condensation nuclei** to the condensation process.
23. What are **supercooled water** droplets?

Adiabatic Processes (p. 138)

24. Which cooling process in the atmosphere is responsible for the formation of most clouds (and nearly all clouds that produce precipitation)?
25. What happens to the relative humidity of an unsaturated parcel of air as it rises? Why?
26. What is the relationship of the dew point temperature of a parcel of air to its **lifting condensation level**?
27. Why does a rising parcel of unsaturated air cool more rapidly than a rising parcel of saturated air (in which condensation is taking place)? In other words, explain why there is a **dry adiabatic rate** and a **saturated adiabatic rate**.
28. Why doesn't descending air form clouds?

Clouds (p. 140)

29. Briefly describe the three main forms of **clouds**: **cirrus clouds**, **stratus clouds**, and **cumulus clouds**.
30. Identify the four families of clouds.
31. Describe the four principal types of **fog**.
32. How and where does **dew** form?

The Buoyancy of Air (p. 143)

33. What is the difference between **stable** air and **unstable** air?
34. What conditions make a parcel of air unstable?
35. Why might the stability of a rising parcel of air change above the lifting condensation level?
36. Are stratus clouds associated with stable or unstable air? Are cumulus clouds associated with stable or unstable air?

Precipitation (p. 147)

37. Briefly describe the following kinds of **precipitation**: **rain**, **snow**, **hail**.
38. How is hail related to atmospheric instability?

Atmospheric Lifting and Precipitation (p. 151)

39. Describe the four main lifting mechanisms of air: **convective**, **orographic**, **frontal**, and **convergent**.
40. What is a **rain shadow** and why does one form? In your answer, explain the role of adiabatic temperature changes, as well as changes in both the relative humidity and the actual water vapor content of the air, in the formation of rain shadows.

Global Distribution of Precipitation (p. 153)

41. What is an **isohyet**?
42. Using the global map of average annual precipitation (Figure 6-34), explain the causes of:
    - Wet regions within the tropics
    - Wet regions along the west coasts of continents in the midlatitudes (between about 40 and 60° N and S)
    - Dry regions along the west coasts of continents in the subtropics (at about 20 to 30° N and S)
    - Dry areas within the midlatitudes
43. Using the maps of average January and July precipitation (Figure 6-35) contrast and explain the seasonal rainfall patterns in central Africa.
44. What is meant by the term **precipitation variability**?
45. What is the general relationship of precipitation variability to average annual precipitation?

Acid Rain (p. 156)

46. What are some of the circumstances that cause **acid rain**?

Log in to **www.mygeoscienceplace.com** for animations, videos, RSS feeds, maps, and self-study quizzes to enhance your study of **Atmospheric Moisture**.

7 Atmospheric Disturbances

In Chapter 5 we explored the general circulation of the atmosphere, mainly the broadscale wind and pressure systems of the troposphere. Now we sharpen our focus a bit.

Over most of Earth, particularly in the midlatitudes, day-to-day weather conditions are accompanied by phenomena that are more limited than general circulation phenomena in both magnitude and permanence. These more limited phenomena include air masses, fronts, as well as a variety of disturbances usually referred to as *storms*. Such disturbances are secondary features of the general circulation of the atmosphere. They move with the general circulation as migratory entities that persist for a relatively short time before dissipating. Although air masses, fronts, and storms are transient and temporary, in some parts of the world they are so frequent and dominating that their interactions are major determinants of weather and, to a lesser extent, climate.

The Impact of Storms on the Landscape

Storms are usually very dramatic phenomena. Under ordinary nonstormy conditions, the atmosphere tends to be quietly inconspicuous, but storms bring excitement. The combination of expansive clouds, swirling winds, and abundant precipitation—often accompanied by thunder and lightning—that characterizes many storms makes us acutely aware of the awesome power latent in the gaseous environment we inhabit.

The landscape may be quickly and significantly transformed by a storm—flooded streets, windblown trees, and darkened skies are prominent examples of the changes that occur. The long-run impact of storms on the landscape is often equally notable. Damage varies with the intensity of the disturbance, but such things as uprooted trees, accelerated erosion, flooded valleys, destroyed buildings, and decimated crops can result. Most storms also have a positive long-term effect on the landscape, however, as they promote diversity in the vegetative cover, increase the size of lakes and ponds, and stimulate plant growth via the moisture they add to the ground.

Air Masses

Although the troposphere is a continuous body of mixed gases that surrounds the planet, it is by no means a uniform blanket of air. Instead, it is composed of many large, variable parcels of air that are distinct from one another. Such parcels are referred to as **air masses**.

◀ Evening thunderstorm near Needles, California.

Characteristics

To be distinguishable, an air mass must meet three requirements:

1. It must be large. A typical air mass is more than 1600 kilometers (1000 miles) across and several kilometers deep (from the surface to its top).
2. It must have uniform properties in the horizontal dimension. This means that at any given altitude in the air mass, its physical characteristics—primarily temperature, humidity, and stability—are relatively homogeneous.
3. It must be a recognizable entity and travel as one. Thus, it must be distinct from the surrounding air, and when it moves, it must retain its original characteristics and not be torn apart by differences in airflow.

Origin

An air mass develops its characteristics by remaining over a uniform land or sea surface long enough to acquire the temperature/humidity/stability characteristics of the surface below. The stagnation needs to last for only a few days if the underlying surface has prominent temperature and moisture characteristics. For the air to stagnate, it needs to be relatively stable, so highly unstable air rarely forms a true air mass because it is constantly moving upward, away from the surface whose characteristics it is to acquire. Most air masses, then, form most commonly in areas with anticyclonic conditions.

The formation of air masses is normally associated with what are called *source regions*: regions of Earth's surface that are particularly well suited to generate air masses. Such regions must be extensive, physically uniform, and associated with air that is stationary or anticyclonic. Ideal source regions are ocean surfaces and extensive flat land areas that have a uniform covering of snow, forest, or desert.

Figure 7-1 portrays the principal recognized source regions for air masses that affect North America. Warm-air masses can form in any season over the waters of the southern North Atlantic, the Gulf of Mexico/Caribbean Sea, and the southern North Pacific and in summer over the deserts of the southwestern United States and northwestern Mexico. Cold-air masses develop over the northern portions of the Atlantic and Pacific Oceans and over the snow-covered lands of north-central Canada.

It may well be that the concept of source regions is of more theoretical value than actual value. A broader view, one subscribed to by many atmospheric scientists, holds that air masses can originate almost anywhere in the low or high latitudes but rarely in the midlatitudes due to the prevailing westerlies where persistent wind would prevent air mass formation.

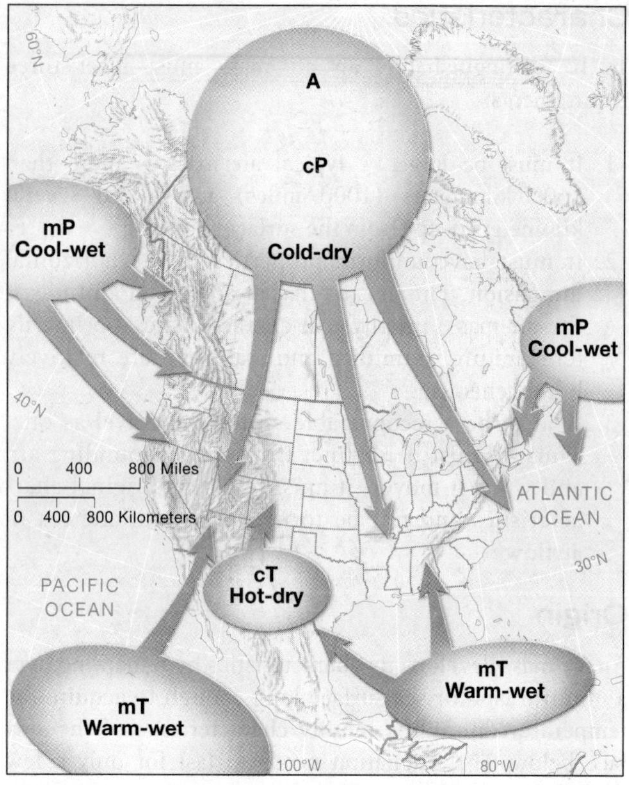

▲ **Figure 7-1** Major air masses that affect North America and their generalized paths. The tropics and subtropics are important source regions as are the high latitudes. Air masses do not originate in the middle latitudes except under unusual circumstances. (For an explanation of the air-mass codes A, cP, mP, cT, and mT, see Table 7-1.)

Classification

Air masses are classified on the basis of source region. The latitude of the source region correlates directly with the temperature of the air mass, and the nature of the surface strongly influences the humidity content of the air mass. Thus, a low-latitude air mass is warm or hot; a high-latitude one is cool or cold. If the air mass develops over a continental surface, it is likely to be dry; if it originates over an ocean, it is usually moist.

A one- or two-letter code is generally used to identify air masses. Although some authorities recognize other categories, the basic classification is sixfold, as shown in Table 7-1.

Movement and Modification

Some air masses remain in their source region for long periods, even indefinitely. In such cases, the weather associated with the air mass persists with little variation. Our interest, however, is in masses that leave their source region and move into other regions, particularly into the midlatitudes.

When an air mass departs from its source region, its structure begins to change, owing in part to thermal modification (heating or cooling from below), in part to dynamic modification (uplift, subsidence, convergence, turbulence), and perhaps also in part to addition or subtraction of moisture.

Once it leaves its source area, an air mass modifies the weather of the regions into which it moves: it takes source-region characteristics into other regions. A classic example of this modification is displayed in Figure 7-2, which diagrams a situation that may occur one or more times every winter. A midwinter outburst of continental polar (cP) air from northern Canada sweeps down across the central part of North America. With a source-region temperature of −46°C (−50°F) around Great Bear Lake, the air mass has warmed to −34°C (−30°F) by the time it reaches Winnipeg, Manitoba, and it continues to warm as it moves southward. Throughout its southward course, the air mass becomes warmer, but it also brings some of the coldest weather that each of these places will receive all winter. Thus, the air mass is modified, but it also modifies the weather in all regions it passes through.

Temperature, of course, is only one of the characteristics modified by a moving air mass. There are also modifications in humidity and stability.

North American Air Masses

The North American continent is a prominent area of air mass interaction. The lack of mountains trending east to west permits polar air to sweep southward and tropical air to flow northward unhindered by terrain, particularly over the eastern two-thirds of the continent (see Figure 7-1). In the western part of the continent, though, air masses moving off the Pacific are impeded by the prominent north–south trending mountain ranges.

Continental polar (cP) air masses develop in central and northern Canada, and *Arctic* (A) air masses originate

TABLE 7-1 Simplified Classification of Air Masses

Type	Code	Source Regions	Source Region Properties
Arctic/Antarctic	A	Antarctica, Arctic Ocean and fringes, and Greenland	Very cold, very dry, very stable
Continental polar	cP	High-latitude plains of Eurasia and North America	Cold, dry, very stable
Maritime polar	mP	Oceans in vicinity of 50°−60° N and S latitude	Cold, moist, relatively unstable
Continental tropical	cT	Low-latitude deserts	Hot, very dry, unstable
Maritime tropical	mT	Tropical and subtropical oceans	Warm, moist, of variable stability
Equatorial	E	Oceans near the equator	Warm, very moist, unstable

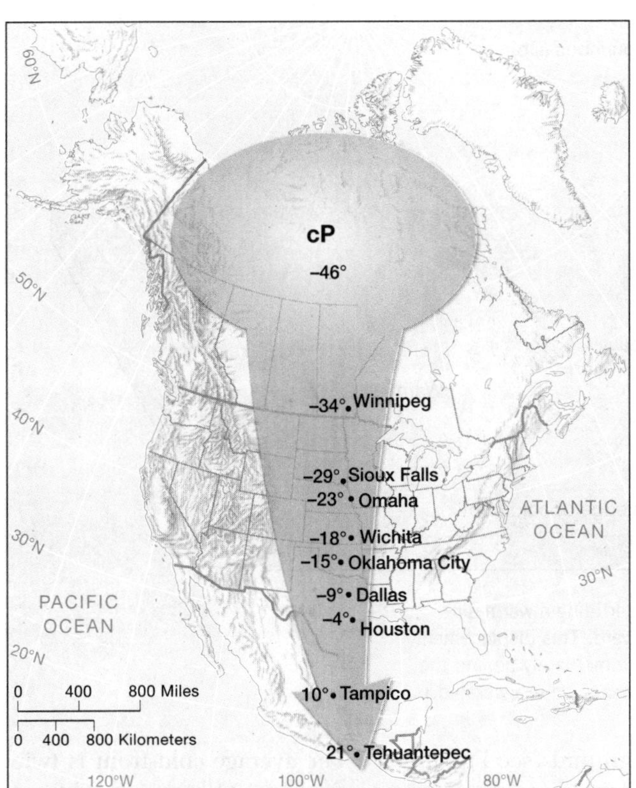

▲ **Figure 7-2** An example of temperatures resulting from a strong midwinter outburst of cP air from Canada. All temperatures are in degrees Celsius.

farther north and so are colder and drier than cP air masses—both are dominant features in winter with their cold, dry, stable nature.

Maritime polar (mP) air from the Pacific in winter can bring cloudiness and heavy precipitation to the mountainous coastal regions, but by the time it reaches midcontinent it may bring moderate temperatures and clear skies. In summer, cool Pacific mP air produces fog and low stratus clouds along the coast. North Atlantic mP air masses are also cool, moist, and unstable, but except for occasional incursions into the mid-Atlantic coastal region, Atlantic mP air does not affect North America because the prevailing circulation of the atmosphere is westerly.

Maritime tropical (mT) air from the Atlantic/Caribbean/Gulf of Mexico is warm, moist, and unstable. It strongly influences weather and climate east of the Rockies in the United States, southern Canada, and much of Mexico, serving as the principal precipitation source in this broad region. It is more prevalent in summer than in winter, bringing periods of uncomfortable humid heat.

Pacific mT air originates over water in areas of anticyclonic subsidence, and so it tends to be cooler, drier, and more stable than Atlantic mT air; it is felt only in the southwestern United States and northwestern Mexico, where it may produce coastal fog and moderate orographic rainfall where forced to ascend mountain slopes. It is also the source of some summer rains in the southwestern interior.

Continental tropical (cT) air is relatively unimportant in North America because its source region is not extensive. In summer hot, very dry, unstable cT air surges into the southern Great Plains area on occasion, bringing heat waves and dry conditions.

Equatorial (E) air affects North America only in association with hurricanes. It is similar to mT air except that E air provides an even more copious source of rain than does mT air because of high humidity and instability.

Fronts

Cold Fronts and Warm Fronts

When unlike air masses meet, they do not mix readily; instead, a boundary zone called a **front** develops between them. A front is not a simple two-dimensional boundary. Rather, it consists of a relatively narrow but nevertheless three-dimensional zone of discontinuity within which the properties of the air change rapidly. A typical front is several kilometers or even tens of kilometers wide.

The frontal concept was developed by Norwegian meteorologists during World War I, and the term "front" was coined because these scientists considered the clash between unlike air masses to be analogous to a confrontation between opposing armies along a battle front. As the more "aggressive" air mass advances at the expense of the other, there is some mixing of the two within the frontal zone, but for the most part the air masses retain their separate identities as one is displaced by the other.

The most conspicuous difference between air masses is usually temperature. A **cold front** forms where an advancing cold air mass meets warmer air (Figure 7-3), while a **warm front** forms where an advancing warm air mass meets colder air (Figure 7-4). In both cases, there is warm air on one side of the front and cool air on the other, with a fairly abrupt temperature gradient between. Air masses may also have different densities, humidity levels, wind patterns, and stability, and so these factors can have a steep gradient through the front as well.

In some cases, a front remains stationary for a few hours or even a few days. More commonly, however, it is in more or less constant motion, shifting the position of the boundary between the air masses but maintaining its function as a barrier between them. Usually one air mass is displacing the other; thus, the front advances in the direction dictated by the movement of the more active air mass.

Regardless of which air mass is advancing, it is always the warmer air that rises over the cooler. The warmer, lighter air is inevitably forced aloft, and the cooler, denser mass functions as a wedge over which the lifting occurs. As you can see in Figures 7-3 and 7-4, fronts "lean" or slope upward from the surface, and it is along this slope that air rises and cools adiabatically to form clouds and often precipitation. Indeed, fronts lean so much that they are much closer to horizontal features than vertical ones. The slope of a typical front averages about 1:150, meaning that 150 kilometers away from the surface position of the front, the height of the front is only 1 kilometer above the ground. Because of this

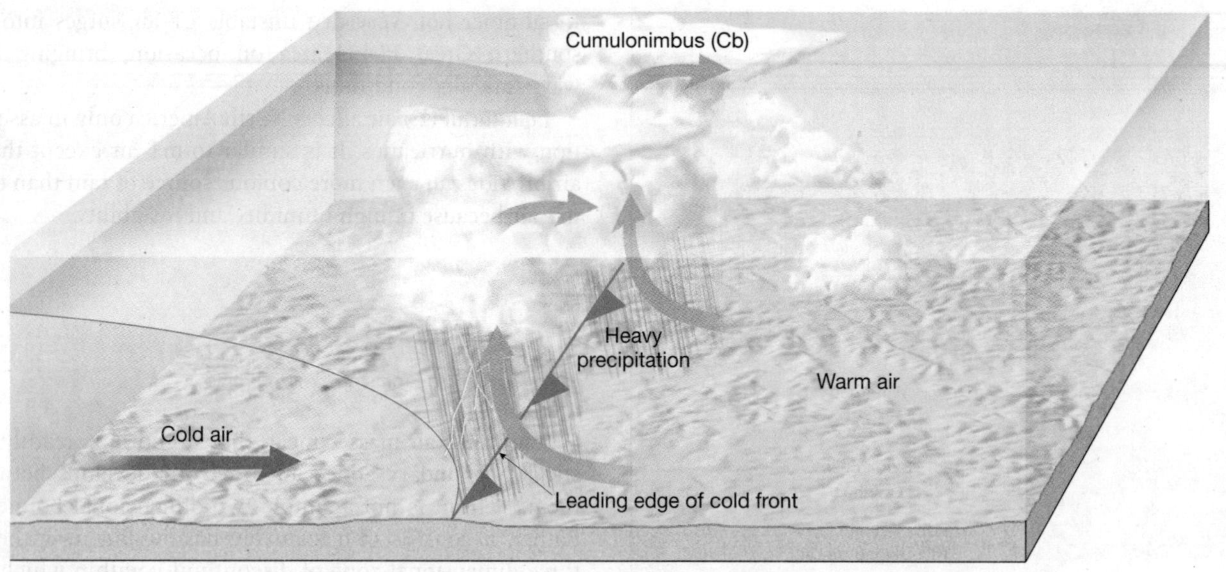

▲ **Figure 7-3** A cold front forms when a cold air mass is actively underriding a warm air mass. As a cold front advances, the warm air ahead of it is forced upward. This displacement often creates cloudiness and relatively heavy precipitation along and immediately behind the ground-level position of the front. (In this diagram, the vertical scale has been exaggerated.)

very low angle of slope (less than 1°), the steepness shown in most diagrams of fronts is greatly exaggerated.

Notice that the "leading edge" of a cold front precedes its higher altitude "trailing edge," whereas a warm front leans "forward" so that the higher altitude part of the front is ahead of its lower altitude "trailing edge."

Cold Fronts

Because of friction with the ground, the advance of the lower portion of a cold air mass is slowed relative to the upper portion. As a result, a cold front tends to become steeper as it moves forward and usually develops a protruding "nose" a few hundred meters above the ground (see Figure 7-3). The average cold front is twice as steep as the average warm front. Moreover, cold fronts normally move faster than warm fronts because the dense, cold air mass easily displaces the lighter, warm air. This combination of steeper slope and faster advance leads to rapid lifting and adiabatic cooling of the warm air ahead of the cold front. The rapid lifting often makes the warm air very unstable, and the result is blustery and violent weather along the cold front. Vertically developed clouds, such as cumulonimbus clouds, are common, with considerable turbulence and showery precipitation. Both clouds and precipitation tend to be concentrated along and immediately behind the ground-level position of the front. Precipitation is usually of higher intensity but shorter duration than that associated with a warm front.

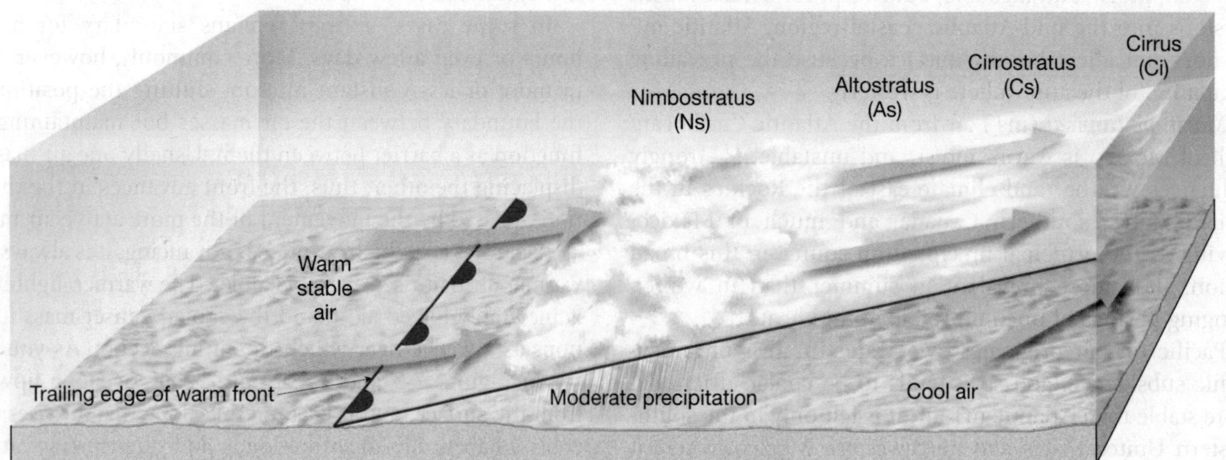

▲ **Figure 7-4** A warm front forms when a warm air mass is actively overriding a cold air mass. As warm air rises above cooler air, widespread cloudiness and precipitation develop along and in advance of the ground-level position of the front. Higher and less dense clouds are often dozens or hundreds of kilometers ahead of the ground-level position of the front. (In this diagram, the vertical scale has been exaggerated.)

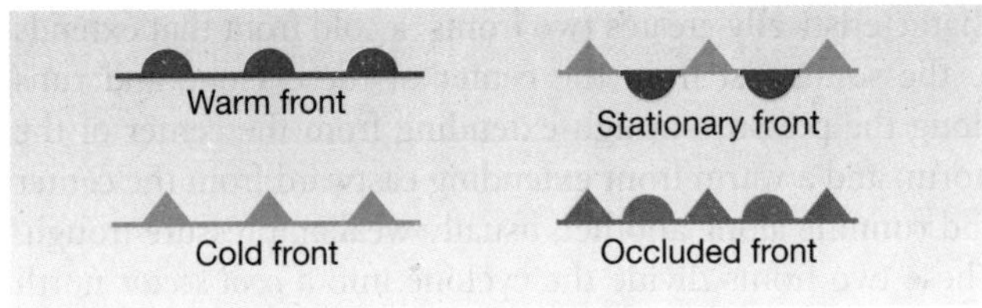

▲ **Figure 7-5** Weather map symbols for fronts.

On a weather map, the ground-level position of a cold front is shown either by a blue line or a solid line studded at intervals with solid triangles that extend in the direction toward which the front is moving (Figure 7-5).

Warm Fronts

The slope of a typical warm front is more gentle than that of a cold front, averaging about 1:200 (see Figure 7-4). As the warm air rises over the retreating cold air, it cools adiabatically, usually resulting in clouds and precipitation. Because the frontal uplift is very gradual, clouds form slowly and there is not much turbulence. High-flying cirrus clouds may signal the approaching front many hours before it arrives. As the front comes closer, the clouds become lower, thicker, and more extensive, typically developing into altocumulus or altostratus. Precipitation usually occurs broadly; it is likely to be protracted and gentle, without much convective activity. If the rising air is inherently unstable, however, precipitation can be showery and even violent. Most precipitation falls ahead of the ground-level position of the moving front.

The ground-level position of a warm front is portrayed on a weather map either by a red line or by a solid line along which solid semicircles are located at regular intervals, with the semicircles extending in the direction toward which the front is moving (see Figure 7-5).

Stationary Fronts

When neither air mass displaces the other, their common boundary is called a **stationary front**. It is difficult to generalize about the weather along such a front, but often gently rising warm air produces limited precipitation similar to that along a warm front. As Figure 7-5 shows, stationary fronts are portrayed on a weather map by a combination of warm and cold front symbols.

Occluded Fronts

A fourth type of front, called an *occluded front*, is formed when a cold front overtakes a warm front. The development of occluded fronts is discussed later in this chapter.

Atmospheric Disturbances

We now turn our attention to the major kinds of atmospheric disturbances that occur within the general circulation. Most of these disturbances involve unsettled and even violent atmospheric conditions and are referred to as *storms*. Some, however, produce calm, clear, quiet weather that is quite the opposite of stormy. The following are common characteristics of atmospheric disturbances in general:

1. They are smaller than the components of the general circulation, although they are extremely variable in size.
2. They are migratory and transient.
3. They have a relatively brief duration, persisting for only a few minutes, a few hours, or a few days.
4. They produce characteristic and relatively predictable weather conditions.

Midlatitude Disturbances The midlatitudes are the principal battleground of tropospheric phenomena: where polar and tropical air masses meet, where most fronts occur, and where weather is most dynamic and changeable from season to season and from day to day. Many kinds of atmospheric disturbances are associated with the midlatitudes, but two of these—*midlatitude cyclones* and *midlatitude anticyclones*—are much more important than the others because of their size and prevalence.

Tropical Disturbances The low latitudes are characterized by monotony—the same weather day after day, week after week, month after month. Almost the only breaks are provided by transient atmospheric disturbances, of which by far the most significant are *tropical cyclones* (locally known as *hurricanes* when they intensify), but also less dramatic disturbances known as *easterly waves*.

Localized Severe Weather Other localized atmospheric disturbances occur in many parts of the world. Short-lived but sometimes severe atmospheric disturbances such as *thunderstorms* and *tornadoes* often develop in conjunction with other kinds of storms.

Midlatitude Cyclones

Midlatitude Cyclones

Probably most significant of all atmospheric disturbances are **midlatitude cyclones**. Throughout the midlatitudes, they dominate weather maps, are basically responsible for most day-to-day weather changes, and bring precipitation to much of the populated portions of the planet. Consisting of large, migratory low-pressure systems, they are usually called *depressions* in Europe and sometimes referred to as *lows*, *wave cyclones*, or *extratropical cyclones* in the United States.

Midlatitude cyclones are associated primarily with airmass convergence and conflict in regions between about 30° and 70° of latitude. Thus, they are found almost entirely within the band of westerly winds. Their general path of movement is toward the east, which explains why weather forecasting in the midlatitudes is essentially a west-facing vocation.

Because each midlatitude cyclone differs from all others in greater or lesser detail, any description must be a general one only. The discussions that follow, then, pertain to "typical" or idealized conditions. Moreover, these conditions are presented as Northern Hemisphere phenomena. For the Southern Hemisphere, the patterns of isobars, fronts, and wind flow should be visualized as mirror images of the Northern Hemisphere patterns (see Figure 7-13).

Characteristics

A typical mature midlatitude cyclone has a diameter of 1600 kilometers (1000 miles) or so. It is essentially a vast cell of low-pressure air, with ground-level pressure in the center typically between 990 and 1000 millibars. The system (shown by closed isobars on a weather map, as in Figure 7-6a) usually tends toward an oval shape, with the long axis trending northeast–southwest. Usually a clear-cut pressure trough extends southwesterly from the center.

Midlatitude cyclones have a converging counterclockwise circulation pattern in the Northern Hemisphere. This wind flow pattern attracts cool air from the north and warm air from the south. The convergence of these unlike air masses characteristically creates two fronts: a cold front that extends to the southwest from the center of the cyclone and runs along the pressure trough extending from the center of the storm, and a warm front extending eastward from the center and running along another, usually weaker, pressure trough. These two fronts divide the cyclone into a *cool sector* north and west of the center and a *warm sector* to the south and east. At ground level, the cool sector is the larger of the two, but aloft the warm sector is more extensive. This size relationship exists because both fronts lean over the cool air. Thus, the cold front slopes upward toward the northwest and the warm front slopes upward toward the northeast, as Figure 7-6b shows.

Clouds and precipitation develop in the zones within a midlatitude cyclone where air is rising. Because warm air rises along both fronts, the typical result is two zones of cloudiness and precipitation that overlap around the center of the storm (where air is rising in the center of the low pressure cell) and extend outward in the general direction of the fronts. Along and immediately behind the ground-level position of the cold front (the steeper of the two fronts), a

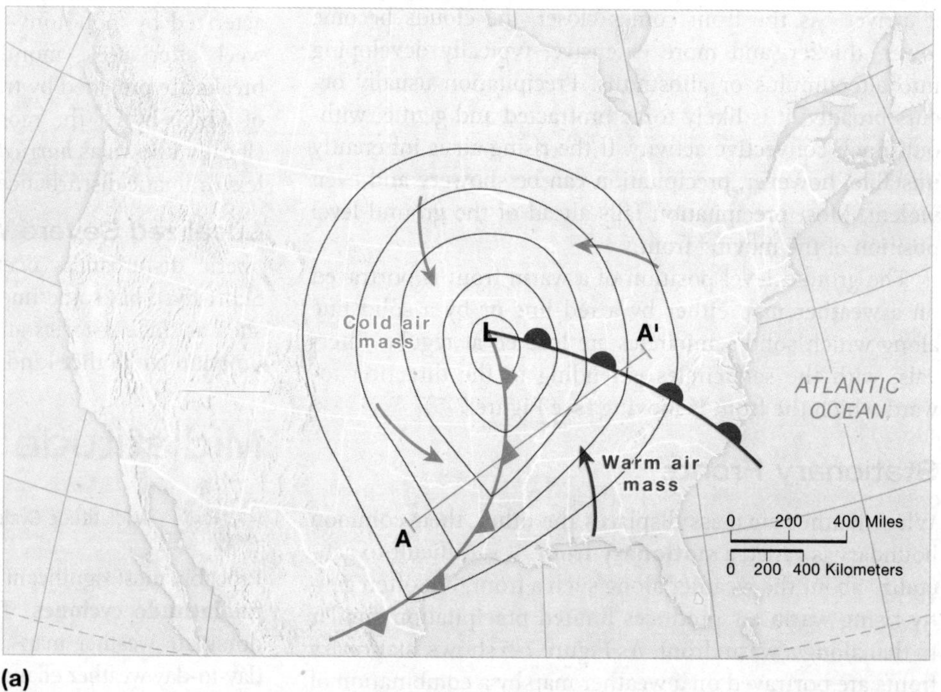

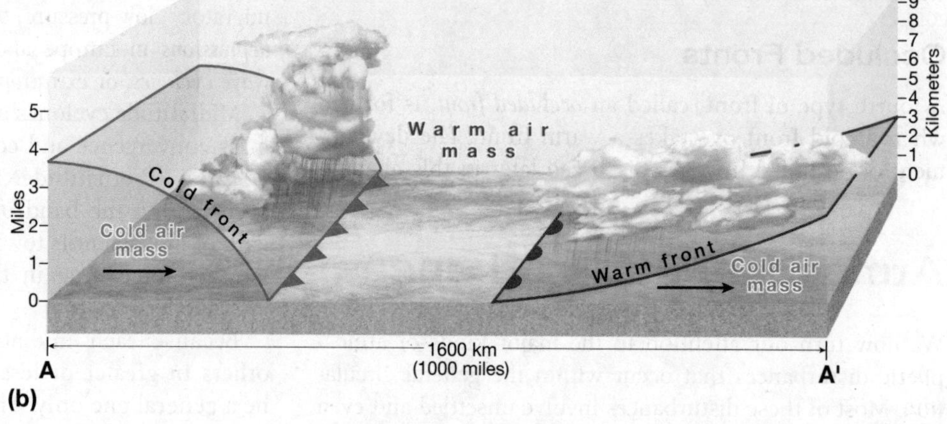

▶ **Figure 7-6** A map (a) and a cross section (b) of a typical mature midlatitude cyclone. In the Northern Hemisphere, there is usually a cold front trailing to the southwest and a warm front extending toward the east. Arrows in (b) indicate the direction of frontal movement.

◄ **Figure 7-7** A large midlatitude cyclone centered near Lake Michigan. The band of clouds extending down across the southern states to the west marks the cold front.

band of cumuliform clouds usually yields showery precipitation. The air rising more gently along the more gradual slope of the warm front produces a more extensive expanse of horizontally developed clouds, perhaps with widespread, protracted, low-intensity precipitation (Figure 7-7). In both cases, most of the precipitation originates in the warm air rising above the fronts and falls down through the front to reach the ground in the cool sector.

This precipitation pattern does not mean that the entire cool sector has unsettled weather and that the warm sector experiences clear conditions throughout. Although most frontal precipitation falls within the cool sector, the general area to the north, northwest, and west of the center of the cyclone is frequently cloudless as soon as the cold front has moved on. Thus, much of the cool sector is typified by clear, cold, stable air. In contrast, the air of the warm sector is often moist and tending toward instability, and so thermal convection and surface-wind convergence may produce sporadic thunderstorms. Also, sometimes one or more *squall lines* of intense thunderstorms develop in the warm sector in advance of the cold front.

Weather Changes with the Passing of a Front

When a front passes through a region, abrupt weather changes normally occur. For example, with the passage of a cold front (see Figures 7-3 and 7-6), we note the following:

1. The temperature decreases sharply.
2. Winds shift from southerly ahead of the front to northwesterly following it (in the Northern Hemisphere).
3. The front is in a pressure trough, so pressure falls as the front approaches and rises after it passes.
4. Generally clear skies ahead of the front are replaced by cloudiness and precipitation at the front—to be replaced again by clear skies after the front moves away.

Similar changes, although of lesser magnitude, occur with the passage of a warm front.

Movements

Midlatitude cyclones are essentially transient features, on the move throughout their existence. There are four kinds of movement involved (Figure 7-8):

1. The whole system moves as a major disturbance in the westerlies, traversing the midlatitudes generally from west to east. The rate of movement averages 30 to 45 kilometers (about 20 to 30 miles) per hour, which means that the storm can cross North America in some three to four days. It typically moves faster in winter than in summer. The route of a cyclone is likely to be undulating and erratic, although it moves generally from west to east, often in association with the path of the jet stream.
2. The system has a cyclonic wind circulation, with wind generally converging counterclockwise (in the Northern Hemisphere) into the center of the storm from all sides.
3. The cold front usually advances faster than the center of the storm.

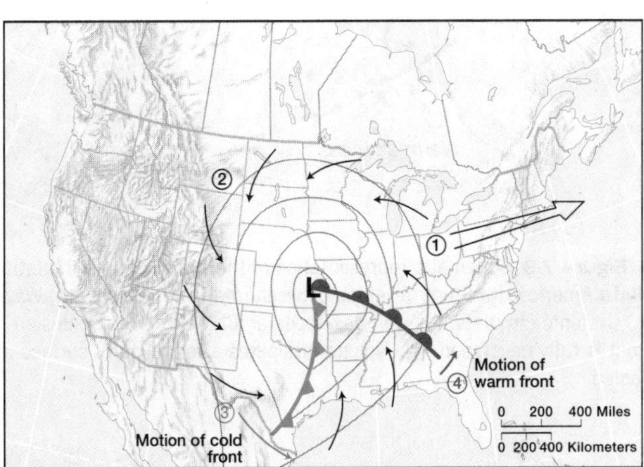

▲ **Figure 7-8** Four varieties of movement occur in a typical midlatitude cyclone: (1) The entire storm moves west to east in the general flow of the westerlies; (2) there is cyclonic converging counterclockwise airflow; (3) the cold front advances; (4) the warm front advances.

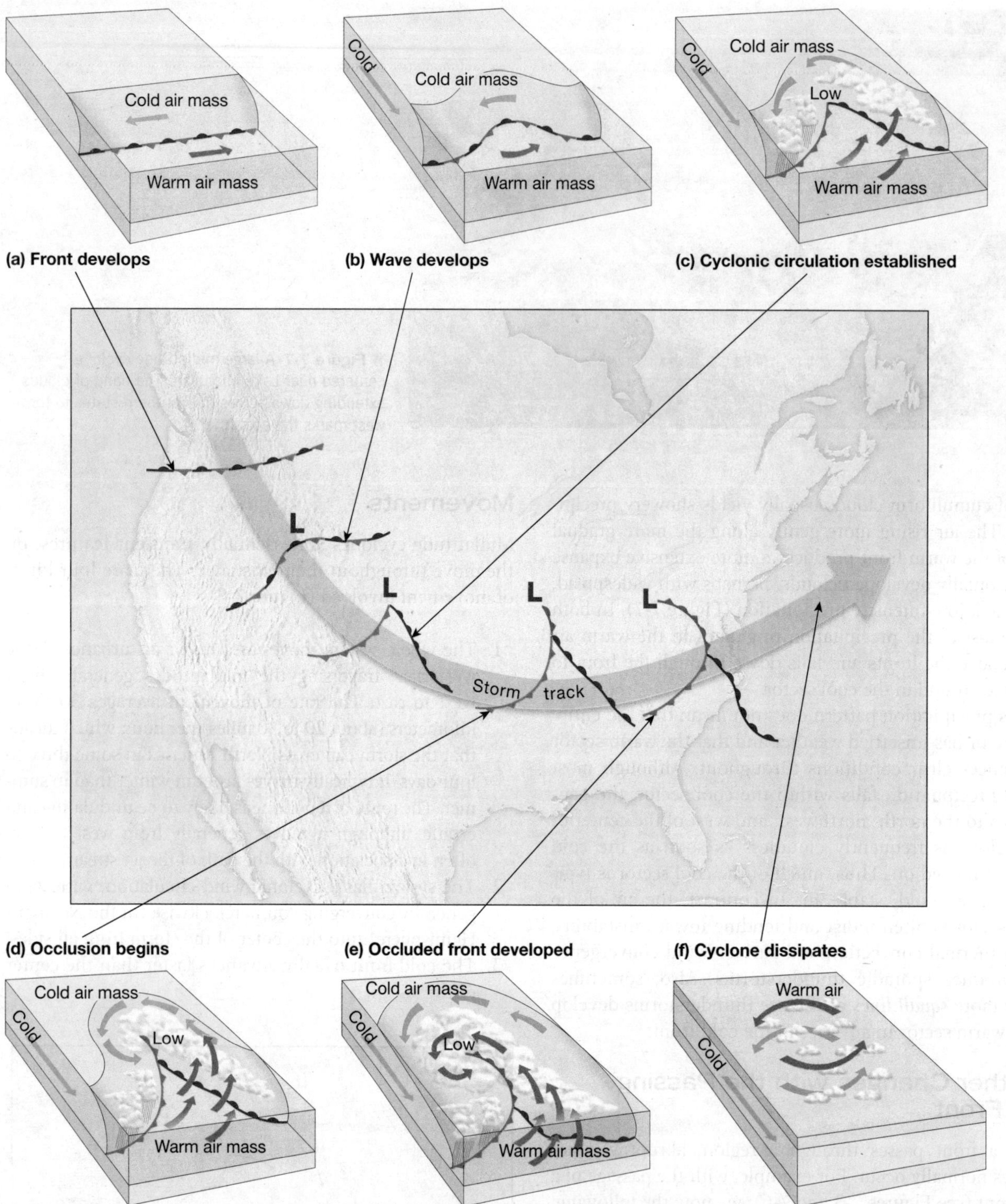

▲ **Figure 7-9** Schematic representation of the life cycle of a midlatitude cyclone passing over North America. (a) Front develops between unlike air masses. (b) Wave appears along front. (c) Cyclonic circulation is well developed around a low. (d) Occlusion begins. (e) Occluded front is fully developed. (f) Cyclone dissipates after all warm surface air has been lifted and cooled.

4. The warm front usually advances more slowly than the center of the storm, causing it to appear to lag behind. (This is only an apparent motion, however. In reality, the warm front is moving west to east, just like every other part of the system.)

Life Cycle

Cyclogenesis A typical midlatitude cyclone develops from origin to maturity in three to six days, and from maturity to dissipation in about the same length of time.

Most begin as "waves" (hence the alternative name *wave cyclone*) along the polar front, which as we saw in Chapter 5 is the contact zone between the relatively cold air of the polar easterlies and the relatively warm air of the westerlies. The opposing airflows normally have a relatively smooth linear motion on either side of the polar front (Figure 7-9a). On occasion, however, the smooth frontal surface may be distorted into a wave shape (Figure 7-9b). Various ground factors—such as topographic irregularities, temperature contrasts between sea and land, or the influence of ocean currents—can apparently initiate a wave along the front, but it is believed that the most common cause of *cyclogenesis* (the birth of cyclones) is upper troposphere conditions in the vicinity of the polar front jet stream.

There appears to be a close relationship between upper-level airflow and ground-level disturbances. When the upper airflow is *zonal*—by which we mean relatively straight from west to east—ground-level cyclonic activity is unlikely. When winds aloft begin to meander north to south in a *meridional airflow* (Figure 7-10), large waves of alternating pressure troughs and ridges are formed and cyclonic activity at ground level is intensified. Most midlatitude cyclones are centered below the polar front jet stream axis and downstream from an upper-level pressure trough.

A cyclone is unlikely to develop at ground level unless there is divergence above it. In other words, the convergence of air near the ground must be supported by divergence aloft. Such divergence can be related to changes in either speed or direction of the wind flow, but it nearly always involves broad north-to-south meanders in the Rossby waves and the jet stream.

Cyclogenesis also occurs on the leeward side of mountains. A low-pressure area drifting with the westerlies becomes weaker when it crosses a mountain range. As it ascends the range, the column of air compresses and spreads, slowing down its counterclockwise spin. When descending the leeward side, the air column stretches vertically and contracts horizontally. This change in shape causes it to spin faster and may initiate cyclonic development even if it were not a full-fledged cyclone before. This chain of events happens with some frequency in winter on the eastern flanks of the Rocky Mountains, particularly in Colorado, and with lesser frequency on the eastern side of the Appalachian Mountains, in North Carolina and Virginia. Cyclones formed in this way typically move toward the east and northeast and often bring heavy rain or snowstorms to the northeastern United States and southeastern Canada.

Occlusion Ultimately, the storm dissipates because the cold front overtakes the warm front. As the two fronts come closer and closer together (Figure 7-9c-e), the warm sector at the ground is increasingly displaced, forcing more and more warm air aloft. When the cold front catches up with the warm front, warm air is no longer in contact with Earth's

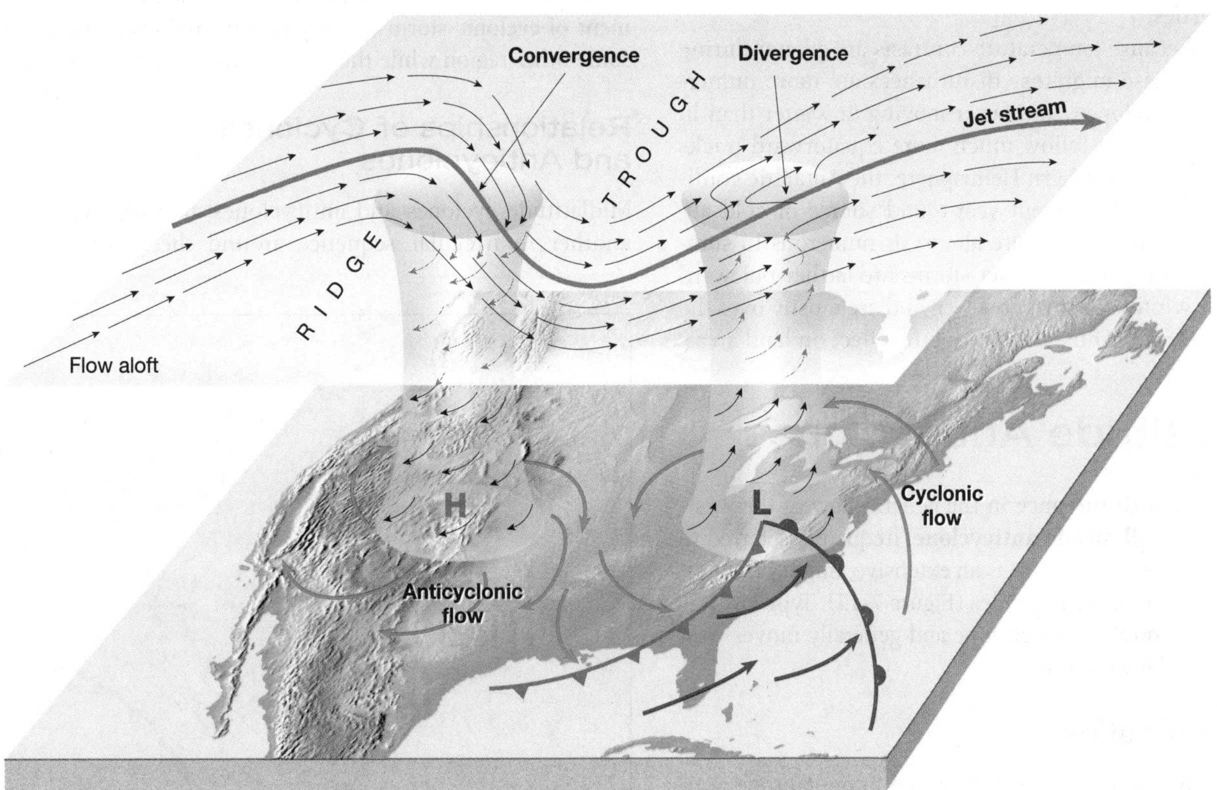

▲ **Figure 7-10** A typical winter situation in which the upper-level airflow is meridional (meanders north and south), creating "standing waves" aloft and cyclonic flow at ground level. Note how convergence and divergence aloft support anticyclonic and cyclonic circulation at ground level.

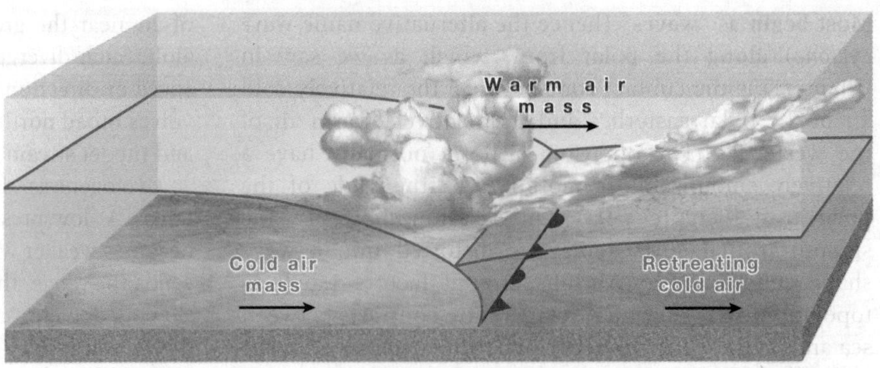

▶ **Figure 7-11** An occluded front develops when the leading edge of a cold front catches up with the trailing edge of a warm front, lifting all of the warm air off the ground. Once lifted, the warm air is much cooler than before.

surface and an **occluded front** is formed (Figure 7-11). This **occlusion** process usually results in a short period of intensified precipitation and wind until eventually all the warm sector is forced aloft and the ground-level low-pressure center is surrounded on all sides by cool air. This sequence of events weakens the pressure gradient and shuts off the storm's energy and air lifting mechanism so that it dies out (Figure 7-9f).

Occurrence and Distribution

At any given time, from 6 to 15 midlatitude cyclones exist in the Northern Hemisphere midlatitudes, and an equal number in the Southern Hemisphere. They occur at scattered but irregular intervals throughout the zone of the westerlies.

In part because temperature contrasts are greater during the winter, these migratory disturbances are more numerous, better developed, and faster moving in winter than in summer. They also follow much more equatorward tracks in winter. In the Southern Hemisphere, the Antarctic continent provides a prominent year-round source of cold air, and so vigorous cyclones are almost as numerous in summer as in winter. The summer storms are farther poleward than their winter cousins, however, and are mostly over the Southern Ocean. Thus, they have little effect on land areas.

Midlatitude Anticyclones

Another major disturbance in the general flow of the westerlies is the **midlatitude anticyclone**, frequently referred to simply as a "high" (H). This is an extensive, migratory high-pressure cell of the midlatitudes (Figure 7-12). Typically it is larger than a midlatitude cyclone and generally moves west to east with the westerlies.

Characteristics

As with any other high-pressure center, a midlatitude anticyclone has air converging into it from above, subsiding, and diverging at the surface, clockwise in the Northern Hemisphere and counterclockwise in the Southern Hemisphere. No air-mass conflict or surface convergence is involved, and so anticyclones contain no fronts (the fronts shown in Figure 7-12 are outside the high-pressure system). The weather is clear and dry with little or no opportunity for cloud formation. Wind movement is very limited near the center of an anticyclone but increases progressively outward. Particularly along the eastern margin (the leading edge) of the system, there may be strong winds. In winter, anticyclones are characterized by very low temperatures.

Anticyclones move toward the east either at the same rate as or a little slower than midlatitude cyclones. Unlike cyclones, however, anticyclones are occasionally prone to stagnate and remain over the same region for several days. This stalling brings clear, stable, dry weather to the affected region, which enhances the likelihood that air pollutants will become concentrated. Such stagnation may block the eastward movement of cyclonic storms, causing protracted precipitation in some other region while the anticyclonic region remains dry.

Relationships of Cyclones and Anticyclones

Midlatitude cyclones and anticyclones alternate with one another in irregular sequence around the world in the

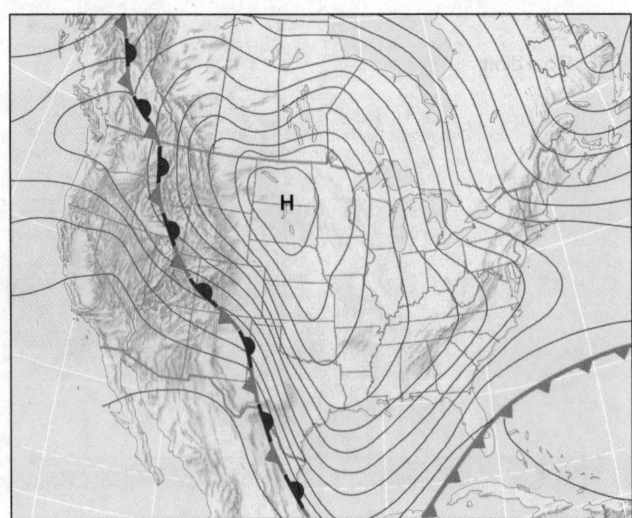

▲ **Figure 7-12** A typical well-developed midlatitude anticyclone centered over the Dakotas. Both fronts shown here are outside the high-pressure system.

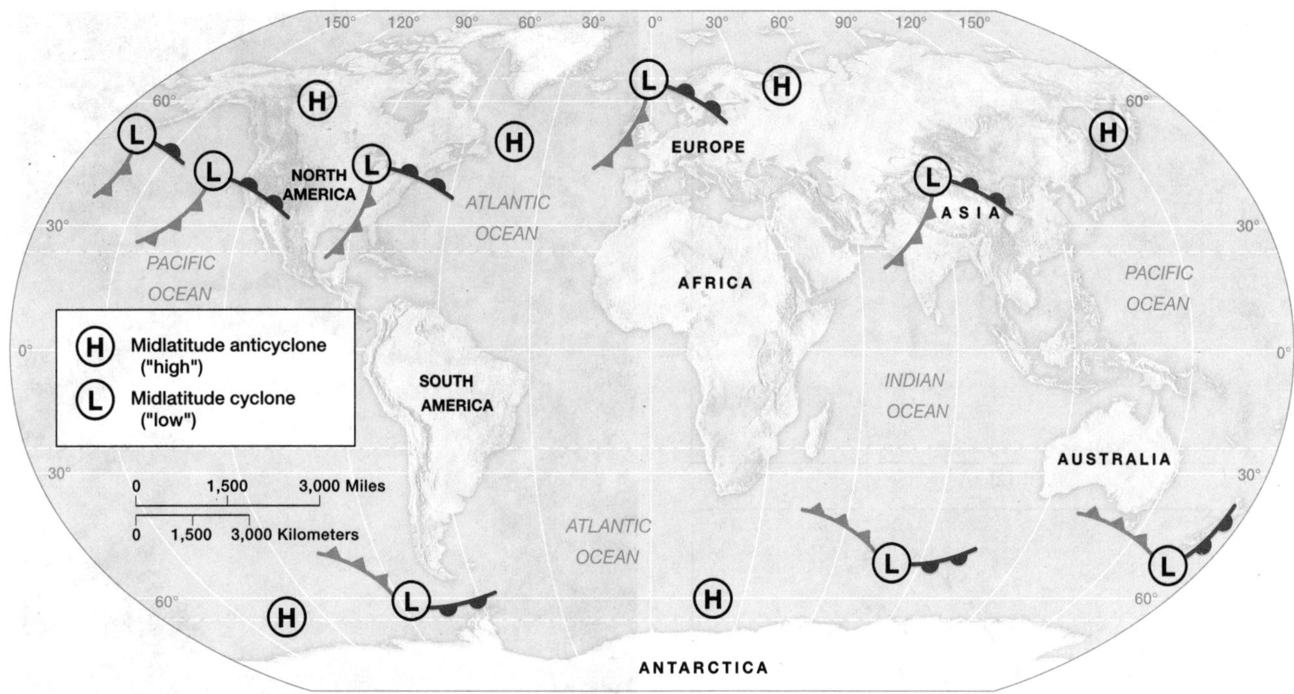

▲ **Figure 7-13** At any given time, the midlatitudes are dotted with midlatitude cyclones and anticyclones. This map depicts a hypothetical situation in January. Note the orientation of fronts in the Southern Hemisphere storms.

midlatitudes (Figure 7-13). Each can occur independently of the other, but there is often a functional relationship between them. This relationship can be seen when an anticyclone closely follows a cyclone, as diagrammed in Figure 7-14. The winds diverging from the eastern margin of the high fit into the flow of air converging into the western side of the low. It is easy to visualize the anticyclone as a polar air mass having the cold front of the cyclone as its leading edge.

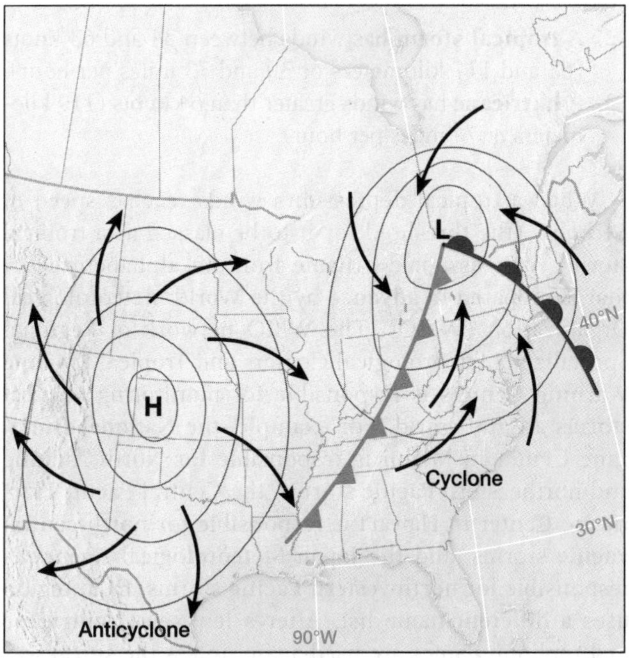

▲ **Figure 7-14** Midlatitude cyclones and anticyclones often occur in juxtaposition in the middle latitudes.

Minor Tropical Disturbances: Easterly Waves

An **easterly wave** is a long but weak migratory, low-pressure system that may occur almost anywhere between 5° and 30° of latitude (Figure 7-15). They are a common kind of tropical disturbance, usually consisting of a band of small thunderstorms with little or no cyclonic rotation. Easterly waves are usually several hundred kilometers long and nearly always oriented north–south. They drift slowly westward in the flow of the trade winds, bringing characteristic weather with them. Ahead of the wave is fair weather with divergent airflow. Behind the wave, convergent conditions prevail, with moist air being uplifted to yield convective thunderstorms and sometimes widespread cloudiness. There is little or no temperature change with the passage of easterly waves.

Most of the easterly waves that move across the North Atlantic originate over North Africa, and then move out over the Atlantic in the trade winds. The vast majority of easterly waves weaken and die out over the ocean, but a small percentage intensify into more powerful tropical cyclones—our next topic.

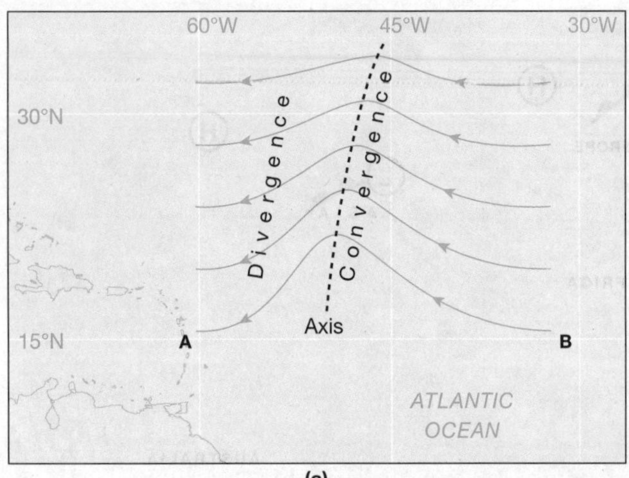

(a)

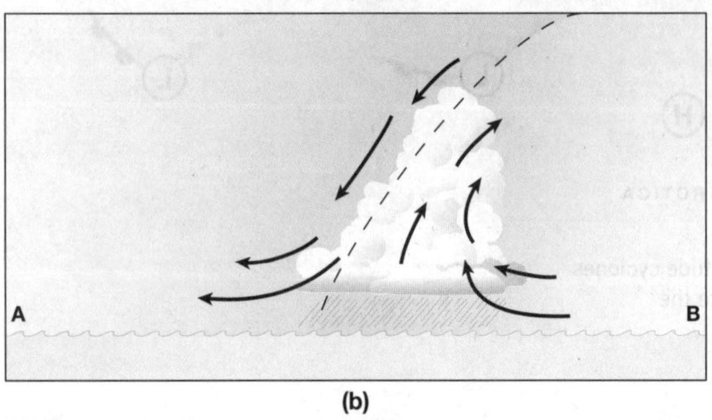

(b)

▲ **Figure 7-15** Diagrammatic map view (a) and cross section (b) of an easterly wave. The arrows indicate general direction of airflow.

▲ **Figure 7-16** Hurricane Ike approaching Cuba as a Category 4 storm on September 7, 2008. This image was captured by the Moderate Resolution Imaging Spectroradiometer (MODIS) on NASA's Aqua satellite.

Major Tropical Disturbances: Hurricanes

Hurricanes

Tropical cyclones are intense, low-pressure disturbances that develop in the tropics and occasionally move poleward into the midlatitudes. Intense tropical cyclones are known by different names in different parts of the world: *hurricanes* in North and Central America, *typhoons* in the western North Pacific, *baguios* in the Philippines, and simply *cyclones* in the Indian Ocean and Australia. Hurricanes are considerably smaller than midlatitude cyclones, typically having a diameter of between 160 and 1000 kilometers (100 and 600 miles; Figure 7-16).

Tropical cyclones develop from incipient low-pressure perturbations in trade-wind flow, generally called *tropical disturbances* by the U.S. National Weather Service. About 100 of these are identified each year over the tropical North Atlantic, but only a few strengthen into hurricanes. Three categories of tropical disturbances are recognized on the basis of wind speed:

1. A **tropical depression** has wind speeds of less than 33 knots (61 kilometers or 38 miles per hour), but has developed a closed wind circulation pattern.
2. A **tropical storm** has winds between 34 and 63 knots (63 and 117 kilometers or 39 and 73 miles per hour).
3. A **hurricane** has winds greater than 64 knots (119 kilometers or 74 miles per hour).

When a tropical depression's winds reach a speed of 34 knots (the threshold for it to be classed as a tropical storm), it is assigned a name from an alphabetical list that is prepared in advance by the World Meteorological Organization (WMO). The WMO network of Regional Specialized Meteorological Centers and Tropical Cyclone Warning Centers is responsible for monitoring tropical storms around world. For example, the National Hurricane Center in Miami is responsible for North Atlantic and northeastern Pacific storms, the Central Pacific Hurricane Center in Hawai'i is responsible for north-central Pacific storms, and the Japan Meteorological Agency is responsible for northwestern Pacific storms. Each region uses a different name list. After a few years, hurricane and typhoon names are used again unless the storm was especially notable—such as Katrina and Andrew—in which case the name is retired from the list.

PEOPLE AND THE ENVIRONMENT

Hurricane Katrina

On August 23, 2005, the 12th tropical depression of the season developed in the southeastern Bahamas. By the next day, the depression had strengthened and was named Tropical Storm Katrina. The storm moved slowly to the northwest, strengthening into a category 1 hurricane just a few hours before making landfall in south Florida on August 25th with winds gusting to 78 knots (145 kph; 90 mph). As Katrina moved westward over Florida, it left more than 40 centimeters (16 inches) of rain in Perrine, Florida.

Although weakened by its passage over land, Katrina quickly intensified again as it moved out over the warm waters of the Gulf of Mexico (Figure 7-A). For the next two days, Katrina gained strength, becoming a category 5 storm on the 28th with winds of more than 150 knots (275 kph; 170 mph), and a central pressure of 902 mb (the 6th lowest of any recorded Atlantic storm; the lowest on record was Hurricane Wilma two months later that year, with a pressure of 882 mb). By the time Katrina made landfall on the Gulf Coast just southeast of New Orleans on the morning of August 29, it had weakened slightly to a strong category 3 storm. By the evening of August 29, Katrina had weakened to tropical storm strength and by the next day had been downgraded to a tropical depression (see Figure 7-20).

Sustained wind speeds at landfall in southeastern Louisiana were more than 108 knots (200 kph; 125 mph); wind gusts of more than 87 knots (160 kph; 100 mph) were recorded in New Orleans. Rainfall storm totals exceeded 25 centimeters (10 inches) along much of Katrina's path, and with a maximum of nearly 38 centimeters (15 inches) recorded in Big Branch, Louisiana. Sixty-two tornadoes were spawned by Katrina.

The greatest devastation, however, came from Katrina's storm surge and flooding from heavy rain. Mobile, Alabama, and large areas of Gulfport and Biloxi, Mississippi, were inundated by Katrina's 8.0 to 8.5 meter (26 to 28 foot) high storm surge. New Orleans, although spared from a direct hit by the storm (the eye passed just east of the city), was first flooded beginning at about 7:00 A.M. A 3.0 to 3.6 meter (10 to 12 foot) storm surge rolled up the Intercoastal Waterway from the Gulf into the city's Industrial Canal, overtopping and scouring levees and flooding the eastern parts of the city. About three hours later—after some news agencies had reported that New Orleans had luckily escaped the worst destruction of Katrina—a second flood hit the city when the London Avenue and 17th Street canals failed, sending water pouring into the central parts of the city from Lake Pontchartrain. By August 31, 80 percent of New Orleans was underwater—in some places 6 meters (20 feet) deep (Figure 7-B). For the next few days, the world watched events in New Orleans on television while government officials seemed unable to coordinate the rescue effort needed to help the thousands of people trapped in the flooded city.

So why was New Orleans so devastated by this storm? Many factors contributed to this disaster. New Orleans sits in a shallow "bowl" alongside the Mississippi River. Draining this once swampy area over the last few hundred years has led to soil compaction and subsidence that has left parts of the city below sea level, protected only by a series of levees. The wetlands of the Mississippi River delta once offered some protection by slowing and absorbing the punch from hurricane storm surges, but flood control efforts upstream have changed the sediment load of the river and over the last few decades these protective marshlands on the lower delta have been sinking and eroding away, now offering little protection from storms. Finally, the levees themselves were inadequate, not being designed to handle such an enormous surge of water.

The loss of life from Hurricane Katrina may never be known with certainty, but is likely to have been more than 1200. More than a year later, tens of thousands of survivors were still displaced—especially citizens from the poorest parts of New Orleans. Some scientists had been warning about such a potential disaster for years, and local authorities were pressing for financial help to strengthen the levees. Even if those planned measures had been taken, it is not clear if it would have been enough to avert the disaster in New Orleans.

▲ **Figure 7-A** Satellite image of Hurricane Katrina at 9:15 A.M. eastern time on August 29, 2005.

▲ **Figure 7-B** New Orleans after Hurricane Katrina on August 30, 2005.

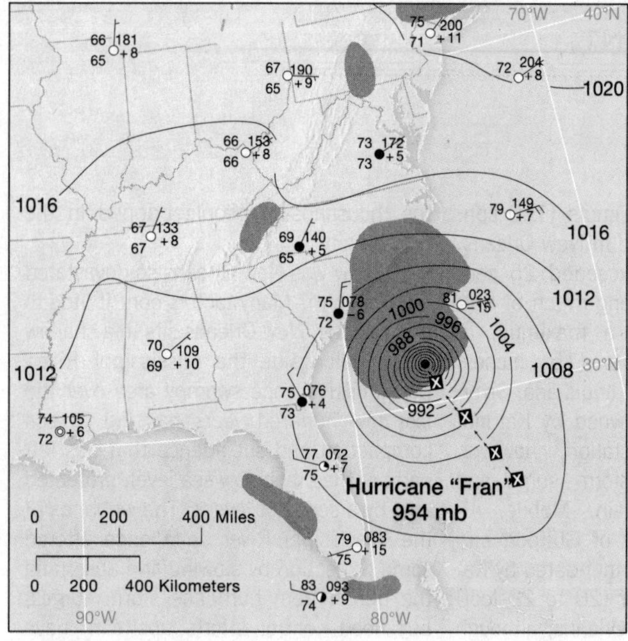

▲ **Figure 7-17** Weather map of Hurricane Fran on September 5, 1996. The isobars show a very steep gradient around the low-pressure center.

Characteristics

Hurricanes (as we generally refer to tropical cyclones in this chapter) consist of prominent low-pressure centers that are essentially circular, with a steep pressure gradient outward from the center (Figure 7-17). As a result, strong winds spiral inward. Winds must reach a speed of 64 knots for the storm to be officially classified as a hurricane, although winds in a well-developed hurricane often double that speed and occasionally triple it.

The converging cyclonic wind pattern of a hurricane pulls in warm, moist air—the "fuel" that powers the storm. As warm, water vapor–laden air spirals into the storm, it rises in intense updrafts within towering cumulonimbus clouds. As the air rises it cools adiabatically, bringing the air to saturation; condensation releases vast amounts of liquid water that builds up the huge clouds and feeds the heavy rain. Condensation also releases latent heat. It is the release of latent heat that powers and strengthens a storm by increasing the instability of the air: in a short period of time and in a relatively small area, a hurricane releases an enormous amount of energy into the atmosphere. An average mature hurricane releases in one day approximately as much energy as generated by all electric utility plants in the United States in one year.

Eye of a Hurricane A remarkable feature of a well-developed hurricane is the nonstormy **eye** in the center of the storm (Figure 7-18). The winds do not converge to a central point, but rather reach their highest speed at the *eye wall,* which is the edge of the eye. The eye has a diameter of from 16 to 40 kilometers (10 to 25 miles) and is a singular area of calmness in the maelstrom that whirls around it.

The weather pattern within a hurricane is relatively symmetrical around the eye. Bands of dense cumulus and cumulonimbus clouds (called *spiral rain bands*) curve in from the edge of the storm to the eye wall, producing heavy rain that generally increases in intensity inward. Updrafts are common throughout the hurricane, becoming most prominent around the eye wall—but within the eye itself a downdraft is found, inhibiting cloud formation. Near the top of the storm, air diverges clockwise out into the upper troposphere.

The clouds of the eye wall tower to heights that may exceed 16 kilometers (10 miles). Within the eye, there is no rain and almost no low clouds; scattered high clouds may part to let in intermittent sunlight. The wall of

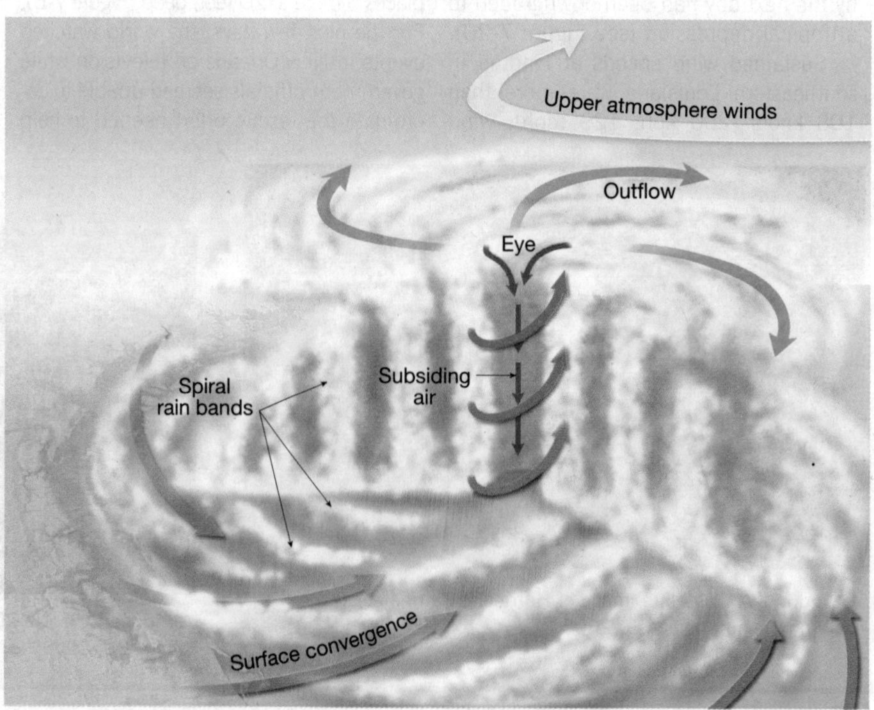

► **Figure 7-18** An idealized cross section through a well-developed hurricane. Air spirals into the storm horizontally and rises rapidly to produce towering cumulus and cumulonimbus clouds that yield torrential rainfall. In the center of the storm is the eye, where air movement is downward.

thunderstorms circling the eye is sometimes surrounded by a new wall of thunderstorms. The inner wall disintegrates and is replaced by the outer wall. The process, called *eye-wall replacement*, usually lasts less than 24 hours and tends to weaken the storm.

Origin

Hurricanes form only over warm oceans in the tropics (the ocean water temperature generally needs to be at least 26.5°C [80°F] to a depth of 50 meters [160 feet] or more), and at least a few degrees north or south of the equator (Figure 7-19). Because the Coriolis effect is so minimal near the equator, no hurricane has ever been observed to form within 3° of it, no hurricane has ever been known to cross it, and the appearance of hurricanes closer than some 8° or 10° of the equator is very rare. More than 80 percent originate in or just on the poleward side of the intertropical convergence zone.

The exact mechanism of formation is not completely understood, but hurricanes always develop out of a preexisting disturbance in the tropical troposphere. Easterly waves, discussed earlier, provide low-level convergence and lifting that catalyze the development of many hurricanes. Even so, fewer than 10 percent of all easterly waves grow into hurricanes. Hurricanes can evolve only when there is no significant **wind shear** with height (*wind shear* refers to a significant change in wind direction or wind speed with increasing elevation), which implies that temperatures at low altitudes are reasonably uniform over a wide area.

Movement

Hurricanes occur in a half-dozen low-latitude regions (Figure 7-19). They are most common in the North Pacific basin, originating largely in two areas: east of the Philippines and west of southern Mexico and Central America. The third most notable region of hurricane development is in the west-central portion of the North Atlantic basin, extending into the Caribbean Sea and Gulf of Mexico. These ferocious storms are also found in the western portion of the South Pacific and all across the South Indian Ocean, as well as in the North Indian Ocean both east and west of the Indian peninsula. They are very rare in the South Atlantic and in the southeastern part of the Pacific, apparently because the water is too cold and because high pressure dominates. The strongest and largest hurricanes are typically those of the western Pacific Ocean. Only a few storms in other parts of the world have attained the size and intensity of the large East Asian "super" typhoons.

Once formed, hurricanes follow irregular tracks within the general flow of the trade winds. A specific path is very difficult to predict many days in advance, but the general pattern of movement is highly predictable. Roughly one-third of all hurricanes travel east to west without much latitudinal change. The rest, however, begin on an east–west path and then curve prominently poleward, where they either dissipate over the adjacent continent or become enmeshed in the general flow of the midlatitude westerlies (Figure 7-20).

Hurricanes sometimes survive (with diminished intensity) off the east coasts of continents in the midlatitudes

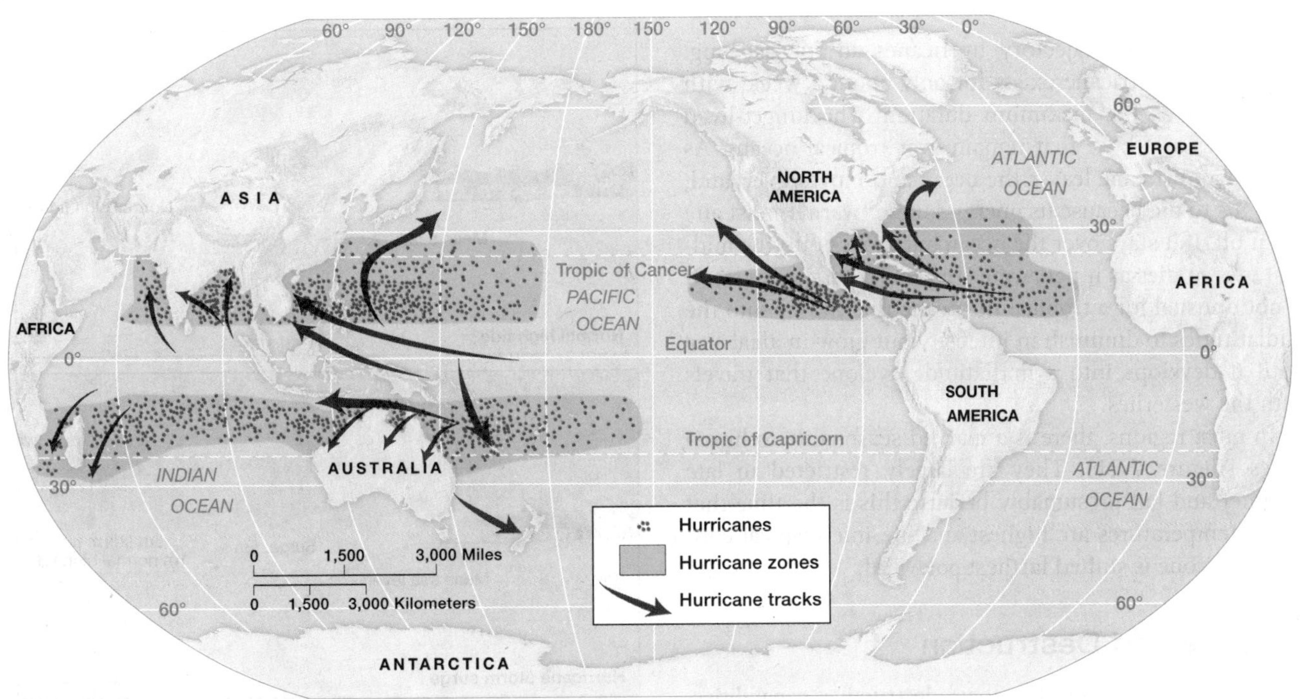

▲ **Figure 7-19** Major generalized hurricane tracks and location of origin points of hurricanes over a 19-year period.

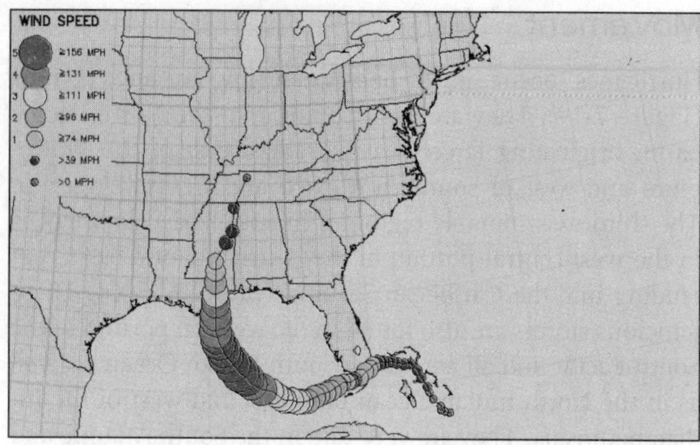

▲ **Figure 7-20** Path of Hurricane Katrina from August 23 to August 30, 2005. Wind speed is indicated by colored dots (light blue = 74 mph; yellow = 111 mph; red = 156 mph). Notice that wind speed increased as Katrina passed over the very warm water in the Gulf of Mexico, but diminished quickly after the storm moved over land.

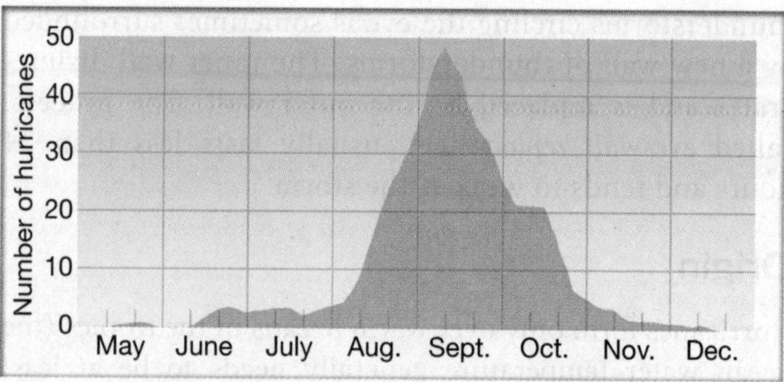

▲ **Figure 7-21** Seasonality of North Atlantic–Caribbean hurricanes over a nine-year period. Hurricane season reaches its peak in the autumn, with a prominent maximum in September.

because of the warm ocean currents there; hurricanes do not survive in the midlatitudes off the west coasts of continents because of the cool ocean currents there.

In one region—the southwestern Pacific Ocean north and northeast of New Zealand—there is a marked variation from this general flow pattern. Hurricanes in this part of the Pacific usually move erratically from northwest to southeast. Therefore, when they strike an island such as Fiji or Tonga, they approach from the west, a situation not replicated anywhere else in the world at such a low latitude. These seemingly aberrant tracks apparently result from the simple fact that the tropospheric westerlies extend quite far equatorward in the Southwest Pacific, and hurricanes there are basically steered by the general circulation pattern.

Whatever their trajectory, hurricanes do not last long. The average hurricane exists for only about a week, with four weeks as the maximum duration. The longer-lived hurricanes are those that remain over tropical oceans. As soon as a hurricane leaves the ocean and moves over land, it begins to die because its energy source (warm, moist air) is cut off. If it stays over the ocean but moves into the midlatitudes, it dies as it penetrates the cooler environment. It is not unusual for a tropical hurricane that moves into the midlatitudes to diminish in intensity but grow in areal size until it develops into a midlatitude cyclone that travels with the westerlies.

In most regions, there is a marked seasonality to hurricanes (Figure 7-21). They are largely restricted to late summer and fall presumably because this is the time that ocean temperatures are highest and the intertropical convergence zone is shifted farthest poleward.

Damage and Destruction

Hurricanes are known for their destructive capabilities. Some of the destruction comes from high winds and torrential rain, and in some cases, tornadoes are spawned by a hurricane. However, the overwhelming cause of damage and loss of life is the flooding brought by high seas. The low pressure in the center of the storm allows the ocean surface to bulge as much as 1 meter (3 feet). To this is added a **storm surge** of wind-driven water as much as 7.5 meters (25 feet) above normal tide level when the hurricane pounds into a shoreline. Thus, a low-lying coastal area can be severely inundated, and 90 percent of hurricane-related deaths are drownings (Figure 7-22).

The greatest hurricane disaster in U.S. history occurred in 1900 when Galveston Island, Texas, was overwhelmed by a 6-meter (20-foot) storm surge that killed as many as 8000 people, nearly one-sixth of Galveston's population. In other regions, hurricane devastation has been much greater. The flat deltas of the Ganges and Brahmaputra rivers in Bangladesh have been subjected to enormous losses of human life from Indian Ocean cyclones: 300,000 deaths in 1737, another 300,000 in 1876, another 500,000

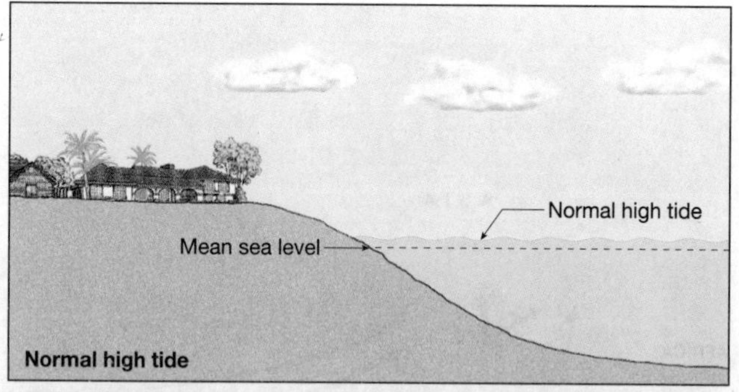

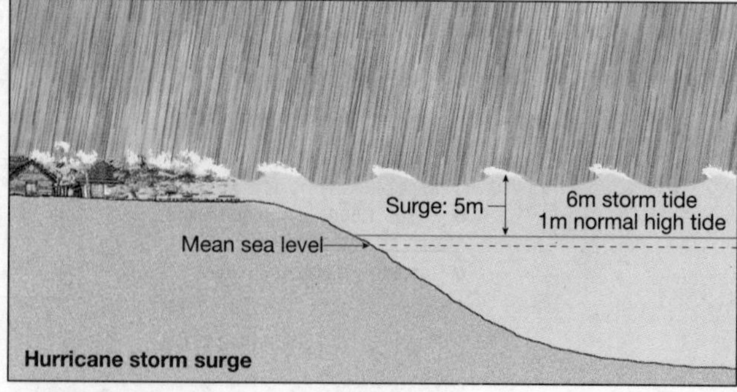

▲ **Figure 7-22** When a hurricane storm surge accompanies a normal high tide, the torrent can overwhelm a coastal area.

PEOPLE AND THE ENVIRONMENT

Hurricanes and Global Warming

The 2005 hurricane season in the North Atlantic was the most active on record, with 28 named storms. Three of the most powerful hurricanes ever measured in terms of minimum atmospheric pressure in the eye occurred that year: Katrina, Rita, and Wilma (Figure 7-C; Table 7-A). The 2008 season was another very active year, tied for fourth overall with 16 named storms. Given the connection between high ocean temperature and hurricane formation in general—as well as evidence that global warming is slightly increasing sea surface temperatures—an obvious question to ask is whether the recent increase in hurricane activity (either the number of storms or their intensity) is tied to global warming?

Over the last 15 years or so, there has been a general increase in the annual number of hurricanes in the North Atlantic. However, many meteorologists think that this increase in frequency is simply part of a multidecadal cycle of hurricane activity that has been well documented since the early 1900s. Known as the Atlantic Multi-Decadal Signal, the pattern includes such factors as higher sea surface temperatures (SST), lower vertical wind shear, and an expanded upper-level westward flow of the atmosphere off North Africa. Although the underlying causes of all the components of the Atlantic Multi-Decadal Signal are not completely understood, the recent upswing in hurricane frequency can likely be explained without tying it to global warming. Meteorologists at NOAA's Climate Prediction Center have generally been forecasting a higher-than-average number of North Atlantic tropical storms since the early 2000s.

On the other hand, based on the computer modeling of a number of greenhouse gas emissions scenarios, the 2007 *Fourth Assessment Report* of the Intergovernmental Panel on Climate Change concluded that it was likely (greater than 66 percent probability) that we will observe an increase in the intensity of tropical cyclones during this century. Several recent studies suggest that there has indeed been an increase in the number of powerful North Atlantic and Indian Ocean hurricanes over the last three decades, consistent with some computer model projections.

Scientists can gauge overall intensity of hurricane activity during a year with the Accumulated Cyclone Energy (ACE) Index (Figure 7-D). The ACE Index describes the total wind energy of all tropical storms during a season. It is calculated by using the maximum sustained wind speed for each named storm every six hours while they are at tropical storm strength or greater—the greater the total period of high wind speed, the greater the total wind energy released during a season.

Some meteorologists think that the evidence is now strong enough to conclude that there is a connection between global warming, higher SST, and an increased frequency of powerful hurricanes in the North Atlantic. For other atmospheric scientists, however, the data set is too limited before regular satellite monitoring of hurricanes began to come to that conclusion just yet.

▲ **Figure 7-C** Hurricane Wilma, the strongest North Atlantic hurricane ever recorded, as it approached Mexico's Yucatan Peninsula on October 20, 2005.

▲ **Figure 7-D** The Accumulated Cyclone Energy (ACE) Index for the North Atlantic from 1950 to 2008. Since the mid-1990s it has been a period of high hurricane activity. The projection for the 2009 season was for average activity.

TABLE 7-A The Ten Most Intense Hurricanes (based on lowest central pressure) in the North Atlantic Basin from 1851 to 2008.

	Hurricane	Year	Minimum Pressure
1	Hurricane Wilma	2005	882 mb
2	Hurricane Gilbert	1988	888 mb
3	1935 Labor Day Hurricane	1935	892 mb
4	Hurricane Rita	2005	895 mb
5	Hurricane Allen	1980	899 mb
6	Hurricane Katrina	2005	902 mb
7	Hurricane Camille	1969	905 mb
	Hurricane Mitch	1998	905 mb
	Hurricane Dean	2007	905 mb
10	1924 Cuba Hurricane	1924	910 mb*
	Hurricane Ivan	2004	910 mb

*The strength of the 1924 Cuba hurricane was recently reevaluated by meteorologists; its minimum central pressure is estimated to have been 910 mb.

(Data source: U.S. Department of Commerce/NOAA)

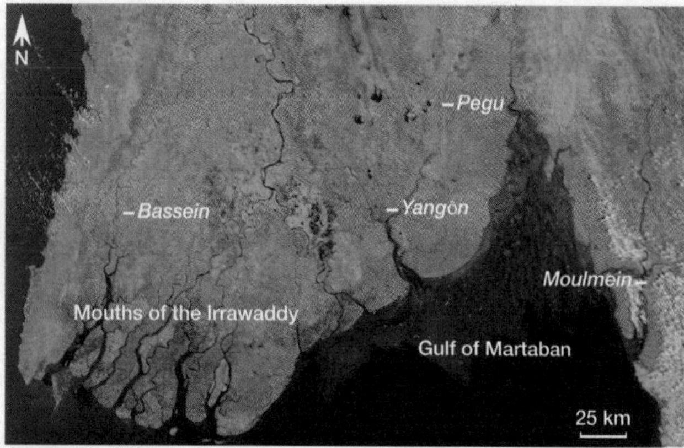

April 15, 2008

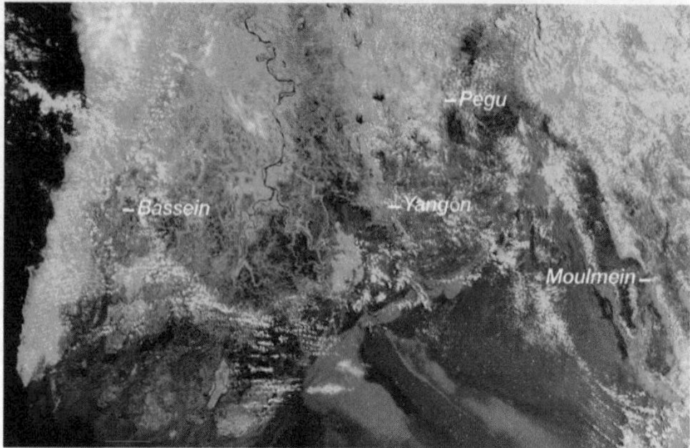

May 5, 2008

▲ **Figure 7-23** In 2008 Cyclone Nargis flooded more than 14,000 square kilometers (5500 square miles) of the low-lying Irrawaddy River delta in Myanmar (Burma), killing more than 130,000 people. The top satellite image was taken on April 15 before the storm hit; the bottom image was taken on May 5 after the storm came ashore.

in 1970, and 175,000 in 1991. In 2008, flooding from Cyclone Nargis in Myanmar (Burma) killed at least 130,000 people (Figure 7-23). Most of the damage along the Gulf coast of the United States in August 2005 from Hurricane Katrina was caused by an eight-meter high (26 foot) storm surge (Figure 7-24).

Although the amount of damage caused by a hurricane depends in part on the physical configuration of the landscape and the population size and density of the affected area, storm strength is the most important factor. In the United States, the **Saffir-Simpson Hurricane Scale** has been established to rank the relative intensity of hurricanes (based on central eye pressure, wind speed, and storm surge height), ranging from 1 to 5, with 5 being the most severe (Table 7-2).

Destruction and tragedy are not the only legacies of hurricanes, however. Such regions as northwestern Mexico, northern Australia, and southeastern Asia rely on tropical storms for much of their water supply. Hurricane-induced rainfall is often a critical source of moisture for agriculture. Although crops within the immediate path of the storm may be devastated by winds and flooding, a much more extensive area may be nurtured by the life-giving rains.

Localized Severe Weather

Several kinds of smaller atmospheric disturbances are common in various parts of the world. Some are locally of great significance, and some are destructive. All occur at a much more localized scale than do tropical cyclones and midlatitude cyclones, although thunderstorms and tornadoes may be associated with both kinds of larger storms.

► **Figure 7-24** Large casino barge washed across Highway 90 in Biloxi, Mississippi, by the storm surge of Hurricane Katrina in August 2005.

TABLE 7-2 Saffir-Simpson Hurricane Scale

Category	Central Pressure (millibars)	Wind Speed: Kilometers per Hour	Wind Speed: Miles per Hour	Storm Surge (meters)	Damage
1	> 979	119–153	74–95	1.2–1.5	Minimal
2	965–979	154–177	96–110	1.6–2.4	Moderate
3	945–964	178–209	111–130	2.5–3.6	Extensive
4	920–944	210–250	131–155	3.7–5.4	Extreme
5	< 920	> 250	> 155	> 5.4	Catastrophic

Thunderstorms

A **thunderstorm**, defined as a violent convective storm accompanied by thunder and lightning, is usually localized and short-lived. It is always associated with vertical air motion, considerable humidity, and instability, a combination that produces a towering cumulonimbus cloud and (nearly always) showery precipitation.

Thunderstorms sometimes occur as individual clouds, produced by nothing more complicated than thermal convection; such developments are commonplace in the tropics and during summer in much of the midlatitudes. Thunderstorms are also frequently found in conjunction with other kinds of storms, however, or are associated with other mechanisms that can trigger unstable uplift. Thus thunderstorms often accompany hurricanes, tornadoes, fronts (especially cold fronts) in midlatitude cyclones, and orographic lifting that may produce instability.

The uplift, by whatever mechanism, of warm, moist air must release enough latent heat of condensation to sustain the continued rise of the air. In the early stage of thunderstorm formation (Figure 7-25), called the *cumulus stage*, updrafts prevail and the cloud grows. Above the freezing level, supercooled water droplets and ice crystals coalesce: when they become too large to be supported by the updrafts, they fall. These falling particles drag air with them, initiating a downdraft. When the downdraft with its accompanying precipitation leaves the bottom of the cloud, the thunderstorm enters the *mature stage*, in which updrafts and downdrafts coexist as the cloud continues to enlarge. The mature stage is the most active time, with heavy rain often accompanied by hail, blustery winds, lightning, thunder, and the growth of an "anvil" top composed of ice crystals on the massive cumulonimbus cloud. Eventually downdrafts dominate and the *dissipating stage* is reached, with light rain ending and turbulence ceasing.

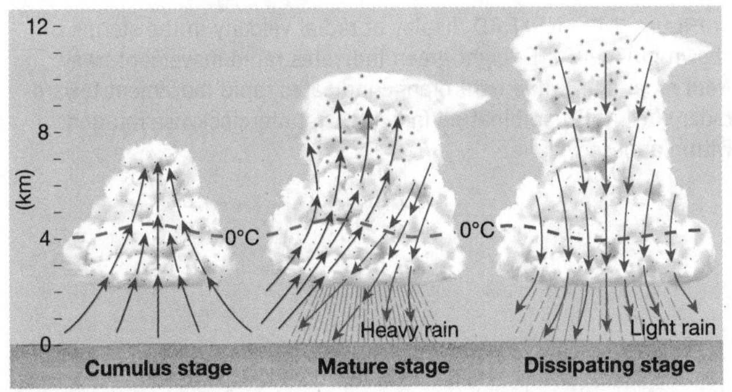

▲ **Figure 7-25** Sequential development of a thunderstorm cell.

Thunderstorms are most common where there are high temperatures, high humidity, and high instability, a combination typical of the intertropical convergence zone. There is a general decrease in thunderstorm frequency away from the equator, and they are virtually unknown poleward of 60° of latitude (Figure 7-26). There is much greater frequency of thunderstorms over land than water because summer temperatures are higher over land and most thunderstorms occur in the summer. In the United States, the greatest annual thunderstorm activity occurs in Florida and the Gulf Coast, where moist, unstable air often prevails in spring and summer (Figure 7-27), while the fewest thunderstorms occur along the Pacific coast where cool water and subsidence from the subtropical high lead to stable conditions.

Lightning At any given moment some 2000 thunderstorms exist over Earth. These storms produce about 6000 flashes of **lightning** every minute, or more than 8.5 million lightning bolts daily. A lightning flash heats the air along its path to as much as 10,000°C (18,000°F), and can develop 100,000 times the amperage used in household electricity (Figure 7-28).

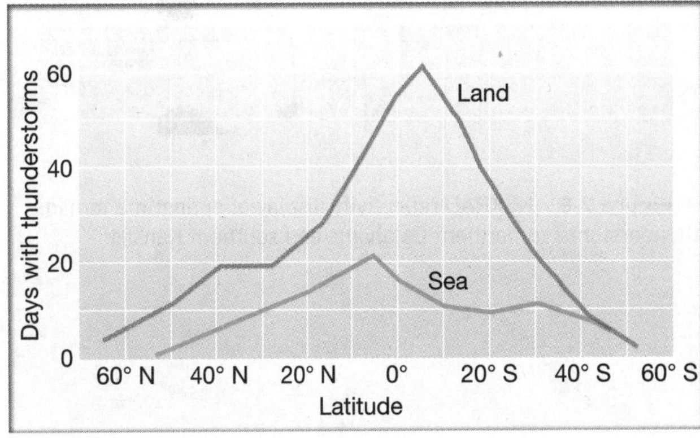

▲ **Figure 7-26** Average number of days per year with thunderstorms, as generalized by latitude. Most thunderstorms are in the tropics. Land areas experience many more thunderstorms than ocean areas because land warms up much more in summer.

FOCUS

Forecasting Severe Storms with NEXRAD

Stephen Stadler, Oklahoma State University

The volatile combination of mT air from the Gulf of Mexico, cP air from Canada, cT air from the desert, and the cold, dry southwesterly flow of the polar front jet stream make the Oklahoma sky the world's most prolific breeder of tornadoes. Technology, such as NEXRAD, now in place in the populated areas of the United States, allows for improved identification of the atmospheric conditions that spawn severe thunderstorms and tornadoes.

NEXRAD (for *next-generation radar)* dissects the internal workings of a severe thunderstorm and thereby determines its severity. The operating principle on which the system works is the Doppler effect, familiar to anyone who has ever been stopped at a railroad track as a train crosses a road. As the train approaches, the pitch of its whistle becomes higher and higher. After the train passes, the whistle pitch becomes lower and lower as the train gets farther and farther away. In reality, the whistle always has the same pitch as the sound is emitted from the train; the apparent change in pitch depends on whether the source of sound waves is approaching or receding from the listener.

Electromagnetic waves (light waves, radio waves, microwaves, and so forth) are also subject to the Doppler effect. NEXRAD transmits microwaves through the atmosphere toward a target storm. Raindrops, ice crystals, and hail reflect some of the microwaves back to the NEXRAD site, and how much of the transmitted signal is reflected gives an estimate of the intensity of the storm.

Figure 7-E shows a NEXRAD reflectivity display of severe thunderstorms in northern Oklahoma and southern Kansas, the type of information also obtainable from any conventional radar system. However, tornado funnels are usually only a few hundred meters across, and it is rare that radar can detect them using reflectivity alone. NEXRAD enjoys an advantage over conventional radar in that NEXRAD can detect motion toward and away from the radar site in increments of 3 kilometers (2 miles) per hour. Figure 7-F is a display of the same storms as shown by Figure 7-E but using NEXRAD's motion-detection capability. In Figure 7-F, notice the southern portion of the large storm at the center. Light green, indicating wind movement in excess of 50 knots away from the radar site, and light orange, indicating wind movement in excess of 50 knots toward the site, exist side by side. This is a strong counterclockwise rotation that is the hallmark of a mesocyclone, the parent cloud of a tornado. A radar operator would issue a tornado warning based on this display.

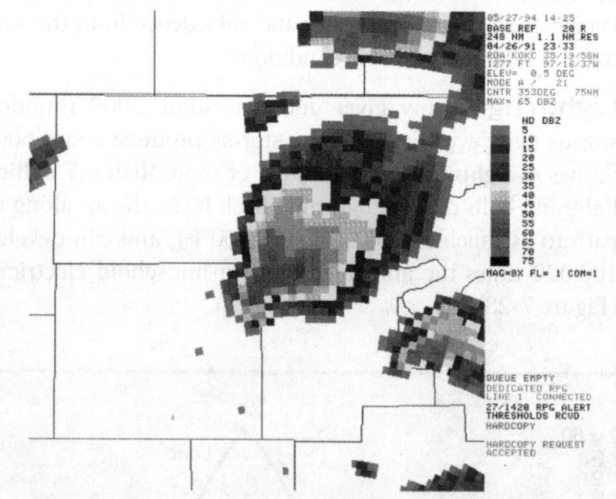

▲ **Figure 7-E** NEXRAD reflectivity display of springtime tornadic thunderstorms in northern Oklahoma and southern Kansas.

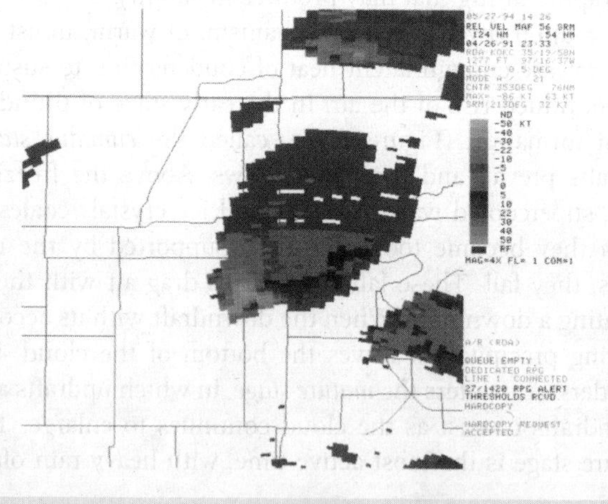

▲ **Figure 7-F** NEXRAD display of radial velocity in the storms shown in Figure 7-E. Light green indicates rapid movement away from radar site, while light orange indicates rapid movement toward radar site—this combination indicates counterclockwise rotation within a mesocyclone.

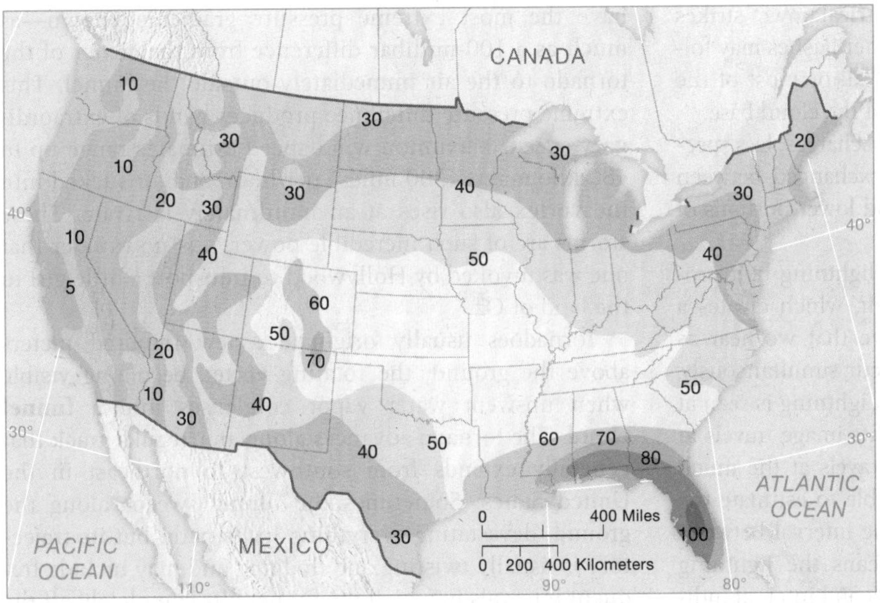

◀ **Figure 7-27** Average number of days per year with thunderstorms in the conterminous United States.

With such frequency and power, lightning clearly poses a significant potential danger for humanity. In the United States, on average about 65 deaths are blamed on lightning annually. The most dangerous places to be during a lightning storm are under a tree, in a boat, on a tractor, and playing golf. However, two out of three people struck by lightning in this country are not killed. It is a massive but brief shock, and quick first aid (mouth-to-mouth resuscitation and cardiopulmonary resuscitation) can save most victims.

▲ **Figure 7-28** Lightning over Tucson, Arizona.

The sequence of events that leads to lightning discharge is known, but the exact mechanism of electrification is not completely understood. Development of a large cumulonimbus cloud causes a separation of electrical charges. Updrafts carrying positively charged water droplets or crystals rise in the icy upper layers of the cloud while falling ice pellets gather negative charges and transport them downward (Figure 7-29). The growing negative charge in the lower part of the cloud attracts a growing positive charge on the Earth's surface immediately below. The contrast between the two (cloud base and ground surface) builds to tens of millions of volts before the insulating barrier of air that separates the charges is overcome.

Finally, a finger of negative current flicks down from the cloud and meets a positive charge darting upward from the ground. This makes an electrical connection of ionized air

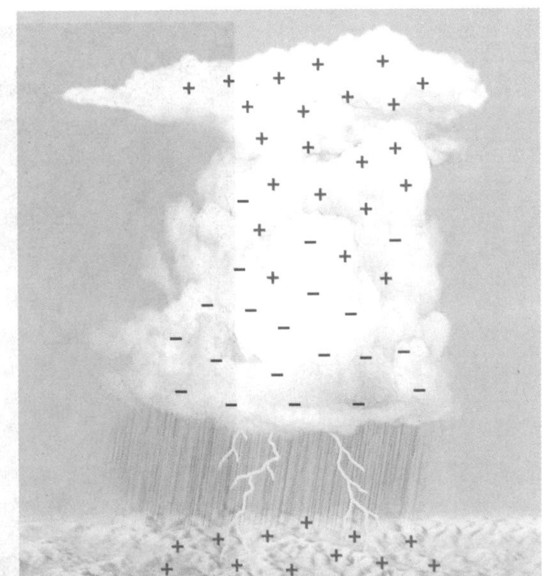

▲ **Figure 7-29** Typical arrangement of electrical charges in a thunderstorm cloud. Positively charged particles are mostly high in the cloud, whereas negatively charged particles tend to be concentrated near the base.

from cloud to ground, and a surge of electrical power strikes downward as the first lightning flash. Other flashes may follow in relatively quick succession, until all or most of the negative charges have been drained out of the cloud base.

In addition to such ground-to-cloud discharges, less spectacular but more frequent lightning is exchanged between adjacent clouds or between the upper and lower portions of the same cloud.

The abrupt heating occasioned by a lightning bolt produces instantaneous expansion of the air, which creates a shock wave that becomes a sound wave that we hear as **thunder**. The lightning and thunder occur simultaneously, but we perceive them at different times. Lightning is seen at essentially the instant it occurs because its image travels at the speed of light. Thunder, however, travels at the much slower speed of sound. Thus, it is possible to estimate the distance of a lightning bolt by timing the interval between sight and sound: A 3-second delay means the lightning strike was about 1 kilometer away; a 5-second interval indicates that the lightning flash was about 1 mile away. Rumbling thunder is indicative of a long lightning trace some distance away, with one portion being nearer than another to the hearer. If no thunder can be heard, the lightning is far away—probably more than 20 kilometers (a dozen miles).

Tornadoes

Tornadoes

Although very small and localized, the **tornado** is one of the most destructive of all atmospheric disturbances. It is the most intense vortex in nature: a deep low-pressure cell surrounded by a violently whirling cylinder of wind (Figure 7-30). These are tiny storms, generally less than 400 meters (a quarter of a mile) in diameter, but they have the most extreme pressure gradients known—as much as a 100-millibar difference from the center of the tornado to the air immediately outside the funnel. This extreme pressure difference produces winds of extraordinary speed. Maximum wind speed estimates range up to 480 kilometers (300 miles) per hour, and air sucked into the vortex also rises at an inordinately fast rate. These storms are of such incredible power, it is no wonder that one was invoked by Hollywood to transport a little girl to the land of Oz.

Tornadoes usually originate a few hundred meters above the ground, the rotating vortex becoming visible when upswept water vapor condenses into a **funnel cloud**. The tornado advances along an irregular track that generally extends from southwest to northeast in the United States. Sometimes the funnel sweeps along the ground, devastating everything in its path, but its trajectory is usually twisting and dodging and may include frequent intervals in which the funnel lifts completely off the ground and then touches down again nearby. Most tornadoes have damage paths 50 meters (about 150 feet) wide, move at about 48 kilometers (30 miles) per hour, and last for only a few minutes. Extremely destructive ones may be more than 1.5 kilometers (1 mile) wide, travel at 95 kilometers (60 miles) per hour, and may be on the ground for more than an hour, with maximum longevity recorded at about eight hours.

The dark, twisting funnel of a tornado contains not only cloud but also sucked-in dust and debris. Damage is caused largely by the strong winds, flying debris, and swirling updraft. The old advice of opening a window when faced with an approaching tornado (supposedly to reduce an abrupt drop in pressure when the center of the storm passes over a closed building) is no longer given—and in fact opening a window may *increase* the chance of injury from flying debris.

▶ **Figure 7-30** This EF-3 tornado in Mulvane, Kansas, on June 12, 2004, severely damaged the barn behind this farmhouse.

FOCUS

Forecasting Severe Storms with the Vertical Wind Profiler and the Oklahoma Mesonetwork

Stephen Stadler, Oklahoma State University

Two kinds of weather data-gathering technology are deployed in Oklahoma to aid in the forecasting of severe storms: the *vertical wind profiler* and the *Oklahoma Mesonetwork.*

Vertical Wind Profiler To estimate the likelihood of storms hours in advance, it is necessary to have some idea of wind conditions far above the ground. Since the 1930s, the United States has had a network of sparsely spaced balloon-launch sites. (Oklahoma, for instance, has one site.) Twice a day, instrumented balloons are launched and data are radioed back from the troposphere and stratosphere. The data are excellent for monitoring movement of Rossby waves and air masses, but smaller features relevant to thunderstorm formation can "slip through the cracks" of the balloon network.

The vertical wind profiler network of the storm-prone central United States solves this problem. A wind profiler is a Doppler radar unit that resembles a chain-link fence turned on its side. It sends radar energy upward to sense wind speed and direction in the first few kilometers of the atmosphere. Not only are profiler sites more closely spaced than balloon sites, but the profilers can also provide a wind profile every few minutes instead of only once every 12 hours. This increased frequency is helpful in severe thunderstorm situations, where atmospheric conditions can change drastically over short amounts of time.

Oklahoma Mesonetwork The word "mesonetwork" is a contraction of "mesoscale" and "network" and implies the observation of the mesoscale events that last from several minutes to several hours. Mesoscale features of 2 to 200 kilometers (1.2 to 125 miles) are sometimes small enough not to be detected by traditional observational networks. Thunderstorms and squall lines are prime examples of mesoscale disturbances. The Oklahoma Mesonetwork is a system of 120 solar-powered, automated weather stations that telemeter their observations every 5 minutes. Before the Mesonetwork, there were only a dozen hourly reporting stations in Oklahoma. Operating since 1994, the Mesonetwork is the oldest mesonetwork in the United States. Others are now in existence and being planned with the vision that the United States will be someday covered with mesonetworks.

Tornado-generating thunderstorms frequently occur along a *dryline*, the boundary between mT and cT air and a zone in which surface air streams are forced to converge and lift. The dryline can generate some of the largest tornadic thunderstorms on Earth. Therefore, its exact position is critical in forecasting where storms will begin. The Oklahoma Mesonetwork, with its average spacing of 30 kilometers (18 miles) between stations, gives forecasters a precise view of dryline location. Figure 7-G shows a dryline in western Oklahoma. Note the relatively high dew point temperatures east of the line in the mT air and the higher air temperatures and lower dew point temperatures in the cT air west of the line. The wind data show air streams are converging and, by implication, rising along the dryline. In this situation, severe thunderstorms are most likely just to the east of the dryline.

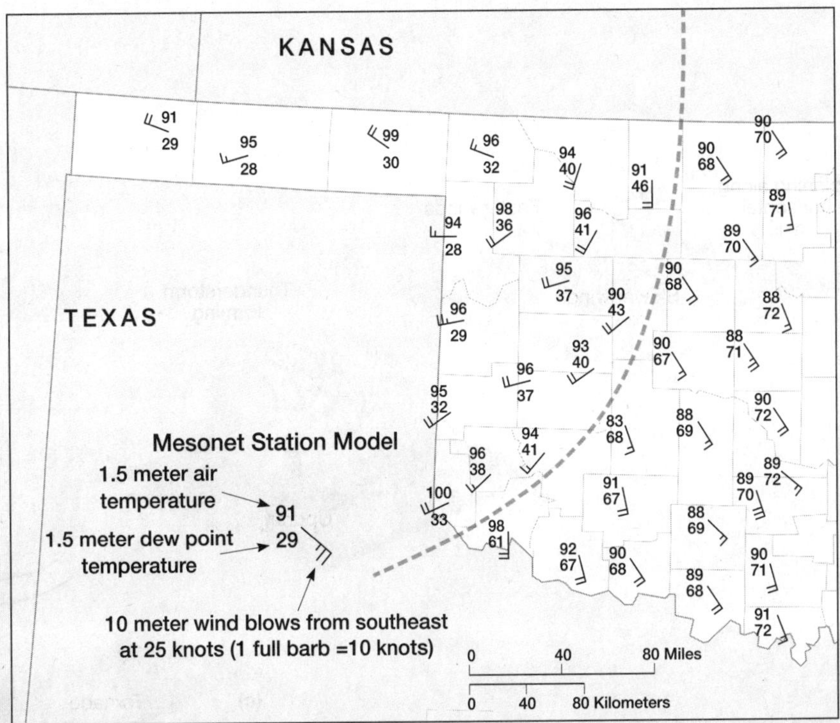

▲ **Figure 7-G** Oklahoma Mesonetwork depiction of a dryline that could generate severe storms.

TABLE 7-3 The Enhanced Fujita Scale (EF Scale) for Tornadoes

Enhanced F Scale			Expected Level of Damage
EF Scale	3-second gust kph	3-second gust mph	
0	105–137	65–85	Light: Broken tree branches; uprooted small trees; billboards and chimneys damaged.
1	138–177	86–110	Moderate: Roof surfaces peeled off; mobile homes overturned or pushed off their foundations.
2	178–217	111–135	Considerable: Mobile homes destroyed; roofs blown off wooden-frame houses; uprooted large trees; light objects become "missiles."
3	218–266	136–165	Severe: Trains derailed or overturned; walls and roofs torn from well-constructed houses; heavy cars thrown off ground.
4	267–322	166–200	Devastating: Structures with weak foundations blown for some distance; well-built houses destroyed; large objects become missiles.
5	Over 322	Over 200	Incredible: Well-built houses lifted off foundations and carried considerable distance before complete destruction; tree bark removed; automobile-sized missiles carried more than 100 meters.

The strength of a tornado is commonly described using the **Enhanced Fujita Scale (EF Scale**; Table 7-3), named after the late University of Chicago meteorologist, Theodore Fujita. The EF Scale is based on estimates of 3-second gust wind speeds as determined by observed damage after a tornado.[1] About 69 percent of all tornadoes in the United States are classified as *light* or *moderate* (EF-0 or EF-1), about 29 percent are classified

[1]The Enhanced Fujita Scale for tornado damage replaced the original F-scale in the United States in 2007.

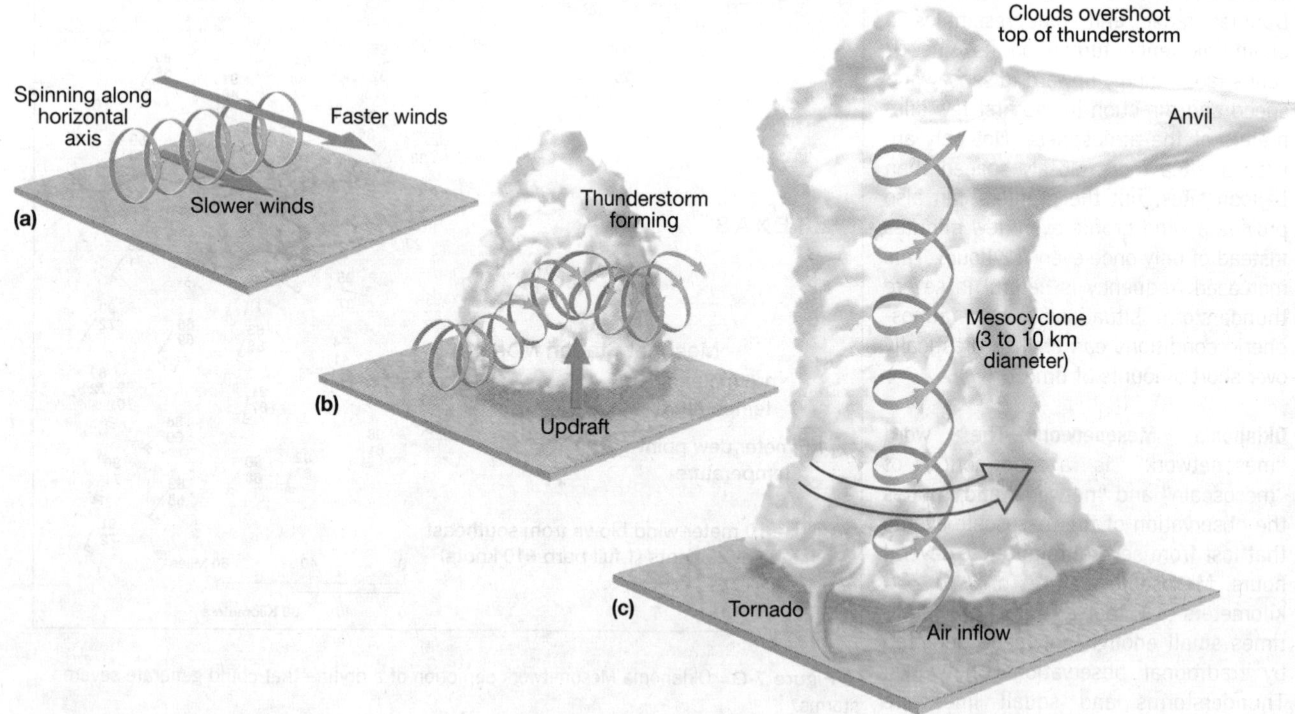

▲ **Figure 7-31** The formation of a mesocyclone often precedes tornado formation. (a) Winds are stronger aloft than at the surface (called *speed wind shear*), producing a rolling motion around a horizontal axis. (b) Strong thunderstorm updrafts tilt the horizontally rotating air to a nearly vertical alignment. (c) The mesocyclone, a vertical cylinder of rotating air, is established. If a tornado develops it will descend from a slowly rotating wall cloud in the lower portion of the mesocyclone.

as *strong* or *severe* (EF-2 or EF-3), and 2 percent are classified as *devastating* or *incredible* (EF-4 or EF-5). The average annual death toll from violent tornadoes in the United States is about 60—a sharp decline from even 35 years ago due to better forecasting and emergency broadcasts.

Formation As with many other storms, the exact mechanism of tornado formation is not well understood. They may develop in the warm, moist, unstable air associated with a midlatitude cyclone, such as along a squall line that precedes a rapidly advancing cold front or along the cold front itself. Virtually all tornadoes are generated by severe thunderstorms. The basic requirement is vertical wind shear—a significant change in wind speed or direction from the bottom to the top of the storm. On a tornado day, low-level winds are southerly and jet stream winds are southwesterly, and this difference in direction causes turbulence on the boundary between the two systems. The conventional updrafts that become thunderstorms reach several kilometers up into the atmosphere, and the wind shear can cause air to roll along a horizontal axis (Figure 7-31). Strong updrafts in such a rapidly maturing *supercell* thunderstorm may then tilt this rotating air vertically, developing into a **mesocyclone**, with a diameter of 3 to 10 kilometers (2 to 6 miles). About half of all mesocyclones formed result in a tornado.

Spring and early summer are favorable for tornado development because of the considerable air-mass contrast present in the midlatitudes at that time; a tornado can form in any month, however (Figure 7-32). Most occur in midafternoon, at the time of maximum heating.

Tornadoes do occur in the midlatitudes and subtropics, but more than 90 percent are reported in the United States, where about 1000 are sighted each year (Figure 7-33). Such concentration in a single area presumably reflects

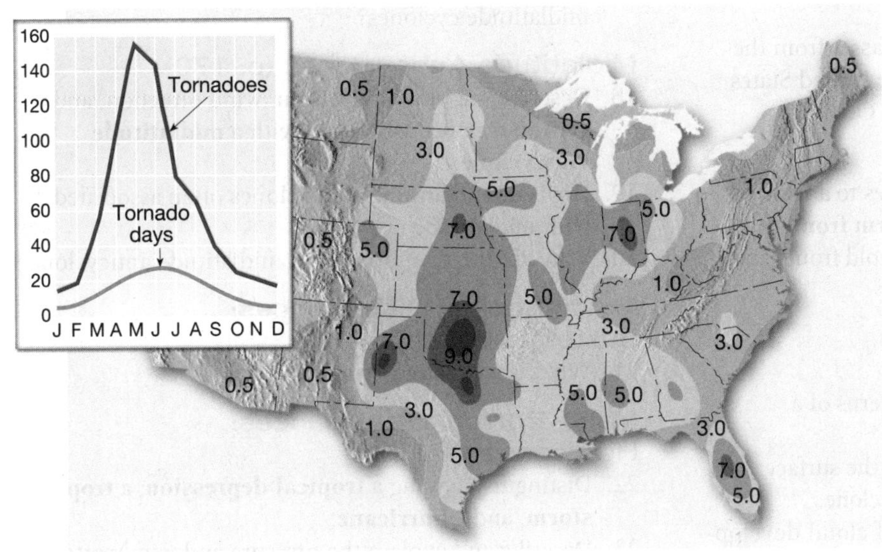

◀ **Figure 7-32** Map showing the average number of tornadoes per 26,000 square kilometers (10,000 square miles) over a 27-year period. The chart shows the seasonality of tornadoes and tornado days in the United States.

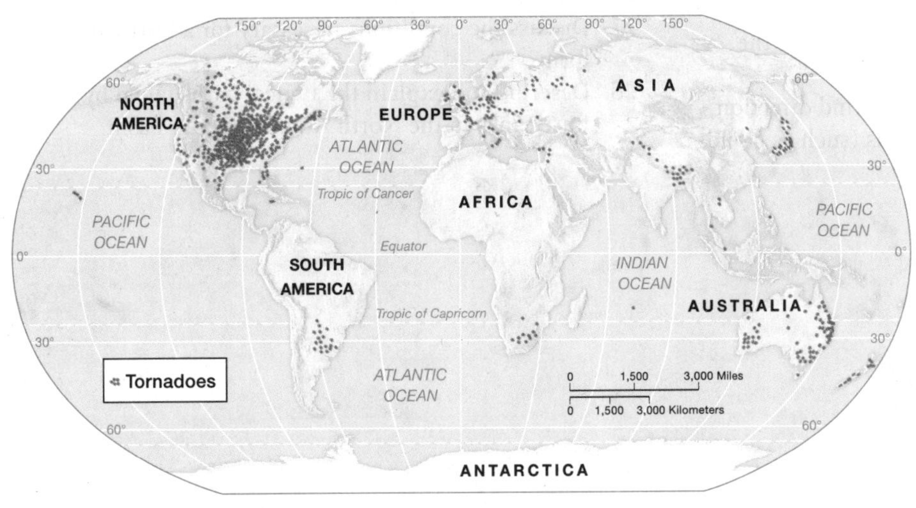

◀ **Figure 7-33** Anticipated distribution of tornadoes over the world in any given five-year period. The Great Plains of North America are clearly a favored area of development.

optimum environmental conditions, with the relatively flat terrain of the central and southeastern United States providing an unhindered zone of interaction between prolific source regions for Canadian cP and Gulf mT air masses. Although between 800 and 1200 tornadoes are recorded annually in the United States, the actual total may be considerably higher than that because many small tornadoes that occur briefly in uninhabited areas are not reported.

True tornadoes are apparently restricted to land areas. Similar-appearing funnels over the ocean, called **waterspouts**, have a lesser pressure gradient, gentler winds, and reduced destructive capability.

CHAPTER 7 LEARNING REVIEW

After studying this chapter, you should be able to answer the following questions. Key terms from each text section are shown in bold type. Definitions for key terms are also found in the glossary at the back of the book.

Air Masses (p. 161)

1. What is an **air mass**, and what conditions are necessary for one to form?
2. What regions of Earth are least likely to produce air masses? Why?
3. Why is an air mass unlikely to form over the Rocky Mountains of North America?
4. Why are maritime polar air (mP) masses from the Atlantic Ocean less important to the United States than mP air masses from the Pacific Ocean?

Fronts (p. 163)

5. What is the relationship of air masses to a **front**?
6. What is a **cold front**? What is a **warm front**?
7. Explain why clouds develop along cold fronts and warm fronts.
8. What is a **stationary front**?

Midlatitude Cyclones (p. 165)

9. Describe the pressure and wind patterns of a **midlatitude cyclone**.
10. Describe the locations of fronts and the surface "sectors" of a mature midlatitude cyclone.
11. Describe and explain the regions of cloud development and precipitation within a midlatitude cyclone.
12. Discuss the four components of movement of a midlatitude cyclone.
13. Explain the process of **occlusion**.
14. Why does an **occluded front** usually indicate the "death" of a midlatitude cyclone?
15. Describe and explain the changes in wind direction, atmospheric pressure, sky conditions (such as clouds and precipitation), and temperature with the passing of a cold front.
16. Discuss the origin of midlatitude cyclones. What is the relationship between upper-level airflow and the formation of surface disturbances in the midlatitudes?
17. What parts of the world are most affected by midlatitude cyclones?

Midlatitude Anticyclones (p. 170)

18. Describe the pressure pattern, wind direction, and general weather associated with a **midlatitude anticyclone**.
19. How are midlatitude anticyclones often associated with midlatitude cyclones?
20. Why are there no fronts in a midlatitude anticyclone?

Minor Tropical Disturbances: Easterly Waves (p. 171)

21. What is an **easterly wave**?

Major Tropical Disturbances: Hurricanes (p. 172)

22. Distinguish among a **tropical depression**, a **tropical storm**, and a **hurricane**.
23. Describe and explain the pressure and wind patterns of a **tropical cyclone (hurricane)**.
24. Discuss the characteristics of the **eye** of a hurricane.
25. Why are there no fronts in a hurricane?
26. What is **wind shear**?
27. Discuss the conditions necessary for a hurricane to form?
28. Describe and explain the typical paths taken by hurricanes in the North Atlantic Ocean basin.

29. Why do hurricanes weaken when they move over land?
30. Why are tropical cyclones common along the east coasts of continents in the midlatitudes, but not along the west coasts?
31. What is a hurricane **storm surge**, and what causes one?
32. Briefly explain the **Saffir-Simpson Hurricane Scale**.
33. Do hurricanes have any beneficial effects? Explain.

Localized Severe Weather (p. 178)

34. Discuss the general sequence of **thunderstorm** development and dissipation.
35. What is the relationship of **thunder** to **lightning**?
36. Why are thunderstorms more common over land than over water?
37. Describe the wind and pressure characteristics of a tornado.
38. What is a **funnel cloud**?
39. Briefly explain the **Enhanced Fujita Scale (EF Scale)** for tornadoes.
40. Discuss the general formation of a tornado from a supercell thunderstorm and **mesocyclone**.
41. What is the difference between a tornado and a **waterspout**?

Log in to **www.mygeoscienceplace.com** for animations, videos, RSS feeds, maps, and self-study quizzes to enhance your study of **Atmospheric Disturbances.**

8 Climate and Climate Change

The basic goal of the geographic study of climate, as with the geographic study of anything else, is to understand its distribution over Earth. Such understanding is exceedingly difficult to achieve for climate, however, because so many variables are involved. It is relatively easy to describe and map the distribution of such uncomplicated phenomena as giraffes, voting patterns, and a host of other concrete entities or simple relationships. Climate, however, is the product of a number of elements that are, for the most part, continuously and independently variable.

Temperature, for instance, is one of the simplest climatic elements to describe, and yet in this textbook more than 40 maps and diagrams are used to convey various temperature patterns around the world: daily, monthly, and annual averages, as well as temperature ranges and extremes. Climate involves almost continuous variation, not only of temperature but also a host of other factors.

Adding to the complexity of understanding climate distribution is that climate can change over time. The geologic record reveals times in both the distant and recent past when Earth's climate was markedly different from today—such as the great ice ages of the Pleistocene Epoch that ended about 10,000 years ago (the Pleistocene is discussed in Chapter 19). More important for us today, however, are the indications of climate change that seems to be underway as a result of human activity.

In this chapter we begin by describing the classification and distribution of global climate types—discussing climate largely as it has been over the last few thousands of years. We conclude the chapter with a discussion of climate change: recognizing, explaining, and anticipating the consequences of a climate system that seems to be in a state of flux.

Climate Classification

To cope with the great diversity of information encompassed by the study of global climate distribution we need a classification scheme to simplify, organize, and generalize a vast array of data.

As an illustration of how a classification system can help us understand global climate patterns, suppose that a geography student in Atlanta, Georgia, is asked to describe the climate of southeastern China. A world map portraying the distribution of climate types (see Figure 8-3 on p. 192) shows that southeastern China has the same climate as the southeastern United States. Thus, the student's familiarity with the home climate in Georgia provides a basis for understanding the general characteristics of the climate of southeastern China.

◀ Hakuba Mountains in Chuba-Sangaku National Park, Japan.

Early Classification Schemes

The earliest known climatic classification scheme originated with the ancient Greeks, perhaps 2200 years ago. Although the "known world" was very small at that time, Greek scholars were aware of the shape and approximate size of Earth. They knew that at the southern limit of their world, along the Nile River and the southern coast of the Mediterranean, the climate was much hotter and drier than on the islands and northern coast of that sea. At the other end of the world known to the Greeks, along the Danube River and the northern coast of the Black Sea, things were much colder, especially in winter. So the Greeks spoke of three climatic zones: the *Temperate Zone* of the midlatitudes, in which they lived (Athens is at 38° N latitude); the *Torrid Zone* of the tropics to the south, and the *Frigid Zone* to the north. Because they knew that Earth is a sphere, they suggested that the Southern Hemisphere has similar Temperate and Frigid zones, making five in all.

For many centuries, this classification scheme was handed down from scholar to scholar. Gradually these five climatic zones were confused with, and eventually their climates ascribed to, the five astronomical zones of the Earth, bounded by the Tropics of Cancer and Capricorn and the Arctic and Antarctic circles (Figure 8-1). This revision put the equatorial rainy zone in with the hot arid region in the Torrid Zone, extended the Temperate Zone to include much of what the Greeks had called Frigid, and moved the Frigid Zone poleward to the polar circles. This simplistic but unrealistic classification scheme persisted for more than a thousand years and was finally discarded only in the twentieth century.

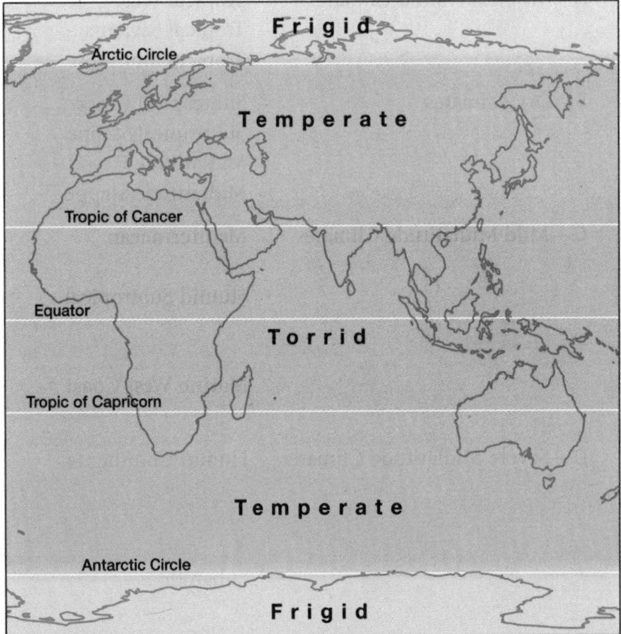

▲ **Figure 8-1** The original Greek climate classification scheme was expanded and correlated with Earth's five astronomical zones to depict a world comprising five climate zones.

The Köppen Climate Classification System

The **Köppen climate classification system** is by far the most widely used modern climate classification system. Wladimir Köppen (1846–1940) was a Russian-born German climatologist who was also an amateur botanist. The first version of his climate classification scheme appeared in 1918, and he continued to modify and refine it for the rest of his life, the last version being published in 1936.

The system uses as a database only the average annual and average monthly values of temperature and precipitation, combined and compared in a variety of ways. Consequently, the necessary statistics are commonly tabulated and easily acquired. Data for any location (called a *station*) on Earth can be used to determine the precise classification of that place, and the areal extent of any recognized climatic type can be determined and mapped. This means that the classification system is functional at both the specific and the general level.

Köppen defined four of his five major climatic groups primarily by temperature characteristics, the fifth (the B group) on the basis of moisture. He then subdivided each group into climate types according to various temperature and precipitation relationships.

Köppen was unsatisfied with his last version and did not consider it a finished product. Thus, many geographers and climatologists have used the Köppen system as a springboard to devise systems of their own or to modify Köppen's classifications. The system of climate classification used in this book is properly called the *modified Köppen system*. It encompasses the basic design of the Köppen system but with a variety of minor modifications. Some of these modifications follow the lead of Glen Trewartha, late geographer and climatologist at the University of Wisconsin.

The modified Köppen system describes five major climate groups (groups A, B, C, D, and E) which are subdivided into a total of 14 individual climate types, along with the special category of highland (H) climate (Table 8-1). In this book no attempt is made to distinguish between pure Köppen and modified Köppen system definitions: our goal is to comprehend the general pattern of world climate, not to learn a specific system or to nitpick about boundaries.

Köppen Letter Code System In the modified Köppen system, each climate type is designated by a descriptive name and by a series of letters defined by specific temperature and/or precipitation values (Figure 8-2 and Table 8-1). The first letter designates the major climate group; the second letter usually describes precipitation patterns, while the third letter (if any) describes temperature patterns. Table 8-2 provides a general description of the meaning of the letters (see Appendix V for the exact definitions used for classification). Figure 8-3 is a map showing the global distribution of all these climates.

TABLE 8-1 Climate Types of the Modified Köppen Classification System

Major Group	Climate Type	Letter Code	General Characteristics
A—Tropical Humid Climates	Tropical Wet	Af	Rain throughout year
	Tropical Savanna	Aw	Winter (low-Sun) dry season
	Tropical Monsoon	Am	Short dry season; heavy rains in other months
B—Dry Climates	Subtropical Desert	BWh	Low-latitude (hot) desert
	Subtropical Steppe	BSh	Low-latitude (hot) semiarid
	Midlatitude Desert	BWk	Midlatitude (cold) desert
	Midlatitude Steppe	BSk	Midlatitude (cold) semiarid
C—Mild Midlatitude Climates	Mediterranean	Csa	Dry, hot summer
		Csb	Dry, warm summer
	Humid Subtropical	Cfa	No dry season; hot summer
		Cwa	Dry winter; hot summer
		Cwb	Dry winter; warm summer
	Marine West Coast	Cfb	No dry season; warm summer
		Cfc	No dry season; cool summer
D—Severe Midlatitude Climates	Humid Continental	Dfa	Severe winter; no dry season; hot summer
		Dfb	Severe winter; no dry season; warm summer
		Dwa	Severe winter; dry winter; hot summer
		Dwb	Severe winter; dry winter; warm summer
	Subarctic	Dfc	Severe winter; no dry season; cool summer
		Dfd	Very cold winter; no dry season
		Dwc	Dry winter; cool summer
		Dwd	Dry winter, very cold winter
E—Polar Climates	Tundra	ET	Polar tundra with no true summer
	Ice Cap	EF	Polar ice cap
H—Highland Climates		H	High elevation climates

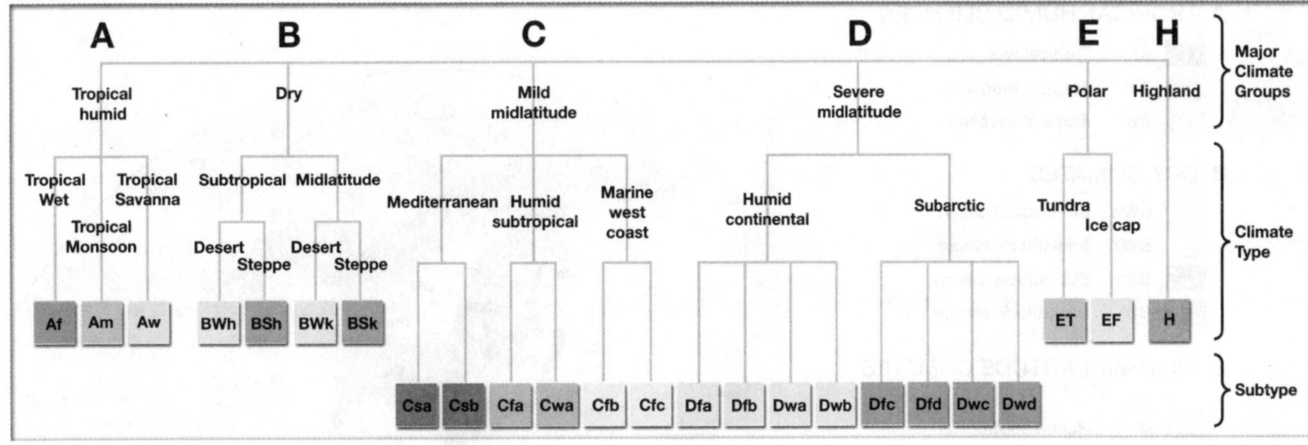

▲ **Figure 8-2** The modified Köppen climatic classification.

TABLE 8-2 Generalized Description of the Letter Codes of the Modified Köppen Climate Classification System

First Letters: Major climate group			Second Letters	Third Letters
A	Tropical Humid	low latitude; warm and wet	A, C, and D Climate Precipitation	C and D Climate Temperature
B	Dry	evaporation exceeds precipitation	f wet all year	a hot summers
C	Mild Midlatitude	mild winters, warm or hot summers	m monsoonal precipitation pattern (very wet summer)	b warm summers c cool summers
D	Severe Midlatitude	severe, cold winters	w winter dry season	d very cold winters
E	Polar	very high latitude, cold climates	s summer dry season	B Climate Temperature
H	Highland	high mountains; elevation is dominant control	B Climate Precipitation W desert S steppe E Climate Temperature T tundra F ice cap	h hot desert or steppe k cold desert or steppe

Although the letter code system seems complicated at first, it provides a shorthand method for summarizing key characteristics of each climate. For example, if we look for the definitions of the letters in *Csa*, one of the letter code combinations for a *mediterranean climate*, we see that:

C = *mild midlatitude climate*

s = *summer dry season*

a = *hot summers*

As we will see later in this chapter, that is a very good synopsis of the distinctive characteristics of this climate type.

Climographs

Probably the most useful tool in a general study of world climatic classification is a simple graphic representation of monthly temperature and precipitation for a specific weather station. Such a graph is called a **climograph** (or *climatic diagram*); a typical one is shown in Figure 8-4. The customary climograph has 12 columns, one for each month, with a temperature scale on the left side and a precipitation scale on the right. Average monthly temperatures are connected by a curved line in the upper portion of the diagram, and average monthly precipitation is represented by bars extending upward from the bottom.

The value of a climograph is twofold: (1) it displays precise details of important aspects of the climate of a specific place, and (2) it can be used to recognize and classify the climate of that place.

World Distribution of Major Climate Types

Seasonal Pressure and Precipitation Patterns

Much of the remainder of this chapter is devoted to a discussion of climatic types. Our attention is focused primarily on three questions:

1. Where are the various climate types located?
2. What are the characteristics of each climate?
3. What are the main controls of these climates?

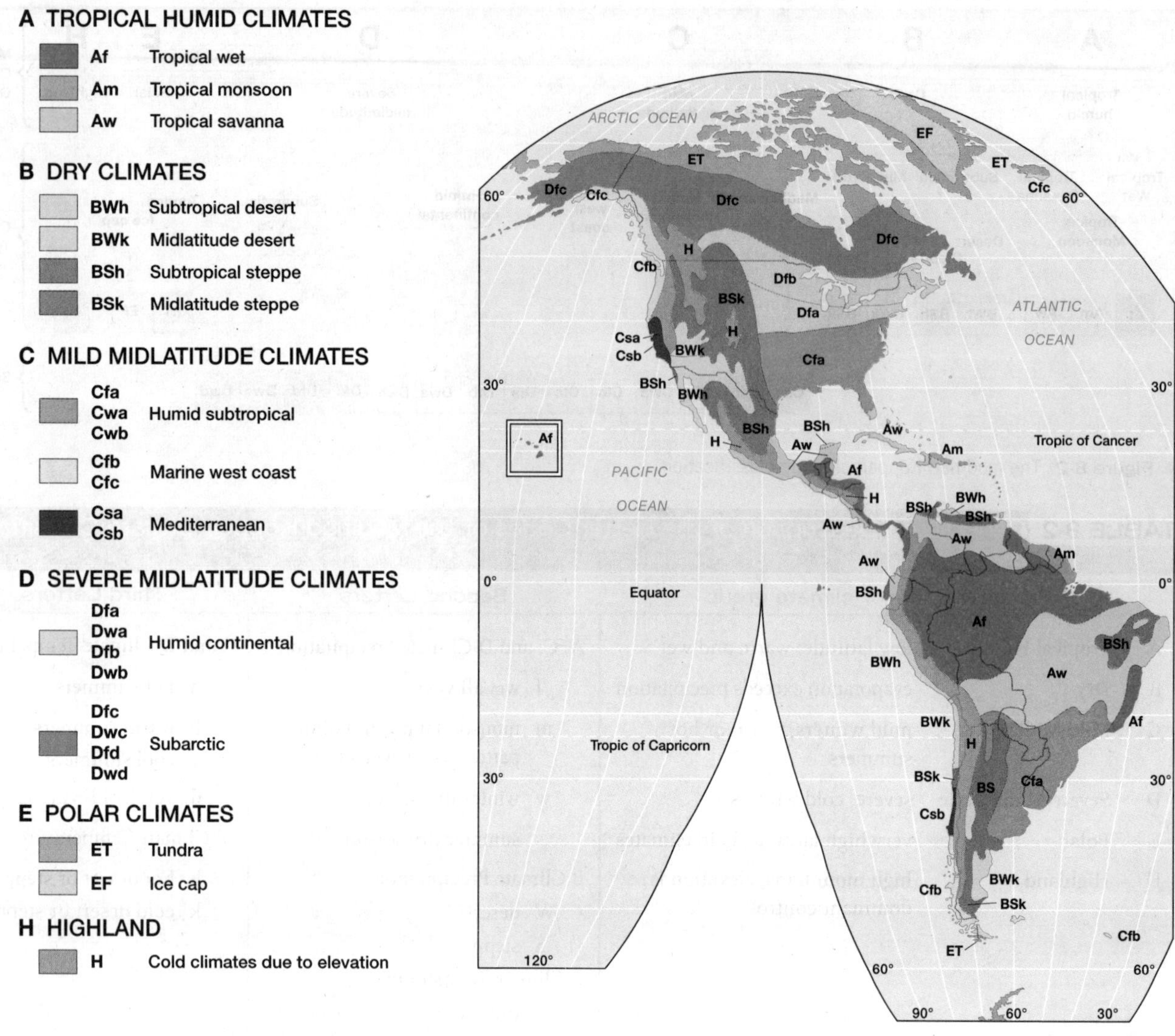

▲ **Figure 8-3** Climatic regions over land areas (modified Köppen system).

As we begin our look at climates of the world—a seemingly daunting task at first glance—keep in mind that very few new concepts are introduced in this section. Rather, this discussion of climate distribution lets us systematically organize what we have *already* learned about weather and climate from previous chapters. Also, take advantage of the climate "summary" tables for each major climate group (such as Table 8-3 on p. 194)—this series of tables concisely describes the general location, temperature and precipitation characteristics, and dominant controls for each climate in the Köppen system.

Tropical Humid Climates (Group A)

The *tropical humid climates* (Group A) occupy almost all the land area of Earth within the tropics. As you can see in Figure 8-5, this globe-girdling belt of A climates is interrupted slightly here and there by mountains or small regions of aridity, but it dominates the tropical regions. The fundamental character of the A climates is molded by this latitudinal location.

The A climates are noted not so much for warmth as for lack of coldness. The Sun is high in the sky every day of the year, and even the shortest days are not appreciably shorter than the longest ones, so these are climates of perpetual warmth (interestingly though, they do not experience the world's highest temperatures for reasons we shall see later).

These are the only truly winterless climates of the world. However, the terms "winter" and "summer" are sometimes used in discussing tropical climates, as in the "winter monsoon" and the "summer monsoon"—in this context, these adjectives mean the time of the year when mid- and high latitude locations are indeed cold or hot. Alternately, we can refer to the "high-Sun season" (summer) and the "low-Sun season" (winter), where "high" and "low" are strictly relative terms.

The second typifying characteristic of the tropical humid climates is the prevalence of moisture. Although

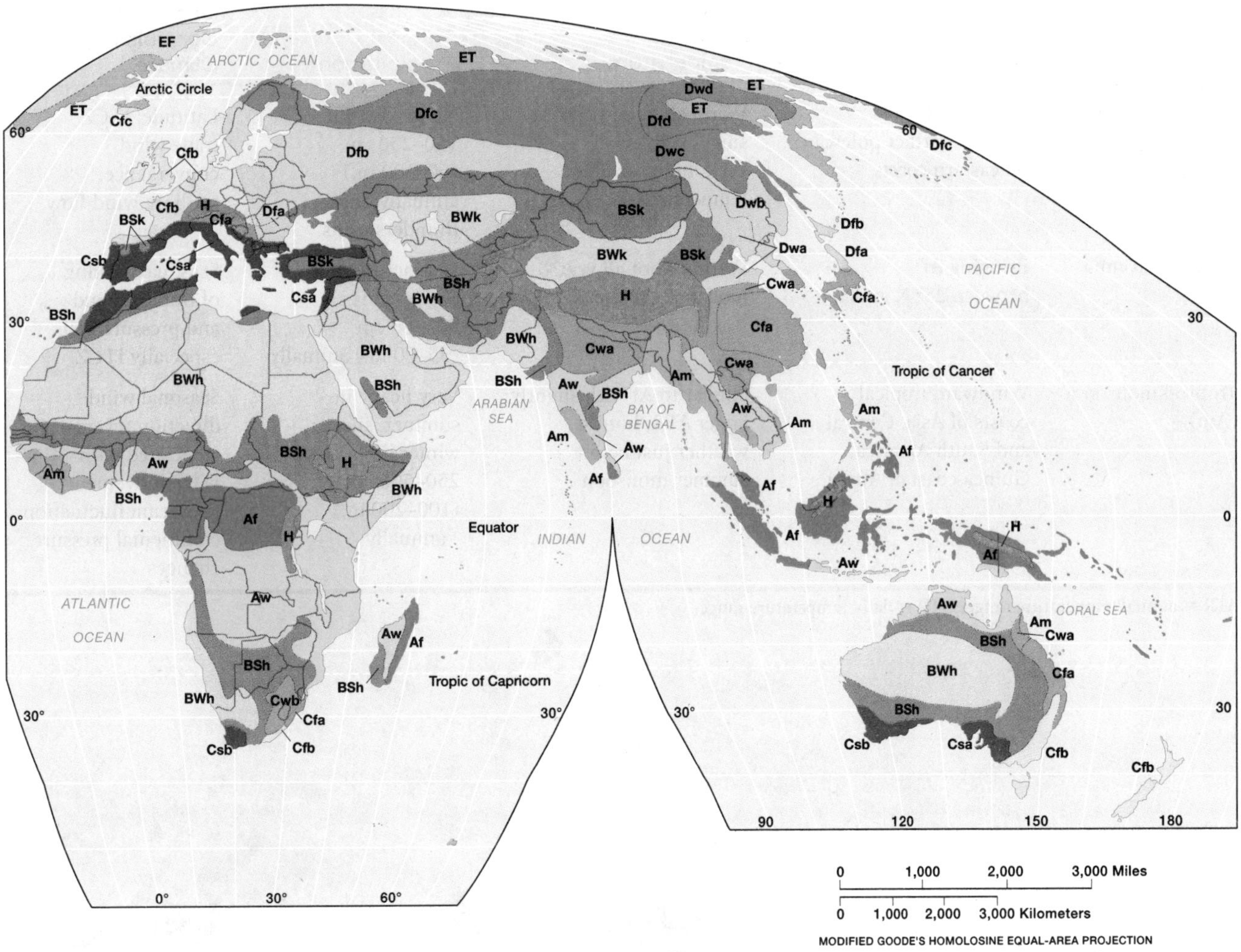

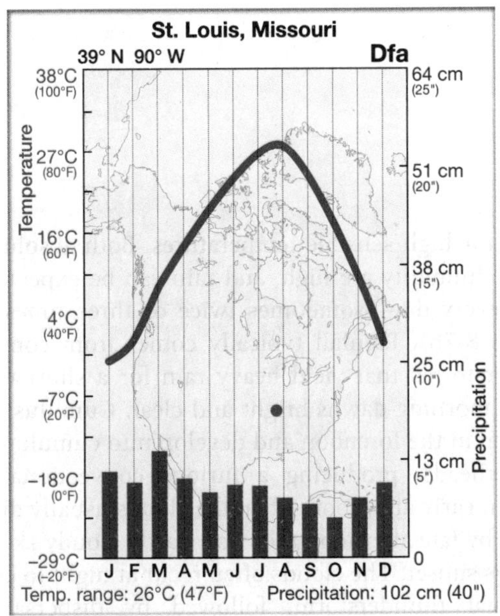

▲ **Figure 8-4** A typical climograph. Average monthly temperature is shown with a solid red line (scale along left side) and average monthly precipitation is shown with blue bars extending from the bottom (scale along the right side).

not universally rainy, much of the tropical humid zone is among the wettest in the world, influenced at least part of the year by the thermal convection and onshore winds associated with the intertropical convergence zone (ITCZ), and regionally from monsoon wind patterns.

The tropical humid climates are classified into three types on the basis of annual rainfall. The *tropical wet* has abundant rainfall every month of the year. The *tropical savanna* is characterized by a low-Sun dry season and a prominent but not extraordinary high-Sun wet season. The *tropical monsoon* has a dry season and a distinct very rainy high-Sun wet season (Table 8-3).

Tropical Wet Climate (Af) Tropical wet climates are predominantly equatorial, found in an east-west sprawl astride the equator—typically extending to about 10° N and S latitude, but in a few east-coast locations as much as 25° poleward of the equator (see Figure 8-5). The largest areas of Af climate occur in the upper Amazon basin of South America, the northern Congo (Zaire) basin of Africa, and the islands of the East Indies.

The single most descriptive word that can be applied to the tropical wet climate is "monotonous." This is a seasonless

TABLE 8-3 Summary of A Climates: Tropical Humid

Type	Location	Temperature	Precipitation	Dominant Controls of Climate
Tropical wet (Af)	Within 5–10° of equator; farther poleward on eastern coasts	Warm all year; very small ATR; small DTR; high sensible temperature	No dry season; 150–250 cm (60–100 in.) annually; many thunderstorms	Latitude; ITCZ; trade wind convergence; onshore wind flow
Tropical savanna (Aw)	Fringing Af between 25° N and S	Warm to hot all year; moderate ATR and DTR	Distinct wet and dry seasons; 90–180 cm (35–70 in.) annually	Seasonal shifting of tropical wind and pressure belts, especially ITCZ
Tropical monsoon (Am)	Windward tropical coasts of Asia, Central and South America, Guinea coast of Africa	Similar to Af with slightly larger ATR; hottest weather just before summer monsoon	Very heavy in summer; short winter dry season; 250–500 cm (100–200 in.) annually	Seasonal wind direction reversal associated with ITCZ movement; jet stream fluctuation; continental pressure changes

ATR = annual temperature range; DTR = daily temperature range.

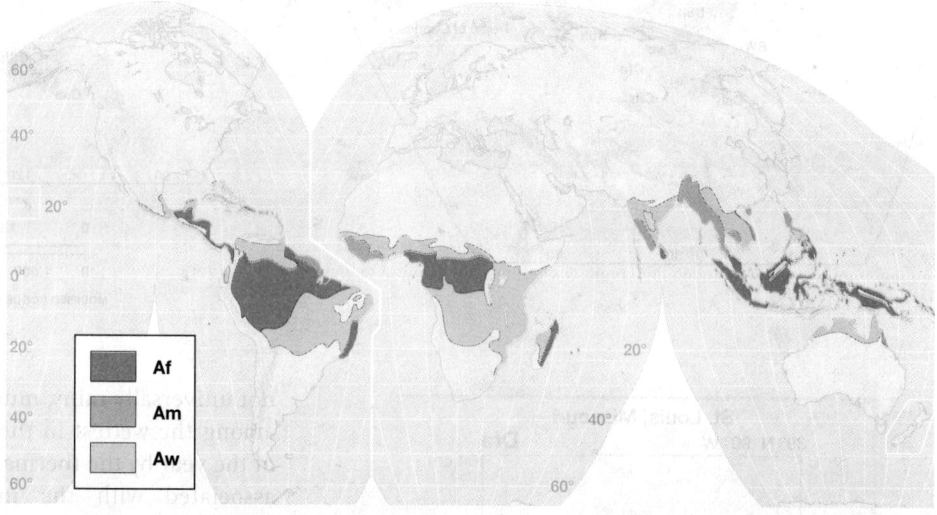

▶ **Figure 8-5** The global distribution of tropical humid (A) climates.

climate, with endless repetition of the same weather day after day after day. Warmth prevails, with every month having an average temperature close to 27°C (80°F), as Figure 8-6 shows, and an average annual temperature range that is typically only 1 or 2°C (2 or 3.6°F)—by far the smallest annual temperature range of any climatic type.

This seasonless condition in Af regions gives rise to the saying, "Night is the winter of the tropics"—this is one of the few climates in which the *average daily temperature range* exceeds the average annual temperature range (Figure 8-7a). On a typical afternoon, the temperature climbs into the low 30s °C (high 80s °F), dropping into the low 20s °C (low 70s °F) in the coolest period just before dawn.

Regardless of the thermometer reading, however, the weather feels warm in this climate because high humidity makes for high sensible temperatures. Both absolute and relative humidity are high, and rain can be expected just about every day—sometimes twice or three times a day (Figure 8-7b). Rainfall typically comes from convective thunderstorms that yield heavy rain for a short time. A typical morning dawns bright and clear. Cumulus clouds build up in the forenoon and develop into cumulonimbus thunderheads, producing a furious convectional rainstorm in early afternoon. Then the clouds usually disperse so that by late afternoon there is a partly cloudy sky and a glorious sunset. The clouds often recur at night to create a nocturnal thunderstorm, followed by dispersal once again. The next day dawns bright and clear, and the sequence repeats.

Each month receives several centimeters of rain, and the annual total normally is between 150 and 250 centimeters

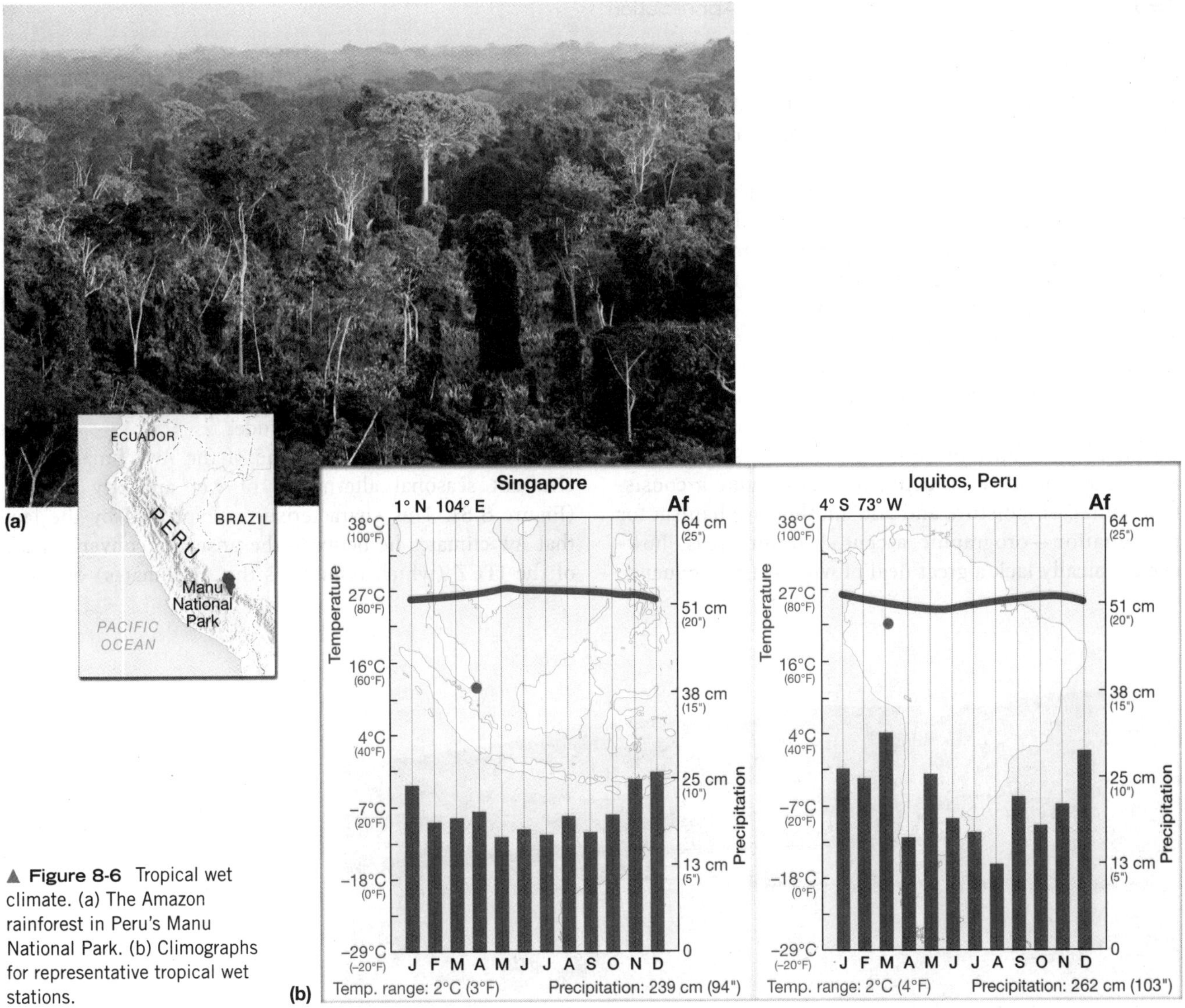

▲ **Figure 8-6** Tropical wet climate. (a) The Amazon rainforest in Peru's Manu National Park. (b) Climographs for representative tropical wet stations.

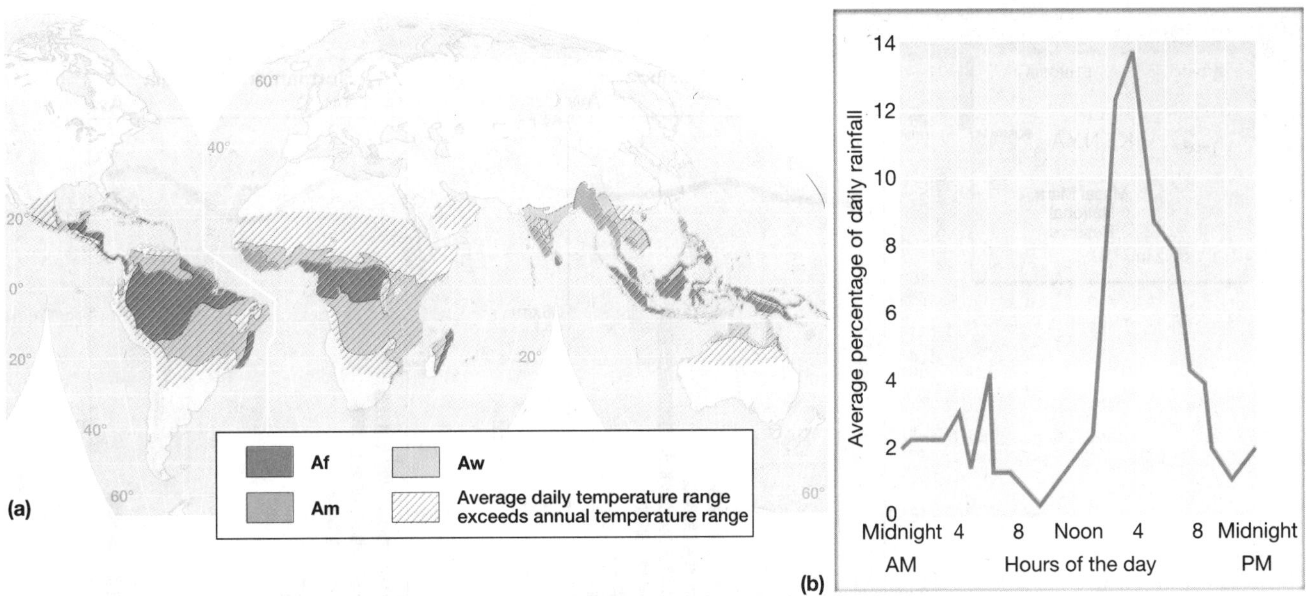

▲ **Figure 8-7** Temperature and rainfall characteristics of tropical humid climates. (a) Land portions of the world where the average daily temperature range exceeds the average annual temperature range. This characteristic is typical of A climates but rare in other climates. (b) A typical daily pattern of rainfall in an Af climate. This example, from Malaysia, shows a heavy concentration in midafternoon with a small secondary peak at or shortly before dawn.

(60 and 100 inches), although in some locations it is considerably greater. Yearly rainfall in the Af climate is exceeded by that of only one other type of climate—*tropical monsoon*.

Why these climatic conditions occur where they do is relatively straightforward. The principal climatic control is latitude. A Sun high in the sky throughout the year makes for relatively uniform insolation, and so there is little opportunity for seasonal temperature variation. This extensive heating produces considerable thermal convection, which accounts for a portion of the raininess. More important, the influence of the ITCZ for most or all of the year leads to widespread uplift of warm, humid, unstable air. Persistent onshore winds along trade-wind (east-facing) coasts provide a consistent source of moisture and add another mechanism for precipitation—orographic ascent. Interior areas, however, typically lack a great deal of wind—a consequence of the persistent influence of the ITCZ and its ongoing convective updrafts (you will recall that the latitudes around the equator are sometimes referred to as the "doldrums" because of the lack of horizontal air movement).

Tropical Savanna Climate (Aw) The most extensive of the A climates, the **tropical savanna climate**, generally lies both to the north and south of Af climates (Figure 8-5). It occurs broadly in South America, Africa, southern Asia, and to a lesser extent in northern Australia, Central America, and the Caribbean islands—extending in a few places to 25° N and S latitude.

The distinctive characteristic of the Aw climate is its clear-cut seasonal alternation of wet and dry periods (Figure 8-8). This characteristic is explained by the fact that Aw climates lie between the unstable, converging air of the ITCZ (which dominates the Af climates) on their

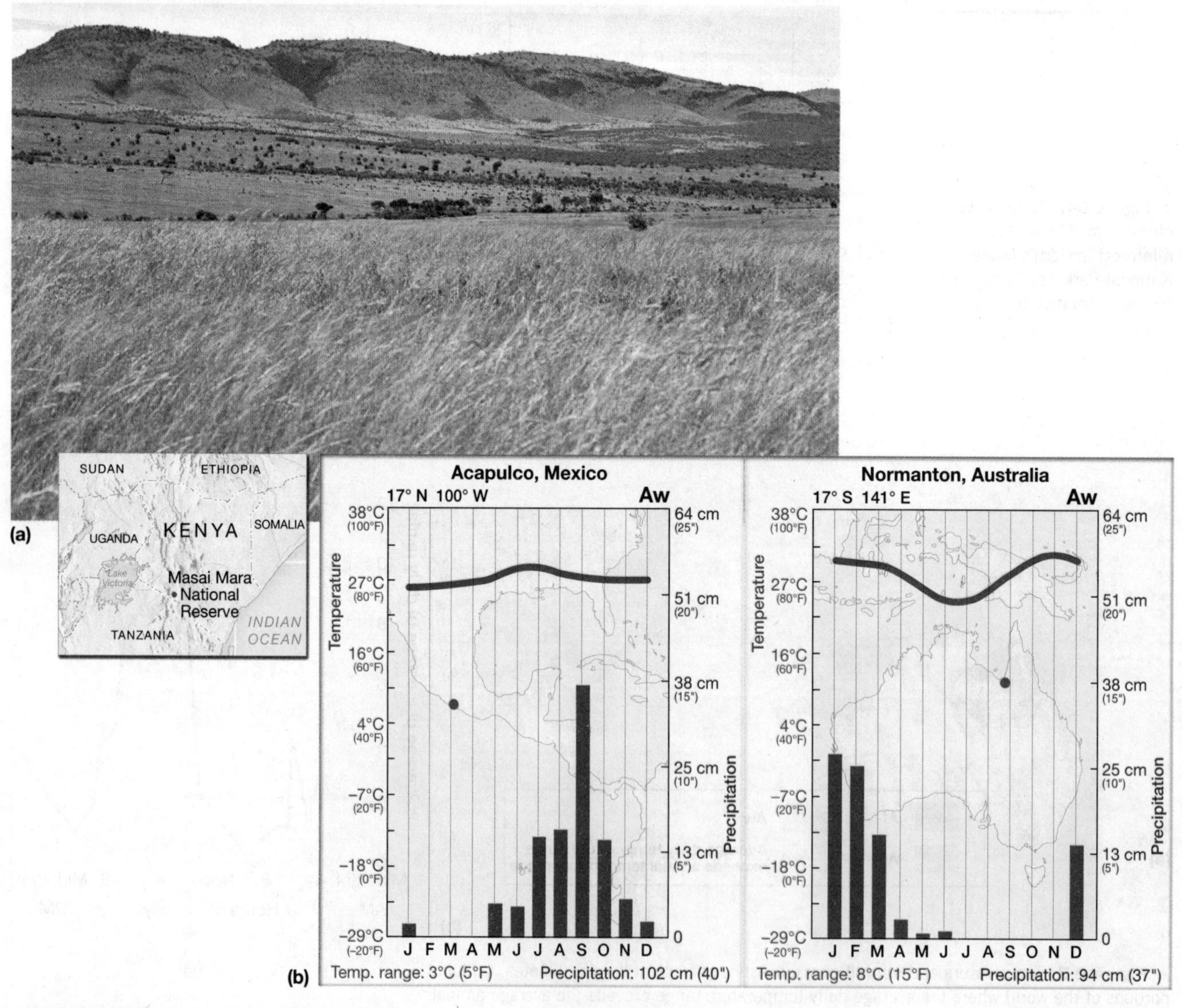

▲ **Figure 8-8** Tropical savanna climate. (a) Tall savanna grassland at the end of the wet season in Masai Mara National Reserve in Kenya. (b) Climographs for representative tropical savanna stations.

equatorial side and stable, subsiding air of the subtropical high on their poleward side. During the low-Sun season (winter), savanna regions are dominated by the dry conditions and clear skies associated with the subtropical highs. In summer, the systems shift as they "follow the Sun" poleward, bringing the thunderstorms and convective rain of the ITCZ into the Aw region (these seasonal precipitation patterns are clearly seen in Figure 6-35 in Chapter 6). The poleward limits of Aw climate are approximately equivalent to the poleward maximum migration of the ITCZ (Figure 8-9).

Annual rainfall totals in the tropical savanna climate are generally less than in the other two A climates. Typical Aw annual rainfall averages are between 90 and 180 centimeters (35 and 70 inches)—almost all coming during the months of the high-Sun season. The average annual temperature range in Aw regions is typically slightly greater than that of Af regions (3 to 8°C [5 to 15°F]), with the higher annual variations occurring in locations farther from the equator. The hottest time of the year is likely to be in late spring just before the onset of the summer rains.

Tropical Monsoon Climate (Am) The **tropical monsoon climate** is found in tropical regions with a prominent monsoon wind pattern. It is most extensive on the windward (west-facing) coasts of southeastern Asia (primarily India, Bangladesh, Myanmar [formerly Burma], and Thailand), but it also occurs in more restricted coastal regions of western Africa, northeastern South America, the Philippines, northeastern Australia, and some islands of the East Indies (Figure 8-5).

The distinctiveness of the Am climate is primarily its rainfall pattern (Figure 8-10). During the high-Sun season, an enormous amount of rain falls in association with the "summer" monsoon. It is not unusual to have more than 75 centimeters (30 inches) of rain in each of two or three months. The annual total for a typical Am station is between 250 and 500 centimeters (100 and 200 inches). An extreme example is Cherrapunji (in the Khasi hills of Assam in eastern India), which has an annual average of 1065 centimeters (425 inches). Cherrapunji has been inundated with 210 centimeters (84 inches) in three days, with 930 centimeters (366 inches) in one month, and with a memorable 2647 centimeters (1042 inches) in its record year (Figure 8-11).

During the low-Sun season, Am climates are dominated by offshore winds. The "winter" monsoon during this season produces little precipitation, and one or two months may be rainless.

A lesser distinction of the Am climate is its annual temperature curve. Although the annual temperature range may be only slightly greater than in a tropical wet climate, the highest Am temperatures normally occur in late spring prior to the onset of the summer monsoon. The heavy cloud cover of the wet monsoon period shields out some of the insolation, resulting in slightly lower temperatures in summer than in spring.

Dry Climates (Group B)

The *dry climates* cover about 30 percent of the land area of the world (Figure 8-12), more than any other climatic zone. Although at first glance their distribution pattern appears erratic and complex, it actually has a considerable degree of predictability. The arid regions of the world develop either as a result of the lack of air uplift necessary for cloud formation, or from the lack of moisture in the air—in some cases, both.

The largest expanses of dry areas are in subtropical latitudes, especially in the western and central portions of continents where subsidence associated with subtropical high-pressure cells suppresses upward vertical motion and cool ocean currents serve to increase atmospheric stability. Desert conditions also occur over extensive ocean areas, and it is quite reasonable to refer to marine deserts.

In the midlatitudes, particularly in central Eurasia, arid climates are found in areas that are cut off from sources of surface moisture either by great distances or by their position in the rain shadow of mountain ranges.

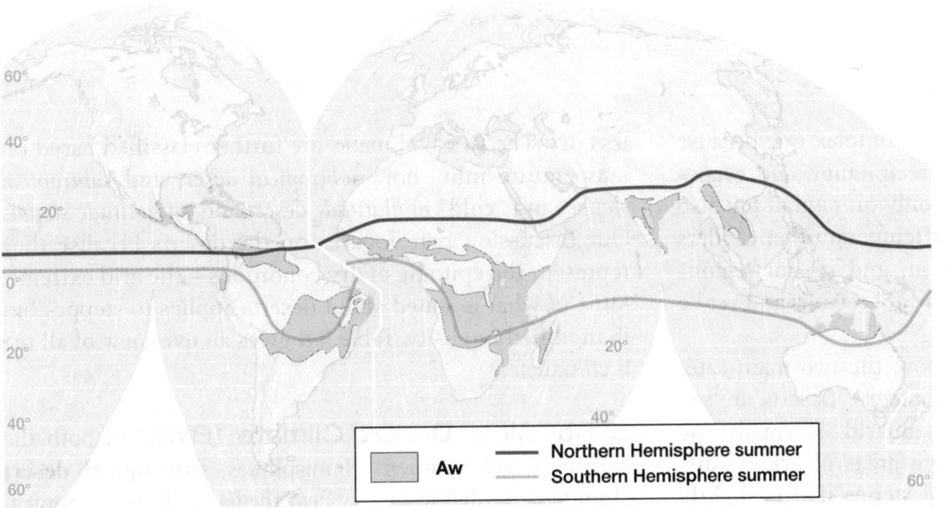

◀ **Figure 8-9** The intertropical convergence zone (ITCZ) migrates widely during the year. The red line shows its typical northern boundary in the Northern Hemisphere summer, and the orange line indicates its southern boundary in the Southern Hemisphere summer. These lines coincide approximately with the poleward limits of Aw climate.

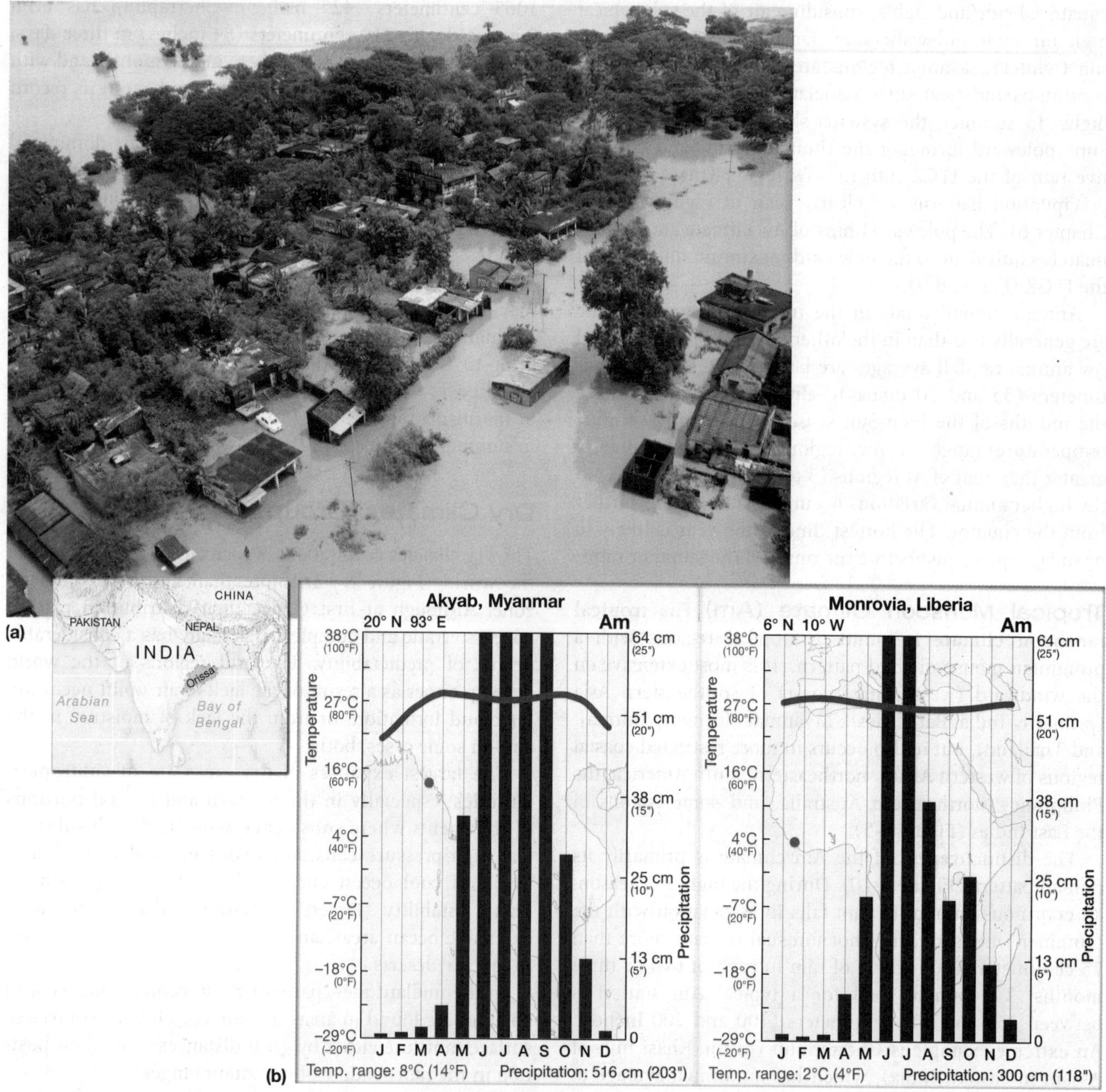

▲ **Figure 8-10** Tropical monsoon climate. (a) Flooding from monsoon rains in Cuttack, Orissa state, India in 2008. (b) Climographs for representative tropical monsoon stations.

The concept of a dry climate is a complex one because it involves the balance between precipitation and evapotranspiration, and so depends not only on rainfall but also on temperature. Generally, higher temperature engenders greater potential evapotranspiration, and so hot regions can receive more precipitation than cool regions and yet be classified as dry.

As Table 8-1 and Figure 8-3 show, the two main categories of B climates are *desert* and *steppe*. Deserts are extremely arid, while steppes are semiarid. Normally the deserts of the world are large core areas of aridity surrounded by a transitional fringe of steppe that is slightly less dry. The two B climates are further classified based on temperature into "hot" *subtropical desert* and *subtropical steppe*, and "cold" *midlatitude desert* and *midlatitude steppe*. Our discussion here focuses on the deserts because they represent the epitome of dry conditions—the arid extreme. Most of what is stated about deserts applies to steppes but in modified intensity. Table 8-4 gives an overview of all the B climates.

Subtropical Desert Climate (BWh) In both the Northern and Southern Hemispheres, **subtropical desert climates** lie either in or very near the band of the subtropical

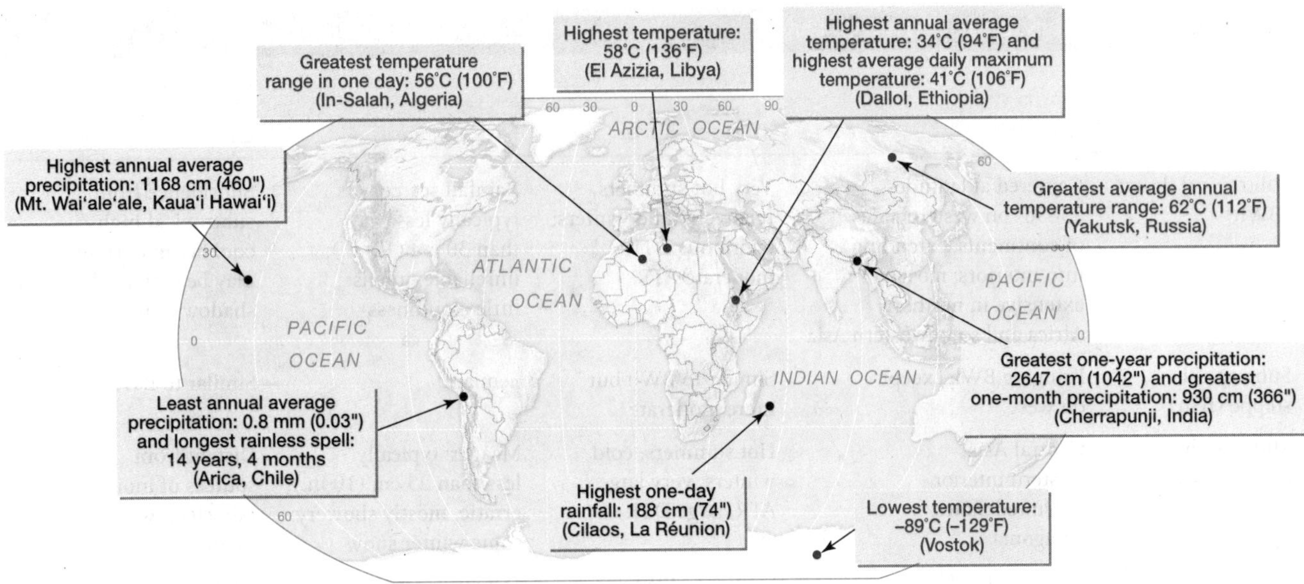

▲ **Figure 8-11** Locations of some extreme weather records.

highs (Figure 8-13), especially along the west coasts of continents where subsidence from the STH is stronger and where cool ocean currents provide added stability. Climographs for two typical BWh locations are shown in Figure 8-14.

The enormous expanse of BWh climate in North Africa (the Sahara) and southwestern Eurasia (the Arabian Desert) represents more desert area than is found in the rest of the world combined—here the adjacency of Eurasia makes Africa a continent without an eastern coast north of 10° N latitude, and so this region lacks maritime moisture sources from the east. Subtropical desert climate is also expansive in Australia (50 percent of the continental area) because the mountains that parallel the continent's east coast are just high enough to prevent Pacific winds from penetrating.

Subtropical deserts are more restricted to coastal regions in southwest Africa, South America, and North America. The greatest north–south elongation occurs along the western side of South America, where the Atacama Desert is not only the "longest" but also the driest of the dry lands. The Atacama is sandwiched in a "double" rain shadow position (Figure 8-15): moist winds from the east are kept out of this region by one of the world's great mountain ranges (the Andes), and Pacific air is thoroughly chilled and stabilized as it passes over the world's most prominent cool ocean current (the Peru or "Humboldt").

The distinctive climatic characteristic of deserts is lack of moisture, and three adjectives are particularly applicable to precipitation conditions in subtropical deserts: scarce, unreliable, and intense.

1. *Scarce*—Subtropical deserts are among the most nearly rainless regions on Earth. According to unofficial records, some have experienced several consecutive years without a single drop of moisture falling from the sky. Most BWh regions, however, are not totally without precipitation. Annual totals of between 5 and 20 centimeters (2 and 8 inches) are characteristic, and some places receive as much as 38 centimeters (15 inches).
2. *Unreliable*—An important climatic axiom is that the lower the average annual precipitation, the greater its variability (see Figure 6-37 in Chapter 6). The very concept of an "average" yearly rainfall in a BWh location is misleading because of year-to-year fluctuations. Yuma, Arizona, for example, has a long-term average rainfall of 7.6 centimeters (3.0 inches), but over a two decade period it has received as little as 1.3 centimeters (0.5 inch) and as much as 18 centimeters (7 inches) in a given year.
3. *Intense*—Most precipitation in these regions falls in vigorous convective showers that are localized and of short duration. Thus, the rare rains may bring brief floods to regions that have been bereft of surface moisture for months.

Temperatures in BWh regions are also distinctive. The combination of low-latitude location (so that the vertical or near-vertical rays of the Sun are striking in summer) and lack of cloudiness permits a great deal of insolation to reach the surface. Summers are long and blisteringly hot, with monthly averages in the middle to high 30s °C (high 90s °F)—significantly hotter than most equatorial regions. Midwinter months have average temperatures in the high teens °C (60s °F), which gives moderate annual temperature ranges of 8 to 14°C (15 to 25°F).

Daily temperature ranges, on the other hand, are sometimes astounding. Summer days are so hot that the nights do not have time to cool off significantly, but during the transition seasons of spring and fall, a 28°C (50°F) fluctuation between the heat of the afternoon and the cool of the

TABLE 8-4 Summary of B Climates: Dry

Type	Location	Temperature	Precipitation	Dominant Controls of Climate
Subtropical desert (BWh)	Centered at latitudes 25–30° on western sides of continents, extending into interiors; most extensive in northern Africa and southwestern Asia	Very hot summers, relatively mild winters; enormous DTR, moderate ATR	Rainfall scarce, typically less than 30 cm (12 in.); unreliable; intense; little cloudiness	Subsidence from subtropical highs; cool ocean currents; may be extended by rain shadow of mountains
Subtropical steppe (BSh)	Fringing BWh except on west	Similar to BWh but more moderate	Semiarid	Similar to BWh
Midlatitude desert (BWk)	Central Asia; western interior of United States; Patagonia	Hot summers, cold winters; very large ATR, large DTR	Meager: typically less than 25 cm (10 in.); erratic, mostly showery; some winter snow	Distant from sources of moisture; rain shadow of mountains
Midlatitude steppe (BSk)	Peripheral to BWk; transitional to more humid climates	Similar to BWk but slightly more moderate	Semiarid; some winter snow	Similar to BWk

ATR = annual temperature range; DTR = daily temperature range.

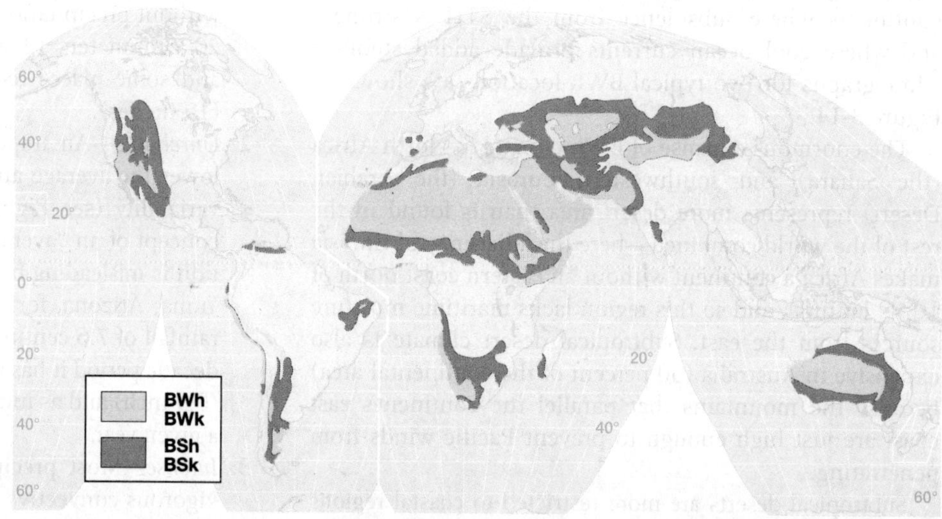

▶ **Figure 8-12** The global distribution of dry (B) climates.

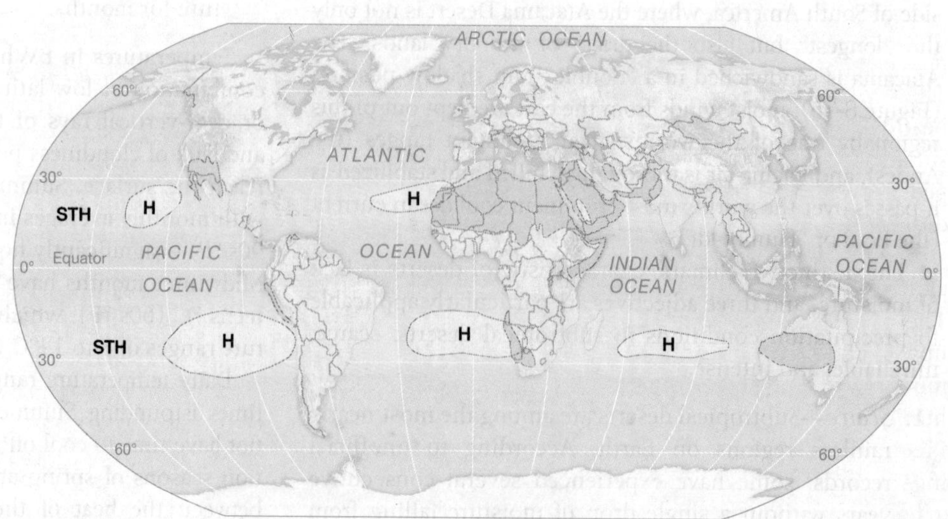

▶ **Figure 8-13** The coincidence of the subtropical high-pressure zones and BWh climates is striking.

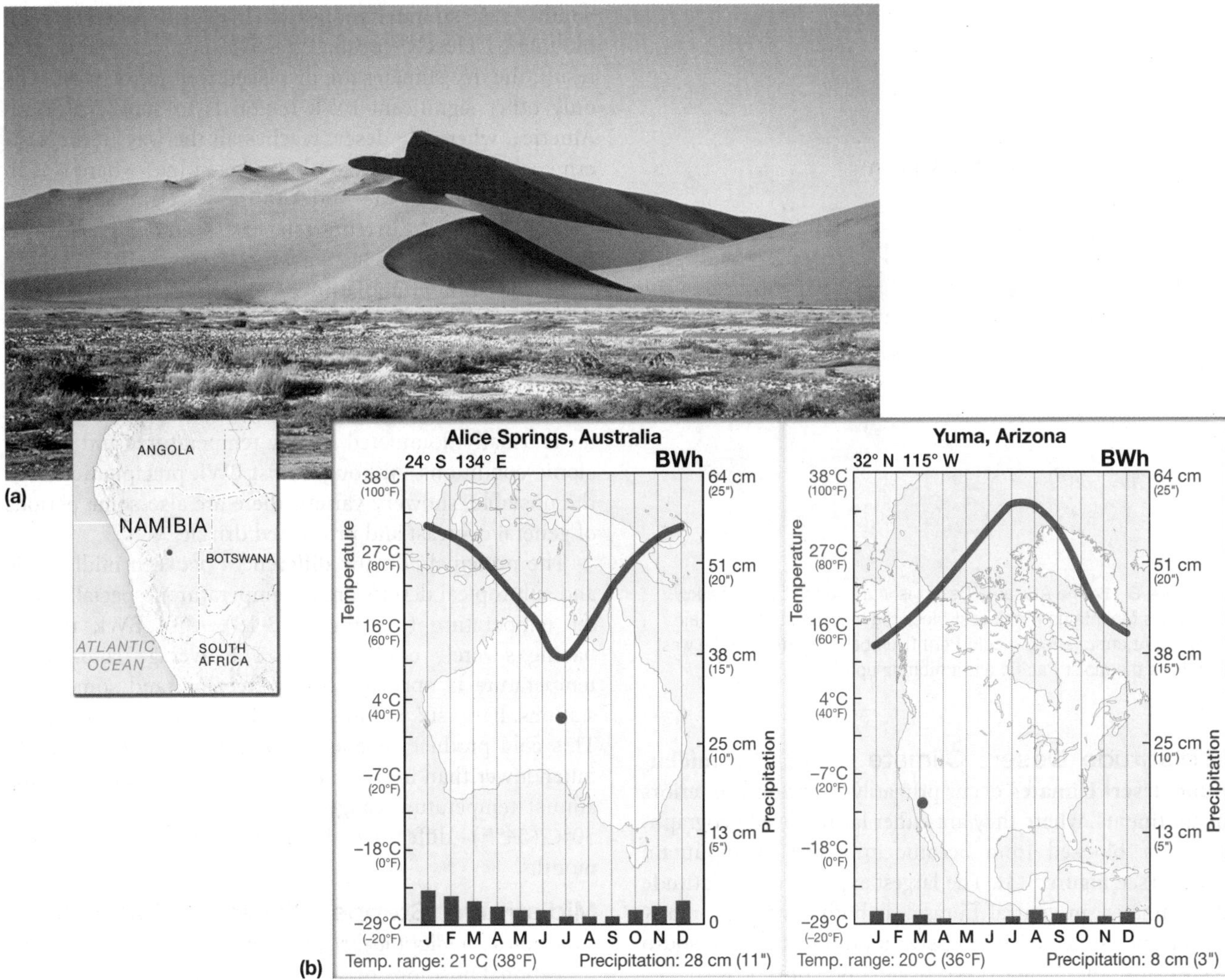

▲ **Figure 8-14** Subtropical desert climate. (a) A subtropical desert landscape sometimes contains an abundance of sand. This scene is from central Namibia. (b) Climographs for representative subtropical desert stations.

following dawn is not unusual—generally clear skies and low water vapor content permit rapid nocturnal cooling from unimpeded longwave radiation transmission through the atmosphere.

Subtropical deserts often experience considerable windiness during daylight, but the air is usually calm at night. The daytime winds are apparently related to rapid daytime heating and strong convective activity, which accelerates surface currents. The persistent winds are largely unimpeded by soil and vegetation, with the result that a great deal of dust and sand is frequently carried along.

Specialized and unusual temperature conditions prevail along western coasts in subtropical deserts (Figure 8-16). The cold waters offshore (the result of cool currents and upwelling of cold deeper ocean water) chill any air that moves across them. This cooling produces high relative humidity as well as frequent fog and low stratus clouds. Precipitation almost never results from this advective cooling, however, and the influence normally extends only a few kilometers inland. The immediate coastal region, however, is characterized by such abnormal desert conditions as relatively low summer temperatures (typical summer averages in the low 20s °C or low 70s °F), continuously high relative humidity, and greatly reduced annual and daily temperature ranges, as indicated by a comparison of Figures 8-14b and 8-16b.

Subtropical Steppe Climate (BSh)

The *subtropical steppe climates* characteristically surround the BWh climates (except on the western side), separating the deserts from the more humid climates beyond. Temperature and precipitation conditions are not significantly unlike those just described for BWh regions except that the extremes are more muted in the steppes (Figure 8-17). Thus, rainfall is somewhat greater and more reliable, and temperatures are slightly moderated. Moreover, the meager precipitation tends to have a seasonal concentration. On the equatorward side of the desert, rain occurs in the high-Sun season; on the poleward side, it is concentrated in the low-Sun season.

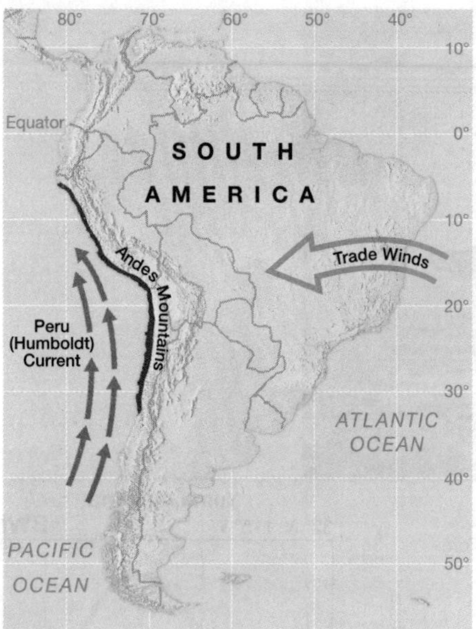

▲ **Figure 8-15** The Atacama Desert is in a "double" rain shadow: the Andes Mountains to the east block the movement of moist air from the Atlantic, and the cold Peru (Humboldt) Current to the west stabilizes the moist Pacific air, inhibiting uplift.

Midlatitude Desert Climate (BWk) The **midlatitude desert climates** occur primarily in the deep interiors of continents, where they are either far removed geographically or blocked from oceanic influence by mountain ranges (see Figure 8-3). The largest expanse of midlatitude dry climates, in central Eurasia, is both distant from any ocean and protected by massive mountains on the south (especially the Himalayas) from any contact with the South Asian summer monsoon. In North America, high mountains closely parallel the western coast and as a result, the dry climates are displaced well to the west. The only other significant BWk region is in southern South America, where the desert reaches all the way to the eastern coast of Patagonia (southern Argentina) where it is in the rain shadow of the Andes to the west and adjacent to the cool Falkland Current to the east. Climographs for two typical BWk locations are shown in Figure 8-18.

Precipitation in midlatitude (BWk) deserts is much like that of subtropical (BWh) deserts—meager and erratic. Differences lie in two aspects: seasonality and intensity. Most BWk regions receive the bulk of their precipitation in summer, when warming and instability are common. Winter is usually dominated by low temperatures and anticyclonic conditions. Although most BWk precipitation is of the unstable, showery variety, there are also some periods of general overcast and protracted drizzle.

The principal climatic differences between midlatitude and subtropical deserts are in temperature, especially winter temperature (see Figure 8-18), with BWk regions having severely cold winters. The average cold-month temperature is normally below freezing, and some BWk stations have six months with below-freezing averages. This cold produces average annual temperatures that are much lower than in BWh regions and greatly increases the annual temperature range—some BWk locations have a 30°C (54°F) difference between winter and summer months.

Midlatitude Steppe Climate (BSk) As in the subtropics, *midlatitude steppe climates* generally occupy transitional positions between deserts and humid climates.

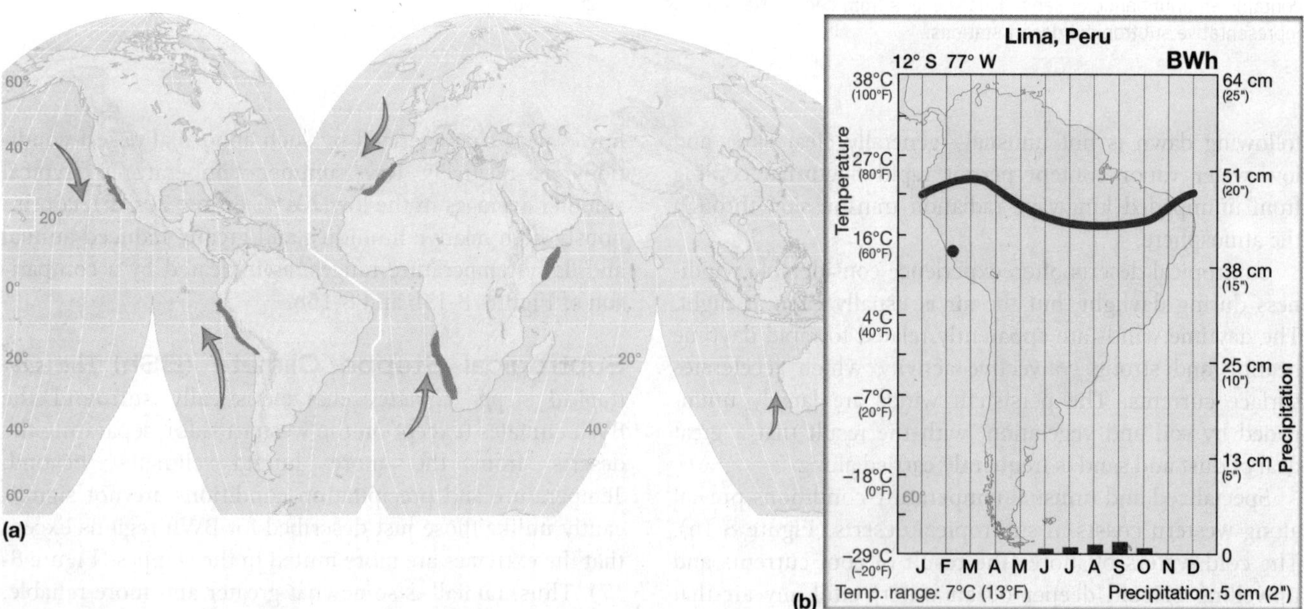

▲ **Figure 8-16** (a) Cool, foggy west coast deserts are found along coasts paralleled by cool ocean currents and cold upwellings. Such deserts are mostly in subtropical west coast locations, with two exceptions: They are absent from the western coast of Australia, and they occur on the eastern coast of the "horn" of Africa (Somalia). (b) Climograph for a cool west coast desert station.

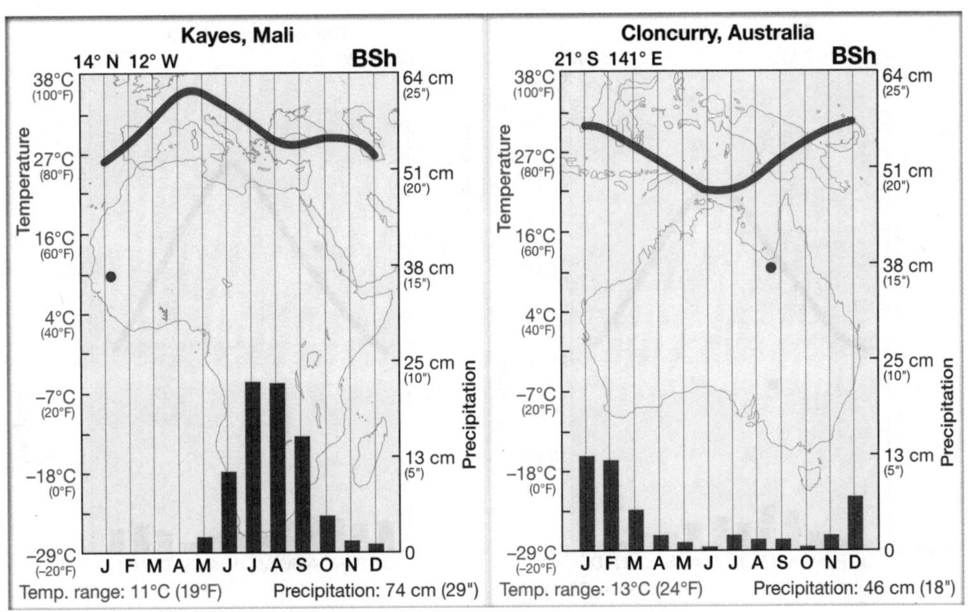

◀ **Figure 8-17** Climographs for representative subtropical steppe stations.

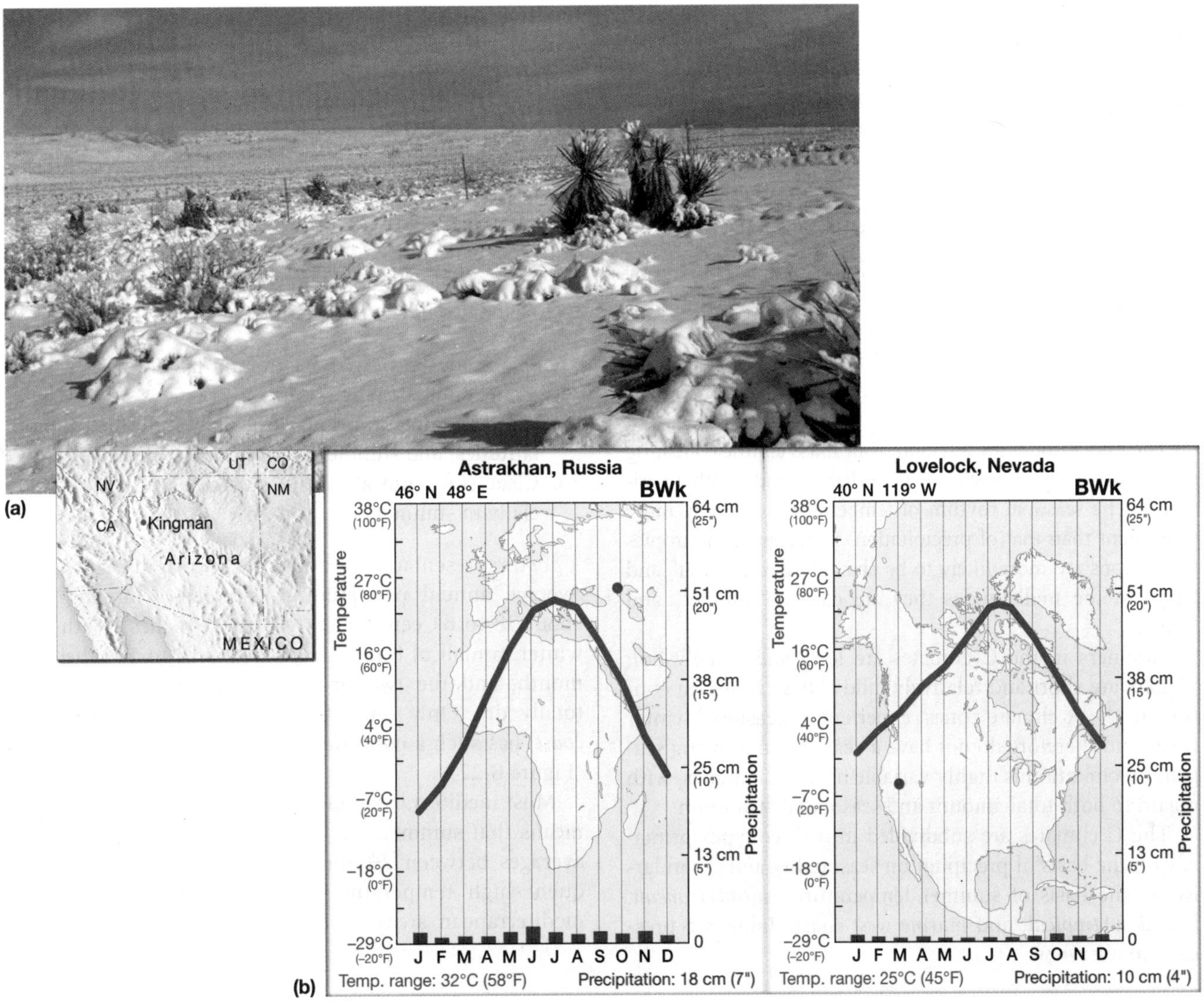

▲ **Figure 8-18** Midlatitude desert climate. (a) Midlatitude deserts are characterized by cold winters. Here a fresh snowfall covers such typical desert plants as creosote bush and yucca near Kingman in northwestern Arizona. (b) Climographs for representative midlatitude desert stations.

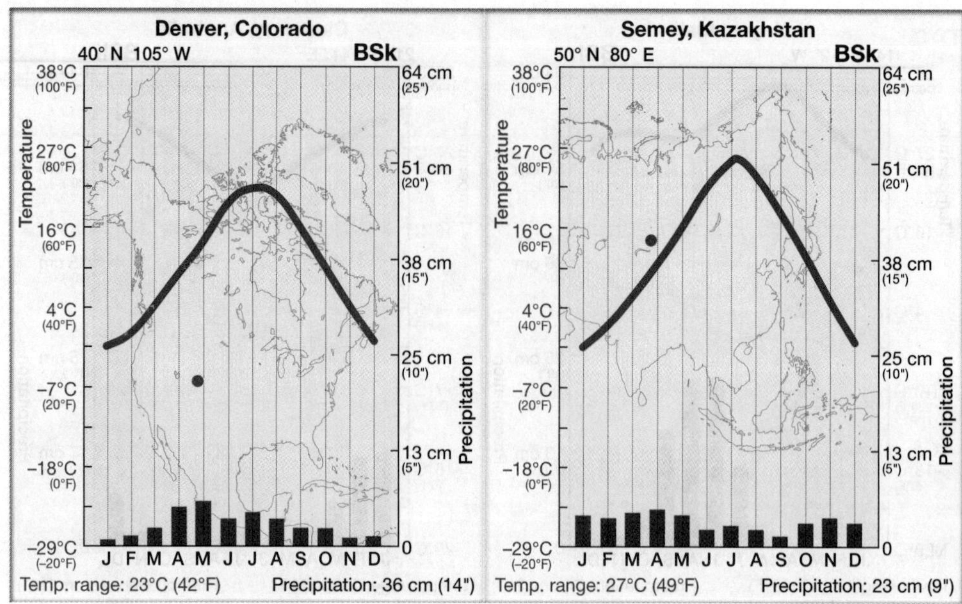

▶ **Figure 8-19** Climographs for representative midlatitude steppe stations.

Typically midlatitude steppes have more precipitation than midlatitude deserts and lesser temperature extreme (Figure 8-19). In western North America, the steppe climate is broadly found, and only in the interior southwest of the United States is the climate sufficiently arid to be classified as desert.

Mild Midlatitude Climates (Group C)

The *mild midlatitude climates* (Group C climates) occupy the equatorward margin of the midlatitudes, extending farther poleward along the west coasts of continents than along the east (Figure 8-20). They constitute a transition between warmer tropical climates and colder severe midlatitude climates.

The midlatitudes are a region of air mass contrast, creating a kaleidoscope of atmospheric disturbances and weather variability. The seasonal rhythm of temperature is usually more prominent than that of precipitation. Whereas in the tropics the seasons are more likely to be characterized as "wet" and "dry," in the midlatitudes they are clearly "summer" and "winter."

Summers in the C climates are long and usually hot; winters are short and relatively mild. These zones, in contrast to the A climate zones, experience occasional winter frosts and therefore do not have a year-round growing season. Precipitation is highly variable in the C climates, with regard to both total amount and seasonal distribution.

The C climates are subdivided into three types primarily on the basis of precipitation seasonality and secondarily on the basis of summer temperatures: *mediterranean*, *humid subtropical*, and *marine west coast*. Table 8-5 presents an overview.

Mediterranean Climate (Csa, Csb) The two Cs climates are sometimes referred to as *dry summer subtropical*, but the more widely used designation is **mediterranean climate.**[1] Cs climates are found on the western side of continents centered at about latitudes 35° N and 35° S. With one exception, all mediterranean regions are small, being restricted mostly to coastal areas either by interior mountains or by limited landmasses. These small regions are in central and southern California (Figure 8-21), central Chile, the southern tip of Africa, and the two southwestern "corners" of Australia. The only extensive area of mediterranean climate is around the borderlands of the Mediterranean Sea.

Cs climates have three distinctive characteristics:

1. The modest annual precipitation falls in winter, summers being virtually rainless.
2. Winter temperatures are unusually mild for the midlatitudes, and summers vary from warm to hot.
3. Clear skies and abundant sunshine are typical especially in summer.

Two representative locations are shown in Figure 8-21. Average annual precipitation is modest, ranging from about 38 to 64 centimeters (15 to 25 inches), with midwinter rainfall of 8 to 13 centimeters (3 to 5 inches) per month, and the two or three midsummer months are totally dry. Only one other climatic type, *marine west coast*, has such a concentration of precipitation in winter (Figure 8-22).

Most mediterranean climate is classified as Csa, which means that summers are hot, with midsummer monthly averages between 24 and 29°C (75 and 85°F) and frequent high temperatures above 38°C (100°F). Coastal mediterranean areas (Csb) have much milder summers

[1]The proper terminology is without capitalization because it is a generic term for a type of climate; Mediterranean with the capital M refers to a specific region around the Mediterranean Sea.

TABLE 8-5 Summary of C Climates: Mild Midlatitude

Type	Location	Temperature	Precipitation	Dominant Controls of Climate
Mediterranean (Csa, Csb)	Centered at 35° latitude on western sides of continents; limited east–west extent except in Mediterranean Sea area	Warm/hot summers; mild winters; year-round mildness in coastal areas	Moderate: 38–64 cm (15–25 in.) annually, nearly all in winter; much sunshine, some coastal fog	Subtropical high subsidence and stability in summer; westerly winds and cyclonic storms in winter
Humid subtropical (Cfa, Cwa, Cwb)	Centered at 30° latitude on eastern sides of continents; considerable east–west extent	Summers warm/hot, sultry; winters mild to cold	Abundant: 100–165 cm (40–65 in.) annually, mostly rain; summer maxima but no true dry season	Westerly winds and storms in winter; moist onshore flow in summer; monsoons in Asia
Marine west coast (Cfb, Cfc)	Latitudes 40–60° on western sides of continents; limited inland extent except in Europe	Very mild winters for the latitude; generally mild summers; moderate ATR	Moderate to abundant: 75–125 cm (30–50 in.) mostly in winter; many days with rain; much cloudiness	Westerly flow and oceanic influence year-round

ATR = annual temperature range.

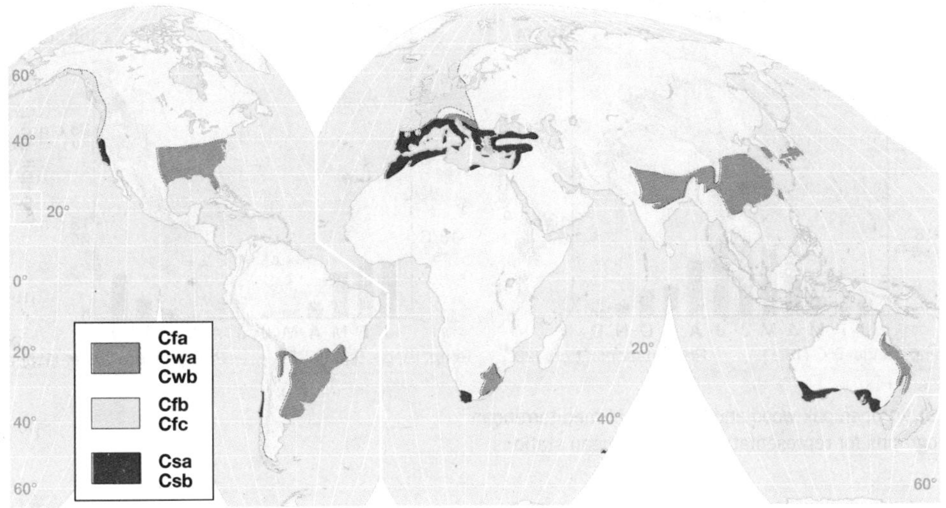

◀ **Figure 8-20** The global distribution of mild midlatitude (C) climates.

than inland mediterranean areas as a result of sea breezes and frequent coastal advection fog (Figure 8-23), with midsummer month temperature averages of 16 to 21°C (60 to 70°F). Csb winters are slightly milder than Csa winters, with midwinter month averages of about 13°C (55°F), and only rarely temperatures falling below freezing.

The origin of mediterranean climates is clear-cut: in summer, these regions are dominated by dry, stable, subsiding air from the eastern portions of subtropical highs; in winter, the wind and pressure belts shift equatorward, and mediterranean regions come under the influence of the westerlies with their migratory midlatitude cyclones and associated fronts. Almost all precipitation comes from these cyclonic storms, except for occasional tropical influences in California.

Humid Subtropical Climate (Cfa, Cwa, Cwb)

Whereas mediterranean climates are found on the western side of continents, **humid subtropical climates** are found on the eastern side of continents centered at about the same latitude. However, humid subtropical climates extend farther inland and over a greater range of latitude than do mediterranean climates—especially in North America, South America, and Eurasia (Figure 8-24).

The humid subtropical climates differ from mediterranean climates in several important respects:

1. Summer temperatures in humid subtropical regions are generally warm to hot, with the highest monthly averages between 24 and 27°C (75 and 80°F). This is not unlike mediterranean climates, but the humid subtropical regions are characterized by much higher

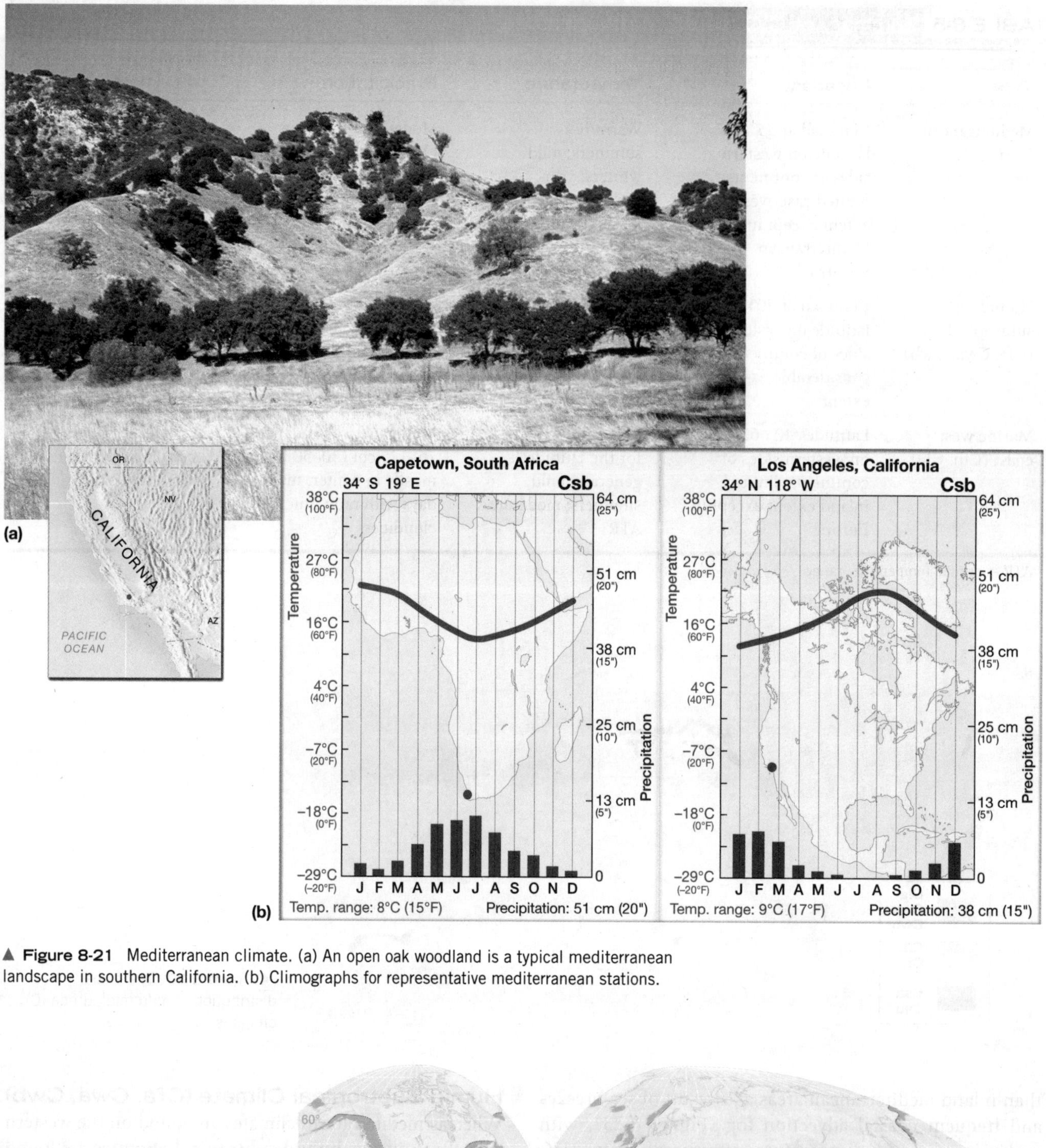

(a)

(b)

Capetown, South Africa	Los Angeles, California
34° S 19° E Csb	34° N 118° W Csb
Temp. range: 8°C (15°F)	Temp. range: 9°C (17°F)
Precipitation: 51 cm (20")	Precipitation: 38 cm (15")

▲ **Figure 8-21** Mediterranean climate. (a) An open oak woodland is a typical mediterranean landscape in southern California. (b) Climographs for representative mediterranean stations.

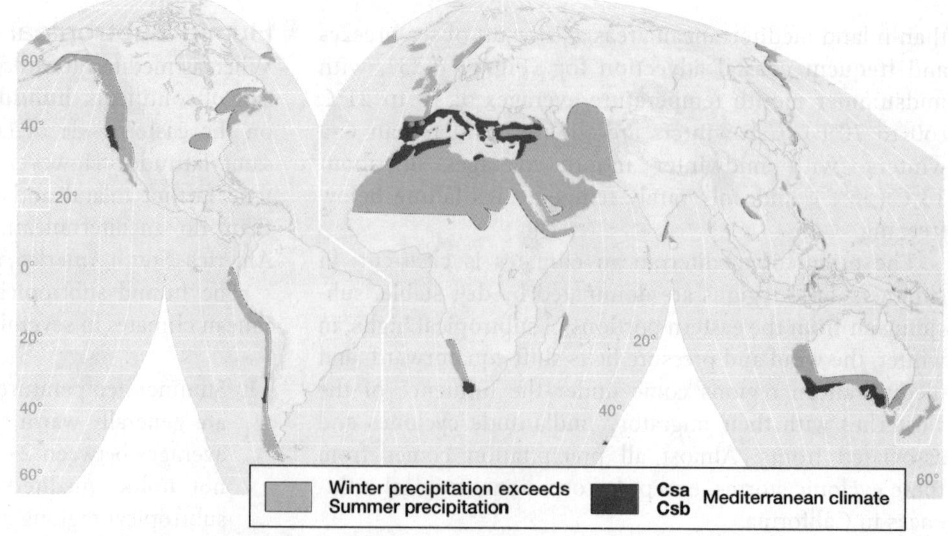

▶ **Figure 8-22** Lands that have precipitation maxima in winter rather than in summer are mostly associated with mediterranean and marine west coast climates and regions adjacent to them.

▲ **Figure 8-23** There are often significant temperature differences between coastal and inland mediterranean areas. This climograph shows the annual temperature curves for three stations in Southern California: Santa Monica (a coastal station; Csb climate) and Pasadena and San Fernando (inland stations having exactly the same average temperature for each month; Csa climate). The physiographic map shows the topographic and coastal relationships of the three stations.

humidity in summer, and so sensible temperatures are higher. Cfa days tend to be hot and sultry, and often night brings little relief.

2. Precipitation in humid subtropical regions tends to have a summer maximum, associated with onshore flow and frequent convection. Winter is a time of diminished precipitation, but it is not really a dry season except in China, where winter monsoon conditions dominate. The rain and occasional snow of winter are the result of midlatitude cyclones passing through the region. In the North American and Asian coastal areas, a late summer-autumn bulge in the precipitation curve is due to rainfall from tropical cyclones. Annual precipitation is generally abundant, averaging 100 to 165 centimeters (40 to 65 inches), with a general decrease inland.
3. Winter temperatures in Cfa regions are mild, but typically 6°C (10°F) lower on average than in mediterranean regions. Winter is punctuated by cold waves that can bring severe weather and killing frosts for a few days at a time. The importance of this fact can be shown by agricultural adjustments—for example, the northernmost limit of commercial citrus production in the eastern United States (much of Florida has a Cfa climate) is at about 29° N latitude, but in the western United States (much of California has a Cs climate) it is at 38° N latitude, 1000 kilometers (625 miles) farther north (Figure 8-25).

To understand the controls of humid subtropical climates, it is again useful to make a comparison with mediterranean regions since both generally lie in the same latitude. During winter, westerly winds bring midlatitude cyclones and precipitation to both regions. In summer, however, in mediterranean regions along west coasts the poleward migration of the subtropical high and relatively cool ocean water stabilizes the atmosphere bringing dry weather, while on the eastern side of continents, where humid subtropical climates are found, no such stability exists—summer rain comes from the onshore flow of maritime air and frequent convective uplift, as well as the influence of tropical cyclones.

Marine West Coast Climate (Cfb, Cfc) As the name implies, **marine west coast climates** are situated on the western side of continents between about 40° and 60°; this is a windward location in the band of the westerlies.

The most extensive area of marine west coast climate is in western and central Europe, where the maritime influence can be carried some distance inland without hindrance from topographic barriers. The North American region is much more restricted by the presence of mountain ranges that run perpendicular to the direction of onshore flow. Only in the Southern Hemisphere, where landmasses are small in these latitudes (New Zealand, southeast Australia, and southernmost South America), does this oceanic climate extend across to eastern coasts (Figure 8-26).

The dominant climate control for these regions is straightforward: the year-round cool maritime influence brought by the onshore flow of the westerlies leads to frequent cloudiness and high proportions of days with some precipitation, as well as an extraordinarily temperate climate considering the latitude. In contrast, mediterranean climates equatorward of Cfb regions come under greater influence of the subtropical high in summer, bringing clear, dry conditions during that time of year. Two typical Cfb climographs are shown in Figure 8-26.

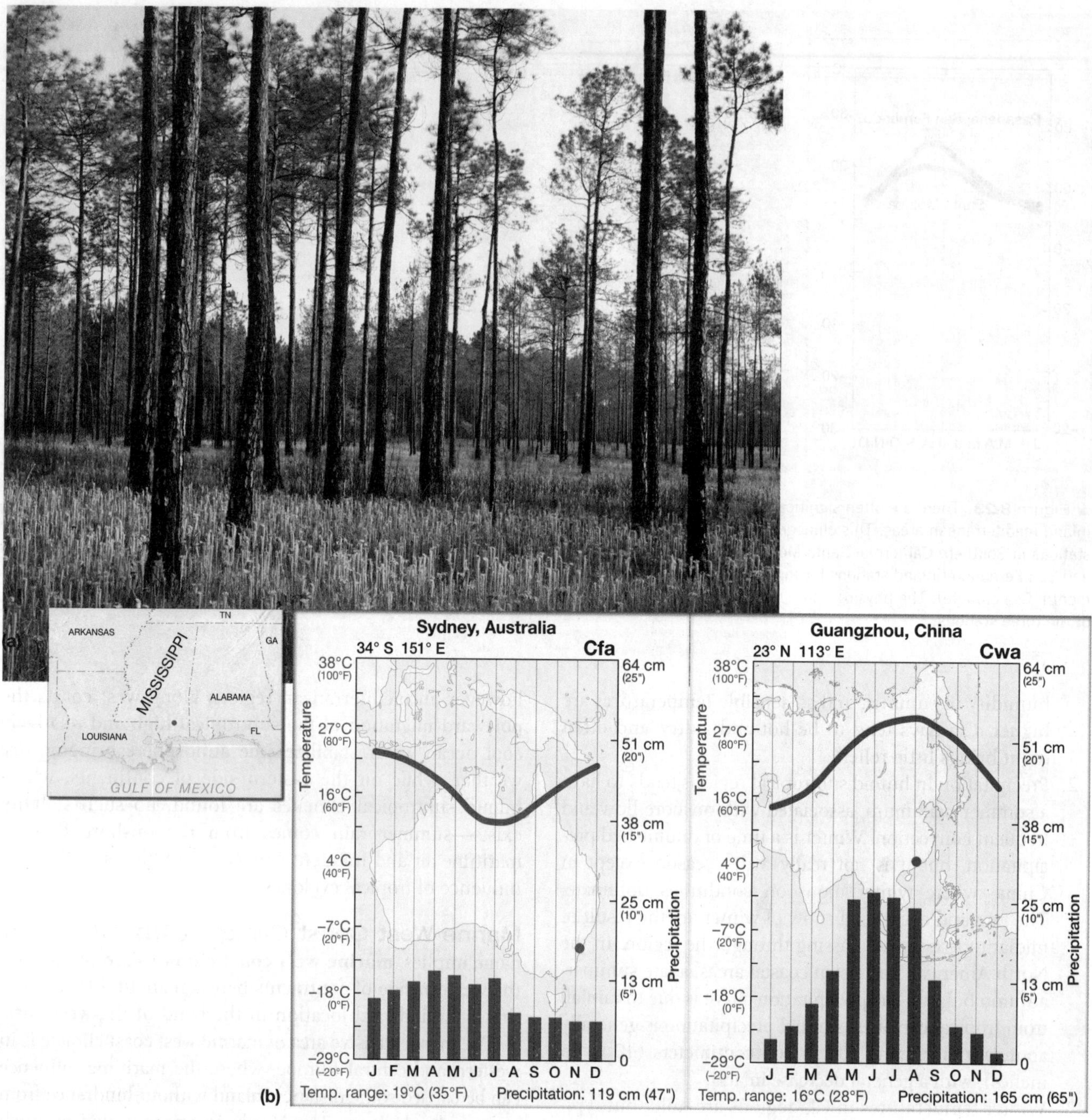

▲ **Figure 8-24** Humid subtropical climate. (a) Pine trees in pitcher plant bog, DeSoto National Forest, Mississippi. (b) Climographs for representative humid subtropical stations.

The oceanic influence moderates temperatures most of the time. Isotherms in marine west coast locations tend to parallel the coastline rather than following their "normal" east–west paths (for example, see Figure 4-30). Average summer month temperatures are generally between 16 and 21°C (60 and 70°F), with winter months averaging between 2 and 7°C (35 and 45°F). There are occasionally very hot days, but prolonged heat waves are rare. Similarly, very cold days occur on occasion, but frosts are relatively infrequent except in interior European locations. London, for example, at 52° N latitude, experiences freezing temperatures on fewer than half the nights in January. There is also an abnormally long growing season for the latitude; around Seattle, for instance, the growing season is a month longer than that around Atlanta, a city lying 14° of latitude closer to the equator.

Marine west coast climates are among the wettest of the midlatitudes. Although some localities receive less, a range of between 75 and 125 centimeters (30 and 50 inches) is typical, with much higher totals recorded on exposed

Citrus production areas
California
Arizona
Texas
Florida
ATLANTIC OCEAN
PACIFIC OCEAN
Gulf of Mexico
0 200 400 Miles
0 200 400 Kilometers
40°
30°
120°
90°
80°
70°

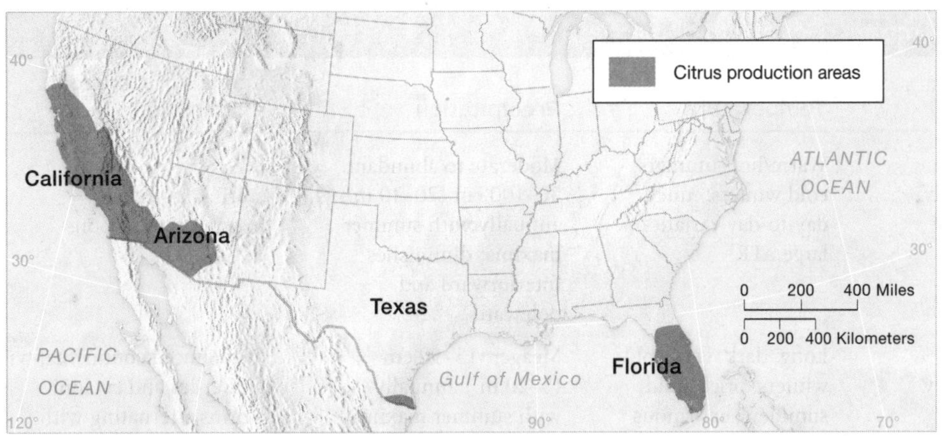

◀ **Figure 8-25** Distribution of citrus production in the conterminous United States. Winter cold spells prohibit commercial citrus growing north of central Florida in the eastern United States; the lack of severe freezes in much of California allows citrus to be grown much farther north.

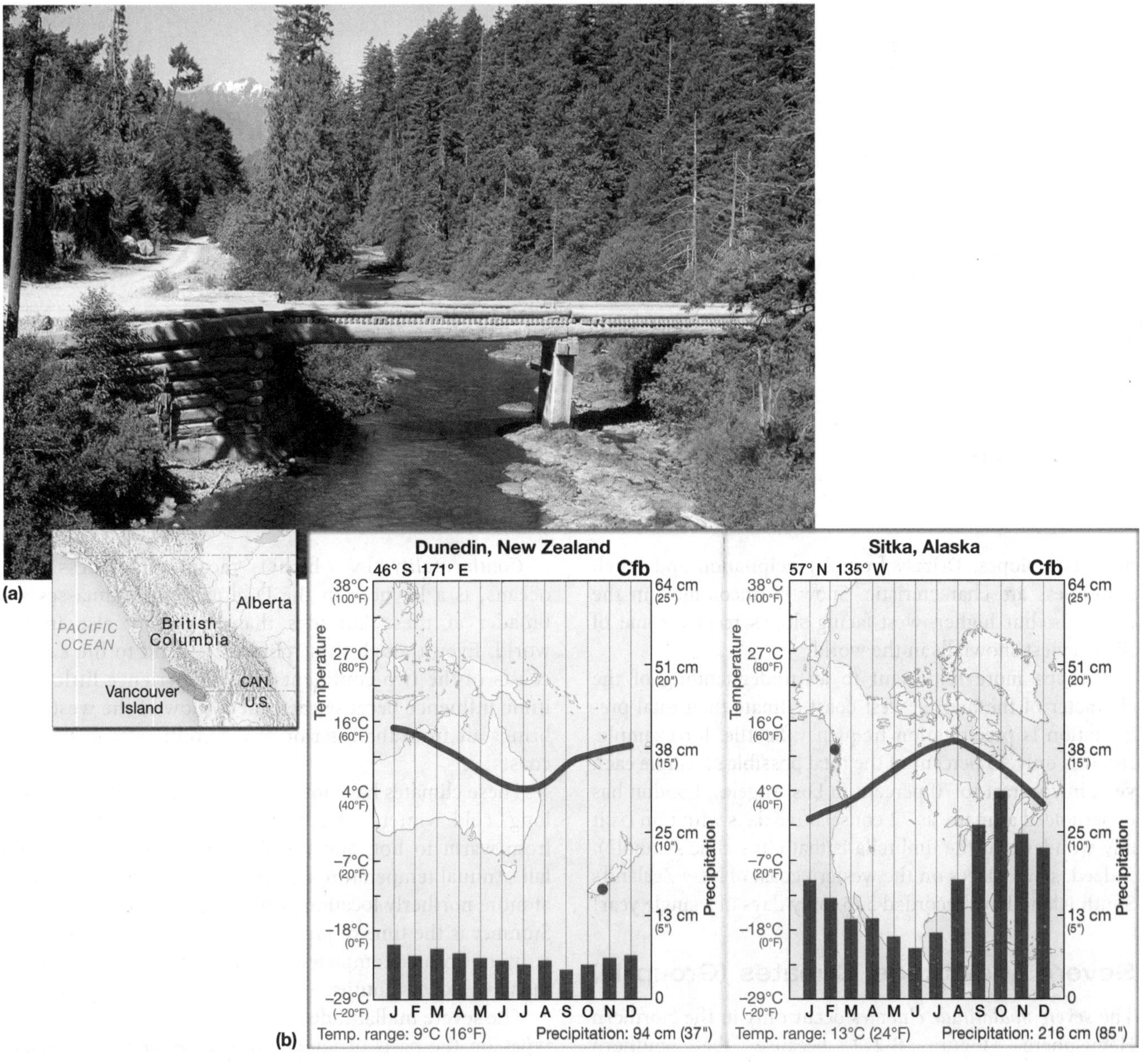

(a)
PACIFIC OCEAN
British Columbia
Alberta
Vancouver Island
CAN.
U.S.
(b)
Dunedin, New Zealand
46° S 171° E
Cfb
Sitka, Alaska
57° N 135° W
Cfb
Temperature
Precipitation
38°C (100°F)
27°C (80°F)
16°C (60°F)
4°C (40°F)
-7°C (20°F)
-18°C (0°F)
-29°C (-20°F)
64 cm (25")
51 cm (20")
38 cm (15")
25 cm (10")
13 cm (5")
0
J F M A M J J A S O N D
Temp. range: 9°C (16°F)
Precipitation: 94 cm (37")
Temp. range: 13°C (24°F)
Precipitation: 216 cm (85")

▲ **Figure 8-26** Marine west coast climate. (a) Big trees dominate this marine west coast landscape on Canada's Vancouver Island. (b) Climographs for representative marine west coast stations.

TABLE 8-6 Summary of D Climates: Severe Midlatitude

Type	Location	Temperature	Precipitation	Dominant Controls of Climate
Humid continental (Dfa, Dfb, Dwa, Dwb)	Northern Hemisphere only; latitudes 35–55° on eastern sides of continents	Warm/hot summers; cold winters; much day-to-day variation; large ATR	Moderate to abundant: 50–100 cm (20–40 in.) annually with summer maxima; diminishes interiorward and poleward	Westerly winds and storms, especially in winter; monsoons in Asia
Subarctic (Dfc, Dfd, Dwc, Dwd)	Northern Hemisphere only, latitudes 50–70° across North America and Eurasia	Long, dark, very cold winters; brief, mild summers; enormous ATR	Meager: 13–50 cm (5–20 in.) annually with summer maxima; light snow in winter but little melting	Pronounced continentality; westerlies and cyclonic storms alternating with prominent anticyclonic conditions

ATR = annual temperature range.

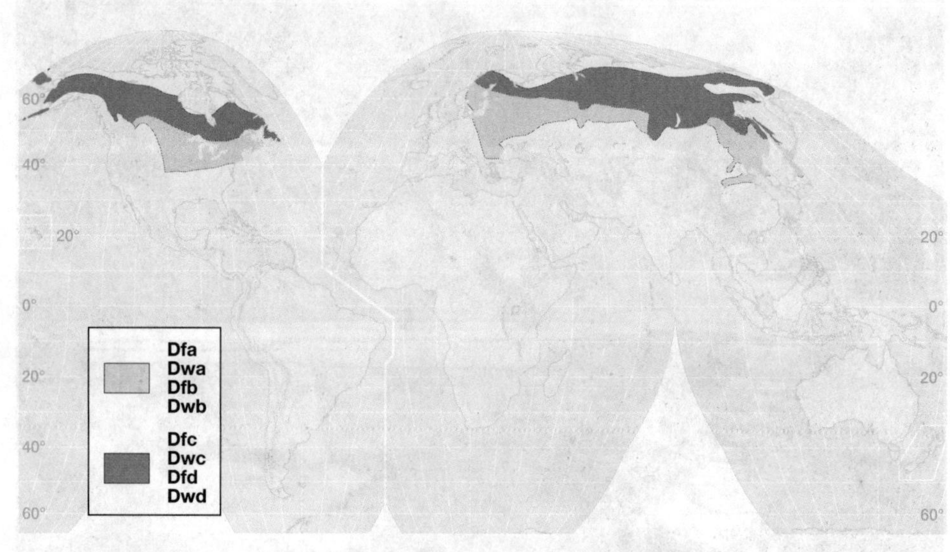

▶ **Figure 8-27** The global distribution of severe midlatitude (D) climates.

mountain slopes. Drizzly frontal precipitation and much cloudiness are characteristic. Snow is uncommon in the lowlands, but higher, west-facing slopes receive some of the heaviest snowfalls in the world.

Perhaps more important to an understanding of the character of the marine west coast climate than total precipitation is precipitation frequency. Seattle, for example, receives only 43 percent of the total possible sunshine each year, in contrast to 70 percent in Los Angeles; London has experienced as many as 72 consecutive days with rain (is it any wonder that the umbrella is that city's civic symbol?). Indeed, some places on the western coast of New Zealand's South Island have recorded 325 rainy days in a single year!

Severe Midlatitude Climates (Group D)

The *severe midlatitude climates* occur only in the Northern Hemisphere (Figure 8-27) because the Southern Hemisphere has limited landmasses at the appropriate latitudes—between 40° and 70°. This climatic group extends broadly across North America and Eurasia.

Continentality, by which is meant remoteness from oceans, is a keynote in the D climates. Landmasses are broader at these latitudes than anywhere else in the world. Even though these climates extend to the eastern coasts of the two continents, they experience little maritime influence because the general flow of the westerlies brings air from the interior of the continents to the east coasts.

These climates have four clearly recognizable seasons: a long, cold winter, a relatively short summer that varies from warm to hot, and transition periods in spring and fall. Annual temperature ranges are very large, particularly at more northerly locations where winters are most severe. Summer is the time of precipitation maximum, but winter is by no means completely dry, and snow cover lasts for many weeks or months.

The severe midlatitude climates are subdivided into two types on the basis of temperature. The *humid continental* type has long, warm summers, while the *subarctic* type is characterized by short summers and very cold winters. Table 8-6 summarizes the D climates.

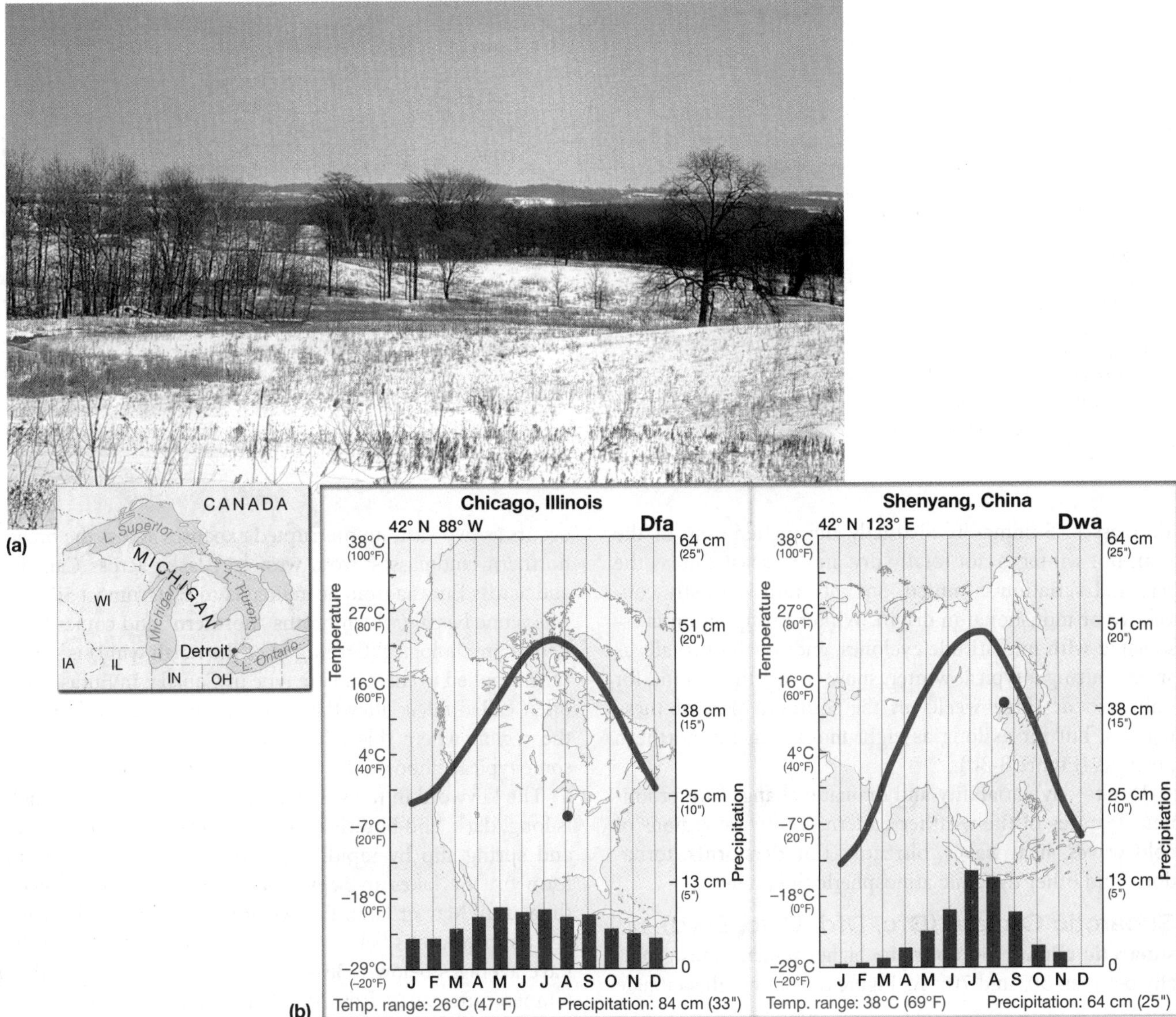

▲ **Figure 8-28** Humid continental climate. (a) The barren look of winter in a humid continental climate. This open woodland of deciduous trees is near Detroit in southern Michigan. (b) Climographs for representative humid continental stations.

Humid Continental Climate (Dfa, Dfb, Dwa, Dwb) The **humid continental climate** (Figure 8-28) is found over a large area of east-central North America and northern and northeastern Eurasia, between 35° and 55° north latitude.

This climatic type is dominated by the westerly wind belt throughout the year, resulting in frequent weather changes associated with the passage of migratory pressure systems, especially in winter. Although large areas of humid continental climate are located along the coastlines of eastern North America and Eurasia—at much the same latitude as marine west coast climates—the temperature pattern is continental. The explanation for this is simple: while westerly winds bring maritime influence to marine west coast locations throughout the year, westerly winds bring continental air to the east coast locations of humid continental climates—especially cold continental air masses during winter.

Summer temperatures are warm, generally averaging in the mid-20s °C (mid-70s °F), and so are comparable although shorter than those of the humid subtropical climate to the south. The average winter month temperature is usually between −12°C and −4°C (10 and 25°F), with from one to five months averaging below freezing. Winter temperatures decrease rapidly northward in the humid continental climates, as Figure 8-29 shows, and the growing season diminishes from about 200 days on the southern margin to about 100 days on the northern edge.

Despite their name, precipitation is not high in humid continental climates. Annual totals average between 50 and 100 centimeters (20 and 40 inches), with the highest values on the coast and a general decrease inland, and from south to north (see Figure 6-36 in Chapter 6). Both of these trends reflect increasing distance from warm moist

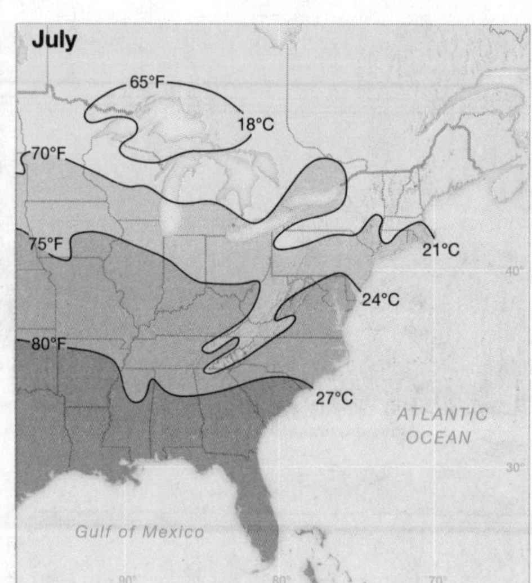

▶ **Figure 8-29** North–south temperature variation in the midlatitudes is much sharper in winter than in summer. These maps of the eastern United States show a very steep north–south average temperature gradient in January (left), with a more gradual gradient in July (right).

air masses. Summer is distinctly the wetter time of the year, but winter is not totally dry, and in coastal areas the seasonal variation is muted. Summer rain is mostly convective or monsoonal in origin. Winter precipitation is associated with midlatitude cyclones, and much of it falls as snow. During a typical winter, snow covers the ground for only two or three weeks in the southern part of these regions, but for as long as eight months in the northern portions (Figure 8-30).

Day-to-day variability and dramatic changes are prominent features of the weather pattern. These are regions of cold waves, heat waves, blizzards, thunderstorms, tornadoes, and other dynamic atmospheric phenomena.

Subarctic Climate (Dfc, Dfd, Dwc, Dwd) The **subarctic climate** occupies the higher midlatitudes generally between 50° and 70°. As Figure 8-3 shows, this climate occurs as two vast, uninterrupted expanses across the broad northern landmasses: from western Alaska across Canada, and across Eurasia from Scandinavia to easternmost Siberia. The name *boreal* (which means "northern" and comes from *Boreas*, mythological Greek god of the north wind) is sometimes applied to this climatic type in Canada; in Eurasia it is often called *taiga*, after the Russian name for the forest in the region where this climate occurs. Figure 8-31 shows some typical climographs.

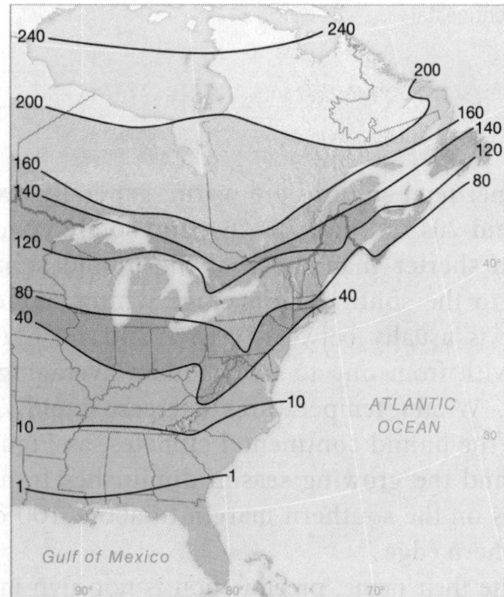

▲ **Figure 8-30** The average duration of snow cover in eastern North America including the Dfa region. The numbers on the isolines represent the average annual number of days with a snow cover of 2.5 centimeters (1 inch) or more.

The keyword in the subarctic climate is "winter," which is long, dark, and bitterly cold. Summers are short, and fall and spring slip by rapidly. In most places, ice begins to form on the lakes in September or October and doesn't thaw until May or later. For six or seven months, the average temperature is below freezing, and the coldest months have averages below −38°C (−36°F). The world's coldest temperatures, apart from the Antarctic and Greenland ice caps, are found in the subarctic climate; the records are −68°C (−90°F) in Siberia and −62°C (−82°F) in Alaska.

Summer warms up remarkably despite its short duration. Although the intensity of the sunlight is low (because of the small angle of incidence), summer days are very long and nights are too short to permit much radiational cooling. Average summer temperatures are typically in the midteens °C, (high 50s °F), but frosts may occur in any month.

Annual temperature ranges in this climate are the largest in the world. Variations from average summer month to average winter month temperatures frequently exceed 45°C (80°F). The *absolute annual temperature variation* (the difference in temperature from the very coldest to the very hottest ever recorded) sometimes reaches unbelievable magnitude—especially in deep inland locations; the world record is −68°C to +37°C (−90°F to +98°F), a range of 105°C (188°F) in Verkhoyansk, Siberia.

Precipitation is usually low in the subarctic climate with annual averages ranging from perhaps 13 to 50 centimeters (5 to 20 inches), with the higher values occurring in coastal

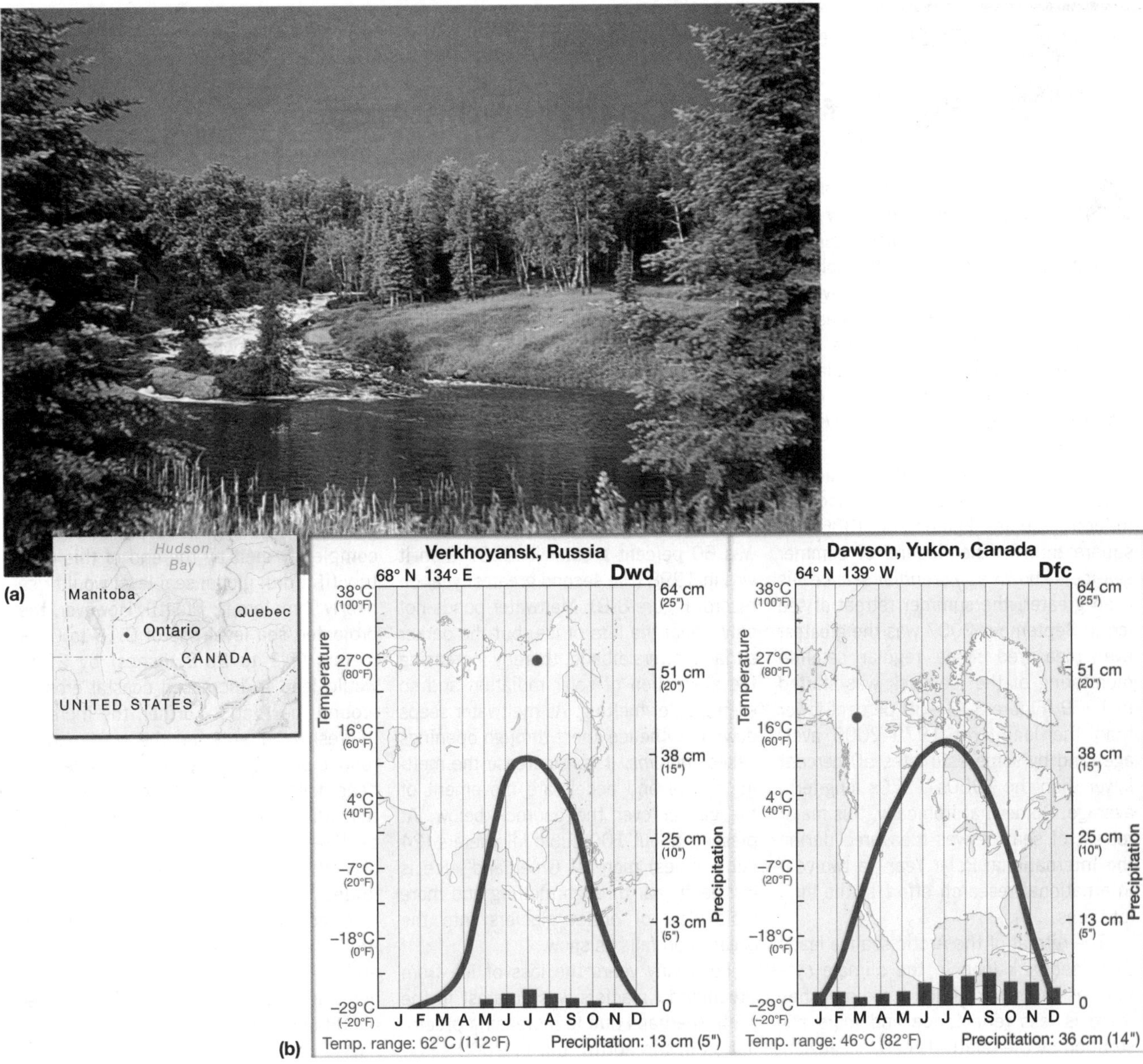

▲ **Figure 8-31** Subarctic climate. (a) Trees, rocks, and water are the principal landscape elements in the subarctic climate. This scene shows the Rushing River in western Ontario, Canada. (b) Climographs for representative subarctic stations.

areas. The low temperatures allow for little moisture in the air, and anticyclonic conditions predominate. Despite these sparse totals, the evaporation rate is low and the soil is frozen for much of the year so that enough moisture is present to support a forest. Summer is the wet season, and most precipitation comes from scattered convective showers. Winter experiences only modest snowfalls (except near the coasts), perhaps accumulating to depths of 60 to 90 centimeters (2 or 3 feet). The snow that falls in October is likely to still be on the ground in May because little melts over the winter. Thus, a continuous thin snow cover exists for many months despite the sparseness of actual snowfall.

Polar Climates (Group E)

Being farthest from the equator, as Figure 8-32 shows, the *polar climates* receive too little insolation for any significant warming. By definition, no month has an average temperature above 10°C (50°F) in a polar climate. If the wet tropics represent conditions of monotonous heat, the polar climates are known for their enduring cold. They have the coldest summers and the lowest annual and absolute temperatures in the world. They are also extraordinarily dry, but evaporation is so minuscule that the group as a whole is classified as nonarid.

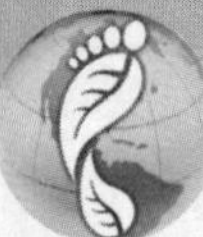

PEOPLE AND THE ENVIRONMENT

Signs of Climate Change in the Arctic

Although signs of global warming have been observed in many parts of the world, among the most dramatic indications have come from the polar regions. During the last 50 years, average temperatures in the Arctic have been increasing at about twice the rate as in lower latitudes. In western Canada and Alaska, winter temperatures are about 3°C (5°F) higher today than they were just 50 years ago.

Since 1979, the winter maximum extent of sea ice decreased by about 1 million square kilometers (386,000 square miles). The decline of summer sea-ice—known as *perennial sea ice*—is even greater: the summer retreat of sea ice in September 2007 was the greatest ever measured since regular satellite monitoring of the ice pack was started in 1979 (Figure 8-A)—39 percent lower than the long-term 1979–2000 average, and by some estimates 50 percent lower than the 1950s–1970s long-term average. Somewhat ironically, this major retreat of sea ice was measured during the International Polar Year—a two-year international research effort in the high latitudes.

The retreat of the Arctic sea ice leads to a feedback loop in the climate system: as the cover of sea ice diminishes there is less reflectance and more absorption by the relatively dark ocean surface now exposed to sunlight—which leads to higher ocean temperatures and so greater melting of the ice pack. Some computer models suggest that with projected temperature increases associated with some greenhouse gas emission scenarios, it is possible that the summer Arctic sea ice cover could actually disappear by the end of this century.

The Greenland Ice Sheet covers nearly all of Greenland. As temperatures have increased over the last few decades, the extent of the seasonal melting of the ice sheet has expanded significantly. The area of the ice sheet exhibiting summer surface melting and *melt pond* formation was 60 percent greater in 2007 than it was in 1998 (the second greatest year on record; Figure 8-B). Meltwater ponds not only signal the loss of ice, but the darker surface lowers albedo, thereby increasing the absorption of solar radiation and so even greater melting. As meltwater seeps down into the ice sheet through openings called *moulins*, it can increase the melting of ice and accelerate movement of the glacier over the bedrock below. At present, about 100 cubic kilometers (24 cubic miles) more of Greenland's ice is lost each year through melting and more rapid flow of outlet glaciers into the ocean than falls as snow.

For many years the loss of ice from Greenland's glaciers was greatest in the south (perhaps due to increasing precipitation in the north), but in 2008 satellite measurements indicated that the northern edges of the Greenland Ice Sheet underwent many more "melting days" than in the 28 years prior. Warmer conditions are also leading to an increase in precipitation falling as rain rather than snow, which contributes to greater amounts of liquid water on the ice surface—which further lowers albedo and increases melting.

Over the last few decades, global sea level has been rising by about 2 mm per year as a result of the melting of glaciers and the thermal expansion of seawater. The latest IPCC projections do not suggest that the Greenland Ice Sheet will completely melt by the end of this century (if it did, global seal level would rise nearly 7 meters [23 feet]). However, the projected sea level rise of 0.18 to 0.59 meters (7.1 to 23.2 inches) by 2100 would lead to increased coastal erosion around the Arctic. Further, the increase in freshwater entering the ocean could reduce salinity and potentially alter the deep ocean global conveyer-belt circulation (discussed in more detail in Chapter 9)—which could alter the exchange of heat between the tropics and high latitudes.

Concern about habitat loss in the Arctic has also been growing among scientists. Satellite images show land areas are becoming "greener" as warmer conditions increase the vegetation cover—parts of the North Slope of Alaska are now changing from tundra to shrub. As the extent of sea ice diminishes decade by decade, the iconic polar bear faces difficult circumstances. Polar bears regularly hunt for seal pups and other prey on the sea ice during the summer, but because the sea ice is breaking up earlier than in the past, the polar bears are facing increasingly challenging times. As a likely consequence of diminishing hunting opportunities, polar bears around western Hudson Bay in Canada now weigh about 15 percent less than they did 30 years ago.

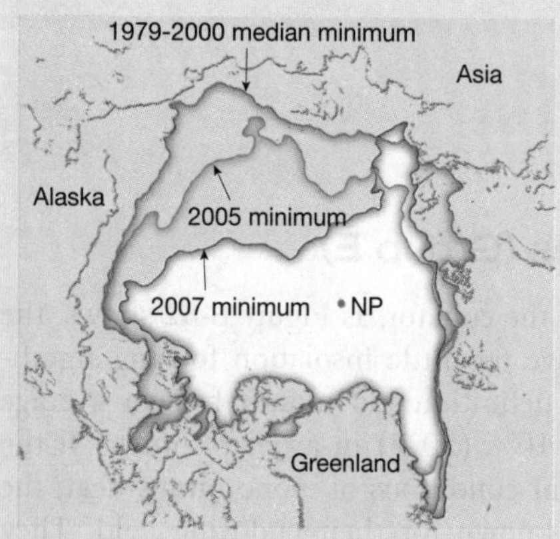

▲ **Figure 8-A** The extent of summer sea ice has been shrinking in the Arctic since regular satellite measurements began in 1979, reaching its lowest measured extent in September 2007.

▲ **Figure 8-B** Melt ponds (shown as blue patches on the white ice) are shown in this false-color infrared satellite image of western Greenland.

TABLE 8-7 Summary of E Climates: Polar

Type	Location	Temperature	Precipitation	Dominant Controls of Climate
Tundra (ET)	Fringes of Arctic Ocean; small coastal areas in Antarctica	Long, cold, dark winters; brief, cool summers; large ATR, small DTR	Sparse: less than 25 cm (10 in.) annually, mostly snow	Latitude; distance from sources of heat and moisture; extreme seasonal contrasts in sunlight and darkness
Ice cap (EF)	Antarctica and Greenland	Long, dark, windy, bitterly cold winters; cold, windy summers; no monthly average above 0°C (32°F); large ATR, small DTR	Very sparse: less than 13 cm (5 in.) annually, all snow	Latitude; distance from sources of heat and moisture; extreme seasonal contrasts in sunlight and darkness; polar anticyclones

ATR = annual temperature range; DTR = daily temperature range.

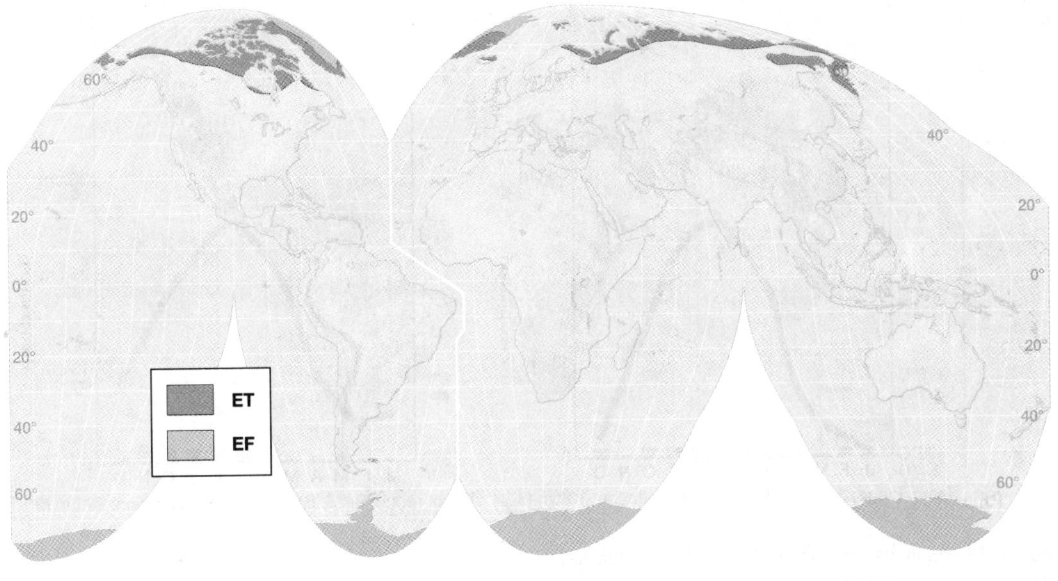

◀ **Figure 8-32** The global distribution of polar (E) climates.

The two types of polar climates are distinguished by summer temperature. The *tundra* climate has at least one month with an average temperature exceeding the freezing point; the *ice cap* climate does not. Table 8-7 summarizes the E climates.

Tundra Climate (ET) The name *tundra* originally referred to the low, ground-hugging vegetation of high-latitude and high-altitude regions, but the term has been adopted to refer to the climate of the high-latitude regions as well. The generally accepted equatorward edge of the **tundra** climate is the 10°C (50°F) isotherm for the average temperature of the warmest month. This same isotherm corresponds approximately with the poleward limit of trees so that the boundary between D and E climates (in other words, the equatorward boundary of the tundra climate) is the " treeline."

At the poleward margin, the ET climate is bounded by the isotherm of 0°C (32°F) for the warmest month, which approximately coincides with the extreme limit for growth of any plant cover. More than for any other climatic type, the delimitation of the tundra climate demonstrates Köppen's contention that climate is best delimited in terms of plant communities (Figure 8-33).

Long, cold, dark winters and brief, cool summers characterize the tundra. Only one to four months experience average temperatures above freezing, and the average summer temperature is between 4 and 10°C (in the 40s °F). Freezing temperatures can occur any time of year, and frosts are likely every night except in midsummer. Although an ET winter is bitterly cold—winter months might average −18°C (0°F) in coastal areas and −35°C (−30°F) in interior regions—it may not be as severe as in the more-continental subarctic climate just equatorward. Annual temperature ranges are fairly large, commonly more than 30°C (54°F). Daily temperature ranges are small because the Sun is above the horizon for most of the time in summer and below the horizon for most of the time in winter; thus nocturnal cooling is limited in summer, and daytime warming is almost nonexistent in winter.

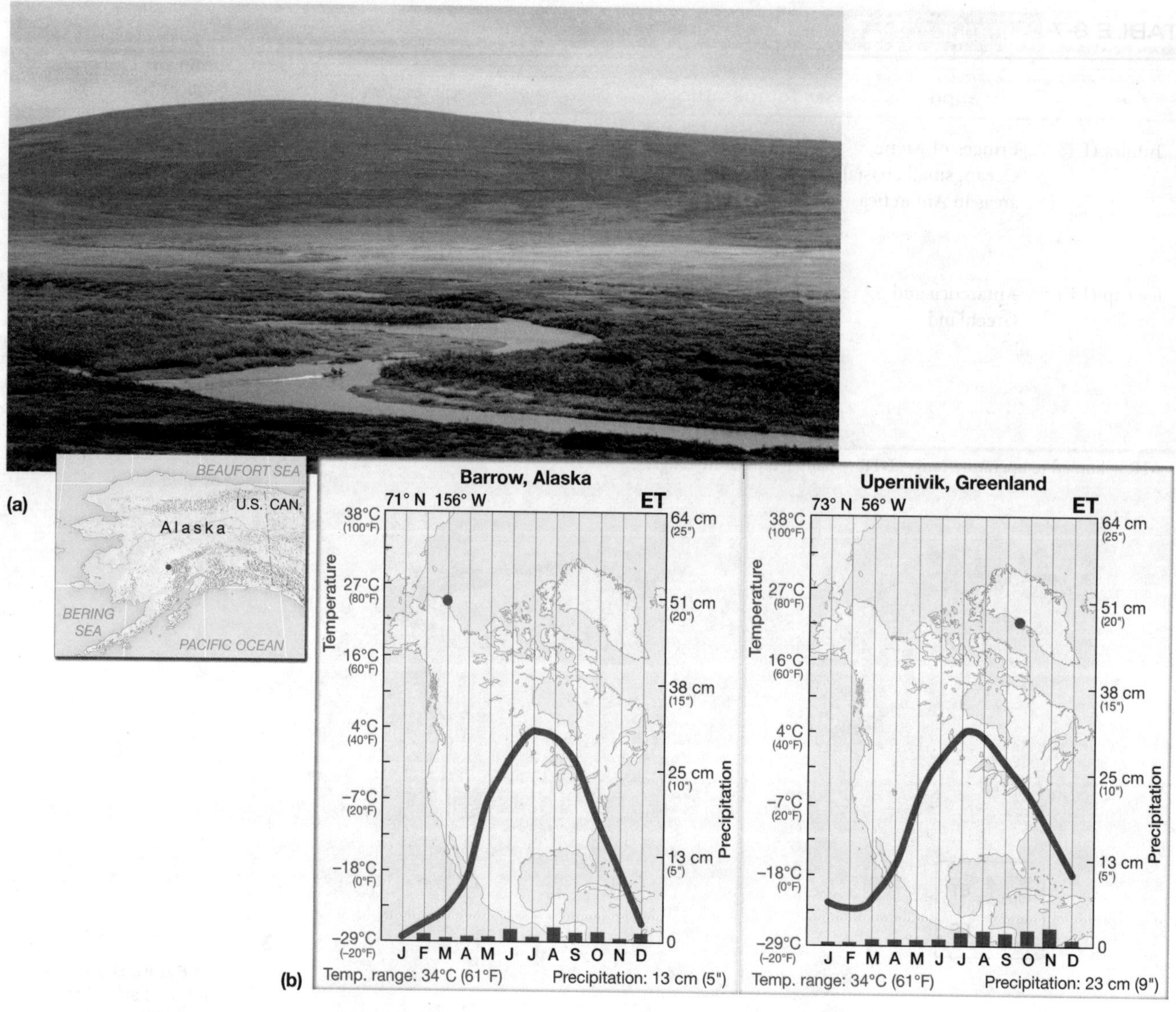

▲ **Figure 8-33** Tundra climate. (a) Tundra in western Denali National Park, Alaska. (b) Climographs for representative tundra stations.

Moisture availability is very restricted in ET regions despite the proximity of an ocean. The air is simply too cold to contain much water vapor, and so the absolute humidity is almost always very low. Moreover, anticyclonic conditions are common, with little uplift to encourage condensation. Annual total precipitation is generally less than 25 centimeters (10 inches), but is somewhat greater in eastern Arctic Canada. Generally, more precipitation falls in the warm season than in winter, although the total amount in any month is small, and the month-to-month variation is minor. Winter snow is often dry and granular; it appears to be more than it actually is because there is no melting and because winds swirl it horizontally even when no snow is falling. Radiation fogs are fairly common throughout ET regions, and sea fogs are sometimes prevalent for days along the coast.

Ice Cap Climate (EF) The most severe of Earth's climates, **ice cap climate**, is restricted to Greenland (all but the coastal fringe) and most of Antarctica, the combined extent of these two regions amounting to more than 9 percent of the world's land area (Figure 8-34). The EF climate is one of perpetual frost where vegetation cannot grow, and the landscape consists of a permanent cover of ice and snow.

The extraordinary severity of EF temperatures is increased because both Antarctica and Greenland are ice plateaus, so that relatively high altitude is added to high latitude as a thermal factor. All months have average temperatures below freezing, and in the most extreme locations the average temperature of even the warmest month is below −18°C (0°F). Winter temperatures average between −34 and −51°C (−30°F and −60°F), and extremes well below −73°C (−100°F) have been recorded at interior Antarctic weather stations.

The air is chilled so intensely from the underlying ice that strong surface temperature inversions prevail most of the time. Heavy, cold air often flows downslope as a vigorous katabatic wind (discussed in Chapter 5). A characteristic feature of the ice cap climate, particularly in Antarctica, is strong winds and blowing snow.

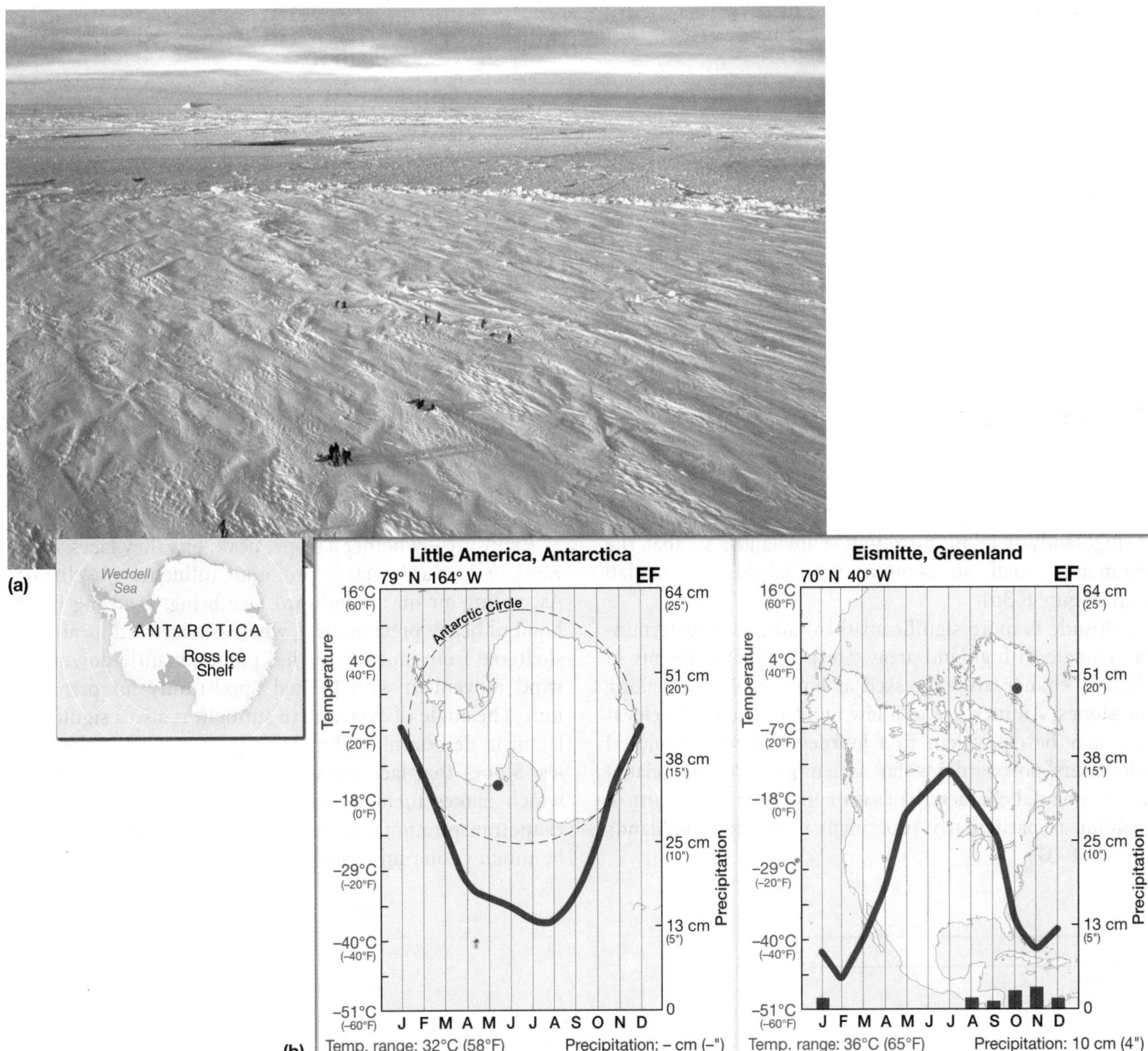

▲ **Figure 8-34** Ice cap climate. (a) The infinite bleakness of an ice cap can be imagined in this view of scientists at work on the Ross Ice Shelf of Antarctica. (b) Climographs for representative ice cap stations. Notice that Little America, Antarctica, receives almost no measurable precipitation in a typical year. (Note that the temperature scales are different from those of the other climographs in the chapter; note also that the climograph for Eismitte, Greenland, is based on less than 30 years of data.)

Precipitation is very limited. These regions are essentially polar deserts; most places receive less than 13 centimeters (5 inches) of moisture annually. The air is too dry and too stable, with too little likelihood of uplift, to permit much precipitation. Evaporation, of course, is minimal, and so an actual moisture surplus may be added to the ice.

Highland Climate (Group H)

Highland climate is not defined in the same sense as all the others we have just studied. Climatic conditions in mountainous areas have almost infinite variations from place to place, and many of the differences extend over very limited horizontal distances. Köppen did not recognize highland climate as a separate group, but most of the researchers who have modified his system have added such a category. Highland climates are delimited in this book to identify relatively high uplands (mountains and plateaus) having complex local climate variation in small areas (Figure 8-35). The climate of any highland location is usually closely related to that of the adjacent lowland, particularly with regard to seasonality of precipitation.

Altitude variations influence all four elements of weather and climate. As we learned in Chapters 4, 5, and 6, with increasing altitude, temperature and pressure generally decrease; wind is less predictable but tends to be brisk and abrupt with many local wind systems. Because of orographic lifting, precipitation is characteristically heavier

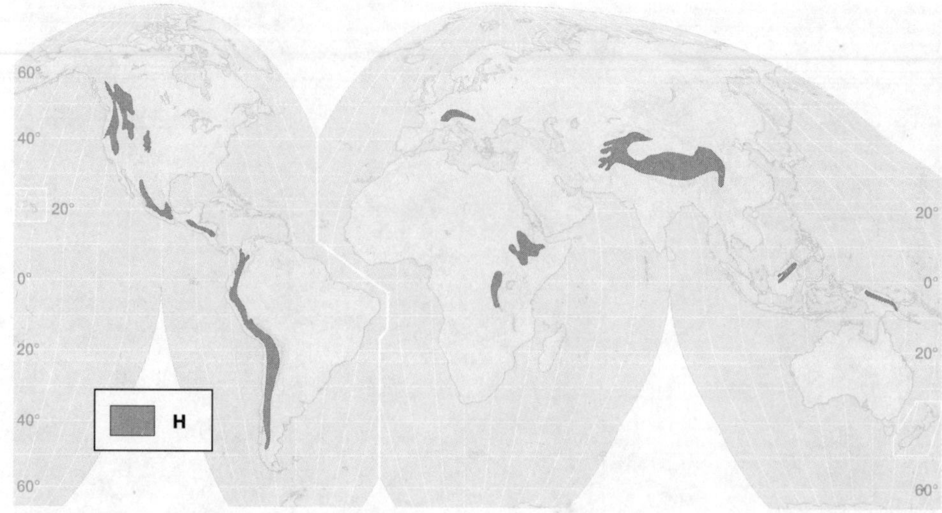

► **Figure 8-35** The global distribution of highland (H) climate.

in highlands than in surrounding lowlands, so that the mountains usually stand out as moist islands on a rainfall map (Figure 8-36).

Altitude is more significant than latitude in determining climate in highland areas. Steep vertical gradients of climatic change are expressed as horizontal bands along the slopes. An increase of a few hundred meters in elevation may be equivalent to a journey of several hundred kilometers poleward insofar as temperature and related environmental characteristics are concerned. Such *vertical zonation* is particularly prominent in tropical highlands (Figure 8-37).

Exposure—whether a slope, peak, or valley faces windward or leeward—has a profound influence on climate. Ascending air on a windward face brings a strong likelihood of heavy precipitation, whereas a leeward location is sheltered from moisture or has predominantly downslope wind movement with limited opportunity for precipitation. The angle of exposure to sunlight is also a significant factor in determining climate, especially outside the tropics. Slopes that face equatorward receive direct sunlight, which makes them warm and dry (through more rapid evapotranspiration); adjacent slopes facing poleward may be much cooler and moister simply because of a smaller

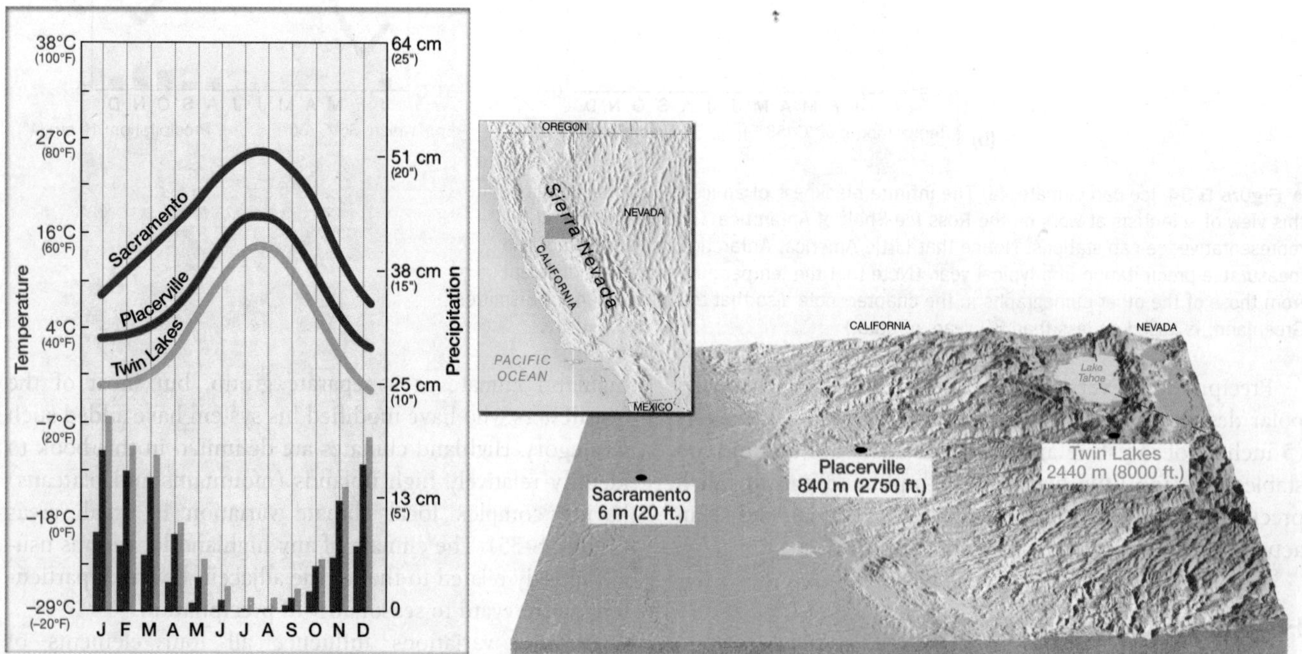

▲ **Figure 8-36** Climate of a lowland station (Sacramento) and two nearby highland localities (Placerville and Twin Lakes) at about the same latitude in California (approximately 38°40′ N). With increasing elevation, we see the expected decrease in temperature and increase in precipitation, although all three exhibit the characteristic summer-dry regime of the lowland mediterranean (Cs) climate.

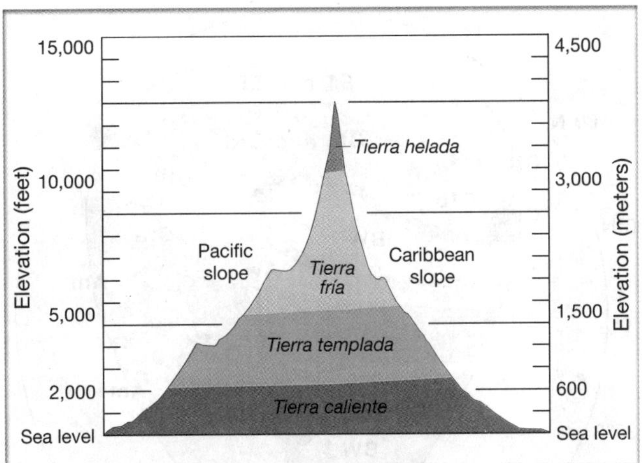

▲ **Figure 8-37** Vertical climate zonation is particularly noticeable in tropical mountainous areas. This diagram idealizes the situation at about 15°N latitude in Guatemala and southern Mexico. *Tierra caliente* ("hot land") is a zone of high temperatures, dense vegetation, and tropical agriculture. *Tierra templada* ("temperate land") is an intermediate zone of slopes and plateaus and temperatures most persons would find comfortable. *Tierra fria* ("cold land") is characterized by warm days and cold nights, and its agriculture is limited to hardy crops. *Tierra helada* ("frozen land") is a zone of cold weather throughout the year.

angle of solar incidence and more shading. Similarly, west-facing slopes receive direct sunlight in the hot afternoon, but east-facing slopes are sunlit during the cooler morning hours.

Changeability is perhaps the single most conspicuous characteristic of highland climate (Figure 8-38). The thin, dry air permits rapid influx of insolation by day and rapid loss of radiant energy at night, and so daily temperature ranges are very large, with frequent and rapid oscillation between freeze and thaw. Daytime upslope winds and convection cause rapid cloud development and abrupt storminess. Travelers in highland areas are well advised to be prepared for sudden changes from hot to cold, from wet to dry, from clear to cloudy, from quiet to windy, and vice versa.

Global Patterns Idealized

It should be clear by now that there is a fairly predictable global pattern of climatic types based primarily on latitude, position on a continent, and the general circulation of the atmosphere and oceans. Figure 8-39 summarizes the idealized distribution of the mild (A, B, and C) climates along the west coasts of continents, where the distribution pattern is slightly more regular than along the east coasts. Notice especially the relationship of climate types to the seasonal shifts of the intertropical convergence zone and the subtropical highs.

A more general model of the climate distribution on a *hypothetical continent* is shown in Figure 8-40. The model portrays the idealized distribution of five of the six groups and most of the types and subtypes; highland climate is not included because its location is determined solely by topography. Groups A, C, D, and E are defined by temperature, which means that their boundaries are strongly latitudinal because they are determined by insolation. The B group is defined by moisture conditions, and its distribution cuts across those of the thermally defined groups. Such a model is a predictive tool. One can state, with some degree of assurance, that at a particular latitude and a general location on a continent, a certain climate is likely to occur. Moreover, the locations of the climatic types relative to one another can be understood more clearly when they are all shown together this way.

◀ **Figure 8-38** Mountain climates are quite variable because of changes in altitude and exposure. This is a summer scene near Rainy Pass in the North Cascade Mountains of Washington.

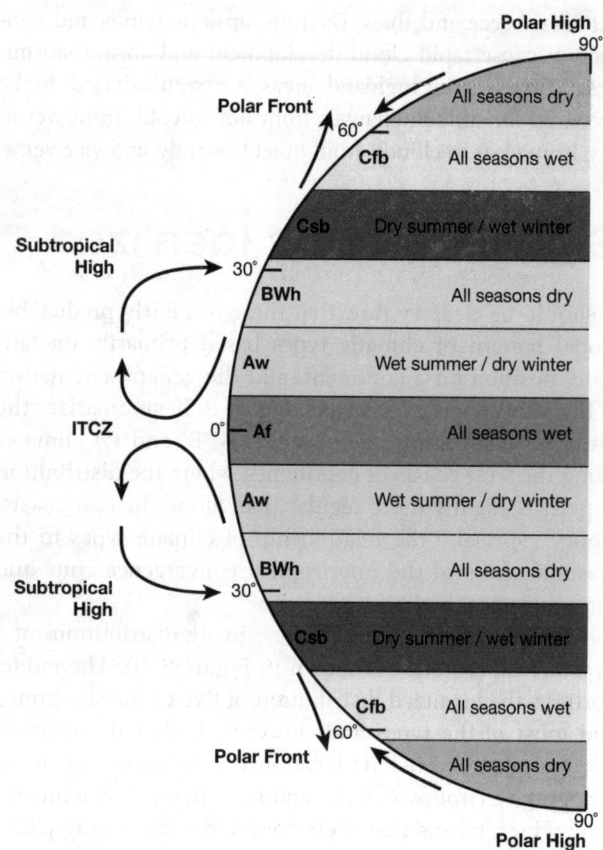

▲ **Figure 8-39** Idealized seasonal precipitation patterns and climates along the west coast of continents. Note that the progressions north and south of the equator are mirror images. Much of this pattern is due to the seasonal shifts of the ITCZ and the subtropical highs.

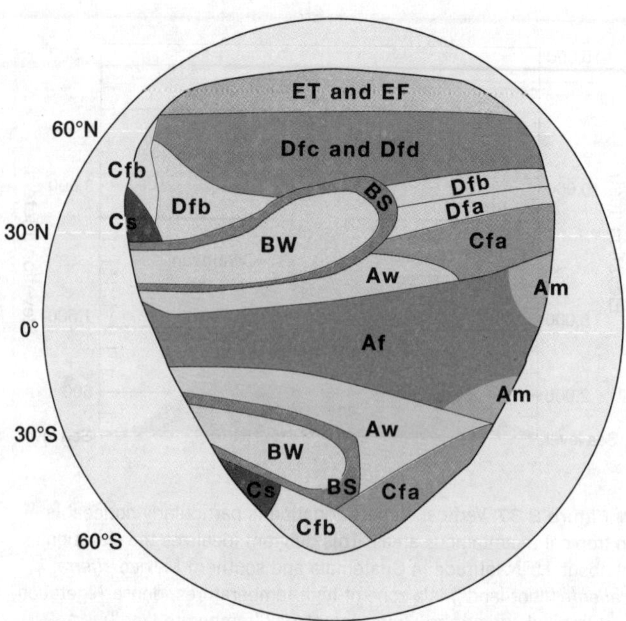

▲ **Figure 8-40** The presumed arrangement of Köppen climatic types on a large hypothetical continent.

Global Climate Change

So far in this chapter, we have discussed the complexity of global climate patterns as we observe them today. Although it is important for us to learn the fundamentals of weather and climate by first studying such "stable" circumstances, a thorough understanding of the atmosphere requires us now to explore how patterns can change over time.

For the remainder of this chapter, we look more specifically at climate change. We especially want to address questions first raised in Chapter 4: is human activity causing global climate change, and if so, what are the likely long-term consequences? In providing a thoughtful answer to these questions, we will look at how climate change is recognized, what kinds of factors may lead to climate change, and what evidence indicates that climate is currently changing.

As we first saw in Chapter 3, within any climate system there is expected fluctuation in weather from year to year. The random variation in weather from year to year can be thought of as the background "noise" in the long-term climate record. Within the climate record, however, other kinds of variation may be present: episodic events and natural cycles of change with periods ranging from a few years to many decades (El Niño and the Pacific Decadal Oscillation discussed in Chapter 5 are examples of this), as well as long-term global climate change (which can also exhibit cycles of its own).

The time scale of observations influences which kinds of patterns stand out in the climate record. For example, in the case of global temperature:

- Over the last 70 million years, a clear global cooling trend is visible (Figure 8-41a).
- Over the last 150,000 years, temperature fluctuated significantly until about 10,000 years ago when temperature increased sharply; temperature has remained warm and fairly steady since. (Figure 8-41b).
- Over the last 150 years a warming trend appears underway—a clear departure from the previous 850 years (Figure 8-41c).

These trends are actually even more complicated than might first appear: these temperature changes weren't necessarily uniform around the world.

Determining Climates of the Past

End of the last Ice Age

Before we can assess the possibility of climate change today, we must have some understanding of past climates and the causes of past climate change, a field of study known as **paleoclimatology**.

Detailed instrument records of weather go back only a few hundred years, so it is necessary to use *proxy* ("substitute") measures of climate to reconstruct conditions in the past. Information about past climate can come from many different sources: tree rings, ice cores, oceanic sediments, coral reefs, relic soils, and pollen, among others. No single method for determining past climate conditions is ideal—each has its own strengths and weaknesses. However, by

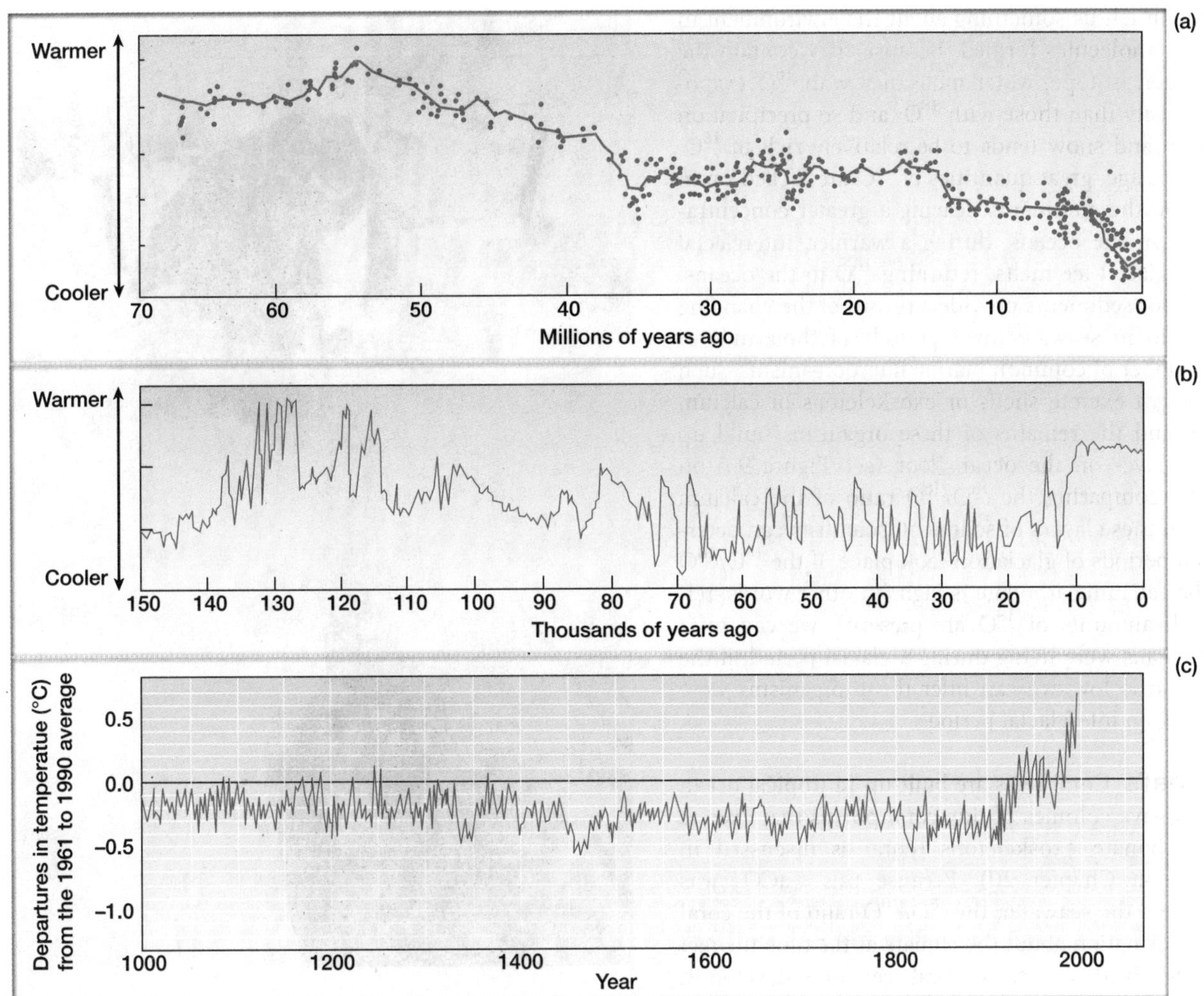

▲ **Figure 8-41** Generalized global temperature trends at three time scales. (a) 70 million years ago to present. (b) 150,000 years ago to present. (c) 1000 years ago to present. The curve in (a) was generalized from ice volume and ocean temperature proxy measures; the curve in (b) was derived from North Atlantic Ocean proxy measures; the curve in (c) was derived from proxy measures and the historical record after 1900.

correlating the results achieved with one method with those from another, paleoclimatologists are able to construct a detailed history of Earth's climate.

Dendrochronology Most trees growing in temperate areas increase trunk diameter by adding one concentric *tree ring* for each year of growth. By counting the number of rings, the age of a tree may be determined. Tree rings also provide information about climate: during years when the growing conditions are favorable (such as mild temperatures and/or ample precipitation), tree rings tend to be wider than in years when the growing conditions are harsher.

Both living and dead trees can be used in **dendrochronology** (the study of past events and past climate through the analysis of tree rings). Small cores are taken so that living trees are not harmed, and fallen trees, buried trunks, and timber from archeological sites all may be used to extend tree ring chronology. By comparing and correlating the tree ring patterns of many trees in an area, dendrochronologists can not only determine the age of the trees, but can establish dates for catastrophic events such as floods and fires that have killed trees, as well as identify periods of drought and lower or higher temperatures.

Oxygen Isotope Analysis of Oceanic Sediments In the nucleus of many atoms, the number of protons and neutrons is the same. For example, the normal oxygen atom contains 8 protons and 8 neutrons, giving it an atomic weight of 16 (^{16}O or "oxygen 16"). However, a small number of oxygen atoms have 2 extra neutrons, giving these atoms an atomic weight of 18 (^{18}O). ^{16}O and ^{18}O are known as *isotopes* of oxygen. Both isotopes of oxygen can be found in common molecules such as water (H_2O) and calcium carbonate ($CaCO_3$).

Through **oxygen isotope analysis**, the ratio of $^{18}O/^{16}O$ in the molecules of substances such as water and calcium carbonate can tell us something about the environment in which those molecules formed. Because they contain the lighter oxygen isotope, water molecules with ^{16}O evaporate more easily than those with ^{18}O, and so precipitation such as rain and snow tends to be relatively rich in ^{16}O. During an ice age, great quantities of ^{16}O are locked up in glacial ice on the continents, leaving a greater concentration of ^{18}O in the oceans; during a warmer, interglacial period, the glacial ice melts, returning ^{16}O to the oceans.

Ocean floor sediments provide a record of the changing $^{18}O/^{16}O$ ratio in seawater over periods of thousands of years. A number of common marine microorganisms, such as *foraminifera* excrete shells or exoskeletons of calcium carbonate, and the remains of these organisms build up layer upon layer on the ocean floor (see Figure 9-6 on p. 240). By comparing the $^{18}O/^{16}O$ ratio of the calcium carbonate in these layers of sediment, scientists can determine when periods of glaciation took place. If the $^{18}O/^{16}O$ ratio in the calcium carbonate is high (in other words, relatively high amounts of ^{18}O are present), we can infer those organisms were living during a glacial period; if the $^{18}O/^{16}O$ ratio is low, we can infer those organisms were living during an interglacial period.

Coral Reefs Coral reefs are built up in tropical ocean areas by massive colonies of tiny coral polyps that excrete calcium carbonate exoskeletons (coral is discussed in more detail in Chapter 20). Because this $CaCO_3$ was extracted from the seawater, the $^{18}O/^{16}O$ ratio of the coral provides information about the climate at the time the reef was forming. Further, because coral reefs only develop in shallow water, the relative height of old reefs may help us determine past fluctuations in sea level.

Ice Cores The $^{18}O/^{16}O$ ratio can also give us climate information through the analysis of ice cores. By drilling down into glaciers, a record of snowfall going back hundreds of thousands of years may be obtained in some locations (Figure 8-42). Such ice cores offer several kinds of information about past conditions. A greater number of water molecules with ^{18}O evaporate from the oceans when temperatures are high than when temperatures are low, thus the $^{18}O/^{16}O$ ratio found in a layer of ice serves as a "thermometer" for the climate at the time that snow fell. Oxygen isotope analysis of ice cores has become one of the most important ways scientists have been able to construct a record of past global temperature—a record that now goes back 800,000 years.

Ice cores also provide direct information about the composition of the atmosphere in the past. As snow accumulates and is compacted into glacial ice, air bubbles from the atmosphere become frozen into the ice. These tiny air bubbles deep in glaciers are preserved samples of the atmosphere in the past, allowing direct measurements of the concentration of CO_2 and other gases to be made.

Ice core analysis shows a close relationship between greenhouse gas concentration (especially carbon dioxide and methane) and temperature over the past 800,000 years. Lower concentrations of greenhouse gases are present during glacial episodes, while higher concentrations of greenhouses gases are present during interglacial warm periods.[2]

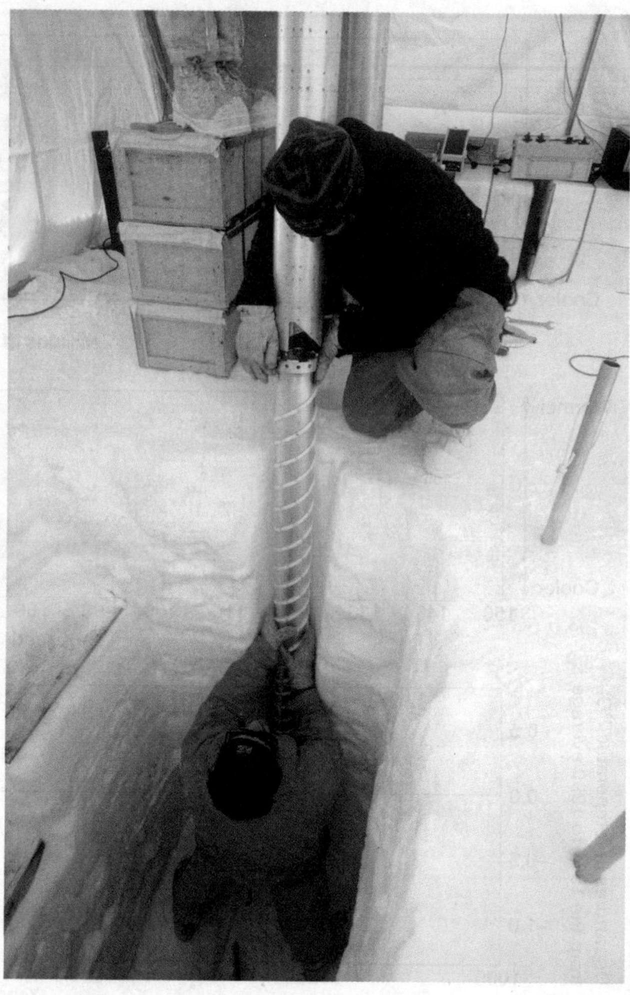

▲ **Figure 8-42** Scientists in Antarctica drilling an ice core as part of the EPIC project (European Project for Ice Coring in Antarctica).

Pollen Analysis Another important method used for determining past climates comes from the field of *palynology*, or *pollen analysis*. Airborne pollen from trees and other plants can be preserved in sediment layers on lake bottoms and in bogs. Cores of these sediment layers are taken and analyzed.

Radiocarbon dating of material in each layer provides an estimate of the age of organic material younger than about 50,000 years. Radiocarbon dating works by comparing the ratio of two isotopes of carbon found in organic material: radioactive ^{14}C decays over time at a known rate into the

[2]Interestingly, some recent evidence shows that it may be several hundred years after an interglacial temperature increase begins before atmospheric CO_2 concentrations increase. This suggests that in the past, an increase in greenhouse gas concentration may not have always been the trigger for an interglacial temperature increase. In other words, increasing CO_2 levels could have been a consequence of a warming climate, and not its cause.

FOCUS

Paleoclimatology at Dome C, Antarctica

Ice cores have yielded a long and detailed proxy record of Earth's past climate. Oxygen isotope analysis, trapped air bubbles, and atmospheric aerosol deposits within the layers of ice have offered information about past temperature and greenhouse gas concentrations, as well as cataclysmic events such as major volcanic eruptions. Of all the locations where ice cores have been extracted from glaciers, the most significant is probably Dome C in Antarctica.

Dome C is located about 1750 kilometers (1090 miles) from the South Pole at a latitude of 75° S and a longitude 123° E—this is where the Antarctic ice cap is at its thickest (Figure 8-C). The coring has been undertaken by the European Project for Ice Coring in Antarctica (EPICA). A multinational team of scientists has extracted a continuous 10 centimeter (4 inch) diameter ice core record to a depth of more than 3 kilometers (2 miles) (Figure 8-D). So far, the analysis has provided climate and atmospheric composition data back 800,000 years—far longer than any other ice core—providing climate information on eight complete glacial/interglacial cycles (Figure 8-E).

The Dome C climate record shows that the present concentration of CO_2 in the atmosphere is greater than at any time in the last 800,000 years, and that increases and decreases in global temperature are closely correlated with changes in CO_2 concentration—with temperatures high when CO_2 concentrations are high, and vice versa. Further, the findings at Dome C closely match the proxy climate record derived from the nearby Vostok, Antarctica, ice core, as well as oxygen isotope analysis of the calcium carbonate in foraminifera (tiny marine creatures) found in oceanic sediments, adding to the scientists' confidence in the soundness of the Dome C data.

Among the most interesting findings of the ice core analysis are from a period between about 400,000 and 650,000 years ago. Two of the warm interglacial periods that occurred during this time span were relatively mild compared to the interglacial periods of the last 400,000 years. Further, carbon dioxide and methane concentrations were lower during those two earlier interglacial periods than during more recent interglacials. This data helps climatologists judge the "sensitivity" of the climate to increased levels of greenhouse gases—which can in turn lead to better computer model predictions of future climate.

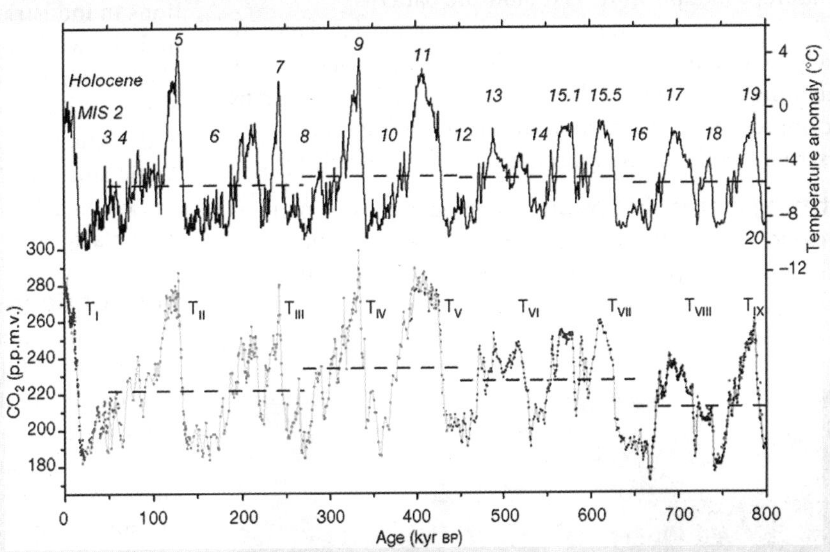

▲ **Figure 8-E** Temperature record over the last 800,000 years from EPICA Dome C ice core, and the corresponding CO_2 record derived from Dome C and other locations. Temperature scale is relative to average temperature over the last 1000 years. The numbers on the temperature graph denote Marine Isotope Stages (MIS) corresponding to temperature data derived from oxygen isotope analysis of oceanic sediments. The end of glacial stages are indicated in the CO_2 graph with roman numerals.

▲ **Figure 8-C** Map of Antarctica showing the location of EPICA Dome C.

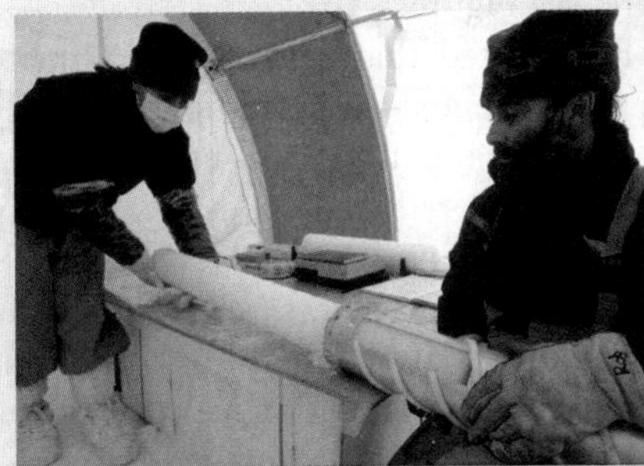

▲ **Figure 8-D** EPICA scientists at Dome C in Antarctica removing a core from the barrel of an ice core drill.

more common "daughter" isotope ^{12}C, so the $^{14}C/^{12}C$ ratio serves as indication of how long ago the plant or animal material was alive. With the age of a sediment layer known, palynologists look at the types of pollen present to determine which plants and trees were living at the time that layer was forming. Since plant species change as the environment changes—some species are better adapted for colder conditions, for example—the prevalence of one plant community over another acts as a proxy for the climate of the time.

Causes of Long-Term Climate Change

Orbital Variations and Climate Change

A quick glance back to Chapter 4 and the diagram showing Earth's solar radiation budget (Figure 4-17) will remind you of the many variables that influence patterns of temperature: the quantity of incoming solar radiation, the albedo of the atmosphere and surface, the concentration of greenhouse gases, the transfer of energy through oceanic and atmospheric circulation, and many others. A change in any one of those variables has the potential to change temperature, wind, pressure, and moisture patterns.

Not surprisingly then, a number of mechanisms are likely to have had a role in past climate change and may be operating in the present day as well. Although by no means a comprehensive description, in the following sections we describe several kinds of mechanisms of climate change.

Atmospheric Aerosols Large quantities of particulates ejected into the atmosphere by volcanic eruptions can alter global temperatures (Figure 8-43). Fine volcanic ash and other aerosols can reach the stratosphere, where they may circle the globe for years, blocking incoming solar radiation and lowering temperature. For example, the 1991 eruption of Mount Pinatubo in the Philippines increased aerosol concentrations (especially sulfur dioxide) in the atmosphere enough to lower global temperatures by about 0.5°C (0.9°F) for the next year.

▲ **Figure 8-43** Eruption of Cleveland Volcano in the Aleutian Islands on May 23, 2006, taken by astronaut Jeffrey N. Williams from the International Space Station. The ash plume from this short-lived eruption was carried to the southwest by the wind.

Large asteroid impacts also can eject enough dust into the atmosphere to significantly alter global climate. It is widely accepted by scientists that the impact of a 10-kilometer wide asteroid 65 million years ago was one factor that contributed to the extinction of the dinosaurs—environmental change associated with massive *flood basalt* eruptions occurring at the same time is likely to have been a more significant factor (flood basalt is discussed in Chapter 14).

As we saw in Chapter 6 ("People and the Environment: Global Dimming"), anthropogenic aerosols may influence climate—especially sulfates and black carbon. Sulfates released by the burning of fossil fuels tend to scatter incoming solar radiation and so have a net cooling effect. On the other hand, black carbon emissions from the burning of diesel and biofuels tend to absorb solar radiation and so lead to warming.

Over the last 30 years stricter environmental regulations in industrialized countries have cut sulfate emissions in half, while over the same time period, emissions of black carbon from industrializing countries in the Northern Hemisphere—especially in Asia—have risen. A 2009 NASA study suggested that much of the warming in the Arctic since 1976 may be linked to changes in aerosols—less cooling from sulfates combined with more warming from black carbon.

Fluctuations in Solar Output A link between past climates and variations in the energy output of the Sun has been pursued for many decades. For example, researchers have looked for a connection between climate and sunspot activity. *Sunspots* appear as dark spots on the surface of the Sun. They are slightly cooler regions of the Sun's photosphere and are associated with intense magnetic storms on the surface of the Sun, as well as the ejection of charged particles (some of the charged particles reach Earth as part of the solar wind where they produce auroral displays in the upper atmosphere; see Figure 3-10).

Some scientists have noted that several past climate events, such as the *Little Ice Age*—a period of unusually cold weather from about 1400 to 1850—corresponded with a period of greatly reduced sunspot activity between 1645 and 1715 (known as the *Maunder Minimum*). Recent satellite measurements show that the energy striking Earth's upper atmosphere, or *total solar irradiance* (TSI), varies by about 0.1 percent over the well-documented 11-year cycle of sunspot activity—with less energy received during periods of low sunspot activity. Ultraviolet radiation seems to vary much more than this, and so fluctuations in TSI may affect the ozone layer and perhaps other aspects of energy flow through the atmosphere.

Although some of the relationships between sunspot cycles and climate appear at first to be strong statistically, many scientists remain unconvinced—in part because not all episodes of unusual sunspot activity correlate with changes in climate. Further, although variation in TSI may well play a role in climate change, the measured fluctuations do not seem to be nearly enough to account for the global warming trend observed over the last century.

Variations in Earth–Sun Relations As we saw in Chapter 1, the change of seasons is largely a consequence of the changing orientation of Earth's rotational axis to the Sun during the year (see Figure 1-22). You will recall that we said that Earth's axis maintains an inclination of 23.5° at all times, that Earth's axis always points toward the North Star Polaris, and that Earth is closest to the Sun on January 3rd. As it turns out, all of those aspects of Earth–Sun relations change over periods of thousands of years in a series of predictable, well-documented cycles (Figure 8-44):

- The "shape" of Earth's elliptical orbit, or *eccentricity*, varies in a series of cycles lasting about 100,000 years. Sometimes Earth's orbit is nearly circular, while at other times the orbit is much more elliptical, thus influencing the distance between the Sun and Earth. The present difference in distance between the Sun and Earth at aphelion and perihelion is about 3 percent (about a 5,000,000 kilometer difference), but over the last 600,000 years or so, the difference has varied from 1 percent to 11 percent. The greater the difference in distance, the greater the difference in insolation at various points in Earth's orbit.
- The inclination of Earth's axis, or *obliquity*, varies from 22.1° to 24.5° in a cycle lasting about 41,000 years. When there is greater inclination, seasonal variation between the low latitudes and high latitudes tends to be greater; when there is less inclination, seasonal contrasts are less significant.
- Earth's axis "wobbles" like a spinning top, and so over time it points in different directions relative to the stars in a 25,800-year cycle called *precession*. Precession alters the timing of the seasons relative to Earth's position in its orbit around the Sun.

As these long-term cycles "overlap" there are periods of time when significantly less radiation reaches Earth surface (especially in the high latitudes), and periods of time when there are greater or smaller seasonal contrasts. For example, during periods when there is less "seasonality"—in other words, when there are smaller contrasts between winter and summer—more snow can accumulate in high latitudes due to greater snowfall from milder winters and less melting will take place due to lower summer temperatures.

These cycles are known as **Milankovitch cycles**, after Milutin Milankovitch, an early twentieth-century Yugoslavian astronomer. Milankovitch looked at the combined effects of these cycles and concluded that they were responsible for an approximately 100,000-year cycle of insolation variation that has taken Earth into and out of periods of glaciation over the last few millions of years. Although his ideas were mostly rejected by other scientists during his lifetime, recent paleoclimatological work has shown that cyclical variation in Earth–Sun relations is one key to the timing of glacial and interglacial events.

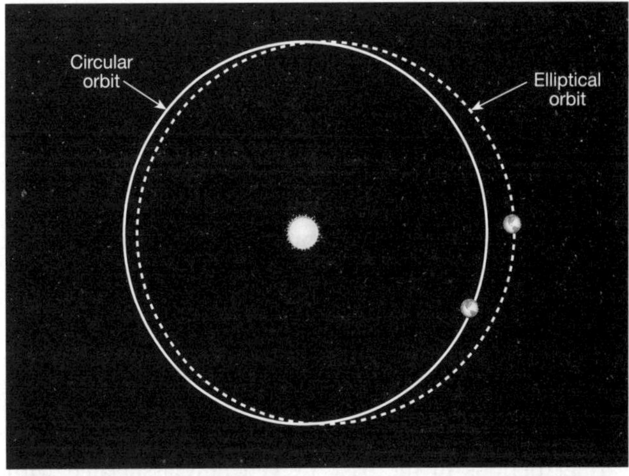

(a)

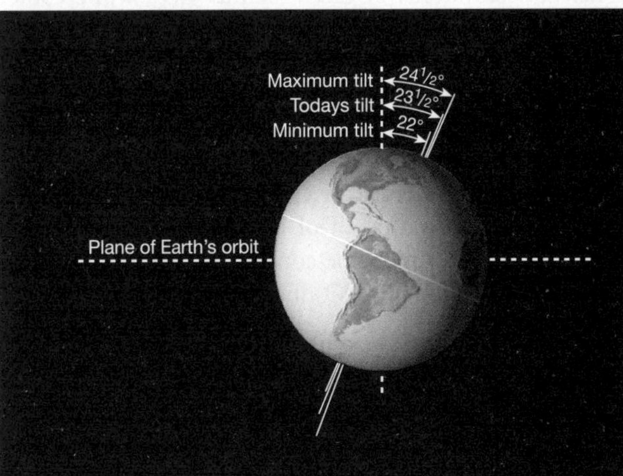

(b)

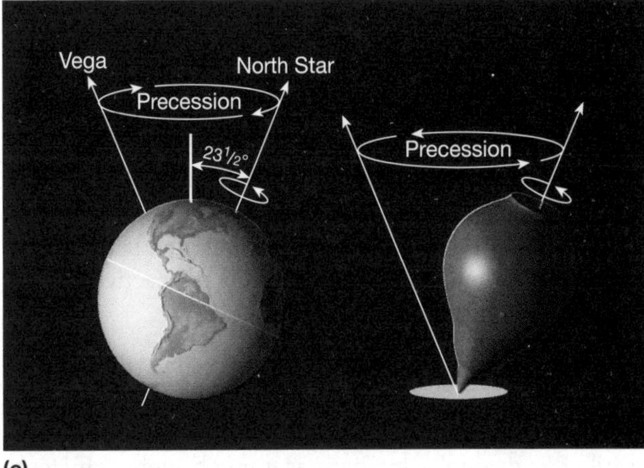

(c)

▲ **Figure 8-44** Orbital variation cycles in Earth-Sun relations. (a) The shape of Earth's elliptical orbit, or eccentricity, varies in a cycle lasting about 100,000 years. (b) The inclination of Earth's axis, or obliquity, varies in a cycle lasting 41,000 years. (c) The orientation of Earth's axis varies in the 25,800-year cycle of precession.

However, these astronomical cycles present some problems as well—for example, the effects of precession that produce cooler conditions in one hemisphere should produce warmer conditions in the other, but this does not appear to be the case. Other factors certainly play a part in the onset and end of glacial periods.

Greenhouse Gases Concentration As we first saw in Chapter 4, greenhouse gases, such as water vapor, carbon dioxide, and methane, play an important role in regulating the temperature of the troposphere. It is also well known—most notably through ice-core analysis—that past fluctuations in global temperature have been accompanied by fluctuations in greenhouse gas concentration: warmer periods in the past are generally associated with higher concentrations of gases such as CO_2, while cooler periods are associated with lower concentrations of CO_2.

The well-documented increase in atmospheric CO_2 that was underway by the early twentieth century (see Figure 4-34), is thought by most atmospheric scientists to be the main cause of contemporary global warming. But how do we know if this increase in CO_2 is the result of human activity or if it is the result of natural processes? Several lines of evidence support the conclusion that human activity is responsible for this CO_2 increase, such as:

1. The increase in atmospheric CO_2 correlates well with fossil fuel use since the industrial revolution—although only about half of the human-released CO_2 seems to have remained in the atmosphere; the balance was either absorbed by the oceans or fixed by plants through photosynthesis.
2. The radioactive isotope ^{14}C forms naturally in the upper atmosphere and exists along with the stable ^{12}C isotope. Measurements of the $^{14}C/^{12}C$ ratio derived from tree cores show that the relative level of the radioactive isotope ^{14}C in the atmosphere has been decreasing since the industrial revolution. This indicates that the additional CO_2 in the atmosphere was released from old "dead" sources of carbon, such as coal and petroleum—formed so long ago that most of the radioactive ^{14}C has decayed to its more stable ^{12}C isotope.

A complication in trying to understand the role of greenhouse gases in climate change is that both the relative potency and the residence time of these gases varies greatly. For example, methane is a much more "efficient" greenhouse gas than carbon dioxide, but it has a shorter "lifetime" in the atmosphere—in other words, the effects of methane in the short term may not be the same as in the long term.

Further adding to the complexity of the relationship between greenhouse gas concentrations and global temperatures are a number of important *feedback mechanisms*.

Feedback Mechanisms Sea ice and continental ice sheets have high albedos—in other words, these surfaces reflect much of the shortwave radiation that strikes them (Figure 8-45). If global climate were to cool, ice would expand and this would in turn reflect more incoming radiation, which would lead to more cooling—which in turn would lead to greater still expansion of ice; this is known as a *positive feedback mechanism*. On the other hand, if climate were to warm, this would reduce the ice cover, which would reduce the albedo and therefore increase absorption by the surface, which would lead to higher temperatures and therefore still less ice cover—another positive feedback mechanism.

▲ **Figure 8-45** Sea ice off the east coast of Greenland during the summer thaw.

One of the most complicated series of feedback mechanisms associated with climate change has to do with water vapor. As climate warms, evaporation increases—thereby increasing the concentration of water vapor in the atmosphere. As we have learned, water vapor is one of the key greenhouse gases, so as warming increases evaporation, the greater concentrations of water vapor in the atmosphere will lead to still more warming—another positive feedback mechanism. However, here is where things become complicated.

An increase in water vapor would also likely lead to increased cloud cover. Clouds can act as efficient reflectors of solar radiation (so leading to cooling), as well as efficient absorbers of terrestrial radiation (and so contributing to warming). In other words, increased cloud cover can lead to both cooling and warming!

The net effect—cooling or warming—evidently depends on the type of clouds that form. For example, a deep layer of cumulus clouds does not increase albedo as much as it increases absorption of longwave radiation from the surface—thus leading to warming. On the other hand, the expansion of stratus cloud cover increases albedo more than it increases longwave absorption—thus leading to cooling. The effects of clouds remain one of the great complications when modeling future climate change.

The Roles of the Ocean Many important variables in climate change involve the oceans. The oceans absorb a great deal of CO_2 from the atmosphere—some estimates suggest that the oceans have absorbed one-third to one-half of the carbon that human activities have released into the atmosphere since the industrial revolution. Once in the ocean, marine plants and animals remove carbon from the ocean water through photosynthesis and the extraction of calcium carbonate to form shells and exoskeletons. Much of this carbon is eventually deposited in sediment layers on the ocean floor.

In addition, vast quantities of carbon in the form of *methane hydrates* (a kind of ice made from methane and water) are found in some oceanic sediments. Should these methane hydrates be destabilized by higher temperatures and the trapped methane (a greenhouse gas) released into the atmosphere, a sudden increase in atmospheric temperature could result (some evidence suggests that this actually happened about 55 million years ago, resulting in a prominent warming period in Earth's history).

Finally, the oceans obviously play an important role in transferring heat from low latitudes to high latitudes. Should patterns of ocean circulation change, climate can be significantly altered (this possibility is considered in Chapter 9).

Climate Models

How do atmospheric scientists predict what climate will be like in the future? Such projections of future temperature and precipitation patterns are mostly the product of sophisticated computer simulations known as *general circulation models (GCMs)*. In short, a GCM is a mathematical model of Earth's climate system. In such a computer-simulated world, assumptions are made about many parameters: the amount of radiation striking the surface, the extent of cloud cover, variations in ocean temperature, changes in wind and pressure patterns, changes in greenhouse gas concentration, variations in surface albedo, as well as many others.

Before trying to predict future climates, a GCM must pass several "tests." First, the GCM must be able to simulate present atmospheric processes and climate—in other words, it must reliably replicate processes and patterns in the present-day atmosphere. Next, the GCM must be able to simulate past climate change, so data on past climate conditions are input into the program and the simulation is run "forward" to the present day. If the model cannot "predict" what has actually happened to climate, adjustments must be made to the program. The model can then be run forward to make projections about future climate.

Because of the enormous complexities in the atmosphere, different GCMs use different assumptions about how the atmosphere works, and so each GCM will come up with slightly different projections about climate in the future. Among the most important uses of GCMs today is to anticipate how climate will change in the future given various scenarios of greenhouse gas increase in the atmosphere. The conclusions of such studies are discussed later in this chapter.

Evidence of Current Global Warming

The paleoclimatological record provides the backdrop with which we can compare both the rate and the magnitude of climate change over the last century. As we saw in Chapter 4, the Intergovernmental Panel on Climate Change (IPCC) is the most authoritative international body providing information about climate change to global leaders. In evaluating evidence of climate change over the last century, the *Fourth Assessment Report of the* IPCC released in 2007 concluded that the warming of global climate is "unequivocal."

The findings of the 2007 *Fourth Assessment Report* include:

- Between 1906 and 2005 global average temperature increased by 0.74°C (1.33°F), with estimates ranging from 0.57 to 0.95°C (1.03 to 1.71°F). Figure 8-46 shows the global *temperature anomalies*, or departure from average, from 1880 to 2008.

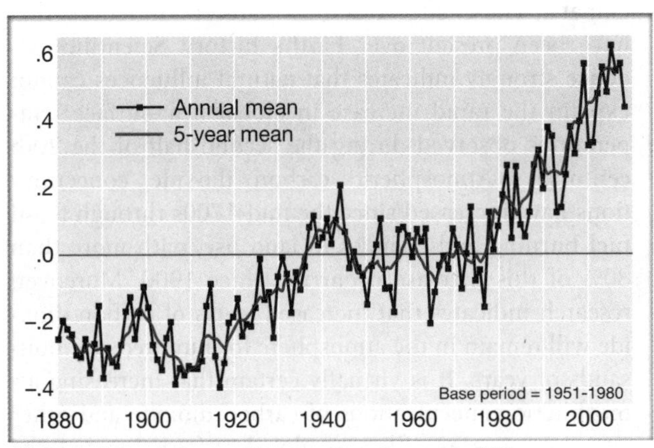

(a) Global Land-Ocean Temperature Anomaly (°C)

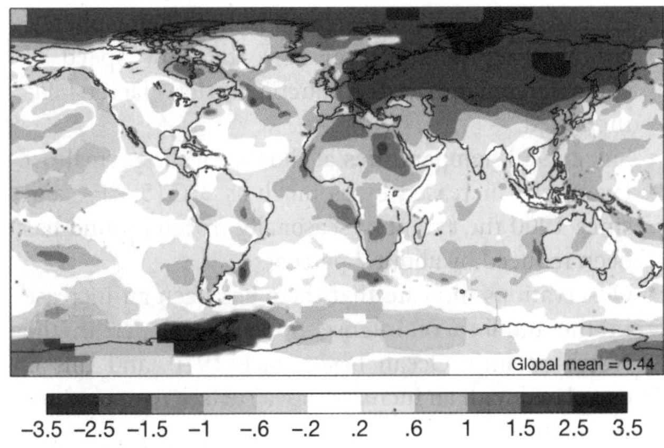

(b) 2008 Surface Temperature Anomaly (°C)

▲ **Figure 8-46** (a) Temperature anomalies of global air temperature from 1880 to 2008. (b) 2008 temperature anomalies; notice that the greatest increase was generally in the high latitudes of the Northern Hemisphere.

- Based on instrument data of temperature since 1850, 11 of the 12 warmest years on record occurred between 1995 and 2006. (By 2008—one year after the IPCC report was released—8 of the 10 warmest years on record had occurred since the year 2000.)
- Over the last 50 years average global temperature has been increasing at a rate of about 0.13°C (0.23°F) per decade, almost twice the rate of the twentieth century as a whole. Average temperatures in the Northern Hemisphere during this time period are likely (greater than 66 percent probability) to be higher than at any time in at least 1300 years.
- Data since 1961 shows that global ocean temperature has increased to depths of at least 3000 meters (9800 feet), and that 80 percent of the heat added to the global climate system has been absorbed by the oceans.
- In part because of thermal expansion of seawater, global sea level has been rising. During the twentieth century, the estimated total global sea level rise was 0.17 meters (6.7 inches). The average rate of global sea level rise between 1961 and 2003 was about 1.8 mm (0.07 inches) per year, although between 1993 and 2003 the rate had increased to about 3.1 mm (0.12 inches) per year; it is not clear if this recent higher rate represents a long-term trend.
- In the Arctic, over the last 100 years average temperatures have been increasing at almost twice the global rate, although in this region there is high observed variability from decade to decade.
- Since 1978, data from satellites shows that average summer sea ice in the Arctic has been decreasing at a rate of about 7.4 percent per decade. (By the end of the summer in 2007 the extent of Arctic sea ice was the smallest measured since regular satellite monitoring of the ice pack began in 1979; 2008 showed a slight increase in the ice pack but was nonetheless well below the long-term average.) Sea ice around Antarctica has shown great annual variation and local changes, but no statistically significant average trend was noted by the IPCC.
- Ice caps and glaciers decreased in both hemispheres, contributing to sea level rise; the flow speed of some Greenland and Antarctic outlet glaciers has increased.
- Since 1980, temperatures at the top of the permafrost layer have increased by as much as 3°C (5.4°F), and since 1900 the extent of seasonally frozen ground has been reduced by about 7 percent.
- Observations indicate that there has been an increase in the number of intense tropical cyclones in the North Atlantic Ocean basin since 1970, and this is correlated with an increase in sea surface temperatures in the tropics.
- The average amount of water vapor in the atmosphere over both land and ocean areas has increased since the 1980s, consistent with the higher water vapor capacity of warmer air.
- Between 1900 and 2005 statistically significant increases in average precipitation were observed in parts of North America, South America, Central and Northern Asia, and Northern Europe, while decreases in average precipitation were observed in Southern Africa, South Asia, around the Mediterranean, and in the Sahel; since the 1970s, longer and more intense droughts have been observed over wide areas.

These observed increases in global temperature and the secondary effects of this warming correlate very closely with an increase in greenhouse gas concentrations tied to human activity. Carbon dioxide in the atmosphere, the most important anthropogenic greenhouse gas, had increased from a preindustrial level of about 280 parts per million (ppm) to 390 ppm by May 2009. Methane, another key anthropogenic greenhouse gas, increased in concentration from a preindustrial level of about 715 parts per billion (ppb) to about 1780 ppb by 2008. Ice-core data from Dome C in Antarctica shows that the current concentrations of both carbon dioxide and methane in the atmosphere are now higher—and that they increased more rapidly in recent decades—than at any time in the past 800,000 years. It is likely that this increase in greenhouse gases would have caused more warming than that observed if not offset by slight cooling from anthropogenic and volcanic aerosols.

So is it possible that the changes in climate observed over the last century have natural causes and are not the result of anthropogenic increases in greenhouses gases? While scientists must always leave open that possibility, the current consensus of atmospheric scientists is that the observed changes in global climate over the last century cannot be explained through natural causes alone.

It may be instructive to provide several examples of the findings of the scientific community about global warming. In 2003 the American Geophysical Union (AGU), an international organization of research scientists, concluded:

> Human activities are increasingly altering the Earth's climate. These effects add to natural influences that have been present over Earth's history. Scientific evidence strongly indicates that natural influences cannot explain the rapid increase in global near-surface temperatures observed during the second half of the 20th century. . . . Atmospheric carbon dioxide concentrations have increased since the mid-1700s through fossil fuel burning and changes in land use, with more than 80% of this increase occurring since 1900. Moreover, research indicates that increased levels of carbon dioxide will remain in the atmosphere for hundreds to thousands of years. It is virtually certain that increasing atmospheric concentrations of carbon dioxide and other greenhouse gases will cause global surface climate to be warmer.
>
> (*Source: American Geophysical Union,* Human Impacts on Climate, *adopted by AGU Council, December 2003.*)

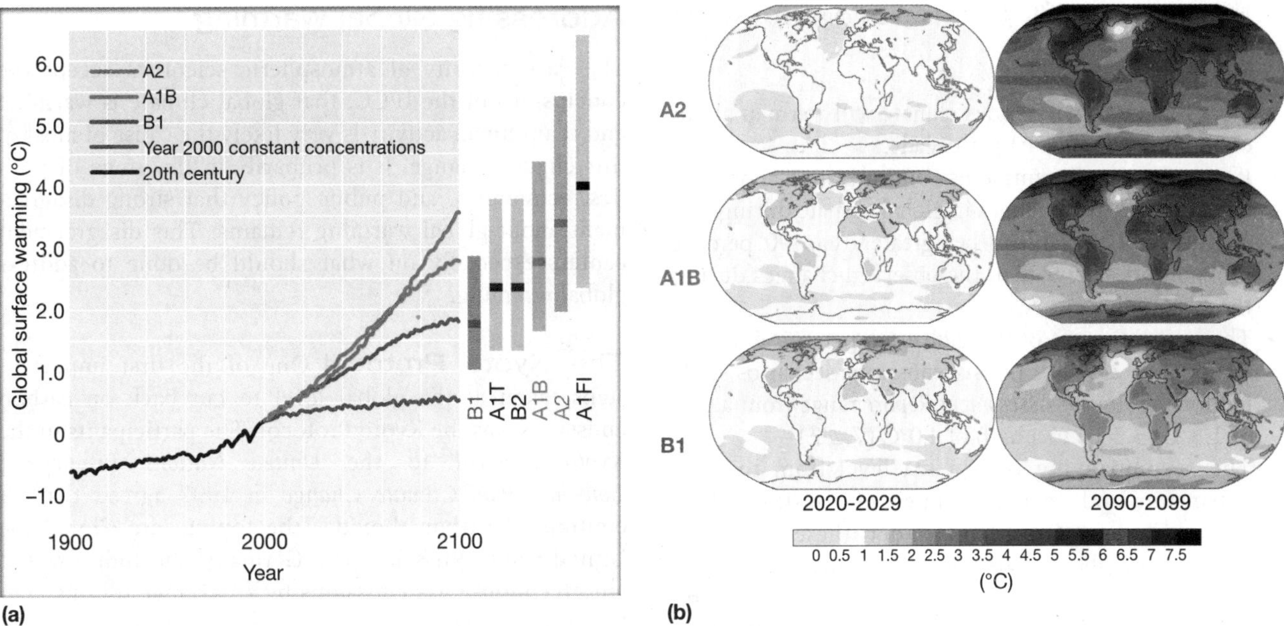

▲ **Figure 8-47** Projected temperature changes from 2000 to 2100 under different IPCC emission scenarios (designated B1, A1T, etc., by the IPCC) in the absence of greenhouse gas reductions. (a) The emission scenarios ranged from a lower emissions scenario (B1) in which greenhouse gas emissions begin to decline after reaching a peak at midcentury, to higher emissions scenarios (A2 and A1FI) in which greenhouse gas emissions continue to rise through the end of the century. For reference, the pink line shows the projected temperature increase if greenhouse gas concentration did not rise above its level in the year 2000—an impossibility. The vertical bars show the range of temperature estimates by century's end. (b) Maps showing projected global surface temperature changes for the years 2020–2029 and 2090–2099 under three different emission scenarios.

As we first saw in Chapter 4, the *Fourth Assessment Report* of the IPCC released in 2007 concluded:

> Most of the observed increase in globally averaged temperatures since the mid-20th century is very likely [greater than 90 percent probability] due to the observed increase in anthropogenic greenhouse gas concentrations. . . . Discernible human influences now extend to other aspects of climate, including ocean warming, continental-average temperatures, temperature extremes and wind patterns.

The report *Global Climate Change Impacts in the United States*, released by the U.S. Global Change Research Program in 2009, reiterated the findings of the IPCC.

In assessing the credibility of these statements and the quality of science behind them, keep in mind that the IPCC reports are the result of work of hundreds of specialists; the conclusions of the report are refereed by a separate group of reviewers, and the final edited work approved by the delegates of more than 100 countries. Because the IPCC's reports are consensus views, the agreed-upon conclusions often involve compromise on the part of scientists. Tim Flannery, climate change researcher and biologist, has described the IPCC report conclusions as "lowest-common-denominator science" and that the conclusions "carry great weight . . . precisely because they represent a consensus view."[3]

Consequences of Global Warming

While the consensus of atmospheric scientists is that human activity is indeed altering global climate, predicting the extent of future climate change and its consequences are less certain. The IPCC's *Fourth Assessment Report* concludes it is likely [a greater than 66 percent probability] that global climate "sensitivity" to a doubling of preindustrial carbon dioxide levels will be a temperature increase of 2.0 to 4.5°C (3.6 to 8.1°F), with a best estimate of about 3.0°C (5.4°F).

However, projections of the temperature increase expected by the middle or end of this century are more complicated to calculate: in addition to the great complexity of the global climate system itself is uncertainty about how levels of greenhouse gases will actually change in coming decades. Six different emission scenarios were modeled by the IPCC. The various scenarios were based on different rates of global population increase, different rates of fossil fuel use, different rates of per capita economic growth around the world, among other factors (Figure 8-47).

[3]Flannery, Tim, *The Weather Makers: How Man Is Changing the Climate and What It Means for Life on Earth*, New York: Atlantic Monthly Press, 2005, p. 246.

The projections of the IPCC in the *Fourth Assessment Report* include:

- Over the next two decades climate will warm at a rate of about 0.2°C (0.4°F) per decade.
- If greenhouse gas emissions continue at or above the present rates, the changes in global climate during this century will very likely be (greater than 90 percent probability) greater than the observed changes during the twentieth century.
- The best estimates of the global temperature increase by the year 2099 for the six emissions scenarios studied for the *Fourth Assessment Report* range from a low of 1.8°C (3.3°F) to a high of 4.0°C (7.2°F).
- The accompanying rise in sea level from thermal expansion and increased rates of ice flow from Antarctica and Greenland under these scenarios ranges from about 0.18 meters (7.1 inches) to 0.59 meters (23.2 inches) by the end of this century. (More recent studies have taken into account a better understanding of ice sheet flows and predict greater sea level increases than the IPCC: from about 0.69 to 0.99 meters [27 to 39 inches] by the end of this century.) Even a modest increase in sea level, however, will cause widespread retreat of shorelines, inundating many heavily populated coastal areas and increasing vulnerability to storms such as hurricanes.
- Warming is expected to be greatest over land and in high northern latitudes, and least over the Southern Ocean; snow cover on land is expected to diminish.
- Sea ice is projected to diminish in the Arctic and Antarctic in all emission scenarios, with summer sea ice disappearing in the Arctic by the end of this century in some scenarios.
- It is likely that tropical cyclones will become more intense in association with projected increases in sea surface temperatures; the storm tracks of midlatitude cyclones are projected to move poleward.
- Precipitation is very likely to increase in high latitudes and likely to decrease in most subtropical areas over land.
- It is very likely (greater than 90 percent probability) that heat waves, heavy precipitation events, and hot extremes will occur more frequently.
- Tropical diseases may become more prevalent in regions beyond their current ranges.
- As climate changes, some plant and animal species will exhibit shifts in their distributions; wildfire risk will increase in areas of decreased rainfall.
- With increased global temperatures, risk of species extinction may increase.

Among the most troubling findings of the IPCC is that global temperatures are projected to continue increasing and sea level to continue rising even if the concentrations of greenhouse gases are stabilized immediately.

Addressing Global Warming

The vast majority of atmospheric scientists accept the conclusions of the IPCC: that global climate is warming and that human activity is very likely the cause of most of this climate change. It is primarily in the arena of politics, economics, and public policy that strong disagreement about global warming remains. This disagreement centers especially on what should be done to address global warming.

The Kyoto Protocol One of the first important steps taken at the global level to cut back on carbon emissions was the Kyoto Protocol. The participants in the *Kyoto Protocol* to the *United Nations Framework Convention on Climate Change* in 1997 agreed that all emitters of carbon dioxide (the top six are China, the United States, Russia, Japan, Germany, and India) would cut greenhouse gas emissions by an average of 5 percent below 1990 levels by the year 2012. As of this writing in 2009, 185 countries have ratified the agreement, but the United States has not.

By even the most optimistic accounts, however, the modest greenhouse gas reductions of the Kyoto Protocol would at best curtail global warming slightly. Nonetheless, the agreement is an historic first step. The commitment period of the Kyoto Protocol is due to expire in 2012, and a new round of international negotiations aimed at much steeper cuts in greenhouse gas emissions is set to be under way by fall 2009.

Newly Industrialized Countries Many developing countries are increasing their greenhouse gas emissions as they expand their industrial infrastructures. Requiring all countries to meet the same emission reductions seems unfair to leaders in some newly industrializing countries. These nations argue that the rich industrialized countries created the problem of global warming through their own unfettered use of fossil fuel and "dirty" technology—should developing countries be asked to cut back emissions as much as rich countries?

The United States makes up less than 5 percent of the world's population, but emits about 19 percent of the world's total carbon dioxide. China recently surpassed the United States as the world leader in CO_2 emissions (about 21 percent of the world total), but on a per capita (per person) basis, the United States still far exceeds China. Clearly, all countries of the world will need to participate in greenhouse gas reductions, but just how the economic needs of all countries can be balanced is not yet clear.

Mitigating and Adapting Because of the long residence time of some greenhouse gases in the atmosphere, even if the goals of carbon dioxide reduction specified under the Kyoto Protocol—or even more ambitious goals proposed for new negotiations—are achieved, we can expect global temperatures to continue rising for some time

PEOPLE AND THE ENVIRONMENT

Strategies for Reducing Greenhouse Gas Emissions

It is likely that many different strategies will need to be employed to address a problem as complex as global climate change. Just a few of the possibilities for reducing greenhouse gas emissions are described below.

Conservation: Perhaps the simplest and most immediate strategy that can be used to reduce greenhouse gas emissions is to burn less fossil fuel. Electrical power generation and transportation are responsible for almost 40 percent of all greenhouse gas emissions worldwide and so conservation in these areas is an obvious goal. Seemingly small steps can add up: driving less, taking public transportation, using energy-efficient lighting, improving home insulation, and simply turning off lights and low-drain electronic devices when not in use (Figure 8-F).

Being aware of our own resource use can be a helpful first step toward conservation. You can calculate your "carbon footprint" by going to one of several Internet sites, such as http://atmospheres.gsfc.nasa.gov/igloo/.

Cap-and-Trade versus Carbon Tax: For the most part, reluctance on the part of the United States and some other industrialized countries to immediately and significantly cut back on greenhouse gas emissions centers on issues of short-term cost. For that reason, many economists argue that any pragmatic approach to reducing carbon emissions must involve market incentives. There are two main approaches to this. One, a cap-and-trade system, places declining limits (a "cap") on carbon emissions from industries, but allows one emitter to sell or trade its excess carbon allotments to another emitter. In other words, industries that emit less than their carbon allotment can sell their remainder to industries still emitting too much—this provides financial incentives to innovate and conserve, while offering more time for industries making a slower transition to cleaner energy technology.

The second method, a carbon tax, on the other hand, simply penalizes industries that emit more than an established emissions cap. To a certain extent, a carbon tax acts as a financial "stick" to reduce emissions, while a cap-and-trade system offers somewhat of a "carrot." There are advocates for both systems, but in either case the key to greenhouse gas reductions is setting meaningful emissions caps and timetables.

In a move suggesting that the United States is beginning to take serious steps to curtail greenhouse gas emissions, after years of delay, in June 2009 the Environmental Protection Agency gave the State of California permission to enforce its tougher mileage and emissions standards for new automobiles—aimed at reducing greenhouse gas emissions by 30 percent by the year 2016. Other states promise to adopt these stricter standards as well.

As of this writing in 2009, the United States Congress is drafting legislation that would regulate greenhouse gas emissions, primarily though a cap-and-trade system, but the outcome of the various proposals is uncertain at this time.

"No Regrets" Opportunities: The IPCC highlights a number of "no regrets" opportunities—practices that reduce greenhouse gas emissions while having a net cost saving for society and therefore need not be postponed for reasons of cost. In other words, the cost of implementing these policies is offset by the cost savings through direct or indirect benefits to individuals or society. Examples of these opportunities include: cost savings from lower energy bills as energy-efficient lighting, building insulation, and heating are adopted; cost savings in health care as air pollution is reduced (in October 2009, a study by the National Academy of Sciences estimated that nearly 20,000 Americans die prematurely each year as a result of air pollution from power plants and vehicles); cost and time savings for commuters as carpooling and expanded public transportation lessen traffic congestion, as well as many others.

The Clean Energy Economy: A final set of economic arguments is also beginning to receive serious attention. As the world slowly but inevitably moves toward less dependence on fossil fuel for its energy needs, an important question remains unanswered: which companies and which countries will lead the way by developing the cost-effective technology and infrastructure to deliver so-called green energy? Many economists argue that those who have the foresight to invest in clean energy technology today may reap enormous financial benefits in the future.

▲ **Figure 8-F** Much of the traffic congestion in American cities during commuting times comes from automobiles with single occupants.

into the future. A 2009 study led by National Oceanic and Atmospheric Administration (NOAA) senior scientist Susan Solomon concluded that temperature increases, sea level increases, and rainfall changes are essentially irreversible for perhaps several centuries after CO_2 emissions are stabilized.

So this raises the question: In addition to efforts to prevent greater global warming, what should be done to adapt to the consequences of climate change? For example, should coastal dikes or seawalls be raised to compensate for higher sea level and storm surge potential? Or should drinking water storage reservoirs be expanded to compensate for reductions in winter snow pack? The costs and environmental consequences of these and many other strategies will need to be thoughtfully weighed.

The findings of the IPCC point to a related problem: the consequences of global climate change may not be shared equally around the world. Many of the populations most vulnerable to sea level rise, changes in precipitation for agriculture and drinking water, and the expanding ranges of insect pests and disease, are found in the poorest parts of the world—populations with the least financial means to adapt. So an appropriate question to ask is: What can and should the richest countries of the world do to assist the poorest to adapt to climate change?

It remains to be seen how the people of the world—as individuals and as countries—will choose to address global climate change, a problem without quick or easy solutions.

CHAPTER 8 LEARNING REVIEW

After studying this chapter, you should be able to answer the following questions. Key terms from each text section are shown in bold type. Definitions for key terms are also found in the glossary at the back of the book.

Climate Classification (p. 189)

1. Explain the basic concept of the **Köppen climate classification system.**
2. In the Köppen climate classification letter code system, what information is given by the first letter, the second letter, and the third letter?
3. What information is conveyed in a **climograph**?

World Distribution of Major Climate Types (p. 191)

4. Briefly describe the major climate groups of the modified Köppen climate classification system: A, B, C, D, E, and H.
5. Describe the general location, temperature characteristics, precipitation characteristics, and main controls of the following 10 climate types. You should also be able to recognize these climates from a climograph:
   - Af **Tropical wet**
   - Aw **Tropical savanna**
   - Am **Tropical monsoon**
   - BWh **Subtropical desert**
   - BWk **Midlatitude desert**
   - Cs **Mediterranean**
   - Cfa **Humid subtropical**
   - Cfb **Marine west coast**
   - Dfa **Humid continental**
   - Dfc **Subarctic**
6. Why do Af (tropical wet) climates receive rain all year while Aw (tropical savanna) climates receive rain only in the summer (the high-Sun season)?
7. What are the main differences in controls of BWh (subtropical desert) and BWk (midlatitude desert) climates?
8. Why are subtropical desert (BWh) climates generally hotter in summer than tropical humid (A) climates?
9. Why are subtropical desert (BWh) climates usually displaced toward the western sides of continents?
10. Why are dry climates much more extensive in North Africa than in any other subtropical location?
11. What is the general difference between a desert climate and a steppe climate?
12. Why do Cs (mediterranean) climates have dry summers and wet winters?
13. What causes the relatively mild temperatures of marine west coast climates?
14. What is meant by the phrase "continentality is a keynote in D climates"?
15. Although both cities are coastal, New York City has a continental climate while Seattle, Washington, has a maritime climate. Why?
16. Why do subarctic (Dfc) climates have such a wide annual temperature range?
17. Why are polar climates so dry?
18. What general temperature characteristic distinguishes **tundra climate** from **ice cap climate**?
19. In what ways is altitude more important in determining a **highland climate** than latitude?

Global Climate Change (p. 220)

20. What is meant by **paleoclimatology**?
21. How does **dendrochronology** provide information about past climate?

22. How does the **oxygen isotope analysis** of ocean floor sediments and glacial ice tell us about past temperatures?
23. How can ice-core analysis provide information about the gas composition of the atmosphere in the past?
24. What are **Milankovitch cycles** and in what ways might they help explain past climate change?
25. Describe and explain at least one feedback mechanism that would further increase global temperatures once a warming trend has started.
26. What kinds of clouds tend to cool the surface of Earth, and what kinds of clouds tend to warm it?
27. What roles do the oceans play in influencing the carbon dioxide concentrations in the atmosphere?

Log in to **www.mygeoscienceplace.com** for animations, videos, RSS feeds, maps, and self-study quizzes to enhance your study of **Climate and Climate Change**.

9 The Hydrosphere

The hydrosphere is at once the most pervasive and the least well defined of the four "spheres" of Earth's physical environment. It includes the surface water in oceans, lakes, rivers, and swamps; all underground water; frozen water in the form of ice, snow, and high-cloud crystals; water vapor in the atmosphere; and the moisture temporarily stored in plants and animals.

The hydrosphere overlaps significantly with the other three spheres. Liquid water, ice, and even water vapor occur in the soil and rocks of the lithosphere. Water vapor and cloud particles composed of liquid water and ice are important constituents of the lower portion of the atmosphere, and water is a critical component of every living organism of the biosphere. Life is impossible without water; every living thing depends on it. Watery solutions in living organisms dissolve or disperse nutrients for nourishment. Most waste products are carried away in solutions. Indeed, the total mass of every living thing is more than half water, the proportion ranging from about 60 percent for some animals to more than 95 percent for some plants.

It is through moisture, then, that the interrelationships of the four spheres are most conspicuous and pervasive. In Chapter 6 we introduced many of the physical properties of water and the roles of water in weather and climate. In this chapter, we examine the geography of water more broadly.

The Hydrologic Cycle

Hydrologic Cycle

This unique substance, water, essential to life and finite in amount, is distributed very unevenly on, in, and above Earth. The great bulk of all moisture, more than 99 percent, is in "storage"—in oceans, lakes, and streams, locked up as glacial ice, or held in rocks below Earth's surface (Figure 9-1). Water frozen as ice in glaciers and continental ice sheets represents about three-fourths of all of the freshwater on the planet.

The proportional amount of moisture in these various storage reservoirs is relatively constant over thousands of years. Only during an ice age is there a notable change in these components: during periods of glaciation, the volume of the oceans becomes smaller as the ice sheets grow and the level of atmospheric water vapor diminishes; then during deglaciation, the ice melts, the volume of the oceans increases as the meltwater flows into them, and there is an increase in atmospheric water vapor.

The remaining small fraction—less than 1 percent—of Earth's total moisture is involved in an almost continuous sequence of movement and change, the effects of which are absolutely critical to life on this planet. This tiny portion of Earth's water supply moves from one storage area to another—from ocean to air, from air to ground, and so on—in the **hydrologic cycle**. We first introduced the hydrologic cycle in Chapter 6. In the present chapter, we examine the complexities of this cycle in more detail.

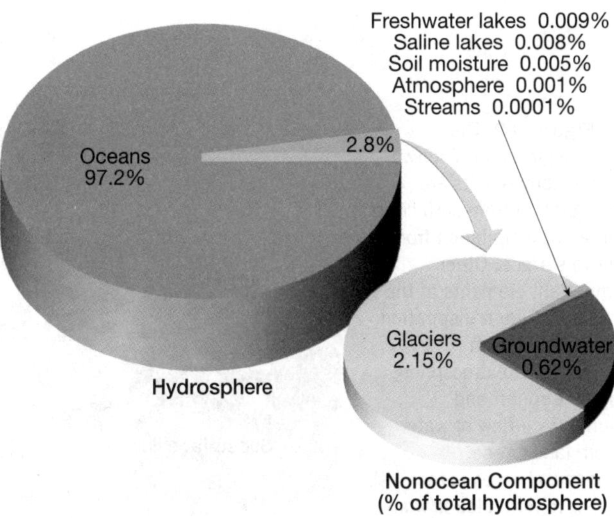

▲ **Figure 9-1** Moisture inventory of Earth. More than 97 percent of all water is contained in the oceans.

The hydrologic cycle can be viewed as a series of storage areas interconnected by various transfer processes, in which there is a ceaseless interchange of moisture in terms of both its geographic location and its physical state (Figure 9-2). Liquid water on Earth's surface evaporates to become water vapor in the atmosphere. That vapor then condenses and precipitates, either as liquid water or as ice, back onto the surface. This precipitated water then runs off into storage areas and later evaporates into the atmosphere once again. Because this is a closed, circular system, we can begin the discussion at any point. It is perhaps clearest to start with the movement of moisture from Earth's surface into the atmosphere.

Surface-to-Air Water Movement

Most of the moisture that enters the atmosphere from Earth's surface does so through evaporation. (Transpiration is the source of the remainder.) The oceans, of course, are the principal source of water for evaporation. They occupy 71 percent of Earth's surface, have unlimited moisture available for evaporation, and are extensive in low latitudes, where considerable heat and wind movement facilitate evaporation. As a result, an estimated 86 percent of all evaporated moisture is derived from ocean surfaces (Table 9-1). (The 14 percent that comes from land surfaces includes the twin processes of evaporation and transpiration.)

Water vapor from evaporation remains in the atmosphere a relatively short time—usually only a few hours or days. During that interval, however, it may move a considerable distance, either vertically through convection or horizontally through advection driven by wind.

◀ Briksdalbreen Glacier near Olden, Norway.

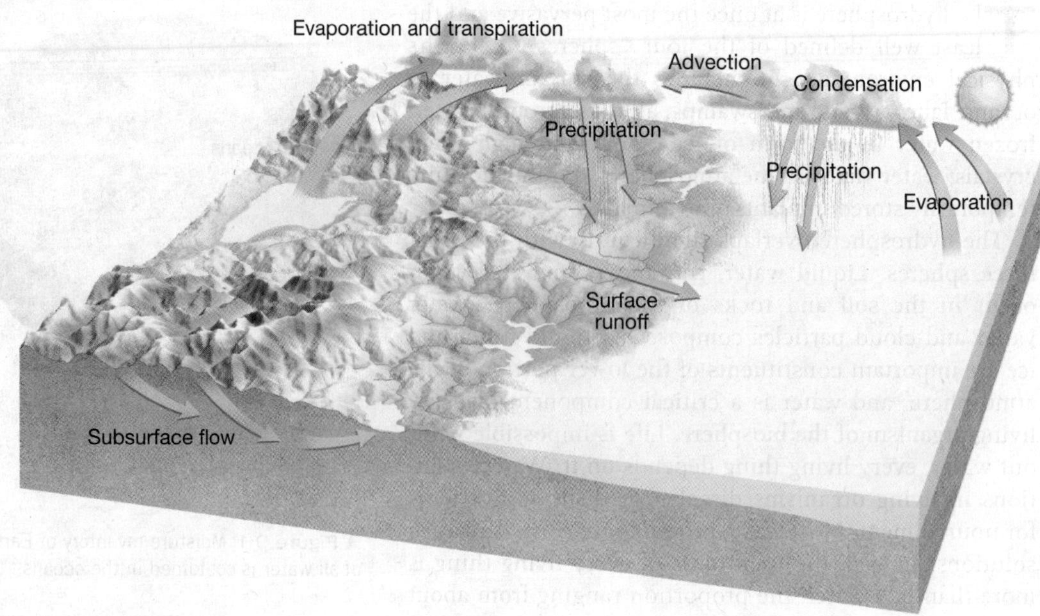

▶ **Figure 9-2** The hydrologic cycle. The two major components are evaporation from surface to air and precipitation from air to surface. Other important elements of the cycle include transpiration of moisture from vegetation to atmosphere, surface runoff and subsurface flow of water from land to sea, condensation of water vapor to form clouds from which precipitation may fall, and advection of moisture from one place to another.

Air-to-Surface Water Movement

Sooner or later, water vapor in the atmosphere condenses to liquid water or sublimates to ice to form cloud particles. As we saw in Chapter 6, under the proper circumstances clouds may drop precipitation in the form of rain, snow, sleet, or hail. As Table 9-1 shows, 78 percent of this precipitation falls into the oceans and 22 percent falls onto land.

Over several years, total worldwide precipitation is approximately equal to total worldwide evaporation/transpiration. Although precipitation and evaporation/transpiration balance in time, they do not balance in place: evaporation exceeds precipitation over the oceans, whereas the opposite is true over the continents. This imbalance is explained by the advection of moist maritime air onto land areas, so that there is less moisture available for precipitation over the ocean.

Except for coastal spray and storm waves, the only route by which moisture moves from sea to land is via the atmosphere.

Movement On and Beneath Earth's Surface

Looking at Table 9-1, you may wonder why the oceans do not dry up and the continents become flooded: there is 8 percent more water leaving the ocean than precipitating back into it, and 8 percent more water falling on land than evaporating off it. The reason such drying up and flooding do not take place is **runoff**, that portion of Earth's circulating moisture that moves in the liquid state from land to sea.

The 78 percent of total global precipitation that falls on the ocean is simply incorporated immediately into the water already there; the 22 percent that falls on land goes through a more complicated series of events. Rain falling on a land surface collects in lakes, runs off if the surface is a slope, or infiltrates the ground. Any water that pools on the surface eventually either evaporates or sinks into the ground, runoff water eventually ends up in the ocean, and infiltrated water is either stored temporarily as soil moisture or percolates farther down to become part of the underground water supply. Much of the soil moisture eventually evaporates or transpires back into the atmosphere, and much of the underground water eventually reappears at the surface via springs. Then sooner or later, and in one way or another, most of the water that reaches the surface evaporates again, and the rest is incorporated into streams and rivers and becomes runoff flowing into the oceans. This runoff water from continents to oceans amounts to 8 percent of all moisture circulating in the global hydrologic cycle. It is this runoff that balances the excess of precipitation over evaporation taking place on the continents and that keeps the oceans from drying up and the land from flooding.

TABLE 9-1 Comparison of Moisture Balance of Continents and Oceans

	Percentage of Total World Surface Area	Percentage of Total World Precipitation Received	Percentage of Total World Evaporation/Transpiration that Occurs from the Surface
Oceans	71	78	86
Continents	29	22	14

Residence Times

Although the hydrologic cycle is a closed system and is believed to have an unvarying total capacity, there is enormous variation in the cycling of individual molecules of water. A particular molecule of water may be stored in oceans and deep lakes, or as glacial ice, for thousands of years without moving through the cycle, and one trapped in rocks buried deep beneath Earth's surface may be excluded from the cycle for thousands or even millions of years.

However, whatever water is moving through the cycle is in almost continuous motion (Figure 9-3). Runoff water can travel hundreds of kilometers to the sea in only a few days, and moisture evaporated into the atmosphere may remain there for only a few minutes or hours before it is precipitated back to Earth. Indeed, at any given moment, the atmosphere contains only a few days' potential precipitation.

▲ **Figure 9-3** Oregon's Lake Trillium with morning fog. Moisture constantly moves through the hydrologic cycle through evaporation, condensation, precipitation, and runoff.

The Oceans

Despite the facts that most of Earth's surface is oceanic and that the vast majority of all water is in the oceans (see Figure 9-1), our knowledge of the seas was fairly limited until recently. The ocean is a hostile environment for most air-breathing creatures, including humans. Thus, only with great care can humans venture beneath the sea's surface, and only within the last five decades or so has sophisticated equipment been available to catalog and measure details of the maritime environment.

How Many Oceans?

From the broadest viewpoint, there is but one interconnected ocean. This "world ocean" has a surface area of 360 million square kilometers (139 million square miles) and contains 1.32 billion cubic kilometers (317 million cubic miles) of saltwater. It spreads over almost three-fourths of Earth's surface, interrupted here and there by continents and islands. Although tens of thousands of bits of land protrude above the blue waters, the world ocean is so vast that half a dozen continent-sized portions of it are totally devoid of islands, without a single piece of land breaking the surface of the water. It is one or more of these large expanses of water we are usually referring to when we use the term *ocean*.

In generally accepted usage, the world ocean is divided into four principal parts—the Pacific, Atlantic, Indian, and Arctic oceans (Figure 9-4). (Some people refer to the waters around Antarctica as the *Antarctic Ocean* or the *Great Southern Ocean*, but this distinction is not universally accepted.) The boundaries of the four oceans are not everywhere precise, and around some of their margins are partly landlocked smaller bodies of water called *seas*, *gulfs*, *bays*, and other related terms. Most of these smaller bodies can be considered as portions of one of the major oceans, although a few are so narrowly connected (Black Sea, Mediterranean Sea, Hudson Bay) to a named ocean as to deserve separate

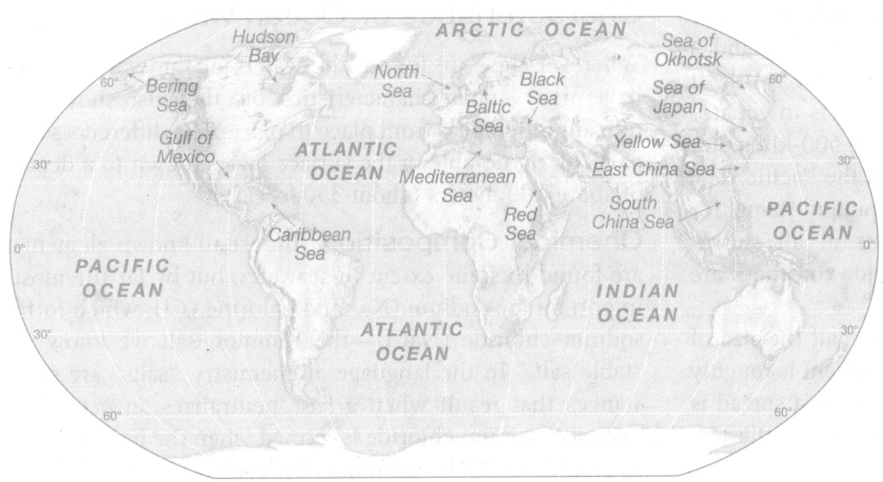

◀ **Figure 9-4** The four principal oceans and the major seas of the world.

TABLE 9-2 Oceans and Major Seas of the World

	Approximate Surface Area		
	Square Kilometers	Square Miles	Percentage of Water Surface on Earth*
Pacific Ocean	166,884,000	64,186,000	46
Atlantic Ocean	82,841,000	31,862,000	23
Indian Ocean	73,710,000	28,350,000	20
Arctic Ocean	14,110,000	5,427,000	4
South China Sea	2,990,000	1,150,000	—
Caribbean Sea	2,525,000	971,000	—
Mediterranean Sea	2,519,000	969,000	—
Bering Sea	2,275,000	875,000	—
Gulf of Mexico	1,560,000	600,000	—
Sea of Okhotsk	1,430,000	550,000	—
East China Sea	1,248,000	480,000	—
Sea of Japan	1,053,000	405,000	—
Hudson Bay	827,000	318,000	—
North Sea	577,000	222,000	—
Yellow Sea	572,000	220,000	—
Black Sea	494,000	190,000	—
Red Sea	455,000	175,000	—
Baltic Sea	416,000	160,000	—

*Values are omitted for the 14 seas because of their minimal size relative to the size of the world's oceans.

consideration, as Figure 9-4 and Table 9-2 show. This nomenclature is further clouded by the term "sea," which is used sometimes synonymously with ocean, sometimes to denote a specific smaller body of water around the edge of an ocean, and occasionally to denote an inland body of water.

The *Pacific Ocean* (Figure 9-5a) is twice as large as any other body of water on Earth. Five continents are on its fringes. It occupies about one-third of the total area of Earth, more than all the world's land surfaces combined. It contains the greatest average depth of any ocean as well as the deepest known oceanic trenches. Although the Pacific extends almost to the Arctic Circle in the north and a few degrees beyond the Antarctic Circle in the south, it is largely a tropical ocean. Its greatest girth is in equatorial regions; almost one-half of the 38,500-kilometer (24,000-mile) length of the equator is in the Pacific. The character of this ocean often belies its tranquil name, for it houses some of the most disastrous of all storms (typhoons) and many of the world's major volcanoes are either in it or around its edge.

The *Atlantic Ocean* is slightly less than half the size of the Pacific (Figure 9-5b). Its north–south extent is roughly the same as that of the Pacific, but its east–west spread is only about half as great. Its average depth is a little less than that of the Pacific.

The *Indian Ocean* (Figure 9-5c) is a little smaller than the Atlantic, and its average depth is slightly less than that of the Atlantic. Nine-tenths of its area is south of the equator.

The *Arctic Ocean* (Figure 9-5d) is much smaller and shallower than the other three and is mostly covered with ice. It is connected to the Pacific by a relatively narrow passageway between Alaska and Siberia, but it has a broad and indefinite connection with the Atlantic between North America and Europe.

Characteristics of Ocean Waters

Wherever they are found, the waters of the world ocean have many similar characteristics, but they also show significant differences from place to place. The differences are particularly notable in the surface layers, down to a depth of about 100 meters (about 350 feet).

Chemical Composition Almost all known elements are found to some extent in seawater, but by far the most important are sodium (Na) and chlorine (Cl), which form sodium chloride (NaCl)—the common salt we know as "table salt." In the language of chemistry, "salts" are substances that result when a *base* neutralizes an *acid*. For instance, sodium chloride is formed when the base sodium hydroxide (NaOH) neutralizes hydrochloric acid (HCl).

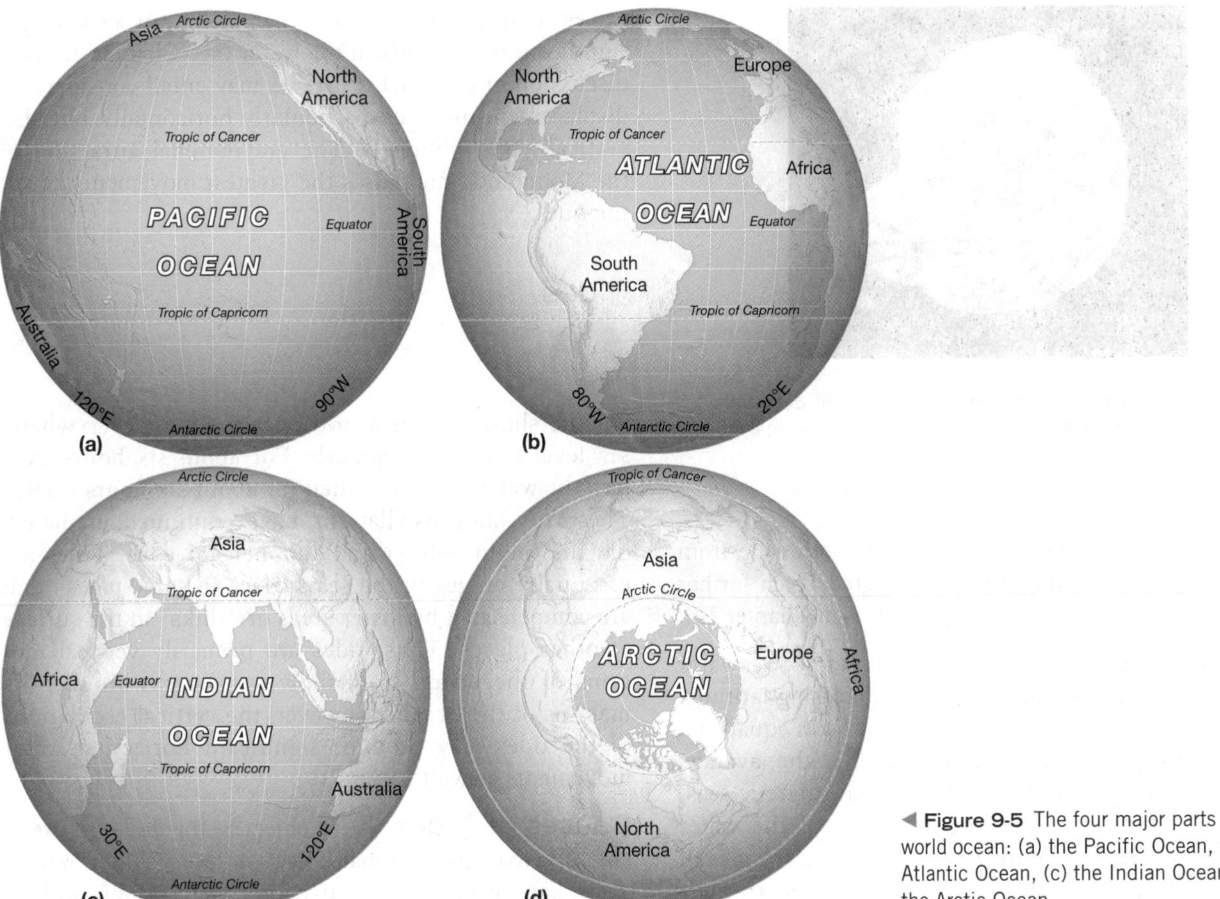

◀ **Figure 9-5** The four major parts of the world ocean: (a) the Pacific Ocean, (b) the Atlantic Ocean, (c) the Indian Ocean, (d) the Arctic Ocean.

The **salinity** of seawater is a measure of the concentration of dissolved salts, which are mostly sodium chloride but also include salts containing magnesium, sulfur, calcium, and potassium. The average salinity of seawater is about 35 parts per thousand, or 3.5 percent of total mass.

The geographic distribution of surface salinity varies. At any given location on the ocean surface, the salinity depends on how much evaporation is taking place and how much freshwater (primarily from rainfall and stream discharge) is being added. Where the evaporation rate is high, so is salinity; where the inflow of freshwater is high, salinity is low. Typically the lowest salinities are found where rainfall is heavy and near the mouths of major rivers. Salinity is highest in partly landlocked seas in dry, hot regions because here the evaporation rate is high and stream discharge is minimal. As a general pattern, salinity is low in equatorial regions because of heavy rainfall, cloudiness, and humidity, all of which inhibit evaporation, and also because of considerable river discharge. Salinity rises to a general maximum in the subtropics, where precipitation is low and evaporation extensive, and decreases to a general minimum in the polar regions, where evaporation is minimal and there is considerable inflow of freshwater from rivers and ice caps.

Increasing Acidity The oceans absorb carbon dioxide from the atmosphere. Perhaps one-third of the excess CO_2 released into the air each year by human activity is absorbed by the oceans. When CO_2 is taken in by the ocean, it forms *carbonic acid* (H_2CO_3), a weak acid. Research now suggests that as a result of the great quantities of CO_2 absorbed since the beginning of the industrial revolution, the ocean is becoming more acidic.

Currently, ocean water is slightly alkaline, with a pH of 8.1 (see "Acid Rain" in Chapter 6 for a description of the pH scale). Although still alkaline, this value is estimated to be about 0.1 lower—in other words more acidic—than it was in the preindustrial era. Given the current rate of fossil fuel use and continued absorption of CO_2 by the oceans, the pH of ocean water could drop to 7.7 by the end of this century.

The consequences of a slightly more acidic ocean are not completely known, but it is likely that it will affect the growth of organisms such as coral polyps and microscopic creatures such as *foraminifera* (Figure 9-6) that build their shells or exoskeletons from calcium carbonate ($CaCO_3$) extracted from seawater. As the oceans become more acidic, there are fewer calcium ions in seawater and so the growth of calcium carbonate shells is inhibited. It is not clear if these creatures will be able to adapt to the changing chemistry of the ocean.

Since foraminifera are at the bottom of the oceanic food web, among the potentially important consequences of a decline in their numbers would be the loss of food for a number of fish, such as mackerel and salmon. If the increased acidity of the oceans reduces the growth of coral

▲ **Figure 9-6** Scanning electron microscope image of *Elphidium crispum*, one of many kinds of single-celled foraminifera, magnified approximately 80 times.

polyps, coral reefs—already under stress worldwide from higher temperatures—might possibly degrade even further (a more complete discussion of coral reefs is in Chapter 20).

Temperature As is to be expected, surface seawater temperatures generally decrease with increasing latitude. The temperature often exceeds 26°C (80°F) in equatorial locations and decreases to −2°C (28°F), the average freezing point for seawater, in Arctic and Antarctic seas. (Dissolved salts lower the freezing point of the water from the 0°C [32°F] of pure water.) The western sides of oceans are nearly always warmer than the eastern margins because of the movement of major ocean currents (see Figure 4-25 in Chapter 4). This pattern of warmer western parts is due to the contrasting effects of poleward-moving warm currents on the west side of ocean basins and equatorward-moving cool currents on the east side of ocean basins.

Density Seawater density varies with temperature, degree of salinity, and depth. High temperature produces low density, and high salinity produces high density. Deep water has high density because of low temperature and because of the pressure of the overlying water.

Surface layers of seawater tend to contract and sink in cold regions, whereas in warmer areas deeper waters tend to rise to the surface. Surface currents also affect this situation, particularly by producing an upwelling of colder, denser water in some localities. As we will see later in this chapter, differences in density are partially responsible for a vast, slow circulation of deep ocean water.

Movement of Ocean Waters

The liquidity of the ocean permits it to be in continuous motion, and this motion can be grouped under three headings: tides, currents, and waves. The movement of almost anything over the surface—the wind, a boat, a swimmer—can set the water surface into motion, and so the ocean surface is almost always ridged with swells and waves. Disturbances of the ocean floor can also trigger significant movements in the water (see Chapter 20 for a discussion of *tsunami*). Currents may entail considerable displacement of water, particularly horizontally, but also vertically and obliquely. The gravitational attraction of the Moon and Sun causes the greatest movements of all: the tides.

Tides

Tides

On the shores of the world ocean, almost everywhere, sea level fluctuates regularly. For about six hours each day, the water rises, and then for about six hours it falls. These rhythmic oscillations have continued unabated, day and night, winter and summer, for eons. **Tides** are essentially bulges in the sea surface in some places that are compensated by lower areas or "sinks" in the surface at other places. Thus, tides are primarily vertical motions of the water. In shallow-water areas around the margins of the oceans, however, the vertical oscillations of the tides may produce significant horizontal water movements as well.

Causes of Tides Every object in the universe exerts an attractive gravitational force on every other object. Thus, Earth exerts an attractive force on the Moon, and the Moon exerts an attractive force on Earth. The same is true for Earth and the Sun. It is the gravitational attraction between the Moon and Earth, and between the Sun and Earth, that cause tides.

The strength of the force of gravity is inversely proportional to the square of the distance between the two bodies, and so, the Sun being 150,000,000 kilometers (93,000,000 miles) from Earth and the Moon 385,000 kilometers (239,000 miles), the Moon produces a greater percentage of Earth's tides than does the Sun. The *lunar tides* are about twice as strong as the *solar tides*. To keep things simple, let us first discuss lunar tides alone, ignoring solar tides for the moment.

Gravitational attraction pulls ocean water toward the Moon. There is more gravitational attraction on the side of Earth facing the Moon than on the opposite side of Earth—where the inward or *centripetal force* needed to keep Earth in its orbit is slightly greater. The difference in forces slightly elongates the shape of the global ocean, so that two bulges of ocean water develop—one on the side of Earth facing the Moon and the other on the opposite side of Earth. As Earth rotates, coastlines move into and out of these bulges, producing simultaneous high tides on the opposite sides of Earth and low tides halfway between.

As Earth rotates eastward, the tidal progression appears to move westward. The tides rise and fall twice in the interval between two "rising" Moons, an interval that is about 50 minutes longer than a 24-hour day. The combination of Earth's rotation and the Moon's revolution around Earth means that Earth makes about 12° more than a full rotation between each rising of the Moon. Thus, two complete tidal cycles have a duration of about 24 hours and 50 minutes. This means that on all oceanic coastlines there are normally two high tides and two low tides about every 25 hours.

The magnitude of tidal fluctuation is quite variable in time and place, but the sequence of the cycle is generally similar everywhere. From its lowest point, the water rises gradually for about 6 hours and 13 minutes, so that there is an actual movement of water toward the coast in what is called a **flood tide**. At the end of the flooding period, the maximum water level, *high tide*, is reached. Soon the water level begins to drop, and for the next 6 hours and 13 minutes there is a gradual movement of water away from the coast, this movement being called an **ebb tide**. When the minimum water level (*low tide*) is reached, the cycle begins again.

Monthly Tidal Cycle The vertical difference in elevation between high and low tide is called the **tidal range**. Changes in the relative positions of Earth, Moon, and Sun induce periodic variations in tidal ranges, as shown in Figure 9-7. The greatest range (in other words, the highest tide) occurs when the three bodies are positioned in a straight line, which usually occurs twice a month near the times of the full and new Moons. When thus aligned, the joint gravitational pull of the Sun and Moon is along the same line, so that the combined pull is at a maximum. This is true both when the Moon is between Earth and the Sun and when Earth is between the Moon and Sun. In either case, this is a time of higher than usual tides, called **spring tides**. (The name has nothing to do with the season; think of water "springing" up to a very high level.)

When the Sun and Moon are located at right angles to one another with respect to Earth, their individual gravitational pulls are diminished because they are now pulling at right angles to each other. This right-angle pulling results in a lower than normal tidal range called a **neap tide**. The Sun–Moon alignment that causes neap tides generally takes place twice a month at about the time of first quarter and third quarter moons.

Tidal range is also affected by the Moon's nearness to Earth. The Moon follows an elliptical orbit in its revolution around Earth, the nearest point (called *perigee*) being about 50,000 kilometers (31,200 miles), or 12 percent, closer than the farthest point (*apogee*). During perigee, tidal ranges are greater than during apogee.

Global Variations in Tidal Range Tidal range fluctuates all over the world at the same times of the month. There are, however, enormous variations in range along different coastlines (Figure 9-8). Midocean islands may experience tides of only one meter (3 feet) or less, whereas continental seacoasts have greater tidal ranges, the amplitude being greatly influenced by the shape of the coastline and the configuration of the sea bottom beneath coastal waters. Along most coasts, there is a moderate tidal range of 1.5–3 meters (5–10 feet). Some partly landlocked seas, such as the Mediterranean, have almost negligible tides. Other places, such as the northwestern coast of Australia, experience enormous tides of 10 meters (35 feet) or so.

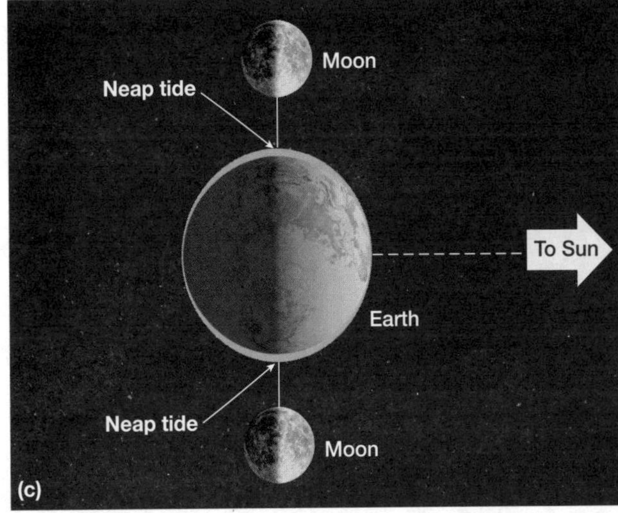

▲ **Figure 9-7** Juxtaposition of the Sun, Moon, and Earth accounts for variations in Earth's tidal range. The three basic positional relationships are illustrated here. (a) When the Moon and Sun are neither aligned nor at right angles to each other, we have normal levels of high tides on both sides of Earth. (b) When the Sun, Earth, and Moon are positioned along the same line, *spring tides* (the highest high tides) are produced. (c) When the line joining Earth and the Moon forms a right angle with the line joining Earth and the Sun, *neap tides* (the lowest high tides) result.

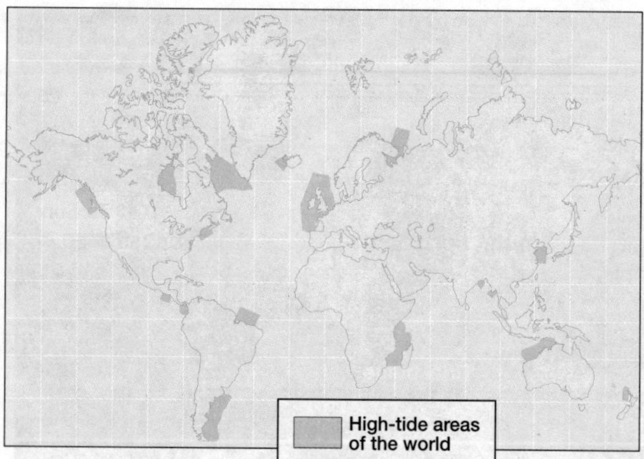

▲ **Figure 9-8** Areas with tidal ranges exceeding 4 meters (13 feet). The pattern is not a predictable one because it depends on a variety of unrelated factors, particularly shoreline and sea bottom configuration.

The greatest tidal range in the world is found at the upper end of the Bay of Fundy in eastern Canada (Figure 9-9), where a 15-meter (50-foot) water-level fluctuation twice a day is not uncommon, and a wall of seawater—called a **tidal bore**—several centimeters to more than a meter in height rushes up the Petitcodiac River in New Brunswick for many kilometers.

Tidal variation is exceedingly small in inland bodies of water. Even the largest lakes usually experience a tidal rise and fall of no more than 5 centimeters (2 inches). Effectively, then, tides are important only in the world ocean, and they are normally noticeable only around its shorelines.

Currents

Ocean Circulation Patterns

As we learned in Chapter 4, the world ocean contains a variety of currents that shift vast quantities of water both horizontally and vertically. Surface currents are caused primarily by wind flow, while other currents are set in motion by contrasts in temperature and salinity. All currents are likely to be influenced by the size and shape of the particular ocean, the configuration and depth of the sea bottom, and the Coriolis effect. Some currents involve subsidence of surface waters downward; other vertical flows bring an upwelling of deeper water to the surface.

Geographically speaking, the most prominent currents are the major horizontal flows that make up the general circulation of the various oceans. The dominant surface currents introduced in Chapter 4 are generally referred to as *subtropical gyres* (see Figure 4-25). They are set up by the action of the dominant surface wind

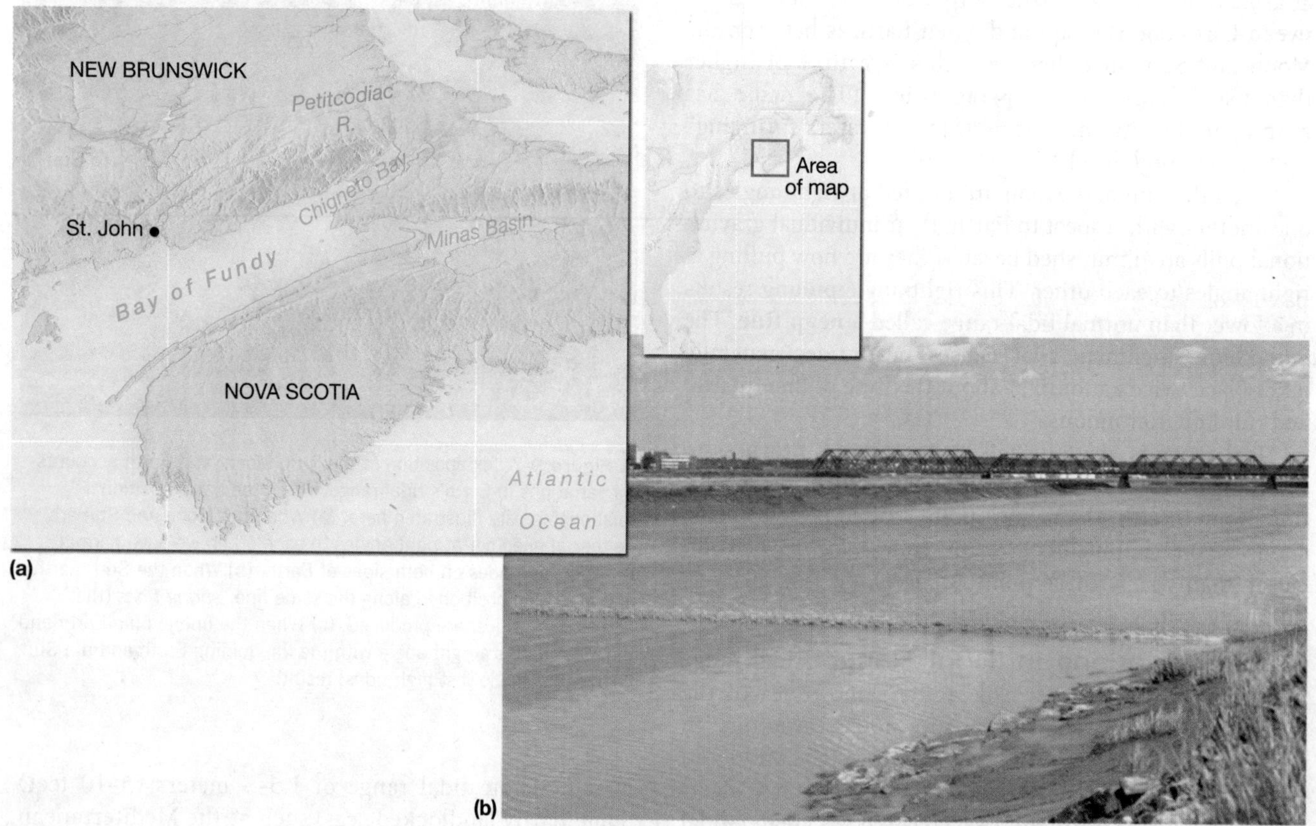

▲ **Figure 9-9** (a) The world's maximum tides are in the Bay of Fundy, where ocean water moves long distances up many of the coastal rivers twice a day. (b) A tidal bore moving up the Petitcodiac River in New Brunswick, Canada.

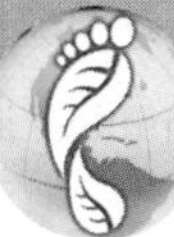

The Great Pacific Garbage Patch

Jennifer Rahn, Samford University, Birmingham, Alabama

Characteristics that make plastic items useful to consumers, their strength and durability, also make them a problem in marine environments. Each year more than 90 billion kilograms (200 billion pounds) of plastic is produced worldwide and about 10 percent of this ends up in the ocean. The trash tends to accumulate in areas where winds and currents are weak in the center of the ocean current *subtropical gyres* (see Figure 4-25)—especially in the area of the subtropical high (Figure 9-A). Some of this trash sinks but most of it floats in the upper 10 meters (30 feet) of the ocean (Figure 9-B). Estimates of the size of floating trash patch now in the Pacific Ocean vary: from the size of Texas to twice the size of Texas. It contains approximately 3.2 billion kilograms (3.5 million tons) of trash, and 80 percent of it is plastic. In the last few years this floating mass of trash has received a vast amount of media attention and is now known as the Great Pacific Garbage Patch.

The Great Pacific Garbage Patch actually consists of two main regions of ever-growing garbage accumulation. The eastern garbage patch—the first to gain attention—floats between Hawai'i and California, generally in the location of the subtropical high. The western garbage patch extends east of Japan to the western archipelago of the Hawaiian Islands, within what is called the *North Pacific Subtropical Convergence Zone.* These patches of garbage and flotsam in the Pacific (and evidently other ocean basins as well) are not new, they have always been there. What is new is that now most of it is nonbiodegradable plastic—formerly it was natural flotsam such as wood, glass bottles, and fishing floats.

There has been increasing use of plastic in our economies in the last few decades. For example, plastic grocery bags came into widespread use in the early 1980s. Environmental groups estimate that 500 billion to 1 trillion plastic bags are used worldwide every year. Because plastic use remains high, because plastic lasts so long in the environment, and because plastic waste from inland areas can end up in the ocean, the size of Great Pacific Garbage Patch is increasing tenfold every decade.

There are a wide variety of environmental hazards associated with the Great Pacific Garbage Patch. Plastic sea trash often floats at the surface and includes bottle caps, bags, and wrappers. Marine animals often mistake the floating plastic for food. To a sea turtle or bird, a plastic bag floating in the water looks like a jellyfish—one of their favorite foods. However, they have no way of digesting plastic. Animals that eat too much plastic die because they cannot pass it—it fills their stomachs and they starve to death.

Marine animals can also become entangled in plastic bags or the plastic rings that hold canned drinks; unable to eat, they starve to death. More than 267 marine species have been harmed by the debris, and about 100,000 whales, seals, turtles, birds, dolphins, and other marine animals are killed each year by plastic induced asphyxiation, strangulation, contamination, and entanglement.

Instead of biodegrading, most plastic slowly breaks up by *photodegrading* into smaller and smaller pieces. Fish mistakenly consume these small plastic particles, thinking they are plankton, their main food source. In the Great Pacific Garbage Patch there is now six times more plastic than plankton. The small particles of plastic ingested by the small fish have many toxic qualities. In the Great Pacific Garbage Patch toxic chemical levels in fish have been as high as a million times the concentration in the surrounding water. The chemicals in the plastic breakdown and enter the system of the fish, and these toxins are concentrated in the top of the food chain where they may be consumed by whales, dolphins—and humans. As people eat the large fish (tuna, shark, king mackerel, marlin and swordfish, to name a few), they are receiving concentrated doses of these toxins because the effects of the pollutants increase as they advance up the food chain.

Pollution of the oceans is arguably the biggest threat to marine life around the planet. The likelihood of harm to both marine life and to humans increases as the size of the Great Pacific Garbage Patch grows, so it is in our best interests to curb—and ideally find a way to remove—these floating pollutants

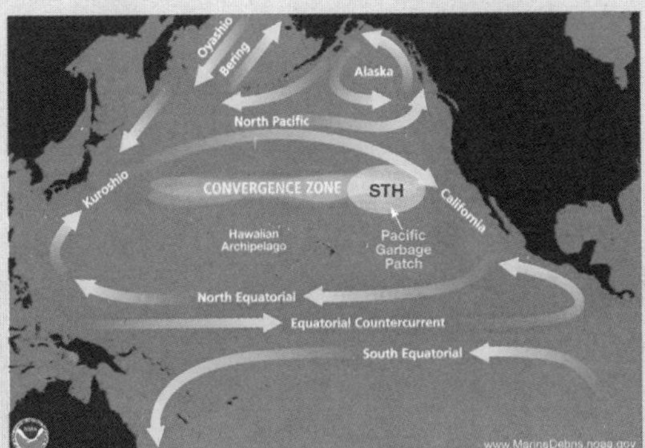

▲ **Figure 9-A** Map showing the location of the Great Pacific Garbage Patch. In addition to the garbage patch generally located in the area of the subtropical high (STH), garbage also accumulates in the western portion of the North Pacific Subtropical Convergence Zone.

▲ **Figure 9-B** Floating debris in the Great Pacific Garbage Patch.

systems in the tropics and midlatitudes: the trade winds and the westerlies.

Deep Ocean Circulation In addition to the major surface ocean currents, there is an important system of deep ocean circulation as well. This circulation of deep ocean water occurs because of differences in water density that arise from differences in salinity and temperature. For this reason, this water movement is sometimes referred to as **thermohaline circulation**. Ocean water will become denser, and thus sink, if its salinity increases or its temperature decreases. This happens predominantly in high-latitude ocean areas where the water is cold and salinity increases when sea ice develops (the dissolved salts are not taken up in the ice when water freezes, so the salinity of the remaining water increases).

The combination of deep ocean water movement through thermohaline circulation, along with influences from surface ocean currents, establishes an overall **global conveyer-belt circulation** pattern (Figure 9-10). Beginning in the North Atlantic, cold, dense water sinks and slowly flows deep below the surface to the south, where it eventually joins the eastward moving deep, cold, high salinity water circulating around Antarctica. Some of this deep water eventually flows north into the Indian and Pacific Oceans, where it rises to form a shallow, warm current that flows back to the North Atlantic Ocean, where it sinks and begins its long journey once again. These deep ocean currents might travel only 15 kilometers (9 miles) in a year—thus requiring many centuries to complete a single circuit.

Recent research suggests that a connection exists between the global conveyer-belt circulation and global climate. For example, if global climate becomes warmer, the freshwater runoff from the melting of Greenland's glaciers could form a pool of lower density water in the North Atlantic; this lower density surface water could disrupt the downwelling of water in the North Atlantic, altering the redistribution of heat—and so climate—around the world.

Waves

To the casual observer, the most conspicuous motion of the ocean is provided by waves. Most of the sea surface is in a state of constant agitation, with wave crests and troughs bobbing up and down most of the time. Moreover, around the margin of the ocean, waves of one size or another lap, break, or pound on the shore in endless procession.

Most of this movement is like running in place from the water's point of view, with little forward progress. Waves in the open ocean are mostly just shapes, and the movement of a wave across the sea surface is a movement of form rather than of substance or, to say the same thing another way, of energy rather than matter. Individual water particles make only small oscillating movements. Only when a wave "breaks" does any significant shifting of water take place. Waves are discussed in detail in Chapter 20.

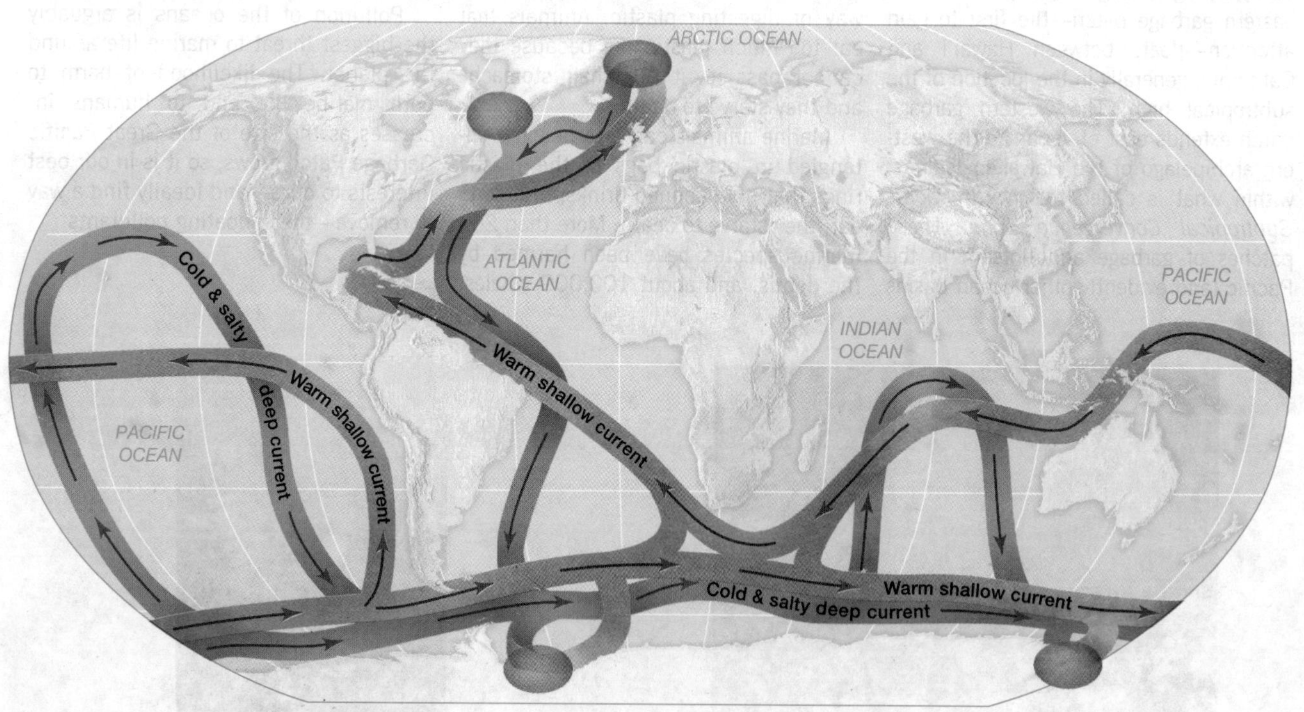

▲ **Figure 9-10** Idealized global conveyor-belt circulation. In the North Atlantic Ocean cool, dense water sinks and moves south as a deep subsurface flow. It joins cold, deep water near Antarctica, eventually moving into the Indian and North Pacific Ocean where the water rises slowly, eventually flowing back into the North Atlantic where it again sinks. The violet ovals show major locations where cool, dense surface water feeds the flow of deep water. One circuit may take many hundreds of years to complete.

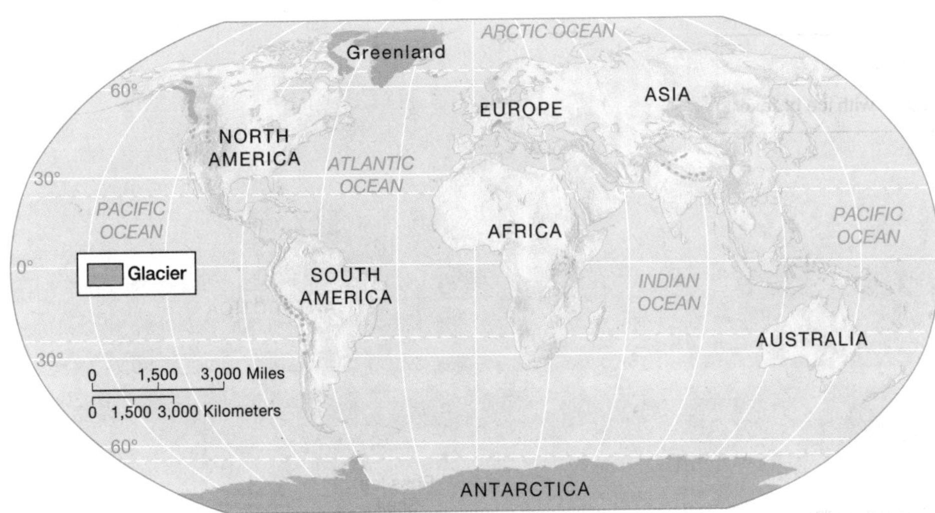

◀ **Figure 9-11** Glacial ice covers about 10 percent of Earth's surface. This ice is confined primarily to Antarctica and Greenland, with small amounts also found at high altitudes in the Canadian Rockies, Andes, Alps, and Himalayas.

Permanent Ice—The Cryosphere

Second only to the world ocean as a storage reservoir for moisture is the solid portion of the hydrosphere—the ice of the world, or *cryosphere*—as we learn from a glance back at Figure 9-1. Although minuscule in comparison with the amount of water in the oceans, the moisture content of ice at any given time is more than twice as large as the combined total of all other types of storage (underground water, surface waters, soil moisture, atmospheric moisture, and biological water).

The ice portion of the hydrosphere is divided between ice on land and ice floating in the ocean, with the land portion being the larger. Ice on land is found as mountain glaciers, ice sheets, and ice caps, all of which are studied in Chapter 19. Approximately 10 percent of the land surface of Earth is covered by ice (Figure 9-11). It is estimated that there is enough water locked up in this ice to feed all the rivers of the world at their present rate of flow for nearly 900 years.

Oceanic ice has various names, depending on size:

- **Ice pack**: An extensive and cohesive mass of floating ice
- **Ice shelf**: A massive portion of a continental ice sheet that projects out over the sea
- **Ice floe**: A large, flattish mass of ice that breaks off from larger ice bodies and floats independently
- **Iceberg**: A chunk of floating ice that breaks off from an ice shelf or glacier (Figure 9-12)

Despite the fact that some oceanic ice freezes directly from seawater, all forms of oceanic ice are composed entirely of freshwater because the salts present in the seawater in its liquid state are not incorporated into ice crystals when that water freezes.

The largest ice pack covers most of the surface of the Arctic Ocean (Figure 9-13); on the other side of the globe, an ice pack fringes most of the Antarctic continent (Figure 9-14). Both of these packs become greatly enlarged during their respective winters, their areas being essentially doubled by increased freezing around their margins. As we saw in Chapter 8, sea ice in the Arctic especially has been diminishing over the last 30 years (see "People and the Environment: Climate Change in the Arctic" in Chapter 8).

There are a few small ice shelves in the Arctic, mostly around Greenland; but several gigantic shelves are attached to the Antarctic ice sheet, most notably the Ross Ice Shelf of some 100,000 square kilometers (40,000 square miles). Some Antarctic ice floes are enormous; the largest ever observed was 10 times as large as the state of Rhode Island.

Over the last two decades, because of increasing temperatures, formerly stable ice shelves in Antarctica have broken apart. Since the early 1990s as much as 8000 square kilometers (over 3000 square miles) of Antarctic

▲ **Figure 9-12** Most of an iceberg is underwater, as shown here in the waters near Svalbard, Norway.

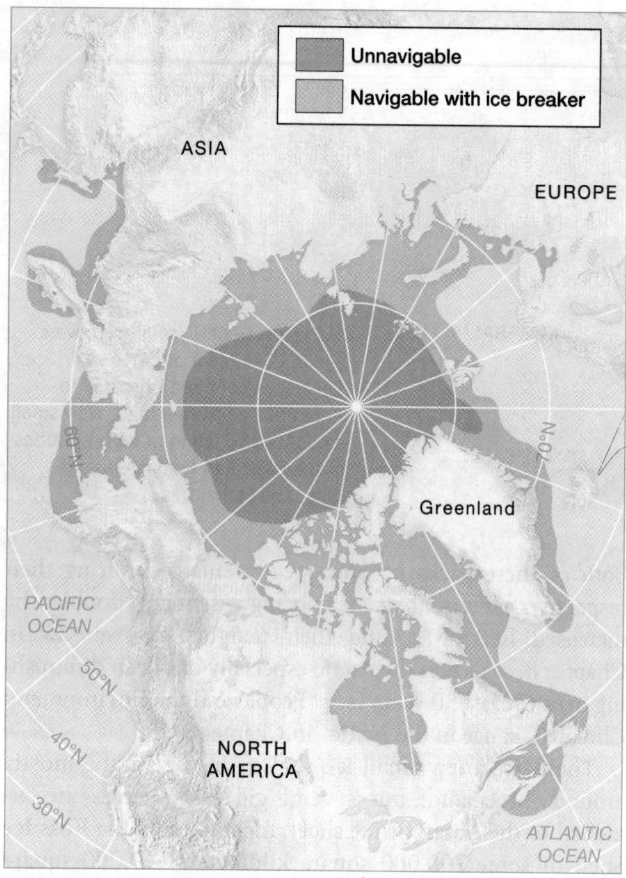

▲ **Figure 9-13** The largest ice pack on Earth covers most of the Arctic Ocean, making that body of water essentially unnavigable. Powerful icebreaker ships allow passage from the Atlantic to the Pacific via this northern route, the fabled "Northwest Passage" of early European explorers. Over the last 30 years the Arctic summer ice pack has diminished by more than 40 percent from its earlier long-term average shown here, and so the Northwest Passage may well become a reality in the future.

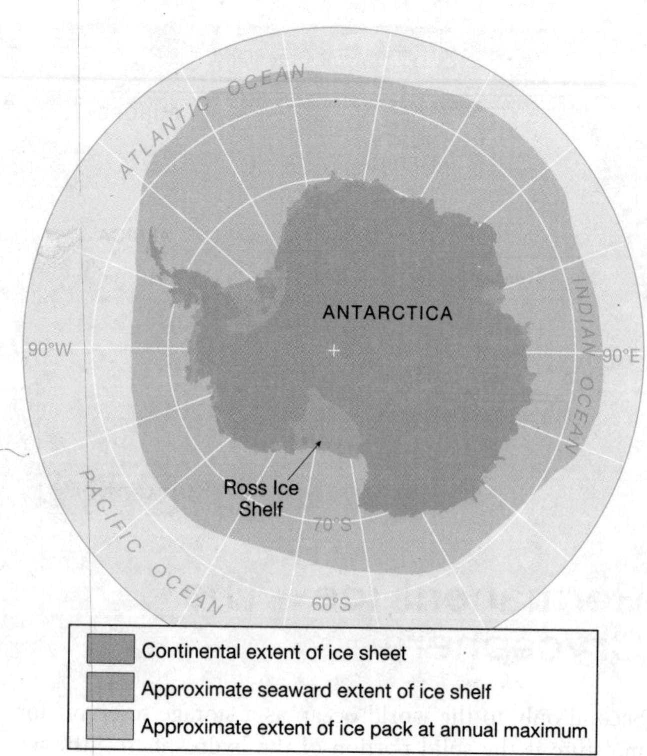

▲ **Figure 9-14** Maximum extent of ice in Antarctica. The ice sheet covers land, and the ice shelf and ice pack are oceanic ice. As is true in the Arctic, over the last three decades the extent of sea ice around Antarctica has diminished from its earlier long-term average shown here.

ice shelves have disintegrated. In 2002, the Larsen-B Ice Shelf on the Antarctic Peninsula disintegrated in less than a month, and the much larger Larsen-C shelf just to the south is showing signs that its mass is being reduced because of increasing water temperatures below it. In 2008, the Wilkins Ice Shelf in Antarctica also began to disintegrate. (See Chapter 19 for further discussion of changes in glaciers and ice sheets around the world.)

Permafrost

A relatively small proportion of the world's ice occurs beneath the land surface as ground ice. This type of ice occurs only in areas where the temperature is continuously below the freezing point, and so it is restricted to high-latitude and high-elevation regions (Figure 9-15). Most permanent ground ice is **permafrost**, which is permanently frozen subsoil. It is widespread in northern Canada, Alaska, and Siberia and found in small patches in many high mountain areas. Some ground ice is aggregated as veins of frozen water, but most of it develops as ice crystals in the spaces between soil particles.

As a response to increasing global temperatures, areas of permafrost in many parts of the world are showing signs of thawing (Figure 9-16). As the ground thaws, buildings, roads, pipelines, and airport runways are increasingly destabilized, and transportation and business are likely to be disrupted as a consequence (see "People and the Environment: Thawing Permafrost in Alaska").

Surface Waters

Surface waters represent only about 0.02 percent of the world's total moisture supply (Figure 9-1), but from the human viewpoint they are of incalculable value. Lakes, swamps, and marshes abound in many parts of the world, and all but the driest parts of the continents are seamed by rivers and streams.

Lakes

Lakes have been called "wide places in rivers." In even simpler terms, a **lake** is a body of water surrounded by land. No minimum or maximum size is attached to this definition, although the word "pond" is often used to designate a very small lake. Well over 90 percent of the nonfrozen surface water of the continents is contained in lakes.

Lake Baykal (often spelled Baikal) in Siberia is by far the world's largest freshwater lake in terms of volume of water, containing considerably more water than the combined contents of all five Great Lakes in the central United States. It is also the world's deepest lake—1742 meters (5715 feet) deep.

▲ **Figure 9-15** Extent of permafrost in the Northern Hemisphere. All the high-latitude land areas and some of the adjacent midlatitude land areas are underlain by permafrost. Climate change is slowly reducing the extent of permafrost.

▲ **Figure 9-16** House in Alaska collapsed as a result of melting permafrost.

Most of the world's lakes contain freshwater, but some of the largest lakes are saline. Indeed, more than 40 percent of the lake water of the planet is salty, with the lake we call the Caspian Sea containing more than three quarters of the total volume of all the world's nonoceanic saline water. (In contrast, Utah's famous Great Salt Lake contains less than 1/2500 the volume of the Caspian.) Any lake that has no natural drainage outlet, either as a surface stream or as a sustained subsurface flow, will become saline.

Most small salt lakes and some large ones are *ephemeral*, which means that they contain water only sporadically and are dry much of the time because they are in dry regions with insufficient inflow to maintain them on a permanent basis. We will discuss ephemeral lakes in desert regions in greater detail in Chapter 18.

Most lakes are fed and drained by streams, but lake origin is usually due to other factors. Two conditions are necessary for the formation and continued existence of a lake: (1) some sort of natural basin having a restricted outlet, and (2) sufficient inflow of water to keep the basin at least partly filled. The water balance of most lakes is maintained by surface inflow, sometimes combined with springs and seeps below the lake surface. A few lakes are fed entirely by springs. Most freshwater lakes have only one stream that serves as a drainage outlet.

Lakes are distributed very unevenly over the land (Figure 9-17 and Table 9-3). They are very common in regions that were glaciated in the recent geologic past

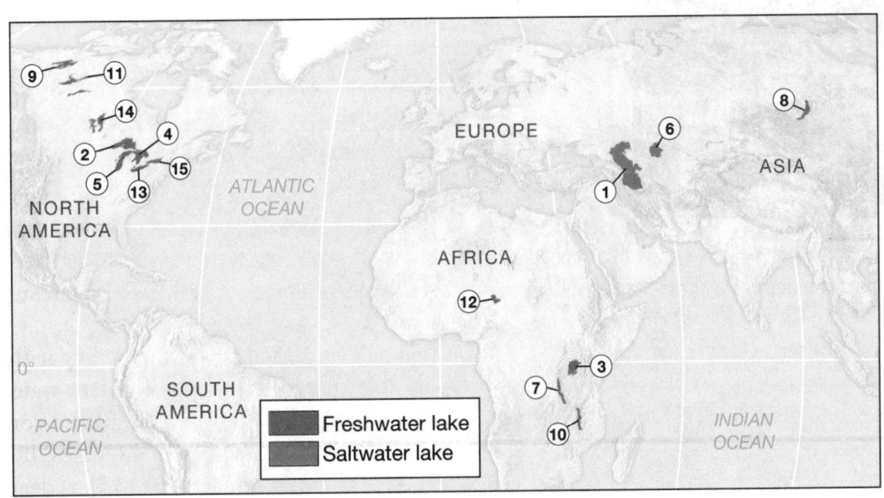

◄ **Figure 9-17** Large lakes of the world: (1) Caspian Sea; (2) Lake Superior; (3) Lake Victoria; (4) Lake Huron; (5) Lake Michigan; (6) Aral Sea (now reduced in size from that shown here); (7) Lake Tanganyika; (8) Lake Baykal; (9) Great Bear Lake; (10) Lake Malawi; (11) Great Slave Lake; (12) Lake Chad (now reduced in size from that shown here); (13) Lake Erie; (14) Lake Winnipeg; (15) Lake Ontario.

PEOPLE AND THE ENVIRONMENT

Thawing Permafrost in Alaska

In locations such as the region around the city of Fairbanks in central Alaska, permanently frozen soil or *permafrost* is widespread just below the surface. During the summer, only the upper 30 to 100 centimeters (12 to 40 inches) of soil thaws in what is called the *active layer*; below that is a layer of permanently frozen ground perhaps 50 meters (165 feet) thick. Much of the permafrost found in the high latitude areas of the world has been frozen for at least the last few thousands of years, but as a response to higher average temperatures, it is beginning to thaw. In just the last 30 years, a warming trend has been observed, bringing the ground temperature in some areas above the melting point of the permafrost. Deep in the permafrost layer where ground still remains frozen, temperatures are rising also (Figure 9-C).

For people used to living in temperate environments, it might seem that having the ground thaw would not be a problem, but such is not the case. As permafrost thaws, it may soften the soil and cause roads and pipelines to deform. In areas with poor surface drainage, the degradation of permafrost can lead to what is called *wet thermokarst* conditions, where the surface subsides and the ground becomes oversaturated with water. In some cases, unpaved roads become impassible. In the last three decades, the number of days that the Alaska Department of Natural Resources permits oil exploration activity in areas of tundra has been cut in half due to the increasingly soft ground.

Along the Beaufort Sea, rising temperatures are thawing permafrost in the coastal bluffs and contributing to more rapid erosion of the coastline. From an average rate of erosion of 6 meters (20 feet) per year between the mid-1950s and 1970s, the rate jumped to nearly 14 meters (45 feet) per year between 2002 and 2007 (Figure 9-D).

As formerly frozen soils thaw, it will likely lead to an increase in the activity of microorganisms in the soil. This could in turn increase the rate of decomposition of organic matter long sequestered in the frozen ground. As this organic matter is decomposed by microorganisms, carbon dioxide or methane can be released, perhaps contributing to increasing greenhouse gas concentrations in the atmosphere.

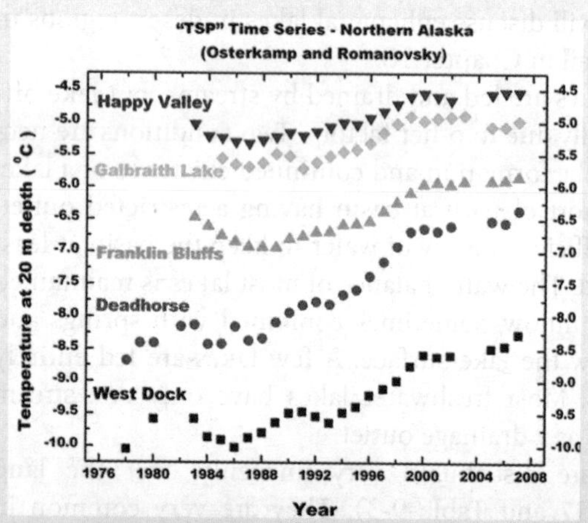

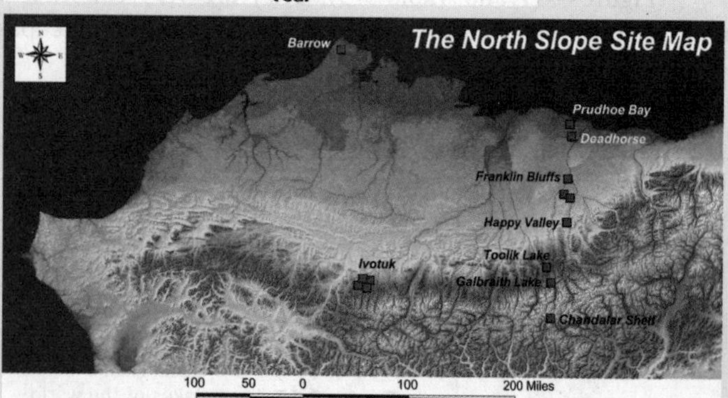

▲ **Figure 9-C** Changes in permafrost temperature at a depth of 20 meters (65 feet) from 1976 to 2008 in Alaska. Sites ranged from the Brooks Range to the North Slope.

▲ **Figure 9-D** Cabin lost to coastal erosion along the Arctic Ocean in Alaska. The increased erosion here is associated with melting permafrost.

because glacial erosion and deposition deranged the normal drainage patterns and created innumerable basins (Figure 9-18; also see Figure 19-28 in Chapter 19). Some parts of the world notable for lakes were not glaciated, however. For example, the remarkable series of large lakes in eastern and central Africa was created by faulting as Earth's crust spread apart tectonically (for example, see Figure 14-14 in Chapter 14); and the many thousands of small lakes in Florida were formed by sinkhole collapse when rainwater dissolved calcium carbonate from the limestone bedrock (for example, see Figure 17-9 in Chapter 17).

Most lakes are relatively temporary features of the landscape. Few have been in existence for more than a few thousand years, a time interval that is momentary in the grand scale of geologic time. Inflowing streams bring sediment to fill lakes up; outflowing streams cut channels progressively deeper to drain them; and as the lake becomes shallower, a continuous increase in plant growth accelerates the infilling (a process discussed in Chapter 10). Thus, the destiny of most lakes is to disappear naturally.

Human Alteration of Natural Lakes Human activity also plays a part in the disappearance of lakes. For example, the diversion of streams flowing into California's Mono Lake (to the east of Yosemite National Park) has reduced its volume by one-half since the 1940s.

More dramatically, the Aral Sea was once the world's fourth largest lake in terms of surface area, but beginning in the 1960s, irrigation projects designed to boost agricultural production in Soviet Central Asia cut off much of the water flowing into the lake. The sea is now only about 10 percent of its original size (Figure 9-19). The once viable commercial fishing industry is gone, and winds now carry away a cloud of choking clay and salt dust lifted from the exposed lake bottom. The sea has split into several pieces. A new dam on the Syr Darya River is allowing the northern remnant of the Aral Sea to recover slightly, but the southern remnants are likely to remain dry.

In some cases, both human and natural changes are responsible for the loss of a lake. Forty years ago, Lake Chad was one of the largest lakes in Africa, but ongoing drought has reduced it to about 10 percent of its original size (Figure 9-20). Nearly all of the lake's water comes from the Chari River flowing into the lake from the south. The lake is shallow and surrounded by an extensive wetlands area—once the second largest in Africa. Because the lake is shallow, it responds quite quickly to changes in inflow. Although water diversion projects along the Chari River have contributed to the reduction of Lake Chad, climate change in the region is likely responsible for much of its ongoing decline.

Reservoirs One of the most notable things people have done to alter the natural landscape is to produce artificial lakes, or *reservoirs*. Such lakes have been created largely by the construction of dams, ranging from small earth mounds heaped across a gully to immense concrete structures blocking the world's major rivers (Figure 9-21). Some reservoirs are as large as medium-sized natural lakes.

Reservoirs are constructed for a number of different reasons, including controlling floods, ensuring a stable agricultural or municipal water supply, and for the generation of hydroelectric power—frequently for all of these reasons. The creation of artificial lakes has had immense ecological and economic consequences, not all of them foreseen at the time of construction (Figure 9-22). In addition to the obvious loss of the land that has been inundated by the waters of the reservoir, downstream ecosystems may be altered by restricted stream flows; and in some locations rapid sedimentation may restrict the useful life of a reservoir.

TABLE 9-3 The World's Largest Lakes Ranked by Surface Area

Rank	Name	Continent	Area		Water
			Square Kilometers	Square Miles	
1	Caspian Sea	Eurasia	372,450	143,250	Salt
2	Lake Superior	North America	82,420	31,700	Fresh
3	Lake Victoria	Africa	69,400	26,700	Fresh
4	Lake Huron	North America	59,800	23,000	Fresh
5	Lake Michigan	North America	58,000	22,300	Fresh
6	Lake Tanganyika	Africa	33,000	12,650	Fresh
7	Lake Baykal	Eurasia	31,700	12,200	Fresh
8	Great Bear Lake	North America	31,500	12,100	Fresh
9	Lake Malawi	Africa	30,000	11,550	Fresh
10	Great Slave Lake	North America	29,400	11,300	Fresh

► **Figure 9-18** Glaciation is responsible for the formation of St. Mary Lake in Montana's Glacier National Park.

▲ **Figure 9-19** In 1960, the Aral Sea was the world's fourth largest lake. However, as irrigation needs have forced farmers to drain more and more water from the rivers feeding the Aral, its size has decreased by 90 percent in the last 40 years. (a) The Aral Sea in 1973 (left) and in 2009 (right). (b) Where fish once swam, camels now wander over the floor of the "sea." If present trends continue, the Aral may cease to exist.

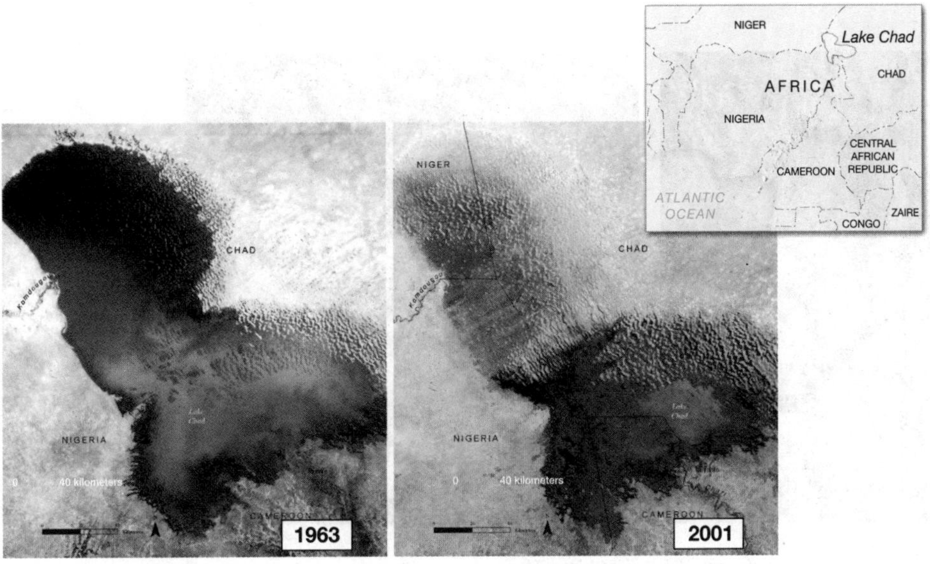

◀ **Figure 9-20** Lake Chad in 1963 and 2001.

▲ **Figure 9-21** Hoover Dam and Lake Mead on the Colorado River.

In Chapter 16, we will consider the implications of flood control through the use of dams and river levees.

Swamps and Marshes

Closely related to lakes but less numerous and containing a much smaller volume of water are swamps and marshes, flattish places that are submerged in water at least part of the time but are shallow enough to permit the growth of water-tolerant plants (Figure 9-23). The conceptual distinction between the terms is that a **swamp** has a plant growth that is dominantly trees, whereas a **marsh** is vegetated primarily with grasses and rushes. Both are usually associated with coastal plains, broad river valleys, or recently glaciated areas. Sometimes they represent an intermediate stage in the infilling of a lake.

Rivers and Streams

Although containing only a small proportion of the world's water at any given time, rivers and streams are an extremely dynamic component of the hydrologic cycle. (Although the terms are basically interchangeable, in common usage a *stream* is smaller than a *river*; geographers, however, call any flowing water a stream, no matter what its size.) Streams provide the means by which the land surface drains and by which water, sediment, and dissolved chemicals are moved ever seaward. The occurrence of rivers and streams is closely, but not absolutely, related to precipitation patterns. Humid lands have many rivers and streams, most of which flow year-round; dry lands have fewer, almost all of which are ephemeral (which means they dry up for part of the year).

Table 9-4 lists the world's largest rivers by discharge volume and length, while Figure 9-24 shows the world's largest river drainage basins (a *drainage basin* is all the land area drained by a river and its tributaries.) A mere two dozen great rivers produce one-half of the total stream discharge of the world. The mighty Amazon yields nearly

▲ **Figure 9-22** The Three Gorges Dam and reservoir on the Yangtze River in China. The Landsat satellite image on the left was taken on April 17, 1987, before the dam was constructed; the image on the right was taken May 9, 2004, after the reservoir behind the 2300-meter-wide dam had started to fill. As many as 1 million people will be displaced by the 600-kilometer-long reservoir.

▲ **Figure 9-23** Marshes are particularly numerous along the poorly drained South Atlantic and Gulf of Mexico coasts of the United States. This is Loxahatchee Marsh in Florida.

20 percent of the world total, more than five times the discharge of the second-ranking river, the Congo River. Indeed, the discharge of the Amazon is three times as great as the total combined discharge of all rivers in the United States. The Mississippi is North America's largest river by far, with a drainage basin that encompasses about 40 percent of the total area of the 48 conterminous states and a flow that amounts to about one-third of the total discharge from all other rivers of the nation.

We will explore the ways in which streams shape the landscape of the continents in Chapter 16.

Underground Water

1. The Water Table
2. Groundwater Cone of Depression

Beneath the land surface is another important component of the hydrosphere—underground water. As Figure 9-1 shows, the total amount of underground water is many times that contained in lakes and streams. Moreover, underground water is much more widely distributed than surface water. Whereas lakes and rivers are found only in restricted locations, underground water is almost ubiquitous, occurring beneath the land surface throughout the world. Its quantity is sometimes limited, its quality is sometimes poor, and its occurrence is sometimes at great depth, but almost anywhere on Earth one can dig deep enough and find water.

More than half of the world's underground water is found within 800 meters (about half a mile) of the surface. Below that depth, the amount of water generally decreases gradually and erratically. Although water has been found at depths below 10 kilometers (6 miles), it is almost immobilized because the pressure exerted by overlying rocks is so great and openings are so few and small.

Almost all underground water comes originally from above. Its source is precipitation that either percolates directly into the soil or eventually seeps downward from lakes and streams.

Once the moisture gets underground, any one of several things can happen to it depending largely on the nature of the soil and rocks it infiltrates. The quantity of water that can be held in subsurface material (rock or soil) depends

TABLE 9-4 The World's Largest Rivers by Discharge and Length

Name	Rank by Discharge Volume	Rank by Length	Approximate Discharge (cubic meters per second)	Approximate Length (kilometers)	Approximate Drainage Area (square kilometers)	Continent
Amazon	1	2	210,000	6400	5,800,000	South America
Congo	2	9	40,000	4700	4,000,000	Africa
Ganges-Brahmaputra	3	23	39,000	2900	1,730,000	Eurasia
Yangtze	4	3	21,000	6300	1,900,000	Eurasia
Paranà-La Plata	5	8	19,000	4900	2,200,000	South America
Yenisey	6	5	17,000	5550	2,600,000	Eurasia
Mississippi-Missouri	7	4	17,000	6000	3,200,000	North America
Orinoco	8	27	17,000	2700	880,000	South America
Nile	25	1	5000	6650	2,870,000	Africa

Note: Estimates of river lengths and discharges are approximate and so the relative rankings of some rivers vary from one data source to another.

on the **porosity** of the material, which is the percentage of the total volume of the material that consists of voids (pore spaces or cracks) that can fill with water. The more porous a material is, the greater the amount of open space it contains and the more water it can hold.

Porosity is not the only factor affecting underground water flow. If water is to move through rock or soil, the pores must be connected to one another and be large enough for the water to move through them. The ability to transmit underground water (as opposed to just hold it) is termed **permeability**, and this property of subsurface matter is determined by the size of pores and by their degree of interconnectedness. The water moves by twisting and turning through these small, interconnected openings.

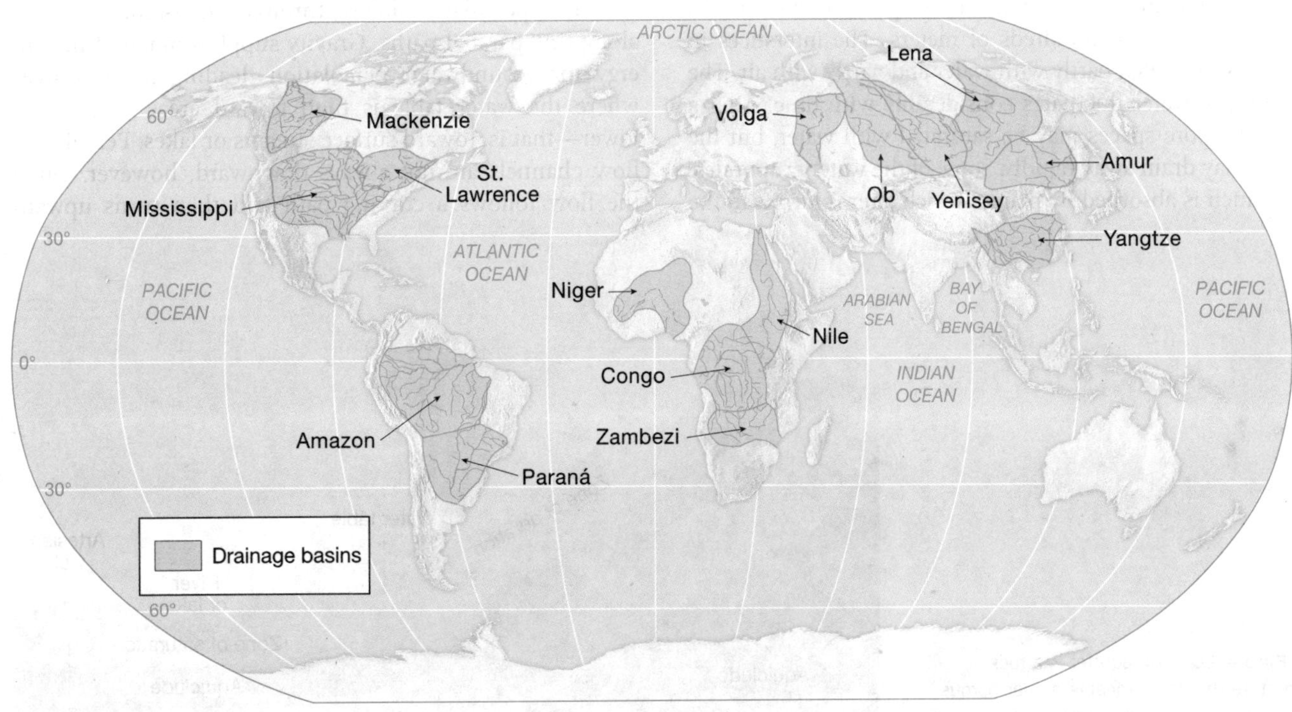

▲ **Figure 9-24** The world's largest drainage basins are scattered over the four largest continents in all latitudes.

The smaller and less connected the pore spaces, the less permeable the material and the slower the water moves.

The rate at which water moves through rock depends on both porosity and permeability. For example, clay is usually of high porosity because it has a great many *interstices* (openings) among the minute flakes that make up the clay, but it generally has low permeability because the interstices are so tiny that the force of molecular attraction binds the water to the clay flakes and holds it in place. Thus, clay is typically very porous but relatively impermeable and consequently can trap large amounts of water and keep it from draining.

Underground water is stored in, and moves slowly through, moderately to highly permeable rocks called **aquifers** (from the Latin, *aqua*, "water," and *ferre*, "to bear"). The rate of movement of the water varies with the situation. In some aquifers, the flow rate is only a few centimeters a day; in others, it may be several hundred meters per day. A "rapid" rate of flow would be 12 to 15 meters (40 to 50 feet) per day.

Impermeable materials composed of components such as clay or very dense unfractured rock, which hinder or prevent water movement, are called **aquicludes** (Figure 9-25).

The general distribution of underground water can probably best be understood by visualizing a vertical subsurface cross section. Usually at least three and often four hydrologic zones are arranged one below another. From top to bottom, these layers are called the *zone of aeration*, the *zone of saturation*, the *zone of confined water*, and the *waterless zone*.

Zone of Aeration

The topmost band, the **zone of aeration**, is a mixture of solids, water, and air. Its depth can be quite variable, from a few centimeters to hundreds of meters. The interstices in this zone are filled partly with water and partly with air. The amount of water fluctuates considerably with time. After a rain, the pore spaces may be saturated with water, but the water may drain away rapidly. Some of the water evaporates, but much is absorbed by plants, which later return it to the atmosphere by transpiration. Water that molecular attraction cannot hold seeps downward into the next zone.

Zone of Saturation

Immediately below the zone of aeration is the **zone of saturation**, in which all pore spaces in the soil and cracks in the rocks are fully saturated with water. The moisture in this zone is called **groundwater**; it seeps slowly through the ground following the pull of gravity and guided by rock structure. The top of the saturated zone is referred to as the **water table**. The orientation and slope of the water table usually conform roughly to the slope of the land surface above, nearly always approaching closer to the surface in valley bottoms and being more distant from it beneath a ridge or hill. Where the water table intersects Earth's surface, water flows out, forming a spring. A lake, swamp, marsh, or permanent stream is almost always an indication that the water table reaches the surface there. In humid regions, the water table is higher than in arid regions, which means that the zone of saturation is nearer the surface in humid regions. Some desert areas have no saturated zone at all.

Sometimes a localized zone of saturation develops above an aquiclude, and this configuration forms a *perched water table*.

A well dug into the zone of saturation fills with water up to the level of the water table. When water is taken from the well faster than it can flow in from the saturated rock, the water table drops in the immediate vicinity of the well in the approximate shape of an inverted cone. This striking feature is called a **cone of depression** (Figure 9-26). If many wells are withdrawing water faster than it is being replenished naturally, the water table may be significantly depressed over a large area.

Water percolates slowly through the saturated zone along tiny parallel paths. Gravity supplies much of the energy for groundwater percolation, leading it from areas where the water table is high toward areas where it is lower—that is, toward surface streams or lakes. Percolation flow channels are not always downward, however. Often the flow follows a curving path and then turns upward

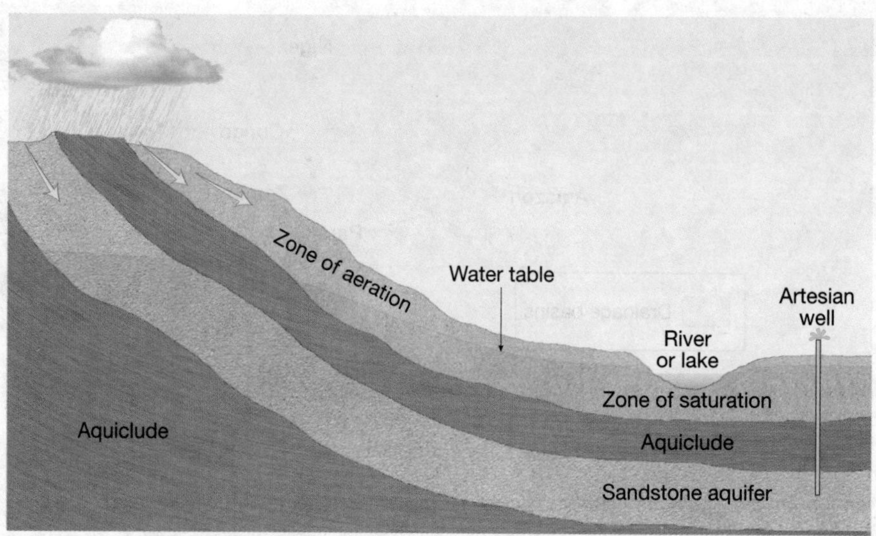

▶ **Figure 9-25** An aquifer is a rock structure that is permeable and/or porous enough to hold water, whereas an aquiclude has a structure that is too dense to allow water to penetrate it.

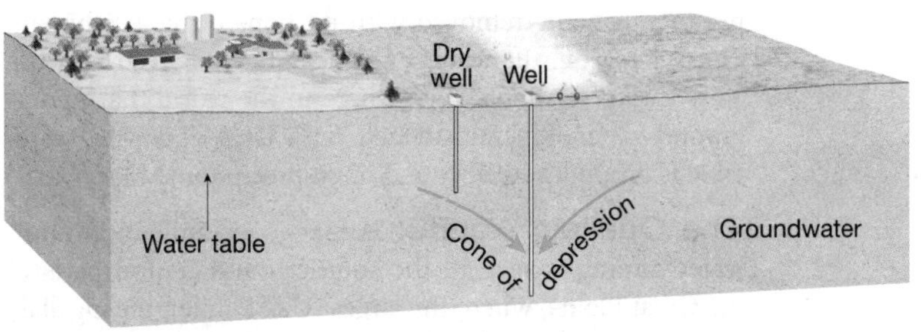

◀ **Figure 9-26** If water is withdrawn from a well faster than it can be replenished, a cone of depression will develop. This can effectively lower the water table over a large area. Nearby shallow wells may run dry because they lie above the lowered water table.

(against the force of gravity) to enter the stream or lake from below. This trajectory is possible because saturated-zone water at any given height is under greater pressure beneath a hill than beneath a stream valley. Thus, the water moves toward points where the pressure is least.

The lower limit of the zone of saturation is marked by the absence of pore spaces and therefore the absence of water. This boundary may be a single layer of impermeable rock, or it may simply be that the increasing depth has created so much pressure that no pore spaces exist in any rocks at that level.

Zone of Confined Water

In many, but not most, parts of the world, a third hydrologic zone lies beneath the zone of saturation, separated from it by impermeable rock. This **zone of confined water** contains one or more aquifers into which water can infiltrate (Figure 9-27). Sometimes aquifers alternate with impermeable layers (aquicludes). Water cannot penetrate an aquifer in this deep zone by infiltration from above because of the impermeable barrier, and so any water the zone contains must have percolated along the aquifer from a more distant area where no aquiclude interfered. Characteristically, then, an aquifer in the confined water zone is a sloping or dipping layer that reaches to, or almost to, the surface at some location, where it can absorb infiltrating water. The water works its way down the sloping aquifer from the catchment area, building up considerable pressure in its confined situation.

If a well is drilled from the surface down into the confined aquifer, which may be at considerable depth, the confining pressure forces water to rise in the well. The elevation to which the water rises is known as the **piezometric surface**. In some cases, the pressure is enough to allow the water to rise above the ground, as shown in Figure 9-28. This free flow of water is called an **artesian well**. If the confining pressure is sufficient to push the water only partway to the surface and it must be pumped the rest of the way, the well is *subartesian*.

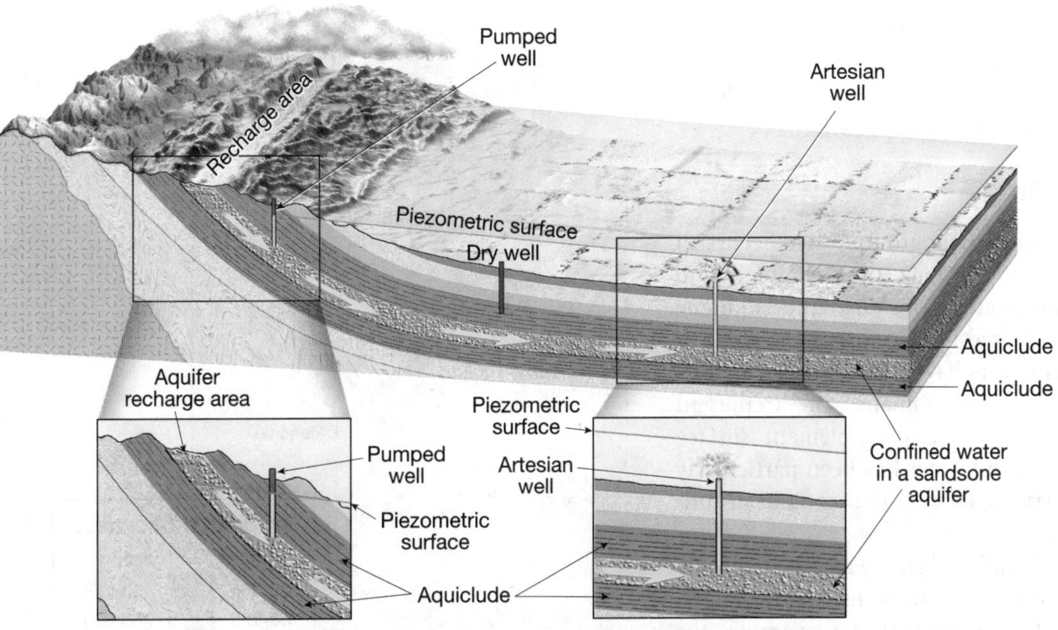

▲ **Figure 9-27** An artesian system. Surface water penetrates the aquifer in the recharge area and infiltrates downward. It is confined to the aquifer by impermeable strata (aquicludes) above and below. If a well is dug through the upper aquiclude into the aquifer, the confining pressure forces the water to rise in the well to the level of the piezometric surface. In an artesian well, the pressure forces the water to the surface; in a subartesian well, the water is forced only part way to the surface and must be pumped the rest of the way.

▲ **Figure 9-28** An artesian well in Australia's Great Artesian Basin, the largest and most productive source of confined water in the world. This scene shows the Pilliga Bore in northwestern New South Wales.

Unlike the distribution of groundwater, which is closely related to precipitation, the distribution of confined water is quite erratic over the world. Confined water underlies many arid or semiarid regions that are poor in surface water or groundwater, thus providing a critical resource for these dry lands.

Waterless Zone

At some depth below the surface, there is no water because the overlying pressure increases the density of the rock and so there are no pores. This *waterless zone* generally begins several kilometers beneath the land surface.

Groundwater Mining

In most parts of the world where groundwater occurs, it has been accumulating for a long time. Rainfall and snowmelt seep and percolate downward into aquifers, where the water may be stored for decades or centuries or millennia. Only in recent years have most of these aquifers been discovered and tapped by humans. They represent valuable sources of water that can supplement surface water resources. Underground water has been particularly utilized by farmers to irrigate in areas that contain insufficient surface water.

The accumulation of underground water is tediously slow. Its use by humans, however, can be distressingly rapid. In many parts of the U.S. Southwest, for example, the recharge (replenishment) rate averages only 0.5 centimeter (0.2 inch) per year, but it is not uncommon for a well to pump 75 centimeters (30 inches) per year. Thus, yearly pumpage is equivalent to 150 years' recharge. This rate of groundwater use can be likened to mining because a finite resource is being removed with no hope of replenishment. For this reason, the water in some aquifers is referred to as "fossil water." Almost everywhere in the world that underground water is being utilized on a large scale, the water table is dropping steadily and often precipitously.

The Ogallala Aquifer A classic example of groundwater mining is seen in the southern and central parts of the Great Plains, where the largest U.S. aquifer, the Ogallala or High Plains Aquifer, underlies 585,000 square kilometers (225,000 square miles) of eight states. The Ogallala formation consists of a series of limey and sandy layers that function as a gigantic underground reservoir ranging in thickness from a few centimeters in parts of Texas to more than 300 meters (1000 feet) under the Nebraska Sandhills (Figure 9-29). Water has been accumulating in this aquifer for some 30,000 years. At the midpoint of the twentieth century, it was estimated to contain 1.4 billion acre-feet (1.7 quadrillion liters or 456 trillion gallons) of water, an amount roughly equivalent to the volume of one of the larger Great Lakes.

Farmers began to tap the Ogallala in the early 1930s. Before the end of that decade, the water table was already dropping. After World War II, the development of high-capacity pumps, sophisticated sprinklers, and other technological innovations encouraged the rapid expansion of irrigation based on Ogallala water. Water use in the region has almost quintupled since 1950. The results of this accelerated usage have been spectacular. Above ground, there has been a rapid spread of high-yield farming into

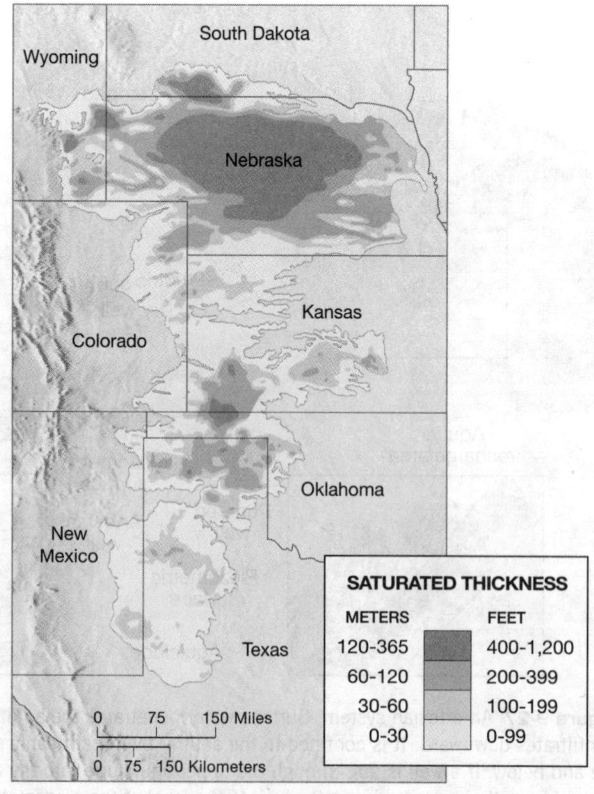

▲ **Figure 9-29** The Ogallala or High Plains Aquifer. Darker areas indicate greater thickness of the water-bearing strata.

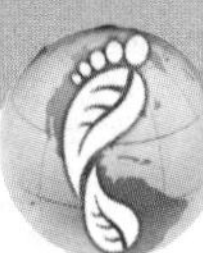

PEOPLE AND THE ENVIRONMENT

Subsidence from Groundwater Extraction

One of the consequences of prolonged groundwater extraction can be the compaction of sediments in an aquifer. If groundwater is extracted at a much greater rate than it is being recharged, not only will the water table drop, but the surface of land itself can subside.

Among the greatest measured amounts of subsidence from groundwater extraction have taken place in California's San Joaquin Valley. Over a 50-year period, from roughly 1925 to 1975, so much groundwater was drawn from the aquifer below—largely for use in crop irrigation—that the water table dropped as much as 120 meters (about 400 feet); the land itself subsided more than 8.5 meters (almost 29 feet) in some places (Figure 9-E). The pace of groundwater pumping and land subsidence eased somewhat after that—in part because of greater availability of surface water from major water projects. However, by June 2009 concerns were raised that groundwater pumping would lead to enough subsidence that the California Aqueduct, which flows across the western San Joaquin Valley and supplies the southern part of the state with some of its municipal water, is itself subsiding and could buckle, disrupting water delivery. During previous dry spells in the San Joaquin Valley when groundwater pumping increased, land under parts of the aqueduct subsided nearly 2 meters (6 feet), slowing the flow of water and forcing repairs.

Las Vegas, Nevada, is also experiencing substantial subsidence from groundwater extraction. Since the 1950s the land surface in the Las Vegas Valley has subsided by as much as 2 meters (about 6 feet). From 1992 to 1997 alone, some areas in the valley subsided 20 centimeters (8 inches). As a result of the subsidence, fissures have developed on the surface and well casings have been damaged.

New technology permits the monitoring of ground subsidence from space using Satellite Interferometric Synthetic Aperture Radar (InSAR). Over a period of time, radar signals from a satellite are bounced off the ground from the same location overhead. By measuring the difference in distance over a period of months or years, ground subsidence or uplift can be measured with a resolution of 5 to 10 millimeters (1/4 to 1/2 inch). Maps showing the relative change in ground level are called interferograms (Figure 9-F).

Such InSAR interferograms have enabled scientists to not only monitor changes in ground level but also to recognize new faults (breaks in a rock structure along which there has been movement). Faults may be revealed in interferograms because in some locations ground subsidence is at least partly controlled by the orientation of subsurface fault planes.

▲ **Figure 9-E** Parts of California's San Joaquin Valley subsided about 8.5 meters (29 feet) between 1925 and 1975 as a consequence of sediment compaction from groundwater extraction. The signs on the utility pole mark the level on the ground from 1925 to 1977.

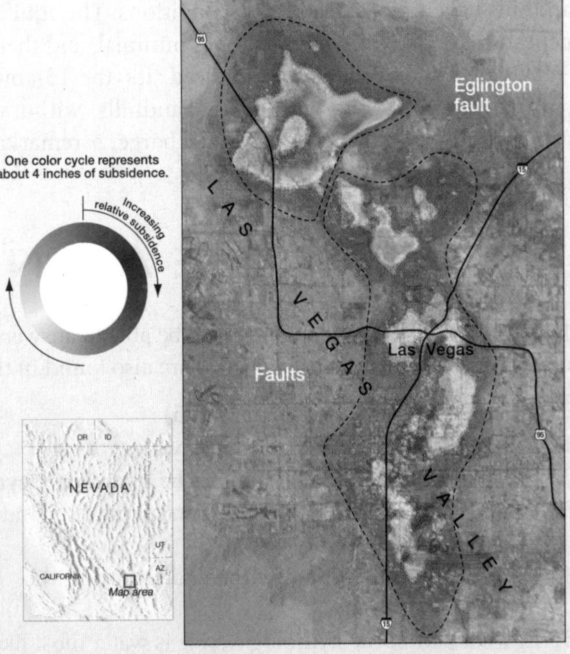

▲ **Figure 9-F** InSAR-derived maps showing ground subsidence due to groundwater pumping in the Las Vegas Valley between April 1992 and December 1997.

areas never before cultivated (especially in Nebraska) and a phenomenal increase in irrigated crops in all eight Ogallala states. Beneath the surface, however, the water table is sinking ever deeper—dropping as much as 30 meters (100 feet) over large areas (Figure 9-30). Farmers who once obtained water from 15-meter (50-foot) wells now must bore to 45 or 75 meters (150 or 250 feet), and as the price of energy increases, the cost of pumping increases operating expenses enormously and so extraction has dropped slightly over the last few years. Some 170,000 wells tapped the Ogallala in the 1970s, but thousands of those have now been abandoned.

Some farmers have been shifting to crops that require less water. Others are adopting water- and energy-conserving measures that range from a simple decision to irrigate less frequently to the installation of sophisticated machinery that uses water in the most efficient fashion. Many farmers have faced or will soon face the prospect of abandoning irrigation entirely. During the next four decades, it is estimated that 2 million hectares (5 million acres) now irrigated will revert to dry-land production. Other farmers concentrate on high-value crops before it is too late, hoping to make a large profit and then get out of farming.

Water conservation is further complicated by the obvious fact that groundwater is no respecter of property boundaries. A farmer who is very conservative in his or her water use must face the reality that less careful neighbors are pumping from the same aquifer and that their profligacy may seriously diminish the water available to everyone.

The situation varies from place to place. The Nebraska Sandhills have the most favorable conditions. The aquifer is deepest there, previous water use was minimal, and there is a relatively rapid recharge rate. Indeed, for the 13-county area that makes up the bulk of the Sandhills, withdrawal averages only about 10 percent of recharge, a remarkable situation. In contrast, the 13 counties of southwestern Kansas have a withdrawal rate more than 20 times the recharge rate—a clearly unsustainable situation.

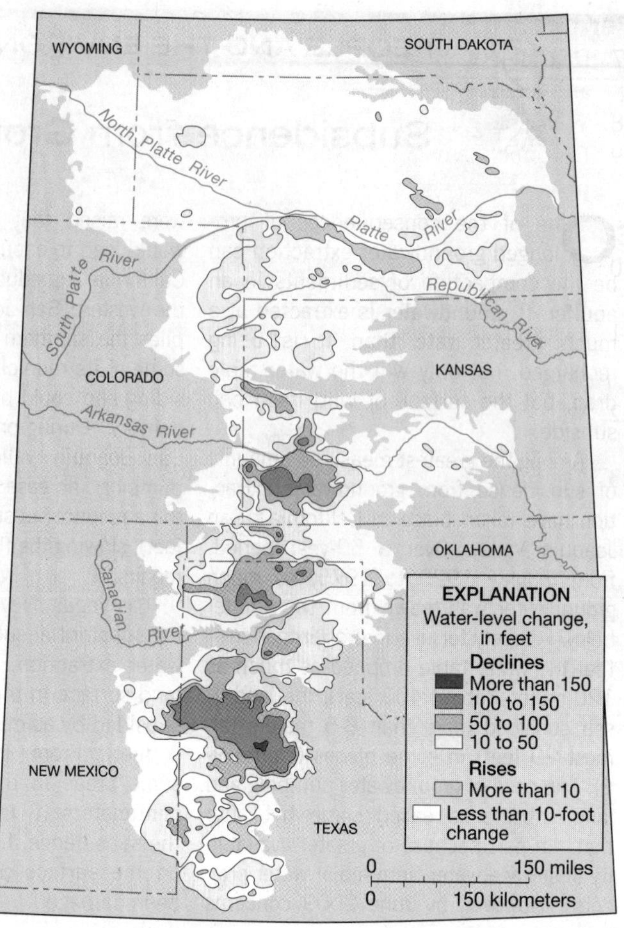

▲ **Figure 9-30** Water table change in the Ogallala Aquifer from the early twentieth century through 1997. Increases in the water table occurred in places where surface water irrigation has been extensive.

CHAPTER 9 LEARNING REVIEW

After studying this chapter, you should be able to answer the following questions. Key terms from each text section are shown in bold type. Definitions for key terms are also found in the glossary at the back of the book.

The Hydrologic Cycle (p. 235)

1. Where is most of the world's freshwater found?
2. Explain the role of evaporation in the **hydrologic cycle**.
3. What is the relationship between transpiration and evaporation?
4. Describe the roles of advection and **runoff** in the hydrologic cycle.
5. In what part of the hydrologic cycle is water most likely to stay for a very short time? A very long time? Why?

The Oceans (p. 237)

6. "How many oceans are there?" Why is this a difficult question to answer?
7. Is the Pacific Ocean significantly different from other oceans? Explain.
8. Why does **salinity** vary in different parts of the world ocean?
9. Why are the oceans becoming slightly more acidic?

Movement of Ocean Waters (p. 240)

10. Why do most oceanic areas experience two high **tides** and two low tides each day?
11. What is meant by the **tidal range** of a coastal location?
12. Distinguish between **flood tide** and **ebb tide**.
13. Describe and explain **spring tides** and **neap tides**.
14. What is a **tidal bore**?
15. What is **thermohaline circulation**?
16. Explain the **global conveyor belt circulation**.

Permanent Ice—The Cryosphere (p. 245)

17. Distinguish among an **ice pack**, **ice shelf**, **ice floe**, and **iceberg**.
18. Why does all sea ice consist of freshwater?
19. Describe the characteristics and global distribution of **permafrost**.

Surface Waters (p. 246)

20. Why are most lakes considered to be "temporary" features of the landscape?
21. Distinguish among a **lake**, a **swamp**, and a **marsh**.

Underground Water (p. 252)

22. What is the difference between **porosity** and **permeability**?
23. Contrast an **aquifer** with an **aquiclude**.
24. Briefly define the following terms: **zone of aeration**, **zone of saturation**, **groundwater**.
25. Explain the concept of a **water table**.
26. Describe and explain the cause of a **cone of depression**.
27. Under what circumstances can a **zone of confined water** develop?
28. What is meant by the **piezometric surface**?
29. Distinguish between an **artesian well** and a subartesian well.
30. Why is the water from some aquifers referred to as fossil water?

Log in to **www.mygeoscienceplace.com** for animations, videos, RSS feeds, maps, and self-study quizzes to enhance your study of **The Hydrosphere**.

10 Cycles and Patterns in the Biosphere

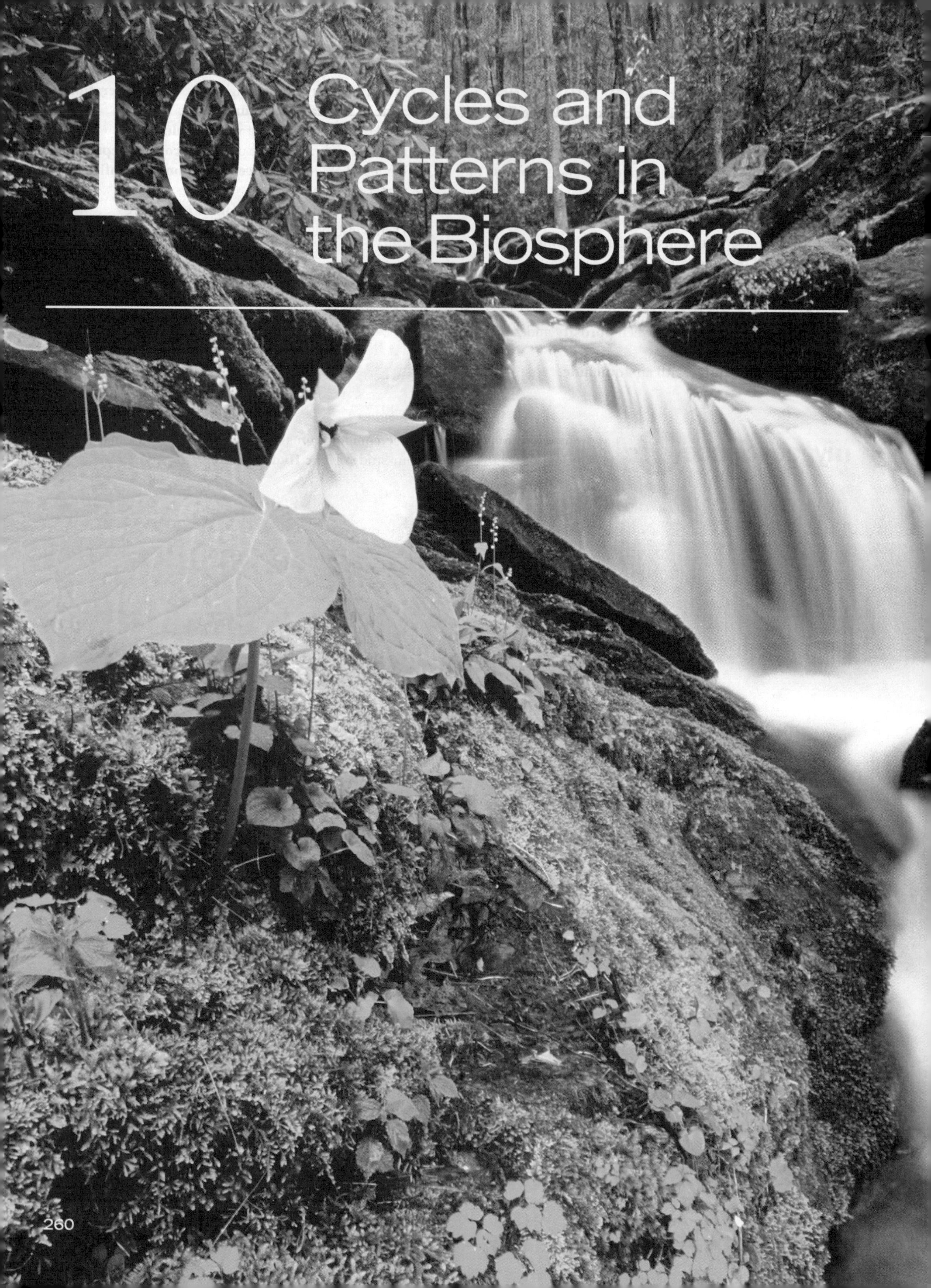

Of the four principal components of our earthly environment, the biosphere has the boundaries that are hardest to pin down. The atmosphere consists of the envelope of air that surrounds the planet, the lithosphere is the solid portion, and the hydrosphere encompasses the various forms of water. These three spheres are distinct from one another and easy to visualize. The biosphere, on the other hand, impinges spatially on the other three. It consists of the incredibly numerous and diverse array of organisms that populate our planet—most obviously plants and animals, but also less obvious types of organisms such as bacteria and fungi.[1] Most of these organisms exist at the interface between atmosphere and lithosphere, but some live largely or entirely within the hydrosphere or the lithosphere, and others move relatively freely from one sphere to another.

The Impact of Plants and Animals on the Landscape

Originally, vegetation grew in profusion over most of the land surface of the planet. Today, native plants are still widespread in parts of the world sparsely populated by humans, as Figure 10-1a shows. However, much of the vegetation in populated areas has been removed, and much that persists has been adulterated and modified by human introduction of crops, weeds, and ornamental plants (Figure 10-1b). Native animal life is much less apparent in the landscape than plant life and is often more conspicuous by sound (especially birds and insects) than by sight. Still we should keep in mind that wildlife usually is shy and reclusive, and its absence may be more apparent than real. Moreover, most species of animals are tiny and therefore less noticeable in our normal scale of vision.

Both plants and animals interact with other components of the natural landscape (such as soil, landforms, and water) and sometimes become important influences in the development and evolution of these components.

The Geographic Approach to the Study of Organisms

Even the simplest living organism is an extraordinarily complex entity. When a student sets out to learn about an organism—be it alga or anteater, tulip or turtle—she or he embarks on a complicated quest for knowledge. An organism differs in many ways from other aspects of the environment, but most significantly in that it is alive and its survival depends on an enormously intricate set of life processes. As beneficial as a complete understanding of the world's organisms might be, as with every other feature of the world, the geographer must focus on certain aspects rather than on a complete comprehension of the whole.

The geographic viewpoint is that of broad understanding, whether we are dealing with plants and animals or with anything else. This does not mean that we ignore the individual organism; rather it means that we seek generalizations and patterns and assess their overall significance. Here, as elsewhere, the geographer is interested in distributions and relationships. A major part of geography's distinctiveness is that it is an integrative discipline that studies phenomena from a broad-scale viewpoint. **Biogeography**, then, is the study of the distribution patterns of living organisms and how these patterns change over time.

The Search for a Meaningful Classification Scheme

When a geographer studies any set of phenomena, he or she attempts to group individual members of the set in some meaningful fashion. In some cases, the geographer borrows classification schemes from specialists in other disciplines. Often, however, schemes devised for other purposes are not particularly useful for geographic studies, and the geographer must develop different ones.

The systematic study of plants and animals is primarily the domain of the biologist, and many biological classification schemes have been devised. By far the most significant and widely used is the *binomial* (two-name) system originally developed by the Swedish botanist Carolus Linnaeus in the eighteenth century. This system focuses primarily on the morphology (structure and form) of organisms and groups them on the basis of structural similarity (see Appendix VI). The Linnaean scheme is generally useful for geographers, but it has certain shortcomings that preclude its total acceptance. Its principal disadvantage for geographic use is that it is based entirely on anatomic similarities, whereas geographers are more interested in distribution patterns and habitat preferences.

It would be nice to be able to say that geographers have come up with a more appropriate classification scheme, but such is not the case, nor is a universally accepted geographic classification of organisms ever likely to be developed. Too many subjective decisions would have to be made, making widespread agreement on any scheme very unlikely.

Seeking Pertinent Patterns

Among the life-forms of our planet are perhaps 600,000 species of plants and more than twice that many species of animals. With such an overwhelming diversity of organisms, how can we study their distributions and relationships

[1]Biologists now generally recognize six kingdoms of living organisms: *Plantae*, *Animalia*, *Archaea*, *Eubacteria*, *Protista*, and *Fungi*. See Appendix VI.

◀ White erect trillium flower in Great Smoky Mountains National Park, Tennessee.

▲ **Figure 10-1** (a) Many parts of Earth are still covered with native vegetation, with little or no human impact in evidence. This forested scene is near Marlinton, West Virginia. (b) A small natural depression of wetland remains, but most of the natural vegetation in this scene has been displaced by crops. This is wheat country in the Canadian province of Alberta.

in any meaningful manner? A logical approach is to decide on some generalizing procedures and useful groupings and then consider the patterns that emerge.

The term **biota** refers to the total complex of plant and animal life. The basic subdivision of biota separates **flora**, or plants, from **fauna**, or animals. In this book, we recognize a further fundamental distinction—between oceanic biota and terrestrial (living on land; *terrenus* is Latin for "earth") biota.

The inhabitants of the oceans are generally divided into three groups—*plankton* (floating plants and animals), *nekton* (animals such as fish and marine mammals that swim freely), and *benthos* (animals and plants that live on or in the ocean bottom). Although these marine life-forms are fascinating, and despite the fact that 70 percent of Earth's surface is oceanic, in this book we pay scant attention to oceanic biota primarily because of constraints of time and space. The terrestrial biota will be our primary focus of interest for most of our study of the biosphere. We begin our study of biogeography with a look at the broadest patterns and cycles in the biosphere—biogeochemical cycles.

Biogeochemical Cycles

The web of life comprises a great variety of organisms coexisting in a diversity of associations. Processes and interactions within the biosphere are exceedingly intricate. Organisms survive only through a bewildering complex of systemic flows of energy, water, and nutrients. These flows are different in different parts of the world, in different seasons of the year, and under various local circumstances (Figure 10-2).

It is generally believed that, for the last billion years or so, Earth's atmosphere and hydrosphere have been composed of approximately the same balance of chemical components we live with today. This constancy implies a planetwide condition in which the various chemical elements have been maintained by cyclic passage through the tissues of plants and animals—first absorbed by an organism and then returned to the air/water/soil through decomposition. These grand cycles—collectively called *biogeochemical cycles*—which sustain all life on our planet, have continued for millennia, at rates and scales almost too vast to conceptualize.

▲ **Figure 10-2** Organisms survive in the biosphere through a complex of systemic flows of energy, water, and nutrients. These wildflowers are in Mount Rainier National Park, Washington.

In recent years, however, the rapid growth of the human population and the accompanying ever-accelerating rate at which we consume Earth's resources have had a deleterious effect on every one of these cycles. Not all of the damage is yet irreparable, but the threat that such disruption will produce irreversible harm to the biosphere is increasing.

Although we describe biogeochemical cycles separately in the following subsections, you will see that many of these cycles are closely interrelated. It is appropriate, then, for us to begin with the most fundamental of these cycles—the flow of energy through the biosphere.

The Flow of Energy

1. Net Primary Productivity
2. Biological Productivity in Midlatitude Oceans

The Sun is the basic energy source on which nearly all life ultimately depends (forms of life utilizing geochemical energy from hydrothermal vents on the ocean floor are a well-known exception). Solar energy can ignite life processes in the biosphere through *photosynthesis*, the production of organic matter by chlorophyll-containing plants and bacteria.

Only about 0.1 percent of the solar energy that reaches Earth is fixed in photosynthesis. More than half of that total is used immediately in the plant's own *respiration*, and the remainder is temporarily stored. Eventually this remainder enters a *food chain*.

If the biosphere is to function properly, its components must be recycled continually. In other words, after one organism uses a component, that component must be converted, at the expense of some energy, to a reusable form. For some components, this conversion can be accomplished in less than a decade; for others, it may require hundreds of millions of years.

Solar energy is of course fundamental to life on Earth. Although readily absorbed by some substances, this energy is also readily reradiated. Thus, it is difficult to store and easy to lose. Happily, most places on Earth receive a daily renewal of the supply.

Photosynthesis and Respiration The biosphere is a temporary recipient of a small fraction of the solar energy that reaches Earth, which is "fixed" (made stable) by green plants through the process of **photosynthesis** (Figure 10-3). The key to photosynthesis is a light-sensitive pigment known as *chlorophyll* that is found within the *chloroplasts* of leaf cells. Chlorophyll absorbs certain wavelengths of visible light, while prominently reflecting green light (this is why leaves look green).

In the presence of sunlight and chlorophyll a photochemical reaction takes carbon dioxide CO_2 from the air and combines it with water (H_2O) to form the energy-rich *carbohydrate* compounds we know as sugars while also releasing molecular oxygen. In this process, the energy from sunlight is stored as chemical energy in the sugars. In simplified form, the chemical equation for photosynthesis is

$$CO_2 + H_2O \xrightarrow{\text{light}} \text{Carbohydrates} + O_2$$

In turn, plants can use simple sugars to build more complex carbohydrates, such as starches.The stored chemical energy in the form of carbohydrates is utilized in the biosphere primarily in two ways: some of the stored energy cycles through the biosphere when animals eat either the photosynthesizing plants or other animals that have eaten plants. The other portion of the stored energy in carbohydrates is consumed directly by the plant itself in a process known as **plant respiration**. In the process of respiration, the stored energy in carbohydrates is *oxidized*, releasing water, carbon dioxide, and heat energy. The simplified chemical equation for plant respiration is

$$\text{Carbohydrates} + O_2 \rightarrow CO_2 + H_2O + \text{Energy (heat)}$$

Net Primary Productivity Plant growth depends on a surplus of carbohydrate production. *Net photosynthesis* is the difference between the amount of carbohydrate produced in photosynthesis and that lost in plant respiration. Annual **net primary productivity** describes the net photosynthesis of a plant community over a period of one year, usually measured in the amount of fixed carbon per unit area (grams of carbon per square meter per year). Monthly or seasonal variations in productivity can also be determined by measuring net photosynthesis. Net primary productivity is, in effect, a measure of the amount of chemical energy stored in a plant community and is reflected in the dry weight of organic material, or **biomass**, of that community.

Net primary productivity varies widely from environment to environment around the world. It tends to be highest on land within the tropics where both high precipitation

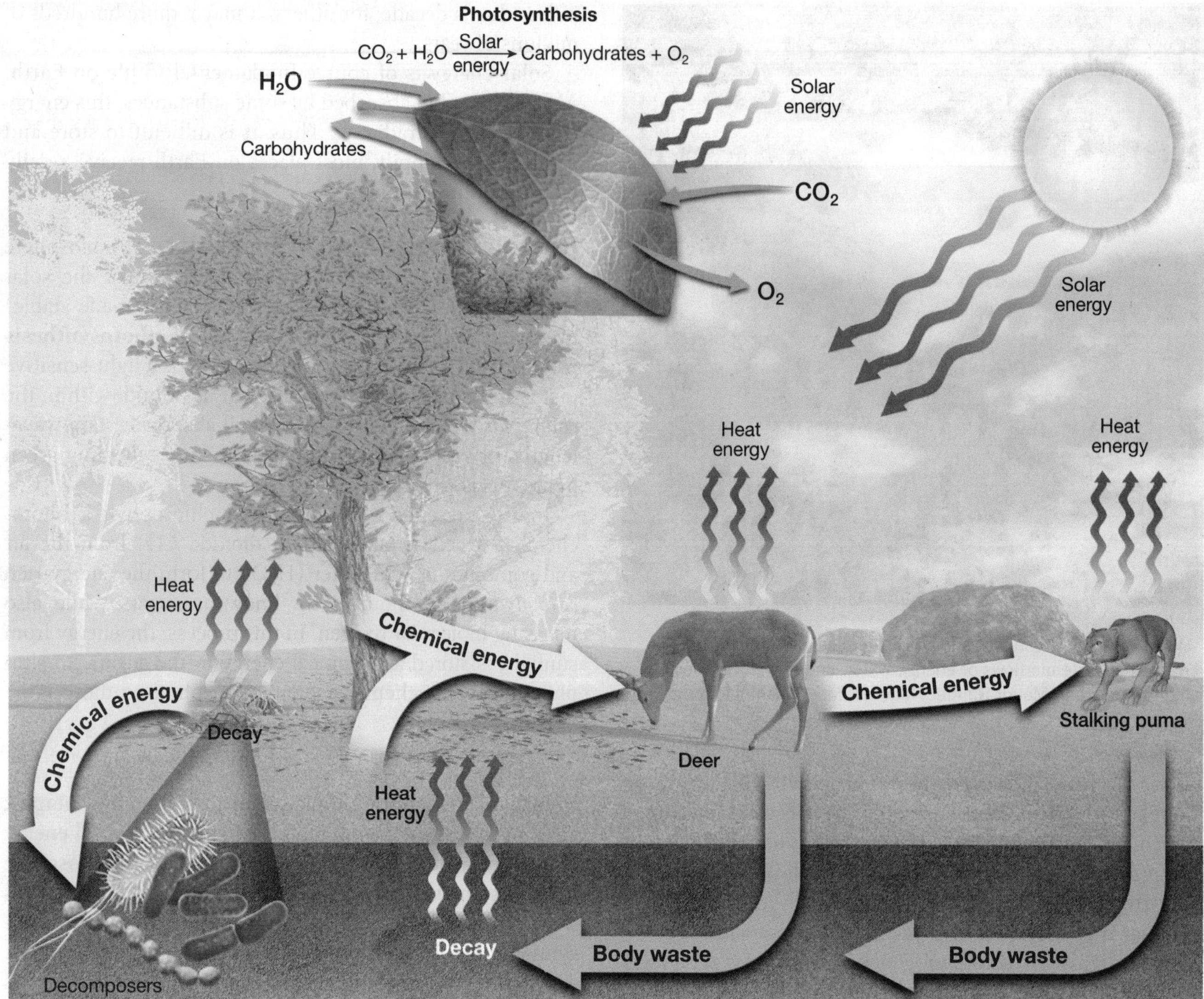

▲ **Figure 10-3** Energy flow in the terrestrial biosphere. Plants use solar energy in photosynthesis, storing that energy in the sugar molecules they manufacture. Grazing and browsing animals then acquire that energy when they eat the plants. Other animals eat the grazers/browsers and thereby acquire some of the energy originally in the plants. Body wastes from all the animals, the bodies of the animals once they die, and dead plant matter that has fallen to the ground all return energy to the soil. As all this waste and dead matter decays, it gives off energy in the form of heat. Thus, the energy originally produced in nuclear reactions in the Sun ends up as random heat energy in the universe.

and high insolation are available for plant growth but generally diminishes poleward, especially in extremely arid and extremely cold environments (Figure 10-4). In the oceans, productivity is strongly influenced by the nutrient content of the water (Figure 10-5). For example, off the west coasts of continents in the midlatitudes, the upwelling of cold, nutrient-rich water results in high net primary productivity.

The carbohydrates incorporated into plant tissue are in turn consumed by animals or decomposed by microorganisms. Plant-eating animals convert some of the consumed carbohydrates back to carbon dioxide and exhale it into the air (animal respiration); the remainder is decomposed by microorganisms after the animal dies. The carbohydrates acted upon by microorganisms are ultimately oxidized into carbon dioxide and returned to the atmosphere (soil respiration).

The Hydrologic Cycle

The most abundant single substance in the biosphere, by far, is water. It is the medium of life processes and the source of their hydrogen. Most organisms contain considerably more water in their mass than anything else, as Table 10-1 shows. Every living thing depends on keeping its water supply within a narrow range. For example, humans can survive without food for two months or more, but they can live without water for only about a week.

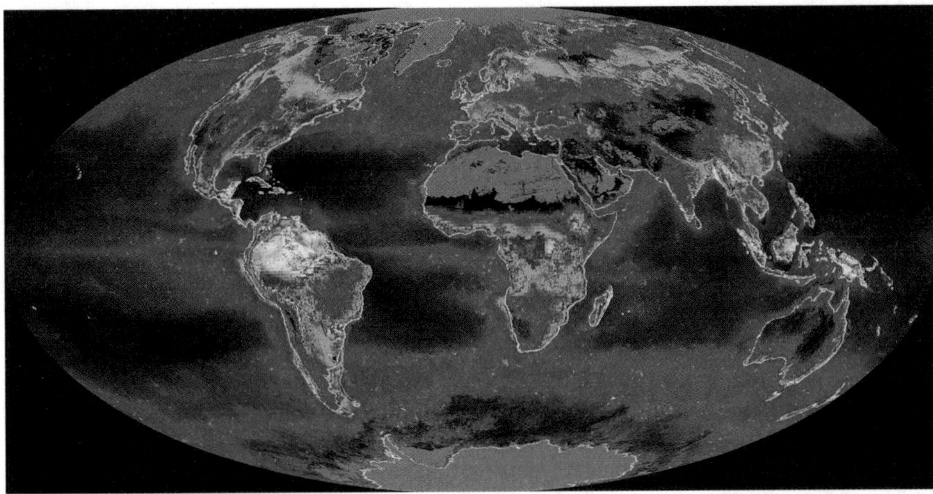

◀ **Figure 10-4** Annual net primary productivity. This composite satellite image shows net primary productivity, based on the rate at which plants absorb carbon dioxide from the air (the mass of carbon absorbed per square meter per year). The areas of highest annual average productivity are shown in yellow and red—most notably in the tropical rainforests of the world; areas of progressively lower productivity are shown in green, blue, and purple.

All living things require water to carry out their life processes. Watery solutions dissolve nutrients and carry them to all parts of an organism. Through chemical reactions that take place in a watery solution, the organism converts nutrients to energy or to materials it needs to grow or to repair itself. In addition, the organism needs water to carry away waste products.

There are two ways in which water is found in the biosphere: (1) *in residence*, with its hydrogen chemically bound into plant and animal tissues; and (2) *in transit*, as part of the transpiration-respiration stream. As we learned in Chapter 9, the movement of water from one sphere to another is called the *hydrologic cycle* (see Figure 9-2).

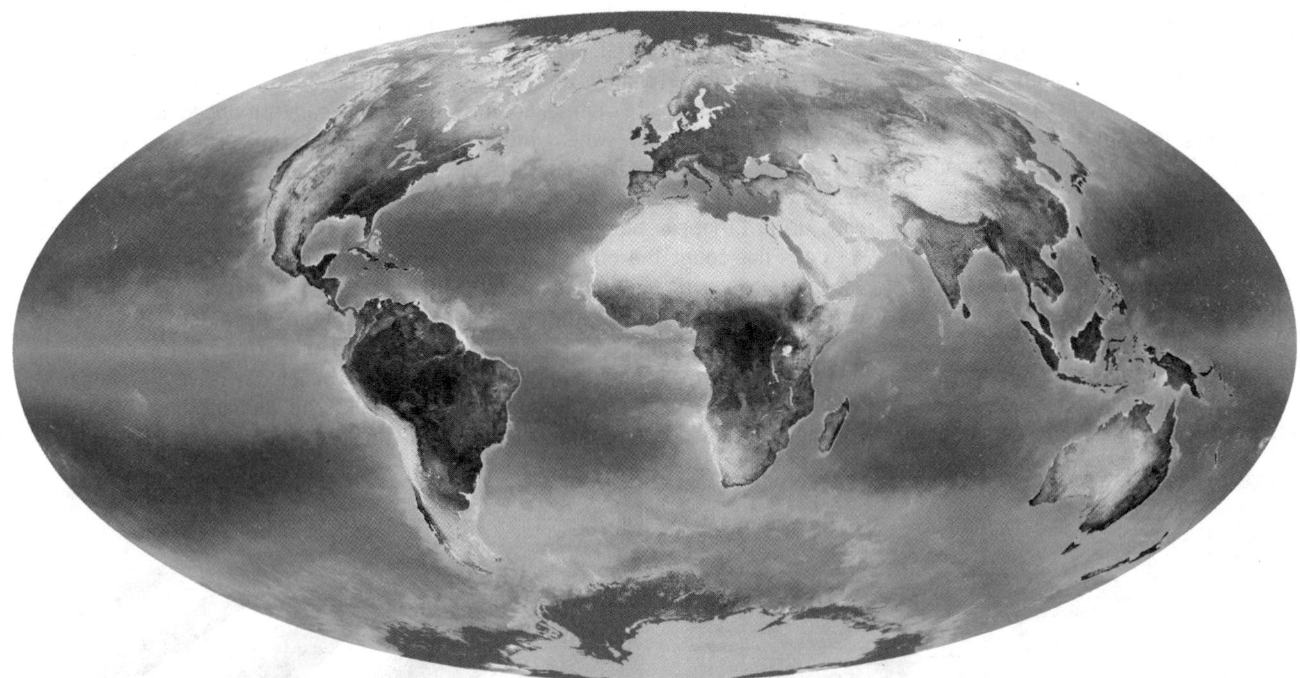

▲ **Figure 10-5** Global vegetation index on land and chlorophyll concentration at the ocean surface for 2008. Relative differences in the amount of vegetation growth on land are depicted with the *vegetation index*: the areas of greatest plant growth are shown in dark green, while areas of the least growth are shown in tan. Chlorophyll concentration in the ocean is an indication of net primary productivity. Values range from 0.01 milligrams of chlorophyll per square meter (dark blue color) to 50 mg/m^2 (yellow). Chlorophyll is generally greatest in areas where phytoplankton is most concentrated such as in the cold, productive waters of the Arctic and Antarctic, and along coastlines where upwelling brings nutrients to the surface. In the center of ocean basins, away from the continental shelves, productivity is generally low. Data were collected by NOAA satellites and the SeaWiFS satellite.

RENEWABLE ENERGY

Biofuels

Valerie Sloan, University of Colorado at Boulder, Ph.D.

As concerns about greenhouse gas emissions affecting climate change have come to the forefront, the search for alternatives to burning fossil fuels is underway. Two fuels of interest are *ethanol*, an alcohol fuel fermented from crops (much like wine or beer), and *biodiesel*, which is made by chemically processing vegetable oil or animal fat. By the early twentieth century, both Rudolph Diesel and Henry Ford had successfully designed engines that would run on such *biofuels*, and many modern internal combustion engines are easily converted to biofuel use.

Burning fossil fuels emits carbon dioxide (CO_2) that has been buried for millions of years, while burning biofuels emits CO_2 which is taken out of the atmosphere during plant or animal growth and so is not considered a new emission. The hope has been that biofuels would emit less greenhouse gas than gasoline and become a new domestic energy source as global oil reserves are depleted. Biofuels have also gained great political and economic momentum, partly because they offer lucrative new markets for the farming industry and because an increase in their production was mandated by the U.S. government in 2007.

However, studies suggest that biofuels are not the cure-all hoped for. Corn ethanol, the biggest source of biofuel in the United States, takes more energy to produce than it provides. In fact, converting most crops, such as soy beans, switchgrass, and palm fruit, into fuel consumes more energy than is produced when all aspects of fossil fuel consumption are accounted for, such as fertilizer production.

Sugarcane ethanol does have a positive energy balance, providing more energy than it takes to produce. This is the primary crop used to produce ethanol in Brazil, the second largest producer and the largest exporter of biofuel in the world. How sugarcane ethanol production and combustion compares to fossil fuels with respect to greenhouse gas emissions depends on whether land use changes were involved to grow the crops. When mature forests are replaced by biofuel crops, trees storing vast quantities of carbon are lost and huge amounts of CO_2 are released to the atmosphere. Critics disagree about whether Brazil's sugarcane ethanol industry creates a net decrease or increase in greenhouse gas emissions compared to fossil fuels.

Recently, carbon accounting problems in international cap-and-trade agreements have surfaced: biodiesel is considered carbon-neutral, and yet the effects of land use change and energy used in production are not taken into account. For example, when Indonesians clear tropical hardwood forests or peat swamps to produce palm oil that ends up in European biodiesel, Indonesia does not count the change in land use emissions or production energy, and Europe does not count the carbon coming out of the tailpipe.

The ethics of using land to grow crops for fuel rather than for food is also becoming a global concern. According to the Director Generals of the United Nations and of the Food and Agriculture Organization, using food crops for fuel is contributing to world hunger, partly because subsistence farmland is replaced by biofuel cash crops. In the United States, the impact is felt through higher food prices that are estimated to have increased 10 to 20 percent because of using food and cattle feed crops for fuel.

Currently, 33 percent of the corn grown in the United States goes into making ethanol (Figure 10-A), but the 34 billion liters (9 billion gallons) of ethanol per year provides the United States with the energy equivalent of only 1.3 percent of total oil consumption. This is a small return, given the environmental costs of growing corn, including the impacts of pesticides, herbicides, groundwater irrigation, and the contribution of nitrogen fertilizer to creating "dead zones," such as the one recently found in the Gulf of Mexico.

For these reasons, innovators and scientists are turning to other sources for biologically produced energy: yeast, algae, and bacteria. The idea is that fats grown inside algae or cyanobacteria can be released by an added chemical solvent; this fat is then refined into biodiesel (Figure 10-B). The U.S. Department of Energy tried to make biodiesel from algae for 18 years but concluded that it was not economically feasible, but today several companies nationally and globally are researching this prospect with interest.

▲ **Figure 10-A** Corn being harvested in Iowa.

▲ **Figure 10-B** Algae growing in research laboratory to make biodiesel.

TABLE 10-1 Water Content in Some Plants and Animals

Organism	Percentage Water in Body Mass
Human	65
Elephant	70
Earthworm	80
Ear of corn	70
Tomato	95

The Carbon Cycle

Carbon is one of the basic elements of life and a part of all living things. The biosphere contains a complex mixture of carbon compounds, more than half a million in total. These compounds are in continuous states of creation, transformation, and decomposition.

The main components of the **carbon cycle** entail the transfer of carbon from carbon dioxide to living matter and then back again to carbon dioxide (Figure 10-6). This conversion is initiated when carbon dioxide from the atmosphere is photosynthesized into carbohydrate compounds (*assimilation*), as shown in the photosynthesis equation we learned earlier. In turn, respiration by plants and soil returns carbon to the atmosphere in the form of carbon dioxide. This is a true cycle or, more precisely, a complex of interlocking cycles; carbon moves constantly from the inorganic reservoir to the living system and back again. A similar cycle takes place in the ocean.

The carbon cycle operates relatively rapidly (the time measured in years or centuries), but only a small proportion (thought to be less than 1 percent) of the total quantity of carbon on or near Earth's surface is part of the cycle at any given moment. The overwhelming bulk of near-surface carbon has been concentrated over millions of years in geologic deposits—such as coal, petroleum, and

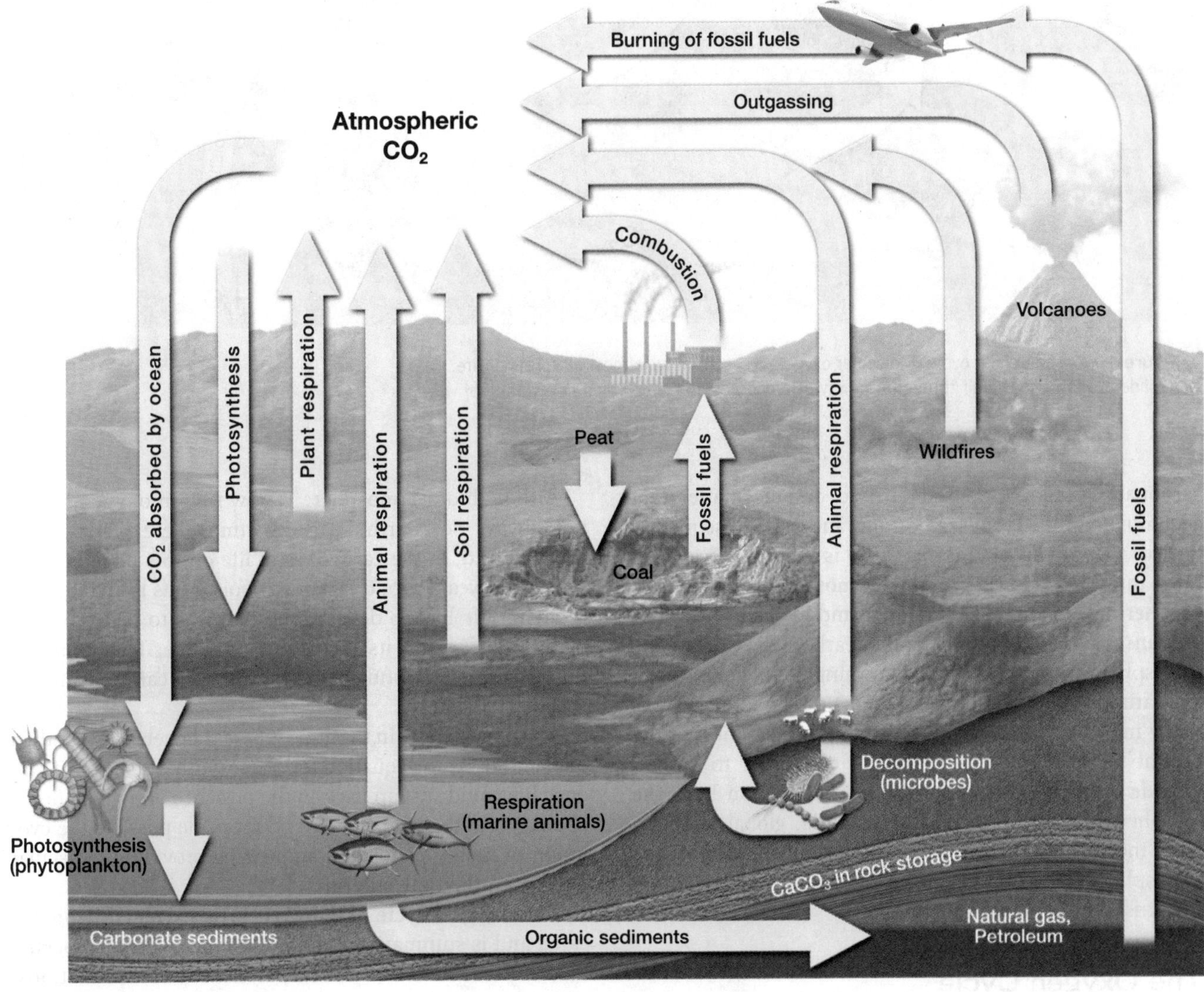

▲ **Figure 10-6** The carbon cycle. Carbon from the carbon dioxide in the atmosphere is used by plants to make the carbon-containing sugars formed during photosynthesis. Through various paths, these compounds are eventually again converted to carbon dioxide and returned to the atmosphere.

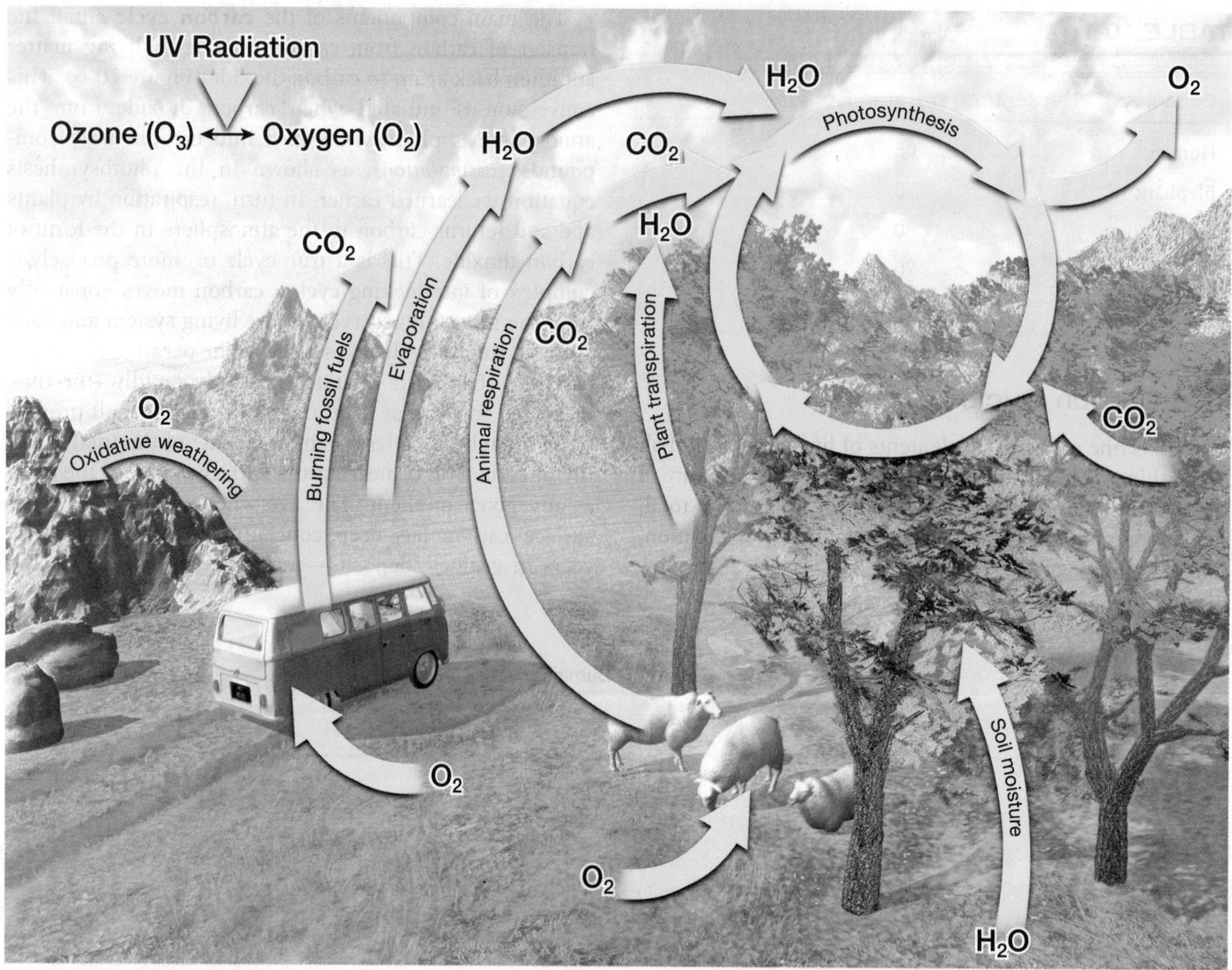

▲ **Figure 10-7** The oxygen cycle. Molecular oxygen is essential for almost all forms of life. It is made available to the air through a variety of processes and is recycled in a variety of ways.

carbonate rocks—composed of dead organic matter that accumulated mostly on sea bottoms and was subsequently buried. Carbon from this reservoir is normally incorporated into the cycle very gradually, mostly by normal rock weathering. In the last century and a half, however, humans have added considerable carbon dioxide to the atmosphere by extracting and burning fossil fuels (coal, oil, natural gas) containing carbon fixed by photosynthesis many millions of years ago. This rapid acceleration of the rate at which carbon is freed and converted to carbon dioxide is likely to have far-reaching effects on both the biosphere and atmosphere. For example, global warming from increasing concentrations of carbon dioxide in the atmosphere may lead to changes in the natural distribution patterns of both plants and animals.

The Oxygen Cycle

Oxygen is a building block in most organic molecules and consequently makes up a significant proportion of the atoms in living matter.

Although our atmosphere is now rich in oxygen, it was not always so. Earth's earliest atmosphere was oxygen poor; indeed, in the early days of life on this planet, about 3.4 billion years ago, oxygen was poisonous to living cells. Evolving life had to develop mechanisms to neutralize or, better still, exploit its poisonous presence. This exploitation was so successful that most life now cannot function without oxygen.

The oxygen now in the atmosphere is largely a by-product of plant life, as the equation for photosynthesis shows. Thus, once life could sustain itself in the presence of high amounts of oxygen in the air, primitive plants made possible the evolution of higher plants and animals by providing molecular oxygen for their metabolism.

The **oxygen cycle** (Figure 10-7) is extremely complicated and is summarized only briefly here. Oxygen occurs in many chemical forms and is released into the atmosphere in a variety of ways. Most of the oxygen in the atmosphere is molecular oxygen (O_2) produced when plants decompose water molecules in photosynthesis. Some atmospheric oxygen is bound up in water molecules

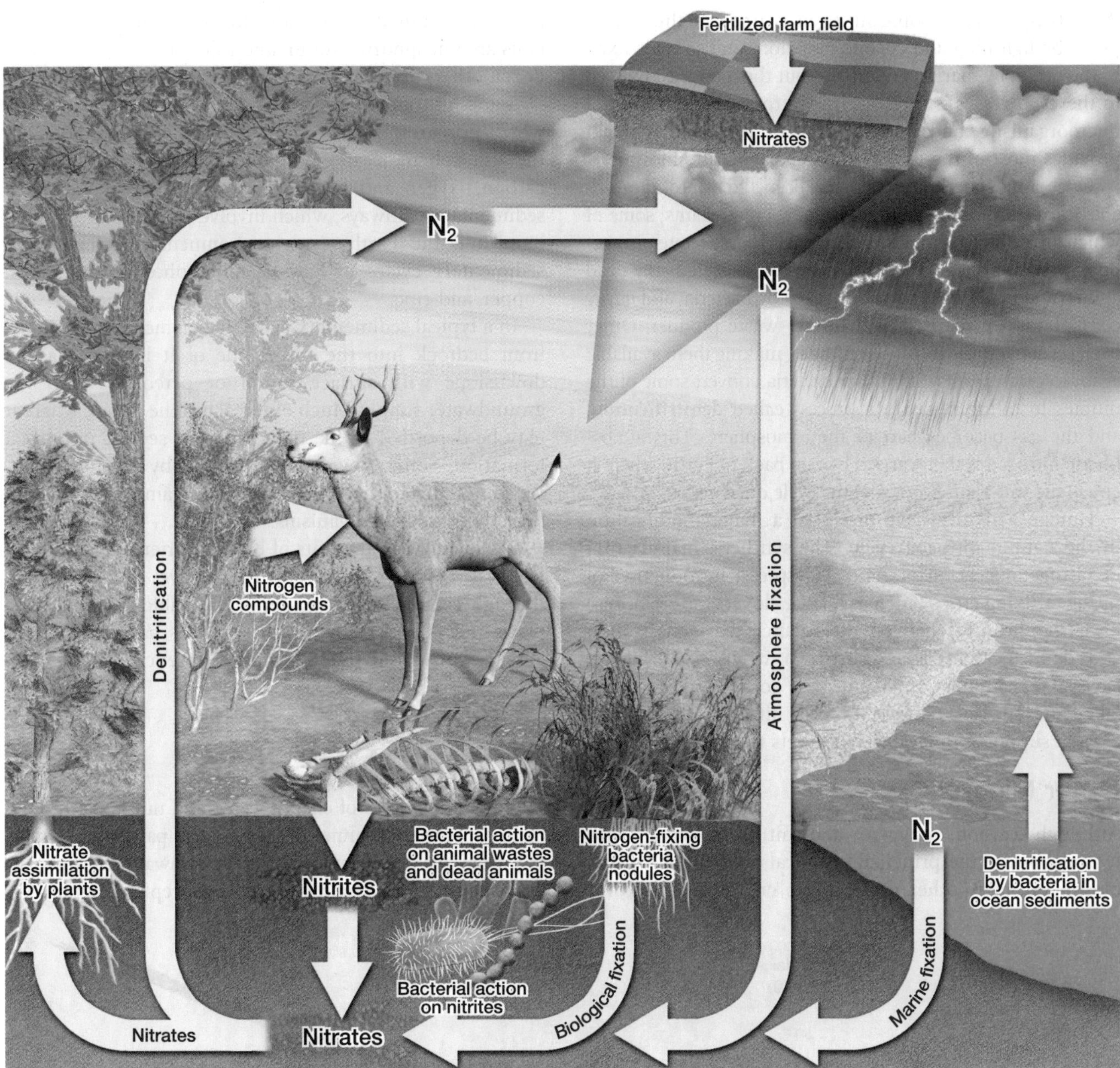

▲ **Figure 10-8** The nitrogen cycle. Atmospheric nitrogen is fixed into nitrates in various ways, and the nitrates are then assimilated by green plants, some of which are eaten by animals. Dead plant and animal materials, as well as animal wastes, contain various nitrogen compounds, and these compounds are acted on by bacteria so that nitrites are produced. The nitrites are then converted by other bacteria to nitrates, and thus the cycle continues. Still other bacteria denitrify some of the nitrates, releasing free nitrogen into the air again.

that came from evaporation or plant transpiration, and some is bound up in the carbon dioxide released during animal respiration. Much of this carbon dioxide and water is eventually recycled through the biosphere via photosynthesis.

Other sources of oxygen for the oxygen cycle include atmospheric ozone, oxygen involved in the oxidative weathering of rocks, oxygen stored in and sometimes released from carbonate rocks, and various other processes, including some (such as the burning of fossil fuels) that are human induced.

The Nitrogen Cycle

Although nitrogen gas (N_2) is an apparently inexhaustible component of the atmosphere (air is about 78 percent N_2), only a limited number of organisms, such as nitrogen-fixing bacteria living in the soil, can use this essential nutrient in its gaseous form. For the vast majority of living organisms, atmospheric nitrogen is usable only after it has been converted to nitrogen compounds (*nitrates*) that can be used by plants (Figure 10-8). This conversion process is called **nitrogen fixation**, and the overall process is called the **nitrogen cycle**. Some nitrogen is fixed in the atmosphere by lightning and cosmic radiation, and some is fixed in the ocean by marine organisms, but the amount involved in these processes is modest. It is nitrogen fixation by soil microorganisms and associated plant root nodules that provides most of the usable nitrogen for Earth's biosphere.

Once atmospheric nitrogen has been fixed into an available form (nitrates), it is assimilated by green plants, some of which are eaten by animals. The animals then excrete nitrogenous wastes in their urine. These wastes, as well as the dead animal and plant material, are attacked by bacteria, and *nitrite* compounds are released as a further waste product. Other bacteria convert the nitrites to nitrates, making them available again to green plants. Still other bacteria convert some of the nitrates to nitrogen gas in a process called **denitrification**, and the gas becomes part of the atmosphere. This atmospheric nitrogen is then carried by rain back to Earth, where it enters the soil-plant portion of the cycle once more.

Human activities have produced a major modification in the natural nitrogen cycle. The synthetic manufacture of nitrogenous fertilizers and widespread introduction of nitrogen-fixing crops (such as alfalfa, clover, and soybeans) have significantly changed the balance between fixation and denitrification. The short-term result has been an excessive accumulation of nitrogen compounds in many lakes and streams. This buildup of nitrogen depletes the oxygen supply of the water and upsets the natural balance.

Other Mineral Cycles

Although carbon, oxygen, and nitrogen—along with hydrogen—are the principal chemical components of the biosphere, many other minerals are critical nutrients for plants and animals. Most notable among these trace minerals are phosphorus, sulfur, and calcium, but more than a dozen others are occasionally significant.

Some nutrients are cycled along gaseous pathways, which primarily involve an interchange between biota and the atmosphere-ocean environment, as we just saw in the carbon, oxygen, and nitrogen cycles. Other nutrients follow sedimentary pathways, which involve interchange between biota and the Earth-ocean environment. Elements with sedimentary cycles include calcium, phosphorus, sulfur, copper, and zinc.

In a typical sedimentary cycle, the element is weathered from bedrock into the soil. Some of it is then washed downslope with surface runoff or percolated into the groundwater supply. Much of it reaches the ocean, where it may be deposited in the next round of sedimentary rock formation. Some, however, is ingested by aquatic organisms and later released into the cycle again through waste products and dead organisms.

In general, the amounts of biotic nutrients available on Earth are finite. These nutrients move over and over through cycles that are extremely variable from place to place; increasingly, some of these cycles have either been damaged or modified by human interference.

Food Chains

The unending flows of energy, water, and nutrients through the biosphere are channeled in significant part by direct passage from one organism to another in pathways referred to as *food chains*. A **food chain** is a simple concept, as Figure 10-9

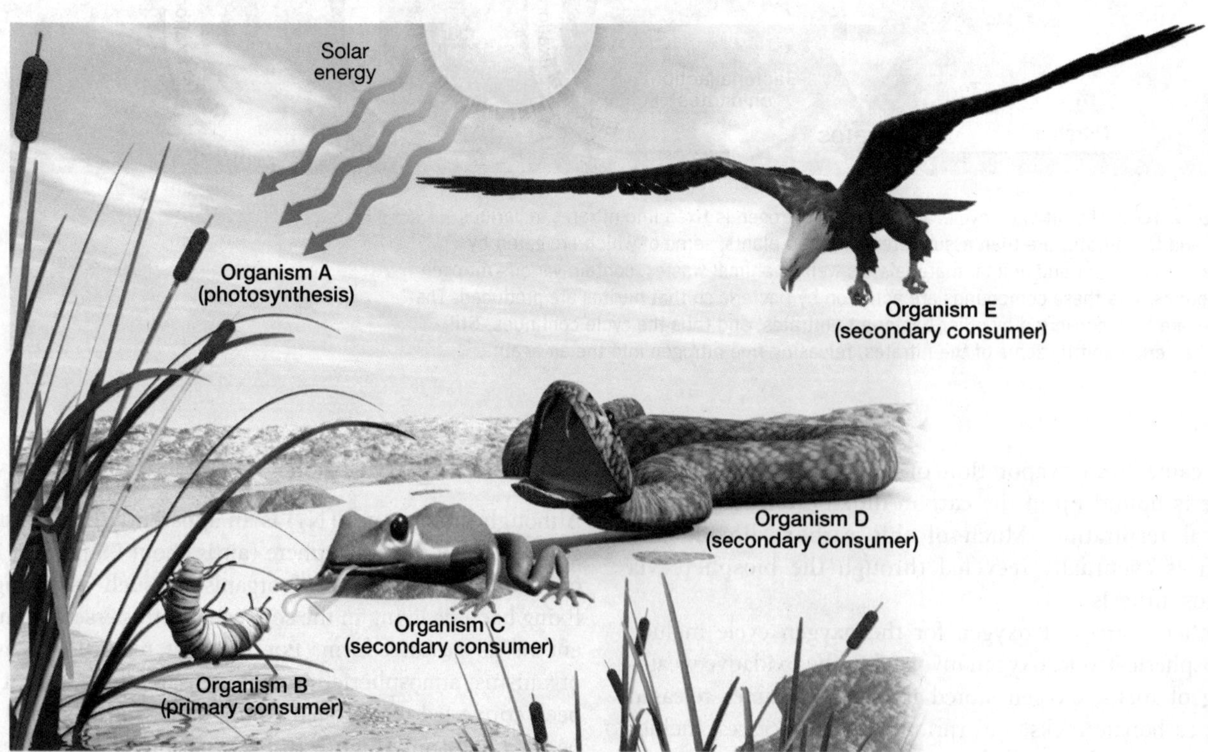

▲ **Figure 10-9** A simple food chain.

shows: Organism A is eaten by Organism B, which thereby absorbs A's energy and nutrients; Organism B is eaten by Organism C, with similar results; Organism C is eaten by D; and so on.

In nature, however, the matter of who eats whom may be extraordinarily complex, with a bewildering number of interlaced strands. Therefore, "chain" probably is a misleading word in this context because it implies an orderly linkage of equivalent units. It is more accurate to think of this energy transfer process as a "web" with interconnected parts or links. Each link acts as an energy transformer that ingests some of the energy of the preceding link, uses some of that energy for its own sustenance, and then passes some of the balance on to the next link.

The fundamental units in any food chain are the **producers**—the *autotrophs*, or "self-feeders"—in other words, plants. Plants fix carbon and effectively store solar energy through photosynthesis. The plants may then be eaten by the **consumers**, or *heterotrophs*. Plant-eating animals are called *herbivores* (*herba* is Latin for "plant"; *vorare*, "to devour"), which are referred to as **primary consumers**. The herbivores then become food for other animals, *carnivores* (*carne*, is Latin for "meat"), which are referred to as **secondary consumers** or *predators*. There may be many levels of secondary consumers in a food chain, as Figure 10-9 shows.

A food chain can also be conceptualized as a **food pyramid** because the number of energy-storing organisms is much, much larger than the number of primary consumers; the number of primary consumers is larger than the number of secondary consumers; and so on up the pyramid (Figure 10-10). There are usually several levels of carnivorous secondary consumers, each succeeding level consisting of fewer and usually larger animals. The final consumers at the top of the pyramid are usually the largest and most powerful predators in the area (Figure 10-11). We say that organisms share the same *trophic level* when they consume the same general types of food in a food pyramid.

The consumers at the apex of the pyramid do not constitute the final link in the food chain, however. When they die, they are fed on by scavenging animals and by tiny (mostly microscopic) organisms that function as **decomposers**, returning the nutrients to the soil to be recycled into yet another food pyramid.

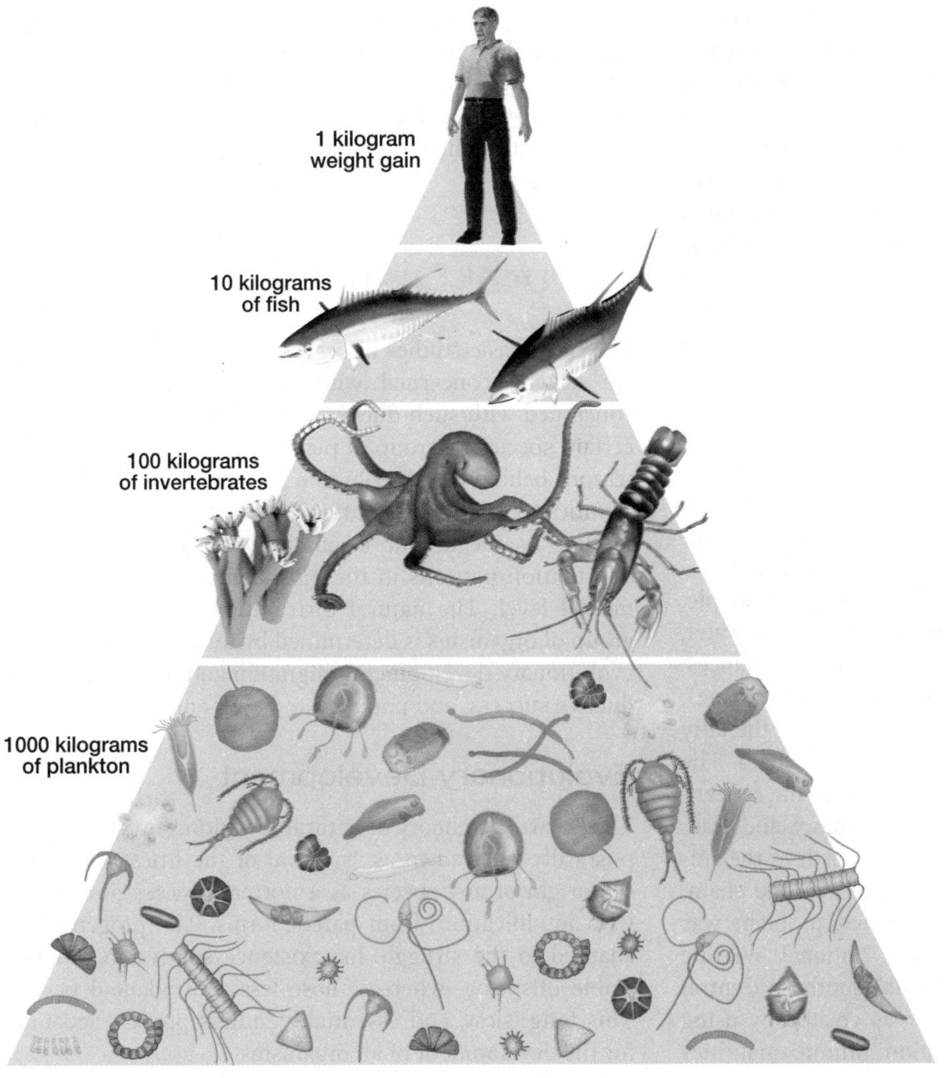

◀ **Figure 10-10** A food pyramid. It takes 1000 kilograms (about 1 ton) of plankton (microscopic marine plants and animals) to provide a 1-kilogram (2-pound) weight gain for a human.

► **Figure 10-11** A lynx about to pounce on a snowshoe hare in a Montana forest.

So why are so many producers needed to support so few top predators in a food pyramid? The answer has to do with the relative inefficiency with which stored energy is transferred from one trophic level to the next. When a primary consumer eats a plant, perhaps only 10 percent of the total energy stored in the plant can be effectively stored by that primary consumer. In turn, when a secondary consumer eats the primary consumer, there is further energy transfer inefficiency. In short, only a portion of the stored energy is transferred from the organisms of one trophic level to those in the next trophic level. (In terms of human diet and food supply, this is why it is much more energy efficient to feed a population directly with grains than with meat from grain-fed animals.)

Pollutants in the Food Chain While stored energy is not efficiently passed on from one organism to the next in a food chain, some chemical pollutants can be. An increasing concern is evidence that some chemical pollutants can become concentrated within a food chain—a process referred to as *biological amplification*. For example, while some chemical pesticides released into the air, water, or soil degrade relatively quickly into harmless substances, others, such as DDT (dichlorodiphenyltrichloroethane), are quite stable and may become concentrated in the fatty tissues of organisms at higher levels of a food chain. The concentration of chemical pesticides, as well as the concentration of some heavy metals such as mercury and lead, have resulted in harmful effects and even death in the animal (and human) consumers at the top of the food chain.

A somewhat similar but unanticipated problem was discovered in 1982 at the Kesterson National Wildlife Refuge amid the farmland of California's southern Central Valley. This artificial wetlands area was created by using runoff from local agricultural irrigation. Minute quantities of natural selenium dissolved in the irrigation water ended up being concentrated through evaporation in the water entering the Wildlife Refuge, resulting in deformities and high mortality for waterfowl using the wetlands. This discovery at Kesterson prompted the U.S. Department of the Interior to create the National Irrigation Water Quality Program to reduce the likelihood of similar problems elsewhere.

Natural Distributions

The most basic studies of organisms made by geographers are usually concerned with distribution. Here we are concerned with such questions as, "What is the range of a certain species or group of plants/animals?" "What are the reasons behind this distribution pattern?" and "What is the significance of the distribution?"

Here we begin our discussion of factors influencing distribution patterns in the biosphere at the most fundamental level. The natural distribution of any species or group of organisms is determined by four basic conditions: evolutionary development, migration and dispersal, reproductive success, and population die-off and extinction.

Evolutionary Development

The Darwinian theory of natural selection, sometimes simplistically referred to as "survival of the fittest," explains the origin of any species as a normal process of descent, with modification, from parent forms. The progeny best adapted to the struggle for existence survive to produce viable offspring, whereas those less well adapted perish. This long, slow, and essentially endless process accounts for the development of all organisms.

◀ **Figure 10-12** Hundreds of species of acacias grow in semiarid and subhumid portions of the tropics. This scene from central Kenya shows acacias in tree form, although lower shrub forms are more common.

To understand the distribution of any species or *genus* (a closely related group of organisms), therefore, we begin with a consideration of where it evolved. In some cases, there was a very localized beginning for a genus; in other cases, similar evolutionary development at several scattered localities led to the same genus. Unfortunately, the evidence on which case applies in a given situation is not always clear. As an example of these two extremes, consider the contrast in apparent origin of two important groups of plants—*acacias* and *eucalyptus* (Figures 10-12 and 10-13). Acacias are an extensive genus of shrubs and low-growing trees represented by numerous species found in low-latitude portions of every continent that extends into the tropics or subtropics. Eucalyptus, on the other hand, are a genus of trees native only to Australia and a few adjacent islands. Acacias apparently evolved prior to the separation of the continents and are now present throughout the Southern Hemisphere, while the genus Eucalyptus evolved after the Australian continent was isolated and only occurs naturally in Australia. (We will discuss the changing locations of the continents over geologic time when we introduce the theory of plate tectonics in Chapter 14.)

Migration and Dispersal

Throughout the millennia of Earth's history, organisms have always moved from one place to another. Animals possess active mechanisms for locomotion—legs, wings, fins, and so on—and their possibilities for migration are obvious. Plants are also mobile, however. Although most individual plants become rooted and therefore fixed in location for most of their life, there is much opportunity for passive migration, particularly in the seed stage. Wind, water, and animals are the principal natural mechanisms of seed dispersal.

◀ **Figure 10-13** The original forests of Australia were composed almost entirely of species of eucalyptus. This scene is from Coolah Tops National Park, New South Wales

The contemporary distribution pattern of many organisms is often the result of natural migration or dispersal from an original center of development. Among thousands of examples that could be used to illustrate this process are the following:

1. **Coconut Palms:** The coconut palm (*Cocos nucifera*) is believed to have originated in southeastern Asia and adjacent Melanesian islands. It is now extraordinarily widespread along the coasts of tropical continents and islands all over the world. Most of this dispersal apparently has come about because coconuts, the large hard-shelled seeds of the plant, can float in the ocean for months or years without losing their fertility. Thus, they wash up on beaches throughout the world and colonize successfully if environmental conditions are right (Figure 10-14). This natural dispersion was significantly augmented by human help in the Atlantic region, particularly by the deliberate transport of coconuts from the Indian and Pacific Ocean areas to the West Indies.
2. **Cattle Egrets:** The cattle egret (*Bubulcus ibis*) apparently originated in southern Asia, but during the last few centuries has spread to other warm areas of the world, particularly Africa (Figure 10-15). In recent decades, a change in land use in South America has caused a dramatic expansion in the cattle egret's range. At least as early as the nineteenth century, some cattle egrets crossed the Atlantic from West Africa to Brazil, but they were unable to find suitable ecological conditions and thus did not become established. The twentieth-century introduction of extensive cattle raising in tropical South America apparently provided the missing ingredient, and egrets quickly adapted to the newly suitable habitat. Their descendants spread northward throughout the subtropics and are now common inhabitants of the Gulf coastal plain in the southeastern United States, are well-established in California, and occur as far north as southeastern Canada. Also within the twentieth century, cattle egrets dispersed at the other end of their "normal" range to enter northern Australia and spread across that continent.

▲ **Figure 10-14** Coconuts have a worldwide distribution in tropical coastal areas, in part because the nut can float long distances and then take root if it finds a favorable environment. This sprouting example is from the island of Vanua Levu in Fiji.

Reproductive Success

A key factor in the continued survival of any biotic population is reproductive success. Poor reproductive success can come about for a number of reasons—heavy predation (a fox eats quail eggs from a nest); climatic change (heavy-furred animals perishing in a climate that was once cold but has warmed up for some reason or other); failure of food supply (a string of unusually cold winters keeps plants from setting seed); and so on.

Changing environmental conditions are also likely to favor one group over another, as when warming waters on the fringes of the Arctic Ocean allowed cods to expand their range at the expense of several other types of fishes. Thus, reproductive success is usually the limiting factor that allows one competing population to flourish while another languishes (Figure 10-16).

Population Die-off and Extinction

The range of a species can be diminished by the dying out of some or all of the population. The history of the biosphere is replete with examples of such range diminution, varying from minor adjustments in a small area to extinction over the entire planet. Evolution is a continuing process. No species is likely to be a permanent inhabitant of Earth, and during the period of its ascendancy there is apt to be a great deal of distributional variation within a species, part of which is caused by local die-offs.

Plant Succession One of the simplest and most localized examples of this process is **plant succession**, in which one type of vegetation is replaced naturally by another. Plant succession is a normal occurrence in a host of situations; a very common one involves the infilling of a lake (Figure 10-17). As the lake gradually fills with sediments and organic debris, the aquatic plants at the bottom are slowly choked out while the sedges, reeds, and mosses of the shallow edge waters become more numerous and extensive. Continued infilling further diminishes the aquatic habitat and allows for the increasing encroachment of low-growing land plants such as grasses and shrubs. As the process continues, trees move in to colonize the site, replacing the grasses and shrubs and completing the transition from lake to marsh to meadow to forest.

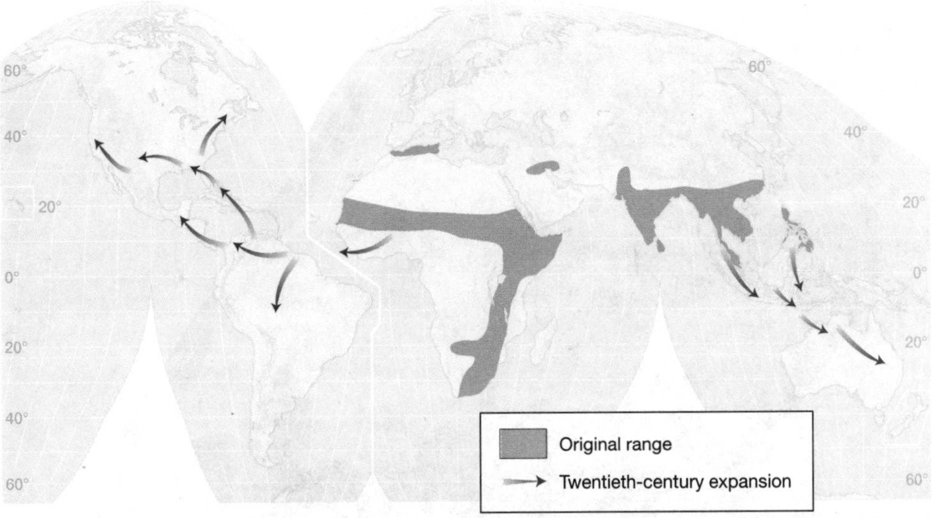

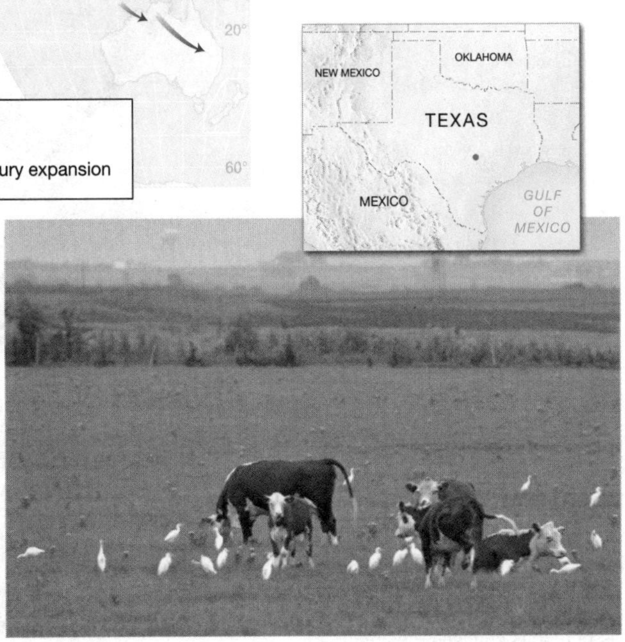

▲ **Figure 10-15** Natural dispersal of an organism. (a) During the twentieth century, cattle egrets expanded from their Asian–African range into new habitats in the Americas and Australia. (b) Cattle egrets have found a happy home among the cattle herds of Texas.

A similar series of local animal replacements would accompany the plant succession because of the significant habitat changes. Lake animals would be replaced by marsh animals, which in turn would be replaced by meadow and forest animals.

▲ **Figure 10-16** Once nearly exterminated, American bison now occur in large numbers in areas of suitable habitat, as here in Elk Island National Park, Alberta, Canada. Under natural conditions they have a high level of reproductive success.

Plant and animal succession occurs after catastrophic natural events as well as gradual ones. For example, as an immediate consequence of the major 1980 eruption of Mount St. Helens, thousands of hectares of forest were covered by volcanic mudflows and pyroclastic flows, or were blown down by the large lateral blast (see "Focus: Mount St. Helens" in Chapter 14). In areas where all remnants of the original ecosystem were effectively covered and the soil completely obliterated, the first species of plants and animals to return comprised what is known as a *pioneer community* in a process called *primary succession.* In other areas—and more common in general—*secondary succession* is taking place, in which remnants of the original plant communities and soil become the starting point for a succession sequence (Figure 10-18).

Plant succession is not to be confused with true extinction. Extinction is permanent, but species succession is not. Although a particular plant species may not be growing at a given time in a given location, it may reappear quickly if environmental conditions change and if there is an available seed source. Extinction means a species is extinct over the entire world, eliminated forever from the landscape. Extinction has taken place many times in Earth's history. Indeed, it is estimated that half a billion species have become extinct during the several-billion-year life of our planet. Probably the most dramatic example is the disappearance of the dinosaurs about 65 million years ago.

▶ **Figure 10-17** A simple example of plant succession: infilling of a small lake with sediment and organic material. Over time, successional colonization by different plant associations changes the area from pond to marsh to meadow to forest.

For many millions of years, those gigantic reptiles were the dominant life-forms of our planet, and yet in a relatively short period of geologic time they were all wiped out. The reasons behind their extermination are imperfectly understood (evidently due in part to the environmental changes associated with massive volcanic eruptions of flood basalt and a catastrophic asteroid impact), but the fact remains that there have been innumerable such natural extinctions of entire species in the history of the world.

▲ **Figure 10-18** The lateral blast from the 1980 eruption of Mount St. Helens in Washington flattened trees many kilometers from the volcano. In this photograph taken 17 years after the eruption, the dead trees are still visible, but a cover of new vegetation has been clearly established.

As we will discuss in Chapter 11, human activities are currently leading to the extinctions of plant and animal species at an alarming rate, primarily through the destruction of natural habitat.

Environmental Relationships

The survival of plants and animals depends on an intimate and sometimes precarious set of relationships with other elements of the environment. The details of these relationships vary with different species, but we can generalize about many of them. They can be discussed at various scales of generalization, however, and you must remember that a generalization that is true at one scale may be quite invalid at another. For example, if we are considering global or continental patterns of biotic distribution, we are concerned primarily with gross generalizations that deal with average conditions, seasonal characteristics, latitudinal extent, zonal winds, and other broad-scale factors. If our interest is

PEOPLE AND THE ENVIRONMENT

Bark Beetle Killing Forests in Western North America

Across western North America, from Canada to New Mexico, millions of acres of pine forest are dead or dying because of an infestation of the mountain pine beetle (*Dendroctonus ponderosae*). Year by year, and acre by acre, green pine forest is being replaced by expanses of brown dead and dying trees (Figure 10-C). By some estimates, the beetles are affecting more than 6 million acres of forest in the United States, and perhaps 34 million acres in British Columbia—and growing rapidly. It may well be the largest insect infestation in North American during historic times.

The black mountain pine beetle is about the size of a fingertip (Figure 10-D). It drills into the wood to lay its eggs, then injects a fungus that stops the tree from excreting sap—which would kill the beetle larvae. The fungus causes a stain that turns the wood blue. The tree responds to the beetle holes by emitting a waxy resin that can plug the holes and kill the beetles, but usually a tree is ultimately overwhelmed as more beetles arrive and drill still more holes.

This extensive mountain pine beetle infestation probably has several causes. First, fire suppression over the last century prevented large sections of the western forests—especially the expanses of lodgepole pines—from burning on a regular basis. This means that a large proportion of the trees are just about the same age—old enough to be susceptible to the bark beetle (younger trees are generally not targeted by the beetles), Second, extended drought over the last 10 years or so has weakened many of the trees, making them more susceptible to the beetle. Finally, climate change in western North America is bringing less frequent severely cold winters, allowing the beetle to extend its range. Today the absolute winter low temperatures are about 3.5 to 5.5°C (6 to 10°F) higher than 50 years ago. With fewer extremely cold days in winter, more beetle larvae are surviving.

All of the consequences of bark beetle infestation are not yet known. Resort towns in the Rocky Mountains are worried about the loss of tourism in areas where forests are dying. One of the biggest concerns is the threat of especially hot wildfires that could burn through the dead forests. Once the trees die or have burned, nearby watersheds may be more susceptible to flashfloods and debris flows. Finally, falling dead trees are becoming a problem along highways—large strips are being clear cut along highways to prevent closure by tree blow downs from storms.

Although the mountain pine beetle infestation is seen as an environmental disaster by many, some scientists see things somewhat differently. The mountain pine beetle is not an exotic species. Rather, it is a native insect that has been around for thousands of years, playing its role in western pine forest ecosystems. Lodgepole pines are well adapted to major stand replacement events during which large expanses of old forest are killed by some event such as a fire—or in this case, an insect infestation. The forest then quickly regenerates as nutrients are cycled through the soil; grasses and other low vegetation return first, to be replaced in time by a new forest.

The wild card in this infestation, however, is climate change. With milder winters, there are fewer checks on the beetles and so they are spreading almost as an exotic species would in some places. At the moment, there is nothing to indicate that the infestation will end any time soon, and the prospect of ever larger expanses of dying pine forests is a real one.

▲ **Figure 10-C** Pine forest in Rocky Mountain National Park, Colorado, dying from the infestation of the mountain pine beetle.

▲ **Figure 10-D** Mountain pine beetle.

▲ **Figure 10-19** In an open stand (left), there is abundant light all around the tree and it responds by broad lateral growth rather than vertical growth. Under crowded conditions (right), less light reaches the tree and it elongates upward rather than spreading laterally.

instead a small area, such as an individual valley or a single hillside, we are more concerned with such localized environmental factors as degree of slope, direction of exposure, and permeability of topsoil.

Whatever the scale, there are nearly always exceptions to the generalizations, and the larger the area, the more numerous the exceptions. Thus, in a region that is generally humid, there are probably many localized sites that are extremely dry, such as cliffs or sand dunes. In even a very dry desert, there are likely to be several places that are always damp, such as an oasis or a spring.

Throughout the following discussion of environmental relationships, keep in mind that both *intraspecific competition* (among members of the same species) and *interspecific competition* (among members of different species) are at work. Both plants and animals compete with one another as they seek light, water, nutrients, and shelter in a dynamic environment. The term **limiting factor** is often used to describe the variable that is most important in determining the survival of an organism.

The Influence of Climate

At almost any scale, the most prominent environmental constraints on biota are exerted by various climatic factors.

Light No green plant can survive without light. We have already discussed the basic process—photosynthesis—whereby plants produce stored chemical energy; this process is activated by light. It is essentially for this reason that photosynthetic vegetation is absent from deeper ocean areas, where light does not penetrate.

Light can have a significant effect on plant shape, as Figure 10-19 shows. In places where the amount of light is restricted, such as in a dense forest, trees are likely to be very tall but have limited lateral growth. In areas that have less dense vegetation, more light is available, and as a result trees are likely to be expansive in lateral spread but truncated vertically.

Another important light relationship involves how much light an organism receives during any 24-hour period. This relationship is called **photoperiodism**. Except in the immediate vicinity of the equator, the seasonal variation in the photoperiod becomes greater with increasing latitude. Fluctuation in the photoperiod stimulates seasonal behavior—such as flowering, leaf fall, mating, and migration—in both plants and animals.

Moisture The broad distribution patterns of the biota are governed more significantly by the availability of moisture than by any other single environmental factor. A prominent trend throughout biotic evolution has been the adaptation of plants and animals to either excesses or deficiencies in moisture availability (Figure 10-20).

◄ **Figure 10-20** Even the most stressful environments often contain distinctive and conspicuous plants, as in this scene from the Sonoran Desert.

▲ **Figure 10-21** Some plants are remarkably adaptable to environmental stress. In this timberline scene from north-central Colorado, there is no question about the prevailing wind direction. Persistent wind from the right has so desiccated this subalpine fir that only branches growing toward the left survive. This preferred growth direction places the trunk of the tree between them and the wind.

Temperature The temperature of the air and the soil is also important to biotic distribution patterns. Fewer species of both plants and animals can survive in cold regions than in areas of more moderate temperatures. Plants, in particular, have a limited tolerance for low temperatures because they are continuously exposed to the weather, and they experience tissue damage and other physical disruption when their cellular water freezes. Animals in some instances are able to avoid the bitterest cold by moving around to seek shelter. Even so, the cold-weather areas of high latitudes and high elevations have a limited variety of animals and plants.

Wind The influence of wind on biotic distributions is more limited than that of the other climatic factors. Where winds are persistent, however, they often serve as a constraint. The principal negative effect of wind is that it causes excessive drying by increasing evaporation from exposed surfaces, thus causing a moisture deficiency (Figure 10-21). In cold regions, wind escalates the rate at which animals lose body heat.

The sheer physical force of wind can also be influential: a strong wind can uproot trees, modify plant forms, and increase the heat intensity of wildfires. On the positive side, wind sometimes aids in the dispersal of biota by carrying pollen, seeds, lightweight organisms, and flying creatures.

Edaphic Influences

Soil characteristics, known as **edaphic factors** (from *edaphos*, Greek for "ground" or "soil"), also influence biotic distributions. These factors are direct and immediate in their effect on flora but usually indirect in their effect on fauna. Soil is a major component of the habitat of any vegetation, of course, and its characteristics significantly affect rooting capabilities and nutrient supply. Especially significant are soil texture, soil structure, humus content, chemical composition, and the relative abundance of soil organisms. Soil is discussed in much greater detail in Chapter 12.

Topographic Influences

In global distribution patterns of plants and animals, general topographic characteristics are the most important factor affecting distribution. For example, the assemblage of plants and animals in a plains region is very different from that in a mountainous region. At a more localized scale, the factors of slope and drainage are likely to be significant, primarily the steepness of the slope, its orientation with regard to sunlight, and the porosity of the soil on the slope.

Wildfire

Most environmental factors that affect the distribution of plants and animals are passive, and their influences are slow and gradual. Occasionally, however, abrupt and catastrophic events, such as floods, earthquakes, volcanic eruptions, landslides, insect infestations, and droughts, also play a significant role. By far the most important of these is wildfire (Figure 10-22). In almost all portions of the continents, except for the always-wet regions where fire simply cannot start and the always-dry regions where there is an insufficiency of combustible vegetation, uncontrolled natural fires have occurred with surprising frequency. Fires generally result in complete or partial devastation of the plant life and the killing or driving away of all or most of the animals. These results, of course, are only temporary; sooner or later, vegetation sprouts and animals return. At least in the short run, however, the composition of the biota is changed, and if the fires occur with sufficient frequency the change may be more than temporary.

Wildfire can be very helpful to the seeding or sprouting of certain plants and the maintenance of certain plant interactions. For example, the forest fires that burned nearly half of Yellowstone National Park in 1988 triggered the extensive regrowth of understory plants and young trees (Figure 10-23)—the lodgepole pine is especially well adapted to rapid regeneration after stand-replacement events such as fire. In some cases, grasslands are sustained by relatively frequent natural fires, which inhibit the encroachment of tree seedlings. Moreover, many plant species, particularly certain trees such as the giant sequoia and the southern yellow pine, scatter their seeds only after the heat of a fire has caused the cones or other types of seedpods to open.

▲ **Figure 10-22** Wildfires are commonplace in many parts of the world. This ground fire in the Top End of the Northern Territory of Australia is almost an annual occurrence under natural conditions.

(a)

(b)

MONTANA
Yellowstone National Park
SD
ID
WYOMING
NE
UT
CO

(c)

▲ **Figure 10-23** Regrowth after the Yellowstone wildfires. (a) Firefighter mopping up some of the last remnants of Yellowstone's fires in 1988. (b) Photograph taken from the same location as (a) five years after the fires showing the regrowth of vegetation on the forest floor. (c) Young stands of lodgepole pine trees 10 years after the fires.

◀ **Figure 10-24** World distribution of tropical rainforest, or *selva*.

Environmental Correlations: Example of Selva

One of the most important themes in physical geography is the intertwining relationships of the various components of the environment. Time after time we note situations in which one aspect of the environment affects another—sometimes conspicuously, sometimes subtly. In terms of broad distribution patterns, climate, vegetation, and soil have a particularly close correlation. Before we go on to discuss some of the details of biogeographical patterns in the following chapter, it may be helpful to provide an example of some of these correlations by examining the distribution of tropical rainforest.

On any map showing the world distribution of major plant associations, one of the conspicuous units is the *selva*, or *tropical rainforest*. A vast extent of selva exists in northern South America (primarily within the watershed of the Amazon River), central Africa (mostly within the watershed of the Congo [Zaire] River), and the East Indies, with more limited patches in Central America, Colombia, West Africa, Madagascar, Southeast Asia, and northeastern Australia (Figure 10-24).

Climate A general explanation of this distribution pattern is simple. With very limited exceptions, the tropical rainforest occurs wherever relatively abundant precipitation and uniformly warm temperatures occur throughout the year, especially in areas of tropical wet (Af) climate (described in Chapter 8). It is tempting to state that the tropical wet climate "creates" the conditions necessary for tropical rainforest, but the cause-and-effect relationship is not that simple—for example, transpiration from the local vegetation is very much a part of the hydrologic cycle in tropical rainforest regions.

Flora Because of the high temperatures and high humidity, regions having a tropical wet climate are normally covered with natural vegetation that is unexcelled in luxuriance and variety. Tropical rainforest is a broadleaf evergreen forest with numerous tree species. Many of the trees are very tall, and their intertwining tops form an essentially continuous canopy that prohibits sunlight from shining on the forest floor. Often shorter trees form a second and even a third partial canopy at lower elevations. Most of the trees are smooth barked and have no low limbs, although there is a profusion of vines and hanging plants that entangle the trunks and dangle from higher limbs (Figure 10-25).

▲ **Figure 10-25** A tropical rainforest scene in Ecuador.

The dimly lit forest floor is relatively clear of growth because lack of sunlight inhibits survival of bushes and shrubs. Where much sunlight reaches the ground, as along the edge of a clearing or banks of a stream, a maze of undergrowth can prosper. This sometimes impenetrable tangle of bushes, shrubs, vines, and small trees is called a *jungle*.

Fauna Regions of tropical rainforest are the realm of flyers, crawlers, creepers, and climbers. Larger species, particularly hoofed animals, are not common. Birds and monkeys inhabit the forest canopy often in great quantity and diversity. Snakes and lizards are common both on the forest floor and in the trees. Rodents are sometimes numerous at ground level, but the sparser population of larger mammals is typically secretive and nocturnal. Aquatic life, particularly fish and amphibians, is usually abundant. Invertebrates, especially insects and arthropods, are characteristically superabundant.

Soil The copious, warm, year-round rains provide an almost continuous infiltration of water downward, with the result that soils are usually deep but highly leached and infertile. Leaves, twigs, flowers, and branches frequently fall from the trees to the ground, where they are rapidly decomposed by the abundant earthworms, ants, bacteria, and microfauna of the soil. The accumulated litter is continuously incorporated into the soil, where some of the nutrients are taken up by plants and the remainder are carried away by the infiltrating water. *Laterization* (rapid weathering of mineral matter and speedy decomposition of organic matter) is the principal soil-forming process; it produces a thin layer of fertile topsoil that is rapidly used by plants and a deep subsoil that is largely an infertile mixture of such insoluble constituents as iron, aluminum, and magnesium compounds. These minerals typically impart a reddish color to the soil. River floodplains tend to develop soils of higher fertility because of flood-time deposition of silt.

Hydrography The abundance of runoff water on the surface feeds well-established drainage systems. There is usually a dense network of streams, most of which carry both a great deal of water and a heavy load of sediment. Lakes are not common because there is enough erosion to drain them naturally. Where the land is very flat, swamps sometimes develop through inadequate drainage and the rapid growth of vegetation.

In the next chapter, we will explore many other correlations among the patterns in the biosphere, lithosphere, atmosphere, and hydrosphere.

CHAPTER 10 LEARNING REVIEW

After studying this chapter, you should be able to answer the following questions. Key terms from each text section are shown in bold type. Definitions for key terms are also found in the glossary at the back of the book.

The Geographic Approach to the Study of Organisms (p. 261)

1. Briefly define the following terms: **biogeography**, **biota**, **flora**, **fauna**.

Biogeochemical Cycles (p. 262)

2. What is the primary source of energy for the biosphere?
3. How is solar energy "stored" in the biosphere?
4. Describe and explain the process of **photosynthesis**.
5. Describe and explain the process of **plant respiration**.
6. What is meant by **net primary productivity**?
7. What is the relationship of **biomass** to net primary productivity?
8. Describe the basic steps in the **carbon cycle**.
9. Most of the carbon at or near Earth's surface is not involved in any short-term cycling. Explain.
10. What is the primary source of carbon that humans have added to the atmosphere?
11. Briefly describe some of the components of the **oxygen cycle**.
12. What is the importance of photosynthesis to the flow of energy, water, oxygen, and carbon through the biosphere?
13. Why is it difficult to integrate nitrogen gas from the atmosphere into the **nitrogen cycle** of the biosphere?
14. Explain the differences between **nitrogen fixation** and **denitrification**.

Food Chains (p. 270)

15. What is the relationship of a **food chain** to a **food pyramid**?
16. Explain the roles of **producers**, **consumers**, and **decomposers** in the food chain.
17. What is the difference between **primary consumers** and **secondary consumers**?
18. Looking at Figure 10-10, explain why it takes 1000 kilograms of plankton to produce only 10 kilograms of fish.

Natural Distributions (p. 272)

19. Describe one mechanism through which plant seeds can be dispersed over great distances.
20. Explain the concept of **plant succession**.

Environmental Relationships (p. 276)

21. What is meant by the term **limiting factor**?
22. Explain how both photosynthesis and **photoperiodism** are dependent on sunlight.
23. Why are trees in dense forests likely to be tall with narrow tops?
24. What is meant by **edaphic factors**?
25. What are the beneficial effects of wildfire?

Log in to **www.mygeoscienceplace.com** for animations, videos, RSS feeds, maps, and self-study quizzes to enhance your study of **Cycles and Patterns in the Biosphere**.

11 Terrestrial Flora and Fauna

Having considered fundamental patterns and processes in the biosphere in the previous chapter, we now turn more specifically to the geographical distribution of plants and animals. We begin with some concepts to help us study groups of organisms.

Ecosystems and Biomes

In our search for organizing principles that help us comprehend the biosphere, two concepts are of particular value—*ecosystem* and *biome*.

Ecosystem: A Concept for All Scales

The term **ecosystem** is a contraction of the phrase *ecological system*. An ecosystem includes all the organisms in a given area, but it is more than simply a community of plants and animals existing together. The ecosystem concept encompasses the totality of interactions among the organisms and between the organisms and the nonliving portion of the environment in the area under consideration. The nonliving portion of the environment includes soil, rocks, water, sunlight, and atmosphere, but it can essentially be considered as nutrients and energy.

An ecosystem, then, is fundamentally an association of plants and animals along with the surrounding nonliving environment and all the interactions in which the organisms take part. The concept is built around the flow of energy among the various components of the ecosystem, which is the essential determinant of how a biological community functions (Figure 11-1).

This functional ecosystem concept is very attractive as an organizing principle for the geographic study of the biosphere. It must be approached with caution, however, because of the various scales at which it can be applied. There is an almost infinite variety in the magnitude of ecosystems we might study. At one extreme of scale, for example, we can conceive of a global ecosystem that encompasses the entire biosphere; at the other end of the scale might be the ecosystem of a fallen log, the underside of a rock, or even a drop of water.

If we are going to identify and understand broad distributional patterns in the biosphere, we must focus only on ecosystems that can be recognized at a useful scale.

Biome: A Scale for All Biogeographers

Among terrestrial ecosystems, the type that provides the most appropriate scale for understanding world distribution patterns is called a **biome**, defined as any large, recognizable assemblage of plants and animals in functional interaction with its environment. A biome is usually identified and named on the basis of its dominant vegetation, which normally constitutes the bulk of the *biomass* (the total weight of all organisms—plant and animal) in the biome, as well as being the most obvious and conspicuous visible component of the landscape.

◀ Juvenile African lions drinking at a waterhole in Serengeti National Park, Tanzania.

▲ **Figure 11-1** The flow of energy in a simple ecosystem. Energy from the Sun is fixed by the grass during photosynthesis. The grass is then eaten by a rabbit, which is eaten by a hawk, which then dies. The energy originally contained in the photosynthetic products made by the grass passes through stages during which it is bound up in the body molecules of the rabbit and the hawk but ultimately becomes heat energy lost from the live animals and from the decaying dead matter.

There is no universally recognized classification system of the world's terrestrial biomes, but scholars commonly accept 10 major types:

- Tropical rainforest
- Tropical deciduous forest
- Tropical scrub
- Tropical savanna
- Desert
- Mediterranean woodland and shrub
- Midlatitude grassland
- Midlatitude deciduous forest
- Boreal forest
- Tundra (Arctic and alpine)

A biome comprises much more than merely the plant association that gives it its name. A variety of other kinds of vegetation usually grows among, under, and occasionally over the dominant plants. Diverse animal species also occupy the area. Often, as we saw in Chapter 10, significant and even predictable relationships exist between the biota (particularly the flora) of a biome and the associated climate and soil types.

On any map showing the major biome types of the world (such as Figure 11-25 later in this chapter), the regional boundaries are somewhat arbitrary. Biomes do not occupy sharply defined areas in nature, no matter how

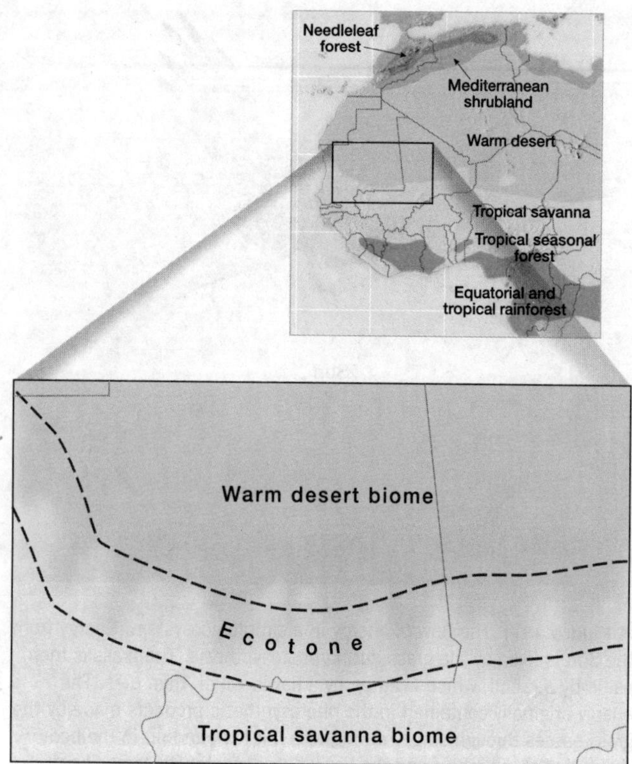

▲ **Figure 11-2** A hypothetical boundary between two biomes of the world. The irregular boundary between the two shows much interfingering, or interdigitation. This transition zone is called an ecotone.

sharp the demarcations may appear on a map. Normally the communities merge more or less imperceptibly with one another through **ecotones**—transition zones of competition in which the typical species of one biome intermingle with those of another (Figure 11-2).

Before we describe the flora and fauna of the world's terrestrial biomes, we need to say something about plants and animals themselves.

Terrestrial Flora

The natural vegetation of the land surfaces of Earth is of interest to the geographer for three reasons:

1. Over much of the planet, the terrestrial (land-dwelling) flora is the most significant visual component of the landscape—topography, soils, animal life, and even water surfaces are often obscured or even completely covered by plants. Only in areas of rugged terrain, harsh climate, or significant human activities are plants not likely to dominate the landscape.
2. Vegetation is a sensitive indicator of other environmental attributes, reflecting subtle variations in sunlight, temperature, precipitation, evaporation, drainage, slope, soil conditions, and other natural parameters. Moreover, the influence of vegetation on soil, animal life, and microclimatic characteristics is frequently pervasive.
3. Vegetation often has a prominent and tangible influence on human settlement and activities. In some cases, it is a barrier or hindrance to human endeavor; in other instances, it provides an important resource to be exploited or developed.

Characteristics of Plants

Despite the fragile appearance of many plants, most varieties are remarkably hardy. They survive, and often flourish, in the wettest, driest, hottest, coldest, and windiest places on Earth. Much of their survival potential is based on a subsurface root system that is capable of sustaining life despite whatever may happen to the above-surface portion of the organism.

The survival capability of a species also depends in part on its reproductive mechanism. Plants that endure seasonal climatic fluctuations from year to year are called **perennials**, whereas those that perish during times of climatic stress (such as winter) but leave behind a reservoir of seeds to germinate during the next favorable period are called **annuals**.

Plant life varies remarkably in form, from microscopic algae to gigantic trees. Most plants, however, have common characteristics—roots to gather nutrients and moisture and anchor the plant; stems and/or branches for support and for nutrient transportation from roots to leaves; leaves to absorb and convert solar energy for sustenance and to exchange gases and transpire water; and reproductive organs for regeneration.

Floristic Terminology

To continue our consideration of plants, we need a specialized vocabulary.

First of all, plants can be divided into two categories: those that reproduce through *spores* and those that reproduce through *seeds*. Plants that reproduce through spores include two major groups:

1. *Bryophytes* include the true mosses, peat mosses, and liverworts. Presumably they have never in geologic history been very dominant among plant communities, except in localized situations.
2. *Pteridophytes* are ferns, horsetails, and club mosses (which are not true mosses). During much of geologic history, great forests of tree ferns, giant horsetails, and tall club mosses dominated continental vegetation, but they are less dominant today.

Plants that reproduce by means of seeds are encompassed in two broad groups:

1. The more primitive of the two groups, the **gymnosperms** ("naked seeds"), carry their seeds in cones, and when the cones open, the seeds fall out. For this reason, gymnosperms are sometimes called **conifers**. Gymnosperms were largely dominant in the geologic past; today the only large surviving gymnosperms are cone-bearing trees such as pines.

2. **Angiosperms** ("vessel seeds") are the flowering plants. Their seeds are encased in some sort of protective body, such as a fruit, nut, or pod. Trees, shrubs, grasses, crops, weeds, and garden flowers are angiosperms. Along with a few conifers, they have dominated the vegetation of the planet for the last 50 or 60 million years.

Several other terms are commonly used to describe vegetation, as summarized in Figure 11-3. Their definitions are not always precise, but their meanings generally are clear.

- One fundamental distinction is made on the basis of stem or trunk composition. *Woody plants* have stems composed of hard fibrous material, whereas *herbaceous plants* have soft stems. Woody plants are mostly trees and shrubs; herbaceous plants are mostly grasses, forbs, and lichens.
- With trees, whether or not a plant loses its leaves sometime during the year is an important distinguishing characteristic. An **evergreen tree** is one that sheds its leaves on a sporadic or successive basis but always appears to be fully leaved. A **deciduous tree** is one that experiences an annual period in which all leaves die and usually fall from the tree, due to either a cold season or a dry season.
- Trees are also often described in terms of leaf shapes. **Broadleaf trees** have leaves that are flat and expansive in shape, whereas **needleleaf trees** are adorned with thin slivers of tough, leathery, waxy needles rather than typical leaves. Almost all needleleaf trees are evergreen, and the great majority of all broadleaf trees are deciduous, except in the rainy tropics, where everything is evergreen.
- *Hardwood* and *softwood* are two of the most unsatisfactory terms in biogeography, but they are widely used in everyday parlance so we must not ignore them. Hardwoods are angiosperm trees that are usually broad leaved and deciduous. Their wood has a relatively complicated structure, but it is not always hard. Softwoods are gymnosperms; nearly all are needleleaf evergreens. Their wood has a simple cellular structure, but it is not always soft.

▲ **Figure 11-3** Terminology used in describing plants. Although the terminology is somewhat confusing, there are conspicuous differences between hardwood (angiosperm) and softwood (gymnosperm) trees. The most obvious difference is in general appearance.

Environmental Adaptations

Despite the hardiness of most plants, there are definite tolerance limits that govern their survival, distribution, and dispersal. During hundreds of millions of years of development, plants have evolved a variety of protective mechanisms to shield against harsh environmental conditions and to enlarge their tolerance limits. Two prominent adaptations to environmental stress are associated with low water availability and high water availability.

Xerophytic Adaptations The descriptive term for plants that are structurally adapted to withstand protracted dry conditions is *xerophytic* (*xero* is Greek for "dry"; *phyt-* comes from phuto-, Greek for "plant"). **Xerophytic adaptations** can be grouped into four general types:

1. Roots are modified in shape or size to enable them to seek widely for moisture. Sometimes *taproots* extend to extraordinary depths to reach subterranean moisture. Also, root modification may involve the growth of a large number of thin hairlike rootlets to penetrate tiny pore spaces in soil (Figure 11-4).
2. Stems are sometimes modified into fleshy, spongy structures that can store moisture. Plants with such fleshy stems are called *succulents*; most cacti are prominent examples.
3. Leaf modification takes many forms; all are designed to decrease transpiration. Sometimes a leaf surface is hard and waxy to inhibit water loss or white and shiny to reflect insolation and thus reduce evaporation. Still more effective is for the plant to have either tiny leaves or no leaves at all. In many types of dry-land shrubs, leaves have been replaced by spines, from which there is virtually no transpiration (Figure 11-5).
4. Perhaps the most remarkable floristic adaptation to aridity is not structural but involves the plant's reproductive cycle. Many xerophytic plants lie dormant for years without perishing. When rain eventually arrives, these plants promptly initiate and pass through an entire annual cycle of germination, flowering, fruiting, and seed dispersal in only a few days, then lapse into dormancy again if the drought resumes.

▲ **Figure 11-4** Desert plants have evolved various mechanisms for survival in an arid climate. Some plants produce long taproots that penetrate deeply in search of the water table. More common are plants that have no deep roots but rather have myriad small roots and rootlets that seek any moisture available near the surface over a broad area.

Hygrophytic Adaptations Plants with **hygrophytic adaptations** are particularly suited to a wet terrestrial environment. Distinction is sometimes made between *hydrophytes* (species living more or less permanently immersed in water, such as the water lilies in Figure 11-6a) and *hygrophytes* (moisture-loving plants that generally require frequent soakings with water, as do many ferns, mosses, and rushes), but both groups are often identified by the latter term.

▲ **Figure 11-5** Cacti mostly or completely lack leaves, but instead have spines from which there is no water loss through transpiration.

Hygrophytes are likely to have extensive root systems to anchor them in the soft ground, and hygrophytic trees often develop a widened, flaring trunk near the ground to provide better support (Figure 11-6b). Many hygrophytic plants that grow in standing or moving water have weak, pliable stems that can withstand the ebb and flow of currents rather than standing erect against them; the buoyancy of the water, rather than the stem, provides support for the plant.

Competition and the Inevitability of Change

As important as climatic, edaphic, and other environmental characteristics are to plant survival, a particular species will not necessarily occupy an area just because all these conditions are favorable. Plants are just as competitive as animals. Of the dozens of plant species that might be suitable for an area, only one or a few are likely to survive. This is not to say that all plants are mutually competitive; indeed, thousands of ecologic niches can be occupied without impinging on one another. However, most plants draw their nutrients from the same soil and their energy from the same Sun, and what one plant obtains cannot be used by another.

The floristic pattern of Earth is impermanent. The plant cover that exists at any given time and place may be in a state of constant change or may be relatively stable for millennia before experiencing significant changes. Sooner or later, however, change is inevitable. Sometimes the change is slow and orderly, as when a lake is filled in, as described in Chapter 10, or when there is a long-term trend toward different climatic conditions over a broad area. On occasion, however, the change is abrupt and chaotic, as in the case of a wildfire.

At some time in Earth's history, all parts of the present landmasses were newly created and therefore unvegetated. Such new land is first occupied temporarily by some plant association that soon gives way to another and then another and another. Many complicated changes occur until some sort of floristic stability is attained. Each succeeding association alters the local environment, making possible the establishment of the next association. The general sequential trend is toward taller plants and greater stability in species composition. The longer plant succession continues the more slowly change takes place because more advanced associations usually contain species that live a relatively long time.

Eventually a plant association of constant composition comes into being. In other words, a point is reached where change is no longer noticeable, and each succeeding generation of the association is much like its predecessor. This stable association is generally referred to as the **climax vegetation**, and the various associations leading up to it are called *seral stages*.

The implication of the term climax vegetation is that the dominant plants of a climax association have demonstrated

(a)

(b)

▲ **Figure 11-6** Some types of plants flourish in a totally aqueous environment. (a) These lily pads virtually cover the surface of a bay of Cold Lake, on the Alberta–Saskatchewan border in western Canada. (b) Many hygrophytic trees, such as these cypress in Louisiana, have wide, flaring trunks near the ground to provide firmer footing in the wet environment.

that, of all possibilities for that particular situation, they can compete the most successfully. Thus, they represent the optimal floristic cover for that environmental context. The climax vegetation presumably persists unchanged for an indefinite period until the next environmental disturbance. The climax vegetation is, then, an association in equilibrium with prevailing environmental conditions. When these conditions change, the climax stage is disturbed and another succession sequence is initiated.

Spatial Associations of Plants

The geographer attempting to recognize spatial groupings of plants faces some significant difficulties. Plant associations that are similar in appearance and in environmental relationships can occur in widely separated localities and are likely to contain totally different species. At the other extreme, exceedingly different plant associations can often be detected within a very small area. As the geographer tries to identify patterns and recognize relationships, generalization invariably is needed. This generalization must accommodate gradations, ecotones, interdigitations, and other irregularities. When associations are portrayed on maps, therefore, their boundaries represent approximations in nearly all cases.

Another problem is that in many areas of the world the natural vegetation has been completely removed or replaced through human interference. Forests have been cut, crops planted, pastures seeded, and urban areas paved. Over extensive areas of Earth's surface, therefore, climax vegetation is the exception rather than the rule. Most world maps that purport to show natural vegetation ignore human interference and are actually maps of theoretical natural vegetation, in which the mapmaker makes assumptions about what the natural vegetation would be if it had not been modified by human activity.

The Major Floristic Associations There are many ways to classify plant associations. For broad geographical purposes, emphasis is usually placed on the structure and appearance of the dominant plants. The major associations generally recognized (Figure 11-7) include the following:

1. **Forests** consist of trees growing so close together that their individual leaf canopies generally overlap. This means that the ground is largely in shade, a condition that usually precludes the development of much undergrowth. Forests require considerable annual precipitation but can survive in widely varying temperature zones. Except where moisture is inadequate or the growing season very short, forests are likely to become the climax vegetation association in any area. Trees depend so much on the availability of moisture primarily because, unlike other plants, they must have a mechanism for transporting mineral nutrients a relatively great distance from their roots to their leaves. Such transport can take place only in a dilute solution; therefore much water is needed by trees throughout the growing season. Other plant forms can flourish in areas of relatively high precipitation, but they rarely become dominant because they are shaded out by trees.
2. **Woodlands** are tree-dominated plant associations in which the trees are spaced more widely apart than in forests and do not have interlacing canopies (Figure 11-8). Ground cover may be either dense or sparse, but it is not inhibited by lack of sunlight.

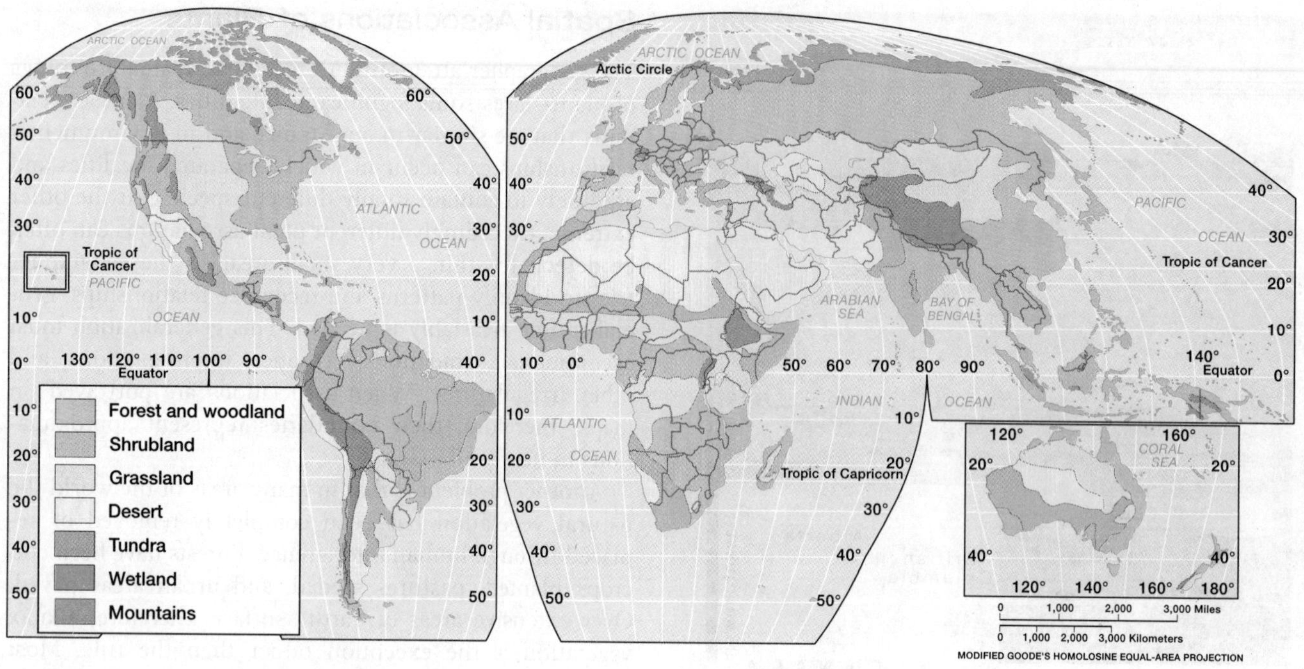

▲ **Figure 11-7** The major natural vegetation associations.

Woodland environments are generally drier than forest environments.

3. **Shrublands** are plant associations dominated by relatively short woody plants generally called *shrubs* or *bushes*. Shrubs take a variety of forms, but most have several stems branching near the ground and leafy foliage that begins very close to ground level. Trees and grasses may be interspersed with the shrubs but are less prominent in the landscape. Shrublands have a wide latitudinal range, but they are generally restricted to semiarid or arid locales.
4. **Grasslands** may contain scattered trees and shrubs, but the landscape is dominated by grasses and forbs (broadleaf herbaceous plants). Prominent types of grassland include *savanna*, low-latitude grassland characterized by tall grasses; *prairie*, midlatitude grassland characterized by tall grasses; and *steppe*, midlatitude grassland characterized by short grasses and bunchgrasses. Grasslands are associated with semiarid and subhumid climates.
5. **Deserts** are typified by widely scattered plants with much bare ground interspersed. Desert is actually a

► **Figure 11-8** A piñon-juniper woodland scene in central Arizona.

◄ **Figure 11-9** Some deserts contain an abundance of plants, but they are usually spindly and always xerophytic. This scene is from the Tonto National Forest near Phoenix, Arizona.

climatic term, and desert areas may have a great variety of vegetation, including grasses, succulent herbs, shrubs, and straggly trees (Figure 11-9). Some extensive desert areas comprise loose sand, bare rock, or extensive gravel, with virtually no plant growth.

6. **Tundra**, as we noted in Chapter 8, consists of a complex mix of very low plants, including grasses, forbs, dwarf shrubs, mosses, and lichens, but no trees. Tundra occurs only in the perennially cold climates of high latitudes or high altitudes.
7. **Wetlands** have a much more limited geographic extent than the associations described above. They are characterized by shallow standing water all or most of the year, with vegetation rising above the water level. The most widely distributed wetlands are swamps (with trees as the dominant plant forms) and marshes (with grasses and other herbaceous plants dominant) (for example, see Figure 11-6).

Vertical Zonation In Figure 8-37 we learned that mountainous areas often have a distinct pattern of **vertical zonation** in vegetation patterns (Figure 11-10). Significant elevational changes in short horizontal distances cause various plant associations to exist in relatively narrow zones on mountain slopes. This zonation is largely due to the effects of elevation on temperature and precipitation (Figure 11-11).

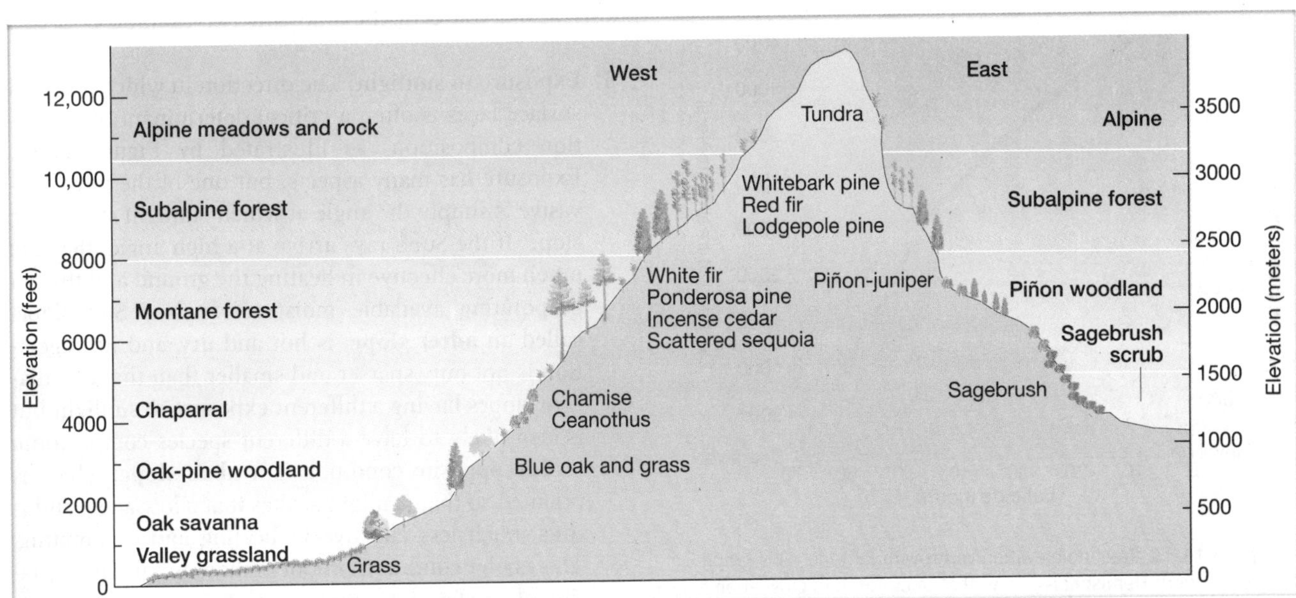

▲ **Figure 11-10** A west-east profile of California's Sierra Nevada, indicating the principal vegetation at different elevations on the western (wet) and eastern (dry) sides of the range. Characteristic plants within each vegetation zone are labeled along the mountain slope.

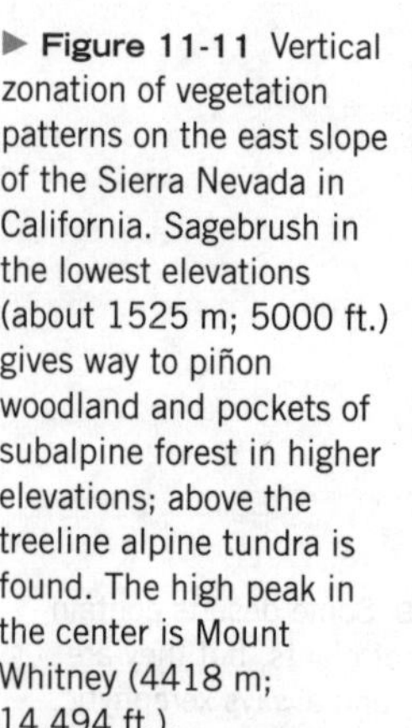

▶ **Figure 11-11** Vertical zonation of vegetation patterns on the east slope of the Sierra Nevada in California. Sagebrush in the lowest elevations (about 1525 m; 5000 ft.) gives way to piñon woodland and pockets of subalpine forest in higher elevations; above the treeline alpine tundra is found. The high peak in the center is Mount Whitney (4418 m; 14,494 ft.).

The essential implication is that elevation changes are the counterpart of latitude changes. In other words, to travel from sea level to the top of a tall tropical peak is roughly equivalent environmentally to a horizontal journey from the equator to the Arctic. This elevation–latitude relationship is shown most clearly by how the elevation of the upper **treeline** (the elevation above which trees are unable to survive, especially due to low summer temperatures and moisture availability) varies with latitude (Figure 11-12).

An interesting detail of vertical zonation is that the elevation–latitude graph for the Southern Hemisphere is different from that of the Northern Hemisphere. For example, between latitudes 35° S and 40° S in Australia and New Zealand, the treeline is below 1800 meters (6000 feet); at comparable latitudes in North America, the treeline is nearly twice as high. The reason for this significant discrepancy is not completely understood.

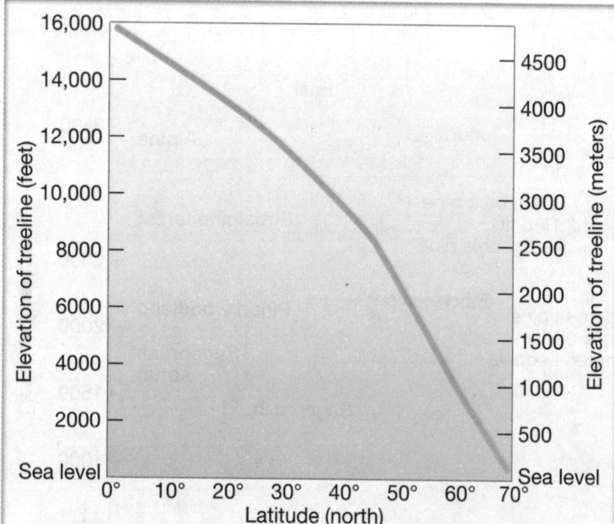

▲ **Figure 11-12** Treeline elevation varies with latitude. This graph for the Northern Hemisphere shows that trees cease to grow at an elevation of about 5000 meters (16,000 feet) in the equatorial Andes of South America but at only 3000 meters (10,000 feet) at 40°N latitude in Colorado. At 70°N latitude in northern Canada, the treeline is at sea level.

Treeline variation represents only one facet of the broader design of vertical zonation in vegetation patterns. All vegetation zones are displaced downward with increasing distance from the equator. This principle accounts for the significant vegetational complexity found in all mountainous areas.

Local Variations Each major vegetation association extends over a large area of Earth's surface. Within a given association, however, there are also significant local variations caused by a variety of local environmental conditions, as illustrated by the following two examples:

1. **Exposure to sunlight**: The direction in which a sloped surface faces is often a critical determinant of vegetation composition, as illustrated by Figure 11-13. Exposure has many aspects, but one of the most pervasive is simply the angle at which sunlight strikes the slope. If the Sun's rays arrive at a high angle, they are much more effective in heating the ground and thus in evaporating available moisture. Such a Sun slope, called an **adret slope**, is hot and dry, and its vegetation is not only sparser and smaller than that on adjacent slopes having a different exposure to sunlight but is also likely to have a different species composition.

   The opposite condition is a **ubac slope**, which is oriented so that sunlight strikes it at a low angle and is thus much less effective in heating and evaporating. This cooler condition produces more luxuriant vegetation of a richer diversity.

   The difference between adret and ubac slopes decreases with increasing latitude, presumably because the ubac flora becomes impoverished under the cooler

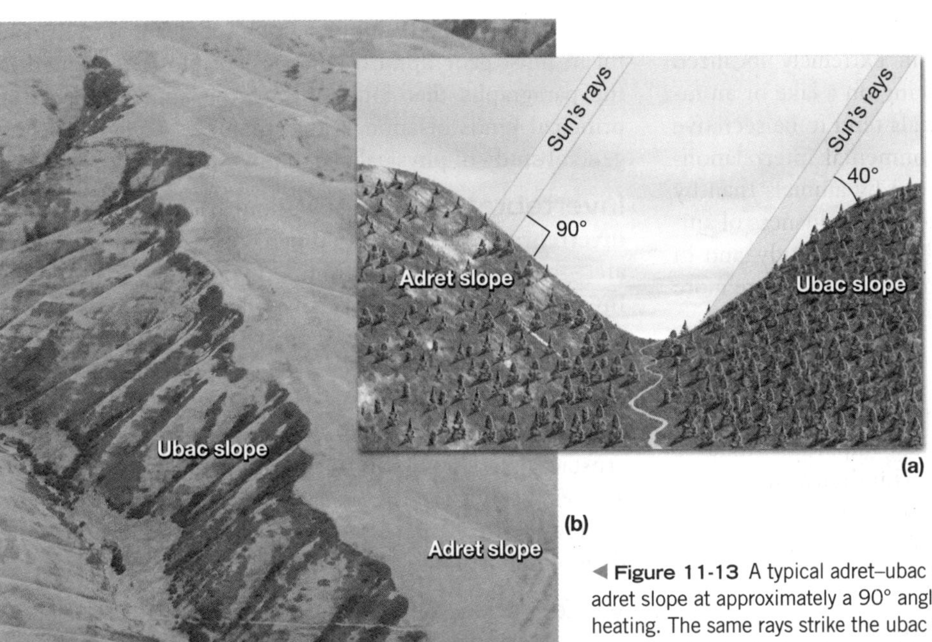

◀ **Figure 11-13** A typical adret–ubac situation. (a) The noon Sun rays strike the adret slope at approximately a 90° angle, a condition that results in maximum heating. The same rays strike the ubac slope at approximately a 40° angle, with the result that the heating is spread over a large area and is therefore less intense. (b) Vegetation differences on south-facing slopes (left and right) and the cooler north-facing slope (center) in the Blue Mountains of eastern Oregon.

conditions that prevail as latitude increases and the relative warmth of the adret surface encourages plant diversity.

2. **Valley-bottom location:** In mountainous areas where a river runs through a valley, the vegetation associations growing on either side of the river have a composition that is significantly different from that found higher up on the slopes forming the valley. This floral gradient is sometimes restricted to immediate streamside locations and sometimes extends more broadly over the valley floor.

The difference is primarily a reflection of the perennial availability of subsurface moisture near the stream and is manifested in a more diversified and more luxuriant flora. This vegetative contrast is particularly prominent in dry regions, where streams may be lined with trees even though no other trees are to be found in the landscape. Such streamside growth is called **riparian vegetation** (Figure 11-14).

Terrestrial Fauna

Animals occur in much greater variety than plants over Earth. As objects of geographical study, however, they have been relegated to a place of lesser prominence for at least

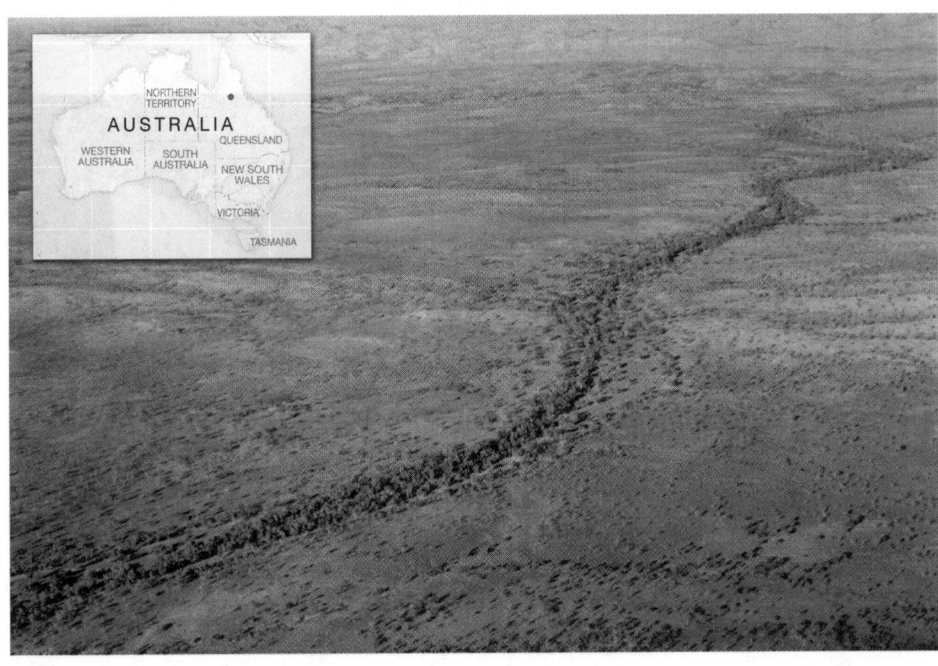

◀ **Figure 11-14** Riparian vegetation is particularly prominent in dry lands. This stream in northern Queensland (Australia) would be inconspicuous in this photograph if it were not for the trees growing along it.

two reasons. First, animals are much less prominent than plants in the landscape; apart from extremely localized situations—such as waterfowl flocking on a lake or an insect plague attacking a crop—animals tend to be secretive and inconspicuous. Second, environmental interrelationships are much less clearly evidenced by animals than by plants. This is due in part to the inconspicuousness of animals, which renders them more difficult to study, and in part to the fact that animals are mobile and therefore more able to adjust to environmental variability.

This is not to say that fauna is inconsequential for students of geography. Under certain circumstances, wildlife is a prominent element of physical geography; and in some regions of the world, it is an important resource for human use and/or a significant hindrance to human activity. Moreover, it is increasingly clear that animals are sometimes more sensitive indicators than plants of the health of a particular ecosystem.

Characteristics of Animals

The diversity of animal life forms is not realized by most people. We commonly think of animals as being relatively large and conspicuous creatures that run across the land or scurry through the trees, seeking to avoid contact with humankind. In actuality, the term animal encompasses not only the larger, more complex forms, but also hundreds of thousands of species of smaller and simpler organisms that may be inconspicuous or even invisible. The variety of animal life is so great that it is difficult to find many unifying characteristics. The contrast, for example, between an enormous elephant and a tiny insect is so extreme as to make their kinship appear ludicrous. Animals really have only two universal characteristics, and even these are so highly modified in some cases as to be almost unrecognizable:

1. Animals are motile, which means that they are capable of self-generated movement.
2. Animals must eat plants and/or other animals for sustenance. They are incapable of manufacturing their food from air, water, and sunlight, which plants can do.

Kinds of Animals

The vast majority of animals are so tiny and/or secretive as to be either invisible or extremely inconspicuous. Their size and habits, however, are not valid indicators of their significance for geographic study. Very minute and seemingly inconsequential organisms sometimes play exaggerated roles in the biosphere—as carriers of disease or as hosts of parasites, sources of infection, or providers of scarce nutrients. For example, no geographic assessment of Africa, however cursory, can afford to ignore the presence and distribution of the small tsetse fly and the tiny protozoan called *Trypanosoma*, which together are responsible for the transmission of trypanosomiasis (sleeping sickness), a widespread and deadly disease for humans and livestock over much of the continent.

Zoological classification is much too detailed to be useful in most geographical studies. Therefore, presented in the paragraphs that follow is a brief summation of the principal kinds of animals that might be recognized in a general study of physical geography.

Invertebrates Animals without backbones are called **invertebrates**. More than 90 percent of all animal species are encompassed within this broad grouping. Invertebrates include worms, sponges, mollusks, various marine animals, and a vast host of creatures of microscopic or near-microscopic size. Very prominent among invertebrates are the *arthropods*, a group that includes insects (Figure 11-15), spiders, centipedes, millipedes, and crustaceans (shellfish). With some 300,000 recognized species, beetles are the most numerous of animals, comprising about 40 percent of all insect species and more than one-fourth of all known animals.

Vertebrates **Vertebrates** are animals that have a backbone that protects the main nerve (or spinal cord). Geographers generally follow biologists in recognizing five principal groups of vertebrates:

1. *Fishes* are the only vertebrates that can breathe under water (a few species are also capable of breathing in air). Most fishes inhabit either freshwater or saltwater only, but some species are capable of living in both environments (such as the bull shark, *Carcharhinus leucas*); and several species, most notably salmon, spawn in freshwater streams but live most of their lives in the ocean.
2. *Amphibians* are semiaquatic animals. When first born, they are fully aquatic and breathe through gills; as adults, they are air-breathers by means of lungs and through their glandular skin. Most amphibians are either frogs or salamanders.
3. Most *reptiles* are totally land based. Ninety-five percent of all reptile species are either snakes or lizards. The remainder are mostly turtles and crocodilians.

▲ **Figure 11-15** Insects, such as this moth, are arthropods, the largest group of invertebrate animals.

4. *Birds* are believed to have evolved from reptiles; indeed, they have so many reptilian characteristics that they have been called "feathered reptiles." There are more than 9000 species of birds, all of which reproduce by means of eggs. Birds are so adaptable that some species can live almost anywhere on Earth's surface. They are **endothermic**, which means that, regardless of the temperature of the air or water in which they live, they maintain a constant body temperature.
5. *Mammals* are distinguished from all other animals by several internal characteristics as well as by two prominent external features. The external features are that only mammals produce milk with which they feed their young, and only mammals possess true hair. (Some mammals have very little hair, but no creatures other than mammals have any hair.) Mammals are also notable for being endothermic. Thus, the body temperature of mammals and birds stays about the same under any climatic conditions, which enables them to live in almost all parts of the world.

The great majority of all mammals are *placentals*, which means that their young grow and develop in the mother's body, nourished by an organ known as the *placenta*, which forms a vital connecting link with the mother's bloodstream (Figure 11-16). A small group of mammals (about 135 species) are *marsupials*, whose females have pouches in which the young, which are born in a very undeveloped condition, live for several weeks or months after birth (Figure 11-17). The most primitive of all mammals are the *monotremes*, of which only two types exist (Figure 11-18); they are egg-laying mammals.

Environmental Adaptations

As with plants, animals have evolved slowly and diversely through eons of time. Evolution has made it possible for animals to diverge remarkably in adjusting to different environments. Just about every existing environmental extreme has been met by some (or many) evolutionary adaptations that make it feasible for some (or many) animal species to survive and even flourish.

▲ **Figure 11-16** Most large land animals are placental mammals, as exemplified by this bull moose in Yellowstone National Park.

▲ **Figure 11-17** A red kangaroo joey (baby) peers out from its mother's pouch. As with other marsupials, baby kangaroos live for a time in their mother's pouch after birth.

Physiological Adaptations The majority of animal adaptations to environmental diversity have been physiological, which is to say that they are anatomical and/or metabolic changes. A classic example is the size of fox ears. Ears are prime conduits for body heat in furred animals as they provide a relatively bare surface for its loss. Arctic foxes (Figure 11-19a) have unusually small ears, which minimizes heat loss; desert foxes (Figure 11-19b) possess remarkably large ears, which are a great advantage during the blistering heat of desert summers.

There are several hundred species of mammals whose skin is covered with a fine, soft, thick, hairy coat referred to as *fur*. These mammals range in size from tiny mice and moles to the largest bears. An examination of their ranges reveals several generalized habitat preferences. Many fur-bearing species live in high-latitude and/or high-elevation locations, where winters are long and cold; a number of fur-bearing species live in aquatic environments; the remainder are widely scattered over the continents, occupying a considerable diversity of habitats. We can conclude tentatively that many fur-bearing species, including all those with the heaviest, thickest fur, live in regions where

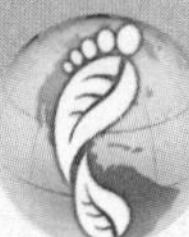

PEOPLE AND THE ENVIRONMENT

Changing Climate Affects Bird Populations

Over the last decade, scientists have documented changes in a number of animal populations that appear to be tied to global warming. Bird populations can be especially sensitive indicators of climate change because many species of birds have specific environmental requirements for feeding, nesting, and migration. Further, since birds are generally quite mobile, a bird population may respond fairly quickly to local environmental changes.

Data suggests that higher global temperatures are already having an effect on some bird populations. The National Audubon Society's annual Christmas Bird Count has gathered information about bird populations for many years. Data from the Christmas Bird Count shows that in warmer years most bird species do not fly as far south for the winter as they do in cooler years; in warmer years, birds may also spend their summers farther north.

In February 2009 the National Audubon Society released a study based on data gathered over the last 40 years with its Christmas Bird Counts. It reports that 177 of 305 North American bird species are wintering an average of 56 kilometers (35 miles) farther north than 40 years ago (Figure 11-A). The "center of abundance" for more than 60 bird species shifted 160 kilometers (100 miles) or more north. The report concludes that higher temperatures are the most likely cause of these shifts—increasing temperatures in winter make areas that were once too cold now habitable for some bird species.

A changing climate may also be affecting the timing of bird migrations. Research has shown that several species of birds that migrate north into Michigan are arriving two or three weeks earlier in the spring than they did 45 years ago. Birds in some locations are nesting earlier because plants are blooming sooner—which results in an earlier proliferation of insects on which the birds feed. The British Trust for Ornithology reports that 20 out of 65 species of birds they studied are laying eggs nine days sooner today than they did just 35 years ago.

Although some bird species may be able to adjust to a warming climate by altering their migration patterns—shifting their ranges north in the Northern Hemisphere, for example—this may not actually occur in some cases. New evidence suggests that habitats may respond to climate change in different ways. Forest communities in North America may not simply shift north of their present locations in response to a warmer climate. Instead, individual plant and animal species within that forest community may respond differently to climate change, perhaps leaving both a dislocated and an altered habitat for birds and other creatures.

One of the earliest studies of the potential effects of global warming on bird populations in North America looked at the songbird the bobolink (*Dolichonyx oryzivorus*; Figure 11-B). Bobolinks have an extraordinarily long migration—a round trip of about 20,000 kilometers (12,400 miles). In summer, they nest in meadows and fields in southern Canada, New England, and south of the Great Lakes to Idaho. In the Northern Hemisphere winter they fly south to the grasslands of northern Argentina, Paraguay, and southwestern Brazil. A study in the mid-1990s modeled the relationship between the distribution of boblinks and climate. The study concluded that with the projected climate and vegetation changes associated with a doubling of atmospheric CO_2, the bobolink might not be able to nest south of the Great Lakes, depriving this region of one of its most beloved songbirds.

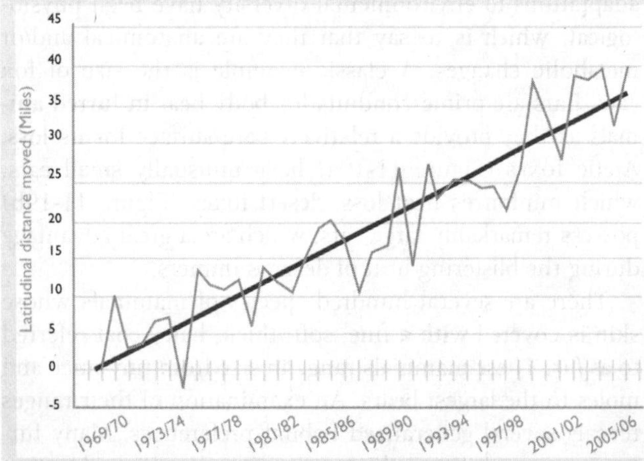

▲ **Figure 11-A** Change in the center of abundance of 305 widespread bird species in North America as determined by National Audubon Society Christmas Bird Counts; as winter temperatures have increased since 1970, the range of many bird species has also shifted north.

▲ **Figure 11-B** Adult male bobolink, *Dolichonyx oryzivorus*.

(a)

(b)

▲ **Figure 11-18** There are only two kinds of monotremes, or egg-laying mammals, in existence—the echidna and the duckbill platypus. (a) The echidna is found only in Australia and New Guinea. (b) The platypus is totally aquatic in its lifestyle.

cold temperatures are common, but the climatic correlation in this case is partial and indistinct.

The catalog of similar adaptations is almost endless: webbing between toes to make swimming easier; broad feet that will not sink in soft snow; increase in size and number of sweat glands to aid in evaporative cooling; and a host of others.

Behavioral Adaptations An important advantage that animals have over plants, in terms of adjustment to environmental stress, is that the former can move about and therefore modify their behavior to minimize the stress. Animals can seek shelter from heat, cold, flood, or fire; they can travel far in search of relief from drought or famine; and they can shift from daytime (diurnal) to nighttime (nocturnal) activities to minimize water loss during hot seasons. Such techniques as migration (periodic movement from one region to another), *hibernation* (spending winter in a dormant condition), and *estivation* (spending a dry–hot period in an inactive state) are behavioral adaptations employed regularly by many species of animals.

(a)

(b)

▲ **Figure 11-19** (a) An Arctic fox near Cape Churchill in the Canadian province of Manitoba. Its tiny ears are an adaptation to conserve body heat in a cold environment. (b) Desert foxes have large ears, the better to radiate away body heat. This is a bat-eared fox in northern Kenya.

Reproductive Adaptations Harsh environmental conditions are particularly destructive to the newly born. As partial compensation for this factor, many species have evolved specialized reproductive cycles or have developed modified techniques of baby care. During lengthy periods of bad weather, for example, some species delay mating or postpone nest building. If fertilization has already taken place, some animal reproductive cycles are capable of almost indefinite delay, resulting in a protracted egg or larval stage or even total suspension of embryo development until the weather improves.

If the young have already been born, they sometimes remain longer than usual in nest, den, or pouch, and the adults may feed them for a longer time. When good weather finally returns, some species are capable of hastened *estrus* (the period of heightened sexual receptivity by the female), nest building, den preparation, and so on, and the progeny produced may be in greater than normal numbers.

Example of Animal Adaptations to Desert Life The desert is a classic illustration of a stressful environment for biotic life, as well as the many adaptations on the part of animals.

The simplest and most obvious adaptation to life in an arid land is for the animals to remain near a perpetual source of water. Thus, the permanent streams and enduring springs of the desert attract a resident faunal population that is often rich and diverse. Birds are more dependent on an open source of water than any other nonaquatic animal, so they often remain near such features throughout the day, or at least visit the water with some frequency, and many other kinds of animals also congregate near open water. Even in areas where open water is not available, it is rare for the desert to be uniformly inhospitable.

Often animals follow the rains in nomadic fashion. Many species demonstrate remarkable instincts for knowing where precipitation has occurred and will travel tens or even hundreds of kilometers to take advantage of locally improved conditions. A high proportion of desert bird species can be classed as nomads, in contrast to the large proportion of sedentary migratory species in other parts of the world. Some of the larger mammals also have rain-following habits. Some African antelopes and Australian kangaroos have been observed to double in numbers almost overnight after a good rain, a feat impossible except by rapid, large-scale movement from one area to another.

(a)

(b)

▲ **Figure 11-20** Desert animals exhibit a number of remarkable adaptations. (a) The Ord kangaroo rat in Arizona can live its entire life without taking a drink, receiving all necessary moisture from the food it eats. (b) Camels are ideal pack animals in arid environments, as here in the Sudan.

Many smaller desert creatures spend a great deal of time underground beneath the level of desiccating heat. Most desert rodents and reptiles live in underground burrows, and many lesser creatures do the same. Desert frogs, while not numerous, are noted burrowers; some have been found as deep as 2 meters (6 feet) underground. Freshwater crayfish and crabs often survive long dry spells by burying themselves. Most desert ants and termites live in underground nests. Most desert animals except birds, ungulates, and many flying insects are almost completely nocturnal. There is clearly much more wildlife activity at night than in the daytime.

A few species of rodents, most notably kangaroo rats (Figure 11-20a), can exist from birth to death without ever taking a drink, surviving exclusively on moisture ingested with their food.

Of all animals, perhaps the *dromedary* or one-humped camel (*Camelus dromedarius*) has developed the most remarkable series of adjustments to the desert environment (Figure 11-20b). The upper lip of the dromedary is deeply cleft, with a groove that extends to each nostril so that any moisture expelled from the nostrils can be caught in the mouth rather than being wasted. The nostrils consist of horizontal slits that can be closed tightly to keep out blowing dust and sand. The eyes are set beneath shaggy brows, which help to shield them from the Sun's glare; the eyes are further protected from blowing sand by a complex double set of eyelids. Dromedaries can tolerate extreme dehydration without their body temperatures rising to a fatal level: whereas humans and most other large animals have body temperatures that fluctuate by only 1° to 2°C (2° to 4°F) on hot days, the range for dromedaries is 7°C (12°F), meaning that the dromedary sweats little except during the very hottest hours, thereby conserving body water. When drinking water is available, they are capable of completely rehydrating in only a short time.

Perhaps the most astounding of all faunal adaptations to arid conditions is one that is overtly very similar to a vegetational adaptation—the ability to delay reproductive processes over long dry periods until more favorable conditions occur, at which time rapid breeding or birth can allow remarkable population regeneration. Australian desert kangaroos, for example, are capable of "delayed implantation," in which a fertilized *blastocyst* (early stage embryo) can remain in an inactive state of development in the uterus during a period of difficult living and then spontaneously resume normal development after conditions have improved. Desert birds may experience enormous die-offs when waters dry up during a prolonged drought, but when rains finally come, the survivors may begin nest construction within a week and ovulation within a fortnight. Moreover, clutch size (the number of eggs laid) may increase, as may the number of clutches produced.

Invertebrates, too, can take advantage of a favorable weather change to proliferate their numbers extraordinarily. Many of them survive the dry period in an egg or larval stage that is extremely resistant to desiccation; when the drought breaks, the egg hatches or the larva develops into an active stage that may not be at all drought resistant. Development continues with great rapidity until the adult form can breed and more drought resistant eggs can be laid. This is a commonplace circumstance for brine shrimp, crayfish, grasshoppers, locusts, flies, mosquitoes, and various other desert arthropods; cases have been known in which more than a quarter of a century passed between successive generations of these tiny creatures in which the adult life cycle lasts for only a few weeks!

Competition among Animals

Competition among animals is even more intense than that among plants because the former involves not only indirect competition in the form of rivalry for space and resources but also the direct antagonism of predation. Animals with similar dietary habits compete for food and occasionally for territory. Animals in the same area also sometimes compete for water. And animals of the same species often compete for territory and for mates. Across this matrix of ecological rivalry is spread a prominent veneer of predator–prey relationships.

Competition among animals is a major part of the general struggle for existence that characterizes natural relationships in the biosphere. Individual animals are concerned either largely or entirely with their own survival (and sometimes with that of their mates) in response to normal primeval instincts. In some species, this concern is broadened to include their own young, although such maternal (and, much more rarely, paternal) instinct is by no means universal among animals. Still fewer species show individual concern for the group, as represented in colonies of ants or prides of lions. For the most part, however, animal survival is a matter of every creature for itself, with no individual helping another and no individual deliberately destroying another apart from normal predatory activities.

Cooperation among Animals

Many animals live together in social groups of varying sizes, generally referred to as herds, flocks, or colonies. This is a common, but by no means universal, behavioral characteristic among animals of the same species (Figure 11-21), and sometimes a social group encompasses several species in a communal relationship, such as zebras, wildebeest, and impalas living together on an East African savanna. Within such groupings, there may be a certain amount of cooperation among unrelated animals, but competition for both space and resources is likely to be prominent as well.

Symbiosis is the arrangement in which two dissimilar organisms live together. There are three principal forms:

1. *Mutualism* involves a mutually beneficial relationship between the two organisms, as exemplified by the tickbirds that are constant companions of many African ungulates (hoofed animals). The birds aid the mammals by removing ticks and other insects that

◀ **Figure 11-21** Seals and sea lions are among the most gregarious of mammals. This "hauling-out" beach near Point Año Nuevo on the central coast of California is crowded with elephant seals.

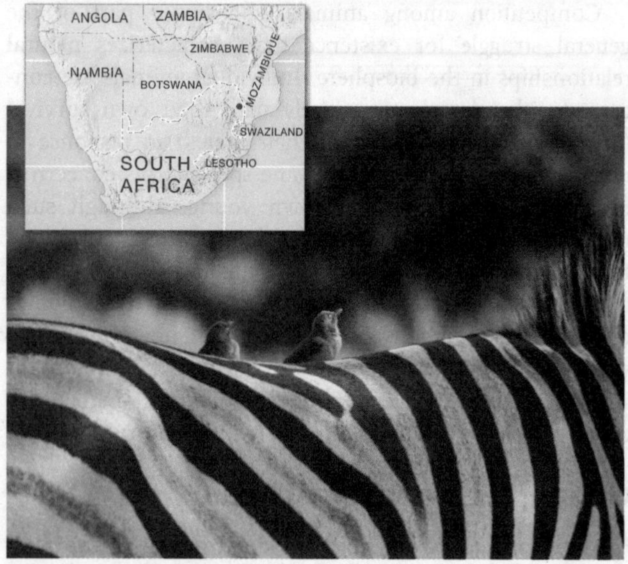

▲ **Figure 11-22** There is often a symbiotic relationship between hoofed animals and insectivorous birds. In this scene from South Africa's Kruger National Park two red-billed oxpeckers are riding on a zebra as they look for tasty morsels.

infest the latter's skin, and the birds benefit by having a readily available supply of food (Figure 11-22).

2. *Commensalism* involves two dissimilar organisms living together with no injury to either, as represented by burrowing owls sharing the underground home of prairie dogs.
3. *Parasitism* involves one organism living on or in another, obtaining nourishment from the host, which is usually weakened and sometimes killed by the actions of the parasite. Mistletoe, for example, is a parasite of forest trees that is widespread in North America and Europe.

Zoogeographic Regions

The distribution of animals over the world is much more complex and irregular than that of plants, primarily because animals are mobile and therefore capable of more rapid dispersal. As with plants, however, the broad distributions of animals are reflective of the general distribution of energy and of food diversity. Thus, the richest faunal assemblages are found in the permissive environment of the humid tropics, and the dry lands and cold lands have the sparsest representations of both species and individuals.

When considering the global patterns of animal geography, most attention is usually paid to the distribution of terrestrial vertebrates, with other animals being given only casual notice. The classical definition of world zoogeographic regions is credited to the nineteenth-century British naturalist A. R. Wallace, whose scheme is based on the work of P. L. Sclater. As shown in Figure 11-23, nine **zoogeographic regions** are generally recognized, but you should understand that this or any other system of faunal regions represents average conditions and cannot portray some common pattern in which different groups of animals fit precisely. It is simply a composite of many diverse distributions of contemporary fauna.

The *Ethiopian Region* is primarily tropical or subtropical and has a rich and diverse fauna. It is separated from other regions by an oceanic barrier on three sides and a broad

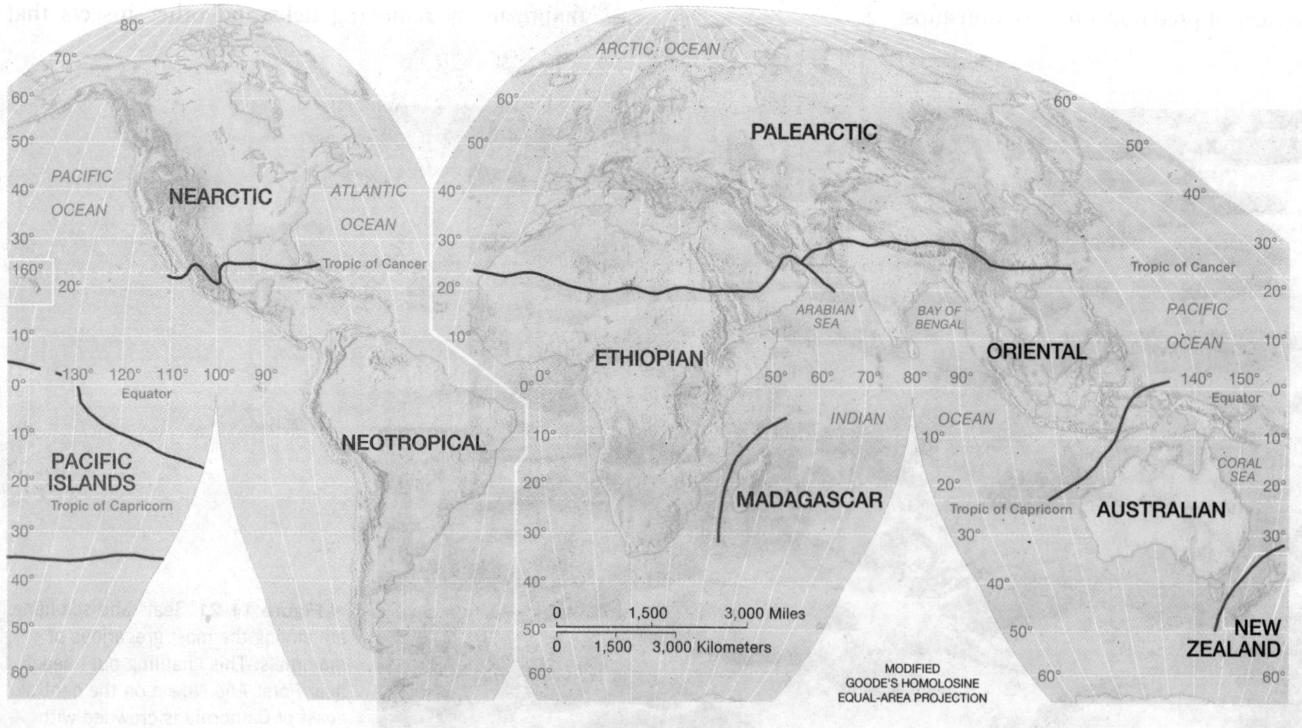

▲ **Figure 11-23** Zoogeographic regions of the world.

desert on the fourth. Despite its isolation, however, the Ethiopian Region has many faunal affinities with the Oriental and Palearctic regions. Its vertebrate fauna is the most diverse of all the zoogeographic regions and includes the greatest number of mammalian families, many of which have no living relatives outside Africa.

The *Oriental Region* is separated from the rest of Eurasia by mountains. Its faunal assemblage is generally similar to that of the Ethiopian Region, with somewhat less diversity. The Oriental Region has some **endemic** groups (*endemic* means these groups are found nowhere else) and a few species that are found only in the Oriental, Palearctic, and Australian regions. Many brilliantly colored birds live in the Oriental Region, and reptiles are numerous, with a particularly large number of venomous snakes.

The *Palearctic Region* includes the rest of Eurasia, all of Europe, and most of North Africa. Its fauna as a whole is much poorer than that of the two regions previously discussed, which is presumably a function of its location in higher latitudes with a more rigorous climate. This region has many affinities with all three bordering regions, particularly the Nearctic. Indeed, the Palearctic has only two minor mammal families (both rodents) that are endemic, and almost all of its birds belong to families that have a very wide distribution.

The *Nearctic Region* consists of the nontropical portions of North America. Its faunal assemblage (apart from reptiles, which are well represented) is relatively poor and is largely a transitional mixture of Palearctic and Neotropical groups. It has few important groups of its own except for freshwater fishes. The considerable similarities between Palearctic and Nearctic fauna have persuaded some zoologists to group them into a single superregion, the *Holarctic*, which had a land connection in the recent geologic past when glaciation lowered global sea level—the Bering land bridge across which considerable faunal dispersal took place (Figure 11-24).

The *Neotropical Region* encompasses all of South America and the tropical portion of North America. Its fauna is rich and distinctive, which reflects both a variety of habitats and a considerable degree of isolation from other regions. Neotropic faunal evolution often followed a path different from that in other regions. It contains a larger number of endemic mammal families than any other region. Moreover, its bird fauna is exceedingly diverse and conspicuous.

The *Madagascar Region*, restricted to the island of that name, has a fauna very different from that of nearby Africa. The Madagascan fauna is dominated by a relic assemblage of unusual forms in which primitive primates (lemurs) are notable.

The *New Zealand Region* has a unique fauna dominated by birds, with a remarkable proportion of flightless types. It has almost no terrestrial vertebrates (no mammals; only a few reptiles and amphibians).

The *Pacific Islands Region* includes a great many far-flung islands, mostly quite small. Its faunal assemblage is very limited.

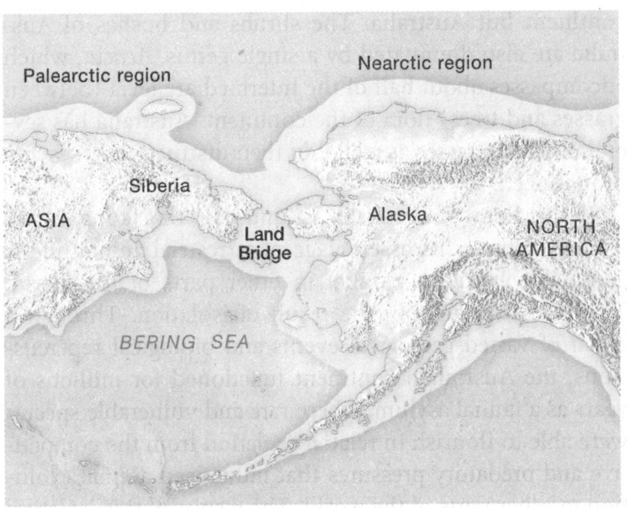

▲ **Figure 11-24** The Bering Land Bridge facilitated the interchange of animals between the Palearctic and Nearctic regions when sea level was lower during recent glaciations. Most of the dispersal was from Asia (Palearctic) to North America (Nearctic).

The *Australian Region* is restricted to the continent of Australia and some adjacent islands, particularly New Guinea. Its fauna is by far the most distinctive of any major region, due primarily to its lengthy isolation from other principal landmasses. The lack of variety of its vertebrate fauna is made up for by uniqueness. There are only nine families of terrestrial mammals, but eight of them are unique to the region. The bird fauna is varied, and both pigeons and parrots reach their greatest diversity here. There is a notable scarcity of freshwater fishes and amphibians. The Australian Region has a moderate amount of reptiles, mostly snakes and lizards, including the largest lizard of all: the Komodo dragon. Within the region, there are many significant differences between the fauna of Australia and that of New Guinea.

So why is the Australian Region so different from the other zoogeographic regions? The answer will become clearer when we discuss plate tectonics in Chapter 14, but for now, a quick explanation.

The Unique Biota of Australia The unusual faunal and floristic developments of Australia are largely the result of isolation. During long periods of time in the geologic past, while most of the continents were connected, the climate was more equitable, allowing for a vast evolution of plant species. As the continents began to separate about two hundred million years ago, climatic changes occurred requiring new adaptation and continued evolution. Australia became isolated from the rest of the continents, and so evolution continued with little genetic influence from the outside. This has engendered specialized evolutionary development among the isolated flora and fauna of Australia.

The most notable vegetation distinction is that nearly all the native trees in Australia—more than 90 percent of the total—are members of a single genus, *Eucalyptus*. Moreover, the eucalyptus (see Figure 10-13), of which there are more than 400 species, are native to no other continent but Australia. The shrubs and bushes of Australia are also dominated by a single genus, *Acacia*, which encompasses about half of the intermediate level (between grasses and trees) flora of the continent. Australia has several unusual grasses as well, but their distinctiveness is less pronounced.

If the flora of Australia is unusual, the fauna is absolutely unique; its assemblage of terrestrial animal life is completely without parallel in other parts of the world. This, too, is primarily the result of isolation. Through a chain of varied geological events and biological repercussions, the Australian continent functioned for millions of years as a faunal asylum, where rare and vulnerable species were able to flourish in relative isolation from the competitive and predatory pressures that influenced animal evolution in other parts of the world. The results of this isolation are bizarre, especially with regard to the highest forms of animal life—the mammals.

Unlike all other continents, the Australian fauna is dominated by a single primitive mammalian order, the marsupials (see Figure 11-17), an order that has long disappeared from most other parts of the world. Australia also provides the only continental home for an even more primitive group, the monotremes (see Figure 11-18). More remarkable, perhaps, is what is lacking in Australia—placental mammals, so notable elsewhere, are limited and inconspicuous. Australia is completely without representatives from such common groups as cats, dogs, bears, monkeys, hoofed animals, and many others. Partially because of the specialized character of the native mammals, *feral* livestock have become established in the wild in much greater profusion in Australia than elsewhere on Earth. Nine feral species exist in numbers that aggregate to several millions. In addition to the remarkable nature of the mammalian fauna, there are also unusual aspects to the bird, reptile, and invertebrate populations of Australia.

The biota of Australia, then, has many singular aspects, due primarily to the circumstances of isolated evolution, with significant recent human-induced modifications.

The Major Biomes

The major biomes of the world (Figure 11-25) are described below. As we learned earlier, most biomes are named for their dominant vegetation association, but the biome concept also encompasses fauna as well as interrelationships with soil, climate, and topography.

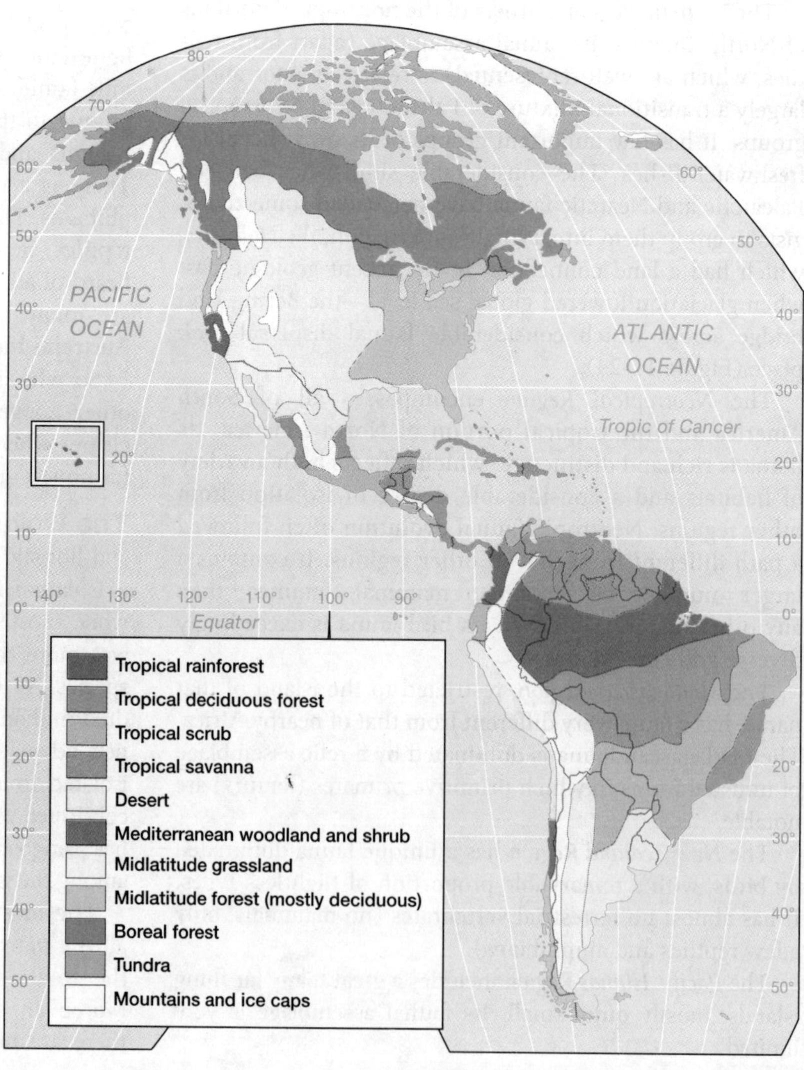

▶ **Figure 11-25** Major biomes of the world.

Tropical Rainforest

As we first saw in Chapter 10, the distribution of the **tropical rainforest** or *selva* biome (selva is the Portuguese–Spanish word for "forest") is closely related to climate—consistent rainfall and relatively high temperatures. Thus, there is an obvious correlation with the location of tropical wet (Af) and some tropical monsoon (Am) climatic regions (Figure 11-26).

The rainforest is probably the most complex of all terrestrial ecosystems, and the one with the greatest *species diversity* (many different species are found in an area, but often relatively few individuals of each species are present). It contains a bewildering variety of trees growing in close conjunction. Mostly they are tall, high-crowned, broadleaf evergreen species that never experience a seasonal leaf fall because seasons are unknown in this environment of continuous warmth and moistness (Figure 11-26a). The selva has a layered structure; the second layer down from the top usually forms a complete canopy of interlaced branches that provides continuous shade to the forest floor. Bursting through the canopy to form the top layer are the forest giants—tall trees that often grow to great heights above the general level. Beneath the canopy is an erratic third layer of lower trees, palms and tree ferns, able to survive in the shade. Sometimes still more layers of increasingly shade-tolerant trees grow at lower levels.

Undergrowth is relatively sparse in the tropical rainforest because the lack of light precludes the survival of most green plants. Only where there are gaps in the canopy, as alongside a river, does light reach the ground, resulting in the dense undergrowth associated with a jungle. *Epiphytes* like orchids and bromeliads hang from or perch on tree trunks and branches. Vines and lianas often dangle from the arching limbs.

The interior of the rainforest, then, is a region of heavy shade, high humidity, windless air, continuous warmth, and an aroma of mold and decomposition. As plant litter accumulates on the forest floor, it is acted on very rapidly by plant and animal decomposers. The upper layers of the forest are areas of high productivity, and there is a much greater concentration of nutrients in

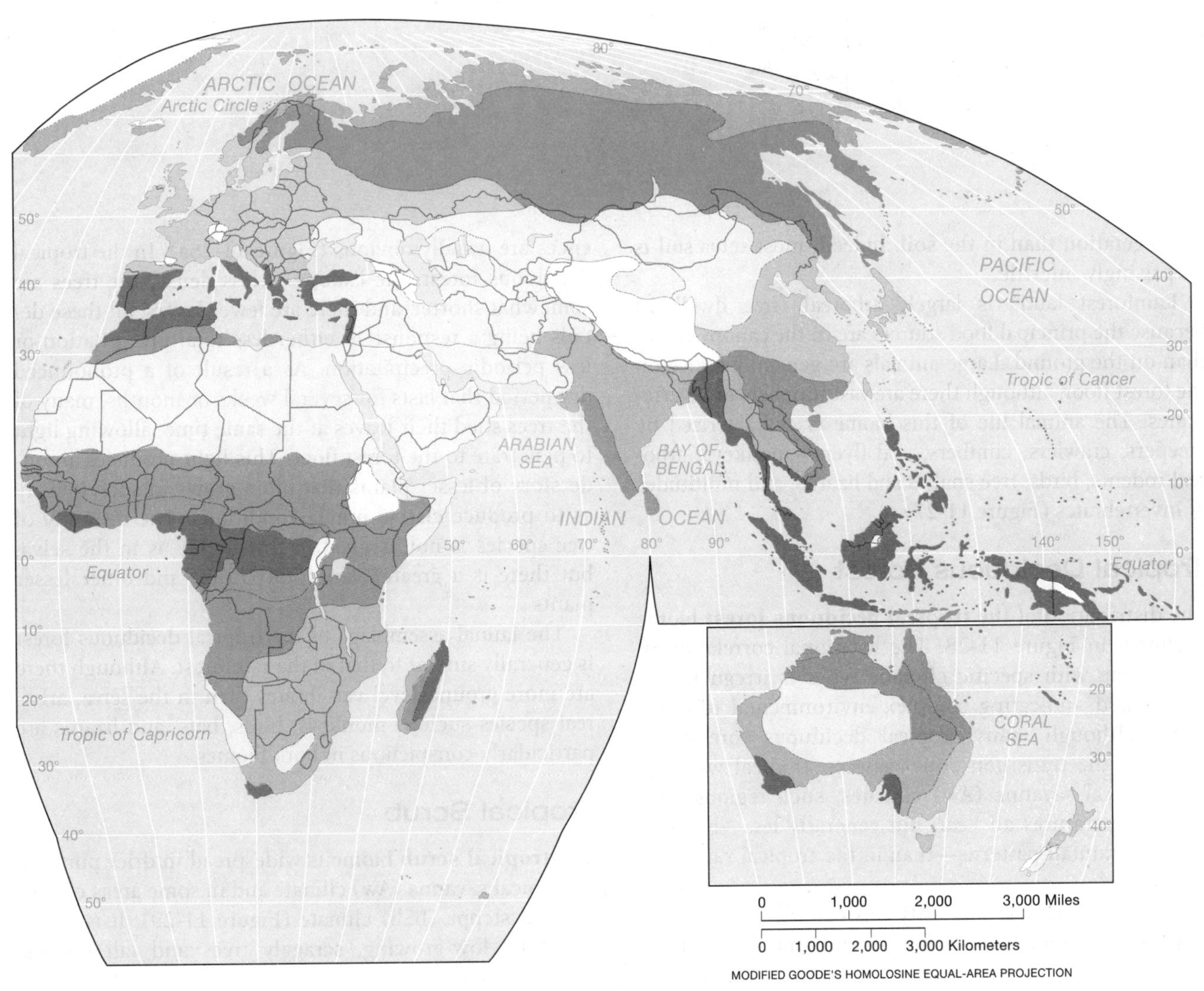

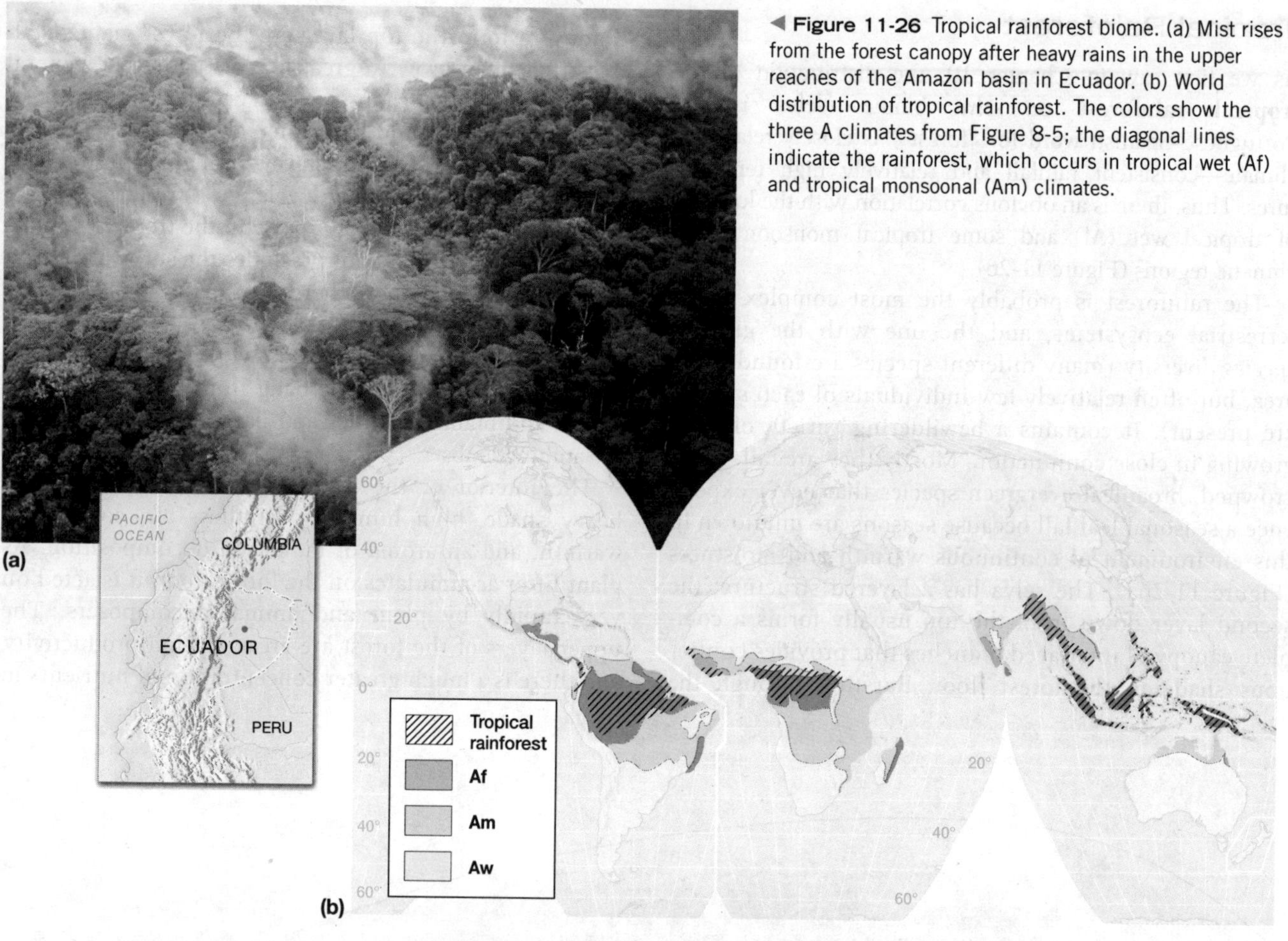

◀ **Figure 11-26** Tropical rainforest biome. (a) Mist rises from the forest canopy after heavy rains in the upper reaches of the Amazon basin in Ecuador. (b) World distribution of tropical rainforest. The colors show the three A climates from Figure 8-5; the diagonal lines indicate the rainforest, which occurs in tropical wet (Af) and tropical monsoonal (Am) climates.

the vegetation than in the soil. Indeed, most selva soil is surprisingly infertile.

Rainforest fauna is largely arboreal (tree dwelling) because the principal food sources are in the canopy rather than on the ground. Large animals are generally scarce on the forest floor, although there are vast numbers of invertebrates. The animal life of this biome is characterized by creepers, crawlers, climbers, and flyers—monkeys, arboreal rodents, birds, tree snakes and lizards, and multitudes of invertebrates (Figure 11-27).

Tropical Deciduous Forest

The distribution of the **tropical deciduous forest** biome is shown in Figure 11-28. The locational correlation of this biome with specific climatic types is irregular and fragmented, indicating complex environmental relationships, although many tropical deciduous forests are found in the transition zone between tropical wet (Af) and tropical savanna (Aw) climates; such regions have high temperatures all year, but generally less—or more seasonal rainfall patterns—than in the tropical rainforest biome.

There is structural similarity between the selva and the tropical deciduous forest, but several important differences are usually obvious (Figure 11-28a). In the tropical deciduous forest, the canopy is less dense, the trees are somewhat shorter, and there are fewer layers, all these details being a response to either less total precipitation or less periodic precipitation. As a result of a pronounced dry period that lasts for several weeks or months, many of the trees shed their leaves at the same time, allowing light to penetrate to the forest floor. This light produces an understory of lesser plants that often grows in such density as to produce classic jungle conditions. The diversity of tree species is not as great in this biome as in the selva, but there is a greater variety of shrubs and other lesser plants.

The faunal assemblage of the tropical deciduous forest is generally similar to that of the rainforest. Although there are more ground-level vertebrates than in the selva, arboreal species such as monkeys, birds, bats, and lizards are particularly conspicuous in both biomes.

Tropical Scrub

The **tropical scrub** biome is widespread in drier portions of tropical savanna (Aw) climate and in some areas of subtropical steppe (BSh) climate (Figure 11-29). It is dominated by low-growing, scraggly trees and tall bushes,

▲ **Figure 11-27** A pair of scarlet macaws in Costa Rica.

usually with an extensive understory of grasses (Figure 11-29a). The trees range from 3 to 9 meters (10 to 30 feet) in height. Their density is quite variable, with the trees sometimes growing in close proximity to one another but often spaced much more openly. Species diversity is much less than in the selva and tropical deciduous forest biomes; frequently just a few species comprise the bulk of the taller growth over vast areas. In the more tropical and wetter portions of the tropical scrub biome, most of the trees and shrubs are evergreen; elsewhere most species are deciduous. In some areas, a high proportion of the shrubs are thorny.

The fauna of tropical scrub regions is notably different from that of the two biomes previously discussed. There is a moderately rich assemblage of ground-dwelling mammals and reptiles, and of birds and insects.

Tropical Savanna

As Figure 11-30 shows, there is an incomplete correlation between the distribution of the **tropical savanna** biome (also called the *tropical grassland* biome) and that of the tropical savanna (Aw) climate. The correlation tends to be most noticeable where seasonal rainfall contrasts are greatest, a condition particularly associated with the broad-scale annual shifting of the intertropical convergence zone (ITCZ).

Savanna lands are dominated by tall grasses (Figure 11-30a). Sometimes the grasses form a complete ground cover, but sometimes there is bare ground among dispersed tufts of grass in what is called a bunchgrass pattern. The name "savanna" without any modifier usually refers to areas that are virtually without shrubs or

(a)

(b)

◀ **Figure 11-28** Tropical deciduous forest biome. (a) A tropical deciduous forest scene at about 2100 meters (7000 ft.) elevation in southeastern Mexico. (b) World distribution of tropical deciduous forest (diagonal lines) compared with distribution of A climates.

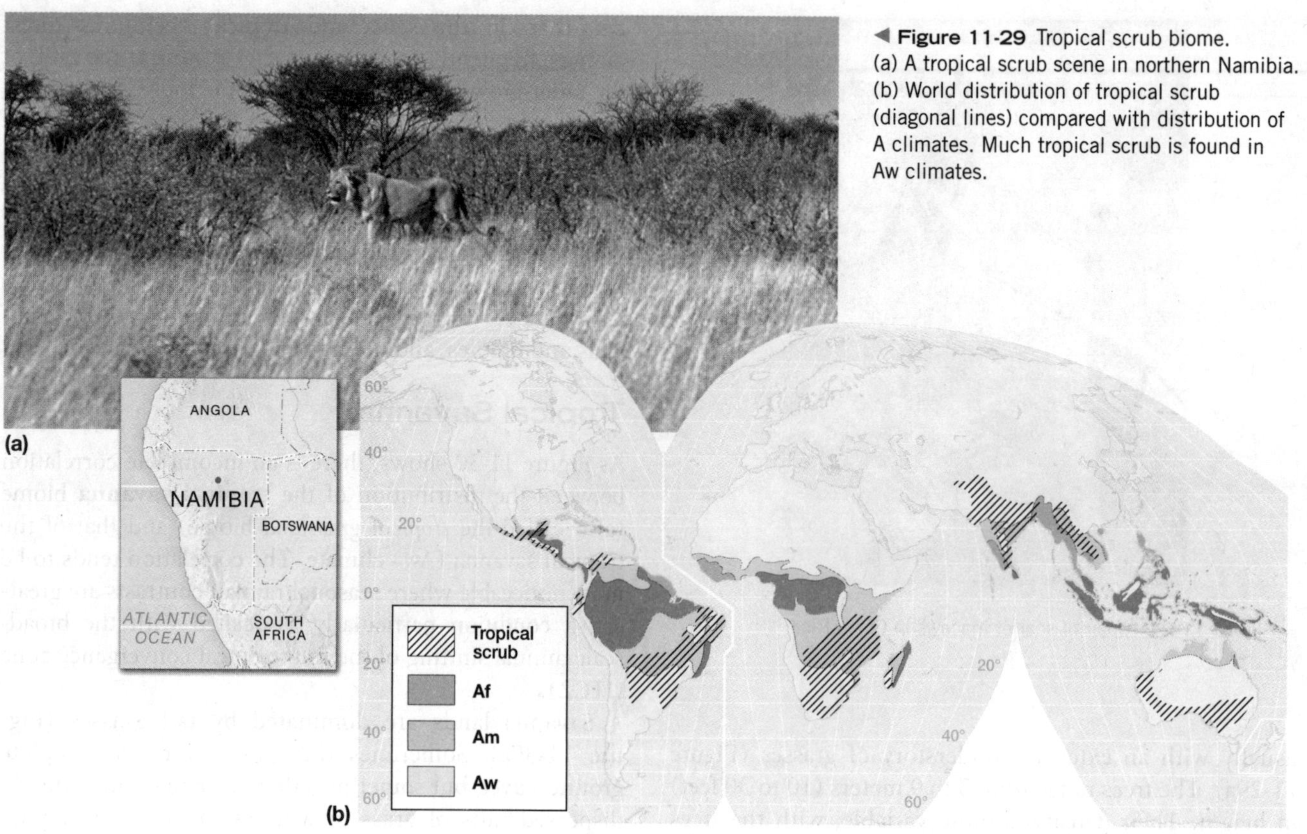

◄ **Figure 11-29** Tropical scrub biome. (a) A tropical scrub scene in northern Namibia. (b) World distribution of tropical scrub (diagonal lines) compared with distribution of A climates. Much tropical scrub is found in Aw climates.

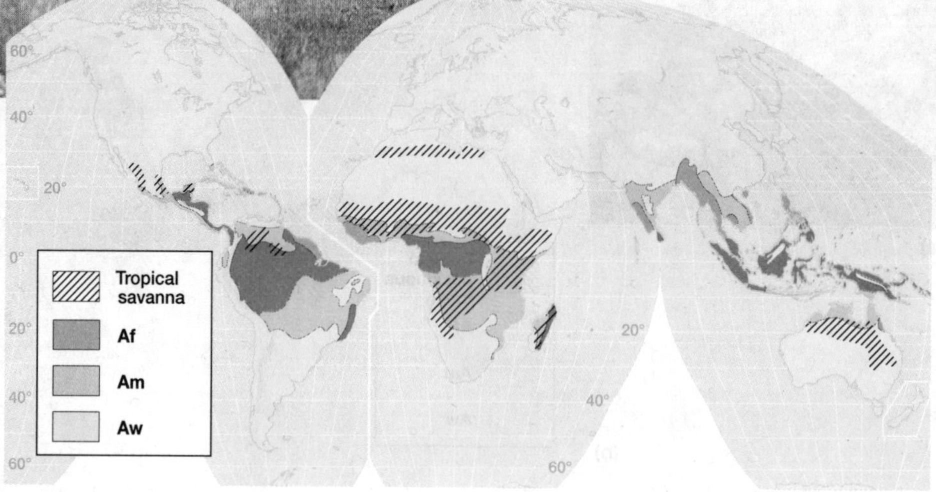

◄ **Figure 11-30** Tropical savanna biome. (a) A mixed array—zebra, kudu, and springbok—of ungulates surround a waterhole in a typical savanna landscape in Namibia's Etosha National Park. (b) World distribution of tropical savanna (diagonal lines) compared with distribution of A climates.

trees, but this type of savanna is not the most common. In most cases, a wide scattering of both types dots the grass-covered terrain, and this mixture of plant forms is often referred to as *parkland* or *park savanna*.

In much of the savanna, the vegetation is degenerate because it has been created by human interference with natural processes. A considerable area of tropical deciduous forest and tropical scrub, and perhaps some tropical rainforest, has been converted to savanna over thousands of years through fires set by humans and through the grazing and browsing of domestic animals.

The savanna biome has a very pronounced seasonal rhythm. During the wet season, the grass grows tall, green, and luxuriant. At the onset of the dry season, the grass begins to wither, and before long the above-ground portion is dead and brown. At this time, too, many of the trees and shrubs shed their leaves. The third "season" is the time of wildfires. The accumulation of dry grass provides abundant fuel, and most parts of the savannas experience natural burning every year or so. The recurrent grass fires are stimulating for the ecosystem, as they burn away the unpalatable portion of the grass without causing significant damage to shrubs and trees. When the rains of the next wet season arrive, the grasses spring into growth with renewed vigor.

Savanna fauna varies from continent to continent. The African savannas are the premier "big game" lands of the world, with an unmatched richness of large animals, particularly *ungulates* (hoofed animals) and carnivores (meat-eaters), but also including a remarkable diversity of other fauna. The Latin American savannas, on the other hand, have only a sparse population of large wildlife, with Asian and Australian areas intermediate between these two extremes.

Desert

In previous chapters, we noted a general decrease in precipitation as one moves away from the equator in the low latitudes. This progression is matched by a gradation from the selva biome of the equator to the **desert** biome of the subtropics. The desert biome also occurs extensively in midlatitude locations in Asia, North America, and South America with a fairly close correlation to subtropical desert (BWh) and midlatitude desert (BWk) climates (Figure 11-31).

◀ **Figure 11-31** Desert biome. (a) Natural vegetation typically is spindly and thorny in deserts, as in this scene from Red Rock Canyon in California's Mojave Desert. (b) World distribution of desert (diagonal lines) compared with distribution of B climates.

Desert vegetation is surprisingly variable (Figure 11-31a). It consists largely of xerophytic plants such as succulents and other drought-resisting plants with structural modifications that allow them to conserve moisture and drought-evading plants capable of hasty reproduction during brief rainy times. The plant cover is usually sparse, with considerable bare ground dotted by a scattering of individual plants. Typically the plants are shrubs, which occur in considerable variety, each with its own mechanisms to combat the stress of limited moisture. Succulents are common in the drier parts of most desert areas, and many desert plants have either tiny leaves or no leaves at all as a moisture-conserving strategy. Grasses and other herbaceous plants are widespread but sparse in desert areas. Despite the dryness, trees can be found sporadically in the desert, especially in Australia.

Animal life is exceedingly inconspicuous in most desert areas, leading to the erroneous idea that animals are nonexistent. Actually, most deserts have a moderately diverse faunal assemblage, although the variety of large mammals is limited. A large proportion of desert animals avoid the principal periods of desiccating heat (daylight in general and the hot season in particular) by resting in burrows or crevices during the day and prowling at night (Figure 11-32).

Generally speaking, life in the desert biome is characterized by an appearance of stillness. In favorable times and in favored places (around water holes and oases), however, there is a great increase in biotic activities, and sometimes the total biomass is of remarkable proportions. Favorable times are at night, and particularly after rains. For example, a heavy rain might trigger the germination of wildflower seeds that had remained dormant for decades.

Mediterranean Woodland and Shrub

As Figure 11-33 shows, the **mediterranean woodland and shrub** biome is found in six widely scattered and relatively small areas of the midlatitudes, all of which experience the pronounced dry summer–wet winter precipitation typical of mediterranean (Cs) climates. In this biome, the dominant vegetation associations are physically very similar to each other but taxonomically quite varied. The biome is dominated mostly by a dense growth of woody shrubs known as a *chaparral* in North America (Figure 11-33a), but having other names in other areas; chaparral includes many species of *sclerophyllous* plants, adapted to the prominent summer dry season by the presence of small, hard leaves that inhibit moisture loss. A second significant plant association of mediterranean regions is an open grassy woodland, in which the ground is almost completely grass covered but has a considerable scattering of trees as well (Figures 11-33b and 11-33c).

▲ **Figure 11-32** This fierce-looking thorny devil lizard in Western Australia is just 10 centimeters (4 inches) long.

The plant species vary from region to region. Oaks of various kinds are by far the most significant genus in the Northern Hemisphere mediterranean lands, sometimes occurring as prominent medium-sized trees but also appearing as a more stunted, shrubby growth. In all areas, the trees and shrubs are primarily broadleaf evergreens. Their leaves are mostly small and have a leathery texture or waxy coating, which inhibits water loss during the long dry season. Moreover, most plants have deep roots.

Summer is a virtually rainless season in mediterranean climates, and so summer fires are relatively common. Many of the plants are adapted to rapid recovery after a wildfire has swept over the area. Indeed, as noted in Chapter 10, some species have seeds that are released for germination only after the heat of a fire has caused their seedpods to open. Part of the seasonal rhythm of this biome is that winter floods sometimes follow summer fires, as slopes left unprotected by the burning away of grass and lower shrubs are susceptible to abrupt erosive runoff if the winter rains arrive before the vegetation has a chance to resprout.

The fauna of this biome is not particularly distinctive. Seed-eating, burrowing rodents are common, as are some bird and reptile groups. There is a general overlap of animals between this biome and adjacent ones.

Midlatitude Grassland

Vast **midlatitude grasslands** occur widely in North America and Eurasia (Figure 11-34). In the Northern Hemisphere, the locational coincidence between this biome and the steppe (BSh and BSk) climatic type is very pronounced. The smaller Southern Hemisphere areas (mostly the *pampa* of Argentina and the *veldt* of South Africa) have less distinct climatic correlations.

The vegetation typical of a grassland biome is a general response either to a lack of precipitation sufficient to support larger plant forms or to the frequency of fires (both natural and human induced) that prevent the growth of tree or shrub seedlings. In the wetter areas of a grassland biome, the grasses grow tall and the term *prairie* is often

◀ **Figure 11-33** Mediterranean woodland and shrub biome. (a) Dense mediterranean shrub biome chaparral vegetation in the San Gabriel Mountains of California. (b) Open mediterranean grassy woodland vegetation in the Mount Lofty Ranges of South Australia. Winter is moist, mild, and green, while early summer (c) is hot, dry, and brown—this is followed by the fire season. (d) World distribution of mediterranean woodland and shrub (diagonal lines) compared with distribution of Cs climates.

applied in North America. In drier regions, the grasses are shorter; such growth is often referred to as *steppe* (Figure 11-34a). Sometimes a continuous ground cover is missing, and the grasses grow in discrete tufts as bunchgrass or tussock grass.

Most of the grass species are perennials, lying dormant during the winter and sprouting anew the following summer. Trees are mostly restricted to riparian locations, whereas shrubs and bushes occur sporadically on rocky sites. Grass fires are fairly common in summer, which helps to explain the relative scarcity of shrubs. The woody plants cannot tolerate fires and can generally survive only on dry slopes where there is little grass cover to fuel a fire.

Grasslands provide extensive pastures for grazing animals, and before encroachment by humans drastically

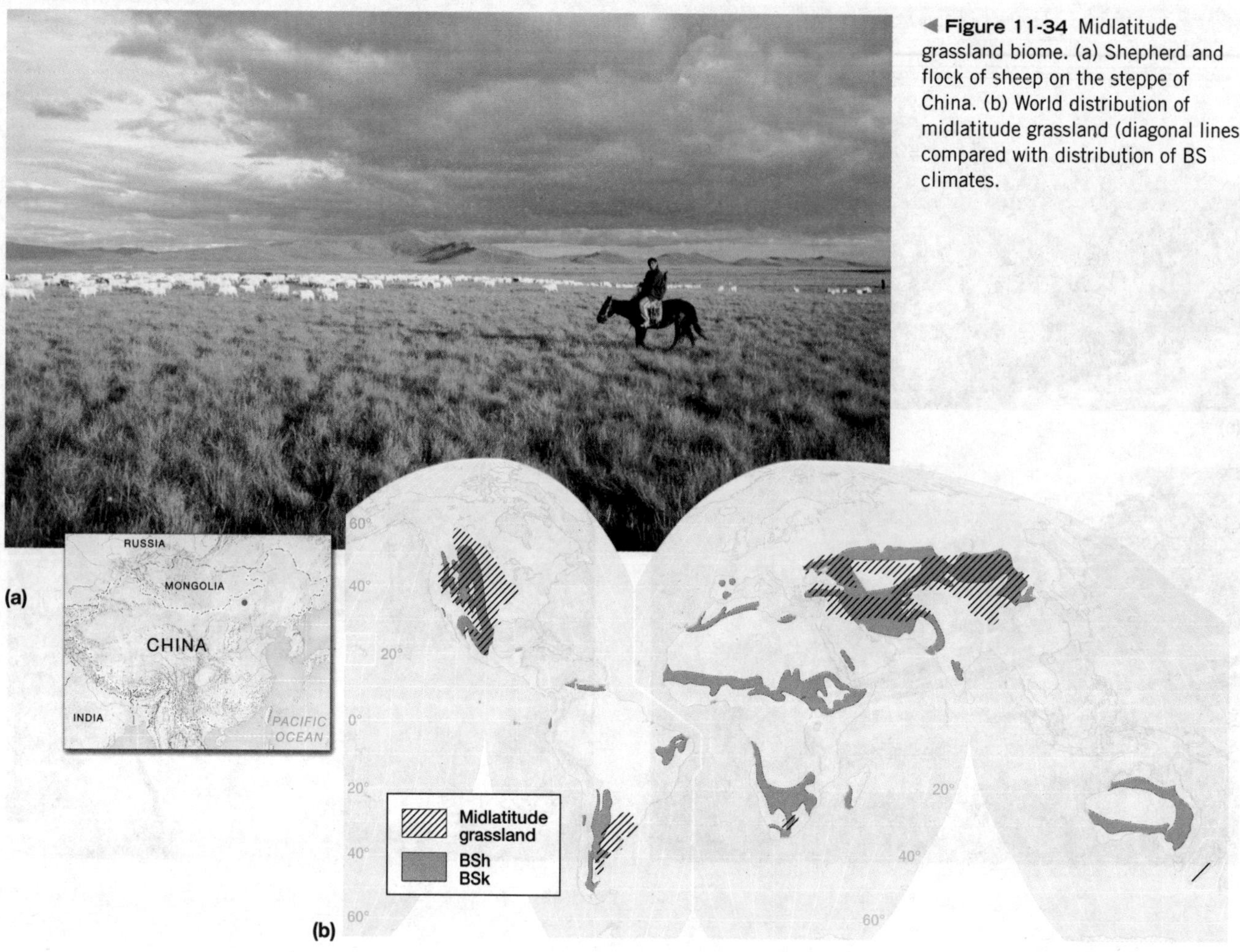

◄ **Figure 11-34** Midlatitude grassland biome. (a) Shepherd and flock of sheep on the steppe of China. (b) World distribution of midlatitude grassland (diagonal lines) compared with distribution of BS climates.

changed population sizes, the grassland fauna comprised large numbers of relatively few species. The larger herbivores were often migratory prior to human settlement. Many of the smaller animals spend all or part of their lives underground, where they find some protection from heat, cold, and fire.

Midlatitude Deciduous Forest

Extensive areas on all Northern Hemisphere continents, as well as more limited tracts in the Southern Hemisphere, were originally covered with a forest of largely broadleaf deciduous trees (Figure 11-35). Except in hilly country, a large proportion of this **midlatitude deciduous forest** has been cleared for agriculture and other types of human use, so that very little of the original natural vegetation remains.

The forest is characterized by a fairly dense growth of tall broadleaf trees with interwoven branches that provide a complete canopy in summer. Some smaller trees and shrubs exist at lower levels, but for the most part, the forest floor is relatively barren of undergrowth. In winter, the appearance of the forest changes dramatically, owing to the seasonal fall of leaves (Figure 11-35a).

Tree species vary considerably from region to region, although most are broadleaf and deciduous. The principal exception is in eastern Australia, where the forest is composed almost entirely of varieties of eucalyptus, which are broadleaf evergreens. Northern Hemisphere regions have a northward gradational mixture with needleleaf evergreen species. An unusual situation in the southeastern United States finds extensive stands of pines (needleleaf evergreens) rather than deciduous species occupying most of the well-drained sites above the valley bottoms. In the Pacific Northwest of the United States, the forest association is primarily evergreen coniferous rather than broadleaf deciduous.

This biome generally has the richest assemblage of fauna to be found in the midlatitudes, although it does not have the diversity to match that of most tropical biomes. It has (or had) a considerable variety of birds and mammals, and in some areas reptiles and amphibians are well represented. Summer brings a diverse and active population of insects and other arthropods. All animal life is

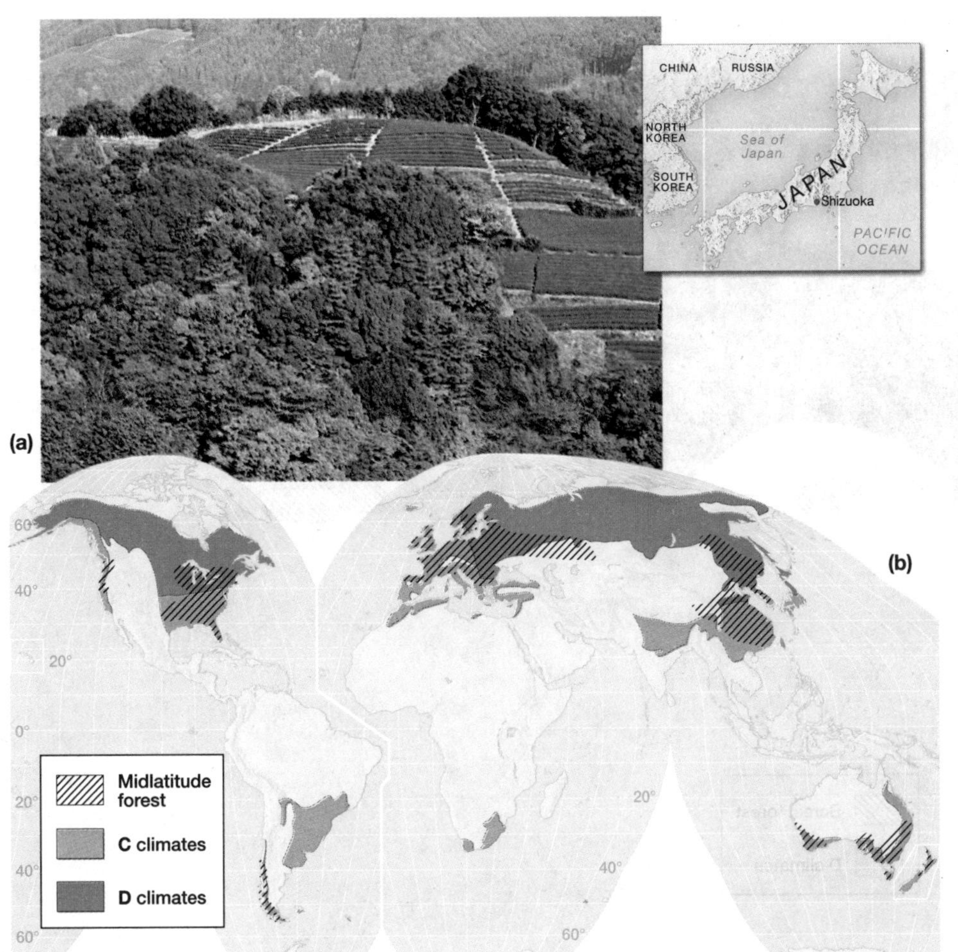

◀ **Figure 11-35** Midlatitude deciduous forest biome. (a) This midlatitude forest scene is in the mountains near Shizuoka, Japan. As is common in some midlatitude locations, the forest here consists of both deciduous and evergreen trees. (b) World distribution of midlatitude forest (diagonal lines) before human activities made significant changes in this biome. The colors indicate distribution of C and D climates.

less numerous (partly due to migrations and hibernation) and less conspicuous in winter.

Boreal Forest

One of the most extensive biomes is the **boreal forest**, sometimes called *taiga* after the Russian word for the northern fringe of the boreal forest in that country (paralleling the way these synonyms are used to describe climates, as mentioned in Chapter 8). The boreal forest occupies a vast expanse of northern North America and Eurasia (Figure 11-36). There is very close correlation between the location of the boreal forest biome and the subarctic (Dfc) climatic type, with a similar correlation between the locations of the tundra climate and the tundra biome.

This great northern forest contains perhaps the simplest assemblage of plants of any biome (Figure 11-36a). Most of the trees are conifers, nearly all needleleaf evergreens, with the important exception of the tamarack or larch which drops its needles in winter. The variety of species is limited to mostly pines, firs, and spruces extending broadly in homogeneous stands. In some places, the coniferous cover is interrupted by areas of deciduous trees. These deciduous stands are also of limited variety (mostly birch, poplar, and aspen) and often represent a seral situation following a forest fire.

The trees grow taller and more densely near the southern margins of this biome, where the summer growing season is longer and warmer. Near the northern margins, the trees are spindly, short, and more openly spaced. Undergrowth is normally not dense beneath the forest canopy, but a layer of deciduous shrubs sometimes grows in profusion. The ground is usually covered with a complete growth of mosses and lichens, with some grasses in the south and a considerable accumulation of decaying needles overall.

Poor drainage is typical in summer, due partially to permanently frozen subsoil, which prevents downward percolation of water, and partially to the derangement of normal surface drainage by the action of glaciers during the recent Pleistocene ice age. Thus, bogs and swamps are numerous, and the ground is generally spongy in summer. During the long winters, of course, all is frozen.

The immensity of the boreal forest gives an impression of biotic productivity, but such is not the case. Harsh climate, floristic homogeneity, and slow plant growth produce only a limited food supply for animals. Faunal

▲ **Figure 11-36** Boreal forest biome. (a) The boreal forest contains trees that generally are short, close growing, and of uniform species composition. This spruce forest surrounds Shady Lake near Prince Albert in the Canadian province of Saskatchewan. (b) World distribution of boreal forest (diagonal lines) compared with distribution of D climates.

species diversity is limited, although the number of individuals of some species is astounding. With relatively few animal species in such a vast biome, populations sometimes fluctuate enormously within the space of only a year or so. Mammals are represented prominently by furbearers (Figure 11-37) and by a few species of ungulates. Birds are numerous and fairly diverse in summer, but nearly all migrate to milder latitudes in winter. Insects are totally absent in winter but superabundant during the brief summer.

Tundra

The **tundra** is essentially a cold desert or grassland in which moisture is scarce and summers so short and cool that trees are unable to survive. This biome is distributed along the northern edge of the Northern Hemisphere continents, correlating closely to the distribution of tundra (ET) climate (Figure 11-38). The plant cover consists of a considerable mixture of species, many of them in dwarf forms (Figure 11-38a). Included are grasses, mosses, lichens, flowering herbs, and a scattering of low shrubs.

▲ **Figure 11-37** Bears live in diverse habitats, but mostly in cold climates. This grizzly bear inhabits Katmai National Park in Alaska.

◀ **Figure 11-38** Tundra biome. (a) Tundra vegetation in Denali National Park, Alaska. (b) World distribution of tundra (diagonal lines) compared with distribution of E climates.

These plants often occur in a dense, ground-hugging arrangement, although some places have a more sporadic cover with considerable bare ground interspersed. The plants complete their annual cycles hastily during the brief summer, when the ground is often moist and waterlogged because of inadequate surface drainage and particularly inadequate subsurface drainage.

▲ **Figure 11-39** Alpine tundra vegetation high in the Rocky Mountains.

Animal life is dominated by birds and insects during the summer. Extraordinary numbers of birds flock to the tundra for summer nesting, migrating southward as winter approaches. Mosquitoes, flies, and other insects proliferate astoundingly during the short warm season, laying eggs that can survive the bitter winter. Other forms of animal life are scarcer—a few species of mammals and freshwater fishes but almost no reptiles or amphibians.

Alpine Tundra An alpine version of tundra is found in many high-elevation areas. Many mountain areas above the timberline exhibit areas with a sparse cover of vegetation consisting of herbaceous plants, grasses, and low shrubs (Figure 11-39).

Human Modification of Natural Distribution Patterns

Thus far in our discussion of the biosphere, most of our attention has been focused on "natural" conditions, that is, events and processes that take place in nature without

▲ **Figure 11-40** An overgrazed range in the Colorado high country illustrates human-induced habitat modification caused by overstocking of livestock in the area to the right of the fence. The locale is near Central City.

the aid or interference of human activities. Such natural processes have been going on for millennia, and their effects on floral and faunal distribution patterns have normally been very slow and gradual. The pristine environment, uninfluenced by human-kind, experiences its share of abrupt and dramatic events, to be sure, but environmental changes generally proceed at a gradual pace. When *Homo sapiens* appear, however, the tempo changes dramatically.

People are capable of exerting extraordinary influences on the distribution of plants and animals. Not only is the magnitude of the changes likely to be great, but also the speed with which they are affected is sometimes exceedingly rapid. In broadest perspective, humankind exerts three types of direct influences on biotic distributions: physical removal of organisms, habitat modification, and artificial translocation of organisms.

Physical Removal of Organisms

One of humankind's most successful skills is in the elimination of other living things. As human population increases and spreads over the globe, there is often a wholesale removal of native plants and animals to make way for the severely modified landscape that is thought necessary for civilization. The natural plant and animal inhabitants are cut down, plowed up, paved over, burned, poisoned, shot, trapped, and otherwise eradicated in actions that have far-reaching effects on overall distribution patterns.

Habitat Modification

Habitat modification is another activity in which humankind excels. The soil environment is changed by farming, grazing, engineering, and construction practices (Figure 11-40); the atmospheric environment is degraded by the introduction of impurities of various kinds; the waters of the planet are impounded, diverted, and polluted. All these deeds influence the native plants and animals in the affected areas.

Among the most dramatic recent human-initiated changes to global habitat have involved the removal of vast areas of tropical rainforest.

Tropical Rainforest Removal Throughout much of history, most rainforests of the world were only modestly populated and as a consequence they were affected by human activities in limited ways. Since the twentieth century, however, rainforests have been exploited and devastated at an accelerating pace; and over the past 30 years or so, tropical deforestation has become one of Earth's most serious environmental problems.

The exact rate of deforestation around the world is not precisely known, but conservative estimates suggest that tropical rainforest is currently being lost at a rate of about

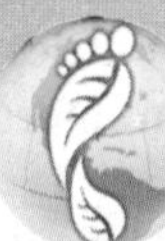

PEOPLE AND THE ENVIRONMENT

Rainforest Loss in Brazil

The country of Brazil contains about one-third of the world's tropical forest, most of it part of the vast rainforest spreading across the Amazon River basin. Although the rainforest has supported a sparse human population for thousands of years, over the last half century large tracts of the forest have been opened up for settlers. As a consequence, the pristine forest is being cleared for settlements, agriculture, ranching, and logging (Figure 11-C).

The Brazilian government completed construction of the Cuiabá-Port Velho highway across the province of Rondônia in 1960. Although the road opened up the region to immigrants and development, the forest was still relatively unchanged 15 years later (Figure 11-D-a). But by the late 1980s, the rate of deforestation had increased substantially and the region had taken on a "fishbone" pattern as forest was cleared along an expanding network of roads. By 2001 the extent of deforestation was extraordinary (Figure 11-D-b) and the pace has eased little since then.

Deforestation of this scale is taking place in a number of areas in the Amazon Basin. In the year 2004, in Brazil alone more than 26,000 square kilometers (10,000 square miles) of rainforest were lost. Such widespread deforestation is leading to a loss of both habitat and species, and in the end, many of the uses of the cleared land are simply not sustainable. In Rondônia, less than 10 percent of agricultural land use has gone toward potentially sustainable perennial crops such as cacao and coffee; much of the rest is either used for cattle grazing or annual crops—and this land must usually be abandoned in a few years after the generally poor tropical soils are depleted of nutrients. Once abandoned for agriculture, it may be decades before the forest can grow back to near its original density and composition.

September 9, 2000

September 26, 2006

▲ **Figure 11-C** Deforestation in Rondônia, Brazil, between September 9, 2000 and September 26, 2006. These large-scale images were obtained using the Advanced Spaceborne Thermal Emissions and Reflection Radiometer (ASTER) on NASA's Terra satellite. Darker green indicates denser vegetation.

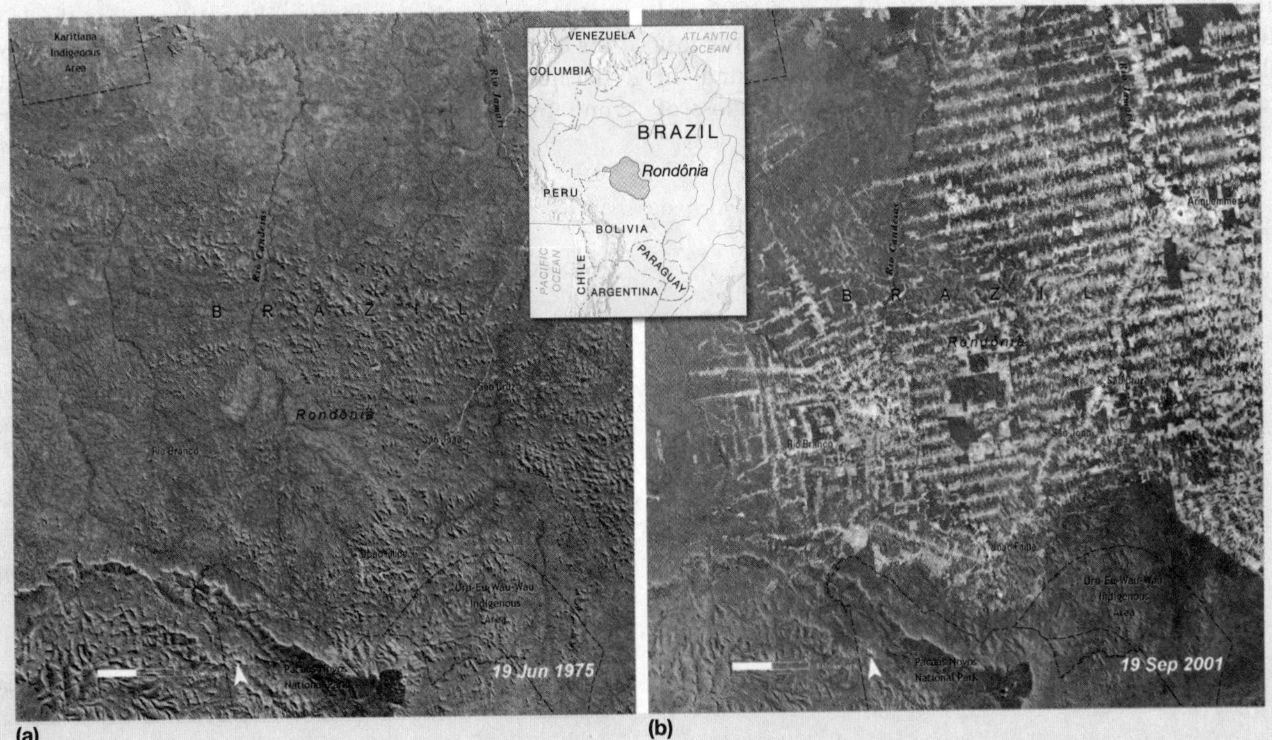

▲ **Figure 11-D** The Brazilian state of Rondônia in (a) 1975 and (b) the same area in 2001.

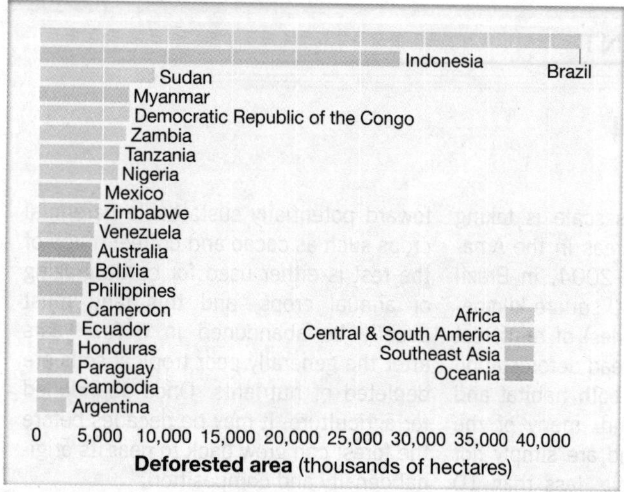

▲ **Figure 11-41** Loss of forest by country between 1990 and 2005 in thousands of hectares.

11 million hectares (27 million acres) per year. The United Nations Food and Agriculture Organization (FAO) estimated that between 1990 and 2005 about 42 million hectares (104 million acres) of rainforest was cleared in Brazil alone—the greatest total of any country during that time period (Figure 11-41) and an area approximately equal to that of California. Indonesia ranked second with more than 25 million hectares (62 million acres) cleared during those years.

In terms of the total percentage of rainforest cleared in each country, the picture is somewhat different (Figure 11-42). The FAO estimates that between 1990 and 2005, the tiny African island country of Comoros lost almost 60 percent of its rainforest, while Burundi in central equatorial Africa ranked second with a loss of about 47 percent. Overall, about half of Africa's original rainforest is now gone, but in some countries, such as Nigeria and Ghana, losses now total more than 80 percent.

The rate of deforestation remains high in South- and Southeast Asia, primarily associated with commercial timber exploration, especially for teak and mahogany—about 45 percent of Asia's rainforest no longer exists. Approximately 40 percent of Latin America's rainforest has been cleared. Much of the very rapid deforestation in Central America has been due to expanded cattle ranching (Figure 11-43). Deforestation of the Amazon region as a percentage of the total area of rainforest has been moderate (perhaps 20 percent of the total has been cleared), but it continues at a rapid pace.

As the forest goes, so goes its habitability for both indigenous peoples and native animal life. In the mid-1980s it was calculated that tropical deforestation was responsible for the extermination of one species per day; by the mid-1990s it was estimated that the rate was two species per hour. Moreover, loss of the forests contributes to accelerated soil erosion, drought, flooding, water quality degradation, declining agricultural productivity, and greater poverty for rural inhabitants. In addition, atmospheric carbon dioxide continues to be increased because burning trees as a way of clearing forest releases carbon to the air.

The irony of tropical deforestation is that the anticipated economic benefits are usually temporary. Much of the forest clearing is in response to the social pressure of overcrowding and poverty in societies where many people are landless. The governments open "new lands" for settlement in the rainforest. The settlers clear the land for crop or livestock. The result is almost always an initial nutrient pulse of high soil productivity, followed in only two or three years by a pronounced fertility decline

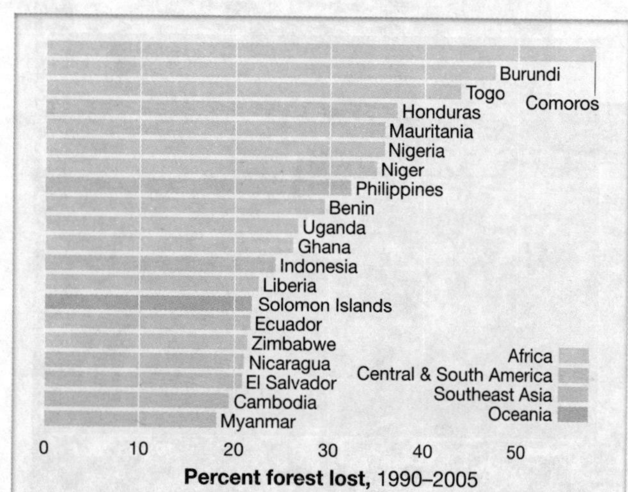

▲ **Figure 11-42** Percentage of forest lost by country between 1990 and 2005.

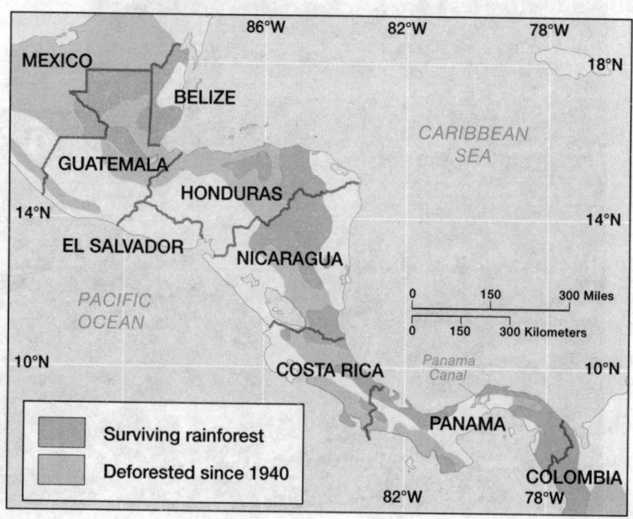

▲ **Figure 11-43** Loss of Central American rainforest since 1940. This region has one of the fastest rates of deforestation in the world.

▲ **Figure 11-44** When rainforests are cleared for agriculture, the results are often small yields and accelerated soil erosion. This scene is from central Thailand.

as the nutrients are quickly leached and cropped out of the soil, weed species rapidly invade, and erosion becomes rampant (Figure 11-44). Ongoing commercial agriculture can generally be expected only with continuous heavy fertilization, a costly procedure, while sustainable land-extensive forms of traditional agriculture are slowly disappearing.

The forests, of course, are renewable. If left alone, they can regenerate, providing there are seed trees in the vicinity and the soil has not been stripped of all its nutrients. The loss of *biodiversity*, however, is a much more serious problem because extinction is irreversible. Valuable potential resources—pharmaceutical products, new food crops, natural insecticides, industrial materials—may disappear before they are even discovered. Wild plants and animals that could be combined with domesticated cousins to impart resistance to disease, insects, parasites, and other environmental stresses may also be lost.

Much concern has been expressed about tropical deforestation, and some concrete steps have been taken. The development of *agroforestry* (planting crops with trees, rather than cutting down the trees and replacing them with crops) is being fostered in many areas. The United Nations Educational, Scientific, and Cultural Organization (UNESCO) administers the *Man and the Biosphere Programme* that coordinates an international effort to establish and protect biosphere reserves. The goal of this project is to set aside tracts of largely pristine land—including regions of tropical rainforest—to preserve biodiversity before it is lost to development. At present, 553 biosphere reserves have been established in 107 countries.

Meanwhile, the sounds of ax, chain saw, and bulldozer continue to be heard throughout the tropical forest lands.

Artificial Translocation of Organisms

People are capable of elaborate rearrangement of the natural complement of plants and animals in almost every part of the world. This is shown most clearly with domesticated species—crops, livestock, pets. There is now, for example, more corn than native grasses in Iowa, more cattle than native gazelles in the Sudan, more canaries than native thrushes in Detroit. More importantly for our discussion here, humans have also accounted for many introductions of wild plants and animals into "new" habitats; such organisms are called **exotic species** in their new homelands.

In some cases, the introduction of exotic species was deliberate. A few examples among a great many include the taking of prickly-pear cactus from Arizona to Australia; crested wheat grass from Russia to Kansas; European boar from Germany to Tennessee; pronghorn antelope from Oregon to Hawai'i; and red fox from England to New Zealand. Frequently, however, the introduction of an exotic species was an accidental result of human carelessness. The European flea, for example, has become one of the most widespread creatures on Earth because it has been an unseen accompaniment to human migrations all over the world. Similarly, the English sparrow and

▶ **Figure 11-45** Burros are prominent feral animals in much of the southwestern desert area of the United States. This group is in the Panamint Range of California.

European brown rat have inadvertently been introduced to all inhabited continents by traveling as stowaways on ships.

One other type of human-induced translocation of animals involves the deliberate release or accidental escape of livestock to become established as a "wild"—properly termed *feral*—population. This has happened in many parts of the world, most notably in North America and Australia (Figure 11-45).

In many parts of the world, natural biotic distribution patterns have been rearranged through deliberate and accidental human efforts to translocate plants and animals. When an exotic (nonnative) species is released in a new area, the results are usually one of two extremes. Either it dies out in a short time because of environmental hazards and/or competitive–predatory pressures, or the introduced species finds both a benign climate and an unfilled ecologic niche and is able to flourish extraordinarily. When the latter situation occurs, the results are occasionally salutary, but in many cases they are unsatisfactory and sometimes absolutely disastrous.

The world abounds with examples that could be cited—sparrows and starlings, rabbits and pigs, mongooses and mynas, lantana and prickly pear, mesquite and broomweed, Australia and New Zealand, Hawaii and Mauritius; the list of species and places is virtually endless. An argument could be made, however, that the recent and contemporary biotic history of Florida represents one of the worst examples of all, made all the more frightening because the cumulative impact will almost surely be much worse in the near future than it is already.

Biotic Rearrangement: The Sad Case of Florida Florida is a very attractive place for human settlement. Particularly in the last three decades, it has experienced one of the highest rates of in-migration of people ever known anywhere in the world. The major inducement is climate. A mild winter more than compensates for a hot summer, particularly when the latter can be ameliorated by artificial air conditioning. Florida's subtropical climate is also very suitable for many plants and animals. The tropical regions of the world contain an incredible variety of biota. Only a small fraction of this total is native to Florida, but a great many alien tropical species are capable of survival and proliferation in the Sunshine State if they can but gain a foothold.

Humankind has been energetically and capriciously willing to help in providing that foothold. In the last few years, Florida has become the major world center for the animal-import industry, and it is almost as important as a focus for plant imports as well. It is inevitable that many of these organisms escape from confinement and try their luck at survival in the wild. Others are deliberately turned loose—by fishers who want new quarry, pet owners who are tired of their pets, gardeners who would like a new shrub in the backyard. Still others are brought in inadvertently, in the holds of freighters or the baggage of travelers. Through these and similar events, which have been accelerating year by year, Florida has become what has been termed a "biological cesspool" of introduced life forms. More than 50 species of exotic animals, not counting invertebrates, have now taken up residence in Florida, and exotic plants are almost too numerous to count.

Not all the introduced species have established viable, breeding populations in the wild, but many of them have, and some occur in plague proportions. Exotics are most likely to prosper in a new environment when the natural ecosystems of the host region are unstable. Florida has experienced massive disruption of its ecosystems in recent years due to the explosive human population increment and the associated accelerated changes in land use. Particularly contributory has been the modification of drainage systems in this flattish state of normally expansive water surfaces (lakes, swamps, marshes, everglades). In contemporary Florida, human interference has destabilized the natural ecosystems, and human-induced introductions provide a large and steady source of exotic plants and animals on a year-round basis.

Dozens of species of exotic plants have become widespread, and almost all of them are continually expanding their ranges. Prominent among them is the melaleuca tree from Australia. Seeds from these "paperbark" trees were deliberately sown by airplane in the 1930s in the hope of developing a timber industry in the swamplands of southwestern Florida. Only recently have they begun to spread extensively, in response to increased artificial drainage and to wildfires. They prosper and flourish in disturbed land much better than the native species do. The spread of melaleuca has changed swamp to forest, radically altering the entire regional ecosystem, but the lumber potential has turned out to be negligible.

Much more extensive and deleterious in impact has been the spread of exotic aquatic weeds, which now infest more than 200,000 hectares (a half million acres) of Florida's waters. The two most significant are the water hyacinth and the hydrilla.

1. The Amazon water hyacinth (*Eichhornia crassipes*) grows incredibly fast. A single plant can double its mass in two weeks under ideal conditions. These plants grow so thickly as to impede boat traffic and shade out other flora, sometimes to the extent that the water receives virtually no oxygen, and biological deserts are created. They are very difficult to cut out, and even when this is done, the decay of displaced portions puts more nutrients into the water, which promotes even lusher growth.
2. The hydrilla (*Hydrilla verticillata*) is a native of tropical Africa and Southeast Asia that was brought to Florida as an aquarium plant but has now spread vastly in the wild (Figure 11-46). It has the form of long green tentacles, growing 2.5 centimeters (an inch) a day, which can become intertwined and form dense mats that will stop an outboard motor propeller dead. It overwhelms native plants and can even thrive in deep water where there is almost total darkness. Tiny pieces, when broken off, can regenerate into a new plant, so it is easily spread by birds, boat propellers, and other things that move from lake to lake. It has already clogged more than 60,000 hectares (150,000 acres) of waterways in Florida and has become established in almost all other southeastern states.

▲ **Figure 11-46** Hydrilla-choked canal near Everglades National Park, Florida.

Exotic animals are less overwhelming in their occurrence, but many species are already well established and some are spreading rapidly. A sampling of nonaquatic species includes Mexican armadillos, Indian rhesus monkeys, Australian parakeets, Cuban lizards, Central American jaguarundi cats, South American giant toads, Great Plains jack rabbits, Amazonian parrots, and many others.

Exotic fish are much more numerous than their terrestrial counterparts and pose even more serious problems, primarily through fierce competition with native species. Most of the drainage systems of Florida are interconnected by numerous irrigation and drainage canals; thus any introduced freshwater species now has access to most of the stream systems of the state. Moreover, Florida has a great abundance of springs that have stable water temperatures throughout the year, providing havens for many tropical fish that would otherwise find winter water temperatures below their tolerance limits.

South American acaras are already the dominant canal fish throughout southern Florida, and African tilapias are the most numerous fish in many lakes in the central part of the state. The greatest present and potential threat, however, is the so-called "walking catfish" (*Clarias batrachus*) from Southeast Asia, which numbers in the millions throughout the state and is considered to be "out of

▲ **Figure 11-47** A walking catfish in Florida.

control," with "no practical method of eradication" (Figure 11-47). These catfish eat insect larvae until the insects are gone; then they eat other fish. They are overwhelming rivals of almost all the native fish, eventually reducing the entire freshwater community to a single species—the walking catfish. They are significantly hardier than other fish because if they do not like the local waters, or if the waters dry up, they can simply hike across land, breathing directly from the air, until they find a new lake or stream.

CHAPTER 11 LEARNING REVIEW

After studying this chapter, you should be able to answer the following questions. Key terms from each text section are shown in bold type. Definitions for key terms are also found in the glossary at the back of the book.

Ecosystems and Biomes (p. 285)

1. Contrast and explain the concepts of **ecosystem** and **biome**.
2. What is an **ecotone**?

Terrestrial Flora (p. 286)

3. What is the difference between a **perennial** and an **annual plant**?
4. Explain the difference between a **gymnosperm (conifer)** and an **angiosperm**. Name trees that are examples of each.
5. Explain the difference between a **deciduous tree** and an **evergreen tree**. Name trees that are examples of each.
6. Explain the difference between a **broadleaf tree** and a **needleleaf tree**. Name trees that are examples of each.
7. Describe some typical **xerophytic adaptations** of plants.
8. Describe some typical **hygrophytic adaptations** of plants.
9. Explain the concept of **climax vegetation.**
10. Why do trees require so much more moisture to survive than grass?
11. What are the differences among **forest**, **woodland**, and **shrubland**?
12. What are the similarities and differences among the **grasslands**: savanna, prairie, and steppe?
13. Briefly describe the **desert**, **tundra**, and **wetlands** plant associations.
14. Explain what is meant by **vertical zonation** of vegetation patterns.
15. In what ways is an increase in altitude similar to an increase in latitude?
16. Define and explain what causes the **treeline**?
17. What generally happens to the elevation of the treeline going from the equator toward the poles? Why?
18. What is the difference between an **adret slope** and a **ubac slope**? How and why is vegetation likely to be different on an adret slope and a ubac slope?
19. What is **riparian vegetation**?

Terrestrial Fauna (p. 293)

20. What are some basic characteristics that distinguish plants from animals?
21. Contrast **invertebrates** with **vertebrates**. Provide one example of each.
22. Describe the distinguishing characteristics of mammals.
23. What are the advantages for an animal to be **endothermic**?
24. Distinguish among the three ways (physiological, behavioral, reproductive) animals adapt to the environment.
25. Describe and explain at least one animal adaptation to desert life.
26. What is meant by **symbiosis**?
27. Distinguish between mutualism and parasitism.

Zoogeographic Regions (p. 300)

28. Explain the concept of **zoogeographic regions**.
29. What is an **endemic** plant or animal?
30. Why are the flora and fauna of the Australian Region so distinctive?

The Major Biomes (p. 302)

31. What climate characteristics are most closely associated with the tropical rainforest biome?
32. Contrast the general characteristics of the **tropical rainforest**, **tropical deciduous forest**, and **tropical scrub** biomes.

33. Describe the seasonal patterns of the **tropical savanna** biome.
34. Describe and explain the global distribution of the **desert** biome.
35. Discuss the general climate characteristics and types of vegetation associated with the **mediterranean woodland and shrub** biome.
36. What are the general differences in climate associated with the tropical savanna and **midlatitude grassland** biomes?
37. What are the general differences in climate associated with the **midlatitude deciduous forest** and **boreal forest** biomes?
38. Refer to the climate map (Figure 8-3) and biomes map in the textbook (Figure 11-25). The boreal forest biome generally corresponds to the distribution of which climate type?
39. Contrast the general species diversity in the tropical rainforest and boreal forest biomes.
40. Describe the general vegetation cover found in the **tundra** biome.

Human Modification of Natural Distribution Patterns (p. 313)

41. Why is it usually difficult to maintain productive agriculture year after year in a cleared area of tropical rainforest?
42. What is an **exotic species**?
43. What is a feral animal population? Provide an example.
44. Describe and explain one example of an exotic species that has disrupted a region's natural ecosystem.

Log in to **www.mygeoscienceplace.com** for animations, videos, RSS feeds, maps, and self-study quizzes to enhance your study of **Terrestrial Flora and Fauna**.

12 Soils

The final major component in our study of Earth's environment is the lithosphere. This fourth sphere is just as complex as the atmosphere, biosphere, or hydrosphere but contrasts with these other realms in its enormity and particularly in its seeming stability.

It is easy to observe change in the three other spheres—clouds forming, flowers blooming, rivers flowing—but the dynamics of the lithosphere, with a few spectacular exceptions, such as earthquakes and volcanic eruptions, operate with such incredible slowness that Earth's crust often appears changeless. Most laypersons consider the phrase "the everlasting hills" a literal expression aptly describing the permanence of Earth's topography. In reality, the phrase is hyperbole that fails to recognize the remarkable alterations that take place over time, largely unrecognizable by the casual observer.

Our goal in the remaining chapters of this book is to understand the contemporary character of Earth's surface and to explain the processes that have caused it to be as it is. In this chapter, we begin with the aspect of Earth that perhaps most dramatically links the lithosphere, the atmosphere, the hydrosphere and the biosphere: the thin veneer of soil.

▲ **Figure 12-1** Farms in Lancaster County, Pennsylvania.

Soil and Regolith

Although the lithosphere encompasses the entire planet, from surface to core,[1] the part that holds our attention here is soil, the topmost layer. Soil is the essential medium in which most terrestrial life is nurtured. Almost all land plants sprout from this precious medium that is spread so thinly across the continental surfaces, with an average worldwide depth of only about 15 centimeters (6 inches).

Despite the implication of the well-known simile "as common as dirt," soil is extremely complex. It is an infinitely varying mixture of weathered mineral particles, decaying organic matter, living organisms, gases, and liquid solutions.

Preeminently, however, soil is a zone of plant growth. Although the concept almost defies definition, **soil** can be conceptualized as a relatively thin surface layer of mineral matter that normally contains a considerable amount of organic material and is capable of supporting living plants. It occupies that part of the outer skin of Earth that extends from the surface down to the maximum depth to which living organisms penetrate, which means basically the area occupied by plant roots. Soil is characterized by its ability to produce and store plant nutrients, an ability made possible by the interactions of such diverse factors as water, air, sunlight, rocks, plants, and animals.

[1] In this chapter we use the word *lithosphere* as a general term for the solid part of Earth. In Chapter 13 we will see that the term lithosphere has a more limited definition when discussing plate tectonics.

◀ Aerial view of a plowed field near Hinton, Oklahoma.

Although thinly distributed over the land surface, soil functions as a fundamental interface where atmosphere, lithosphere, hydrosphere, and biosphere meet. The bulk of most soil is inorganic material, so soil is usually classified as part of the lithosphere, but its relationship to the other three spheres is both intimate and complex (Figure 12-1).

Soil development begins with the physical and chemical disintegration of rock exposed to the atmosphere and to the action of water percolating down from the surface. This disintegration is called *weathering*. As we shall learn in Chapter 15, the basic result of weathering is the weakening and breakdown of solid rock, the fragmentation of coherent rock masses, the making of little rocks from big ones. The principal product is a layer of loose inorganic material called **regolith** ("blanket rock") because it lies like a blanket over the unfragmented rock below (Figure 12-2). Typically then, the regolith consists of material that has weathered from the underlying rock and that has a crude gradation of particle sizes, with the largest and least fragmented pieces at the bottom, immediately adjacent to the bedrock. Sometimes, however, the regolith consists of material that was transported from elsewhere by the action of wind, water, or ice. Thus, the regolith may vary significantly in composition from place to place.

The upper half meter or so of the regolith normally differs from the material below in several ways, most notably in the intensity of biochemical weathering. This upper portion is soil, which is composed largely of finely fragmented mineral

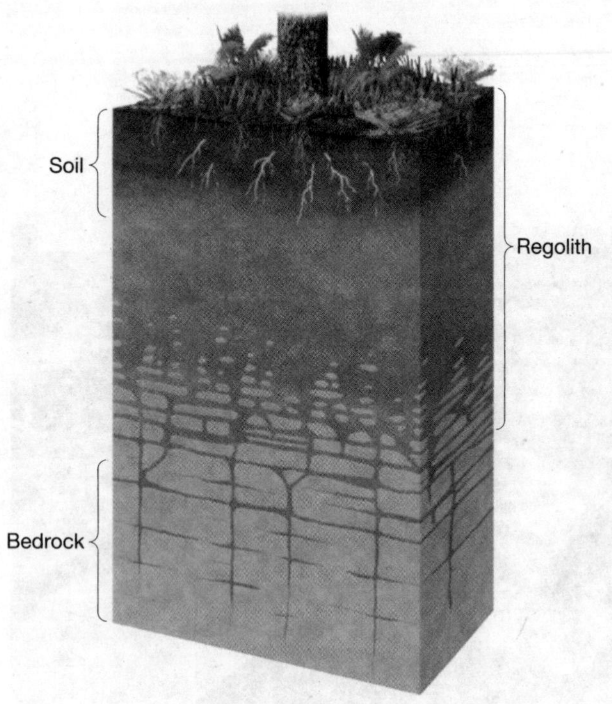

▲ **Figure 12-2** Vertical cross section from surface to bedrock, showing the relationship between soil and regolith.

particles, the ultimate product of weathering. It normally also contains an abundance of living plant roots, dead and rotting plant parts, microscopic plants and animals both living and dead, and a variable amount of air and water. Soil is not the end product of a process, but rather a stage in a never-ending continuum of physical–chemical–biotic activities (Figure 12-3).

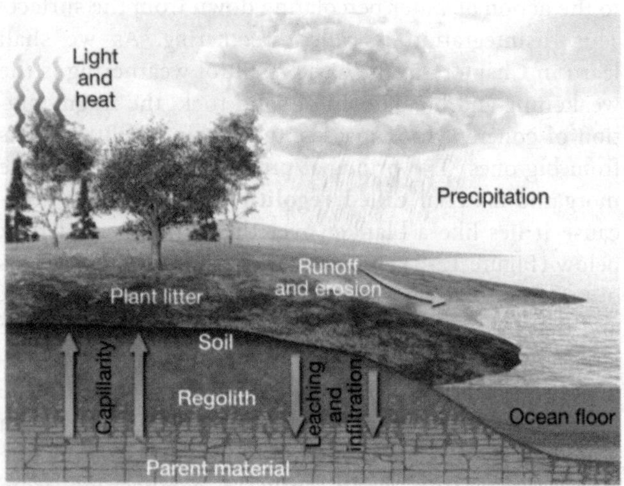

▲ **Figure 12-3** Soil develops through a complex interaction of physical, chemical, and biological processes. Parent-material bedrock weathers to regolith, and then plant litter combines with the regolith to form soil. Some of that soil washes to the ocean floor, where, over the expanse of geologic time, it is transformed to sedimentary rock. Some day that ocean floor may be uplifted above sea level and the exposed sedimentary rock will again be weathered into soil.

Soil as a Component of the Landscape

The surface of the lithosphere is usually, but not always, represented by soil. Although extremely pervasive, soil is an inconspicuous component of the landscape because it is beneath our feet, and its presence is normally masked by either vegetation or human constructed features.

We recognize soil in the landscape mostly by its color, which can often be seen through the filigree of plant life that covers it. A second main aspect of soil—its depth—becomes obvious only where some of its vertical dimension is exposed by gully erosion, road cuts, or some other type of excavation.

Soil-Forming Factors

Soil is an ever-evolving material. Metaphorically, soil acts like a sponge—taking in inputs and being acted upon by the local environment—changing over time and when the inputs or local environment change. Five principal soil-forming factors are responsible for soil development: geology, climate, topography, biology, and time.

The Geologic Factor

The source of the rock fragments that make up soil is **parent material**, which may be either bedrock or loose sediments transported from elsewhere by water, wind, or ice. Whatever the parent material, it is sooner or later disintegrated and decomposed at and near Earth's surface, providing the raw material for soil formation.

The nature of the parent material often influences the characteristics of the soil that develops from it, and, particularly in the early stages of soil formation, this factor sometimes dominates all others. The chemical composition of parent material is obviously reflected in the resulting soil, and parent-material physical characteristics may also be influential in soil development, particularly in terms of texture and structure. Bedrock that weathers into large particles (as does sandstone, for example) normally produces a coarse-textured soil, one easily penetrated by air and water to some depth. Bedrock that weathers into minute particles (shale, for example) yields fine-textured soils with a great number of pores but of very small size, which inhibits air and water from easily penetrating the surface.

Young soils are likely to be very reflective of the rocks or sediments from which they were derived. With the passage of time, however, other soil-forming factors become increasingly important, and the significance of the parent material diminishes. Eventually the influence of the parent material may be completely obliterated, so that it is sometimes impossible to ascertain the nature of the rock from which the soil evolved.

The Climatic Factor

Temperature and moisture are the climatic variables of greatest significance to soil formation. As a basic generalization, both the chemical and biological processes in soil are usually accelerated by high temperatures and abundant moisture and are slowed by low temperatures and lack of moisture. One predictable result is that soils tend to be deepest in warm, humid regions and shallowest in cold, dry regions.

It is difficult to overemphasize the role of moisture moving through the soil. The flow is mostly downward because of the pull of gravity, but it is sometimes lateral in response to drainage opportunities and sometimes, in special circumstances, even upward. Wherever and however water moves, it always carries dissolved chemicals in solution and usually also carries tiny particles of matter in suspension. Thus, moving water is ever engaged in rearranging the chemical and physical components of the soil, as well as contributing to the variety and availability of plant nutrients.

In terms of general soil characteristics, climate is likely to be the most influential factor in the long run. This generalization has many exceptions, however, and when soils are considered on a local scale, climate is likely to be less prominent as a determinant.

The Topographic Factor

Slope and drainage are the two main features of topography that influence soil characteristics. Wherever soil develops, its vertical extent undergoes continuous, if usually very slow, change. This change comes about through a lowering of both the bottom and top of the soil layer (Figure 12-4). The bottom slowly gets deeper as weathering penetrates into the regolith and parent material and as plant roots extend to greater depths. At the same time, the soil surface is being lowered by sporadic removal of its uppermost layer through normal erosion, which is the removal of individual soil particles by running water, wind, and gravity.

Where the land is flat, soil tends to develop at the bottom more rapidly than it is eroded away at the top. This does not mean that the downward development is speedy; rather it means that surface erosion is extraordinarily slow. Thus,

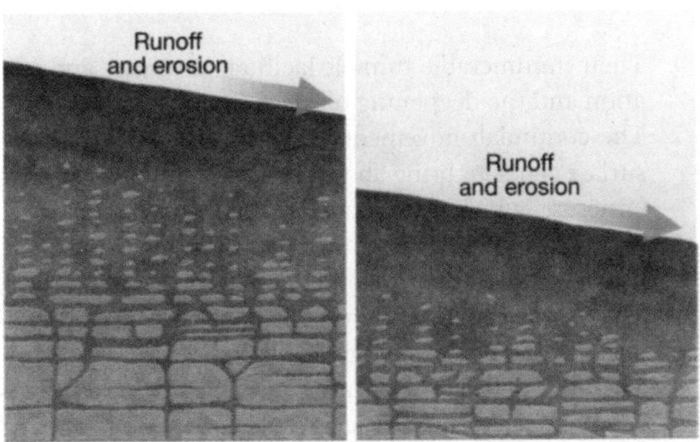

▲ **Figure 12-4** Over time, the extent of soil undergoes slow, continuous change. The bottom of the soil layer is lowered by the break-up of parent material as weathering processes extend deeper into the regolith and bedrock. At the same time, the top of the soil layer can be lowered through erosional processes.

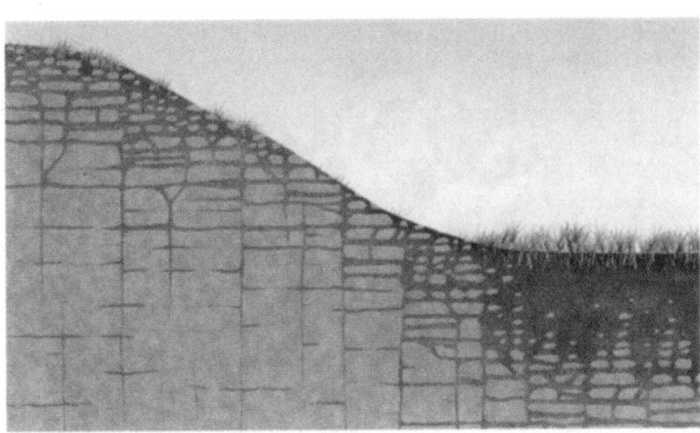

▲ **Figure 12-5** Slope is a critical determinant of soil depth. On flat land, soil normally develops more deeply with the passage of time because there is very little erosion washing away the topmost soil. On a slope, the rate of erosion is equal to or greater than the rate at which soil is formed at the bottom of the soil layer, with the result that the soil remains shallow.

the deepest soils are usually on flat land. Where slopes are relatively steep, surface erosion is more rapid than soil deepening, with the result that such soils are nearly always thin and immaturely developed (Figure 12-5).

If soils are well drained, moisture relationships may be relatively unremarkable factors in soil development. If soils have inefficient natural drainage, however, significantly different characteristics may develop. For example, a waterlogged soil tends to contain a high proportion of organic matter, and the biological and chemical processes that require free oxygen are impeded (because air is the source of the needed oxygen and a waterlogged soil contains essentially no air). Most poorly drained soils are in valley bottoms or in some other flat locale because soil drainage is usually related to slope.

In some cases, such subsurface factors as permeability and the presence or absence of impermeable layers are more influential than slope.

The Biological Factor

From a volume standpoint, soil is about half mineral matter and about half air and water, with only a small fraction of organic matter. However, the organic fraction, consisting of both living and dead plants and animals, is of utmost importance. The biological factor in particular gives life to the soil and makes it more than just "dirt." Every soil contains a quantity (sometimes an enormous quantity) of living organisms, and every soil incorporates some (sometimes a vast amount of) dead and decaying organic matter.

Vegetation of various kinds growing in soil performs certain vital functions. Plant roots, for instance, work their way down and around, providing passageways for drainage and aeration, as well as being the vital link between soil nutrients and the growing plants.

Many kinds of animals contribute to soil development as well. Even such large surface-dwelling creatures as elephants and bison affect soil formation by compaction with their hooves, rolling in the dirt, grazing the vegetation,

▲ **Figure 12-6** Like other burrowing animals, these prairie dogs in South Dakota contribute to soil development by bringing subsoil to the surface and providing passageways for air and moisture to get underground.

and dropping excreta. Ants, worms, and all other land animals fertilize the soil with their waste products and contribute their carcasses for eventual decomposition and incorporation into the soil.

Many smaller animals spend most or all of their lives in the soil layer, tunneling here and there, moving soil particles upward and downward, and providing passageways for water and air (Figure 12-6). Mixing and plowing by soil fauna is sometimes remarkably extensive. Ants and termites, as they build mounds, also transport soil materials from one layer to another. The mixing activities of animals in the soil, generalized under the term *bioturbation*, tend to counteract the tendency of other soil-forming processes to accentuate the vertical differences among soil layers.

The abundance and variety of animal life connected with the soil are quite surprising. Such organisms vary in size from the gigantic to the microscopic, and in numbers from a few per hectare to billions per gram. The organic life of the soil ranges from almost impalpable protozoans existing in colloidal slime to larger animals that may accidentally alter certain soil characteristics. Of all creatures, however, it is probable that the earthworm is the most important to soil formation and development (Figure 12-7).

Earthworms The cultivating and mixing activities of earthworms are of great value in improving the structure, increasing the fertility, lessening the danger of accelerated erosion, and deepening the profile of the soil. The distinctive evidence of this value is that the presence of many well-nourished earthworms is almost always a sign of productive, or potentially productive, soil.

The mere presence of earthworms does not guarantee that a soil will be highly productive, as there may be other kinds of inhibiting factors such as a high water table. Nevertheless, an earthworm-infested soil has a higher

▲ **Figure 12-7** The common earthworm (*Lumbricus terrestris*) in an underground burrow.

potential productivity than similar soils lacking earthworms. In various controlled experiments, the addition of earthworms to wormless soil has enhanced plant productivity by several hundred percent.

At least seven beneficial functions have been attributed to earthworms:

1. Their innumerable tunnels facilitate drainage and aeration and the deepening of the soil profile.
2. The continual movement of the creatures beneath the surface tends to bring about the formation of a crumbly structure, which is generally favorable for plant growth.
3. The soil is further mixed by material being carried and washed downward into their holes from the surface. This is notably in the form of leaf litter dragged downward by the worms, which fertilizes the subsoil.
4. The digestive actions and tunneling of earthworms form aggregate soil particles that increase porosity and resist the impact of raindrops, helping to deter erosion.
5. Nutrients in the soil are increased by the addition of *casts* excreted by earthworms (casts are expelled by earthworms and consist of mineral material bound together with decomposed organic material), which have

been shown to be 5 times richer in available nitrogen, 7 times richer in available phosphates, and 11 times richer in available potash than the surrounding soil.
6. They rearrange material in the soil, especially by bringing deeper matter to the surface, where it can be weathered more rapidly. Where earthworms are numerous, they may deposit as much as 9000 kg/hectare (25 tons/acre) of casts on the surface in a year.
7. Nitrification is also promoted by the presence of earthworms, due to increased aeration, alkaline fluids in their digestive tracts, and the decomposition of earthworm carcasses.

In many parts of the world, of course, earthworms are lacking. They are, for example almost totally absent from arid and semiarid regions. In these dry lands, some of the earthworm's soil-enhancing functions are carried out by ants and earth-dwelling termites, but much less effectively.

Microorganisms in the Soil Another important component of the biological factor are microorganisms, both plant and animal, that occur in uncountable billions. An estimated three quarters of a soil's metabolic activity is generated by microorganisms. These microbes help release nutrients from dead organisms for use by live ones by decomposing organic matter into *humus*, a dark adhesive of minute particles, and by converting nutrients to forms usable by plants. Algae, fungi, protozoans, actinomycetes, and other minuscule organisms all play a role in soil development, but bacteria probably make the greatest contribution overall. This is because certain types of bacteria are responsible for the decomposition and decay of dead plant and animal material and the consequent release of nutrients into the soil.

The Time Factor

For soil to develop on a newly exposed land surface requires time, with the length of time needed varying according to the nature of the exposed parent material and the characteristics of the environment. Soil-forming processes are generally very slow, and many centuries may be required for a thin layer of soil to form on a newly exposed surface. A warm, moist environment is conducive to soil development. Normally of much greater importance, however, are the attributes of the parent material. For example, soil develops from sediments relatively quickly and from bedrock relatively slowly.

Most soil develops with geologic slowness—so slowly that changes are almost imperceptible within a human life span. It is possible, however, for a soil to be degraded, either through the physical removal associated with accelerated erosion or through depletion of nutrients, in only a few years (Figure 12-8). In the grand scale of geologic time, then, soil can be formed and reformed, but in the dimension of human time, it is a nonrenewable resource.

▲ **Figure 12-8** Accelerated erosion cutting a deep gully in the Coast Ranges of central California.

Soil Components

Soil is made up of a variety of natural components existing together in myriad combinations. All these components can be classified, however, into just a few main groups: inorganics, organics, air, and water.

Inorganic Materials

As mentioned earlier, the bulk of most soils is mineral matter, mostly in the form of small but macroscopic particles. Inorganic material also occurs as microscopic clay particles and as dissolved minerals in solution.

About half the volume of an average soil is small, granular mineral matter called *sand* and *silt*. These particles may consist of a great variety of minerals, depending on the nature of the parent material from which they were derived, and are simply fragments of the wasting rock. Most common are bits of quartz, which are composed of silica (SiO_2) and appear in the soil as very resistant grains of sand. Other prominent minerals making up sand and silt are some of the feldspars and micas.

The smallest particles in the soil are **clay**, which is usually a combination of silica and of oxides of aluminum and iron found only in the soil and not in the parent material. Clay has properties significantly different from those of larger (sand and silt) fragments. Most clay particles are *colloidal* in size, which means they are larger than molecules but too small to be seen with the naked eye. They are usually flat platelets, as Figure 12-9 illustrates, and therefore have a relatively large surface area. For this reason, clay has an important influence on chemical activity in the soil because many chemical reactions occur at the surfaces of soil particles. The platelets group together in loose, sheetlike assemblages, and

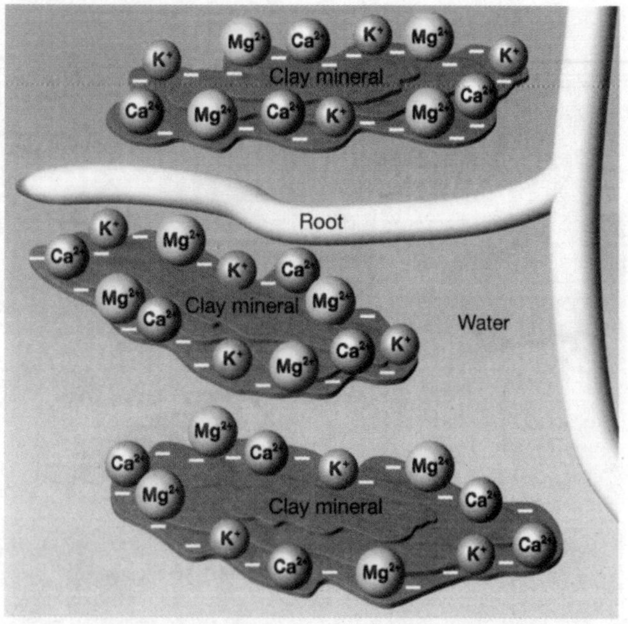

▲ **Figure 12-9** Clay particles offer a large surface area on which substances dissolved in soil water can cling. The particles are negatively charged and therefore attract cations (positively charged ions) from the water. These cations held by the clay are then absorbed by plant roots and become nutrients for the plant.

water moves easily between these sheets. Substances dissolved in the water are attracted to and held by the sheets. Since the sheets are negatively charged, they attract positively charged ions called **cations**. Many essential plant nutrients occur in soil solutions as cations, with the result that clay is an important reservoir for plant nutrients, just as it is for soil water. We discuss the role of cations in soils in greater detail later in this chapter.

Organic Matter

Although organic matter generally constitutes less than 5 percent of total soil volume, it has an enormous influence on soil characteristics and plays a fundamental role in the biochemical processes that make soil an effective medium of plant growth. Some of the organic matter is living organisms, some is dead but undecomposed plant parts and animal carcasses, some is totally decomposed and so has become humus, and some is in an intermediate stage of decomposition.

Apart from plant roots, evidence of the variety and bounty of organisms living in the soil may be inconspicuous or invisible, but most soils are seething with life. A half hectare (about one acre) may contain a million earthworms, and the total number of organisms in 28 grams (an ounce) of soil is likely to exceed 100 trillion. Microorganisms far exceed more complex life forms, both in total numbers and in cumulative mass. They are active in rearranging and aerating the soil and in yielding waste products that are links in the chain of nutrient cycling. Some make major contributions to the decay and decomposition of dead organic matter, and others make nitrogen available for plant use.

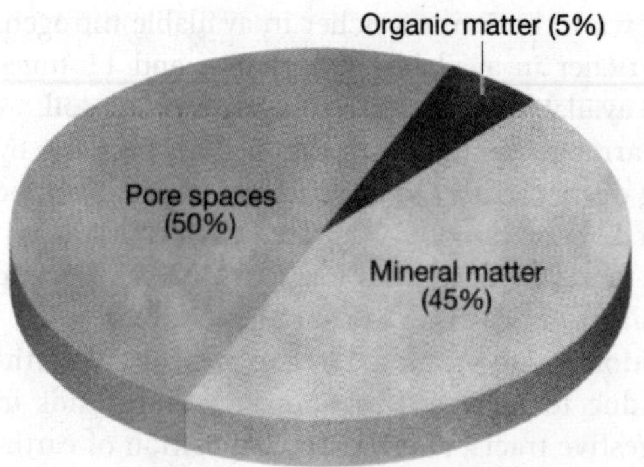

▲ **Figure 12-10** The best soil for plant growth is about half solid material (by volume) and about half pore spaces. Most of the solids are mineral matter, with only a small amount being organic. On average, about half of the pore spaces in an ideal soil are filled with air and the other half are filled with water.

Leaves, twigs, stalks, and other dead plant parts accumulate at the soil surface, where they are referred to collectively as **litter**. The eventual fate of most litter is decomposition, in which the solid parts are broken down into chemical components, which are then either absorbed into the soil or washed away. In cold, dry areas, litter may remain undecomposed for a very long time; where the climate is warm and moist, however, decomposition may take place almost as rapidly as litter accumulates.

After most of the residues have been decomposed, a brown or black, gelatinous, chemically stable organic matter remains; this is referred to as **humus**. This "black gold" is of utmost importance to agriculture because it loosens the structure and lessens the density of the soil, thereby facilitating root development. Moreover, humus, like clay, is a catalyst for chemical reactions and a reservoir for plant nutrients and soil water.

Soil Air

Nearly half the volume of an average soil is made up of pore spaces (Figure 12-10). These spaces provide a labyrinth of interconnecting passageways, called *interstices*, among the soil particles. This labyrinth lets air and water penetrate into the soil. On the average, the pore spaces are about half filled with air and half with water, but at any given time and place, the amounts of air and water are quite variable, the quantity of one varying inversely with that of the other (Figure 12-11).

The characteristics of air in the soil are significantly different from those of atmospheric air. Soil air is found in openings generally lined with a film of water, and since this air exists in such close contact with water and is not exposed to moving air currents, it is saturated with water vapor. Soil air is also very rich in carbon dioxide and poor in oxygen because plant roots and soil organisms remove oxygen from, and respire carbon dioxide into, the pore spaces. The carbon dioxide then slowly escapes into the atmosphere.

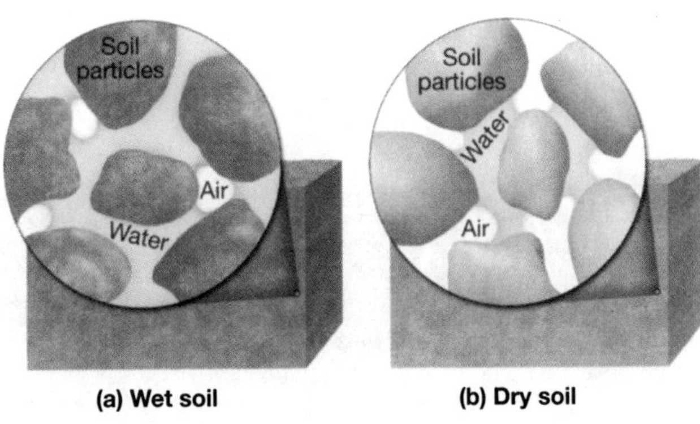

▲ **Figure 12-11** The relative amounts of water and air in soil pores vary from place to place and from time to time. (a) The interstices of wet soil contain much water and little air. (b) In dry soil there is much air and little water.

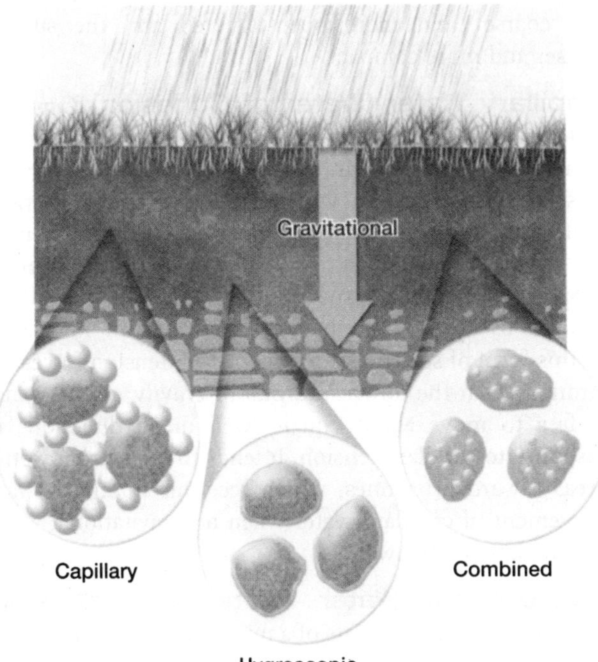

▲ **Figure 12-13** The four forms of soil moisture.

Soil Water

Water comes into the soil largely by percolation of rainfall and snowmelt, but some is also added from below when groundwater is pulled up above the water table by *capillary action* (Figure 12-12; see Chapter 6 for a discussion of capillarity). Once water has penetrated the soil, it envelops in a film of water each solid particle that it contacts, and it either wholly or partially fills the pore spaces. Water can be lost from the soil by percolation down into the groundwater, by upward capillary movement to the surface followed by evaporation, or by plant use (transpiration).

Four forms of soil moisture are generally recognized (Figure 12-13) and are discussed below.

Gravitational Water (Free Water) *Gravitational water* is temporary in that it results from prolonged infiltration from above (usually due to prolonged precipitation) and is pulled downward by gravity, through the interstices toward the groundwater zone. Thus, this water stays in the soil only for a short time and is not very effective in supplying plants because it drains away rapidly once the external supply ceases.

Gravitational water accomplishes significant functions during its passage through the soil, however. It is the principal agent of *eluviation* and *illuviation* (discussed below) and is therefore a translocation agent that makes the top

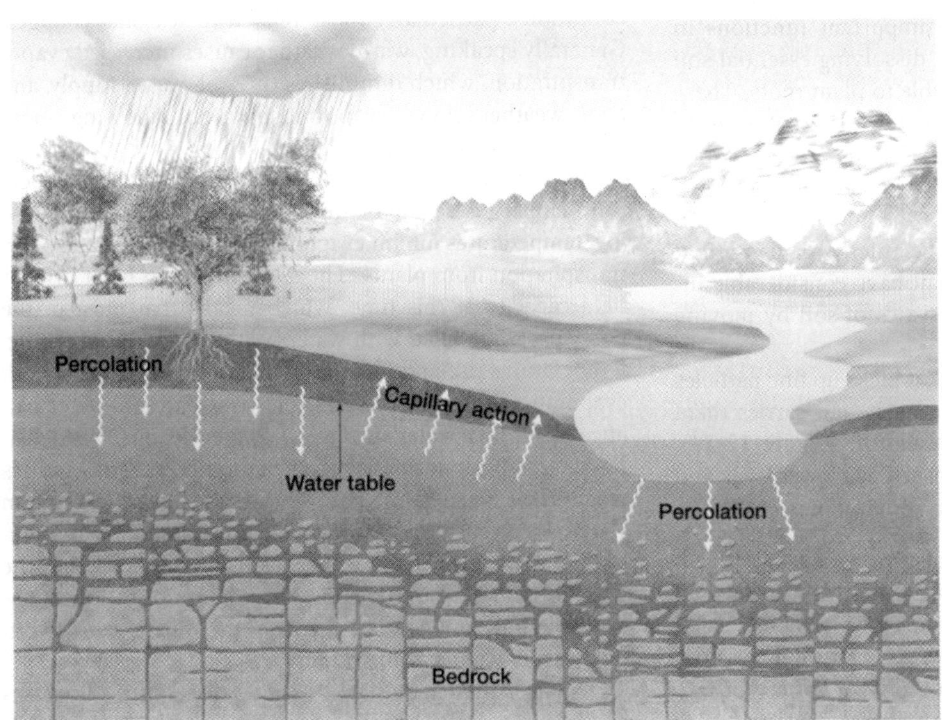

◀ **Figure 12-12** Water is added to the soil layer by the percolation of rainwater and snowmelt from above. Additional moisture enters the soil from below as groundwater is pulled up above the water table by capillary action.

soil coarser and more open-textured and the subsoil denser and more compact.

Capillary Water (Water of Cohesion) *Capillary water*, which remains after gravitational water has drained away, consists of moisture held at the surface of soil particles by *surface tension*, which is the attraction of water molecules for each other (discussed in Chapter 6, this is the same property that causes water to form rounded droplets rather than dispersing in a thin film). Capillary water is by far the principal source of moisture for plants. In this form of soil moisture, the surface-tension forces are stronger than the downward pull of gravity, so this water is free to move about equally well in all directions in response to capillary tension. It tends to move from wetter areas toward drier ones, which accounts for the upward movement of capillary water when no gravitational water is percolating downward.

Hygroscopic Water (Water of Adhesion) *Hygroscopic water* consists of a microscopically thin film of moisture bound rigidly to all soil particles by *adhesion*, which is the attraction of water molecules to solid surfaces. Hygroscopic water adheres so tightly to the particles that it is normally unavailable to plants.

Combined Water *Combined water* is least available of all. It is held in chemical combination with various soil minerals and is freed only if the chemical is altered.

For plants, capillary water is the most important and gravitational water is largely superfluous. After gravitational water has drained away, the remaining volume of water represents the **field capacity** of the soil. If drought conditions prevail and the capillary water is all used up by plants or evaporated, the plants are no longer able to extract moisture from the soil, and the **wilting point** is reached.

Water performs a number of important functions in the soil. It is an effective solvent, dissolving essential soil nutrients and making them available to plant roots. These dissolved nutrients are carried downward in solution, to be partly redeposited at lower levels. This process, called **leaching**, tends to deplete the topsoil of soluble nutrients. Water is also required for many of the chemical reactions of clay and for the actions of the microorganisms that produce humus. In addition, it can have considerable influence on the physical characteristics of soil by moving particles around.

As water percolates into the soil, it picks up fine particles of mineral matter from the upper layers and carries them downward in a process called **eluviation** (Figure 12-14). These particles are eventually deposited at a lower level, and this deposition process is known as **illuviation**.

Soil-Water Budget The moisture added to the soil by percolation of rainfall or snowmelt is diminished largely through evapotranspiration. The dynamic relationship between these two processes is referred to as the **soil–water balance**. It is influenced by a variety of factors, including soil and vegetation characteristics, but is primarily determined by temperature and humidity.

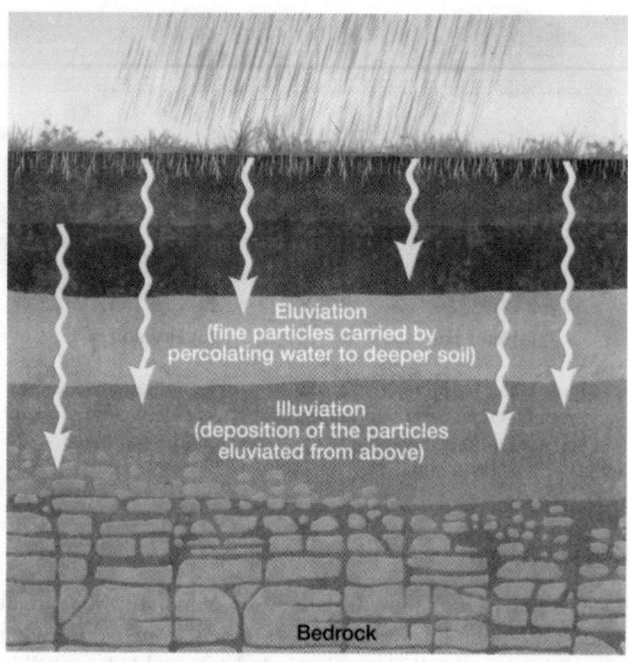

▲ **Figure 12-14** In the process of eluviation, fine particles in upper soil layers are picked up by percolating water and carried deeper into the soil. In the process of illuviation, these particles are deposited in a lower soil layer.

How much water is available to plants is much more important to an ecosystem than is the amount of precipitation. Much water derived from rainfall or snowmelt becomes unavailable to plants because of evaporation, runoff, deep infiltration, or other processes. At any given time and place, there is likely to be either a surplus or a deficit of water in the soil. Such a condition is only temporary, and it varies in response to changing weather conditions, particularly those related to seasonal changes. Generally speaking, warm weather causes increased evapotranspiration, which diminishes the soil–water supply, and cool weather slows evapotranspiration, allowing more moisture to be retained in the soil.

In a hypothetical Northern Hemisphere midlatitude location, January is a time of surplus water in the soil because low temperatures inhibit evaporation and there is little or no transpiration from plants. The soil is likely to be at or near field capacity at this time, which means that most of the pore spaces are filled with water. With the arrival of spring, temperatures rise and plant growth accelerates, so that both evaporation and transpiration increase. The soil–water balance tips from a water surplus to a water deficit. This deficit builds to a peak in middle or late summer, as temperatures reach their greatest heights and plants need maximum water. Heavy use and diminished precipitation may combine to deplete all the moisture available to plants, and the wilting point is reached. Thus, the amount of soil moisture available for plant use is essentially the difference between field capacity and wilting point.

In late summer and fall, as air temperature decreases and plant growth slackens, evapotranspiration diminishes

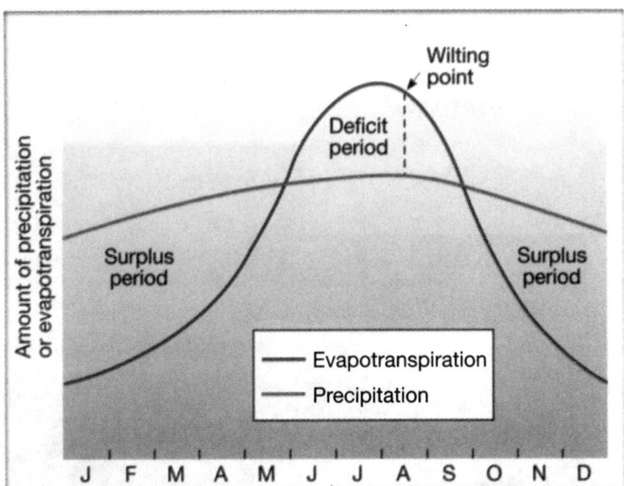

▲ **Figure 12-15** Hypothetical soil–water budget for a Northern Hemisphere midlatitude location. From January through May, there is more precipitation than evapotranspiration, and consequently the soil contains a surplus of water, more than sufficient for any plant needs. From mid-May to mid-September, the evapotranspiration curve rises above the precipitation curve, indicating that more water leaves the soil than enters it. About the first of August, so much water has been removed from the soil that plants begin to wilt. After mid-September, the evapotranspiration curve again dips below the precipitation curve and there is again a surplus of water in the soil.

rapidly. At this time, the soil–water balance shifts once again to a water surplus, which continues through the winter. Then the cycle begins again. Figure 12-15 illustrates the annual sequence just described. Such variation in the soil–water balance through time is called a **soil–water budget**.

Soil Properties

As one looks at, feels, smells, tastes, and otherwise examines soils, various physical and chemical characteristics appear useful in describing, differentiating, and classifying them. Some soil properties are easily recognized, but most can be ascertained only by precise measurement.

Color

The most conspicuous property of a soil is usually its color, but color is by no means the most definitive property. Soil color can provide clues about the nature and capabilities of the soil, but the clues are sometimes misleading. Soil scientists recognize 175 gradations of color. The standard colors are generally shades of black, brown, red, yellow, gray, and white. Soil color occasionally reflects the color of the unstained mineral grains, but in most cases, color is imparted by stains on the surface of the particles; stains caused by either metallic oxides or organic matter.

Black or dark brown usually indicates a considerable humus content; the blacker the soil, the more humus it contains. Color gives a strong hint about fertility, therefore, because humus is an important catalyst in releasing nutrients to plants. Dark color is not invariably a sign of fertility, however, because it may be due to other factors, such as poor drainage or high carbonate content.

Reddish and yellowish colors generally indicate iron oxide stains on the outside of soil particles. These colors are most common in tropical and subtropical regions, where many minerals are leached away by water moving under the pull of gravity, leaving insoluble iron compounds behind. In such situations, a red color bespeaks good drainage, and a yellowish hue suggests imperfect drainage. Red soils are also common in desert and semidesert environments, where the color is carried over intact from reddish parent materials rather than representing a surface stain (Figure 12-16).

Gray and bluish colors typically indicate poor drainage, whereas mottling indicates saturated conditions for part of the year. In humid areas, a light color implies so much leaching that even the iron has been removed, but in dry climates, it indicates an accumulation of salts. It may also indicate simply a lack of organic matter.

Texture

All soils are composed of myriad particles of various sizes, as Figure 12-17 shows, although smaller particles usually predominate. Rolling a sample of soil about between the fingers can provide a feel for the principal particle sizes. Table 12-1 shows the standard classification scheme for particle sizes; in this scheme, the size groups are called **separates**. The gravel, sand, and silt separates are fragments of the weathered parent material and are mostly the grains of minerals found commonly in rocks, especially quartz, feldspars, and micas. These coarser particles are the inert materials of the soil mass, its skeletal framework. As noted above, only the clay particles take part in the intricate chemical activities that occur in the soil.

Because no soil is made up of particles of uniform size, the texture of any soil is determined by the relative amounts of the various separates present. The *texture triangle* (Figure 12-18) shows the standard classification scheme for soil texture; this scheme is based on the percentage of each separate by weight. Near the center of the triangle is **loam**, the name given to a texture in which none of the three principal separates dominates the other two. This fairly even-textured mix is generally the most productive for plants.

Structure

The individual particles of most soils tend to aggregate into clumps called **peds**, and it is these clumps that determine soil structure. The size, shape, and stability of peds have a marked influence on how easily water, air, and organisms (including plant roots) move through the soil, and consequently on soil fertility. Peds are classified on the basis of shape as spheroidal, platy, blocky, or prismatic, with these four shapes giving rise to seven generally recognized soil

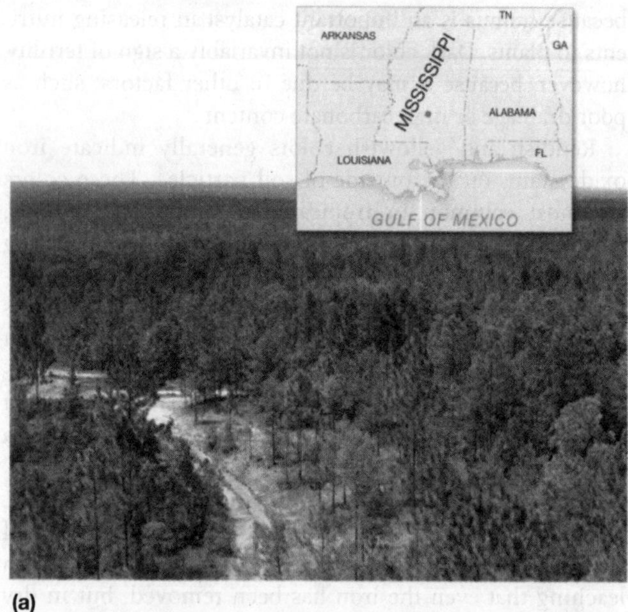

▲ **Figure 12-16** A red soil means different things in different places. (a) In this Mississippi forest, the soluble minerals have been leached away, leaving the insoluble iron to impart its reddish color to the soil. (b) In this Australian desert, the reddish color reflects the iron content of the underlying bedrock.

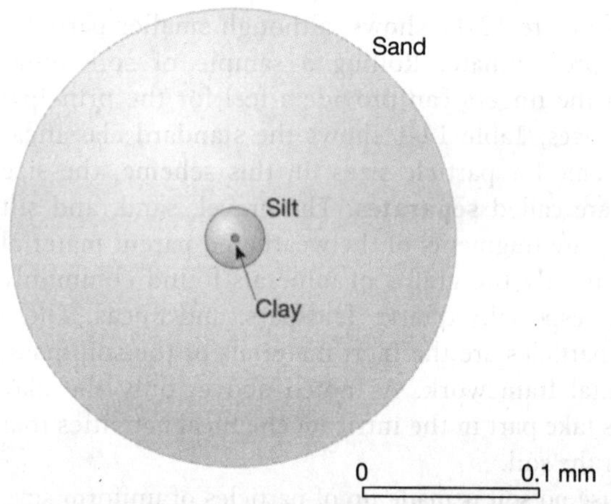

▲ **Figure 12-17** The relative sizes of sand, silt, and clay particles.

TABLE 12-1 Standard U.S. Classification of Soil Particle Size

Separate	Diameter
Gravel	Greater than 2 mm (0.08 in.)
Very coarse sand	1–2 mm (0.04–0.08 in.)
Coarse sand	0.5–1 mm (0.02–0.04 in.)
Medium sand	0.25–0.5 mm (0.01–0.02 in.)
Fine sand	0.1–0.25 mm (0.004–0.01 in.)
Very fine sand	0.05–0.1 mm (0.002–0.004 in.)
Coarse silt	0.02–0.05 mm (0.0008–0.002 in.)
Medium silt	0.006–0.02 mm (0.00024–0.0008 in.)
Fine silt	0.002–0.006 mm (0.00008–0.00024 in.)
Clay	0.002 mm (less than 0.00008 in.)

structure types (Figure 12-19). Aeration and drainage are usually facilitated by peds of intermediate size; both massive and fine structures tend to inhibit these processes.

Some soils, particularly those composed largely of sand, do not develop a true structure, which is to say that the individual grains do not aggregate into peds. Silt and clay particles readily aggregate in most instances. Other things being equal, aggregation is usually greatest in moist soils and least in dry ones.

Structure is an important determinant of a soil's *porosity* and *permeability*. As we learned in Chapter 9, porosity

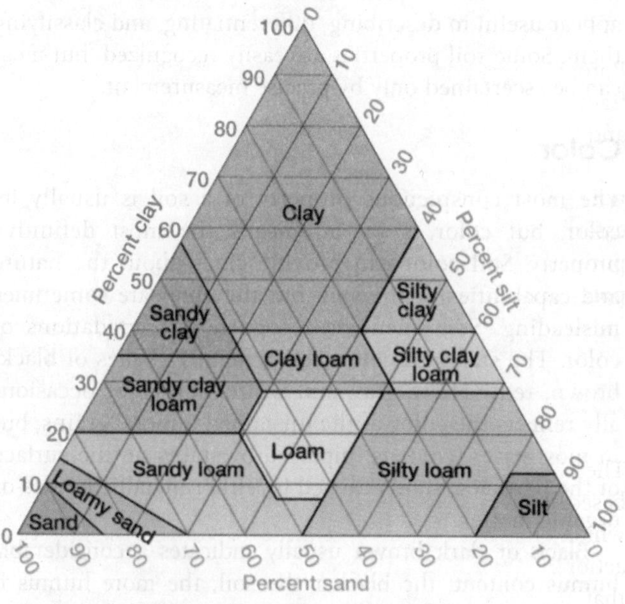

▲ **Figure 12-18** The standard soil-texture triangle.

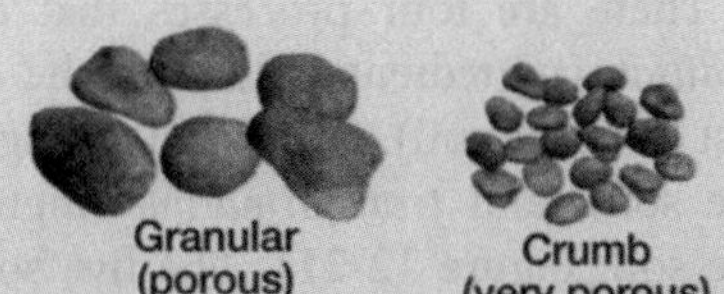

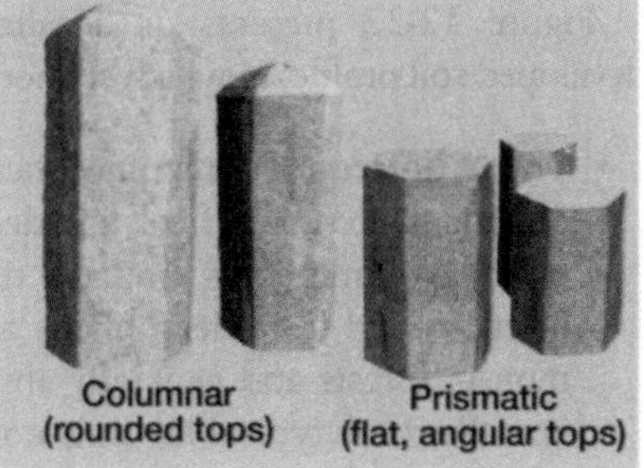

▲ **Figure 12-19** Various structure types of soil particle clumps, or peds, found in mineral soils.

refers to the amount of pore space between soil particles or between peds. We can define it as

$$\text{porosity} = \frac{\text{volume of voids}}{\text{total volume}}$$

Porosity is usually expressed as a percentage or a decimal fraction. It is a measure of a soil's capacity to hold water and air.

The relationship between porosity and permeability is not simple; that is, the most porous materials are not necessarily the most permeable. Clay, for example, is the most porous separate, but it is the least permeable because the pores are too small for water to easily pass through.

Soil Chemistry

The effectiveness of soil as a growth medium for plants is based largely on the presence and availability of nutrients, which are determined by an intricate series of chemical reactions. Soil chemistry is an extraordinarily complex subject that revolves primarily around microscopic particles and electrically charged atoms or groups of atoms called *ions*.

Colloids

Particles smaller than about 0.1 micrometer in diameter are called **colloids**. Inorganic colloids consist of clay in thin, crystalline, platelike forms created by the chemical alteration of larger particles; organic colloids represent decomposed organic matter in the form of humus; and both types are the chemically active soil particles. When mixed with water, colloids remain suspended indefinitely as a homogeneous, murky solution. Some have remarkable storage capacities, and consequently colloids are major determinants of the water-holding capacity of a soil. They function as a virtual sponge, soaking up water, whereas the soil particles that are too large to be classified as colloids can maintain only a surface film of water.

Both inorganic and organic colloids attract and hold great quantities of ions.

Cation Exchange

As was noted earlier, *cations* are positively charged ions. Elements that form them include calcium, potassium, and magnesium, which are all essential for soil fertility and plant growth. Colloids carry mostly negative electrical charges on their surfaces, and these charges attract swarms of nutrient cations that would be leached from the soil if their ions were not retained by the colloids.

The combination of colloid and attached cations is called the *colloidal complex*, and it is a delicate mechanism. If it holds the nutrients strongly, they will not be leached away, yet if the bond is too strong, they cannot be absorbed by plants. Thus, a fertile soil is likely to be one in which the cation–colloid attraction is intricately balanced.

Adding to the complexity of the situation is the fact that some types of cations are bound more tightly than others. Cations that tend to bond strongly in the colloid complex may replace those that bond less strongly. For example, basic ions are fairly easily replaced by metal ions or hydrogen ions, and this process is called *cation exchange*. The capability of a soil to attract and exchange cations is known as its **cation exchange capacity (CEC)**. As a generalization, the higher the CEC, the more fertile the soil. Soils with a high clay content have a higher CEC than more coarsely grained soils because the former have more colloids. Humus is a particularly rich source of high-CEC activity because humus colloids have a much higher CEC than inorganic clay minerals. The most fertile soils, then, tend to be those with a notable clay and humus content.

Acidity/Alkalinity

An *acid* is a chemical compound that produces hydrogen ions (H^+) or hydronium ions (H_3O^+) when dissolved in water, whereas a *base* is a chemical compound that produces hydroxide ions (OH^-) when dissolved in water. An acid reacts with a base to form a *salt*. Solutions that contain dissolved acids are described as being *acidic*. Those that contain dissolved bases are called either *basic* or *alkaline* solutions. Any chemical solution can be characterized on the basis of its acidity or alkalinity.

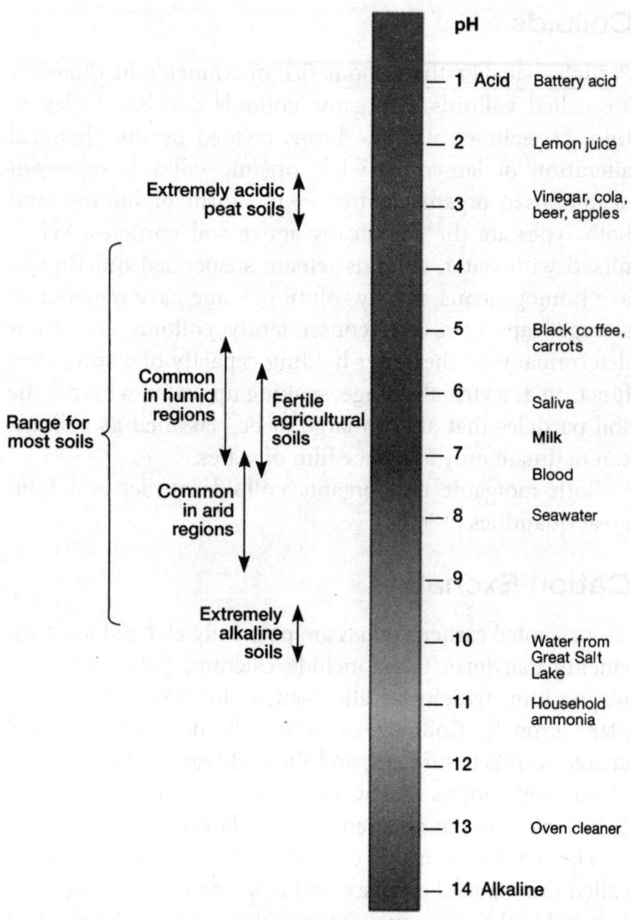

▲ **Figure 12-20** The standard pH scale. The most fertile soils tend to be neither too acidic nor too alkaline.

Nearly all nutrients are provided to plants in solution. An overly alkaline soil solution is inefficient in dissolving minerals and releasing their nutrients. However, if the solution is highly acidic, the nutrients are likely to be dissolved and leached away too rapidly for plant roots to absorb them. The optimum situation, then, is for the soil solution to be neutral, neither too alkaline nor too acidic. The acidity/alkalinity of a soil is determined to a large extent by its CEC.

As we first saw in Chapter 6, the chemist's symbol for the measure of the acidity/alkalinity of a solution is pH, which is based on the relative concentration of hydrogen ions (H^+) in the solution. The scale ranges from 0 to 14. The lower end represents acidic conditions; higher numbers indicate alkaline conditions (Figure 12-20). Neutral conditions are represented by a value of 7, and it is soil having a pH of about 7 that is most suitable for the great majority of plants and microorganisms.

Soil Profiles

The development of any soil is expressed in two dimensions: depth and time. There is no straight-line relationship between depth and age, however; some soils deepen and develop much more rapidly than others.

There are four processes that deepen and age soils: *addition* (ingredients added to the soil), *loss* (ingredients lost from the soil), *translocation* (ingredients moved within the soil), and *transformation* (ingredients altered within the soil; Figure 12-21). The five soil-forming factors discussed earlier—geologic, climatic, topographic, biological, and time—influence the rate of these four processes, the result being the development of various soil *horizons* and the *soil profile*.

The vertical variation of soil properties is not random but rather an ordered layering with depth. Soil tends to have more or less distinctly recognizable layers, called **horizons**, each with different characteristics. The horizons are positioned approximately parallel with the land surface, one above the other, normally, but not always, separated by a transition zone rather than a sharp line. A vertical cross section (as might be seen in a road cut or the side of a trench dug in a field) from the Earth's surface down through the soil layers and into the parent material is referred to as a **soil profile**. The almost infinite variety of soils in the world are usually grouped and classified on the basis of differences exhibited in their profiles.

Figure 12-22 presents an idealized sketch of a well-developed soil profile, in which six horizons are differentiated:

- The **O horizon** is sometimes the surface layer, and in it organic matter, both fresh and decaying, makes up most of the volume. This horizon results essentially from litter derived from dead plants and animals. It is common in forests and generally absent in grasslands. It is actually more typical for soils not to possess an O horizon; the surface horizon of most soils is the A horizon.
- The **A horizon**, colloquially referred to as *topsoil*, is a mineral horizon that also contains considerable organic matter. It is formed either at the surface or immediately below an O horizon. A horizons generally contain enough partially decomposed organic matter to give the soil a darker color than underlying horizons. They are also normally coarser in texture, having lost some of the finer materials by erosion and eluviation. Seeds germinate mostly in the A horizon.
- The **E horizon** is normally lighter in color than either the overlying A or the underlying B horizon. It is essentially an eluvial layer from which clay, iron, and aluminum have been removed, leaving a concentration of abrasion-resistant sand or silt particles.
- The **B horizon**, usually called *subsoil*, is a mineral horizon of illuviation where most of the materials removed from above have been deposited. A collecting zone for clay, iron, and aluminum, this horizon is usually of heavier texture, greater density, and relatively greater clay content than the A horizon.
- The **C horizon** is unconsolidated parent material (regolith) beyond the reach of plant roots and most soil-forming processes except weathering. It is lacking in organic matter.
- The **R horizon** is bedrock, with little evidence of weathering.

True soil, which is called **solum**, only extends down through the B horizon.

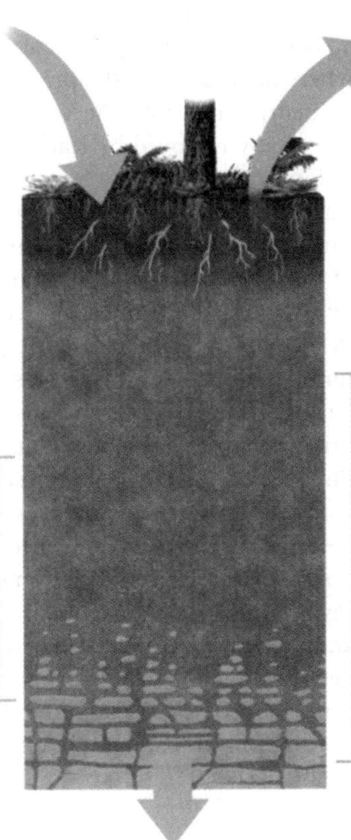

◀ **Figure 12-21** The four soil-forming processes: addition, loss, translocation, transformation. Geologic, climatic, topographic, biological, and chronological (time) soil-forming factors influence the rate at which these four processes occur and therefore the rate at which soil is formed.

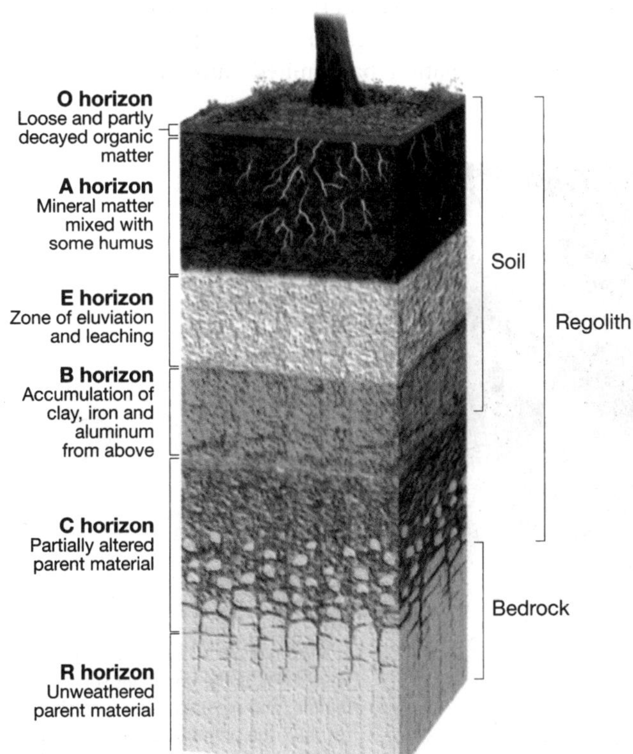

▲ **Figure 12-22** Idealized soil profile. The true soil, or solum, consists of the O, A, E, and B horizons.

As we learned earlier in the chapter, time is a critical passive factor in profile development, but the vital active factor is surface water. If there is no surface water, from rainfall or snowmelt or some other source, to infiltrate the soil, there can be no profile development. Descending water carries material from the surface downward, from topsoil into subsoil, by eluviation and leaching. This transported material is mostly deposited a few tens of centimeters (a few feet) below the surface. In the usual pattern, topsoil (A) becomes a somewhat depleted horizon through eluviation and leaching, and subsoil (B) develops as a layer of accumulation due to illuviation.

A profile that contains all horizons is typical of a humid area on well-drained but gentle slopes in an environment that has been undisturbed for a long time. In many parts of the world, however, such idealized conditions do not pertain, and the soil profile may have one horizon particularly well developed, one missing altogether, a *fossil horizon* formed under a different past climate, an accumulation of a hardpan (a very dense and impermeable layer), surface layers removed through accelerated erosion, or some other variation. Moreover, many soils are too young to have evolved a normal profile. A soil containing only an A horizon atop partially altered parent material (C horizon) is said to be *immature*. The formation of an illuvial B horizon is normally an indication of a *mature* soil.

Pedogenic Regimes

Soil-forming factors and processes interact in almost limitless variations to produce soils of all descriptions. Fundamental to an understanding of soil classification and distribution is the realization that only five major **pedogenic regimes** ("soil-forming" regimes) exist: *laterization*, *podzolization*, *gleization*, *calcification*, and *salinization*. These regimes can be thought of as environmental settings in which certain physical–chemical–biological processes prevail.

Laterization

Laterization is named for the brick-red color of the soil it produces (*Later*: Latin, "brick"). The processes associated with this regime are typical of the warm, moist regions of the world, and a significant annual moisture surplus is a requisite condition. The soil formed by laterization is most prominent, then, in the tropics and subtropics, in regions dominated by forest, shrub, and savanna vegetation.

A laterization regime is characterized by rapid weathering of parent material, dissolution of nearly all minerals, and speedy decomposition of organic matter. Probably the most distinctive feature of laterization is the leaching away of silica, the most common constituent of most rock and soil and a constituent that is usually highly resistant to being dissolved. That silica is indeed removed during laterization indicates the extreme effectiveness of chemical weathering and leaching under this regime. Most other minerals are also leached out rapidly, leaving behind primarily iron and aluminum oxides and barren grains of quartz sand. This residue normally imparts to the resulting soil the reddish color that gives this regime its name (Figure 12-23). The A horizon is highly eluviated and leached, whereas the B horizon has a considerable concentration of illuviated materials.

Because plant litter is rapidly decomposed in places where laterization is the predominant regime, little humus is incorporated into the soil. Even so, plant nutrients are not totally removed by leaching because the natural vegetation, particularly in a forest, quickly absorbs many of the nutrients in solution. If the vegetation is relatively undisturbed by human activities, this regime has the most rapid of nutrient cycles, and the soil is not totally impoverished by the speed of mineral decomposition and leaching. Where the forest is cleared for agriculture or some other human purpose, however, most base nutrients are likely to be lost from the cycle because the tree roots that would bring them up are gone. The soil then rapidly becomes impoverished, and hard crusts of iron and aluminum compounds are likely to form.

The general term applied to soils produced by laterization is *latosols*. These soils sometimes develop to depths of several meters because of the strong weathering activities and the fact that laterization continues year-round in these benign climates. Most latosols have little to offer as agricultural soils, but laterization often produces such concentrations of iron and aluminum oxides that mining them can be profitable.

Podzolization

Podzolization is another regime named after the color of the soil it produces; in this case, gray (*podzol*: Russian for "like ashes"). It also occurs in regions having a positive moisture balance and involves considerable leaching, but beyond those two characteristics, it bears little similarity to laterization. Podzolization occurs primarily in areas where the vegetation has limited nutrient requirements

◀ **Figure 12-23** In the wet tropics, laterization is the dominant soil-forming regime, and most soils are reddish as a result of the prominence of iron and aluminum compounds. Lateritic soils are exposed in this road cut near Savusavu on the Fijian island of Vanua Levu.

and where the plant litter is acidic. These conditions are most prominent in mid- and high-latitude locales having a coniferous forest cover. Thus, podzolization is largely a Northern Hemisphere phenomenon because there is not much land in the higher midlatitudes south of the equator. The typical location for podzolization is under a boreal forest in subarctic climates, which is found only in the Northern Hemisphere.

In these cool regions, chemical weathering is slow, but the abundance of acids, plus adequate precipitation, makes leaching very effective. Mechanical weathering from frost action is relatively rapid during the unfrozen part of the year. Moreover, much of the land was bulldozed by Pleistocene glaciers, leaving an abundance of broken rock debris at the surface. Bedrock here consists mostly of ancient crystalline rocks rich in quartz and aluminum silicates and poor in the alkaline mineral cations important in plant nutrition. The boreal forests, dominated by conifers, require little in the way of soil nutrients, and their litter returns few nutrient minerals when it decays. The litter is largely needles and twigs, which accumulate on the surface of the soil and decompose slowly. Microorganisms do not thrive in this environment, and so humus production is retarded. Moisture is relatively abundant in summer, so that leaching of whatever nutrient cations are present in the topsoil, along with iron oxides, aluminum oxides, and colloidal clays, is relatively complete.

Podzolization, then, produces soils that are shallow and acidic and have a fairly distinctive profile. There is usually an O horizon. The upper part of the A horizon is eluviated to a silty or sandy texture and is so leached as to appear bleached. It is usually the ashy, light gray color that gives this regime its name, a color imparted by its high silica content. The illuviated B horizon is a receptacle for the iron-aluminum oxides and clay minerals leached from above and has a sharply contrasting darker color (sometimes with an orange or yellow tinge). Soil fertility is generally low, and a crumbly structure makes the soil very susceptible to accelerated erosion if the vegetation cover is disturbed, whether by human activities or by such natural agencies as wildfire. Soils produced by podzolization often are referred to collectively as *podzols*.

Gleization

Gleization is a regime restricted to waterlogged areas, normally in a cool climate. (The name comes from *glej*, Polish for "muddy ground.") Although occasionally widespread, it is generally much more limited in occurrence than laterization and podzolization. The poor drainage that produces a waterlogged environment can be associated with flat land, but it can also result from a topographic depression, a high water table, or various other conditions. In North America, gleization is particularly prominent in areas around the Great Lakes, where recent glacial deposition has interrupted preglacial drainage patterns.

The general term for soils produced by gleization is *gley soils*. They characteristically have a highly organic A horizon, where decomposition proceeds slowly because bacteria are inhibited by the lack of oxygen in a waterlogged situation. In such *anaerobic* conditions, chemical reduction takes place—notably ferric iron (Fe^{3+}) compounds are reduced to ferrous iron (Fe^{2+}) compounds (*reduction* occurs when a substance gains an electron—the opposite of *oxidation*). Reduced iron is more easily carried away than oxidized iron, and so over time the soil tends to become poor in iron and gray in color.

Gley soils are usually too acidic and oxygen poor to be productive for anything but water-tolerant vegetation. If drained artificially, however, and fertilized with lime to counteract the acid, their fertility can be greatly enhanced.

Calcification

In semiarid and arid climates, where precipitation is less than potential evapotranspiration, leaching is either absent or transitory. Natural vegetation in such areas consists of grasses or shrubs. **Calcification** (so called because many calcium salts are produced in this regime) is the dominant pedogenic process in these regions, as typified by the drier prairies of North America, the steppes of Eurasia, and the savannas and steppes of the subtropics.

Both eluviation and leaching are restricted by the absence of percolating water, and so materials that would be carried downward in other regimes become concentrated in the soil where calcification is at work. Moreover, there is considerable upward movement of water by capillary action in dry periods. Calcium carbonate ($CaCO_3$) is the most important chemical compound active in a calcification regime. It is carried downward by limited leaching after a rain and is often concentrated in the B horizon to form a dense layer of *hardpan*, then brought upward by capillary water and by grass roots, and finally returned to the soil when the grass dies. Little clay is formed because of the limited amount of chemical weathering. Organic colloidal material, however, is often present in considerable quantity.

Where calcification takes place under undisturbed grassland, the resulting soils are likely to have remarkable agricultural productivity. Humus from decaying grass yields abundant organic colloidal material, imparting a dark color to the soil and contributing to a structure that can retain both nutrients and soil moisture. Grass roots tend to bring calcium up from the B horizon sufficiently to inhibit or delay the formation of calcic hardpans. Where shrubs are the dominant vegetation, roots are fewer but deeper, so that nutrients are brought up from deeper layers, and little accumulates at the surface with less humus being incorporated into the soil.

Where a calcification regime is operative in true deserts, the soils tend to be shallower and sandier, calcic hardpans may form near the surface, little organic matter accumulates either on or in the soil, and the soils are not very different from the parent material.

Salinization

In arid and semiarid regions, it is fairly common to find areas with inadequate drainage, particularly in enclosed valleys and basins. Moisture is drawn upward and into the atmosphere by intense evaporation. The evaporating water leaves behind various salts in or on the surface of the soil, sometimes in such quantity as to impart a brilliant white surface color to the land, and the pedogenic regime is called **salinization**. These salts, which are mostly chlorides and sulfates of calcium and sodium, are toxic to most plants and soil organisms, and the resulting soil is able to support very little life apart from a few salt-tolerant grasses and shrubs.

Soils developed in a salinization (*salin*: Latin for "salt") regime can be sometimes made productive through careful water management, but artificial drainage is equally necessary or else salt accumulation will be intensified. Indeed, human-induced salinization has ruined good agricultural land in various parts of the world many times in the past.

Climate and Pedogenic Regimes

The pedogenic regimes are distinguished primarily on the basis of climate as reflected in temperature and moisture availability (Figure 12-24) and secondarily on the basis of vegetation cover. In regions where there is normally a surplus of moisture—which is to say annual precipitation exceeds annual evapotranspiration—water movement in the soil is predominantly downward and leaching is a prominent process. In such areas where temperatures are relatively high throughout the year, laterization is the dominant regime; where winters are long and cold, podzolization predominates; and where the soil is saturated most of the time due to poor drainage, gleization is notable. A broadly valid generalization is that gleization can occur in any of the pedogenic regimes, as it is more dependent on local topography than on macroclimate. In regions having a moisture deficit, the principal soil moisture movement is upward (through capillarity) and leaching is limited. Calcification and salinization are the principal pedogenic regimes under these conditions.

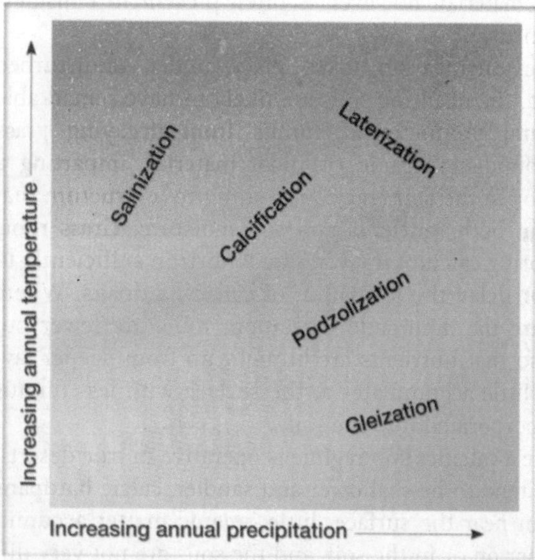

▲ **Figure 12-24** Temperature–moisture relationships for the five principal pedogenic regimes. Where temperature is high but precipitation low, salinization is the main method of soil formation. Where temperature is low but precipitation high, gleization predominates. Laterization occurs where both temperature and precipitation are high. Calcification and podzolization take place in less extreme environments.

Soil Classification

Some of the most significant products of scholarly studies are classification systems. If phenomena can be classified meaningfully, it becomes easier to remember them and to understand the relationships among them. Our consideration thus far has included various classifications (for example, climate and biomes). In no other subdiscipline of physical geography, however, is the matter of classification more complicated than with soil.

The Soil Taxonomy

Over the past century, various soil classifications have been devised in the United States and other countries. As the knowledge of soil characteristics and processes has become greater, so have the efforts at soil classification become more sophisticated. Several different systems have been developed in other countries, particularly in Canada, the United Kingdom, Russia, France, and Australia. Moreover, United Nations agencies have their own classification schemes. The system that is presently in use in the United States is called simply **Soil Taxonomy**.

The basic characteristic that sets Soil Taxonomy apart from previous systems is that it is *generic*, which means it is organized on the basis of observable soil characteristics. The focus is on the existing properties of a soil rather than on the environment, genesis, or properties it would possess under virgin conditions. The logic of such a generic system is theoretically impeccable: soil has certain properties that can be observed, measured, and at least partly quantified.

Like other logical generic systems, the Soil Taxonomy is a hierarchical system, which means that it has several levels of generalization, with each higher level encompassing several members of the level immediately below it. There are only a few similarities among all the members of the highest level category, but the number of similarities increases with each step downward in the hierarchy, so that in the lowest-level category, all members have mostly the same properties. (See Appendix VII for details.)

At the highest level (the smallest scale of generalization) of the Soil Taxonomy is **soil order**, of which only 12 are recognized worldwide (Table 12-2). The soil orders, and many of the lower-level categories as well, are distinguished from one another largely on the basis of certain diagnostic properties, which are often expressed in combination to form *diagnostic horizons*. The two basic types of diagnostic horizon are the *epipedon* (based on the Greek word *epi*, meaning "over" or "upon"), which is essentially the A horizon or the combined O/A horizon, and the *subsurface horizon*, which is roughly equivalent to the B horizon. (Note that all A and B

FOCUS

Using Soil Profiles to Decipher Past Environments

Randall Schaetzl, Michigan State University

The term *palimpsest* dates back to ancient Egypt where papyrus parchments were used like we use paper today. Papyrus parchments were often used and reused many times. Palimpsest refers to this papyrus manuscript after it has been written on more than once; earlier writings on papyrus were never quite completely erased, and so were often still faintly legible. This happens even today with modern paper. And it happens with soils, so to speak.

Soils develop their characteristics—be they physical morphologies or chemical properties—depending on the soil-forming processes that operate on and in them. Climate and vegetation largely govern the types of processes that soils experience, and given enough time, these processes change raw parent material into soil with definite physical (morphology) and chemical signatures. Like reading words on paper, if properly trained, one can "read" a soil's morphology and determine the various environmental inputs that have factored into its formation. And like palimpsests, many of these characteristics are difficult to erase. Below we discuss one such example to illustrate how soils can provide evidence of environmental change.

Many soils in the midwestern United States have developed entirely under forest or entirely under grassland vegetation. Some, however, have had periods of development under both. Under grassland vegetation, soils develop thick, dark, humus-rich A horizons (mollic epipedons), as grass roots continually form and decay in the upper 20 to 40 centimeters of the soil. Because the grasses biocycle certain base cations (especially Ca^{2+} and Mg^{2+}), the pH of the soil remains high. The high pH inhibits strong weathering of soil minerals, as well as the movement of clay to the lower profile, and thus, these Mollisols often lack a reddish, clay-rich B horizon (Figure 12-A). They have relatively simple profiles: a dark A horizon, a weak B horizon, and a C horizon (parent material). When one sees this kind of "soil writing" it is easy to infer a history of grassland influence.

Forest soils have a very different formative history. Their main source of organic matter comes from leaf decay on the soil surface, and thus the little organic matter they have is concentrated in a much thinner A horizon (an ochric epipedon). Most trees do not cycle bases, and thus the soil can acidify over time, which frees up clay for translocation to the lower profile, forming a reddish, clay-rich B horizon (known as an argillic horizon). Between the A and B horizons often exists a zone of clay depletion, the "albic" E horizon, which is light in color. These morphological characteristics are evident in most forest soils (Figure 12-B).

Soil characteristics develop slowly, over many centuries, and are therefore difficult to "erase." Thus, a soil that has experienced major environmental shifts in climate and vegetation will retain evidence of that past environment. Such soils are called "polygenetic," because they show clear evidence of different pathways (or influences) of soil genesis. Particularly, grassland soils that had a period of forest in their past will often have a clay-rich B horizon (the forest "signature") below the thick A horizon—the signature of a former period of grassland influence (Figure 12-C).

▲ **Figure 12-A** Profile of grassland soil in a soybean field near Markesan, southeastern Wisconsin. The soil has a thick, dark A horizon, but because of little clay illuviation, it lacks a reddish, clay-rich B horizon, probably because it has been continually forming under grassland vegetation (the slight reddening between 60 and 90 centimeters is mainly due to weak, in situ weathering).

▲ **Figure 12-B** Profile of forest soil in a maple-basswood forest near Withee, central Wisconsin. The soil shows clear evidence of clay illuviation—the reddish-brown B horizon. Note the thin A horizon and the light-colored E horizon between 18 and 50 centimeters.

▲ **Figure 12-C** Profile of grassland soil near Fox Lake, southeastern Wisconsin that also shows evidence of forest cover in the past. Early settlers to this area found tall prairie vegetation here, but the soil clearly shows that forest cover was present in the past. The thick A horizon is the signature of grassland, while the reddish, clay-rich B horizon (below 50 centimeters) tells of the forest.

TABLE 12-2 Name Derivations of Soil Orders

Order	Derivation
Alfisols	"al" for aluminum, "f" for iron (chemical symbol Fe), two prominent elements in these soils
Andisols	andesite, rock formed from type of magma in Andes Mountains volcanoes; soils high in volcanic ash
Aridisols	Latin *aridus*, "dry"; dry soils
Entisols	last three letters in "recent"; these are recently formed soils
Gelisols	Latin *gelatio*, "freezing"; soils in areas of permafrost
Histosols	Greek *histos*, "living tissue"; these soils contain mostly organic matter
Inceptisols	Latin *inceptum*, "beginning"; young soils at the beginning of their "life"
Mollisols	Latin *mollis*, "soft"; soft soils
Oxisols	soils with large amounts of oxygen-containing compounds
Spodosols	Greek *spodos*, "wood ash"; ashy soils
Ultisols	Latin *ultimus*, "last"; soils that have had the last of their nutrient bases leached out
Vertisols	Latin *verto*, "turn"; soils in which material from O and A horizons falls through surface cracks and ends up below deeper horizons; the usual horizon order is inverted

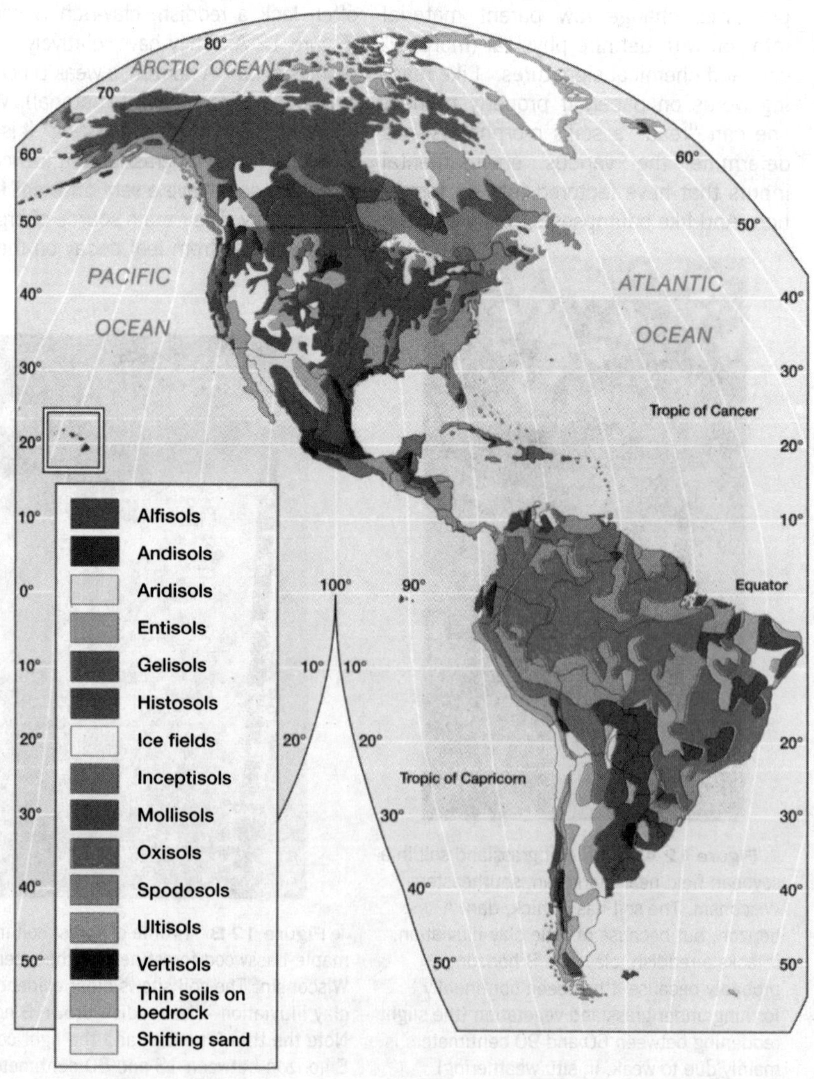

▶ **Figure 12-25** Soils of the world.

horizons are not necessarily diagnostic, and so the terms and concepts are not synonymous.)

Soil orders are subdivided into suborders, of which about 50 are recognized in the United States. The third level consists of *great groups*, which number about 250 in the United States. Successively lower levels in the classification are subgroups, families, and series. About 19,000 soil series have been identified in the United States to date, and the list will undoubtedly be expanded in the future. For the purpose of comprehending general world distribution patterns, however, we need to concern ourselves only with orders (Figure 12-25).

The Mapping Question

Maps are a basic tool of geographic study, and one of the fundamental problems that confronts any geographic inquiry is how the phenomena under study should be mapped. Because the higher levels of the Soil Taxonomy do not represent phenomena that actually exist but rather are generalized abstractions of average or typical conditions over broad areas, the selection of an appropriate mapping technique can significantly influence our understanding of soil distribution.

Most soil maps use the same timeworn technique of areal expression. If one soil type (at whatever level of generalization is being studied) is more common in an area than any other, that area is classified by the prevailing type and colored or shaded appropriately. Such a map is effective in indicating the principal type of soil in each region and is useful in portraying the general distribution of the major soil types.

Maps of this type are compiled through generalization of data, and the smaller the scale, the greater the generalization needed. Thus, more intricate patterns can be shown on the larger scale map of the United States (Figure 12-39 on p. 349) than on the smaller scale world map (Figure 12-25). In either case, the map is only as good as its generalizations are meticulous.

Global Distribution of Major Soils

There are 12 orders of soils, which are distinguished largely on the basis of properties that reflect a major course of development, with considerable emphasis on the presence or absence of notable diagnostic horizons. We consider each

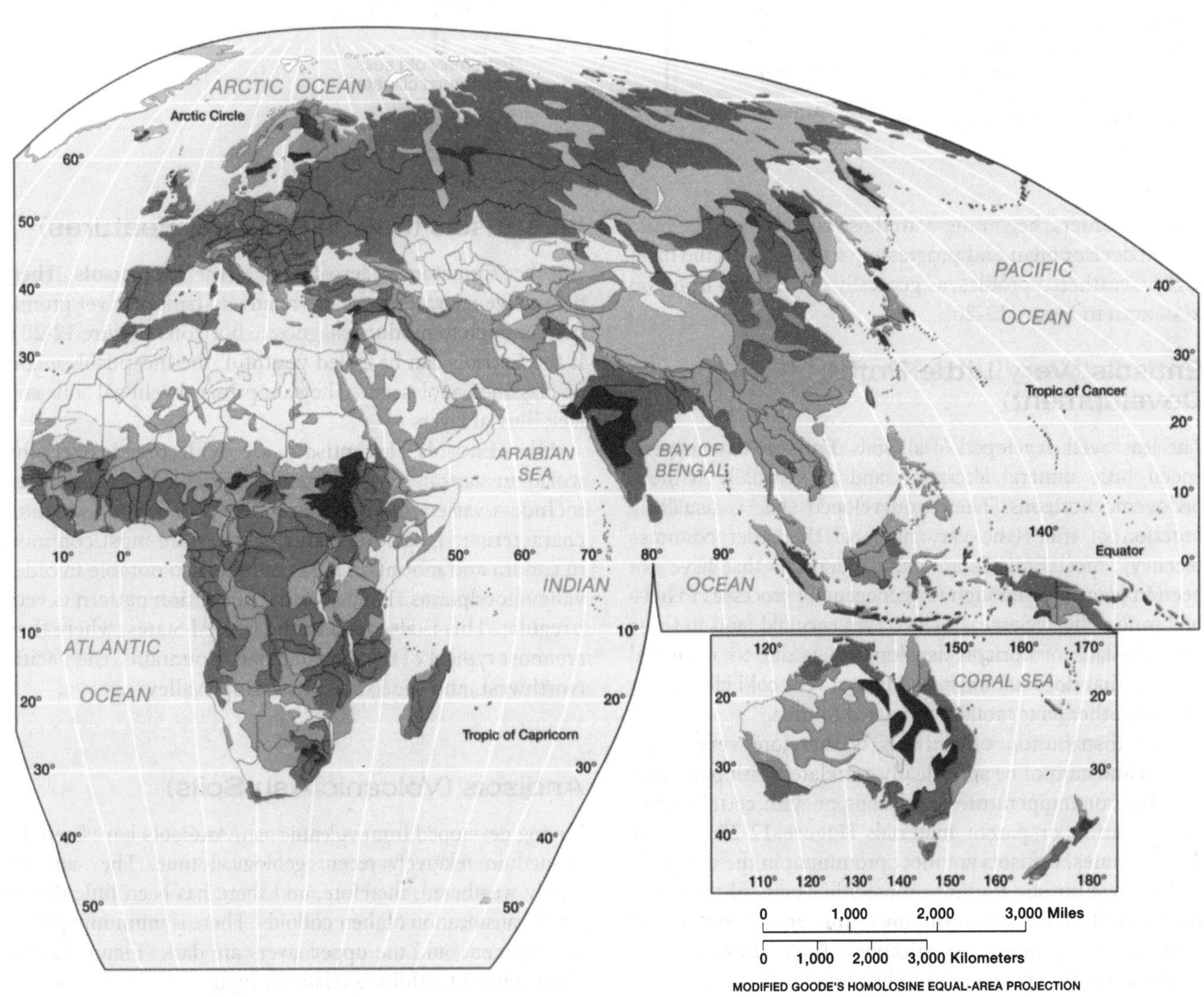

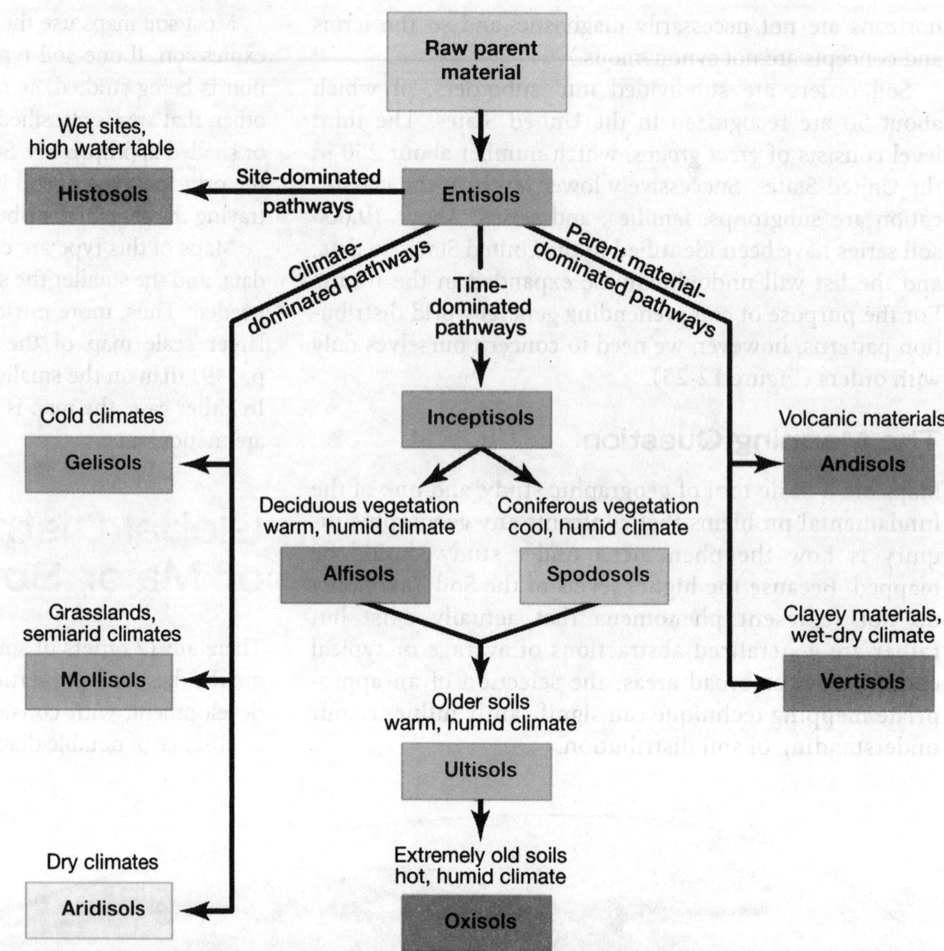

▶ **Figure 12-26** Theoretical soil order development pathways. Soils evolve along "pathways" in which different factors—such as parent material, climate, local site conditions, or length of time—may dominate.

of the 12 orders, beginning with those characterized by little profile development and progressing to those with the most highly weathered profiles, as generally represented from top to bottom in Figure 12-26.

Entisols (Very Little Profile Development)

The least well developed of all soils, **Entisols** have experienced little mineral alteration and are virtually without pedogenic horizons. Their undeveloped state is usually a function of time (the very name of the order connotes recency); most Entisols are surface deposits that have not been in place long enough for pedogenetic processes to have had much effect. Some, however, are very old, and in these soils the lack of horizon development is due to a mineral content that does not alter readily, to a very cold climate, or to some other factor totally unrelated to time.

The distribution of Entisols is therefore very widespread and cannot be specifically correlated with particular moisture or temperature conditions or with certain types of vegetation or parent materials (Figure 12-27). In the United States, Entisols are most prominent in the dry lands of the West but are found in most other parts of the country as well. They are commonly thin and/or sandy and have limited productivity, although those developed on recent alluvial deposits tend to be quite fertile.

Inceptisols (Few Diagnostic Features)

Another immature order of soils is the **Inceptisols**. Their distinctive characteristics are relatively faint, not yet prominent enough to produce diagnostic horizons (Figure 12-28). If the Entisols can be called youthful, the Inceptisols might be classified adolescent. They are primarily eluvial soils and lack illuvial layers.

Like Entisols, Inceptisols are widespread over the world in various environments. Also like Entisols, they include a variety of fairly dissimilar soils whose common characteristic is lack of maturity. They are most common in tundra and mountain areas but are also notable in older valley floodplains. Their world distribution pattern is very irregular. This is also true in the United States, where they are most typical of the Appalachian Mountains, the Pacific Northwest, and the lower Mississippi Valley.

Andisols (Volcanic Ash Soils)

Having developed from volcanic ash, **Andisols** have been deposited in relatively recent geological time. They are not highly weathered, therefore, and there has been little downward translocation of their colloids. There is minimum profile development, and the upper layers are dark (Figure 12-29). Their inherent fertility is relatively high.

▲ **Figure 12-27** Entisols. (a) World distribution of Entisols. (b) Profile of an Entisol in northern Michigan. This weakly developed soil evolved from sandy parent material in an area with only moderate precipitation.

▲ **Figure 12-28** Inceptisols. (a) World distribution of Inceptisols. (b) Profile of a New Zealand Inceptisol with a distinctive B horizon of white pebbly material. The scale is in meters.

Andisols are found primarily in volcanic regions of Japan, Indonesia, and South America, as well as in the very productive wheat lands of Washington, Oregon, and Idaho.

Gelisols (Cold Soils with Permafrost)

Gelisols are young soils with minimal profile development (Figure 12-30). They develop only slowly because of cold temperatures and frozen conditions. These soils typically have a permafrost layer that is a defining characteristic. Also commonly found in Gelisols is *cryoturbation* or frost churning, which is the physical disruption and displacement of soil material by freeze–thaw action in the soil. Most of the soil-forming processes in Gelisols take place above the permafrost in the active layer that thaws every year or so.

Gelisols are the dominant soils of Arctic and subarctic regions. They occur in association with boreal forest and tundra vegetation; thus, they are primarily found in Russia, Canada, and Alaska and are prominent in the Himalaya Mountain country of central Asia. Altogether nearly 9 percent of Earth's land area has a Gelisol soil cover.

Histosols (Organic Soils on Very Wet Sites)

Least important among the soil orders are the **Histosols**, which occupy only a small fraction of Earth's land surface, a much smaller area than any other order. These are organic rather than mineral soils, and they are invariably saturated with water all or most of the time. They may occur in any waterlogged environment but are most characteristic in

▲ **Figure 12-29** Andisols. (a) World distribution of Andisols. (b) Profile of an Andisol in Washington state. The various horizons are very clearly delineated.

▲ **Figure 12-30** Gelisols. (a) World distribution of Gelisols. (b) Profile of a Gelisol in Alaska. The scale is in centimeters.

mid- and high-latitude regions that experienced Pleistocene glaciation. In the United States, they are most common around the Great Lakes, but they also occur in southern Florida and Louisiana. Nowhere, however, is their occurrence extensive (Figure 12-31).

Some Histosols are composed largely of undecayed or only partly decayed plant material, whereas others consist of a thoroughly decomposed mass of muck. The lack of oxygen in the waterlogged soil slows down the rate of bacterial action, and the soil becomes deeper mostly by growing upward, that is, by more organic material being added from above.

Histosols are usually black, acidic, and fertile only for water-tolerant plants. If drained, they can be very productive agriculturally for a short while. Before long, however, they are likely to dry out, shrink, and oxidize, a series of steps that leads to compaction, susceptibility to wind erosion, and danger of fire.

Aridisols (Soils of Dry Climates)

Nearly one-eighth of Earth's land surface is covered with **Aridisols**, one of the most extensive spreads of any soil order (Figure 12-32). They are preeminently soils of the dry lands, occupying environments that do not have enough water to remove soluble minerals from the soil. Thus, their distribution pattern is largely correlated with that of desert and semidesert climate.

Aridisols are typified by a thin profile that is sandy and lacking in organic matter, characteristics clearly associated

▲ **Figure 12-31** Histosols. (a) World distribution of Histosols. (b) Histosols are characteristically dark in color and composed principally of organic matter.

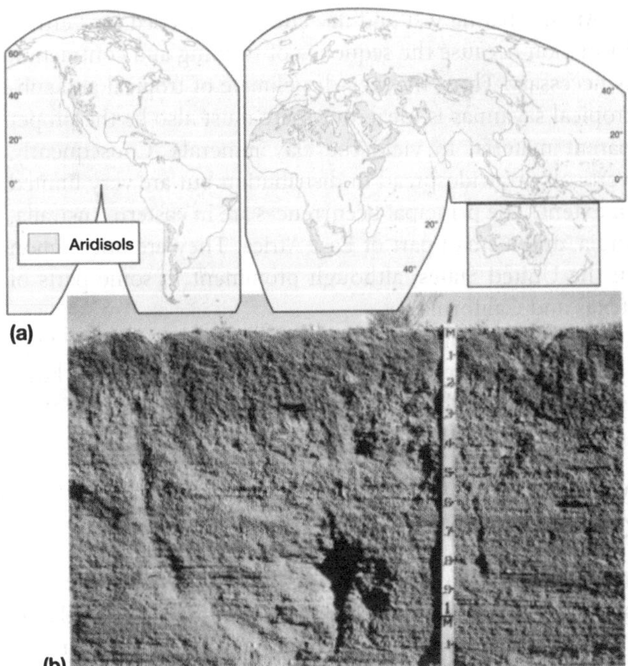

▲ **Figure 12-32** Aridisols. (a) World distribution of Aridisols. (b) The typical sandy profile of an Aridisol, in this case from New Mexico.

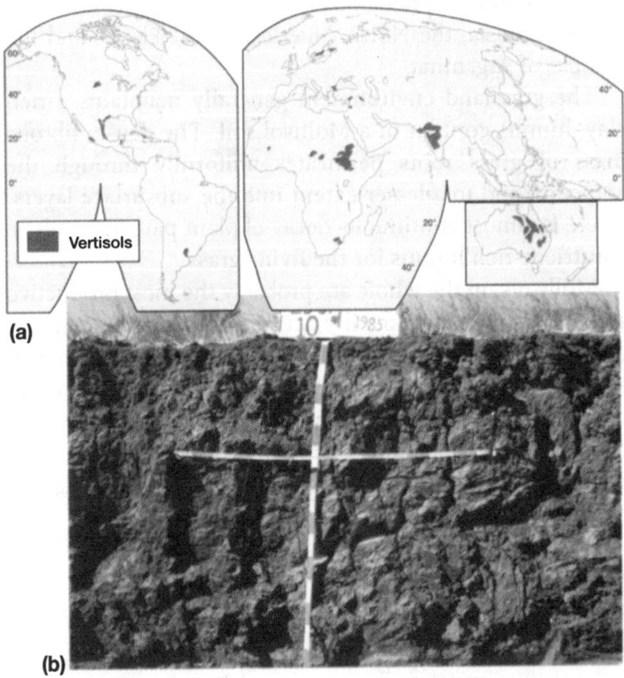

▲ **Figure 12-33** Vertisols. (a) World distribution of Vertisols. (b) A dark Vertisol profile from Zambia. Many cracks are typically found in Vertisols.

with a dry climate and a scarcity of penetrating moisture. The epipedon is almost invariably light in color. There are various kinds of diagnostic subsurface horizons, nearly all distinctly alkaline. Most Aridisols are unproductive, particularly because of lack of moisture; if irrigated, however, some display remarkable fertility. The threat of salt accumulation is ever present, however.

Vertisols (Swelling and Cracking Clays)

Vertisols contain a large quantity of clay that becomes a dominant factor in the soil's development. The clay of Vertisols is described as "swelling" or "cracking" clay. This clay-type soil has an exceptional capacity for absorbing water: when moistened, it swells and expands; as it dries, deep, wide cracks form, sometimes 2.5 centimeters (an inch) wide and as much as 1 meter (a yard) deep. Some surface material falls into the cracks, and more is washed in when it rains. When the soil is wetted again, more swelling takes place and the cracks close. This alternation of wetting and drying and expansion and contraction produces a churning effect that mixes the soil constituents (the name Vertisol connotes an inverted condition), inhibits the development of horizons, and may even cause minor irregularities in the land surface (Figure 12-33).

An alternating wet and dry climate is needed for Vertisol formation because the sequence of swelling and contraction is necessary. Thus, the wet–dry climate of tropical and subtropical savannas is ideal, but there must also be the proper parent material to yield the clay minerals. Consequently, Vertisols are widespread in distribution but are very limited in extent. The principal occurrences are in eastern Australia, India, and a small part of East Africa. They are uncommon in the United States, although prominent in some parts of Texas and California.

The fertility of Vertisols is relatively high, as they tend to be rich in nutrient bases. They are difficult to till, however, because of their sticky plasticity, and so they are often left uncultivated.

Mollisols (Dark, Soft Soils of Grasslands)

The distinctive characteristic of **Mollisols** is the presence of a mollic epipedon, which is a mineral surface horizon that is dark and thick, contains abundant humus and basic cations, and retains a soft character (rather than becoming hard and crusty) when it dries out (Figure 12-34). Mollisols can be thought of as transition soils that evolve in regions not dominated by either humid or arid conditions. They are typical of the midlatitude grasslands and are thus most common in central Eurasia, the North American Great Plains, and the pampas of Argentina.

The grassland environment generally maintains a rich clay–humus content in a Mollisol soil. The dense, fibrous mass of grass roots permeates uniformly through the epipedon and to a lesser extent into the subsurface layers. There is almost continuous decay of plant parts to produce a nutrient-rich humus for the living grass.

Mollisols on the whole are probably the most productive soil order. They are generally derived from loose parent material rather than from bedrock and tend to have favorable structure and texture for cultivation. Because they are not overly leached, nutrients are generally retained within reach of plant roots. Moreover, Mollisols provide a favored habitat for earthworms, which contribute to softening and mixing the soil.

▲**Figure 12-34** Mollisols. (a) World distribution of Mollisols. (b) A Mollisol profile from Nebraska. It has a typical mollic epipedon, a surface horizon that is dark and replete with humus.

Alfisols (Clay-Rich B Horizons, High Base Status)

The most wide ranging of the mature soils, Alfisols occur extensively in low and middle latitudes, as Figure 12-35 shows. They are found in a variety of temperature and moisture conditions and under diverse vegetation associations. By and large, they tend to be associated with transitional environments and are less characteristic of regions that are particularly hot or cold or wet or dry. Their global distribution is extremely varied. They are also widespread in the United States, with particular concentrations in the Midwest.

Alfisols are distinguished by a subsurface clay horizon and a medium to generous supply of basic cations, plant nutrients, and water. The epipedon is ochric (light-colored), as Figure 12-35b shows, but beyond that, it has no characteristics that are particularly diagnostic and can be considered an ordinary eluviated horizon. The relatively moderate conditions under which Alfisols develop tend to produce balanced soils that are reasonably fertile. Alfisols rank second only to Mollisols in agricultural productivity.

Ultisols (Clay-Rich B Horizons, Low Base Status)

Ultisols are roughly similar to Alfisols except that Ultisols are more thoroughly weathered and more completely leached of nutrient bases. They have experienced greater mineral

▲ **Figure 12-35** Alfisols. (a) World distribution of Alfisols. (b) Profile of an Alfisol in east-central Illinois. The soil has reddened, clay-rich B horizon. It formed under forest vegetation and is heavily cropped with corn and soybeans.

▲ **Figure 12-36** Ultisols. (a) World distribution of Ultisols. (b) A tropical Ultisol from Thailand. It is reddish throughout its profile, indicative of much leaching and weathering.

alteration than any other soil in the midlatitudes, although they also occur in the low latitudes. Many pedologists believe that the ultimate fate of Alfisols is to degenerate into Ultisols.

Typically, Ultisols are reddish as a result of the significant proportion of iron and aluminum in the A horizon. They usually have a fairly distinct layer of subsurface clay accumulation. The principal properties of Ultisols have been imparted by a great deal of weathering and leaching (Figure 12-36). Indeed, the connotation of the name (derived from the Latin *ultimos*) is that these soils represent the ultimate stage of weathering in the conterminous United States. The result is a fairly deep soil that is acidic, lacks humus, and has a relatively low fertility due to the lack of bases.

Ultisols have a fairly simple world distribution pattern. They are mostly confined to humid subtropical climates and to some relatively youthful tropical land surfaces. In the United States, they are restricted largely to the southeastern quarter of the country and to a narrow strip along the northern Pacific Coast.

Spodosols (Soils of Cool, Forested Zones)

The key diagnostic feature of a **Spodosol** is a spodic subsurface horizon, an illuvial dark or reddish layer where organic matter, iron, and aluminum accumulate. The upper layers are light-colored and heavily leached (Figure 12-37). At the top of the profile is usually an O horizon of organic litter. Such a soil is a typical result of podzolization.

▲ **Figure 12-37** Spodosols. (a) World distribution of Spodosols. (b) Profile of a Spodosol in northern Michigan. This weakly developed soil was formed in sandy material under coniferous forest vegetation and contains few nutrients. The scale is in centimeters.

▲ **Figure 12-38** Oxisols. (a) World distribution of Oxisols. (b) Oxisols are impoverished tropical soils that usually are heavily leached. The horizons typically are indistinct. This sample is from Hawai'i. The scale is in centimeters.

Spodosols are notoriously infertile. They have been leached of useful nutrients and are acidic throughout. They do not retain moisture well and are lacking in humus and often in clay. Spodosols are most widespread in areas of coniferous forest where there is a subarctic climate. Alfisols, Histosols, and Inceptisols also occupy these regions, however, and Spodosols are sometimes found in other environments, such as poorly drained portions of Florida.

Oxisols (Highly Weathered and Leached)

The most thoroughly weathered and leached of all soils are the **Oxisols**, which invariably display a high degree of mineral alteration and profile development. They occur mostly on ancient landscapes in the humid tropics, particularly in Brazil and equatorial Africa, and to a lesser extent in Southeast Asia (Figure 12-38). The distribution pattern is often spotty, with Oxisols mixed with less-developed Entisols, Vertisols, and Ultisols. Oxisols are totally absent from the United States, except for Hawai'i, where they are common.

Oxisols are essentially the products of laterization (and in fact were called Latosols in the older classification systems). They have evolved in warm, moist climates, although some are now found in drier regions, an indication of climatic change since the soils developed. The diagnostic horizon for Oxisols is a subsurface dominated by oxides of iron and aluminum and with a minimal supply of nutrient bases (this is called an *oxic horizon*). These are deep soils

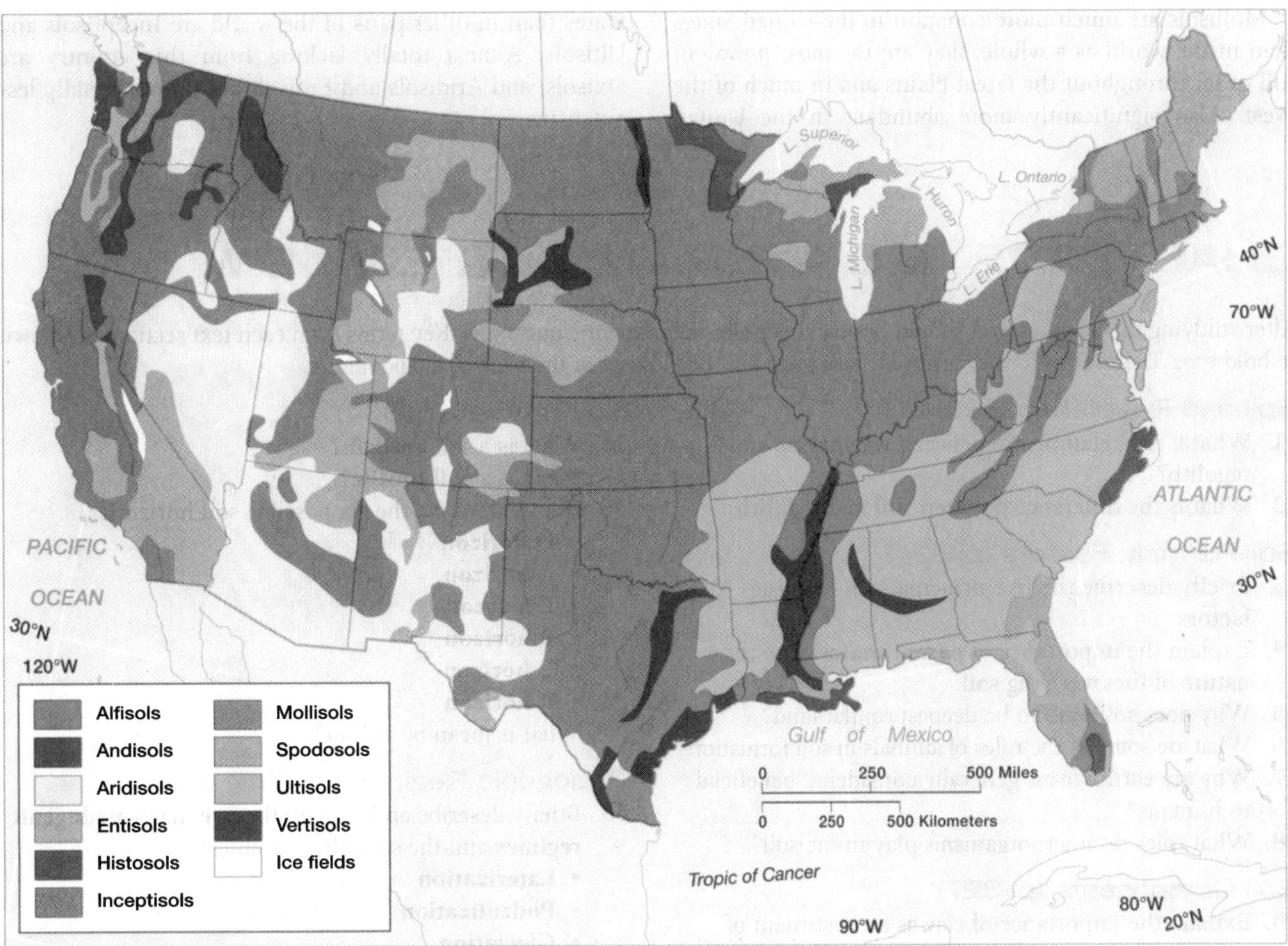

▲ **Figure 12-39** Distribution of predominant soil orders in the conterminous United States.

but not inherently fertile. The natural vegetation is efficient in cycling the limited nutrient supply, but if the flora is cleared (to attempt agriculture, for example), the nutrients are rapidly leached out, and the soil becomes impoverished.

Distribution of Soils in the United States

The distribution of the various soils in the United States is quite different from that of the world as a whole. This difference is due to many factors, the most important being that the United States is essentially a midlatitude country, and it lacks significant expanses of area in the low and high latitudes (Figure 12-39). Table 12-3 compares the relative areas occupied by the 12 soil orders, nationally and globally. The statistics are generalized estimates prepared by the Soil Conservation Service of the U.S. Department of Agriculture and should not be considered definitive.

TABLE 12-3 Approximate Proportional Extent of Soil Orders

Order	Percentage of Land Area Occupied	
	United States	World
Alfisols	14.5	9.7
Andisols	1.7	0.7
Aridisols	8.8	12.1
Entisols	12.2	16.3
Gelisols	7.5	8.6
Histosols	1.3	1.2
Inceptisols	9.1	9.9
Mollisols	22.4	6.9
Oxisols	—	7.6
Spodosols	3.3	2.6
Ultisols	9.6	8.5
Vertisols	1.7	2.4

Source: Brady, Nyle C., and Ray R. Weil, *The Nature and Properties of Soils*, Upper Saddle River, NJ: Prentice Hall, 1999, p.86.

Mollisols are much more common in the United States than in the world as a whole; they are the most prevalent soil order throughout the Great Plains and in much of the West. Also significantly more abundant in the United States than in other parts of the world are Inceptisols and Ultisols. Almost totally lacking from this country are Oxisols, and Aridisols and Entisols are proportionally less extensive.

CHAPTER 12 LEARNING REVIEW

After studying this chapter, you should be able to answer the following questions. Key terms from each text section are shown in bold type. Definitions for key terms are also found in the glossary at the back of the book.

Soil and Regolith (p. 323)

1. What is the relationship between weathering and regolith?
2. What is the difference between **soil** and **regolith**?

Soil-Forming Factors (p. 324)

3. Briefly describe the five principal soil-forming factors.
4. Explain the importance of **parent material** to the nature of the overlying soil.
5. Why does soil tend to be deepest on flat land?
6. What are some of the roles of animals in soil formation?
7. Why are earthworms generally considered beneficial to humans?
8. What roles do microorganisms play in the soil?

Soil Components (p. 327)

9. Explain the importance of clay as a constituent of soil.
10. What are the importance of **clay** and **cations** to plant nutrients in the soil?
11. What is the difference between **litter** and **humus**?
12. Describe and explain the four forms of soil moisture.
13. Distinguish between **field capacity** and **wilting point**.
14. What is meant by **leaching**?
15. Explain the processes of **eluviation** and **illuviation**.
16. What is the **soil–water balance**?
17. What is the role of temperature in the **soil–water budget**?

Soil Properties (p. 331)

18. What can you learn about a soil from its color?
19. Distinguish between soil texture and soil structure.
20. What are soil **separates**?
21. What is meant by **loam**?
22. Explain the difference between porosity and permeability.
23. How do **peds** influence soil porosity?

Soil Chemistry (p. 333)

24. What is a **colloid**?
25. Explain what is meant by the **cation exchange capacity (CEC)** of a soil.

Soil Profiles (p. 334)

26. What is a soil **horizon**?
27. What is a **soil profile**?
28. Briefly describe the six possible soil horizons:
    - **O horizon**
    - **A horizon**
    - **E horizon**
    - **B horizon**
    - **C horizon**
    - **R horizon**
29. What is mean by **solum**?

Pedogenic Regimes (p. 336)

30. Briefly describe and explain the five major **pedogenic regimes** and the soils they produce:
    - **Laterization**
    - **Podzolization**
    - **Gleization**
    - **Calcification**
    - **Salinization**
31. Why do tropical rainforests usually have poor soils?

Soil Classification (p. 338)

32. How does the **Soil Taxonomy** differ from previous soil classification schemes?
33. What is meant by a **soil order**?
34. Why is it so difficult to portray soil distribution with reasonable accuracy on a small-scale map?

Global Distribution of Major Soils (p. 341)

35. Briefly describe the most distinguishing characteristics for each of the 12 soil orders:
    - **Entisols**
    - **Inceptisols**
    - **Andisols**
    - **Gelisols**
    - **Histosols**
    - **Aridisols**
    - **Vertisols**
    - **Mollisols**
    - **Alfisols**
    - **Ultisols**
    - **Spodosols**
    - **Oxisols**

36. Using the map of soil distribution in the United States (Figure 12-39) or the map showing the world distribution of soils (Figure 12-25), choose one of the soil orders that is found nearby your home and describe that order's general distribution and characteristics.

37. For each of the following circumstances, indicate which soil order is mostly likely to be found and the reasons for this:
    - Regions of permafrost
    - Areas with coniferous forest
    - Deserts

Log in to **www.mygeoscienceplace.com** for animations, videos, RSS feeds, maps, and self-study quizzes to enhance your study of **Soils**.

13 Introduction to Landform Study

In this chapter, we turn our attention more directly to the study of the solid portion of Earth. Thoughtful beginning students of physical geography may be daunted by the magnitude of their object of study because Earth is enormous from the human viewpoint. Its diameter of about 12,800 kilometers (about 8000 miles) and circumference of 40,000 kilometers (about 25,000 miles) are distances well beyond our normal scale of living and thinking. Our endeavor is greatly simplified, however, because we can largely concentrate our attention on the surface—as geographers, we want to know about Earth's interior because it helps us to comprehend the nature and characteristics of the processes shaping Earth's surface features.

We begin our task with a description of the structure of Earth as a whole. We then turn to a discussion of rocks—the solid material from which the planet is made. We conclude the chapter by introducing some fundamental concepts that we will use in our study of Earth's surface features.

The Structure of Earth

Our knowledge of the interior of Earth is based largely on indirect evidence. No human activity has explored more than a minute fraction of the vastness beneath the surface. No person has penetrated as much as one-thousandth of the radial distance from the surface to the center of Earth; the deepest existing mine shaft extends a mere 3.8 kilometers (2.4 miles). Nor have probes extended much deeper; the deepest drill holes from which sample cores have been brought up have penetrated only a modest 12 kilometers (8 miles) into Earth. Earth scientists, in the colorful imagery of writer John McPhee, "are like dermatologists: they study, for the most part, the outermost two per cent of the earth. They crawl around like fleas on the world's tough hide, exploring every wrinkle and crease, and try to figure out what makes the animal move."[1]

Even so, a considerable body of inferential knowledge concerning Earth's interior has been amassed by geophysical means, primarily by monitoring patterns of the shock waves transmitted through Earth from earthquakes or from human-made explosions. Such seismic waves change their speed and direction whenever they cross a boundary from one type of material to another. Analysis of these changes, augmented by related data on Earth's magnetism and gravitational attraction, has enabled Earth scientists to develop a model of Earth's internal structure.

[1]McPhee, John, *Assembling California*, New York: Farrar, Straus and Giroux, 1993, p. 36.

◀ Spider Rock in Canyon de Chelly National Monument, Arizona.

Figure 13-1 shows the internal structure of Earth. It has been deduced that Earth has a dense inner core surrounded by three concentric layers of various compositions and density. Starting at the surface and moving inward, these four regions are called the *crust*, the *mantle*, the *outer core*, and the *inner core*.

The Crust

The **crust**, the outermost shell, consists of a broad mixture of rock types. Beneath the oceans the crust has an average thickness of only about 7 kilometers (4 miles), whereas beneath the continents the thickness averages more than five times as much, and in places exceeds 70 kilometers (40 miles). There is a gradual increase in density with depth in this rigid outer shell. Altogether, the crust makes up less than 1 percent of Earth's volume and about 0.4 percent of Earth's mass.

At the base of the crust there is a significant change in mineral composition. This relatively narrow zone of change is called the **Mohorovičić discontinuity**, or simply the **Moho** for short, named for the Yugoslavian seismologist Andrija Mohorovičić (1857–1936) who discovered it.

The Mantle

Beneath the Moho is the **mantle**, which extends downward to a depth of approximately 2900 kilometers (1800 miles). Volumetrically, the mantle is by far the largest of the four shells. Although its depth is only about one-half the distance from the surface to the center of the Earth, its location on the periphery of the sphere gives it a vast breadth. It makes up 84 percent of the total volume of Earth and about two-thirds of Earth's total mass.

Earth scientists now think that there are three sublayers within the mantle, as Figure 13-1a shows. The uppermost zone is relatively thin but hard and rigid, extending down to a depth of 65 to 100 kilometers (40 to 60 miles)—somewhat deeper under the continents than under the ocean floors. This uppermost mantle zone together with the overlying crust is called the **lithosphere**.[2]

Beneath this rigid zone, and extending downward to a depth of as much as 350 kilometers (200 miles), is a zone in which the rocks are so hot that they lose much of their strength and become "plastic"—they are easily deformed, somewhat like tar. This is called the **asthenosphere** ("weak sphere"). Below the asthenosphere is the *lower mantle*, where the rocks are believed to be largely rigid again.

[2]For the remainder of our study of landforms, the term *lithosphere* will have a more restricted meaning than earlier in the book when we introduced the four "spheres." For our purposes now, lithosphere refers specifically to the crust and upper rigid mantle, or simply the "plates" of plate tectonics.

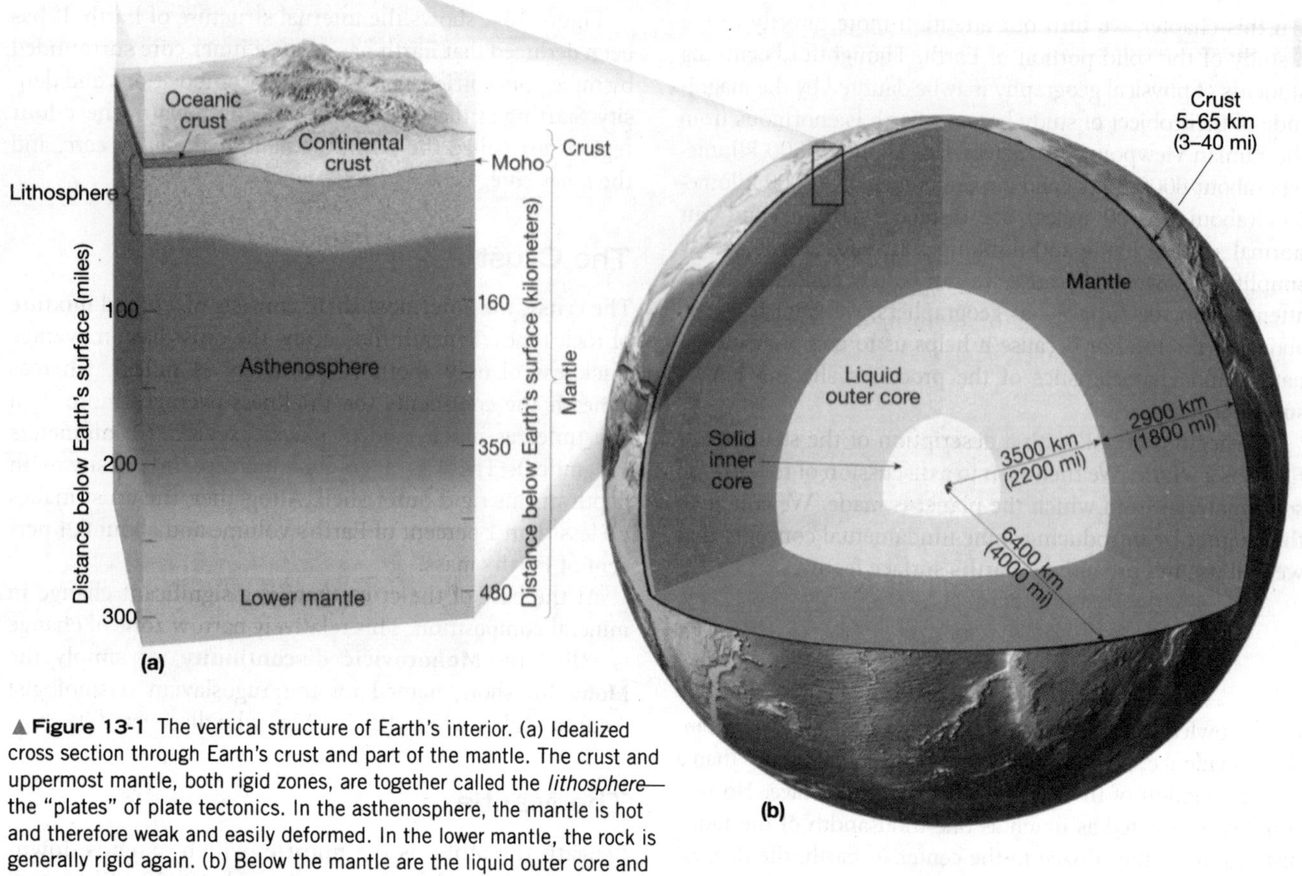

▲ **Figure 13-1** The vertical structure of Earth's interior. (a) Idealized cross section through Earth's crust and part of the mantle. The crust and uppermost mantle, both rigid zones, are together called the *lithosphere*—the "plates" of plate tectonics. In the asthenosphere, the mantle is hot and therefore weak and easily deformed. In the lower mantle, the rock is generally rigid again. (b) Below the mantle are the liquid outer core and the solid inner core of Earth.

The Inner and Outer Cores

Beneath the mantle is the **outer core** (Figure 13-1b), thought to be molten (liquid) and extending to a depth of about 5000 kilometers (3100 miles).

The innermost portion of Earth is the **inner core**, an evidently solid and very dense mass having a radius of about 1450 kilometers (900 miles). Both the inner and outer cores are thought to be made of iron/nickel or iron/silicate. These two zones together make up about 15 percent of Earth's volume and 32 percent of its mass.

Earth's *magnetic field* is generated primarily in the outer core—circulation within the liquid outer core induces the magnetic field of our planet. Interestingly, the magnetic field of Earth changes over time, and the location of the north magnetic pole does not exactly match the true geographic North Pole. The position of the north magnetic pole slowly but continually shifts by several tens of kilometers each year—it is currently located at about 83° N, 114° W—but the position of the north magnetic pole can even change slightly during a single day! In addition, for reasons that are not completely understood, at irregular intervals of thousands of years, the polarity of Earth's magnetic field reverses, with the north magnetic pole becoming the south. As we will see in Chapter 14, a record of these magnetic polarity reversals has been recorded in the iron-rich rocks of the ocean floor.

Plate Tectonics and the Structure of Earth

The generalized model of Earth's interior shown in Figure 13-1 is a useful starting point for understanding physical geography. However, many details of the processes taking place within the interior of Earth are still being worked out by scientists. Our understanding of the interior of Earth was dramatically transformed in the 1960s when the notion of "continental drift," first propounded in the early 1900s but held in disdain by most scientists for the next half-century, was revived and expanded into the present-day theory of *plate tectonics*—a theory that is now accepted by virtually all Earth scientists. Geological, paleontological, seismic, and magnetic evidence makes it clear that the lithosphere of Earth is broken into large, sometimes continent-sized slabs, commonly called "plates," that are slowly moving over the warm, soft asthenosphere below. These enormous plates are quite literally pulling apart, colliding, or sliding past each other. Many internal processes such as faulting, folding, and volcanic activity are directly linked to the action taking place along the boundaries of these plates.

We will explore the dynamics of plate tectonics in much greater detail in Chapter 14. For the remainder of this chapter, however, we set the stage for that discussion by introducing some key concepts in the study of the surface features of Earth, such as rocks, geologic time, and the doctrine of uniformitarianism.

The Composition of Earth

About 100 natural chemical elements are found in Earth's crust, mantle, and core, occasionally as discrete elements but usually bonded with one or more other elements to form compounds. These naturally formed compounds and elements of Earth are called *minerals*—the building blocks of *rocks*, which are in turn the building blocks of the landscape itself.

Minerals

For a substance to be considered a **mineral**, it must:

- Be solid
- Be naturally found in nature
- Be inorganic
- Have a specific chemical composition wherever found that varies only within certain limits
- Contain atoms arranged in a regular pattern to form solid crystals

Only about one-fourth of the elements are involved in the formation of minerals to any significant magnitude. Approximately 4400 minerals have been identified, with new types identified almost every year. The majority of known minerals are found only in the crust; a more limited number of minerals are found within the mantle or have been identified in extraterrestrial rocks such as meteorites or those brought back from the Moon.

Of the more than 4000 recognized minerals, only a few dozen are of much importance as constituents of the rocks of Earth's crust. Some of the most common rock-forming minerals are listed in Table 13-1. Mineral nomenclature is very unsystematic. Some names reflect the mineral's chemical composition or a physical property, some names are

TABLE 13-1 Some Common Rock Forming Minerals

Group	Mineral	Chemical Composition	Common Characteristics
Ferromagnesian ("dark") Silicate	Olivine	$(Mg, Fe)_2SiO_4$	Green to brown color; glassy luster; gem peridot
	Pyroxene group (Augite)	$(Mg, Fe)SiO_3$	Green to black color; commonly has two cleavage planes at 90°
	Amphibole group (Hornblende)	$Ca_2(Fe, Mg)_5Si_8O_{22}(OH)_2$	Black or dark green color; often appear as elongated rod-shaped crystals
	Biotite mica	$K(Mg, Fe)_3AlSi_3O_{10}(OH)_2$	Black or brown; often appear as hexagonal crystals in thin sheets
Nonferromagnesian ("light") Silicate	Muscovite mica	$KAl_2(AlSi_3O_{10})(OH)_2$	Colorless or brown; splits into thin, translucent sheets
	Potassium feldspar (Orthoclase)	$KAlSi_3O_8$	White to gray or pink color; good cleavage in two planes at 90°
	Plagioclase feldspar	$(Ca, Na)AlSi_3O_8$	White to gray color; striations often appear along cleavage planes
	Quartz	SiO_2	Commonly colorless; forms six-sided elongated crystals
Oxide	Hematite	Fe_2O_3	Red to silver-gray color; type of iron ore
	Magnetite	Fe_3O_4	Black with metallic luster; magnetic
	Corundum	Al_2O_3	Brown or blue color; gems sapphires and rubies
Sulfide	Galena	PbS	Black to silver metallic luster; lead ore
	Pyrite	FeS_2	Brassy or golden yellow color; often seen in well formed cubes
	Chalcopyrite	$CuFeS_2$	Brassy or yellow metallic luster; copper ore
Sulfate	Gypsum	$CaSO_4 \cdot 2H_2O$	White or transparent; used in plaster and wallboard
Carbonate	Calcite	$CaCO_3$	White or colorless; cleavage in 3 planes forms rhombohedra; fizzes in dilute acid
	Dolomite	$CaMg(CO_3)_2$	White or transparent; cleaves into rhombohedra; powdered form fizzes in dilute acid
Halide	Halite	$NaCl$	Transparent or white; table salt
	Fluorite	CaF_2	Transparent purple, green or yellow color; fluorine ore

based on a person or a place, and some appear to have been simply chosen at random.

The rock-forming minerals can be grouped into seven principal categories or families based on their chemical properties and internal structure.

Silicates The largest and most important mineral family consists of the **silicates**. The bulk of the rocks of the crust are composed of silicate minerals, which combine the two most abundant chemical elements in the lithosphere: oxygen (O) and silicon (Si). Most silicates are hard and durable. The major subcategories of this group are *ferromagnesian silicates* (sometimes called the "dark" silicates) and *nonferromagnesian silicates* (sometimes called the "light" silicates), which are distinguished from one another by the presence or absence of iron and magnesium in their composition. Feldspars and quartz (Figure 13-2) are the most abundant of the silicate minerals. Quartz itself is composed of pure *silica* (SiO_2).

Oxides An *oxide* is an element combined with oxygen. Only a relatively few elements are able to do so. The most widespread of the oxides are those that combine with iron, particularly hematite, magnetite, and limonite, all three of which are major sources of iron ore as well as being common rock-forming minerals. (Although quartz has the chemical composition of an oxide, it is classified as a silicate because of its internal structure.)

Sulfides *Sulfides* are composed of sulfur in some combination with one or more other elements. Pyrite (Figure 13-3), for example, is a combination of iron and sulfur (FeS_2). Many of the most important ore minerals—such as galena (lead), sphalerite (zinc), and chalcopyrite (copper)—are sulfides. This group is common in many types of rock and may be abundant in veins.

Sulfates The *sulfate* group includes minerals such as gypsum that contain sulfur and oxygen in combination with some other element. Calcium is the principal combining element. The sulfate minerals are usually light-colored and are mostly found in sedimentary rocks.

Carbonates *Carbonates* are also light-colored (or colorless) minerals that are common constituents of sedimentary rocks such as limestone (largely made from calcium carbonate [$CaCO_3$] in its mineral form of *calcite*). Carbonates in general are composed of one or more elements in combination with carbon and oxygen.

Halides The *halide* group is the least widespread of any. The name is derived from a word meaning "salt." Halide minerals are notably salty—such as halite, or common table salt (NaCl).

Native Elements A few minerals may occur as discrete elements (not combined chemically with another element) in nature. These are referred to as native elements. Included are such precious metals as gold and silver.

Rocks

Rocks are composed of mineral material—sometimes just one kind of mineral, but usually several different minerals. Rocks occur in bewildering variety and complexity in the lithosphere, although fewer than 20 minerals account for more than 95 percent of the composition of all continental and oceanic crustal rocks.

Solid rock is sometimes found right at the surface, in which case it is called an **outcrop** (Figure 13-4). Over most of Earth's land area, though, solid rock exists as a buried layer of *bedrock* and covered by a layer of broken rock called *regolith*. Soil, when present, comprises the upper portion of the regolith (see Chapter 12).

The enormous variety of rocks can be classified systematically. A detailed knowledge of *petrology* (the characteristics of different kinds of rocks) is unnecessary for our purposes

◀ **Figure 13-2** Quartz crystal.

▲ **Figure 13-3** Pyrite crystal, a common sulfide mineral.

▲ **Figure 13-4** Bedrock is usually buried under a layer of soil and regolith but occasionally appears as an outcrop.

here, however, so we restrict coverage to a survey of the three major rock classes—*igneous*, *sedimentary*, and *metamorphic*—and their basic attributes (Figure 13-5 and Table 13-2).

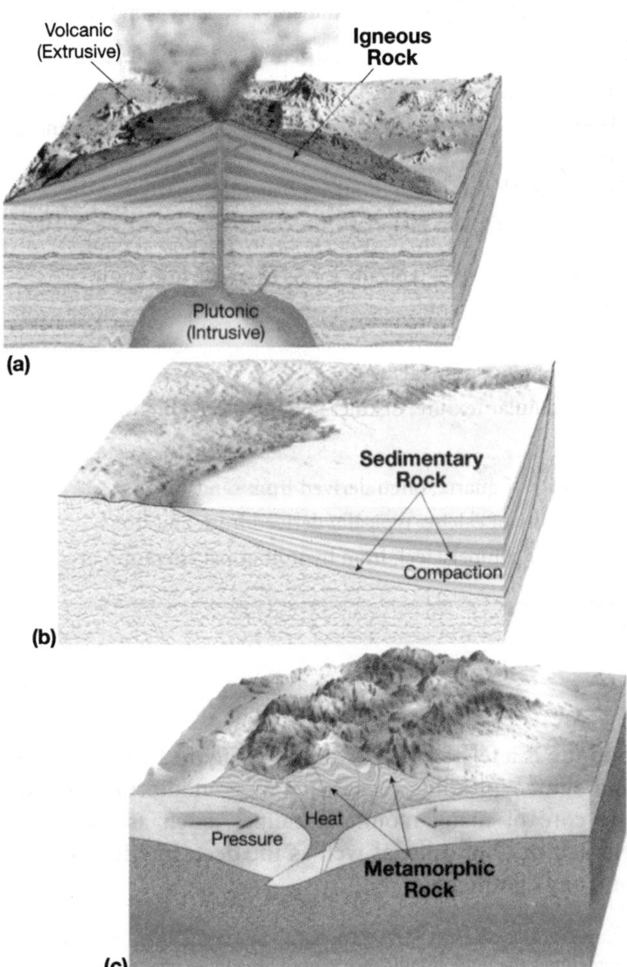

▲ **Figure 13-5** The three rock classes. (a) Igneous rocks are formed when magma cools. (b) Sedimentary rocks result from consolidation of deposited particles. (c) Metamorphic rocks are produced when heat and/or pressure act on existing igneous and sedimentary rocks.

Igneous Rocks

The word igneous is derived from the Latin, *igneus* ("fire"). **Igneous rocks** are formed by the cooling and solidification of molten rock (Figure 13-5a). **Magma** is a general term for molten rock beneath the surface of Earth, while the term **lava** refers to molten rock when it flows out on, or is squeezed up onto, the surface. Most igneous rocks form directly from the cooling of magma or lava; however, some igneous rocks develop from the "welding" of tiny pieces of solid volcanic rock, called **pyroclastics**, that have been explosively ejected out onto the surface by a volcanic eruption.

The classification of igneous rocks is based largely on mineral composition and texture. The mineral composition of an igneous rock is determined by the "chemistry" of the magma—in other words, by the particular combination of molten mineral material in the magma. As we will see in Chapter 14, one of the most important variables of magma composition is the relative amount of *silica* (SiO_2) present. Magmas with relatively large amounts of silica generally cool to form *felsic* igneous rocks that contain large portions of light-colored silicate minerals such as quartz and feldspar (the word felsic comes from *fel*dspar and *si*lica [quartz]); these minerals tend to have relatively lower densities and lower melting temperatures than *mafic* minerals. Magmas with relatively low amounts of silica generally cool to form mafic igneous rocks that contain large portions of dark-colored, magnesium- and iron-rich silicate minerals such as olivine and pyroxene (mafic comes from *ma*gnesium and *f*errum [Latin for iron]).

The texture of an igneous rock is determined primarily by where and how the molten material cools; for example, the slow cooling of magma beneath the surface leads to a coarse-grained texture, while the rapid cooling of lava on the surface leads to a fine-grained texture. Thus, one particular kind of magma can produce a number of quite different igneous rocks depending on whether the material cools below the surface, on the surface as a lava flow, or on the surface as an accumulation of pyroclastics.

Igneous rocks are generally subdivided into two main categories based on where the rocks form: **volcanic**, or **extrusive igneous rocks**, form from the cooling of lava or the bonding of pyroclastic materials on the surface, whereas **plutonic**, or **intrusive igneous rocks**, form from the cooling of magma below the surface. Some common plutonic and volcanic rocks are shown in Figure 13-6.

Plutonic (Intrusive) Rocks

Plutonic rocks cool and solidify beneath Earth's surface, where surrounding rocks serve as insulation around the intrusion of magma that greatly slows the rate of cooling. Because the magma may require many thousands of years for complete cooling, the individual mineral crystals in a plutonic rock can grow to a relatively large size—large enough to see with

TABLE 13-2 Rock Classification

Class	Subclass	Rock	General Characteristics
Igneous	Plutonic (Intrusive)	Granite	Coarse-grained: "salt & pepper" appearance; from high-silica felsic magma; plutonic equivalent of rhyolite.
		Diorite	Coarse-grained; from intermediate silica magma; plutonic equivalent of andesite.
		Gabbro	Coarse-grained; black or dark gray color; from low-silica mafic magma; plutonic equivalent of basalt.
		Peridotite	Common mantle rock consisting primarily of olivine and/or pyroxene.
	Volcanic (Extrusive)	Rhyolite	Light color; from high-silica felsic magma; volcanic equivalent of granite.
		Andesite	Typically gray in color; from intermediate silica magma; volcanic equivalent of diorite.
		Basalt	Usually black in color; from low-silica mafic magma; volcanic equivalent of gabbro.
		Obsidian	Volcanic glass; typically black in color and rhyolitic in composition.
		Pumice	Volcanic glass with frothy texture; often rhyolitic in composition.
		Tuff	Rock made from volcanic ash or pyroclastic flow deposits.
Sedimentary	Clastic (Detrital)	Shale	Composed of very fine-grained sediments; typically thin bedded.
		Sandstone	Composed of sand-sized sediments.
		Conglomerate	Composed of rounded, pebble-sized sediments in a fine-textured matrix.
		Breccia	Composed of coarse-grained, angular, sediments; typically thin bedded poorly-sorted fragments.
	Chemical & Organic	Limestone	Composed of calcite; may contain shells or shell fragments.
		Travertine	Limestone deposited in caves or around hot springs; often deposited in banded layers.
		Chert	Common chemical rock composed of microcrystalline quartz.
		Coal	Composed of compacted plant material.
Metamorphic	Foliated	Slate	Fine-grained; smooth surfaces; typically forms from the low-grade metamorphism of shale.
		Phyllite	Fine-grained; glossy, wavy surfaces.
		Schist	Thin, flaky layers of platy minerals.
		Gneiss	Coarse-grained; granular texture; distinct mineral layers; high-grade metamorphism.
	Nonfoliated	Quartzite	Composed primarily of quartz; often derived from sandstone.
		Marble	Composed principally of calcite; typically derived from limestone.
		Serpentinite	Greenish-black color; slippery feel; from hydrothermal alteration of peridotite.

the naked eye—giving the rock a very coarse-grained texture. Although originally buried, plutonic rocks may subsequently become important to topographic development by being pushed upward to the surface or by being exposed by erosion.

The most common and well-known plutonic rock is **granite**, a generally light-colored, coarse-grained igneous rock (Figure 13-7). Granite is made of a combination of light- and dark-colored minerals such as quartz, plagioclase feldspar, potassium feldspar, hornblende, and biotite (granite is a felsic igneous rock; see Figure 13-6).[3] Granite and similar plutonic rocks such as granodiorite make up the core of many mountain ranges, such as the Sierra Nevada in California, as well as the deep interior "shield" of many continents.

[3]Sometimes the term *granitic rock* is used broadly to refer to granite as well as closely related plutonic rocks such as granodiorite and tonalite that have slightly different combinations of potassium feldspar and plagioclase feldspar from true granite.

▲ **Figure 13-6** Common igneous rocks. The chemistry of the magma and the rate of cooling help determine the final igneous rock formed. Felsic magmas contain relatively high amounts of silica and cool to form a plutonic rock such as granite or the volcanic rock rhyolite. Mafic magmas contain relatively low amounts of silica and cool to form a plutonic rock such as gabbro or the volcanic rock basalt. Intermediate composition magmas cool to form a plutonic rock such as diorite or the volcanic rock andesite.

Volcanic (Extrusive) Rocks Volcanic rocks form on the surface of Earth, either from the cooling of lava or from the accumulation of pyroclastic material such as volcanic ash and cinders. When lava cools rapidly on Earth's surface, the solidification may be complete within hours, and so the mineral crystals in many volcanic rocks are so small as to be invisible without microscopic inspection. On the other hand, volcanic rocks that form from the accumulation of pyroclastics may clearly show tiny fragments of shattered rock that was explosively ejected from a volcano.

Of the many kinds of volcanic rocks, by far the most common is the black or dark gray, fine-grained rock called **basalt** (Figure 13-8). Basalt forms from the cooling of lava and is comprised only of dark-colored minerals such as plagioclase feldspar, pyroxene, and olivine (basalt is a mafic igneous rock; see Figure 13-6). Basalt is the most common volcanic rock in Hawai'i and is widespread in parts of some continents—such as in the Columbia Plateau of the northwestern United States. Basalt also makes up the bulk of the ocean floor crust.

There are many other well-known volcanic rocks. *Obsidian* is a type of volcanic glass (meaning that there is no organized mineral crystal structure)—typically black in color—that forms from extremely rapid cooling of lava; *pumice* forms from the rapid cooling of frothy, gas-rich molten material (a piece of pumice is sometimes light enough to float on water!); *Tuff* is a volcanic rock consisting of welded pyroclastic fragments (see Figure 13-14b).

In Chapter 14 we will discuss volcanism and some of the topographic features associated with plutonic and volcanic activity.

▲ **Figure 13-7** (a) Massive outcrops of granite at Sylvan Lake reservoir in the Black Hills of South Dakota. Granite is a plutonic (intrusive) igneous rock that solidified beneath Earth's surface and was subsequently exposed by erosion. (b) A sample of Vermont granite, showing glassy quartz grains and black biotite mica.

Sedimentary Rocks

External processes, mechanical and chemical, operating on rocks cause them to disintegrate. This disintegration produces fragmented mineral material, some of which is removed by water, wind, ice, gravity, or a combination of these agents. Much of this material is transported by water moving in rivers or streams as sediment. Eventually the sediment is deposited somewhere in a quiet body of water, particularly on the floor of an ocean (Figure 13-5b). Over a long period of time, sedimentary deposits can build to a remarkable thickness—many thousands of meters. The sheer weight of this massive overburden exerts enormous pressure,

▲ **Figure 13-8** (a) Several horizontal layers of basalt are exposed on the wall of the Snake River Canyon near Twin Falls, Idaho. (b) A porous chunk of black basalt.

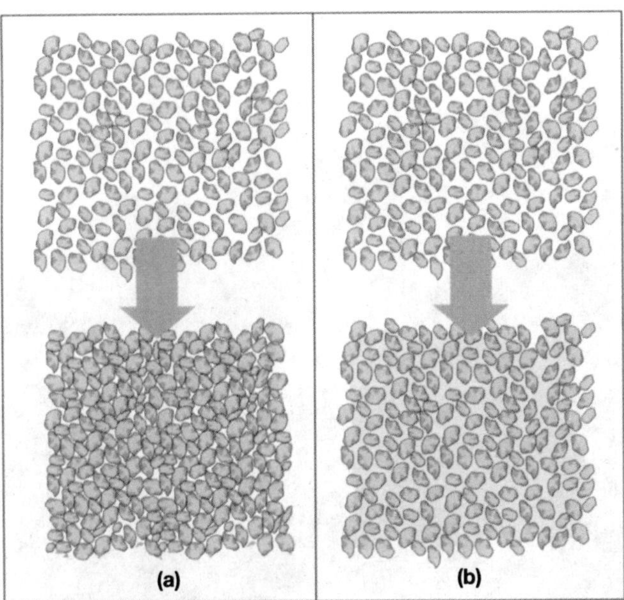

▲ **Figure 13-9** Sedimentary rocks are composed of small particles of matter deposited by water or wind in layers and then consolidated by compaction and/or cementation. (a) Compaction consists of the packing of the particles as a result of the weight of overlying material. (b) Cementation involves the infilling of pore spaces among the particles by a cementing agent, such as silica, calcium carbonate, or iron oxide.

which causes individual particles in the sediment to adhere to each other and to interlock. In addition, chemical cementation normally takes place. Various cementing agents—especially silica, calcium carbonate, and iron oxide—precipitate from the water into the pore spaces in the sediment (Figure 13-9). This combination of pressure and cementation consolidates and transforms the sediments to **sedimentary rock**.

As we will see in Chapter 16, during transportation and deposition, sediments may be sorted roughly by size and so many sedimentary rocks are comprised of particles of fairly uniform size. Other variations in the composition of the sediments may be due to factors such as different rates of deposition, changes in climatic conditions, or patterns of sediment movement in the oceans. Most sedimentary deposits are built up in more or less distinct horizontal layers called **strata** (Figure 13-10), which vary in thickness and composition (wind-deposited sediments are a notable exception to the horizontal layers of most sediments). The resulting parallel structure, or *stratification*, is a characteristic feature of most sedimentary rocks. Although originally deposited and formed in horizontal orientation, the strata may later be uplifted, tilted, and deformed by pressures from within Earth (Figure 13-11).

Sedimentary rocks are generally classified into two subcategories based on how they were formed.

Clastic Sedimentary Rocks Sedimentary rocks composed of fragments of preexisting rocks in the form of cobbles, gravel, sand, silt, or clay are known as *clastic* or *detrital* sedimentary rocks. By far the most common of these rocks are *shale* (or *mudstone*) that is composed of very fine silt and clay particles, and *sandstone* that is made up of compacted, sand-size grains (Figure 13-12). When the rock is composed of rounded, pebble-size fragments it is called *conglomerate*.

Chemical and Organic Sedimentary Rocks Chemically accumulated sedimentary rocks are usually formed by the precipitation of soluble materials or sometimes by more complicated chemical reactions. Calcium carbonate ($CaCO_3$) is a common component of such rocks, and *limestone* is the most widespread result (see Figure 13-10). *Chert* forms in a similar way to

▲ **Figure 13-10** (a) Nearly horizontal strata of limestone and shale in a road cut near Lyons, Colorado. (b) A sample of limestone containing an abundance of small fossil mollusks such as snails.

▶ **Figure 13-11** Sedimentary strata (mostly limestone and shale) that have been folded and tilted into an almost vertical orientation on Mount Angeles in the Olympic Mountains of Washington.

limestone but is composed of silica (SiO_2) instead of calcium carbonate.

Organically accumulated sedimentary rocks, including *coal*, are formed from the compacted remains of dead plant material. Limestone can also be formed from the accumulated skeletal remains of coral and other lime-secreting sea animals.

There is considerable overlap among these formation methods, with the result that, in addition to there being many different kinds of sedimentary rocks, there also are many gradations among them. Taken together, however, the vast majority of all sedimentary rocks are shale, sandstone, or limestone (Figure 13-13).

▲ **Figure 13-12** (a) The structure of these abrupt sandstone cliffs in southeastern Utah is easy to see. (b) A typical light-colored piece of sandstone.

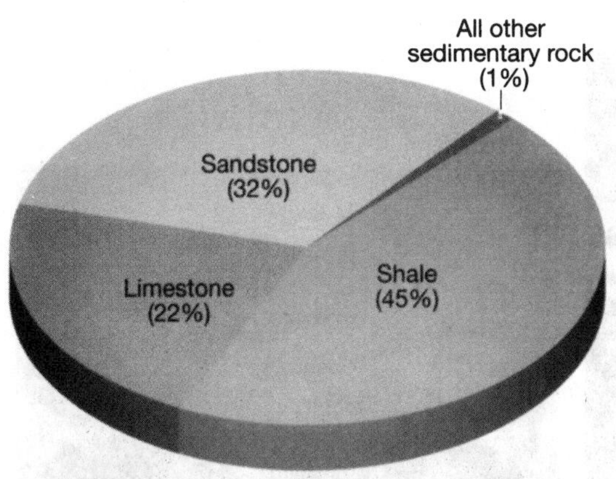

▲ **Figure 13-13** The estimated relative abundance of various types of sedimentary rocks.

Metamorphic Rocks

Metamorphic Rock Foliation

Originally either igneous or sedimentary, **metamorphic rocks** are those that have been drastically changed by heat and/or pressure. The effects of heat and pressure on rocks are complex, being strongly influenced by such things as the presence or absence of fluids in the rocks and the length of time the rocks are heated and/or subjected to high pressure. Metamorphism can be a virtual "cooking" process that heats the rock, causing its mineral components to be recrystallized and rearranged (Figure 13-14). The metamorphic result is often quite different from the original rock; the rocks are changed in structure, texture, composition, and appearance.

Metamorphism can take place in a number of different circumstances. For example, metamorphism can occur beneath the surface of Earth where magma comes in contact with surrounding rocks, altering the surrounding rocks through heat and pressure. Because of such **contact metamorphism**, it is common to find exposed metamorphic rocks adjacent to plutonic rocks, such as granite. **Regional metamorphism** takes place where large volumes of rock deep within the crust are subjected to heat and/or pressure over long periods of time (see Figure 13-5c). As we will see in Chapter 14, metamorphism is common along many of the boundaries between lithospheric plates.

If the minerals in a metamorphic rock show a prominent alignment or orientation, we say that the rock is *foliated*. Such rocks may have a platy, wavy, or banded texture. If the original rock was dominated by a single mineral (as in some sandstones or limestones), such foliation is less common.

Some rocks, when metamorphosed, change in predictable fashion. Thus, limestone usually becomes *marble*, sandstone normally is changed to *quartzite*, and shale to *slate*. In many cases, however, the metamorphosis is so great that it is difficult to ascertain the nature of the original rock; in this case, metamorphic rocks are identified according to their physical characteristics and appearance. Among the most common foliated metamorphic rocks—ranging from the least metamorphism to the most—are fine-grained slate and *phyllite* (Figure 13-15), narrowly-foliated/medium-grained *schist*, and broadly banded *gneiss* (Figure 13-16).

▲ **Figure 13-14** (a) This bedrock exposure in northeastern California, near Alturas, shows a light brown basalt overlying a colorful layer of tuff (a volcanic rock formed by consolidation of volcanic ash). The basalt was extruded onto the tuff in molten form, and its great heat "cooked," or metamorphosed, the upper portion of the tuff. Visual evidence of the metamorphosis is seen in the difference in color between the metamorphosed and unmetamorphosed part of the tuff stratum. (b) A representative sample of tuff, which is formed by the consolidation of pyroclastic material.

(a)

(b)

▲ **Figure 13-15** (a) A large outcropping of the foliated metamorphic rock phyllite in California's Merced River Canyon. Thin parallel layers of minerals are characteristic. (b) A close-up of a sample of phyllite.

The Rock Cycle

As the preceding description suggests, over long periods of time, the minerals in one rock might well end up in a different rock: igneous rocks can be broken down into sediments that might then form a sedimentary rock, which in turn might undergo metamorphism, only to be worn back again into sediments. This ongoing "recycling" of lithospheric material is sometimes referred to as the **rock cycle** (Figure 13-17).

Continental and Ocean Floor Rocks

The lithosphere has a very uneven distribution of the three principal rock classes. Sedimentary rocks compose the most commonly exposed bedrock (perhaps as much as 75 percent) of the continents (Figure 13-18). The sedimentary cover is not thick, however, averaging less than 2.5 kilometers (1.5 miles), and sedimentary rocks accordingly constitute only a very small proportion (perhaps 5 percent) of the total volume of the crust (Figure 13-19). The bulk of the continents consist of granite (along with an unknown proportion of metamorphic rocks such as gneiss and schist). On the other hand, the ocean floor crust is composed almost entirely of basalt and gabbro (covered by a relatively thin veneer of oceanic sediments).

The distinction between the dominant rocks of the ocean floors and continents is important. Basalt is a denser rock than granite (the density of basalt is about 3.0 g/cm^3 while that of granite is about 2.7 g/cm^3), and for this reason, continental lithosphere is less dense than oceanic lithosphere. Continental lithosphere "floats" quite easily on the denser asthenosphere below, while oceanic lithosphere is dense enough that it can actually be pushed down, or *subducted*, into the asthenosphere—the consequences of this are discussed at length in the following chapter.

(a)

(b)

▲ **Figure 13-16** (a) An outcrop of banded gneiss in Greenland. This particular formation is one of the oldest on Earth; its age is 3.8 billion years. (b) A typical piece of gneiss, showing contorted foliation.

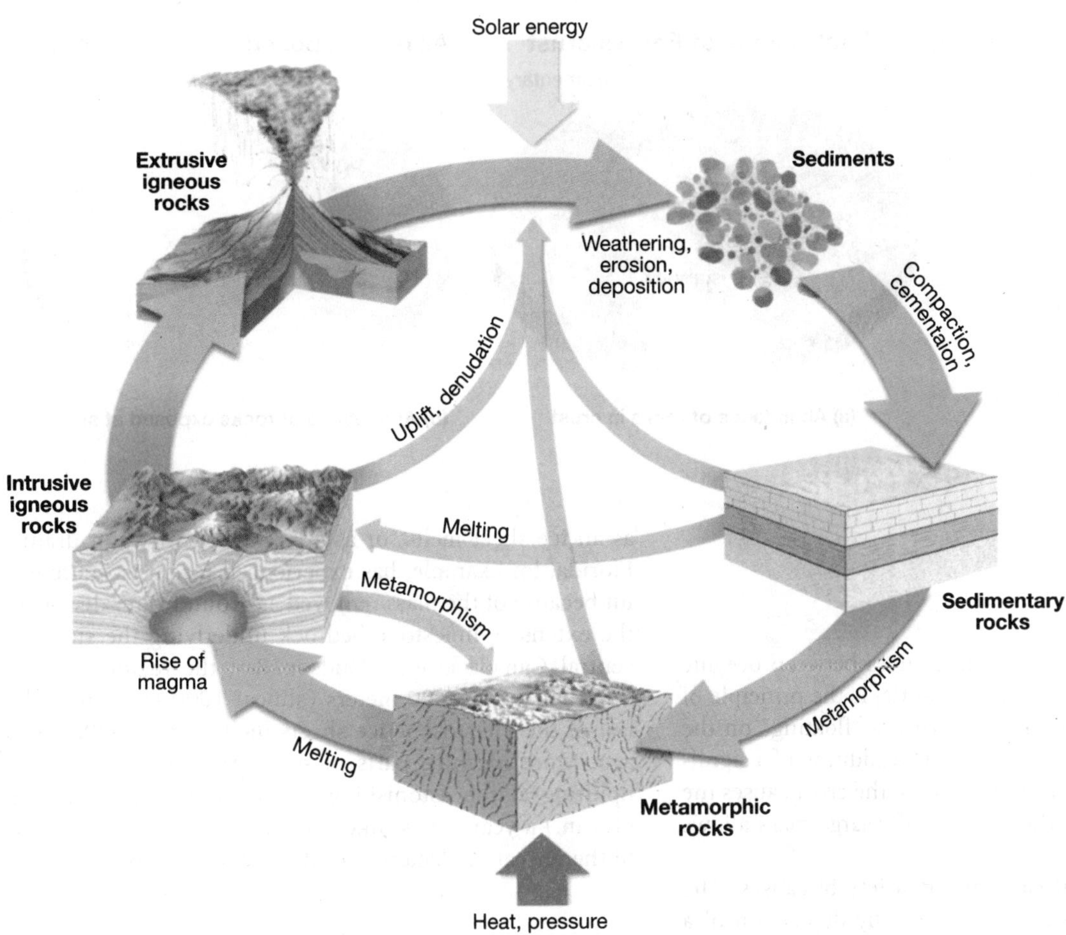

▲ **Figure 13-17** The rock cycle, showing the relationships among the three classes of rocks.

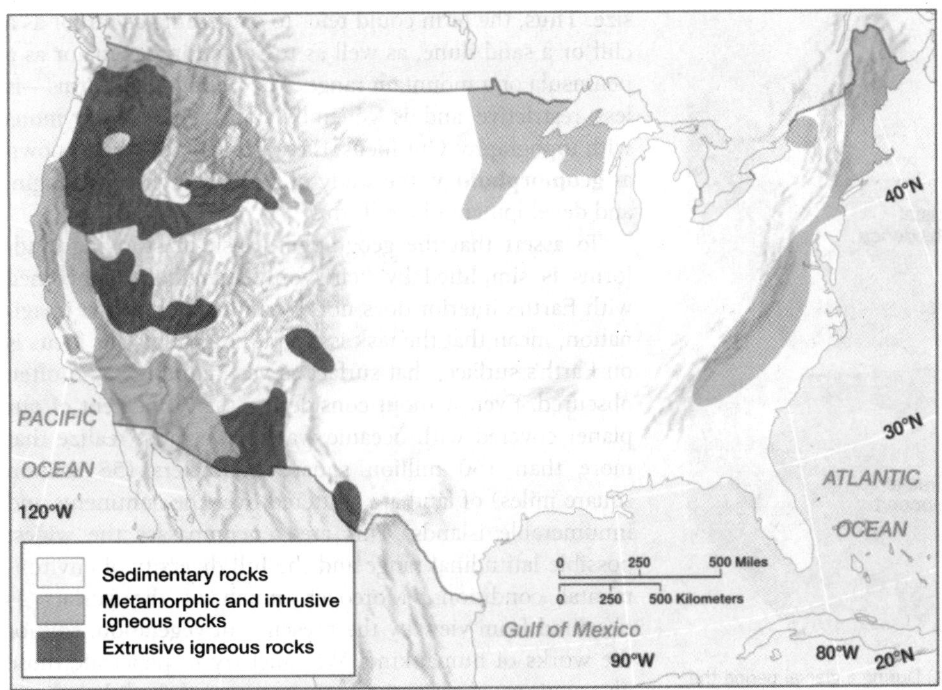

▲ **Figure 13-18** Surface distribution of rock classes in the conterminous United States. Sedimentary rocks clearly dominate, and this trend also holds true for the rest of the world.

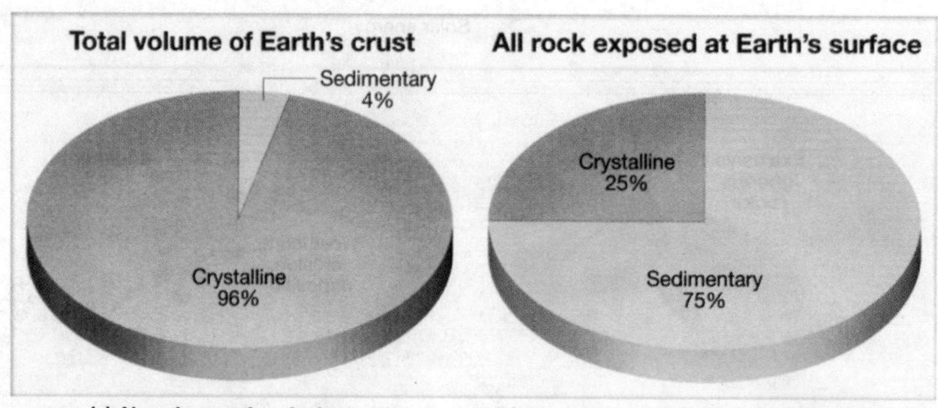

▶ **Figure 13-19** Relative abundances of sedimentary and crystalline (igneous and metamorphic) rocks (a) in Earth's crust and (b) exposed at Earth's surface.

Isostasy

Isostasy

Related to the recognition of differences between oceanic crust, continental crust, and the mantle, is the principle of **isostasy**. In simplest terms, the crust is "floating" on the denser, deformable, mantle below. The addition of a significant amount of mass onto a portion of the crust causes the crust to sink, whereas the removal of a large mass allows the crust to rise.

Isostatic adjustment can have a variety of causes. The crust may be depressed, for example, by deposition of a large amount of sediment on a continental shelf, or by the accumulation of a great mass of glacial ice on a landmass, as illustrated in Figure 13-20, or even by the weight of water trapped behind a large dam. Depressed crust may rebound to a higher elevation as material is eroded away, as an ice sheet melts, or as a large body of water drains. Florida, for example, has experienced recent isostatic uplift because of the mass removed as groundwater dissolved the extensive limestone bedrock underlying the state. In central Canada around Hudson Bay, the region has uplifted more than 300 meters (almost 1000 feet) since the last of the Pleistocene ice sheets melted away 8000 years ago. More recently, parts of southern Alaska have been uplifting at the astonishing rate of as much as 36 mm (1.4 in.) a year—most likely at least in part as a response to the retreat of glaciers over the last century or so.

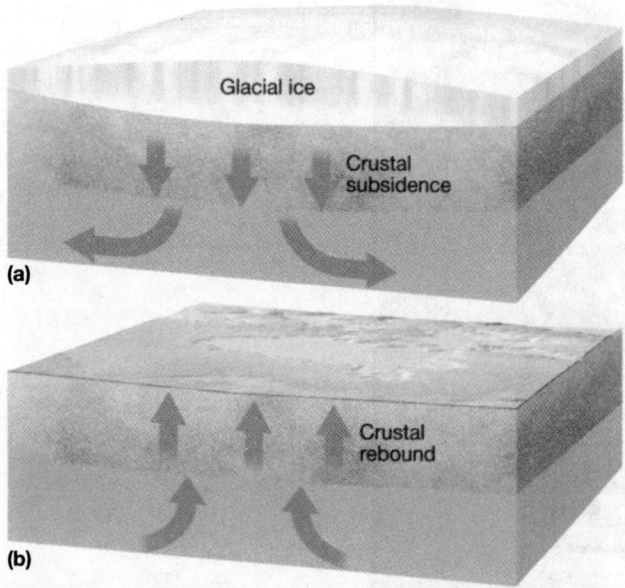

▲ **Figure 13-20** Isostatic adjustment. (a) During a glacial period the heavy weight of accumulated ice depresses the crust. (b) Deglaciation removes the weighty overburden and as the ice melts the crust rises, or "rebounds."

The Study of Landforms

Our attention for the remainder of this book is directed primarily to **topography**—the surface configuration of Earth. A **landform** is an individual topographic feature, of any size. Thus, the term could refer to something as minor as a cliff or a sand dune, as well as to something as major as a peninsula or a mountain range. The plural—landforms—is less restrictive and is generally considered synonymous with topography. Our focus, then, is a field of study known as **geomorphology**, the study of the characteristics, origin, and development of landforms.

To assert that the geographer's task in studying landforms is simplified by being only marginally concerned with Earth's interior does not, by any stretch of the imagination, mean that the task is simple. Although our focus is on Earth's surface, that surface is vast, complex, and often obscured. Even without considering the 70 percent of our planet covered with oceanic waters, we must realize that more than 150 million square kilometers (58 million square miles) of land are scattered over the continents and innumerable islands. This area encompasses the widest possible latitudinal range and the full diversity of environmental conditions. Moreover, much of the surface is obscured from view by the presence of vegetation, soil, or the works of humankind. We must try to penetrate those obstructions, observe the characteristics of the lithospheric surface, and encompass the immensity and diversity of a worldwide landscape. This is far from a simple task.

◀ **Figure 13-21** The spectacular terrain of Montana's Glacier National Park was produced by a variety of interacting processes including glaciation. This view is across St. Mary Lake toward Mount Jackson.

To organize our thinking for such a complex endeavor, it may be helpful to isolate certain basic elements for an analytic approach:

Structure *Structure* refers to the nature, arrangement, and orientation of the materials making up the landform being studied. Structure is essentially the geologic underpinning of the landform. Is it composed of bedrock or not? If so, what kind of bedrock and in what configuration? If not, what are the nature and orientation of the material? With a structure as clearly visible as that shown in Figures 13-11 and 13-12, these questions are easily answered, but such is not always the case, of course.

Process *Process* considers the actions that have combined to produce the landform. A variety of processes—usually geologic, hydrologic, atmospheric, and biotic—are always at work shaping the features of the lithospheric surface, and their interaction is critical to the formation of the feature(s). For example, a landscape resulting from the processes of *glaciation* is shown in Figure 13-21.

Slope *Slope* is the fundamental aspect of shape for any landform. The angular relationship of the surface is essentially a reflection of the contemporary balance among the various components of structure and process (Figure 13-22). The inclinations and lengths of the slopes provide details that are important both in describing and analyzing the feature.

Drainage *Drainage* refers to the movement of water (from rainfall and snowmelt), either over Earth's surface or

◀ **Figure 13-22** Slope is a conspicuous visual element of some landscapes. The abrupt cliffs of the volcanic backbone of the island of Bora-Bora (in French Polynesia) dominate this scene.

down into the soil and bedrock. Although moving water is an outstanding force under the "process" heading, the ramifications of slope wash, streamflow, stream patterns, and other aspects of drainage are so significant that the general topic of drainage is considered a basic element in landform analysis (a topic discussed in detail in Chapter 16).

Once these basic elements have been recognized and identified, the geographer is prepared to analyze the topography by answering the fundamental questions at the heart of any geographic inquiry:

- What? The *form* of the feature or features
- Where? The *distribution* and pattern of the landform assemblage
- Why? An explanation of *origin* and *development*
- So what? The *significance* of the topography in relationship to other elements of the environment and to human life and activities.

Some Critical Concepts

One term used frequently in the following pages is **relief**, which refers to the difference in elevation between the highest and lowest points in an area. The term can be used at any scale. Thus, as we saw in Figure 1-6, the maximum world relief is approximately 20 kilometers (12 miles), which is the difference in elevation between the top of Mount Everest and the bottom of the Mariana Trench. At the other extreme, the local relief in some places like Florida can be a matter of merely a few meters.

Internal and External Geomorphic Processes

The relief we see in a landscape is temporary. It represents the momentary balance of two largely opposing sets of processes that are operating to shape and reshape the surface of Earth: *internal processes* and *external processes* (Figure 13-23). These processes are relatively few in number but extremely varied in nature and operation. The great variety of Earth's topography reflects the complexity of interactions between these processes and the underlying structure of the surface.

Internal Processes The **internal processes** originate from within Earth, energized by internal heat that generates forces that apparently operate outside of any surface or atmospheric influences. These processes result in crustal movements through folding, faulting, and volcanic activity of various kinds. In general, they are constructive, uplifting, building processes that tend to increase the relief of the land surface. These processes are considered in some detail in the chapter that follows.

External Processes In contrast, the **external processes** are largely subaerial (meaning that they operate at the base of the atmosphere) and draw their energy mostly from sources above the lithosphere, either in the atmosphere or in the oceans. Unlike some of the internal processes, external processes are generally well understood and their behavior is often predictable. Moreover, their behavior may be significantly influenced by the existing topography, particularly its shape and the nature of the surface materials. The external processes may be thought of generally as wearing-down or destructive processes—broadly called *denudation*—that eventually tend to diminish topographic irregularities and decrease the relief of Earth's surface.

In succeeding chapters, we consider these various processes—their nature, dynamics, and effects—in detail, but it is useful here to summarize them so that they can be glimpsed in totality before we treat them one at a time. Table 13-3 presents such an overview. Note, however, that our classification scheme is a great simplification of reality; while some items are clearly separate and discrete, others overlap with each other. The table, then, represents a simple, logical way to begin a study of the processes, but is not necessarily the only or ultimate framework.

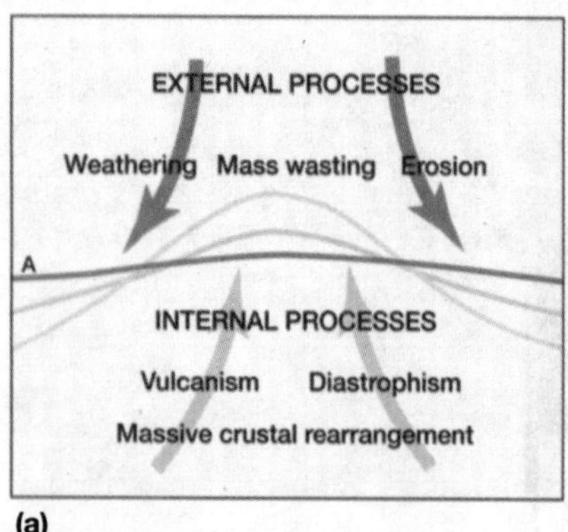

(a)

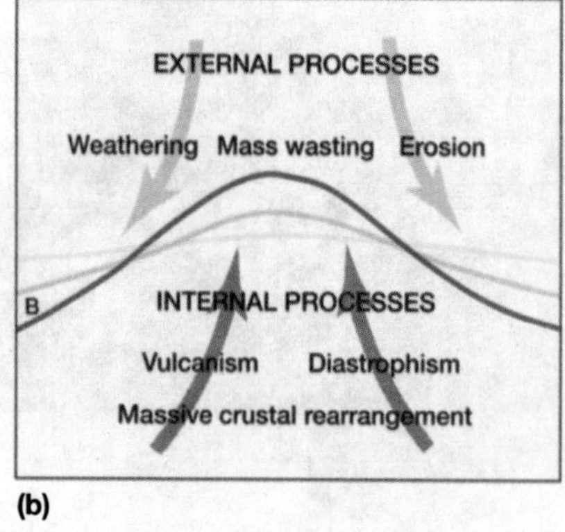

(b)

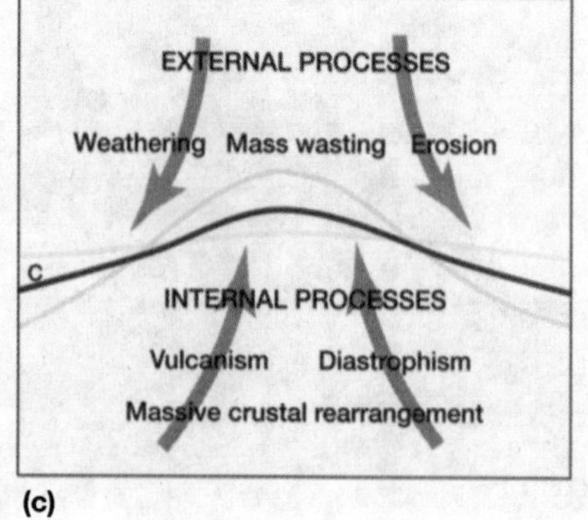

(c)

▲ **Figure 13-23** Schematic relationship between external and internal geomorphic processes. (a) A surface that has been worn down by external forces but not uplifted by internal forces. (b) A surface that has been uplifted by internal forces but not worn away by external forces. (c) A "normal" surface that has been both uplifted by internal forces and worn away by external forces.

TABLE 13-3 A Summary of Geomorphic Processes

Internal
Lithospheric Rearrangement (plate tectonics)
Volcanism
Extrusive
Intrusive
Tectonism (Diastrophism)
Folding
Faulting
External (Denudation)
Weathering
Mass Wasting
Erosion/deposition
Fluvial (running water)
Aeolian (wind)
Glacial (moving ice)
Solution (ground water)
Waves and currents (oceans/lakes)

Thus, internal and external processes work in more or less direct opposition to one another—in some landscapes, both operating at the same time; we see the momentary balance between them. Their battleground is Earth's surface, the interface between lithosphere and atmosphere, where this remarkable struggle has persisted for billions of years and may continue endlessly into the future.

Uniformitarianism

Fundamental to an understanding of internal and external processes and topographic development is familiarity with the doctrine of **uniformitarianism**, which holds that "the present is the key to the past." This concept—first put forth by James Hutton in 1795—means that the processes that are shaping the contemporary landscape are the same processes that formed the topography of the past—and are the same processes that will shape the topography of the future.[4] Therefore, by understanding the geomorphic processes we see working today, we can study a landscape and begin to comprehend the history of its development. The development of landforms is a virtually endless event, with the topography at any given time simply representing one moment in a continuum of change. The processes involved are mostly ongoing and slow acting, although some of these processes reveal change rather episodically—an earthquake, a volcanic eruption, a flood.

Because many of the internal and external processes operate very slowly by human standards, they may be difficult to grasp. For example, the enormous lithospheric plates that we will discuss in Chapter 14 are moving, on average, about as fast as your fingernails grow. To understand how sometimes imperceptibly slow processes work to alter the landscape, we must stretch our concept of time.

Geologic Time

Probably the most mind-boggling concept in all physical geography is the vastness of geologic time (Figure 13-24). In our daily lives, we deal with such brief intervals of time as hours, months, years, and sometimes centuries, which does nothing to prepare us for the scale of Earth's history. The colossal sweep of geologic time encompasses epochs of millions and hundreds of millions of years.

The concept of *geologic time* refers to the vast periods of time over which geologic processes operate. For uniformitarianism to work as a basic premise for interpreting Earth history, that history must have occurred over a span of time vast enough to allow feats of considerable magnitude to be accomplished by means of processes infinitesimally slow when measured on a human time scale.

How can we grasp the enormous sweep of geologic time? To state that Earth has existed for about 4.6 billion years, or that the Age of Dinosaurs persisted for some 160 million years, or that the Rocky Mountains were initially uplifted approximately 65 million years ago is to identify expanses of time of almost unfathomable scope.

An analogy may help us begin to comprehend the almost incomprehensible length of geologic time. If we could envision the entire 4.6-billion-year history of Earth compressed into a single calendar year, each day would be equivalent to 12.6 million years, each hour to 525,000 years, each minute to 8750 years, and each second to 146 years. On such a time scale the planet was lifeless for the first four months, with primal forms of one-celled life not appearing until early May. These primitive algae and bacteria had the world to themselves until early November, when the first multicelled organisms began to evolve. The first vertebrate animals, fishes, appeared about November 21; and before the end of the month, amphibians began establishing themselves as the first terrestrial vertebrates. Vascular plants, mostly tree ferns, club mosses, and horsetails, appeared about November 27, and reptiles began their era of dominance about December 7. Mammals arrived about December 14, with birds on the following day. Flowering plants would first bloom about December 21, and on Christmas Eve would appear both the first grasses and the first primates. The first hominids walked upright in midafternoon on New Year's Eve, and *Homo sapiens* appeared on the scene about an hour before midnight. The age of written history encompassed the last minute of the year—just the last 60 ticks of the clock!

Only with such an extraordinary time scale as this can one give credence to the doctrine of uniformitarianism or accept that the Grand Canyon is a youthful feature carved by that relatively small river seen deep in its inner gorge or accept that Africa and South America were once joined together and have drifted 3200 kilometers (2000 miles) apart. Indeed, the remainder of this

[4] We do know, however, that the relative balance or combination of processes operating has not been identical throughout Earth's history. For example, there have been periods of time in the past when glaciation was more important than it is today.

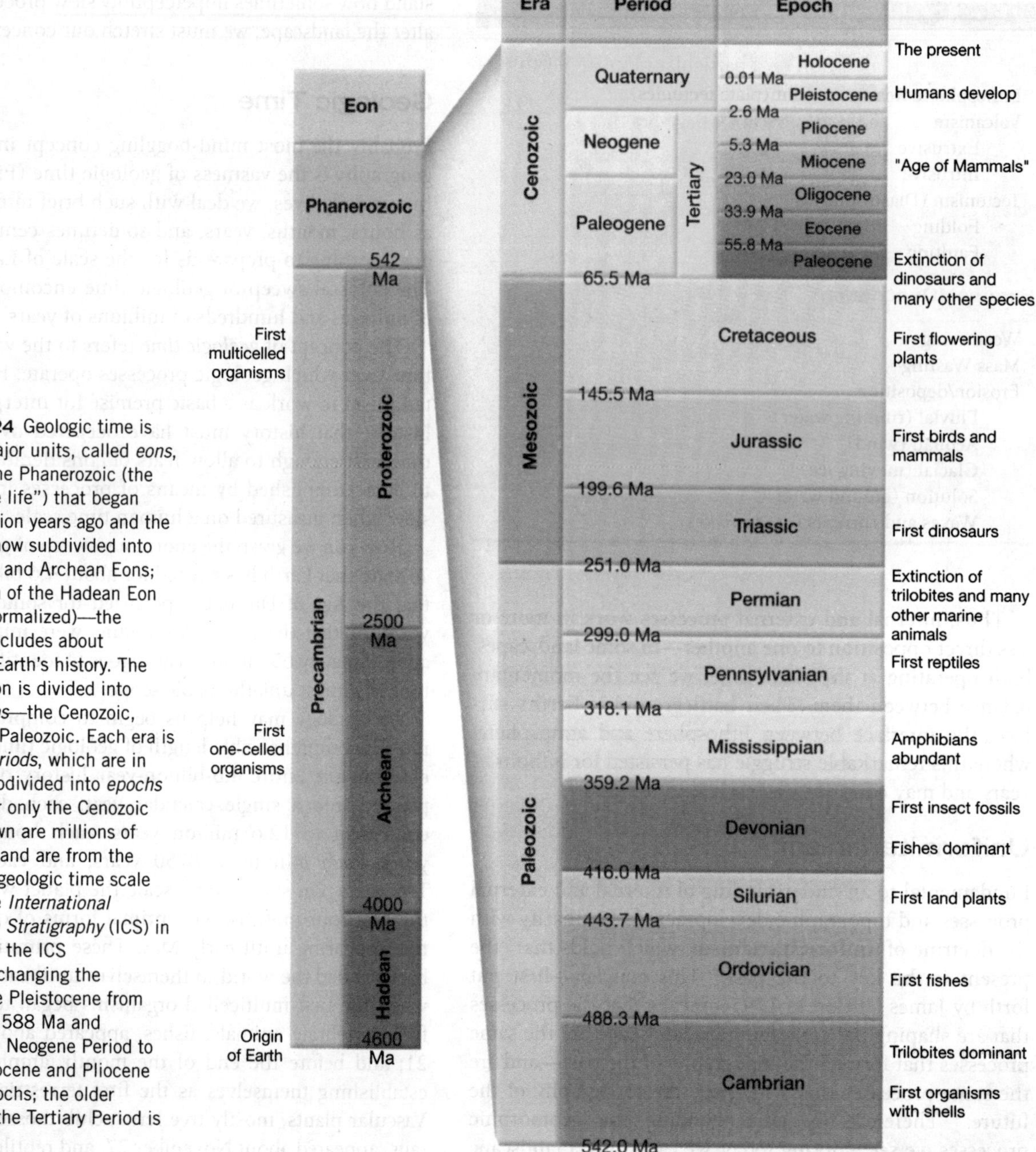

▶ **Figure 13-24** Geologic time is divided into major units, called *eons*, most notably the Phanerozoic (the time of "visible life") that began about 542 million years ago and the Precambrian (now subdivided into the Proterozoic and Archean Eons; the designation of the Hadean Eon has not been formalized)—the Precambrian includes about 88 percent of Earth's history. The Phanerozoic Eon is divided into three major *eras*—the Cenozoic, Mesozoic, and Paleozoic. Each era is divided into *periods*, which are in turn further subdivided into *epochs* (shown here for only the Cenozoic Era). Ages shown are millions of years ago (Ma) and are from the revision of the geologic time scale provided by the *International Commission on Stratigraphy* (ICS) in 2008. In 2009 the ICS recommended changing the boundary of the Pleistocene from 1.806 Ma to 2.588 Ma and designating the Neogene Period to include the Miocene and Pliocene but no later epochs; the older designation of the Tertiary Period is still in use.

book can be considered only as fanciful fiction unless one can rely on the concept of geologic time. The geomorphic processes generally operate with excruciating slowness, but the vastness of geologic history provides a suitable time frame for their accomplishments.

If daunted by the noxious stench
Exhaled from Time's abyss,
Retreat into some lesser trench
Where ignorance is bliss.

Claude C. Albritton Jr.,
The Abyss of Time

Scale and Pattern

Before we proceed with our systematic study of geomorphic processes, two final concepts should be kept in mind—scale and pattern.

An Example of Scale

The question of scale is fundamental in geography. Regardless of the subject of geographic inquiry, recognizable features and associations are likely to vary considerably depending on the scale of observation. This simply means that the aspects of the landscape one observes in a close-up view are different from those observed from a more distant view.

As an example of the complexity and significance of scale, let us focus our attention on a particular place on Earth's surface and view it from different perspectives. The location is north-central Colorado within the boundaries of Rocky Mountain National Park, some 13 kilometers (8 miles) due west of the town of Estes Park. It encompasses a small valley called Horseshoe Park, through which flows a clear mountain stream named Fall River and adjacent to which is the steep slope of Bighorn Mountain (Figure 13-25).

1. To illustrate the largest scale of ordinary human experience, we will hike northward from the center of Horseshoe Park up the side of Bighorn Mountain. At this level of observation, the first topographic feature of note is a smooth stretch of Fall River we must cross. We walk over a small sandbar at the south edge of the river, wade for a few steps in the river, and step up a half-meter (20-inch) bank onto the mountainside, noting the dry bed of a small intermittent pond on our left. After 20 minutes or so of steep uphill scrambling, we reach a rugged granite outcrop (locally called Hazel Cone), which presents us with an almost vertical cliff face to climb.
2. At a significantly different scale of observation, we might travel for 20 minutes by car in this same area. The road through Horseshoe Park is part of U.S.

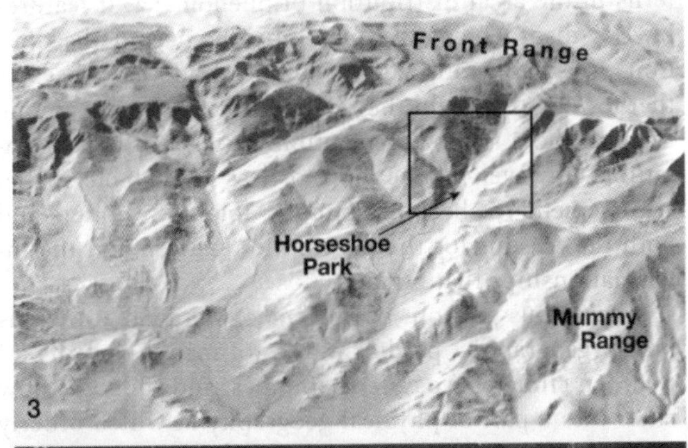

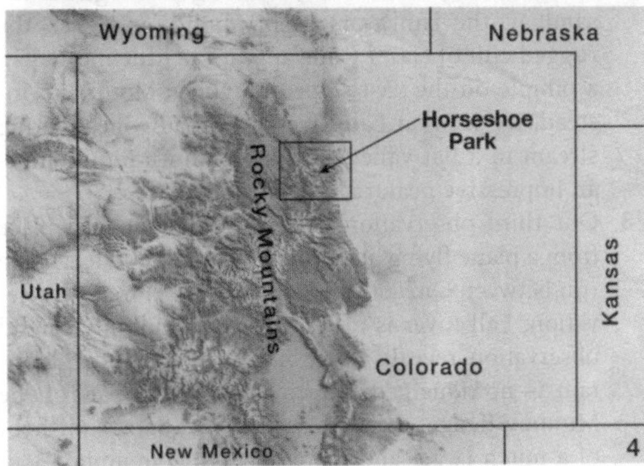

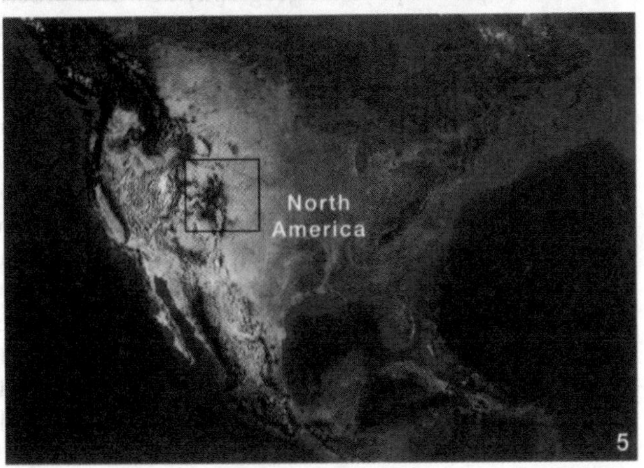

◄ **Figure 13-25** An experience with change of scale: (1) A close view of Horseshoe Park, Colorado, from the west. (2) Looking down on Horseshoe Park from Trail Ridge Road. (3) An aerial view of the Mummy Range and part of the Front Range. (4) A high-altitude look at Colorado. (5) North America as it might be seen from a distant spacecraft.

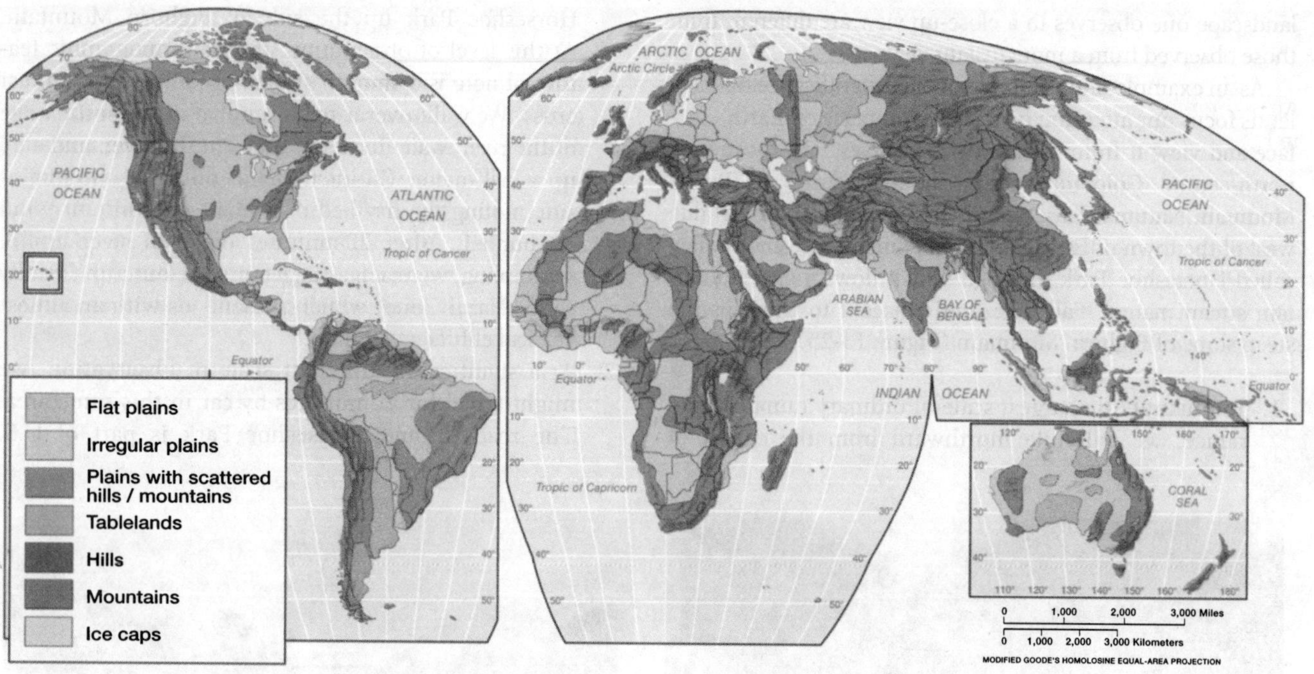

▲ **Figure 13-26** Major landform assemblages of the world.

Highway 34, which here is called the Trail Ridge Road. After 20 minutes, we reach a magnificent viewpoint high on the mountain to the southwest of Horseshoe Park. From this vantage point, our view of the country through which we have hiked is significantly enlarged. We can no longer recognize the sandbar, the bank, or the dry pond, and even the rugged cliff of Hazel Cone appears as little more than a pimple on the vast slope of Bighorn Mountain. Instead, we see that Fall River is a broadly meandering stream in a flat valley and that Bighorn Mountain is an impressive peak rising high above.

3. Our third observation of this area might take place from a plane flying at 12 kilometers (39,000 feet) on a run between Omaha and Salt Lake City. From this elevation, Fall River is nearly invisible, and only careful observation reveals Horseshoe Park. Bighorn Mountain is now merged indistinguishably as part of the Mummy Range, which is seen to be a minor offshoot of a much larger and more impressive mountain system called the Front Range.
4. A fourth level of observation would be available to us if we could hitch a ride on the Space Shuttle orbiting 340 kilometers (210 miles) above Earth. Our brief glimpse of northern Colorado would probably be inadequate to distinguish the Mummy Range, and even the 400-kilometer (250-mile) long Front Range would appear only as a component of the mighty Rocky Mountain cordillera, which extends from New Mexico to northern Canada.
5. At the smallest scale, the final viewpoint possible to humans could come from a spacecraft rocketing toward some distant heavenly body. Looking back in the direction of Horseshoe Park from a near-space position, one might possibly recognize the Rocky Mountains, but the only conspicuous feature in this small-scale view would be the North American continent.

The Pursuit of Pattern

A prime goal of any geographic study is to detect patterns in the areal distribution of phenomena. If features have a disordered and apparently haphazard distribution, it is more difficult to comprehend the processes that formed them and the relationships that exist among them. If there is some perceptible pattern to their distribution, however—some more or less predictable spatial arrangement—it becomes simpler for us to understand both the reasons for the distribution pattern and the interrelationships that pertain.

In previous portions of this book, we saw that there is broad geographic predictability to many patterns of weather, climate, ecosystems, biomes, and soils. We now enter into a part of physical geography in which orderly patterns of distribution are more difficult to discern, as the landform distribution pattern in Figure 13-26 so readily shows. There are a few aspects of predictability; for example, one can anticipate that in desert areas, certain geomorphic processes are more conspicuous than others and certain landform features are likely to be found. Overall, however, the global distribution of topography appears somewhat irregular. Largely for this reason, the geomorphology portion of this book concentrates less on distribution and more on process. Comprehending the dynamics of topographic development is more important to an understanding of systematic physical geography than any amount of detailed study of landform distribution.

CHAPTER 13 LEARNING REVIEW

After studying this chapter, you should be able to answer the following questions. Key terms from each text section are shown in bold type. Definitions for key terms are also found in the glossary at the back of the book.

The Structure of Earth (p. 353)

1. Briefly describe the overall structure of Earth, noting the four main layers: **crust**, **mantle**, **outer core**, and **inner core**.
2. What are the differences between the **lithosphere** ("plates") and the **asthenosphere**?
3. What is the **Moho (Mohorovičić discontinuity)**?

The Composition of Earth (p. 355)

4. How is a **mineral** different from a **rock**?
5. Provide a general description of the **silicate** mineral family. Name at least one common silicate mineral.
6. What is an **outcrop**?
7. Describe the general differences among **igneous**, **sedimentary**, and **metamorphic** rocks.
8. Briefly define the following terms: **magma**, **lava**, **pyroclastics**.
9. What is the difference between a **plutonic (intrusive) igneous rock** and a **volcanic (extrusive) igneous rock**?
10. How can one kind of magma (one kind of magma chemistry) produce two or more different kinds of igneous rocks?
11. What are the main differences between **granite** and **basalt**?
12. Why do most sedimentary rocks form in flat, horizontal layers **(strata)**?
13. Briefly contrast **regional metamorphism** and **contact metamorphism**.
14. Why are metamorphic rocks often found in contact with plutonic rocks such as granite?
15. Over long periods of time, how can the minerals in one rock end up in a different rock (or in a different kind of rock)? In other words, explain the **rock cycle**.
16. Why are sedimentary rocks so common on the surface of the continents?
17. Contrast the composition and characteristics of oceanic lithosphere with that of continental lithosphere.
18. Explain the concept of **isostasy**.

The Study of Landforms (p. 366)

19. Briefly define the terms **topography**, **landform**, and **geomorphology**.
20. What is meant by the structure of a landform?

Some Critical Concepts (p. 368)

21. In the context of geomorphology, what is meant by the term **relief**?
22. Contrast the concepts of **internal processes** and **external processes** in geomorphology.
23. How does the doctrine of **uniformitarianism** help us understand the history of Earth?
24. What is generally meant by the term geologic time?
25. What is the importance of geologic time to the doctrine of uniformitarianism?

Scale and Pattern (p. 370)

26. In the study of geomorphology, why do we primarily concentrate on processes rather than on distribution patterns by latitude?

Log in to **www.mygeoscienceplace.com** for animations, videos, RSS feeds, maps, and self-study quizzes to enhance your study of **Introduction to Landform Study**.

14 The Internal Processes

The geomorphic processes originating from within Earth are our starting points for understanding processes shaping the surface features of Earth. We begin our study of these internal processes with a look at the theory of plate tectonics, and finish by looking more specifically at volcanism, folding and faulting.

The Impact of Internal Processes on the Landscape

In our endeavor to understand the development of the earthly landscape, no pursuit is more rewarding than a consideration of the internal processes, for they are the supreme builders of terrain. Energized by awesome forces within Earth, the internal processes actively reshape the crustal surface. The crust is buckled and bent, land is raised and lowered, rocks are fractured and folded, solid material is melted, and molten material is solidified. These actions have been going on for billions of years and are fundamentally responsible for the gross shape of the lithospheric landscape at any given time. They do not always act independently and separately, but in this chapter we isolate them in order to simplify our analysis.

From Rigid Earth to Plate Tectonics

The shapes and positions of the continents may seem fixed at the time scale of human experience, but at the geologic time scale, measured in millions or tens of millions of years, continents are quite mobile. Continents have moved, collided and merged, and then been torn apart again; ocean basins have formed, widened, only to be eventually closed off. These changes on the surface of Earth continue today, so that the contemporary configuration of the ocean basins and continents is by no means the ultimate one. It is only in the last half century, however, that Earth scientists have come to understand how all of this could actually happen.

Until the mid-twentieth century, most Earth scientists assumed that the planet's crust was rigid, with continents and ocean basins fixed in position and significantly modified only by changes in sea level and periods of mountain building. The uneven shapes and irregular distribution of the continents were puzzling, but it was generally accepted that the present arrangement was emplaced in some ancient age when Earth's crust cooled from its original molten state.

◀ Eruption of Mount Yasur on Tanna, Vanuatu.

The idea that the continents had changed position over time, or that a single "supercontinent" once existed before separating into large fragments, has been around for a long time. Various naturalists, physicists, astronomers, geologists, botanists, and geographers from a number of countries have been putting forth this idea since the days of geographer Abraham Ortelius in the 1590s and philosopher Francis Bacon in 1620. Until fairly recently, however, the idea was generally unacceptable to the scientific community at large.

Wegener's Continental Drift

During the second and third decades of the twentieth century, the notion of **continental drift** was revived, most notably by the German meteorologist and geophysicist Alfred Wegener. Wegener put together the first comprehensive theory to describe and partially explain the phenomenon, publishing his landmark book, *Die Entstehung der Kontinente und Ozeane* (The Origin of Continents and Oceans), in 1915. Wegener postulated a massive supercontinent, which he called **Pangaea** (Greek for "whole land"), as having existed about 225 million years ago and then breaking apart into several large sections—the present-day continents—that have continued to move away from one another to this day (Figure 14-1).

Wegener accumulated a great deal of evidence to support his hypothesis, most notably the remarkable number of close affinities of geologic features on both sides of the Atlantic Ocean. He found the continental margins of the subequatorial portions of Africa and South America fit together with jigsaw-puzzle-like precision (Figure 14-2). He also determined that the petrologic records on both sides of the Atlantic show many distributions—such as ancient coal deposits—that would be continuous if the ocean did not intervene. Moreover, when the continents are placed back in their Pangaean configuration, mountain belts in Scandinavia and the British Isles match up with the Appalachian Mountains in eastern North America (Figure 14-3).

Supporting evidence came from paleontology: the fossils of some dinosaur and other reptile species, such as the freshwater swimming reptile the *Mesosaurus*, are found on both sides of the southern Atlantic Ocean, but nowhere else in the world (Figure 14-4). Fossilized plants, such as the fern-like *Glossopteris*, are found in similar-aged rocks in South America, South Africa, Australia, India, and Antarctica—its seeds too large and heavy to have been carried across the expanse of the present-day oceans by wind.

Wegener worked with climatologist Wladimir Köppen to study the past climate patterns of Earth. For example, they studied glacial deposits that indicated that large portions of the southern continents and India were extensively glaciated about 300 million years ago. The pattern of deposits made sense if the continents had been together in Pangaea when this glaciation took place (Figure 14-5).

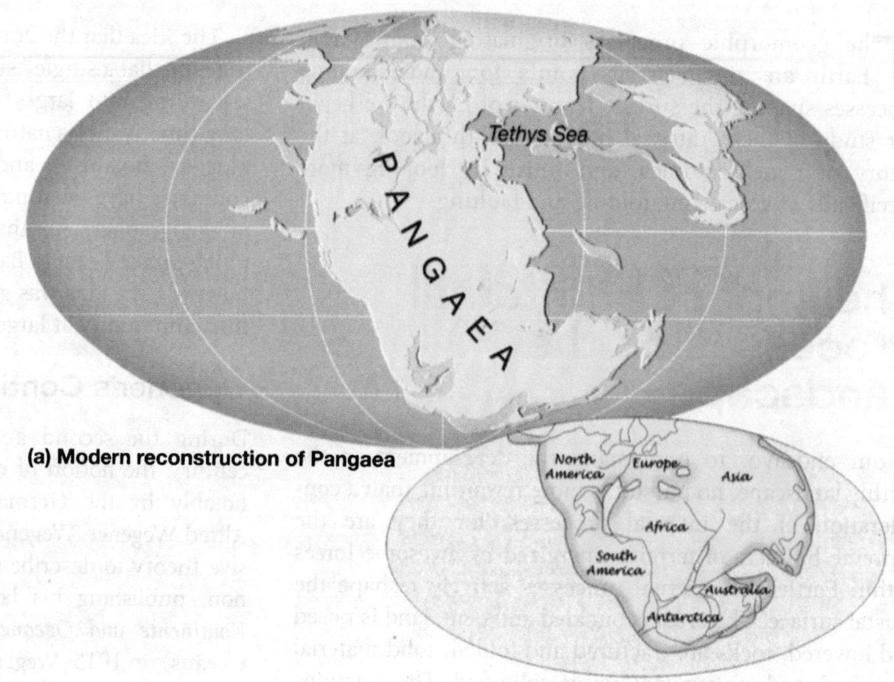

▶ **Figure 14-1** The supercontinent Pangaea. (a) Contemporary reconstruction of Pangaea as it probably appeared about 200 million years ago. (b) Wegener's reconstruction from 1915.

Wegener's accumulated evidence could be most logically explained by continental drift. His ideas attracted much attention in the 1920s—and generated much controversy. Some Southern Hemisphere geologists, particularly in South Africa, responded with enthusiasm. The general response to Wegener's hypothesis, however, was disbelief.

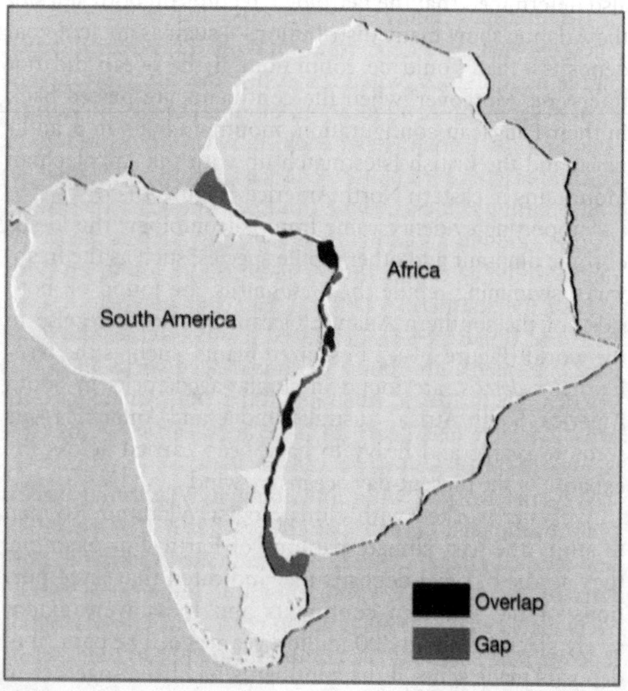

▲ **Figure 14-2** The matching of the coastlines of Africa and South America. When the continental shapes are compared at a depth of about 900 meters (500 fathoms) along the continental slope, there are few large gaps or overlaps.

Despite the vast number of distributional coincidences, most scientists felt that two difficulties made the theory improbable if not impossible: (1) Earth's crust was believed to be too rigid to permit such large-scale motions—after all, how could solid rock plow through solid rock? (2) Further, Wegener did not offer a suitable mechanism that could displace such large masses for a long journey. For these reasons, most Earth scientists ignored or even debunked the idea of continental drift for the better part of half a century after Wegener's theory was presented.

Although certainly discouraged that his ideas on continental drift were rejected by most scientists, Wegener continued his other scientific work—most notably in meteorology and polar research, where his contributions are widely acknowledged.[1] In 1930, Wegener was leading a meteorological expedition to the ice cap of Greenland. After delivering supplies to scientists stationed in the remote research outpost of Eismitte in the middle of the ice cap (see Figure 8-34b for a climograph of this station), on November 1 Wegener and a fellow expedition member, Rasmus Villumsen, set out by skis and dogsled to return to their base camp near the coast, but neither arrived. Wegener's body was found six months later buried in the snow—he died decades before his ideas on continental drift would receive serious attention by the majority of Earth scientists.

[1]Before Wegener developed his ideas on continental drift, his 1911 textbook, *The Thermodynamics of the Atmosphere*, had become a standard in German universities. The Swedish meteorologist Tor Bergeron openly acknowledged Wegener's contribution to our understanding of the raindrop formation process known today as the *Bergeron process* (see Chapter 6).

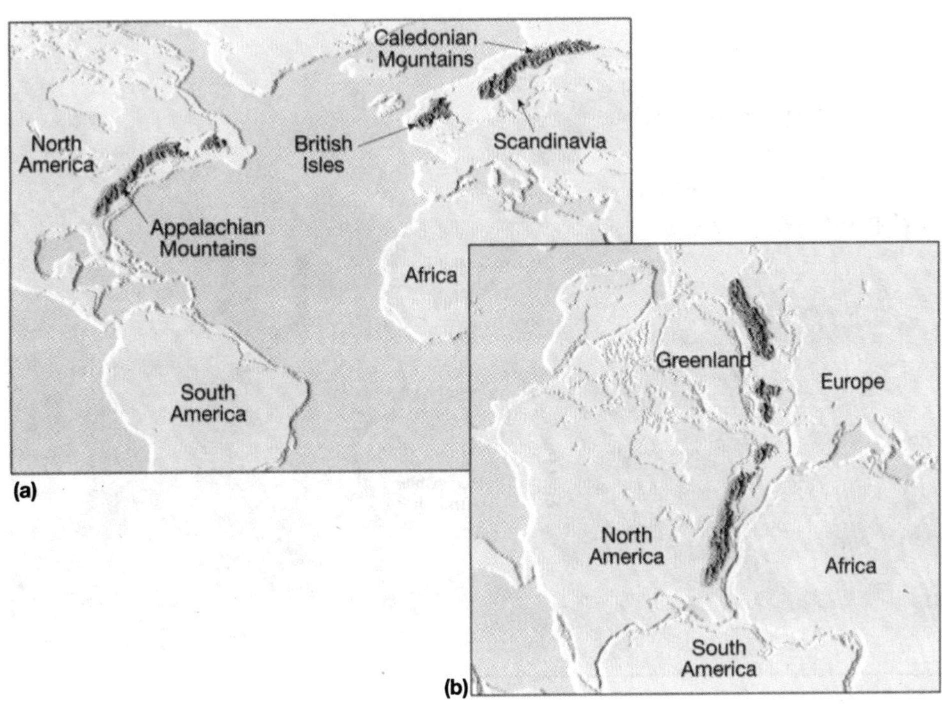

◀ **Figure 14-3** Mountain ranges match on both sides of the North Atlantic Ocean. (a) The Appalachian Mountains are of similar age and structure to those found in the British Isles and Scandinavia. (b) Before Pangaea rifted apart, these mountain belts formed a nearly continuous chain.

Plate Tectonics

Despite the questions about the validity of continental drift, continuing research revealed more and more about crustal mechanics throughout the middle of the twentieth century.

The Evidence

1. Plate Boundaries
2. Seafloor Spreading
3. Paleomagnetism
4. Convection and Plate Tectonics

Among the many gaps in scientific knowledge at the time of Alfred Wegener was an understanding of the dynamics of the ocean floors. By the 1950s, geologists, geophysicists, seismologists, oceanographers, and physicists had accumulated a large body of data about the ocean floor and the underlying crust.

One of the most intriguing early findings came when thousands of depth soundings from the oceans of the world were used to construct a detailed map of ocean floor topography (Figure 14-6). The result was remarkable: vast abyssal plains were seen dotted with chains of undersea

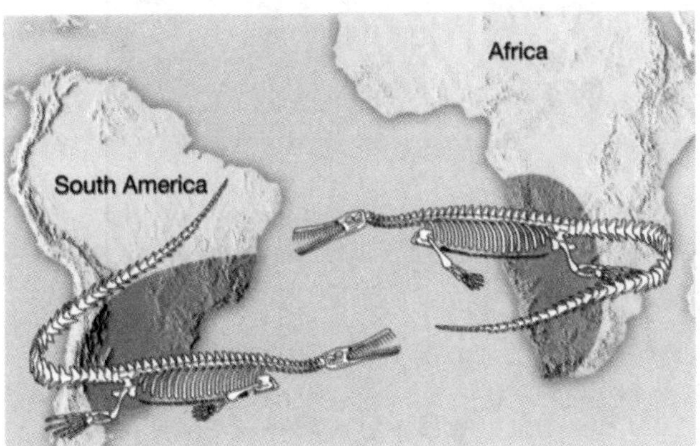

▲ **Figure 14-4** *Mesosaurus* fossils have been found in southeastern South America and southwestern Africa. These animals were alive at the time that present-day South America and Africa were together as part of a larger landmass in the geologic past.

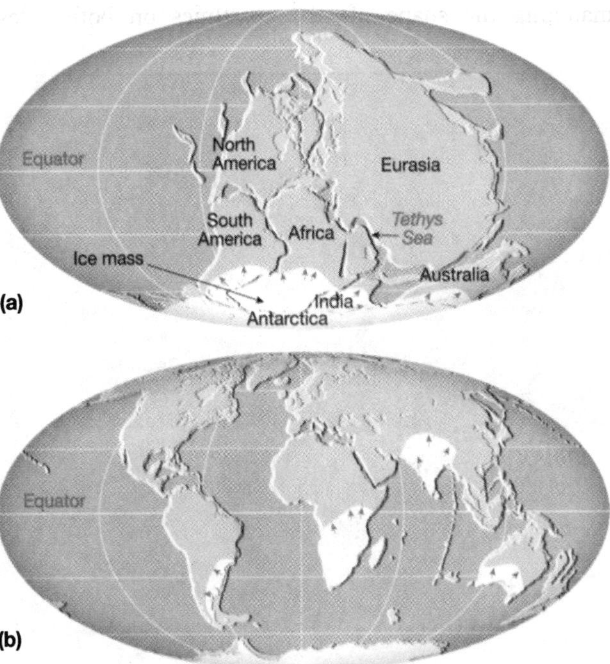

▲ **Figure 14-5** About 300 million years ago extensive areas of Pangaea were glaciated. When the continents are placed back in their former positions, the patterns of ice movement and glacial deposition make sense.

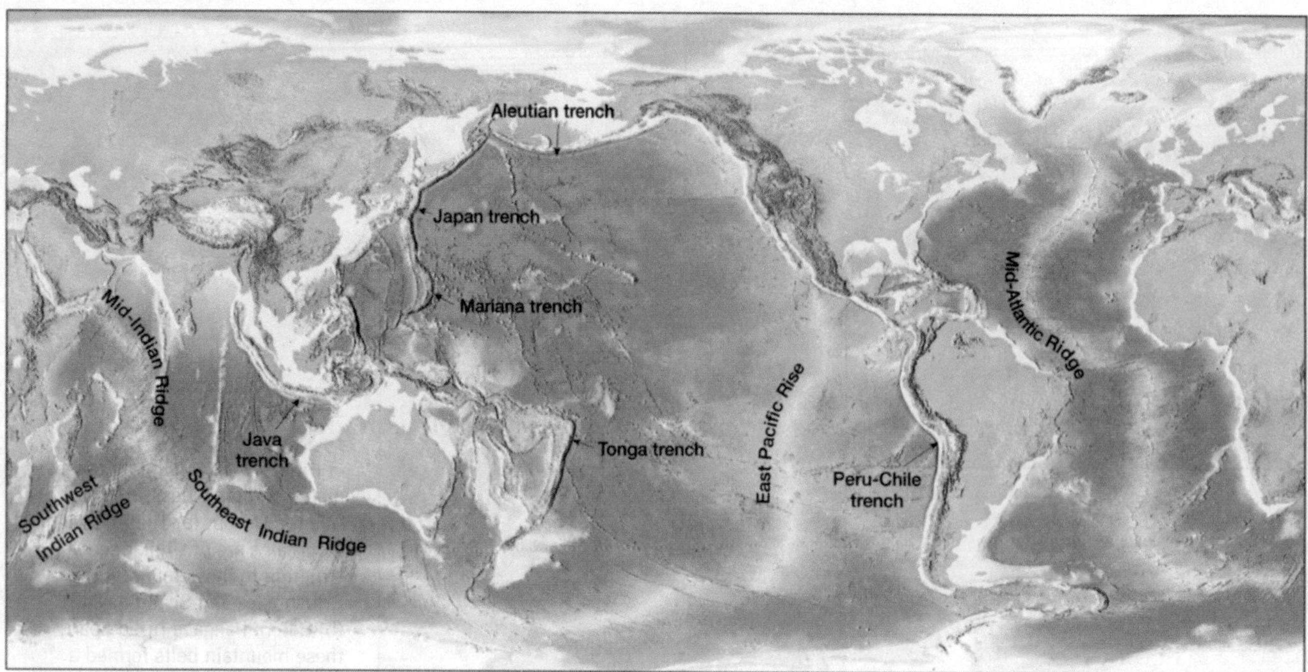

▲ **Figure 14-6** A continuous system of ridges—including the Mid-Atlantic Ridge, the East Pacific Rise, and the Southwest- and Southeast Indian Ridges—runs across the floor of the world ocean. In addition to the ridges, this map also shows the major oceanic trenches.

volcanoes known as *seamounts*. Narrow, deep *oceanic trenches* occurred in many places, often around the margins of the ocean basins. Perhaps most stunning of all was a continuous ridge system running across the floors of all the oceans for 64,000 kilometers (40,000 miles)—wrapping around the globe like the stitching on a baseball. The mid-Atlantic segment of this *midocean ridge system* is especially striking, running exactly halfway between—and matching the shape of—the coastlines on both sides, almost as if a giant seam had opened up in the ocean floor between the continents.

By the 1960s a world network of seismographs was able to pinpoint the location of every significant earthquake in the world. When earthquake locations were mapped, it was clear that earthquakes do not occur randomly around the world; instead, most earthquakes occur in bands, often coinciding with the pattern of the midocean ridge system and oceanic trenches (Figure 14-7).

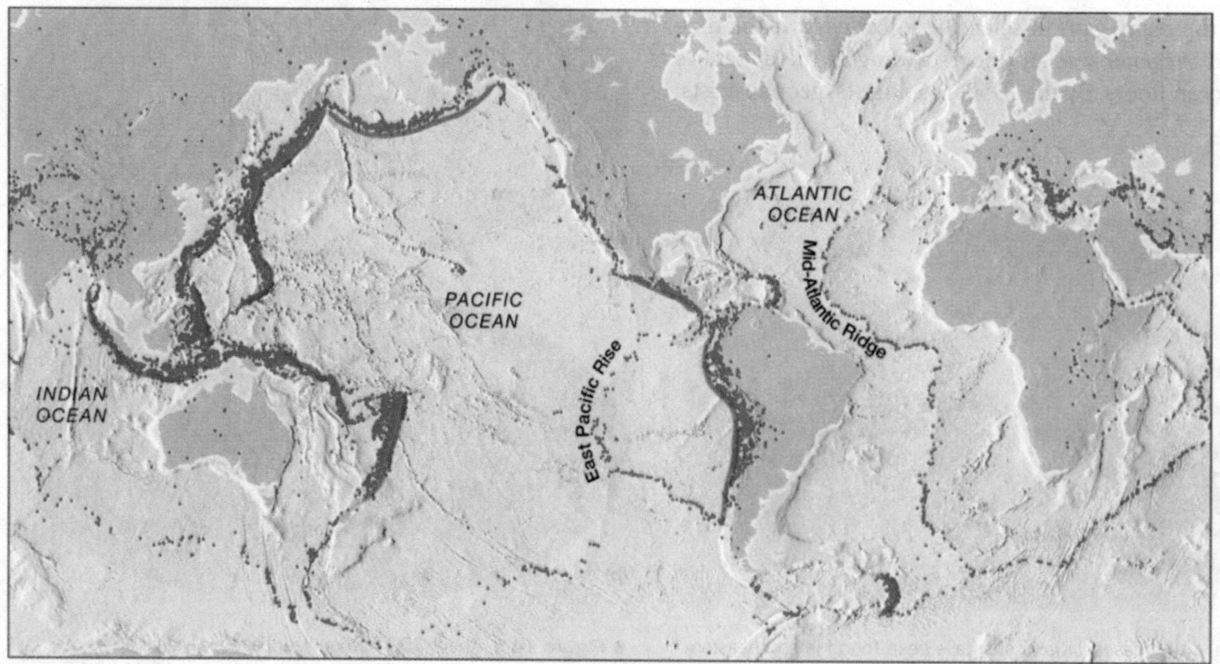

▲ **Figure 14-7** The distributions of epicenters for all earthquakes of at least 5.0 magnitude over a 10-year period. Their relationship to midocean ridges and oceanic trenches is striking.

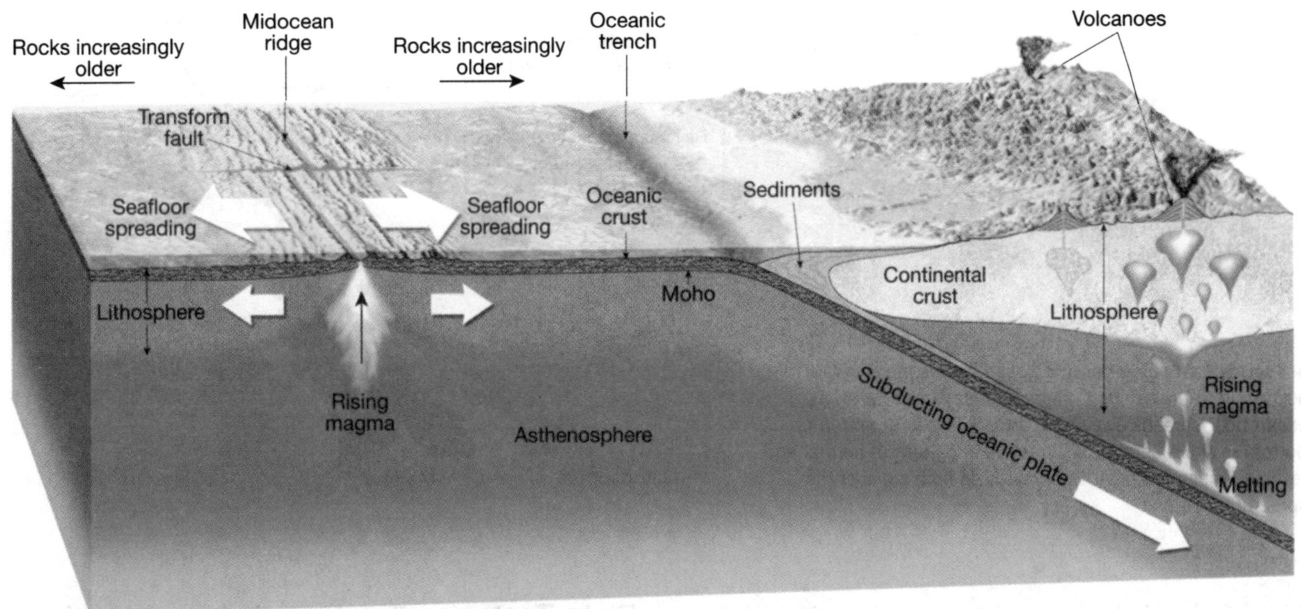

▲ **Figure 14-8** Seafloor spreading. Convection currents bring magma from the asthenosphere up through fissures in the oceanic lithosphere at the midocean ridge. The cooled and therefore solidified magma becomes a new portion of ridge along the ocean floor and the two sides of the ridge spread away from each other, as indicated by the arrows. Where denser ocean lithosphere converges with less-dense continental lithosphere, the oceanic plate slides under the continental plate in a process called subduction. Magma produced by this subduction rises to form volcanoes and igneous intrusions.

Seafloor Spreading In the early 1960s, a new theory was propounded, most notably by the American oceanographer Harry Hess and geologist Robert S. Dietz, that could explain the significance of the midocean ridges, the oceanic trenches, the pattern of earthquakes—and could provide a possible mechanism for Wegener's continental drift. Known as **seafloor spreading**, this theory stated that midocean ridges are formed by currents of magma rising up from the mantle; volcanic eruptions create new basaltic ocean floor, that then spreads away laterally from the ridge (Figure 14-8). Thus, the midocean ridges contain the newest crust formed on the planet. At other places in the ocean basin—at the oceanic trenches—older lithosphere descends into the asthenosphere in a process called *subduction*, where it is ultimately "recycled." The amount of new seafloor created is compensated for by the amount lost at subduction zones.

Verification of Seafloor Spreading The validity of seafloor spreading was confirmed most notably by two lines of evidence: paleomagnetism and ocean floor core sampling.

When any rock containing iron is formed—such as iron-rich ocean floor basalt—it is magnetized so that the iron-rich grains become aligned with Earth's magnetic field. This orientation then becomes a permanent record of the polarity of Earth's magnetic field at the time the rock solidified. Over the last 100 million years, for reasons that are not fully understood, the polarity of Earth's magnetic field has reversed itself more than 170 times—with the north magnetic pole becoming the south magnetic pole.

In 1963, Fred Vine and D.H. Matthews used **paleomagnetism** to test the theory of seafloor spreading by studying paleomagnetic data from a portion of the midocean ridge system. If the seafloor has spread laterally by the addition of new crust at the oceanic ridges, there should be a relatively symmetrical pattern of magnetic orientation—normal polarity, reversed polarity, normal polarity, and so on—on both sides of the ridges (Figure 14-9). Such was found to be the case, as Figure 14-10 clearly shows.

Final confirmation of seafloor spreading was obtained from core holes drilled into the ocean floor by the research ship, the *Glomar Challenger* in the late 1960s. Several thousand **ocean floor cores** of sea-bottom sediments were analyzed, and it was evident from this work that, almost invariably, sediment thickness and age increase with increasing distance from the midocean ridges, indicating that sediments farthest from the ridges are oldest. At the ridges, ocean floor material is almost all igneous, with little accumulation of sediment—any sediment near the ridges is thin and young.

Thus, the seafloors can be likened to gigantic conveyor belts, moving ever outward from the midocean ridges toward the trenches. Oceanic lithosphere has a relatively short life at Earth's surface. New crust is formed at the oceanic ridges, and within 200 million years is returned to the mantle by subduction. Because lower density continental lithosphere cannot be subducted, once it forms it is virtually permanent. The continual recycling of oceanic crust means that its average age is only about 100 million years, whereas the average age of continental crust is

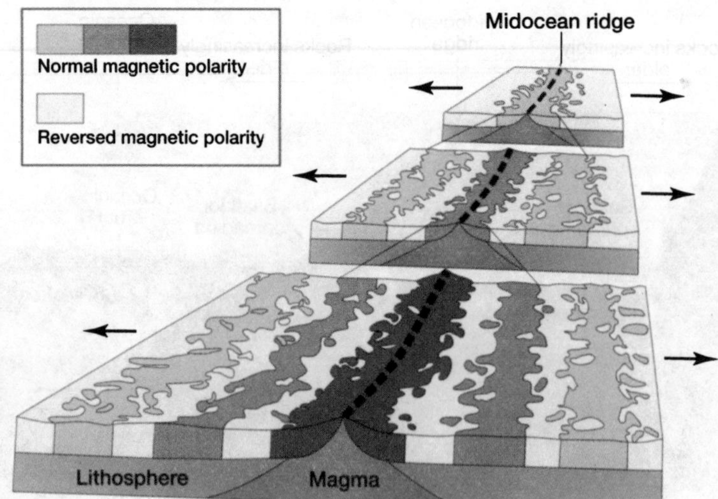

▶ **Figure 14-9** New basaltic ocean floor is magnetized according to the existing magnetic field of the Earth. As ocean floor spreads away from a ridge (shown from top to bottom in the diagram), a symmetrical pattern of normal and reversed magnetic polarity develops on both sides of the spreading center.

20 times that. Indeed, some fragments of continental crust have been discovered that are more than 4 billion years old—nearly nine-tenths of the age of Earth!

So, as it turns out, Alfred Wegener was wrong about one important detail in his theory of continental drift: it is not just the continents that are drifting. The continents are embedded in the thicker lithospheric plates, carried along by the action of seafloor spreading.

Plate Tectonics By 1968, on the basis of these details and a variety of other evidence, the theory of **plate tectonics**, as it had become known, was being accepted by the scientific community. Plate tectonics provides a framework with which we can understand and relate a wide range of internal processes and topographic patterns around the world.

The lithosphere is a mosaic of rigid lithospheric plates floating over the underlying plastic asthenosphere

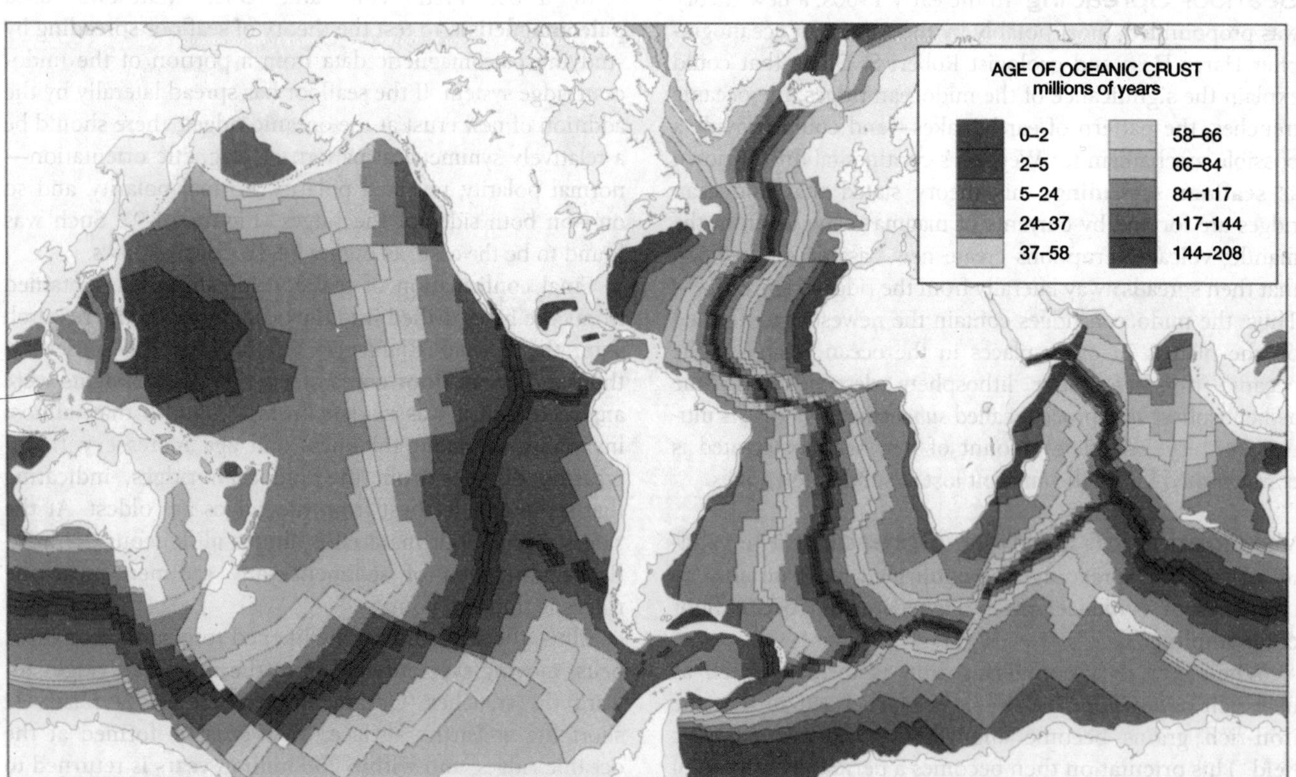

▲ **Figure 14-10** The age of the ocean floors. The patterns of magnetic reversals recorded in the ocean floor as it spreads away from midocean ridges have helped establish the age of oceanic crust. The youngest ocean floor is found at midocean ridges, while ocean floor farther from the ridges is the oldest. (*Note:* The length of time shown is not the same in each age category.)

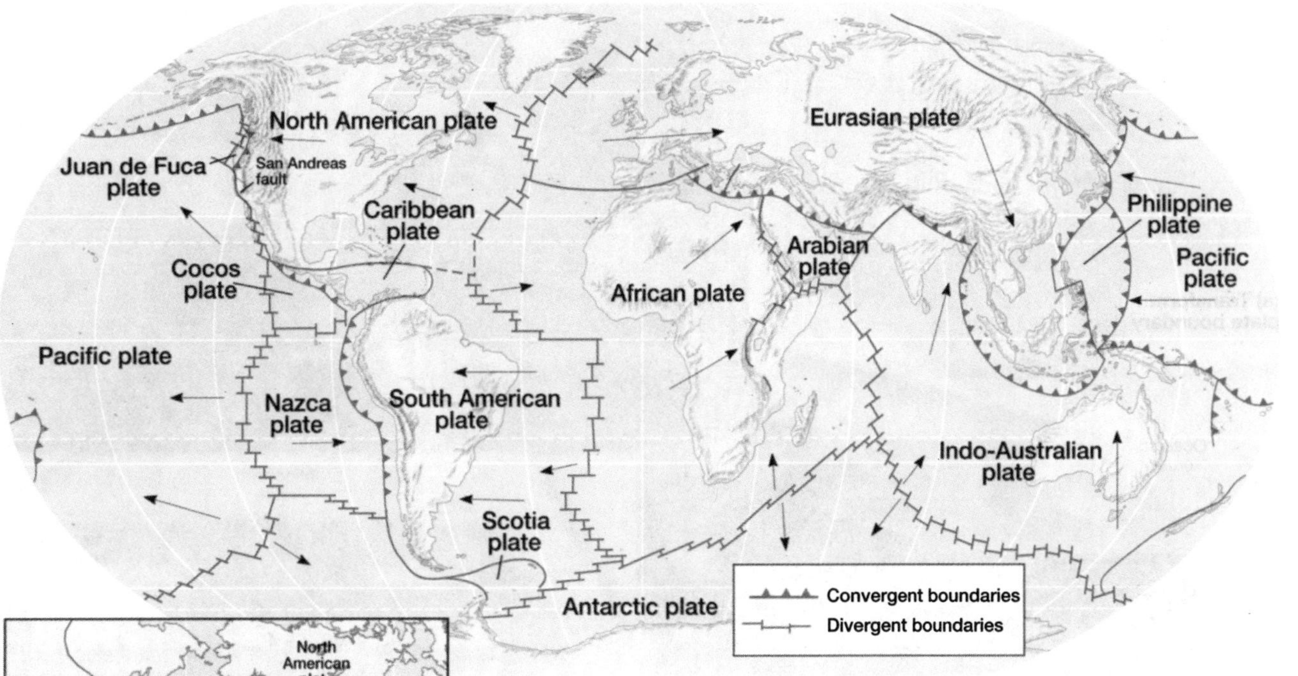

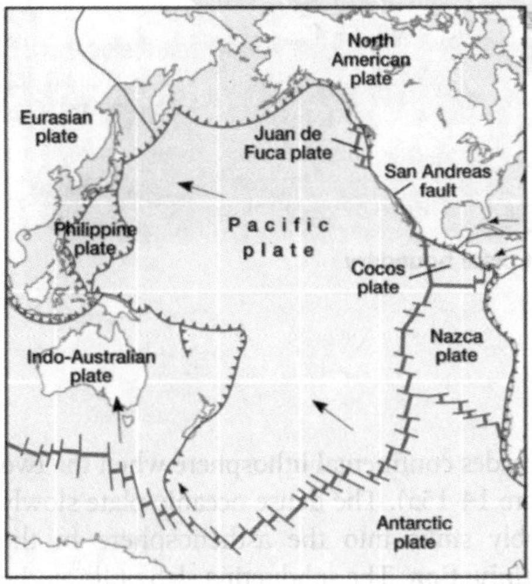

◀ **Figure 14-11** The major contemporary tectonic plates and their generalized direction of movement.

(Figure 14-11). The plates vary considerably in area: some are almost hemispheric in size, whereas others are much smaller. The number of plates and some of their boundaries are not completely clear. Seven major plates, an equal number of intermediate-sized plates, and perhaps a dozen smaller plates, are recognized. Many of the smaller plates are remnants of once larger plates that are now being subducted. These plates are about 65 to 100 kilometers (40 to 60 miles) thick, and most consist of both oceanic and continental crust.

The driving mechanism for plate tectonics is thought to be convection within Earth's mantle (the process of convection was discussed in Chapter 4 in the context of atmospheric processes). A very sluggish thermal convection system appears to be operating within the planet, bringing deep-seated heated rock slowly to the surface. Plates may be "pushed" away from midocean ridges to a certain extent, but it appears that much of the motion is a result of the plates being "pulled" along by the subduction of dense oceanic lithosphere down into the asthenosphere. The complete details of thermal convection within the mantle and the ultimate fate of subducted plates remain to be worked out.

These plates move slowly over the asthenosphere. The rates of seafloor spreading vary from less than 1 cm (0.4 in.) per year in parts of the Mid-Atlantic Ridge to as much as 10 cm (4 in.) per year in the Pacific-Antarctic Ridge.

Plate Boundaries

1. Divergent Boundaries
2. Subduction Zones
3. Collision of India with Eurasia
4. Transform Faults and Boundaries

The plates are rigid and therefore deformed significantly only at the edges and only where one plate impinges on another. Most of the "action" in plate tectonics takes place along such plate boundaries. Only three types of contact between plates are possible: two plates may diverge from one another, converge toward one another, or slide laterally past one another (Figure 14-12).

Divergent Boundaries At a **divergent boundary**, magma from the asthenosphere wells up in the opening between plates. This upward flow of molten material produces a line of volcanic vents that spill out basaltic lava onto the ocean floor, with the plutonic rock gabbro solidifying deeper below. A divergent boundary is usually represented by a **midocean ridge** (Figure 14-13). Most of the midocean ridges of the world are either active or extinct spreading ridges. Such *spreading centers* are associated with *shallow-focus earthquakes* (meaning that the ruptures that generate

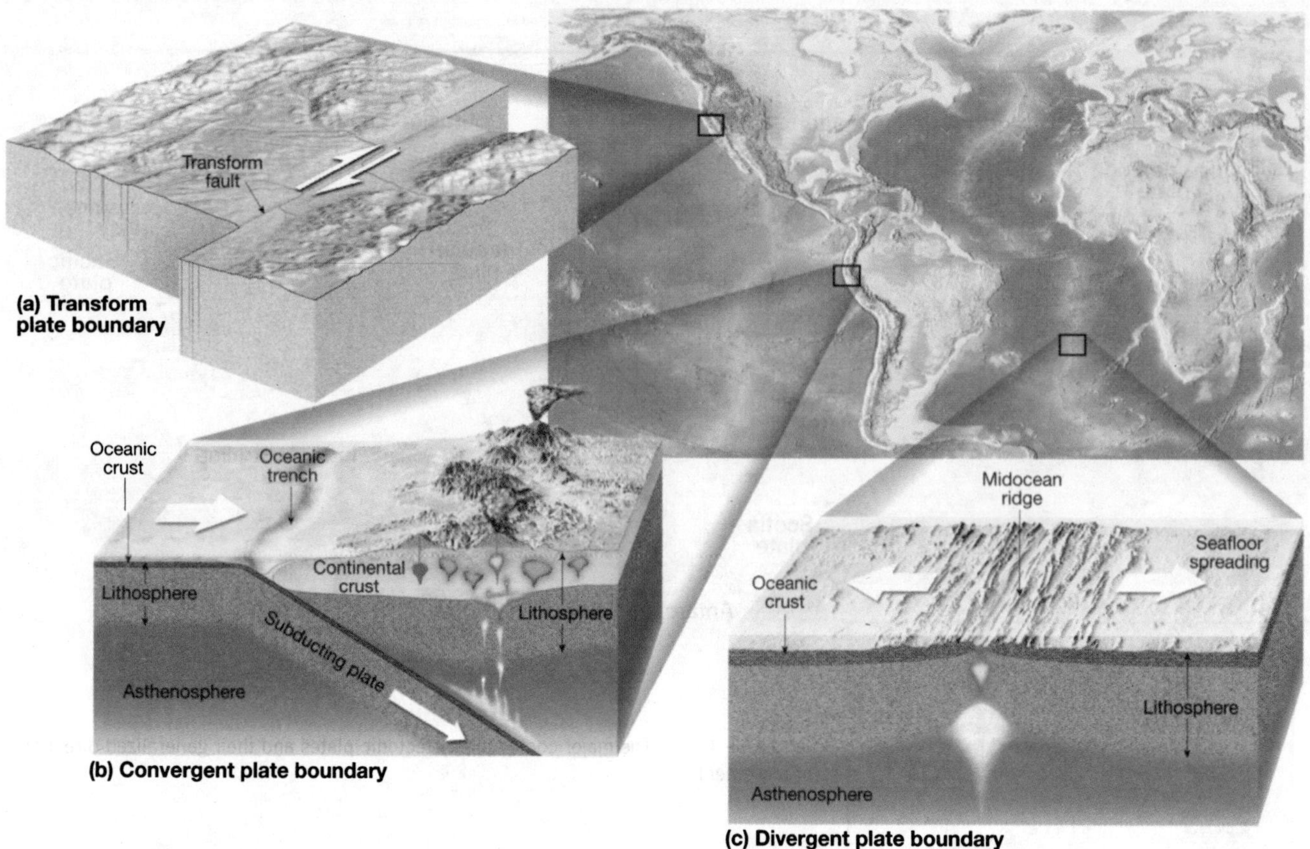

▲ **Figure 14-12** Three kinds of plate boundaries. The edges of lithospheric plates move apart at divergent boundaries, come together at convergent boundaries, and slide past each other along transform boundaries.

the earthquakes are within about 70 kilometers [45 miles] of the surface), volcanic activity, and hydrothermal metamorphism. Divergent boundaries are said to be "constructive" because material is being added to the crustal surface at such locations.

Divergent boundaries can also develop within a continent (Figure 14-14), resulting in a **continental rift valley** such as the great East African Rift Valley that extends from Ethiopia southward through Mozambique. The Red Sea is also the outcome of spreading taking place within a continent—in this case the spreading has been great enough to form a "proto-ocean."

Convergent Boundaries At a **convergent boundary**, plates collide and as such are sometimes called "destructive" boundaries because they result in removal or compression of the surface crust (Figure 14-15). Convergent plate boundaries are responsible for some of the most massive and spectacular of earthly landforms: major mountain ranges, volcanoes, and oceanic trenches. There are three types of convergent boundary: *oceanic–continental convergence*, *oceanic–oceanic convergence*, and *continental–continental convergence.*

1. **Oceanic–continental convergence:** Because oceanic lithosphere includes dense basaltic crust, it is denser than continental lithosphere, and so oceanic lithosphere always underrides continental lithosphere when the two collide (Figure 14-15a). The dense oceanic plate slowly and inexorably sinks into the asthenosphere in the process of **subduction**. The subducting slab pulls on the rest of the plate—such "slab pull" is probably the main cause of most plate movement, pulling the rest of the plate in after itself, as it were. Wherever such an oceanic–continental convergent boundary exists, a mountain range is formed on land (the Andes range of South America is one notable example; the Cascades in northwestern North America is another) and a parallel **oceanic trench** develops as the seafloor is pulled down by the subducting plate.

   Earthquakes take place along the margin of a subducting plate. Shallow-focus earthquakes are common at the trench, but as the subducting plate descends into the asthenosphere, the earthquakes become progressively deeper, with some subduction zones generating earthquakes as deep as 600 kilometers (375 miles) below the surface.

   Volcanoes develop from magma generated in the subduction zone. Early researchers thought that a subducted plate would melt when pushed down into the hot asthenosphere. However, more recent research indicates that such a result is unlikely. Oceanic crust is relatively cold when it approaches a subduction zone and would take a long time to become hot

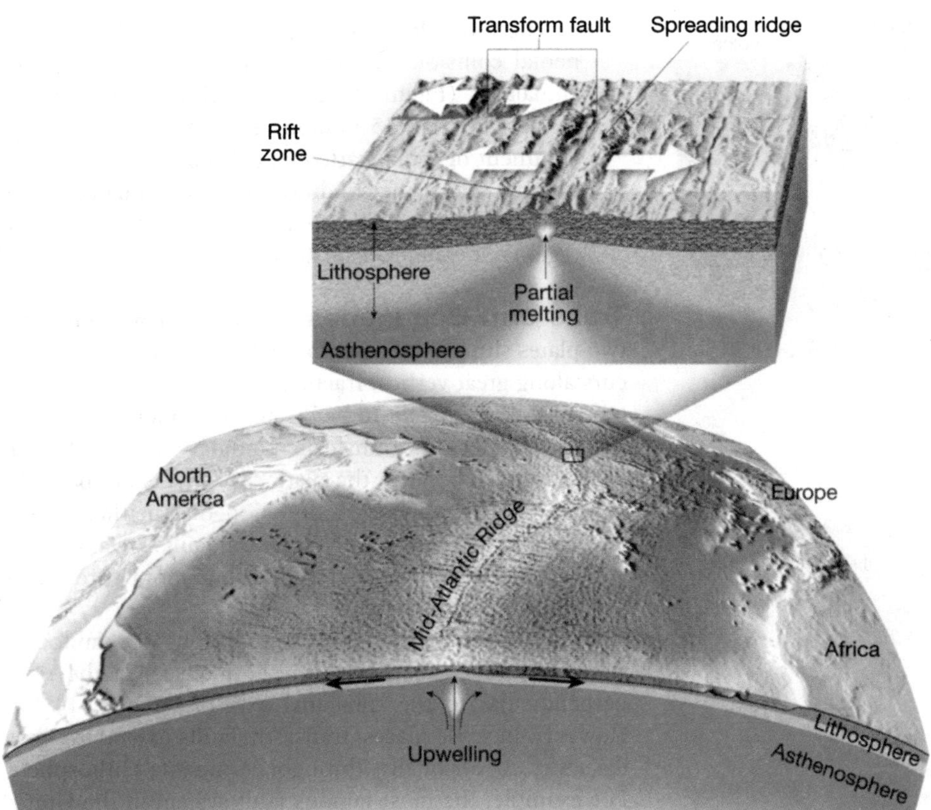

◄ **Figure 14-13** Midocean ridge spreading center. Seafloor spreading involves the rise of magma from within Earth and the lateral movement of new ocean floor away from the zone of upwelling. This gradual process moves the older material farther away from the spreading center as it is replaced by newer material from below. Transform faults are found along the short offsets associated with slight bends in the ridge system.

enough to melt. A more likely explanation is that beginning at a depth of about 100 kilometers (about 60 miles) water is driven off from the oceanic crust as it is subducted, and this water lowers the melting point of the mantle rock above, causing it to melt. This magma rises through the overriding plate, producing both extrusive and intrusive igneous rocks. The chain of volcanoes that develops in association with an oceanic-continental plate subduction zone is sometimes referred to as a *continental volcanic arc*. As we will see later in this chapter, such subduction zone volcanoes frequently erupt explosively.

Metamorphic rocks often develop in association with subduction zones. The margin of a subducting oceanic plate is subjected to increasing pressure, although relatively modest heating, as it begins to descend—this can lead to the formation of high-pressure, low-temperature metamorphic rocks, such as *blueschist*. In addition, the magma generated in the subduction zone may cause contact metamorphism as it rises through the overlying continental rocks.

2. **Oceanic–oceanic convergence**: If the convergent boundary is between two oceanic plates, subduction also takes place, as in Figure 14-15b. As one of the

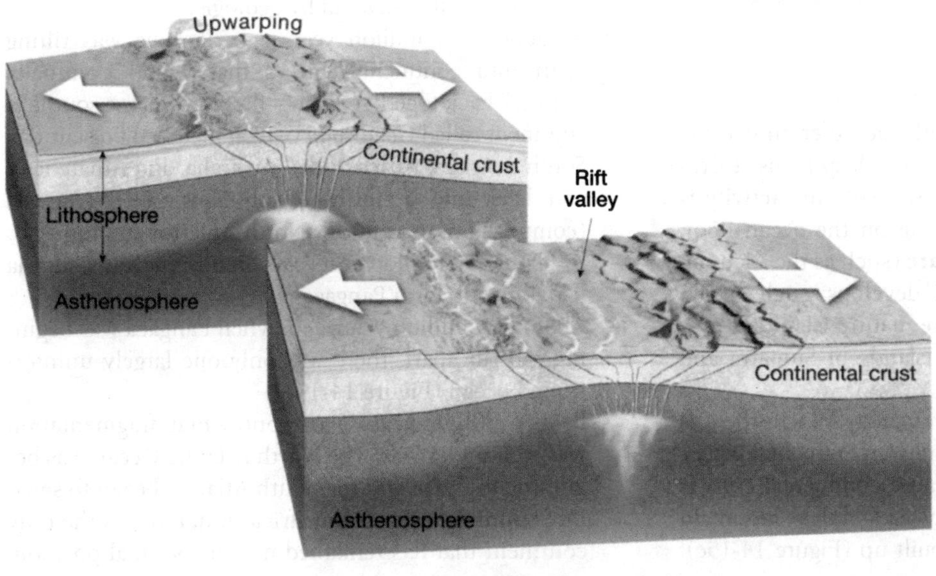

◄ **Figure 14-14** A continental rift valley develops where divergence takes place within a continent. As spreading proceeds, blocks of crust drop down to form a rift valley.

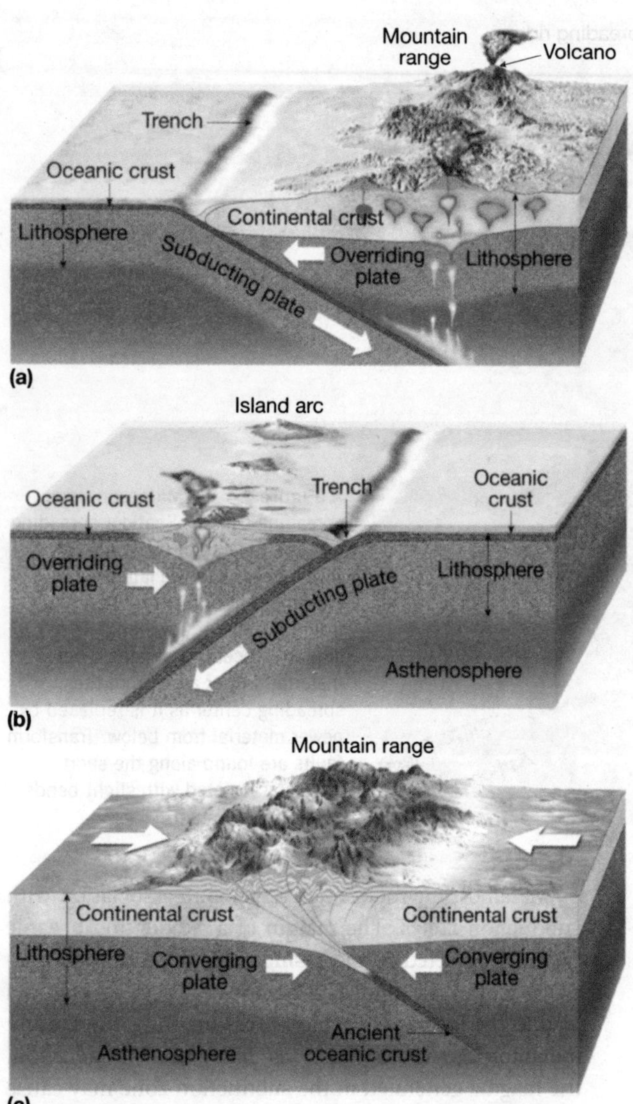

▲**Figure 14-15** Idealized portrayals of convergent plate boundaries: (a) Where an oceanic plate converges with a continental plate, the oceanic plate is subducted and a oceanic trench and coastal mountains with volcanoes are usually created. (b) Where an oceanic plate converges with another oceanic plate, an oceanic trench and volcanic island arc result. (c) Where a continental plate collides with a continental plate no subduction takes place, but mountains are generally thrust upward.

oceanic plates subducts beneath the other an oceanic trench is formed, shallow- and deep-focus earthquakes occur (Figure 14-16), and volcanic activity is initiated with volcanoes forming on the ocean floor. With time, a **volcanic island arc** (such as the Aleutian Islands and Mariana Islands) develops; such an arc may eventually become a more mature island arc system (such as Japan and the islands of Sumatra and Java in Indonesia are today).

3. **Continental–continental convergence**: Where there is a convergent boundary between two continental plates, no subduction takes place because continental crust is too buoyant to subduct. Instead, huge mountain ranges, such as the Alps, are built up (Figure 14-15c).

The most dramatic present-day example of continental collision has resulted in the formation of the Himalayas (Figure 14-17). The Himalayas began to form more than 45 million years ago, when the subcontinent of India started its collision with the rest of Eurasia. Under the conditions of continental collision, volcanoes are rare, but shallow-focus earthquakes and regional metamorphism are common.

Transform Boundaries At a **transform boundary**, two plates slip past one another laterally. This slippage occurs along great vertical fractures called *transform faults* (a type of *strike-slip fault* to be discussed in more detail later in this chapter). A transform boundary is classified as "conservative" because the plate movements are basically parallel to the boundary, a situation that neither creates new crust nor destroys old. Transform faults are associated with a great deal of seismic activity, commonly producing shallow-focus earthquakes.

Most transform faults are found along the midocean ridge system where they form short offsets in the ridge perpendicular to the spreading axis (see Figure 14-13). However, in some places, transform faults extend for great distances, occasionally through continental lithosphere. For example, the most famous fault system in the United States, the San Andreas Fault in California, is on a transform boundary between the Pacific and North American plates (Figure 14-18).

The Rearrangement

Assembly and Breakup of Pangaea

Plate tectonics provides us with a grand framework for understanding the extensive lithospheric rearrangement that has taken place during the history of Earth. A brief summary of major events might highlight the following:

- Between about 1.1 billion and 800 million years ago—before Pangaea existed—there was an earlier supercontinent, called *Rodinia* by geologists.
- By about 700 million years ago, Rodinia was rifting apart into continental pieces that would eventually "suture" back together again—first into a large southern continent called *Gondwana* (which included present-day South America, Africa, India, Australia, and Antarctica), and later into a northern continent called *Laurasia* (comprised of present-day North America and Eurasia). By about 250 million years ago, Gondwana and Laurasia had joined to form Pangaea.
- About 200 million years ago, when Pangaea was beginning to rift apart, there was only one largely uninterrupted ocean (Figure 14-19).
- By 90 million years ago, continental fragmentation was well under way. The North Atlantic Ocean was beginning to open, and the South Atlantic began to separate South America from Africa. Antarctica is the only continent that has remained near its original position.

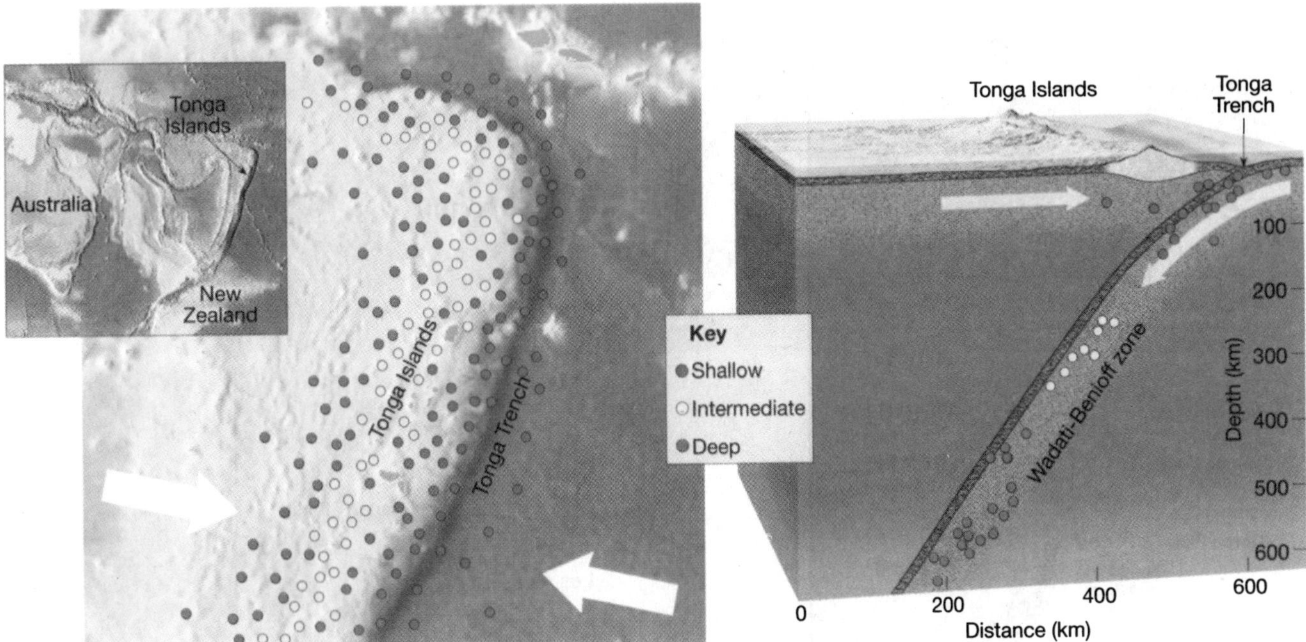

▲ **Figure 14-16** Earthquake patterns associated with the Tonga Trench subduction zone (shown in a map view on the left and a side view on the right). Shallow-focus earthquakes occur where the Pacific Plate begins to subduct at the trench. Intermediate- and deep-focus earthquakes occur as the subducting oceanic plate goes deeper into the asthenosphere below. The *Wadati-Benioff zone* is named for seismologists Kigoo Wadati and Hugo Benioff who were the first scientists to describe these inclined zones of earthquakes.

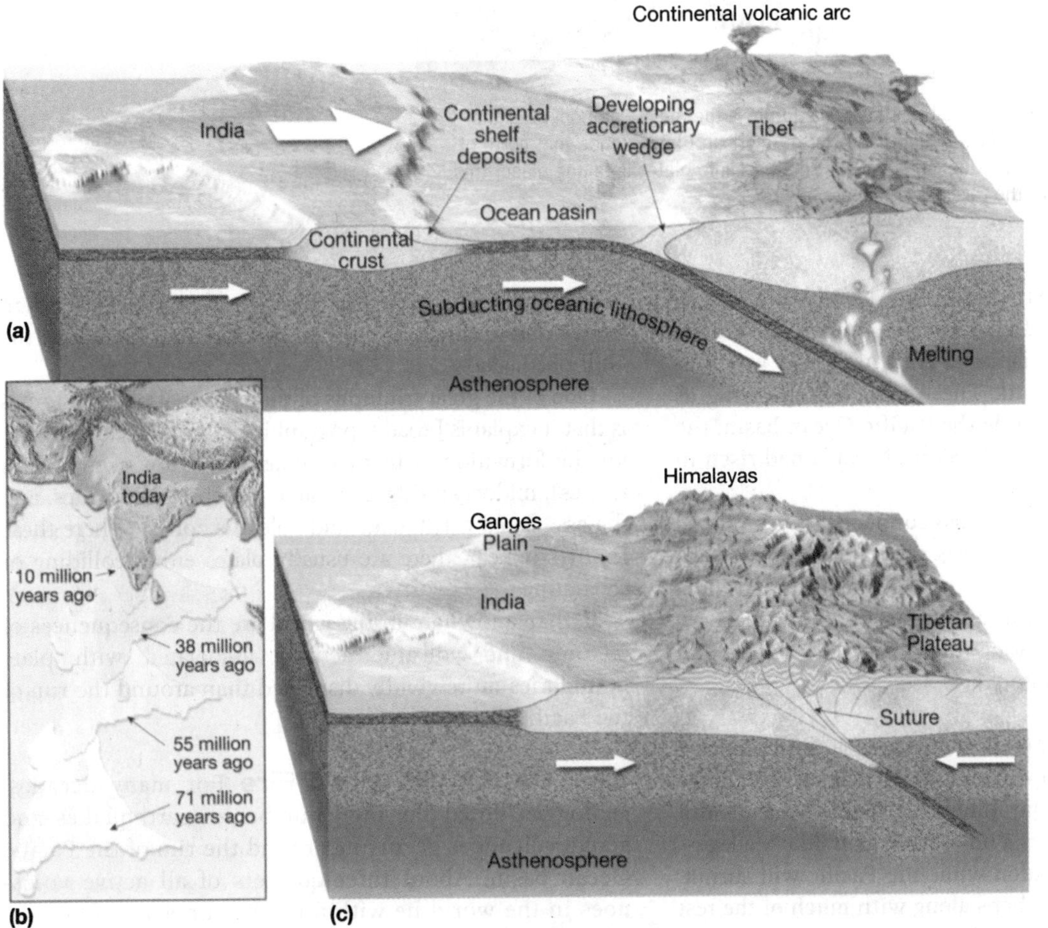

◀ **Figure 14-17** The collision of the subcontinent of India with Eurasia began about 45 million years ago. This collision and continental "suture" has uplifted the Himalayas and the Tibetan Plateau.

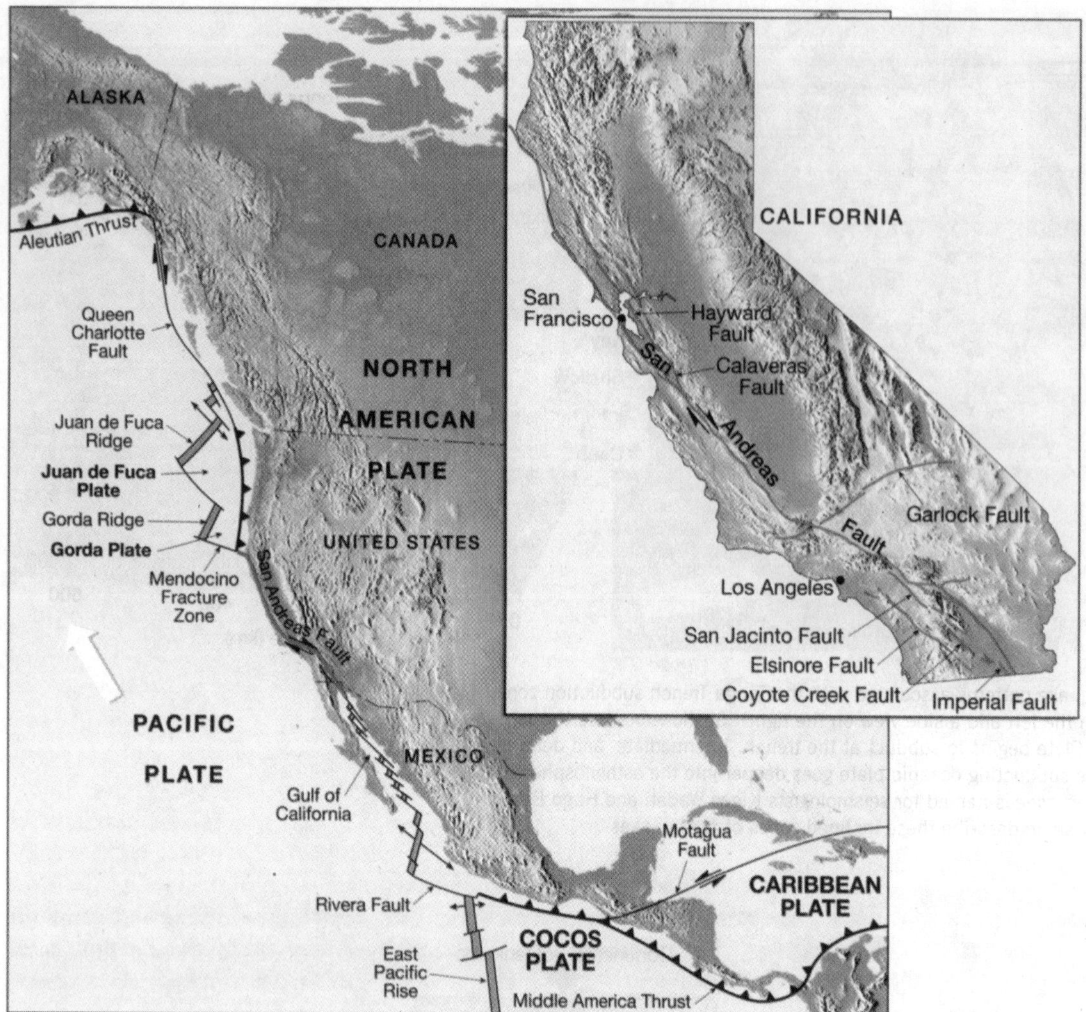

▲ **Figure 14-18** The San Andreas Fault system of California is the major component of the transform boundary between the Pacific Plate and North American Plate. Notice that north of the San Andreas Fault system, subduction of several small oceanic plates is taking place; this subduction is associated with the Cascade volcanoes.

- By 50 million years ago, the North and South Atlantic Oceans had both opened, and South America was a new and isolated continent that was rapidly moving westward. The Andes were growing as South America overrode the Pacific Ocean basin; the Rockies and the ancestral Sierra Nevada had risen in North America.
- Today, South America has connected with North America. North America has separated from western Eurasia, Australia has split from Antarctica, and India has collided with Eurasia to thrust up the Himalayas. Africa is splitting along the Great Rift Valley and slowly rotating counterclockwise.

If current plate movement continues, 50 million years into the future Australia will straddle the equator as a huge tropical island. Africa may pinch the Mediterranean shut while East Africa becomes a new large island like Madagascar. The Atlantic will widen while the Pacific will shrink. Southern California—perhaps along with much of the rest of the state—will slide past the rest of North America en route to its ultimate destiny in the Aleutian Trench in the Gulf of Alaska.

One of the great triumphs of the theory of plate tectonics is that it explains broad topographic patterns. It can account for the formation of many *cordilleras* (groups of mountain ranges), midocean ridges, oceanic trenches, island arcs, and the associated earthquake and volcanic zones. Where these features appear, there are usually plates either colliding or separating.

Perhaps nowhere in the world are the consequences of tectonic and volcanic activity associated with plate boundaries more vividly displayed than around the rim of the Pacific Ocean.

The Pacific Ring of Fire For many decades, geologists noted the high number of earthquakes and active volcanoes occurring around the rim of the Pacific Ocean basin. About three-quarters of all active volcanoes in the world lie within the Pacific Rim, but it was

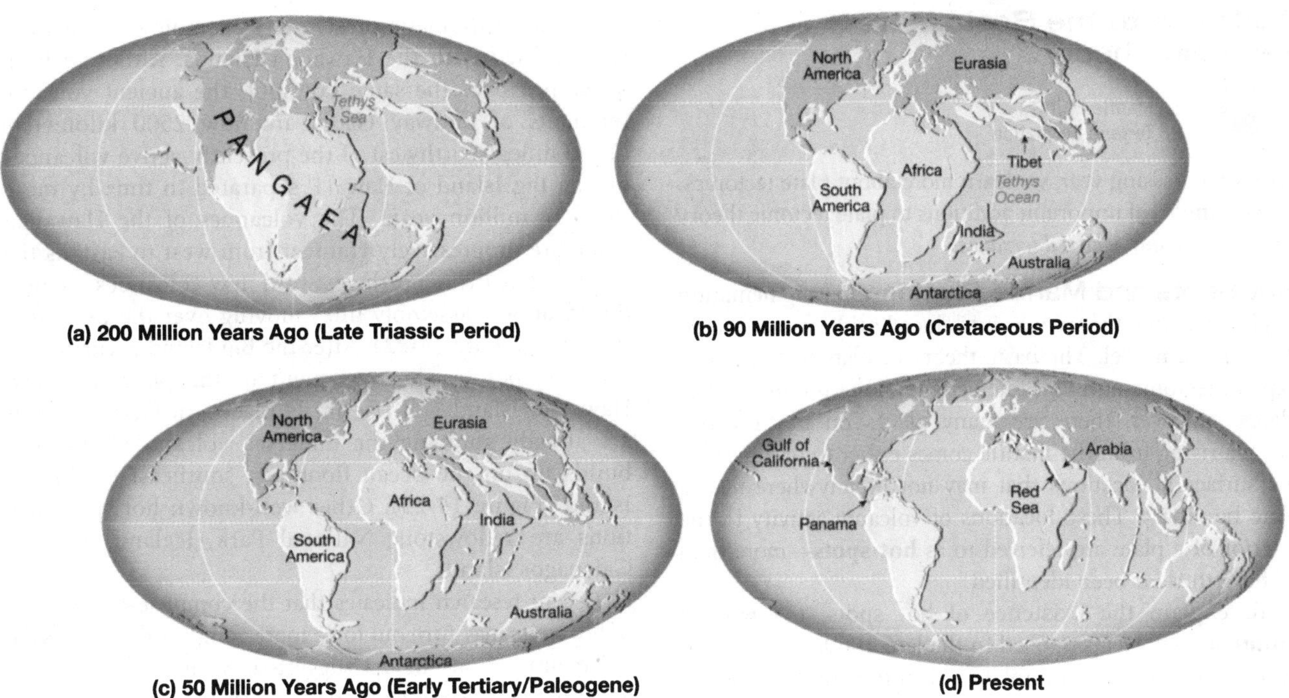

▲ **Figure 14-19** The break-up of Pangaea beginning about 200 million years ago.

only in the late 1960s that the theory of plate tectonics provided an explanation for this pattern. Plate boundaries are found all of the way around the Pacific basin—primarily subduction zones, along with segments of transform and divergent boundaries. It is along these plate boundaries that the many volcanoes and earthquakes take place in what is now called the **Pacific Ring of Fire** (Figure 14-20).

The Pacific Rim is home to millions of people. Active or potentially active volcanoes and major faults systems are within sight of some of the largest metropolitan regions in the world, such as Mexico City, Los Angeles, and Tokyo. In recent decades we have had many reminders of the ever-active Ring of Fire: the 1980 Mount St. Helens eruption; the 1985 Nevado del Ruiz volcano tragedy in Colombia; the 1991 eruption of Mount Pinatubo in the Philippines; the 1994 Northridge earthquake in California; the 1995 Kobe earthquake in Japan; and the December 2004 Sumatra, Indonesia, earthquake and tsunami that killed more than 270,000 people.

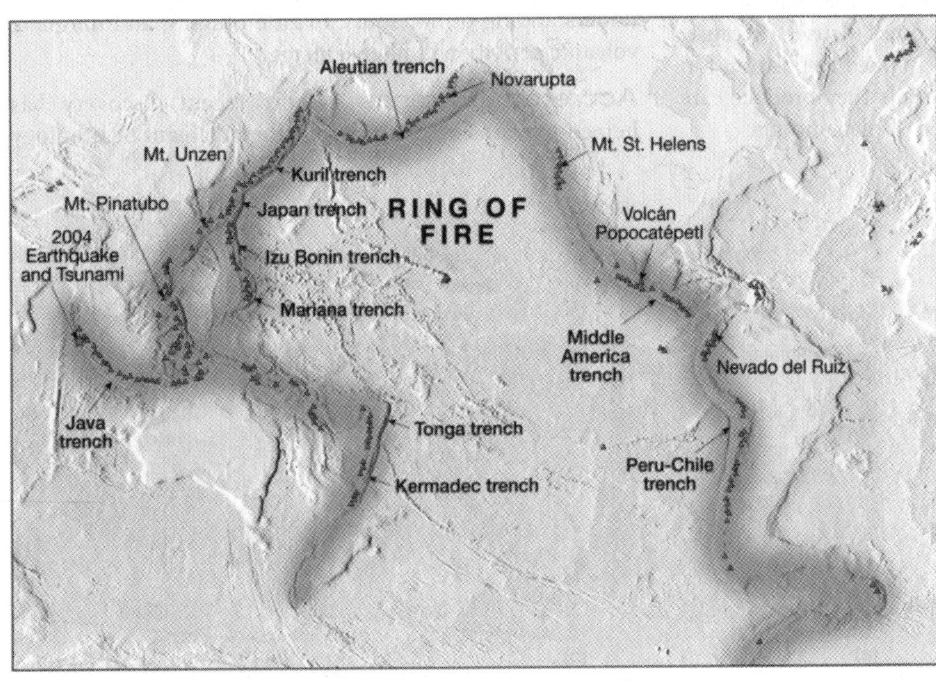

◀ **Figure 14-20** The Pacific Ring of Fire.

Additions to the Basic Plate Tectonic Theory

1. Mantle Plumes
2. Terrane Formation

With each passing year, we learn more about plate tectonics. Two examples of important additions to plate tectonic theory are *hot spots* and *accreted terranes*.

Hot Spots and Mantle Plumes One augmentation to plate tectonic theory was introduced at the same time as the original model. The basic theory of plate tectonics can explain tectonic and volcanic activity along the margins of plates, however, there are many places on Earth where magma rising from the mantle comes either to or almost to the surface at locations that may not be anywhere near a plate boundary. These locations of volcanic activity in the interior of a plate are referred to as **hot spots**—more than 50 have thus far been identified.

To explain the existence of hot spots, the **mantle plume** model was proposed in the late 1960s. This explanation suggests that midplate volcanic activity develops over narrow plumes of heated material rising through the mantle—perhaps originating as deep as the core–mantle boundary (Figure 14-21). Such mantle plumes are believed to be relatively stationary over long periods of time (in some cases, as long as tens of millions of years). As the magma rises through the plate above, it creates hot spot volcanoes and/or hydrothermal (hot water) features on the surface—often after an initial large outpouring of lava known as *flood basalt* (discussed later in this chapter).

The plate above the hot spot is moving, so the volcanoes or other hot spot features are eventually carried off the plume and become inactive, while in turn new volcanic features develop over the plume, so generating a straight-line *hot spot trail*. Volcanic islands carried off the hot spot may eventually subside to form underwater *seamounts* as the oceanic lithosphere cools and becomes denser. Because many hot spots seem to be effectively fixed in position for long periods of time, the hot spot trails they produce can indicate both the direction and speed of plate motion.

The most dramatic present-day example of a hot spot is associated with the Hawaiian Islands. Although both developed over the same hot spot, the ancient volcanic remnants of Midway Island are now 2500 kilometers (1600 miles) northwest of the presently-active volcanoes on the Big Island of Hawai'i, separated in time by more than 27 million years. The volcanoes of the Hawaiian chain are progressively younger from west to east; as the Pacific Plate drifts northwestward new volcanoes are produced on an "assembly line" moving over the persistent hot spot (Figure 14-22). After the Big Island is carried off the hot spot by the movement of the plate, the next Hawaiian island will rise in its place—in fact, scientists are already studying the undersea volcano Lō'ihi as it builds up on the ocean floor just southeast of the Big Island (Figure 14-23). Other well-known hot spot locations are Yellowstone National Park, Iceland, and the Galapagos Islands.

Recent research indicates that the complete explanation of hot spots may turn out to be more complex than the original mantle plume model suggested. *Seismic tomography*—a technique that uses earthquake waves to produce a kind of "ultrasound" of Earth—suggests that the magma source of at least some hot spots is quite shallow, while the source for others are mantle plumes originating deep from within the mantle. Further, some researchers cite evidence suggesting that several mantle plumes may have changed location in the geologic past. For example, the Emperor Seamounts—a chain of seamounts to the northwest of Midway Island—are part of the Hawaiian hot spot trail, but they appear to divert quite significantly in direction from the straight line of the rest of the Hawaiian chain (see Figure 14-22). This "bend" in the hot spot trail is due either to a significant change in direction of the Pacific Plate about 43 million years ago or to the migration of the hot spot itself—perhaps both.

As additional information is gathered, a more complete understanding of hot spots, mantle plumes, and midplate volcanic activity will likely emerge.

Accreted Terranes A more recent discovery has helped explain the often perplexing amalgam of lithology

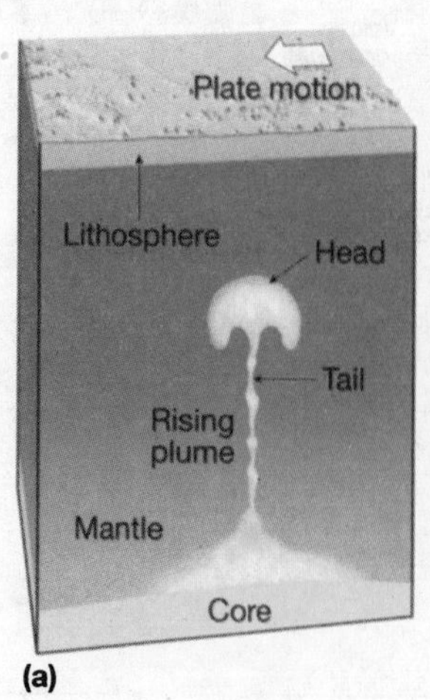

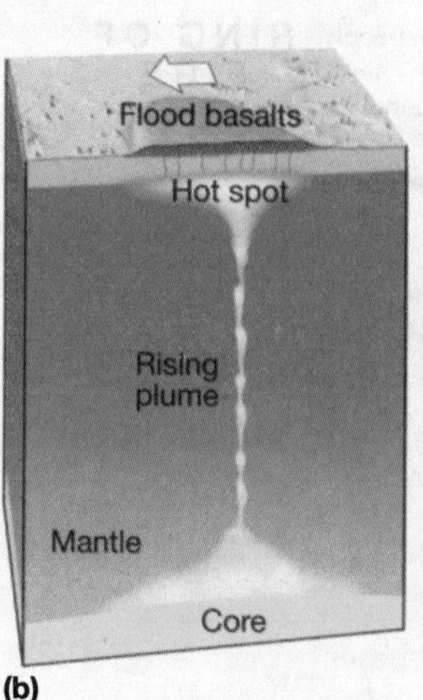

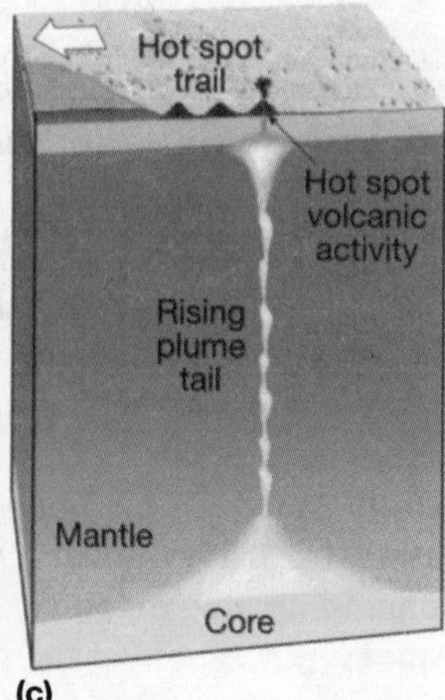

▶ **Figure 14-21** The idealized mantle plume model of hot spot origin. A plume of heated material rises from deep within the mantle. When the large head of the plume reaches the surface, an outpouring of flood basalt results. Plate motion carries the flood basalts off the stationary plume and a new volcano or volcanic island forms. As the moving plate carries each volcano off the hot spot, it becomes extinct, resulting in a straight-line "hot spot trail." As volcanic islands move off the hot spot, the plate cools, becomes denser, and subsides; some islands may eventually sink below the surface to become *seamounts*.

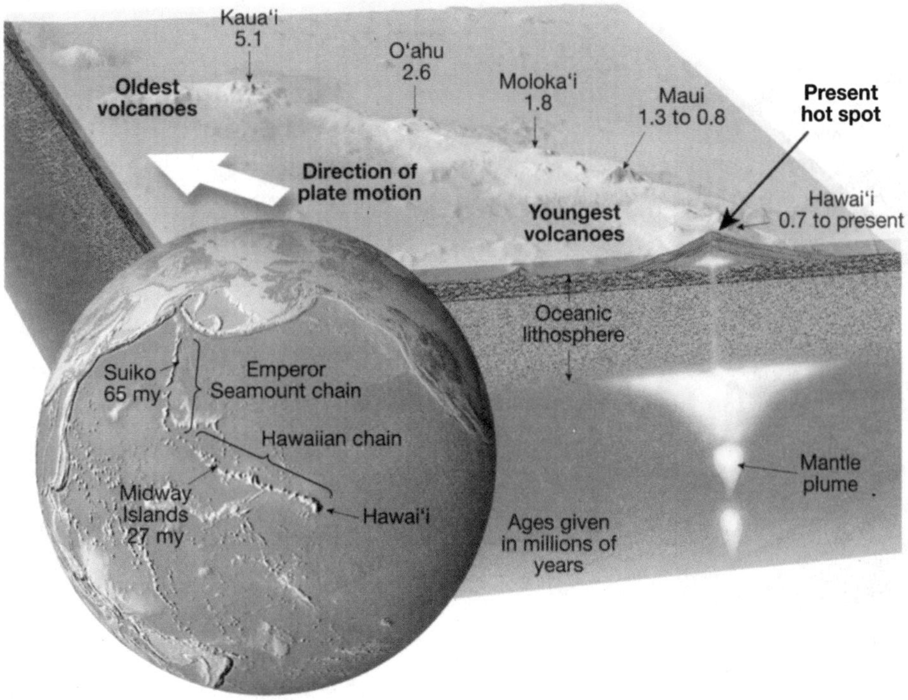

◀ **Figure 14-22** The Hawaiian hot spot. A hot spot has persisted here for many millions of years. As the Pacific Plate moved northwest, a progression of volcanoes was created and then died as their source of magma was shut off. Among the oldest is Midway Island. Later volcanoes developed down the chain. The numbers on the main islands indicate the age of the basalt that formed the volcanoes, in millions of years before the present.

seen along the margins of some continents. A **terrane** is a small-to-medium mass of lithosphere—bounded on all sides by faults—that may have been carried a long distance by a drifting plate, eventually to converge with the edge of another plate. The terrane is too buoyant to be subducted in the collision and instead is fused ("accreted") to the other plate, often being fragmented in the process. In some cases, slices of oceanic lithosphere have accreted in terranes (including the accumulated sediment in what is called the *accretionary wedge* of a subduction zone); in other cases, it appears that entire old island arcs have fused with the margin of a continent (Figure 14-24).

Terranes are distinctive geologically because their lithologic complement (types of rock) is generally quite different from that of the plate to which they are accreted. It is generally believed that every continent has grown outward by the accumulation of accreted terranes on one or more of its margins. North America is a prominent example (Figure 14-25): most of Alaska and much of western Canada and the western United States consist of a mosaic of several dozen accreted terranes, some of which have been traced to origins south of the equator.

Remaining Questions

Plate tectonic theory has advanced our understanding of the internal processes of Earth dramatically. However, a number of questions remain unanswered for the time being. For example, several major mountain ranges in North America and Eurasia are in the middle of plates rather than in boundary zones. Although the genesis of some midplate ranges, such as the Appalachians in North America and the Ural mountains in Eurasia, can be traced to continental collisions in the geologic past, other midplate mountain ranges or regions of seismic activity are not yet fully understood. Further, while convection of heated material within the mantle provides the general mechanism for plate movement, the details of heat flow within Earth and the possible relationships of mantle plumes to these overall patterns are still being worked out.

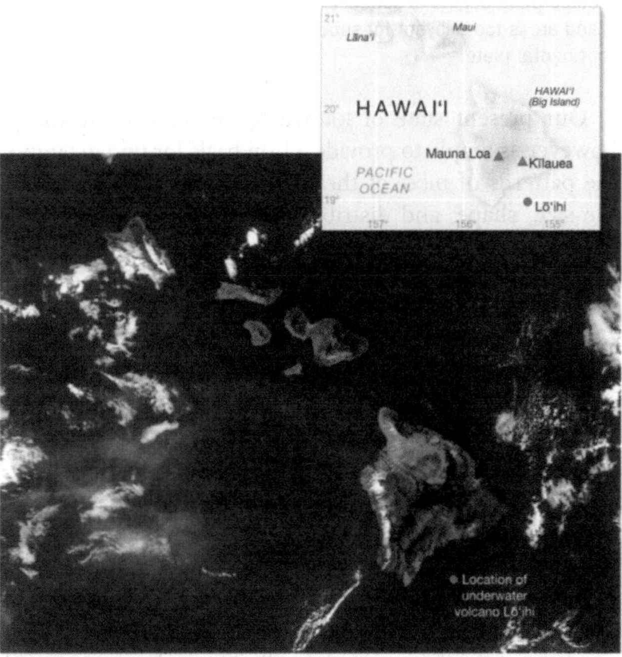

▲ **Figure 14-23** The only recently active volcanoes in the Hawaiian chain are on the Big Island of Hawai'i—Mauna Loa and Kīlauea have been repeatedly active during historic times. Lō'ihi, an underwater volcano southeast of the Big Island, is the next volcanic island being built.

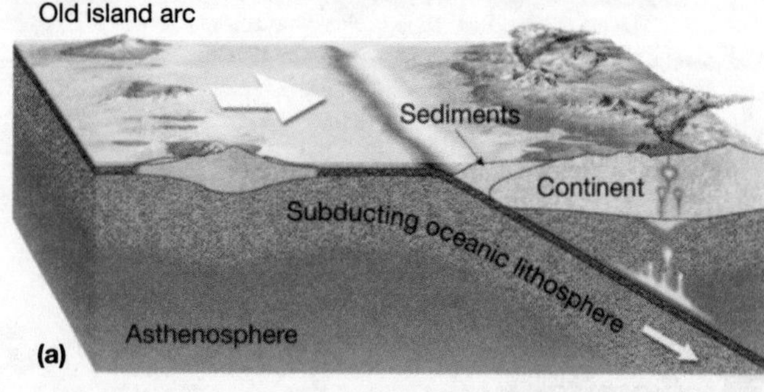

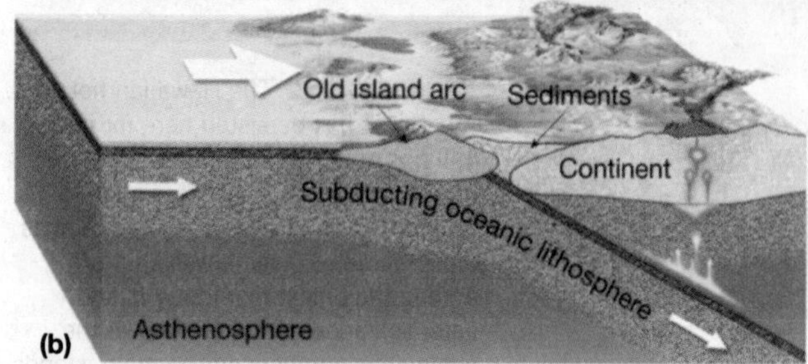

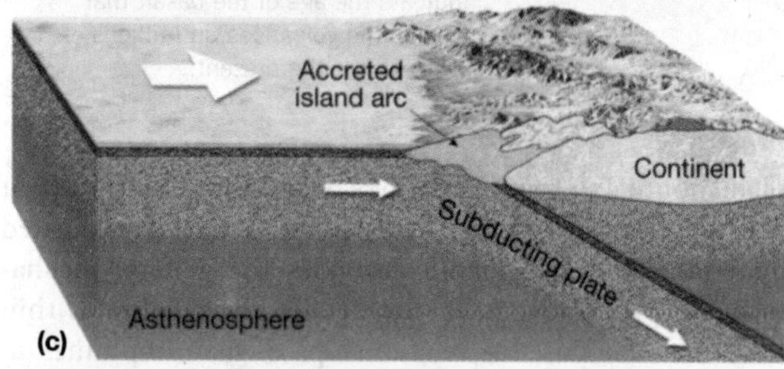

▲**Figure 14-24** The origin of an accreted terrane in a convergent boundary. (a) A moving oceanic plate carries along an old island arc. (b) The oceanic plate converges with a continental plate. (c) The oceanic plate begins to subduct under the continental plate, but the island arc is too buoyant for subduction and so is accreted to the continental plate.

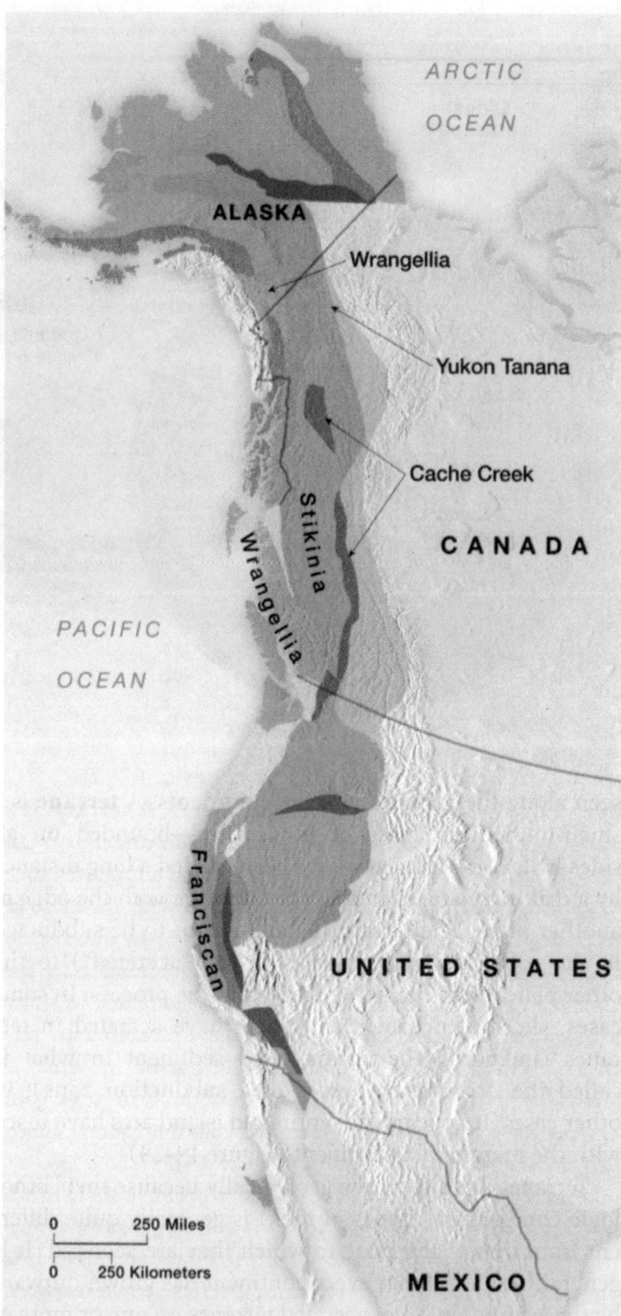

▲**Figure 14-25** The western part of North America consists of a complicated mixture of terranes that have been accreted to the North American Plate.

Our present state of knowledge about plate tectonics, however, is ample to provide a firm basis for understanding the patterns of most of the world's major relief features—the size, shape, and distribution of the continents, major mountain ranges, and ocean basins. To understand more detailed topographic features, however, we must now turn to less spectacular, but no less fundamental, internal processes that are often directly associated with tectonic movement.

Volcanism

Volcanism (or *vulcanism*) is a general term that refers to all the phenomena connected with the origin and movement of molten rock. These phenomena include the well-known explosive volcanic eruptions that are among the most spectacular and terrifying events in all nature, along with much more quiescent events, such as the slow solidification of molten material below the surface.

We noted in Chapter 13 the distinction between *volcanic* (extrusive) and *plutonic* (intrusive) igneous rocks; a similar differentiation is made between extrusive and intrusive volcanism. When magma is expelled onto Earth's surface while still molten, the activity is extrusive and is called *volcanism*; when magma solidifies below the surface it is referred to as *intrusive volcanism* or *plutonic activity* and results in intrusive igneous features.

Volcanism

Magma (molten mineral material below the surface) extruded onto Earth's surface is called **lava** (Figure 14-26). The ejection of lava into the open air is sometimes volatile and explosive, devastating the area for many kilometers around; in other cases, it is gentle and quiet, affecting the landscape more gradually. All eruptions, however, alter the landscape because the fiery lava is an inexorable force until it cools, even if it is expelled only slowly and in small quantities.

The explosive eruption of a volcano is an awesome spectacle. In addition to an outward flow of lava, such solid matter as rock fragments, solidified lava blobs, ashes, and dust—collectively called **pyroclastic** material—as well as gas and steam, may be hurled upward in extraordinary quantities. In some cases, the volcano literally explodes, disintegrating in an enormous self-destructive blast. The supreme example of such self-destruction within historic times was the final eruption of Krakatau, a volcano that occupied a small island in Indonesia between Sumatra and Java. When it exploded in 1883, the noise was heard 2400 kilometers (1500 miles) away in Australia, and 9 cubic kilometers (2.2 cubic miles) of material was blasted into the air. The island disappeared, leaving only open sea where it had been. The *tsunamis* (great seismic sea waves; discussed in Chapter 20) it generated drowned more than 30,000 people, and sunsets in various parts of the world were colored by fine volcanic dust for many months afterward.

▲ **Figure 14-26** Eruption of Kīlauea Volcano, Hawai'i. Note the streams of very fluid basaltic lava flowing away from the vent toward the left.

Volcano Distribution A volcano is considered active if it has erupted at least once within historical times and is considered likely to do so again. There are about 550 active volcanoes in the world. On average, about 15 of them will erupt this week, 55 this year, and perhaps 160 this decade. Moreover, there will be one or two eruptions per year from volcanoes with no historic activity. In addition to surface eruptions on continents and islands, there is a great deal of underwater volcanic activity; indeed, it is estimated that more than three-fourths of all volcanic activity is undersea activity such as at midocean ridge spreading centers.

Within the conterminous 48 United States prior to the 1980 eruption of Mount St. Helens, there was only one volcano classified as active—Lassen Peak in California, which last erupted in 1917 but still occasionally produces gas and steam. A number of other volcanoes, notably California's Mount Shasta and Long Valley Caldera, Washington's Mount Baker and Mount Rainier, and the Yellowstone Caldera show signs of potential activity but have not erupted in recorded time, and there are hundreds of extinct volcanoes, primarily in the West Coast states. Both Alaska and Hawai'i have many volcanoes, both active and inactive.

Areas of volcanism are widespread over the world, but as Figure 14-27 shows, their distribution is uneven around the world, with many volcanoes in some regions and none in others. Volcanic activity is primarily associated with plate boundaries. At a divergent boundary, magma wells up from the interior both by eruption from active volcanoes and by flooding out of fissures. At convergent boundaries where subduction of oceanic lithosphere is taking place, volcanoes are formed in association with the generation of magma. Hot spots are responsible for volcanic and hydrothermal activity in many places such as Yellowstone, Hawai'i, and the Galapagos Islands.

It is apparent from Figure 14-27 that the most notable area of volcanism in the world is around the margin of the Pacific Ocean in the Pacific Ring of Fire (see Figure 14-20)—also called the *Andesite Line* because the volcanoes consist primarily of the volcanic rock andesite. About 75 percent of the world's volcanoes, both active and inactive, are associated with the Pacific Rim.

Magma Chemistry and Styles of Eruption

The nature of a volcanic eruption apparently is determined largely by the chemistry of the magma that feeds it, although the relative strength of the surface crust and the degree of confining pressure to which the magma is subjected may also be important. The chemical relationships are complex, but the critical component seems to be the relative amount of silica (SiO_2) in the magma. Recall that common magmas include relatively high-silica *felsic* magma (such as produces the volcanic rock rhyolite and the plutonic rock granite), intermediate-silica andesitic magma (such as produces the volcanic rock andesite and the plutonic rock diorite), and relatively low-silica *mafic* magma (such as produces the volcanic rock basalt and the plutonic rock gabbro). See Figure 13-6 in Chapter 13 for a review of igneous rocks resulting from different magma chemistries.

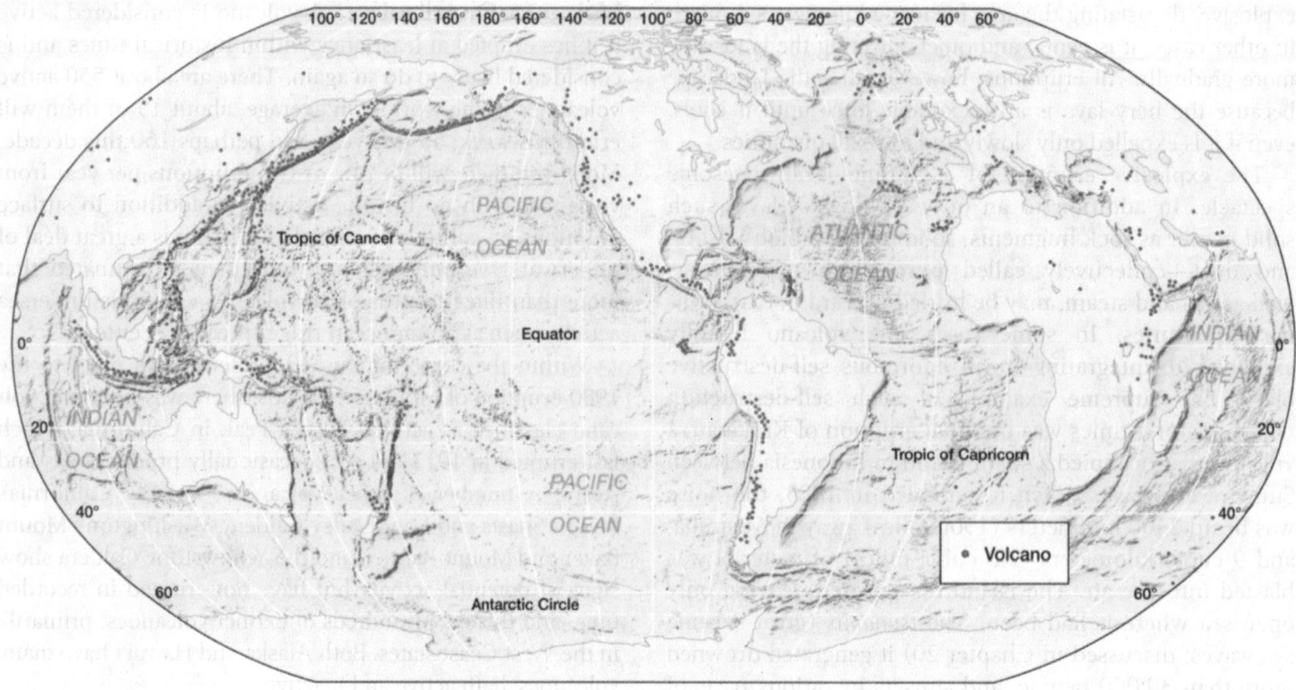

▲ **Figure 14-27** Distribution of volcanoes known to have erupted at some time in the recent geological past. The Pacific Ring of Fire is quite conspicuous.

In high-silica felsic magmas, long chains consisting of silicate structures can develop even before crystallization of minerals begins, greatly increasing the *viscosity* (thickness or "stickiness") of the magma. A high silica content also usually indicates cooler magma in which some of the heavier minerals have already crystallized and a considerable amount of gas has already separated. Some of this gas is trapped in pockets in the magma under great pressure. Unlike the more fluid lavas, gas bubbles can rise only slowly through viscous felsic magma. As the magma approaches the surface, the confining pressure is diminished and the pent-up gases are released explosively, generating an eruption in which large quantities of pyroclastic material are ejected from the volcano. Any lava flows are likely to be very thick and slow moving.

On the other hand, mafic magma is likely to be hotter and considerably more fluid because of its lower silica content. Dissolved gases can bubble out of very fluid mafic magma much more easily than from viscous felsic magma. The resulting eruptions usually yield a great outpouring of lava, quietly and without explosions or large quantities of pyroclastic material. ("Quietly" is a descriptive term that is relative and refers to the nonexplosive flow of fluid lava.) The highly active volcanoes of Hawai'i erupt in this fashion.

Volcanoes with intermediate-silica content andesitic magmas erupt in a style somewhat between that of felsic and mafic magmas: periodically venting fairly fluid andesitic lava flows and periodically having explosive eruptions of pyroclastic material. Many of the major volcanoes associated with subduction zones are of this type.

Volcanic Activity All volcanoes are relatively temporary features of the landscape. Some may have an active life of only a few years, whereas others are sporadically active for thousands of centuries. At the other end of the scale, new volcanoes are spawned from time to time. Two of the more spectacular recent developments were the birth of Surtsey, which rose out of the sea as a new island above a hot spot off the coast of Iceland in 1964, and the eruption of an undersea volcano near Tonga in 2009 (Figure 14-28).

Despite the destruction they cause, volcanoes do provide vital services to the planet. Much of the water on Earth today was originally released as water vapor during volcanic eruptions during the early history of our planet, and just as our blood carries nutrients that nourish our bodies, so volcanoes do the same for the "skin" of Earth.

▲ **Figure 14-28** Underwater volcano erupting to the surface near Tongo on March 18, 2009.

▲ **Figure 14-29** Young basalt of the Big Island of Hawai'i.

Magma contains the major elements—phosphorus, potassium, calcium, magnesium, sulfur—required for plant growth. When this magma is extruded as lava that hardens into rock, the weathering that releases the nutrients into soil may require decades or centuries (Figure 14-29). When the magma is blasted out as ash, however, nutrients can be leached into the soil within months. It is no coincidence that Java, one of the most volcanically active parts of the planet, is also one of the world's most fertile areas.

Lava Flows Whether originating from a volcanic crater or a crustal fissure, a lava flow spreads outward approximately parallel with the surface over which it is flowing, and this parallelism is maintained as the lava cools and solidifies. Although some viscous flows cling to relatively steep slopes, the vast majority eventually solidify in a horizontal orientation that may resemble the stratification of sedimentary rock, particularly if several flows have accumulated on top of one another.

The topographic expression of a lava flow, then, is often a flattish plain or plateau. The strata of sequential flows may be exposed by erosion as streams usually incise very steep-sided gullies into lava flows. The character of the flow surface varies with the nature of the lava and with the extent of erosion, but as a general rule the surface of relatively recent lava flows tends to be extremely irregular and fragmented. When flows of fluid lava such as basalt cool uniformly, a distinctive pattern of joints frequently develops resulting in hexagonal columns (Figure 14-30).

Flood Basalt Many of the world's most extensive lava flows were not extruded from volcanic peaks but rather issued from fissures associated with hot spots. The lava that flows out of these vents is nearly always basaltic and frequently comes forth in great volume. As we saw in Figure 14-21, many scientists think that the initial consequence of a large mantle plume reaching the surface can be a huge outpouring of lava. The term **flood basalt** is applied to the vast accumulations of lava that build up, layer upon layer, sometimes covering tens of thousands of square kilometers to depths of many hundreds of meters. A prominent example of flood basalt in the United States is the Columbia Plateau, which covers 130,000 square kilometers (50,000 square miles) in Washington, Oregon, and Idaho (Figure 14-31). Larger outpourings are evidenced on other continents, most notably the Deccan Traps of India (520,000 square kilometers [200,000 square miles]; *trap* is derived from the

▲ **Figure 14-30** Basalt cliffs in Devils Postpile National Monument, California. Note the hexagonal columns—a common result when a fluid lava flow cools uniformly.

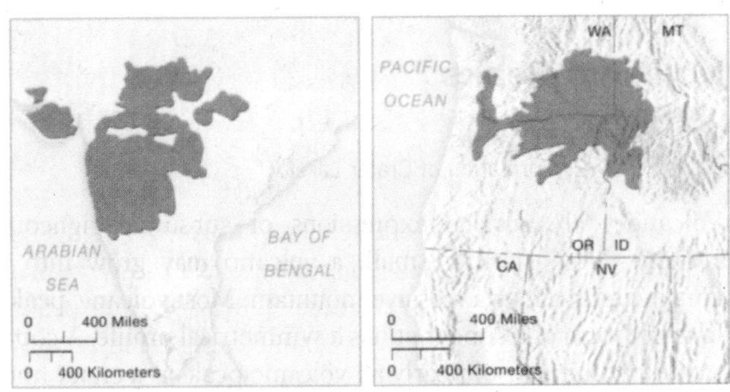

▲ **Figure 14-31** Two extensive outpourings of flood basalt: the Deccan Traps (also called the Deccan Plateau) in India and the Columbia Plateau in the northwestern United States.

TABLE 14-1 Principal Types of Volcanoes

Volcano Type	Shape and Size	Structure	Magma and Eruption Style	Examples
Shield	Broad, gently sloping mountain; much broader than high; size varies greatly.	Layers of solidified lava flows.	Magma usually basaltic; characterized by quiet eruptions of fluid lava.	Hawaiian Islands; Tahiti
Composite (Stratovolcano)	Steep-sided symmetrical cone; heights to over 3700 m (12,000 ft.).	Layers of lava flows, pyroclastics, and hardened mudflow deposits.	Magma usually intermediate in chemistry, often andesitic; long life span; characterized by both explosive eruptions of pyroclastics and quiet eruptions of lava.	Mt. Fuji, Japan; Mt. Rainier, Washington; Mt. Shasta, California; Mt. Vesuvius, Italy; Mt. St. Helens, Washington
Lava Dome (Plug Dome)	Usually small, typically less than 600 m (2000 ft.) high; sometimes irregular shape.	Solidified lava that was thick and viscous when molten; plug of lava often covered by pyroclastics; frequently occur within the crater of composite volcano.	Magma usually high in silica, often rhyolitic; dome grows by expansion of viscous lava from within; explosive eruptions common.	Lassen Peak, California; Mono Craters, California
Cinder Cone	Small, steep-sided cone; maximum height 500 m (1500 ft.).	Loose pyroclastic material; may be composed of ash or cinder-size pieces.	Chemistry of magma varies, often basaltic; short life span; pyroclastics ejected from central vent; occasionally produce lava flows.	Paricutin, Mexico; Sunset Crater, Arizona

Sanskrit word for "step" in reference to the layers of lava flows found here). Over the world as a whole, more lava has issued quietly from fissures than from the combined outpourings of all volcanoes.

Research indicates that the timing of several major flood basalt eruptions in the geologic past correlate with mass extinctions of plants and animals—perhaps caused by the environmental disruption brought by the massive lava flows and outgassing from the eruptions. For example, many scientists now think that the major extinctions about 65 million years ago that ended the reign of the dinosaurs were as much, or more, a consequence of the flood basalt eruptions of the Deccan Traps than of the asteroid impact that occurred at the same time.

Volcanic Peaks

1. Volcanoes
2. Formation of Crater Lake

Volcanoes are surface expressions of subsurface igneous activity. Often starting small, a volcano may grow into a conspicuous hill or a massive mountain. Most volcanic peaks take the form of a cone that has a symmetrical profile. A common denominator of nearly all volcanic peaks is a crater normally set conspicuously at the apex of the cone. Frequently, smaller subsidiary cones develop around the base or on the side of a principal peak, or even in the crater. Generally, differences in magma, and therefore eruption style, result in different types of volcanic peaks (Table 14-1).

Shield Volcanoes Basaltic lava tends to flow quite easily over the surrounding surface, forming broad, low-lying **shield volcanoes**, built up of layer upon layer of solidified lava flows with relatively little pyroclastic material. Some shield volcanoes are massive and very high, but they are never steep-sided (Figure 14-32).

The Hawaiian Islands are composed of numerous shield volcanoes. Produced by the Hawaiian "hot spot," Mauna Loa on the Big Island of Hawai'i is the world's largest

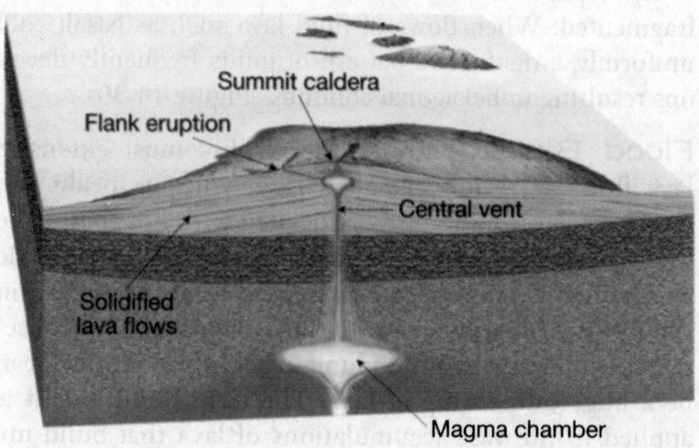

▲ **Figure 14-32** Shield volcanoes, such as those on the Big Island of Hawai'i, have gentle slopes and consist of layer after layer of solidified lava flows with little pyroclastic material.

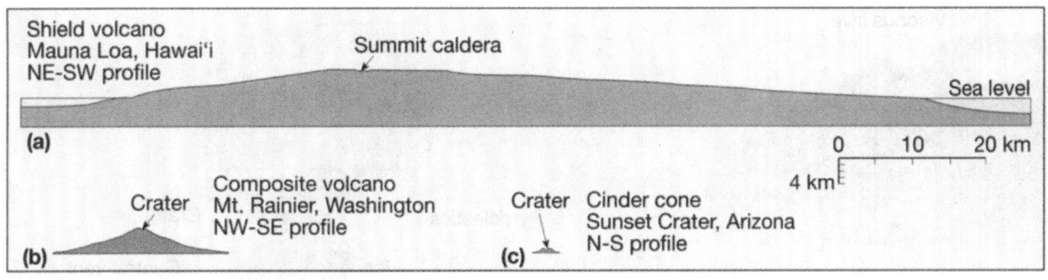

◀ **Figure 14-33** Profiles of volcanoes drawn at identical scales. (a) Mauna Loa, Hawai'i—a shield volcano. (b) Mt. Rainier, Washington—a composite volcano. (c) Sunset Crater, Arizona—a cinder cone.

volcano—it is more than 9 kilometers (6 miles) high from its base on the floor of the ocean to the top of its summit (Figure 14-33). Kīlauea, currently the most active of the Hawaiian shield volcanoes, is on the southeast flank of Mauna Loa (see Figure 14-23).

Composite Volcanoes Volcanoes that emit higher silica "intermediate" lavas such as andesite often erupt explosively and tend to develop into symmetrical, steep-sided volcanoes known as **composite volcanoes** or *stratovolcanoes* (Figure 14-34). These mountains build up steep sides by having layers of ejected pyroclastics (ash and cinders) from explosive eruptions alternate with lava flows from nonexplosive eruptions. The pyroclastic material tends to produce the steep slopes, while the solidified lava flows hold the pyroclastics together. Famous examples of composite volcanoes include Mt. Fuji in Japan, Mt. Rainier in Washington, and Volcán Popocatépetl near Mexico City (Figure 14-35).

Lava Domes **Lava domes**—also called **plug domes**—have masses of very viscous lava such as high-silica rhyolite that are too thick and pasty to flow very far. Instead, lava bulges up from the vent, and the dome grows largely by expansion from below and within (Figure 14-36). The Mono Craters are a chain of young rhyolitic plug domes just to the east of the Sierra Nevada and Yosemite National Park in California—the most recent activity taking place just a few hundred years ago (Figure 14-37).

Lava domes may also develop within the craters of composite volcanoes when viscous lava moves up into the vent. Shortly after the large eruption of Mount St. Helens in 1980, such a lava dome began to develop.

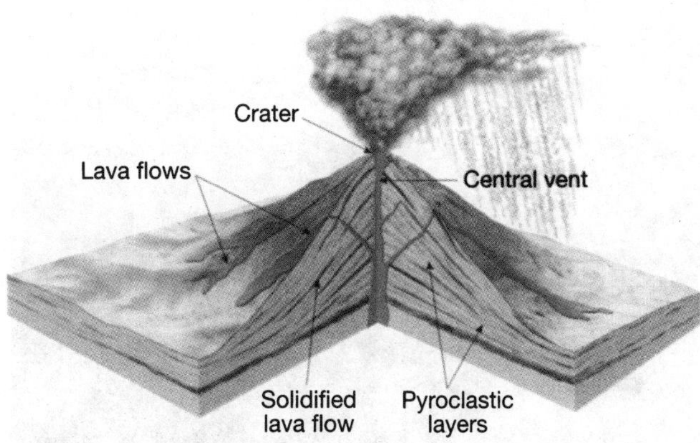

▲ **Figure 14-34** Composite volcanoes consist of layers of pyroclastic material and solidified lava flows.

Cinder Cones **Cinder cones** are the smallest of the volcanic peaks. Their magma chemistry varies, but basaltic magma is most common. They are cone-shape peaks built by the pyroclastic materials that are ejected from the volcanic vent (Figure 14-38). The size of the particles being ejected determines the steepness of the slopes. Tiny particles ("ash") can support slopes as steep as 35 degrees, whereas the larger ejecta ("cinders") will produce slopes up to about 25 degrees. Cinder cones are generally less than 450 meters

▲ **Figure 14-35** Volcán Popocatépetl, a composite volcano near Mexico City, erupting in November 1998. The Church of Our Lady of Remedies is in the foreground.

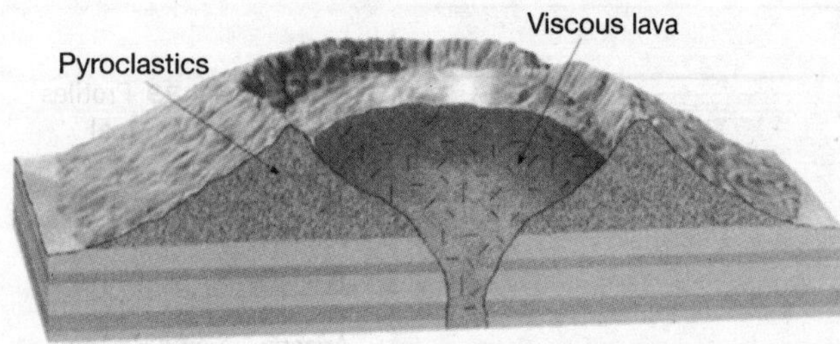

▲ **Figure 14-36** Lava domes (plug domes) develop when viscous lava (commonly rhyolite or dacite) is "squeezed" up into a volcanic vent. The plug of lava may be mantled or surrounded by explosively ejected pyroclastic material.

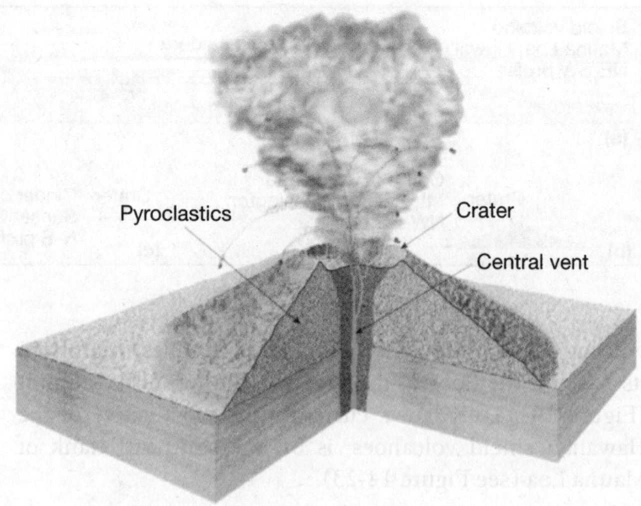

▲ **Figure 14-38** Cinder cones are small volcanoes consisting of pyroclastic material.

(1500 feet) high and are often found in association with other volcanoes (Figure 14-39). Lava flows occasionally issue from the same vent that produces a cinder cone.

Calderas Uncommon in occurrence but spectacular in result is the formation of a **caldera**, which is produced when a volcano explodes, collapses, or does both. The result is an immense basin-shaped depression, generally circular, that has a diameter many times larger than that of the original volcanic vent or vents. Some calderas are tens of kilometers in diameter.

North America's most famous caldera is Oregon's misnamed Crater Lake (Figure 14-40). Mount Mazama was a composite volcano that reached an estimated elevation of 3660 meters (12,000 feet) above sea level. During a major eruption about 7700 years ago, the walls of Mount Mazama weakened and collapsed as enormous volumes of pyroclastic material were ejected from the volcano (Figure 14-41). The partial emptying of magma chamber below Mount Mazama may have contributed to this collapse. The final, cataclysmic eruption removed—by explosion and collapse—the upper 1220 meters (4000 feet) of the peak and produced a caldera whose bottom is 1220 meters (4000 feet) below the crest of the remaining rim. Later, half this depth filled with water, creating one of the deepest lakes in North America. A subsidiary volcanic cone has subsequently built up from the bottom of the caldera and now breaks the surface of the lake as Wizard Island (Figure 14-42).

Shield volcanoes may develop *summit calderas* in a different way. When large quantities of fluid lava are vented from rift zones along the sides of a volcano, the magma chamber below the summit can empty and collapse, forming a relatively shallow caldera. Both Mauna Loa and Kīlauea on the Big Island of Hawai'i have calderas that formed in this way.

Volcanic Necks More limited still but very prominent where it does occur is a *volcanic neck*, a small, sharp spire that rises abruptly above the surrounding land. It represents the pipe, or throat, of an old volcano that filled with solidified lava after its final eruption. Millennia of erosion have removed the less resistant material that made

► **Figure 14-37** Crater Mountain is the highest peak in the Mono Craters (also called the Mono Domes) chain of rhyolitic plug domes east of Yosemite National Park in California. Note the irregular summits of these volcanoes, formed by the bulging up of viscous lava.

◀ **Figure 14-39** Sunset Crater, in northern Arizona near Flagstaff, is a classic example of a cinder cone volcano.

up the cone, leaving the harder, lava-choked neck as a conspicuous remnant (Figure 14-43).

Volcanic Hazards

Millions of people around the world live in close proximity to active or potentially active volcanoes. In the United States alone, there are more than 50 volcanoes that have erupted within the last 200 years, and others that could well become active in the near geologic future. Some of these volcanoes are located near population centers in Washington, Oregon, California, Alaska, Hawai'i, and the region around Yellowstone National Park. Future eruptions could expose large numbers of people to a wide range of volcanic hazards (Figure 14-44).

Volcanic Gases A volcano emits large quantities of gas during an eruption. Water vapor makes up the bulk of the gas emitted, but other volcanic gases include

▲ **Figure 14-40** Oregon's Crater Lake occupies an immense caldera. Wizard Island represents a more recent subsidiary volcanic cone.

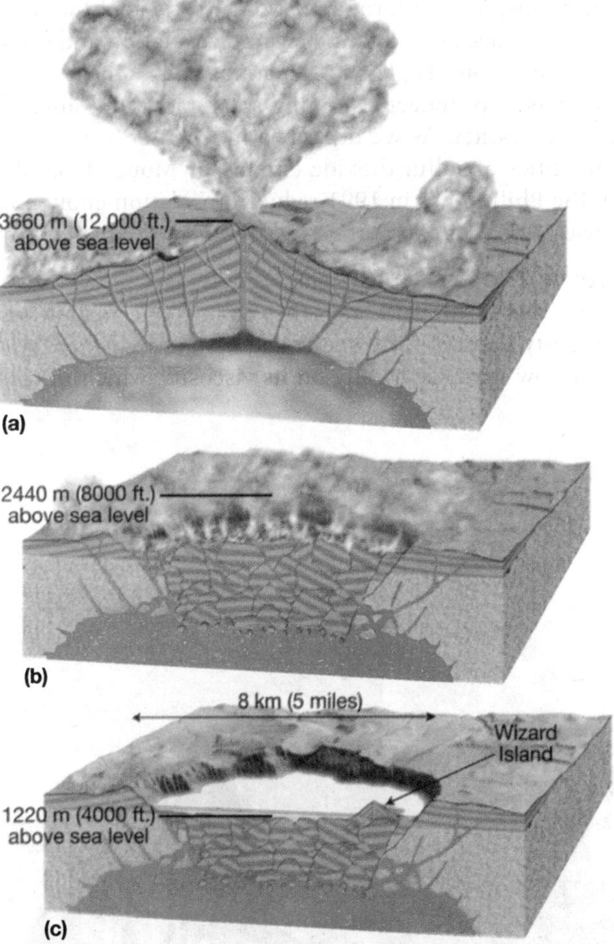

▲ **Figure 14-41** Formation of Crater Lake. (a) Mount Mazama about 7700 years ago. (b) During the cataclysmic eruption, an enormous volume of pyroclastic material was erupted from the magma chamber and the volcano collapsed, forming a caldera 1220 meters (4000 feet) deep. (c) The caldera partially filled with water for form a lake; a new fissure formed the volcano known as Wizard Island.

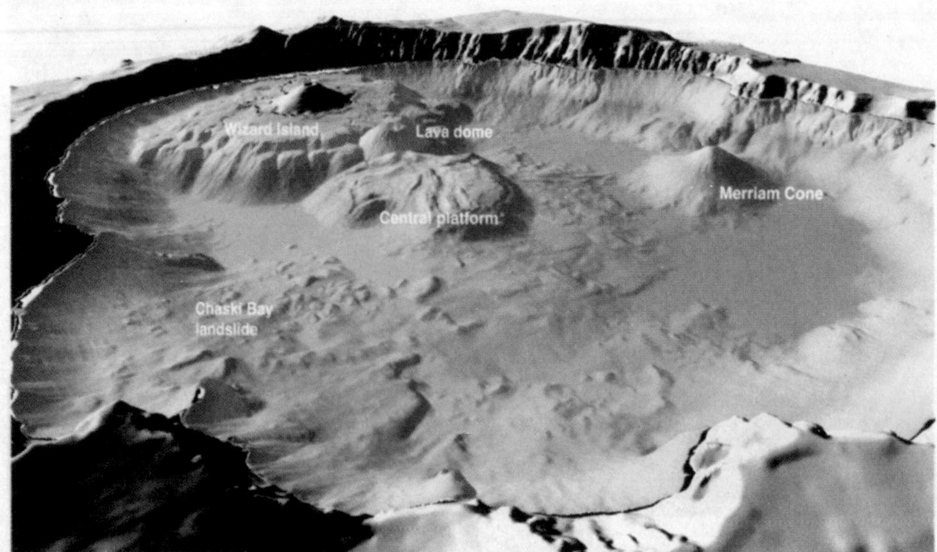

▶ **Figure 14-42** Bathymetry of Crater Lake. This detailed map of the floor of Crater Lake in Oregon shows evidence of postcollapse volcanic activity and massive landslides.

carbon dioxide, sulfur dioxide, hydrogen sulfide, and fluorine. The hazards resulting from these gases vary. For example, sulfur dioxide may combine with water in the atmosphere to form a mist of sulfuric acid that falls to the surface as acid rain, harming vegetation and causing corrosion. High in the atmosphere, these same droplets may reflect incoming solar radiation, altering global weather. As we described in Chapter 8, the large quantities of sulfur dioxide emitted by Mount Pinatubo in the Philippines in 1991 reduced insolation enough to lower global temperatures slightly for more than a year.

Lava Flows Perhaps surprisingly, lava flows rarely cause loss of life, although they can produce significant property damage. The speed and distance covered by a lava flow depends mostly on its viscosity, which in turn depends on its silica content. Low-silica basaltic lava associated with shield volcanoes, such as those on Hawai'i, tends to be quite fluid and fast moving. Although most basaltic lava flows move more slowly than a person can walk, some flows travel at speeds of over 25 kilometers (15 miles) per hour and cover distances of more than 120 kilometers (75 miles) before congealing. Since the paths taken by lava flows tend to be predictable they cause few injuries, although many dozens of homes have been destroyed by lava flows from Kīlauea Volcano in Hawai'i over the last few decades.

Higher-silica lava, such as the andesitic lava associated with the many composite volcanoes around the rim of the Pacific, tends to be thicker than basaltic lava, usually moving only short distances down the slopes of the volcano. Very viscous rhyolitic lava often does little more than squeeze out of a vent, bulging up to form a lava dome. While lava flows from intermediate- and high-silica magmas rarely represent a direct danger to people, the explosive nature of these volcanoes does often produce a number of significant hazards.

▲ **Figure 14-43** Shiprock in northwestern New Mexico is a prime example of a volcanic neck. It rises 490 meters (1600 feet) above the surrounding landscape.

Eruption Column and Clouds It is common for composite volcanoes and lava domes to erupt explosively. The violent ejection of pyroclastic material and gases from a volcano can form an *eruption column* reaching altitudes of 16 kilometers (10 miles) or more. Large fragments of solid rock, called volcanic "bombs," drop to the ground immediately around the volcano, while smaller fragments of volcanic ash and dust form an enormous *eruption cloud* from which great quantities of ash may fall. A heavy covering of ash can damage crops and even cause the collapse of buildings.

Pyroclastic Flows The collapse of a lava dome or the rapid subsidence of an eruption column during an explosive eruption of a volcano can lead to a terrifying high-speed avalanche of searing hot gases, ash, and rock fragments known as a **pyroclastic flow**. A pyroclastic flow

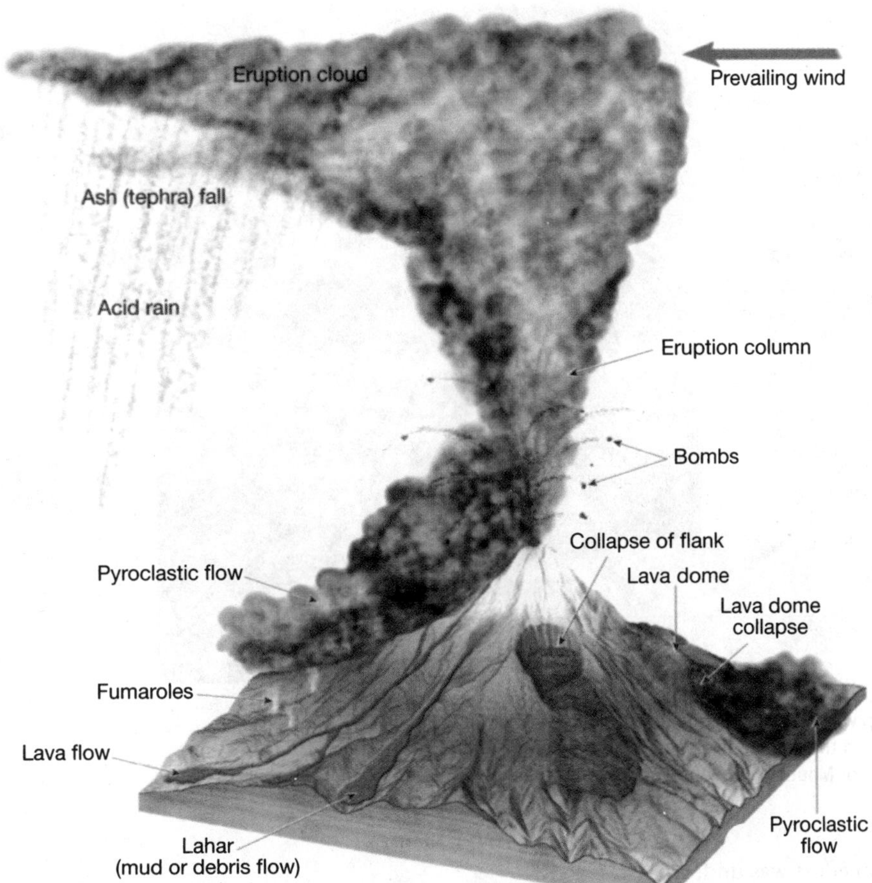

◀ **Figure 14-44** The hazards associated with a typical composite volcano. Some of these hazards may also occur with other kinds of volcanoes.

(sometimes called a *nuée ardente*) can travel down a volcano at speeds of more than 160 kilometers (100 miles) per hour, burning and burying everything in its path (Figure 14-45). Probably the most famous example of a pyroclastic flow took place in 1902 on the Caribbean island of Martinique. An explosive eruption of Mont Pelée sent a massive pyroclastic flow down onto the port city of St. Pierre, destroying the town and killing nearly all of its 28,000 inhabitants in a matter of moments.

More recently, the Unzen volcano complex on the Japanese island of Kyushu began a series of eruptions in 1990. After a period of escalating seismic activity, a lava dome formed in Fugen-dake, a peak adjacent to the large Mayu-yama dome. By the spring of 1991, lava blocks were collapsing off the dome, generating hundreds of pyroclastic flows, some of which traveled more than 5 kilometers (3 miles) off the flanks of the volcano (Figure 14-46). By 1993, pyroclastic flows and volcanic mudflows had destroyed 2000 structures, forced the evacuation of as many as 12,000 local residents, and taken the lives of 43 people, including the well-known French filmmakers and volcanologists Maurice and Katia Krafft.

Volcanic Mudflows (Lahars) One of the most common hazards associated with composite volcanoes are **volcanic mudflows**—also known by their Indonesian name, **lahar**. A loose mantle of ash and pyroclastic flow deposits on the slopes of a volcano can be mobilized easily by heavy rain or by the melting of snow and glaciers during an eruption. The water mixes with unconsolidated pyroclastic material to produce a fast-moving—and sometimes hot—slurry of mud and boulders, flowing with the consistency of wet concrete. Lahars typically flow down stream valleys off the slopes of a volcano, leaving the valley floor buried in thick mud and debris. They can reach speeds of over 50 kilometers (30 miles) per hour and travel distances of over 80 kilometers (50 miles).

One of the most tragic examples of a lahar took place in 1985 when the Nevado del Ruiz volcano in Colombia produced a mudflow that inundated the town of Armero nearly 50 kilometers (30 miles) away, killing more than 20,000 people.

Monitoring Volcanic Hazards Volcanoes may lie dormant for hundreds of years between major eruptions, and so the volcanic hazards in the surrounding regions are not always evident. In North America, the Cascade Range alone contains more than a dozen potentially active volcanoes, including seven that have exhibited activity since the late 1700s (Figure 14-47). Because of rapidly expanding populations here and elsewhere around the Pacific Rim, increasing numbers of people are now exposed to the same volcanic hazards that have killed thousands in the past.

▲ **Figure 14-45** A pyroclastic flow can develop when an eruption column collapses, sending a surge of hot gases and pyroclastic material down the side of a volcano. The photograph shows a pyroclastic flow heading down the flank of Mount St. Helens during an eruption on August 7, 1980.

By the 1970s, a concerted international effort was underway to both assess volcanic hazards and to gain enough knowledge to anticipate the onset of major volcanic eruptions. In the United States, the U.S. Geological Survey and research universities are looking at past historical eruptions as well as evidence from the geological record to map out the most likely paths of pyroclastic flows and lahars from volcanoes such as Mt. Rainier in Washington, Mt. Hood in

▲ **Figure 14-46** Aerial view of Unzen volcano in Japan showing the path of pyroclastic flows that moved down toward the town of Shimabara in 1993.

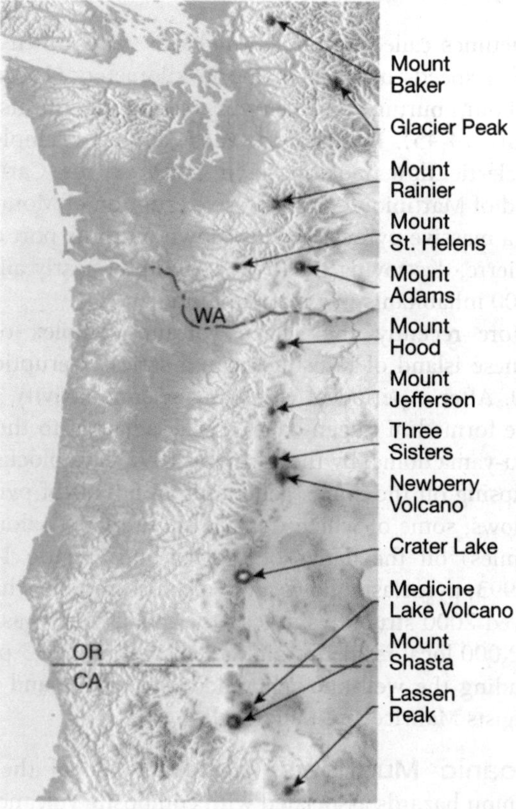

▲ **Figure 14-47** Prominent volcanoes (active, dormant, and extinct) of the Cascade Range.

FOCUS

Eruption of Mount St. Helens

In the spring of 1980, Mount St. Helens in Washington, one of the most beautiful volcanic peaks of the Cascades, suddenly came back to life after lying dormant for 123 years. On the morning of May 18, after several months of seismic activity, swelling, and sporadic eruptions of steam and ash, the volcano unleashed a devastating eruption (Figure 14-A).

With little warning, the entire north slope of the mountain unhinged during an earthquake in an enormous landslide. The landslide depressurized the magma within the volcano, triggering a powerful lateral blast that leveled trees more than 24 kilometers (15 miles) north of the volcano (see Figure 10-18), completely devastating an area of more than 500 square kilometers (200 square miles). At the foot of the volcano, Spirit Lake, once one of the most scenic bodies of water in North America, was instantly laden with debris and dead trees.

Immediately following the lateral blast, a strong vertical explosion of ash and steam began, forming an eruption column that rose to an altitude of 19 kilometers (12 miles) in less than 10 minutes. Over the course of the day, more than 470 billion kilograms (520 million tons) of ash were carried to the east by the prevailing winds. Spokane, Washington, 400 kilometers (250 miles) from the volcano, was in darkness at midday. During the course of the eruption, a series of pyroclastic flows rolled down the mountain, reaching as far as 8 kilometers (5 miles) north of the volcano. Volcanic mudflows, or lahars, some created when the eruption melted snow and ice that capped the volcano, poured down nearby river valleys. The largest of these lahars carried tons of mud and debris into the Toutle River, where the raging mudflow knocked out every bridge for 48 kilometers (30 miles), overwhelmed the Cowlitz River valley, and clogged the Columbia River shipping channel to less than half of its normal depth of 12 meters (40 feet).

The cataclysmic eruption of May 18, 1980, reduced the elevation of Mount St. Helens by more than 390 meters (1300 feet), removed about 2.8 cubic kilometers (0.67 cubic miles) of rock from the volcano, spread volcanic ash over an area of 56,000 square kilometers (22,000 square miles), caused more than $1 billion in property and economic losses, and killed 57 people.

It is sobering to realize that the 1980 eruption of Mount St. Helens was not an extraordinarily large volcanic eruption. The 1991 eruption of Mount Pinatubo in the Philippines was more than 10 times larger, and the 1912 eruption of Novarupta in Alaska 30 times larger. The geologic record shows that Mount St. Helens has had much larger eruptions in the past, as have many other volcanoes in the Cascades—the catastrophic eruption of 7700 years ago that formed present-day Crater Lake in Oregon was 20 times larger than the 1980 eruption of Mount St. Helens.

Now 30 years after its last major eruption, Mount St. Helens continues to show activity. By late 2004, the volcano was having small eruptions. A new lava dome was building up inside the old crater, indicating that a plug of near-solid rock was being pushed up from the magma chamber below. Glaciers inside the crater that had grown back since the 1980 eruption were being pushed aside by the bulging new dome. Whether another major eruption comparable to the one in 1980 will occur anytime soon is unclear, although in the long run, Mount St. Helens is quite likely to erupt vigorously again.

▲ **Figure 14-A** (a) Mount St. Helens prior to the 1980 eruption. The peak in the background is Mount Adams. (b) Mount St. Helens after the 1980 eruption. Its posteruption elevation is 400 meters (1300 feet) less than the pre-eruption elevation.

Oregon, the Yellowstone Caldera, and Mt. Shasta, Lassen Peak, and the Long Valley Caldera in California.

The monitoring of active volcanoes includes measuring slight changes in the slope of a mountain using sensitive "tiltmeters" that can detect swelling of a volcano with magma, measuring variations in gas composition and quantity vented from a volcano that may indicate changes in magma, and monitoring earthquake activity below a volcano—swarms of small earthquakes may indicate the filling of the magma chamber below a volcanic peak. Remote cameras can produce time-lapse movie sequences that document changes to lava domes and other crater features. Especially promising research is pointing toward distinctive seismic patterns that often accompany the onset of major eruptions.

It is hoped that this growing knowledge will enable authorities to make more informed choices about when and where to evacuate local populations when the next volcano erupts.

Intrusive Igneous Features

Igneous Features

When magma solidifies below Earth's surface in association with intrusive volcanism or plutonic activity, it produces igneous rock. If this rock is pushed upward into the crust either before or after solidification, it is called an **igneous intrusion**. Most such intrusions have no immediate effect on the surface landscape, but sometimes the igneous mass is raised high enough to deform the overlying material and change the shape of the surface. In many cases, the intrusion is eventually exposed at the surface through the action of the external processes, often becoming conspicuous because they are usually resistant to erosion and with the passage of time stand up relatively higher than the surrounding land.

Pluton is a general term used to refer to intrusive igneous bodies of nearly any size. Intrusions come in all shapes, sizes, and compositions. Moreover, their relationship to overlying or surrounding rock is also quite variable. The intrusive process usually disturbs the preexisting rock. Rising magma makes room for itself by a process called *stoping* (a mining term for ore removal by working upward). The molten, invading magma can assimilate the rock being invaded or heat it enough to make it flow out of the way or either split or bow it upward. Adjacent to the area the magma has invaded, the invaded rock usually experiences contact metamorphism from being exposed to the heat and pressure of the rising intrusion.

Although there is almost infinite variety in the forms assumed by igneous intrusions, most can be broadly classified according to a scheme that contains only a few types (Figure 14-48).

Batholiths By far the largest and most amorphous intrusion is the **batholith**, which is a subterranean igneous body of enormous size (typically with a surface area of at least 100 square kilometers [40 square miles]) and perhaps of unknown depth. A large batholith may be comprised of dozens of plutons, perhaps emplaced as a series of "pulses" of magma over periods of millions of years. The term *stock* is sometimes used to refer to similar but smaller igneous intrusions of indefinite depth.

Batholiths form the core of many major mountain ranges. Such notable ranges as the Sierra Nevada in California, Idaho's Sawtooth Mountains, and Colorado's Front Range were created at least partially by the uplift of massive batholiths (Figure 14-49). Almost all the plutonic bedrock exposed in these ranges consists of granite or granodiorite (a felsic plutonic rock with a mineral composition between that of granite and diorite [see Figure 13-6]), now exposed after the extensive overburden of other rocks eroded away.

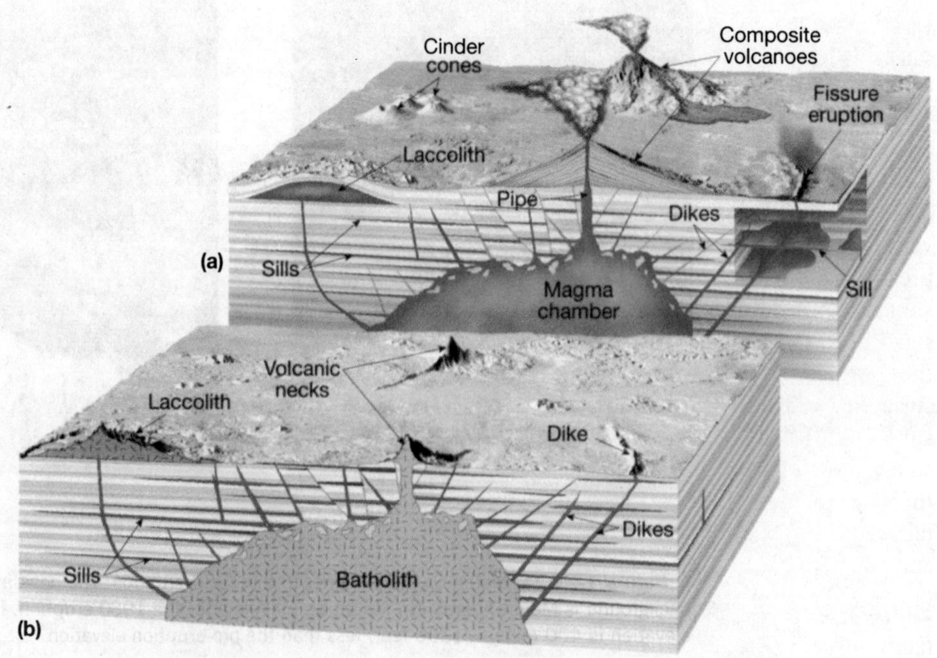

▶ **Figure 14-48** The formation of common types of igneous intrusions. (a) Volcanic eruptions and intrusion of magma. (b) After erosion.

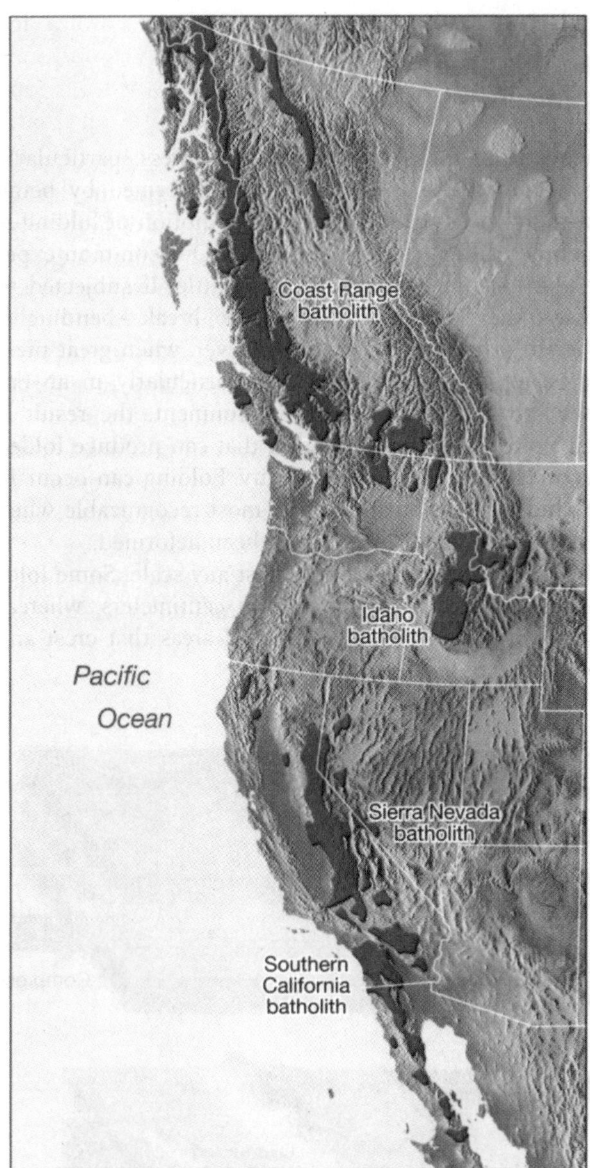

▲ **Figure 14-49** The major batholiths of western North America. These extensive intrusions are mostly associated with the former subduction boundary along the western edge of the North American Plate.

Laccoliths A specialized form of intrusion is the *laccolith*, which is produced when slow-flowing, usually viscous felsic magma is forced between horizontal layers of preexisting rock. Although continuing to be fed from some subterranean source, the magma resists flowing and instead builds up into a mushroom-shaped mass that domes the overlying strata. If this dome is near enough to Earth's surface, a rounded hill will rise, like a blister, above the surrounding area. Many laccoliths are small, but some are so large as to form the cores of hills or mountains in much the same fashion as batholiths. Several well-known mountain groups in southeastern Utah such as the Henry, Abajo, and La Sal Mountains have laccolithic cores.

Sills A *sill* is also a long, thin intrusive body, but its orientation is determined by the structure of the preexisting rocks. It is formed when magma, typically basaltic, is forced between strata that are already in place; the result is often a horizontal igneous sheet between horizontal sedimentary layers.

Dikes One of the most widespread of all intrusive forms is the **dike**, formed by the intrusion of a vertical or nearly vertical sheet of magma into preexisting rock, sometimes forcing its way into vertical fractures and sometimes melting its way upward. Dikes are notable because they are vertical, narrow (a few centimeters to a few meters wide), and usually quite resistant to erosion. As with most other igneous intrusions, their depth is often indeterminate, having served as a conduit through which deep-seated magma reached the surface. In some cases, dikes are quite long, extending for kilometers or even tens of kilometers in one direction. When exposed at the surface by erosion, they commonly form sheer-sided walls that rise above the surrounding terrain, as Figure 14-50 shows.

Dikes are often found in association with volcanoes, occurring as radial walls extending outward from the volcano like spokes of a wheel. Notable examples of radial dike development can be seen around Shiprock in northwestern New Mexico (see Figure 14-43) and around the Spanish Peaks in south-central Colorado.

A special kind of dike complex is found below the layers of basaltic lava on the ocean floor. A sequence of *sheeted dikes*—one vertical dike right next to another—develops in a midocean ridge spreading center as basaltic magma is injected into the ever-spreading gap produced by the diverging plates.

Veins Least prominent among igneous intrusions but widespread in occurrence are thin *veins* of igneous rock that may occur individually or in profusion. They are commonly formed when magma or hydrothermal fluids are forced into small fractures in the preexisting rocks, but they can also result from melting by an upward surge of magma. Intrusive veins may take very irregular shapes, but they normally have a generally vertical orientation.

▲ **Figure 14-50** The erosion of less-resistant surrounding material has left this dike standing as an abrupt natural wall. This scene is in southern Colorado near La Veta.

Tectonism

Tectonism (or *diastrophism*) is a general term that refers to the deformation of Earth's crust. The term covers various kinds of crust movement and implies that the material is solid and not molten. The rocks may be bent or broken in a variety of ways in response to great pressure exerted either in the mantle or crust. Sometimes stress is exerted by the rise of molten material from below. In some cases the stress is clearly the result of plate movement, but in other cases it appears to be unrelated directly to plate boundaries.

Whatever the causes of tectonism, the results are often conspicuous in the landscape. Tectonic movements are particularly obvious in sedimentary rocks because nearly all sedimentary strata are initially deposited in a horizontal or near-horizontal attitude—so if they now are bent, broken, or nonhorizontal, we know that they have clearly experienced some sort of diastrophic deformation.

Two types of tectonism are described below—*folding* and *faulting*—although the separation in nature is not always discrete and clear-cut.

Folding

Folding

When crustal rocks are subjected to stress, particularly lateral compression, they are often deformed by being bent in a process called **folding**. The notion of folding is sometimes difficult to conceptualize. Our common experience is that rocks are hard and brittle. If subjected to stresses, they might be expected to break—bending is harder to visualize. In nature, however, when great pressure is applied for long periods, particularly in an enclosed, buried, subterranean environment, the result is often a slow plastic deformation that can produce folded structures of incredible complexity. Folding can occur in any kind of rock, but it is often most recognizable when once-flat sedimentary strata have been deformed.

Folding can take place at almost any scale. Some folds can be measured in no more than centimeters, whereas others can develop over such broad areas that crest and trough are tens of kilometers apart.

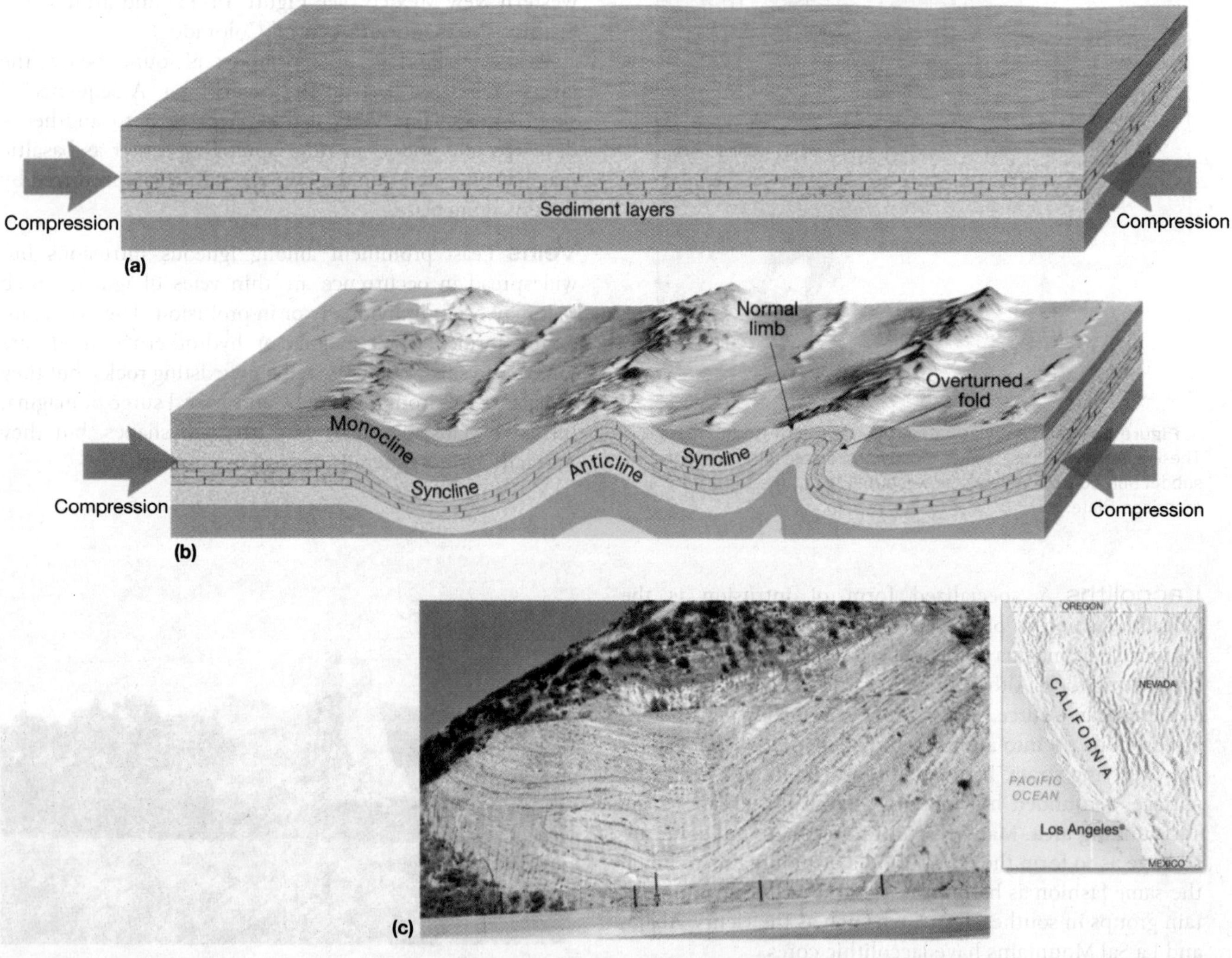

▲ **Figure 14-51** The development of folded structures. (a) Compressive stresses cause sedimentary strata that are initially horizontal to fold. (b) The basic types of folds. (c) Tightly folded sedimentary strata in a road cut near Los Angeles.

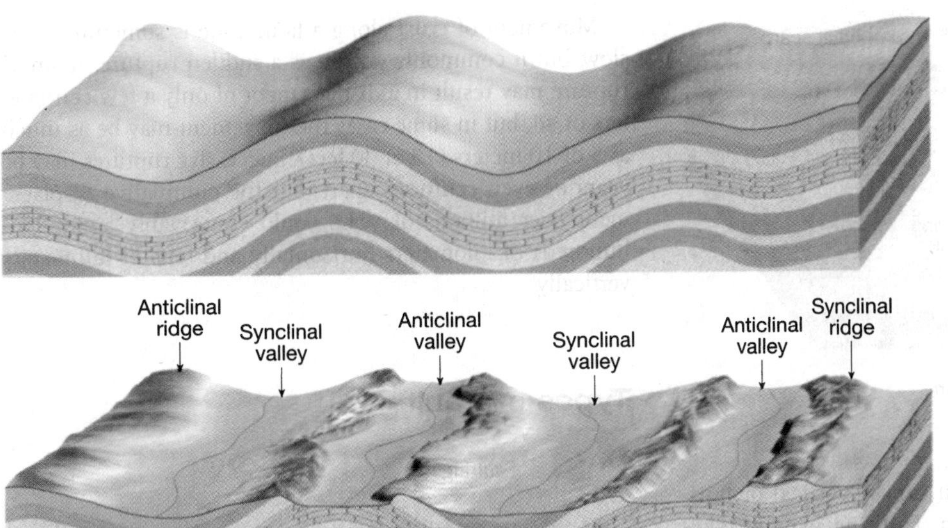

◀ **Figure 14-52** Formation of anticlinal valleys and synclinal ridges.

The configuration of the folds can be equally variable. In some cases, the folding is simple and symmetrical; elsewhere it may be of extraordinary complexity and totally without symmetry. Moreover, the structure may become even more complex when fracturing and displacement of the rock structure (faulting) accompanies folding.

Types of Folds

Structural geologists recognize many kinds of folds. For introductory physical geography, however, only a few terms are necessary (Figure 14-51). A *monocline* is a one-sided fold—a slope connecting two horizontal or gently inclined strata. A simple symmetrical upfold is an **anticline**, and a simple downfold is a **syncline**. Also relatively common is an upfold that has been pushed so extensively from one side that it becomes oversteepened enough to have a reverse orientation on the other side; such a structure is referred to as an *overturned fold*. If the pressure is enough to break the oversteepened limb and cause a shearing movement, the result is an *overthrust fold*, which causes older rock to ride above younger rock.

Topographic Features Associated with Folding

The simplest relationship between structure and topography, and one that often occurs in nature, finds the upfolded anticlines producing ridges and the downfolded synclines forming valleys. The converse relationship is also possible, however, with valleys developing on the anticlines and ridges on the synclines (Figure 14-52). This inverted topography is most easily explained by the effects of tension (pulling apart) and compression (pushing together) on the folded strata. Where a layer is arched over an upfold, tension cracks can form and provide easy footholds for erosional forces to remove materials and incise downward into the underlying strata. Conversely, the compression that acts on the downfolded beds increases their density and therefore their resistance to erosion. Thus, over a long period of time, the upfolds may be eroded away faster than the downfolds, producing *anticlinal valleys* and *synclinal ridges*.

All these types of folding, as well as many variations, are found in what is called the Ridge-and-Valley section of the Appalachian Mountains, a world-famous area noted for its remarkably parallel sequence of mountains and valleys developed on folds (Figure 14-53). The section extends for about 1600 kilometers (1000 miles) in a northeast-southwest direction across parts of nine states, with a width that varies from 40 to 120 kilometers (25 to 75 miles).

Faulting

Another prominent result of crustal stresses is the breaking apart of rock structures. When rock is broken with accompanying displacement (movement of the crust on one or both sides of the break), the action is called **faulting**

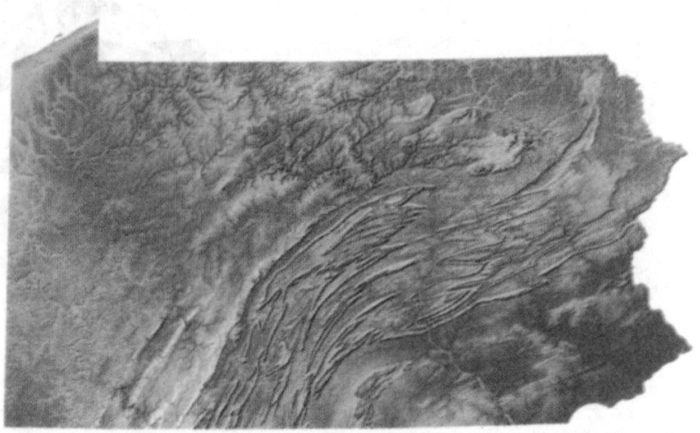

▲ **Figure 14-53** The prominently folded Appalachian Mountains are clearly shown in this shaded relief map of Pennsylvania.

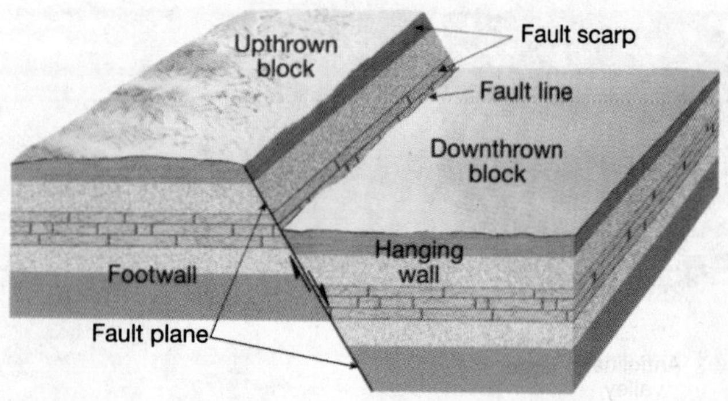

▲ **Figure 14-54** A simple fault structure.

(Figure 14-54). The displacement can be vertical or horizontal or a combination of both. Faulting usually takes place along zones of weakness in the crust; such an area is referred to as a *fault zone*, and the intersection of that zone with Earth's surface is called a *fault line*.

Major faults penetrate many kilometers into Earth's crust. Indeed, the deeper fault zones can serve as conduits to allow both water and heat from inside Earth to approach the surface. Frequently springs are found along fault lines, sometimes with hot water gushing forth. Volcanic activity is also associated with some fault zones as magma forces its way upward in the zone of weakness.

Movement of crust along a fault zone is sometimes very slow, but it commonly occurs as a sudden rupture. A single rupture may result in a displacement of only a few centimeters or so, but in some cases the movement may be as much as 5 or 10 meters (15 or 30 feet). Successive ruptures may be years or even centuries apart, but the cumulative displacement over millions of years could conceivably amount to hundreds of kilometers horizontally and tens of kilometers vertically.

Types of Faults

Although structural geologists recognize more than a dozen kinds of faults, they can be generalized into four principal types on the basis of direction and angle of movement (Figure 14-55). Two types involve displacement that is mostly vertical, a third encompasses only horizontal movement, and the fourth includes both horizontal and vertical offsets:

1. **Normal faults:** A **normal fault** results from *tension stresses* (pulling apart or extension) in the crust. It produces a very steeply inclined fault zone, with the block of land on one side being pushed up, or upthrown,

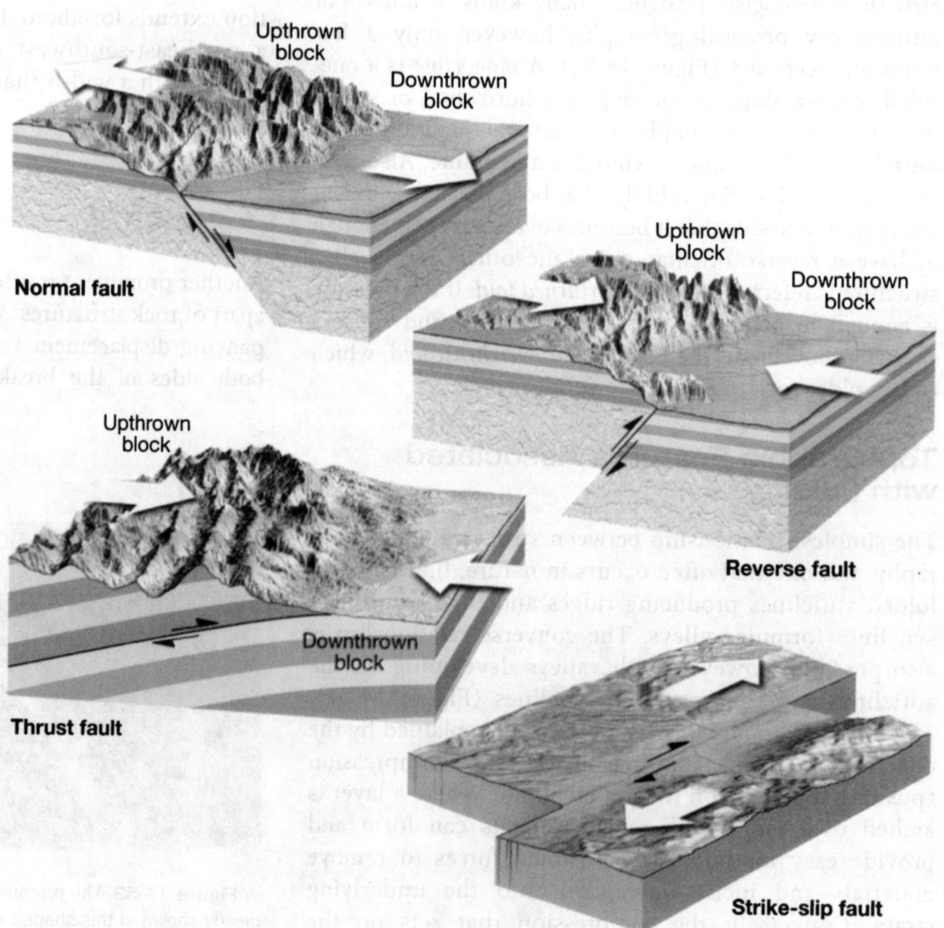

▲ **Figure 14-55** The principal types of faults. The large arrows show the direction of stress.

▲ **Figure 14-56** The Calaveras Fault, one of the strike-slip faults in the San Andreas Fault system, runs through the town of Hollister, California. Dramatic evidence of its activity is shown by this offset wall and sidewalk.

relative to the downthrown block on the other side. A prominent *fault scarp* is usually formed.

2. **Reverse faults:** A **reverse fault** is produced from *compression stresses*, with the upthrown block rising steeply above the downthrown block, so that the fault scarp would be severely oversteepened if erosion did not act to smooth the slope somewhat. Landslides often accompany reverse faulting.
3. **Thrust faults:** More complicated in structure and more impressive in their dynamics are **thrust faults** (or *overthrust faults*), in which compression forces the upthrown block to override the downthrown block at a relatively low angle, sometimes for many kilometers. Overthrusting occurs frequently in mountain building, resulting in unusual geologic relationships such as older strata being piled on top of younger rocks. Both thrust faults and reverse faults are commonly associated with subduction zones and continental collision zones.
4. **Strike-slip faults:** In a **strike-slip fault** the movement is horizontal, with the adjacent blocks being displaced laterally relative to each other (Figure 14-56). Strike-slip faults are a consequence of *shear stresses*. Transform faults described earlier in this chapter are one variety of strike-slip fault.

Fault-Produced Landforms

Fault lines are often marked by prominent topographic features. Most obvious perhaps are **fault scarps**—steep cliffs that represent the edge of a vertically displaced block (see Figure 14-54). Although most commonly associated with normal faulting, slight amounts of vertical displacement along strike-slip faults may also leave a scarp. Some fault scarps are as much as 3 kilometers (2 miles) high and extend for more than 150 kilometers (100 miles) in virtually a straight line. The abruptness of their rise, the steepness of their slope, and the linearity of their orientation combine to make some fault scarps extremely spectacular features in the landscape.

Landforms Associated with Normal Faulting

Under extensional stresses, a surface block may be faulted and relatively upthrown on one side without any faulting or uplift on the other. When this happens, the block is tilted asymmetrically, producing a steep slope along the fault scarp and a relatively gentle slope on the other side of the block. The classic example of such a **tilted fault-block mountain** range (also simply called a *fault-block mountain*) is California's Sierra Nevada (Figure 14-57), an immense block nearly 640 kilometers (400 miles) north–south and about 96 kilometers (60 miles) east–west. The spectacular eastern face is a fault scarp that has a vertical relief of about 3 kilometers (about 2 miles) in a horizontal distance of only about 20 kilometers (12 miles; Figure 14-58). In contrast, the general slope of the western flank of the range, from crest toward its "hinge line" (the line along which the gentle side begins to rise), covers a horizontal distance of more than 80 kilometers (50 miles). The shape of the range has of course been modified by other processes, but its general configuration was determined by block faulting.

Another occurrence is the relative uplift of a block of land between two parallel faults, an action that produces a structure called a **horst** (Figure 14-59). Frequently such horsts are the result of the land on both sides being downthrown rather than the block itself being uplifted. In either case, the horst may take the form of a plateau or a mountain mass with two steep, straight sides.

At the other extreme is a **graben**—a block of land bounded by parallel faults in which the block has been relatively downthrown, producing a distinctive structural valley with straight, steep-sided fault scarps on both sides.

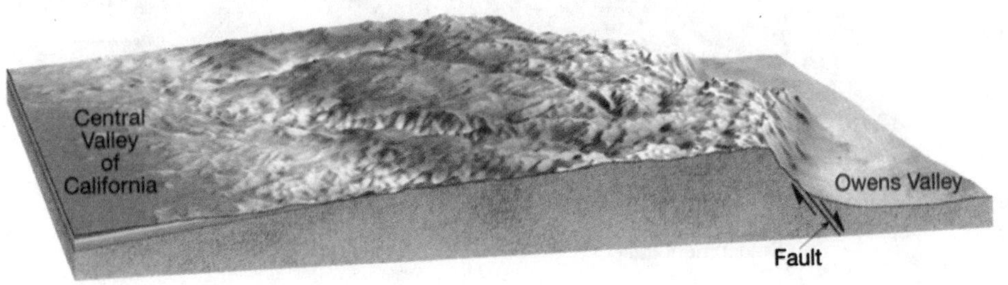

◀ **Figure 14-57** The western slope of the Sierra Nevada is long and gentle, whereas the eastern slope is short and steep. This gentle-steep combination is the result of enormous block faulting on the eastern side.

▲ **Figure 14-58** The steep east side of the Sierra Nevada mountains in California is a prominent fault scarp.

Tilted fault-block mountains, grabens, and horsts may occur side by side. The Basin-and-Range region of the western United States (covering most of Nevada and portions of surrounding states; see Figure 18-26) is a vast sequence of tilted fault-block mountains and *half-grabens* (basins formed by blocks tilting down along just one fault), along with a few true grabens and horsts. As the Basin-and-Range region undergoes sustained extensional faulting, the landscape continues to spread and develop along normal faults and *detachment faults*. Detachment faults form where steeply dipping normal faults tie into low angle (or nearly horizontal) faults below the surface that develop along the boundary between brittle crustal rocks above and more ductile crustal rocks below.

Rift Valleys Where a divergent plate boundary develops within a continent, the resulting downfaulted grabens occasionally extend for extraordinary distances as linear valleys enclosed between steep fault scarps. Such lengthy troughs are called *rift valleys* (see Figure 14-14), and they comprise some of Earth's most notable structural lineaments, particularly the Great Rift Valley in East Africa.

Landforms Associated with Strike-Slip Faulting

A wide variety of landforms can result from strike-slip faulting (Figure 14-60). The surface trace of a large strike-slip fault may be marked by a **linear fault trough** or valley, formed by repeated movement and fracturing of rock within the fault zone. Small depressions known as *sags* develop through the settling of rock within the fault zone, and may become filled with water to form **sag ponds**. Linear ridges and scarps trending parallel to the trace of the fault may develop from slight vertical displacement that may occur along a strike-slip fault.

Perhaps the most conspicuous landform produced by strike-slip faulting is an **offset stream** (Figure 14-61). Streams flowing across the fault are displaced by periodic fault movement or diverted when a *shutter ridge* is faulted in front of a drainage channel.

Earthquakes

1. Earthquake Waves
2. Seismographs

Usually, but not exclusively, associated with faulting is the abrupt shaking of the crust known as an *earthquake*. The movement varies from mild, imperceptible tremors to wild shaking that persists for many seconds.

An **earthquake** is essentially a vibration in Earth produced by shock waves resulting from a sudden displacement along a fault (earthquakes may also develop from the movement of magma or sudden ground subsidence). The fault movement allows an abrupt release of energy, usually after a long, slow accumulation of strain. While a fault rupture can take place at the surface, displacement can also take place at considerable depth—as deep as 600 kilometers (375 miles) beneath the surface in the case of subduction zones.

Earthquake Waves The energy released in an earthquake moves through Earth in several different kinds of seismic waves that originate at the center of fault motion, called the *focus* of the earthquake (Figure 14-62). These waves travel outward in widening circles, like the ripples produced when a rock is thrown into a pond, gradually diminishing in amplitude with increasing distance from the focus. The strongest shocks, and greatest crustal vibration, are often felt on the ground directly above the focus, at the location known as the **epicenter** of the earthquake.

The fastest moving earthquake waves, and the first to be felt during an earthquake, are known as *P* or *primary waves*. P waves move through Earth in the same fashion as sound waves, by alternately compressing and relaxing the medium through which they are traveling. The initial jolt of the P waves is followed by the strong side-to-side and

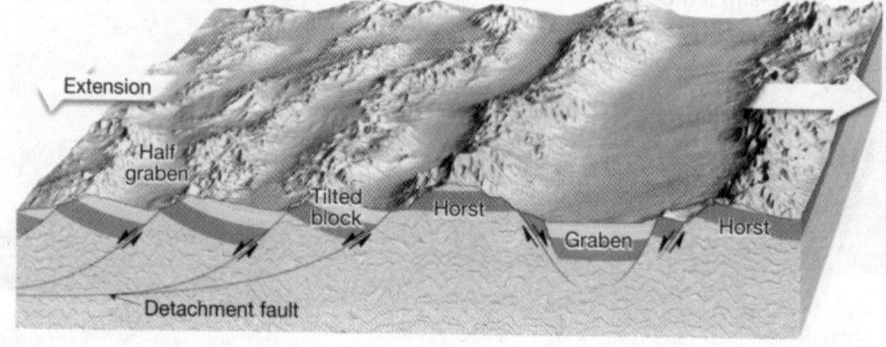

▶ **Figure 14-59** Landforms associated with tensional stresses and normal faulting. Horsts and grabens are bound by faults on both sides, while tilted blocks have been faulted along just one side, forming fault block mountains and "half grabens." Extension may also result in detachment faults, where a steep normal fault ties in with a nearly horizontal fault below.

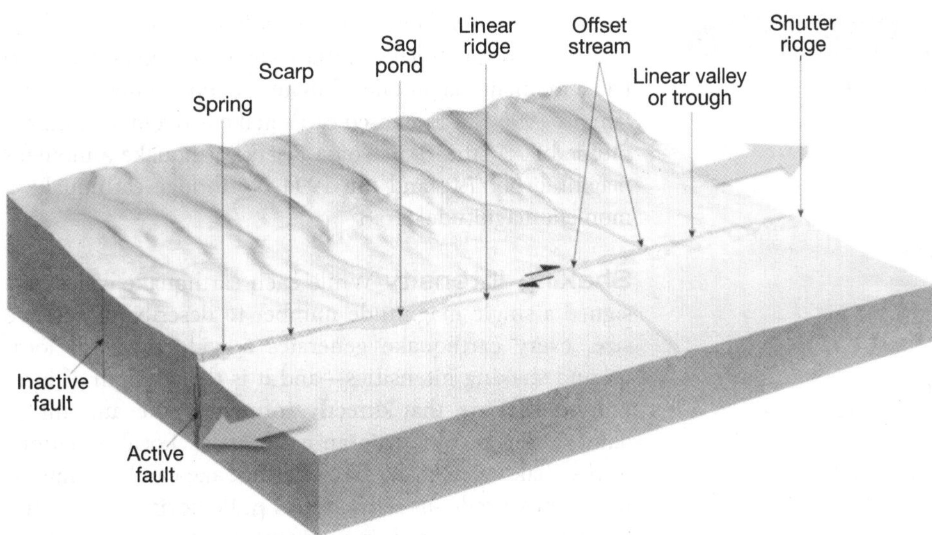

◀ **Figure 14-60** Common landforms produced by strike-slip faulting.

up-and-down shearing motion of the slower-moving *S* or *secondary waves*. Both P waves and S waves travel through the body of Earth (so they are also known as *body waves*), while the motion of a third type of earthquake wave is limited to the surface. These *surface waves* typically arrive immediately after the S waves and produce strong side-to-side movement as well as the up-and-down "rolling" motion often experienced during a large earthquake.

Because P waves and S waves travel at different speeds—at the surface P waves travel through bedrock at about 6 kilometers (3.7 miles) per second, while S waves travel at about 3.5 kilometers (2.2 miles) per second—it is possible to determine the distance to an earthquake's focus. The farther away an earthquake, the greater the time lag between the arrival of the P waves and the S waves. By using a network of *seismographs* (instruments used to record earthquakes) and comparing arrival times of the seismic waves, seismologists can pinpoint the focus of an earthquake with great precision.

Earthquake Magnitude Perhaps the most widely mentioned—and most widely misunderstood—aspect of an earthquake is its **magnitude**. Magnitude describes the relative amount of energy released during an earthquake. Magnitudes are calculated on a logarithmic scale, with an energy increase from one magnitude to the next of about 32 times. The difference in size between small earthquakes (magnitudes less than 3) and large earthquakes (magnitude 7 and higher) is enormous. A magnitude 4 earthquake releases about 32 times more energy than a magnitude 3; a magnitude 5 releases 1000 times more energy than a magnitude 3; and a magnitude 7 releases 1,000,000 times more energy than a magnitude 3.

The most commonly quoted magnitude is the *Richter scale*, devised in 1935 by Charles Richter at the California Institute of Technology in Pasadena. While the Richter scale magnitude is relatively easy for seismologists to calculate, it is not ideal for comparing the sizes of very large earthquakes (magnitude 7 and higher). The more recently

▲ **Figure 14-61** Offset stream along the San Andreas Fault in the Temblor Range of California.

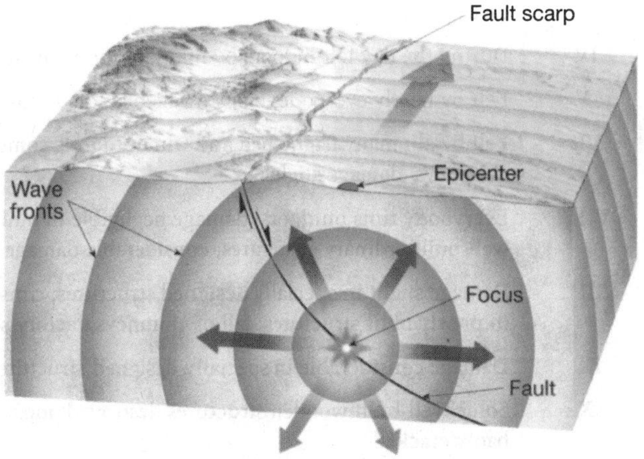

▲ **Figure 14-62** Relationship among focus, epicenter, and seismic waves of an earthquake. The earthquake waves are indicated by the concentric circles.

TABLE 14-2 Worldwide Earthquake Frequency

Magnitude	Number per Year
<3.4	800,000
3.5–4.2	30,000
4.3–4.8	4800
4.9–5.4	1400
5.5–6.1	500
6.2–6.9	100
7.0–7.3	15
7.4–7.9	4
8.0	1 every 1 or 2 years

devised *moment magnitude* is now the most commonly used scale to describe the size of large quakes, although the Richter scale (also known as the *local magnitude* scale) and other scales are still in use. Because the method of calculation varies among the magnitude scales, slightly different magnitude numbers are often reported for the same earthquake.

In any given year, tens of thousands of earthquakes occur around the world (Table 14-2). The vast majority of these are too small to be felt by people, but perhaps 60 or 70 are large enough to cause damage or loss of life (commonly magnitude 6 or higher). Great earthquakes—those with magnitude 8 or higher—might occur only once every few years. The two largest earthquakes yet recorded were the 1960 Chile earthquake with a moment magnitude of 9.5, and the 1964 Alaska earthquake with a moment magnitude of 9.2. The great Sumatra earthquake of December 2004 also had a moment magnitude of about 9.2. For comparison, the famous 1906 San Francisco earthquake had a moment magnitude of 7.7, the 1989 Loma Prieta earthquake a moment magnitude of 7.0; and the 1994 Northridge earthquake a moment magnitude of 6.8.

Shaking Intensity While each earthquake can be assigned a single magnitude number to describe its relative size, every earthquake generates a wide range of local ground shaking intensities—and it is the strength of local ground shaking that directly influences the amount of damage that results from an earthquake. Local variations in the shaking intensity of an earthquake can be quantitatively measured—usually as the peak horizontal ground acceleration expressed as a percentage of gravity—but the most widely used intensity scale was originally devised in 1902 by the Italian geologist Giuseppe Mercalli. Updated by the U.S. Coast and Geodetic Survey, the **modified Mercalli intensity scale** assigns the strength of local shaking to one of 12 categories, based on the observed effects and damage (Table 14-3). The modified Mercalli scale is also used by seismologists to describe the anticipated intensity of ground shaking that may occur during future earthquakes in a region.

Earthquake Hazards Most of the damage from an earthquake is due to ground shaking. Generally, the strength of ground shaking decreases with increasing distance from the epicenter of the earthquake, but this pattern can be significantly modified by the local geology. Loose, unstable regolith, sediments, and soil tend to amplify ground shaking—buildings on such "soft"

TABLE 14-3 Modified Mercalli Intensity Scale

I.	Not felt except by very few people under especially favorable circumstances
II.	Felt only by a few persons at rest, especially on upper floors of buildings
III.	Felt quite noticeably indoors, especially on upper floors of buildings, but many people do not recognize it as an earthquake
IV.	During the day felt indoors by many, outdoors by few; sensation like heavy truck striking building
V.	Felt by nearly everyone, many awakened; disturbances of trees, poles, and other tall objects sometimes noticed
VI.	Felt by all; many frightened and run outdoors; some heavy furniture moved; few instances of fallen plaster or damaged chimneys; damage slight
VII.	Everybody runs outdoors; damage negligible in buildings of good design and construction; slight to moderate in well-built ordinary structures; considerable damage in poorly built or badly designed structures
VIII.	Damage slight in specially designed structures; considerable in ordinary substantial buildings, with partial collapse; great in poorly built structures; fall of chimneys, factory stacks, columns, monuments, and other vertical features
IX.	Damage considerable in specially designed structures; buildings shifted off foundations; ground cracked conspicuously
X.	Some well-built wooden structures destroyed; most masonry and frame structures destroyed with foundations; ground badly cracked
XI.	Few, if any, masonry structures remain standing; bridges destroyed; broad fissures in ground
XII.	Damage total; waves seen on ground surfaces; objects thrown upward into the air

ground will shake much more strongly than those located on bedrock the same distance from the epicenter. In addition, loose, water-saturated sediments, such as coastal landfill, may undergo **liquefaction**—during the shaking of an earthquake, the water-saturated material turns fluid, resulting in subsidence, fracturing, and horizontal sliding of the ground surface. A great deal of the damage in the 1989 Loma Prieta earthquake and in the 1995 Kobe earthquake was due to amplified ground shaking and liquefaction in artificial fill (Figure 14-63).

Landslides are often triggered by earthquakes. During the 1964 Alaska earthquake, for example, dozens of homes in the Turnagain Heights residential area outside of Anchorage were destroyed when a 2.6-kilometer (1.6-mile) wide section of the development moved toward the sea in a series of massive landslides.

Another kind of hazard associated with earthquakes involves water movements in lakes and oceans. Abrupt crustal movement can set great waves (known as *seiches*) in motion in lakes and reservoirs, causing them to overflow shorelines or dams in the same fashion that water can be sloshed out of a dishpan. Much more significant, however, are great seismic sea waves called *tsunami* (the Japanese term for these large waves), which are sometimes generated by undersea earthquakes or landslides. These waves, often occurring in a sequential train, move quickly across the ocean. They are barely perceptible in deep water, but when they reach shallow coastal waters, they sometimes build up to heights of more than 15 meters (50 feet) and crash onto the shoreline with devastating effect (where they are often incorrectly called "tidal waves"). Most of the deaths from the 1964 Alaska earthquake were the result of tsunami—even as far away as Crescent City, California, 12 people were killed when a 6-meter (19-foot) high tsunami came on shore. In December 2004, about 227,000 people were killed when a tsunami generated by a subduction zone earthquake off the Indonesian island of Sumatra spread through the Indian Ocean. A detailed description of tsunami formation and the 2004 Indian Ocean tsunami disaster is found in Chapter 20, where we discuss the dynamics of ocean waves.

The Complexities of Crustal Configuration

Considering each internal process in turn as we have just done is a helpful way to systematize knowledge. Doing so presents an artificial and misleading picture, however, because in nature these processes are interrelated. To attempt a more balanced assessment, let us consider a highly simplified statement of the origin of the gross contemporary topographic features of a small part of Earth's surface—a mountainous section of northwestern Montana that encompasses the spectacular scenery of Glacier National Park.

This area, now part of the northern Rocky Mountains, was below sea level for many millions of years. Most of the rocks in the region were formed in the Precambrian Era (more than 542 million years ago), when much of the area now occupied by the Rocky Mountain Cordillera consisted of a large, shallow, seawater-filled trough. Muds and sands were washed into this Precambrian sea for millennia, and a

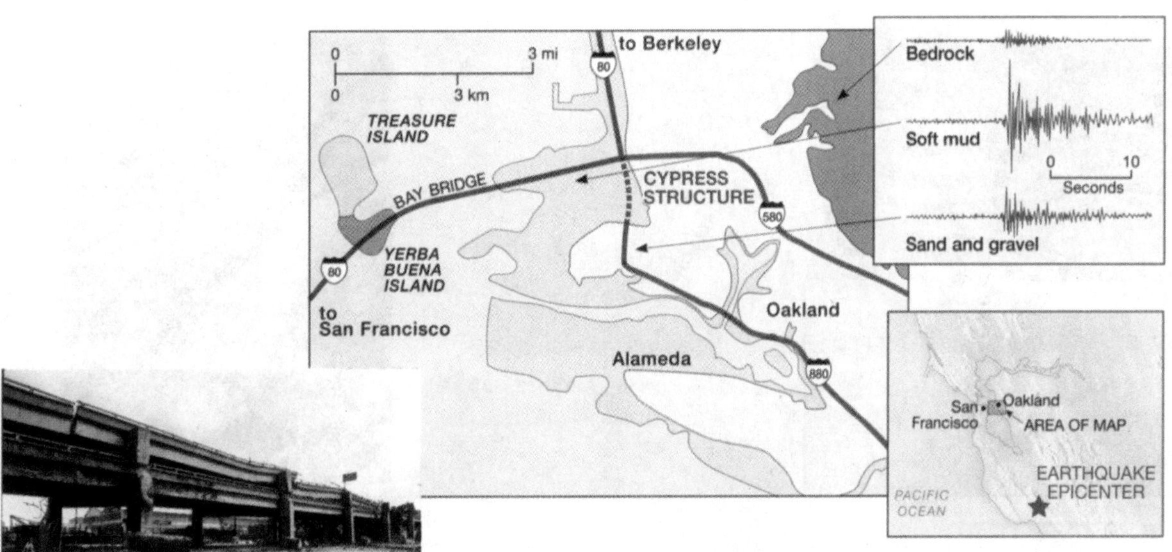

▲ **Figure 14-63** The portion of the Cypress freeway structure in Oakland, California that collapsed during the 1989 magnitude 7.0 Loma Prieta earthquake stood on soft mud (dashed red line). Adjacent parts of the structure built on firmer ground (solid red line) remained standing. Seismograms in the upper right show that ground shaking was more severe on the soft mud than on the nearby bedrock.

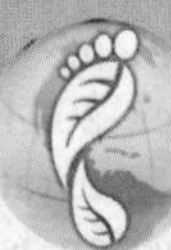

PEOPLE AND THE ENVIRONMENT

Earthquake Forecasting

In recent years considerable attention has been focused on earthquake prediction and forecasting.

The goal of earthquake prediction is to provide warnings far enough in advance to minimize loss of life and property. Such short-term predictions hinge on the existence of "precursors" or warning signs of an impending earthquake. Many such possible warning signs are being studied, including patterns of small earthquakes that sometimes occur before major quakes ("foreshocks"), changes in water-well levels, changes in the amount of dissolved radon gas in groundwater, variations in electrical conductivity of rocks, the bulging of the ground surface, and changes in the behavior of animals.

One of the world's most elaborate earthquake research projects has been underway since 1985 in the tiny California town of Parkfield, which sits atop the San Andreas Fault in the central part of the state. Since the mid-1800s this segment of the fault has ruptured at fairly regular intervals—1857, 1881, 1901, 1922, 1934, and 1966—each time producing an earthquake of about magnitude 6. The U.S. Geological Survey (USGS) and dozens of university and industry researchers began intensive monitoring of the fault here in hopes of "capturing" the next earthquake as it happens. They especially wanted to have enough monitoring equipment in place to detect any precursors before the quake; such precursors might lead seismologists closer to short-term predictions of earthquakes. The Parkfield earthquake came on September 28, 2004—a number of years later than expected, but it yielded a great deal of data for the scientists to analyze.

Just a few months before the quake, drilling was started on the San Andreas Fault Observatory at Depth (SAFOD), a project that is installing instruments 2 to 3 kilometers (1.2–1.9 miles) deep in the active fault zone of the San Andreas (Figure 14-B). The Phase 3 drilling of the project was completed in September 2007—after the September 2004 Parkfield earthquake, but SAFOD is already providing seismologists with valuable data on fault-zone rocks and fluids.

A new tool currently being used by NASA scientists to study the San Andreas Fault is airborne Synthetic Aperture Radar (SAR; Figure 14-C). By repeatedly gathering SAR data over periods of time, very slight differences in the surface can be detected, enabling scientists to measure slight amounts of creep or strain developing along the fault that might signal an impending earthquake.

Currently, there is no reliable method of predicting earthquakes, and some seismologists think it unlikely that the short-term prediction of earthquakes will ever exist. However, seismologists have made great strides toward long-term forecasts of earthquakes: the probability of an earthquake occurring along a segment of a fault within a period of several decades.

Long-term probabilities are based on the balance of two processes: the loading of strain onto faults, and the slip along the faults (which should, over time, release the strain). By recognizing seismic "gaps" where little strain has been released in recent decades, seismologists can identify which fault segments are most likely to rupture in the future. For example, the USGS recently concluded that in the San Francisco Bay region, there is a 63 percent probability of at least one magnitude 6.7 or greater earthquake occurring before the year 2036. Although such forecasts do not state the likelihood of a quake coming in any particular month or year, long-term forecasts can help local governments, businesses, and the general public to establish priorities for retrofitting structures and for disaster preparation.

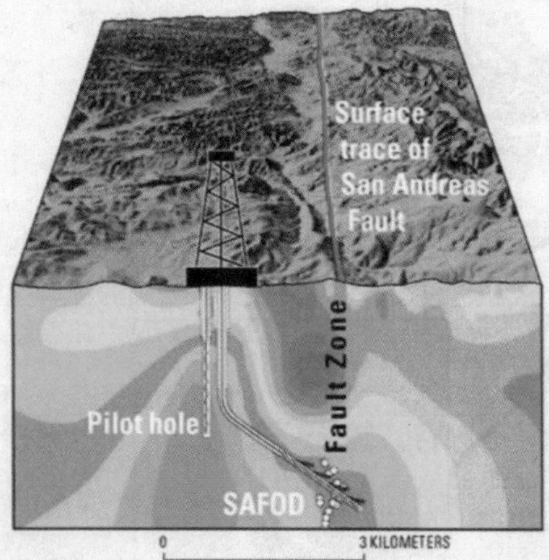

▲**Figure 14-B** The San Andreas Fault Observatory at Depth instrument bore at Parkfield, California, extends diagonally into the active fault zone. The colors below the surface represent differences in electrical resistivity in the rocks. The lowest resistivity is shown with red—the zone above the SAFOD instruments may represent a fluid-rich zone in the fault.

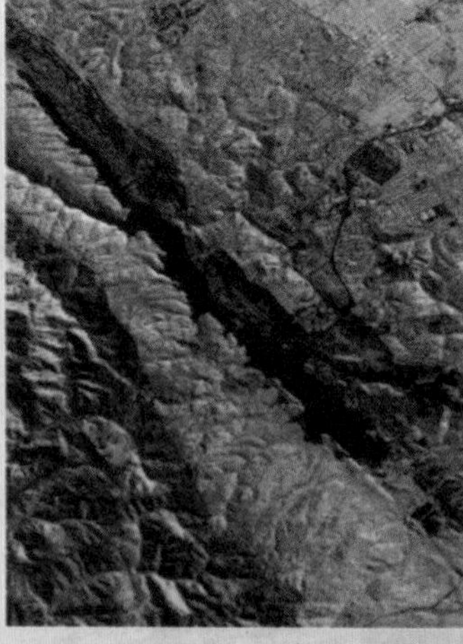

▲**Figure 14-C** Radar image of the San Andreas Fault zone south of San Francisco. The fault runs diagonally from upper left to lower right. The area shown in black is Crystal Springs Reservoir, a key water storage facility for San Francisco that is built right on the fault; pink shows urbanized areas.

vast thickness of sedimentary strata built up. These limestones, shales, and sandstones accumulated as six distinct formations, each with a conspicuous color variation known collectively as the Belt Series.

Occasionally during this lengthy epoch of sedimentary accumulation, igneous activity added variety. Most notable was a vast outpouring of flood basalt that issued from fissures in the ocean floor and was extruded in the form of a submarine lava flow. Further sediments were then deposited on top of the lava flow.

Igneous intrusions were also injected from time to time, including one large sill and a number of dikes. The igneous rocks—both the flood basalt and the various intrusions—initiated contact metamorphism, whereby the tremendous heat of the igneous material converted some of the adjacent sedimentary rocks into metamorphic rocks (mostly changing limestone to marble and sandstone to quartzite).

After a long gap in the geologic record, during which the Rocky Mountain region was mostly above sea level, the land once again sank below the ocean during the Cretaceous Period (145.5 to 65.5 million years ago) and another thick series of sediments was deposited. This was followed by a period of mountain building so significant that it has been named the Rocky Mountain Revolution. In the Glacier National Park area, the rocks were compressed and uplifted, converting the site of the former sea into a mountainous region.

Along with uplift came extreme lateral pressure from the west, convoluting the gently downfolded strata into a prominent anticline. Continuing pressure then overturned the anticline toward the east, with a lengthening western limb and a truncating eastern limb. This additional strain on the rock and the persistent crustal pressure eventually caused a vast rupture and faulting. The entire block was then pushed eastward by one of the greatest thrust faults known, the Lewis Overthrust (Figure 14-64). This remarkable fault forced the Precambrian sedimentary rocks out over the Cretaceous strata that underlie the plains to the east by as much as 30 kilometers (about 20 miles). The plane of the thrust fault was only slightly above the horizontal, nowhere exceeding a dip of 10°. This had the peculiar effect of placing older rock layers on top of much younger strata. The terrain thus produced is referred to as "mountains without roots." Chief Mountain is world famous as a rootless mountain because of its conspicuous location as an erosional outlier east of the main range (Figure 14-65).

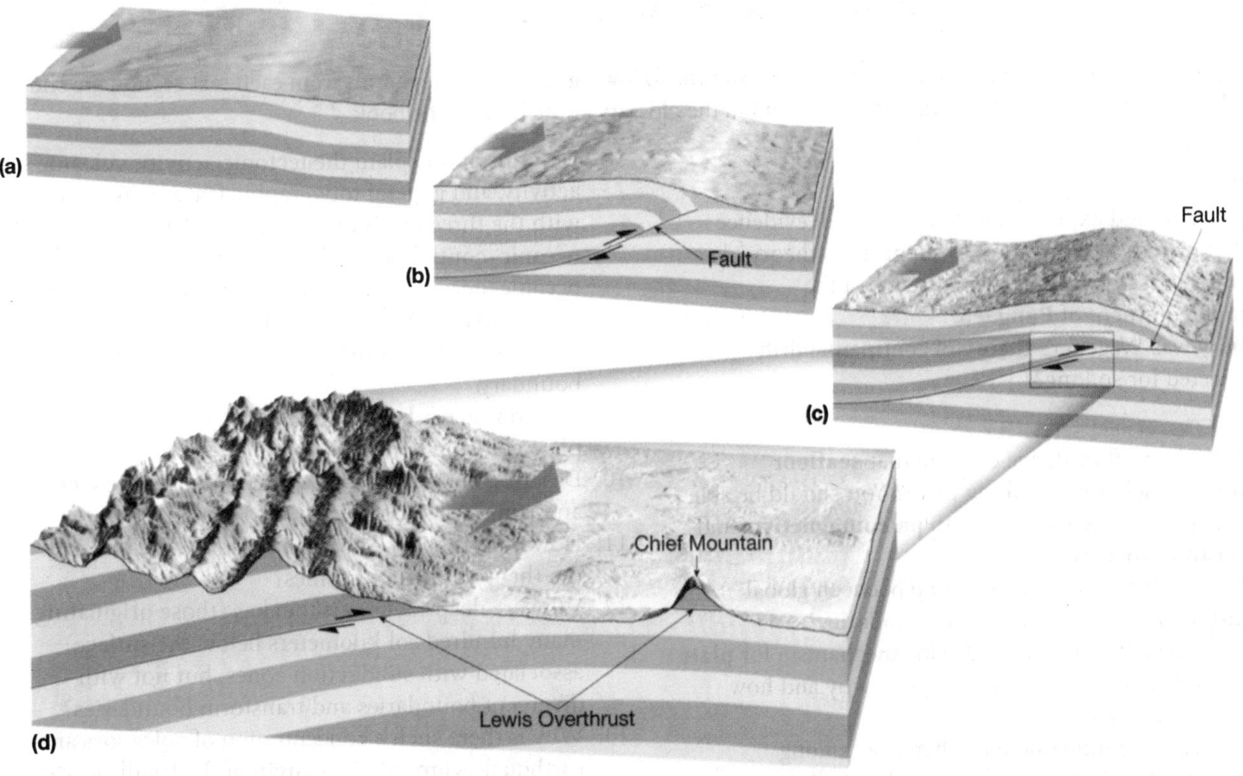

▲ **Figure 14-64** Sequential development of the Lewis Overthrust. (a) Compression of original sedimentary rocks begins the process of uplift and folding. (b and c) Continued pressure from the west causes overturning of the fold and faulting along the eastern limb of the anticline. (d) Subsequent erosion produces the present topography, with Chief Mountain as a residual outlier to the east of the range.

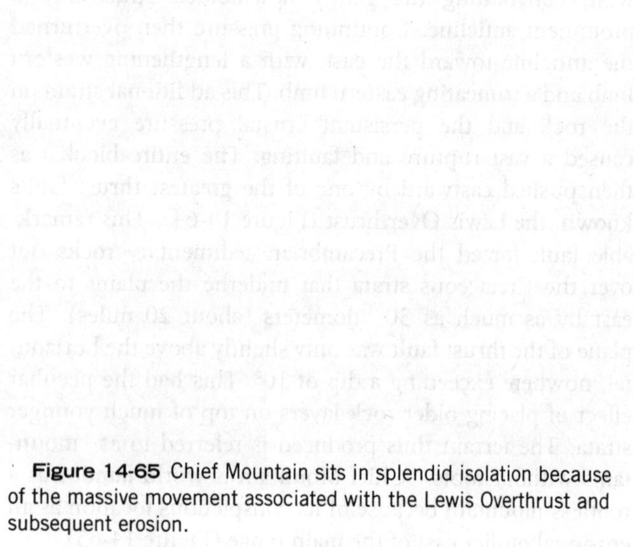

Figure 14-65 Chief Mountain sits in splendid isolation because of the massive movement associated with the Lewis Overthrust and subsequent erosion.

CHAPTER 14 LEARNING REVIEW

After studying this chapter, you should be able to answer the following questions. Key terms from each text section are shown in bold type. Definitions for key terms are also found in the glossary at the back of the book.

From Rigid Earth to Plate Tectonics (p. 375)

1. Describe and explain at least two lines of evidence that Alfred Wegener used to support his theory of **continental drift** and the existence of the supercontinent of **Pangaea**.
2. Why was Wegener's theory of continental drift rejected for so long?

Plate Tectonics (p. 377)

3. What lines of evidence confirm that **seafloor spreading** has been taking place? You should be able to explain evidence from both **paleomagnetism** and **ocean floor cores**.
4. What is the general relationship between global earthquake activity and plate boundaries?
5. Describe and explain the driving mechanism for **plate tectonics**—in other words, explain why and how lithospheric plates move.
6. Describe the fundamental differences among **divergent**, **convergent**, and **transform plate boundaries**.
7. Describe and explain the tectonic activity, volcanic activity, and general topographic features associated with the two kinds of divergent plate boundary: **midocean ridges** and **continental rift valleys**. Mention at least one present-day example of each of these kinds of divergent boundary.
8. Describe and explain the tectonic activity, volcanic activity, and general topographic features associated with the three kinds of convergent plate boundary: oceanic–continental plate subduction, oceanic–oceanic plate subduction, and continental plate collision. Mention at least one present-day example of each of these kinds of convergent boundary.
9. Why doesn't **subduction** take place in a continental plate collision zone?
10. Describe the differences between an **oceanic trench** and a midocean ridge.
11. How does the San Andreas Fault system fit in with the theory of plate tectonics?
12. Why are deep-focus earthquakes (those originating many hundreds of kilometers below the surface) associated with subduction zones, but not with divergent boundaries and transform boundaries?
13. Why is there such a concentration of volcanoes and earthquakes around the margin of the Pacific Ocean—the region referred to as the **Pacific Ring of Fire**?
14. What is a **hot spot**? Name at least one present-day example of a hot spot.
15. How is a hot spot trail different from a **volcanic island arc**?
16. How does the **mantle plume** model explain the existence of hot spots?

17. In what ways don't hot spots fit in with the basic theory of plate tectonics and plate boundaries?
18. In what ways have hot spots been used to verify that plate motion is taking place?
19. What is a **terrane** and how does one form?

Volcanism (p. 390)

20. What is meant by the term **volcanism**?
21. Define and contrast the following: **magma**, **lava**, and **pyroclastics**.
22. Explain the general differences in silica content and style of volcanic eruption (i.e., quiet lava flows versus explosive eruptions of pyroclastics) associated with basaltic magma, andesitic magma, and rhyolitic magma.
23. What is **flood basalt**?
24. Describe and explain the general formation, shape, and structure of the following kinds of volcanic peaks: **shield volcanoes**, **composite volcanoes**, **lava domes (plug domes)**, and **cinder cones**.
25. Explain the formation of a **caldera** such as Crater Lake in Oregon.
26. Describe and explain the origin and characteristics of **pyroclastic flows** and **volcanic mudflows (lahars)**.
27. Define the following terms: **igneous intrusion**, **pluton**, **batholith**, and **dike**.

Folding (p. 404)

28. How is **folding** different from **faulting**?
29. What is a **syncline**? What is an **anticline**?
30. How is it possible for a syncline to be associated with a topographic ridge and for an anticline to be associated with a topographic valley?

Faulting (p. 405)

31. Explain the differences in stress direction and displacement among the four basic kinds of faults: **normal faults**, **reverse faults**, **thrust faults**, and **strike-slip faults**.
32. What is a **fault scarp**?
33. Describe and explain the formation of landforms that result from normal faulting (such as **grabens**, **horsts**, and **tilted fault-block mountains**).
34. Describe and explain the formation of landforms that result from strike-slip faulting (such as **linear fault troughs**, **sag ponds**, and **offset streams**).
35. What is an **earthquake**?
36. What is the difference between the focus of an earthquake and the **epicenter** of an earthquake?
37. What is the difference between earthquake **magnitude** and earthquake shaking intensity?
38. About how much more energy is released during a magnitude 6 earthquake than during a magnitude 5 earthquake?
39. What does the **modified Mercalli intensity scale** convey?
40. Briefly describe the differences among the P waves, S waves, and surface waves of an earthquake.
41. Under what conditions does **liquefaction** occur?
42. Why do areas far away from the epicenter sometimes experience greater damage during an earthquake than areas closer to the epicenter? (Consider factors other than differences in building construction.)

Log in to **www.mygeoscienceplace.com** for animations, videos, RSS feeds, maps, and self-study quizzes to enhance your study of **The Internal Processes**.

15 Preliminaries to Erosion: Weathering and Mass Wasting

If the internal processes of landscape formation are overwhelming, the external processes are inexorable. During all the time that lithospheric plates are moving, the crust is bending and breaking, and volcanoes are erupting and intrusions forming, another suite of natural forces is simultaneously at work. These are the external forces, often mundane and even minuscule in contrast to those described in the preceding chapter. Yet the cumulative effect of the external processes is awesome; they are capable of wearing down anything the internal forces can erect. No rock is too resistant and no mountain too massive to withstand their unrelenting power. Ultimately, the detailed configuration of peaks, slopes, valleys, and plains is molded by the work of gravity, water, wind, and ice.

Denudation

The overall effect of the disintegration, wearing away, and removal of rock material is generally referred to as **denudation**, a term that implies a lowering of continental surfaces. Denudation is accomplished by the interaction of three types of activities:

- **Weathering** is the breaking down of rock into smaller components by atmospheric and biotic agencies.
- **Mass wasting** involves the relatively short-distance downslope movement of broken rock material under the direct influence of gravity.
- **Erosion** consists of more extensive and generally more distant removal, transportation, and eventual deposition of fragmented rock material.

All three processes are illustrated conceptually in Figure 15-1.

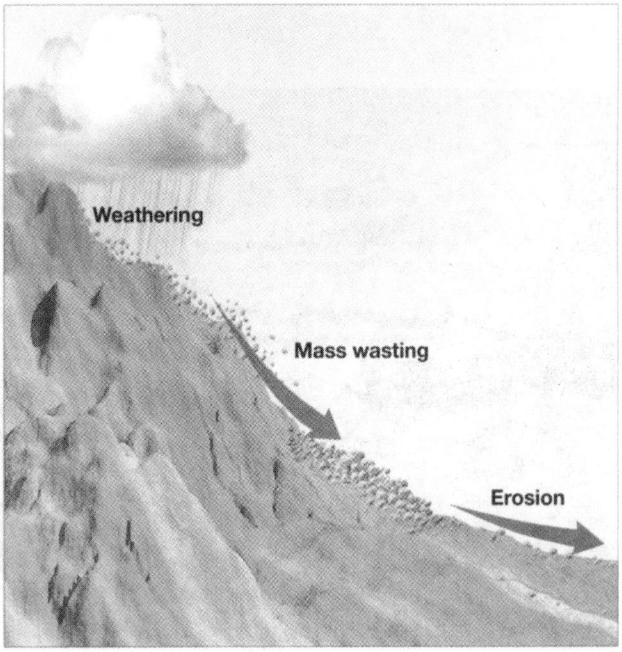

▲ **Figure 15-1** Denudation, the lowering of continental surfaces, is accomplished by a combination of weathering, mass wasting, and erosion.

The Impact of Weathering and Mass Wasting on the Landscape

The most readily observable landscape effect of weathering is the fragmentation of bedrock—the reduction of large rock units into more numerous and less cohesive smaller units.

Mass wasting always involves the downslope movement of rock material. Its impact on the landscape is normally twofold: an open scar may be left on the surface that was vacated, and an accumulation of debris is deposited somewhere downslope from the scar.

Although weathering and mass wasting are often deemphasized as processes of landform modification and shaping, the role they play is critical. Take the Grand Canyon of the Colorado River, for example. One of the most impressive landscape features of the world, the canyon is usually attributed only to the erosive powers of the river. However, most of what we see in the Grand Canyon are forms produced by weathering and mass wasting (Figure 15-2). Besides deepening its channel, the river's main role is to transport the sediment loosened by weathering and pulled into the stream by gravity and mass wasting. The Grand Canyon also exhibits the mark of *differential weathering*: some kinds of rock are easily weathered, while others are much more resistant (this important topic will be explored further in Chapter 18).

◀ Badlands topography in Theodore Roosevelt National Park in North Dakota.

Weathering and Rock Openings

The first step in the shaping of Earth's surface by external processes is *weathering*, the mechanical disintegration and/or chemical decomposition that destroys the coherence of bedrock and begins to fragment rock masses into progressively smaller components. Fast or slow, mechanical and chemical weathering occurs wherever the lithosphere and atmosphere meet. It occurs with great subtlety, however, breaking chemical bonds, separating grain from grain, pitting the smooth, and fracturing the solid. It is the aging process of rock surfaces, the process that must precede the other, more active, forms of denudation.

Whenever bedrock is exposed, it weathers. It often has a different color or texture than neighboring unexposed bedrock. Most significant from a topographic standpoint,

◀ **Figure 15-2** The Colorado River in Arizona's Grand Canyon is the principal erosive agent that transports rock out of the canyon, but nearly all of the material removed to form the massive canyon is the result of weathering and mass wasting.

exposed bedrock is likely to be looser and less coherent than the underlying rock. Blocks or chips may be so loose that they can be detached with little effort. Sometimes pieces are so "rotten" that they can be crumbled by finger pressure. Slightly deeper in the bedrock, there is firmer, more solid rock, although along cracks or crevices weathering may extend to considerable depths. In some cases, the weathering may reach as much as several hundred meters beneath the surface. This penetration is made possible by open spaces in the rock bodies and even between the mineral grains. Subsurface weathering is initiated along these openings, which can be penetrated by such weathering agents as water, air, and plant roots. As time passes, the weathering effects spread from the immediate vicinity of the openings into the denser rock beyond (Figure 15-3).

Openings in the surface and near-surface bedrock are frequently microscopic, but they may also be large enough to be conspicuous and are sometimes huge. In any case, they occur in vast numbers and provide avenues along which weathering agents can attack the bedrock and break it apart.

Broadly speaking, five types of openings are common:

1. **Microscopic openings**: Microscopic openings in the rock surface occur in profusion. Although tiny, they are so numerous that they can be responsible for extensive weathering. They may consist of spaces between crystals of igneous or metamorphic rocks, pores between grains of sedimentary rocks, or minute fractures within or alongside mineral grains in any kind of rock.
2. **Joints**: The most common structural features of the rocks of the lithosphere are **joints**—cracks that develop as a result of stress—but the rocks do not show appreciable displacement along these breaks parallel

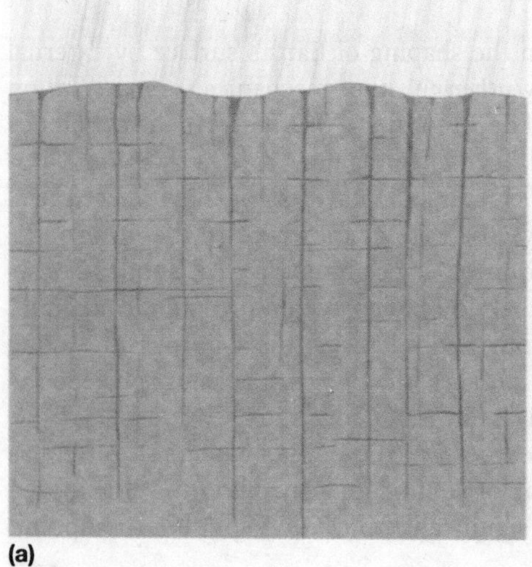
(a)

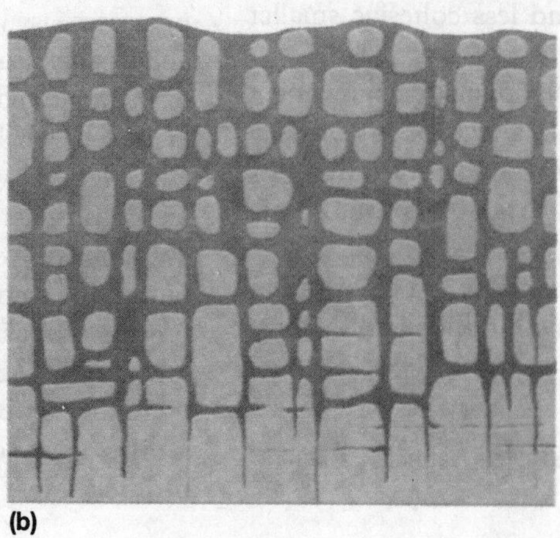
(b)

◀ **Figure 15-3** Weathering below the surface can take place in bedrock containing many cracks (a) before weathering; (b) after weathering.

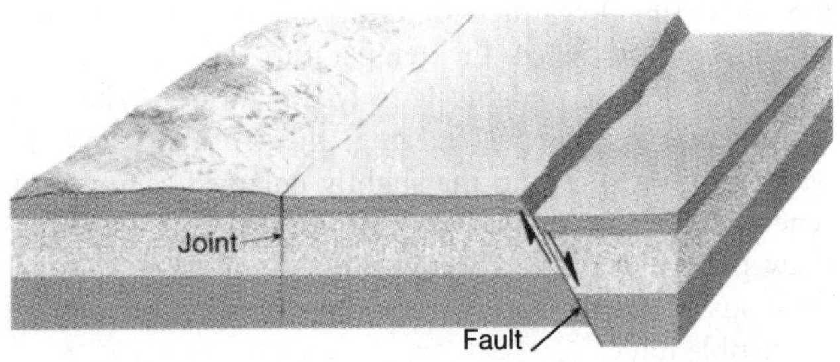

▲ **Figure 15-4** The essential difference between joints and faults is that joints exhibit no displacement along either side of the crack.

to the joint walls. Joints are innumerable in most rock masses, dividing them into blocks of various sizes. Because of their ubiquity, joints are the most important of all rock openings in facilitating weathering.

3. **Faults:** As we saw in Chapter 14, *faults* are breaks in bedrock along which there is relative displacement of the walls of the crack (Figure 15-4). Faults are generally individual or occur only in small numbers, whereas joints normally are numerous. A further difference between faults and joints is that faults sometimes appear as major landscape features, extending for tens or even hundreds of kilometers, whereas joints are normally minor structures extending only a few meters. Faults allow easy penetration of weathering agents into subsurface areas because not only fracturing but also displacement is involved.
4. **Lava vesicles:** Lava *vesicles* are holes of various sizes, usually small, that develop in cooling lava when gas is unable to escape as the lava solidifies.
5. **Solution cavities:** *Solution cavities* are holes formed in calcareous rocks (particularly limestone) as the soluble minerals are dissolved and carried away by percolating water. Most solution cavities are small, but sometimes huge holes and even massive caverns are created when large amounts of solubles are removed. (A more lengthy description of solution processes and topography appears in Chapter 17.)

The Importance of Jointing

Almost all lithospheric bedrock is jointed, resulting sometimes from contractive cooling of molten material, sometimes from contraction of sedimentary strata as they dry, and sometimes from tectonic stresses. At Earth's surface, the separation between blocks on either side of a joint may be conspicuous because weathering emphasizes the fracture. Below the surface, however, the visible separation is minimal.

Joints are relatively common in most rock, but they are clearly more abundant in some places than in others. Where numerous, they are usually arranged in sets, each set comprising a series of approximately parallel fractures. Frequently, two prominent sets intersect almost at right angles; such a combination constitutes a *joint system* (Figure 15-5). A well-developed joint system, particularly in sedimentary rock having prominent natural bedding

▲ **Figure 15-5** The badlands topography of Bryce Canyon in southern Utah from Inspiration Point. The closely spaced joints and bedding planes contribute to intricate sculpturing by weathering and erosion.

planes, can divide stratified rock into a remarkably regular series of close-fitting blocks. Generally speaking, jointing is more regularly patterned, and the resulting blocks are more sharply defined in fine-grained rocks than in coarse-grained ones.

In some places, large joints or joint sets extend for long distances and through a considerable thickness of rocks; these are termed **master joints** (Figure 15-6). Master joints play a role in topographic development by functioning as a plane of weakness, a plane more susceptible to weathering and erosion than the rock around it. Thus, the location of large features of the landscape, such as valleys and cliffs, may be influenced by the position of master joints.

Weathering Agents

Most weathering agents are atmospheric. Because it is gaseous, the atmosphere is able to penetrate readily into all cracks and crevices in bedrock. From a chemical standpoint, oxygen, carbon dioxide, and water vapor are the three

▲ **Figure 15-6** In Utah's Zion National Park master joints are widely spaced, allowing for the development of massive blocks and precipitous cliffs.

atmospheric components of greatest importance in rock weathering.

Temperature changes are a second important weathering agent. Most notable, however, is water, which can penetrate downward effectively into openings in the bedrock. Biotic agents also contribute to weathering, in part through the burrowing activities of animals and the rooting effects of plants, but especially through the production of chemical substances that attack and decompose the rock.

The total effect of these agents is complicated and is influenced by a variety of factors: the nature and structure of the bedrock, the abundance and size of openings in it, the surface configuration, prevailing climatic conditions, the vegetative cover, and the variety and abundance of digging animals. For analytical purposes, however, it is convenient to recognize three principal categories of weathering—*mechanical*, *chemical*, and *biotic*. Although we now consider each of them in turn, we should bear in mind that they often act together.

Mechanical Weathering

Mechanical Weathering

Mechanical weathering (or *physical weathering*) is the physical disintegration of rock material without any change in its chemical composition. In essence, big rocks are mechanically weathered into little ones by various stresses that cause the rock to fracture into smaller fragments. Most mechanical weathering occurs at or very near the surface, but under certain conditions it may occur at considerable depth.

Frost Wedging Probably the most important single agent of mechanical weathering is the freeze-thaw action of water. When water freezes, it expands by almost 10 percent. Moreover, the upper surface of the water freezes first, which means that the principal force of expansion is exerted against the wall of the confining rock rather than upward. This expanding wedge of ice splits the rock, as shown in Figure 15-7.

Even the strongest rocks cannot withstand frequent alternation of freezing and thawing. Repetition is the key to understanding the inexorable force of **frost wedging** or *frost shattering*. Regardless of its size, if an opening in rock contains water, when the temperature falls below 0°C (32°F) ice forms, wedging its way downward. When the temperature rises above freezing, the ice melts and the water sinks farther into the slightly enlarged cavity. With renewed freezing, the wedging is repeated. Such a freeze-thaw pattern may be repeated millions of times through the eons of Earth history, providing what is literally an irresistible force.

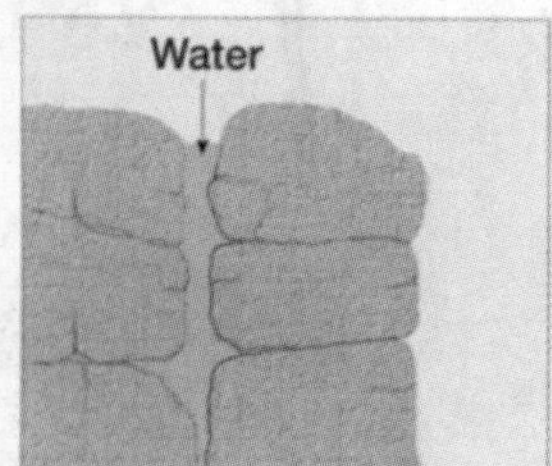

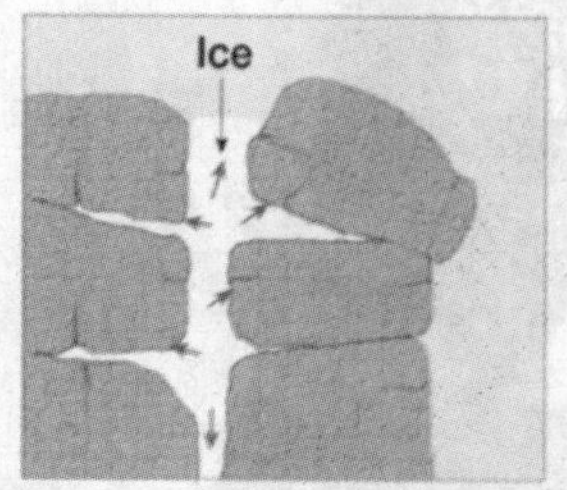

▲ **Figure 15-7** Schematic illustration of frost wedging. When water in a rock crack freezes, the ice expansion exerts a force that can deepen and widen the crack, especially if the process is repeated many times.

Frost wedging in large openings may produce large boulders, whereas that occurring in small openings may granulate the rock into sand and dust particles, with every size gradation in between. A common form of breakup in coarse-grained crystalline rocks is a shattering caused by frost wedging between grains. This type of shattering produces gravel or coarse sand in a process termed *granular disintegration*.

The physical characteristics of the rock are important determinants of the rate and magnitude of mechanical weathering, as are temperature and moisture variations. The process is most effective where freezing is prolonged and intense—in high latitudes, in midlatitudes during winters, and at high elevations. It is most conspicuous above the treeline of mountainous areas (Figure 15-8), where broken blocks of rock are likely to be found in profusion everywhere except on slopes that are too steep to allow them to lie without sliding downhill.

Salt Wedging Related to frost wedging but much less significant is **salt wedging**, which happens when salts crystallize out of solution as water evaporates. In areas of dry climate, water is often drawn upward in rock openings by capillary action (*capillarity* was discussed in Chapter 6). This water nearly always carries dissolved salts. When the water evaporates, as it commonly does, the salts are left behind as tiny crystals. With time, the crystals grow, prying apart the rock grain by grain, much in the fashion previously described for freezing water, although less intensely (Figure 15-9).

Along ocean coastlines, salt wedging may also be a weathering factor. Above the tideline, seawater from ocean spray gets between mineral grains; after the water evaporates the growth of salt crystals can slowly pry off mineral grains.

Temperature Changes Temperature changes not accompanied by freeze–thaw cycles may also weather rock mechanically, but they do so much more gradually than the processes just described. The fluctuation of temperature from day to night and from summer to winter can cause minute changes in the volume of most mineral particles, forcing expansion when heated and contraction when cooled. This volumetric variation weakens the coherence of the mineral grains and tends to break them apart. Millions of repetitions are normally required for much weakening or fracturing to occur. This factor is most significant in arid areas and near mountain summits, where direct solar radiation is intense during the day and radiational cooling is prominent at night.

◄ **Figure 15-8** Frost wedging is an especially pervasive force on mountaintops above the treeline, as with these granite boulders on Australia's highest peak, Mount Kosciusko, in New South Wales.

Probably more important but less widespread than simple solar heating and cooling is heat from forest fires or brushfires. The intense heating of a fire can cause a rock to expand and break apart.

Exfoliation One of the most striking of all weathering processes is **exfoliation**, in which curved layers peel off bedrock. (We saw in Chapter 13 that the term *foliation* refers to the parallel alignment of textural and structural features of a rock; *exfoliation* involves the stripping away of roughly parallel, concentric rock slabs.) Curved and concentric sets of joints, usually minor and inconspicuous, develop in the bedrock, and parallel shells of rock break away in succession, somewhat analogous to the way layers of an onion separate. The sheets that split off are sometimes only a millimeter or so thick; in other cases, however, they may be several meters thick.

The results of exfoliation are conspicuous. If the rock mass is a large one, such as Half Dome or one of the other granitic monoliths overlooking Yosemite Valley, California, its surface consists of imperfect curves punctuated by several partially fractured shells of the surface layers, and the mass is referred to as an **exfoliation dome** (Figure 15-10). Exfoliation may also be seen occurring on boulders in which case the result is usually a rounder shape, with each layer of shelling revealing a smaller spherical mass. Overall, then, especially in regions of exposed plutonic bedrock, exfoliation tends to gently smooth the landscape.

▲ **Figure 15-9** Salt wedging. This crystalline rock is being shattered by salt wedging near the floor of Death Valley in California.

The dynamics of exfoliation are not fully understood. The most widely accepted explanation of massive exfoliation is that the rock cracks after an overlying weight has been removed, a process called *unloading* or pressure release (Figure 15-11). The intrusive bedrock may originally have been deeply buried beneath a heavy overburden—perhaps several kilometers deep. When the overlying material is stripped away by erosion, the release of pressure allows expansion in the rock. The outer layers are unable to contain the expanding mass, and the expansion can be absorbed only by cracking along the sets of joints. Exfoliation occurs mainly in granite and related intrusive rocks, but under certain circumstances it is also seen in sandstone and other sedimentary strata.

Other processes, however, are likely at work to produce the very thin exfoliation layers seen on boulders (Figure 15-12). Fracturing resulting from temperature variations may be a component of the mechanism, although exfoliation is also found in areas where temperature fluctuations are not marked. Volumetric changes in minerals, which set up strains in the rock, may also be involved. Such changes are most notably produced by *hydration*, in which water molecules become attached to other substances and then the added water causes the original substance to swell without any change in its chemical composition. This swelling weakens the rock mass and is usually sufficient to produce some fracturing.

◀ **Figure 15-10** Two large exfoliation domes in Yosemite National Park, California.

Other Mechanical Weathering Processes

Chemical changes (discussed in more detail in the next subsection) may also contribute to mechanical weathering. Various chemical actions can cause an increase in volume of the affected mineral grains. This swelling sets up strain that weakens the coherence of the rock and causes fractures.

Some biotic activities also contribute to mechanical weathering. Most notable is the penetration of growing plant roots into cracks and crevices, which exerts an expansive force that widens the openings. This factor is especially conspicuous where trees grow out of joint or fault planes, with their large roots showing amazing tenacity and persistence as wedging devices (Figure 15-13). Additionally, burrowing animals sometimes are factors in rock disintegration. The total effect of these biotic actions is probably significant, but it is difficult to assess because it is obscured by subsequent chemical weathering.

When acting alone, mechanical weathering breaks up rock masses into ever smaller (and often angular) pieces, producing boulders, cobbles, pebbles, sand, silt, and dust. As more and more rock surface area is exposed over time, the process proceeds at an accelerating rate (Figure 15-14).

Chemical Weathering

Mechanical weathering is usually, but not always, accompanied by **chemical weathering**, which is the decomposition of rock by the chemical alteration of its minerals. Almost all minerals are subject to chemical alteration when exposed to atmospheric and biotic agents. Some minerals, such as quartz, are extremely resistant to chemical change, but many others are very susceptible. There are very few rocks that cannot be significantly affected by chemical weathering because the alteration of even a single significant mineral constituent can lead to the eventual disintegration of an entire rock mass.

One important effect of mechanical weathering is to expose bedrock to the forces of chemical weathering.

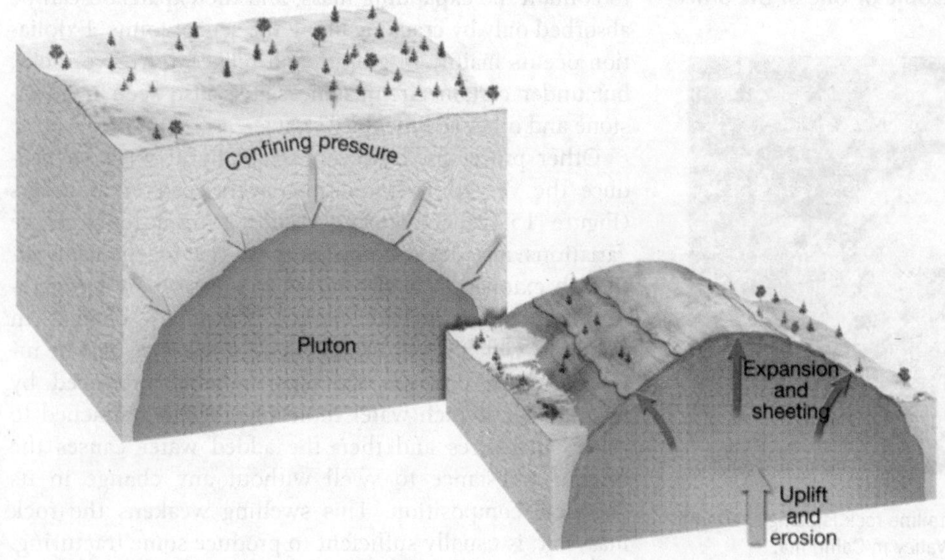

◀ **Figure 15-11** When formerly buried rocks such as granite are exposed at the surface, the unloading of pressure causes slight expansion of the rock mass. As a consequence, curved layers of rock peel off in the process of exfoliation.

◄ **Figure 15-12** An exfoliating granite boulder in California's Joshua Tree National Park.

The greater the surface area exposed, the more effective the chemical weathering. Thus, finer-grained materials decompose more rapidly than coarser-grained materials of identical composition because in fine-grained materials there is more exposed surface area.

Virtually all chemical weathering requires moisture. Thus, an abundance of water enhances the effectiveness of chemical weathering, and chemical processes operate more rapidly in humid climates than in arid areas. Moreover, chemical reactions are more rapid under high-temperature conditions than in cooler regions. Consequently, chemical weathering is most efficient and conspicuous in warm, moist climates. In cold or dry lands, there is less chemical weathering and so mechanical weathering tends to dominate.

Some of the chemical reactions that affect rocks are very complex, but others are simple and predictable. The principal reacting agents are oxygen, water, and carbon dioxide, and the most significant processes are *oxidation*, *hydrolysis*, and *carbonation*. These processes often take place more or less simultaneously, largely because they all involve water that contains dissolved atmospheric gases. Water percolating into the ground acts as a weak acid because of the presence of these gases and of decay products from the local vegetation; the presence of these impurities increases the water's capacity to drive chemical reactions.

Oxidation When the oxygen dissolved in water comes into contact with certain rock minerals, the minerals undergo **oxidation**, in which the oxygen atoms combine with atoms of various metallic elements making up the minerals in the rock and form new products. The new substances

◄ **Figure 15-13** Roots can serve as formidable natural tools to enlarge cracks and crevices in bedrock. (a) A small plant growing out of a rock crack in California's Desolation Valley. (b) A full-sized lodgepole pine tree growing out of a joint in the granitic bedrock of Desolation Valley.

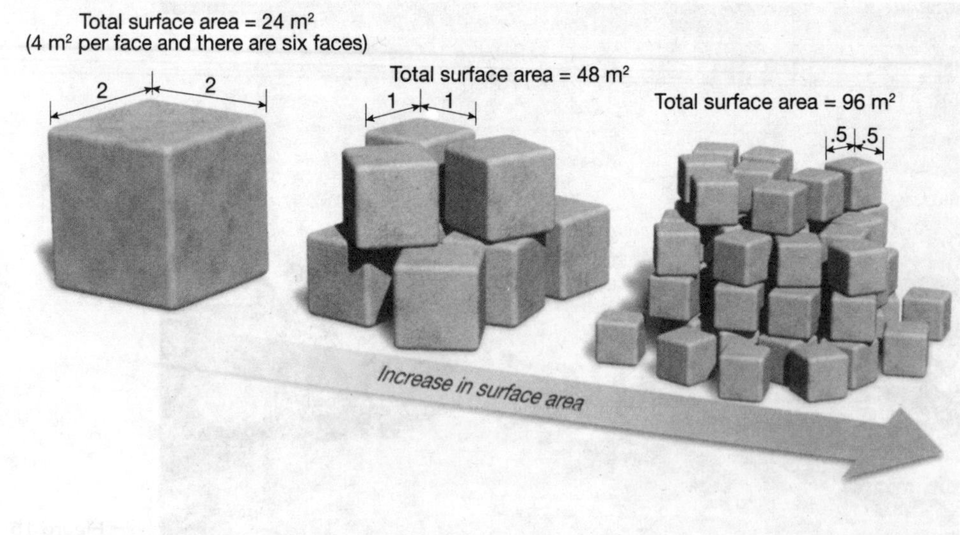

▶ **Figure 15-14** As mechanical weathering fragments rock, the amount of surface area exposed to further weathering is increased. Each successive step shown here doubles the surface area of the preceding step.

are usually more voluminous, softer, and more easily removed than the original compounds.

When iron-bearing minerals react with oxygen (in other words, become oxidized), *iron oxide* is produced:

$4Fe$ +	$3O_2$	$\rightarrow$	$2Fe_2O_3$
Iron	Oxygen		Iron Oxide (Hematite)

This reaction, probably the most common oxidation in the lithosphere, is called *rusting*, and the prevalence of rusty red stains on the surface of many rocks attests to its widespread occurrence (Figure 15-15). Similar effects are produced by the oxidation of aluminum. Since iron and aluminum are very common in Earth's crust, a reddish brown color is seen in many rocks and soils, particularly in tropical areas because there oxidation is the most notable chemical weathering process. Rusting contributes significantly to weathering because oxides are usually softer and more easily removed than the original iron and aluminum compounds from which the oxides were formed.

Hydrolysis **Hydrolysis** is the chemical union of water with another substance to produce a new compound that is nearly always softer and weaker than the original. Igneous rocks are particularly susceptible to hydrolysis because their silicate minerals combine readily with water. Hydrolysis invariably increases the volume of the mineral,

▲ **Figure 15-15** Iron oxide (rust) stains on a sandstone cliff in Capitol Reef National Park in Utah.

and this expansion can contribute to mechanical disintegration. In tropical areas, where water frequently percolates to considerable depth, hydrolysis often occurs far below the surface.

Carbonation **Carbonation** is the reaction between the carbon dioxide in water (which forms *carbonic acid* [H_2CO_3], a mild acid) and carbonate rocks such as limestone to produce a very soluble product (calcium bicarbonate) that can readily be removed by runoff or percolation and can also be deposited in crystalline form if the water is evaporated. We will discuss this process in greater detail in Chapter 17.

These and other less common chemical weathering processes are continuously at work at and beneath Earth's surface. Most chemically weathered rocks are changed physically, too. Their coherence is weakened, and the loose particles produced at Earth's surface are quite unlike the parent material. Beneath the surface, the rock holds together—but in a chemically altered condition. The major eventual products of chemical weathering are clays.

Biological Weathering

As already mentioned, plants frequently and animals occasionally contribute to weathering, and such processes involving living organisms are called **biological weathering**. Most notable is the penetration of growing plant roots into cracks and crevices (see Figure 15-13).

Lichens are primitive organisms that consist of algae and fungi living as a single unit. Typically they live on bare rock, bare soil, or tree bark (Figure 15-16). They draw minerals from the rock by ion exchange, and this leaching can weaken the rock. Moreover, expansion and contraction of lichens as they get alternately wet and dry flake off tiny particles of rock.

Burrowing by animals mixes soil effectively and is sometimes a factor in rock disintegration (for example, see Figure 12-6 in Chapter 12).

The total effect of all these biotic actions is probably significant but difficult to evaluate because it may be obscured by subsequent chemical or mechanical weathering.

Climate and Weathering

Geomorphic research tells us more and more every year about how climate influences weathering. The basic generalization is that weathering, particularly chemical weathering, is enhanced by a combination of high temperatures and abundant precipitation. Of these two factors, the moisture is usually more important than temperature. For example, we will see in Chapter 18 (The Topography of Arid Lands) that in most desert regions, because of a general lack of precipitation, mechanical weathering may be more conspicuous than chemical weathering. As we saw in Chapter 12 (Soils), the climatic regime significantly influences patterns of soil development.

There are many variations in the connection between weathering and climate; Figure 15-17 represents a generalized model of relevant relationships. Notice that, all else being equal, the depth of active weathering tends to be relatively shallow in regions of tundra and desert but relatively deep in regions of tropical rainforest.

▲ **Figure 15-16** Rocks covered with multicolored lichens in the Lake District of England.

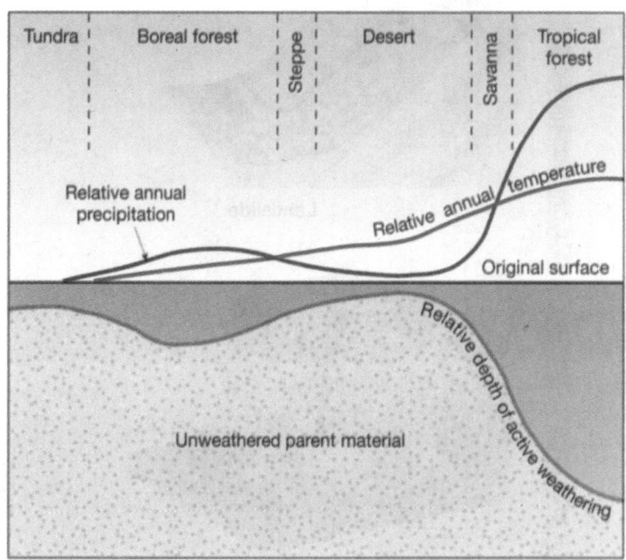

▲ **Figure 15-17** Relationship between important climatic elements and depth of weathering. The extreme effectiveness of weathering in the humid tropics is clearly shown. Humid midlatitude regions are omitted from this diagram because the relationships are more complex there.

Mass Wasting

Mass Wasting

As we learned above, the denudation of Earth's surface is accomplished by weathering followed by mass wasting followed by erosion. The ultimate destiny of all weathered material is to be carried away by erosion, a topic we cover in the remaining chapters of this book. The remainder of this chapter is concerned with mass wasting (also called *mass movement*), the process whereby weathered material is moved a relatively short distance downslope under the direct influence of gravity. Although it is sometimes circumvented when erosive agents act on weathered material directly, mass wasting is normally the second step in a three-step denudation process.

Gravity is inescapable; everywhere on the surface it pulls objects toward the center of Earth. Where the land is flat, the influence of gravity on topographic development is minimal. Even on gentle slopes, however, minute effects are likely to be significant in the long run, and on steep slopes the results are often immediate and conspicuous. Any loosened material is impelled downslope by gravity—in some cases falling abruptly or rolling rapidly, in others flowing or creeping with imperceptible gradualness.

The materials involved in these movements are all the varied products of weathering. Gigantic boulders respond to the pull of gravity in much the same fashion as do particles of dust, although the larger the object, the more immediate and pronounced the effect. Of particular importance, however, is the implication of "mass" in mass wasting; the accumulations of material moved—fragmented rock, regolith, and soil—are often extremely large and contain enormous amounts of mass.

Factors Influencing Mass Wasting All rock materials, from individual fragments to cohesive layers of soil, lie at rest on a slope if undisturbed unless the slope attains a critical steepness. The steepest angle that can be assumed by loose fragments on a slope without downslope movement is called the **angle of repose**. This angle, which varies with the nature and internal cohesion of the material, represents a fine balance between the pull of gravity and the

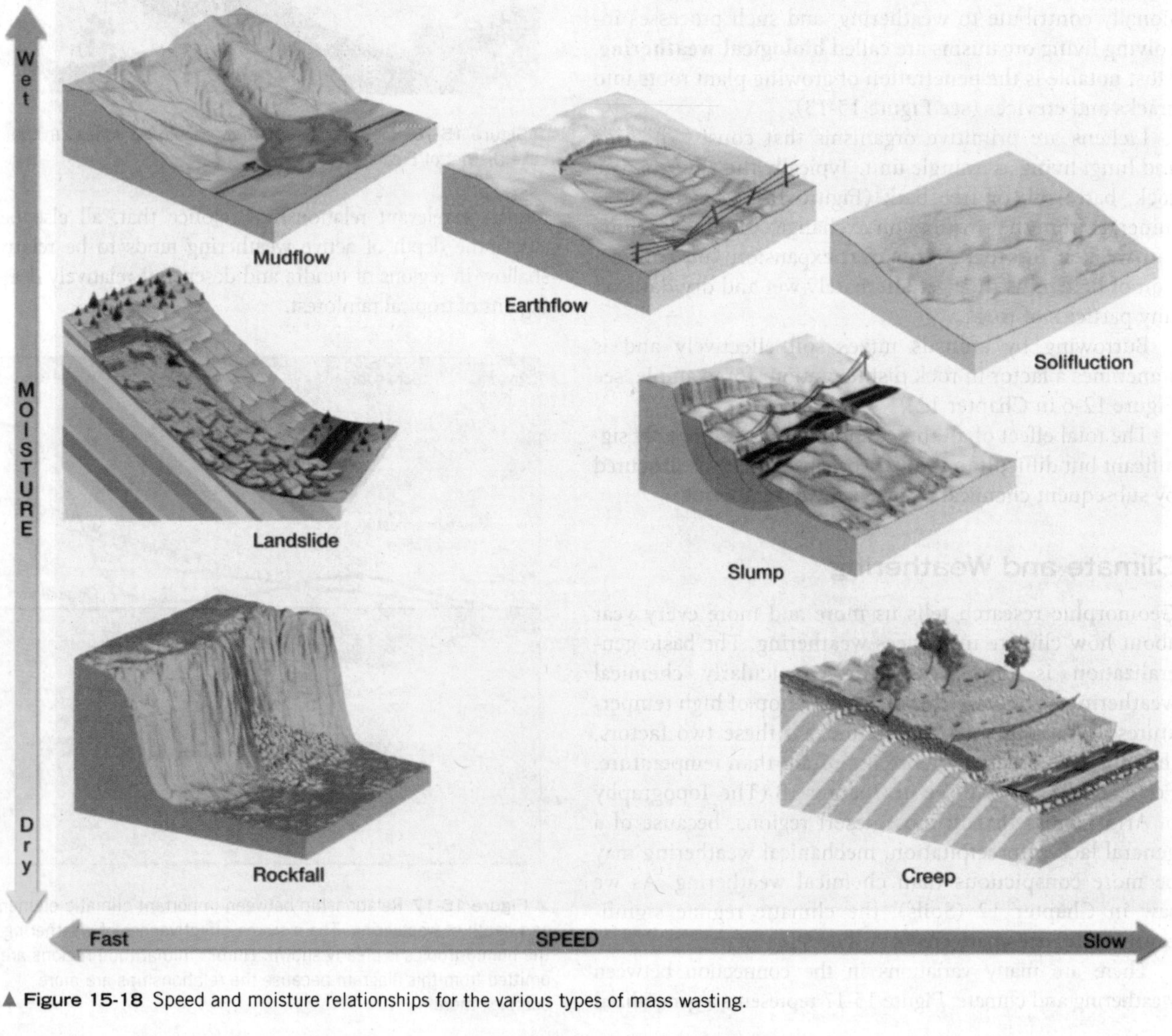

▲ **Figure 15-18** Speed and moisture relationships for the various types of mass wasting.

cohesion and friction of the rock material. If additional material accumulates on a debris pile lying on a slope that is near the angle of repose, the newly added material may upset the balance (because the added weight overcomes the friction force that is keeping the pile from sliding) and may cause all or part of the material to slide downward.

If water is added to the rock material through rainfall, snowmelt, or subsurface flow, the mobility is usually increased, particularly if the rock fragments are small. Water is a "lubricating" medium, and it diminishes friction between particles so that they can slide past one another more readily. Water also adds to the buoyancy and weight of the weathered material, which makes for a lower angle of repose and adds momentum once movement is under way. For this reason, mass wasting is particularly likely during and after heavy rains.

Another facilitator of mass wasting is *clay*. As noted in Chapter 12, clays readily absorb water. This absorbed water combined with the fine-grained texture of the material makes clay a very slippery and mobile substance. Any material resting on clay can often be set in motion by rainfall or an earthquake shock, even on very gentle slopes. Indeed, some clay formations are called *quick clays* because they spontaneously change from a relatively solid mass to a near-liquid condition as the result of a sudden disturbance or shock.

In subarctic regions and at high latitudes, mass wasting is often initiated by the heaving action of frozen groundwater. The presence of thawed, water-saturated ground in summer overlying permanently frozen subsoil (*permafrost*) also contributes to mass wasting in such regions. Some geomorphologists assert that, in the subarctic, mass wasting is the single most important means of transport of weathered material.

Although some types of mass wasting are rapid and conspicuous, others are slow and gradual (Figure 15-18). The principles involved are generally similar, but the extent of the activity and particularly the rate of movement are quite variable. In our consideration here, we proceed from the most rapid to the slowest, discussing the characteristics of each as if they were discrete movements, although in nature the various types often overlap.

Fall

The simplest and most obvious form of mass wasting is **rockfall**, or simply **fall**, the falling of pieces of rock downslope. When loosened by weathering on a very steep slope, a rock fragment may simply be dislodged and fall, roll, or bounce down to the bottom of that slope. This is a very characteristic event in mountainous areas, particularly as a result of frost wedging. Normally the fragments do not travel far before they become lodged, although the lodging may be unstable and temporary.

Pieces of unsorted, angular rock that fall in this fashion are referred to collectively as **talus** or **scree** (some geomorphologists use the term *talus* when referring to larger blocks and *scree* when the material is smaller, but the terms are often used interchangeably). Sometimes the fragments accumulate relatively uniformly along the base of the slope, in which case the resultant landform is called a *talus slope* or *talus apron*. More characteristically, however, the dislodged rocks collect in sloping, cone-shaped heaps called **talus cones** (Figure 15-19). This cone pattern is commonplace because most steep bedrock slopes and cliffs are seamed by vertical ravines and gullies that funnel the falling rock fragments into piles directly beneath the ravines, usually producing a series of talus cones side by side along the base of the slope or cliff. Some falling fragments, especially larger ones with their greater momentum, tumble and roll to the base of the cone. Most of the new talus, however, comes to rest at the upper end of

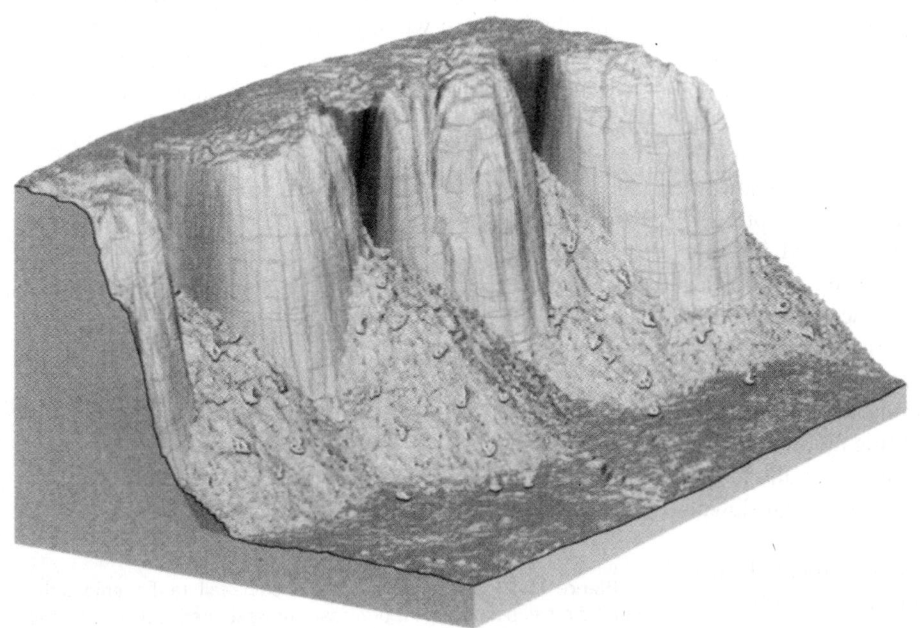

◀ **Figure 15-19** Talus cones at the foot of a steep slope develop as a result of rockfall.

▲ **Figure 15-20** An accumulation of talus at the base of a steep slope in Bloody Canyon, near Tioga Pass in the Sierra Nevada of California.

the cone. The cone thereby grows up the mountainside (Figure 15-20).

The angle of repose for talus is very high, generally about 35° and sometimes as great as 40°. The slope of a talus accumulation is gently concave upward, with the steepest angle near the apex of the cone. New material is frequently added to the top of the cone, where the fragments are invariably in delicate equilibrium, and each new piece that bombards down from above may cause a readjustment, with considerable downhill sliding. The freshest blocks, then, tend to be at the upper end of the cone. There is sometimes a rough sorting of fragments according to size, however, with the larger pieces rolling farther downslope and the smaller bits lodging higher up. With the passage of time, all talus slowly migrates downslope, encouraged by the freeze-thaw action of water in the many open spaces of the cones and aprons. All may be further reduced in size by more weathering.

In some rugged mountain areas, talus accumulates in great masses, and these masses may move slowly downslope under their own weight. As we shall see in Chapter 19, glaciers move in a somewhat similar way, and for this reason these extremely slow flows of talus are called **rock glaciers**. The flow in rock glaciers is caused primarily by the pull of gravity, aided by freeze–thaw temperature changes and may largely be independent of any lubricating effect of water. Rock glaciers occur primarily in glacial environments; many are found in midlatitude mountains that are relics of periods of glaciation. They are normally found on relatively steep slopes but sometimes extend far downvalley and even out onto an adjacent plain.

In some parts of the world it appears that rockfalls may be becoming more common as climate warms. Perhaps in part due to the melting of ice deep within talus cones or rock glaciers, talus slopes may become more unstable and susceptible to mass movement.

Slide

Eruption of Mount St. Helens

In mountainous terrain, landslides carry large masses of rock and soil downslope abruptly and often catastrophically. A **landslide** is a general term referring to an essentially instantaneous collapse of a slope (a type of *slope failure*) along a generally flat sliding plane that may not involve the lubricating effects of water or clay (Figure 15-21). In other words, the sliding material represents a rigid mass that is suddenly displaced without any fluid flow. The presence of water may contribute to the action, however; many slides are triggered by rains that add weight to already overloaded

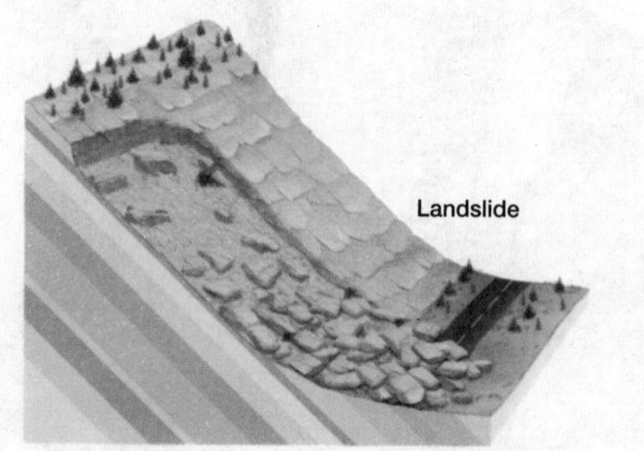

▲ **Figure 15-21** Landslide is a type of slope failure that entails the rapid downslope movement of masses of weathered rock along a flat sliding plane.

▲ **Figure 15-22** In August 1959 an earthquake triggered a large landslide in the Madison River Valley in Montana near Yellowstone National Park. The landslide dammed the river, forming "Quake Lake."

slopes. Landslides may be activated by other stimuli as well, most notably by earthquakes (Figure 15-22). Slides are also sometimes initiated simply by lateral erosion of a stream that undercuts its bank and thus oversteepens the slope above.

Some slides move only regolith, but many large slides also involve masses of bedrock detached along joint planes and fracture zones. The term *rock avalanche* (or *debris avalanche*) is often used to describe slides consisting primarily of rock without large quantities of soil or other fine-grained material.

Landslide action is not only abrupt but also rapid. Precise measurement of the rate of movement is obviously impossible, but eyewitness accounts affirm speeds of 160 kilometers (100 miles) per hour. Thunderous noise accompanies the slide, and the blasts of air that the slide creates can strip leaves, twigs, and even branches from nearby trees.

The immediate topographic result of a landslide may be threefold. First, on the hill where the slide originated, there is a deep and extensive *scar*, usually exposing a mixture of bedrock and scattered debris.

Second, because most landslides occur in steep, mountainous terrain, the great mass of material (displaced volume is sometimes measured in cubic kilometers) that roars downslope may choke the valley at the bottom with debris, usually in the form of either a broad ridge or low-lying cone. Its surface consists of a jumble of unsorted material, ranging from immense boulders to fine dust. Moreover, the momentum of the slide may push material several hundred meters up the slope on the other side of the valley.

Finally, in the valley bottom where the slide material comes to rest, a natural dam across the width of the valley may form, blocking the valley-bottom stream and producing a new lake, which becomes larger and larger until it either overtops the dam or cuts a path through it.

Slump An extremely common form of mass wasting is the type of slide called a **slump**. Slumping involves slope failure in which the rock or regolith moves downward and at the same time rotates outward along a curved slide plane that has its concave side facing upward (for this reason, a slump is sometimes called a *rotational slide*). The upper portion of the moving material tilts down and back, and the lower portion moves up and out. The top of the slump is usually marked by a crescent-shaped scarp face, sometimes with a steplike arrangement of smaller scarps and terraces below. The bottom of the slumping block consists of a bulging lobe of saturated debris protruding downslope or into the valley bottom (Figure 15-23).

Flow

In another form of mass wasting, a section of a slope becomes unstable, owing to the addition of water, and flows downhill. In some cases, the flow is fairly rapid, but it

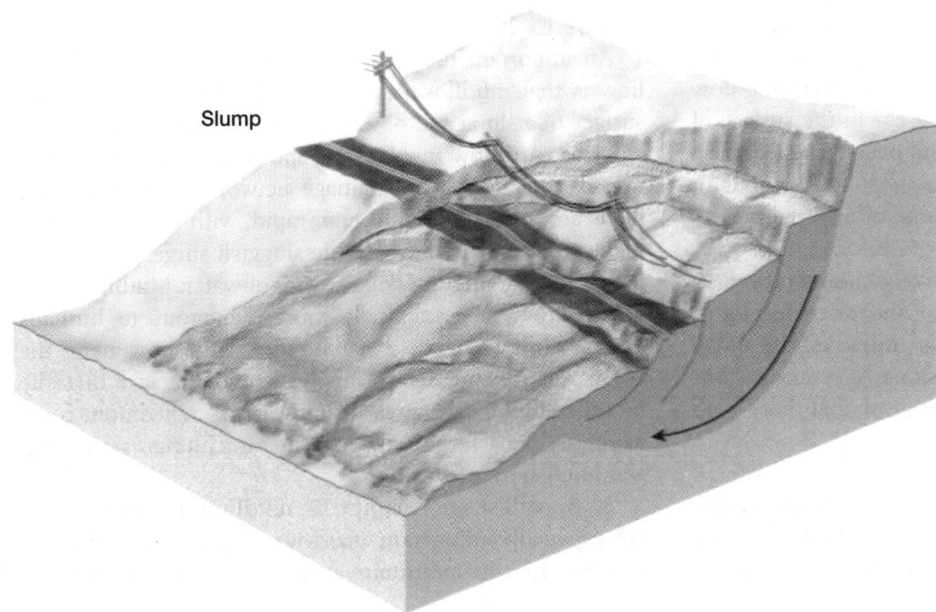

◀ **Figure 15-23** Slump involves movement along a curved sliding plane.

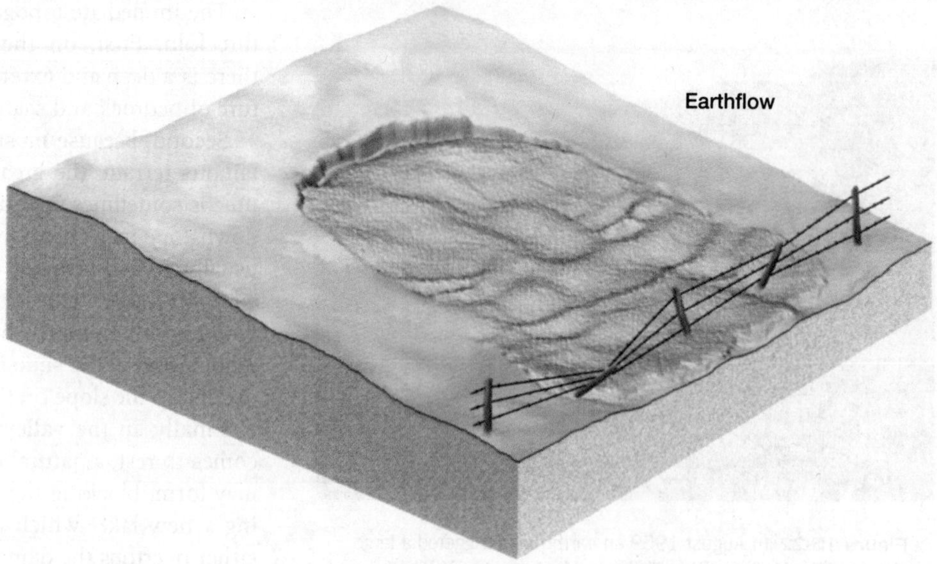

► **Figure 15-24** Earthflow takes place on hillsides when wet surface material begins to flow downslope.

is in others gradual and sluggish. Usually the center of the mass moves more rapidly than the base and sides, which are retarded by friction.

Many flows are relatively small, often encompassing an area of only a few square meters. More characteristically, however, they cover several tens or hundreds of hectares. Normally, they are relatively shallow phenomena, including only soil and regolith, but under certain conditions a considerable amount of bedrock may be involved.

As with other forms of mass movement, gravity is the impelling force. Water is an important catalyst to the movement; the surface materials become unstable with the added weight of water, and their cohesion is diminished by waterlogging, so that they are more responsive to the pull of gravity. The presence of clay also promotes flow, as some clay minerals become very slippery when lubricated with any sort of moisture.

Earthflow The most common flow movement is **earthflow**, in which a portion of a water-saturated slope moves a limited distance downhill, normally during or immediately after a heavy rain. At the place where the flow originates, there is usually a distinct scar in the surface of the slope, either cracks or a prominent oversteepened scarp face (Figure 15-24). An earthflow is most conspicuous in its lower portion, where a bulging lobe of material pushes out onto the valley floor.

This type of slope failure is relatively common on hillsides that are not densely vegetated and often results in blocked transportation lines (roads, railways) in valley bottoms. Property damage is occasionally extensive, but the rate of movement is usually so sluggish that there is no threat to life.

Mudflow A **mudflow** originates in drainage basins in arid and semiarid country when a heavy rain following a long dry spell produces a cascading runoff too voluminous to be absorbed into the soil. Fine debris is picked up from the hillsides by the runoff and concentrated in the valley bottoms, where it flows down-valley with the consistency of wet concrete. The leading edge of the mudflow continues to accumulate load, becoming increasingly stiff and retarding the flow of the more liquid upstream portions, so that the entire mudflow moves haltingly down the valley. When a mudflow reaches the mouth of the valley and abruptly leaves its confining walls for the more open slopes of the piedmont zone, the pent-up liquid behind the glutinous leading edge breaks through with a rush, spreading muddy debris into a wide sheet (Figure 15-25). A *volcanic mudflow* (*lahar*) described in Chapter 14 is a special kind of mudflow that develops on the slopes of volcanoes.

Mudflows often pick up large rocks, including huge boulders, and carry them along as part of their load. In some cases, the large pieces are so numerous that the term **debris flow** is used in preference to mudflow (Figure 15-26).

An important distinction between earthflow and mudflow is that mudflows (and debris flows) move along the surface of a slope and down established drainage channels, whereas earthflow involves a slope failure and has little or no relationship to the drainage network. Moreover, mud flows are normally much more rapid, with a rate of movement intermediate between the sluggish surge of an earthflow and the rapid flow of a stream of water. Mudflows and debris flows are potentially more dangerous to humans than earthflows because of the more rapid movement, the larger quantity of debris involved, and the fact that the mudflow often discharges abruptly across a piedmont zone, which is likely to be a favored area for human settlement and intensive agriculture.

As we will see in Chapter 18, mudflow and debris flow are especially important mass wasting processes in arid regions. In the mountain areas of deserts, an intense,

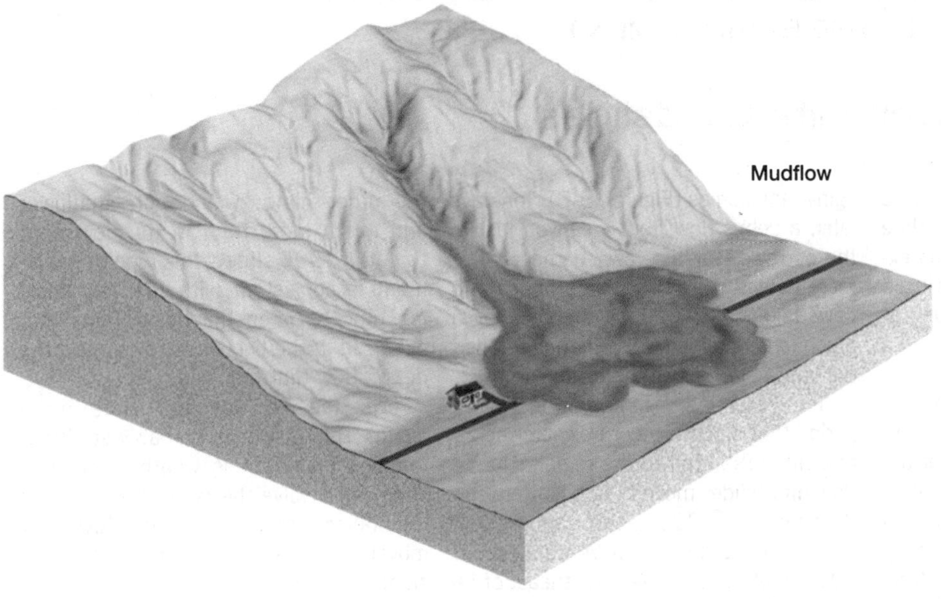

Mudflow

◀ **Figure 15-25** Mudflows entail the rapid movement of very wet material through a canyon or valley. The mud and debris typically comes to rest at the mouth of the valley in a fan-shaped deposit.

local thunderstorm can quickly mobilize loose, weathered material; the flow will move rapidly down a desert canyon, depositing the mud and debris at the foot of the mountains.

The media often incorrectly refers to almost all kinds of mass wasting with the rather ambiguous term, "mudslide." As we have just seen, there are significant differences between the various kinds of slides and flows.

Creep

The slowest and least perceptible form of mass wasting, **creep**—or **soil creep**—consists of a very gradual downhill movement of soil and regolith so unobtrusive that it can normally be recognized only by indirect evidence. Generally the entire slope is involved. Creep is such a pervasive phenomenon that it occurs all over the world on sloping land. Although most notable on steep, lightly vegetated slopes, it also occurs on gentle slopes that have a dense plant cover. Wherever weathered materials are available for movement on land that is not flat, creep is a persistent form of mass wasting.

▲ **Figure 15-26** The city of Caraballeda in Venezuela after a series of debris flows in 1999 that killed 19,000 people.

Creep is universal. Infinite numbers of tiny bits of lithospheric material, as well as many larger pieces, march slowly and sporadically downslope from the places where they were produced by weathering or deposited after erosion. When water is present in the surface material, creep is usually accelerated because the lubricating effect allows individual particles to move more easily and because water adds to the weight of the mass.

Creep is caused by the interaction of various factors, the most significant being alternation of freeze/thaw or wet/dry conditions. When water in the soil freezes, soil particles tend to be displaced upward and in the direction perpendicular to the ground surface due to ice expansion, as Figure 15-27 shows. After thawing, however, the particles settle downward, not directly into their original position but rather pulled slightly downslope by gravity. With countless repetitions, this process can result in downhill movement of the entire slope.

Many other agents also contribute to creep. Indeed, any activity that disturbs soil and regolith on a sloping surface is a contributor because gravity affects every rearrangement of particles, attracting them downslope. For example, burrowing animals pile most of their excavated material downslope, and subsequent infilling of burrows is mostly by material from the upslope side. As plant roots grow, they also tend to displace particles downslope. Animals that walk on the surface exert a downslope movement as well. Even the shaking of earthquakes or thunder produces disturbances that stimulate creep.

Whenever it occurs, creep is a very slow process, but its rate of movement is faster under some circumstances than

PEOPLE AND THE ENVIRONMENT

The La Conchita Landslides

Once a landslide has occurred on a hillside, in coming years that slide is quite likely to move again. A good example of the propensity for destabilized slopes to fail multiple times is found in the coastal California community of La Conchita (Figure 15-A).

La Conchita is a small town just southeast of Santa Barbara. Houses here are built on a narrow coastal strip at the foot of a steep bluff about 180 meters (600 feet) high. The bluff is comprised of poorly cemented marine sedimentary rocks and shows evidence of prehistoric landslides. The recent sequence of slope failures at La Conchita began in 1995.

Evidence of cracks and slight movement at the top of the bluff was noticeable by December 1994. That winter was exceptionally wet, with the region receiving twice its average amount of rainfall. In January 1995 alone 62.3 centimeters (24.53 inches) of rain fell—more than five times the monthly average. Then, after a dry February, a modest-size storm arrived on March 2 and 3. On the afternoon of March 4, 1995, a 120 meter (400 foot) wide section of the slope above La Conchita failed, moving about 330 meters (1100 feet) downhill as a slump and earthflow, destroying or damaging nine homes (Figure 15-B). Ten days later, a debris flow from a canyon just to the northwest of the town damaged five more homes. The 1995 slide appears to have taken place within the area of a much larger prehistoric slide that entailed the entire bluff above the town.

Ten years later, following a 15-day period that brought record or near-record amounts of rainfall to the region, the La Conchita slide moved again. Between December 27, 2004, and January 10, 2005, the nearby city of Ventura (about 20 kilometers [12 miles] southeast of La Conchita) received 37.8 centimeters (14.9 inches) of rain—the equivalent of nearly a year's average rainfall. In the afternoon of January 10, 2005, the slope above La Conchita failed again, this time remobilizing a 90 meter (300 foot) wide, 350 meter (1150 foot) long section of the 1995 landslide deposit as a fast-moving debris flow (Figure 15-C). The 2005 event destroyed 13 houses, damaged 23 more, and killed 10 people.

Analysis showed that the 1995 slide entailed fairly deep movement—consistent with deep infiltration of water into the slope, and therefore the long lag time between the heavy rains of January and the slide itself. The 2005 event, on the other hand, was much shallower and much more fluid in its movement than the 1995 slide—this type of movement is consistent with the very short lag time between the heavy rains that occurred in the two weeks prior, and the debris flow that followed.

Will the La Conchita slide move again? The answer is probably so. What remains uncertain is whether the next slide will involve the rapid movement of remobilized material as in 2005, or movement originating deeper within the slope as in 1995.

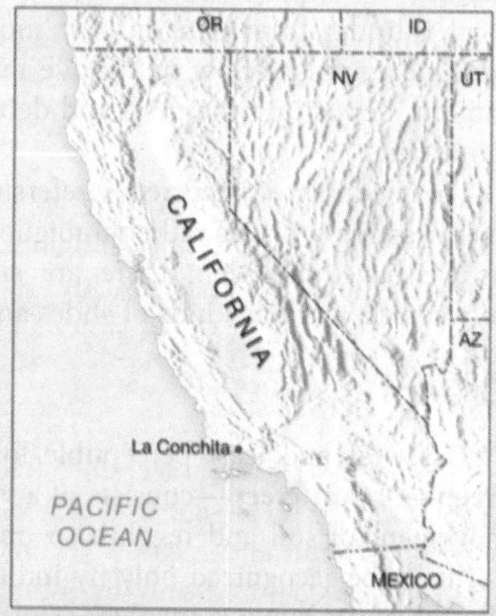

▲ **Figure 15-A** Regional map of La Conchita, California.

▲ **Figure 15-B** Aerial photograph of the 1995 La Conchita landslide.

▲ **Figure 15-C** Aerial photograph of the 2005 La Conchita debris flow.

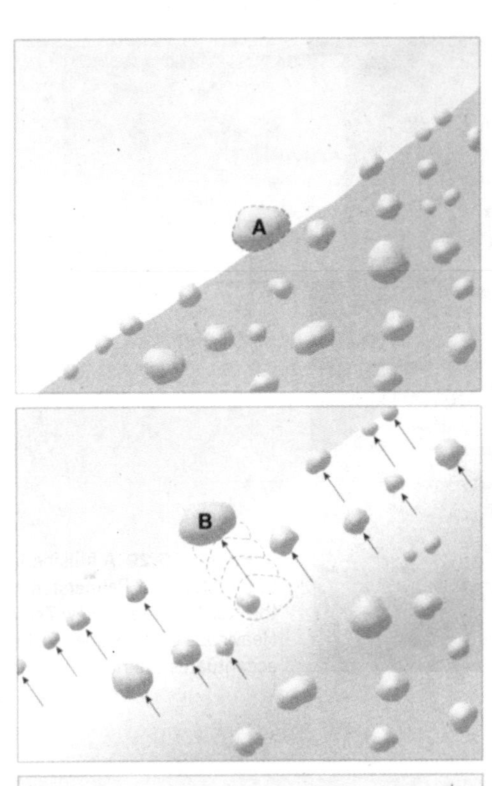

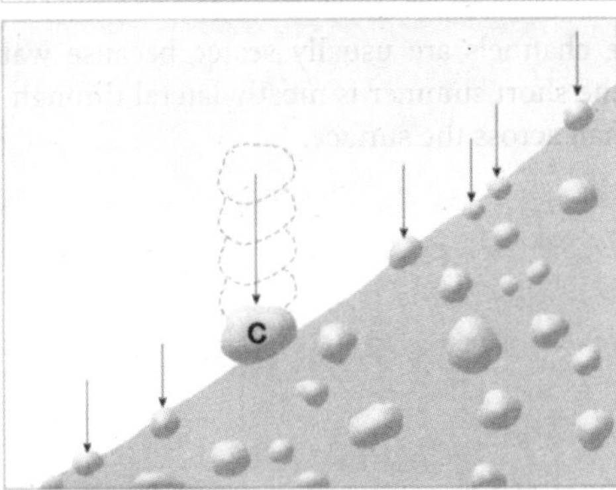

▲ **Figure 15-27** The movement of a typical rock particle in a freeze–thaw situation. Freezing and ice expansion lifts the particle perpendicular to the slope (from A to B); upon thawing, the particle settles slightly downslope (from B to C).

others. The principal variables are slope angle, vegetative cover, and moisture supply. Creep operates faster on steep slopes than on gentle ones, for obvious reasons. Deep-rooted and dense-growing vegetation inhibits creep because the roots bind the soil. Also, as mentioned previously, creep is generally faster on water-saturated slopes than on dry ones. Although extreme creep rates of up to 5 centimeters (2 inches) per year have been reported, much more common are rates of just a fraction of a centimeter per year.

Whatever the creep rate, it is much too slow for the eye to perceive and the results are all but invisible. Creep is usually recognized only through the displacement of human-built structures—most commonly when fence posts and utility poles are tilted downhill. Retaining walls may be broken or displaced, and even roadbeds may be disturbed (Figure 15-28). Unlike the other forms of mass wasting, creep produces few distinctive landforms. Rather it induces an imperceptible diminishing of slope angles and gradual lowering of hilltops—in other words, a widespread but minute smoothing of the land surface.

Under certain conditions, and usually on steep grassy slopes, grazing animals accentuate soil creep through the formation of a network of hillside ridges known as *terracettes*. Over time, the entire hillside may be covered with a maze of these terracettes (Figure 15-29).

Solifluction A special form of creep that produces a distinctive surface appearance is **solifluction** (meaning "soil flowage"), a process largely restricted to high latitude and high altitude tundra landscapes beyond the treeline (Figure 15-30). During the summer, the near-surface portion of the ground (called the *active layer*) thaws, but the meltwater cannot percolate deeper because of the permafrost below. The spaces between the soil particles become saturated, and the heavy surface material sags slowly downslope. Movement is erratic and irregular, with lobes

◀ **Figure 15-28** Visual evidence of soil creep: displacement and/or bending of fences, utility poles, and retaining walls.

◀ **Figure 15-29** A hillside laced with terracettes near Palmerston North, on the North Island of New Zealand. Heavy use of the slope by sheep accentuates the ridges.

overlapping one another in a haphazard, fish-scale pattern. The lobes move only a few centimeters per year, but they remain very obvious in the landscape, in part because of the scarcity of vegetation. Where solifluction occurs, drainage channels are usually scarce because water flow during the short summer is mostly lateral through the soil rather than across the surface.

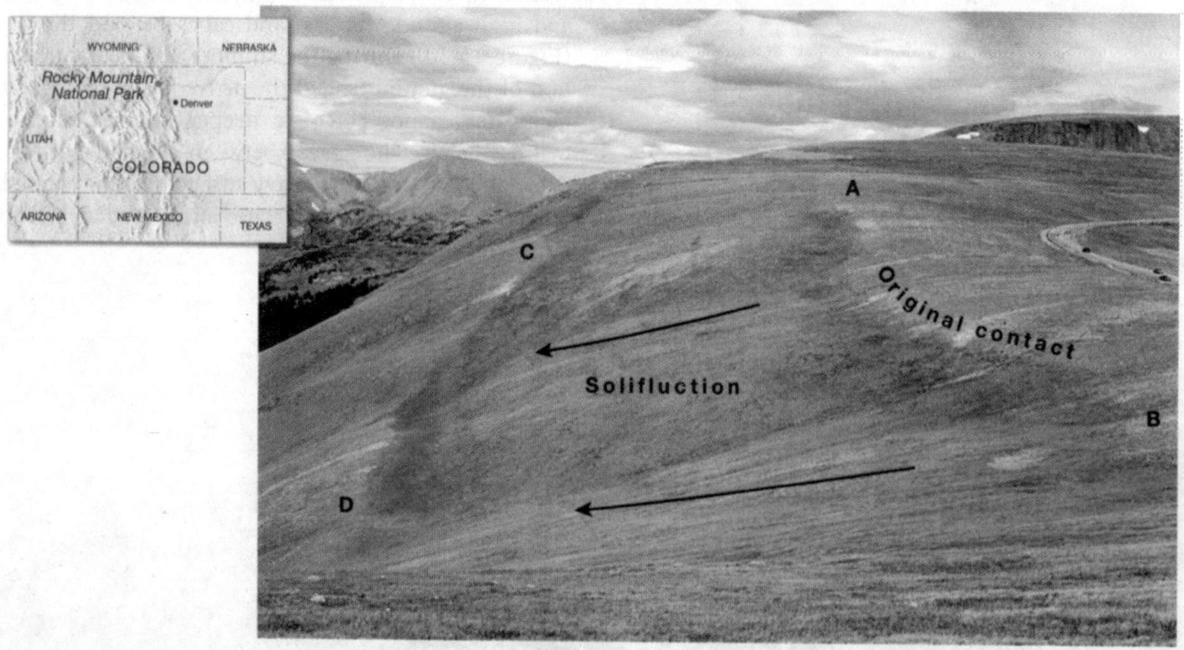

▲ **Figure 15-30** Solifluction in the high country of Colorado's Rocky Mountain National Park. The surface has gradually slipped downslope about 60 meters (200 feet). A–B and C–D were originally in contact with each other.

CHAPTER 15 LEARNING REVIEW

After studying this chapter, you should be able to answer the following questions. Key terms from each text section are shown in bold type. Definitions for key terms are also found in the glossary at the back of the book.

Denudation (p. 417)

1. What is meant by **denudation**?
2. Distinguish among **weathering**, **mass wasting**, and **erosion**.

Weathering and Rock Openings (p. 417)

3. What roles do rock openings play in weathering processes?
4. What is the difference between a **joint** and a fault?
5. What are **master joints** and how can they influence the topography of a landscape?
6. Explain how it is possible for weathering to take place beneath the surface of bedrock.

Weathering Agents (p. 419)

7. What are the general differences between **mechanical weathering** and **chemical weathering**?
8. Explain the mechanics of **frost wedging**.
9. Explain the process of **salt wedging**.
10. Explain the weathering process of **exfoliation** ("unloading") that is responsible for features such as **exfoliation domes**.
11. What is the relationship between **oxidation** and rust?
12. Briefly describe the chemical weathering processes of **hydrolysis** and **carbonation**.
13. Describe and explain one example of **biological weathering**.
14. Why is chemical weathering more effective in humid climates than in arid climates?

Mass Wasting (p. 426)

15. What is the relationship between gravity and mass wasting?
16. How does the **angle of repose** affect mass wasting?
17. What is the role of clay in mass wasting?
18. In what ways can moisture expedite mass wasting?
19. Describe the process of **rockfall (fall)**.
20. Describe the origin and general characteristics of **talus (scree)**.
21. What is a **talus cone** and where does one typically develop?
22. What is a **rock glacier**?
23. What roles may heavy rain play in triggering a **landslide**?
24. How is a **slump** different from other kinds of landslide?
25. What are the differences between a landslide and a **mudflow**?
26. How is **earthflow** different from mudflow and **debris flow**?
27. Explain the process of **soil creep**.
28. In what kinds of environments is **solifluction** common?

Log in to **www.mygeoscienceplace.com** for animations, videos, RSS feeds, maps, and self-study quizzes to enhance your study of **Preliminaries to Erosion: Weathering and Mass Wasting**.

16 Fluvial Processes

The landscape is shaped by a variety of processes functioning in concert. In Chapter 14, we considered internal processes, and in Chapter 15, we studied the external processes of weathering and mass wasting. In Chapters 16 through 20, we concentrate on the external land-shaping agents that work by erosion and deposition: running water, subsurface waters, wind, moving ice, and coastal waters. By far the most important of these external agents in transporting and depositing material is running water moving over the land—*fluvial processes*—the topic of this chapter.

Running water probably contributes more to shaping landforms than all other external agents combined. This is true not because running water is more forceful than other agents (moving ice and pounding waves often apply much greater amounts of energy per unit area), but rather because running water is nearly ubiquitous: it exists everywhere except in Antarctica. (Wind is ubiquitous, too, but trivial in its power as a terrain sculptor.) So it is appropriate for us to begin our study of erosional and depositional processes with the work of running water.

▲ **Figure 16-1** The work of some rivers is very conspicuous. The Green River in northeastern Utah helped carve Whirlpool Canyon through fluvial erosion and by transporting away the rock deposited into the valley bottom by side streams and mass wasting.

The Impact of Fluvial Processes on the Landscape

Moving water is so widespread and effective as an agent of erosion and deposition that its influence on the landscape is usually prominent—and in many cases, dominant. The shapes of most valleys are strongly influenced by the water that runs through them—especially through the transportation and deposition of sediment (Figure 16-1). Areas above the valleys are less affected by running water, but even there flowing water may significantly influence the shape of the land.

The basic landscape-sculpturing effect of running water is to smooth irregularities—in simplest terms, to wear down the hills by erosion and transportation of debris, and fill up the valleys by deposition.

Streams and Stream Systems

As we learned in Chapter 9, in the study of geomorphology we call any channeled flow of water a "stream"—whether it is a tiny creek or an enormous river. Streams and their valleys dissect the land in a multitude of patterns. However, there are some important threads of similarity, and the study of individual streams can produce additional generalities to aid in our understanding of topographic development. Before we get into the specifics of fluvial processes and their topographic results, we need to introduce a few basic concepts.

Streamflow and Overland Flow

Fluvial processes, defined as those that involve running water, encompass both the unchanneled downslope movement of surface water, called **overland flow**, and the channeled movement of water along a valley bottom, called **streamflow**.

Valleys and Interfluves

All the surfaces of the continents can be considered to consist of two topographic elements: *valleys* and *interfluves*. The distinction between the two is not always obvious in nature, but the conceptual difference is clear. A **valley** is that portion of the terrain in which a drainage system is clearly established (Figure 16-2). It includes the valley bottom that is partially or totally occupied by the channel of a stream, as well as the valley walls that rise above the valley bottom on both sides. Some valley bottoms are narrow and limited in area, while others are broad and extensive. Valley walls are also variable, with slopes that are steep or gentle, and an extent that is limited or expansive. The upper limit of a valley is not always readily apparent, but it can be clearly conceptualized as a lip or rim at the top of the valley

◀ The Tamul Waterfall forms where the Rio Gallinas enters the Rio Santa Maria in Mexico.

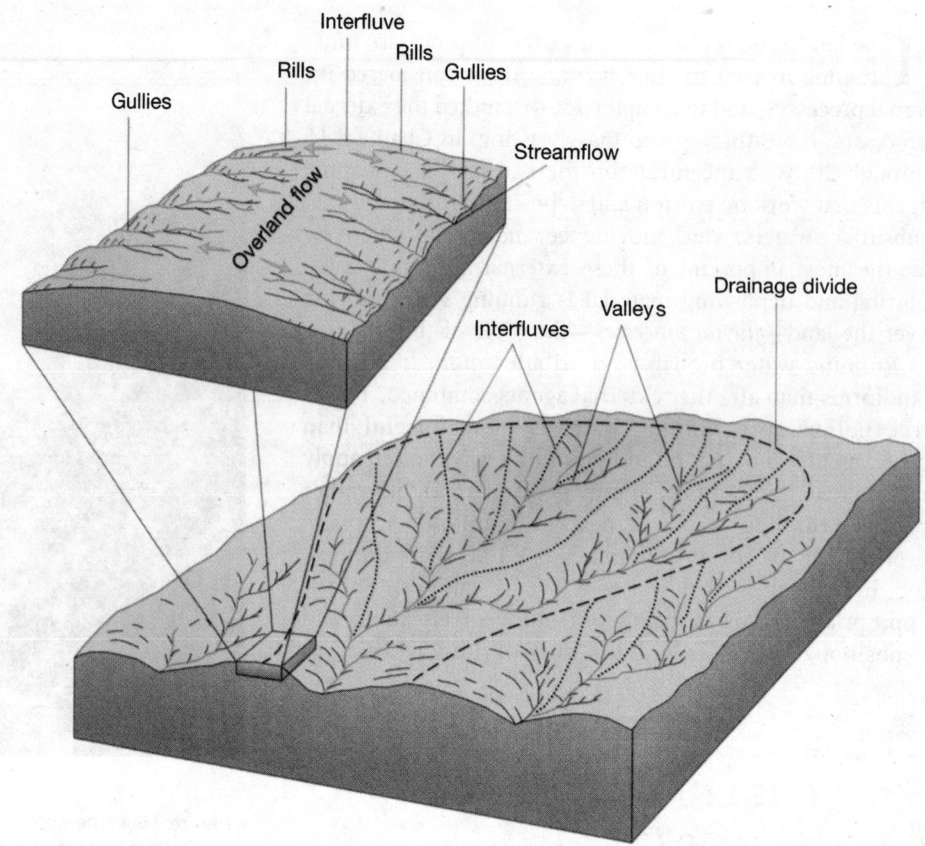

▶ **Figure 16-2** Valleys and interfluves. Valleys normally have clear-cut drainage systems; interfluves do not.

walls above which drainage channels are either absent or indistinct.

An **interfluve** (from Latin, *inter*, "between," and *fluvia*, "rivers") is the higher land above the valley walls that separates adjacent valleys. Some interfluves consist of ridgetops or mountain crests with precipitous slopes, but others are simply broad and flattish divides between drainage systems. Conceptually, all parts of the terrain not in a valley are part of an interfluve. We can envision that on an interfluve water will move downslope through unchanneled overland flow until it reaches the lip of the interfluve; there, as the water drops off the lip of the interfluve into the first small gullies of the valley system, streamflow begins.

These simplistic definitions are not always applicable in nature because some terrain elements defy classification. Swamps and marshes, for example, may be found on interfluves, but more typically are found in sections of valleys where there is not a clearly established drainage system. Such exceptional cases, however, should not inhibit our acceptance of the valley–interfluve concept.

Drainage Basins

The **drainage basin** or **watershed** of a particular stream is all the area that contributes overland flow, streamflow, and groundwater to that stream. In other words, the drainage basin consists of a stream's valley bottom, valley sides, and those portions of the surrounding interfluves that drain toward the valley. Conceptually, the drainage basin terminates at a **drainage divide**, which is the line of separation between runoff that descends in the direction of one drainage basin, and runoff that goes toward an adjacent basin.

Every stream of any size has its own drainage basin, but for practical purposes the term is often reserved for major streams. The drainage basin of a principal river encompasses the smaller drainage basins of all its tributaries; consequently, larger basins include a hierarchy of smaller tributary basins (Figure 16-3).

Stream Orders

In every drainage basin, small streams come together to form successively larger ones, and small valleys join more extensive ones. This relationship, although variable in detail, holds true for drainage basins of any size. This systematic characteristic makes it possible to recognize a natural organization within a watershed, and the concept of **stream order** has been devised to describe the arrangement (Figure 16-4).

A *first-order stream*, the smallest unit in the system, is a stream without tributaries. Where two first-order streams unite, a *second-order stream* is formed. At the confluence of two second-order streams, a *third-order stream* begins, and so on through successively higher orders.

Note from Figure 16-4 that the joining of a lower-order stream with a higher-order stream does not increase the order below that junction. For example, the confluence of a first-order stream and a second-order one does not produce

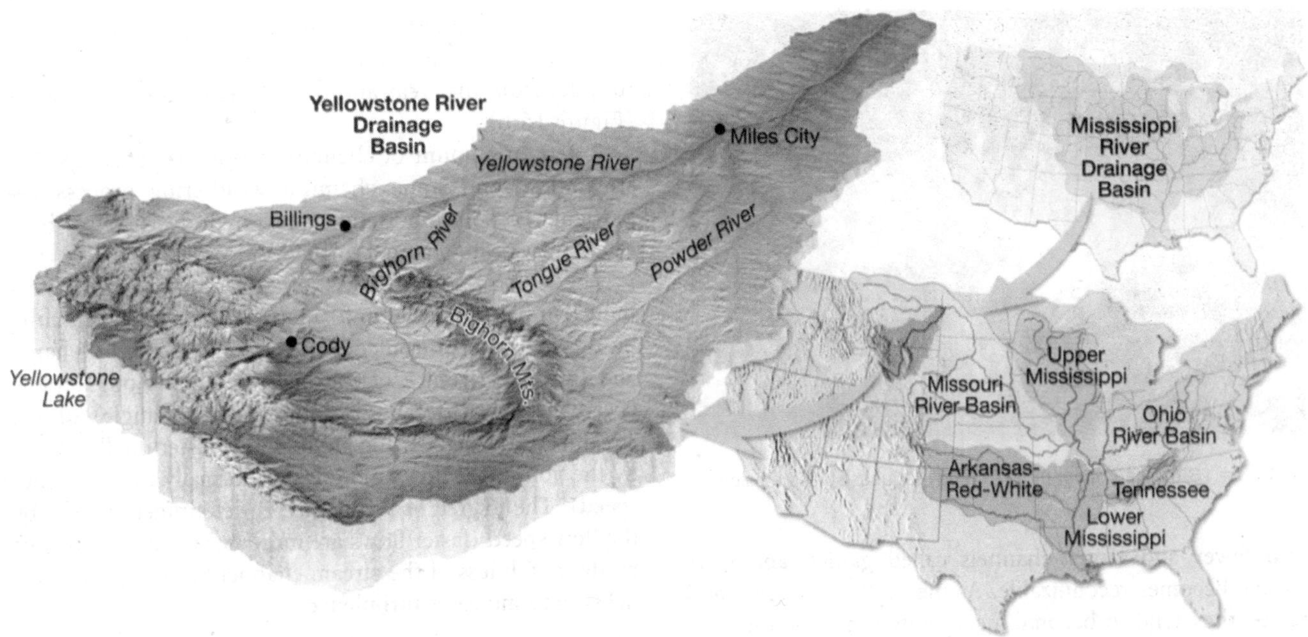

▲ **Figure 16-3** A nested hierarchy of drainage basins. Many sub-basins are found within the overall Mississippi River drainage basin. For example, beginning in the Yellowstone River drainage basin, the Bighorn River flows into the Yellowstone River, which flows into the Missouri River, which flows into the Mississippi River; finally, the Mississippi empties into the Gulf of Mexico.

a third-order stream. A third-order stream is formed only by the joining of two second-order streams.

The concept of stream order is more than simply a numbers game because several significant relationships are involved. In a well-developed drainage system, for example, one can predict with some certainty that first-order streams and valleys are more numerous than all others combined and that each succeeding higher order is represented by fewer and fewer streams. Some other predictable relationships are that average stream length generally increases with increasing order, that average watershed area generally increases with increasing order, and that average stream gradient decreases with increasing order.

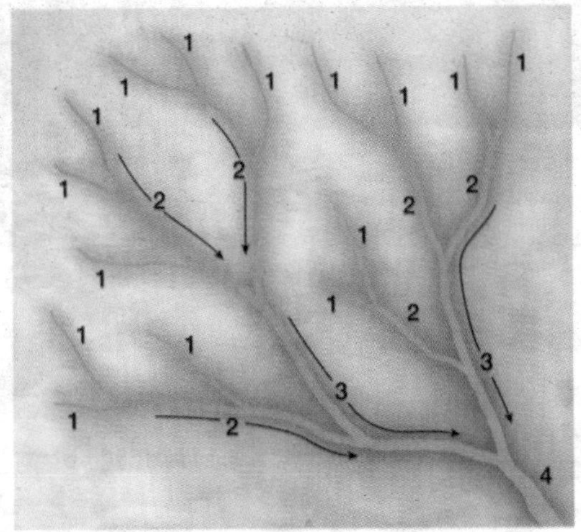

▲ **Figure 16-4** Stream orders. The branching component of a stream and its tributaries can be classified into a hierarchy of segments, ranging from smallest to largest (in this case, from 1 to 4).

Fluvial Erosion and Deposition

Stream Sediment Movement

All external processes remove fragments of bedrock, regolith, and soil from their original positions (erosion), transport them to another location, and set them down (deposition). The fluvial processes, our concern in this chapter, produce one set of landforms by erosion and another quite different set by deposition.

Erosion by Overland Flow On the interfluve, fluvial erosion begins when rain starts to fall. Unless the impact of rain is absorbed by vegetation or some other protective covering, the collision of raindrops with the ground is strong enough to blast fine soil particles upward and outward, shifting them a few millimeters laterally. On sloping ground, most particles move downhill by this *splash erosion* (Figure 16-5).

In the first few minutes of a rain, much of the water infiltrates the soil, and consequently there is little runoff. During heavy or continued rain, however, particularly if the land is sloping and has only a sparse vegetative cover, infiltration is greatly diminished and most of the water moves downslope as overland flow. The water flows across the surface as a thin sheet, transporting material already loosened by splash erosion, in a process termed *sheet erosion.* As overland flow moves downslope and its volume increases, the resulting turbulence tends to break up the sheet flow into multitudinous tiny channels called *rills*. This more concentrated flow picks up additional material and scores the slope with numerous parallel seams; this sequence of events is termed *rill erosion*. If the process continues, the rills begin to coalesce

▲ **Figure 16-5** The impact of a raindrop produces splash erosion.

into fewer and larger channels called *gullies*, and *gully erosion* becomes recognizable. As the gullies get larger and larger, they tend to become incorporated into the drainage system of the adjacent valley, and the flow changes from overland flow to streamflow.

Erosion by Streamflow Once surface flow is channeled, its ability to erode and transport material is greatly enhanced by the increased volume of water. Erosion is accomplished in part by the direct hydraulic power of the moving water, which can excavate and transport material at the bottom and sides of the stream. Banks can also be undermined by streamflow, particularly at times of high water, dumping more loose material into the water to be swept downstream.

The erosive capability of streamflow is also increased by the abrasive "tools" it picks up and carries along with it. All sizes of rock fragments, from silt to boulders, chip and grind the stream bed as they travel downstream in the moving water. These rock fragments break off more fragments from the bottom and sides of the channel, and they collide with one another, becoming both smaller and rounder from the wear and tear of the frequent collisions (Figure 16-6).

A certain amount of chemical action also accompanies streamflow, and some chemical weathering processes—particularly solution action and hydrolysis—also help erode the stream channel through *corrosion*.

The erosive effectiveness of streamflow varies enormously from one situation to another, determined primarily by the speed and turbulence of the flow on the one hand, and by the resistance of the bedrock on the other. Flow speed is governed by the gradient (slope angle) of the streambed (the steeper the gradient, the faster the flow) and by the volume of flow (more water normally means higher speed). The degree of turbulence is determined in part by the flow speed (faster flows are more turbulent) and in part by the roughness of the stream channel (an irregular channel surface increases turbulence).

Transportation Any water moving downslope, whether moving as overland flow or as streamflow, can transport rock material. At any given time and place, the load carried by overland flow is likely to be small in comparison with what a stream can transport. Eventually, most of this material reaches the streams in the valley bottoms, where it is added to the stream-eroded debris and material contributed by mass wasting to constitute the **stream load**.

Essentially, the stream load contains three fractions (Figure 16-7):

1. Some minerals, mostly salts, are dissolved in the water and carried in solution as the **dissolved load**.
2. Very fine particles of clay and silt are carried in suspension, moving along with the water without ever touching the streambed. These tiny particles, called the **suspended load**, have a very slow settling speed, even

► **Figure 16-6** Rock transported by streams becomes rounded by frequent collisions. These streamworn rocks are in the bed of the Aluna River in the Southern Highlands of Papua New Guinea.

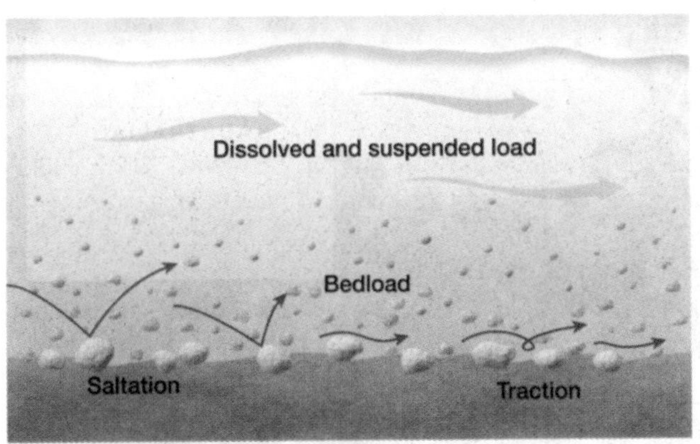

▲ **Figure 16-7** A stream moves its load in three ways. The dissolved and suspended loads are carried in the general water flow. The bedload is moved by traction (dragging) and saltation (bouncing).

in still water. (Fine clay may require as much as a year to sink 30 meters [100 feet] in perfectly quiet water.)

3. Sand, gravel, and larger rock fragments constitute the **bedload**. The smaller particles are moved along with the general streamflow in a series of jumps or bounces collectively referred to as *saltation*. Coarser pieces are moved by *traction*, which is defined as rolling or sliding along the streambed. The bedload is normally moved spasmodically, especially during floods: debris is transported some distance, dropped, then picked up later and carried farther.

Geomorphologists employ two concepts—*competence* and *capacity*—in describing the load a stream can transport. **Competence** is a measure of the particle size a stream can transport, expressed by the diameter of the largest particles that can be moved. Competence depends mainly on flow speed, with the power of the water generally increasing by the square of its speed. In other words, if the flow speed is doubled, the size of the largest movable particle is increased fourfold (2^2); if the flow speed triples, the force increases ninefold (3^2)—as stream speed increases even moderately, the largest moveable particle increases significantly. Thus, a stream that normally can transport only sand grains, might easily be able to move large boulders during a flood.

Capacity is a measure of the amount of solid material a stream has the potential to transport, normally expressed as the volume of material passing a given point in the stream channel during a given time interval. Capacity may vary tremendously over time, depending mostly on fluctuations in volume and flow speed but also on the characteristics of the load (particularly the mix of coarse and fine sediments). It is difficult to overemphasize the significance of the greatly expanded capacity of a stream to transport material at flood time.

Deposition Whatever is picked up must eventually be set down, which means that erosion is inevitably followed by deposition. Moving water, whether moving as overland flow or as streamflow, carries its load downslope or downvalley toward an ultimate destination—either ocean, lake, or interior drainage basin. A large volume of fast-moving water can carry its debris a great distance, but sooner or later deposition will take place as either flow speed or water volume decreases. Diminished flow is often the result of a change in gradient, but it may also occur when a channel either widens or changes direction. Therefore, stream deposits are found at the mouths of canyons, on floodplains, and along the inside bank of river bends. Eventually, however, much of a stream's load is dumped by moving water (stream) into quiet water (ocean or lake).

The general term for stream-deposited sediment is **alluvium**. Deposits of alluvium may include all sizes of rock debris, but smaller particles constitute by far the bulk of the total. Alluvial deposits often—but not always—exhibit one or more of the following characteristics:

1. Alluvial material is typically smooth and round due to the battering the rocks receive from each other as they flow downstream. In addition, over the great distances of many stream systems, this battering and buffeting (along with chemical weathering) eventually reduces boulders, cobbles, and pebbles to sand, silt, and clay.
2. Alluvial deposits often display distinct strata or layers, due to episodes of deposition following periodic floods.
3. Finally, alluvium is often "sorted"—that is, an alluvial deposit is often comprised of rocks of just about the same size. Sorting can occur when the speed of water flow diminishes—as stream competence drops, the heaviest (and therefore largest) rocks are deposited first, while smaller rocks are carried away to be deposited elsewhere.

Perennial and Intermittent Streams

We tend to think of rivers as permanent, but in fact many of the world's streams do not flow year-round. In humid regions, the large rivers and most tributaries are **perennial streams**—that is, permanent—but in less well-watered parts of the world, many of the major streams and most tributaries carry water only part of the time, either during the wet season or during and immediately after rains. These impermanent flows are called **intermittent streams** or *seasonal streams* if they flow for only part of the year and **ephemeral streams** if they carry water only during and immediately after a rain, although the term intermittent is sometimes applied to both cases (Figure 16-8).

Even in humid regions, many first-order and second-order streams are intermittent. These are generally short streams with relatively steep gradients and small watersheds. If rain is not frequent, or snowmelt not continuously available, these low-order streams simply run out of water. High-order streams in the same regions are likely to have permanent flow because of their larger drainage areas and because previous rainfall or snowmelt that sank into the ground can emerge in the valleys as groundwater runoff long after the rains have ceased.

▶ **Figure 16-8** Most desert streams are either intermittent or ephemeral, carrying water only rarely. This dry bed is near Wiluna in Western Australia.

The Role of Floods

Our discussion of factors associated with effective stream erosion and deposition leads us to an important concept in fluvial geomorphology—the role of floods. Most of us have stood on a bridge or on the brink of a canyon looking at a small stream far below and wondered how that meager flow could have carved the enormous gorge we see beneath us. On such occasions, we might be tempted to discount uniformitarianism and consider some brief, catastrophic origin as a more logical linking of what we can see with what we can imagine. Such reasoning, however, is erroneous, for uniformitarianism easily accounts for the existence of tiny streams in extensive valleys by invoking three geomorphic principles: (1) the extraordinary length of geologic time, (2) the role of mass wasting in moving weathered rock down into valleys where it can be transported away, and (3) the remarkable effectiveness of floods.

We have already discussed the incredible expanse of time involved in the evolution of topography (Chapter 13). This vast temporal sweep provides the opportunity for countless repetitions of an action, and the repetitive movement of even a tiny flow of water can wear away the strongest rock. In Chapter 15 we discussed the range of processes that can move masses of rock downslope where later it can be picked up and carried away.

Of perhaps equal importance is that the amount of water flowing in a stream varies. Most streams have fairly erratic flow regimes, with great fluctuation in **discharge**, or volume of flow. Many of the world's streams carry a relatively modest amount of water during most of the year but a relatively large volume during flood time. In some streams, flood flow occurs for several weeks or even months each year; but in the vast majority, the duration of high-water flow is much more restricted. In many cases, the "wet season" may be only one or two days a year. Yet the amount of denudation that can be accomplished during high-water flow is supremely greater than that done during periods of normal discharge. As stream discharge increases during a flood, so does the speed of the stream, and as the speed of flow increases, the competence and capacity of the stream increase exponentially. Thus, the epic work of almost all streams—the excavating of great valleys, the forming of vast floodplains—is primarily accomplished during flood flow.

Measuring Streamflow Each stream system has its own characteristic flow regime, influenced by local factors such as the climate and area of drainage. To understand the flow characteristics of a particular stream, data are gathered from a network of *stream gages*. Some 14,000 such stream gaging stations are operating across the United States and Canada, with many thousands more in operation around the world. On an ongoing basis, stream gages log information about the height of a stream (or "gage height" above a local reference point) as well as the discharge of a stream, commonly measured in cubic meters of water flow per second (cms).

By correlating stream discharge and gage height to local weather events, it is possible to estimate when floods will arrive or crest at difference places along the course of a river. The amount of urbanization in a region significantly influences this relationship. For example, in heavily urbanized areas, where there tends to be less infiltration of rain because much of the region is covered with impermeable surfaces such as pavement and buildings, a heavy rain tends to produce greater and faster runoff into local streams than in areas with more permeable surfaces (Figure 16-9).

Flood Recurrence Intervals Streamflow data gathered for a stream over many decades can also provide information about the relative frequency and size of floods. The probability of a given-size flood occurring is usually described as a **recurrence interval** (or *return period*). For example, a "100-year flood" refers to stream discharge that has a 1 in 100 (1 percent) probability of being exceeded in any single year. Such terminology is somewhat unfortunate, since it may leave the impression that such a flood is expected only once every 100 years, but this is not the case—it

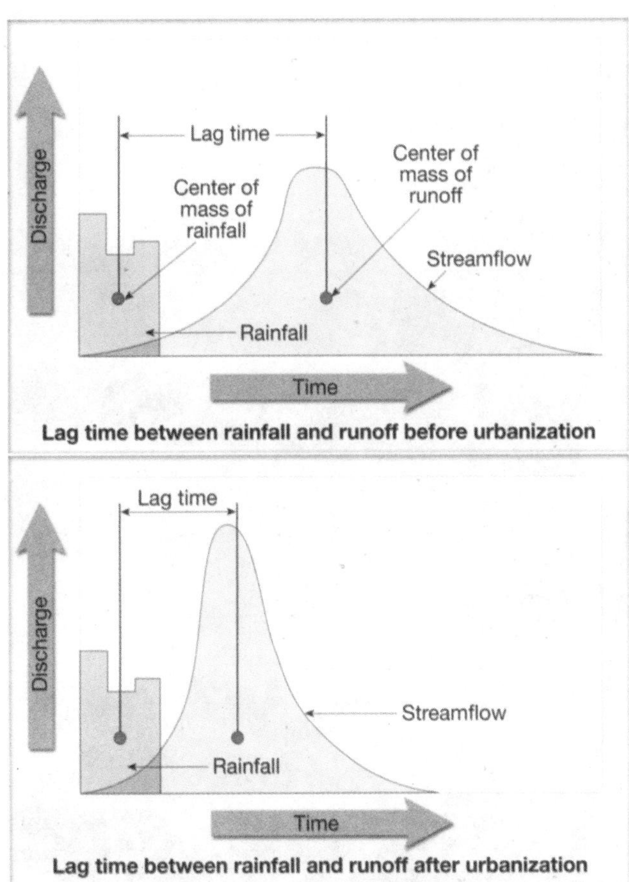

▲ **Figure 16-9** The lag time between the peak of rainfall and peak stream discharge during a flood is usually reduced in urbanized areas where there is less infiltration of runoff.

is possible to have more than one 100-year flood each century, or no such floods (Figure 16-10).

Flood recurrence intervals are calculated based on available streamflow data—the longer the record, the more accurate the probabilities. But changing circumstances—such as greater urbanization or climate change—can, of course, alter these probabilities.

Stream Channels

Overland flow is a relatively simple process. It is affected by such factors as rainfall intensity and duration, vegetative cover, surface characteristics, and slope shape, but its general characteristics are straightforward and easily understood. Streamflow is much more complicated, in part because streams represent not only a process of denudation but also an element of the landscape—an active force as well as an object of study.

Channel Flow

A basic characteristic of streamflow, and one that distinguishes it from the randomness of overland flow, is that it is normally confined to channels, which gives it a three-dimensional nature with scope for considerable complexity. In any channel having even a slight gradient, gravitational pull overcomes friction forces to move the water down-channel. Except under unusual circumstances, however, this movement is not straight, smooth, and regular. Rather it tends to be unsystematic and irregular, with many directional components and with different speeds in different parts of the channel.

Channel Cross Section A principal cause of flow irregularity is the retarding effect of friction along the bottom and sides of the channel, which causes the water to move most slowly there and fastest in the center of the stream (Figure 16-11). The amount of friction is determined by the width and depth of the channel and the roughness of its surface. A narrow, shallow channel with a rough bottom has a much greater retarding effect on streamflow than a wide, deep, smooth-bottomed one.

Turbulence In turbulent flow, the general downstream movement is interrupted by irregularities in the direction and speed of the water. Such irregularities produce momentary currents that can move in any direction, including upward. Turbulence in streamflow is caused partly by

◄ **Figure 16-10** Flooding along the Red River near Fargo, North Dakota, on March 25, 2009.

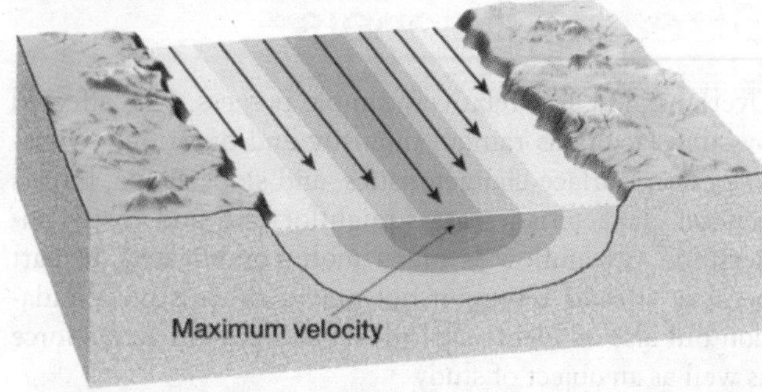

▲ **Figure 16-11** Friction between the stream bank and liquid surfaces cause a stream to flow fastest in the middle and at the surface. The arrow lengths indicate that water speed is greatest at the surface, where the water is farthest from any solid surfaces, and least at the bottom, where the water drags along the streambed.

friction, partly by internal shearing stresses between currents within the flow, and partly by surface irregularities in the channel. Stream speed also contributes to the development of turbulence, with faster streams more turbulent than slow-moving ones.

Eddies and whirlpools are conspicuous results of turbulence, as is the roiling whitewater of rapids. Even streams that appear very placid and smooth on the surface, however, are often turbulent at lower levels.

Turbulent flow creates a great deal of frictional stress as the numerous internal currents interfere with one another. This stress dissipates much of the stream's energy, decreasing the amount available to erode the channel and transport sediment. On the other hand, turbulence contributes to erosion by creating flow patterns that pry and lift rock materials from the streambed.

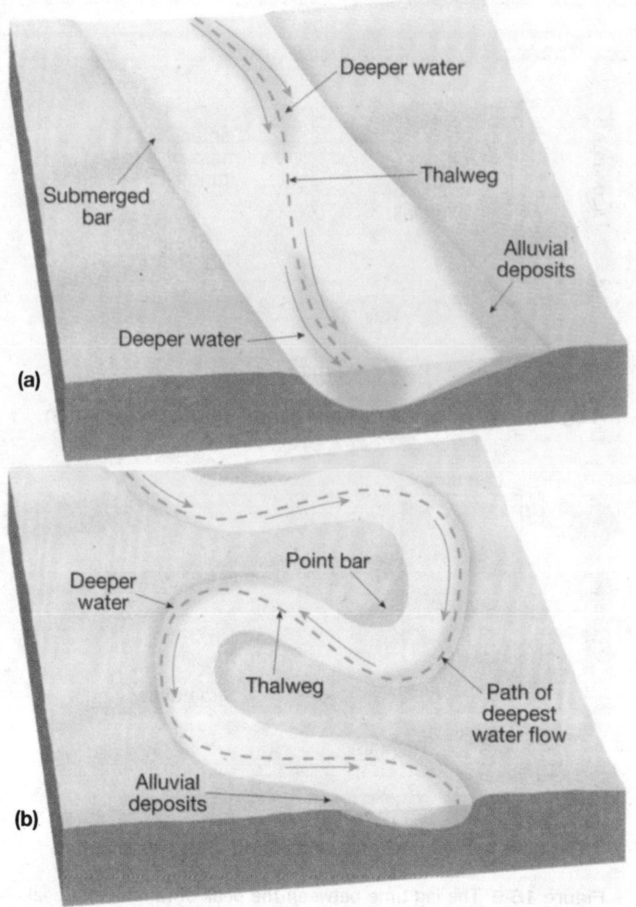

▲ **Figure 16-12** (a) Straight and (b) meandering channels. Erosion usually takes place along the outside of a turn where the stream is deepest and fastest, while alluvial deposition takes place along the inside of a turn where the water speed is slowest.

Stream Channel Patterns

Irregularities in streamflow are manifested in various ways, but perhaps most conspicuously by variations in channel patterns. If streamflow were smooth and regular, one might expect stream channels to be straight and direct. Few natural stream channels are straight and uniform for any appreciable distance, however. Instead, they wind about to a greater or lesser extent, sometimes developing remarkable sinuosity. In some instances, this winding is a response to the underlying geologic structure, but even in areas of perfectly uniform structure, stream channels are rarely straight.

Stream channel patterns are generally grouped into four categories—straight, sinuous, meandering, and braided—each with variations:

1. Straight channels are short and uncommon and usually indicative of strong control by the underlying geologic structure. A straight channel does not necessarily mean straight flow, however. A line running in the direction of the water and indicating the deepest parts of the channel, called the *thalweg* (German: *thal*, "valley," and *weg*, "way"), rarely follows a straight path midway between the stream banks; rather it wanders back and forth across the channel, as Figure 16-12 shows. Opposite the place where the thalweg approaches one bank, a deposit of alluvium is likely to be found. Thus, straight channels are likely to have many of the characteristics of sinuous channels.
2. **Sinuous channels** are much more common than straight ones. They are winding channels and are found in almost every type of topographic setting (Figure 16-13). Their curvature is usually gentle and irregular. Stream channels are likely to be sinuous even when the gradient of the stream is high (in other words, when flowing down a steep slope); where gradients are low (over flatter land), many stream channels will tend to develop greater sinuosity and begin to meander.
3. **Meandering channels** exhibit an extraordinarily intricate pattern of smooth curves in which the stream follows a serpentine course, twisting and contorting and turning back on itself, forming tightly curved loops and then abandoning them, cutting a new, different, and equally twisting course. Meandering generally occurs when the land is flat and the gradient is low, especially

▲ **Figure 16-13** The sinuous channel of the Merced River flowing out of Yosemite Valley in California.

▲ **Figure 16-14** Floodplain of the Trinity River in Big Thicket National Preserve, Texas.

when most of the rock being transported by the stream is fine-grained suspended load (Figure 16-14). A meander shifts its location almost continuously. This is accomplished by erosion on the outside of curves and deposition on the inside. In this fashion, meanders migrate across the floodplain and also tend to shift downvalley, producing rapid and sometimes abrupt changes in the channel. (Landforms that develop along a meandering stream channel will be discussed later in this chapter.)

4. **Braided channels** consist of a multiplicity of interwoven and interconnected channels separated by low bars or islands of sand, gravel, and other loose debris (Figure 16-15). Braiding often takes place when a very flat stream channel has a heavy load of sediment (such as the coarse-grained bedload supplied by glaciers to a meltwater stream) and in regions with prominent dry seasons and periods of low stream discharge (such as is common in arid environments). At any given time, the active channels of a braided stream may cover less than one-tenth of the width of the entire channel system, but in a single year most or all of the surface sediments may be reworked by the flow of the laterally shifting channels.

▲ **Figure 16-15** A typical braided stream on the South Island of New Zealand. This is the valley of the Lower Matukituki River.

RENEWABLE ENERGY

Hydroelectric Power

Nancy Lee Wilkinson, San Francisco State University

Moving water has the power to do work, and people have captured stream power since ancient times. Water wheels supzplied power for early mills and for the looms of the Industrial Revolution. Water wheels worked wherever the volume and velocity of flowing water were sufficient—most famously at "fall lines" where terrain dropped sharply.

Hydroelectric generation on a commercial scale began in the United States in the early 1880s. Plants are usually built below dams that store water to increase its volume, reliability, and potential energy. Water flows through large pipes (penstocks) to turn turbines many meters across. The turbines spin giant shafts topped by generators. As the turbines spin, the generator produces electric current (Figure 16-A).

Hydroelectric plants convert the potential energy of water stored behind a dam into kinetic energy; the amount of electricity they generate depends on the volume of water available and the elevation difference between the water supply and the turbine. The best hydropower sites are deep, narrow canyons with solid bedrock where elevation changes rapidly and a tall dam can be built with little material.

Hydroelectricity has important advantages. Dams generate minimal air and water pollution, providing an appealing alternative to coal, gas, or nuclear power plants. They require no fuel and only a small labor force to operate once built. Unlike generators that use steam to turn turbines and can take many hours to reach operating temperature, hydroelectric generators can be turned on or off in response to power demand. Finally, dams built for hydropower production can store water for irrigation or other uses, facilitate navigation, and provide recreational opportunities and flood protection. Revenues from these "multiple uses" can offset construction and operating costs.

Dams are an attractive tool for economic development because their construction employs large numbers of laborers. They symbolize progress and permanence and can continue generating electricity for a hundred years or more. Yet the social and environmental impacts of dams are substantial. Dams flood forests, fertile fields, and cities along streams, displacing millions of people worldwide. They block the upstream migration of fish that spawn in freshwater and flood upstream habitats. They dry up wetlands and block the downstream transport of water, sand, and gravel, increasing erosion of riverbanks and beaches. The failure of aging or poorly-built dams also poses a hazard to downstream communities.

The consequences of dam building became apparent as hydroelectric development boomed in the southeastern and western United States during the twentieth century (Figure 16-B) and as dams became a major tool in international development. By the year 2000, there were large dams on more than half of Earth's major rivers and new hydro-projects were encountering angry resistance in many parts of the world. Environmentalists had even begun to call for the destruction of dams they felt had outlived their usefulness and to initiate restoration efforts for damaged streams and watersheds.

Hydropower's controversial environmental and social impacts make it difficult to build large dams anywhere in the United States today. Off-stream reservoirs that store water but do not disrupt the stream's flow are one way to reduce the social and environmental costs of hydroelectricity. Pump-back facilities below existing dams also increase hydroelectric generation without further disruption to streams: although energy is lost each time, water can be pumped back during the night and reused to generate electricity again when it is in high demand. Finally, "small hydro" facilities can utilize stream flows to generate electricity at a smaller scale, without dams.

Hydroelectric dams provide less than 20 percent of the world's energy today, with the largest facilities in China, Canada, Brazil, the United States, and Russia. Dam building has declined from a peak in the 1970s, but many massive projects are still under construction and more are proposed. China recently completed the world's largest dam at Three Gorges (see Figure 9-22); international interests continue to support dam construction in the Global South.

While hydroelectric power is often thought of as a "green" energy source, social and environmental impacts suggest that dam projects must be planned and sited carefully to serve long-term interests. Changing precipitation patterns associated with global warming may also call for reassessment of hydroelectricity's role in regional energy production.

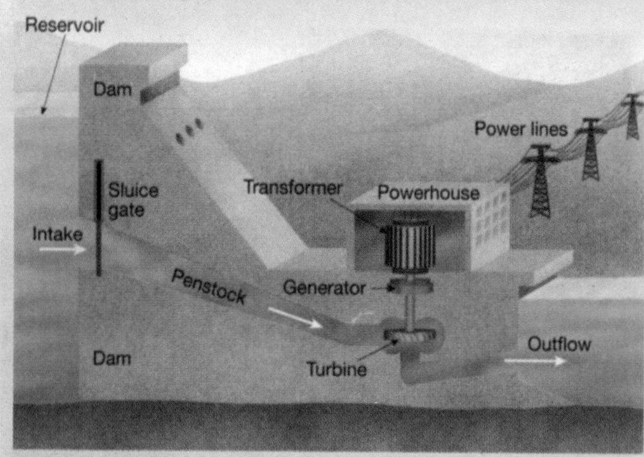

▲ **Figure 16-A** Hydroelectric generator.

▲ **Figure 16-B** Grand Coulee Dam in Washington.

Structural Relationships

Many factors affect stream development, and perhaps the most important is the geologic–topographic structure over or through which the stream must make its way and carve its valley. Each stream faces particular structural obstacles as it seeks the path of least resistance in its descending course to the sea. Most streams respond directly and conspicuously to structural controls, which is to say that their courses are guided and shaped by the nature and arrangement of the underlying bedrock.

Although streams change in pattern and flow characteristics, they may persist through eons of time, outlasting mountain ranges and other topographic assemblages that are more temporary occupants of continental surfaces. Thus, sometimes a relationship can be seen between the location of a stream and the contemporary structure of the land over which it flows, although sometimes it is necessary to delve into the geomorphic history of a region before the location of a drainage channel can be comprehended.

Consequent and Subsequent Streams The simplest and most common relationship between underlying structure and channel development is one in which the stream follows the initial slope of the land. Such a *consequent stream* is normally the first to develop on newly uplifted land, and many streams remain consequent throughout their evolutionary development. Streams that develop along zones of structural weakness are termed *subsequent streams*—they may excavate their channels along an outcrop of weak bedrock, or perhaps follow a fault zone or a master joint. Subsequent streams often trend at right angles to other drainage channels.

Antecedent and Superimposed Streams Some streams seem to "defy" the structure with courses that cut through ridges or other significant structures. One way this can occur is when an established stream is interrupted by an uplift of land that is so slow that the stream is able to maintain its previously established course by downward erosion, leaving a deep gorge carved through hills or mountains. Because such a stream antedates (predates) the existence of the uplift, it is called an **antecedent stream** (Figure 16-16).

A *superimposed stream* may also seem to ignore the local structure. Such streams originally existed on a higher sequence of land that has been entirely or largely eroded away, so that the original drainage pattern becomes incised into an underlying sequence of rocks of quite different structure. The result may be a drainage system that seems to bear no relation at all to the present surface structure.

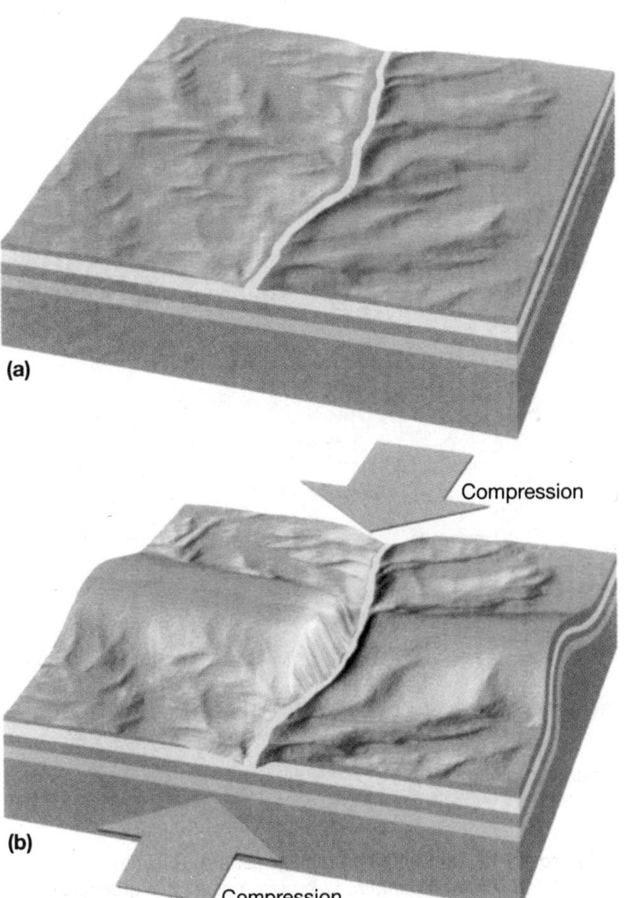

▲ **Figure 16-16** The development of an antecedent stream. (a) Stream course established before uplift begins. (b) Stream maintains its course and erodes through slowly rising anticlinal ridge.

Stream Drainage Patterns

In addition to the response of individual streams to the underlying structure, entire stream systems often form conspicuous drainage patterns in the landscape. These patterns develop largely in response to the underlying structure and slope of the land surface. Geologic-topographic structure can often be deduced from a drainage pattern, and, conversely, drainage patterns often can be predicted from structure.

Dendritic Pattern The most common drainage pattern over the world is a treelike, branching one called a **dendritic drainage pattern** (Figure 16-17). It consists of a random merging of streams, with tributaries joining larger streams irregularly but always at an angle smaller than 90°. The pattern resembles branches on a tree or veins on a leaf. The relationship between drainage pattern and land structure is negative, which is to say that the underlying structure does not control the evolution of the drainage pattern—the underlying rocks are more or less equally resistant to erosion. Dendritic patterns are more numerous than all others combined and can be found almost anywhere.

Trellis Pattern A **trellis drainage pattern** usually develops as a response to an underlying structure consisting of alternating bands of tilted hard and soft strata, with long, parallel streams linked by short, right-angled segments (Figure 16-18). Two regions of the United States are particularly noted for their trellis drainage patterns: the ridge-and-valley section of the Appalachian Mountains and the

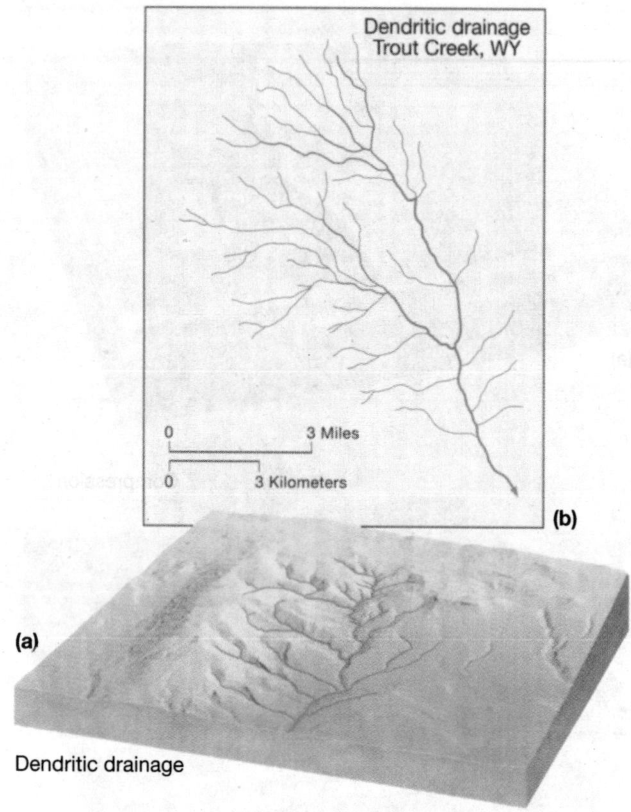

▲ **Figure 16-17** (a) Dendritic drainage pattern. (b) Trout Creek, from the *Pat O'Hara Mountain, Wyoming,* topographic quadrangle.

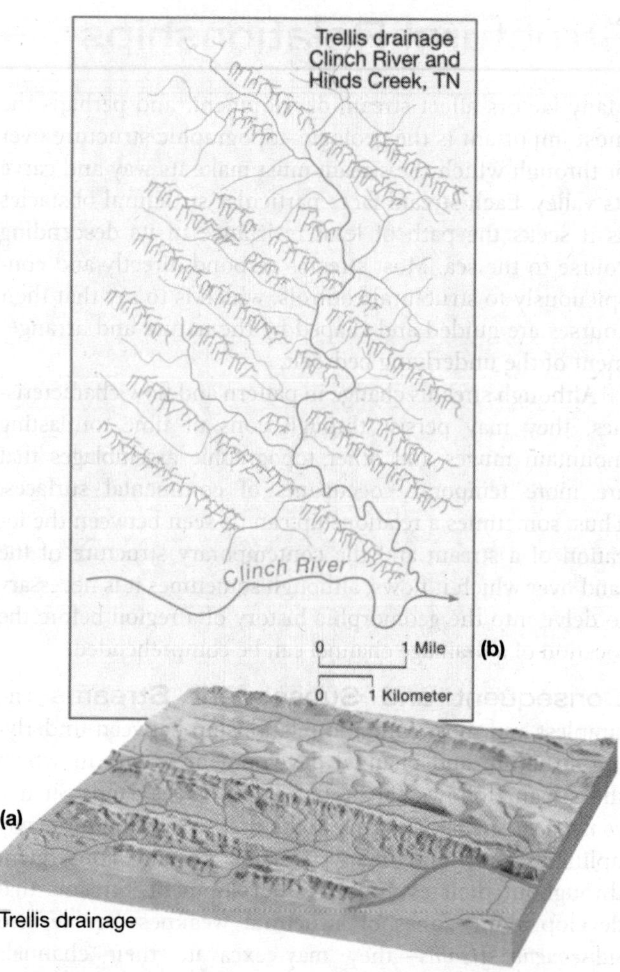

▲ **Figure 16-18** (a) Trellis drainage pattern. (b) Clinch River and Hinds Creek from the *Norris, Tennessee,* topographic quadrangle.

Ouachita Mountains of western Arkansas and southeastern Oklahoma.

In the ridge-and-valley section, which extends northeast–southwest for more than 1280 kilometers (800 miles) from New York to Alabama, the drainage pattern developed in response to tightly folded Paleozoic sedimentary strata forming a world-famous series of parallel ridges and valleys (see Figure 14-53). Parallel streams flow in the valleys between the ridges, with short, right-angled connections here and there cutting through the ridges.

The marked contrast between trellis and dendritic patterns is shown dramatically by the principal streams of West Virginia (Figure 16-19). The folded structures of the eastern part of the state produce trellising, whereas the nearly horizontal strata of the rest of the state are characterized by dendritic patterns.

Radial Pattern A *radial drainage pattern* is usually found when streams descend from some sort of concentric uplift, such as an isolated volcano. Figure 16-20 shows one example: Mount Egmont on the North Island of New Zealand.

Centripetal Pattern A *centripetal drainage pattern,* essentially the opposite of a radial one, is usually associated with streams converging in a basin. Occasionally, however, centripetal drainage develops on a much grander scale. Shown in Figure 16-21 is the northeastern part of Australia, where rivers from hundreds of kilometers away converge toward the Gulf of Carpentaria, a basin partially inundated by the sea.

Annular Pattern More complex is an *annular drainage pattern,* which can develop either on a dome or in a basin where dissection has exposed alternating concentric bands of tilted hard and soft rock. The principal streams follow curving courses on the softer material, occasionally breaking through the harder layers in short, right-angled segments. The Maverick Spring Dome of Wyoming portrays a prominent example of annular drainage (Figure 16-22). This dome of ancient crystalline rocks was pushed up through a sedimentary overlay and has been deeply eroded, thus exposing crystallines in the higher part of the hills, with upturned concentric sedimentary ridges (called *hogbacks*) around the margin. The streams are mostly incised into the softer layers.

The Shaping and Reshaping of Valleys

Running water shapes terrain partly by overland flow on interfluves but mostly by streamflow in the valleys. Thus, by focusing our attention on the processes through which streams shape and reshape their valleys, we can understand

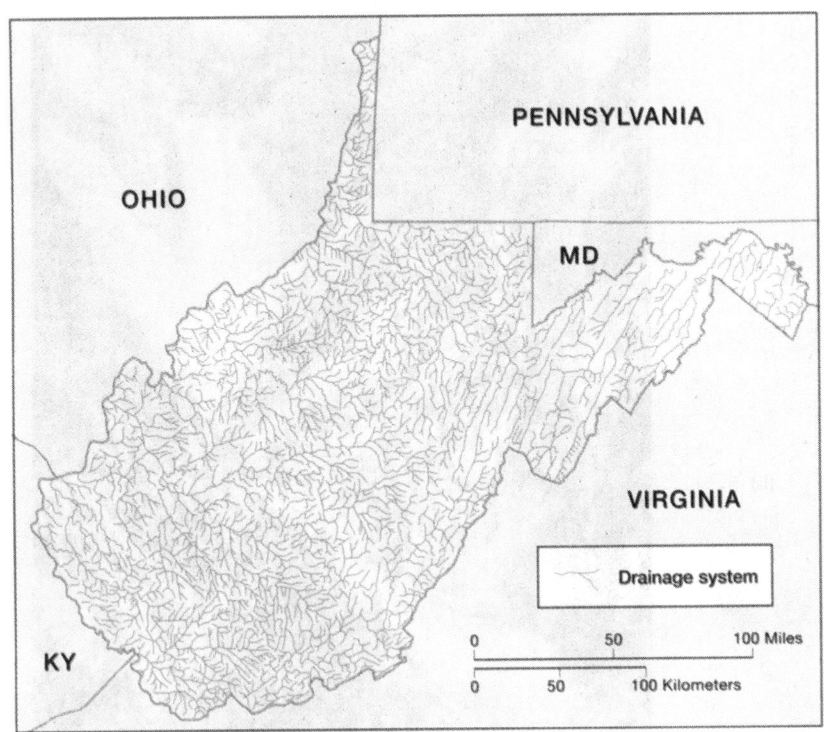

◀ **Figure 16-19** Drainage pattern contrasts in West Virginia. The trellis systems in the east are a response to parallel folding; the dendritic drainages in the west have developed because there are no prominent structural controls.

the development of fluvial landforms in general. The shaping of valleys and their almost continuous modification through time produce a changing balance between valleys and interfluves and consequently an ongoing dynamism in the configuration of most parts of the continental surfaces.

Valley Deepening

Wherever it has either a relatively rapid speed or a relatively large volume, a stream expends most of its energy in **downcutting**. This lowering of the streambed involves the hydraulic power of the moving water, the prying and lifting capabilities of turbulent flow, and the abrasive effect of the stream's bedload as it rolls, slides, and bounces along the channel. Downcutting is most frequent in the upper reaches of a stream, where the gradient is usually steep and the valley narrow. The general effect of downcutting is to produce a deep valley with steep sides and often a V-shaped cross section (Figure 16-23).

Base Level A stream excavates its valley by eroding the channel bed. If only downcutting were involved, the resulting valley would be a narrow, steep-sided gorge. Such gorges sometimes occur, but usually other factors are at work also and the result is a wider valley. In either case, there is a lower limit to how much downcutting a stream

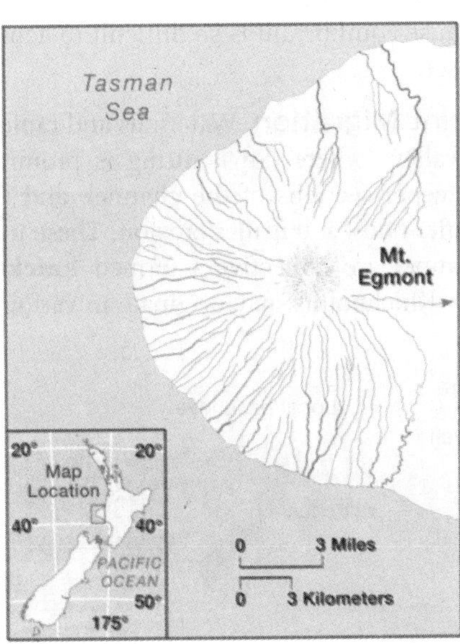

▲ **Figure 16-20** The extraordinary radial drainage pattern of Mount Egmont in New Zealand.

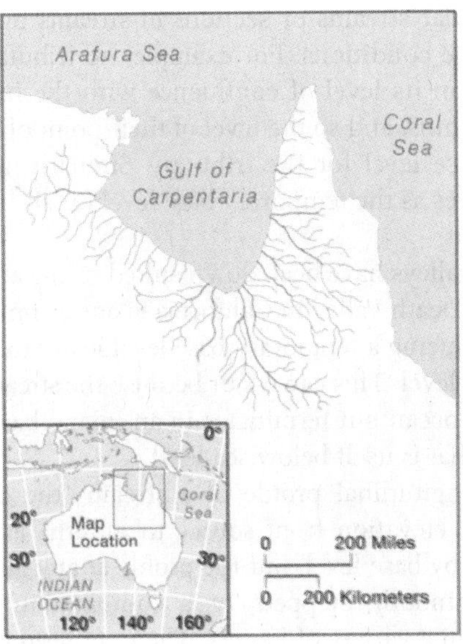

▲ **Figure 16-21** The centripetal drainage pattern of the region around the Gulf of Carpentaria in northeastern Australia.

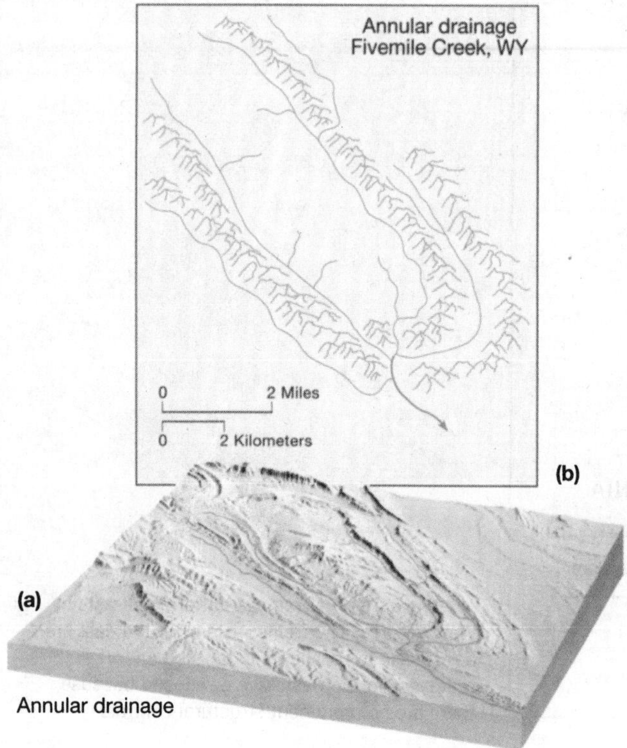

▲ **Figure 16-22** (a) Annular drainage pattern. (b) Fivemile Creek from the *Maverick Spring Dome, Wyoming,* topographic quadrangle.

▲ **Figure 16-23** The Yellowstone River occupies a conspicuous V-shaped valley, shaped in part through downcutting.

can do, and this limit is called the **base level** of the stream. Base level is an imaginary surface extending underneath the continents from sea level at the coasts (Figure 16-24). This imaginary surface is not simply a horizontal extension of sea level, however; inland it is gently inclined at a gradient that allows streams to maintain some flow. Sea level, then, is the absolute, or *ultimate base level*, or lower limit of downcutting for most streams.

As Figure 16-24 shows, there are also local, or temporary, base levels, which are limits to downcutting imposed on particular streams or sections of streams by structural or drainage conditions. For example, no tributary can cut deeper than its level of confluence with the higher-order stream it joins, and so the level of their point of junction is a local base level for the tributary. Similarly, a lake normally serves as the temporary base level for all streams that flow into it.

Some valleys have been downfaulted to elevations below sea level (Death Valley in California is one example), a situation producing a temporary base level lower than the ultimate base level. This can occur because the stream does not reach the ocean but terminates in an inland basin or body of water that is itself below sea level.

The longitudinal profile of a stream (the downvalley change in elevation from source to mouth) is ultimately restricted by base level, but the profile at any given time—whether smooth, stepped, or a combination of both—depends on a variety of factors. The long-term tendency is toward a profile in which the amount of sediment entering a stretch of a stream is equal to the amount leaving it. This hypothetical condition is called a *graded stream*, defined as one in which the gradient just allows the stream to transport its load. A graded stream is more theoretical than actual because equilibrium is so difficult to achieve and so easy to upset.

Knickpoint Migration Waterfalls and rapids are often found in valleys where downcutting is prominent. They occur in steeper sections of the channel, and their faster, more turbulent flow intensifies erosion. These irregularities in the channel are collectively termed **knickpoints** (or *nickpoints*). Knickpoints may originate in various ways but

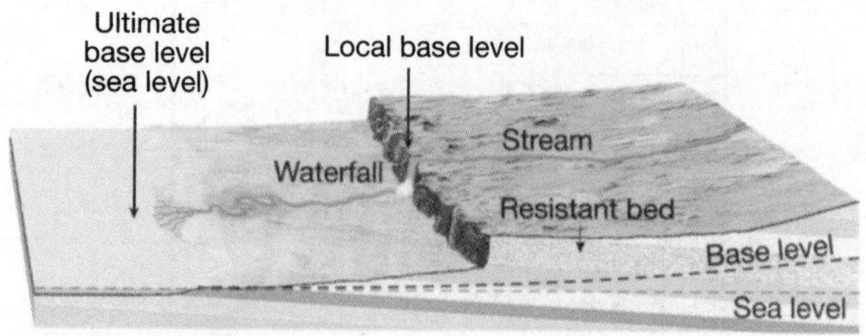

▲ **Figure 16-24** Comparison of sea level, base level, and local base level.

are commonly the result of abrupt changes in bedrock resistance. The more resistant material inhibits downcutting, and as the water plunges over the waterfall or rapids with accelerated vigor, it tends to scour the channel above and along the knickpoint and fill the channel immediately downstream. This intensified action eventually erodes away the harder material, so that the knickpoint migrates upstream with a successively lower profile until it finally disappears and the channel gradient is smoothed (Figure 16-25). Knickpoint migration can be relatively rapid when the bedrock consists of soft sedimentary rocks, but relatively slow when the rock is resistant plutonic or metamorphic rock.

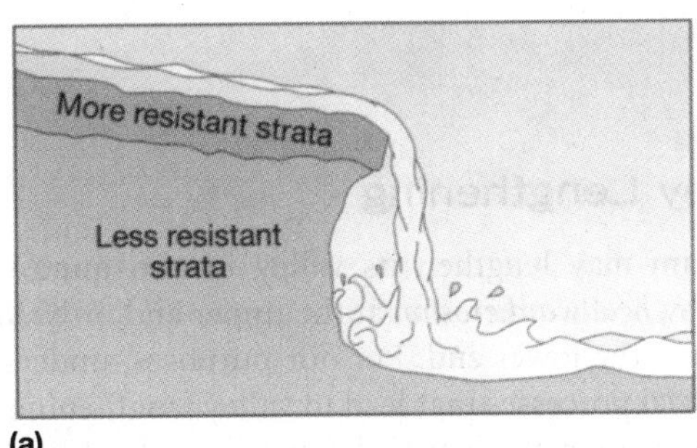

(a)

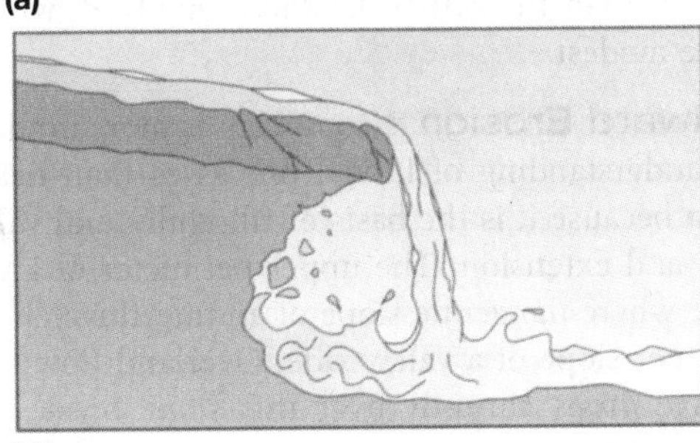

(b)

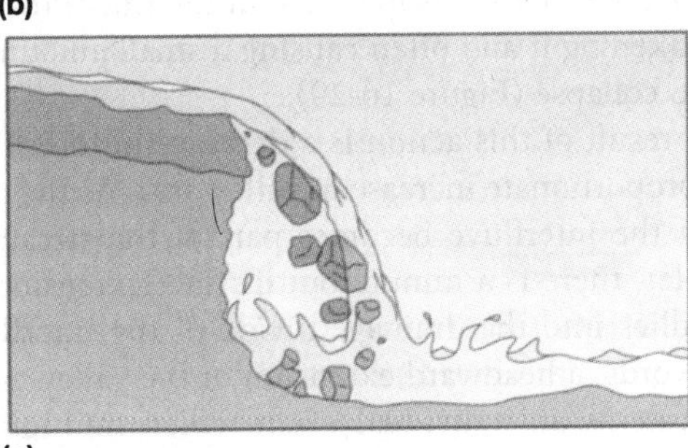

(c)

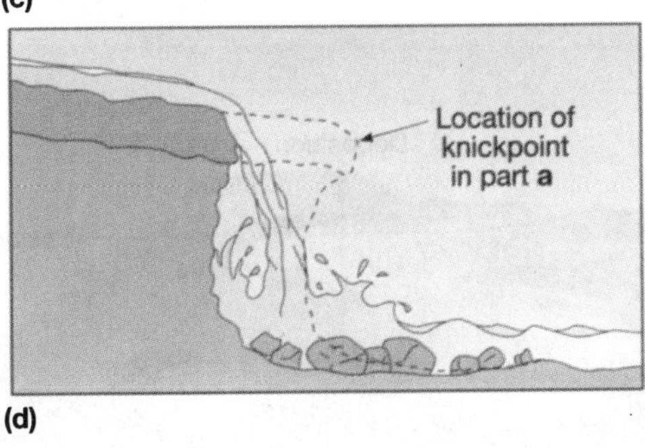

(d)

▲ **Figure 16-25** (a) Knickpoint formed where a stream flows over a resistant layer of rock. (b) The water flow undercuts the lip and (c) causes it to collapse. (d) Position of knickpoint has migrated upstream.

▲ **Figure 16-26** Niagara Falls from the American side. The Niagara River drops about 55 meters (180 feet) over a knickpoint formed by the Niagara Escarpment.

Niagara Falls is a particularly good example of knickpoint migration on a massive scale (Figure 16-26). The Niagara River forms the connecting link between Lake Erie and Lake Ontario. As the contemporary drainage system became established in this area following the last retreat of Pleistocene ice sheets about 12,000 years ago, Lake Ontario was about 50 meters (about 150 feet) higher than its present level, and the Niagara River had no falls. However, an easterly outlet, the Mohawk Valley, developed for Lake Ontario, and the lake drained down to approximately its present level, exposing a prominent escarpment directly across the course of the river. The Niagara Escarpment is formed by a massive bed of resistant limestone that dips gently toward Lake Erie and is underlain by similarly dipping but softer strata of shale, sandstone, and limestone (Figure 16-27).

As the river pours over the escarpment, the swirling water undermines the hard limestone by erosion of the weaker beds beneath, leaving a lip of resistant rock projecting without support. Through the years, the lip has collapsed, with block after block of limestone tumbling into the gorge below. After each collapse, rapid undermining takes place again, leading to further collapse. In this fashion the falls has gradually retreated upstream, moving southward a distance of about 11 kilometers (7 miles) from its original position along the trend of the escarpment, the retreat being marked by a deep gorge.

This principle of **knickpoint migration** is important in understanding fluvial erosion because it illustrates dramatically the manner in which valley shape often develops first in the lower reaches and then proceeds progressively upstream, even though the water obviously flows downstream.

Valley Widening

Where a stream gradient is steep and the channel well above the local base level, downcutting is usually the dominant activity; as a result, valley widening is likely to be

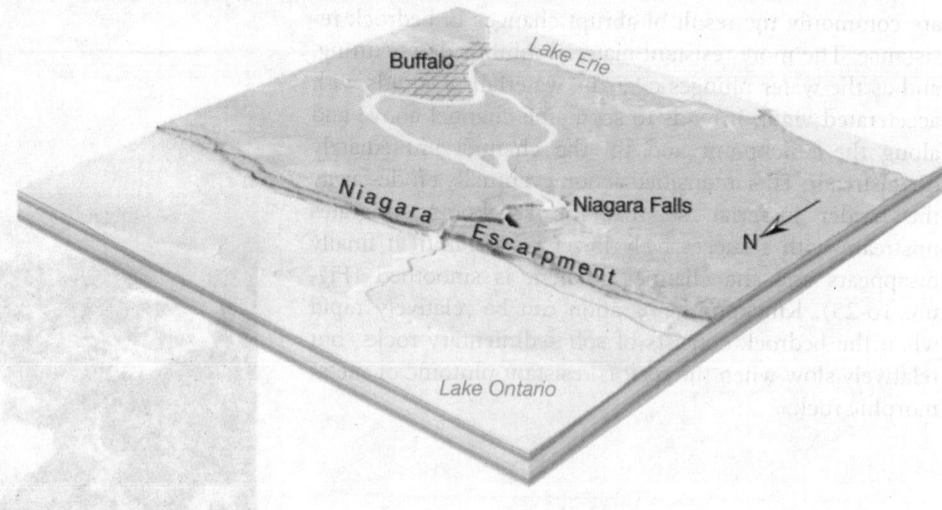

▶ **Figure 16-27** The situation of Niagara Falls. The falls was originally located where the Niagara River crosses the Niagara Escarpment, but it has retreated upstream to its present location.

slow. Even at this stage, however, some widening takes place as the combined action of weathering, mass wasting, and overland flow removes material from the valley sides. Downcutting diminishes with time as the stream reduces its gradient, or wherever the stream course flows down a gentle slope. The stream's energy is then increasingly diverted into a meandering flow pattern. As the stream sways from side to side, **lateral erosion** begins: the main flow of the current swings from one bank to the other, eroding where the water speed is greatest and depositing where it is least. The water moves fastest on the outside of curves (the *cut bank*), and there it undercuts the bank. On the inside of a curve where water is moving most slowly, alluvium is likely to accumulate, forming a *point bar* along the inside bank of the stream (Figure 16-28). The channel shifts position often, so that undercutting is not concentrated in just a few locations. Rather, over a long period of time, most or all parts of the valley sides are undercut. (Landforms associated with such meandering streams will be described later in this chapter.)

All the while a valley floor is being widened through lateral erosion, mass wasting is usually helping wear back the valley walls—and in some valleys, mass wasting is the dominant process responsible for widening. In addition, similar processes along tributary streams also contribute to the general widening of the main valley.

Valley Lengthening

A stream may lengthen its valley in two quite different ways: by *headward erosion* at the upper end or by *delta formation* at the lower end. For our purposes, understanding the fluvial processes that lead to valley lengthening is more important than the actual distances involved, which may be quite modest.

Headward Erosion No concept is more fundamental to an understanding of fluvial processes than **headward erosion** because it is the basis of rill, gully, and valley formation and extension. The upper perimeter of a valley is the line where the gentle slope of an interfluve changes to the steeper slope of a valley side. Overland flow from the interfluve drops abruptly over this slope break, and the fast-moving water tends to undercut the rim of the perimeter, weakening it and often causing a small amount of material to collapse (Figure 16-29).

The result of this action is a decrease in interfluve area and a proportionate increase in valley area. As the overland flow of the interfluve becomes part of the streamflow of the valley, there is a minute but distinct extension of rills and gullies into the drainage divide of the interfluve—in other words, a headward extension of the valley. Although minuscule as an individual event, when multiplied by a

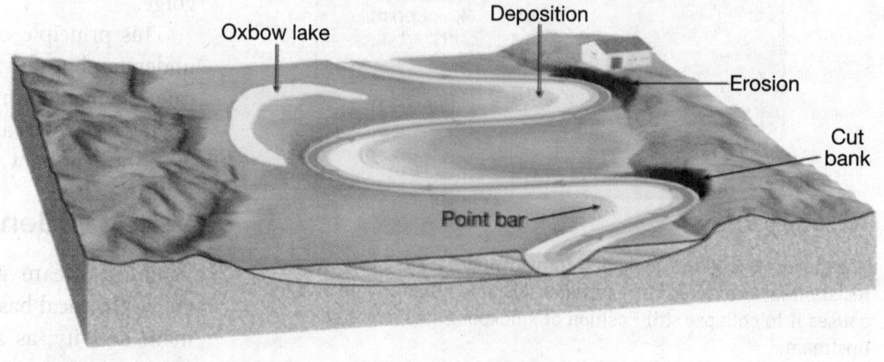

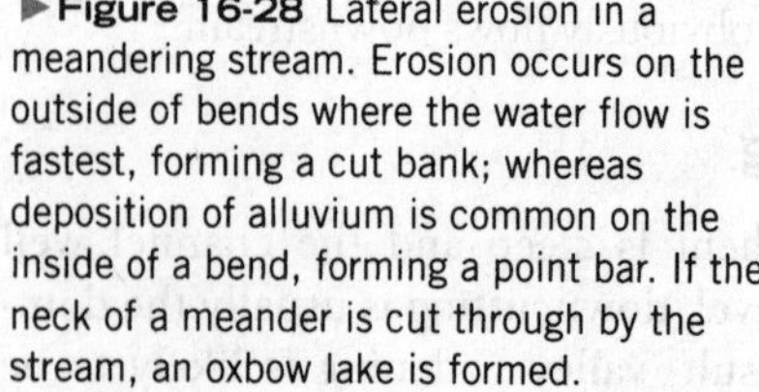

▶ **Figure 16-28** Lateral erosion in a meandering stream. Erosion occurs on the outside of bends where the water flow is fastest, forming a cut bank; whereas deposition of alluvium is common on the inside of a bend, forming a point bar. If the neck of a meander is cut through by the stream, an oxbow lake is formed.

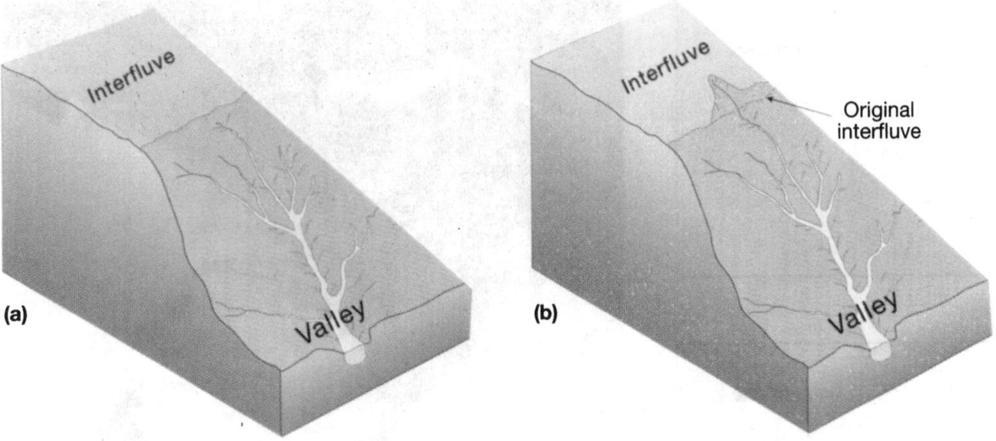

◀ **Figure 16-29** Headward erosion occurs at the upper end of a stream where overland flow pours off the lip of the interfluve into the valley. (b) The channeled streamflow wears back the lip of the interfluve. Over time, this erosion extends the valley headward at the expense of the interfluve.

thousand gullies and a million years, this action can lengthen a valley by tens of kilometers and the expansion of a drainage basin by hundreds of square kilometers. Thus, the valley lengthens at the expense of the interfluve (Figure 16-30).

Stream Capture Headward erosion is illustrated dramatically when a portion of the drainage basin of one stream is diverted into the basin of another stream by natural processes. This event, called **stream capture** or **stream piracy**, is relatively uncommon in nature, but evidence that it does sometimes occur is found in many places.

As a hypothetical example, let us consider two streams flowing across a coastal plain, as shown in Figure 16-31.

▲ **Figure 16-30** Headward erosion of tributary streams into the formerly undissected landscape along the San Rafael River in Utah.

Their valleys are separated by an interfluve, which for this example can be thought of as an undulating area of low relief. Stream A is shorter than stream B but is also more powerful, and A's valley is aligned so that headward extension will project it in the direction of B's valley.

As stream A lengthens its valley through headward erosion, the drainage divide between the two valleys is reduced. As the process continues, the headwaters of stream A eventually extend completely into the valley bottom of stream B and the flow from the upper reaches of B is diverted into A. In the parlance of the geographer, stream A has "captured" part of stream B. Stream A is called the *captor stream*, the lower part of B is the *beheaded stream*, the upper part of B is the *captured stream*, and the abrupt bend in the stream channel where the capture took place is called an *elbow of capture*.

Stream capture on a grand scale can be detected on a map of West Africa. The mighty Niger River has its headwaters relatively near the Atlantic Ocean, but it flows inward rather than seaward. After flowing northeast for nearly 1600 kilometers (1000 miles), it makes an abrupt turn to the southeast and then continues in that direction for another 1600 kilometers before finally emptying into the Atlantic. At some time in the past, the upper reaches of what is now the Niger was a separate river, one that did not change course but rather flowed northeast until it reached a great inland lake in what is now

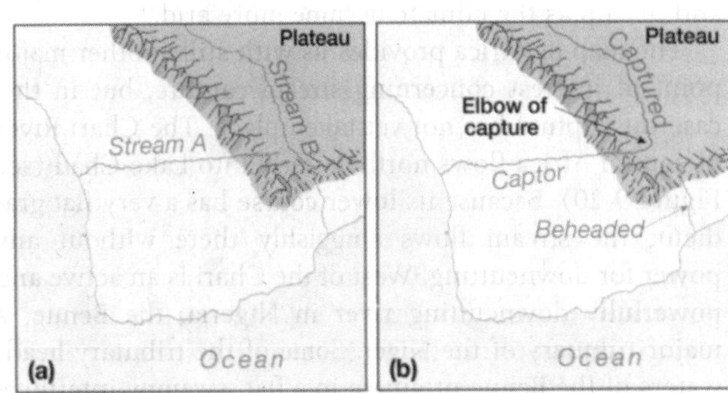

▲ **Figure 16-31** A hypothetical stream-capture sequence. The valley of stream A is extended by headward erosion until stream A captures and beheads stream B.

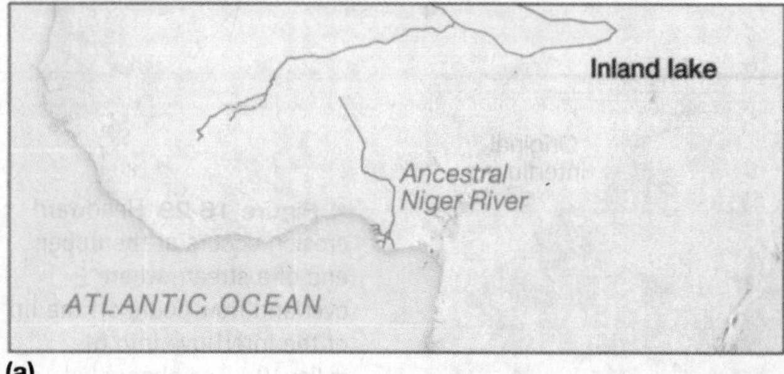

(a)

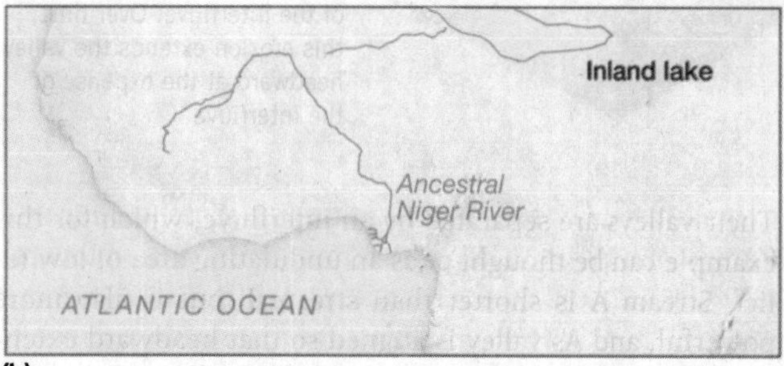

(b)

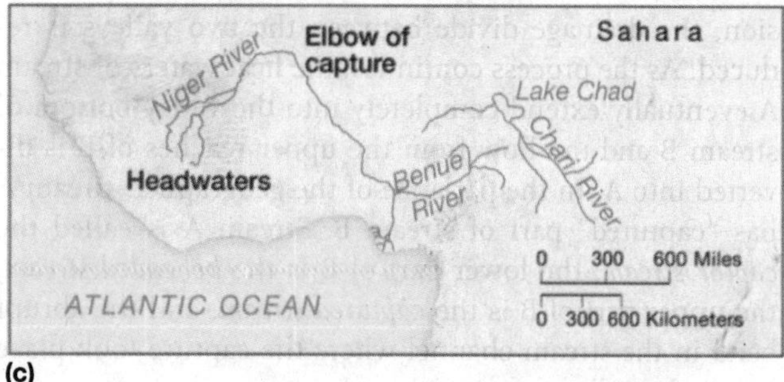

(c)

▲ **Figure 16-32** Stream capture, actual and anticipated in West Africa. (a) The upper course of the Niger River was once part of an unnamed stream that flowed into a large lake in what is now the Sahara. (b) This ancient stream was captured by headward erosion of the ancestral Niger. (c) At present, headward erosion on the Benue River gives promise of capturing the Chari River just a few tens or hundreds of centuries from now.

▲ **Figure 16-33** The Nile River delta captured by a NASA satellite using the Multi-angle Imaging Spectroradiometer's (MISR's) nadir camera in January 2001. The view is northward, with the Nile entering from the bottom and the Mediterranean Sea at the top.

the central Sahara (Figure 16-32). This river was beheaded by the ancestral Niger, producing a great elbow of capture and leaving the beheaded stream to wither and dry up as the climate became more arid.

The map of Africa provides us with still another major point of interest concerning stream capture, but in this case the capture has not yet taken place. The Chari River of central Africa flows northwesterly into Lake Chad (see Figure 9-20). Because its lower course has a very flat gradient, the stream flows sluggishly there without any power for downcutting. West of the Chari is an active and powerfully downcutting river in Nigeria, the Benue, a major tributary of the Niger. Some of the tributary headwaters of the Benue originate in a flat, swampy interfluve only a short distance from the floodplain of the Chari. Since the Benue is more active than the Chari and its alignment is such that headward erosion cuts directly into the Chari drainage, the Benue is likely to behead the Chari before our very eyes, so to speak, provided we can wait a few thousand years.

Delta Formation A valley can also be lengthened at its seaward end—in this case, by deposition. Flowing water slows down whenever it enters the quiet water of a lake or ocean and deposits its load. Most of this debris is dropped right at the mouth of the river in a landform called a **delta**, after a fancied resemblance to the Greek capital letter delta, Δ (Figure 16-33). The classic triangular shape is maintained in some deltas, but it is severely modified in others because of imbalances between the amount of sediment deposited by rivers and the removal of those sediments by ocean waves and currents.

The stream slows down, losing both competence and capacity, and drops much of its load, which partially blocks the channel and forces the stream to seek another path (Figure 16-34). Later this new path is likely to become clogged, and the pattern is repeated. As a result, deltas usually consist of a maze of roughly parallel channels called *distributaries* through which the water flows slowly toward the sea. Continued deposition builds up the surface of the delta so that it is at least partially exposed above sea level.

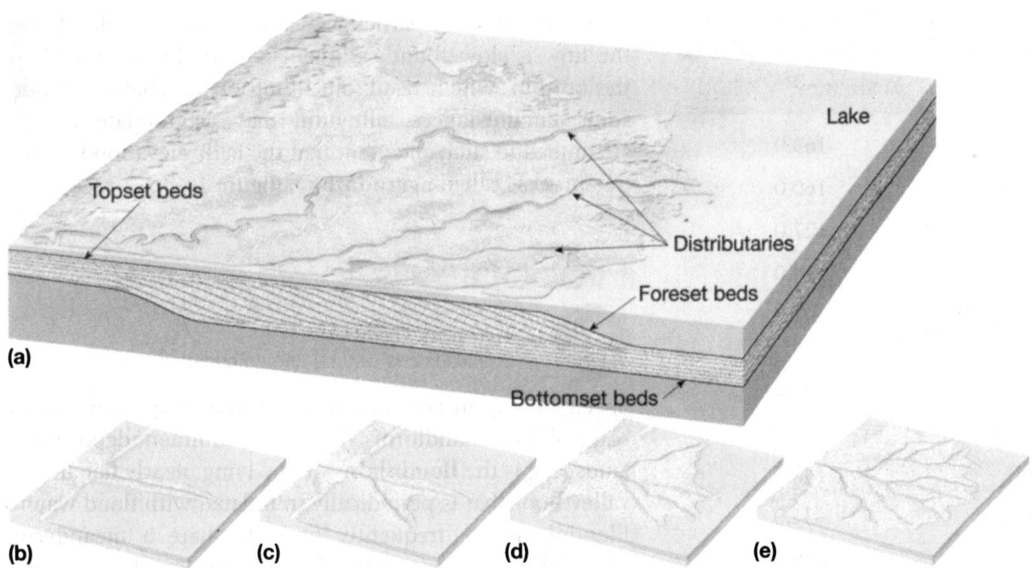

▲ **Figure 16-34** Hypothetical sequence showing the formation of a simple delta in a quiet body of water. *Foreset bed* deposits are comprised of coarser sediments dropped immediately by the stream; *topset beds* are deposited during floods over the foreset beds; *bottomset beds* are composed of fine sediments that settle some distance from the mouth of the stream.

Rich alluvial sediments and an abundance of water favor the establishment of vegetation, which provides a base for further expansion of the delta. In this fashion, the stream valley is extended downstream.

At some river mouths, the local coastal currents are so vigorous that no delta is formed—the stream sediment is simply swept away to be deposited elsewhere along the coast or offshore (Figure 16-35). Often, however, a prominent and clear-cut delta does form (Table 16-1).

As we will see later in this chapter when we discuss flood control strategies, many of the world's major deltas, such as the Nile and Mississippi, are degrading as a consequence of direct human manipulation of the rivers that feed them, as well as slowly rising sea levels.

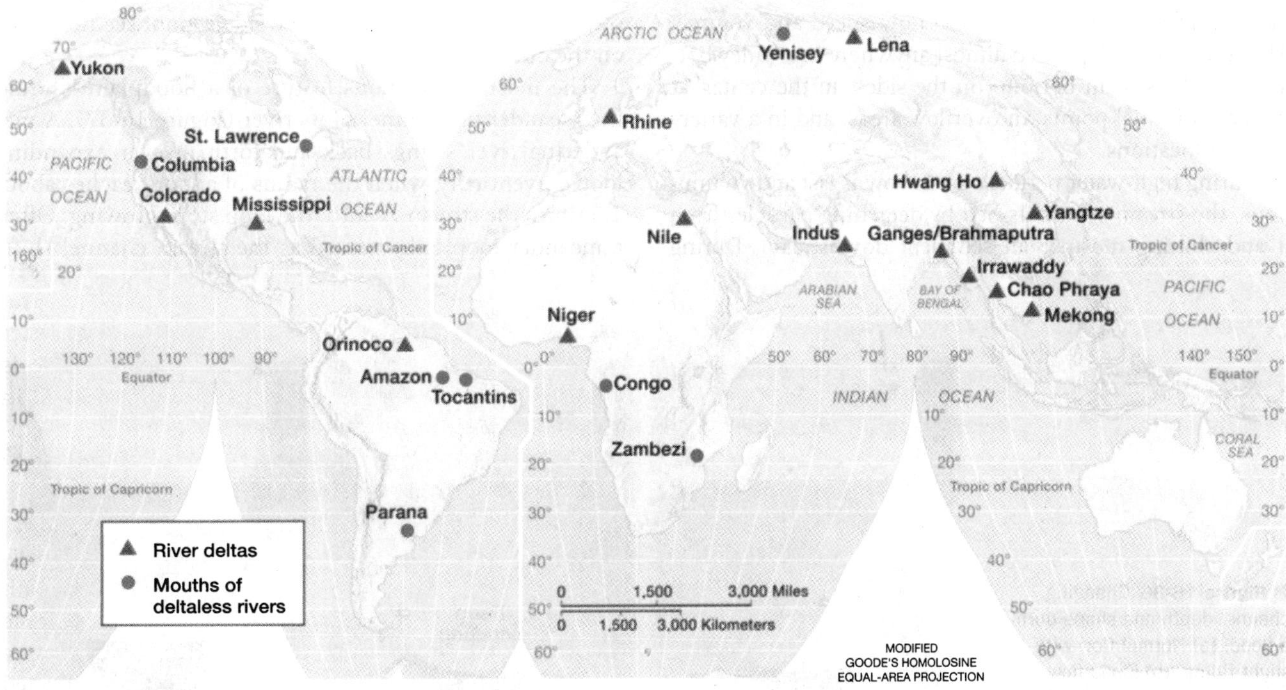

▲ **Figure 16-35** Locations of the world's largest deltas and deltaless rivers. Not all large rivers form deltas. Major rivers such as the Amazon, Congo, Yenisey, and the Paraná do not have deltas.

TABLE 16-1 The World's Largest Deltas

Rank	River (Country)	Area ($km^2 \times 1000$)
1	Indus (Pakistan)	163.0
2	Nile (Egypt)	160.0
3	Hwang Ho (China)	127.0
4	Yangtze (China)	124.0
5	Ganges/Brahmaputra (Bangladesh)	91.0
6	Orinoco (Venezuela)	57.0
7	Yukon (United States)	54.0
8	Mekong (Vietnam)	52.0
9	Irrawaddy (Myanmar)	31.0
10	Lena (Russia)	28.5
11	Mississippi (United States)	28.0
12	Chao Phraya (Thailand)	24.6
13	Rhine (Netherlands)	22.0
14	Colorado (Mexico)	19.8
15	Niger (Nigeria)	19.4

Deposition in Valleys

Thus far, we have emphasized the prominence of the removal and transportation of material in the formation and shaping of valleys, but deposition, too, has a role in these processes.

Nearly every stream continuously rearranges its sediment in response to variations in flow speed and volume. Alluvium can be deposited almost anywhere in a valley bottom: on the stream bottom, on the sides, in the center, at the base of knickpoints, in overflow areas, and in a variety of other locations.

During high-water periods, when flow is fast and voluminous, the stream scours its bed by detaching particles from it and shifting most or all sediment downstream. During low-water periods, particularly after a period of flood flow, the flow is slowed and sediment is more likely to settle to the bottom, which results in filling of the channel. Under some circumstances, alluvium may accumulate on the streambed to such an extent that the bed's elevation is raised in a process called **aggradation** (Figure 16-36).

Floodplains

1. Meandering Streams
2. Floods and Natural Levee Formation

Special attention needs to be given to an important assemblage of fluvial landforms. The most prominent depositional landscape is the **floodplain**—a low-lying, nearly flat alluvial valley floor that is periodically inundated with flood waters. Floodplains are frequently formed where a meandering stream flows across a wide, nearly level valley floor.

Floodplain Landforms

The frequent shifting of stream meanders produces an increasingly broader, flattish valley floor largely or completely covered with deposits of alluvium left by periodic floods. At any given time, a stream is likely to occupy only a small portion of the flatland, although during periods of flood flow, the entire floor may be flooded. For this reason, the valley bottom is properly termed a floodplain. The outer edges of the floodplain are usually bounded by an increase in slope, sometimes marking the outer limit of lateral erosion and undercutting where the flat terrain abruptly changes to a line of *bluffs*. Valley widening and floodplain development can extend for great distances; the floodplains of many of the world's largest rivers are so broad that a person standing on the bluffs at one side cannot see the bluffs on the other.

The most conspicuous feature of a floodplain is often the meandering channel of its river (Figure 16-37). A meandering river swings back and forth in ever expanding loops. Eventually, when the radius of a loop reaches about 2.5 times the stream's width, the loop stops growing. Often a meander loop is by-passed as the stream channel shifts

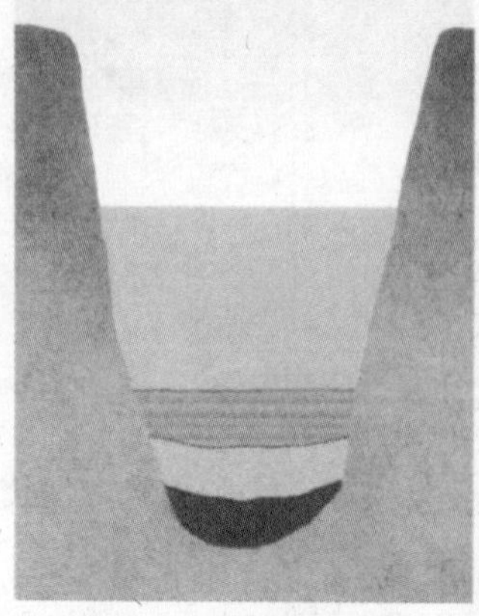
(a) Normal flow—slight filling

(b) Flood flow—deep scouring

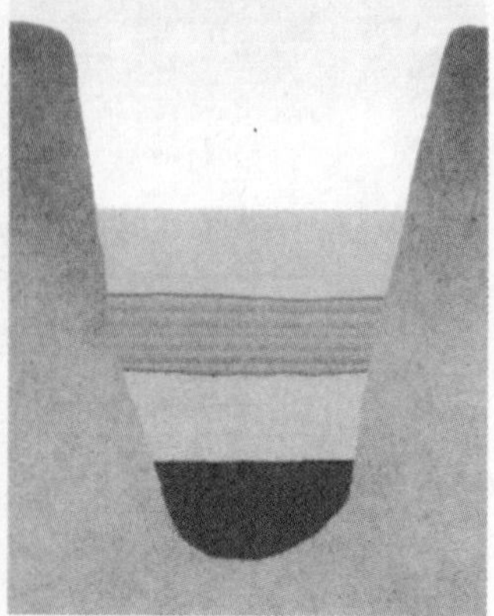
(c) Receding water after flood-filling

▶ **Figure 16-36** Changing channel depth and shape during a flood: (a) Normal flow with slight filling. (b) Flood flow significantly deepens the channel by scouring. (c) As the flood recedes, considerable filling raises the channel bed again.

◀ **Figure 16-37** The floodplain of the Bear River near Brigham City, Utah. As with all other floodplains, the meandering river channel is the most conspicuous feature. In this case, snow cover highlights the pattern.

through lateral erosion and cuts a new channel across its neck and starts meandering again, leaving the old meander loop as a **cutoff meander**. The cutoff portion of the channel may remain for a period of time as an **oxbow lake**, so named because its rounded shape resembles the bow part of yokes used on teams of oxen. Oxbow lakes gradually fill with sediment and vegetation to become oxbow swamps and eventually retain their identity only as **meander scars** (Figure 16-38).

A floodplain is slightly higher along the banks of the stream channel. As the stream overflows at flood time, the current, as it leaves the normal channel, is abruptly slowed by friction with the floodplain surface. This slowdown causes the principal deposition to take place along the margins of the main channel, producing **natural levees** (from the Old French word *levée*, "act of raising," derived from the Latin *levare*, "to raise") on each side of the stream (Figure 16-39). The natural levees merge outwardly and almost imperceptibly with the less well-drained and lower portions of the floodplain, generally referred to as *backswamps*.

Sometimes a tributary stream entering a floodplain that has prominent natural levees cannot flow directly into the main channel and so flows downvalley in the backswamp zone, running parallel to the main stream for some distance before finding an entrance. A tributary stream with such a pattern is referred to as a **yazoo stream**, after Mississippi's Yazoo River, which flows parallel to the Mississippi River for about 280 kilometers (175 miles) before joining it.

Although the landforms we have just described are most prominently displayed on the floodplains of large streams, nearly identical processes and landforms may also be found where a stream is meandering over a nearly flat surface and down a gentle slope. Thus, examples of cutoff meanders and oxbow lakes might well be seen on the valley floor of a small creek meandering across a flat alpine meadow.

Modifying Rivers for Flood Control

There are obvious attributes—flat land, abundant water, productive soils—that attract humans to valley bottoms, and therefore such areas are often places of intensive agriculture, transportation routes, and urban development. However, fluvial processes are always ongoing in valley bottoms, with the result that nature and humans coexist in an uneasy juxtaposition shrouded by the specter of flood. We have seen that every river is subject to at least occasional flooding, and thus the existence of a floodplain—so very attractive for human settlement—is incontrovertible evidence that floods do occur from time to time.

Accordingly, wherever humans have settled in considerable concentrations in river valleys, they have gone to extraordinary lengths to mitigate potential flood damage. The principal means for averting disaster are sizable earthwork or concrete water-containment and diversion structures in the form of dams, artificial levees, and overflow floodways. As an example of the remarkable efforts that go into such endeavors, we might consider the major river system of North America: the Mississippi.

Flood Control on the Mississippi River The Mississippi originates in Minnesota and flows more or less directly southward to the Gulf of Mexico below New Orleans (Figure 16-40). It is joined by a number of right-bank tributaries along the way, of which the Missouri is by far the most important. There are also many left-bank tributaries, of which the Ohio, with its tributary the Tennessee, is the most notable. (Tributary streams are designated as "right bank" or "left bank" from the perspective of an observer looking downstream.)

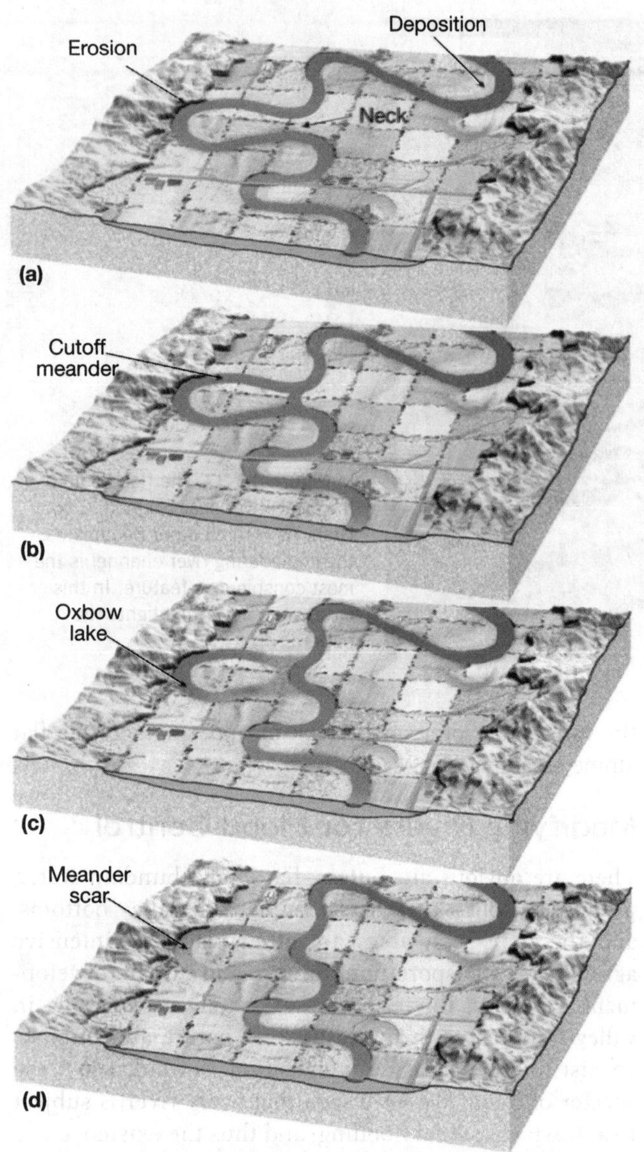

▲ **Figure 16-38** The formation of a cutoff meander on a floodplain. As the river cuts across the narrow neck of a meander, the river bend becomes an oxbow lake, which becomes an oxbow swamp, which becomes a meander scar.

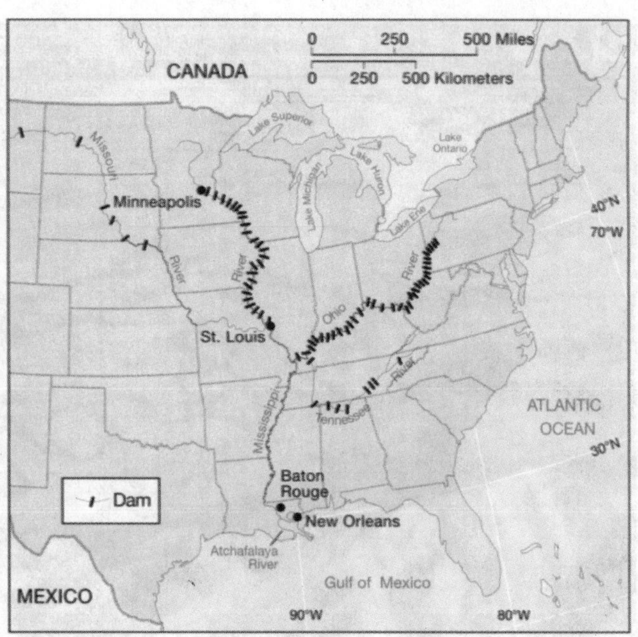

▲ **Figure 16-40** Dams on the Mississippi, Missouri, Ohio, and Tennessee Rivers.

All four of these rivers have been thoroughly dammed, largely for flood control but also for such other benefits as hydroelectricity production, navigation stabilization, and recreation. On the Mississippi, the Army Corps of Engineers operates and maintains 27 low dams between St. Louis and the head of navigation at Minneapolis, most equipped with hydroelectricity facilities and each with locks to allow barges and other shallow-draft boats to pass (one of the purposes of the dams is to maintain a water depth of at least 2.7 meters [9 ft.] throughout the navigable course of the river). The Missouri has fewer but larger dams, six in all, widely scattered from Montana to Nebraska. They are primarily flood-control dams. The Ohio is punctuated by more than three dozen low dams, whose primary purpose is maintenance of pools deep enough for barge navigation, with flood control as a

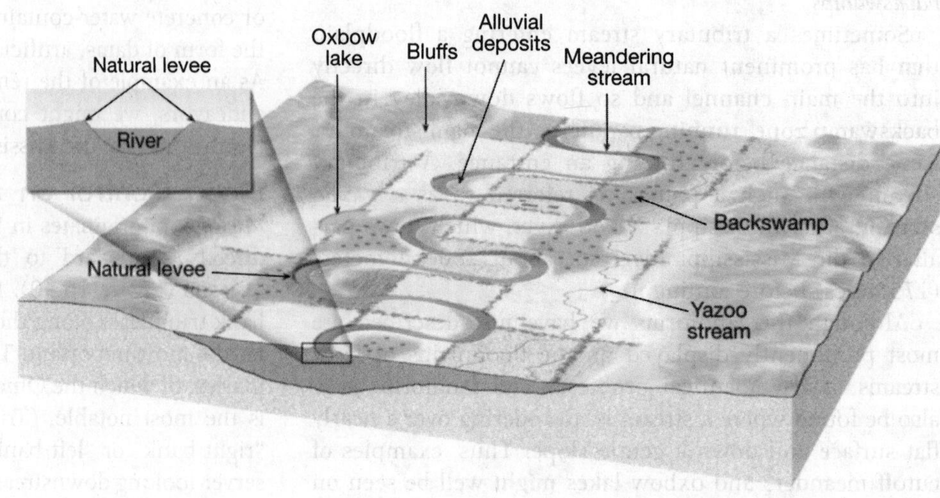

▲ **Figure 16-39** Typical landforms in a floodplain. Natural levees build up from alluvium deposited along the sides of a stream during floods.

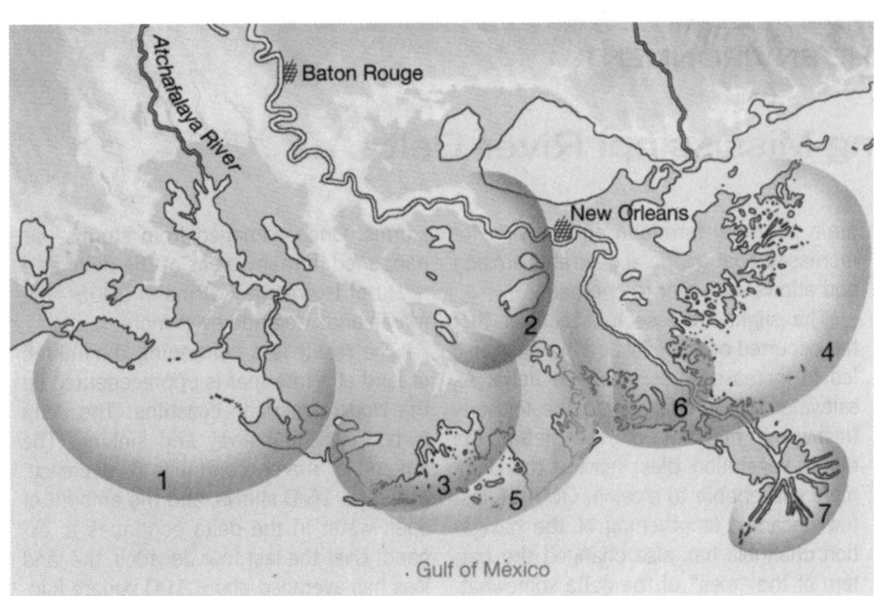

◀ **Figure 16-41** Over the last 5000 years, seven subdeltas were built by the Mississippi River (on the map, "1" is the oldest and "7" is the youngest). The present bird's foot delta ("7") began to form about 600 years ago.

secondary consideration. Most thoroughly dammed is the Tennessee, whose nine mainstream dams have reduced it to a series of quiet reservoirs for its entire length, apart from the upper headwaters. These are Tennessee Valley Authority (TVA) dams built during the 1930s particularly for flood control and other subsidiary benefits.

These four river valleys contain an extensive series of artificial levees designed to protect the local floodplain and move floodwaters downstream. There is a vicious-circle aspect to levee building: anytime a levee is raised in an upstream area most downstream locales require higher levees to pass the floodwaters on without overflow. These levees are usually the highest parts of the landscape, particularly in the ever-flatter and more extensive floodplains downstream. In addition, while the system of artificial levees may work to contain flooding in most years, if the levees should fail, the results can be catastrophic for the population living on the adjacent floodplain. (Many experienced hydrologists and geologists say, somewhat cynically, that there are only two kinds of flood control levees: "those that have failed, and those that will fail.")

At the lower end of the Mississippi system, in southern Louisiana, river control is extremely complicated and the results of human efforts are somewhat ambiguous. This region, which receives the full flow of the continent's mightiest river system, is exceedingly flat and thus has poor natural drainage. During the past 5000 years, the lower course of the river has shifted several times, producing at least seven subdeltas that are the principal elements of the present complicated *bird's-foot delta* of the Mississippi—a series of narrow, sediment-lined distributary channels (Figure 16-41). The main flow of the river during the past 600 years or so has been along its present course, southeast from New Orleans. This portion of the delta was built out into the Gulf of Mexico at a rate of more than 10 kilometers (6 miles) per century in that period.

Under the normal pattern of deltaic fluctuations, the main flow of the river would now be shifting to the shorter and slightly steeper channel of the Atchafalaya River, a prominent distributary of the ancestral delta a few kilometers west of the present main channel of the Mississippi. However, enormous flood-control structures—part of the Old River Control Project—were erected above Baton Rouge in an effort (thus far successful) to prevent the river from abandoning its present channel and delta (Figure 16-42).

In part because of human-built flood control and navigation measures, the natural processes operating on the floodplain and delta of the Mississippi River are changing—the delta is sinking and natural marshlands are retreating (see "People and the Environment: The Changing Mississippi River Delta" in this chapter, and "People and the Environment: Hurricane Katrina" in Chapter 7).

So, can people live on a floodplain without incurring untenable economic and ecological costs? Part of the answer

▲ **Figure 16-42** Old River Control Project on the Mississippi River near Baton Rouge, Louisiana. Diversion dams keep most of the Mississippi River (in the upper left) from draining into the Atchafalaya River.

PEOPLE AND THE ENVIRONMENT

The Changing Mississippi River Delta

Since the end of the last ice age, the "bird's-foot" delta of the Mississippi River has changed its form many times and occupied different locations along a 320 kilometer (200 mile) stretch of the gulf coastline (see Figure 16-41). It has been in its current position for about the last 600 years.

The size, shape, and configuration of the delta are the outcome of a number of different processes. The Mississippi River supplies sediment to the delta, deposited when the velocity of the river decreases as it flows into the Gulf of Mexico (Figure 16-C). At the same time, however, the delta slowly subsides into the gulf. Before dams and other kinds of human manipulation of the Mississippi–Missouri River system began decreasing the sediment load of the river, the delta was growing faster than it was sinking. Today the delta is sinking faster than sediments can replenish it (Figure 16-D).

The extensive artificial-drainage and river-channeling efforts to provide a dry surface for human settlement in southeastern Louisiana have restricted the Mississippi and its distributaries to relatively narrow channels, thus keeping both silt and freshwater from getting to the extensive surrounding marshlands. In addition, a maze of canals dredged to drain potential farmland and provide for increased boat traffic accelerates erosion and allows saltwater to encroach.

The slight increase in sea level that has occurred over the last century has also led to increased erosion of the delta. As saltwater moves farther into the formerly freshwater marshes of the delta, the marsh vegetation dies, leaving the delta more susceptible to erosion. Over the last few decades, lengthening of the navigation channels has also changed the pattern of the "toes" of the delta somewhat.

Before the shipping channels were as extensively lined with artificial levees as they are today, floodwaters would regularly deposit sediment into the marshes of the delta. With time, the sediment compacted around the roots of the marsh plants—the delta would subside, but a continuing supply of sediment would replenish the marshes (Figure 16-E). Once the artificial levees cut off much of the sediment supply to the marshes just outside the shipping channels, the marsh vegetation became discontinuous and much more susceptible to damage from storms. Once damaged from storms—as happened in many areas of the delta as a result of Hurricane Katrina in 2005—the marshlands degrade even more.

The result is a continuing diminution of land at a rate that is unprecedented on the North American coastline. The delta is both washing away and sinking. The amount of marsh continues to decrease, as Figure 16-D shows, and the amount of open water in the delta continues to expand. Over the last four decades, the land loss has averaged about 100 square kilometers (40 square miles) annually. The freshwater and brackish ecosystems suffer, and some of the higher land containing human settlements is sinking.

Scientists and policy makers have been struggling for years to come up with a practical way to restore the wetlands of the delta, but no approach can be agreed upon by all. One promising strategy involves controlled diversions of water into the marshes, but it is too early to tell if that will be enough to restore some of the marshlands—or even stop the ongoing decline.

▲ **Figure 16-C** Landsat image of the Mississippi River Delta in 2003.

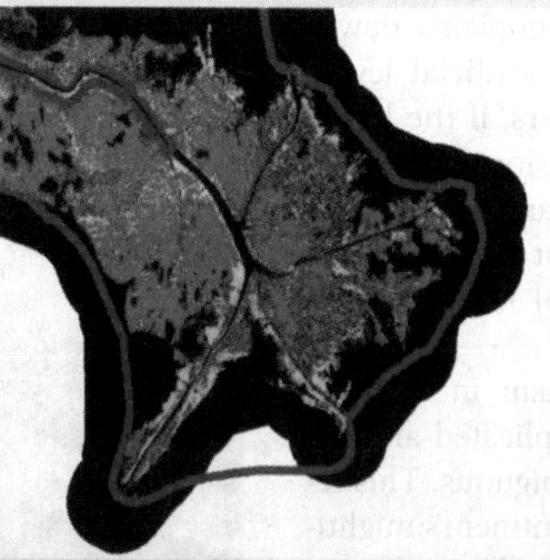

▲ **Figure 16-D** Land loss in the Mississippi River Delta. Red shows land loss from 1932 to 2000, yellow shows predicted land loss from 2000 to 2050, light green shows land gain from 1932 to 2000, and dark green the predicted land gain from 2000 to 2050. The blue line marks the limits of the study conducted by the U.S. Geological Survey National Wetlands Research Center in cooperation with the U.S. Army Corps of Engineers and the Louisiana Department of Natural Resources.

▲ **Figure 16-E** Freshwater marsh within the Mississippi River delta, Louisiana.

rests in sensible land use practices—the local governments of most urban areas adjacent to streams have drafted maps showing the extent of floodplain inundation for a *design flood*, such as the 100-year flood; zoning regulations can then restrict land use in such a way as to reduce the likelihood of costly damage. Diversion or bypass channels can be constructed to divert flood waters out onto an "artificial" floodplain—such diversion areas can be used profitably for other activities, such as farming, most of the year.

In some locations levees are actually being removed in an effort to give streams back at least part of their original floodplains. This approach not only restores valuable wetlands habitat, but it may also serve to dampen the surge of a flood downstream by increasing the "storage capacity" of a floodplain upstream. One such experiment is currently underway in the Upper Ouachita National Wildlife Refuge along the Ouachita River in northern Louisiana. A proposal to open levees south of New Orleans is currently under review by state officials in Louisiana.

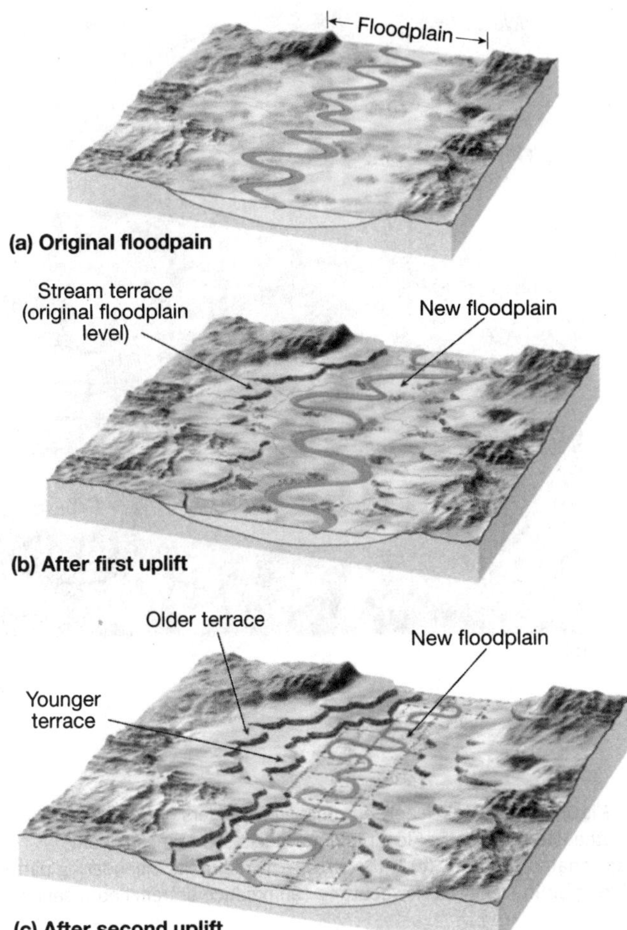

▲ **Figure 16-43** Stream terraces normally represent sequences of uplift and rejuvenation. (a) Before uplift a river meanders across an alluvial floodplain. (b) After uplift, the stream downcuts and then widens a new floodplain, leaving the remnants of the original floodplain as a pair of stream terraces. (c) Following a second episode of uplift, the stream leaves a second pair of stream terraces.

Stream Rejuvenation

All parts of the continental surfaces experience episodes, sometimes frequent and sometimes rare, when their elevation relative to sea level changes. This change is occasionally caused by a drop in sea level (as occurred throughout the world during the various ice ages when frozen water accumulated on the land, diminishing the amount of water in the oceans), but it is much more commonly the result of tectonic uplift of the land surface. When such uplift occurs, it "rejuvenates" the streams in the area. The increased gradient causes the streams to flow faster, which provides renewed energy for downcutting (a dramatic, sustained increase in discharge may also rejuvenate a stream). Vertical incision, which may have long been dormant, is initiated or intensified by such **stream rejuvenation**.

Stream Terraces If a stream occupied a broad floodplain prior to rejuvenation, renewed downcutting enables the stream to carve a new valley floor (Figure 16-43). This means that the old floodplain can no longer function as an overflow area and instead becomes an abandoned stretch of flat land overlooking the new valley. This remnant of the previous valley floor is called a **stream terrace**. Terraces often, but not always, occur in pairs, one on either side of the newly incised stream channel.

Entrenched Meanders Under certain circumstances, rejuvenative uplift has a different and even more conspicuous topographic effect: **entrenched meanders** (Figure 16-44). These topographic features are formed when an area containing a meandering stream is uplifted slowly and the stream incises downward while still retaining the meandering course—often with the entrenchment extending itself headward upstream. In some cases, such meanders may become entrenched in narrow gorges hundreds of meters deep.

Theories of Landform Development

The various internal and external processes operating on Earth's surface produce an infinite variety of landscapes. Systematizing this vast array of facts and relationships into a coherent body of knowledge has been the goal of many students of geomorphology. As we finish this chapter describing the often dominant role of running water in all of these processes, it may be helpful to discuss how comprehensive theories of landform development were devised and how they have changed over time.

Davis' Geomorphic Cycle

The first, and in many ways most influential, model of landscape development was propounded by William Morris Davis, an American geographer and geomorphologist active in the 1890s and early 1900s. He called it the *geographical cycle*, but many of his students considered the adjective too

▲ **Figure 16-44** (a) Deeply entrenched meanders of the Green River in southeastern Utah. (b) Floodplain meanders sometimes become rejuvenated by uplift. (c) If the stream maintains its meandering pattern during uplift, renewed downcutting can produce entrenched meanders.

generalized, and so the theory came to be known as the *cycle of erosion*. This term in turn was considered too restrictive because weathering and deposition are also involved, and so the Davisian theory is now usually referred to as the *geomorphic cycle*.

Davis envisioned a continuous sequence of terrain evolution in which a relatively flat surface is uplifted, then incised by fluvial erosion into a landscape of slopes and valleys, and finally denuded until it is once again a flat surface at low elevation (Figure 16-45). He metaphorically likened this sequence to the life cycle of an organism, recognizing stages of development he called "youth," "maturity," and "old age."

Davis stressed that any landscape can be comprehended by analyzing structure, process, and stage. Structure refers to the type and arrangement of the underlying rocks and surface materials, process is concerned with the internal and external forces that shape the landforms, and stage is the length of time during which the processes have been at work.

Youth Davis postulated that the initial surface is uplifted rapidly, so that erosion has little time to act until the uplift is complete. Thus, the initial surface consists of relatively flat land far above sea level. He further assumed no significant subsequent crustal movement or deformation for the duration of the cycle as well as a stable base level during all stages. During the youthful stage of development, streams become established and a drainage pattern begins to take shape. These streams, which incise deep, narrow, steep-sided, V-shaped valleys, flow rapidly and have irregular gradients marked by waterfalls and rapids. During this stage, most of the initial surface is broad, flattish interfluves, largely unaffected by stream erosion and encompassing shallow lakes and swamps because of the incomplete drainage system.

Maturity In the mature stage, the main streams approach an equilibrium condition, having worn away the falls and rapids and developed smooth profiles. Vertical erosion ceases in the main valleys, the streams begin to meander, floodplains are formed, and the drainage system is more extensive than in the youth stage. The interfluves are thoroughly dissected during this stage, their lakes and swamps drained by headward erosion, and their remnants existing only as narrow drainage divides between valleys. Whereas youth is characterized by the presence of a vast area of initial surface, maturity is marked by the absence of initial surface.

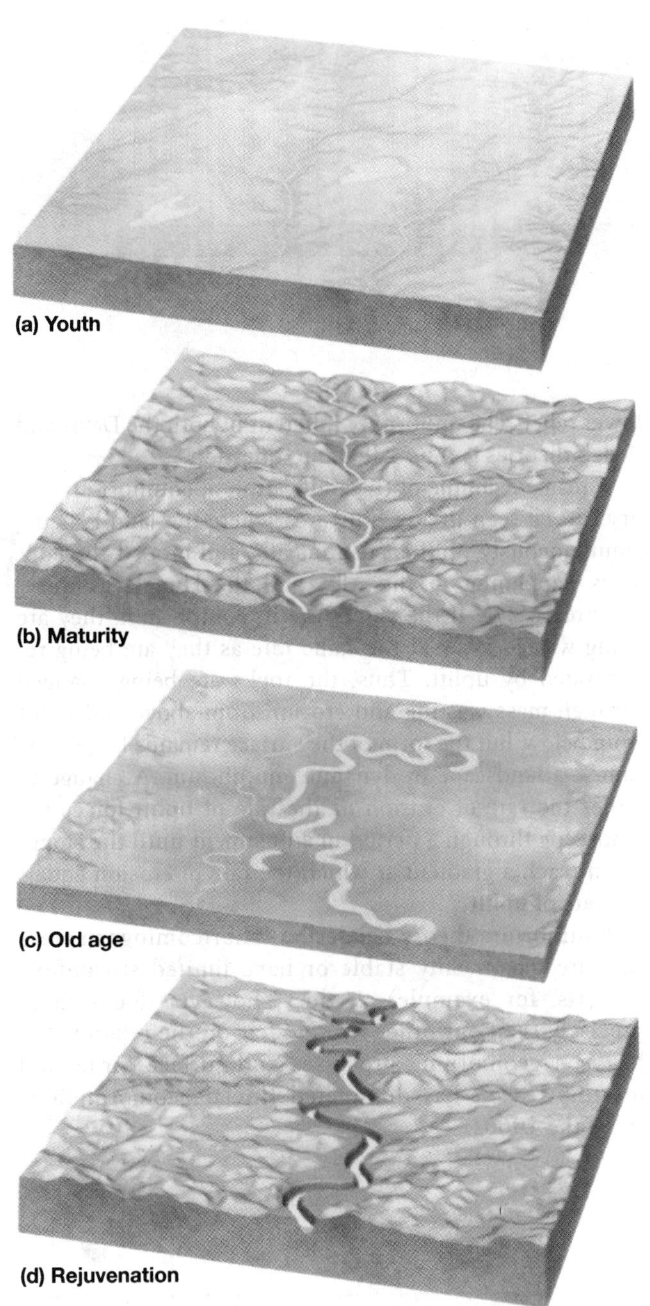

▲ **Figure 16-45** The idealized Davisian geomorphic cycle.

Old Age With the passage of a vast amount of time, erosion reduces the entire landscape to near base level. Sloping land is virtually absent, and the entire region is dominated by extensive floodplains over which a few major streams meander broadly and slowly. The end product of the geomorphic cycle is a flat, featureless landscape with minimal relief. Davis called this a *peneplain* (*paene* is Latin for "almost"; hence, "almost a plain"). He envisioned occasional remnants of exceptionally resistant rock rising slightly above the peneplain surface; such erosional remnants are dubbed *monadnocks* after a mountain of this name in New Hampshire.

Rejuvenation Because Davis recognized that the extraordinarily long time without crustal deformation required by his model is unlikely, his theory also covered rejuvenation, whereby regional uplift could raise the land and interrupt the cycle at any stage. This tectonic activity would reenergize the system, initiate a new period of downcutting, and restart the cycle.

Davis was both a prolific writer and a persuasive teacher, and his theory had a profound influence for many decades, especially in the United States. Even in the early days, however, there were strong dissenters. Other geomorphologists recognized imperfections in some of his assumptions and questioned some of his conclusions. For example, apparently no intact peneplains exist; remnants of peneplain surfaces are recognized in some areas, but nowhere does an actual peneplain occur. A more important difficulty with Davis' model concerns his idea that little erosion takes place while the initial surface is being uplifted, a notion unacceptable to most geomorphologists. Moreover, the causal interplay between uplift and erosion is open to varying interpretations. Finally, there are serious doubts about sequential development, and some people feel that the biological analogy may be more misleading than helpful.

The sequential aspect of the geomorphic cycle is very appealing because it provides an orderly, evolutionary, and predictable train of development. Moreover, many areas in nature have the appearance of youthful, mature, or old-age topography. However, no proof has ever been found that one stage commonly precedes another in regular fashion, and even in a single valley the terrain characteristic of the various stages is often jumbled. For landform analysis, therefore, it is probably better to use the terms youth, maturity, and old age as descriptive summaries of regional topography rather than as distinct implications of sequential development.

Penck's Theory of Crustal Change and Slope Development

Another theory of terrain evolution grew out of critical analysis of Davis' geomorphic cycle. In the Davisian cycle, drainage divides waste away in a steady and predictable pattern. As a slope retreats, it becomes less steep and more rounded, always maintaining a convex form. In nature, however, not all slopes are convex; some are straight and others are concave. Walther Penck, a young German geomorphologist and a prominent early critic of Davis, pointed out in the 1920s that slopes assume various shapes as they erode. Penck stressed that uplift stimulates erosion immediately and that slope form is significantly influenced by the rate of uplift or other crustal deformation. He argued that steep slopes, particularly, maintain a constant angle as they erode, retaining their steepness as they diminish in a sort of "parallel retreat" rather than being worn down at a continually lower slope angle—meaning that

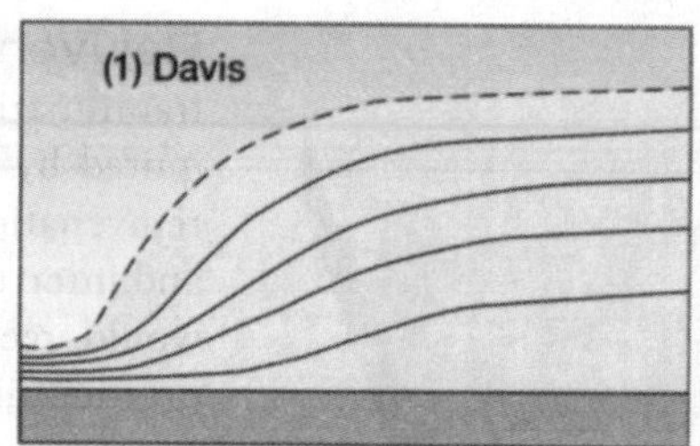

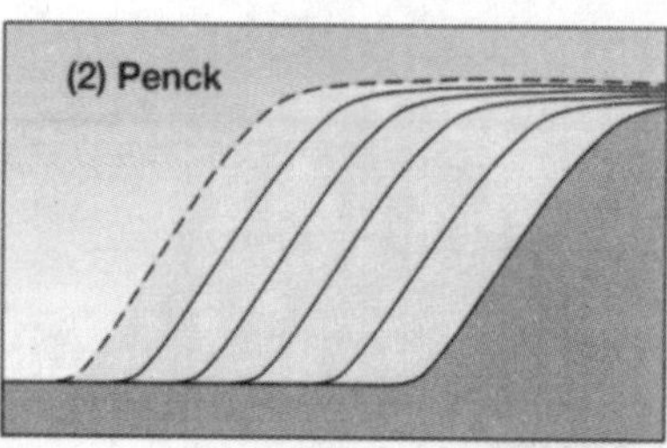

▶ **Figure 16-46** Slope retreat in (1) the Davis model and (2) the Penck model. The Davis concept proposes a continually diminishing angle of slope, whereas Penck theorized parallel retreat in which the slope angle remains approximately the same over time.

the shape of many initial surfaces is retained long after they would have been worn away in the Davisian concept (Figure 16-46). Many, but not all, of Penck's ideas have been substantiated by subsequent workers, and his ideas have come to be called the *theory of crustal change and slope development*.

Equilibrium Theory

A third model of landform development became known as *equilibrium theory*. In the last four decades or so, many geomorphologists have been studying the physics of landform development. This approach emphasizes the delicate balance between form and process in the landscape. It is believed that the influence of crustal movement and the resistance of the underlying rock vary significantly from place to place and that these variations are as significant as differences in process in determining terrain. Thus, equilibrium theory suggests that slope forms are adjusted to geomorphic processes so that there is a balance of energy—the energy provided is just adequate for the work to be done. For example, harder rock develops steeper slopes and higher relief, and softer rock has gentler slopes and lower relief. The uniformity inherent in both the Davis and Penck theories is thus called into question.

A prime example of the application of equilibrium theory can be seen in any hilly area where the land is being simultaneously uplifted tectonically and eroded fluvially, as is happening in the Alps and the Himalayas today (Figure 16-47). If the slopes are in equilibrium, they are being wasted away at the same rate as they are being regenerated by uplift. Thus, the rocks are being changed through mass wasting and erosion from above and uplift from below, but the form of the surface remains largely the same—a landscape in dynamic equilibrium. A change in either the rate of erosion or the rate of uplift forces the landscape through a period of adjustment until the slopes again reach a gradient at which the rate of erosion equals the rate of uplift.

Equilibrium theory has serious shortcomings in areas that are tectonically stable or have limited streamflow (deserts, for example). It does, however, focus more precisely than our other two models on the relationship between geomorphic processes and surface forms and for this reason has dominated fluvial geomorphology since the 1960s.

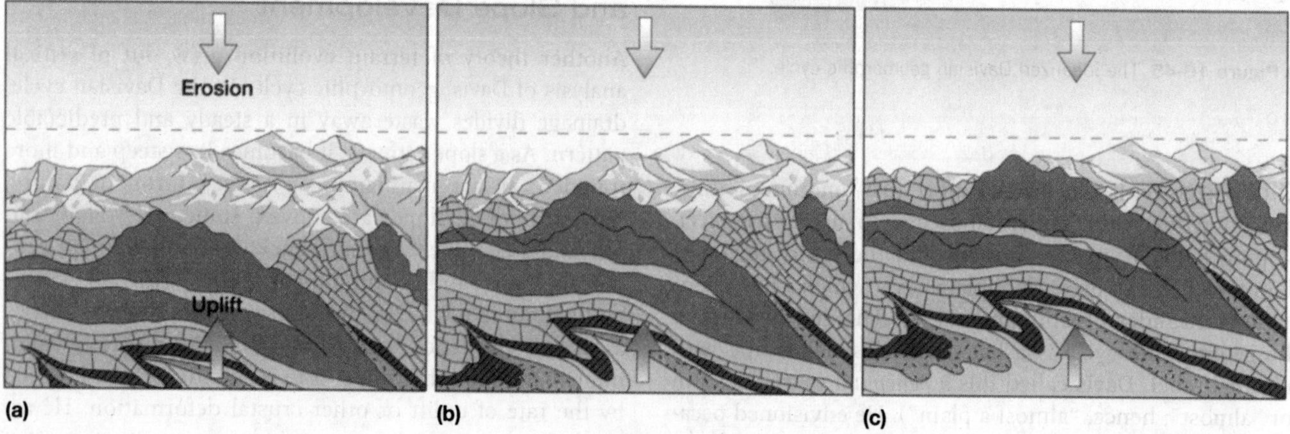

▲ **Figure 16-47** The dynamic equilibrium concept. This vertical cross section through an area of the Swiss Alps (a) shows that erosion reduces relief just about as rapidly as uplift raises the land, with the result that the elevation of the mountains remains essentially constant over time (b and c). The surface rocks are continually changing through uplift and erosion, but the shape of the surface remains approximately the same because removal and replacement are in balance.

CHAPTER 16 LEARNING REVIEW

After studying this chapter, you should be able to answer the following questions. Key terms from each text section are shown in bold type. Definitions for key terms are also found in the glossary at the back of the book.

Streams and Stream Systems (p. 437)

1. What are **fluvial processes**?
2. Describe the difference between **streamflow** and **overland flow**.
3. What are the differences between an **interfluve** and a **valley**?
4. What is a **drainage basin (watershed)**? A **drainage divide**?
5. Using the concept of **stream order**, explain the difference between a first-order stream and a second-order stream.
6. What factors influence the erosional effectiveness of a stream?
7. What are the abrasive tools used by a stream in its erosive activities?
8. Describe the different components of **stream load**: **dissolved load**, **suspended load**, and **bedload**.
9. What is the difference between stream **competence** and stream **capacity**?
10. What factor determines the competence of a stream?
11. How does a stream sort **alluvium** by size?
12. Why are deposits of alluvium often rounded and stratified?
13. Explain how talus from rockfall (Chapter 15) is likely to look different from a deposit of alluvium.
14. Contrast **perennial streams** with **intermittent streams** and **ephemeral streams**.
15. What is meant by stream **discharge**?
16. Why are floods so important in the development of fluvial landforms?
17. What is meant by a flood **recurrence interval** of 50 years—in other words, what is meant by a 50-year flood?

Stream Channels (p. 443)

18. In what way is the pattern of water flow in a straight stream channel likely to be similar to that of a **sinuous channel**?
19. Under what circumstances is a stream likely to have a **meandering channel** pattern?
20. Under what circumstances is a stream likely to have a **braided channel** pattern?

Structural Relationships (p. 447)

21. What is an **antecedent stream** and how might one form?
22. Describe and explain the circumstances under which **dendritic drainage patterns** develop, and the conditions under which **trellis drainage patterns** develop.
23. Describe radial drainage patterns and one kind of location where such a drainage pattern might be found.

The Shaping and Reshaping of Valleys (p. 448)

24. Explain the process of valley deepening through **downcutting**.
25. What is meant by **base level**?
26. What prevents a stream from downcutting to sea level throughout its entire course?
27. Why would few depositional features be found in most V-shaped stream valleys?
28. What is a **knickpoint**?
29. Describe and explain the process of **knickpoint migration**.
30. Explain how a meandering stream widens its valley through the process of **lateral erosion**.
31. Describe and explain the process of **headward erosion**.
32. Explain the process of **stream capture (stream piracy)**.
33. Describe the formation of a **delta**.
34. Why don't all large rivers form deltas where they enter the ocean?
35. Under what conditions does **aggradation** of a stream channel take place?

Floodplains (p. 456)

36. Describe the general characteristics of a **floodplain**.
37. Describe and explain the formation process of a **cutoff meander**.
38. Explain the relationships among a cutoff meander, an **oxbow lake**, and a **meander scar**.
39. What are **natural levees** and how do they form?
40. What explains the presence of a **yazoo stream** on a floodplain?
41. Why might meandering rivers make poor political boundaries?

Stream Rejuvenation (p. 461)

42. Explain the circumstances that can lead to **stream rejuvenation**.
43. Describe and explain the formation of **stream terraces**.
44. How is it possible for **entrenched meanders** to form?

Theories of Landform Development (p. 461)

45. Describe and explain at least one problem with Davis' geomorphic cycle model of fluvial landform development.
46. How does equilibrium theory differ from earlier theories of topographic development?

Log in to **www.mygeoscienceplace.com** for animations, videos, RSS feeds, maps, and self-study quizzes to enhance your study of **Fluvial Processes**.

17 Karst and Hydrothermal Processes

In studying topographic development, we pay a great deal of attention to the role of water. We noted that water running across the ground is a significant external shaper of terrain, and we will see that coastal waters produce distinctive landforms around the margins of oceans and lakes. In both cases, the water moves rapidly, and much energy is expended in erosion, transportation, and deposition.

In this brief chapter, we focus on underground water, which, because it is confined, functions in a much more restricted fashion than surface water. Underground water is largely unchanneled and therefore generally diffused, and it moves very slowly for the most part. Consequently, it is almost totally ineffective in terms of hydraulic power and other kinds of mechanical erosion. However, underground water can leave a distinctive mark on the surface landscape through a variety of solution processes associated with the development of karst and hydrothermal features—the topic of this chapter.

The Impact of Solution Processes on the Landscape

The mechanical effects of underground water have only a limited influence on topographic development. Some subsurface mechanical weathering does take place, but the surface landscape is rarely directly affected by it, although certain forms of mass wasting (such as earthflows and slumps) are facilitated when loose materials are lubricated by underground water.

Through its chemical action, however, underground water is an effective shaper of the topographic landscape. Water is a solvent for certain rock-forming chemicals, dissolving them from rock and then carrying them away in solution and depositing them elsewhere. Under particular circumstances, the aboveground results of this dissolution are widespread and distinctive.

Underground water also affects surface topography via the creation of such hydrothermal features as hot springs and geysers, formed when hot water from underground is discharged at the ground surface.

Solution and Precipitation

The chemical reactions involving underground water are relatively simple. Although pure water is a relatively poor solvent, almost all underground water is laced with enough chemical impurities to make it a good solvent for the compounds that make up a few common minerals (Figure 17-1).

◀ Geyser in Lake Bogoria National Park, Kenya.

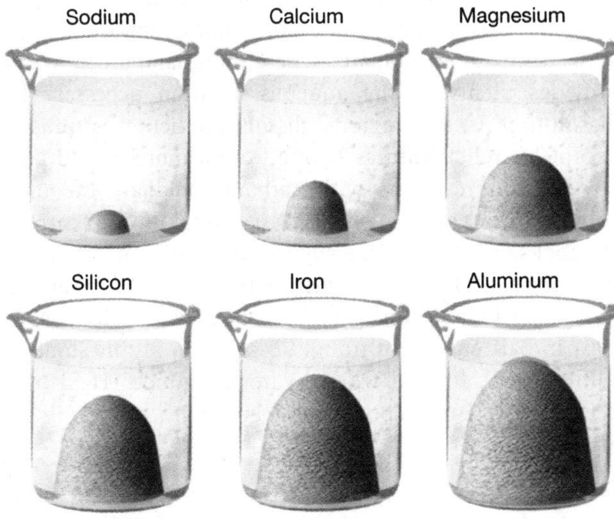

▲ **Figure 17-1** Relative solubility in water of some common rock-forming elements. The darker portion of each diagram indicates the proportion of the element that is insoluble in water. Sodium and calcium can be almost completely dissolved, for instance, whereas iron and aluminum are essentially insoluble.

Basically, underground water is a weak solution of **carbonic acid** (H_2CO_3) because it contains dissolved carbon dioxide gas—as we first saw in Chapter 15, the resulting *carbonation* can lead to the dissolution of bedrock.

Dissolution Processes **Dissolution** is the removal of bedrock through the chemical action of water. Dissolution is an important weathering/erosion process for all rocks, but it is particularly effective on carbonate sedimentary rocks, especially limestone. A common sedimentary rock, limestone is composed largely of calcium carbonate, which reacts strongly with carbonic acid solution to yield calcium bicarbonate, a compound that is very soluble in (and thus easily removed by) water. Other carbonate rocks, such as gypsum, chalk, and even the metamorphic rock marble, undergo similar reactions. Dolomite is a calcium magnesium carbonate rock that dissolves almost as quickly as limestone. The pertinent chemical equations (in simplified form) for the dissolving of limestone and dolomite are as follows:

$$\underset{\text{limestone (calcium carbonate)}}{CaCO_3} + \underset{\text{water}}{H_2O} + \underset{\text{carbon dioxide}}{CO_2} \rightarrow \underset{\text{calcium bicarbonate}}{Ca(HCO_3)_2}$$

$$\underset{\text{dolomite}}{CaMg(CO_3)_2} + \underset{\text{water}}{2H_2O} + \underset{\text{carbon dioxide}}{2CO_2} \rightarrow \underset{\text{calcium bicarbonate}}{Ca(HCO_3)_2} + \underset{\text{magnesium bicarbonate}}{Mg(HCO_3)_2}$$

These reactions are the most notable dissolution processes. Water percolating down into carbonate bedrock dissolves and carries away a part of the rock mass. Since limestone and related rocks are largely composed of soluble minerals, great volumes of rock are sometimes dissolved and removed, leaving conspicuous voids in the bedrock. This action occurs more rapidly and on a larger scale in a humid climate, where abundant precipitation provides plenty of the aqueous medium necessary for dissolution. In arid regions, dissolution action is unusual except for relict features dating from a more humid past.

Although reactions with carbonic acid are the most common processes involved in the dissolution of carbonate rocks, recent studies suggest that in some locations sulfuric acid (H_2SO_4) may also be important. For example, it appears that Lechuguilla Cave in New Mexico was at least in part enlarged through dissolution of limestone by sulfuric acid, formed when hydrogen sulfide (H_2S) from deeper petroleum deposits combined with oxygen in the groundwater.

Role of the Bedrock Structure Bedrock structure is also a factor in dissolution. A profusion of joints and bedding planes permits groundwater to penetrate the rock readily. That the water is moving also helps because, as a given volume of water becomes saturated with dissolved calcium bicarbonate, it can drain away and be replaced by fresh unsaturated water that can dissolve more rock. Such drainage is enhanced by some outlet at a lower level, such as a deep subsurface stream.

Most limestone is resistant to mechanical erosion and often produces rugged topography. Thus, its ready solubility contrasts notably with its mechanical durability—a vulnerable interior beneath a durable surface.

Precipitation Processes Complementing the removal of calcium carbonate is its precipitation from solution. Mineralized water may trickle in along a cave roof or wall. The reduced air pressure in the open cave induces precipitation of whatever minerals the water is carrying.

One other type of precipitation is worth mentioning despite its scarcity, because of its dramatic distinctiveness. Hot springs and geysers nearly always provide an accumulation of precipitated minerals, frequently brilliant white but sometimes orange, green, or some other color due to associated algae. Wherever it comes in contact with magma, underground water becomes heated, and this water sometimes finds its way back to the surface through a natural opening so rapidly that it is still hot when it reaches the open air. Hot water is generally a much better solvent than cold, and so a hot spring or geyser usually contains a significant quantity of dissolved minerals. When exposed to the open air, the hot water precipitates much of its mineral content as its temperature and the pressure on it decrease, as the dissolved gases that helped keep the minerals in solution dissipate, and as algae and other organisms living in it secrete mineral matter. These deposits (such as *travertine, tufa*, and *sinter*) contain a variety of calcareous minerals and take the form of mounds, terraces, walls, and peripheral rims (see Figure 17-17).

It should be noted, however, that the solubility of carbon dioxide increases as water temperature declines. Thus, cool water often is more potent than hot water as a solvent for calcium carbonate.

Caverns and Related Features

Some of the most spectacular landforms produced by dissolution are not visible at Earth's surface. Solution along joints and bedding planes in limestone beneath the surface often creates large open areas called **caverns**. The largest of these openings are usually more expansive horizontally than vertically, indicating a development along bedding planes. In many cases, however, the cavern pattern has a rectangularity that demonstrates a relationship to the joint system (Figure 17-2).

Caverns are found almost anywhere there is a massive limestone deposit at or near the surface. The state of Missouri, for example, has more than 6000. Caverns often are difficult to find because their connection to the surface may be extremely small and obscure, or nonexistent. Beneath the surface, however, some caverns are very extensive (Mammoth Cave in Kentucky has more than 560 kilometers [350 miles] of known passages), with an elaborate system of galleries and passageways, usually very irregular in shape and

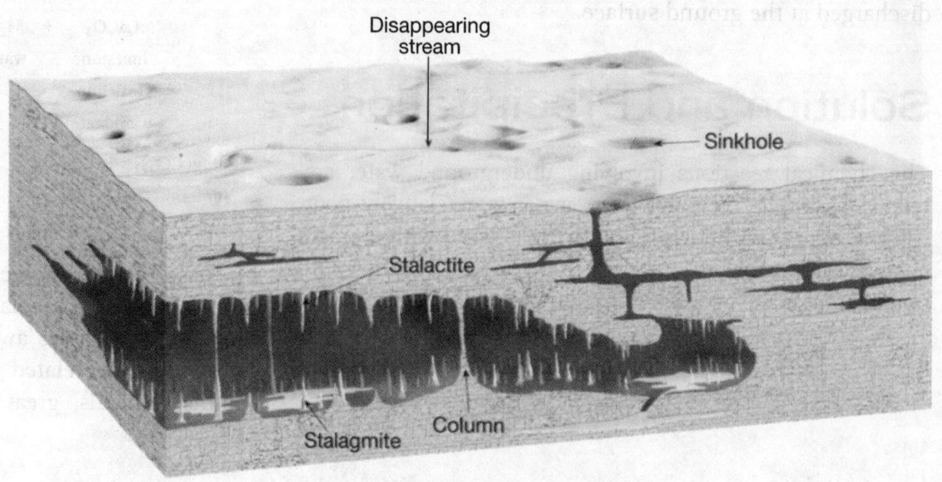

▲ **Figure 17-2** Caverns are formed by solution action of underground water as it trickles along bedding planes and joint systems.

sometimes including massive openings ("rooms") scattered here and there along the galleries. A stream may flow along the floor of a large cavern, adding another dimension to erosion/deposition.

There are two principal stages in cavern formation. First there is the initial excavation, wherein percolating water dissolves the carbonate bedrock and leaves voids. This dissolution is followed, often after a drop in the water table, by a "decoration stage" in which ceilings, walls, and floors are decorated with a wondrous variety of **speleothems** (Figure 17-3). These forms are deposited when water leaves behind the compounds (principally carbon dioxide and calcite) it was carrying in solution. Once out of solution, the carbon dioxide gas diffuses into the cave atmosphere, and calcite is deposited. Much of the deposition occurs on the sides of the cavern, but the most striking features are formed on the roof and floor. Where water drips from the roof, a pendant structure grows slowly downward like an icicle—a

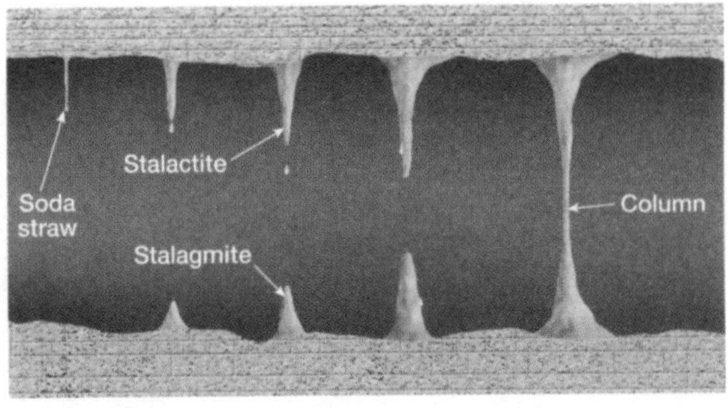

▲ **Figure 17-4** The development of stalagmites, stalactites, and columns.

stalactite. Where the drip hits the floor, a companion feature, a **stalagmite**, grows upward. Stalactites and stalagmites may extend until they meet to form a *column* (Figure 17-4). In some caverns, long, slender *soda straws* hang down from the ceiling—little more than one water drop wide, these delicate hollow tubes may eventually grow into stalactites.

▲ **Figure 17-3** A multitude of stalactites hang from the ceiling of this room in New Mexico's Carlsbad Caverns. Some of them have joined with stalagmites to form columns.

Karst Topography

In many areas where the bedrock is limestone or similarly soluble rock, dissolution has been so widespread and effective that a distinctive landform assemblage has developed at the surface, in addition to whatever caves may exist underground. The term **karst** (a Germanized form of an ancient Slavic word meaning "barren land") is applied to this topography. The name derives from the Kras or Krš Plateau region of Slovenia (formerly part of Yugoslavia), a rugged hilly area that has been shaped almost entirely by solution action in limestone formations (Figure 17-5).

The term karst connotes both a set of processes and an assemblage of landforms. The word is used as the catchall name of a cornerstone concept that describes the special

▲ **Figure 17-5** Irregular topography is prominent in the namesake karst region of the former Yugoslavia.

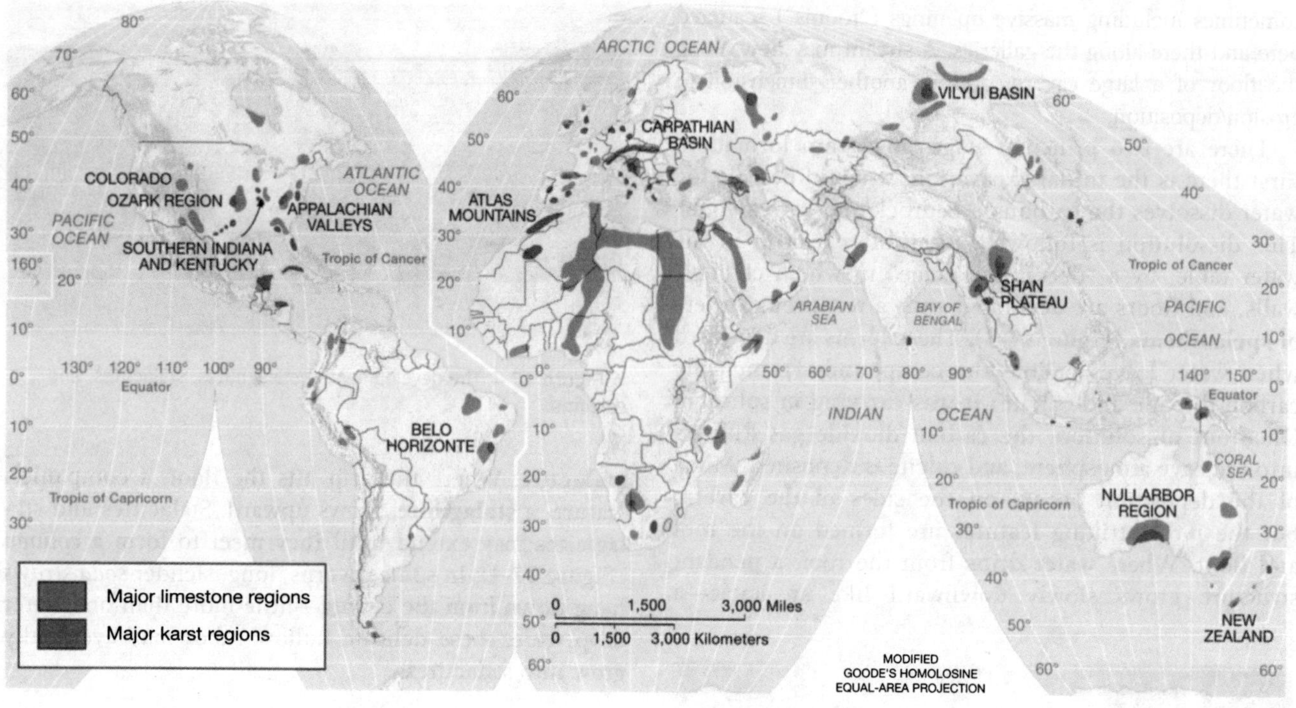

▲ **Figure 17-6** Major limestone and karst regions of the world.

landforms that develop on exceptionally soluble rocks, although there is a broad international vocabulary to refer to specific features in specific regions.

Typical landforms in karst regions include sinkholes, disrupted surface drainage, and underground drainage networks that have openings formed from solution action. The openings range in size from enlarged cracks to huge caverns.

Karst landscapes usually evolve where there is massive limestone bedrock. However, karstic features may also occur where other highly soluble rocks—dolomite, gypsum, or halite—predominate. Karst landforms are worthy of study not only because of their dramatic appearance but also because of their abundance. It is estimated that about 10 percent of Earth's land area has soluble carbonate rocks at or near the surface; in the conterminous United States, this total rises to 15 percent (Figure 17-6).

The most common surface features of karst landscapes are **sinkholes** (also called *dolines*), which occur by the hundreds and sometimes by the thousands. Sinkholes are rounded depressions formed by the dissolution of surface carbonate rocks, typically (but by no means always) at joint intersections. The sinkholes erode more rapidly than the surrounding area, forming closed depressions. Their sides generally slope inward at the angle of repose (usually 20° to 30°) of the adjacent material, although some have more gentle side slopes. A sinkhole that results from the collapse of the roof of a subsurface cavern is called a **collapse sinkhole** (or *collapse doline*); these may have vertical walls or even overhanging cliffs (Figure 17-7a/b).

Sinkholes range in size from shallow depressions a few meters in diameter and a few centimeters deep to major features kilometers in diameter and hundreds of meters deep. The largest are associated with tropical regions, where they develop rapidly and where adjacent holes often intersect.

The bottom of a sinkhole may lead into a subterranean passage, down which water pours during a rainstorm. More commonly, however, the subsurface entrance is blocked by rock rubble, soil, or vegetation, and rains form temporary lakes until the water percolates away. Indeed, sinkholes are the karst analog of river valleys, in that they are the fundamental unit of both erosion and weathering.

Where sinkholes occur in profusion, they often channel surface runoff into the groundwater circulation, leaving networks of dry valleys as relict surface forms. The Serbo-Croatian term **uvala** refers to such a chain of intersecting sinkholes. In many cases, sinkholes evolve into uvala over time.

In many karst locations, depressions dominate the landscape. Sinkholes are commonplace in central Florida and parts of the Midwest, particularly Kentucky, Illinois, and Missouri. For example, at the University of Missouri in Columbia, both the football stadium and the basketball fieldhouse are built in sinkholes, and parking lots around them occasionally sink.

Apart from the ubiquity of sinkholes, karst areas show considerable topographic diversity. Where the relief is slight, as in central Florida, sinkholes are the dominant features. Where the relief is greater, however, cliffs and steep slopes alternate with flat-floored, streamless valleys. Limestone bedrock exposed at the surface tends to be pitted, grooved, etched, and fluted with a great intricacy of erosive detail.

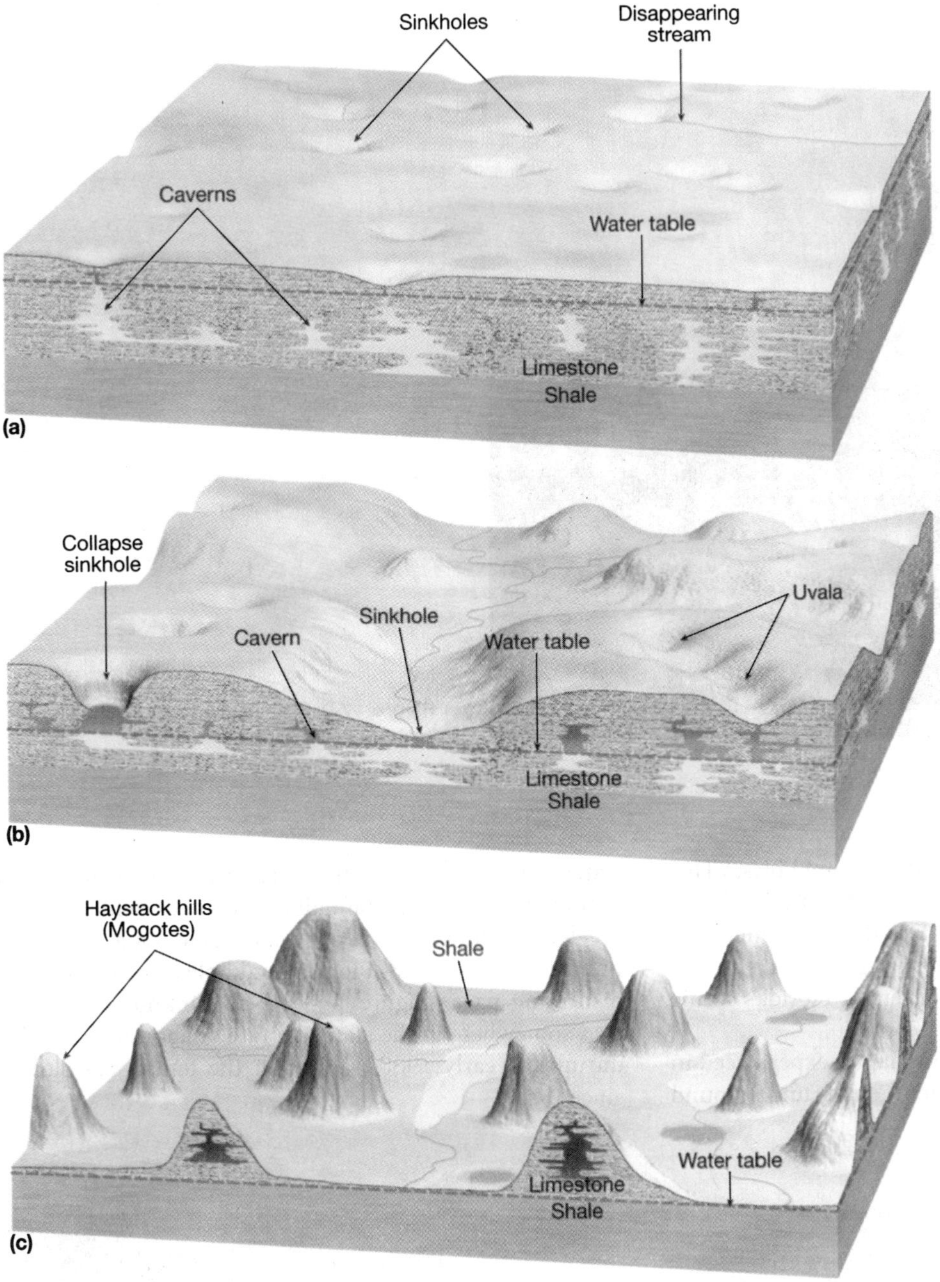

◄ **Figure 17-7** The development of karst topography. (a) Landscape dominated by sinkholes and disappearing streams. Dissolution of bedrock below the water table may leave openings that develop into caverns. (b) Where underground caverns collapse, a collapse sinkhole (collapse doline) forms. Streams may disappear into sinkholes through swallow holes. If the water table drops enough to leave open caverns, speleothems may gradually develop. (c) Tower karst topography consists of residual towers of limestone (haystack hills or mogotes).

Residual karst features, in the form of very steep-sided hills, dominate some parts of the world (Figure 17-7c). These formations are sometimes referred to as **tower karst** because of their almost vertical sides and conical or hemispheric shapes. Such towers are sometimes riddled with caves. The tower karst of southeastern China and adjacent parts of northern Vietnam is world famous for its spectacular scenery (Figure 17-8), as are the *mogotes* (haystack hills) of western Cuba.

In many ways, the most notable feature of karst regions is what is missing—surface drainage. Most rainfall and snowmelt seep downward along joints and bedding planes, enlarging them by dissolution. Surface runoff that does become channeled does not usually go far before it disappears into a sinkhole or joint crack—such streams are often termed **disappearing streams** (see Figure 17-7a). The water that collects in sinkholes generally percolates downward, but some sinkholes have distinct openings at their bottom, called **swallow holes**, through which surface drainage can pour directly into an underground channel, often to reappear at the surface through another hole some distance away. Where dissolution has been effective for a long time, there may be a complex underground drainage system that has superseded any sort of surface drainage net. An appropriate generalization concerning surface drainage in karst regions is that valleys are relatively scarce and mostly dry.

Groundwater Extraction and Sinkhole Formation Human activities can have direct and immediate consequences in some areas of karst topography. For example, most of central Florida is underlain by massive limestone bedrock, which is particularly susceptible to the

◀ **Figure 17-8** Spectacular tower karst hills (magotes) in Guilin, China.

formation of sinkholes and collapse sinkholes. This process is accelerated when the water table drops. Sinkholes have been forming in Florida for a long time, with several thousand of them having appeared in the twentieth century. Indeed, most of central Florida's scenic lakes began as sinkholes.

The population growth that Florida has experienced in recent decades has put a heavy drain on its underground water supply. This depletion was exacerbated by several years of below-normal rainfall in the 1970s, which caused accelerated drawdown of the water table. As a result, the number and size of Florida sinkholes increased at a disturbing pace (Figure 17-9). New sinkholes materialized somewhere in the state at a rate of about one per day during the early 1980s, although the tempo has slowed since then.

▶ **Figure 17-9** The sinkhole that swallowed a house in central Florida. The neighboring house also had to be abandoned.

Hydrothermal Features

In many parts of the world, there are small areas where hot water comes to the surface through natural openings. Such outpouring of hot water, often accompanied by steam, is known as **hydrothermal activity** and usually takes the form of either a hot spring or a geyser.

Hot Springs

The appearance of hot water at Earth's surface usually indicates that the underground water has come in contact with heated rocks or magma and has been forced upward through a fissure by the pressures that develop when water is heated (Figure 17-10). The usual result at the surface is a **hot spring**, with water bubbling out either continuously or intermittently. The hot water invariably contains a large amount of dissolved mineral matter, and a considerable proportion of this load is precipitated out as soon as the water reaches the surface and its temperature and the pressure on it both decrease.

The deposits around and downslope from hot springs can take many forms. If the opening is on sloping land, terraces are usually formed. Where the springs emerge onto flat land, there may be cones, domes, or irregular concentric deposits. Since calcium carbonate is so readily soluble in water containing carbonic acid, the deposits of most springs are composed largely of massive (*travertine*) or porous (*tufa* or *sinter*) accumulations of calcium carbonate. Various other minerals are also contained in the deposits on occasion, especially silica, but are much less common than calcium compounds.

Sometimes the water bubbling out of a hot spring builds a continually enlarging mound or terrace. As the structure is built higher, the opening through which the hot water comes to the surface also rises, so that the water is always emerging above the highest point. As the water flows down the sides of the structure, more deposition takes place there, thus broadening the structure as well, often with brilliantly colored algae, which add to the striking appearance as well as contribute mineral secretions to the deposit (Figure 17-11).

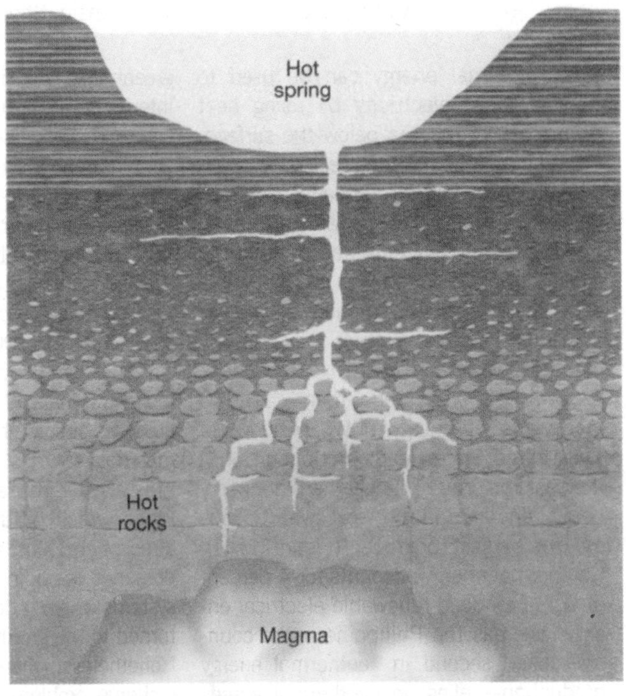

▲ **Figure 17-10** Cross section through a hot spring.

◀ **Figure 17-11** The sides and bottoms of hot springs are often brilliantly colored by algal growth. This is Beauty Pool in Yellowstone National Park. The black area at the bottom of the pool is the opening to the fissure that brings hot water to the surface. The water temperature of this hot spring is typically about 77°C (170°F).

RENEWABLE ENERGY

Geothermal Energy

Karl Byrand, University of Wisconsin, Sheboygan

Geothermal energy can be used to generate electricity by using heat from rocks or magma below the surface. Although geothermal heat exists throughout Earth's interior because of the radioactive decay of certain elements, existing technology limits accessibility to sites near plate boundaries and hot spots where Earth's crust is thinner.

Currently, 24 countries have implemented geothermal energy to produce a capacity of greater than 10,000 megawatts of energy (MWe) worldwide. The United States leads production, with the vast majority of its geothermal energy being generated in the West, with California the largest producer (Figure 17-A). Geothermal energy accounts for 4 percent of U.S.-produced renewable electrical energy, whereas the Philippines—the country ranked second in geothermal energy production—relies on geothermal energy to produce 23 percent of all its electricity.

The most common geothermal energy facility uses a "dry steam" process, directly extracting steam from inside Earth to turn a turbine. Once used, the steam is then condensed into water and returned to the ground, where it will be reheated naturally and used again. Another type of facility uses a "closed loop system" with subterranean pipes filled with fluids such as isobutane and pentafluropropane, which have a lower boiling point than water, thus requiring less heat to create steam (Figure 17-B).

Geothermal energy generation is self-sustaining, virtually inexhaustible, and free of combustion-based emissions. In other words, there is no burning of fossil fuels. While carbon dioxide (CO_2), a known greenhouse gas, is released from Earth's interior during extraction of the steam, the amount is considerably smaller than the amount generated by coal-fired power facilities. According to the U.S. Department of Energy, a geothermal facility releases 27 to 40 kilograms (60 to 90 pounds) of CO_2 on average to generate one megawatt of electricity, whereas a coal-fired facility releases 900 kilograms (2000 pounds).

Other byproducts of geothermal energy include what is considered negligible amounts of some harmful gasses and other materials such as hydrogen sulfide and arsenic. By far, the largest byproduct is water vapor, which diffuses harmlessly into the atmosphere if lost as steam. This, however, does not occur in closed-loop systems; in systems where it does, far more water is returned to the ground than is lost as steam. Nonetheless, there can be groundwater recharge problems and a potential reduction in steam pressure if water extraction rates exceed the rate of recharge.

The other major disadvantage associated with geothermal energy generation is the possibility of induced seismicity because of well drilling, water extraction, and the injection of water into the ground. For instance, in 2006, operations at a geothermal power plant in Switzerland were halted after it triggered several earthquakes. However, because existing technology only permits geothermal facilities to operate within already seismically active regions, it can be difficult to distinguish seismic activity that occurs naturally from that created because of a geothermal power plant.

Although the potential for these induced microearthquakes exists, studies at facilities in California have determined the risk there is negligible. The greatest risk is associated with Enhanced Geothermal Systems (EGS), a process currently being developed and tested because of its capability to capture geothermal energy in regions without adequate groundwater flow. In EGS, cold water is pumped directly into subterranean regions to intentionally fragment geothermally active, but dry, rock to create paths of subsurface water flow. A study from Australia, which has the world's largest EGS project, concluded that seismic risks from EGS would be low.

Continued investigations into EGS and other geothermal technologies are expected to make this power source more common and allow many nations to decrease their reliance on fossil fuels. For example, in Indonesia, $3 billion is financing 1068 MWe of geothermal projects. Likewise, a small Caribbean nation, St. Kitts and Nevis, is in the process of developing geothermal facilities that will generate more than four times its energy needs.

In the United States, studies have identified between 3000 and 16,000 MWe of geothermal energy that is available but not being implemented. The National Renewable Energy Laboratory estimates that U.S. production of geothermal energy could equal 26,000 MWe by 2015, and a Massachusetts Institute of Technology study estimated that domestic production could increase to 100 gigawatts within the next 50 years, with the employment of proper research and development. In 2009 President Obama committed $350 million in stimulus funding to geothermal energy research.

▲ **Figure 17-A** Geothermal power generation facilities at The Geysers in northern California.

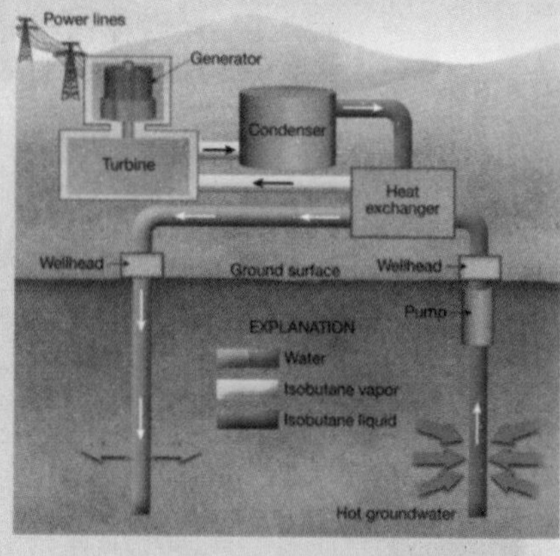

▲ **Figure 17-B** Geothermal electrical generation using a heat exchanger to increase efficiency.

Geysers

A special form of intermittent hot spring is the **geyser**. Hot water usually issues from a geyser only sporadically, and most or all of the flow is a temporary ejection (called an *eruption*) in which hot water and steam spout upward. Then the geyser subsides into apparent inactivity until the next eruption.

The basic principle of geyser activity involves the building up of pressure within a restricted subterranean tube until that pressure is relieved by an eruption. The process begins when underground water seeps into subterranean openings that are connected in a series of narrow caverns and shafts. Heated rocks and/or magma are close enough to these storage reservoirs to provide a constant source of heat. As the water accumulates in the reservoirs, it is heated to 200°C (400°F) or higher, which is much above the boiling point at sea level and normal pressure. (Such superheating is possible because of the high underground water pressure.) At these high temperatures, much of the water becomes steam. The accumulation of steam deep in the tube and of boiling water higher in the tube eventually causes a great upward surge that sends water and steam showering out of the geyser vent. This eruption releases the pressure, and when the eruption subsides, underground water again begins to collect in the reservoirs in preparation for a repetition of the process.

A tremendous supply of heat is essential for geyser activity. Recent studies in Yellowstone Park's Upper Geyser Basin indicate that the heat emanating from that basin is at least 800 times greater than the heat flowing from a nongeyser area of the same size.

Some geysers erupt continuously, indicating that they are really hot springs that have a constant supply of water through which steam is escaping. Most geysers are only sporadically active, however, apparently depending on the accumulation of sufficient water to force an eruption. Some eruptions are very brief, whereas others continue for many minutes. The interval between eruptions for most geysers is variable. Most erupt at intervals of a few hours or a few days, but some wait years or even decades between eruptions. The temperature of the erupting water generally is near the boiling point for pure water (100°C or 212°F at sea level). In some geysers, the erupting water column goes up only a few centimeters in the air, whereas in others the column rises to more than 45 meters (150 feet).

Geyser comes from the Icelandic word *geysir* ("to gush" or "to rage"), the Great Geysir in southern Iceland being the namesake origin for this term. The most famous of all geysers is Old Faithful in Yellowstone National Park (Figure 17-12). Its reputation is based partly on the force of its eruptions (the column goes more than 30 meters or 100 feet high) but primarily on its regularity. Since first timed by scientists more than a century ago, Old Faithful has maintained an average interval of 65 minutes between eruptions (ranging from about 30 minutes to 120 minutes), day and night, winter and summer, year after year.

▲ **Figure 17-12** Yellowstone's Old Faithful geyser in all its eruptive grandeur.

In the early 1980s, however, several consecutive earthquakes on the Yellowstone plateau apparently upset the geyser's internal plumbing. Over the last decade, Old Faithful's average interval between eruptions has been about 90 minutes, typically varying between 44 and 125 minutes. Thus, Old Faithful has become slightly more erratic, which is to say, more like other geysers.

The deposits resulting from geyser activity are usually much less notable than those associated with hot springs. Some geysers erupt from open pools of hot water, throwing tremendous sheets of water and steam into the air but usually producing relatively minor depositional features. Other geysers are of the "nozzle" type and consequently build up a depositional cone and erupt through a small opening in it (Figure 17-13). Most deposits resulting from geyser activity are simply sheets of precipitated mineral matter spread irregularly over the ground.

Fumaroles

A third hydrothermal feature is the **fumarole**, a surface crack directly connected to a deep-seated heat source (Figure 17-14). For some reason, very little water drains into the tube of a fumarole. The water that does drain in is instantly converted to steam by the heat, and a cloud of steam is then expelled from the opening—often with an accompanying roaring or hissing sound. Thus, a fumarole

▶ **Figure 17-13** Most geysers erupt from vents or hot pools, but some build up prominent depositional "nozzles" through which water is expelled. This is Castle Geyser in Yellowstone National Park.

is marked by steam issuing either continuously or sporadically from a surface vent; in essence, a fumarole is simply a hot spring that lacks water.

Hydrothermal Features in Yellowstone

Hydrothermal features are found in many volcanic areas, being particularly notable in Iceland, New Zealand, Chile, and Siberia's Kamchatka Peninsula. By far the largest concentration, however, occurs in Yellowstone National Park, located mostly in northwestern Wyoming, which contains about 225 of the world's 425 geysers as well as more than half of the world's other hydrothermal phenomena. The area consists of a broad, flattish plateau bordered by extensive mountains (the Absaroka Range) on the east and by more limited highlands (particularly the Gallatin Mountains) on the west. The bedrock surface of the plateau is almost entirely volcanic materials,

▶ **Figure 17-14** A fumarole is like a geyser except that it erupts no water; it sends out only steam. This scene is from Iceland.

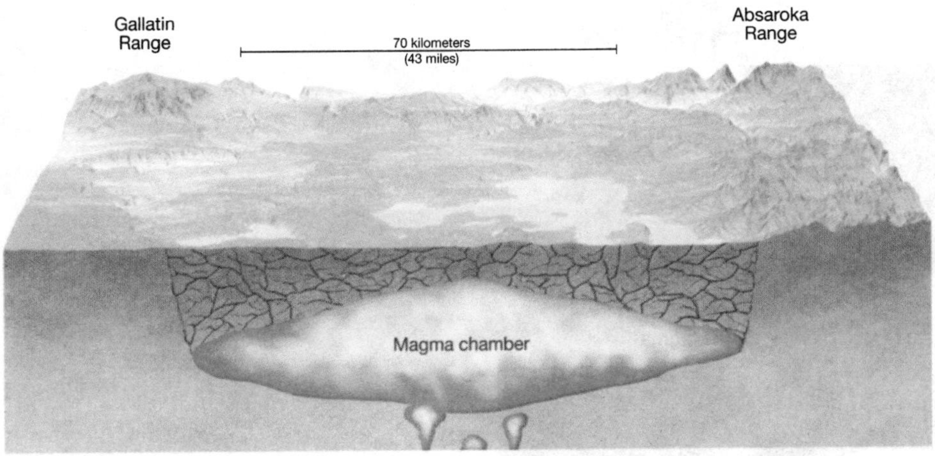

◀ **Figure 17-15** Schematic west-east cross section through the Yellowstone Plateau showing the extent of the Yellowstone Caldera and the magma chamber below.

although no volcanic cones are in evidence. About 640,000 years ago, a catastrophic volcanic eruption here ejected about 1000 cubic kilometers (240 cubic miles) of pyroclastics—about 1000 times more material than the 1980 Mount St. Helens eruption—covering the surrounding region with thick deposits of volcanic ash, and resulting in the formation of a 70 kilometer (43 mile) diameter caldera (Figure 17-15). An even larger eruption took place about 2.1 million years ago, and geologists think that future eruptions of the Yellowstone volcano are a distinct possibility.

The uniqueness of Yellowstone's geologic setting stems from the presence of a large, shallow magma chamber beneath the plateau—thought to be the result of a hot spot formed by a mantle plume rising up from the mantle. Test boreholes and geophysical studies reveal a high thermal gradient, in which the temperature increases with depth at a rate of about 67°C per 100 meters (36°F per 100 feet), and indicate that the top of the magma chamber is perhaps only 8 kilometers (5 miles) below the surface. This shallow magma pool provides the heat source—the most important of the three conditions necessary for the development of hydrothermal features.

The second requisite is an abundance of water that can seep downward and become heated. Yellowstone receives copious summer rain and a deep winter snowpack (averaging more than 250 centimeters or 100 inches).

The third necessity for hydrothermal development is a weak or broken ground surface that allows water to move up and down easily. Here, too, Yellowstone fits the bill, for the ground surface there is very unstable and subject to frequent earthquakes, faulting, and volcanic activity. Consequently, many fractures and weak zones provide easy avenues for vertical water movement.

The park contains about 225 geysers, more than 3000 hot springs, and 7000 other thermal features (fumaroles, steam vents, hot-water terraces, hot-mud cauldrons, and so forth). There are five major geyser basins, a half dozen minor ones, and an extensive scattering of individual or small groups of thermal features.

The principal geyser basins are all in the same watershed on the western side of the park (Figure 17-16). The Gibbon River from the north and the Firehole River from the south unite to form the Madison River, which flows westward into Montana, eventually to join two other rivers in forming the Missouri (a 1959 earthquake in the Yellowstone area triggered a major landslide in the Madison River valley just to the west of the park; see Figure 15-22). The Gibbon River drains the Norris and Gibbon geyser basins, whereas the Firehole drains the Upper, Midway, and Lower basins. The Firehole River derives its name from the great quantity of hot water fed into it from the hot springs and geysers along its way. Approximately two-thirds of the hydrothermal features of Yellowstone are in the drainage area of the Firehole River.

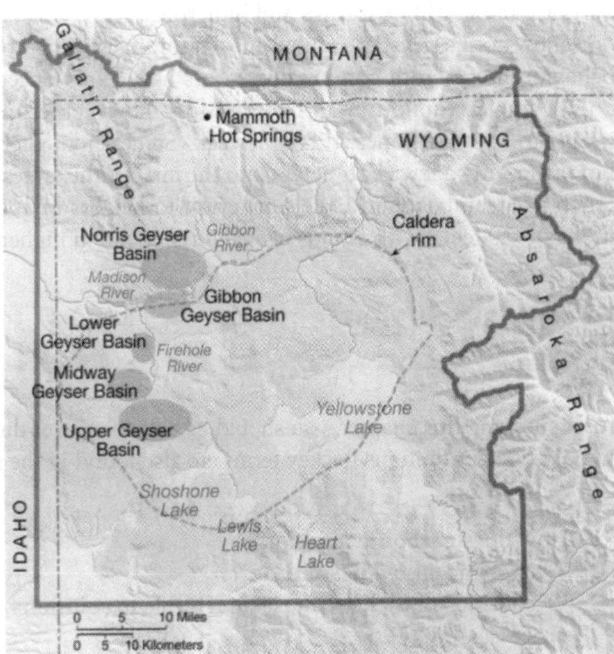

▲ **Figure 17-16** Yellowstone National Park and its major geyser basins. The approximate outline of the Yellowstone Caldera is shown.

▲ **Figure 17-17** Yellowstone's Mammoth Hot Springs. (a) Although the white material looks like snow or ice, it is actually travertine. (b) Water from rainfall and snowmelt percolates down into the underlying limestone, where it is heated and flows downslope. Hot water issues onto the surface and precipitates travertine deposits when exposed to the air.

All the major geyser basins consist of gently undulating plains or valleys covered mostly with glacial sediments and large expanses of whitish siliceous material called *geyserite*. Each basin contains from a few to several dozen geysers, some of which are inconspicuous holes in the geyserite, but others of which are built-up cones that rise a few meters above the basin. In addition, each basin contains a number of hot springs and fumaroles.

Yellowstone's geysers exhibit an extraordinary range of behavior. Some erupt continually; others have experienced only a single eruption in all history. Most, however, erupt irregularly several times a day or week. Some shoot their hot water only a few centimeters into the air, but the largest (such as Steamboat and Excelsior) erupt to heights of 100 meters (330 feet), with clouds of steam rising much higher.

In the northwestern portion of the park is the most remarkable aggregation of hot-water terraces in the world—the Mammoth Hot Springs Terraces (Figure 17-17). There, groundwater percolates down from surrounding hills into thick layers of limestone. Hot water, carbon dioxide, and other gases rise from the heated magma to mingle with the groundwater and produce a mild carbonic acid solution that rapidly dissolves great quantities of the limestone. Saturated with dissolved minerals, the temporarily carbonated water seeps downslope until it gushes forth near the base of the hills as the Mammoth Hot Springs. The carbon dioxide escapes into the air, and the calcium carbonate is precipitated as massive deposits of travertine in the form of flat-topped, steep-sided terraces.

CHAPTER 17 LEARNING REVIEW

After studying this chapter, you should be able to answer the following questions. Key terms from each text section are shown in bold type. Definitions for key terms are also found in the glossary at the back of the book.

Solution and Precipitation (p. 467)

1. How does **carbonic acid** form?
2. What is meant by **dissolution**?
3. What kinds of rock are most susceptible to solution processes? Why?
4. How does the underground structure of the bedrock influence the dissolution process?
5. How is it possible for percolating groundwater to both remove mineral material and deposit it?

Caverns and Related Features (p. 468)

6. What is the importance of jointing and bedding planes to the underground structure of **caverns**?
7. Describe and explain the formation of **speleothems** such as **stalactites**, **stalagmites**, and columns.

Karst Topography (p. 469)

8. In what kinds of rocks does **karst** topography usually develop?
9. Explain how a **sinkhole** is formed.
10. Describe the formation of a **collapse sinkhole** and a **uvala**?
11. Describe the characteristics of **tower karst**.
12. Why is there a scarcity of surface drainage in karst areas?
13. What is a **swallow hole**? A **disappearing stream**?

Hydrothermal Features (p. 473)

14. What is **hydrothermal activity**?
15. What are the differences among a **hot spring**, a **geyser**, and a **fumarole**? What causes these differences?
16. Briefly explain the eruption sequence of a typical geyser.
17. What three conditions are necessary for hydrothermal features to develop?
18. What is the importance of jointing and bedding planes to the development of hot springs and geysers?

Log in to **www.mygeoscienceplace.com** for animations, videos, RSS feeds, maps, and self-study quizzes to enhance your study of **Karst and Hydrothermal Processes**.

18 The Topography of Arid Lands

Arid lands are in many ways distinctive from humid ones, but there are no obvious boundaries to separate the two. In this chapter, we focus on the dry lands of the world without attempting to establish precise definitions or borders. We are concerned not with where such borders lie, but rather with the processes that shape desert landscapes. It should be understood, however, that both the processes and the landforms of desert landscapes occur more widely than the term *desert* might imply, and some of the landforms we describe may even be seen in humid regions.

It also important to understand that some of today's deserts had quite different climates in the geologic past. Parts of today's Sahara Desert, for example, were much wetter just a few thousand years ago than they are today. Thus, in addition to processes operating today, some desert landscapes we see have also been shaped by processes that were at work in the past.

A Specialized Environment

As we learned in Chapter 11, desert terrain is usually stark and abrupt, unsoftened by regolith, soil, or vegetation. Despite the great difference in appearance between arid lands and humid, however, most of the terrain-forming processes active in humid areas are also at work in desert areas. There are, however, special conditions found in deserts that do significantly influence landform development.

Special Conditions in Deserts

Desert landforms are often conspicuously different from those found in wetter locations. These differences are largely the result of a variety of factors and special conditions found in arid regions. The most important of these special conditions include:

Weathering Because moisture is required for nearly all kinds of chemical weathering, in many desert regions mechanical weathering is dominant—although chemical weathering is likely to be absent in only the driest of deserts. Mechanical weathering processes such as *salt wedging* are more common in arid regions than in humid ones (see Figure 15-9). This predominance of mechanical weathering results not only in a generally slower rate of total weathering in deserts, but also in the production of more angular particles of weathered rock.

Soil and Regolith In deserts, the covering of soil and regolith is either thin or absent in most places, a condition that exposes the bedrock to weathering and erosion, and contributes to the stark, rugged, rocky terrain (Figure 18-1).

Soil Creep Soil creep is a relatively minor phenomenon on most desert slopes. This is due partly to the lack of soil but primarily to the lack of the lubricating effects of water. Creep is a smoothing phenomenon in more humid climates, and its lack in deserts accounts in part for the angularity of desert slopes.

◀ Dried mudflow deposits amid sand dunes in Death Valley National Park, California.

▲ **Figure 18-1** Slope comparisons in dry and moist climates. The steep relief is more easily seen in a dry climate because there is very little obscuring vegetation cover. In a humid climate, the steep faces are to some degree obscured by vegetation.

Impermeable Surfaces A relatively large proportion of the desert surface is impermeable to percolating water, permitting little moisture to seep into the ground. *Caprocks* (resistant bedrock surfaces) and *hardpans* (hardened and generally water-impermeable subsurface soil layers) of various types are widespread, and what soil has formed is usually thoroughly compacted and often does not readily absorb water. Such impermeable surfaces lead to high runoff when it rains.

Sand Some deserts have an abundance of sand in comparison with other parts of the world (Figure 18-2). This is not to say that deserts are mostly sand covered; indeed, the notion that deserts consist of great seas of sand is quite erroneous. Nevertheless, the relatively high proportion of sand in deserts has three important influences on topographic development: (1) A sandy cover allows water to infiltrate the ground and inhibits drainage via streams and overland flow, (2) sand is readily moved by heavy rains, and (3) it can be transported and redeposited by the wind (the development of desert sand dunes is discussed later in this chapter).

Rainfall Although rainfall is limited in desert areas, much of the rain that does fall comes from intense convective thunderstorms—which result in very high and rapid runoff. Floods, although often brief and covering only a limited area, are the rule rather than the exception in deserts. Thus, fluvial erosion and deposition, however sporadic and rare, are remarkably effective and conspicuous.

Fluvial Deposition Almost all streams in desert areas are *ephemeral*, flowing only during and immediately after a rain. Such streams are effective agents of erosion, shifting enormous amounts of material in a short time. This is mostly short-distance transportation, however. A large volume of unconsolidated debris is moved to a nearby location, and as the stream dries up, the debris is dumped on slopes or in valleys, where it is readily available for the next rain. As a consequence, depositional features of alluvium are unusually common in desert areas.

▲ **Figure 18-2** Many deserts contain considerable accumulations of sand, which sometimes builds up to form conspicuous dunes. Here, tourists explore dunes in the extreme southeastern corner of California.

Wind Another fallacy associated with deserts is that their landforms are produced largely by wind action. This is not true, even though high winds are characteristic of most deserts and even though sand and dust particles are easily shifted.

Basins of Interior Drainage Desert areas contain many watersheds that do not drain ultimately into any

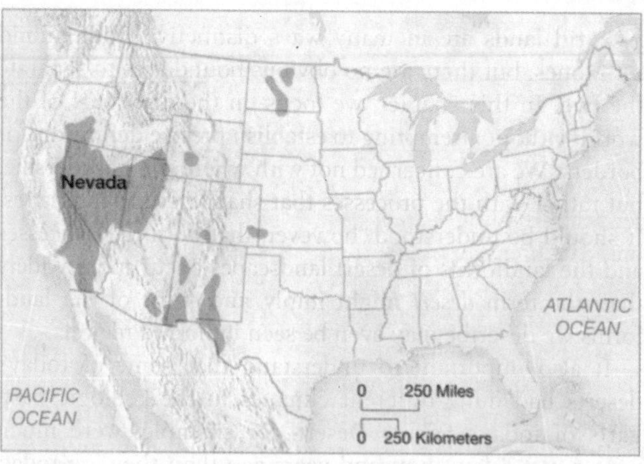

▲ **Figure 18-3** The basins of interior drainage in the United States are all in the western part of the nation. By far the largest area of interior drainage is in the "Great Basin" centered in Nevada and the adjoining states.

ocean. For most continental surfaces, rainfall has the potential of flowing all the way to the sea. In dry lands, however, drainage networks are frequently underdeveloped, and the terminus of a drainage system is often a basin or valley with no external outlet, as Figure 18-3 shows for the United States. Any rain that falls in Nevada, for example, except in the extreme southeastern and northeastern corners of the state, has no hope of flowing in a stream to the sea.

Vegetation All the previous environmental factors have important effects on topographic development, but perhaps the single most obvious feature of dry lands is the lack of a continuous vegetative cover. The plant cover consists mostly of widely spaced shrubs or sparse grass, which provide little protection from the force of raindrops and function inadequately to bind the surface material with roots (Figure 18-4).

◀ **Figure 18-4** In most arid regions, the cover of vegetation is sparse. The creosote bush is the green shrub in this Death Valley, California, scene.

Running Water in Waterless Regions

Probably the most fundamental fact of desert geomorphology is that running water is by far the most important external agent of landform development. The erosional and depositional work of running water influences the shape of the terrain surface almost everywhere outside areas of extensive sand accumulation. The lightly vegetated ground is defenseless to whatever rainfall may occur, and erosion by rain splash, sheetwash, rilling, and streamflow is enormously effective. Despite the rarity of precipitation, its intensity and the presence of impermeable surfaces produce abrupt runoff, and great volumes of sediment can be moved in a very short time.

The steeper gradients of mountain streams increase the capacity of these streams for carrying large loads, of course, but the sporadic flow of mountain streams in arid lands results in an unpredictable imbalance between erosion and deposition. At any given time, therefore, much transportable rock debris and alluvium sit at rest in the dry stream bed of a desert mountain, awaiting the next flow. Loose surface material is either thin or absent on the slopes, and bedrock is often clearly exposed, with the more resistant strata standing out as caprocks and cliff faces.

Where slopes are gentle in an arid land, the streams rapidly become choked with sediment as the brief flood time subsides. Here stream channels are readily subdivided by braiding, and main channels often break up into distributaries in the basins. Much silt and sand are thus left on the surface for the next flood to move, unless wind moves them first.

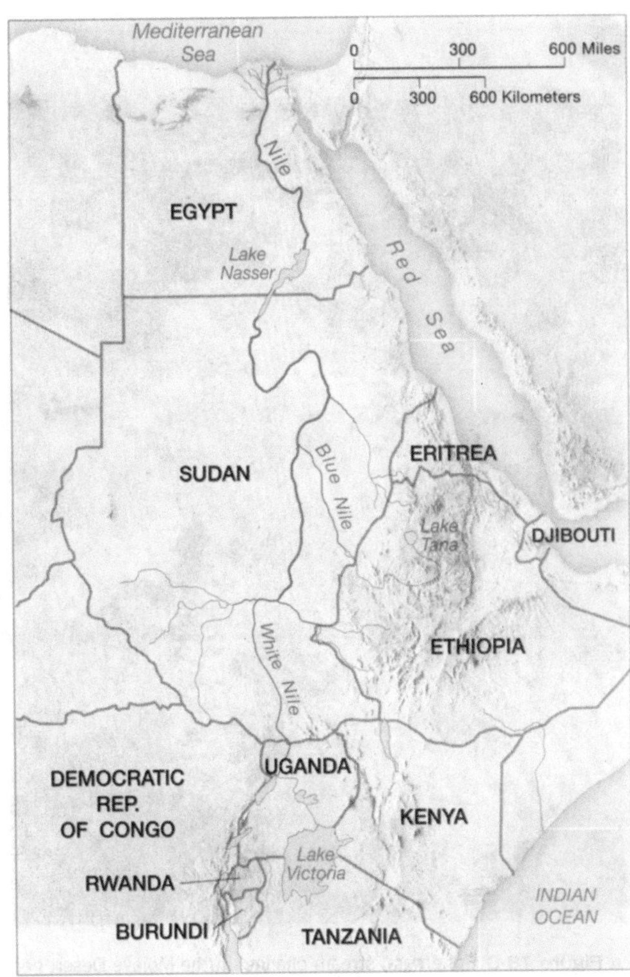

▲ **Figure 18-5** The world's preeminent example of an exotic stream, the Nile River, flows for many hundreds of kilometers without being joined by a tributary.

Surface Water in the Desert

Surface water in deserts is conspicuous by its absence. These are lands of sandy streams and dusty lakes, in which the presence of surface water is usually episodic and brief.

Exotic Streams Permanent streams in dry lands are few, far between, and, with scarce exceptions, *exotic*, meaning they are sustained by water that originates outside the desert. This water that feeds **exotic streams** comes from an adjacent wetter area or a higher mountain area in the desert and has sufficient volume to survive passage across the dry lands. The Nile River of North Africa is the classic example of an exotic stream (Figure 18-5). Its water comes from the mountains and lakes of central Africa and Ethiopia in sufficient quantity to survive a 3200-kilometer (2000-mile) journey across the desert without benefit of tributaries. In North America, the Colorado River is a prominent example of an exotic stream.

In humid regions, a river becomes larger as it flows downstream, nourished by tributaries and groundwater inflow. In dry lands, however, the flow of exotic rivers diminishes downstream because the water seeps into the riverbed, evaporates—or is diverted for irrigation.

Ephemeral Streams Although almost every desert has a few prominent exotic rivers, more than 99 percent of all desert streams are **ephemeral streams** (Figure 18-6). The brief periods during which these streams flow are times of intense erosion, transportation, and deposition, however. Most ephemeral desert streamflow eventually dissipates through seepage and evaporation, although sometimes such a stream is able to reach the sea, a lake, or an exotic river.

Desert Lakes Although lakes are uncommon in desert areas, dry lake beds are not (Figure 18-7). We have already noted the prevalence of basins of interior drainage in dry lands; most of them have a lake bed occupying their area of lowest elevation, which functions as the local base level for that basin. These dry lake beds are called **playas** (Figure 18-8), although the term **salina** may be used if there is an unusually heavy concentration of salt in the lake-bed sediments. If a playa surface is heavily impregnated with clay, the formation is called a *claypan*. On rare occasions, the intermittent streams may have sufficient flow to bring water to the playa, forming a temporary *playa lake*.

Playas are among the flattest and most level of all landforms—in some cases they are several kilometers across, but with a local relief of only a few centimeters. A playa develops such a flat surface when it is periodically covered with water as a playa lake. The suspended silt in the shallow lake eventually settles out and the water evaporates, leaving a flat

▲ **Figure 18-6** Ephemeral stream channel in the Mojave Desert near Baker, California.

▲ **Figure 18-8** Desert playa in central Nevada. Playas are among the flattest of all landforms.

layer of dried mud. Through repeated episodes such as this, over the centuries a playa attains its flat surface.

A few desert lakes are permanent. The smaller ones are nearly always the product of either subsurface structural conditions that provide water from a permanent spring or exotic streams or from streams flowing down from nearby mountains. Many permanent lakes in desert areas are **saline lakes**—high rates of evaporation relative to the inflow and/or basins of interior drainage lead to the accumulation of dissolved salts in such lake waters.

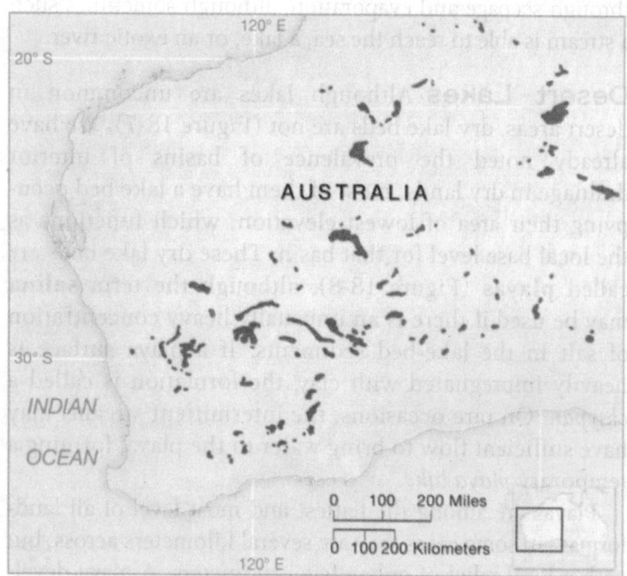

▲ **Figure 18-7** Dry lake beds are often numerous in desert areas. This map shows the principal playas and salinas in Western Australia.

Many of the largest natural desert lakes are remnants of still larger bodies of water that formed in a previously wetter climate. Utah's Great Salt Lake is the outstanding example in the United States. Although in terms of surface area it is the second largest lake wholly in the United States (after Lake Michigan), it is a mere shadow of the former Lake Bonneville, which was formed during the wetter conditions of the Pleistocene Epoch (see Figure 19-6).

Fluvial Erosion in Arid Lands

Although fluvial erosion takes place in desert areas only during a small portion of each year, it does its work rapidly and effectively, and the results are conspicuous. In desert areas of any significant relief, large expanses of exposed bedrock are common because of the lack of soil and vegetation. During the rare rains, this bedrock is both mechanically weathered and eroded by running water, and the result of the latter process is steep, rugged, rocky surfaces.

A typical circumstance leading to fluvial erosion in a desert might first entail a brief but intense thunderstorm, perhaps over a mountain drainage basin. The localized thunderstorm puts lots of water on the ground, but because the desert surface is either exposed bedrock or some other kind of relatively impermeable surface, most of the water

◀ **Figure 18-9** The effects of differential weathering and erosion are conspicuous on the Red Cliffs near Gateway in western Colorado. The more resistant layers near the top form an abrupt escarpment, whereas the softer layers below are weathered and eroded into gentler slopes.

quickly runs off into a nearby dry ephemeral stream channel. This stream channel quickly fills up with water, perhaps developing into a *flash flood* or a debris flow (described in Chapter 15) that travels for many kilometers down out of the desert mountains onto a basin floor. Such flash floods and debris flows can move a remarkable amount of material in just a few minutes. It is through such localized but infrequent events that most change takes place in the desert.

Flashfloods and debris flows also pose a significant hazard to humans in arid lands—travelers are cautioned against camping or parking in dry stream washes, since such floods can arrive with little local warning.

Differential Weathering and Erosion Whenever a land surface weathers or erodes variations in rock type and structure produce differences in the slope and shape of the resulting landform—processes called **differential weathering and erosion** (Figure 18-9). In many cases, rocks resistant to weathering and erosion form the cliffs, pinnacles, spires, and other sharp crests, while softer rocks wear away more rapidly to produce gentler slopes. Differential erosion is very common in sedimentary landscapes because there are significant differences in resistance from one stratum to the next; such areas often have vertical escarpments (steep, clifflike slopes) and abrupt changes in slope angle. In areas dominated by igneous or metamorphic bedrock, however, differential weathering and erosion may be less obvious because there is not much difference in resistance from one part of the bedrock to another.

The steep rock faces that are evidence of differential weathering and erosion are more noticeable in dry lands than in humid lands. The reason for this difference is that in humid areas the lush vegetation overgrows the faces and obscures their steepness (see Figure 18-1).

Residual Surfaces and Features Scattered throughout the arid and semiarid lands of the world are isolated landforms that rise abruptly from the surrounding plains. Such steep-sided mountains, hills, or ridges are referred to as **inselbergs** ("island mountains") because they resemble rocky islands standing above the surface of a broad sea. One notable type of inselberg is the *bornhardt*, which is composed of highly resistant rock and has a rounded form (Figure 18-10). Differential weathering and

▶ **Figure 18-10** Kata Tjuta (the Olgas) is a massive bornhardt in the desert of central Australia.

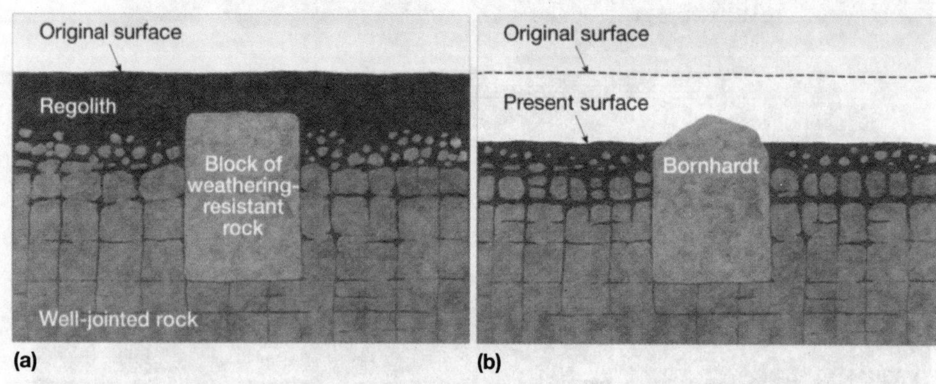

▶ **Figure 18-11** The development of a bornhardt. (a) The well-jointed rock is more susceptible to weathering and erosion than the resistant block in the center. (b) As a result, the bornhardt remains as a conspicuous result of differential weathering and erosion.

erosion lower the surrounding terrain, leaving the resistant bornhardt standing high (Figure 18-11). Bornhardts are very stable and may persist for tens of millions of years.

Along the lower slopes of desert mountains and hills, a distinctive kind of surface is often found. This gently inclined bedrock platform, called a **pediment**, is a "residual" surface (in other words, a surface left by weathering and the removal of rock rather than by deposition) extending outward from the mountain front (Figure 18-12).

Pediments have a complicated origin that is still incompletely understood. Long thought to develop as desert mountains are worn down and back by weathering and erosion during previous periods of wetter climate, recent analysis suggests that pediments may form in other ways. Some pediments may be the result of deep subsurface weathering during a period of wetter climate—after the overlying weathered material is removed by erosion, a flat or gently sloping bedrock surface remains. Other pediments may have formed as a consequence of detachment faulting in areas undergoing tectonic extension, whereby blocks of crust are displaced along nearly horizontal fault planes—the bedrock surface of the pediment representing the fault plane.

Pediments can be found in many deserts and are sometimes the dominant terrain feature. They are not easily recognizable, however, because almost invariably they are covered with a veneer of debris deposited by water and wind.

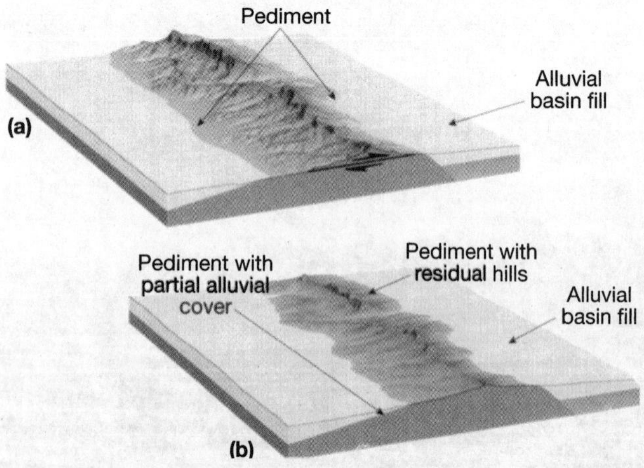

▲ **Figure 18-12** Desert pediments. (a) Some pediments may develop as a consequence of detachment faulting. (b) Other pediments may represent the top of a now exhumed weathered bedrock surface. The residual desert hills are sometimes called inselbergs.

Desert Stream Channels The normally dry beds of ephemeral streams typically have flat floors, sandy bottoms, and steep sides (see Figure 18-6). In the United States, they are variously referred to as *arroyos*, *gullies*, *washes*, or *coulees*. In North Africa and Arabia, the name *wadi* is common; in South Africa, *donga;* in India, *nullah*.

Stream channels in desert areas tend to be deep. Both mountain canyons and flatland arroyos typically have flat, narrow bottoms and steep, often near-vertical sides. Deep accumulations of sand and other loose debris usually cover the channel bed, although flash floods sometimes scour away all alluvial fill right down to the bedrock.

Fluvial Deposition in Arid Lands

Except in hills and mountains, depositional features are more notable than erosional ones in a desert landscape. Depositional features consist mostly of talus accumulations at the foot of steep slopes and deposits of alluvium and other fragmented debris in ephemeral stream channels, the latter representing bedload left behind with the subsidence of the last flood.

"Piedmont" is a generic term meaning any zone at the foot of a mountain range (the term *pediment* comes from the same Latin root but refers to a specific landform). The **piedmont zone** of a desert mountain range is a prominent area of fluvial deposition. There is normally a pronounced change in slope at the mountain base, with a steep slope giving way abruptly to a gentle one (Figure 18-13). This is an area of significant accumulation of rock debris because the break in slope greatly reduces the speed of any sheetwash, streamflow, or debris flow that travels into the piedmont zone. Moreover, the streams issue more or less abruptly from canyons onto the more open piedmont and so are freed from lateral constraint. The resulting fluvial deposition in the piedmont often reaches depths of hundreds of meters.

The flatter portions of desert areas also often exhibit a prominent accumulation of alluvium, in part because there is not usually enough streamflow to carry sediments very far from a mountain front—and no place to go if it is a basin of interior drainage. Any sheetwash or streamflow

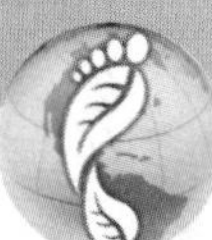

PEOPLE AND THE ENVIRONMENT

Desertification

Desertification is sure to appear on almost any short list of major world environmental problems. Although coined much earlier, the term desertification has come into prominent use in the last four decades to refer to the expanding of desert conditions into areas previously not desert. Desertification is occurring today all over the world: in southern Africa, the Middle East, India, western China, southern Australia, Chile and Peru, northeastern Brazil, the southwestern United States, Mexico, and other places. The United Nations Environment Program (UNEP) estimates that perhaps 21 million hectares (8100 square miles) of land is desertified each year.

Although the concept of deserts progressively and clearly moving over productive land—usually in the form of a readily recognizable front of mobile sand dunes—may be appealing in its simplicity and its potential for raising political and public awareness, but this is not the form that desertification commonly takes. The United Nations (UN) describes desertification as a type of land degradation: "the loss of the land's biological productivity, caused by human-induced factors and climate change."

Deserts can enlarge through entirely natural causes, primarily through climate change that leads to recurrent drought. However, deserts can also expand as a result of human activity: overgrazing of livestock, imprudent agricultural practices, deforestation, and improvident use of water resources.

Desertification is normally associated with the margins of existing deserts and implies an expansion of an already-existing desert, but this is not always the case. For example, "dust bowl" conditions in the North American Great Plains in the 1930s—brought on by drought made worse by imprudent land use practices and poor economic conditions—were not peripheral to any existing desert, yet they represented a clear and spectacular case of desertification long before the term was in common use.

The specter of desertification was brought to worldwide attention in the late 1960s by the onset of a cruel six-year drought in the African Sahel. This subhumid to semiarid region on the southern margin of the Sahara occupies parts of 10 countries, from the Atlantic Ocean on the west to the Ethiopian highlands on the east (Figure 18-A). It has an east–west expanse of more than 4800 kilometers (3000 miles), and a north–south extent that varies from 480 to 800 kilometers (300 to 500 miles).

As the Sahelian drought intensified from 1968 to 1974, vegetation disappeared, millions of livestock perished, tens of thousands of people died, and there was a prominent outflow of surviving humans to an uncertain future in already overcrowded lands, particularly cities to the south of the drought-stricken region.

Since the 1970s, the problem of drought and desertification in the Sahel has remained a problem. In 2009, the March-to-June rainy season—generally associated with the shift of the ITCZ poleward in summer—was drier than average. As a result, crops did not grow well and the grass cover of the Sahel was much sparser than in average years (Figure 18-B), stressing food supplies in many countries.

In an effort to gain international recognition for the problem of desertification, as well as to seek cooperative long-term solutions to help the millions of mostly impoverished people displaced by it, in 1994 the United Nations adopted the United Nations Convention to Combat Desertification. More recently, the General Assembly of the UN designated 2006 as the International Year of Deserts and Desertification (IYDD). The goals of the IYDD include raising awareness of desertification so that the urgency of the problem is more generally recognized around the world.

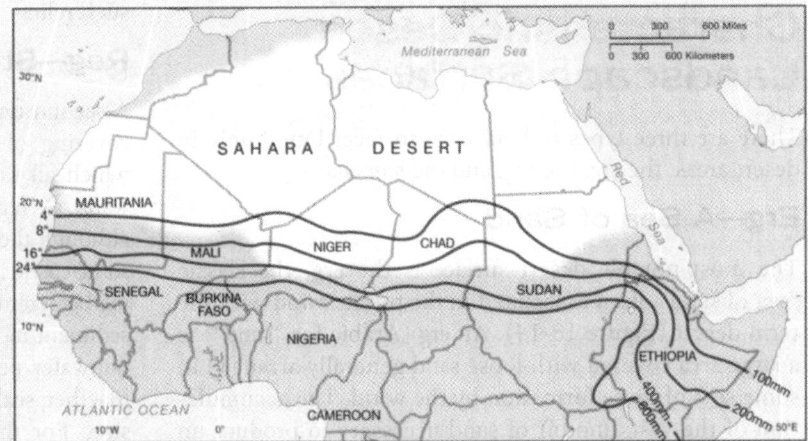

▲ **Figure 18-A** The Sahel Region of Africa. The isolines are isohyets of average annual rainfall. Only the Sahelian countries are named.

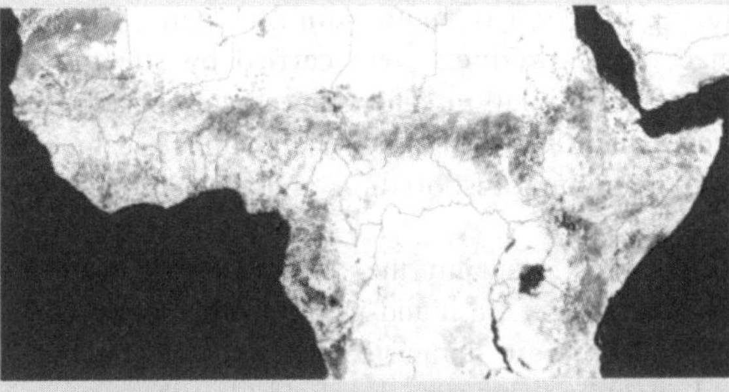

▲ **Figure 18-B** The recent decline in vegetation growing in the Sahel is clearly shown in this image comparing the density of vegetation in 2009 with the average density of vegetation between 2003 and 2008. Brown areas show less than average vegetation growth while green areas show greater than average growth. The data were gathered by using the AVHRR instrument aboard the NOAA-17 satellite.

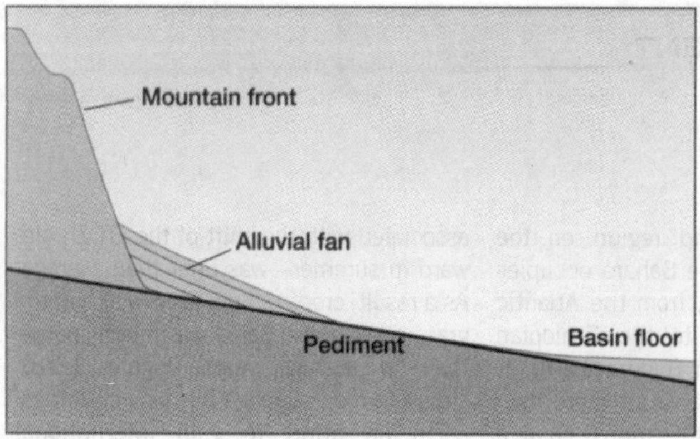

▲ **Figure 18-13** An idealized cross section of a desert piedmont zone. The pediment surface (in this case associated with a low-angle fault) is often covered with alluvial deposits at the mountain front and along the basin floor.

that reaches into such low-lying flatlands usually had to travel a considerable distance over low-angle slopes, which means that both flow volume and flow speed are likely to be limited. Consequently, larger rock fragments are rarely transported into middle of the basins; instead, they are covered with fine particles of sand, silt, and clay, sometimes to a considerable depth.

▲ **Figure 18-14** Sand dunes in an erg area of the Namib Desert, Sossusvlei National Park, Namibia.

Characteristic Desert Landscape Surfaces

There are three types of landscape surfaces found only in desert areas: the *erg*, the *reg*, and the *hamada*.

Erg—A Sea of Sand

The most notable desert surface is the **erg**, the classic "sea of sand" often associated in the public mind with the term desert (Figure 18-14). An erg (Arabic for "sand") is a large area covered with loose sand generally arranged in some sort of dune formation by the wind. The accumulation of the vast amount of sand necessary to produce an erg probably cannot be explained only by processes operating in deserts today—since desert weathering processes are very slow. Rather, much of this sand probably accumulated during a period of more humid climate. After being formed, these products were carried by streams into an area of accumulation. Then the climate became drier and consequently wind, rather than water, became the principal agent of transportation and deposition of the sand.

Several large ergs occur in the Sahara and Arabian deserts, and smaller ergs are found in most other deserts. The Australian deserts are dominated by large accumulations of sand, including extensive dunefields, but these are not true ergs because most of the sand is anchored by vegetation and therefore not free to move with the wind. Relict ergs (usually in the form of sand dunes covered with vegetation) are sometimes found in nonarid areas, indicative of a drier climate in the past. Much of western Nebraska has such relicts.

Reg—Stony Deserts

A second type of desert landscape surface is the **reg**, a tight covering of coarse gravel, pebbles, and/or boulders from which all sand and dust have been removed by wind and water. A reg (Arabic for "stone"), then, is a stony desert, although the surface covering of stones may be very thin (in some cases, it is just one pebble deep). The finer material having been removed through surface erosion—perhaps aided by sediment movement below the surface through the action of rainwater percolation—the surface pebbles often fit closely together, sealing whatever material is below from further erosion. For this reason, a reg is often referred to as **desert pavement** or *desert armor* (Figure 18-15). In Australia, where regs are widespread, they are called *gibber plains*.

A striking feature of some deserts, one particularly but not exclusively associated with regs, is **desert varnish**. This is a dark, shiny coating, consisting mostly of iron and manganese oxides, that forms on the surface of pebbles, stones, and larger outcrops after long exposure to the desert air (see the larger rocks shown in Figure 18-15). Desert varnish is characterized by a high content of iron and manganese oxides along with wind-delivered clay. The relatively high concentrations of manganese in desert varnish seem to be a consequence of a biochemical process involving bacteria. Desert varnish can be used as a relative dating tool for geomorphologists because the longer a rock surface has been exposed to weathering, the greater is the concentration of the oxide coating and thus the darker the color.

▲ **Figure 18-15** Desert pavement in California's Panamint Valley. Note the dark brown coating of desert varnish on the larger rocks.

Hamada—Barren Bedrock

A third desert landscape surface is the **hamada** (Arabic for "rock"), a barren surface of consolidated material. A hamada surface usually consists of exposed bedrock, but it is sometimes composed of sedimentary material that has been cemented together by salts evaporated from groundwater. In either case, fragments formed by weathering are quickly swept away by the wind, so that little loose material remains.

Ergs, regs, and hamadas are all limited to plains areas. Regs and hamadas are exceedingly flat, whereas ergs are as high as the sand dunes built by the wind. The boundaries of these landscapes are often sharp because of the abrupt change in friction-layer speed as the wind moves from a sandy surface to a nonsandy surface. For example, wind passing from an erg to a reg or hamada is no longer slowed by the drag of loose sand and so can speed up and sweep the hard surface clean. On the other hand, wind moving from a barren reg or hamada to a sandy erg is slowed down perceptibly, and deposition results.

Although the extent of ergs, regs, and hamadas is significant in some desert areas, the majority of the arid land in the world contains only a limited number of these surfaces. For example, only one-third of the Arabian Desert, the sandiest desert of all, is covered with sand, and much of that is not in the form of an erg.

The Work of the Wind

The irrepressible winds of the desert create spectacular sand and dust storms and continuously reshape minor details of the landscape (Figure 18-16). However, the effect of wind as a sculptor of terrain is very limited, with the important exception of such relatively impermanent features as sand dunes.

The term wind is theoretically restricted to horizontal air movement. Some turbulence is nearly always involved, however, so there is usually a vertical component of flow as well. In general, the motion of air passing over the ground is similar to that of water flowing over a streambed, and that similarity is the cause of the turbulence. In a thin layer right at the ground surface, wind speed is zero, just as the speed of the water layer touching the banks and bed of a stream is zero, but wind speed

▲ **Figure 18-16** The wind is sometimes a prominent force in rearranging loose particles. This scene is near Barrow Creek in the Northern Territory of Australia.

increases with distance above the ground. The shear developed between different layers of air moving at different speeds causes turbulence similar to that in a stream of water. Wind turbulence can also be caused by heating from below, which causes the air to expand and move upward.

Aeolian processes are those related to wind action (Aeolus was the Greek god of the winds). They are most pronounced, widespread, and effective wherever fine-grained unconsolidated sedimentary material is exposed to the atmosphere, without benefit of vegetation, moisture, or some other form of protection—in other words, in deserts and along sandy beaches. Our focus here is wind action in desert regions.

Aeolian Erosion

The erosive effect of wind can be divided into two categories: deflation and abrasion.

Deflation **Deflation** is the shifting of loose particles as a result of their being blown either through the air or along the ground. Except under extraordinary circumstances, the wind is not strong or buoyant enough to move anything more than dust and small sand grains, and therefore no significant landforms are created by deflation. Sometimes a **blowout**, or *deflation hollow*, may be formed; this is a shallow depression from which an abundance of fine material has been deflated. Most blowouts are small, but some exceed 1.5 kilometers (1 mile) in diameter. Along with fluvial erosion, deflation is also a factor in the formation of a reg surface.

Abrasion *Aeolian abrasion* is analogous to fluvial abrasion, except that the aeolian variety is much less effective. Whereas deflation is accomplished entirely by air currents, abrasion requires "tools" in the form of airborne sand and dust particles. The wind drives these particles against rock and soil surfaces in a form of natural sandblasting. Wind abrasion does not construct or even significantly shape a landform; it merely sculptures those already in existence. The principal results of aeolian abrasion are the pitting, etching, faceting, and polishing of exposed rock surfaces and the further fragmenting of rock fragments. Rocks so faceted by wind sandblasting are called **ventifacts** (Figure 18-17).

▲ **Figure 18-17** A sand-blasted rock, or ventifact. This piece of basalt is in Death Valley, near Badwater.

Aeolian Transportation

Wind Transportation of Sediment

Rock materials are transported by wind in much the same fashion as they are moved by water, but less effectively. The finest particles are carried in suspension as dust. Strong, turbulent winds can lift and carry thousands of tons of suspended dust. Some dust storms extend for hundreds of meters above Earth's surface and may move material through more than 1600 kilometers (1000 miles) of horizontal distance.

Particles larger than dust are moved by wind through *saltation* and *traction*, just as in streamflow (Figure 18-18). Wind is unable to lift particles larger than the size of sand grains, and even these are likely to be carried less than one meter above the surface. Indeed, most sand, even when propelled by a strong wind, leaps along in the low, curved trajectory typical of saltation, striking the ground at a low angle and bouncing onward. Larger particles move by traction, being rolled or pushed along the ground by the wind. It is estimated that three-fourths of the total volume of all wind-moved material in dry lands is shifted by saltation and traction, particularly the former. At the same time, the entire surface layer of sand moves slowly downwind as a result of the frequent impact of the saltating grains; this process is called *creep*, but it should not be confused with soil creep.

Because the wind can lift particles only so high, a true sandstorm is a cloud of horizontally moving sand that extends for only a few centimeters or feet above the ground surface. Persons standing in its path have their legs peppered by sand grains, but their heads are most likely above the sand cloud. The abrasive impact of a sandstorm, while having little erosive effect on the terrain, may be quite significant for the works of humans near ground level; unprotected wooden poles and posts can be rapidly cut down by the sandblasting, and cars traveling through a sandstorm are likely to suffer etched windshields and chipped paint.

Aeolian Deposition

Desert Sand Dunes

Sand and dust moved by the wind are eventually deposited when the wind dies down. The finer material, which may be carried long distances, is usually laid down as a thin coating of silt and has little or no landform significance. The coarser sand, however, is normally deposited locally. Sometimes it is spread across the landscape as an amorphous sheet called a *sandplain*. The most notable of all aeolian deposits, however, is the **sand dune**, in which loose, windblown sand is heaped into a mound or low hill.

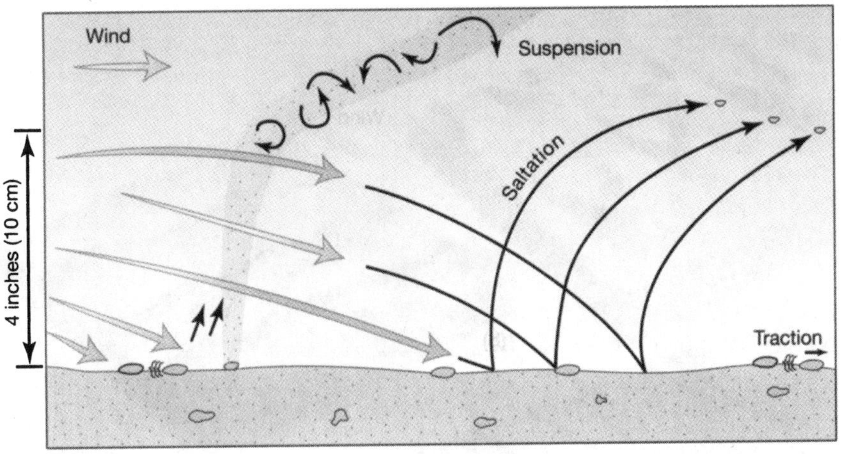

◀ **Figure 18-18** Wind carries tiny dust particles in suspension; larger particles are moved by saltation and traction.

Desert Sand Dunes In some instances, dunefields are composed entirely of unanchored sand that is mostly uniform grains of quartz (occasionally gypsum, rarely some other minerals) and usually brownish-gray colored, although sometimes a brilliant white. Unanchored dunes are deformable obstructions to airflow. Because they are unanchored, they can move, divide, grow, or shrink. They develop sheltered air pockets on their leeward sides that slow down and baffle the wind, so that deposition is promoted there.

Unanchored dunes are normally moved by local winds. The wind erodes the windward slope of the dune, forcing the sand grains up and over the crest to be deposited on the steeper leeward side, or **slip face** (Figure 18-19). The slip face of a dune typically maintains an angle of 32° to 34°—the angle of repose of dry sand. If the wind prevails from one direction for many days, the dune may migrate downwind without changing shape. Such migration is usually slow, but in some cases dunes can move tens of meters in a year.

Not all dunes are unanchored, however, and another characteristic dunefield arrangement is one in which the dunes are mostly or entirely anchored and therefore no longer shifting with the wind. Various agents can anchor dunes, the primary one being vegetation. Dunes provide little nourishment or moisture for plant growth, but desert vegetation is remarkably hardy and persistent and often able to survive in a dune environment. Where vegetation manages to gain a foothold, it may proliferate and anchor the dunes.

Dune patterns are almost infinite in their variety. Several characteristic dune forms are widespread in the world's deserts, their configuration largely a consequence of the relative amount of sand present and the persistence of the wind direction. Here we consider four of the most common:

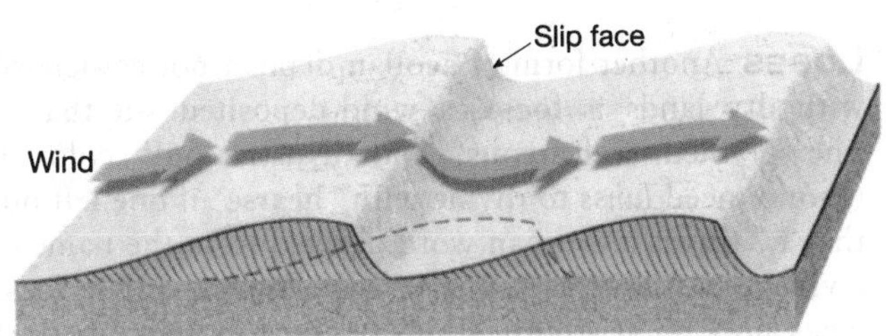

▲ **Figure 18-19** Sand dunes migrate downwind as sand grains move up the gentle windward slope and are deposited on the steep slip face.

1. Best known of all dune forms is the **barchan**, which usually occurs as an individual dune migrating across a nonsandy surface, although barchans may also be found in groups. A barchan is crescent-shaped, with the horns of the crescent pointing downwind (Figures 18-20a and 18-21). Sand movement in a barchan is not only over the crest, from windward side to slip face, but also around the edges of the crescent to extend the horns. Barchans form where strong winds blow consistently from one direction. They tend to be the fastest moving of all dunes and are found in all deserts except those of Australia. They are most widespread in the deserts of central Asia (Thar and Takla Makan) and in parts of the Sahara.
2. **Transverse dunes**, which are also crescent-shaped but less uniformly so than barchans (Figure 18-20b), occur where the supply of sand is much greater than that found in locations that have barchans; normally the entire landscape leading to transverse dune formation is sand covered. As with a barchan, the convex side of a transverse dune faces the prevailing wind direction. In a formation of transverse dunes, all the crests are perpendicular to the wind direction, and the dunes are aligned in parallel waves across the land. They migrate downwind just as barchans do, and if the sand supply decreases, they are likely to break up into barchans.
3. **Seifs** are a type of linear or *longitudinal dune*. They are long, narrow dunes that usually occur in multiplicity and in a generally parallel arrangement (Figures 18-20c and 18-22). They are typically a few dozen to a few hundred meters high, a few tens of meters wide, and kilometers or even tens of kilometers long. Their lengthy, parallel orientation apparently represents an intermediate direction between two dominant wind directions—blowing from one direction part of the year, and from another direction the rest of the year. Seifs are

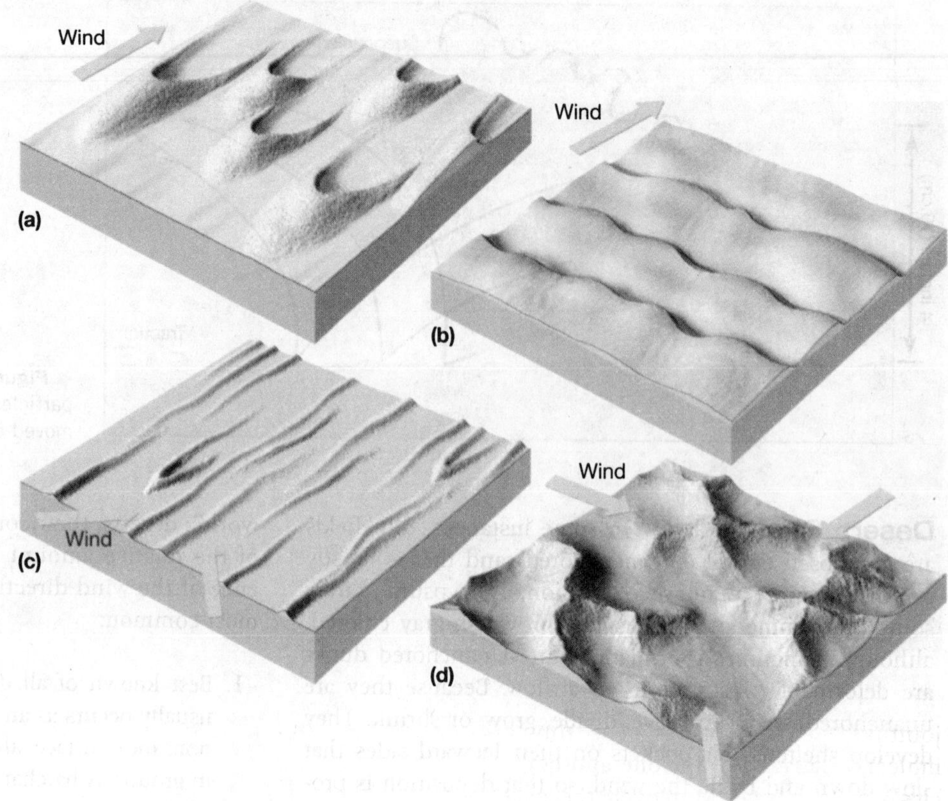

▶ **Figure 18-20** Common desert sand dune types: (a) barchans and (b) transverse dunes develop where there is a consistent wind direction; (c) seifs or longitudinal dunes generally develop where the wind blows from one direction part of the year, and another direction the rest of the year; (d) star dunes develop where the wind direction varies throughout the year.

rare in American deserts but may be the most common dune forms in other parts of the world.

4. **Star dunes** are large pyramid-shaped dunes with arms radiating out in three or more directions (Figure 18-20d). Star dunes develop in areas where the wind frequently varies in direction.

▲ **Figure 18-21** The characteristic crescent shape of barchan dunes in Namibia's Namib-Naukluft National Park. The dominant wind direction is from left to right.

In some locations what are sometimes called "fossil" sand dunes are found. For example, in some regions of the southwestern United States vast deposits of sandstone exhibit the characteristic "cross-bedding" of wind-deposited sand, rather than the more typical horizontal strata of sediments accumulated in bodies of water (Figure 18-23).

Coastal Dunes Winds are also active in dune formation along many stretches of ocean and lake coasts, whether the climate is dry or otherwise. On almost all flattish coastlines, ocean waves deposit sand along the beach. A prominent onshore wind can blow some of the sand inland, often forming dunes. In areas where vegetation becomes established on the sand, *parabolic dunes* develop—they look much like a barchan dune, but with the "horns" pointing into the wind offshore. Most coastal dune aggregations are small, but they sometimes cover extensive areas. The largest area is probably along the Atlantic coastline of southern France, where dunes extend for 240 kilometers (150 miles) along the shore and reach inland for 3 to 10 kilometers (2 to 6 miles).

Loess Another form of aeolian deposit not associated with dry lands is **loess**, a wind-deposited silt that is fine grained, calcareous, and usually buff colored. (Pronounced *luhss* to rhyme with "hearse" if one left out the "r," this is a German word derived from the name of a village in Alsace.) Despite its depositional origin, loess lacks horizontal stratification. Perhaps its most distinctive characteristic is its great vertical durability, which results from its fine grain size, high porosity, and vertical

◀ **Figure 18-22** The parallel linearity of seifs is characteristic of many desert areas. This scene is from the Simpson Desert of central Australia.

jointlike cleavage planes. The tiny grains have great molecular attraction for one another, making the particles very cohesive. Moreover, the particles are angular, which increases porosity. Thus, loess accepts and holds large amounts of water. Although relatively soft and unconsolidated, when exposed to erosion loess maintains almost vertical slopes because of its structural characteristics, as though it were firmly cemented rock (Figure 18-24). Prominent bluffs are often produced as erosional surfaces in loess deposits.

The formational history of loess is complex, although much of the silt was produced in association with Pleistocene glaciation (discussed in Chapter 19). During both glacial and interglacial periods, rivers carried large amounts of debris-laden meltwater from the glaciers, producing many broad floodplains. During periods of low water, winds whipped the smaller, dust-sized particles from the floodplains and dropped them in some places in very thick deposits. Some loess also seems to have been generated by deflation of dust from desert areas, especially in central Asia.

Most deposits of loess are in the midlatitudes, where some are very extensive, particularly in the United States, Russia, China, and Argentina (Figure 18-25). Indeed, some 10 percent of Earth's land surface is covered with loess, and in the conterminous United States the total approaches 30 percent. Loess deposits provide fertile possibilities for agriculture, for they serve as the parent material for some of the world's most productive soils, especially for growing grain.

▲ **Figure 18-23** Cross-bedded sandstone in "fossil" sand dunes near Zion National Park in Utah.

▲ **Figure 18-24** Loess has a remarkable capability for standing in vertical cliffs, as seen in this road cut near Maryville, Missouri.

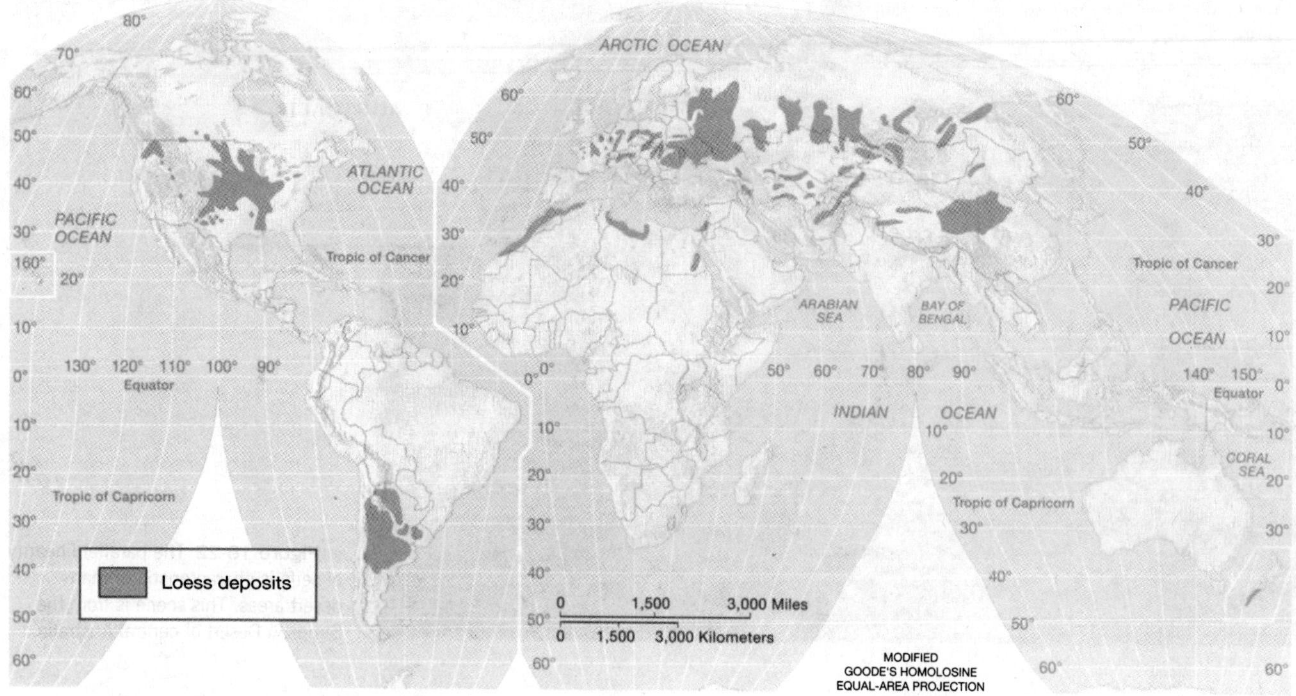

▲ **Figure 18-25** Major loess deposits of the world.

The loess areas have been particularly significant in China because of their agricultural productivity. The Yellow River (Hwang Ho) received its name from the vast amount of buff-colored sediment it carries, as did its destination, the Yellow Sea. Also, numerous cave dwellings have been excavated in the Chinese loess because of its remarkable capability for standing in vertical walls. Unfortunately, however, this region is also prone to earthquakes, and the cave homes collapse readily when tremors occur; thus, some of the world's greatest earthquake disasters in terms of loss of life have taken place there.

Two Representative Desert Landform Assemblages

Surface features vary considerably from one desert to the next. Overall, however, the most common desert landform is a mountain or mountain range, and the most frequently recurring desert profile is a mountain flanked by plains.

Two particular assemblages of landforms are the most common in the deserts of the United States: basin-and-range terrain and mesa-and-scarp terrain. Their pattern of development is repeated time and again over thousands of square kilometers of the American Southwest (Figure 18-26). Although desert landscapes vary greatly around the world, these two landform assemblages are good representative examples of the outcome of the special conditions and the external desert landform-shaping processes we have just described.

Basin-and-Range Terrain

As Figure 18-26 shows, most of the southwestern interior of the United States is characterized by basin-and-range topography. This is a land largely without external drainage, with only a few exotic rivers (notably the Colorado and Rio Grande) flowing through or out of the region. This region of North America has undergone

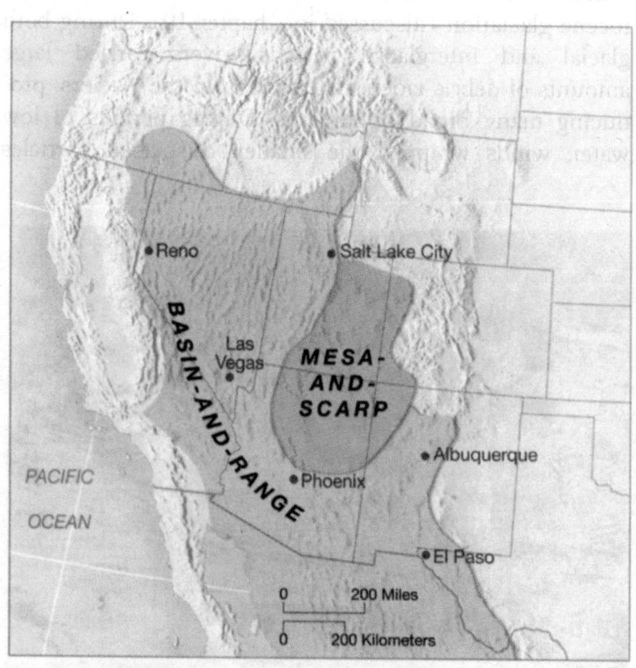

▲ **Figure 18-26** The southwestern interior of the United States contains two principal assemblages of landforms: basin-and-range and mesa-and-scarp.

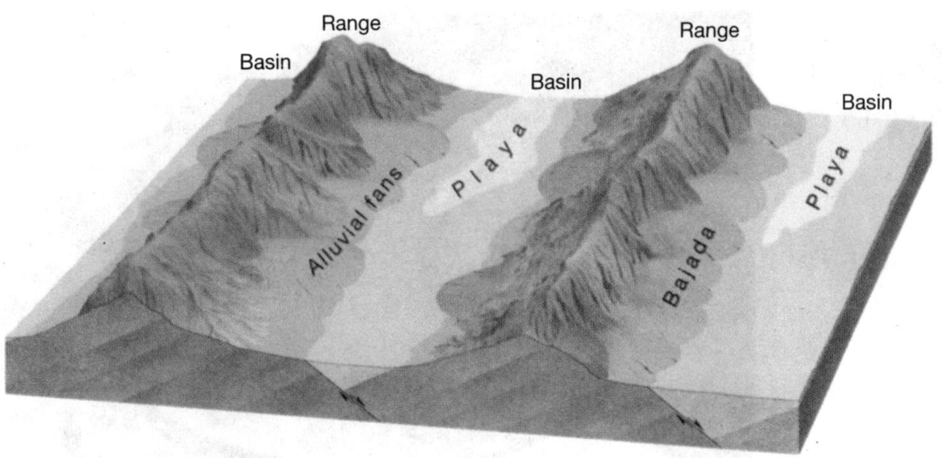

◀ **Figure 18-27** A typical basin-and-range desert landscape.

extensive normal faulting—including movement along low-angle detachment faults—leaving a landscape consisting of numerous fault-block mountain ranges surrounding a series of interior drainage basins, including many down-dropped grabens and down-tilted "half-grabens" (see Figure 14-59).

Basin-and-range terrain has three principal features: ranges, piedmont zones, and basins (Figure 18-27).

The Ranges If we stand in the basin of any basin-and-range landscape, the mountain ranges dominate the horizon in all directions. Some of the ranges are high and some quite low, but the prevalence of steep and rocky slopes presents an aura of ruggedness. Although the tectonic origins of these mountains vary (most have been tilted by faulting, but others were formed by folding, by volcanism, or in a more complex fashion), their surface features have been largely shaped by weathering, mass wasting, and fluvial processes.

Ridge crests and peaks are usually sharp, steep cliffs are common, and rocky outcrops protrude at all elevations. The mountain ranges of a basin-and-range formation are usually long, narrow, and parallel to one another. Most of them are seamed by numerous gullies, gorges, and canyons that rarely have flowing streams. These dry drainage channels are usually narrow and steep sided and have a V-shaped cross section. Typically the channel bottoms are filled with sand and other loose debris.

In some areas, the ranges may have been eroding for a long time, but in other places the ranges are just a few million years old, and bare rock outcrops are still prominent. If the range stands in isolation and the alluvial plains and basins roundabout are extensive, the term inselberg is applied to the mountain remnant (see Figure 18-12).

The Piedmont Zone At the base of the ranges, there is usually a sharp break in slope that marks the change from range to basin floor (see Figure 18-13). This piedmont zone is a transition area from the steep slopes of the ranges to the near-flatness of the basins.

Much of the piedmont zone may be underlain by a pediment, although the pediment is rarely visible. It is normally covered with several meters of unconsolidated sediment because the piedmont zone is an area of fluvial deposition. During the occasional rainfall, flash floods and debris flows come roaring out of the gullies and gorges of the surrounding ranges, heavily laden with sedimentary material. As they burst out of the mouths of the confining canyons onto the piedmont zone, their speed and load capacity drop abruptly. Significant deposition is the result.

One of the most prominent and widespread topographic features to be found in any desert area is the **alluvial fan**, particularly characteristic of the basin-and-range region (although a dominant feature in many deserts, alluvial fans also can develop in humid areas). As a stream leaves the narrow confines of a mountain gorge and emerges onto the open piedmont zone, it abruptly loses both capacity and competence and breaks into distributaries that wend their way down the piedmont slope, sometimes cutting shallow new channels in the loose alluvium but frequently depositing more debris atop the old (Figure 18-28). Channels become choked and overflow, developing new ones. In this fashion, a moderately sloping, fan-shaped landform is constructed at the mouth of the canyon. When one part of the fan is built up, the channeled flow shifts to another section and builds that up. This means that the entire fan is eventually covered more or less symmetrically with alluvium. As deposition continues, the fan is extended outward across the piedmont zone and onto the basin floor.

The material in an alluvial fan is not well sorted. Instead, the fan may have a heterogeneous mixture of particle sizes because volumes of water flow vary, often considerably, from year to year. Thus, there is not the neat sorting that occurs in a delta. In general, however, large boulders are dropped near the apex of the fan and finer material around the margins, with a considerable mixture of particle sizes throughout. The dry drainage channels across the fan surface frequently shift their positions as well as their balance between erosion and deposition.

As alluvial fans become larger, neighboring ones often overlap and are called *coalescing alluvial fans*. Continued growth and more complete overlap may eventually result in a continuous alluvial surface all across the piedmont zone, in which case it is difficult to distinguish between

▶ **Figure 18-28** (a) An alluvial fan develops at the mouth of a canyon. (b) Alluvial fan in Death Valley, California.

individual fans. This feature is known as a **bajada**. Near the mountain front, a bajada surface is undulating, with convex sections near the canyon mouths and concave sections in the overlap areas between the canyons. In the portion of the bajada away from the range and out on the basin floor, however, no undulations occur because the component fans have coalesced so thoroughly.

The Basin Beyond the mountain front is the flattish floor of the basin, which has a very gentle slope from all sides toward some low point. A playa is usually found in this low point (see Figure 18-8). Drainage channels across the basin floor are sometimes clear cut, but more often shallow and ill defined, frequently disappearing before reaching the low point. This low point thus functions as the theoretical drainage terminal for all overland and streamflow from the near sides of the surrounding ranges, but only sometimes does much water reach it. Most is lost by evaporation and seepage long before.

Salt accumulations are commonplace on the playa surrounding the low point of a desert basin because of all the water-soluble minerals washed out of the surrounding watershed. Once out of the mountains and in the basin, the water evaporates or seeps away, but the salts cannot evaporate and are only marginally involved in seepage. There is usually sufficient water to allow flow across the outer rim of the basin floor and into the playa area, however, and so salts become increasingly concentrated in the playa, which is then more properly called a salina. The presence of the salt usually gives a brilliant whitish color to the playa or salina surface. Many different salts can be involved, and their accumulations are sometimes large enough to support a prosperous mining enterprise.

In the rare occasions when water does flow into a playa, the formation becomes a playa lake. Such lakes may be extensive, but they are usually very shallow and normally persist for only a few days or weeks. Saline lakes are marked by clear water and a salty froth around the edges, whereas shallow freshwater lakes have muddy water because they lack salt to make the silt settle.

The basin floor is covered with very fine-grained material because the contributory streams are too weak to transport large particles. Silt and sand predominate and sometimes accumulate to remarkable depths. Indeed, the normal denudation processes in basin-and-range country tend to raise the floor of the basins. Debris from the surrounding ranges has nowhere to go but the basin of interior drainage. Thus, as the mountains are being worn down, the basin is gradually filling up.

The fine material of basin floors is very susceptible to the wind, with the result that small concentrations of sand dunes are often found in some corner of the basin. Free-moving sand is rarely found in the center of the basin because winds push it to one side or another.

Death Valley: A Remarkable Example of Basin-and-Range Terrain California's Death Valley is a vast topographic museum—a veritable primer of basin-and-range terrain. Located in east-central California, close to the Nevada border, the "valley" is actually a down-dropped basin—the result of the crustal extension taking place throughout the basin-and range—with extensive and complex fault zones both east and west of the valley floor (Figure 18-29). The trough is about 225 kilometers (140 miles) long, in a general northwest–southeast orientation; its width ranges from 6 to 26 kilometers (4 to 16 miles). Not a classic graben, the basin floor of Death Valley has tilted down more along its eastern side than its western side—created at least in part as a "pull-apart" basin, formed where land drops down between two parallel strike-slip faults. The downfaulting has been so pronounced that nearly 1425 square kilometers

▲ **Figure 18-29** The setting of Death Valley. The mountains and basins are a consequence of extensional faulting. Alluvial fans and bajadas are found all along the mountain fronts.

(550 square miles) of the valley floor is below sea level, reaching a depth of 86 meters below sea level (−282 feet). Lengthy, tilted fault-block mountain ranges border the valley on either side. The Panamint Range on the west is the most prominent; the high point at Telescope Peak 3368 meters or 11,049 feet above sea level is only 29 kilometers (18 miles) due west from the low point of the valley. The Amargosa Range on the east is equally steep and rugged, but a bit lower overall.

The most conspicuous topography associated with Death Valley is the surrounding mountains. The ranges have all the characteristic features of desert mountains, being rugged, rocky, and generally barren. Their erosional slopes are steep and their escarpments steeper. The canyons that seam the ranges are invariably deep, narrow, V-shaped gorges. Many of those in the Amargosa Range and some in the Panamints are *wineglass canyons*. The cup of the glass is the open area of dispersed headwater tributaries high in the range, the stem is the narrow gorge cut through the mountain front, and the base is the fan that opens out onto the piedmont zone.

The piedmont zone at the foot of the Panamints and the Amargosas is almost completely alluviated in one of the most extensive fan complexes imaginable. Every canyon mouth is the apex either of a fan or a fan-shaped debris flow deposit, and most of the fans overlap with neighbors to the north and south. The fans on the western side of the valley (those formed by debris from the Panamints) are much more extensive than those on the eastern side. For the most part, the Panamint fans are thoroughly coalesced into a conspicuous bajada that averages about 8 kilometers (5 miles) wide, with the outer margin of the bajada as much as 610 meters (2000 feet) lower than the canyon mouths.

The Amargosa fans are much smaller, primarily because faulting has tilted down the eastern side basin floor so that the thickest sediments are nestled close to the base of the Amargosa Range. Thus, the west-side fans have been able to extend outward onto the valley floor, whereas tilting has reduced the size of the east-side fans by creating shorter slopes (and therefore, smaller drainage basins to supply sediment), and by facilitating their partial burial by valley-floor deposits. The Amargosa fans, then, are mostly short, steep, and discrete so there is no bajada on the eastern side of Death Valley, although some of the fans do overlap with one another.

The floor of Death Valley is also of great topographic interest, although its flatness makes the features less easy to see and understand. The valley is filled with an incredible depth of alluvium, most of which has been washed down from the surrounding mountains; in places the fill is estimated to consist of 900 meters (3000 feet) of young alluvium resting atop another 1800 meters (6000 feet) of Tertiary sediment. The surface of the valley floor has little relief and slopes gently toward the low point near Badwater, a permanent saltwater pond. Drainage channels appear irregularly on the valley floor, trending toward Badwater. In some places, distinct braided channels appear; in other locations, the channels disappear in sand or playa.

There are several extensive crusty-white *salt pans*, particularly in the middle of the valley (Figure 18-30), along with several sand accumulations, with one area of mobile dunes covering 36 square kilometers (14 square miles).

During the most recent ice age, Death Valley was occupied by an immense lake. Lake Manly, as it is now called, was more than 160 kilometers (100 miles) long and 180 meters (600 feet) deep. It was fed by three rivers that flowed into the valley from the west, carrying meltwater from Sierra

▲ **Figure 18-30** The Death Valley "salt pan" here consists mostly of ordinary table salt (NaCl). The polygonal ridges form as salt crystals grow through the evaporation of water from a salty "slush" just below the surface.

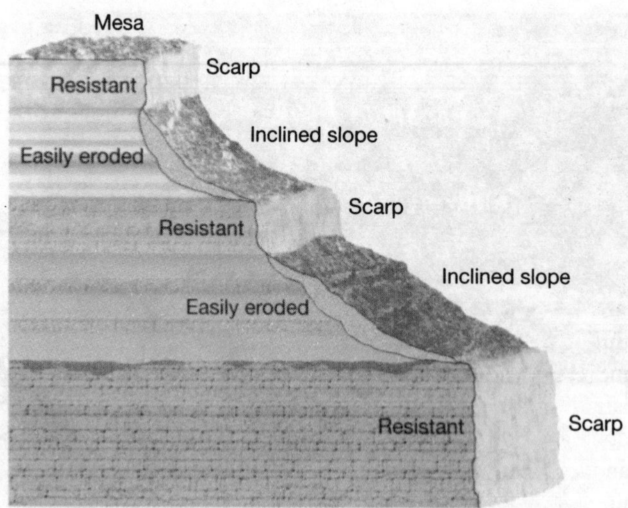

▲ **Figure 18-32** Cross section of a mesa-and-scarp formation. Differential weathering and erosion shows up prominently: The resistant strata weather and erode into mesas or scarps, whereas the more easily eroded strata yield gentler inclined slopes.

Nevada glaciers. As the climate became drier and warmer, the lake eventually disappeared through evaporation and seepage, but traces of its various shoreline levels can still be seen at several places on the lower slopes. Much of the salt accumulated in the valley is due to the evaporating waters of Lake Manly as well as lakes that existed more recently.

Mesa-and-Scarp Terrain

The other major landform assemblage of the American Southwest is mesa-and-scarp terrain. It is most prominent in Four Corners country—the place where Colorado, Utah, Arizona, and New Mexico come together (Figure 18-31). **Mesa** is Spanish for "table" and implies a flat-topped surface. *Scarp* is short for "escarpment" and pertains to steep, more or less vertical cliffs.

Mesa-and-scarp terrain is normally associated with horizontal sedimentary strata—shown off prominently in this arid region because it lacks a continuous cover of trees and vegetation. Such strata invariably offer different degrees of resistance to weathering and erosion, and so abrupt changes in slope angle are characteristic of this terrain. The most resistant layers, typically limestone or sandstone, often play a dual role in the development of a mesa-and-scarp terrain. They form an extensive caprock, which becomes the mesa; and at the eroded edge of the caprock, the hard layer protects underlying strata and produces an escarpment. Thus, it is the resistant layers that are responsible for both elements of slope (mesa and scarp) that describe this terrain type.

Often mesa-and-scarp topography has a broad and irregular stair-step pattern. Figure 18-32 shows a cross section

◀ **Figure 18-31** The stair-step pattern of mesa-and-scarp terrain is shown on an imposing scale in Arizona's Grand Canyon.

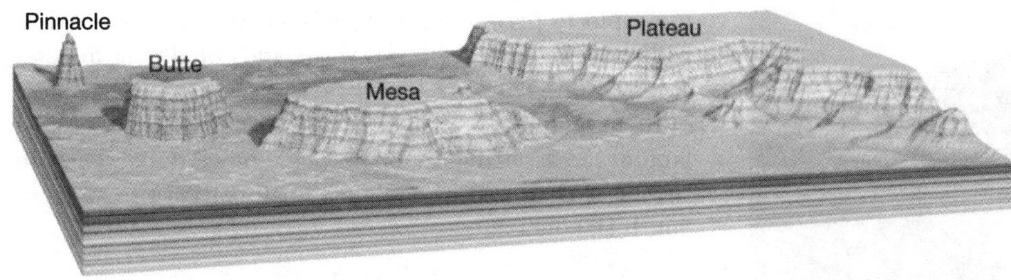

◀ **Figure 18-33** Typical development of residual landforms in horizontal sedimentary strata with a hard caprock. With the passage of time, larger features are eroded into smaller features.

through a typical formation. An extensive, flat platform (the mesa) in the topmost resistant layer of a sedimentary accumulation terminates in an escarpment (the scarp) that extends downward to the bottom of the resistant layer(s). From here, another slope, steep but not as steep as the escarpment, continues down through softer strata. This inclined slope extends downward as far as the next resistant layer, which forms either another escarpment or another mesa ending in an escarpment.

The top platforms are properly referred to as **plateaus** if they are bounded on one or more sides by a prominent escarpment. If a scarp edge is absent or relatively inconspicuous, the platform is called a *stripped plain*.

The escarpment edge is worn back by weathering, mass wasting, and fluvial erosion. The cliffs retreat, maintaining their perpendicular faces, as they are undermined by the more rapid removal of the less resistant strata (often shale) beneath the caprock. Much of the undermining is accomplished by a process called *sapping*, in which groundwater seeps and trickles out of the scarp face, eroding fine particles and weakening the cohesion of the face. When thus undermined, blocks of the caprock break off, usually along vertical joint lines. Throughout this process, the harder rocks are the cliff-formers, and the less resistant beds develop more gently inclined slopes. Talus often accumulates at the base of the slope.

Although the term mesa is applied generally to many flat surfaces in dry environments, it properly refers to a particular landform: a flat-topped, steep-sided hill with a limited summit area. It is a remnant of a formerly more extensive surface, most of which has been worn away (Figure 18-33). Sometimes it stands in splendid isolation as a final remnant in an area where most of the previous surface has been removed, but more commonly it occurs as an outlying mass not very distant from the retreating escarpment face to which it was once connected. A mesa is invariably capped by some sort of resistant material that helps keep the summit flat even as the bulk of the rock mass is reduced by continuing mass wasting and erosion of its rimming cliffs.

A related but smaller topographic feature is the **butte**, an erosional remnant having a very small surface area and cliffs that rise conspicuously above their surroundings. Some buttes have other origins, but most are formed by the mass wasting of mesas (Figure 18-34). With further denudation, a still smaller residual feature, usually referred to as a **pinnacle** or *pillar*, may be all that is left—a final spire of resistant caprock protecting weaker underlying beds. Buttes, mesas, and pinnacles are typically found not far from some retreating escarpment face (see Figure 18-33).

▲ **Figure 18-34** The spectacular starkness of arid-land topography is demonstrated dramatically in the view of the Mitten Buttes in Arizona's Monument Valley.

▲ **Figure 18-35** Badlands are characterized by innumerable ravines and gullies dissecting the land and forming a maze of low but very steep slopes. This scene is from Zabriskie Point, in Death Valley, California.

Badlands One of the most striking topographic features of arid and semiarid regions is the intricately rilled and barren terrain known as **badlands** (Figure 18-35). In areas underlain by horizontal strata of shale and other clay formations that are poorly consolidated, overland flow after the occasional rains is an extremely effective erosive agent. Innumerable tiny rills that develop over the surface evolve into ravines and gullies that dissect the land in an extraordinarily detailed manner. A maze of short but very steep slopes is etched in a filigree of rills, gullies, and gorges, with a great many ridges, ledges, and other erosional remnants scattered throughout. Erosion is too rapid to permit soil to form or plants to grow, and so badlands are barren, lifeless wastelands of almost impassable terrain. They are found in scattered locations (most of them mercifully small) in every western state, the most famous areas being in Bryce Canyon National Park in southern Utah and in Badlands National Park in western South Dakota.

Arches and Natural Bridges Mesa-and-scarp terrain is also famous for numerous minor erosional features, most produced by a combination of weathering, mass wasting, and fluvial erosion. These features are not confined to arid regions, but the mesa-and-scarp region has many examples. An *arch* (Figure 18-36) can form when the lower portions of a narrow "fin" of sedimentary rock—formed by prominent, closely-spaced vertical joints—weaken and collapse, leaving an arch of more resistant rock above. A *natural bridge* can form anytime the rock over which water flows changes from an erosion-resistant type to a less resistant type. One place a natural bridge frequently forms is where an entrenched meander wears away the rock in a narrow neck between meander loops (see Figure 16-44 in Chapter 16).

Pedestals and pillars, sometimes larger at the top than at the bottom, rise abruptly above their surroundings, their caps resistant material but their narrow bases continuously weathered by rainwater trickling down the surface. This water dissolves the cementing material that holds the sand grains together, and the loosened grains are easily blown or washed away.

One other notable characteristic of mesa-and-scarp terrain is vivid colors. The sedimentary outcrops and sandy debris of these regions are often resplendent in various shades of red, brown, yellow, and gray, due mostly to iron compounds.

▲ **Figure 18-36** Delicate Arch in Utah's Arches National Park.

CHAPTER 18 LEARNING REVIEW

After studying this chapter, you should be able to answer the following questions. Key terms from each text section are shown in bold type. Definitions for key terms are also found in the glossary at the back of the book.

A Specialized Environment (p. 481)

1. List several ways in which topographic development in arid lands is different from that in humid regions.
2. What is meant by an impermeable surface and how does such a surface influence the results of rainfall in a desert?
3. What is a basin of interior drainage?

Running Water in Waterless Regions (p. 483)

4. What is the difference between an **ephemeral stream** and an **exotic stream** in a desert?
5. Although there is very little rainfall in deserts, running water is still the most important process of erosion and deposition in arid environments. Describe and explain at least two special conditions in deserts that tend to increase the likelihood of fluvial erosion whenever it does rain.
6. What is a **playa** and why does one form?
7. What is the difference between a playa and a **salina**?
8. Why are playas so flat and level?
9. Why is a desert lake in a basin of interior drainage likely to be a **saline lake**?
10. Explain the concept of **differential weathering and erosion**.
11. Describe the formation of an **inselberg**.
12. What is the difference between **pediment** and **piedmont zone**?
13. Why are depositional features of alluvium so prominent in many desert regions?

Characteristic Desert Landscape Surfaces (p. 488)

14. Distinguish among **erg**, **reg**, and **hamada**.
15. Describe and explain the formation of **desert pavement** and **desert varnish**.

The Work of the Wind (p. 489)

16. Contrast the **aeolian processes** of **deflation** and abrasion.
17. How does a **blowout** form?
18. What is a **ventifact** and how does one form?
19. Overall, how important is wind in the erosion of desert landforms?
20. Describe and explain the general cross section of an unanchored desert **sand dune**. Be sure to contrast the windward side with the **slip face** of the dune.
21. Explain the general shape and movement of **barchan** sand dunes.
22. How are **transverse dunes** different from **seif** dunes?
23. Under what circumstances do **star dunes** form?
24. Most **loess** is not found in arid regions; why is it discussed in this chapter?

Two Representative Desert Landform Assemblages (p. 494)

25. What is an **alluvial fan** and why are they so prevalent in the basin-and-range desert area of North America?
26. How does an alluvial fan differ from a delta?
27. How is a **bajada** different from an alluvial fan?
28. Why are some playas in the basin-and-range desert so salty?
29. Why are there few deep stream channels cutting across basin floors in the basin-and-range desert?
30. How does a **plateau** differ from a **mesa**?
31. Describe and explain the processes involved that can change a mesa into a **butte** or **pinnacle**.
32. What is distinctive about **badlands** terrain? How did it come to be that way?

Log in to **www.mygeoscienceplace.com** for animations, videos, RSS feeds, maps, and self-study quizzes to enhance your study of **The Topography of Arid Lands**.

19 Glacial Modification of Terrain

In the long history of our planet, ice ages have occurred an unknown number of times. The cause or causes of the climate changes that led to these ice ages are still incompletely understood, a topic we discuss further at the end of the chapter. With one outstanding exception, however, much of the evidence of past glacial periods has been eradicated by subsequent geomorphic events, with the result that only the most recent ice age has influenced contemporary topography. Consequently, when referring to the Ice Age, capitalized, we usually mean this most recent ice age, which is the main feature of the geologic epoch known as the *Pleistocene*, a period that began about 2.5 million years ago and ended about 10,000 years ago.

In this chapter, we are concerned with Pleistocene events both because they significantly modified pre-Pleistocene topography and because their results are so thoroughly imprinted on many parts of the continental terrain today. Glacial processes are still at work, to be sure, but their importance is much less now than it was just a few thousand years ago simply because so much less glacial ice is present today.

The Impact of Glaciers on the Landscape

Wherever glaciers have developed, they have had a significant impact on the landscape simply because moving ice grinds away almost anything in its path: most soil is carried away, and bedrock is polished, scraped, gouged, plucked, and abraded. Moreover, the rock that is picked up is eventually deposited in a new location, further changing the shape of the terrain. In short, preglacial topography is significantly reshaped.

Perhaps 7 percent of all contemporary erosion and transportation of rock debris on the continents is accomplished by glaciers. This is a paltry total in comparison with fluvial erosion, to be sure, but considering how small a land area is covered by glacial ice today, it is clear that glaciers make a respectable contribution to continental denudation.

Glaciation modifies flat landscapes greatly, with the result that postglacial slope, drainage, and surficial material are likely to be totally different from what they were before the glacier passed by. In mountainous areas, the metamorphosis of the landscape may be less complete, but the topography is sharpened, and often deepened and steepened throughout.

Types of Glaciers

A glacier is more than a block of ice filling up a mountain valley or resting on a continental plain; it is a finely tuned environment with a delicately balanced nourishment budget. Although glacial ice behaves in similar fashion wherever it accumulates, its pattern of movement and its effect on topographic shaping can vary considerably depending on the quantity of ice involved and particularly on the environment. These variations are best understood by first considering the different types of glaciers: *mountain glaciers* and *continental ice sheets*.

Mountain Glaciers

In a few high-mountain areas, ice accumulates in an unconfined sheet that may cover a few hundred or few thousand square kilometers, submerging all the underlying topography except perhaps for some protruding pinnacles called *nunataks*. Such **highland icefields** are notable in parts of the high country of western Canada and southern Alaska and on various Arctic islands (particularly Iceland). Their outlets are often tongues of ice that travel down valleys in the mountains and so are called **valley glaciers** (Figure 19-1). If the leading edge of a valley glacier reaches a flat area and

◀ **Figure 19-1** Mountain glaciers in southeastern Alaska near Skagway. A half-dozen small valley glaciers in the high country have united to form the very large Davidson Glacier, which extends off the bottom of the photo.

◀ The McBride Glacier in Muir Inlet, Glacier Bay National Park, Alaska.

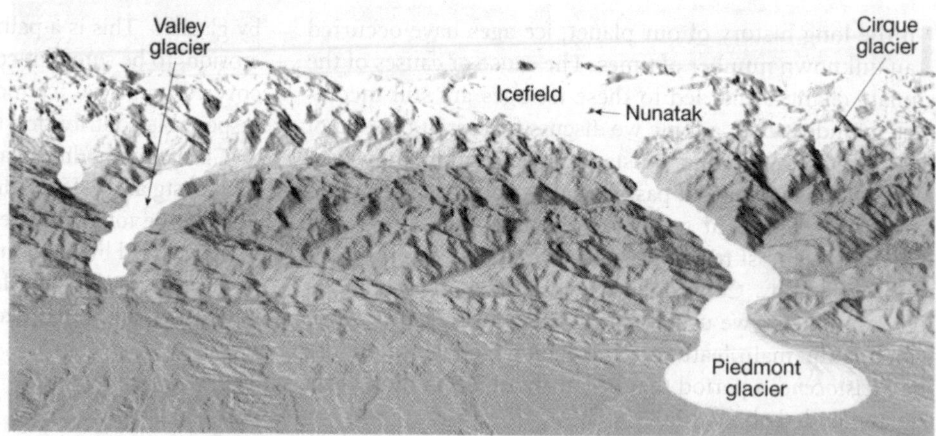

▶ **Figure 19-2** Types of mountain glaciers. Valley and piedmont glaciers can originate in a highland icefield, or when an alpine glacier overflows its cirque and flows downvalley. Nunataks are pinnacles rising above the ice of a highland icefield.

so escapes from the confines of its valley walls, it is called a *piedmont glacier* (Figure 19-2).

Sometimes the term **alpine glacier** is used to describe glaciers that develop individually high in the mountains rather than as part of a broad icefield, usually at the heads of valleys. Very small alpine glaciers confined to the basins where they originate are called **cirque glaciers** (because the basin is called a *cirque*, as we shall see later in the chapter). Normally, however, alpine glaciers spill out of their originating basins and flow downvalley as long, narrow valley glaciers. Occasionally they extend to the mouth of the valley and become piedmont glaciers.

Continental Ice Sheets

Glaciers that form in nonmountainous areas of the continents are called **continental ice sheets**. During the Pleistocene these were vast blankets of ice that completely inundated the underlying terrain to depths of hundreds or thousands of meters. Because of their immense size, ice sheets have been the most significant agents of glaciation across the land surface. Only two true ones exist today, in Antarctica and Greenland (Figure 19-3). The ice in an ice sheet accumulates to great depths in the interior of the sheet but is much thinner at the outer edges. Around the margin of the sheet, some long tongues of ice, called *outlet glaciers*, extend between rimming hills to the sea. In other places, the ice reaches the ocean along a massive front, where it sometimes projects out over the sea as an *ice shelf* (Figure 19-4). As we saw in Chapter 9, great chunks of ice frequently break off, both from the ice shelves and from the ends of outlet glaciers, fall into the sea, and float away. These huge floating ice masses are *icebergs*.

Glaciations Past and Present

The amount of glacial ice on Earth's surface has varied remarkably over the last few million years, with periods of accumulation interspersed with periods of melting and

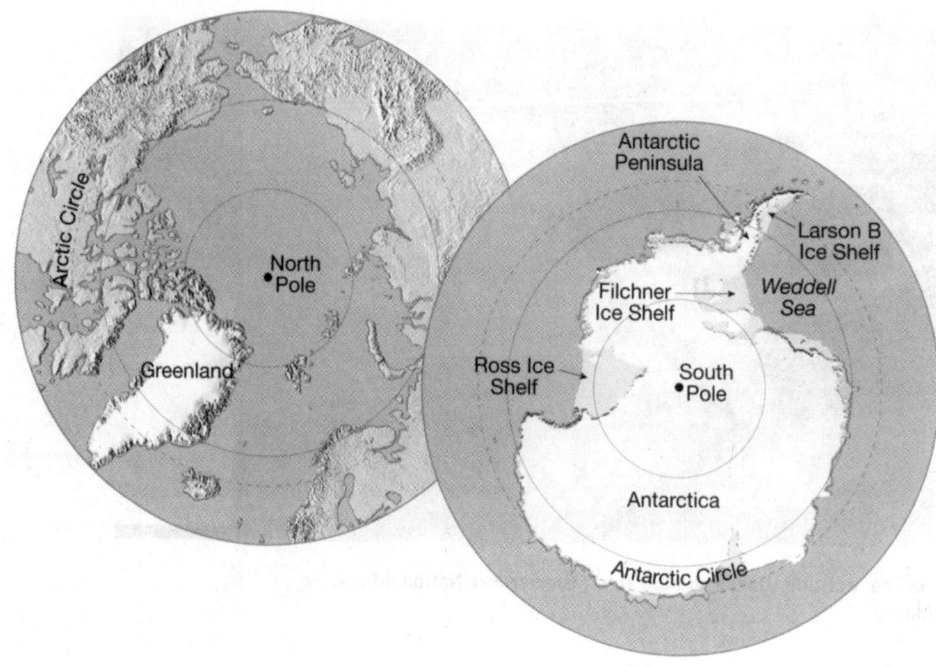

▶ **Figure 19-3** Current continental ice sheets in the Arctic and Antarctic.

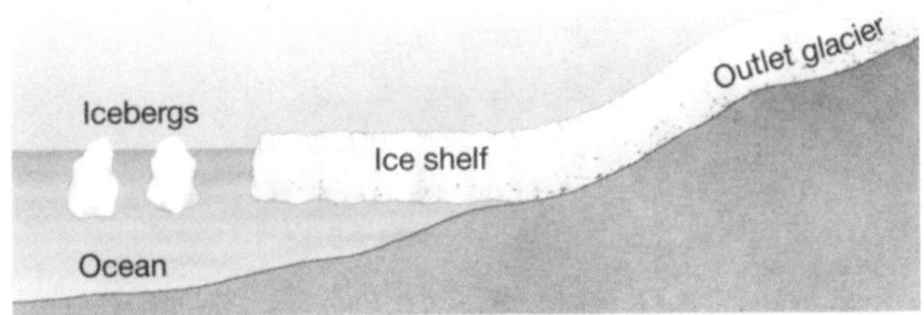

▲ **Figure 19-4** When either an ice sheet or an outlet glacier reaches the ocean, some of the ice may extend out over the water as an ice shelf. Icebergs form when the hanging ice of the shelf breaks off and floats away, a process called calving.

times of ice advance alternating with times of ice retreat. A great deal of secondary evidence was left behind by the moving and melting ice, and scientists have been remarkably insightful in piecing together the chronology of past glaciations. Nevertheless, the record is incomplete and often approximate. As is to be expected, the more recent events are best documented; the further one delves into the past, the murkier the evidence becomes.

Pleistocene Glaciation

1. End of the Last Ice Age
2. Isostasy

The precise boundaries of the **Pleistocene Epoch** (see Figure 13-24) are still subject to debate. Current estimates define the start of the Pleistocene as 2.59 million years ago, but geochronologists now recognize that the glaciations began even earlier. Evidence suggests, for example, that Antarctica was covered by an ice cap similar in size to today's as long ago as perhaps 10 million years. The most recent findings tell us that by the start of the Pleistocene, the "amplitude" of climate fluctuations—from glacial period to interglacial period—had increased and some parts of the Northern Hemisphere were covered by glaciers.

New evidence has also changed the date of the close of the Pleistocene. Although the end of the Pleistocene is now set at 11,700 years ago, evidence suggests that some of the Pleistocene glaciers were still retreating as recently as 9000 years ago. Even this most recent estimate for the close date cannot be cast in stone, however, because the Ice Age may not yet have ended at all, a possibility we consider later in the chapter. For now, let us just say that, to the best of present knowledge, the Pleistocene Epoch occupied almost all of the most recent two and a half million years of Earth's history.

The dominant environmental characteristic of the Pleistocene was the cooling of high-latitude and high-elevation areas, so that a vast amount of ice accumulated in many places. However, the epoch was by no means universally icy. During several lengthy periods, most or all of the ice melted, only to be followed by intervals of ice accumulation. In broad terms, the Pleistocene consisted of an alternation of *glacial* periods (times of ice accumulation) and *interglacial* periods (times of ice retreat). Current evidence suggests as many as 18 or 19 glacial episodes took place during the Pleistocene.

The end of the Pleistocene Epoch coincided with the conclusion of what is known in North America as the "Wisconsin" glacial stage (known as the "Würm" in the Alps), approximately 11,000 years ago. The period since then is identified as the *Holocene Epoch*. Conceptually, then, the Holocene is either a postglacial epoch or the latest in a series of interglacial interludes.

At its maximum Pleistocene extent, ice covered one-third of the total land area of Earth—nearly 47,000,000 square kilometers (19 million square miles), as Figure 19-5 shows. Ice thickness varied and can be estimated only roughly, but we do know that in some areas it reached a depth of several thousands of meters.

The greatest total area of ice-covered land was in North America. The *Laurentide* ice, which covered most of Canada and a considerable portion of the northeastern United States, was the most extensive Pleistocene ice mass; its area was slightly larger than that of the present glacier covering Antarctica. It extended southward into the United States to approximately the present location of Long Island, the Ohio River, and the Missouri River. Most of western Canada and much of Alaska were covered by an interconnecting network of smaller glaciers. For reasons we do not fully understand, however, a small area in northwestern Canada as well as extensive portions of northern and western Alaska were never glaciated during the Pleistocene. Moreover, a small area (29,000 square kilometers, 11,200 square miles) in southwestern Wisconsin and parts of three adjoining states was also left uncovered, as Figure 19-5b shows. This area, referred to as the *Driftless Area*, apparently never existed as an island completely surrounded by ice; rather, ice encroached first on one side and then on another side during different glacial advances.

More than half of Europe was overlain by ice during the Pleistocene (Figure 19-5c). Asia was less extensively covered, presumably because in much of its subarctic portion there was not enough precipitation for the ice to last. Nevertheless, ice covered much of Siberia, and extensive glaciation occurred in most Asian mountain ranges.

In the Southern Hemisphere, Antarctic ice was only slightly more extensive than it is today, a large ice complex covered southernmost South America, and the South Island of New Zealand was largely covered with ice. Other high mountain areas all over the world—in central Africa, New Guinea, Hawai'i—experienced more limited glaciation.

The accumulation of ice and the movement and melting of the resulting glaciers had an enormous effect on topography and drainage, a topic we discuss in detail later in the chapter. In addition, however, there were several indirect effects of Pleistocene glaciations:

1. **Periglacial processes:** Beyond the outermost extent of ice advance is an area of indefinite size called the **periglacial zone**, which was never touched by glacial ice but where indirect influence of the ice was felt. The most important periglacial processes were the erosion and

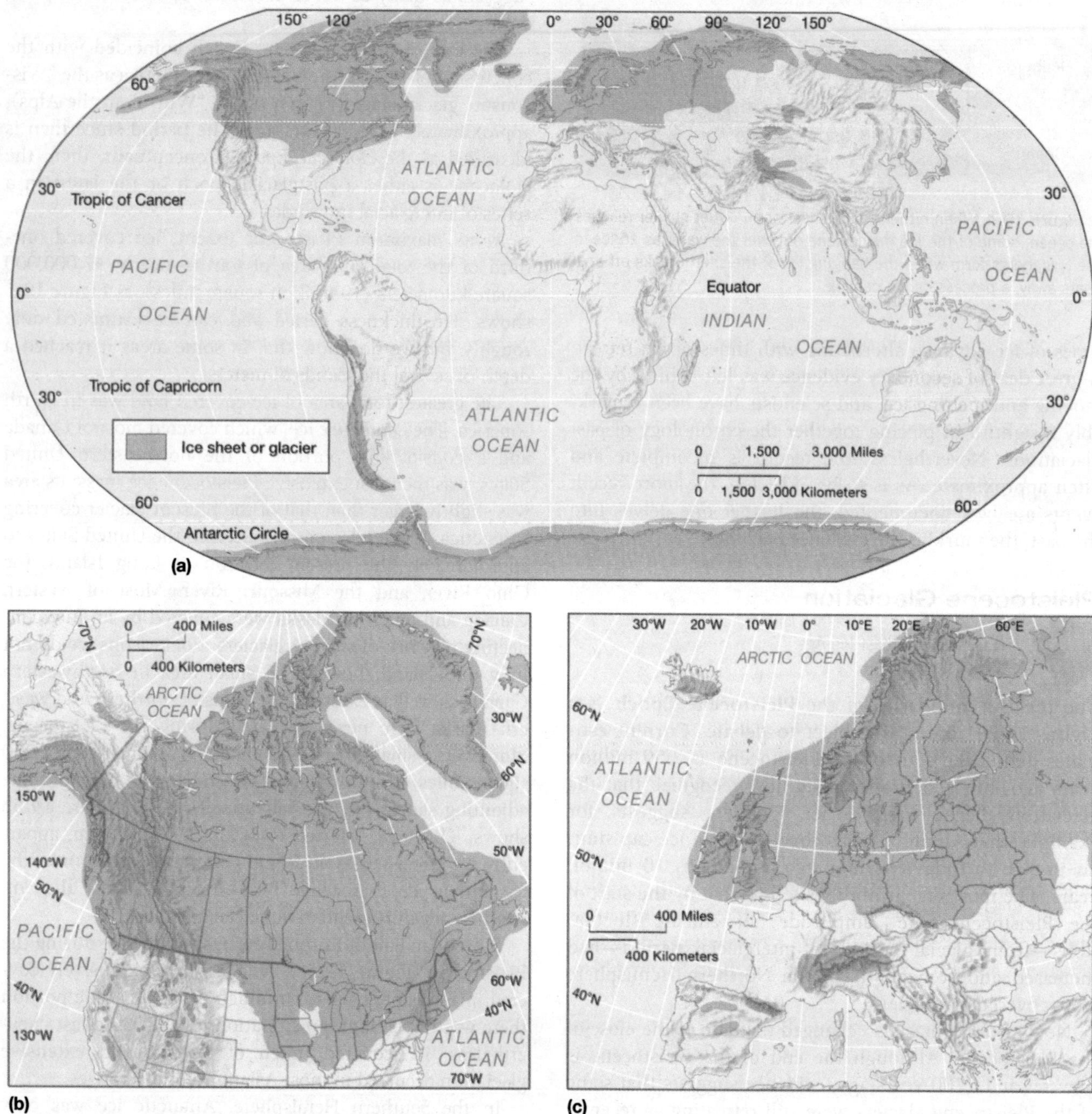

▲ **Figure 19-5** The maximum extent of Pleistocene glaciation: (a) worldwide; (b) in North America; (c) in Western Eurasia.

deposition done by the prodigious amounts of meltwater released as the glaciers melted. Also important were frost weathering caused by the low temperatures in the periglacial zone and the associated *solifluction* of frozen subsoil (see Chapter 15 for a review of solifluction). It is estimated that periglacial conditions extended over more than 20 percent of Earth's land area. (Periglacial landforms are discussed later in this chapter.)

2. **Sea-level changes**: The buildup of ice on the continents meant that less water was available to drain from the continents into the oceans, a condition that resulted in a worldwide lowering of sea level during every episode of glacial advance; when the glaciers retreated, sea level would again rise as meltwater returned to the oceans. At the peak of the Pleistocene glaciations, global sea level was about 130 meters (430 feet) lower than it is today. These fluctuations in the amount of ocean water caused a significant difference in drainage patterns and topographic development on seashores and coastal plains (the influence of Pleistocene sea level changes on coastal topography will be discussed in Chapter 20). At the peak of the Pleistocene glaciations, the Bering Strait between present-day Alaska and Russia was a dry land bridge,

allowing the migration of both animals and humans (see Figure 11-24).

3. **Crustal depression:** The enormous weight of accumulated ice on the continents caused portions of Earth's crust to sink, in some cases by as much as 1200 meters (4000 feet). After the ice melted, the crust slowly began to rebound. This *isostatic adjustment* has not yet been completed, and some portions of Canada and northern Europe are still rising as much as 20 centimeters (8 inches) per decade. Isostasy is discussed in detail in Chapter 13 (for example, see Figure 13-20).
4. **Pluvial (increased rain) developments:** During the Pleistocene glaciations, there was, on almost all areas of the continents, a considerable increase in the amount of moisture available. This increase was caused by a combination of meltwater runoff, increased precipitation, and decreased evaporation. A prominent result of these **pluvial effects** was the creation of many lakes in areas where none had previously existed. Most of these lakes have subsequently been drained or significantly reduced in size, but they have left lasting imprints on the landscape. **Pleistocene lakes** in the western part of the United States are shown in Figure 19-6. The present-day Great Salt Lake in Utah is a tiny remnant of a much larger Pleistocene lake known as Lake Bonneville, and today's Bonneville Salt Flats were once the floor of this enormous lake.

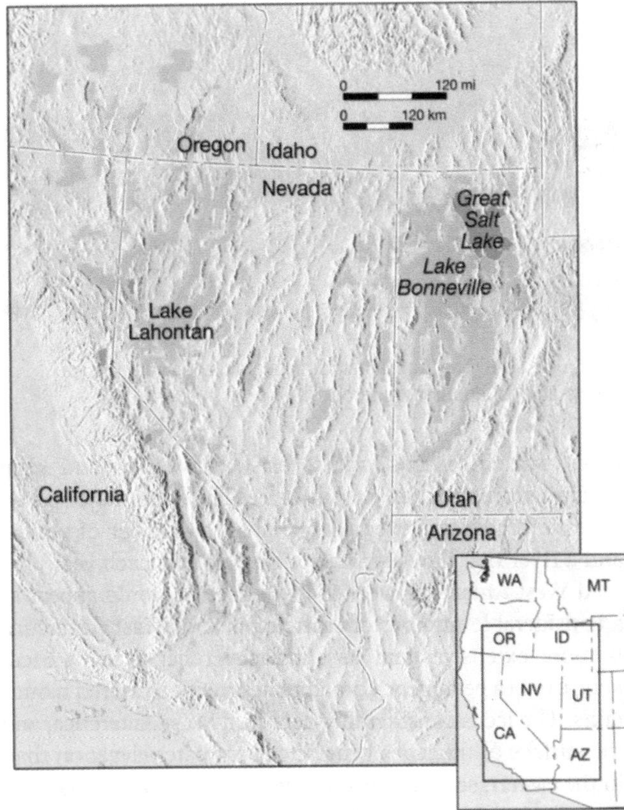

▲ **Figure 19-6** Pleistocene lakes of the intermontane region of the United States. Today's Great Salt Lake (Utah) is shown outlined in blue inside the boundaries of the ancestral Lake Bonneville.

Contemporary Glaciation

In marked contrast to Pleistocene glaciation, the extent of ice covering the continental surfaces today is very limited (Figure 19-7). About 10 percent of Earth's land surface—some 15 million square kilometers (6 million square miles)—is covered with ice today, but more than 96 percent of that total is in Antarctica and Greenland. Something more than two-thirds of all the world's freshwater is at this moment frozen into glacial ice.

Antarctic Ice Cap Antarctic ice is by far the most extensive ice cap on Earth (see Figure 19-3). At present, about 98 percent of its surface is covered with glacial ice, representing about 85 percent of the world's land–ice total. This ice is more than 4000 meters (13,000 feet) thick in some places and more than 1500 meters (5000 feet) thick over most of the continent.

Physically the continent and its ice sheets can be thought of as consisting of two unequal sections separated by the wide upland belt of the Transantarctic Mountains, which extend for some 4000 kilometers (2500 miles; Figure 19-8). West Antarctica, the smaller of the two sections, is generally mountainous. It contains, however, a few interior valleys that are curiously ice free. The "Dry Valleys" area consists of about 3900 square kilometers (1500 square

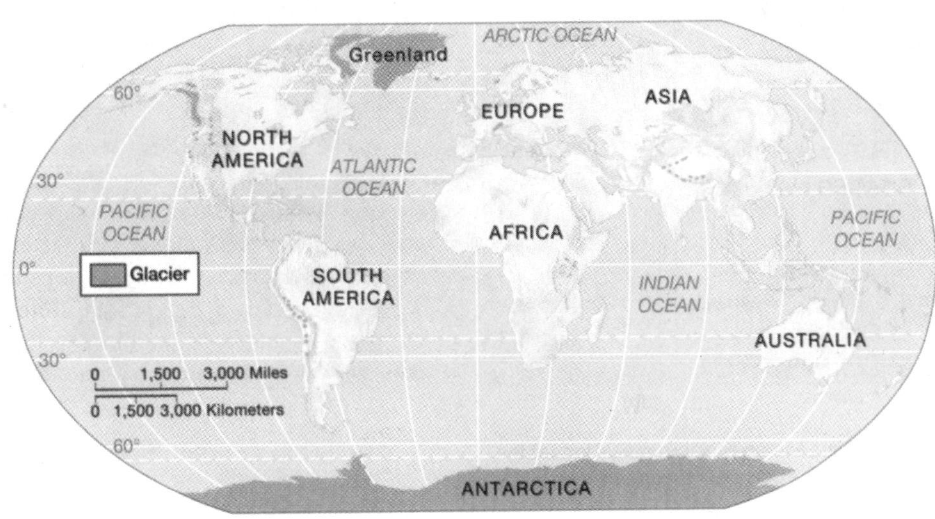

◀ **Figure 19-7** World distribution of glacial ice today.

▲ **Figure 19-8** The Transantarctic Mountains separate the ice sheets of West and East Antarctica.

miles) which, because winds blast away snow and keep precipitation out, does not build ice. The three major parallel valleys contain several large lakes, a number of ponds, and a river that flows for one or two months each year.

If West Antarctica were to lose its ice, it would appear as a considerable number of scattered islands. East Antarctica is more extensive, and its subglacial relief is less varied, appearing to be largely a broad plateau with scattered mountains. The ice is considerably deeper in West Antarctica, and the surface of the ice is generally at a greater elevation than in the eastern section. Most of the surface of West Antarctica exceeds 2.4 kilometers (8000 feet) above sea level, and a considerable portion is more than 3 kilometers (10,000 feet) in altitude.

Greenland Ice Cap Greenland ice is much less extensive—1,740,000 square kilometers (670,000 square miles)—but still of impressive size (see Figure 19-3). Elsewhere there are only relatively small ice masses on certain islands in the Canadian Arctic, Iceland, and some of the islands north of Europe.

Mountain Glaciers Other than the two major ice caps, the remainder of the world's present-day glaciers are concentrated in the high mountain areas. In the conterminous United States, most glaciers are in the Pacific Northwest, and more than half of these are in the North Cascade Mountains of Washington (Figure 19-9). In Alaska, there are 75,000 square kilometers (29,000 square miles) of glacial ice, amounting to about 4 percent of the total area of the state. The largest Alaskan glacier is the Bering Glacier, near Cordova, which covers 5830 square kilometers (2250 square miles) and is more than twice the size of Rhode Island.

Climate Change and Contemporary Glaciation

As we saw in earlier chapters, global climate change is significantly influencing the extent of contemporary glaciation. The retreat of the Arctic sea ice pack and the loss of mass of Greenland's ice sheets is indicative of the higher temperatures experienced in the high latitudes of the Northern Hemisphere over the last 30 years (see "People and the Environment: Climate Change in the Arctic" in Chapter 8). As we will see later in this chapter ("Focus: Shrinking Glaciers"), glaciers are sensitive indicators of environmental change—reflecting variations in both temperature and precipitation.

Perhaps nowhere today is the attention of scientists focused more keenly on contemporary ice sheets than in Antarctica. Not only does the Antarctic ice sheet provide invaluable information about Earth's past climate (see "Focus: Paleoclimatology at Dome C, Antarctica" in Chapter 8), but changes in the ice sheet today are pointing to a contemporary climate shift. A NASA study released in January 2009 showed that Antarctica has been warming at a rate of about 0.12°C (0.22°F) per decade over the last 50 years—much faster than scientists thought even a few

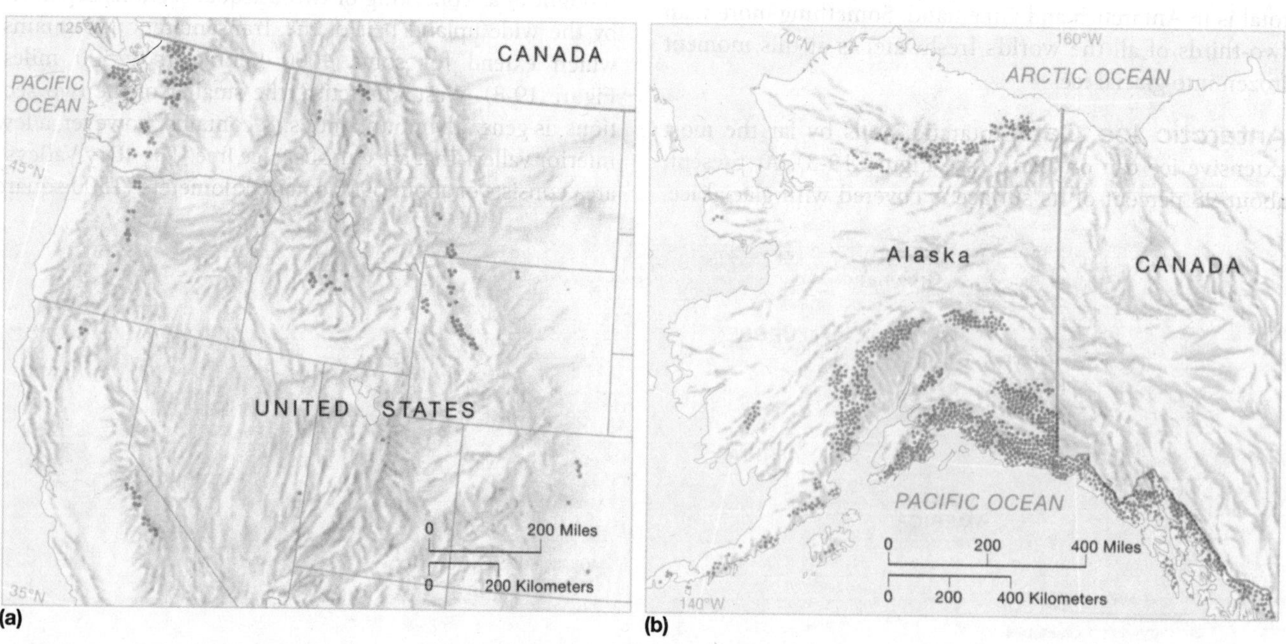

▲ **Figure 19-9** The location of contemporary glaciers and perennial snowfields in (a) the conterminous western United States and (b) Alaska. The dots on the map considerably exaggerate the size of each glacier.

PEOPLE AND THE ENVIRONMENT

Disintegration of Antarctic Ice Shelves

Despite its remoteness, Antarctica exerts a prominent influence on the world's environment—global sea level, oceanic temperature, ocean nutrient content, and patterns of atmospheric circulation all are affected by conditions in Antarctica. At the same time, the global environment exerts its own influence on the conditions in Antarctica. As is true in the Arctic, the ice sheets of Antarctica appear to be sensitive indicators of changes in global climate. Over the last half century, average temperatures in the Antarctic have increased by about 2.5°C (4.5°F)—much more than the average increase in temperature worldwide. As a consequence, Antarctic ice has been undergoing changes.

If all 30,000,000 cubic kilometers (7,200,000 cubic miles) of Antarctic ice were to melt, the sea level around the world would rise by about 73 meters (240 feet)—up to the nose of the Statue of Liberty. While the complete melting of the Antarctic ice cap is unlikely, the long-term equilibrium between accumulation and wastage of the ice does appear to be changing significantly as a response to higher global temperatures. Some of the most dramatic changes are occurring in Antarctica's ice shelves.

The Antarctic ice sheets flow outward from the interior of the continent in nearly all directions toward the sea, which means that icebergs are being calved (broken off) into the sea more or less continuously around the perimeter of the continent. Some of these icebergs originate through outlet glaciers, but many are broken off from ice shelves. There are several of these great plates of floating ice, particularly in West Antarctica, with the Ross Ice Shelf being the largest (520,000 square kilometers or 200,000 square miles; see Figure 19-3). On the Antarctic Peninsula, a number of smaller ice shelves are found, such as the Larsen Ice Shelves on the eastern side of the peninsula (Figure 19-A).

Over the last few years, large sections of the ice shelves along the Antarctic Peninsula have disintegrated—more than 8000 square kilometers (3100 square miles) of ice shelf have disappeared since 1993. After slowly retreating for many years, in 1995 the Larsen-A Ice Shelf simply collapsed and disappeared. In 2002, the Larsen-B Ice Shelf collapsed in little more than a month (Figure 19-B). The larger Larsen-C Ice Shelf is currently losing mass as a result of both higher air temperatures above and higher water temperatures below. Farther to the south on the opposite side of the Antarctic Peninsula, the Wilkins Ice Shelf began to disintegrate in February 2008, losing a total area of about 3600 square kilometers (1400 square miles) by the end of the year (Figure 19-C).

Although the loss of ice shelves does not raise global sea level (for the same reason that a floating ice cube does not raise the level of the water in a glass as it melts), changing ice shelves can trigger a change in the flow of land-based ice off the continent. To a certain extent, an intact ice shelf holds back the flow of continental ice into the ocean. Once an ice shelf is gone, the continental ice may flow faster into the ocean, and with land-based ice entering the ocean at a greater rate than before, sea level will rise. For example, since the break up of the Larsen-B shelf, the flow rate of glaciers feeding it accelerated by as much as six times their previous rate.

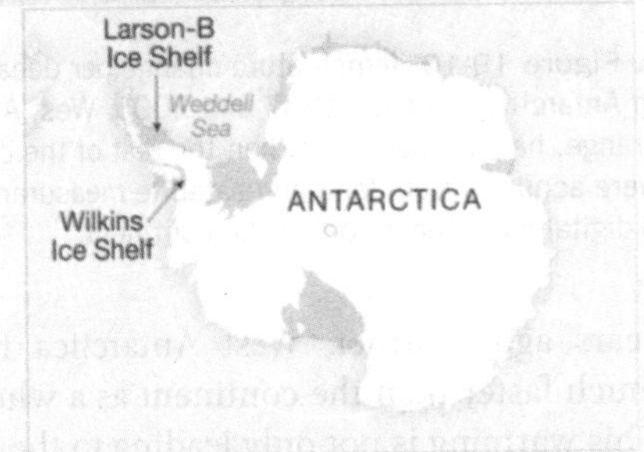

▲ **Figure 19-A** Portions of Antarctic ice shelves, such as the Larsen-B and Wilkins, have disintegrated in recent years.

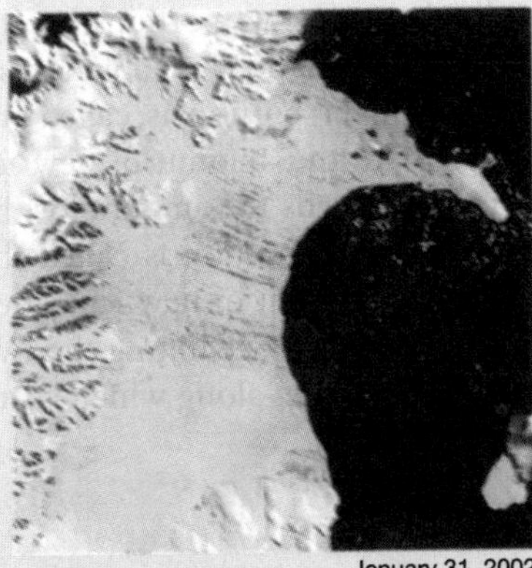

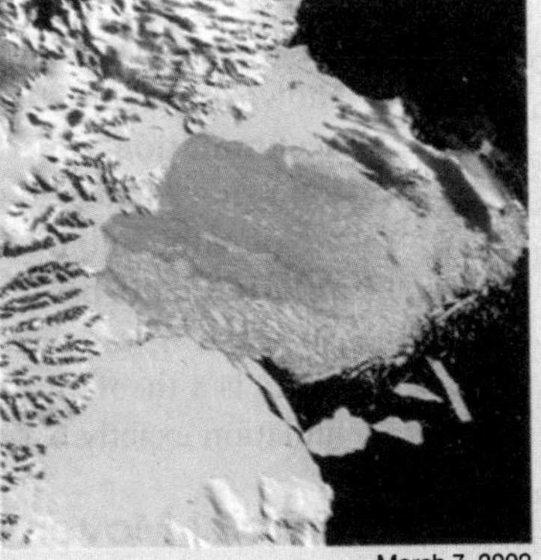

▲ **Figure 19-B** The Larsen-B Ice Shelf in Antarctica began to collapse in January 2002, as shown in these images taken by NASA's Terra satellite. On January 31, 2002, blue "melt ponds" are seen on top of the ice shelf. By March 7, 2002, the ice shelf had disintegrated. The blue area shows the shattered floating ice from the shelf.

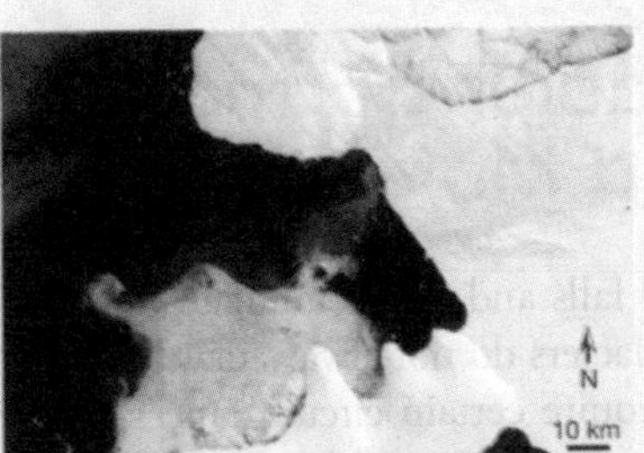

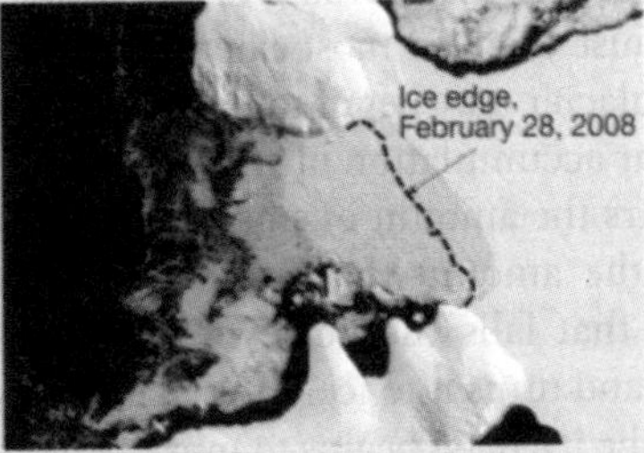

▲ **Figure 19-C** The Wilkins Ice Shelf began to disintegrate in 2008 as recorded by the MODIS sensors on NASA's Terra and Aqua satellites. The top image was taken on February 28; the bottom image was taken on March 17 after a large portion of the ice shelf collapsed, leaving broken masses of ice floating in the ocean.

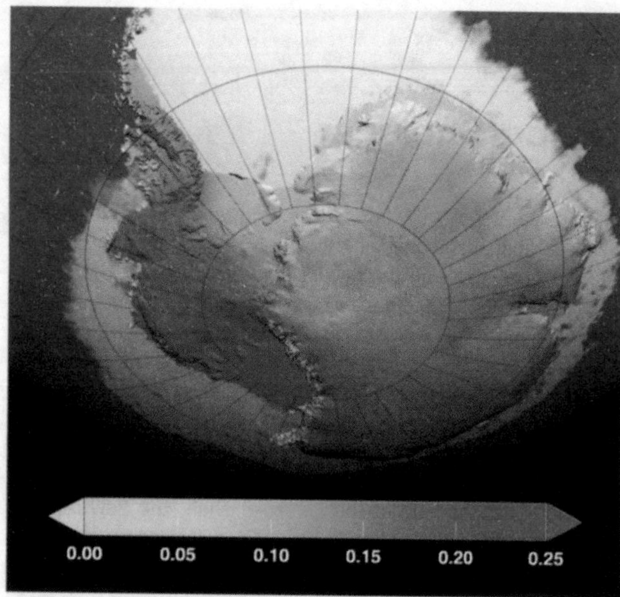

▲ **Figure 19-10** Temperature change per decade in degrees Celsius in Antarctica between 1957 and 2006. West Antarctica, shown in orange, has warmed more than the rest of the continent. The data were acquired from land and satellite measurements, and overlain on a digital elevation model of the continent.

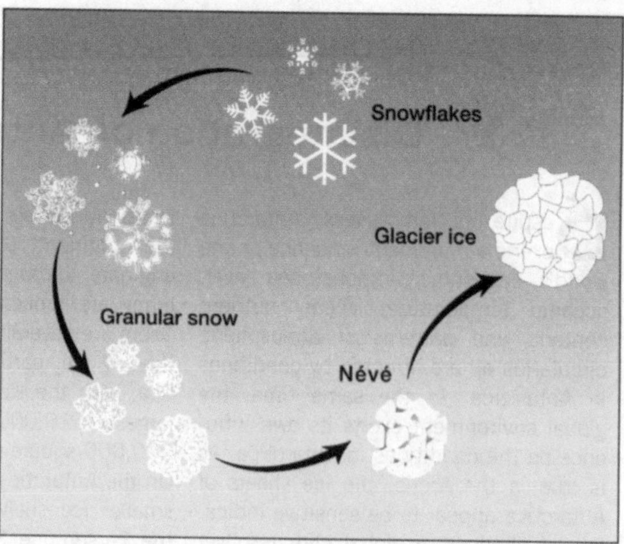

▲ **Figure 19-11** Snow is changed to ice by compression and coalescence, following a sequence from snowflake to granular snow to névé to glacial ice.

years ago. Further, West Antarctica has been warming much faster than the continent as a whole (Figure 19-10). This warming is not only leading to the breakup of Antarctic ice shelves (see "People and the Environment: Disintegration of Antarctic Ice Shelves") but to higher flow rates of outlet glaciers as well.

Glacier Formation and Movement

Snow falls and ice accumulates in many parts of the world, but glaciers do not always develop from these events. Glaciers require certain circumstances to form and then depend on just the right combination of temperature and moisture to survive. A slight warming or drying trend for a few decades can cause even the most extensive ice sheet to disappear.

A glacier may begin to develop when there is a net year-to-year accumulation of snow—that is, when over a period of years the amount of snow that falls in a winter is greater than the amount that melts the following summer. The snow that falls the next winter weighs down on the old snow and turns it to ice. After many years of such accumulation, the ice mass begins to move under the pull of gravity—and a glacier is formed. The persistence of any glacier depends on the balance between **accumulation** (addition of ice by incorporation of snow) and **ablation** (wastage of ice through melting and sublimation).

Changing Snow to Ice

Snow is not merely frozen water; rather, it is a substance that has crystallized directly from water vapor in the atmosphere and floats to Earth as lacy, hexagonal crystals that are only about one-tenth as dense as liquid water. Sooner or later (within a few hours if the temperature is near freezing, but only over a period of years in very cold situations), crystalline snow is compressed by overlying snow into granular form, and in the process its density is approximately doubled. With more time and further compression, the granules are packed more closely and begin to coalesce, the density increasing steadily until it is about half the density of water (Figure 19-11). This material is called **névé** or **firn**. As time passes, the pore spaces with their trapped air among the whitish névé crystals gradually diminish as the air is squeezed out by the weight of the overlying snow; the density approaches 90 percent of that of liquid water, and the material takes on the bluish tinge of glacial ice. This ice continues to change, although very slowly, with more air being forced out, the density increasing slightly, and the crystals increasing in size.

Every glacier can be divided into two portions on the basis of the balance between accumulation and ablation, as shown in Figure 19-12. The upper portion is called the **accumulation zone** because here the amount of new ice from snowfall added each year exceeds the amount lost by melting and sublimation. The lower portion is called the **ablation zone** because here the amount of new ice added each year is less than the amount lost. Separating the two zones is a theoretical **equilibrium line**, along which accumulation exactly balances ablation.

Glacial Movement

Flow of Ice within a Glacier

Despite the fact that glaciers are often likened to rivers of ice, there is very little similarity between liquid flow and glacial movement.

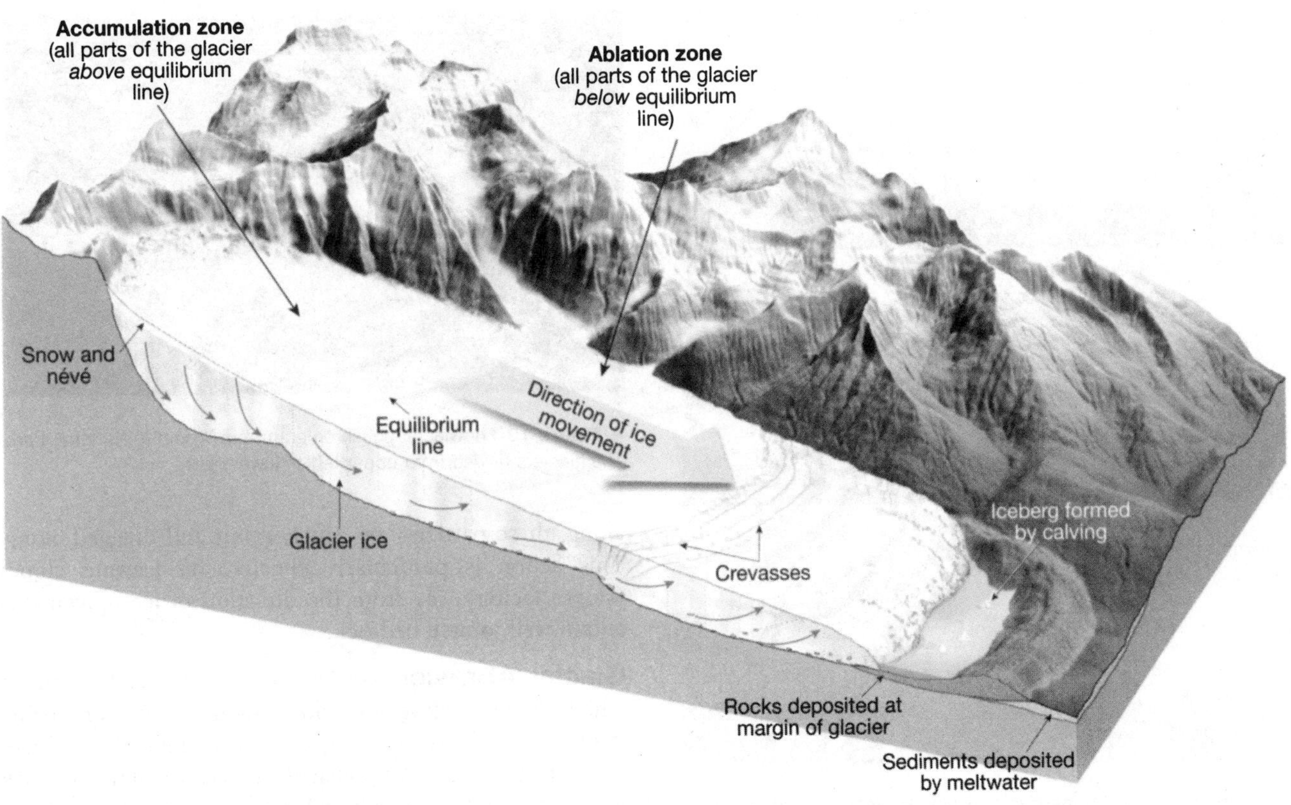

▲ **Figure 19-12** Cross section through an alpine glacier. The upper portion is an area of net ice accumulation. Below the equilibrium line there is more ablation than accumulation.

We usually think of ice as being a brittle substance that breaks rather than bends and resists any sort of deformation. This is generally true for surface ice, as evidenced by the cracks and *crevasses* that often appear at the surface of a glacier. However, ice under considerable confining pressure, as below the surface of a glacier, behaves quite differently: it deforms rather than breaks. Moreover, partial melting, due to the stresses within and the pressure at the bottom of the glacier, aids movement because the meltwater sinks to the bottom of the glacier and becomes a slippery layer on which the glacier can slide.

When a mass of ice attains a thickness of about 50 meters (165 feet)—less on steep slopes—the **plastic flow of ice** begins in response to the overlying weight. The entire mass does not move uniformly; rather, there is an oozing outward from around the edge of an ice sheet or downvalley from the toe of an alpine glacier. A second kind of movement is **basal slip** at the bottom of the glacier, in which the entire mass slides over its bed on a lubricating film of water. The glacier more or less molds itself to the shape of the terrain over which it is riding.

Glaciers usually move very slowly; indeed, the adjective "glacial" is synonymous with "exceedingly slow." The movement of most glaciers can be measured in a few centimeters per day, although an advance of several meters per day would not be unusual, and extreme examples of nearly 30 meters (100 feet) in a 24-hour period have been recorded. Also, the flow is often erratic, with irregular pulsations and surges over a short span of time.

As might be expected, all parts of a glacier do not move at the same rate. The fastest-moving ice is that at and near the surface, with speed generally decreasing with depth. If the glacier is confined, the way a valley glacier is, for instance, the center of the surface ice moves faster than the sides, which is similar to streamflow patterns.

Glacier Flow versus Glacier Advance In discussing glacier movement, it is important to distinguish between glacier *flow* and glacier *advance*. As long as a glacier exists, the ice in it is flowing, either laterally outward or downhill. This does not necessarily mean that the outer edge of the ice is advancing, however. The ice in a glacier always moves forward, but the outer margin of the glacier may or may not be advancing, depending on the balance between accumulation and ablation (Figure 19-13). Even in a retreating glacier (one whose outer margin is retracting toward its point of origin due to heavy ablation), the ice is flowing forward.

During wetter or cooler periods when there is a great accumulation of ice, a glacier can flow farther before it finally wastes away, and so the outer margin of the glacier advances. During warmer or drier periods when the rate of ablation is increased, the glacier continues to flow, but it wastes away sooner, and so the end or terminus of the glacier retreats.

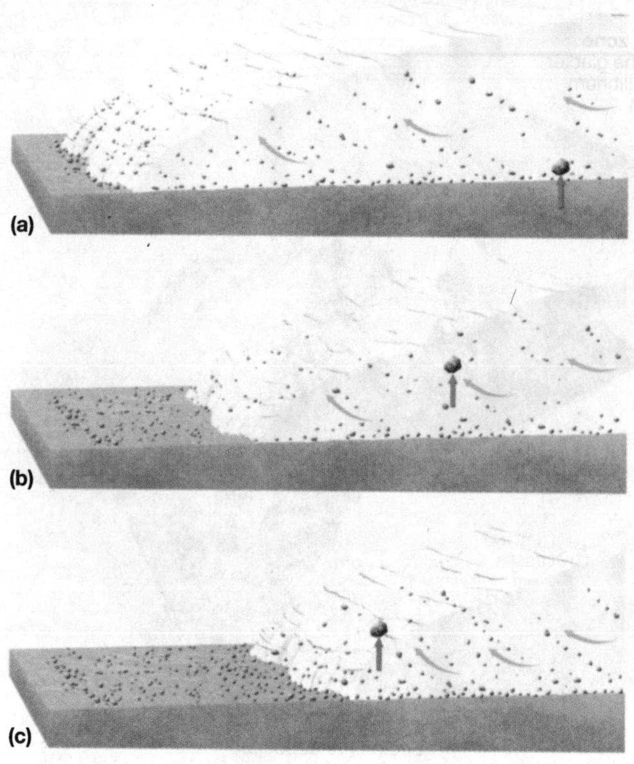

▲ **Figure 19-13** A flowing glacier is not necessarily an advancing glacier. In this sequential illustration, the front of the glacier is clearly retreating, but the ice continues to flow forward. The boulder marked by the red arrow illustrates the principle.

The Effects of Glaciers

Glacial Processes

As glaciers move across a landscape, they can reshape the topography through the erosion, transportation, and deposition of rock.

Erosion by Glaciers

As with streams, volume and speed determine the effectiveness of glacial erosion. The amount of erosion caused by a glacier is roughly proportional to the thickness of the ice and its rate of flow. The depth of the erosion is limited in part by the structure and texture of the bedrock and in part by the relief of the terrain.

Glacial Plucking The direct erosive power of moving ice is greater than that of flowing water but not remarkably so. As the slowly moving ice scrapes against bedrock, friction between rock and ice causes the lowermost ice to melt, and the layer of water created reduces the pressure on the rock. This water can refreeze around rocky protrusions, however, and the refrozen ice can exert a significant force as it is pushed by the ice behind it. Probably the most significant erosive work of glacial ice is accomplished by this **glacial plucking**. Rock particles beneath the ice are grasped as meltwater refreezes in bedrock joints and fractures where frost wedging further loosens the rock. As the ice moves along, these particles are plucked out and dragged along. This action is particularly effective on leeward slopes (slopes facing away from the direction of ice movement) and in well-jointed bedrock.

▲ **Figure 19-14** Glacial polish on granite in Yosemite National Park. The glaciers flowed from upper left to lower right.

Glacial Abrasion Glaciers also erode by abrasion, in which the bedrock is worn down by the rock debris being dragged along in the moving ice. Abrasion mostly produces minor features, such as polished surfaces when the bedrock is of highly resistant material and striations (fine parallel indentations) and grooves (indentations deeper and larger than striations) in less resistant bedrock (Figure 19-14). Whereas plucking tends to roughen the underlying surface, abrasion tends to polish it and dig striations and grooves.

Subglacial Meltwater Erosion A third process also contributes to glacial erosion: Meltwater streams flowing below the glacier not only transport rock, they can erode smooth grooves and channels into the bedrock (Figure 19-15).

In plains areas, the topography produced by glacial erosion may be inconspicuous. Prominences are smoothed and small hollows may be excavated, but the general appearance of the terrain changes little. In hilly areas, however, the effects of glacial erosion are much more notable. Mountains and ridges are sharpened, valleys are deepened and steepened and made more linear, and the entire landscape becomes more angular and rugged.

Transportation by Glaciers

Glaciers are extremely competent, as well as indiscriminate, in their ability to transport rock debris. They can move immense blocks of rock—the size of houses in some cases—for dozens or even hundreds of kilometers. Most of a glacier's load, however, is not such huge blocks but rather a heterogeneous collection of particles of all sizes, including finely ground rock material known as **glacial flour**.

Most of the material transported by continental glaciers is plucked or abraded from the underlying surface and so is carried along at the base of the ice. Thus, there is a narrow zone at the bottom of the glacier that is likely to be well armored by miscellaneous rock debris frozen into it, with most of the rest

◀ **Figure 19-15** Grooves and striae near Cuzco, Peru, caused by meltwater erosion below a glacier.

of the glacial ice relatively free of rock fragments. With mountain glaciers, in addition to rock transported within the ice, some of the material is also transported on top of the ice, deposited by rockfall or other forms of mass wasting from the surrounding slopes. With at least some mountain glaciers, mass wasting may supply more rock debris to a glacier than that added by glacial erosion.

A glacier transports its load outward or downvalley at a variable speed. The rate of flow usually increases in summer and slows in winter but also depends on variations in ice accumulation and in the gradient of the underlying slopes.

Melt Streams Another important aspect of transportation by glaciers is the role of flowing water on, in, and under the ice. During the warmer months, streams of meltwater normally flow along with the moving ice (Figure 19-16). Such streams may run along the surface of the glacier until they find cracks or crevasses into which to plunge—including steep drainage shafts in the ice called *moulins*—continuing their flow as a subglacial stream either within the ice or along the interface between glacier and bedrock. Wherever such streams flow, they transport rock debris, particularly smaller particles and glacial flour, providing an effective mechanism for shifting debris from the ice surface to a position within or at the bottom of the glacier. Further, where subglacial streams encounter finely ground rock along the bottom of a glacier, the lubricating effect can accelerate the flow of ice.

Even if a glacier is retreating, the debris inside it is still carried forward because of the ice flow taking place all through the glacier. The transport function of a glacier persists indefinitely unless and until the ice becomes so thin that subglacial obstacles, such as a hill, prevent further flow.

Deposition by Glaciers

Probably the major role of glaciers in landscape modification is to pick up lithospheric material from one area and take it to some distant region, where it is left in a fragmented and vastly changed form. This is clearly displayed in North America, where an extensive portion of central Canada has been glacially scoured of its soil, regolith, and much of its surface bedrock, leaving a relatively barren, rocky, gently undulating surface dotted with bodies of water. Much of the removed material was taken southward and deposited in the midwestern part of the United States,

◀ **Figure 19-16** A meltwater stream flowing over the surface of Athabaska Glacier in the Canadian Rockies.

◀ **Figure 19-17** Unsorted glacial till in a roadcut near Bridgeport, California.

producing an extensive plains area of remarkably fertile soil. Thus, the legacy of Pleistocene ice sheets for the Midwest was the evolution of one of the largest areas of productive soils ever known, at the expense of central Canada, which was left impoverished of soil by those same glaciers. (On the other hand, many valuable Canadian mineral deposits were exposed when the glaciers removed soil and regolith.)

The general term for all material moved by glaciers is **drift**, a misnomer coined in the eighteenth century when it was believed that the vast debris deposits of the Northern Hemisphere were leftovers from biblical floods. Despite the erroneous thinking that led to the name, the term is still used today to refer to any material deposited by glaciers and/or their meltwater.

Direct Deposition by Glacial Ice Rock debris deposited directly by moving or melting ice, with no meltwater redeposition involved, is given the more distinctive name **till** (Figure 19-17). Direct deposition by ice is usually the result of melting around the margin of an ice sheet or near the lower end of an alpine glacier, but it is also accomplished whenever debris is dropped on the ground beneath the ice, especially in the ablation area. In either case, the result is an unsorted and unstratified agglomeration of fragmented rock material. Most of the fragments are angular because they have been held in position while carried in the ice (and so had little opportunity to become rounded by frequent impact the way pebbles in a stream would) or were deposited on top of the ice by mass wasting.

Sometimes outsized boulders are deposited by a retreating glacier. Such enormous fragments, which may be very different from the local bedrock, are called **glacial erratics** (Figure 19-18).

◀ **Figure 19-18** A glacial erratic carried many kilometers before being deposited by a Pleistocene glacier in what is now Yellowstone National Park. The ice has long since melted, and a forest has grown up around the erratic.

Secondary Deposition by Meltwater Glacial runoff has several peculiarities—peak flows in midsummer, distinct day-and-night differences in volume, large silt content, and occasional floods—that set meltwater streams apart from other kinds of natural waterways. Much of the debris carried by glaciers is eventually deposited or redeposited by meltwater. In some cases, this is accomplished by subglacial streams issuing directly from the ice and carrying sedimentary material washed from positions in, on, or beneath the glacier. Much meltwater deposition, however, involves debris that was originally deposited by ice and subsequently picked up and redeposited by the meltwater well beyond the outer margin of the ice. Such **glaciofluvial deposition** occurs around the margins of all glaciers, as well as far out in some periglacial zones.

Continental Ice Sheets

Apart from the oceans and continents, continental ice sheets are the most extensive features ever to appear on the face of the planet. Their actions during the Pleistocene significantly reshaped both the terrain and the drainage of nearly one-fifth of the total surface area of the continents.

Development and Flow

Pleistocene ice sheets, with the exception of the one covering Antarctica, did not originate in the polar regions. Rather, they developed in subpolar and midlatitude locations and then spread outward in all directions, including poleward. Several (perhaps several dozen) centers of original ice accumulation have been identified. The accumulated snow/névé/ice eventually produced such a heavy weight that the ice began to flow outward from each center of accumulation.

The initial flow was channeled by the preexisting terrain along valleys and other low-lying areas, but in time the ice developed to such depths that it overrode almost all preglacial topography. In many places, it submerged even the highest points under thousands of meters of ice. Eventually the various ice sheets coalesced into only one, two, or three massive sheets on each continent. These vast ice sheets flowed and ebbed as the climate changed, always modifying the landscape with their enormous erosive power and the great masses of debris they deposited. The elaborate result was nothing less than a total reshaping of the land surface and a total rearrangement of the drainage pattern.

Erosion by Ice Sheets

Except in mountainous areas of great initial relief, the principal topography resulting from the erosion caused by an ice sheet is a gently undulating surface. The most conspicuous features are valley bottoms gouged and deepened by the moving ice. Such troughs are deepest where the preglacial valleys were oriented parallel to the direction of ice movement, particularly in areas of softer bedrock. A prime example of such development is the Finger Lakes District of central New York, where a set of parallel stream valleys was accentuated by glaciation into a group of long, narrow, deep lakes (Figure 19-19). Even where the preglacial valley was not oriented parallel to the direction of ice flow, however, glacial gouging and scooping normally produced a large number of shallow excavations that became lakes after the ice disappeared. Indeed, the postglacial landscape in areas of ice-sheet erosion is notable for its profusion of lakes.

Hills are generally sheared off and rounded by the moving ice. A characteristic shape produced by both continental ice sheets and mountain glaciers is the **roche moutonnée**, which is often produced when a bedrock hill is overridden by moving ice (Figure 19-20).[1] The *stoss side*

[1]The origin of the French term roche moutonnée is unclear. It is often translated as "sheep's back," but some authorities believe it is based on a fancied resemblance to wavy wigs that were fashionable in France in the late 1700s and were known as *moutonnées* because they were pomaded with mutton tallow.

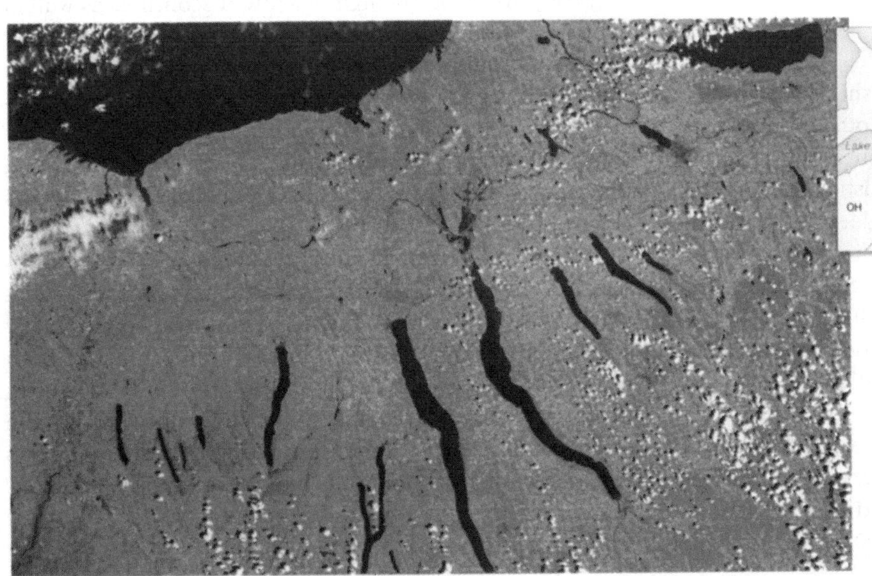

◄ **Figure 19-19** The Finger Lakes of upstate New York occupy glacial valleys that were accentuated by ice sheets moving from north–northwest to south–southeast. The large body of water at upper left is Lake Ontario. This false-color composite image was taken from an altitude of about 920 kilometers (570 miles) by the ERTS-1 satellite.

▲ **Figure 19-20** (a) Lembert Dome in Yosemite National Park is an example of a roche moutonnée. The glaciers moved from right to left (from east to west) across this granite dome. (b) The formation of a roche moutonnée. The glacier rides over a resistant bedrock surface, smoothing the stoss side by abrasion and steepening the lee side by plucking. When the ice has melted, an asymmetrical hill is the result.

(facing in the direction from which the ice came) of a roche moutonnée is smoothly rounded and streamlined by grinding abrasion as the ice rides up the slope, but the lee side (facing away from the direction from which the ice came) is shaped largely by plucking, which produces a steeper and more irregular slope.

The postglacial landscape produced by ice sheets is one of relatively low relief but not absolute flatness. The principal terrain elements are ice-scoured rocky knobs and scooped-out depressions. Soil and weathered materials are largely absent, with bare rock and lakes dominating the surface. Stream patterns are erratic and inadequately developed because the preglacial drainage system was deranged by ice erosion. Once eroded by the passing ice sheet, however, most of this landscape was subjected to further modification by glacial deposition. Thus, the starkness of the erosional landscape is modified by depositional debris.

Deposition by Ice Sheets

In some cases, the till transported by ice sheets is deposited heterogeneously and extensively, without forming any identifiable topographic features; a veneer of unsorted debris is simply laid down over the preexisting terrain. This veneer is sometimes quite shallow and does not mask the original topography. In other cases, till is deposited to a depth of several hundred meters, completely obliterating the shape of the preglacial landscape. In either case, deposition tends to be uneven, producing an irregularly undulating surface of broad, low rises and shallow depressions. Such a surface is referred to as a *till plain*.

In many instances, glacial sediments are laid down in more defined patterns, creating characteristic and identifiable landforms (Figure 19-21). **Moraine** is a general term for glacier-deposited landforms composed entirely or largely of till. They typically consist of irregular rolling topography rising some small height above the surrounding terrain. Moraines are usually much longer than they are wide, although the width can vary from a few tens of meters to as much as several kilometers. Some moraines are distinct ridges, whereas others are much more irregular in shape. Their relief is not great, varying from a few meters to a few hundred meters. When originally formed, moraines tend to have relatively smooth and gentle slopes, which become more uneven with the passage of time, as the blocks of stagnant ice, both large and small, included within the till eventually melt, leading to the collapse of the surface of the moraine.

Three types of moraines are particularly associated with deposition from continental ice sheets, although all three may be produced by mountain glaciers as well. A **terminal moraine** is a ridge of till that marks the outermost limit of glacial advance. It can vary in size from a conspicuous rampart tens of meters high to a low, discontinuous wall of debris. A terminal moraine is formed when a glacier reaches its equilibrium point and so is wasting at the same rate that it is being nourished. Although the toe of the glacier is not advancing, the interior continues to flow forward, delivering a supply of till. As the ice melts around the margin, the till is deposited, and the moraine grows (Figure 19-22).

Behind the terminal moraine, **recessional moraines** may develop as the glacier recedes. These are ridges that mark positions where the ice front was temporarily stabilized during the final retreat of the glacier. Both terminal and recessional moraines normally occur in the form of concave arcs that bulge outward in the direction of ice movement, indicating that the ice sheets advanced not along an even line but rather as a connecting series of great tongues of ice, each with a curved front (Figure 19-23).

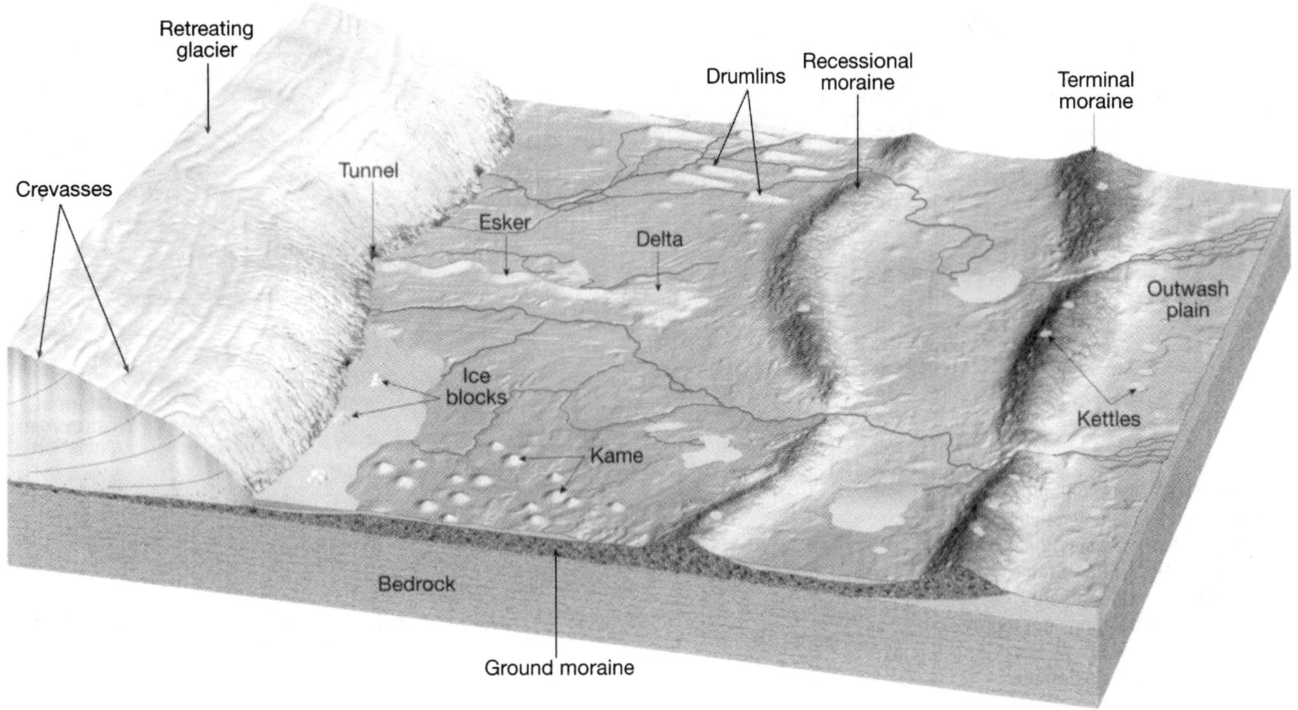

▲ **Figure 19-21** Glacier-deposited and glaciofluvially deposited features of a landscape as a continental ice sheet retreats.

The third type of moraine is the **ground moraine**, formed when large quantities of till are laid down from underneath the glacier rather than from its edge. A ground moraine usually means gently rolling plains across the landscape. It may be shallow or deep and often consists of low knolls and shallow *kettles*. **Kettles** can form when large blocks of ice left by a retreating glacier become surrounded or even covered by glacial drift; after the ice block melts, the morainal surface collapses, leaving an irregular depression (Figure 19-24).

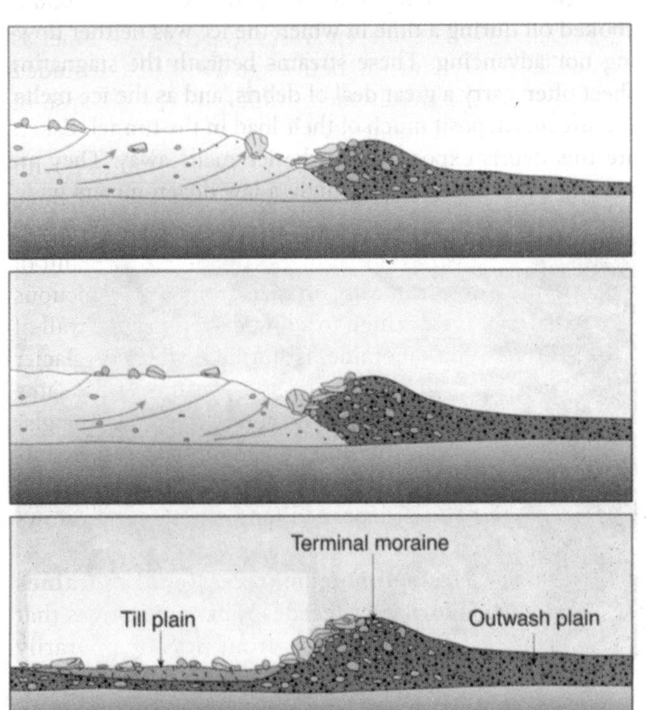

▲ **Figure 19-22** Moraine growth at the terminus of a glacier. Rock is carried within the ice, emerging at the end of the glacier where it is deposited in a moraine. Some debris also moves on top of the ice. The final diagram represents the situation after the ice has melted.

Another prominent feature deposited by ice sheets is a low, elongated hill called a **drumlin**, a term that comes from *druim*, an old Irish word for "ridge." Drumlins are much smaller than moraines but composed of similarly unsorted till. The long axis of the drumlin is aligned parallel with the direction of ice movement. The end of the drumlin facing the direction from which the ice came is blunt and slightly steeper than the opposite end. Thus, the configuration is the reverse of that of a roche moutonnée. The origin of drumlins is complex, but most of them are apparently the result of ice readvance into an area of previous glacial deposition. In other words, they are depositional features subsequently shaped by erosion. Drumlins usually occur in groups, sometimes numbering in the hundreds, with all drumlins in a group oriented parallel to each other (Figure 19-25). The greatest concentrations of drumlins in the United States are found in central New York and eastern Wisconsin.

Glaciofluvial Features

The deposition or redeposition of debris by ice-sheet meltwater produces certain features both where the sheet covered the ground and in the periglacial region. These features are composed of **stratified drift**, which means that there has been some sorting of the debris as it was

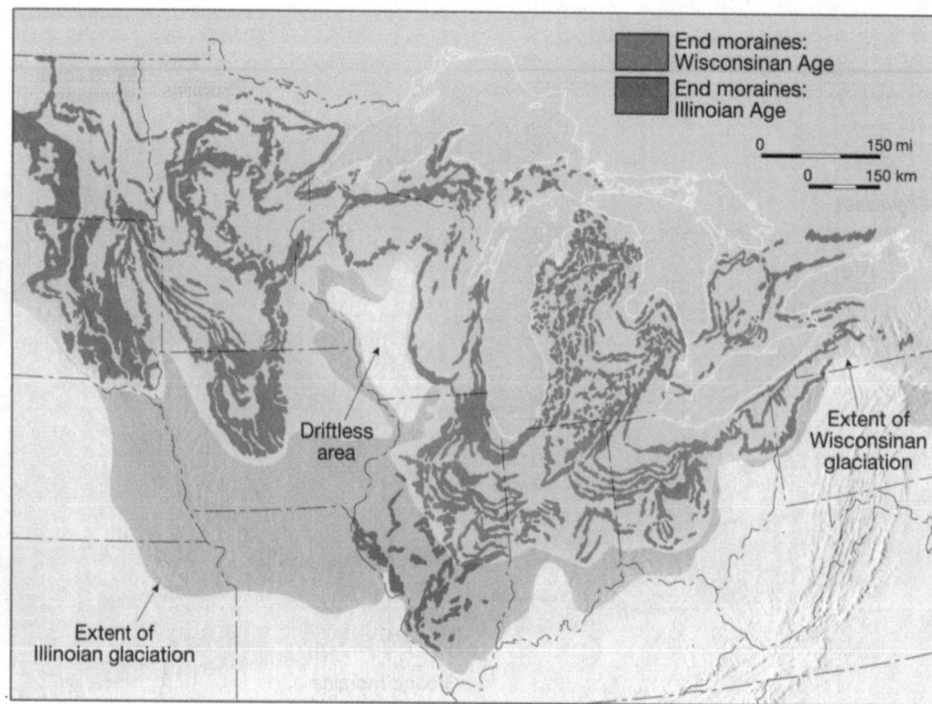

▶ **Figure 19-23** Terminal and recessional moraines (collectively referred to as end moraines) near the Great Lakes in the United States resulting from the more-recent Wisconsin stage glaciations are more prominent than those left by the earlier Illinoian stage glaciations.

carried along by the meltwater. Glaciofluvial features are composed largely or entirely of gravel, sand, and silt because meltwater is incapable of moving larger material.

The most extensive glaciofluvial features are **outwash plains**, which are smooth, flat alluvial aprons deposited beyond recessional or terminal moraines by streams issuing from the ice (see Figure 19-21). Streams of water, heavily loaded with reworked till or with debris washed directly from the ice, issue from the melting glacier to form a braided pattern of channels across the area beyond the glacial front. As they flow away from the ice, these braided streams, choked with debris, rapidly lose their speed and deposit their load. Such outwash deposits sometimes cover many hundreds of square kilometers. They are occasionally pitted by kettles that often become ponds or small lakes. Beyond the outwash plain, there is sometimes a lengthy deposit of glaciofluvial alluvium confined to a valley bottom; such a deposit is termed a **valley train**.

Less common than outwash plains, but more conspicuous, are long sinuous ridges of stratified drift called **eskers**, named from *eiscir*, another Irish word for "ridge." These landforms are composed largely of glaciofluvial gravel and are thought to have originated when streams flowing through tunnels in the interior of the ice sheet became choked off during a time in which the ice was neither flowing nor advancing. These streams beneath the stagnating sheet often carry a great deal of debris, and as the ice melts, the streams deposit much of their load in the tunnel. Eskers are this debris exposed once the ice melts away. They are usually a few dozen meters high, a few dozen meters wide,

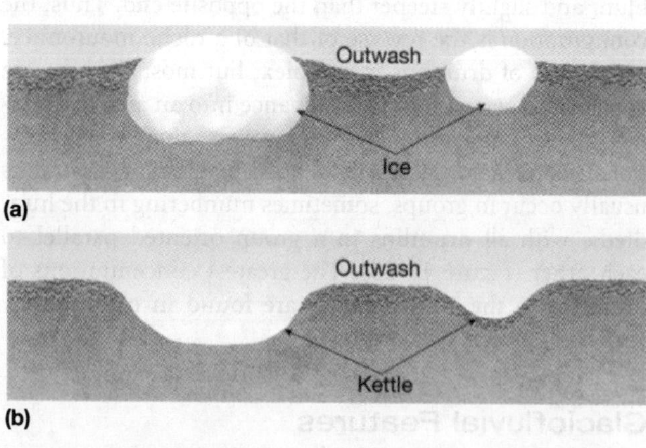

▲ **Figure 19-24** The formation of kettles. During deglaciation isolated masses of ice are often mixed in with the glacial debris and outwash and melt slowly because of the insulation provided by the surrounding debris. When the ice does eventually melt, sizable depressions known as kettles may pit the surface of the outwash. Today, many kettles contain water as lakes.

▲ **Figure 19-25** Drumlin west of Rochester, New York. The glacier flowed from right to left.

FOCUS

Shrinking Glaciers

Glaciers are sensitive indicators of environmental change. As we have seen, the size of a glacier is a delicate balance between the accumulation of ice and the ablation of ice. A relatively small increase in summer temperature or decrease in winter precipitation for a number of years may lead to the retreat of a glacier. But the relationships are sometimes counterintuitive: a glacier might also advance if slightly higher temperatures lead to an increase in winter precipitation.

Throughout most of the world glaciers are retreating, and some of the most striking evidence for this comes not from specialized instruments or satellite measurements, but from old-fashioned photographs. Bruce Molina of the U.S. Geological Survey (USGS) has compared photographs of Alaskan glaciers taken from the 1890s to the 1970s with new photographs he has taken from the exact same locations. The contrast between pairs of images taken just decades apart in some cases is remarkable (Figure 19-D).

Overall, Molina estimates that since the most recent peak of glacial ice cover in the 1700s, Alaska has lost about 15 percent of its total glacier cover—a loss of perhaps 10,000 square kilometers (3800 square miles) of ice surface area.

Not only have glaciers been shrinking, but the photographs also reveal that vegetation patterns have been changing. In some areas where little vegetation was visible in the past, an extensive plant cover is now obvious.

In other parts of the world, the retreat of glaciers is also being observed: in the Alps, the Himalayas, the Cascades, the Andes, and the Sierra Nevada. Greenland's largest outlet glacier, Jakobshavn's Isbrae, near the town of Ilulissat along the west coast, has retreated about 40 kilometers (25 miles) over the last 150 years—retreating half a kilometer in 2007 alone (Figure 19-E).

The causes of the observed changes in glaciers and vegetation are certainly linked to global climate change. Although some scientists are reluctant to conclude that the changes are solely a result of human-induced global warming, many researchers see the findings as some of the most convincing evidence yet of the human impact on global climate.

▲ **Figure 19-D** Change in the Muir Glacier and Riggs Glacier in Alaska's Glacier Bay National Park and Preserve. In 1941 (left) the Muir and Riggs Glaciers filled Muir Inlet. By 2004 (right), the Muir Glacier had retreated out of view and lush vegetation is now growing on the hillsides in the foreground. The "trimline" along the mountain front of the left side of the photograph indicates that the Muir Glacier was at least 600 meters (2000 feet) thick in 1941.

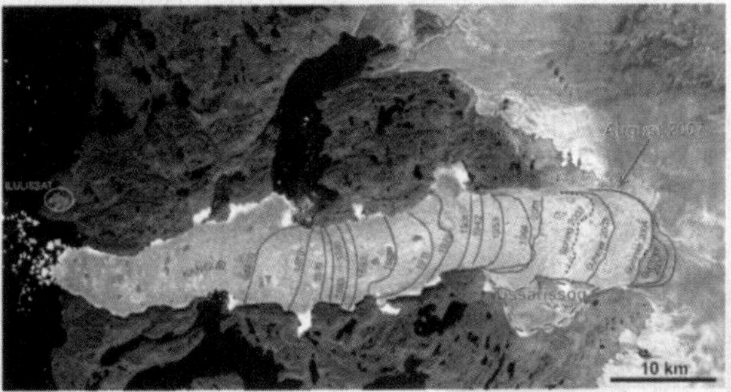

▲ **Figure 19-E** Jakobshavn's Isbrae glacier in Greenland has retreated 40 kilometers over the last 150 years. The red line marks the terminus of the glacier in 2007; the blue color in the inlet is floating ice that has calved off the glacier. This satellite image is a mosaic of Landsat and ASTER images taken in June 2003.

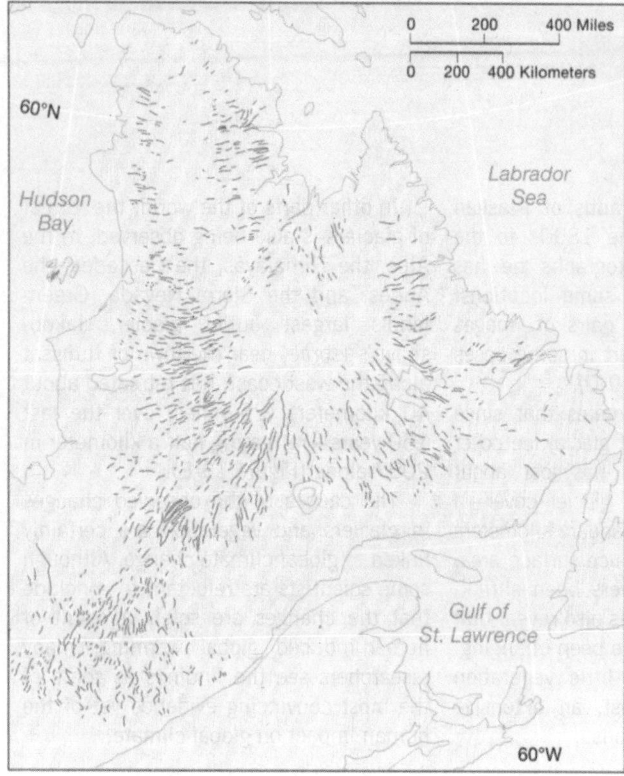

▲ **Figure 19-26** The distribution of eskers in eastern Canada.

and may be a few dozen kilometers (or up to 160 kilometers [a hundred miles]) long (Figure 19-26).

Small, steep mounds or conical hills of stratified drift are found sporadically in areas of ice-sheet deposition. These *kames* (the word derives from *comb*, an old Scottish word referring to an elongated steep ridge) appear to be of diverse origin, but they are clearly associated with meltwater deposition in stagnant ice. They are mounds of poorly sorted sand and gravel that probably formed within glacial fissures or between the glacier and the land surface (Figure 19-27).

▲ **Figure 19-27** A kame in southeastern Wisconsin, near Dundee.

Many seem to have been built as steep fans or deltas against the edge of the ice that later collapsed partially when the ice melted. Morainal surfaces containing a number of mounds and depressions are called *kame-and-kettle topography*.

Lakes are very common in areas that were glaciated during the Pleistocene. The old stream systems were obliterated or deranged by the ice sheets, and water remains ponded in the many erosional basins and kettles, and behind morainal dams. One has only to compare the northern and southern parts of the United States to recognize this fact (Figure 19-28). Most of Europe and the northern part of Asia demonstrate a similar correlation between past glaciation and present-day lakes.

Mountain Glaciers

Most of the world's high-mountain regions experienced extensive Pleistocene glaciation, and many mountain glaciers exist today. Mountain glaciers do not usually reshape the terrain as completely as the ice sheets did, partly because some portions of the mountains protrude above the ice and partly because the movement of mountain glaciers is channeled by the mountains. However, the effect of glacial action, particularly erosion, on mountainous topography is to create slopes that are often steeper and relief that is greater than the preglacial slope and relief (Figure 19-29). This action, of course, is in contrast to ice-sheet action, which tends to smooth and round the terrain.

Development and Flow

A highland icefield can extend broadly across the high country, submerging all but the uppermost peaks and finding its outlet in a series of lobes that issue from the icefield and move down adjacent drainage channels (Figure 19-30). Individual alpine glaciers usually form in sheltered depressions near the heads of stream valleys (often far below the level of the peaks). Glaciers from either source advance downslope, pulled by gravity and normally finding the path of least resistance along a preexisting stream valley. A system of merging glaciers usually develops, with a trunk glacier in the main valley joined by tributary glaciers from smaller valleys.

Erosion by Mountain Glaciers

Erosion and transportation of rock debris by highland icefields and alpine glaciers can reshape the topography in dramatic fashion. It remodels the peaks and ridges and transforms the valleys leading down from the high country (Figure 19-31).

In the High Country The basic landform feature in glaciated mountains is the **cirque**, a broad amphitheater hollowed out at the head of a glacial valley (Figure 19-32). It has very steep, often perpendicular, head and side walls and a floor that is either flat or gently sloping or else gouged enough to form a basin. A cirque marks the place where an alpine glacier originated. It is the first landform feature

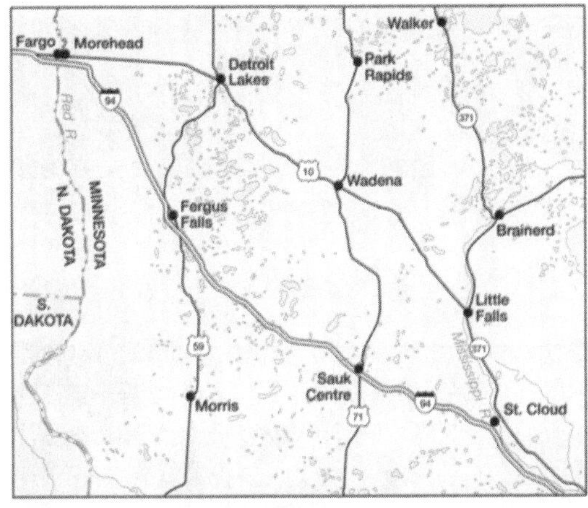

(a)

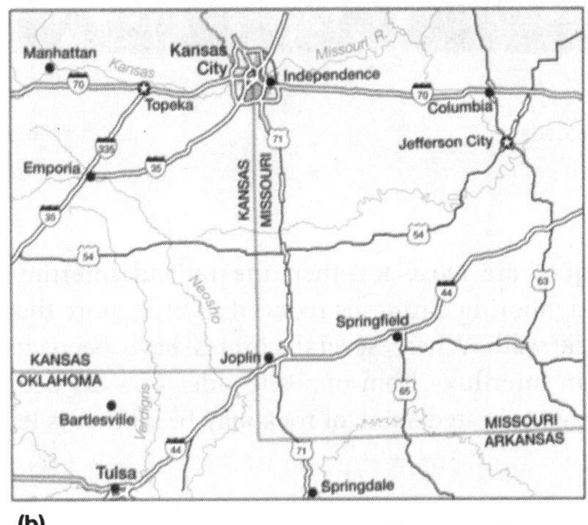

(b)

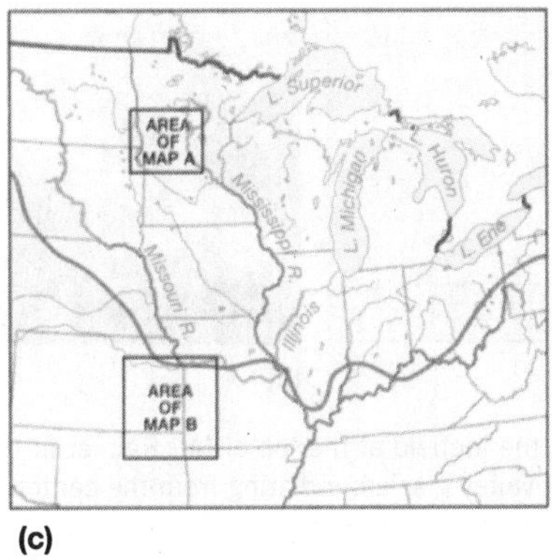

(c)

◀ **Figure 19-28** There is an abundance of natural lakes in the region north of the Ohio and Missouri rivers, a result of glacial action. South of these rivers, however, there was no glaciation and consequently natural lakes are almost unknown.

produced by alpine glaciation, essentially being quarried out of the mountainside. The shifting of the equilibrium line back and forth as a result of minor climatic changes may generate much of this quarrying action, along with plucking, mass wasting, and frost wedging. By the middle of summer, a large crevice known as the *bergschrund* opens at the top of the glacier, exposing part of the headwall to frost wedging. The shattered rock from the headwall is eventually incorporated into the glacial ice. As the glacier grows, its erosive effectiveness within the cirque increases, and when the glacier begins to extend itself down-valley out of the cirque, quarried fragments from the cirque are carried away with the flowing ice (Figure 19-33). Cirques vary considerably in size, ranging from a few hectares to a few square kilometers in extent. Many large cirques apparently owe their development to repeated episodes of glaciation.

▲ **Figure 19-29** The Teton Range of Wyoming has been sculpted in part by the action of mountain glaciers.

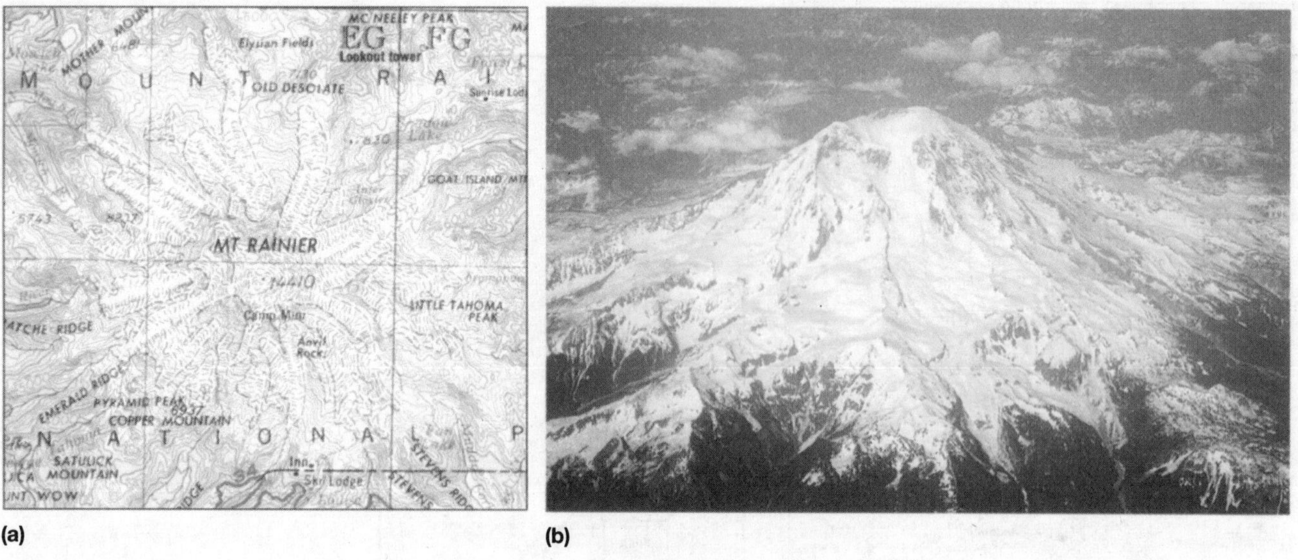

▲ **Figure 19-30** (a) Topographic map of the icefield at the top of Mt. Rainier in the Cascade Range in Washington. Note the tongues of valley glaciers radiating from the central field. (b) The ice-covered peak of Mt. Rainier, the highest volcano in the Cascade Range.

When the glacial ice in a cirque has melted away, there is often enough of a depression formed to hold water. Such a cirque lake is called a **tarn**.

A cirque grows steeper, in part through frost wedging, as its glacier plucks rock from the head and side walls. Where cirques are close together, the upland interfluve between neighboring cirques is reduced to little more than a steep rock wall. Where several cirques have been cut back into an interfluve from opposite sides of a divide, a narrow, jagged, serrated spine of rock may be all that is left

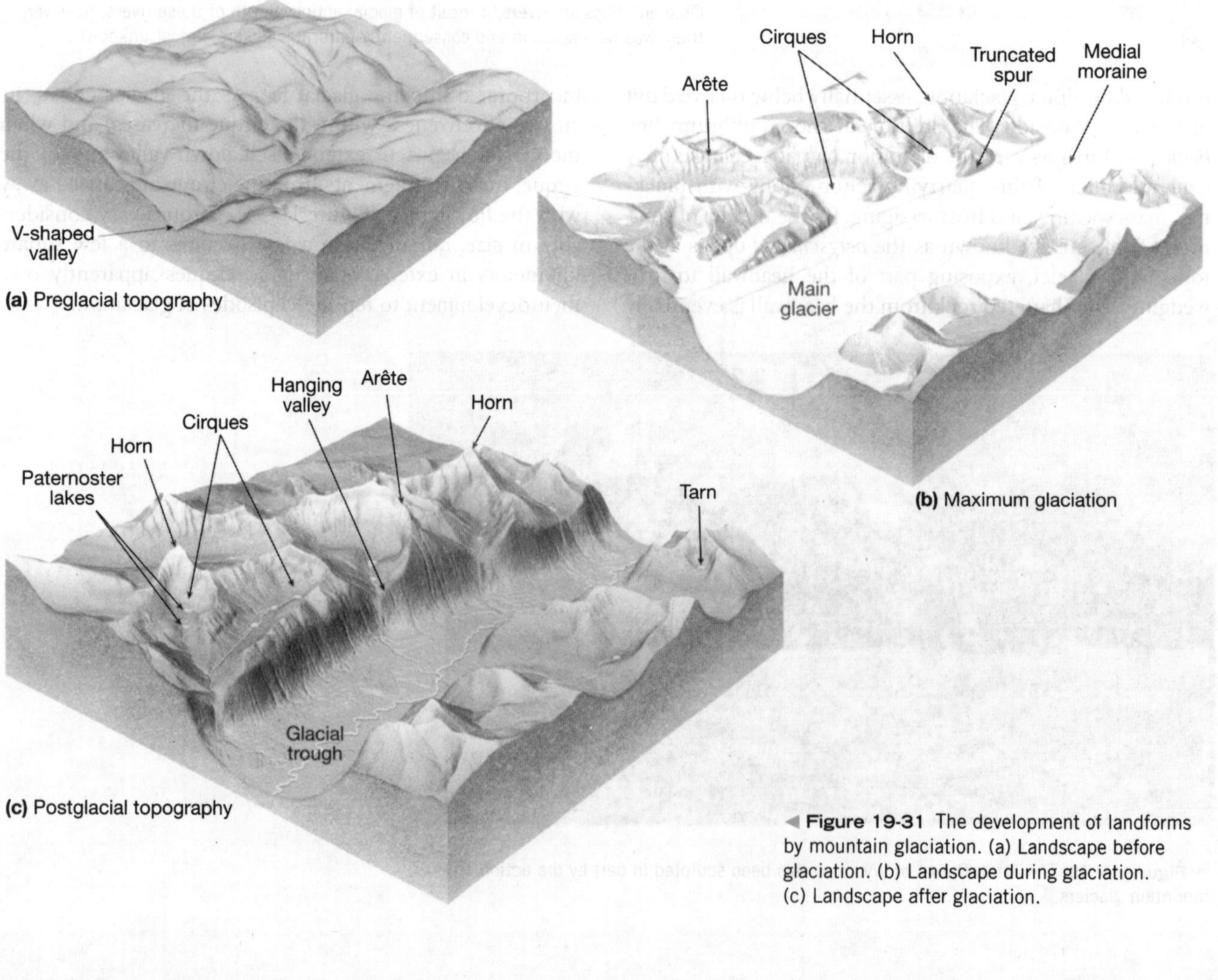

◀ **Figure 19-31** The development of landforms by mountain glaciation. (a) Landscape before glaciation. (b) Landscape during glaciation. (c) Landscape after glaciation.

▲ **Figure 19-32** Only a small remnant glacier with a small terminal moraine remains in this cirque on the north side of Wheeler Peak in Nevada's Great Basin National Park.

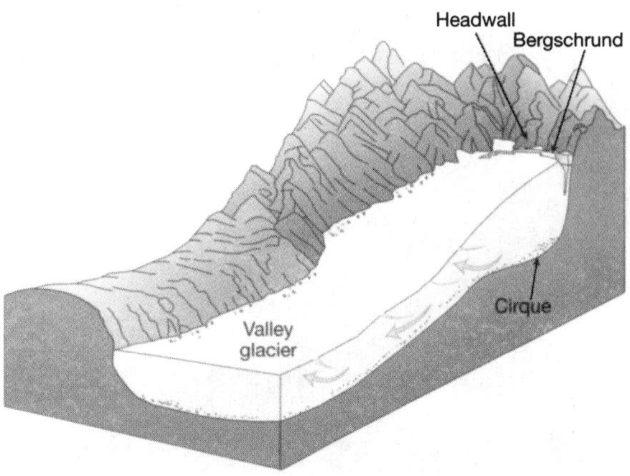

◄ **Figure 19-33** The development of a cirque at the head of a valley glacier.

of the ridge crest; this is called an **arête** (French for "fishbone"; derived from the Latin *arista*, "spine"). If two adjacent cirques on opposite sides of a divide are being cut back enough to remove part of the arête between them, the sharp-edged pass or saddle through the ridge is referred to as a **col** (*collum* is Latin for "neck"; Figure 19-34). An even more prominent feature of glaciated highland summits is a **horn**, a steep-sided, pyramid-shaped mountain peak formed by expansive quarrying of the headwalls where three or more cirques intersect (Figure 19-35). The name is derived from Switzerland's Matterhorn, the most famous example of such a glaciated spire.

In the Valleys Some alpine glaciers, such as those shown in Figure 19-36, did little more than erode cirques, presumably because of insufficient accumulation of ice to force a downvalley movement, and as a result the valleys below are unmodified by glacial erosion. Most alpine glaciers, however, as well as those issuing from highland icefields, flow down preexisting valleys and reshape them (Figure 19-37).

A glacier moves down a mountain valley with greater erosive effectiveness than a stream. The glacier is denser, carries more abrasive tools, and has an enormously greater

► **Figure 19-34** A glaciated section of the Front Range in north-central Colorado. A *col* is seen in the center of the photograph between the sharp peaks on either side in this scene from Rocky Mountain National Park.

▲ **Figure 19-35** Mount Aspiring is a prominent horn in the Southern Alps of New Zealand.

volume. It erodes by both abrasion and plucking. The lower layers of the ice can even flow uphill for some distance if blocked by resistant rock on the valley floor, permitting rock fragments to be dragged out of depressions in the valley floor.

An alpine glacier can deepen, steepen, and widen its valley. Abrasion and plucking take place not only on the valley floor but along the sides as well. The cross-sectional profile is often changed from a stream-cut "V" shape to an ice-eroded "U" shape flared at the top (Figure 19-38). Moreover, the general course of the valley may be straightened somewhat because the ice does not meander like a stream; rather it can grind away the protruding spurs that separate side canyons, creating what are called *truncated spurs*, and thereby replacing the sinuous course of the stream with a somewhat straighter, U-shaped **glacial trough** (see Figure 19-31). Not all glacial valleys are U-shaped, however; the resistance of the bedrock, jointing patterns, and processes such as exfoliation and rockfall may also influence the final shape of a glacial trough.

As might be expected, a glacier grinding along the floor of a glacial trough does not produce a very smooth surface. Valley glaciers do not erode a continuously sloping channel because differential erosion works with ice as well as with water. Therefore, resistant rock on the valley floor is gouged less deeply than weaker or more fractured rock. As a result, the down-valley profile of a glacial trough is often marked by an irregular series of rock steps or benches, separated by steep (although usually short) cliffs on the down-valley side. Such landforms are known as **glacial steps** (Figure 19-39).

A postglacial stream that flows out of a cirque and down a glacial trough usually has a relatively straight course, but typically includes rapids and waterfalls—particularly down the cliffs that mark the series of glacial steps. Small lakes may remain in shallow excavated depressions on the benches of the glacial steps, forming a sequence called **paternoster lakes** after a fancied resemblance to beads on a rosary.

As we will discuss in Chapter 20, some of the most spectacular glacial troughs occur along coastlines where valleys have been partly drowned by the sea to create *fjords*.

Our focus in the last few paragraphs has been on major valleys occupied by trunk glaciers. The same processes are at work and the same features produced in tributary valleys, although usually on a lesser scale. However, one important distinction between main valleys and tributary valleys is the amount of glacial erosion. Erosive effectiveness is determined largely by the amount of ice that passes through the valley; thus, the smaller ice streams of tributary valleys cannot widen and deepen as much as the main valley glaciers.

When occupied by glaciers and thus covered by a relatively level field of ice, main and tributary valleys may appear equally deep (see Figure 19-1). When the ice melts, however, the valleys are of different depths because

▶ **Figure 19-36** Three small cirques lined up in a row at about the 3300-meter (11,000-foot) level on Mount Nebo in central Utah. The original glaciers may never have been large enough to move out of their cirques to become valley glaciers.

▲ **Figure 19-37** The magnificent glacial landscape of Yosemite National Park, looking past Half Dome up Tenaya Canyon. During the Pleistocene, glaciers flowed down Tenaya Canyon, past Half Dome and into Yosemite Valley. The largest glaciers filled the valley, but did not cover Half Dome—the shape of Half Dome is largely due to exfoliation and jointing in the granite.

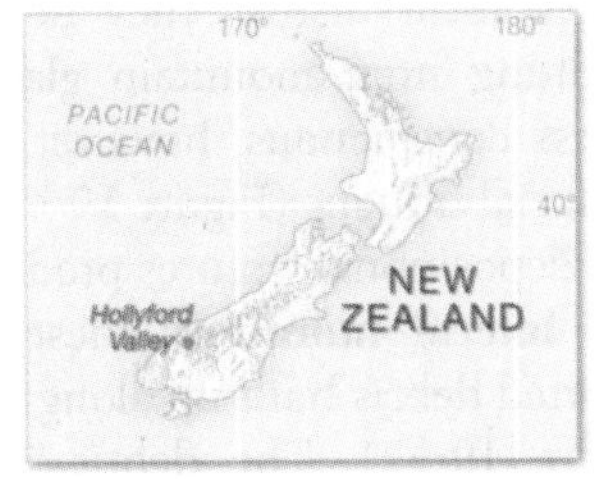

▶ **Figure 19-38** The Hollyford Valley in New Zealand's South Island has been shaped by glacial erosion. The U-shaped cross section is conspicuous.

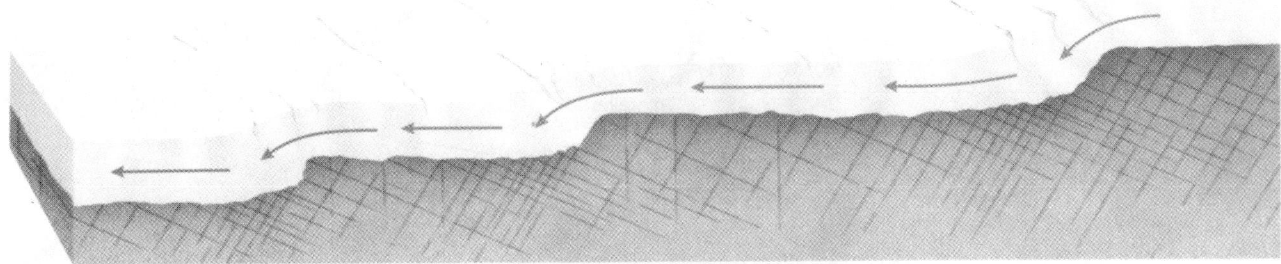

▲ **Figure 19-39** A longitudinal cross section of a glaciated valley in hilly or mountainous terrain showing a sequence of glacial steps.

▲ **Figure 19-40** Bridalveil Creek plunges out of a hanging valley in California's Yosemite Valley.

the smaller tributary glaciers could not erode as deeply as the larger trunk glaciers. The mouths of the tributary valleys are characteristically perched high along the sides of the major troughs, forming **hanging valleys** (see Figure 19-31). Typically, streams that drain the tributary valleys must plunge over waterfalls to reach the floor of the main trough. Several of the world-famous falls of Yosemite National Park are of this type (Figure 19-40).

Deposition by Mountain Glaciers

Depositional features are less significant in areas of mountain glaciation than in areas where continental ice sheets have been at work. The high country is almost totally devoid of till; only in the middle and lower courses of glacial valleys can much deposition be found.

The principal depositional landforms associated with mountain glaciation are moraines. Terminal and recessional moraines form just as they do with ice sheets. Moraines resulting from mountain glaciation are much smaller and less conspicuous, however, because they are restricted to glacial troughs (Figure 19-41).

The largest depositional features produced by mountain glaciation are **lateral moraines**; these are well-defined ridges of unsorted debris built up along the sides of valley glaciers (Figure 19-42). The debris is partly material deposited by the glacier and partly rock that falls or is washed down the valley walls (Figure 19-43).

Where a tributary glacier joins a trunk glacier, their lateral moraines (and debris carried on top of the ice along the sides of the glaciers) become united at the intersection and often continue together down the middle of the combined

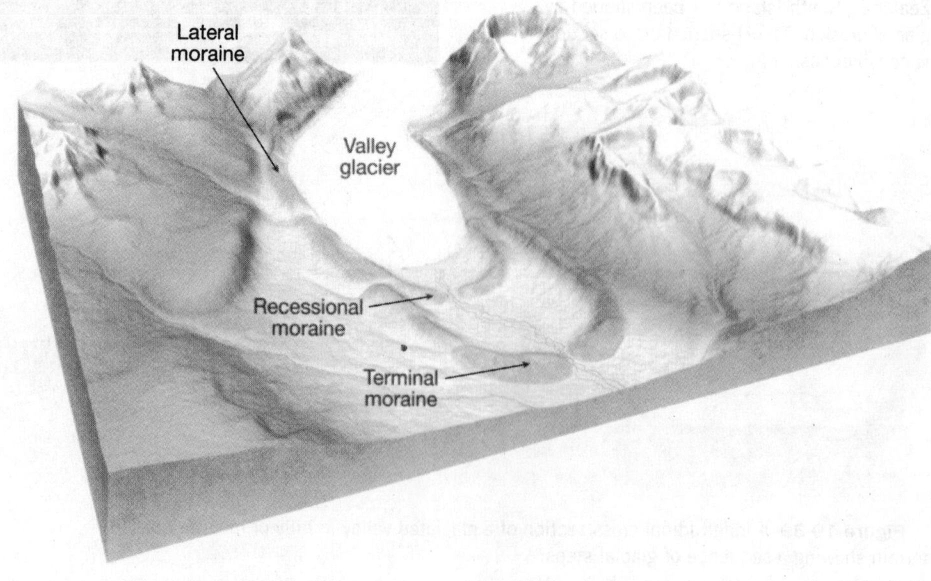

▶ **Figure 19-41** Common types of moraines in mountainous areas.

◀ **Figure 19-42** The Athabaska valley glacier issuing from a highland icefield in the Canadian Rockies. The distance from the highest icefall to the toe is about 11 kilometers (7 miles). The glacier has conspicuous lateral moraines along both sides.

glacier as a dark band of rocky debris known as a **medial moraine** (Figure 19-44). Medial moraines are sometimes found in groups of three or four running together, indicating that several glaciers have joined to produce a candy-cane effect of black (moraine) and white (ice) bands extending down the valley.

The debris left by meltwater below mountain glaciers is similar to that bordering ice sheets because similar outwash is produced.

The Periglacial Environment

The term *periglacial* means "on the perimeter of glaciation." More than 20 percent of the world's land area is presently periglacial, but most of this was covered by ice on one or more occasions during the Pleistocene Epoch.

Periglacial lands are found either in high latitudes or in high elevations. Almost all are in the Northern Hemisphere because the Southern Hemisphere continents either do not extend far enough into the high latitudes to be significantly affected by contemporary glaciation (Africa, South America, Australia) or are mostly ice covered (Antarctica).

Nonglacial land-shaping processes function in periglacial areas, but in addition the pervasive cold imparts some distinctive characteristics, the most notable of which is *permafrost* (discussed in Chapter 9). Either continuous or discontinuous permafrost occurs over most of Alaska and more than half of Canada and Russia. There are also extensive high-altitude areas of permafrost in Asia, Scandinavia, and the western United States. In some cases, the frozen ground extends to extraordinary depths; a thickness of 1000 meters (3000 feet) has been found in Canada's

◀ **Figure 19-43** Emerald Bay on the California side of Lake Tahoe is nearly closed off from the rest of the lake by the arms of two lateral moraines left by Pleistocene glaciers.

◀ **Figure 19-44** Prominent medial moraines on McBride Glacier, near Glacier Bay in southeastern Alaska.

Northwest Territories and 1500 meters (4500 feet) in north-central Siberia.

The most unique and eye-catching periglacial terrain is **patterned ground**, the generic name applied to various geometric patterns that repeatedly appear over large areas in the Arctic (Figure 19-45). The patterns are varied, with their formation related to the freeze-thaw cycle that slowly disrupts uniform surfaces of soil and regolith, forming rough polygons. The principal significance of patterned ground is that it demonstrates the mobility of periglacial surfaces, emphasizing the role of soil ice in producing geomorphic activities largely unknown in warmer regions.

Proglacial Lakes Another sometimes conspicuous development in periglacial regions is a **proglacial lake** (*pro* here means "marginal to" or "in advance of"). Where ice flows across a land surface, the natural drainage is either impeded or blocked, and meltwater from the ice can become impounded against the ice front, forming a proglacial lake. Such an event sometimes occurs in alpine glaciation but is much more common along the margin of continental ice sheets, particularly when the ice stagnates.

Most proglacial lakes are small and quite temporary because subsequent ice movements cause drainage changes and because normal fluvial processes, accelerated by the growing accumulation of meltwater in the lake, cut spillways

▲ **Figure 19-45** Polygonal ground patterns near Prudhoe Bay, Alaska, formed by the action of the freeze-thaw cycle on soil and regolith.

▲ **Figure 19-46** The channeled scablands of eastern Washington near the city of Spokane were scoured by a series of enormous floods during the Pleistocene.

or channels to drain the impounded waters. Sometimes, however, proglacial lakes are large and relatively long lived. Such major lakes are characterized by considerable fluctuations in size due to the changing location of the receding or advancing ice front. Several huge proglacial lakes were impounded along the margins of the ice sheets as they advanced and retreated during the Pleistocene Epoch in North America, Europe, and Siberia.

Among the most dramatic consequences of lakes once impounded by continental glaciers is found in eastern Washington near Spokane. A series of great floods during the Pleistocene scoured eastern Washington's "channeled scablands" when enormous volumes of water were periodically let loose from an ice sheet-dammed lake near present-day Missoula, Montana. Dozens of times during the Pleistocene, a lobe of the continental ice sheets blocked the Clark Fork River, forming Pleistocene Lake Missoula, which at times was more than 300 meters (1000 feet) deep. When the ice dam failed, a wall of water surged across the landscape, scouring deep channels and leaving enormous ripple marks before the water eventually drained into the Columbia River (Figure 19-46).

Causes of the Pleistocene Glaciations

Ice ages are fascinating not only because of the landscape changes they bring about but also because of the questions they raise. What initiates massive accumulations of ice on the continental surfaces, stimulates their advances and retreats, and finally causes them to disappear? Scientists and other scholars have pondered these questions for decades. Any complete theory of the causes of the Pleistocene glaciations must be able to account for four main glacial characteristics:

1. The accumulation of ice masses more or less simultaneously at various latitudes in both hemispheres but without uniformity (e.g., much less in Siberia and Alaska than in similar latitudes in Canada and Scandinavia)
2. The apparently concurrent development of pluvial conditions in dryland areas
3. Multiple cycles of ice advance and retreat, including both minor fluctuations over decades and centuries and major glaciations and deglaciations over tens of thousands of years
4. Eventual total (or near total) deglaciation

We know that glaciers grow when there is a net accumulation of snow over a period of time and that glaciers waste away when summer melting exceeds winter snowfall. Beyond that simplistic statement, however, in some cases it is not always clear whether a colder climate would be more conducive to glaciation than a warmer but wetter one! Although colder conditions would inhibit summer wastage and thus enhance the longevity of the winter accumulation, cold air cannot contain much water vapor. Hence, warmer winters would favor increased snowfall, whereas cooler summers are needed for decreased melting. Even a theory that accommodated either significantly increased snowfall or significantly decreased melting, or a combination of both, would still have to take into account the timing of glacial advances and retreats.

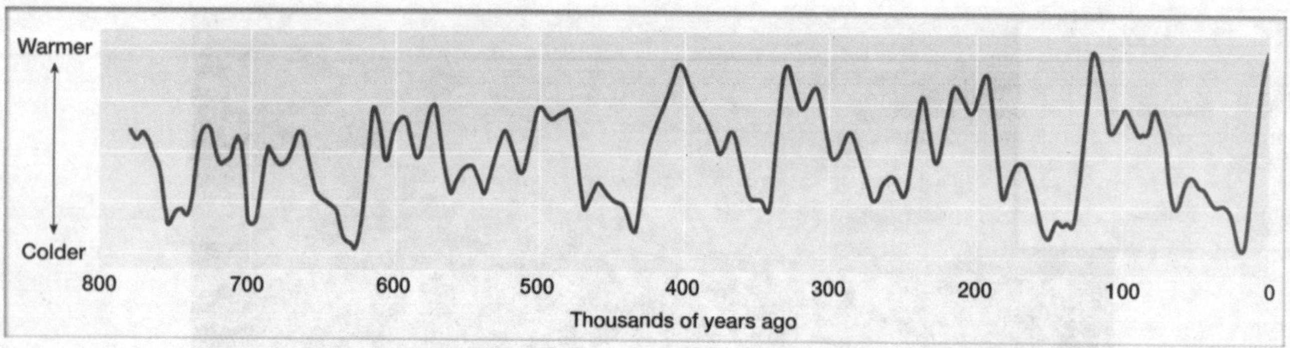

▲ **Figure 19-47** Global temperature fluctuations from 800,000 years ago to the present. Many glacial and interglacial periods are easily recognizable.

Our understanding of the sequence of glacial and interglacial episodes during the Pleistocene has been dramatically enhanced in recent years through oxygen isotope analysis. As we saw in Chapter 8, the ratio of $^{18}O/^{16}O$ derived from carbonate sediments on the ocean floor and from ice cores in glaciers can be used as a proxy for temperature. The climate record that emerged from these studies shows a sequence of major glacial and interglacial episodes, along with a great many minor advances—some of which ended quite abruptly (Figure 19-47).

A lot is known about many of the factors associated with the Pleistocene glaciations, but certainly not all of the factors. As we discussed in Chapter 8, cyclical variations in Earth–Sun relations, often referred to as *Milankovitch cycles*, played a part. The combination of slight variations in the inclination of Earth's axis and the eccentricity of Earth's orbit, as well as the changing orientation of Earth's axis relative to the stars (known as the "precession of the equinoxes"), seem to correlate fairly well with some—but not all—of the major glacial advances and retreats during the Pleistocene.

Other factors that have been postulated as triggers or components of glacial episodes include variations in the energy output of the Sun; variations in the level of carbon dioxide and other greenhouse gases in the atmosphere; the changing positions of continents and the configuration of ocean basins and ocean circulation patterns; atmospheric circulation changes due to increased elevation of continental masses after a period of tectonic uplift; and reductions in insolation reaching the surface due to particulates released during massive volcanic eruptions.

We make no attempt here to offer details of these many theories, primarily because none of them alone or in combination yet provides a complete explanation of the climate changes during the Pleistocene. The search for a complete explanation continues.

Are We Still in an Ice Age?

One significant outgrowth of our study of what causes a glacial age is the further puzzle of whether or not Earth's most recent ice age has actually ended. Are we now living in a postglacial period or in an interglacial period? Is the cycle of glaciation and deglaciation that marked the Pleistocene really over, or have the glaciers merely "gone back to get some more rocks"? Based on the climate patterns of the last two and a half million years, it is quite possible that Earth will enter another period of glaciation within a few tens of thousands of years. However, might human-enhanced greenhouse effect and global warming "postpone" the onset of this glacial period?

We are, of course, much too close to the event in a temporal sense to obtain a definitive answer.

CHAPTER 19 LEARNING REVIEW

After studying this chapter, you should be able to answer the following questions. Key terms from each text section are shown in bold type. Definitions for key terms are also found in the glossary at the back of the book.

Types of Glaciers (p. 503)

1. Describe and contrast **continental ice sheets**, **highland icefields**, **alpine glaciers**, **valley glaciers**, and **cirque glaciers**.

Glaciations Past and Present (p. 504)

2. Why is the **Pleistocene Epoch** so important to physical geography, whereas other ice ages are not?
3. Briefly describe the world-wide extent of ice cover during the peak of the Pleistocene glaciations.
4. What is meant by a **periglacial zone**?
5. Explain how and why global sea level fluctuated during the Pleistocene.
6. What is the relationship of large continental ice sheets to the crustal depression associated with isostatic adjustment (isostasy)?
7. What is meant by the **pluvial effects** of the Pleistocene?
8. Describe and explain the formation of large **Pleistocene lakes** in western North America.
9. Describe the global extent of glaciers today.

Glacier Formation and Movement (p. 510)

10. Describe and contrast the processes of glacial ice **accumulation** and **ablation**.
11. Describe the metamorphosis from snow to **névé (firn)** to glacial ice.
12. Explain how the balance between ice accumulation and ablation influences the "advance" or "retreat" of a glacier. (In other words, explain what causes glaciers to grow larger or to become smaller.)
13. What is the relationship of the **equilibrium line** of a glacier to its **accumulation zone** and **ablation zone**?
14. Discuss the different components of glacial movement: **plastic flow of ice** and **basal slip**.
15. Why can a glacier continue to erode and transport rock even while it is retreating?

The Effects of Glaciers (p. 512)

16. Contrast the erosional processes of **glacial plucking** and glacial abrasion.
17. What roles does meltwater below a glacier play in the transportation and erosion of rock?
18. What is **glacial flour**? Glacial **drift**?
19. Describe the characteristics of glacial **till**.
20. How is a deposit of till likely to look different from a deposit of alluvium?
21. What is a **glacial erratic**?
22. How does **glaciofluvial deposition** differ from deposition directly by glacial ice?

Continental Ice Sheets (p. 515)

23. Describe and explain the formation of a **roche moutonnée**.
24. A **moraine** is made from what kind of glacial material?
25. Explain the formation of **terminal moraines** and **recessional moraines**.
26. Describe the general appearance of **ground moraine**.
27. Describe and explain the formation of a **kettle**.
28. Describe and explain the formation and orientation of a **drumlin**.
29. What is meant by **stratified drift**?
30. Describe the formation of a glacial **outwash plain** and a **valley train**.
31. How does an **esker** form?
32. Why are there so many lakes in areas that were glaciated by continental ice sheets during the Pleistocene?

Mountain Glaciers (p. 520)

33. Why are mountainous areas that have experienced glaciation usually quite rugged?
34. Describe and explain the formation of a **cirque**, **horn**, **arête**, and **col**.
35. Where in a glaciated landscape is a **tarn** found?
36. Contrast the general cross-sectional shape of a typical stream valley in the high mountains with that of a typical **glacial trough** (glacial valley).
37. Describe the general downvalley profile of a glacial trough and the formation of **glacial steps**.
38. What are **paternoster lakes** and where do they form?
39. Explain the formation of a **hanging valley**.
40. What is a **lateral moraine**?
41. How does a **medial moraine** form?

The Periglacial Environment (p. 527)

42. Briefly describe **patterned ground** in periglacial areas.
43. What is a **proglacial lake**?

Log in to **www.mygeoscienceplace.com** for animations, videos, RSS feeds, maps, and self-study quizzes to enhance your study of **Glacial Modification of Terrain**.

20 Coastal Processes and Terrain

In addition to running water, two other forms of moving water—waves and ocean currents—shape the land. Coastlines are shaped by the agitated waters crashing or lapping against them, with the result that coastal terrain is often quite different from the landscape just a short distance from shore.

The Impact of Waves and Currents on the Landscape

Coastal processes affect only a tiny fraction of the total area of Earth's surface, but they create a landscape that is almost totally different from any other on the planet. Although there is some overlap, in essence waves are agents of erosion, and currents are agents of transportation and deposition. The most notable land features created by waves are rocky cliffs and headlands (promontories of sloping land projecting into the sea). Depositional features are diverse in form, but by far the most common are beaches and sandbars.

Beaches along the shorelines of both oceans and lakes are the most distinctive aspect of coastal landscapes. They provide a transition from land to water and are usually impermanent features of the landscape, built up during times of "normal" weather and eroded during storms.

Coastal Processes

The coastlines of the world's oceans and lakes extend for hundreds of thousands of kilometers. Every conceivable variety of structure, relief, and topography can be found somewhere along these coasts. The distinctiveness of the coastal milieu, however, is that it is at the interface of three major components of Earth's environment: hydrosphere, lithosphere, and atmosphere. This interface is dynamic and highly energetic, primarily because of the restless motions of the waters (Figure 20-1).

We saw in Chapter 18 that wind can be an important shaper of landforms on the continents—particularly through the deposition of sand. Along coastlines, the wind has an even greater influence on topography because the surface of a large body of water can be deformed abruptly and rapidly by wind action. This deformation of the water surface is what creates waves and ocean currents, both of which shape coastlines, producing topographic features found throughout the world in all latitudes and in a variety of climates.

▲ **Figure 20-1** A coastline is the place where hydrosphere, lithosphere, and atmosphere meet. It is often an interface of ceaseless movement and energy transfer. Pounding waves do not always erode evenly and smoothly. Sea stacks and cliffs testify to the principle of differential erosion in this scene from Port Campbell National Park, Victoria, Australia.

◀ Keoneloa Bay on the island of Kaua'i in Hawai'i.

The wind is not the only force causing water to move, of course, but from the standpoint of geomorphic effects it is the most important. Oceanic coastlines also experience daily tidal fluctuations which often move enormous quantities of water. Tectonic events, particularly earthquakes, contribute to water motion, as does volcanic activity upon occasion. Even more fundamental are long-term variations in sea or lake level caused by tectonic forces and *eustatic sea-level change* (sea-level change entirely due to an increase or decrease in the amount of water in the world ocean—especially as a result of variations in the size of continental ice sheets during glacial and interglacial periods).

The processes that shape the topography of oceanic coastlines are similar to the processes acting on lakeshores, with three important exceptions:

1. Along lakeshores, the range of tides is so small that they are insignificant to landform development.
2. The causes of sea-level fluctuations are quite different from the causes of lake-level fluctuations.
3. Coral reefs are built only in tropical and subtropical oceans, not in lakes.

With these exceptions, the topographic forms produced on seacoasts and lakeshores are generally similar. Even so, the larger the body of water, the greater the effects of the coastal processes. Thus, topographic features developed along seacoasts are normally larger, more conspicuous, and more distinctive than those found along lakeshores.

There are many processes that contribute to the shaping of coastal features. Virtually all of the internal and external processes we have discussed so far may be at work along a shoreline. In addition, a number of processes largely confined to coastal areas may be at work, and of these, by far the most important is the work of waves.

Waves

1. Wave Motion and Wave Refraction
2. Tsunami

Waves entail the transfer of energy through a cyclical rising and falling motion in a substance. Our interest here is in water waves, which are simply undulations in the surface layers of a water body. Although water waves appear to move water horizontally, this appearance is misleading: in open water, the form of the wave (and therefore its energy) moves along through the water, although the water itself is shifted only very slightly; as we will see, this motion changes in shallow water, where waves crest and break.

Most water waves are wind generated, set in motion largely by the friction of air blowing across the water. This transfer of energy from wind to water initiates wave motion. Some water waves (called *forced waves*) are generated directly by wind stress on the water surface; they can develop to considerable size if the wind is strong and turbulent but usually last for only a limited time and do not travel far. Water waves become **swells** when they escape the influence of the generating wind, and in so doing can travel enormous distances. A small number of all water waves are generated by something other than the wind, such as a tidal surge, volcanic activity, or undersea tectonic movement (discussed later in this chapter).

Waves of Oscillation and Translation As a wave passes a given point on the water surface, the water at that point makes a small circular or oscillatory movement, with very little forward motion (to oscillate means to move back and forth over the same space again and again). Waves that cause water to move this way are called **waves of oscillation**. As the wave passes, the water moves upward, producing a *wave crest*, as shown in Figure 20-2. Then crest formation is followed by a sinking of the surface that creates a *wave trough*. The horizontal distance from crest to crest or from trough to trough is called the **wavelength**. The vertical dimension of wave development is determined by the circular orbit of the surface water as the wave form passes; the vertical distance from crest to trough is equivalent to the diameter of this orbit and is called the **wave height**. The height of any wave depends on wind speed, wind duration, water depth, and *fetch* (the area of open water over which the wind blows).

The passage of a wave of oscillation normally moves water only very slightly in the direction of flow. Thus an object floating on the surface simply bobs up and down without advancing, except as it may be pushed by the wind. The influence of wave movement diminishes rapidly

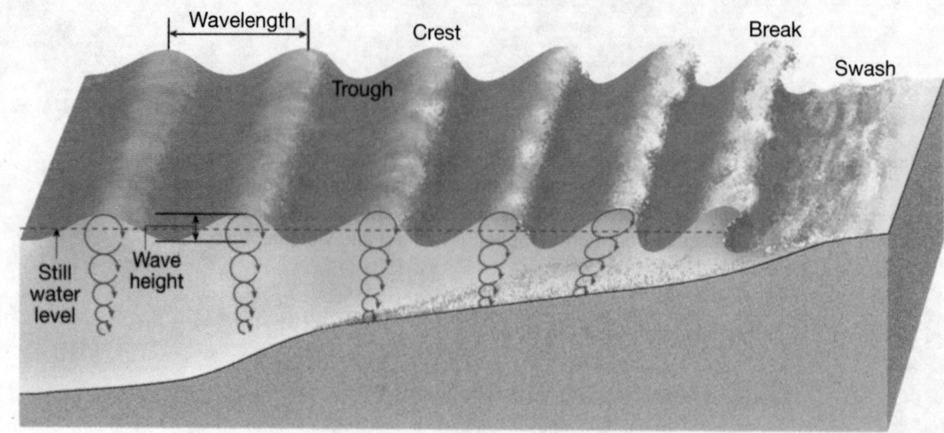

▶ **Figure 20-2** In deep water, the passage of a wave involves almost circular movements of the water. Agitation diminishes rapidly with increasing depth, as shown here by decreasing orbital diameter downward. As the wave moves into shallow water, the revolving orbits of the water become more elliptical, the wavelength becomes shorter, and the wave becomes steeper. Eventually the wave "breaks" and dissipates its remaining energy as it washes up onto the beach.

with depth; even very high waves stir the subsurface water to a depth of only a few tens of meters.

Waves often travel great distances across deep water with relatively little change in speed or shape. As they roll into shallow water, however, a significant metamorphosis occurs. When the water depth becomes equal to about half the wavelength, the wave motion begins to be affected by frictional drag on the sea bottom. The waves of oscillation then rapidly become changed into **waves of translation**, and the result is significant horizontal movement of the surface water. Friction retards the progress of the waves so that they are slowed and bunch together, marking a decrease in wavelength, while at the same time their height is increased. As the wave becomes higher and steeper, frictional drag becomes even greater, which causes the wave to tilt forward and become more and more unstable. Soon and abruptly the wave *breaks* (Figure 20-3), collapsing into whitewater surf or plunging forward as a breaker or, if the height is small, perhaps simply surging up the beach without cresting.

When a wave breaks, the motion of the water instantly becomes turbulent, like that of a swift river. The breaking wave rushes toward shore or up the beach as **swash**. This surge can carry sand and rock particles onto the beach, or it can pound onto rocky headlands and sea cliffs with considerable force (Figure 20-4). The momentum of the surging swash is soon overcome by friction and gravity, and a return flow, called **backwash**, drains much of the water seaward again carrying loose material with it, usually to meet the oncoming swash of the next wave.

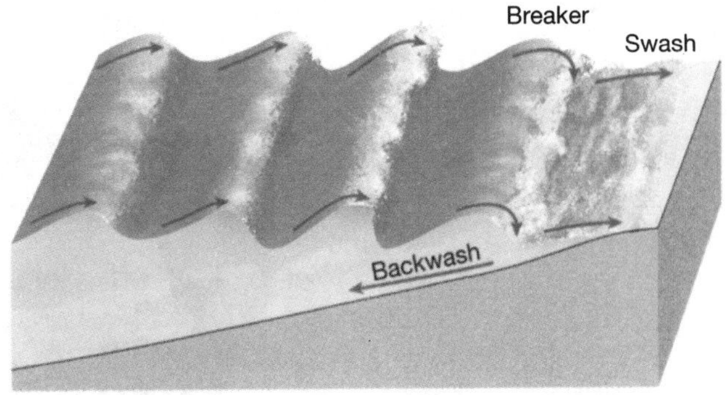

▲ **Figure 20-3** A breaking wave. In shallow water the ocean bottom impedes oscillation, causing the wave to become increasingly steeper until it is so oversteepened that it collapses and tumbles forward as a breaker. The surging water then rushes up the beach as swash, and then drains off the beach below the waves as backwash.

Wave Refraction Waves often change direction as they approach the shore, a phenomenon known as **wave refraction**. It occurs when a line of waves does not approach exactly parallel to the shore, or where the coastline is uneven or there are irregularities in water depth in the near-shore zone. For one or more of these reasons, one portion of a wave reaches shallow water sooner than other portions and is thus slowed down. This slowing down causes the wave line to bend (*refract*) as it pivots toward the obstructing area, finally breaking roughly parallel to shore. Thus wave energy tends to be concentrated in the vicinity of an obstruction and diminished in other areas (Figure 20-5).

The most conspicuous geomorphic result of wave refraction is the focusing of wave action on headlands (Figure 20-6), subjecting them to the direct onslaught of pounding waves, whereas an adjacent bay experiences much gentler, low-energy wave action. Other things being equal (such as the resistance of the bedrock), the differential effect of wave refraction tends to smooth the coastal outline by wearing back the headlands and increasing sediment accumulation in the bays.

▲ **Figure 20-4** The continuous pounding of waves can erode even the most resistant coastal rocks. This scene is near Port Edward on the Natal coast of South Africa.

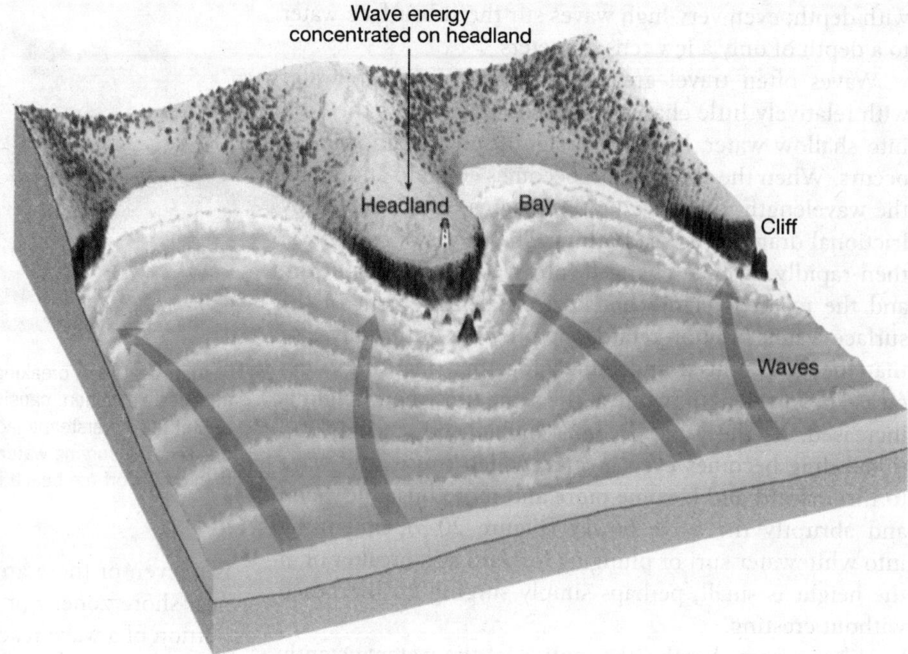

▲ **Figure 20-5** Refraction of waves on an irregular coastline. The waves reach the headland first and then "wrap" around it, breaking nearly parallel to the coastline as a result of wave refraction. Thus wave energy is concentrated on the headlands and is diminished in the bays.

Wave Erosion The most notable erosion along coastlines is accomplished by wave action. The incessant pounding of even small waves is a potent force in wearing away the shore, and the enormous power of storm waves almost defies comprehension—it is often large storm waves that accomplish most of the erosion along a shoreline. Waves break with abrupt and dramatic impact, hurling water, debris, and air in a thunderous crash onto the shore. Spray from breaking waves commonly moves as fast as 115 kilometers (70 miles) per hour, and small jets have been measured at more than twice that speed. This speed, coupled with the sheer mass of the water involved in such hydraulic pounding, is responsible for much coastal erosion, which is made much more effective by the abrasive rock particles carried by the waves.

Wherever the land along a shore is rocks or cliffs rather than sand, there is another dimension to wave erosion: the air forced into cracks in the rock as the wave hits the shore. The resulting compression is abruptly released as the water recedes, allowing instant expansion of the air. This pneumatic action is often very effective in loosening rock particles of various sizes.

▲ **Figure 20-6** The incessant pounding of waves on this soft-rock headland on the southern coast of the state of Victoria, Australia, produced a double arch that eventually eroded into a single arch. The double-arch photograph was taken in 1985 and the single-arch one in 1992. The view is in Port Campbell National Park.

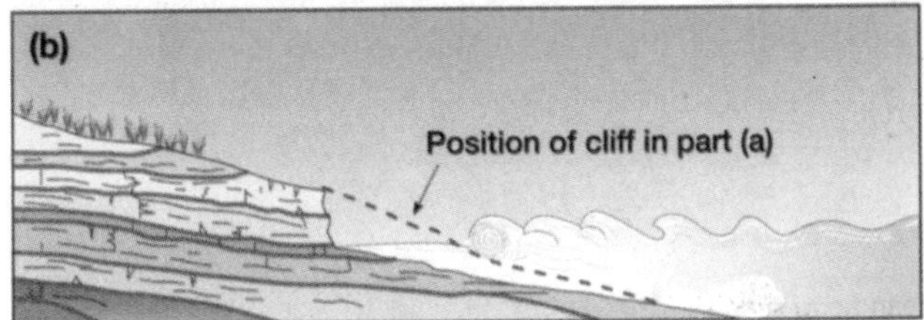

▲ **Figure 20-7** (a) Waves pounding an exposed rocky shoreline erode the rock most effectively at water level, with the result that a notch may be cut in the face of the headland. (b) The presence of this notch undermines the higher portion of the headland, which may subsequently collapse (c), producing a steep cliff. (d) The notching/undercutting/collapse sequence may be repeated many times, causing the cliff face to retreat.

Chemical action also plays a part in the erosion of rocks and cliffs because most rocks are to some extent soluble in seawater. In another form of weathering action, salts from seawater crystallize in the crevices and pores of onshore rocks and cliffs, and this deposition is a further mechanism for weakening and breaking up the rock (such *salt wedging* is discussed in Chapter 15).

On shorelines made up of cliffs, the most effective erosion takes place just at or slightly above sea level, so that a notch is cut in the base of the cliff. The cliff face then retreats as the slope above the undercutting collapses (Figure 20-7). The resulting debris is broken, smoothed, and made smaller by further wave action, and eventually most of it is carried seaward.

Where a shoreline is composed of sand or other unconsolidated material, currents and tides may also cause rapid erosion. Storms greatly accelerate the erosion of sandy shores; a violent storm can remove an entire beach in just a few hours, cutting it right down to bedrock.

Tsunami Occasionally, major oceanic wave systems are triggered by a sudden disruption of the ocean floor. These waves are called **tsunami** (from the Japanese *tsu* for harbor and *nami* for wave) or *seismic sea waves* (improperly called "tidal waves"). Most tsunami are a consequence of abrupt movement along an ocean floor fault—especially from the vertical displacement caused by reverse or thrust faulting along a subduction zone. Tsunami may also result from underwater volcanic eruptions and major underwater and coastal landslides.

The great destructive power unleashed by some tsunami comes from the way in which the ocean is disrupted to form such a wave. Recall that with wind-generated waves, only the surface of the ocean exhibits significant movement—the orbital movement of the water only extends down to a depth of about one-half the wavelength of the wave (rarely more than a few tens of meters). On the other hand, when fault rupture on the ocean floor generates a tsunami, the entire water column—from the floor of the ocean to the surface—is disrupted, displacing an enormous volume of water (Figure 20-8).

Out in the open ocean, tsunami are usually quite inconspicuous because they are low and have very long wavelengths (in the open sea a tsunami might have a wave height of only 0.5 meter [1.5 feet] with a wavelength of perhaps 200 kilometers [125 miles]), although they can travel at speeds exceeding 700 kilometers an hour (435 miles per hour). When tsunami reach shallow water, however, they change considerably. As a tsunami approaches a coast, it slows—as do all waves—causing the wavelength to decrease and the wave height to increase.

When they strike a shoreline, tsunami rarely form towering breaking waves; instead, most tsunami arrive as a very rapidly advancing surge of water, sometimes up to 40 meters (130 feet) high—unlike many large wind-generated waves, however, immediately behind the wave crest of a tsunami is an enormous volume of water that can surge great distances inland before receding. In many cases, before a tsunami arrives, the water withdraws from the coast, appearing like a very sudden, very low tide—this happens when the trough of the tsunami arrives. Unfortunately, people sometimes venture out on the freshly exposed subtidal areas to collect shellfish or stranded fish, only to be caught a few minutes later by the rapid surge of water when the crest of the tsunami comes onshore. Frequently, there will be a series of surges and withdrawals, with the largest surge not necessarily the first to arrive.

Tsunami can usually be anticipated long enough in advance to allow time for evacuation of the impact area. Since they often originate from sudden fault displacement in a subduction zone, the resulting earthquakes are readily detectable by seismographs. However, as the great northern Sumatra earthquake and tsunami disaster of December 2004 revealed, if local warning systems are not in place,

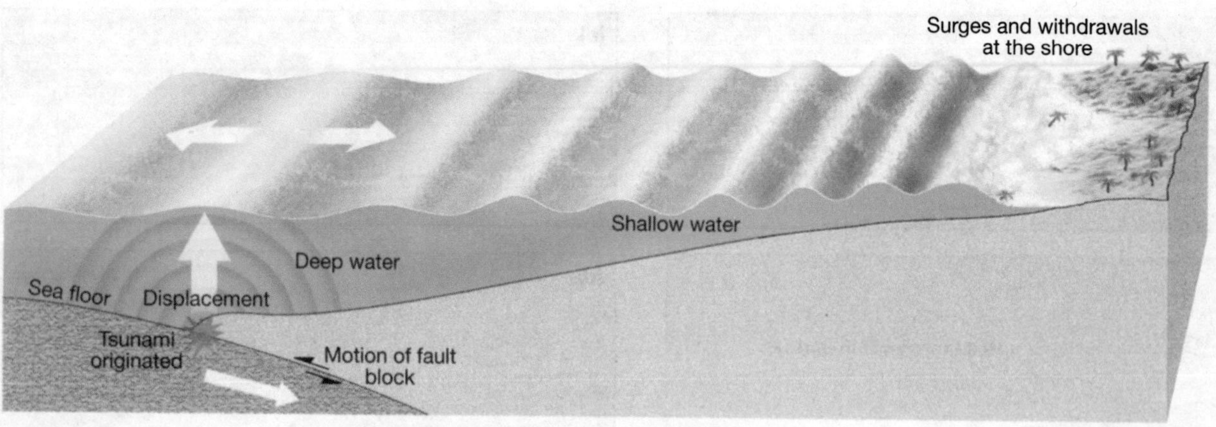

▲ **Figure 20-8** The formation of a tsunami. A vertical disruption of the ocean floor, such as from faulting, displaces the entire water column from the ocean floor to the surface. In the open ocean, the tsunami may be almost indistinguishable because of its great wavelength. Once it reaches shallow water, however, wave height increases and the tsunami comes onshore as a series of surges and withdrawals.

there may be no way to get evacuation orders to coastal populations in time (see "People and the Environment: The Sumatra-Andaman Earthquake and Tsunami of 2004").

Whether they are awesome tsunami or mild swells, the peculiar contradiction of water waves is that they normally pass harmlessly under such fragile things as boats or swimmers in open water but can wreak devastation on even the hardest rocks of a shoreline. In other words, a wave of oscillation is a relatively gentle phenomenon, but a wave of translation can be a powerful force of destruction.

Tides

Tides

As we learned in Chapter 9, the waters of the world ocean oscillate in a regular and predictable pattern called *tides*, resulting from the gravitational influence of the Sun and Moon (see Figure 9-7). The tides rise and fall in a cycle that takes about 12 hours, producing two high tides and two low tides a day on most (but not all) seacoasts.

Despite the enormous amount of water moved by tides and despite the frequency of this movement, the topographic effects are surprisingly small. Tides are significant agents of erosion only in narrow bays, around the margin of shallow seas, and in passages between islands, where they produce currents strong enough to scour the bottom and erode cliffs and shorelines (Figure 20-9).

Changes in Sea Level and Lake Level

Sea-level changes can result either from the uplift or sinking of a landmass (tectonic cause), or from an increase or decrease in the amount of water in the oceans—**eustatic sea-level change**. During Earth's recent history, there have been many changes in sea level, sometimes worldwide and

◀ **Figure 20-9** Under certain circumstances, a large tidal range can influence the shaping of coastal landforms. These gigantic pedestal rocks on the edge of the Bay of Fundy in New Brunswick, Canada, were carved by waves in this region which has the greatest tidal range in the world of up to 15 meters. For scale, the spruce trees on top of the rocks are about 9 meters (30 feet) tall.

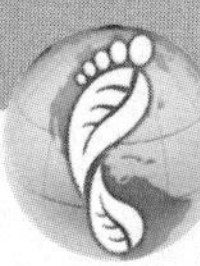

PEOPLE AND THE ENVIRONMENT

The Sumatra-Andaman Earthquake and Tsunami of 2004

On December 26, 2004, one of the greatest natural disasters in recent history was triggered after a magnitude 9.2 earthquake—the largest since the 1964 Alaska earthquake—shook the northern coast of Sumatra in Indonesia. A 1200 kilometer- (750 mile-) long section of the interplate thrust fault (or "megathrust"), formed where the Indo-Australian Plate is subducting beneath the Burma Plate, ruptured, uplifting the ocean floor by as much as 4.9 meters (16 feet). The sharp movement of the ocean floor displaced a massive column of ocean water all of the way up to the surface, generating a tsunami that spread away in all directions (Figure 20-A).

About 28 minutes after the earthquake struck, a 24 meter- (80 foot-) high wave rushed onshore at the city of Banda Aceh on the northern tip Sumatra, just 100 kilometers (60 miles) from the epicenter of the earthquake. Along a 100 kilometer (60 mile) stretch of coastline in northwestern Sumatra, wave heights of 20 to 30 meters (65 to 100 feet) were documented.

The tsunami also raced away from Sumatra to the west across the Indian Ocean at about 800 kilometers an hour (500 mph). Although almost unnoticeable in the open water because of its great wavelength, whenever the tsunami reached shallow water wave height grew (Figure 20-B). Two hours after the earthquake, waves as high as 14 meters (45 feet) struck Sri Lanka. Three and a half hours after the earthquake, waves as high as 4 meters (13 feet) reached the Maldives Islands southwest of India. Seven hours after the earthquake, 1.2 meter (4 foot) waves reached the coast of Somalia in northeast Africa. In most places, at least three waves came on shore; in some locations, there were more than three.

The results were astonishing. The exact death toll will likely never be known, but estimates now suggest that nearly 227,000 people died and many tens of thousands more were seriously injured. In a few locations, entire villages were quite literally washed away. Devastation in some cities extended inland from the coast for more than a kilometer (Figure 20-C). The recovery will take years.

Adding to the tragedy is that some scientists were aware that a large tsunami was likely moving across the Indian Ocean shortly after the earthquake occurred—but there was no way to issue a tsunami warning in the region. In the Pacific basin, where about 85 percent of the world's tsunami occur, a well-organized tsunami warning system is in place: earthquakes are carefully monitored; data buoys are in place to detect changes in ocean movement; warnings can be relayed to government officials around the Pacific Rim. After the 2004 tsunami in the Indian Ocean, an international effort was started to establish a tsunami warning system for the region—had such a system been in place in December of 2004, thousands of lives would likely have been saved.

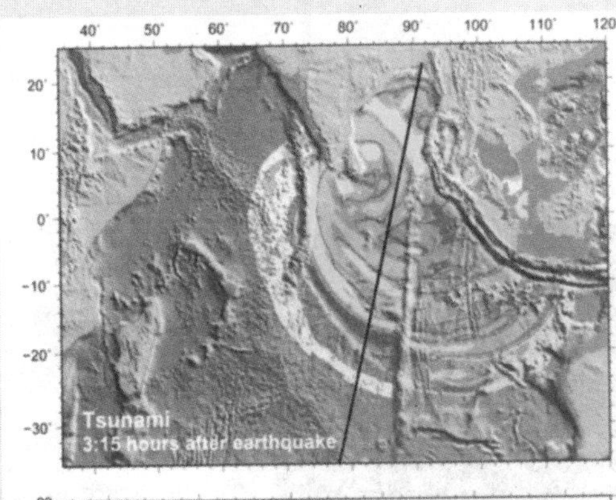

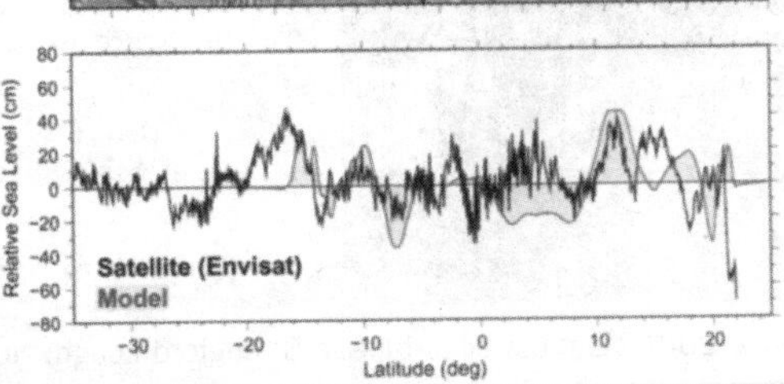

▲ **Figure 20-A** The pattern of tsunami waves 3 hours and 15 minutes after the Sumatra-Andaman earthquake of 2004. The wave height shown in the bottom chart was measured by satellite.

▲ **Figure 20-B** Tsunami coming onshore at Hat Rai Lay Beach, near Krabi, Thailand, on December 26, 2004.

▲ **Figure 20-C** Damage from the December 26, 2004, tsunami in Banda Aceh, Indonesia.

RENEWABLE ENERGY

Tidal Power

Jennifer Rahn, Samford University, Birmingham, Alabama

Before large-scale commercial electrical power generation, many civilizations used the force of moving water to power machines such as textile mills and lumber mills. Just as wind turns the blades of a windmill, recent technological advances utilize the inflow and outflow of tidal water during flood tides and ebb tides to turn a turbine in order to generate electricity (Figure 20-D).

Tides are more predictable than solar power and wind energy—tides change twice a day (in some places just once a day) like clockwork—and the higher the tidal range the greater the potential for tidal energy generation. The ideal locations for tidal power are areas with high flow volumes and/or high tidal ranges (see Figure 9-8).

The first tidal power plant was built in 1966 in France, and the first one in North America was installed in the 1980s in an inlet of the Bay of Fundy in eastern Canada. In the Bay of Fundy, Nova Scotia Power owns and operates one of three tidal power plants in the world, and the only one in the western hemisphere. Currently, one of the largest tidal power installations is being completed in Strangford Lough in Northern Ireland (Figure 20-E). In the United States, no commercial tidal power plants exist.

Internationally, some places under consideration for tidal power plants include the River Severn between Wales and England, and the Cook Strait and Kaipara Harbour in New Zealand. In the United States, tidal power has been suggested for the Puget Sound in the Pacific Northwest, San Francisco Bay, New York's East River, Maine, Alaska, Hawai'i, and off Atlantic City, New Jersey.

Currently, tidal power is not widely used, in part because a limited number of places in the world have high flow tidal regimes. Tidal-power projects are also extremely expensive because massive structures must be built in difficult saltwater environments. However, there are many long-term advantages for using this technology. Tidal power is a renewable energy resource with a virtually unlimited supply and no greenhouse gas emissions.

Estimates suggest that tidal power could meet 10 percent of the United States' and 20 percent of the United Kingdom's electricity demand within the next few decades. Globally, in about 10 years tidal energy could supply 10 percent of the world's energy if full commercialization of this technology materializes.

The technology needed to efficiently and economically turn the tides onto electricity is still being developed. Several new models for generating tidal power have been developed in recent years but none are operating commercially. Other problems include damage to the local aquatic environment, mostly with small juvenile fish getting caught in the rotors. Many people are hesitant to adopt large scale tidal power plants because of the unknown harm they could cause the environment. As we are now seeing with dammed rivers, when humans alter the flow of energy and the surrounding environment, many times there are unintended consequences.

The growth of tidal power generation will depend on the public and political desire for governments to make the large investment needed for this type of alternative energy and the innovation of green energy entrepreneurs to find better and more cost-effective infrastructure, as well as the ability of scientists to predict the major environmental effects of individual projects ahead of time in order to protect local ecosystems.

▲ **Figure 20-D** Tidal power turbines generate electric power when tidal currents turn underwater propellers that in turn power generators.

▲ **Figure 20-E** Tidal power turbine in Strangford Lough, Northern Ireland.

sometimes only around one or a few continents or islands. The eustatic changes of greatest magnitude and most extensive effect are those associated with seawater volume before, during, and after the Pleistocene glaciations. As we saw in Chapter 19, at the peak of the Pleistocene glaciations, sea level around the world was as much as 130 meters (430 feet) lower than today.

As a result of both tectonic and eustatic sea level changes, many present-day ocean coastlines show submergence characteristics, with a portion of a previous landscape now underwater, while others show emergent characteristics, in which shoreline topography of the past is now situated well above the contemporary sea level (Figure 20-10). We will consider the topographic consequences of these circumstances later in this chapter.

Most water-level changes in lakes are less extensive and less notable than those along ocean shorelines. These changes are usually the result of the total or partial drainage of a lake, and their principal topographic expression is exposed ancestral beach lines and wave-cut cliffs above present lake levels.

Global Warming and Sea-Level Change In Chapters 4 and 8, we discussed the consequences of global warming, including the rise in sea level associated with the thermal expansion of ocean water and the increase in volume from the melting of continental glaciers and ice caps. If worldwide temperature trends continue to warm, we can anticipate an ongoing period of deglaciation, with the ice sheets of Antarctica and Greenland slowly melting. Such a situation would cause a global eustatic rise in sea level that would inundate many islands and coastal plains of the world, and would expose coastal populations to greater risks from storm waves such as those generated by hurricanes.

Should the ice caps of Antarctic and Greenland melt completely, global sea level would rise by about 80 meters (260 feet). However, even a modest increase in global sea level will be potentially devastating to populations now living in low-lying coastal areas. Given the anticipated levels of greenhouse gases and the associated temperature rise due to global warming, in 2007 in the *Fourth Assessment Report* of the Intergovernmental Panel on Climate Change projected a sea level rise of 0.18 to 0.59 meters (7.1 to 23.2 inches) by the end of this century. More recent studies suggest that the rise could be as much as 0.69 to 0.99 meters (27 to 39 inches). Such an increase in sea level would cause shorelines around the world to retreat on average more than 30 meters (about 100 feet)—eliminating thousands of square kilometers of coastal land in North America alone. With such an increase in global sea level, some island countries would literally disappear.

Ice Push

The shores of bodies of water that freeze over in winter are sometimes significantly affected by *ice push*, which is usually the result of the contraction and expansion that occurs when the water freezes and thaws as the weather changes. As more and more water turns to ice—and therefore expands in volume (recall the discussion of *frost wedging* in Chapter 15)—near-shore ice is shoved onto the land, where it can deform the shoreline by pushing against it, more or less in the fashion of a small glacial advance.

Ice push is usually unimportant on seashores outside the Arctic and Antarctic, but it can be responsible for numerous minor alterations of the shorelines of high-latitude or high-elevation lakes.

▲ **Figure 20-10** A tectonically active coastline has helped produce the sharp cliffs in this scene along the north coast of California.

Organic Secretions

Several aquatic animals and plants produce solid masses of rocklike material by secreting calcium carbonate. By far the most significant of these organisms is the coral *polyp*, a tiny animal that builds a hard external skeleton of calcium carbonate and then lives inside it. Coral polyps are of many species, and they cluster together in social colonies of uncounted billions of individuals. Under favorable conditions (clear, shallow, salty warm water), the coral can accumulate into enormous masses, forming reefs, platforms, and atolls, all commonplace features in tropical and subtropical oceans. (Coral reef structures will be discussed later in this chapter.)

Stream Outflow

The outflow of streams and rivers into oceans and lakes is included here in our discussion of shore-shaping forces because these outflows are important feeders of sediments to oceans and lakes. They provide much of the sand and other sedimentary material that is moved around and deposited by coastal waters (Figure 20-11). As we will discuss later in this chapter, should the source of sediment from stream outflow be disrupted, beaches along a coastline may begin to shrink.

Currents and Coastal Sediment Transport

Coastal Sediment Transport

Currents consist of large volumes of water moving horizontally. Many kinds of currents flow in the oceans and lakes of the world, usually confined to surface areas but sometimes deeper. Nearly all movement of sediment along coastlines is accomplished by wave action and local currents.

Longshore Currents Coastal topography is affected most by **longshore currents**, in which the water moves roughly parallel to the shoreline. (Think of *longshore* as a contraction for "along the shore.") Longshore currents develop just offshore and are set up by the action of the waves striking the coast at a slight angle (Figure 20-12). Since most waves are generated by wind, the direction of longshore current typically reflects the local wind direction. Longshore currents are prominent transporters of sand and other sediment along many shorelines.

▲ **Figure 20-11** The sediment plume of the Betsiboka River of Madagascar spreads broadly into the blue waters of the Mozambique Channel, which separates the island of Madagascar from the African coast.

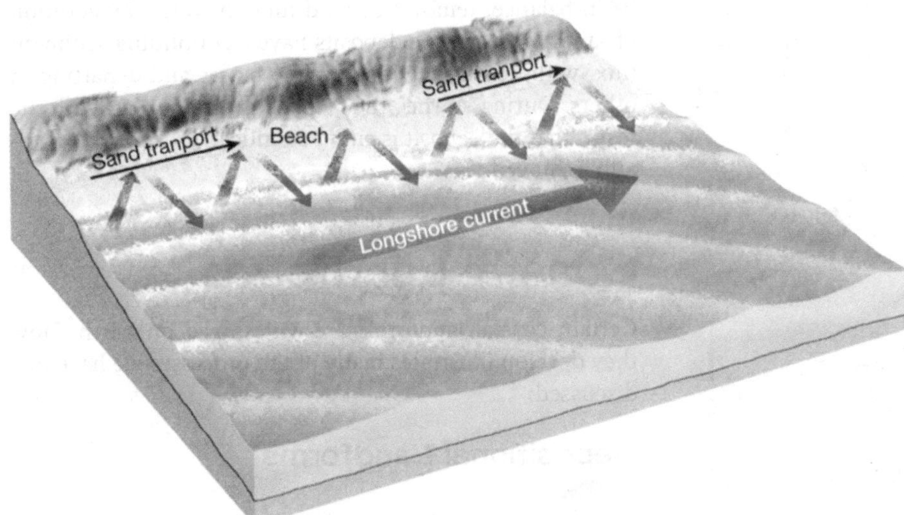

Figure 20-12 Beach drifting involves a zigzag movement of sand along the coast. Sand is brought obliquely onto the beach by the wave and is then returned seaward by the backwash. Longshore currents develop just offshore and move sediment parallel to the coastline. Since most waves develop in response to the wind, longshore currents and beach drifting typically move sediment in a general downwind direction along a shore.

Beach Drifting Another significant mechanism of coastal sediment transport involves the short-distance shifting of sand directly onshore by breaking waves and directly offshore by the retreating water. The debris (mostly sand) that advances and retreats with the repeated surging and ebbing of the waves abrades and smoothes the beach, while the continuous abrasion reduces the size of the particles.

This movement may take the form of **beach drifting** along a coastline, a zigzag movement of sediment that results in a general downwind displacement parallel to the coast (see Figure 20-12). Nearly all waves approach the coast obliquely rather than at a right angle (Figure 20-13), and therefore the sand and other debris carried onshore by the breaking wave move up the beach at an oblique angle. Some of the water soaks into the beach, but much of it returns seaward directly downslope, which is normally at a right angle to the shoreline. This return flow takes some of the sand with it, much of which is picked up by the next surging wave and carried shoreward again along an oblique path. This infinitely repetitious pattern of movement shifts the debris farther and farther along the coastline longitudinally. Because wind is the driving force for wave motion, the strength, direction, and duration of the wind are the principal determinants of beach drifting.

Tides are not very effective in moving coarser debris, but where channels are narrow and relatively shallow, tides can move fine material.

Some sediment transport along shorelines is accomplished directly by the wind. Wherever waves have carried or hurled sand and finer-grained particles to positions above the water level, these particles can be picked up by a breeze and moved overland. This type of movement frequently results in dune formation and sometimes moves sand a considerable distance inland (Figure 20-14; see Chapter 18 for a more lengthy discussion of sand dunes).

Figure 20-13 Here a series of waves rolls in obliquely to the beach at Cape Byron, on the eastern coast of Australia. The direction of beach drifting would be from the bottom of the photograph toward the top.

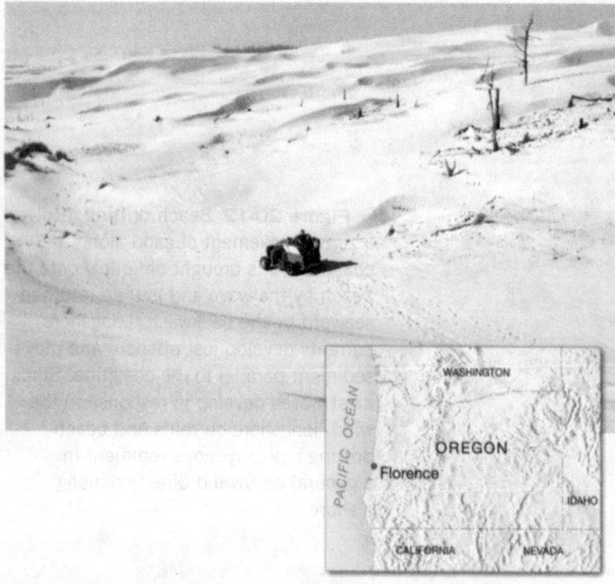

▲ **Figure 20-14** Sand is sometimes heaped into dunes that cover extensive areas. One of the largest sand accumulations in North America is along the central coast of Oregon, near Florence.

Coastal Deposition

Although the restless waters of coastal areas accomplish notable erosion and transportation, in many cases the most conspicuous topographic features of a shoreline are formed by deposition. Just as in streamflow over the surface of a continent, marine deposition occurs wherever the energy of moving water is diminished.

Maritime deposits along coastlines tend to be more ephemeral than noncoastal deposits. This is due primarily to the composition of marine deposits, which typically consist of relatively small particles (sand and gravel), and to the fact that the sand is not stabilized by a vegetation cover. Most coastal deposits are under constant onslaught by agitated waters which can rapidly wash away portions of the sediment. Consequently, the **sediment budget** must be in balance if the deposit is to persist; for the budget to be in balance, removal of sand must be offset by addition of sand. Most marine deposits have a continuing sediment flux, with debris arriving in some places and departing in others. During storms, the balance is often upset, with the result that the deposit is either significantly reshaped or totally removed.

Coastal Landforms

Certain coastal landforms are widespread on Earth. How they develop illustrates many of the processes we have just discussed.

Depositional Landforms

Coastal Stabilization Structures

Beaches The most widespread marine depositional feature is the **beach**, which is an exposed deposit of loose sediment adjacent to a body of water. Although the sediment can range in size from fine sand to large cobbles, it is usually relatively homogeneous in size on a given section of beach. Beaches composed of smaller particles (which is to say sand, because silt and clay get carried away in suspension and do not form beaches) are normally broad and slope gently seaward, whereas those formed of larger particles (gravel, cobbles) generally slope more steeply.

Beaches occupy the transition zone between land and water, sometimes extending well above the normal sea level into elevations reached only by the highest storm waves. On the seaward side, they generally extend down to the level of the lowest tides and can often be found at still lower levels, where they merge with muddy bottom deposits. Figure 20-15 portrays an idealized beach profile. The *backshore* is the upper part of the beach, landward of the high-water line. It is usually dry, being covered by waves only during severe storms. It contains one or more *berms*, which are flattish wave-deposited sediment platforms. The *foreshore* is the zone that is regularly covered

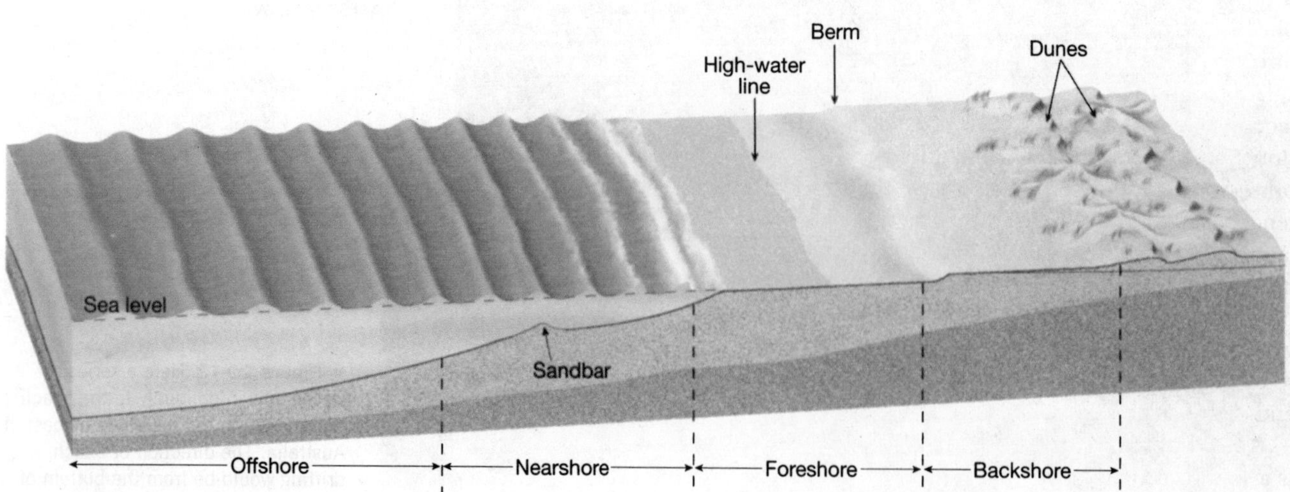

▲ **Figure 20-15** An idealized beach profile.

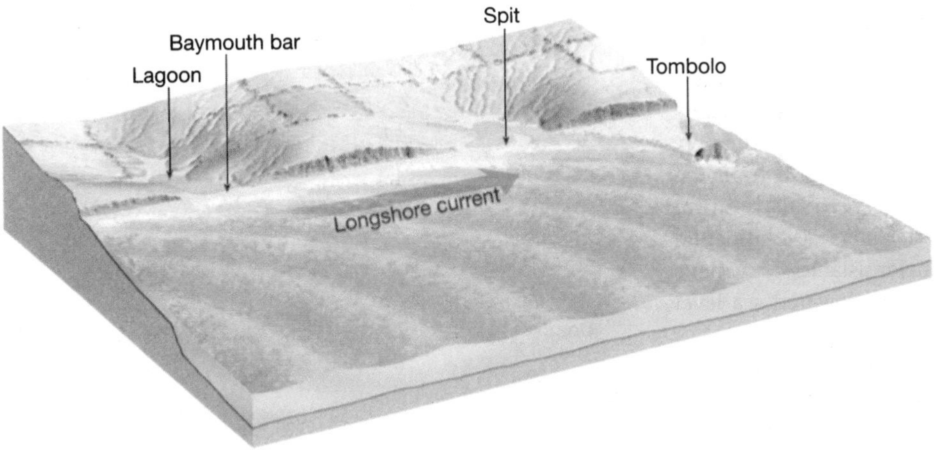

◀ **Figure 20-16** Common depositional landforms along a coastline include spits, baymouth bars, and tombolos. Note the orientation of the spit to the direction of the longshore current.

and uncovered by the rise and fall of tides. The *nearshore* extends from the low-tide mark, seaward to where the low-tide breakers begin to form—the nearshore is not exposed to the atmosphere, but it is the place where waves break and where surf action is greatest. The *offshore* zone is permanently underwater and deep enough that wave action rarely influences the bottom.

Beaches sometimes extend for dozens of kilometers along straight coastlines, particularly if the relief of the land is slight and the bedrock unresistant. Along irregular shorelines, beach development may be restricted largely or entirely to bays, with the bays frequently alternating with rocky headlands.

Beach shape may change greatly from day to day and even from hour to hour. Normally beaches are built up during quiet weather and removed rapidly during storms. Most midlatitude beaches are longer and wider in summer and greatly worn away by the storminess of winter.

Spits Near-shore ocean waters normally contain a considerable amount of fine-grained sediment that is shifted about by waves, currents, and tides. At the mouth of a bay, sediment may be moved by longshore currents into deeper water. There the flow speed is slowed and the sediment is deposited. The growing bank of land guides the current farther into the deep water, where still more material is dropped. Any such linear deposit attached to the land at one end and extending into open water in a down-current direction is called a **spit** (Figure 20-16).

Although most spits are straight, sandy peninsulas projecting out into a bay or other coastal indentation, variations in local currents, winds, and waves often give them other configurations. In some cases, the spit becomes extended clear across the mouth of a bay to connect with land on the other side, producing what is called a **baymouth bar** and transforming the bay to a lagoon. Another common modification of spit shape is caused by water movements in the bay which can cause the deposits to curve toward the mainland, forming a *hook* at the outer end of the spit (Figure 20-17).

A less common but even more distinctive development is a **tombolo**—a depositional feature that connects a nearshore island with the mainland (Figure 20-18). While some tombolos form as a type of spit where waves deposit sediment on the landward side of the island, other tombolos develop where a bedrock structure at, or just below, sea level connects the island with the mainland and serves to trap sand.

▲ **Figure 20-17** Farewell spit on the South Island of New Zealand has developed a slight hook that curves back toward shore.

▲ **Figure 20-18** This tombolo connects the nearshore island of Point Stephens to the mainland in Tomaree National Park, New South Wales, Australia.

Barrier Islands Another prominent coastal deposition is the **barrier island** (Figure 20-19), a long, narrow sandbar built up in shallow offshore waters, sometimes only a few hundred meters from the coast but often several kilometers at sea. Barrier islands are always oriented approximately parallel to the shore. They are believed to result from the heaping up of debris where large waves (particularly storm waves) begin to break in the shallow waters of continental shelves. However, many larger barrier islands may have more complicated histories linked to the lowered sea level during the Pleistocene.

Barrier islands often become the dominant element of a coastal terrain. Although they usually rise at most only a few meters above sea level and are typically only a few hundred meters wide, they may extend many kilometers in length. Most of the Atlantic and Gulf of Mexico coastline of the United States, for instance, is paralleled by lengthy barrier islands, several more than 50 kilometers (30 miles) long (Figure 20-20).

An extensive barrier island isolates the water between itself and the mainland, forming a body of quiet salt- or brackish water called a **lagoon**. Over time, a lagoon becomes increasingly filled with water-deposited sediment from coastal streams, wind-deposited sand from the barrier island, and tidal deposits if the lagoon has an opening to the sea. All three of these sources contribute to the buildup of mudflats on the edges of the lagoon. Unless tidal inlets across the barrier island permit vigorous tides or currents to carry lagoon debris seaward, therefore, the ultimate destiny of most lagoons is to slowly be transformed, first to mudflats and then to coastal marshes (Figure 20-21).

In addition to infilling by sediment, another factor contributes to a lagoon's disappearance. After a barrier island becomes a certain size, it often begins to migrate slowly shoreward as waves wear away its seaward shore and sediments accumulate to build up its landward shore. Eventually, if the pattern is not interrupted by such things as changing sea level, the island and the mainland shore will merge.

Because most barrier islands rise just a few meters above sea level and are largely built up of coastal sediment,

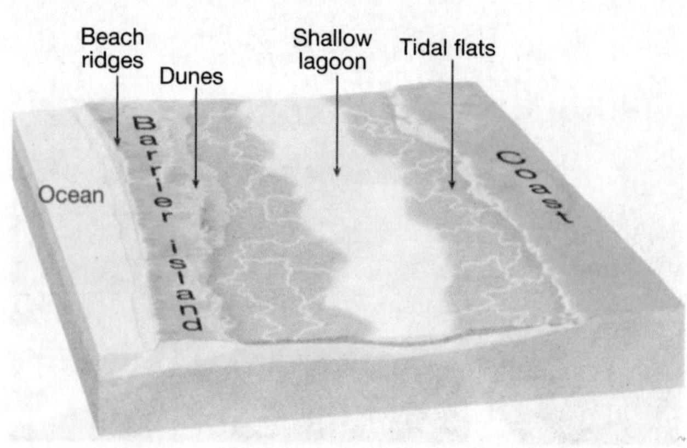

▲ **Figure 20-19** A typical relationship between ocean, barrier island, and lagoon.

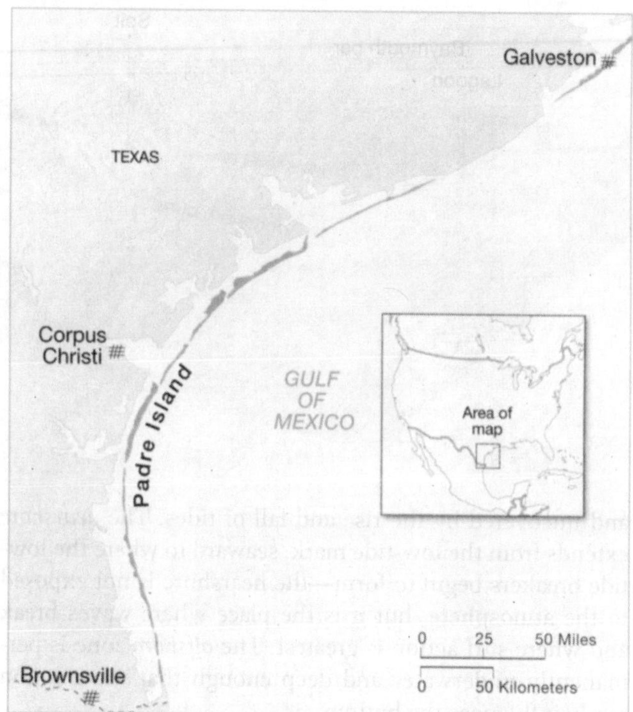

▲ **Figure 20-20** The longest of the barrier islands off the Gulf Coast of the United States is Texas' Padre Island.

they can be quite susceptible to damage from large storms. A number of inhabited barrier islands along the Gulf Coast of the United States experienced severe erosion during hurricanes such as Lili in 2002, Katrina in 2005, and Ike in 2008 (Figure 20-22). With even the slight rise in global sea level anticipated this century by the Intergovernmental Panel on Climate Change, barrier islands will become even more vulnerable to storm waves.

Human Alteration of Coastal Sediment Budgets Over the last half-century, human activity has disrupted the sediment budgets of beaches along many shorelines of the world—this has been especially true in many coastal areas of North America. For example, dams built along rivers for flood control or hydroelectric power generation effectively act as sediment traps. With less sediment reaching the mouths of rivers, there is less sediment to be transported along the shoreline by longshore currents

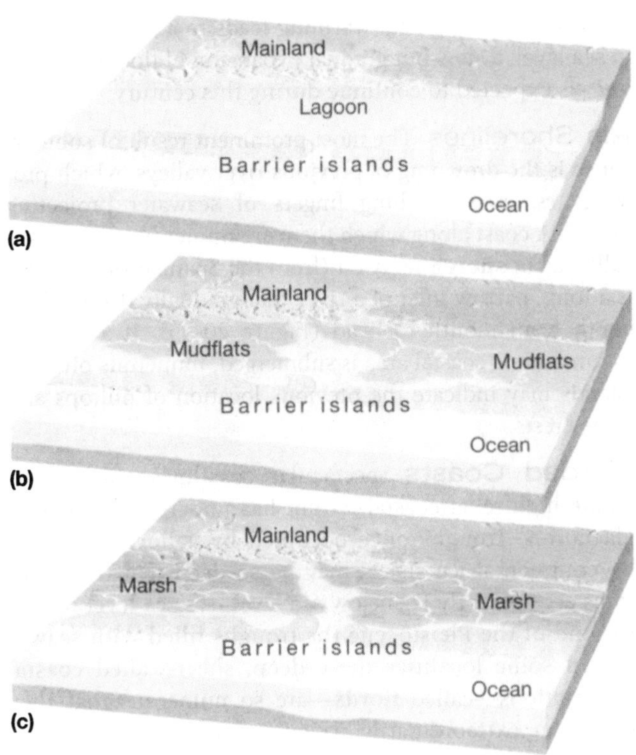

▲ **Figure 20-21** Barrier islands are separated from the mainland by lagoons (a). With the passage of time, the lagoons often become choked with sediment and become converted to mudflats (b) or marshes (c).

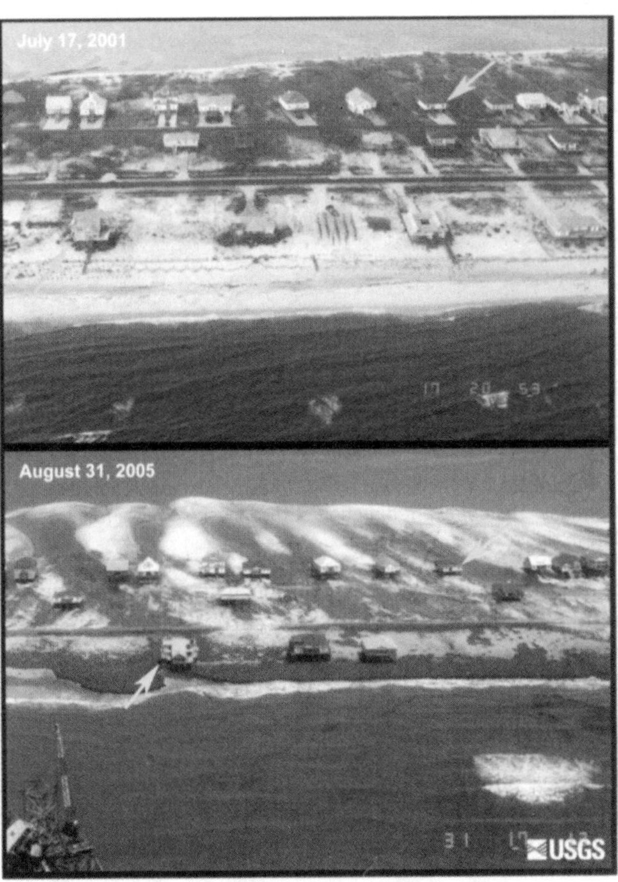

▲ **Figure 20-22** Dauphin Island south of Mobile, Alabama, experienced extensive erosion from storm waves during Hurricane Katrina in 2005. The top photograph was taken on July 21, 2001, before Katrina. The bottom photograph was taken on August 31, 2005, after the passage of Hurricane Katrina. The oil rig in the foreground was washed onto the island during the storm.

and beach drifting, and so the downcurrent beaches begin to shrink. In addition, artificial structures built by one community to increase or stabilize their beaches may reduce the amount of sediment transported farther down a shoreline, thus causing downcurrent beaches to be reduced in size.

Local communities have taken a number of different approaches to solving the problem of shrinking beaches. One direct, but relatively expensive, approach is to "nourish" a beach by dumping tons of sand just slightly upcurrent of the beach. Unfortunately, since longshore currents and beach drifting will eventually transport the sand away, such nourishment must be undertaken repeatedly in order to maintain the beach at a desired size.

Another approach to maintaining beaches is the use of "hard" stabilization structures. For example, a **groin** is a short wall or dam built out from a beach to impede the longshore current and force sand deposition on the upcurrent side of the structure (Figure 20-23). While groins do trap sediment on their upcurrent side, erosion tends to take place on their downcurrent side; to reduce this erosion, another groin can be built just downcurrent. In some locations, a series of groins, known as a *groin field*, has been built (Figure 20-24).

Jetties are usually built in pairs on either side of a river or harbor entrance. The idea is to confine the flow of water to a narrow zone, thereby keeping the sand in motion and

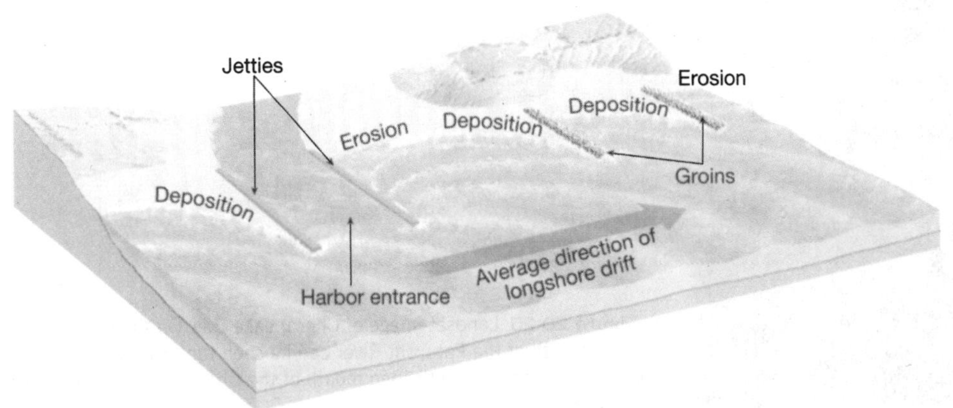

◀ **Figure 20-23** Jetties and groins along a shoreline trap sediment on their upcurrent side, while erosion tends to remove sediment on their downcurrent side.

▲ **Figure 20-24** Groin field along Norderney Island in northern Germany. The dominant direction of the longshore current is from left to right.

inhibiting its deposition in the navigation channel. Jetties tend to interfere with longshore currents in the same way as groins, trapping sand on the upcurrent side while causing erosion on the downcurrent side.

Even after undertaking such expensive projects as building groin fields and carrying out regular beach nourishment, some communities are finding that it is a losing proposition—the beaches continue to shrink and there seems to be no clear solution available to them.

Shorelines of Submergence

In the recent geological past, sea level has fluctuated sharply. For example, during a warmer climatic interval in an interglacial period, 125,000 years ago, sea level was about 6 meters (20 feet) higher than it is today. During the last glacial peak (about 20,000 years ago), sea level is estimated to have been about 120 meters (400 feet) lower than it is at present.

Almost all the world's oceanic coastlines show evidence of submergence during the last 15,000 years or so, a result of the melting of the Pleistocene ice sheets. As water from melting glaciers returned to the oceans, rising sea level caused widespread submergence of coastal zones. Further, as contemporary global warming leads to a slight increase in sea level, a slow but gradual expansion of flooded shorelines is expected to continue during this century.

Ria Shorelines The most prominent result of submergence is the drowning of previous river valleys, which produces *estuaries*, or long fingers of seawater projecting inland. A coast along which there are numerous estuaries is called a **ria shoreline**. A ria (from the Spanish *ría*, "river") is a long, narrow inlet of a river that gradually decreases in depth from mouth to head (Figure 20-25). If a hilly or mountainous coastal area is submerged, numerous offshore islands may indicate the previous location of hilltops and ridge crests.

Fjorded Coasts Spectacular coastlines often occur where high-relief coastal terrain has undergone extensive glaciation. Troughs once occupied by valley glaciers or by continental ice sheets may be so deep that their bottoms are presently far below sea level—as sea level rose at the end of the Pleistocene the troughs filled with seawater. In some localities these deep, sheer-walled coastal indentations—called **fjords**—are so numerous that they create an extraordinarily irregular coastline, often with long, narrow fingers of saltwater reaching more than 160 kilometers (100 miles) inland.

The most extensive and spectacular fjorded coasts are in Norway, western Canada, Alaska, southern Chile, the South Island of New Zealand, Greenland, and Antarctica (Figure 20-26).

Shorelines of Emergence and Erosion

Evidence of previously higher sea levels is sometimes related to ice melting during past interglacial ages but is more often associated with tectonic uplift. The clearest topographic results of coastal emergence are shoreline features raised well above the present water level. Often the emerged portion of a continental shelf appears as a broad, flat coastal plain.

Wave-Cut Cliffs and Platforms One of the most common coastal landform complexes comprises wave-cut cliffs, sea stacks, and wave-cut platforms (see Figure 20-10).

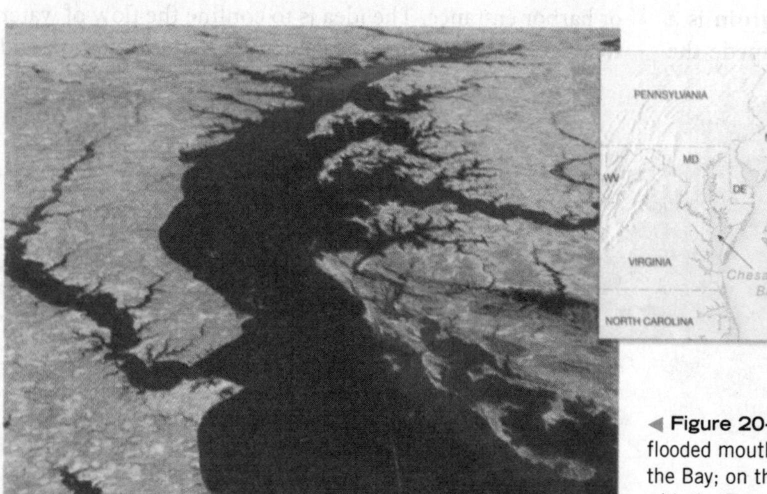

◀ **Figure 20-25** Landsat image of Chesapeake Bay. On the left, the flooded mouth of the Patuxent River can be seen running parallel to the Bay; on the right the flooded mouth of the Choptank River connects with the Bay.

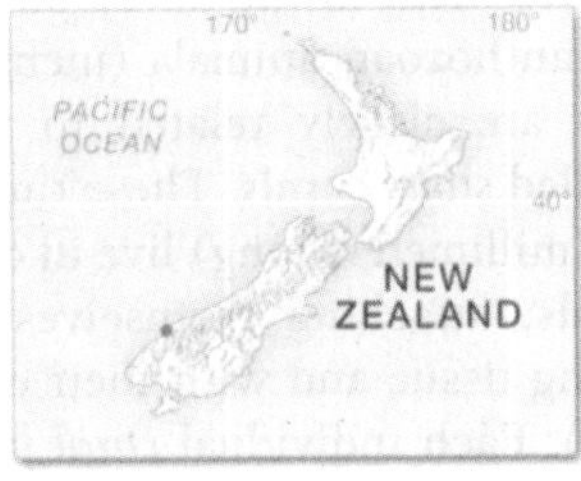

▶ **Figure 20-26** Entrance to fjords at Milford Sound, New Zealand.

As we discussed earlier, as waves erode away at a rocky headland, steep wave-cut cliffs are formed, and these cliffs receive the greatest pounding at their base, where the power of the waves is concentrated.

A combination of hydraulic pounding, abrasion, pneumatic push, and chemical solution at the cliff base frequently cuts a notch at the high-water level (Figure 20-27a).

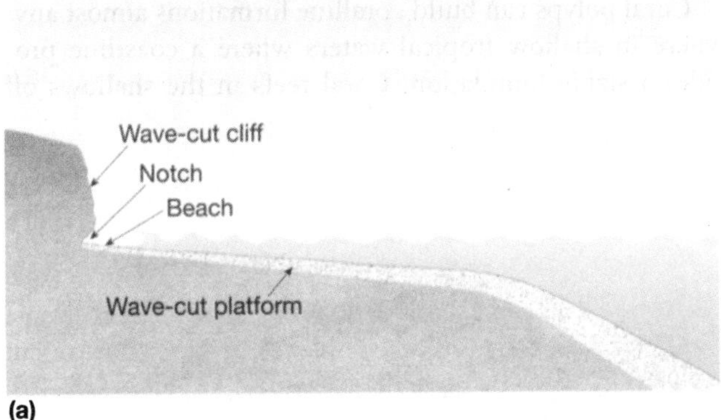

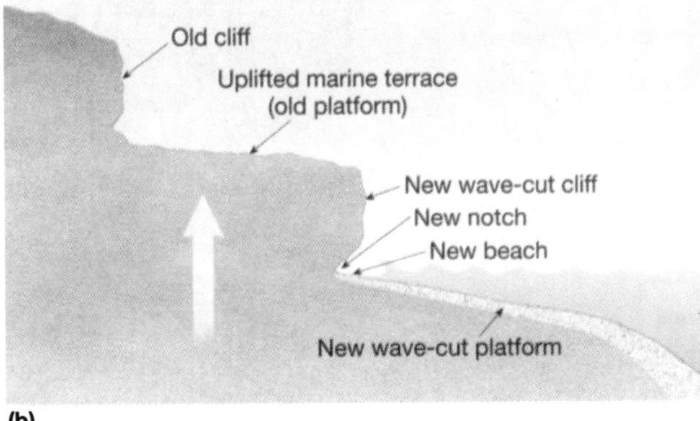

▲ **Figure 20-27** (a) A wave-cut platform develops where a coastal cliff is worn back by wave erosion. (b) A marine terrace develops when a wave-cut platform is tectonically uplifted above sea level.

As the notch is enlarged, the overhang sporadically collapses, and the cliff recedes as the ocean advances. Where wave action cuts through the bottom of a cliff-topped headland, a *sea arch* may be formed (see Figure 20-6), while *sea stacks* develop where wave erosion leaves towers of rock isolated just offshore from the coastal cliff (see Figure 20-1). Seaward of the cliff face, the pounding and abrasion of the waves create a broad erosional surface called a **wave-cut platform** (or *wave-cut bench*) usually slightly below water level. The combination of wave-cut cliff and wave-cut platform produces a profile that resembles a letter "L," with the steep vertical cliff descending to a notched base and the flat horizontal platform extending seaward.

The debris eroded from cliff and platform is mostly removed by the swirling waters. The larger fragments are battered into smaller and smaller pieces until they are small enough to be transportable. Some of the sand and gravel produced in this fashion may be washed into an adjacent bay to become, at least temporarily, a part of the beach. Much of the debris, however, is shifted directly seaward where it may be deposited. With the passage of time and the wearing away of the cliff by weathering and erosion, these deposits may eventually cover the wave-cut platform entirely, resulting in a beach that extends to the base of the cliff.

Marine Terraces When a wave-cut platform is uplifted along a tectonically rising coast, a **marine terrace** is formed (Figure 20-27b). It appears that fluctuations of sea level during the Pleistocene played a part in the formation of at least some marine terraces: when sea level drops during a glacial period, the wave-cut platform is left well above sea level; gradual tectonic uplift during the period of low sea level leaves the terrace high enough to be preserved after sea level rises again during the subsequent interglacial period.

Along some shorelines of the world, a series of marine terraces is present, reflecting several episodes of terrace formation (Figure 20-28).

▲ **Figure 20-28** Uplifted marine terraces, well above the present shoreline. This scene is on the north coast of California near Fort Ross.

Coral Coasts

In tropical oceans, nearly all continents and islands are fringed with either *coral reefs* or some other type of coralline formation (Figure 20-29). Coralline structures are built by a complicated series of events that involve animals, algae, and various physical and chemical processes.

The critical element in the development of coral reefs is a group of anthozoan animals (members of the class *Anthozoa* that are closely related to jellyfish and sea anemones) called *stony corals*. These tiny creatures (most are only a few millimeters long) live in colonies of countless individuals, attaching themselves to one another both with living tissue and with their external skeletons (Figure 20-30). Each individual *coral polyp* extracts calcium carbonate from the seawater and secretes a limy skeleton around the lower half of its body. Most polyps withdraw into their skeletal cups during the day and extend their armlike feeding structures at night. At the top of the body is a mouth surrounded by rings of tentacles, which gives them a blossomlike appearance that for centuries caused biologists to believe that they were plants rather than animals. They feed on minute animal and plant plankton. Although the coral polyp is an animal, reef-building hard corals are hosts to symbiotic algae that provide additional food for the coral polyp through photosynthesis.

The ubiquity of coral reefs in shallow tropical waters is a tribute to the remarkable productiveness of the polyps because they are not actually very hardy creatures. They cannot survive in water that is very cool or very fresh or very dirty. Moreover, they require considerable sunlight, so they cannot live more than a few tens of meters below the surface of the ocean. Many coral reefs around the world show signs of degradation—from both natural and human-generated causes (see "Focus: Imperiled Coral Reefs").

Coral polyps can build coralline formations almost anywhere in shallow tropical waters where a coastline provides a stable foundation. Coral reefs in the shallows off

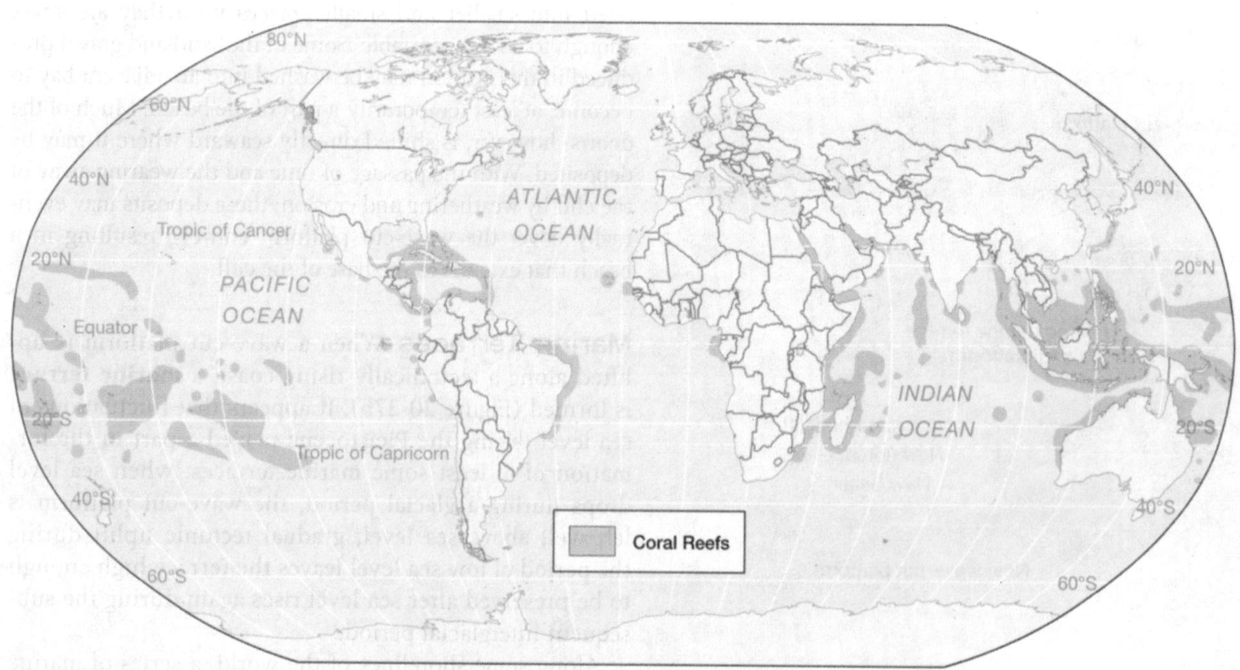

▲ **Figure 20-29** Distribution of coral reefs and other coralline structures in the oceans of the world.

FOCUS

Imperiled Coral Reefs

All species of reef-building coral contain *zooxanthellae*, a type of algae that lives in a symbiotic relationship with the coral polyp. Through photosynthesis, the algae provides nutrients to the coral—it is also the algae that gives coral its beautiful color.

For reasons that are not completely understood, when coral is stressed it expels the algae, leaving its exoskeleton of calcium carbonate a translucent white color. This phenomenon is known as *coral bleaching* (Figure 20-F). In some cases, bleached corals die within just a few weeks if the zooxanthellae are not replaced.

Bleaching events have been observed for decades, caused by such factors as a sharp decrease in water salinity, sedimentation, pollution, and abrupt changes in temperature; but it became clear to researchers in the 1980s that stress from high water temperature was the most common cause. The warming of coastal waters during the 1982–83 El Niño was observed to cause bleaching of coral in Panama; the 1997–98 El Niño was even stronger and caused bleaching in reefs around the world. In recent years, some researchers have started pointing to higher sea surface temperatures (SST) associated with global warming as at least one of the causes of the rising number of coral bleaching incidents. In 2005 a major bleaching event occurred in the Caribbean—at one monitoring station south of St. John in the U.S. Virgin Islands, 30 percent of the coral died within a six-month period.

The National Oceanic and Atmospheric Administration (NOAA) monitors stress on coral reefs through its Coral Reef Watch program. NOAA is currently testing a Bleaching Outlook system that issues predictions of bleaching potential for periods of up to three months based on experimental SST forecasts (Figure 20-G)

In addition to bleaching, coral reefs are being stressed by the slight acidification of the ocean waters caused by the absorption of carbon dioxide (acidification of the oceans is discussed in Chapter 9). In October 2008, an international panel of 155 marine scientists issued the Monaco Declaration, stating that damage from ocean acidification is already detectable, and that with the projected increase in atmospheric carbon dioxide—and the associated increased acidification of the ocean—many regions of the world will become "chemically inhospitable" to coral reefs by mid century.

The loss of coral reefs through bleaching and other natural and human-produced causes is alarming many researchers. When coral dies, an entire ecosystem is at risk: the fish and other creatures that depend on coral for survival are stressed, local fisheries can decline, the protection from storm waves offered to low-lying islands fringed with reefs is diminished, and, of course, the loss of species and biodiversity may be irreparable.

▲ **Figure 20-F** Bleached coral in the Great Barrier Reef, Australia.

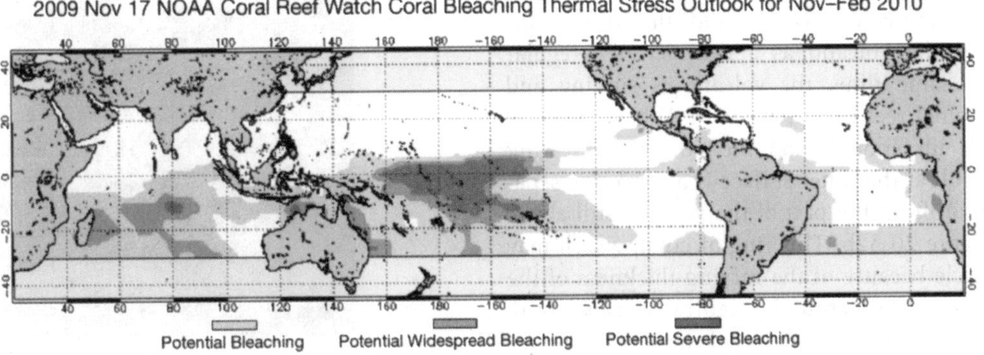

▲ **Figure 20-G** Coral Reef Watch experimental Coral Bleaching Thermal Stress Outlook for winter 2009-2010. The darker the shade of brown, the higher the potential for coral bleaching.

▲**Figure 20-30** Coral comes in an extraordinary variety of sizes, shapes, forms, and species. This close-up reef scene in the Bahamas shows a "Christmas tree" worm in a colony of coral.

the coasts of Florida, for instance, are built on such stable bases. The famous Great Barrier Reef off the northeastern coast of Australia is an immense shallow-water platform of bedrock largely, but not entirely, covered with coral. Its enormously complex structure includes many individual reefs, irregular coral masses, and a number of islands (Figure 20-31).

One favored location for coral reefs is around a volcanic island in tropical waters; as the volcano forms and then subsides, the following different types of reefs grow upward. When the volcano first forms, as, for example, over the Hawaiian hot spot described in Chapter 14, coral accumulates on the part of the mountain flank just below sea level because it is in these shallow waters that polyps live. The result is a reef built right onto the volcano, as shown in Figure 20-32a; such an attached reef is called a **fringing reef** (Figure 20-33).

As new layers are laid down over old, the coral builds upward around the volcano as a cylinder of irregular height. At the same time, the volcano is sinking and pulling the original reef base downward. When the coral has been built up enough and the volcano has sunk enough, the result at the water surface is a coral ring separated by a lagoon from the part of the volcano still above water, as in Figure 20-32b. This ring of coral, which may be a broken circle because of the varying thickness of the coral, appearing to float around a central volcanic peak (but is actually attached to the flanks of the sinking mountain far below the water surface), is called a **barrier reef**.

▲**Figure 20-31** A view of Australia's Great Barrier Reef.

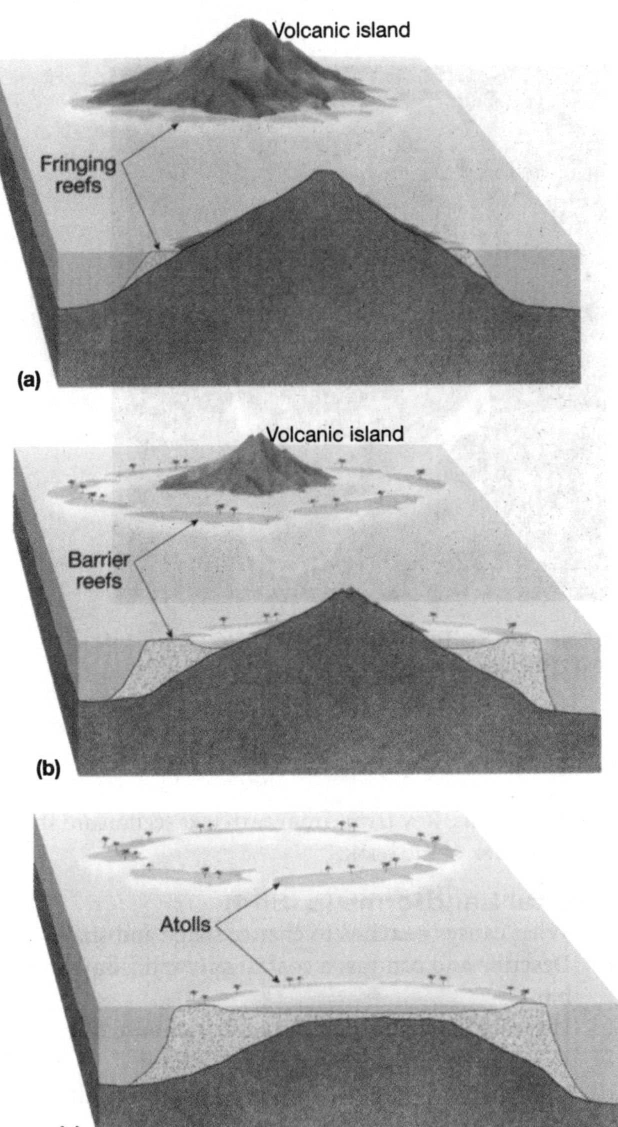

◀ **Figure 20-32** Coral reef formation around a sinking volcano. (a) Around a newly formed volcano rising above the water of a tropical ocean, secretions from coral polyps living along the shallow-water flanks of the volcano accumulate into a fringing reef attached to the mountain. (b) As the volcano becomes dormant and begins to sink, the coral continues to grow upward over the original base, essentially a cylinder surrounding the mountain. Such a reef separated from its mainland by a lagoon is a barrier reef. (c) Once the volcano is completely submerged, the coral surrounding a landless lagoon is called an atoll reef.

▲ **Figure 20-33** A part of the fringing reef on the island of Moorea in French Polynesia.

The surface of a barrier reef is usually right at sea level, with some portions projecting upward into the air.

Coral polyps continue to live in the upper, shallow-water portions of a barrier reef, and so the reef continues to grow upward. Once the top of the volcano sinks below the water surface, the reef surrounding a now landless lagoon is called an **atoll** (Figure 20-32c). The term atoll implies a ring-shaped structure. In actuality, however, the ring is rarely unbroken; rather it consists of a string of closely spaced coral islets separated by narrow channels of water. Each individual islet is called a *motu*. Since coral cannot live above sea level, much of the coral debris that makes up the above-water portion of an atoll has been deposited there by storm waves (Figure 20-34).

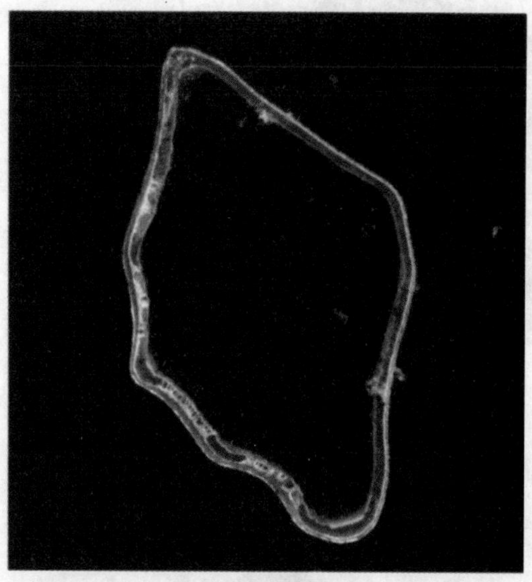

▲ **Figure 20-34** The island of Tureia in French Polynesia in the South Pacific is a coral atoll.

CHAPTER 20 LEARNING REVIEW

After studying this chapter, you should be able to answer the following questions. Key terms from each text section are shown in bold type. Definitions for key terms are also found in the glossary at the back of the book.

Coastal Processes (p. 533)

1. How are most ocean waves generated?
2. What are **swells** in the ocean?
3. How is a **wave of oscillation** different from a **wave of translation**?
4. Contrast the characteristics of ocean waves in deep water with waves in shallow water, especially note how **wavelength** and **wave height** change as a wave comes onshore.
5. Contrast **swash** and **backwash** on a beach.
6. Describe and explain the process of **wave refraction**.
7. What factors influence the erosional power of waves striking a coastline?
8. How does air serve as a tool of erosion in wave action?
9. Explain the formation and characteristics of a **tsunami**.
10. Why did **eustatic sea-level changes** take place during the Pleistocene?
11. Describe the formation and characteristics of **longshore currents**.
12. Describe the process of **beach drifting**.
13. Explain the concept of the **sediment budget** of a coastal depositional landform such as a beach.

Coastal Landforms (p. 544)

14. What causes **beaches** to change shape and size?
15. Describe and contrast a coastal **spit** and a **baymouth bar**.
16. Under what circumstances does a **tombolo** form?
17. Describe the features of a **barrier island**.
18. What happens to most coastal **lagoons** with the passage of time?
19. What will likely happen to a downcurrent beach when a major river flowing into the ocean is dammed? Why?
20. How do **groins** and **jetties** typically affect the beaches around them?
21. Why are shorelines of submergence so common today?
22. Explain the formation of **ria shorelines** and **fjords**.
23. Explain how wave-cut cliffs and **wave-cut platforms** develop.
24. Describe the formation of a **marine terrace** along a shoreline of emergence.
25. Explain how a coral **fringing reef** forms and how it can subsequently become transformed first to a **barrier reef** and then to an **atoll**.

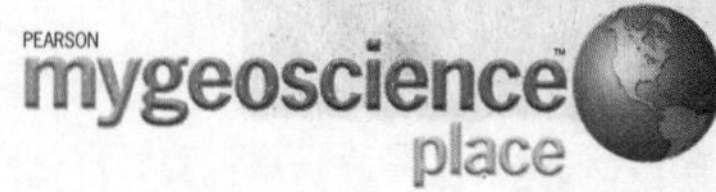

Log in to **www.mygeoscienceplace.com** for animations, videos, RSS feeds, maps, and self-study quizzes to enhance your study of **Coastal Processes and Terrain**.

Appendix I
The International System of Units (SI)

With the major exception of the United States, the system of weights and measures used worldwide both in scientific work and in everyday life is the International System, usually abbreviated SI from its French name, *Système Internationale*. The system has seven base units and supplementary units for angles (Table I-1). The beauty of the system lies in its reliance on multiples of the number 10, with the prefixes shown in Table I-2 being used to cover a magnitude range from the astronomically large to the infinitesimally small. Table I-3 lists the most frequently used conversion factors.

TABLE I-1 SI Units

Quantity	Unit	Symbol
	Base Units	
Length	Meter	m
Mass	Kilogram	kg
Time	Second	s
Electric current	Ampere	A
Temperature	Kelvin	K
Amount of substance	Mole	mol
Luminous intensity	Candela	Cd
	Supplementary Units	
Plane angle	Radian	rad
Solid angle	Steradian	Sr

TABLE I-2 Common Multiples and SI Prefixes

Multiple	Value	Prefix	Symbol
1 000 000 000 000	10^{12}	tera	T
1 000 000 000	10^{9}	giga	G
1 000 000	10^{6}	mega	M
1000	10^{3}	kilo	k
100	10^{2}	hecto	h
10	10^{1}	deka	da
0.1	10^{-1}	deci	d
0.01	10^{-2}	centi	c
0.001	10^{-3}	milli	m
0.000001	10^{-6}	micro	μ
0.000000001	10^{-9}	nano	n
0.000000000001	10^{-12}	pico	P

TABLE I-3 SI-English Conversion Units

Multiply	By	To Get
	Length	
Inches	2.540	Centimeters
Foot	0.3048	Meters
Yards	0.9144	Meters
Miles	1.6093	Kilometers
Millimeters	0.039	Inches
Centimeters	0.3937	Inches
Meters	3.2808	Foot
Kilometers	0.6214	Miles

TABLE I-3 SI-English Conversion Units (Continued)

Multiply	By	To Get
	Area	
Square inches	6.452	Square centimeters
Square feet	0.0929	Square meters
Square yards	0.8361	Square meters
Square miles	2.590	Square kilometers
Acres	0.4047	Hectares
Square centimeters	0.155	Square inches
Square meters	10.764	Square feet
Square meters	1.196	Square yards
Square kilometers	0.3861	Square miles
Hectares	2.471	Acres
	Volume	
Cubic inches	16.387	Cubic centimeters
Cubic feet	0.028	Cubic meters
Cubic yards	0.7646	Cubic meters
Fluid ounces	29.57	Milliliters
Pints	0.47	Liters
Quarts	0.946	Liters
Gallons	3.785	Liters
Cubic centimeters	0.061	Cubic inches
Cubic meters	35.3	Cubic feet
Cubic meters	1.3079	Cubic yards
Milliliters	0.034	Fluid ounces
Liters	1.0567	Quarts
Liters	0.264	Gallons
	Mass (Weight)	
Ounces	28.3495	Grams
Pounds	0.4536	Kilograms
Tons (2000 lb)	907.18	Kilograms
Tons (2000 lb)	0.90718	Tonnes
Grams	0.03527	Ounces
Kilograms	2.2046	Pounds
Kilograms	0.0011	Tons (2000 lb)
Tonnes	1.1023	Tons (2000 lb)

Appendix II
U.S. Geological Survey Topographic Maps

The U.S. Geological Survey (USGS) is one of the world's largest mapping agencies and is primarily responsible for the country's National Mapping Program. The USGS produces a broad assortment of maps, but its topographic "quadrangles" are among the most widely used by geographers. The quadrangles come in various sizes and scales (from 1:24,000 to 1:1,000,000), but all are rectangles bordered by parallels and meridians rather than by political boundaries. For many years the quadrangles were produced by surveys on the ground. Today, however, they are created from aerial photographs, satellite imagery, and computer rectification.

Topographic maps convey both human-built features such as roads and buildings and natural features such as rivers, glaciers, and areas of forest cover. Topographic maps also depict the topography of a landscape with *elevation contour lines* (Figure II-1). From these two-dimensional maps, it is possible to envision the three-dimensional topography, making "topo" maps especially useful when studying landform patterns.

The USGS has produced more than 55,000 different printed topographic quadrangles, as well as about 220,000 digital orthorectified aerial images. In 2001, the USGS began an ambitious multiyear plan to develop a seamless, continuously maintained, nationally consistent set of online, public-domain, geographic based information for the United States known as *The National Map*. The National Map incorporates topographic maps, aerial and satellite imagery, along with a wide range of other map layers, and is available online at http://nationalmap.gov.

Elevation Contour Lines

USGS topographic maps use elevation contour lines to portray the shape, slope, elevation, and relief (the difference in elevation between the highest and lowest locations) of a landscape (Figure II-2). The following guidelines are followed when interpreting contour lines:

1. A contour line connects points of equal elevation.
2. The difference in elevation between two contour lines is called the *contour interval*.
3. Usually every fourth or fifth contour line is a darker *index contour*.
4. Elevations on one side of a contour line are higher than on the other side.
5. Contour lines never touch or cross one another (except at some cliffs).
6. Contour lines have no beginning or end—every line closes on itself, either on or off the map.
7. Uniformly spaced contours indicate a uniform slope.
8. If spaced far apart, contour lines indicate a relatively gentle slope; if spaced close together, they represent a steep slope.
9. Contour lines bend upstream in a "V" shape when crossing a stream valley, gully, or "draw" (the "V" points uphill).
10. Along a spur or ridge running down a hillside, a contour line forms a "V" pointing downhill.
11. A contour line that closes within the limits of the map represents a hill or rise. The land inside that contour is higher than the elevation of the closed contour itself.
12. A depression may be represented by a closed contour that is hachured on the side leading into the depression—the elevation of such a *depression contour* is the same as that of the adjacent lower regular contour (unless otherwise marked).

Topographic Map Symbols

Topographic quadrangles also portray a variety of other features using symbols and colors. Standard colors are used to distinguish various kinds of map features:

- **Brown:** contour lines and other topographic features
- **Blue:** hydrographic (water) features
- **Black:** features constructed or designated by humans, such as buildings, roads, boundary lines, and names
- **Green:** areas of vegetation, such as woodlands, forests, orchards, and vineyards
- **Red:** important roads and lines of the public land survey system
- **Gray or red tint:** urban areas
- **Purple:** features added from aerial photos during map revision

The principal standard symbols used on USGS topographic maps are shown in Figure II-3. Note that in a few cases more than one kind of symbol is used to show the same kind of feature.

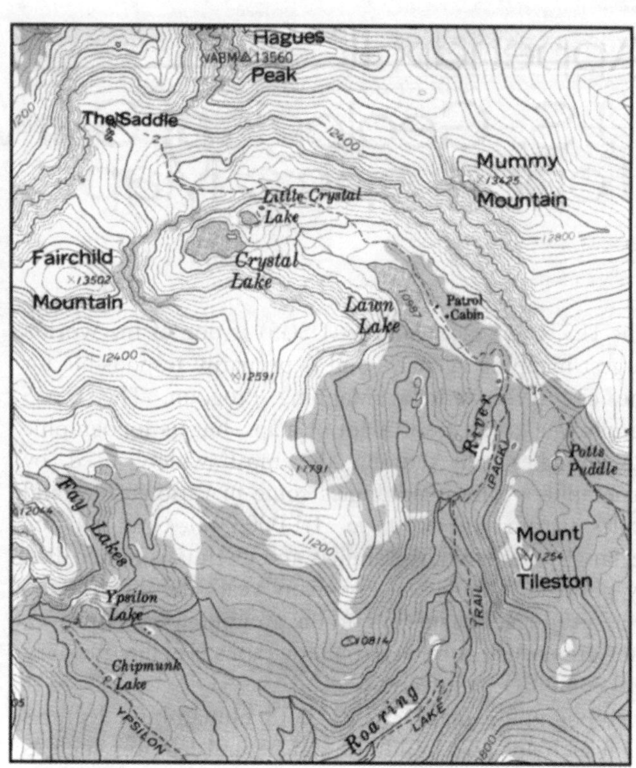

► **Figure II-1** Portrayal of terrain by means of elevation contour lines. Here is a photo of Hagues Peak and Mummy Mountain in Colorado's Rocky Mountain National Park along with a standard topographic map of the same area. (Original map scale 1:125,000; contour interval is 80 feet [27 meters].)

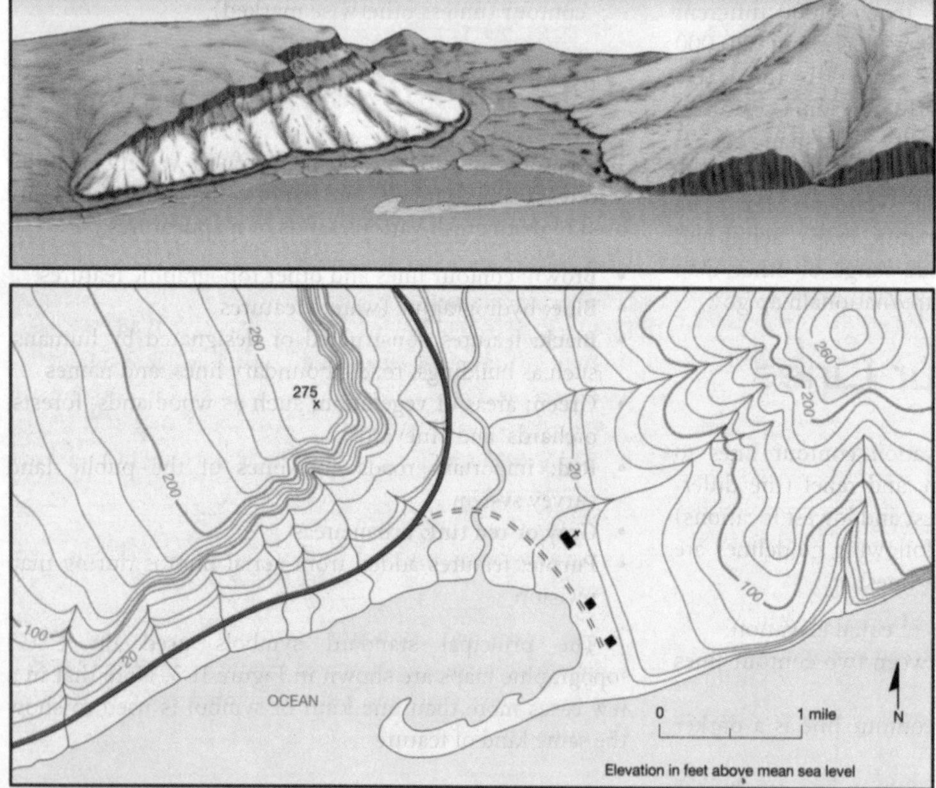

▲ **Figure II-2** A fictitious landscape and a contour line map of the same area. Where the contours are close together, the slope is steep; where the lines are far apart, the slope is gentle.

BATHYMETRIC FEATURES	
Area exposed at mean low tide; sounding datum line***	
Channel***	
Sunken rock***	+

BOUNDARIES	
National	
State or territorial	
County or equivalent	
Civil township or equivalent	
Incorporated city or equivalent	
Federally administered park, reservation, or monument (external)	
Federally administered park, reservation, or monument (internal)	
State forest, park, reservation, or monument and large county park	
Forest Service administrative area*	
Forest Service ranger district*	
National Forest System land status, Forest Service lands*	
National Forest System land status, non-Forest Service lands*	
Small park (county or city)	

BUILDINGS AND RELATED FEATURES	
Building	
School; house of worship	
Athletic field	
Built-up area	
Forest headquarters*	
Ranger district office*	
Guard station or work center*	
Racetrack or raceway	
Airport, paved landing strip, runway, taxiway, or apron	
Unpaved landing strip	
Well (other than water), windmill or wind generator	
Tanks	
Covered reservoir	
Gaging station	
Located or landmark object (feature as labeled)	
Boat ramp or boat access*	
Roadside park or rest area	
Picnic area	
Campground	
Winter recreation area*	
Cemetery	Cem

COASTAL FEATURES	
Foreshore flat	Mud
Coral or rock reef	Reef
Rock, bare or awash; dangerous to navigation	
Group of rocks, bare or awash	
Exposed wreck	
Depth curve; sounding	18 23
Breakwater, pier, jetty, or wharf	
Seawall	
Oil or gas well; platform	

CONTOURS	
Topographic	
Index	6000
Approximate or indefinite	
Intermediate	
Approximate or indefinite	
Supplementary	
Depression	
Cut	
Fill	
Continental divide	
Bathymetric	
Index***	
Intermediate***	
Index primary***	
Primary***	
Supplementary***	

CONTROL DATA AND MONUMENTS	
Principal point**	3-20
U.S. mineral or location monument	USMM 438
River mileage marker	Mile 69
Boundary monument	
Third-order or better elevation, with tablet	BM 9134 BM 277
Third-order or better elevation, recoverable mark, no tablet	5628
With number and elevation	67 4567
Horizontal control	
Third-order or better, permanent mark	Neace Neace
With third-order or better elevation	BM 52 Pike BM393
With checked spot elevation	1012
Coincident with found section corner	Cactus Cactus
Unmonumented**	

▲ **Figure II-3** Topographic map symbols used on USGS quadrangles.

CONTROL DATE AND MONUMENTS — *continued*	
Vertical control	
Third-order or better elevation, with tablet	BM × 5280
Third-order or better elevation, recoverable mark, no tablet	× 528
Bench mark coincident with found section corner	BM + 5280
Spot elevation	× 7523
GLACIERS AND PERMANENT SNOWFIELDS	
Contours and limits	
Formlines	
Glacial advance	
Glacial retreat	
LAND SURVEYS	
Public land survey system	
Range or Township line	
Location approximate	
Location doubtful	
Protracted	
Protracted (AK 1:63,360-scale)	
Range or Township labels	R1E T2N R3W T4S
Section line	
Location approximate	
Location doubtful	
Protracted	
Protracted (AK 1:63,360-scale)	
Section numbers	1 - 36 1 - 36
Found section corner	
Found closing corner	
Witness corner	WC
Meander corner	MC
Weak corner*	
Other land surveys	
Range or Township line	
Section line	
Land grant, mining claim, donation land claim, or tract	
Land grant, homestead, mineral, or other special survey monument	
Fence or field lines	
MARINE SHORELINES	
Shoreline	
Apparent (edge of vegetation)***	
Indefinite or unsurveyed	
MINES AND CAVES	
Quarry or open pit mine	⚒
Gravel, sand, clay, or borrow pit	×
Mine tunnel or cave entrance	
Mine shaft	
Prospect	X
Tailings	Tailings
Mine dump	
Former disposal site or mine	

PROJECTION AND GRIDS	
Neatline	3915 ′ 9037 30 ″
Graticule tick	55′
Graticule intersection	
Datum shift tick	
State plane coordinate systems	
Primary zone tick	640 000 FEET
Secondary zone tick	247 500 METERS
Tertiary zone tick	260 000 FEET
Quaternary zone tick	98 500 METERS
Quintary zone tick	320 000 FEET
Universal transverse metcator grid	
UTM grid (full grid)	273
UTM grid ticks*	269
RAILROADS AND RELATED FEATURES	
Standard guage railroad, single track	
Standard guage railroad, multiple track	
Narrow guage railroad, single track	
Narrow guage railroad, multiple track	
Railroad siding	
Railroad in highway	
Railroad in road	
Railroad in light duty road*	
Railroad underpass; overpass	
Railroad bridge; drawbridge	
Railroad tunnel	
Railroad yard	
Railroad turntable; roundhouse	
RIVERS, LAKES, AND CANALS	
Perennial stream	
Perennial river	
Intermittent stream	
Intermittent river	
Disappearing stream	
Falls, small	
Falls, large	
Rapids, small	
Rapids, large	
Masonry dam	
Dam with lock	
Dam carrying load	

▲ **Figure II-3** (continued)

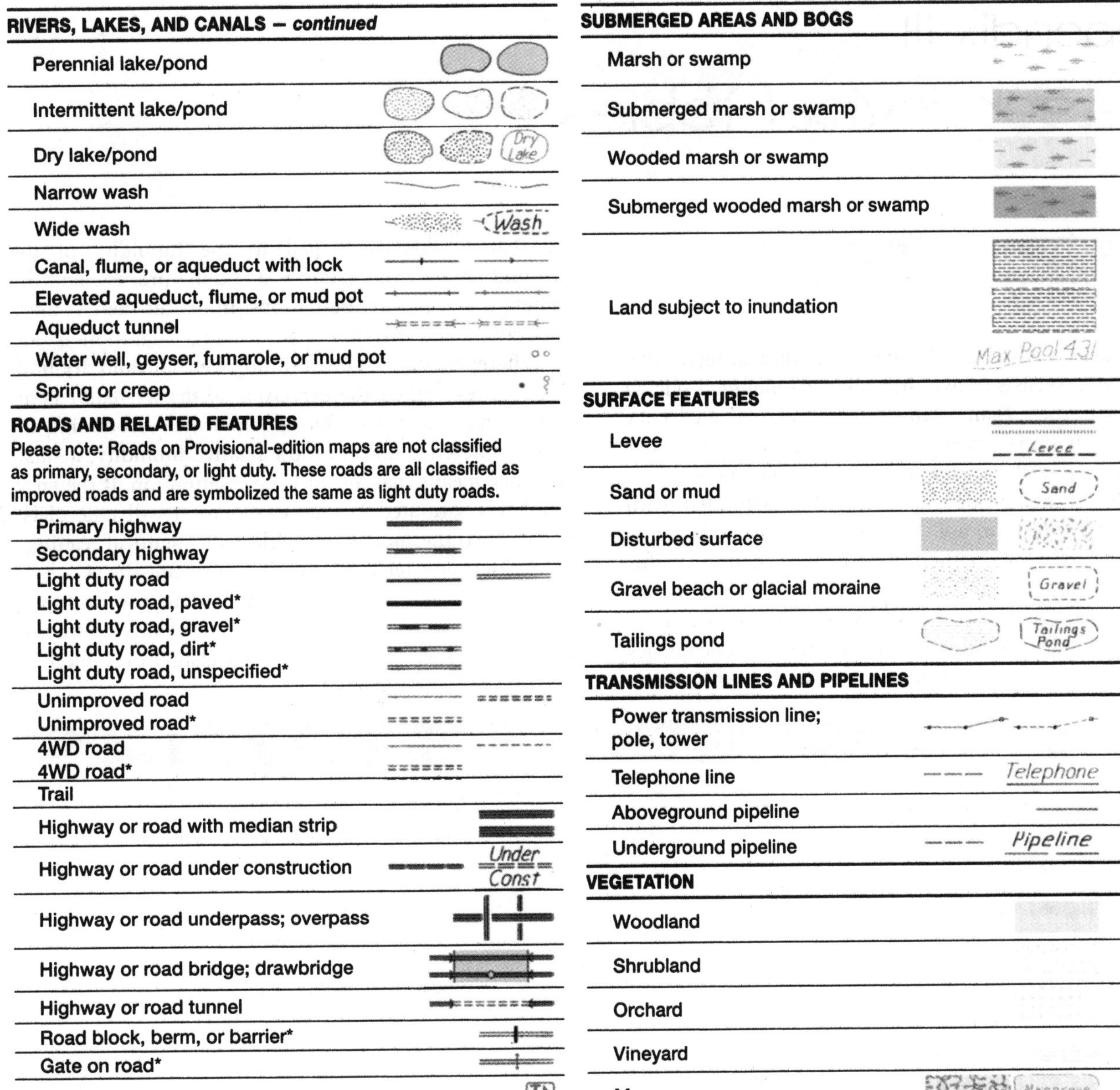

* USGS-USDA Forest Service Single-Edition
** Provisional-Edition maps only.
*** Topographic Bathymetric maps only.

▲ **Figure II-3** (continued)

Appendix III

Meteorological Tables

Determining Relative Humidity

A *psychrometer* is an instrument used for measuring relative humidity. It consists of two thermometers mounted side by side. One of these is an ordinary thermometer (called a *dry bulb*), which simply measures air temperature. The other thermometer (called a *wet bulb*) has its bulb encased in a covering of muslin or gauze which is saturated with distilled water prior to use. The two thermometers are then thoroughly ventilated either by being whirled around (this instrument has a handle around which the thermometers can be whirled and is referred to as a *sling psychrometer*) or by fanning air past them. This ventilation encourages evaporation of water from the covering of the wet bulb at a rate that is directly related to the humidity of the surrounding air. Evaporation is a cooling process, and the temperature of the wet bulb drops. In dry air there is more evaporation, and therefore more cooling, than in moist air. The difference between the resulting wet-bulb and dry-bulb temperatures (called the *depression of the wet bulb*) is an expression of the relative saturation of the surrounding air. A large difference indicates low relative humidity; a small difference means that the air is near saturation. If the air is completely saturated, no net evaporation will take place; thus the two thermometers would have identical readings.

To determine relative humidity, measure the wet-bulb and dry-bulb temperatures using a sling psychrometer. Then use these two values to read the relative humidity from Table Table III-1 (for degrees Celsius) or Table III-2 (for degrees Fahrenheit).

TABLE III-1 Relative Humidity Psychrometer Tables (°C)

Relative Humidity (%)

Air Temp. °C	Depression of Wet-Bulb Thermometer (°C)																					
	1	*2*	*3*	*4*	*5*	*6*	*7*	*8*	*9*	*10*	*11*	*12*	*13*	*14*	*15*	*16*	*17*	*18*	*19*	*20*	*21*	*22*
−4	77	54	32	11																		
−2	79	58	37	20	1																	
0	81	63	45	28	11																	
2	83	67	51	36	20	6																
4	85	70	56	42	27	14																
6	86	72	59	46	35	22	10	0														
8	87	74	62	51	39	28	17	6														
10	88	76	65	54	43	33	24	13	4													
12	88	78	67	57	48	38	28	19	10	2												
14	89	79	69	60	50	41	33	25	16	8	1											
16	90	80	71	62	54	45	37	29	21	14	7	1										
18	91	81	72	64	56	48	40	33	26	19	12	6	0									
20	91	82	74	66	58	51	44	36	30	23	17	11	5									
22	92	83	75	68	60	53	46	40	33	27	21	15	10	4	0							
24	92	84	76	69	62	55	49	42	36	30	25	20	14	9	4	0						
26	92	85	77	70	64	57	51	45	39	34	28	23	18	13	9	5						
28	93	86	78	71	65	59	53	45	42	36	31	26	21	17	12	8	4					
30	93	86	79	72	66	61	55	49	44	39	34	29	25	20	16	12	8	4				
32	93	86	80	73	68	62	56	51	46	41	36	32	27	22	19	14	11	8	4			
34	93	86	81	74	69	63	58	52	48	43	38	34	30	26	22	18	14	11	8	5		
36	94	87	81	75	69	64	59	54	50	44	40	36	32	28	24	21	17	13	10	7	4	
38	94	87	82	76	70	66	60	55	51	46	42	38	34	30	26	23	20	16	13	10	7	5

TABLE III-2 Relative Humidity Psychrometer Tables (°F)

Air temp. °F	Depression of Wet-Bulb Thermometer (°F)																													
	1	2	3	4	5	6	7	8	9	10	11	12	13	14	15	16	17	18	19	20	21	22	23	24	25	26	27	28	29	30
0	67	33	1																											
5	73	46	20																											
10	78	56	34	13	15																									
15	82	64	46	29	11																									
20	85	70	55	40	26	12																								
25	87	74	62	49	37	25	13	1																						
30	89	78	67	56	46	36	26	16	6												Relative Humidity (%)									
35	91	81	72	63	54	45	36	27	19	10	2																			
40	92	83	75	68	60	52	45	37	29	22	15	7																		
45	93	86	78	71	64	57	51	44	38	31	25	18	12	6																
50	93	87	74	67	61	55	49	43	38	32	27	21	16	10	5															
55	94	88	82	76	70	65	59	54	49	43	38	33	28	23	19	11	9	5												
60	94	89	83	78	73	68	63	58	53	48	43	39	34	30	26	21	17	13	9	5	1									
65	95	90	85	80	75	70	66	61	56	52	48	44	39	35	31	27	24	20	16	12	9	5	2							
70	95	90	86	81	77	72	68	64	59	55	51	48	44	40	36	33	29	25	22	19	15	12	9	6	3					
75	96	91	86	82	78	74	70	66	62	58	54	51	47	44	40	37	34	30	27	24	21	18	15	12	9	7	4	1		
80	96	91	87	83	79	75	72	68	64	61	57	54	50	47	44	41	38	35	32	29	26	23	20	18	15	12	10	7	5	3
85	96	92	88	84	81	77	73	70	66	63	59	57	53	50	47	44	41	38	36	33	30	27	25	22	20	17	15	13	10	8
90	96	92	89	85	81	78	74	71	68	65	61	58	55	52	49	47	44	41	39	36	34	31	29	26	24	22	19	17	15	13
95	96	93	89	86	82	79	76	73	69	66	63	61	58	55	52	50	47	44	42	39	37	34	32	30	28	25	23	21	19	17
100	96	93	89	86	83	80	77	73	70	68	65	62	59	56	54	51	49	46	44	41	39	37	35	33	30	28	26	24	22	21
105	97	93	90	87	84	81	78	75	72	69	66	64	61	58	56	53	51	49	46	44	42	40	38	36	34	32	30	28	26	24

For example, if the dry-bulb temperature is 20°C and the wet-bulb temperature is 14°C, the depression of the wet bulb is 6°C. In Table III-1, find 20°C in the "Air Temperature" column; move across to the 6° column under "Depression of the Wet-Bulb Thermometer," at the point of intersection, read the relative humidity, 51 percent.

The Beaufort Scale of Wind Speed

Early in the nineteenth century Admiral Beaufort of the British Navy developed a scale of wind speed widely used in the English-speaking world. It has been modified through the years, but the essentials have not changed. The scale is shown in Table III-3.

Wind Chill

Wind chill is the popular name used to describe what cold weather feels like at various combinations of low temperature and high wind. On a cold, windless day, one's body heat is conducted sluggishly to a thin layer of atmospheric molecules near the skin. These heated molecules diffuse away slowly, to be replaced by other, cooler molecules. One's body thus comes into contact with a relatively small number of cool molecules, and one's body heat dissipates slowly.

With an increase in wind speed, however, body heat dissipates much more rapidly, as the protective layer of warmer molecules is speedily removed and supplanted by a continually renewing supply of cold air molecules against the skin. Up to a certain speed, the greater the wind velocity, the greater the cooling effect on one's body.

TABLE III-3 Beaufort Scale

Beaufort Force	Speed Kilometers per Hour	Miles per Hour	Knots	Description
0	<1	<1	<1	Calm
1	1–5	1–3	1–3	Light air
2	6–11	4–7	4–6	Light breeze
3	12–19	8–12	7–10	Gentle breeze
4	20–29	13–18	11–16	Moderate breeze
5	30–38	19–24	17–21	Fresh breeze
6	39–49	25–31	22–27	Strong breeze
7	50–61	32–38	28–33	Near gale
8	62–74	39–46	34–40	Gale
9	75–87	47–54	41–47	Strong gale
10	88–101	55–63	48–55	Storm
11	102–116	64–72	56–63	Violent storm
12	117–132	73–82	64–71	Hurricane
13	133–148	83–92	72–80	Hurricane
14	149–166	93–103	81–89	Hurricane
15	167–183	104–114	90–99	Hurricane
16	184–201	115–125	100–108	Hurricane
17	202–219	126–136	109–118	Hurricane

Wind chill affects only organisms that generate heat; inanimate objects have no such heat to lose, so their temperatures are not affected at all by wind movement.

The term wind chill was apparently first used in 1939 by Paul Siple, a geographer and polar explorer. Meteorologists from the National Weather Service of the United States and from the Meteorological Services of Canada have refined the calculations several times, most recently in 2002. The revised wind-chill index currently in use accounts for the wind effects at face level (the "old" system relied on observed winds 10 meters [33 feet] above the ground) and a better calculation for body heat loss.

The updated index also includes a new frostbite chart, which shows how long skin can safely be exposed to the air given varying temperatures and wind (Tables Table III-4 and III-5).

The Heat Index

Sensible temperatures may be significantly influenced by humidity, which can make the weather seem either colder or warmer than it actually is—although humidity is more likely to impinge on our lives in hot weather. Quite simply, high humidity makes hot weather seem hotter.

The National Weather Service has developed a *heat index* that combines temperature and relative humidity to produce an "apparent temperature" that quantifies how hot the air feels to one's skin. A sample heat index chart is given in Table III-6.

To this index has been added a general heat stress index to indicate heat-related dangers at various apparent temperatures, as shown in Table III-7.

TABLE III-4 Wind Chill (°C)

Wind Speed (kilometers per hour)	Actual Air Temperature °C									
Calm	5	0	−5	−10	−15	−20	−25	−30	−35	−40
5	4	−2	−7	−13	−19	−24	−30	−36	−41	−47
10	3	−3	−9	−15	−21	−27	−33	−39	−45	−51
15	2	−4	−11	−17	−23	−29	−35	−41	−48	−54
20	1	−5	−12	−18	−24	−31	−37	−43	−49	−56
25	1	−6	−12	−19	−25	−32	−38	−45	−51	−57
30	0	−7	−13	−20	−26	−33	−39	−46	−52	−59
35	0	−7	−14	−20	−27	−33	−40	−47	−53	−60
40	−1	−7	−14	−21	−27	−34	−41	−48	−54	−61
45	−1	−8	−15	−21	−28	−35	−42	−48	−55	−62
50	−1	−8	−15	−22	−29	−35	−42	−49	−56	−63
55	−2	−9	−15	−22	−29	−36	−43	−50	−57	−63
60	−2	−9	−16	−23	−30	−37	−43	−50	−57	−64
65	−2	−9	−16	−23	−30	−37	−44	−51	−58	−65
70	−2	−9	−16	−23	−30	−37	−44	−51	−59	−66
75	−3	−10	−17	−24	−31	−38	−45	−52	−59	−66
80	−3	−10	−17	−24	−31	−38	−45	−52	−60	−67

Frostbite within 30 minutes | Frostbite within 10 minutes | Frostbite within 5 minutes

TABLE III-5 Wind Chill (°F)

Wind Speed (miles per hour)	Actual Air Temperature °F															
Calm	40	35	30	25	20	15	10	5	0	−5	−10	−15	−20	−25	−30	−35	−40
5	36	31	25	19	13	7	1	−5	−11	−16	−22	−28	−34	−40	−46	−52	−57
10	34	27	21	15	9	3	−4	−10	−16	−22	−28	−35	−41	−47	−53	−59	−66
15	32	25	19	13	6	0	−7	−13	−19	−26	−32	−39	−45	−51	−58	−64	−71
20	30	24	17	11	4	−2	−9	−15	−22	−29	−35	−42	−48	−55	−61	−68	−74
25	29	23	16	9	3	−4	−11	−17	−24	−31	−37	−44	−51	−58	−64	−71	−78
30	28	22	15	8	1	−5	−12	−19	−26	−33	−39	−46	−53	−60	−67	−73	−80
35	28	21	14	7	0	−7	−14	−21	−27	−34	−41	−48	−55	−62	−69	−76	−82
40	27	20	13	6	−1	−8	−15	−22	−29	−36	−43	−50	−57	−64	−71	−78	−84
45	26	19	12	5	−2	−9	−16	−23	−30	−37	−44	−51	−58	−65	−72	−79	−86
50	26	19	12	4	−3	−10	−17	−24	−31	−38	−45	−52	−60	−67	−74	−81	−88
55	25	18	11	4	−3	−11	−18	−25	−32	−39	−45	−54	−61	−68	−75	−82	−89
60	25	17	10	3	−4	−11	−19	−26	−33	−40	−48	−55	−62	−69	−76	−84	−91

Frostbite within 30 minutes | Frostbite within 10 minutes | Frostbite within 5 minutes

TABLE III-6 Heat Index (Apparent Temperature)

Temperature	10%	20%	30%	40%	50%	60%	70%	80%	90%
	Relative Humidity								
46°C 115°F	44°C 111°F	49°C 120°F	57°C 135°F	66°C 151°F	*	*	*	*	*
43°C 110°F	41°C 105°F	44°C 112°F	51°C 123°F	58°C 137°F	66°C 150°F	*	*	*	*
41°C 105°F	38°C 100°F	41°C 105°F	45°C 113°F	51°C 123°F	57°C 135°F	65°C 149°F	*	*	*
38°C 100°F	35°C 95°F	37°C 99°F	40°C 104°F	43°C 110°F	49°C 120°F	56°C 132°F	62°C 144°F	*	*
35°C 95°F	32°C 90°F	34°C 93°F	36°C 96°F	38°C 101°F	42°C 107°F	46°C 114°F	51°C 124°F	58°C 136°F	*
32°C 90°F	29°C 85°F	31°C 87°F	32°C 90°F	34°C 93°F	36°C 96°F	38°C 100°F	41°C 106°F	45°C 113°F	50°C 122°F
29°C 85°F	27°C 80°F	28°C 82°F	29°C 84°F	30°C 86°F	31°C 88°F	32°C 90°F	34°C 93°F	36°C 97°F	39°C 102°F
27°C 80°F	24°C 75°F	25°C 77°F	26°C 78°F	26°C 79°F	27°C 81°F	28°C 82°F	29°C 85°F	30°C 86°F	31°C 88°F

*Temperature–relative humidity conditions rarely observed in the atmosphere.

TABLE III-7 General Heat Stress Index

Danger Category	Heat Index	Heat Syndrome
IV. Extreme Danger	Above 54°C (130°F)	Heat/sunstroke highly likely with continued exposure.
III. Danger	40–54°C (105–130°F)	Sunstroke, heat cramps, or heat exhaustion likely; heatstroke possible with prolonged exposure and/or physical activity.
II. Extreme Caution	32–40°C (90–105°F)	Sunstroke, heat cramps, or heat exhaustion possible with prolonged exposure and/or physical activity.
I. Caution	27–32°C (80–90°F)	Fatigue possible with prolonged exposure and/or physical activity.

Appendix IV
The Weather Station Model

Weather data are recorded at regular intervals for a great many locations on Earth, each location being called a *weather station*. These data are then plotted on weather maps according to a standard format and code. The format for a standard *station model* is shown in Figure IV-1, along with an explanation of the code. Figure IV-2 shows the same model but with the codes replaced by sample data, and Figures IV-3 and IV-4 list some of the codes and symbols used by meteorologists. Tables IV-1 through IV-3 give additional codes and symbols, and Figure IV-5 is a sample weather map.

The wind symbol (Table IV-3) gives two pieces of information. Wind direction is indicated by an arrow shaft entering the station circle from the direction in which the wind is blowing, as at the one-o'clock position in Figure IV-2, and wind speed is indicated by the number of "feathers" and half-feathers protruding from the shaft; each half-feather represents a 5-knot increase in speed (a knot is one nautical mile per hour, which is the same as 1.15 statute miles per hour or 1.85 kilometers per hour); each full feather represents a 10-knot increase; a triangular pendant represents a 50-knot increase.

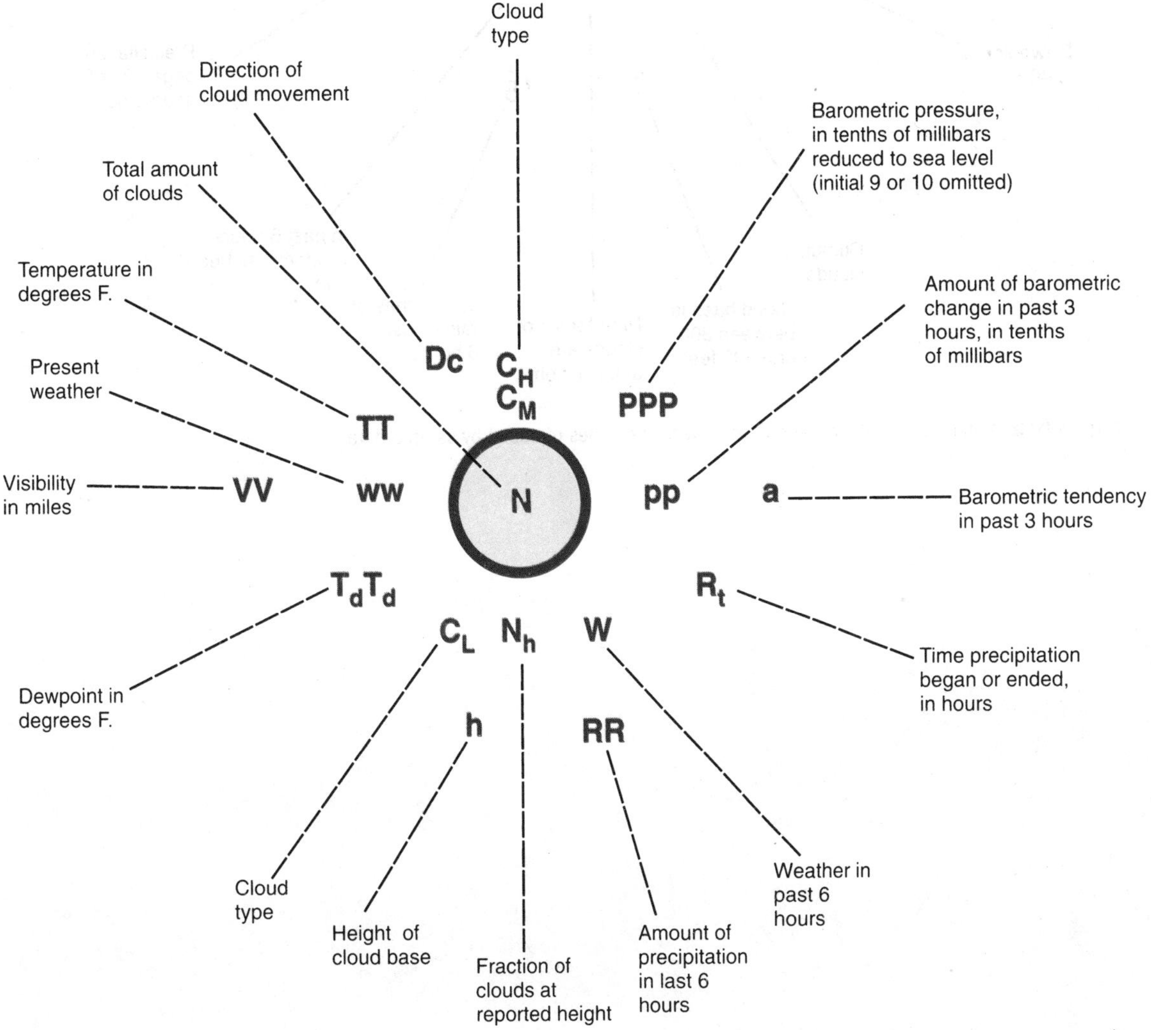

▲ **Figure IV-1** A standard weather station model. (No symbol for wind speed and direction is shown here because there is no one assigned place for this symbol; instead, its position on the model depends on wind direction.)

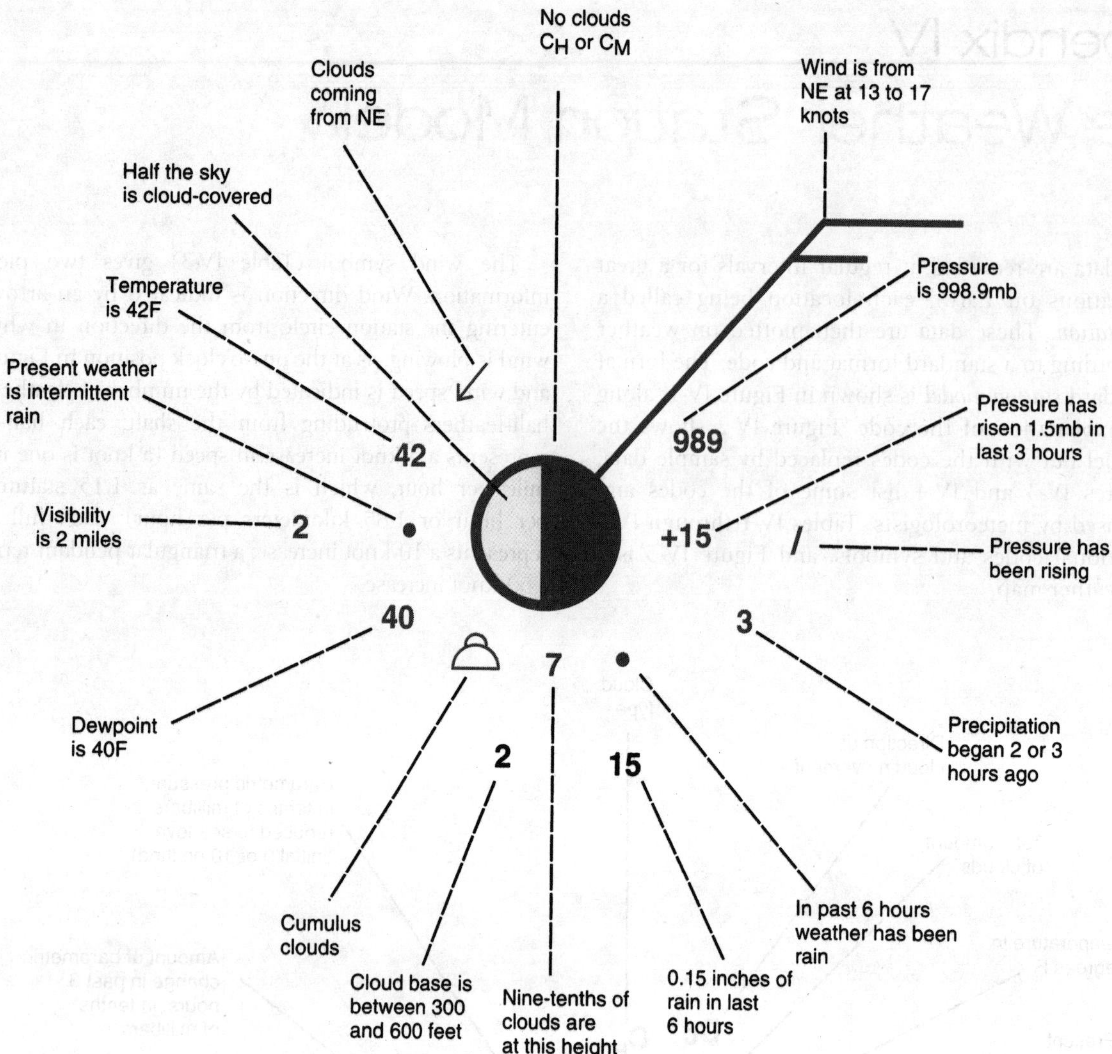

▲**Figure IV-2** A standard weather station model with the codes replaced by sample data.

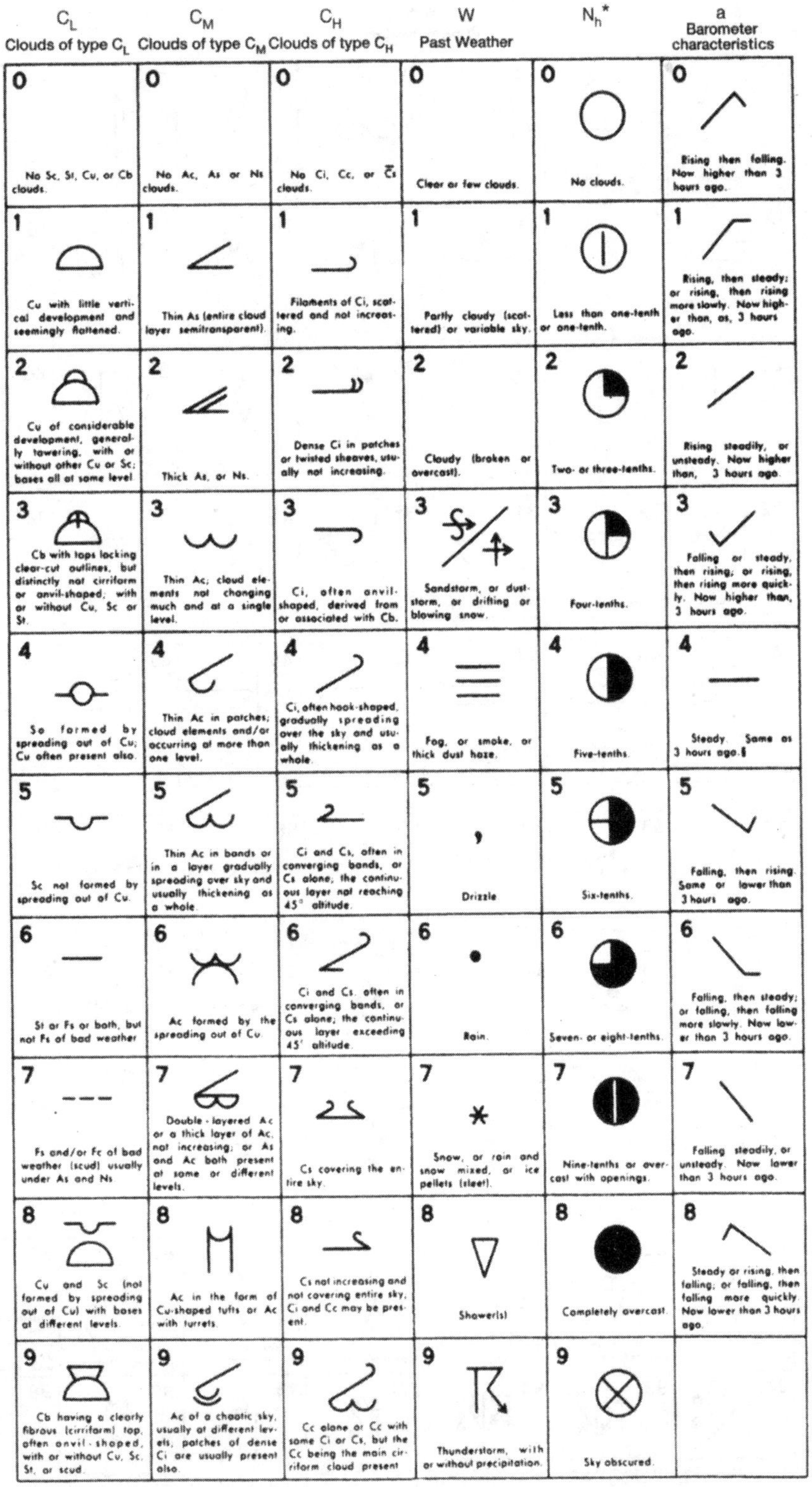

	C_L Clouds of type C_L	C_M Clouds of type C_M	C_H Clouds of type C_H	W Past Weather	N_h*	a Barometer characteristics
0	No Sc, St, Cu, or Cb clouds.	No Ac, As or Ns clouds.	No Ci, Cc, or Cs clouds.	Clear or few clouds.	No clouds.	Rising then falling. Now higher than 3 hours ago.
1	Cu with little vertical development and seemingly flattened.	Thin As (entire cloud layer semitransparent).	Filaments of Ci, scattered and not increasing.	Partly cloudy (scattered) or variable sky.	Less than one-tenth or one-tenth.	Rising, then steady; or rising, then rising more slowly. Now higher than, as, 3 hours ago.
2	Cu of considerable development, generally towering, with or without other Cu or Sc; bases all at same level	Thick As, or Ns.	Dense Ci in patches or twisted sheaves, usually not increasing.	Cloudy (broken or overcast).	Two- or three-tenths.	Rising steadily, or unsteady. Now higher than, 3 hours ago.
3	Cb with tops lacking clear-cut outlines, but distinctly not cirriform or anvil-shaped; with or without Cu, Sc or St.	Thin Ac; cloud elements not changing much and at a single level.	Ci, often anvil-shaped, derived from or associated with Cb.	Sandstorm, or duststorm, or drifting or blowing snow.	Four-tenths.	Falling or steady, then rising; or rising, then rising more quickly. Now higher than, 3 hours ago.
4	So formed by spreading out of Cu; Cu often present also.	Thin Ac in patches; cloud elements and/or occurring at more than one level.	Ci, often hook-shaped, gradually spreading over the sky and usually thickening as a whole.	Fog, or smoke, or thick dust haze.	Five-tenths.	Steady. Same as 3 hours ago.§
5	Sc not formed by spreading out of Cu.	Thin Ac in bands or in a layer gradually spreading over sky and usually thickening as a whole.	Ci and Cs, often in converging bands, or Cs alone; the continuous layer not reaching 45° altitude.	Drizzle	Six-tenths.	Falling, then rising. Same or lower than 3 hours ago.
6	St or Fs or both, but not Fs of bad weather	Ac formed by the spreading out of Cu.	Ci and Cs. often in converging bands, or Cs alone; the continuous layer exceeding 45° altitude.	Rain.	Seven- or eight-tenths.	Falling, then steady; or falling, then falling more slowly. Now lower than 3 hours ago.
7	Fs and/or Fc of bad weather (scud) usually under As and Ns.	Double-layered Ac or a thick layer of Ac, not increasing; or As and Ac both present at some or different levels.	Cs covering the entire sky.	Snow, or rain and snow mixed, or ice pellets (sleet).	Nine-tenths or overcast with openings.	Falling steadily, or unsteady. Now lower than 3 hours ago.
8	Cu and Sc (not formed by spreading out of Cu) with bases at different levels.	Ac in the form of Cu-shaped tufts or Ac with turrets.	Cs not increasing and not covering entire sky, Ci and Cc may be present.	Shower(s)	Completely overcast.	Steady or rising, then falling; or falling, then falling more quickly. Now lower than 3 hours ago.
9	Cb having a clearly fibrous (cirriform) top, often anvil-shaped, with or without Cu, Sc, St, or scud.	Ac of a chaotic sky, usually at different levels; patches of dense Ci are usually present also.	Cc alone or Cc with some Ci or Cs, but the Cc being the main cirriform cloud present	Thunderstorm, with or without precipitation.	Sky obscured.	

*Fraction representing how much of the total cloud cover is at the reported base height.

▲ **Figure IV-3** Standard symbols used to indicate cloud conditions, past weather, and barometer characteristics. The numbers in the upper-left-hand corner of the cells are used in a standard model and the icons are used on weather maps.

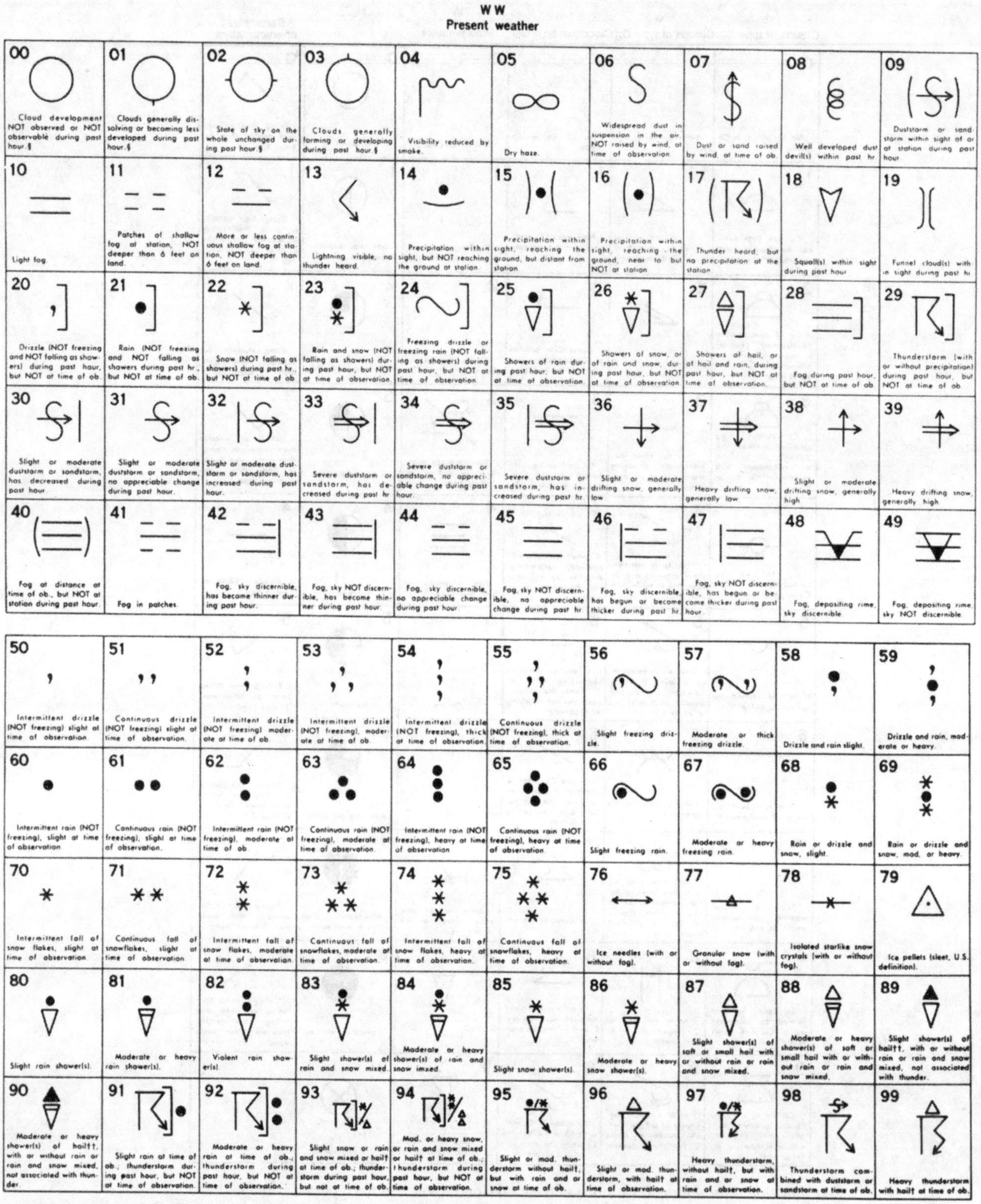

WW
Present weather

00	Cloud development NOT observed or NOT observable during past hour.§
01	Clouds generally dissolving or becoming less developed during past hour.§
02	State of sky on the whole unchanged during past hour.§
03	Clouds generally forming or developing during past hour.§
04	Visibility reduced by smoke.
05	Dry haze.
06	Widespread dust in suspension in the air, NOT raised by wind, at time of observation
07	Dust or sand raised by wind, at time of ob.
08	Well developed dust devil(s) within past hr.
09	Duststorm or sandstorm within sight of or at station during past hour.
10	Light fog.
11	Patches of shallow fog at station, NOT deeper than 6 feet on land.
12	More or less continuous shallow fog at station, NOT deeper than 6 feet on land.
13	Lightning visible, no thunder heard.
14	Precipitation within sight, but NOT reaching the ground at station.
15	Precipitation within sight, reaching the ground, but distant from station.
16	Precipitation within sight, reaching the ground, near to but NOT at station
17	Thunder heard, but no precipitation at the station.
18	Squall(s) within sight during past hour
19	Funnel cloud(s) within sight during past hr
20	Drizzle (NOT freezing and NOT falling as showers) during past hour, but NOT at time of ob.
21	Rain (NOT freezing and NOT falling as showers during past hr., but NOT at time of ob.
22	Snow (NOT falling as showers) during past hr., but NOT at time of ob
23	Rain and snow (NOT falling as showers) during past hour, but NOT at time of observation.
24	Freezing drizzle or freezing rain (NOT falling as showers) during past hour, but NOT at time of observation.
25	Showers of rain during past hour, but NOT at time of observation.
26	Showers of snow, or of rain and snow, during past hour, but NOT at time of observation
27	Showers of hail, or of hail and rain, during past hour, but NOT at time of observation.
28	Fog during past hour, but NOT at time of ob.
29	Thunderstorm (with or without precipitation) during past hour, but NOT at time of ob.
30	Slight or moderate duststorm or sandstorm, has decreased during past hour.
31	Slight or moderate duststorm or sandstorm, no appreciable change during past hour.
32	Slight or moderate duststorm or sandstorm, has increased during past hour.
33	Severe duststorm or sandstorm, has decreased during past hr
34	Severe duststorm or sandstorm, no appreciable change during past hour.
35	Severe duststorm or sandstorm, has increased during past hr.
36	Slight or moderate drifting snow, generally low
37	Heavy drifting snow, generally low.
38	Slight or moderate drifting snow, generally high
39	Heavy drifting snow, generally high.
40	Fog at distance at time of ob., but NOT at station during past hour.
41	Fog in patches.
42	Fog, sky discernible, has become thinner during past hour.
43	Fog, sky NOT discernible, has become thinner during past hour.
44	Fog, sky discernible, no appreciable change during past hour.
45	Fog, sky NOT discernible, no appreciable change during past hr.
46	Fog, sky discernible, has begun or become thicker during past hr.
47	Fog, sky NOT discernible, has begun or become thicker during past hour.
48	Fog, depositing rime, sky discernible.
49	Fog, depositing rime, sky NOT discernible.
50	Intermittent drizzle (NOT freezing) slight at time of observation.
51	Continuous drizzle (NOT freezing) slight at time of observation.
52	Intermittent drizzle (NOT freezing) moderate at time of ob.
53	Intermittent drizzle (NOT freezing), moderate at time of ob.
54	Intermittent drizzle (NOT freezing), thick at time of observation.
55	Continuous drizzle (NOT freezing), thick at time of observation.
56	Slight freezing drizzle.
57	Moderate or thick freezing drizzle.
58	Drizzle and rain slight.
59	Drizzle and rain, moderate or heavy.
60	Intermittent rain (NOT freezing), slight at time of observation
61	Continuous rain (NOT freezing), slight at time of observation.
62	Intermittent rain (NOT freezing), moderate at time of ob.
63	Continuous rain (NOT freezing), moderate at time of observation.
64	Intermittent rain (NOT freezing), heavy at time of observation
65	Continuous rain (NOT freezing), heavy at time of observation.
66	Slight freezing rain.
67	Moderate or heavy freezing rain.
68	Rain or drizzle and snow, slight.
69	Rain or drizzle and snow, mod. or heavy.
70	Intermittent fall of snow flakes, slight at time of observation
71	Continuous fall of snowflakes, slight at time of observation
72	Intermittent fall of snow flakes, moderate at time of observation.
73	Continuous fall of snowflakes, moderate at time of observation.
74	Intermittent fall of snow flakes, heavy at time of observation.
75	Continuous fall of snowflakes, heavy at time of observation.
76	Ice needles (with or without fog).
77	Granular snow (with or without fog).
78	Isolated starlike snow crystals (with or without fog).
79	Ice pellets (sleet, U.S. definition).
80	Slight rain shower(s).
81	Moderate or heavy rain shower(s).
82	Violent rain shower(s).
83	Slight shower(s) of rain and snow mixed.
84	Moderate or heavy shower(s) of rain and snow imxed.
85	Slight snow shower(s).
86	Moderate or heavy snow shower(s).
87	Slight shower(s) of soft or small hail with or without rain or rain and snow mixed.
88	Moderate or heavy shower(s) of soft or small hail with or without rain or rain and snow mixed.
89	Slight shower(s) of hail††, with or without rain or rain and snow mixed, not associated with thunder.
90	Moderate or heavy shower(s) of hail††, with or without rain or rain and snow mixed, not associated with thunder.
91	Slight rain at time of ob.; thunderstorm during past hour, but NOT at time of observation.
92	Moderate or heavy rain at time of ob.; thunderstorm during past hour, but NOT at time of observation.
93	Slight snow or rain and snow mixed or hail† at time of ob.; thunderstorm during past hour, but not at time of ob.
94	Mod. or heavy snow, or rain and snow mixed or hail† at time of ob.; thunderstorm during past hour, but NOT at time of observation.
95	Slight or mod. thunderstorm without hail†, but with rain and or snow at time of ob.
96	Slight or mod. thunderstorm, with hail† at time of observation.
97	Heavy thunderstorm, without hail†, but with rain and or snow at time of observation.
98	Thunderstorm combined with duststorm or sandstorm at time of ob.
99	Heavy thunderstorm with hail† at time of ob.

§ The symbol is not plotted for "ww" when "00" is reported. When "01, 02, or 03" is reported for "ww," the symbol is plotted on the station circle. Symbols are not plotted for "a" when "3 or 8" is reported.

† Refers to "hail" only.

††Refers to "soft hail," "small hail," and "hail."

▲ **Figure IV-4** Standard weather map symbols used to indicate present weather.

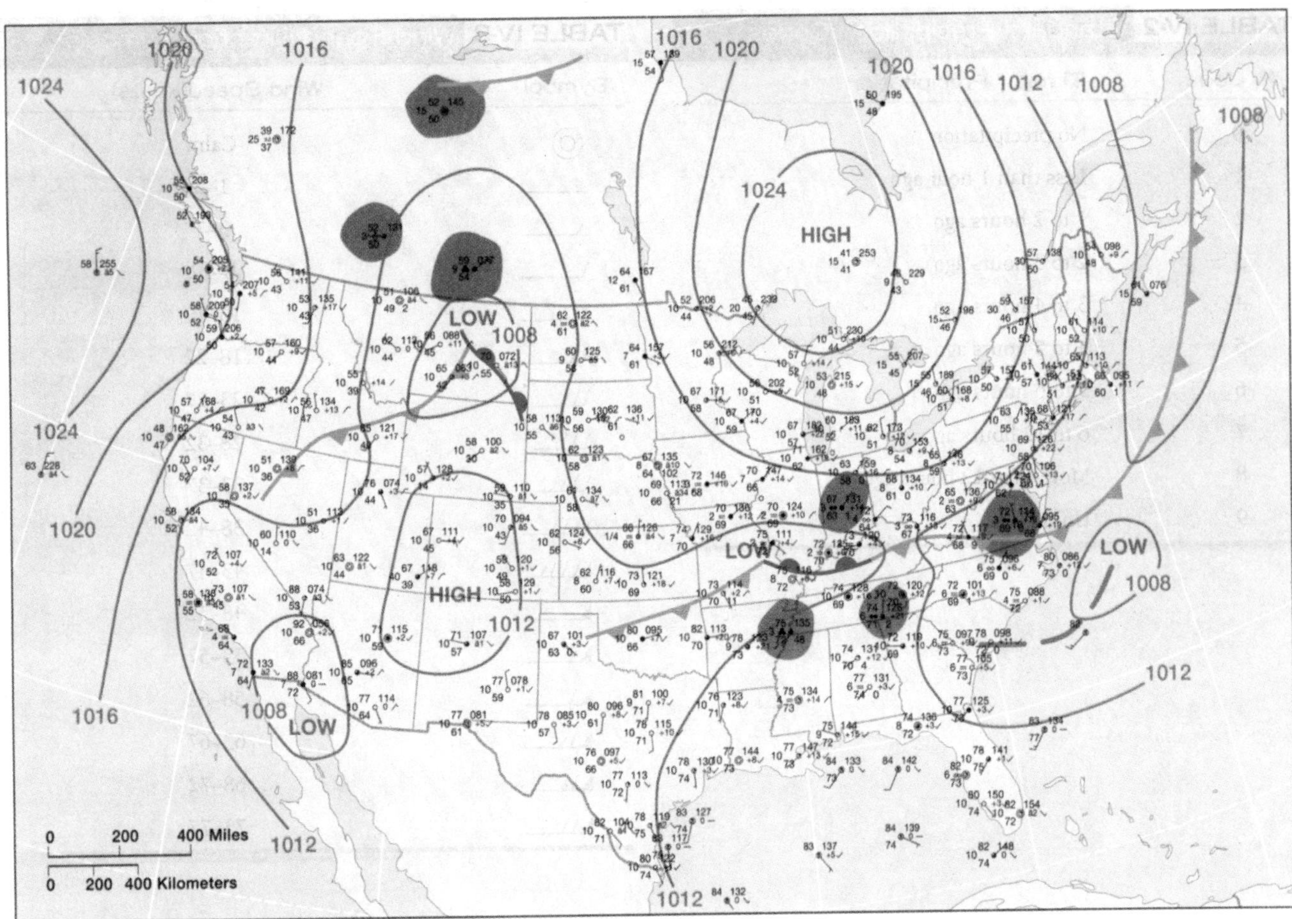

▲ **Figure IV-5** A sample weather map.

TABLE IV-1 Standard Cloud Height Codes

h (height of cloud base)	Approximate Cloud Height	
	Meters	Feet
0	0–49	0–149
1	50–99	150–299
2	100–199	300–599
3	200–299	600–999
4	300–599	1000–1999
5	600–999	2000–3499
6	1000–1499	3500–4999
7	1500–1999	5000–4699
8	2000–2499	6500–7999
9	> 2500 or no clouds	> 8000 or no clouds

TABLE IV-2 Standard Precipitation Codes

R_t Code	Time of Precipitation
0	No precipitation
1	Less than 1 hour ago
2	1 to 2 hours ago
3	2 to 3 hours ago
4	3 to 4 hours ago
5	4 to 5 hours ago
6	5 to 6 hours ago
7	6 to 12 hours ago
8	More than 12 hours ago
9	Unknown

TABLE IV-3 Wind Speed/Direction Symbols

Symbol	Wind Speed(knots)
◎	Calm
	1–2
	3–7
	8–12
	13–17
	18–22
	23–27
	28–32
	33–37
	38–42
	43–47
	48–52
	53–57
	58–62
	63–67
	68–72
	73–77

Appendix V
Köppen Climate Classification

Table V-1 provides the definitions of the code letters used in the modified Köppen climate classification system. The exact classification of a climate using the Köppen system may require additional calculations, and so the definitions of each letter given here may be approximate. For more detailed directions for classifying climates with the Köppen system, see the *Laboratory Manual for Physical Geography: A Landscape Appreciation*, 10th edition, by Darrel Hess.

TABLE V-1 Code Letters of the Modified Köppen Classification System

Letters				
1st	2nd	3rd	Description	Definitions
A			Low-latitude humid climates	Average temperature of each month above 18°C (64°F)
	f		No dry season [German: *feucht* ("moist")]	Average rainfall of each month at least 6 cm (2.4 in.)
	m		Monsoon; short dry season compensated by heavy rains in other months	1 to 3 months with average rainfall less than 6 cm (2.4 in.)
	w		Winter dry season (low-Sun season)	3 to 6 months with average rainfall less than 6 cm (2.4 in.)
B			Dry climates	Evaporation exceeds precipitation
	W		Desert [German: *wüste* ("desert")]	Average annual precipitation generally less than 38 cm (15 in.) in low latitudes; 25 cm (10 in.) in midlatitudes
	S		Steppe (semiarid)	Average annual precipitation generally between 38 cm (15 in.) and 76 cm (30 in.) in low latitudes; between about 25 cm (10 in.) and 64 cm (25 in.) in midlatitudes; without pronounced seasonal concentration
		h	Low-latitude (subtropical) dry climate [German: *heiss* ("hot")]	Average annual temperature more than 18°C (64°F)
		k	Midlatitude dry climate [German: *kalt* ("cold")]	Average annual temperature less than 18°C (64°F)
C			Mild midlatitude climates	Average temperature of coldest month between 18°C (64°F) and −3°C (27°F); average temperature of warmest month above 10°C (50°F)
	s		Summer dry season	Wettest winter month has at least 3× precipitation of driest summer month
	w		Winter dry season	Wettest summer month has at least 10× precipitation of driest winter month
	f		No dry season [German: *feucht* ("moist")]	Does not fit either s or w above
		a	Hot summers	Average temperature of warmest month more than 22°C (72°F)
		b	Warm summers	Average temperature of warmest month below 22°C (72°F); at least 4 months with average temperature above 10°C (50°F)
		c	Cool summers	Average temperature of warmest month below 22°C (72°F); less than 4 months with average temperature above 10°C (50°F); coldest month above −38°C(−36°F)
D			Humid midlatitude climates with severe winters (2nd and 3rd letters same as in C climates)	Warmest month above 10°C (50°F); coldest month below −3°C (27°F)
		d	Very cold winters	Average temperature of coldest month less than −38°C(−36°F)

TABLE V-1 Code Letters of the Modified Köppen Classification System (*Continued*)

Letters 1st	2nd	3rd	Description	Definitions
E			Polar climates; no true summer	No month with average temperature more than 10°C (50°F)
	T		Tundra climates	At least one month with average temperature more than 0°C (32°F) but less than 10°C (50°F)
	F		Ice cap climates ("frost")	No month with average temperature more than 0°C (32°F)
H			Highland climates	Significant climatic changes within short horizontal distances due to altitudinal variations

Appendix VI
Biological Taxonomy

Taxonomy is the science of classification. As a term it was originally applied to the classification of plants and animals, although its meaning has been broadened to encompass any sort of systematic classification. Our concern here is only with biological taxonomy. People have attempted to devise meaningful classifications of plants and animals for thousands of years. One of the most useful of the early classifications was designed by Aristotle 2300 years ago, and this system was in general use for nearly 20 centuries. In the late 1700s, the Aristotelian classification was finally replaced by a much more comprehensive and systematic one developed by the Swedish naturalist Carolus Linnaeus. Linnaeus made use of ideas from other biologists, but the system is largely his own work.

The *Linnaean system of classification* is generic, hierarchical, comprehensive, and binomial. *Generic* means that it is based on observable characteristics of the organisms it classifies, primarily their anatomy, structures, and details of reproduction. *Hierarchical* means that the organisms are grouped on the basis of similar characteristics, with each lower level of grouping having a larger number of similar characteristics and therefore containing fewer individuals in the group. *Comprehensive* means that all plants and animals, existing and extinct, can be encompassed within the system. *Binomial* means that every kind of plant and animal is identified by two names.

The binomial naming of organisms is highly systematized. Each type of living thing has a name with two parts. The first part, in which the first letter is capitalized, designates the *genus*, or group; the second part, which is not capitalized, indicates the *species*, or specific kind of organism. The combination of genus and species is referred to as the *scientific name*; it is always in Latin, although many of the words have Greek derivations.

Each type of organism, then, has a scientific name that distinguishes it from all other organisms. Although the popular name may be variable, or even indefinite, the scientific name is unvarying. Thus, in different parts of the Western Hemisphere, the large native cat may be called a mountain lion, cougar, puma, panther, painter, or leon, but its scientific name is always *Felis concolor*.

The intellectual beauty of the Linnaean system is twofold: (1) every organism that has ever existed can be fitted into the scheme in a logical and orderly manner for the system is capable of indefinite expansion, and (2) the various hierarchical levels in the system provide a conceptual framework for understanding the relationships among different organisms or groups of organisms. This is not to say the system is perfect. Linnaeus believed that species were unchanging entities, and his original system had no provision for variations. The concept of subspecies was a major modification of the system that was introduced subsequently to accommodate observed conditions of evolution.

Nor is the system even completely objective. Whereas its concept and general organization are accepted by scientists throughout the world, details of the classification depend on judgments and opinions made by biologists. These judgments and opinions are based on careful measurements and observations of plant and animal specimens, but there is often room for differing interpretations of the relevant data. Consequently, some details of biological taxonomy are disputed and even controversial. Despite confusion about some of the details, the generally accepted Linnaean system provides a magnificent framework for biological classification. The seven main levels of the system, in order from largest to smallest, are (1) kingdom, (2) phylum, (3) class, (4) order, (5) family, (6) genus, and (7) species.

Kingdom is the broadest category and contains the largest number of organisms. Until recently, the system recognized only two kingdoms: one encompassing all plants and the other including all animals. Increasingly, however, taxonomists encountered difficulty in accommodating the many varieties of one-celled and other simple microscopic organisms into such a two-kingdom system. It is now widely, but by no means universally, accepted that six kingdoms exist at the highest level of taxonomic distinction.[1]

1. *Archaea* are simple organisms that live in harsh environments such as underwater hydrothermal vents, hot sulfur springs, and hypersaline water. Some may also live in the open ocean and can be either autotrophs or heterotrophs.
2. *Eubacteria* are true, one-celled bacteria.
3. *Protista* consist of other one-celled organisms and some simple multicelled algae, most of which were formerly classed as plants.
4. *Fungi* were also previously classified as plants, but it is now recognized that they differ in origin, direction of evolution, and primary nutrition from plants and therefore deserve separation.

[1]Until a few years ago only five kingdoms were recognized: *Monera*, *Protista*, *Fungi*, *Plantae*, and *Animalia*. The new six-kingdom classification is not recognized by all biologists.

5. *Plantae* include the multicelled green plants and higher algae.
6. *Animalia* consist of the multicelled animals.

Phylum is the second major level of the system. Of the two or three dozen phyla within the animal kingdom, one, *Chordata*, includes all animals with backbones, which means almost every animal more than a few centimeters in length (Figure VI-1). In the plant kingdom, the term *division* is often used in place of phylum. Most large plants belong to the division *Tracheophyta*, or vascular plants, which have efficient internal systems for transporting water and sugars and a complex differentiation of organs into leaves, stem, and roots.

The third principal level is *class*. Among the several dozen animal classes, the most important are *Mammalia* (mammals), *Aves* (birds), *Reptilia* (reptiles), *Amphibia* (amphibians), and two classes that encompass fishes. There are somewhat fewer classes of plants, of which the most notable is *Angiospermae*, the flowering plants.

The fourth level of the classification is called *order*, and the three lower levels are *family*, *genus*, and *species*. As in all levels of the hierarchy, each succeeding lower level contains organisms that are increasingly alike. Species is the basic unit of the classification. In theory, only members of the same species are capable of breeding with one another. In practice, however, interbreeding is possible among just a few species (always within the same genus), although the offspring of such interspecific breeding are nearly always infertile (i.e., totally incapable of reproducing). In some cases, species are further subdivided into subspecies, also called varieties or races (Figure VI-2).

The relative diversity of living and extinct species is worthy of note. About 1,500,000 species of living organisms have been identified and described. From 3 to 10 million additional species, mostly microscopic in size, may not yet be identified. An estimated 500 million species have become extinct in Earth's history. Thus, more than 95 percent of all evolutionary lines have already completely disappeared.

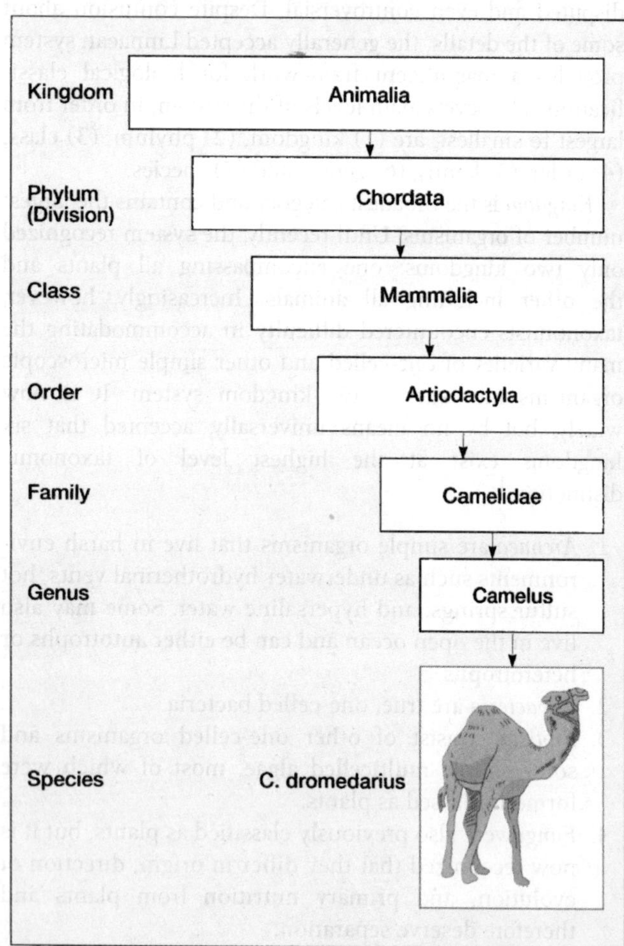

▲ **Figure VI-1** The taxonomic classification of an animal, as illustrated by the Arabian camel, or dromedary.

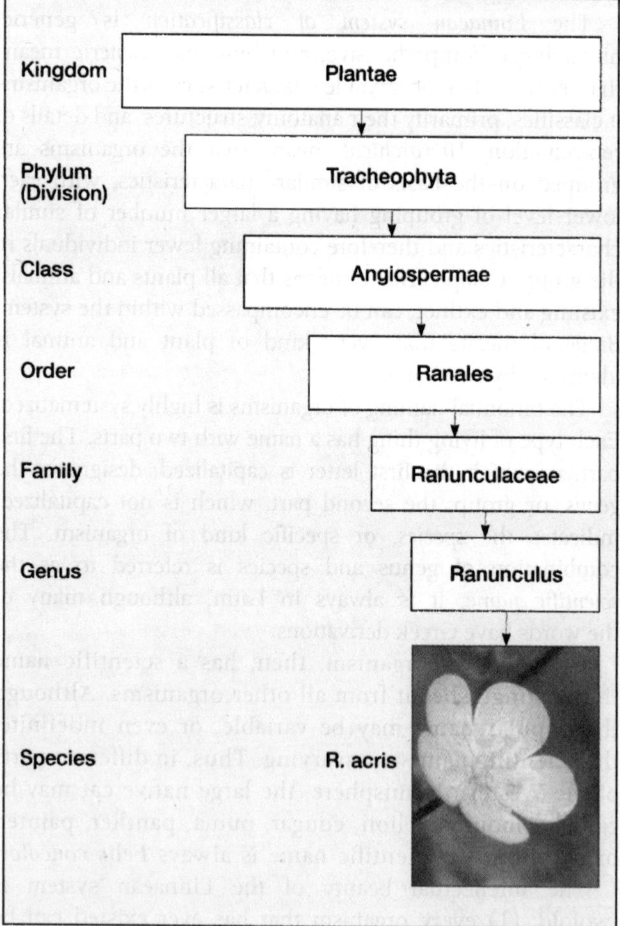

▲ **Figure VI-2** The taxonomic classification of a plant, as illustrated by the buttercup.

Appendix VII
The Soil Taxonomy

The Soil Taxonomy, described in Chapter 12, utilizes a nomenclature "invented" for the purpose. This nomenclature consists of "synthetic" names, which means that syllables from existing words are rearranged and combined to produce new words for the names of the various soil types. The beauty of the system is that each newly coined name is highly descriptive of the soil it represents.

The awkwardness of the nomenclature is threefold:

1. Most of the terms are new and have never before appeared in print or in conversation. Thus they look strange, and the words do not easily roll off the tongue. It is almost a new language.
2. Many of the words are difficult to write, and great care must be taken in the spelling of seemingly bizarre combinations of letters.
3. Many of the syllables sound so much alike that differences in pronunciation are often slight, although enunciating these slight differences is essential if the new nomenclature is to serve its purpose.

Nevertheless, the Soil Taxonomy has a sound theoretical base, and once the user is familiar with the vocabulary, every syllable of every word gives important information about a soil. Almost all the syllables are derived from Greek or Latin roots, in contrast to the English and Russian terms used in previous systems. In some cases, the appropriateness of the classical derivatives may be open to question, but the uniformity and logic of the terminology extend throughout the system.

The Hierarchy

The top level in the Soil Taxonomy hierarchy is *soil order*. The names of orders are made up of three or four syllables, the last of which is always *sol* (from the Latin word *solum*, "soil"). The next-to-last syllable consists of a single linking vowel, either *i* or *o*. The syllable (or two) that begins the word contains the formative element of the name and gives information regarding some distinctive characteristic of the order. Thus each of the 12 soil order names contains a distinctive syllable that (1) identifies the formative element of that soil order and (2) appears in all the names in the next three levels of the hierarchy for that soil order.

Suborder is the second level in the hierarchy. All suborder names contain two syllables: The first indicates some distinctive characteristic of the suborder; the second identifies the order to which the suborder belongs. For example, *aqu* is derived from the Latin word for "water," and so *Aquent* is the suborder of wet soils of the order Entisols. The names of the four dozen suborders are constructed from the two dozen root elements shown in Table VII-1.

The third level of the hierarchy contains *great groups*, the names of which are constructed by grafting one or more syllables to the beginning of a suborder name. Hence, a *Cryaquent* is a cold soil that is a member of the Aquent suborder of the Entisol order (*cry* comes from the Greek word for "coldness"). The formative prefixes for the great group names are derived from about 50 root words, samples of which are presented in Table VII-2.

The next level is called *subgroup*, and there are more than 1000 subgroups recognized in the United States. Each subgroup name consists of two words: the first derived from a formative element higher up in the hierarchy (with a few exceptions), the second the same as that of the relevant great group. Thus, a *Sphagnic Cryaquent* is of the order Entisol, the suborder Aquent, the great group *Cryaquent*, and it contains sphagnum moss (derived from the Greek word *sphagnos*, "bog").

The *family* is the penultimate level in the hierarchy. It is not given a proper name but is simply described by one or more lowercase adjectives, as "a skeletal, mixed, acidic family."

Finally, the lowest level of the hierarchy is the *series*, named for geographic location. Table VII-3 gives a typical naming sequence.

Summary of Soil Suborders

Alfisols

Alfisols have five suborders. *Aqualfs* have characteristics associated with wetness. *Boralfs* are associated with cold boreal forests. *Udalfs* are brownish or reddish soils of moist midlatitude regions. *Ustalfs* are similar to Udalfs in color but subtropical in location and usually have a hard surface layer in the dry season. *Xeralfs* are found in mediterranean climates and are characterized by a thick, hard surface horizon in the dry season.

Andisols

Seven suborders are recognized in the Andisols, five of which are distinguished by moisture content. *Aquands* have abundant moisture, often with poor drainage.

TABLE VII-1 Name Derivations of Soil Suborders

Root	Derivation	Connotation	Example of Suborder Name
alb	Latin *albus*, "white"	Presence of bleached eluvial horizon	Alboll
and	Japanese *ando*, a volcanic soil	Derived from pyroclastic material	Andept
aqu	Latin *aqua*, "water"	Associated with wetness	Aquent
ar	Latin *arare*, "to plow"	Horizons are mixed	Arent
arg	Latin *argilla*, "white clay"	Presence of a horizon containing illuvial clay	Argid
bor	Greek *boreas*, "northern"	Associated with cool conditions	Boroll
ferr	Latin *ferrum*, "iron"	Presence of iron	Ferrod
fibr	Latin *fibra*, "fiber"	Presence of undecomposed organic matter	Fibrist
fluv	Latin *fluvius*, "river"	Associated with floodplains	Fluvent
fol	Latin *folia*, "leaf"	Mass of leaves	Folist
hem	Greek *hemi*, "half"	Intermediate stage of decomposition	Hemist
hum	Latin *humus*, "earth"	Presence of organic matter	Humult
ochr	Greek *ochros*, "pale"	Presence of a light-colored surface horizon	Ochrept
orth	Greek *orthos*, "true"	Most common or typical group	Orthent
plag	German *plaggen*, "sod"	Presence of a human-induced surface horizon	Plaggept
psamm	Greek *psammos*, "sand"	Sandy texture	Psamment
rend	Polish *rendzino*, a type of soil	Significant calcareous content	Rendoll
sapr	Greek *sapros*, "rotten"	Most decomposed stage	Saprist
torr	Latin *torridus*, "hot and dry"	Usually dry	Torrox
trop	Greek *tropikos*, "of the solstice"	Continuously warm	Tropert
ud	Latin *udud*, "humid"	Of humid climates	Udoll
umbr	Latin *umbro*, "shade"	Presence of a dark surface horizon	Umbrept
ust	Latin *ustus*, "burnt"	Of dry climates	Ustert
xer	Greek *xeros*, "dry"	Annual dry season	Xeralf

TABLE VII-2 Name Derivations for Great Groups

Root	Derivation	Connotation	Example of Great Group Name
calc	Latin *calcis*, "lime"	Presence of calcic horizon	Calciorthid
ferr	Latin *ferrum*, "iron"	Presence of iron	Ferrudalf
natr	Latin *natrium*, "sodium"	Presence of a natric horizon	Natraboll
pale	Greek *paleos*, "old"	An old development	Paleargid
plinth	Greek *plinthos*, "brick"	Presence of plinthite	Plenthoxeralf
quartz	The German name	High quartz content	quartzipsamment
verm	Latin *vermes*, "worm"	Notable presence of worms	Vermudoll

TABLE VII-3 A Typical Soil Taxonomy Naming Sequence

Order	Entisol
Suborder	Aquent
Great group	Cryaquent
Subgroup	Sphagnic Cryaquent
Family	Skeletal, mixed, acidic, Sphagnic Cryaquent
Series	Aberdeen

Torrands are associated with a hot, dry regime. *Udands* are found in humid climates, *Ustands* in dry climates with hot summers. *Xerands* have a pronounced annual dry season. In addition, *Cryands* are found in cold climates, and *Vitrands* are distinguished by the presence of glass.

Aridisols

Two suborders are generally recognized on the basis of degree of weathering. *Argids* have a distinctive subsurface horizon with clay accumulation, whereas *Orthids* do not.

Entisols

There are five suborders of Entisols. *Aquents* occupy wet environments where the soil is more or less continuously saturated with water; they may be found in any temperature regime. *Arents* lack horizons because of human interference, particularly that involving large agricultural or engineering machinery. *Fluvents* form on recent water-deposited sediments that have satisfactory drainage. *Orthents* develop on recent erosional surfaces. *Psamments* occur in sandy situations, where the sand is either shifting or stabilized by vegetation.

Gelisols

The three suborders of Gelisols—*Histels*, *Orthels*, and *Turbels*—are distinguished largely on the basis of quantity and distribution of organic material.

Histosols

The four suborders of Histosols—*Fibrists*, *Folists*, *Hemists*, and *Saprists*—are differentiated on the basis of degree of plant-material decomposition.

Inceptisols

The six suborders of Inceptisols—*Andepts*, *Aquepts*, *Ochrepts*, *Plaggepts*, *Tropepts*, and *Umbrepts*—have relatively complicated distinguishing characteristics.

Mollisols

The seven suborders of Mollisols—*Albolls*, *Aquolls*, *Borolls*, *Rendolls*, *Udolls*, *Ustolls*, and *Xerolls*—are distinguished largely, but not entirely, on the basis of relative wetness/dryness.

Oxisols

The five suborders of Oxisols—*Aquox*, *Humox*, *Orthox*, *Torrox*, and *Ustox*—are distinguished from one another primarily by what effect varying amounts and seasonality of rainfall have on the profile.

Spodosols

Of the four suborders of Spodosols, most widespread are the *Orthods*, which represent the typical Spodosols. *Aquods*, *Ferrods*, and *Humods* are differentiated on the basis of the amount of iron in the spodic horizon.

Ultisols

Five suborders of Ultisols—*Aquults*, *Humults*, *Udults*, *Ustults*, and *Xerults*—are recognized. The distinction among them is largely on the basis of temperature and moisture conditions and on how these parameters influence the epipedon.

Vertisols

The four principal suborders of Vertisols are distinguished largely on the frequency of "cracking," which is a function of climate. *Torrerts* are found in arid regions, and the cracks in these soils remain open most of the time. *Uderts* are found in humid areas, and in these soils cracking is irregular. *Usterts* are associated with monsoon climates and have a relatively complicated cracking pattern. *Xererts* occur in mediterranean climates and have cracks that open and close regularly once each year.

Glossary

ablation Wastage of glacial ice through melting and sublimation.

ablation zone The lower portion of a glacier where there is a net annual loss of ice due to melting and sublimation.

absolute humidity One measure of the actual water vapor content of air, expressed as the mass of water vapor in a given volume of air, usually as grams of water per cubic meter of air.

absorption The ability of an object to assimilate energy from electromagnetic waves that strike it.

accumulation (glacial ice accumulation) Addition of ice into a glacier by incorporation of snow.

accumulation zone The upper portion of a glacier where there is a greater annual accumulation of ice than there is wastage.

acid rain Precipitation with a pH less than 5.6. It may involve dry deposition without moisture.

adiabatic cooling Cooling by expansion, such as in rising air.

adiabatic warming Warming by compression, such as in descending air.

adret slope A slope oriented so that the Sun's rays arrive at a relatively high angle. Such a slope tends to be relatively warm and dry.

advection Horizontal transfer of heat, such as through the movement of wind across Earth's surface.

aeolian processes Processes related to wind action that are most pronounced, widespread, and effective in dry lands.

aerial photograph A photograph taken from an elevated "platform," such as a balloon, airplane, rocket, or satellite.

aerosols Solid or liquid particles suspended in the atmosphere; also called *particulates*.

aggradation The process in which a stream bed is raised as a result of the deposition of sediment.

A horizon Upper soil layer in which humus and other organic materials are mixed with mineral particles.

air mass An extensive body of air that has relatively uniform properties in the horizontal dimension and moves as an entity.

albedo The reflectivity of a surface. The fraction of total solar radiation that is reflected back, unchanged, into space.

Alfisol A widely distributed soil order distinguished by a subsurface clay horizon and a medium-to-generous supply of plant nutrients and water.

alluvial fan A fan-shaped depositional feature of alluvium laid down by a stream issuing from a mountain canyon.

alluvium Any stream-deposited sedimentary material.

alpine glacier Individual glacier that develops near a mountain crest line and normally moves down-valley for some distance.

Andisol Soil order derived from volcanic ash.

angiosperms Plants that have seeds encased in some sort of protective body, such as a fruit, a nut, or a seedpod.

angle of incidence The angle at which the Sun's rays strike Earth's surface.

angle of repose Steepest angle that can be assumed by loose fragments on a slope without downslope movement.

annual plants (annuals) Plants that perish during times of climatic stress but leave behind a reservoir of seeds to germinate during the next favorable period.

Antarctic Circle The parallel of 66.5° south latitude.

antecedent stream Stream that predates the existence of the hill or mountain through which it flows.

anticline A simple symmetrical upfold in the rock structure.

anticyclone A high-pressure center.

antitrade winds Tropical upper-atmosphere westerly winds at the top of the Hadley cells that blow toward the northeast in the Northern Hemisphere and toward the southeast in the Southern Hemisphere.

aphelion The point in Earth's elliptical orbit at which Earth is farthest from the Sun (about 152,171,500 kilometers or 94,555,000 miles).

aquiclude An impermeable rock layer that is so dense as to exclude water.

aquifer A permeable subsurface rock layer that can store, transmit, and supply water.

Arctic Circle The parallel of 66.5° north latitude.

arête A narrow, jagged, serrated spine of rock; remainder of a ridge crest after several glacial cirques have been cut back into an interfluve from opposite sides of a divide.

Aridisol A soil order occupying dry environments that do not have enough water to remove soluble minerals from the soil; typified by a thin profile that is sandy and lacking in organic matter.

artesian well The free flow that results when a well is drilled from the surface down into the aquifer and the confining pressure is sufficient to force the water to the surface without artificial pumping.

asthenosphere Plastic layer of the upper mantle that underlies the lithosphere. Its rock is very hot and therefore weak and easily deformed.

atmosphere The gaseous envelope surrounding Earth.

atmospheric pressure The force exerted by the atmosphere on a surface.

atoll Coral reef in the general shape of a ring or partial ring that encloses a lagoon.

average annual temperature range Difference in temperature between the average temperature of the hottest and coldest months for a location.

average lapse rate The average rate of temperature decrease with height in the troposphere—about 6.5°C per 1000 meters (3.6°F per 1000 feet).

backwash Water moving seaward after the momentum of the wave swash is overcome by gravity and friction.

badlands Intricately rilled and barren terrain of arid and semiarid regions, characterized by a multiplicity of short, steep slopes.

bajada A continual alluvial surface that extends across the piedmont zone, slanting from the range toward the basin, in which it is difficult to distinguish between individual alluvial fans.

barchan dune A crescent-shaped sand dune with cusps of the crescent pointing downwind.

barometer Instrument used to measure atmospheric pressure.

barrier island Narrow offshore island composed of sediment; generally oriented parallel to shore.

barrier reef A prominent ridge of coral that roughly parallels the coastline but lies offshore, with a shallow lagoon between the reefs and the coast.

basal slip The term used to describe the sliding of the bottom of a glacier over its bed on a lubricating film of water.

basalt Fine-grained, dark (usually black) volcanic rock; forms from mafic (relatively low silica content) lava.

base level An imaginary surface extending underneath the continents from sea level at the coasts and indicating the lowest level to which land can be eroded.

batholith The largest and most amorphous of igneous intrusions.

baymouth bar A spit that extends entirely across the mouth of a bay, transforming the bay into a lagoon.

beach An exposed deposit of loose sediment, normally composed of sand and/or gravel, and occupying the coastal transition zone between land and water.

beach drifting The zigzag movement of sediment caused by waves washing particles onto a beach at a slight angle; the net result is the movement of sediment along the coast in a general downwind direction.

bedload Sand, gravel, and larger rock fragments moving in a stream by saltation and traction.

B horizon Mineral soil horizon located beneath the A horizon.

biogeography The study of the distribution patterns of plants and animals, and how these patterns change over time.

biological weathering Rock weathering processes involving the action of plants or animals.

biomass The total mass (or weight) of all living organisms in an ecosystem or per unit area.

biome A large, recognizable assemblage of plants and animals in functional interaction with its environment.

biosphere The living organisms of Earth.

biota The total complex of plant and animal life.

blowout (deflation hollow) A shallow depression from which an abundance of fine material has been deflated by wind.

boreal forest (taiga) An extensive needleleaf forest in the subarctic regions of North America and Eurasia.

braided channel pattern (braided stream) A stream that consists of a multiplicity of interwoven and interconnected shallow channels separated by low islands of sand, gravel, and other loose debris.

broadleaf trees Trees that have flat and expansive leaves.

butte An erosional remnant of very small surface area and clifflike sides that rises conspicuously above the surroundings.

calcification One of the dominant pedogenic regimes in areas where the principal soil moisture movement is upward because of a moisture deficit. This regime is characterized by a concentration of calcium carbonate ($CaCO_3$) in the B horizon forming a hardpan.

caldera Large, steep-sided, roughly circular depression resulting from the explosion and/or collapse of a large volcano.

capacity (stream capacity) The maximum load that a stream can transport under given conditions.

capacity (water vapor capacity) Maximum amount of water vapor that can be present in the air at a given temperature.

capillarity The action by which water can climb upward in restricted confinement as a result of its high surface tension, and thus the ability of its molecules to stick closely together.

carbonation A process in which carbon dioxide in water reacts with carbonate rocks to produce a very soluble product (calcium bicarbonate), which can readily be removed by runoff or percolation, and which can also be deposited in crystalline form if the water is evaporated.

carbon cycle The change from carbon dioxide to living matter and back to carbon dioxide.

carbon dioxide CO_2; minor gas in the atmosphere; one of the greenhouse gases; by-product of combustion and respiration.

carbonic acid Mild acid formed when carbon dioxide dissolves in water; H_2CO_3.

cation An atom or group of atoms with a positive electrical charge.

cation exchange capacity (CEC) Capability of soil to attract and exchange cations.

cavern Large opening or cave, especially in limestone; often decorated with *speleothems*.

chemical weathering The chemical decomposition of rock by the alteration of rock-forming minerals.

chinook A localized downslope wind of relatively dry and warm air, which is further warmed adiabatically as it moves down the leeward slope of the Rocky Mountains.

chlorofluorocarbons (CFCs) Synthetic chemicals commonly used as refrigerants and in aerosol spray cans; destroy ozone in the upper atmosphere.

C horizon Lower soil layer composed of weathered parent material that has not been significantly affected by translocation or leaching.

cinder cone Small, common volcano that is composed primarily of pyroclastic material blasted out from a vent in small but intense explosions. The structure of the volcano is usually a conical hill of loose material.

circle of illumination The edge of the sunlit hemisphere that is a great circle separating Earth into a light half and a dark half.

cirque A broad amphitheater hollowed out at the head of a glacial valley by glacial erosion and frost wedging.

cirque glacier A small glacier confined to its cirque and not moving down-valley.

cirrus cloud High cirriform clouds of feathery appearance.

clay Very small inorganic particles produced by chemical alteration of silicate minerals.

climate An aggregate of day-to-day weather conditions and weather extremes over a long period of time.

climax vegetation A stable plant association of relatively constant composition that develops at the end of a long succession of changes.

climograph (climatic diagram) Chart showing the average monthly temperature and precipitation for a weather station.

cloud Visible accumulation of tiny liquid water droplets or ice crystals suspended in the atmosphere.

col A pass or saddle through a ridge produced when two adjacent glacial *cirques* on opposite sides of a divide are cut back enough to remove part of the *arête* between them.

cold front The leading edge of a cool air mass actively displacing warm air.

collapse sinkhole A sinkhole produced by the collapse of the roof of a subsurface cavern; a collapse doline.

colloids Organic and inorganic microscopic particles of soil that represent the chemically active portion of particles in the soil.

competence The size of the largest particle that can be transported by a stream.

composite volcano Volcanoes with the classic symmetrical, cone-shaped peak, produced by a mixture of lava outpouring and pyroclastic explosion.

condensation Process by which water vapor is converted to liquid water; a warming process because latent heat is released.

condensation nuclei Tiny atmospheric particles of dust, bacteria, smoke, and salt that serve as collection centers for water molecules.

conduction The movement of energy from one molecule to another without changing the relative positions of the molecules. It enables the transfer of heat between different parts of a stationary body.

cone of depression The phenomenon whereby the water table has sunk into the approximate shape of an inverted cone in the immediate vicinity of a well as the result of the removal of a considerable amount of groundwater.

conformal map projection A projection that maintains proper angular relationships over the entire map; over limited areas shows the correct shapes of features shown on a map.

conic projection A family of maps in which one or more cones is set tangent to, or intersecting, a portion of the globe and the geographic grid is projected onto the cone(s).

conifer See *gymnosperm*.

consumer Animal that consumes plants or other animals; heterotrophs.

contact metamorphism Metamorphism of surrounding rocks by contact with magma.

continental drift Theory that proposed that the present continents were originally connected as one or two large landmasses that have broken up and drifted apart over the last several hundred million years.

continental ice sheet Large ice sheet covering a portion of a continental area.

continental rift valley Fault-produced valley resulting from spreading or rifting of continent.

controls of weather and climate The most important influences acting upon the elements of weather and climate.

convection Vertical circulation and movement of fluids, such as air, due to density differences.

convection cell Closed pattern of convective circulation.

convective precipitation Showery precipitation resulting from the convective lifting of air.

convergent [plate] boundary Location where two lithospheric plates collide.

convergent precipitation Showery precipitation that occurs as a result of the forced uplift of air in areas of air convergence.

Coriolis effect (Coriolis force) The apparent deflection of free-moving objects to the right in the Northern Hemisphere and to the left in the Southern Hemisphere, in response to the rotation of Earth.

creep (soil creep) The slowest and least perceptible form of mass wasting, which consists of a very gradual downhill movement of soil and regolith.

crust The outermost solid layer of Earth.

cryosphere Subsphere of the hydrosphere that encompasses water frozen as snow or ice.

cultural geography The study of the human and/or cultural elements of geography.

cumulus cloud Puffy white cloud that forms from rising columns of air.

cutoff meander A portion of an old meandering stream course left isolated from the present stream channel because the narrow meander neck has been cut through by stream erosion.

cyclone Low-pressure center.

cylindrical projection A family of maps derived from the concept of projection onto a paper cylinder that is tangential to, or intersecting with, a globe.

daylight-saving time Shifting of clocks forward one hour.

debris flow Streamlike flow of muddy water heavily laden with sediments of various sizes; a *mudflow* containing large boulders.

December solstice Day of the year when the vertical rays of the Sun strike the Tropic of Capricorn; on or about December 21; winter solstice in the Northern Hemisphere.

deciduous tree A tree that experiences an annual period in which all leaves die and usually fall from the tree, due either to a cold season or a dry season.

decomposers Mainly microscopic organisms such as bacteria that decompose dead plant and animal matter.

deflation The shifting of loose particles by wind blowing them into the air or rolling them along the ground.

delta A landform comprised of alluvium at the mouth of a river produced by the sudden reduction of a stream's velocity and the resulting deposition of the stream's load.

dendritic drainage pattern A treelike, branching pattern that consists of a random merging of streams, with tributaries joining larger streams irregularly, but always at acute angles; generally develops in regions where the underlying structure does not significantly control the drainage pattern.

dendrochronology Study of past events and past climate through the analysis of tree rings.

denitrification The conversion of nitrates into free nitrogen in the air.

denudation The total effect of all actions (weathering, mass wasting, and erosion) that lower the surface of the continents.

desert Climate, landscape, or biome associated with extremely arid conditions.

desert pavement Hard and relatively impermeable desert surface of tightly packed small rocks.

desert varnish A dark shiny coating of iron and manganese oxides that forms on rock surfaces exposed to desert air for a long time.

dew The condensation of beads of water on relatively cold surfaces.

dew point temperature (dew point) The critical air temperature at which water vapor saturation is reached.

differential weathering and erosion The process whereby different rocks or parts of the same rock weather and/or erode at different rates.

dike A vertical or nearly vertical sheet of magma that is thrust upward into preexisting rock.

disappearing stream Stream that abruptly disappears from the surface where it flows into an underground cavity; common in karst regions.

discharge Volume of flow of a stream.

dissolution Removal of bedrock through chemical action of water; includes removal of subsurface rock through action of groundwater.

dissolved load The minerals, largely salts, that are dissolved in water and carried invisibly in solution.

divergent [plate] boundary Location where two lithospheric plates spread apart.

doldrums Belt of calm air associated with the region between the trade winds of the Northern and Southern hemispheres, generally in the vicinity of the equator. The region of the intertropical convergence zone (ITCZ).

downcutting Action of stream to erode a deeper channel; occurs when stream is flowing swiftly and/or flowing down a steep slope.

drainage basin An area that contributes overland flow and groundwater to a specific stream (also called a *watershed* or *catchment*).

drainage divide The line of separation between runoff that descends into two different drainage basins.

drift (glacial drift) All material carried and deposited by glaciers.

dripstone Features formed by precipitated deposits of minerals that decorate the walls, floor, and roof of a cavern.

drumlin A low, elongated hill formed by ice-sheet deposition and erosion. The long axis is aligned parallel with the direction of ice movements, with the blunt, steeper end facing the direction from which the ice came.

dry adiabatic rate (dry adiabatic lapse rate) The rate at which a parcel of unsaturated air cools as it rises (10°C per 1000 meters [5.5°F per 1000 feet]).

dynamic high High pressure cell associated with prominently descending air.

dynamic low Low pressure cell associated with prominently rising air.

earthflow Mass wasting process in which a portion of a water-saturated slope moves a short distance downhill.

earthquake Vibrations generated by abrupt movement of Earth's crust.

easterly wave A long but weak migratory low-pressure trough in the tropics.

ebb tide A periodic falling of sea level during a tidal cycle.

ecosystem The totality of interactions among organisms and the environment in the area of consideration.

ecotone The transition zone between biotic communities in which the typical species of one community intermingle with those of another.

edaphic factors Having to do with soil.

E horizon A light-colored, eluvial layer that usually occurs between the A and B horizons.

electromagnetic radiation Flow of energy in the form of electromagnetic waves; radiant energy.

electromagnetic spectrum Electromagnetic radiation, arranged according to wavelength.

elements of weather and climate The basic ingredients of weather and climate—temperature, pressure, wind, and moisture.

elevation contour line (contour line) A line on a map joining points of equal elevation.

El Niño Periodic atmospheric and oceanic phenomenon of the tropical Pacific that typically involves the weakening or reversal of the trade winds and the warming of surface water off the west coast of South America.

eluviation The process by which gravitational water picks up fine particles of soil from the upper layers and carries them downward.

emission See *radiation*.

endemic Organism found only in a particular area.

endothermic [animal] Warm-blooded animal.

Enhanced Fujita Scale Classification scale of tornado strength, with EF-0 being the weakest tornadoes and EF-5 being the most powerful.

ENSO (El Niño/Southern Oscillation) Linked atmospheric and oceanic phenomenon of pressure and water temperature. *Southern Oscillation* refers to a periodic seesaw of atmospheric pressure in the tropical southern Pacific Ocean basin. Also see *El Niño*.

Entisol The least developed of all soil orders, with little mineral alteration and no pedogenic horizons.

entrenched meanders A winding, sinuous stream valley with abrupt sides; possible outcome of the rejuvenation of a meandering stream.

environmental lapse rate The observed vertical temperature gradient of the troposphere.

ephemeral stream A stream that carries water only during the "wet season" or during and immediately after rains.

epicenter Location on the surface directly above the center of fault rupture during an earthquake.

equal area projection See *equivalent map projection*.

equator The parallel of 0° latitude.

equilibrium line A theoretical line separating the ablation zone and accumulation zone of a glacier along which accumulation exactly balances ablation.

equivalent map projection A projection that maintains constant area (size) relationships over the entire map; also called an *equal area projection*.

erg "Sea of sand." A large area covered with loose sand, generally arranged in some sort of dune formation by the wind.

erosion Detachment and removal of fragmented rock material.

esker Long, sinuous ridge of stratified glacial drift composed largely of glaciofluvial gravel and formed by the choking of subglacial streams during a time of glacial stagnation.

eustatic sea-level change Change in sea level due to an increase or decrease in the amount of water in the world ocean; also known as *eustasy*.

evaporation Process by which liquid water is converted to gaseous water vapor; a cooling process because latent heat is stored.

evapotranspiration The transfer of moisture to the atmosphere by transpiration from plants and evaporation from soil and plants.

evergreen A tree or shrub that sheds its leaves on a sporadic or successive basis but at any given time appears to be fully leaved.

exfoliation Weathering process in which curved layers peel off bedrock in sheets. This process commonly occurs in granite and related intrusive rocks after overlying rock has been removed, allowing the body to expand slightly. Also referred to as *unloading*.

exfoliation dome A large rock mass with a surface configuration that consists of imperfect curves punctuated by several partially fractured shells of the surface layers; result of exfoliation.

exotic species (exotics) Organisms that are introduced into "new" habitats in which they did not naturally occur.

exotic stream A stream that flows into a dry region, bringing its water from somewhere else.

external geomorphic processes Destructive processes that serve to denude or wear down the landscape. Includes weathering, mass wasting, and erosion.

eye (eye of tropical cyclone) The nonstormy center of a tropical cyclone, which has a diameter of 16 to 40 kilometers (10 to 25 miles) and is a singular area of calmness in the maelstrom that whirls around it.

fall Mass wasting process in which pieces of weathered rock fragments fall to the bottom of a cliff or steep slope; also called *rockfall*.

fault A fracture or zone of fracture where the rock structure is forcefully broken and one side is displaced relative to the other. The movement can be horizontal or vertical, or a combination of both.

fault-block mountain (tilted-fault-block mountain) A mountain formed where a surface block is faulted and relatively upthrown on one side without any faulting or uplift on the other side. The block is tilted asymmetrically, producing a steep slope along the fault scarp and a relatively gentle slope on the other side of the block.

faulting See *fault*.

fault scarp Cliff formed by faulting.

fauna Animals.

feral A domesticated creature that has reverted to a wild existence, or the progeny thereof.

field capacity The maximum amount of water that can be retained in the soil after the gravitational water has drained away.

firn (névé) Snow granules that have become packed and begin to coalesce due to compression, achieving a density about half as great as that of water.

fix (the verb) To make firm or stable; example: to fix nitrogen.

fjord A glacial trough that has been partly drowned by the sea.

flood basalt A large-scale outpouring of basaltic lava that may cover an extensive area of Earth's surface.

floodplain A flattish valley floor covered with stream-deposited sediments (alluvium) and subject to periodic or episodic inundation by overflow from the stream.

flood tide The movement of ocean water toward the coast in a tidal cycle—from the ocean's lowest surface level the water rises gradually for about 6 hours and 13 minutes.

flora Plants.

fluvial processes Processes involving the work of running water on the surface of Earth.

foehn See *chinook*. The word foehn is used particularly in Europe.

fog A cloud whose base is at or very near ground level.

folding The bending of crustal rocks by compression and/or uplift.

foliation Wavy or layered appearance of minerals in metamorphic rock.

food chain Sequential predation in which organisms feed upon one another, with organisms at one level providing food for organisms at the next level, and so on. Energy is thus transferred through the ecosystem.

food pyramid A conceptualization of energy transfer through the ecosystem from large numbers of "lower" forms of life through succeedingly smaller numbers of "higher" forms, as the organisms at one level are eaten by the organisms at the next higher level. Also see *food chain*.

forest An assemblage of trees growing closely together so that their individual leaf canopies generally overlap.

fractional scale (fractional map scale) Ratio of distance measured on a map and the actual distance that represents on Earth's surface, expressed as a ratio or fraction; assumes that the same units of measure are used on the map and on Earth's surface.

friction layer Zone of the atmosphere, between Earth's surface and an altitude of about 1000 meters (3300 feet), where most frictional resistance to air flow is found.

fringing reef A coral reef built out laterally from the shore, forming a broad bench that is only slightly below sea level, often with the tops of individual coral "heads" exposed to the open air at low tide.

front A sharp zone of discontinuity between unlike air masses.

frontal precipitation Precipitation associated with the lifting of air along a front.

frost wedging Fragmentation of rock due to expansion of water that freezes into ice within rock openings.

fumarole A hydrothermal feature consisting of a surface crack that is directly connected with a deep-seated source of heat. The little water that drains into this tube is instantly converted to steam by heat and gases, and a cloud of steam is then expelled from the opening.

funnel cloud Funnel-shaped cloud extending down from a cumulonimbus cloud; a tornado is formed when the funnel cloud touches the surface.

Gelisol Soil order that develops in areas of permafrost.

geographic information systems (GIS) Computerized systems for the capture, storage, retrieval, analysis, and display of spatial (geographic) data.

geomorphology The study of the characteristics, origin, and development of landforms.

geostrophic wind A wind that moves parallel to the isobars as a result of the balance between the pressure gradient force and the Coriolis effect.

geyser A specialized form of intermittent hot spring with water issuing only sporadically as a temporary ejection, in which hot water and steam are spouted upward for some distance.

glacial erratic Outsize boulder included in the glacial till, which may be very different from the local bedrock.

glacial flour Rock material that has been ground to the texture of very fine talcum powder by glacial action.

glacial plucking Action in which rock fragments beneath the ice are loosened and grasped by the freezing of meltwater in joints and fractures, and then pried out and dragged along in the general flow of a glacier. Also called *glacial quarrying*.

glacial steps Series of level or gently sloping bedrock benches alternating with steep drops in the down-valley profile of a glacial trough.

glacial trough A valley reshaped by an alpine glacier, usually U-shaped.

glaciofluvial deposition The action whereby rock debris that is carried along by glaciers is eventually deposited or redeposited by glacial meltwater.

gleization The dominant pedogenic regime in areas where the soil is saturated with water most of the time due to poor drainage.

global conveyer-belt circulation Slowly moving circulation of deep ocean water that forms a continuous loop from the North Atlantic to the Antarctic, into the Indian and Pacific Oceans, and back into the North Atlantic.

Global Positioning System (GPS) A satellite-based system for determining accurate positions on or near Earth's surface.

global warming Popular name given to the recent warming of Earth's climate due to human-released greenhouse gases.

graben A block of land bounded by parallel faults in which the block has been downthrown, producing a distinctive structural valley with a straight, steep-sided fault scarp on either side.

granite The most common and well-known plutonic (intrusive) rock; coarse-grained rock consisting of both dark- and light-colored minerals; forms from felsic (relatively high silica content) magma.

graphic scale (graphic map scale) The use of a line marked off in graduated distances as a map scale.

grassland Plant association dominated by grasses and forbs.

great circle Circle on a globe formed by the intersection of Earth's surface with any plane that passes through Earth's center.

greenhouse effect The warming in the lower troposphere because of differential transmissivity for shortwave and longwave radiation through the gases in the atmosphere; the atmosphere easily transmits *shortwave radiation* from the Sun but inhibits the transmission of *longwave radiation* from the surface.

greenhouse gases Gases with the ability to transmit incoming shortwave radiation from the Sun but absorb outgoing longwave terrestrial radiation. The most important natural greenhouse gases are water vapor and carbon dioxide.

Greenwich Mean Time (GMT) Time in the Greenwich time zone. Today more commonly called *UTC* or *Universal Time Coordinated.*

groin A short wall built perpendicularly from the beach into the shore zone to interrupt the longshore current and trap sand.

ground moraine A moraine consisting of glacial till deposited widely over a land surface beneath an ice sheet.

groundwater Water found underground in the zone of saturation.

gymnosperms Seed-reproducing plants that carry their seeds in cones; "naked seeds."

Hadley cells Two complete vertical convective circulation cells between the equator, where warm air rises in the *ITCZ*, and 25° to 30° of latitude, where much of the air subsides into the *subtropical highs*.

hail Rounded or irregular pellets or lumps of ice produced in cumulonimbus clouds as a result of active turbulence and vertical air currents. Small ice particles grow by collecting moisture from supercooled cloud droplets.

hamada A barren desert surface of consolidated material that usually consists of exposed bedrock but is sometimes composed of sedimentary material that has been cemented together by salts evaporated from groundwater.

hanging valley A tributary glacial trough, the bottom of which is considerably higher than the bottom of the principal trough that it joins.

headward erosion Erosion that cuts into the interfluve at the upper end of a gully or valley.

heat Energy that transfers from one object or substance to another because of a difference in temperature. Sometimes the term *thermal energy* is used interchangeably with the term heat. A form of energy associated with the random motion of molecules. Things can be made hotter by the collision of the moving molecules.

high [pressure cell] Area of relatively high atmospheric pressure.

highland climate High mountain climate where altitude is dominant control. Designated H in Köppen system.

highland ice field Largely unconfined ice sheet in high mountain area.

Histosol A soil order characterized by organic, rather than mineral, soils, which is invariably saturated with water all or most of the time.

horizon (soil horizon) The more or less distinctly recognizable layer of soil, distinguished from one another by differing characteristics and forming a vertical zonation of the soil.

horn A steep-sided, pyramidal rock pinnacle formed by expansive glacial plucking and frost wedging of the headwalls where three or more cirques intersect.

horse latitudes Areas in the *subtropical highs* characterized by warm, tropical sunshine and an absence of wind.

horst A relatively uplifted block of land between two parallel faults.

hot spot An area of volcanic activity within the interior of a lithospheric plate associated with magma rising up from the mantle below.

hot spring Hot water at Earth's surface that has been forced upward through fissures or cracks by the pressures that develop when underground water has come in contact with heated rocks or magma beneath the surface.

humid continental climate Severe midlatitude climate characterized by hot summers, cold winters, and precipitation throughout the year.

humid subtropical climate Mild midlatitude climate characterized by hot summers and precipitation throughout the year.

humidity General term for water vapor in the air.

humus A dark-colored, gelatinous, chemically stable fraction of organic matter on or in the soil.

hurricane A tropical cyclone with wind speeds of 64 knots or greater affecting North or Central America.

hydrogen bond Attraction between water molecules in which the negatively charged oxygen side of one water molecule is attracted to the positively charged hydrogen side of another water molecule.

hydrologic cycle A series of storage areas interconnected by various transfer processes, in which there is a ceaseless interchange of moisture in terms of its geographical location and its physical state.

hydrolysis A chemical union of water with another substance to produce a new compound that is nearly always softer and weaker than the original.

hydrophytic adaptations Terrestrial plants adapted to living in very wet environments.

hydrosphere Total water realm of Earth, including the oceans, surface waters of the lands, groundwater, and water held in the atmosphere.

hydrothermal activity The outpouring or ejection of hot water, often accompanied by steam, which usually takes the form of either a hot spring or a geyser.

iceberg A great chunk of floating ice that breaks off an ice shelf or the end of an outlet glacier.

ice cap climate Polar climate characterized by temperatures below freezing throughout the year.

ice floe A mass of ice that breaks off from larger ice bodies (ice sheets, glaciers, ice packs, and ice shelves) and floats independently in the sea. This term is generally used with large, flattish, tabular masses.

ice pack The extensive and cohesive mass of floating ice that is found in the Arctic and Antarctic oceans.

ice shelf A massive portion of an ice sheet that projects out over the sea.

igneous intrusion Features formed by the emplacement and cooling of magma below the surface.

igneous rock Rock formed by solidification of molten magma.

illuviation The process by which fine particles of soil from the upper layers are deposited at a lower level.

Inceptisol An immature order of soils that has relatively faint characteristics; not yet prominent enough to produce diagnostic horizons.

inclination [of Earth's axis] The tilt of Earth's rotational axis relative to its orbital plane (the *plane of the ecliptic*).

infrared [radiation] Electromagnetic radiation in the wavelength range of about 0.7 to 1000 micrometers; wavelengths just longer than visible light.

inner core The solid, dense, innermost portion of Earth, believed to consist largely of iron and nickel.

inselberg "Island mountain"; isolated summit rising abruptly from a low-relief surface.

insolation Incoming solar radiation.

interfluve The higher land or ridge above the valley sides that separates adjacent valleys.

intermittent stream A stream that carries water only part of the time, during the "wet season" or during and immediately after rains.

internal geomorphic processes Geomorphic processes originating below the surface; include volcanism, folding, and faulting.

International Date Line The line marking a time difference of an entire day from one side of the line to the other. Generally, this line falls on the 180th meridian except where it deviates to avoid separating an island group.

International System of measurement (SI) Popularly known as the "metric system" of measurement.

intertropical convergence zone (ITCZ) The region near or on the equator where the northeast trades and the southeast trades converge; associated with rising air of the Hadley cells and frequent thunderstorms.

invertebrates Animals without backbones.

island arc See *volcanic island arc*.

isohyet A line joining points of equal numerical value of precipitation.

isoline A line on a map connecting points that have the same quality or intensity of a given phenomenon.

isostasy Maintenance of the hydrostatic equilibrium of Earth's crust; the sinking of the crust as weight is applied and the rising of crust as weight is removed.

isotherm A line joining points of equal temperature.

ITCZ See *intertropical convergence zone*.

jet stream A rapidly moving current of wind in the upper troposphere; jet streams can be thought of as the high-speed "cores" of the high altitude westerly wind flow that frequently meander in a north-south direction over the midlatitudes.

jetty A wall built into the ocean at the entrance of a river or harbor to protect against sediment deposition, storm waves, and currents.

joints Cracks that develop in bedrock due to stress, but in which there is no appreciable movement parallel to the walls of the joint.

June solstice Day of the year when the vertical rays of the Sun strike the Tropic of Cancer; on or about June 21; summer solstice in the Northern Hemisphere.

karst Topography developed as a consequence of subsurface solution.

katabatic wind A wind that originates in cold upland areas and cascades toward lower elevations under the influence of gravity.

kettle An irregular depression in a morainal surface created when blocks of stagnant ice eventually melt.

kinetic energy The energy of movement.

knickpoint A sharp irregularity (such as a waterfall, rapid, or cascade) in a stream-channel profile; also known as a *nickpoint*.

knickpoint migration Upstream shift in location of a knickpoint due to erosion.

Köppen climate classification system A climatic classification of the world devised by Wladimir Köppen.

lagoon A body of quiet salt or brackish water in an area between a barrier island or a barrier reef and the mainland.

lahar Volcanic mudflow; a fast-moving muddy flow of volcanic ash and rock fragments.

lake A body of water surrounded by land.

land breeze Local wind blowing from land to water, usually at night.

landform An individual topographic feature, of any size.

landforms Topography.

landslide An abrupt and often catastrophic event in which a large mass of rock and/or soil slides bodily downslope in only a few seconds or minutes. An instantaneous collapse of a slope.

La Niña Atmospheric and oceanic phenomenon associated with cooler than usual water off the west coast of South America. Sometimes described as the opposite of *El Niño*.

large-scale map A map with a scale that is a relatively large representative fraction and therefore portrays only a small portion of Earth's surface, but in considerable detail.

latent heat Energy stored or released when a substance changes state. For example, evaporation is a cooling process because latent heat is stored and condensation is a warming process because latent heat is released.

latent heat of condensation Heat released when water vapor condenses back to liquid form.

latent heat of evaporation Energy stored when liquid water evaporates to form water vapor.

latent heat of vaporization Stored energy absorbed by escaping molecules during evaporation or boiling.

lateral erosion Erosion that occurs when the principal current of a stream swings laterally from one bank to the other, eroding where the velocity is greatest on the outside bank and depositing alluvium where it is least on the inside bank.

lateral moraine Well-defined ridge of unsorted debris built up along the sides of valley glaciers, parallel to the valley walls.

laterization The dominant pedogenic regime in areas where temperatures are relatively high throughout the year and which is characterized by rapid weathering of parent material, dissolution of nearly all minerals, and the speedy decomposition of organic matter.

latitude Location described as an angle measured north and south of the equator.

lava Molten magma that is extruded onto the surface of Earth, where it cools and solidifies.

lava dome (plug dome) Dome or bulge formed by the pushing up of viscous magma in a volcanic vent.

lifting condensation level (LCL) The altitude at which rising air cools sufficiently to reach 100 percent relative humidity at the dew point temperature, and condensation begins.

lightning A luminous electric discharge in the atmosphere caused by the separation of positive and negative charges associated with cumulonimbus clouds.

limiting factor Variable that is important or most important in determining the survival of an organism.

linear fault trough Straight-line valley that marks the surface position of a fault, especially a strike-slip fault; formed by the erosion or settling of crushed rock along the trace of a fault.

liquefaction Phenomenon observed during an earthquake when water-saturated soil or sediments become soft or even fluid during the time of strong ground shaking.

lithosphere Tectonic plates consisting of the crust and upper rigid mantle. Also used as a general term for the entire solid Earth.

litter The collection of dead plant parts that accumulate at the surface of the soil.

loam A soil texture in which none of the three principal soil separates—sand, silt, and clay—dominates the other two.

loess A fine-grained, wind-deposited silt. Loess lacks horizontal stratification, and its most distinctive characteristic is its ability to stand in vertical cliffs.

longitude Location described as an angle measured (in degrees, minutes, and seconds) east and west from the prime meridian on Earth's surface.

longshore current A current in which water moves roughly parallel to the shoreline in a generally downwind direction; also called a littoral current.

longwave radiation Wavelengths of thermal infrared radiation emitted by Earth and the atmosphere; also referred to as *terrestrial radiation*.

low [pressure cell] Area of relatively low atmospheric pressure.

loxodrome (rhumb line) A true compass heading; a line of constant compass direction.

magma Molten material below Earth's surface.

magnitude [of an earthquake] Scale used to describe the relative amount of energy released during an earthquake. Several different magnitude scales are in current use, such as the *moment magnitude* and the *Richter scale*.

mantle The portion of Earth beneath the crust and surrounding the core.

mantle plume A plume of mantle magma that rises to, or almost to, Earth's surface; not directly associated with most lithospheric plate boundaries, but associated with many *hot spots*.

map A flat representation of Earth at a reduced scale, showing only selected detail.

map projection A systematic representation of all or part of the three-dimensional Earth surface on a two-dimensional flat surface.

map scale Relationship between distance measured on a map and the actual distance on Earth's surface.

March equinox One of two days of the year when the vertical rays of the Sun strike the equator; every location on Earth has equal day and night; occurs on or about March 20 each year.

marine west coast climate Mild midlatitude climate characterized by mild temperatures and precipitation throughout the year.

marine terrace A platform formed by marine erosion that has been uplifted above sea level.

marsh Flattish surface area that is submerged in water at least part of the time but is shallow enough to permit the growth of water-tolerant plants, primarily grasses and sedges.

mass wasting The downslope movement of broken rock material by gravity, sometimes lubricated by the presence of water; also called *mass movement*.

master joints Major joints that run for great distances through a bedrock structure.

meandering channel pattern (meandering stream channel) Highly twisting or looped stream channel pattern.

meander scar A dry former stream channel meander through which the stream no longer flows.

mechanical weathering The physical disintegration of rock material without any change in its chemical composition.

medial moraine A dark band of rocky debris down the middle of a glacier created by the union of the *lateral moraines* of two adjacent glaciers.

mediterranean climate Mild midlatitude climate characterized by dry summers and wet winters.

mediterranean woodland and shrub Woodland and shrub plant association found in regions of mediterranean climate.

Mercator projection A cylindrical projection mathematically adjusted to attain complete conformality which has a rapidly increasing scale with increasing latitude; straight lines on a Mercator projection are lines of constant compass heading.

meridian An imaginary line of longitude extending from pole to pole, crossing all parallels at right angles, and being aligned in true north–south directions.

mesa A flat-topped, steep-sided hill with a limited summit area.

mesocyclone Cyclonic circulation of air within a severe thunderstorm; diameter of about 10 kilometers (6 miles).

metamorphic rock Rock that was originally something else but has been drastically changed by massive forces of heat and/or pressure working on it from within Earth.

midlatitude anticyclone An extensive migratory high-pressure cell of the midlatitudes that moves generally with the westerlies.

midlatitude cyclone Large migratory low-pressure system that occurs within the middle latitudes and moves generally with the westerlies. Also known as *extratropical cyclones* and *wave cyclones*.

midlatitude deciduous forest Broadleaf forest plant assemblage comprised of mostly deciduous trees.

midlatitude desert climate Desert climate characterized by warm summers but cold winters.

midlatitude grassland Grassland plant assemblage in semiarid regions of the midlatitudes; regionally called *steppe*, *prairie*, *pampa*, and *veldt*.

midocean ridge A lengthy system of deep-sea mountain ranges, generally located at some distance from any continent; formed by divergent plate boundaries on the ocean floor.

Milankovitch cycles Combination of long-term astronomical cycles involving Earth's inclination, precession, and eccentricity of orbit; believed at least partially responsible for major periods of glaciation and deglaciation. Named for Milutin Milankovitch, an early twentieth century Yugoslavian astronomer, who studied these cycles.

millibar A measure of pressure, consisting of one-thousandth part of a bar, or 1000 dynes per square centimeter (1 dyne is the force needed to accelerate 1 gram of mass 1 centimeter per second per second).

mineral A naturally formed solid inorganic substance that has a specified chemical composition and crystal structure.

mistral A cold, high-velocity wind that sometimes surges down France's Rhone Valley, from the Alps to the Mediterranean Sea.

modified Mercalli intensity scale Qualitative scale from I to XII used to describe the relative strength of ground shaking during an earthquake.

Mohorovičić discontinuity The boundary between Earth's crust and mantle. Also known simply as the *Moho*.

Mollisol A soil order characterized by the presence of a mollic epipedon, which is a mineral surface horizon that is dark, thick, contains abundant humus and base nutrients, and retains a soft character when it dries out.

monsoon A seasonal reversal of winds; a general onshore movement in summer and a general offshore flow in winter, with a very distinctive seasonal precipitation regime.

moraine The largest and generally most conspicuous landform feature produced by glacial deposition of *till*, which consists of irregular rolling topography that rises somewhat above the level of the surrounding terrain.

mountain breeze Downslope breeze from a mountain due to chilling of air on its slopes at night.

mudflow Downslope movement of a thick mixture of soil and water.

multispectral [remote sensing] A remote sensing instrument that collects multiple digital images simultaneously in different bands.

natural levee An embankment of slightly higher ground fringing a stream channel in a floodplain; formed by deposition during floodtime.

neap tides The lower-than-normal tidal variations that occur twice a month as the result of the alignment of the Sun and Moon at a right angle to one another.

needleleaf trees Trees adorned with thin slivers of tough, leathery, waxy needles rather than typical leaves.

net primary productivity The net photosynthesis of a plant community over a period of one year, usually measured in the amount of fixed carbon per unit area (kilograms of carbon per square meter per year).

névé See *firn*.

nitrogen cycle An endless series of processes in which nitrogen moves through the environment.

nitrogen fixation Conversion of gaseous nitrogen into forms that can be used by plant life.

normal fault The result of tension (extension) producing a steeply inclined fault plane, with the block of land on one side being pushed up, or upthrown, in relation to the block on the other side, which is downthrown.

North Pole Latitude of 90° north.

occluded front A complex front formed when a cold front overtakes a warm front, lifting all of the warm air mass off the ground.

occlusion Process of cold front overtaking a warm front to form an occluded front.

ocean floor core samples Rock and sediment samples removed from ocean floor.

oceanic trench (deep oceanic trench) Deep linear depression in the ocean floor where subduction is taking place.

offset stream A stream course displaced by lateral movement along a fault.

O horizon The immediate surface layer of a soil profile, consisting mostly of organic material.

orographic precipitation Precipitation that occurs when air, forced to ascend over topographic barriers, cools to the dew point.

orthophoto map A map produced through computerized rectification of aerial imagery.

outcrop Surface exposure of bedrock.

outer core The liquid shell beneath the mantle that encloses Earth's inner core.

outwash plain Extensive glaciofluvial feature that is a relatively smooth, flattish alluvial apron deposited beyond recessional or terminal moraines by streams issuing from ice.

overland flow The general movement of unchanneled surface water down the slope of the land surface.

oxbow lake A cutoff meander that initially holds water.

oxidation The chemical union of oxygen atoms with atoms from various metallic elements to form new products, which are usually more voluminous, softer, and more easily eroded than the original compounds.

Oxisol The most thoroughly weathered and leached of all soils. This soil order invariably displays a high degree of mineral alteration and profile development.

oxygen cycle The movement of oxygen by various processes through the environment.

oxygen isotope analysis Using the ratio of ^{16}O (oxygen 16) and ^{18}O (oxygen 18) isotopes in compounds such as water and calcium carbonate to infer temperature and other conditions in the past.

ozone A gas composed of molecules consisting of three atoms of oxygen, O_3.

ozone layer A layer of ozone between 16 and 40 kilometers (10 and 25 miles) high, which absorbs ultraviolet solar radiation.

Pacific ring of fire Name given to the rim of the Pacific Ocean basin due to widespread volcanic and seismic activity; associated with lithospheric plate boundaries.

paleoclimatology The study of past climates.

paleomagnetism Past magnetic orientation.

Pangaea The massive supercontinent that Alfred Wegener postulated to have existed about 200 million years ago. He visualized Pangaea as breaking up into several large sections that have continually moved away from one another and that now comprise the present continents.

parallel A circle resulting from an isoline connecting all points of equal latitude.

parallelism See *polarity*.

parent material The source of the weathered fragments of rock from which soil is made; solid bedrock or loose sediments that have been transported from elsewhere by the action of water, wind, or ice.

particulate Composed of distinct tiny particles or droplets suspended in the atmosphere; also known as *aerosols*.

paternoster lakes A sequence of small lakes found in the shallow excavated depressions of a glacial trough.

patterned ground Polygonal patterns in the ground that develop in areas of seasonally frozen soil and permafrost.

pediment A gently inclined bedrock platform that extends outward from a mountain front, usually in an arid region.

pedogenic regimes Soil-forming regimes that can be thought of as environmental settings in which certain physical/chemical/biological processes prevail.

ped A larger mass or clump that individual soil particles tend to aggregate into and that determines the structure of the soil.

perennial plants (perennials) Plants that can live more than a single year despite seasonal climatic variations.

perennial stream A permanent stream that contains water the year-round.

periglacial zone An area of indefinite size beyond the outermost extent of ice advance that was indirectly influenced by glaciation.

perihelion The point in its orbit where Earth is nearest to the Sun.

permafrost Permanent ground ice or permanently frozen subsoil.

permeability A soil or rock characteristic in which there are interconnected pore spaces through which water can move.

photochemical smog Form of secondary air pollution caused by the reaction of nitrogen compounds and hydrocarbons to ultraviolet radiation in strong sunlight.

photogrammetry The science of obtaining reliable measurements from photographs and, by extension, mapping from aerial photos.

photoperiodism The response of an organism to the length of exposure to light in a 24-hour period.

photosynthesis The basic process whereby plants produce stored chemical energy from water and carbon dioxide and which is activated by sunlight.

physical geography Study of the physical elements of geography.

piedmont Zone at the "foot of the mountains."

piezometric surface The elevation to which groundwater will rise under natural confining pressure in a well.

pinnacle An erosional remnant in the form of a steep-sided spire that has a resistant caprock; normally found in an arid or semiarid environment; also speleothem column.

plane of the ecliptic The imaginary plane that passes through the Sun and through Earth at every position in its orbit around the Sun; the orbital plane of Earth.

plane projection A family of maps derived by the perspective extension of the geographic grid from a globe to a plane that is tangent to the globe at some point.

plant respiration Stored energy in carbohydrates consumed directly by the plant itself; carbohydrates are oxidized, releasing water, carbon dioxide, and heat energy.

plant succession The process whereby one type of vegetation is replaced naturally by another.

plastic flow [of glacial ice] Slow, nonbrittle flow and movement of ice under pressure.

plateau Flattish erosional platform bounded on at least one side by a prominent escarpment.

plate tectonics A coherent theory of massive crustal rearrangement based on the movement of continent-sized lithospheric plates.

playa Dry lake bed in a basin of interior drainage.

Pleistocene Epoch An epoch of the Cenozoic era between the Pliocene and the Holocene; from about 2.6 million to about 10,000 years ago.

Pleistocene lakes Large freshwater lakes that formed in basins of interior drainage because of higher rainfall and/or lower evaporation during the Pleistocene.

plucking See *glacial plucking*.

pluton A large, intrusive igneous body.

plutonic rock Igneous rock formed below ground from the cooling and solidification of magma.

pluvial (pluvial effects) Pertaining to rain; often used in connection with a past rainy period.

podzolization The dominant pedogenic regime in areas where winters are long and cold, and which is characterized by slow chemical weathering of soils and rapid mechanical weathering from frost action, resulting in soils that are shallow, acidic, and with a fairly distinctive profile.

polar easterlies A global wind system that occupies most of the area between the *polar highs* and about 60° of latitude. The winds move generally from east to west and are typically cold and dry.

polar front The contact between unlike air masses in the subpolar low-pressure zone.

polar high A high-pressure cell situated over either polar region.

polarity [of rotation axis] A characteristic of Earth's axis wherein it always points toward Polaris (the North Star) at every position in Earth's orbit around the Sun. Also called *parallelism*.

porosity The amount of pore space between the soil particles and between the *peds*, which is a measure of the capacity of the soil to hold water and air.

precipitation Drops of liquid or solid water falling from clouds.

precipitation variability Expected departure from average annual precipitation in any given year.

pressure gradient Change in atmospheric pressure over some horizontal distance.

primary consumer Animals that eat plants as the first stage in a food pyramid or chain.

primary pollutants Contaminants released directly into the air.

prime meridian The meridian passing through the Royal Observatory at Greenwich (England), just east of London, and from which longitude is measured.

producers Organisms that produce their own food through photosynthesis; plants.

proglacial lake A lake formed when ice flows across or against the general slope of the land and the natural drainage is impeded or completely blocked so that meltwater from the ice becomes impounded against the ice front.

pseudocylindrical projection (elliptical projection) A family of map projections in which the entire world is displayed in an oval shape.

pyroclastic flow High-speed avalanche of hot gases, ash, and rock fragments emitted from a volcano during an explosive eruption; also known as a *nuée ardente*.

pyroclastics (pyroclastic material) Solid rock fragments thrown into the air by volcanic explosions.

radar Radio detection and ranging.

radiation The process in which electromagnetic energy is emitted from a body; the flow of energy in the form of electromagnetic waves.

rain The most common and widespread form of precipitation, consisting of drops of liquid water.

rain shadow Area of low rainfall on the leeward side of a mountain range or topographic barrier.

recessional moraine A glacial deposit of till formed during a pause in the retreat of the ice margin.

recurrence interval [of a flood] The probability of a given-size flood occurring in a year; also called the *return period*.

reflection The ability of an object to repel waves without altering either the object or the waves.

reg A desert surface of coarse material from which all sand and dust have been removed by wind and water erosion. Often referred to as *desert pavement* or *desert armor*.

regional metamorphism Widespread subsurface metamorphism of rock as a result of prolonged exposure to heat and high pressure.

regolith A layer of broken and partly decomposed rock particles that covers bedrock.

relative humidity An expression of the amount of water vapor in the air (the water vapor content) in comparison with the maximum amount that could be there if the air were saturated (the *capacity*). This is a ratio that is expressed as a percentage.

relief The difference in elevation between the highest and lowest points in an area; the vertical variation from mountaintop to valley bottom.

remote sensing Study of an object or surface from a distance by using various instruments.

representative fraction (r.f.) The ratio that is an expression of a *fractional map scale* that compares map distance with ground distance (using the same units of measurement).

reverse fault A fault produced from compression, with the upthrown block rising steeply above the downthrown block.

revolution [around the Sun] The orbital movement of Earth around the Sun over the year.

R horizon The consolidated bedrock at the base of a soil profile.

ria shoreline An embayed coast with numerous estuaries; formed by the flooding of stream valleys by the sea.

ridge [of atmospheric pressure] Linear or elongated area of relatively high atmospheric pressure.

rift valley See *continental rift valley*.

riparian vegetation Anomalous streamside growth, particularly prominent in relatively dry regions, where stream courses may be lined with trees, although no other trees are to be found in the landscape.

roche moutonnée A characteristic glacial landform produced when a bedrock hill or knob is overridden by moving ice. The *stoss* side is smoothly rounded and streamlined by grinding abrasion as the ice rides up the slope, but the lee side is shaped largely by plucking, which produces a steeper and more irregular slope.

rock Solid material composed of aggregated mineral material.

rock cycle Term given to the long-term "recycling" of mineral material from one kind of rock to another.

rockfall (fall) Mass wasting process in which weathered rock drops to the foot of a cliff or steep slope.

rock glacier An accumulated *talus* mass that moves slowly but distinctly downslope under its own weight.

Rossby wave A very large north–south undulation of the upper-air westerlies and jet stream.

rotation [of Earth] The spinning of Earth around its imaginary north-south axis.

runoff Flow of water from land to oceans by overland flow, streamflow, and groundwater flow.

Saffir-Simpson Hurricane Scale Classification system of hurricane strength with category 1 the weakest and category 5 the strongest.

sag pond A pond caused by the collection of water from springs and/or runoff into sunken ground, resulting from the crushing of rock in an area of fault movement.

salina Dry lake bed that contains an unusually heavy concentration of salt in the lake-bed sediment.

saline lake Salt lake; commonly caused by interior stream drainage in an arid environment.

salinity A measure of the concentration of dissolved salts.

salinization One of the dominant pedogenic regimes in areas where principal soil moisture movement is upward because of a moisture deficit.

salt wedging Rock disintegration caused by the crystallization of salts from evaporating water.

sand dune A mound, ridge, or low hill of loose, windblown sand.

Santa Ana winds Name given to dry, usually warm, and often very powerful winds blowing offshore in southern California region.

saturated adiabatic rate (saturated adiabatic lapse rate) The diminished rate of cooling, averaging about 6°C per 1000 meters (3.3°F per 1000 feet) of rising air above the lifting condensation level; a result of the latent heat of condensation counteracting some of the adiabatic cooling of rising air.

saturation vapor pressure The maximum pressure that can be exerted by water vapor at a given temperature; the pressure exerted by water vapor when the air is saturated.

scattering A change in direction, but not in wavelength, of light waves.

sea breeze A wind that blows from the sea toward the land, usually during the day.

seafloor spreading The pulling apart of lithospheric plates to permit the rise of deep-seated magma to Earth's surface in midocean ridges.

secondary consumer Animals that eat other animals, as the second and further stages in a food pyramid or chain.

secondary pollutant Pollutants formed in the atmosphere as a consequence of chemical reactions or other processes.

sediment Small particles of rock debris or organic material deposited by water, wind, or ice.

sediment budget [of a beach] The balance between the sediment being deposited on a beach and the sediment that is being transported away from a beach.

sedimentary rock Rock formed of sediment that is consolidated by the combination of pressure and cementation.

seif (longitudinal dune) Long, narrow desert dunes that usually occur in multiplicity and in parallel arrangement.

sensible temperature The relative apparent temperature that is sensed by a person's body.

separates The size groups within the standard classification of soil particle sizes.

September equinox One of two days of the year when the vertical rays of the Sun strike the equator; every location on Earth has equal day and night; occurs on or about September 22 each year.

shield volcanoes Volcanoes built up in a lengthy outpouring of very fluid basaltic lava. Shield volcanoes are broad mountains with gentle slopes.

shortwave radiation Wavelengths of radiation emitted by the Sun, especially ultraviolet, visible, and short infrared radiation.

shrubland Plant association dominated by relatively short woody plants.

silicate mineral (silicates) A category of minerals composed of silicon and oxygen combined with another element or elements.

sinkhole (doline) A small, rounded depression that is formed by the dissolution of surface limestone, typically at joint intersections.

sinuous channel pattern (sinuous stream channel) Gently curving or winding stream channel pattern.

slip face Steeper leeward side of a sand dune.

slump A slope collapse with rotation along a curved sliding plane.

small-scale map A map whose scale is a relatively small representative fraction and therefore shows a large portion of Earth's surface in limited detail.

snow Solid precipitation in the form of ice crystals, small pellets, or flakes, which is formed by the direct conversion of water vapor into ice.

soil An infinitely varying mixture of weathered mineral particles, decaying organic matter, living organisms, gases, and liquid solutions. Soil is that part of the outer "skin" of Earth occupied by plant roots.

soil order The highest (most general) level of soil classification in the *Soil Taxonomy*.

soil profile A vertical cross section from Earth's surface down through the soil layers into the parent material beneath.

Soil Taxonomy The system of soil classification currently in use in the United States. It is genetic in nature and focuses on the existing properties of the soil rather than on environment, genesis, or the properties it would possess under virgin conditions.

soil–water balance The relationship between gain, loss, and storage of soil water.

soil–water budget An accounting that demonstrates the variation of the soil–water balance over a period of time.

solar altitude Angle of the Sun above the horizon.

solifluction A special form of soil creep in tundra areas that produces a distinctive surface appearance. During the summer the near-surface portion of the ground thaws, but the meltwater cannot percolate deeper because of the permafrost below. The spaces between the soil particles become saturated, and the heavy surface material sags slowly downslope.

solum The true soil that includes only the top four horizons: O, the organic surface layer; A, the topsoil; E, the eluvial layer; and B, the subsoil.

sonar Sound navigation and ranging.

South Pole Latitude of 90° south.

species The major subdivision of a genus, regarded as the basic category of biological taxonomic classification.

specific heat The amount of energy required to raise the temperature of 1 gram of a substance by 1°C.

specific humidity A direct measure of water-vapor content expressed as the mass of water vapor in a given mass of air (grams of vapor/ kilograms of air).

speleothem A feature formed by precipitated deposits of minerals on the wall, floor, or roof of a cave.

spit A linear deposit of marine sediment that is attached to the land at one or both ends.

Spodosol A soil order characterized by the occurrence of a spodic subsurface horizon, which is an illuvial layer where organic matter and aluminum accumulate, and which has a dark, sometimes reddish, color.

spring tide A time of maximum tide that occurs as a result of the alignment of Sun, Moon, and Earth.

stable [air] Air that rises only if forced.

stalactite A pendant structure hanging downward from a cavern's roof.

stalagmite A projecting structure growing upward from a cavern's floor.

star dune Pyramid-shaped sand dune with arms radiating out in three or more directions.

stationary front The common boundary between two air masses in a situation in which neither air mass displaces the other.

storm surge A surge of wind-driven water as much as 8 meters (25 feet) above normal tide level, which occurs when a hurricane pounds into a shoreline.

strata Distinct layers of sediment.

stratified drift Drift that was sorted as it was carried along by the flowing glacial meltwater.

stratosphere Atmospheric layer directly above the troposphere.

stratus clouds Layered, horizontal clouds, often below altitudes of 2 kilometers (6500 feet), which sometimes occur as individual clouds but more often appear as a general overcast.

stream capture (stream piracy) An event where a portion of the flow of one stream is diverted into that of another by natural processes.

streamflow Channeled movement of water along a valley bottom.

stream load Solid matter carried by a stream.

stream order Concept that describes the hierarchy of a drainage network.

stream piracy See *stream capture.*

stream rejuvenation When a stream gains downcutting ability, usually through regional tectonic uplift.

stream terrace Remnant of a previous valley floodplain of a rejuvenated stream.

strike-slip fault A fault produced by shearing, with adjacent blocks being displaced laterally with respect to one another. The movement is mostly or entirely horizontal.

subarctic climate Severe midlatitude climate found in high latitude continental interiors, characterized by very cold winters and an extreme annual temperature range.

subduction Descent of the edge of an oceanic lithospheric plate under the edge of an adjoining plate.

sublimation The process by which water vapor is converted directly to ice, or vice versa.

subpolar low A zone of low pressure that is situated at about 50° to 60° of latitude in both Northern and Southern hemispheres (also referred to as the *polar front*).

subtropical Bordering on the tropics; between tropical and temperate regions.

subtropical gyres The closed-loop pattern of surface ocean currents around the margins of the major ocean basins; the flow is clockwise in the Northern Hemisphere and counterclockwise in the Southern Hemisphere.

subtropical desert climate A hot desert climate; generally found in subtropical latitudes, especially on the western sides of continents.

subtropical high (STH) Large, semipermanent, high-pressure cells centered at about 30° latitude over the oceans, which have average diameters of 3200 kilometers (2000 miles) and are usually elongated east–west.

supercooled water Water that persists in liquid form at temperatures below freezing.

supersaturated [air] Air in which the relative humidity is greater than 100 percent but condensation is not taking place.

surface tension Because of electrical polarity, liquid water molecules tend to stick together—a thin "skin" of molecules forms on the surface of liquid water causing it to "bead."

suspended load The very fine particles of clay and silt that are in suspension and move along with the flow of water without ever touching the streambed.

swallow hole The distinct opening at the bottom of some sinkholes through which surface drainage can pour directly into an underground channel.

swamp A flattish surface area that is submerged in water at least part of the time but is shallow enough to permit the growth of water-tolerant plants—predominantly trees.

swash The cascading forward motion of a breaking wave that rushes up the beach.

swell An ocean wave, usually produced by stormy conditions, that can travel enormous distances away from the source of the disturbance.

symbiosis A mutually beneficial relationship between two organisms.

syncline A simple downfold in the rock structure.

talus Pieces of weathered rock, of various sizes, that fall directly downslope; also called *scree.*

talus cone Sloping, cone-shaped heaps of dislodged talus.

tarn Small lake in the shallow excavated depression of a glacial cirque.

teleconnection The coupling or relationship of weather and/or oceanic events in one part of the world with those in another.

temperature Description of the average kinetic energy of the molecules in a substance; the more vigorous the "jiggling" of the molecules (and therefore the greater the internal kinetic energy), the higher the temperature of a substance; in popular terms, a measure of the degree of hotness or coldness of a substance.

temperature inversion A situation in which temperature increases with higher altitude and so the normal condition is inverted.

terminal moraine A glacial deposit of till that builds up at the outermost extent of ice advance.

terrane A mass of lithosphere, bounded on all sides by faults, that has become accreted to a lithospheric plate margin with different lithologic characteristics from those of the terrane; often comprised of lithosphere that is too buoyant to subduct.

terrestrial radiation Longwave radiation emitted by Earth's surface or atmosphere.

thermal high High pressure cell associated with cold surface conditions.

thermal infrared radiation (thermal IR) The middle and far infrared part of the electromagnetic spectrum.

thermal low Low pressure cell associated with warm surface conditions.

thermohaline circulation Slow circulation of deep ocean water because of differences in water density that arise from differences in salinity and temperature.

thrust fault A fault created by compression forcing the upthrown block to override the downthrown block at a relatively low angle; also called an *overthrust fault.*

thunder The sound that results from the shock wave produced by the instantaneous expansion of air that is abruptly heated by a lightning bolt.

thunderstorm A relatively violent convective storm accompanied by thunder and lightning.

tidal bore A wall of seawater several centimeters to several meters in height that rushes up a river as the result of enormous tidal inflow.

tidal range The vertical difference in elevation between high and low tide.

tides The rise and fall of the coastal water levels caused by the alternate increasing and decreasing gravitational pull of the Moon and the Sun on varying parts of Earth's surface.

till Rock debris that is deposited directly by moving or melting ice, with no meltwater flow or redeposition involved.

tilted fault block mountain See *fault-block mountain.*

time zone Region on Earth (generally a north-south band defined by longitude) within which the agreed-upon local time is the same.

tombolo **A spit formed by sand deposition that connects an island to** the mainland.

topography Surface configuration of Earth.

tornado A localized cyclonic low-pressure cell surrounded by a whirling cylinder of violent wind; characterized by a funnel cloud extending below a cumulonimbus cloud.

tower karst Tall, steep-sided hills in an area of karst topography.

trade winds The major easterly wind system of the tropics, issuing from the equatorward sides of the subtropical highs and diverging toward the west and toward the equator.

transform [plate] boundary Two plates slipping past one another laterally.

transmission The ability of a medium to allow electromagnetic waves to pass through it.

transverse dune Crescent-shaped sand dune ridges that have convex sides facing the prevailing direction of wind and which occur where the supply of sand is great. The crest is perpendicular to the wind direction vector and aligned in parallel waves across the land.

travertine Massive accumulation of calcium carbonate generally associated with water seeps and springs.

treeline The elevation above (or below) which trees do not grow.

trellis drainage pattern A drainage pattern that is usually developed on alternating bands of hard and soft strata, with long parallel subsequent streams linked by short, right-angled segments and joined by short tributaries.

tropical cyclone A storm most significantly affecting the tropics and subtropics, which is intense, revolving, rain-drenched, migratory, destructive, and erratic. Such a storm system consists of a prominent low-pressure center that is essentially circular in shape and has a steep pressure gradient outward from the center. When wind speed reaches 64 knots, they are called *hurricanes* in North America and the Caribbean.

tropical deciduous forest Tropical forest found in regions with a pronounced dry period during which many of the trees shed their leaves.

tropical depression By international agreement, an incipient tropical cyclone with winds not exceeding 33 knots.

tropical monsoon climate Tropical humid climate with a pronounced winter dry season and a very wet summer rainy season; associated with monsoon wind pattern.

tropical rainforest A distinctive assemblage of tropical vegetation that is dominated by a great variety of tall, high-crowned trees; also called *selva*.

tropical savanna A tropical grassland dominated by tall grasses.

tropical savanna climate Tropical humid climate with a dry winter season and a moderately wet summer season; associated with the seasonal migration of the ITCZ.

tropical scrub A widespread tropical assemblage of low trees and bushes.

tropical storm By international agreement, an incipient tropical cyclone with winds between 34 and 63 knots.

tropical wet climate Tropical humid climate that is wet all year; usually under the influence of the ITCZ all year.

Tropic of Cancer The parallel of 23.5° north latitude, which marks the northernmost location reached by the vertical rays of the Sun in the annual cycle of Earth's revolution.

Tropic of Capricorn The parallel of 23.5° south latitude, which marks the southernmost location reached by the vertical rays of the Sun in the annual cycle of Earth's revolution.

troposphere The lowest thermal layer of the atmosphere, in which temperature decreases with height.

trough [of atmospheric pressure] Linear or elongated band of relatively low atmospheric pressure.

tsunami Very long wavelength oceanic wave generated by submarine earthquake, landslide, or volcanic eruption; also called *seismic sea wave*.

tundra A complex mix of very low-growing plants, including grasses, forbs, dwarf shrubs, mosses, and lichens, but no trees. Tundra occurs only in the perennially cold climates of high latitudes or high altitudes.

tundra climate Polar climate in which no month of the year has an average temperature above 10°C (50°F).

ubac slope A slope oriented so that sunlight strikes it at a low angle and hence is much less effective in heating and evaporating than on the *adret slope*, thus producing more luxuriant vegetation of a richer diversity.

Ultisol A soil order similar to Alfisols, but more thoroughly weathered and more completely leached of bases.

ultraviolet (UV) radiation Electromagnetic radiation in the wavelength range of 0.1 to 0.4 micrometers.

uniformitarianism The concept that the "present is the key to the past" in geomorphic processes. The processes now operating have also operated in the past.

Universal Time Coordinated (UTC)or Coordinated Universal Time The world time standard reference; previously known as *Greenwich mean time (GMT)*.

unstable [air] Air that rises without being forced.

upwelling Cold, deep ocean water that rises to the surface where wind patterns deflect surface water away from the coast; especially common along the west coasts of continents in the subtropics and midlatitudes.

uvala A compound sinkhole (doline) or chain of intersecting sinkholes.

valley That portion of the total terrain in which a drainage system is clearly established.

valley breeze Upslope breeze up a mountain due to heating of air on its slopes during the day.

valley glacier A long, narrow feature resembling a river of ice, which spills out of its originating basins and flows down-valley.

valley train A lengthy deposit of glaciofluvial alluvium confined to a valley bottom beyond the outwash plain.

vapor pressure The pressure exerted by water vapor in the atmosphere.

ventifact Rock that has been sandblasted by the wind.

verbal scale [of a map] Scale of a map stated in words; also called a word scale.

vertebrates Animals that have a backbone that protects their spinal cord—fishes, amphibians, reptiles, birds, and mammals.

vertical zonation The horizontal layering of different plant associations on a mountainside or hillside.

Vertisol A soil order comprising a specialized type of soil that contains a large quantity of clay and has an exceptional capacity for absorbing water. An alternation of wetting and drying, expansion and contraction, produces a churning effect that mixes the soil constituents, inhibits the development of horizons, and may even cause minor irregularities in the surface of the land.

visible light Waves in the electromagnetic spectrum in the narrow band between about 0.4 and 0.7 micrometers in length; wavelengths of electromagnetic radiation to which the human eye is sensitive.

volcanic island arc Chain of volcanic islands associated with an oceanic plate–oceanic plate subduction zone.

volcanic mudflow A fast-moving, muddy flow of volcanic ash and rock fragments; also called a *lahar*.

volcanic rock Igneous rock formed on the surface of Earth; also called *extrusive rock*.

volcano A conical mountain or hill from which extrusive material is ejected.

volcanism General term that refers to movement of magma from the interior of Earth to or near the surface.

Walker Circulation General circuit of air flow in the southern tropical Pacific Ocean; warm air rises in the western side of the basin (in the updrafts of the ITCZ), flows aloft to the east where it descends into the subtropical high off the west coast of South America; the air then flows back to the west in the surface trade winds. Named for the British meteorologist Gilbert Walker (1868–1958) who first described this circumstance.

warm front The leading edge of an advancing warm air mass.

waterspout A funnel cloud in contact with the ocean or a large lake; similar to a weak tornado over water.

watershed See *drainage basin*.

water table The top of the saturated zone within the ground.

water vapor The gaseous state of moisture.

wave-cut platform Gently sloping, wave-eroded bedrock platform that develops just below sea level; common where coastal cliff is being worn back by wave action; also called *wave-cut bench*.

wave height The vertical distance from wave crest to trough.

wavelength The horizontal distance from wave crest to crest or from trough to trough.

wave of oscillation Motion of wave in which the individual particles of the medium (such as water) make a circular orbit as the wave form passes through.

wave of translation The horizontal motion produced when a wave reaches shallow water and finally "breaks" on the shore.

wave refraction Phenomenon whereby waves change their directional trend as they approach a shoreline; results in ocean waves generally breaking parallel with the shoreline.

weather The short-term atmospheric conditions for a given time and a specific area.

weathering The physical and chemical disintegration of rock that is exposed to the atmosphere.

westerlies The great wind system of the midlatitudes that flows basically from west to east around the world in the latitudinal zone between about 30° and 60° both north and south of the equator.

wetland Landscape characterized by shallow, standing water all or most of the year, with vegetation rising above the water level.

wilting point The point at which plants are no longer able to extract moisture from the soil because the capillary water is all used up or evaporated.

wind Horizontal air movement.

wind shear (vertical wind shear) Significant change in wind direction or speed in the vertical dimension.

woodland Tree-dominated plant association in which the trees are spaced more widely apart than those of forests and do not have interlacing canopies.

xerophytic adaptations Plants that are structurally adapted to withstand protracted dry conditions.

yazoo stream A tributary unable to enter the main stream because of natural levees along the main stream.

zone of aeration (vadose zone) The topmost hydrologic zone within the ground, which contains a fluctuating amount of moisture (soil water) in the pore spaces of the soil (or soil and rock).

zone of confined water The third hydrologic zone below the surface of the ground, which contains one or more permeable rock layers (*aquifers*) into which water can infiltrate and is separated from the zone of saturation by impermeable layers.

zone of saturation (phreatic zone) The second hydrologic zone below the surface of the ground, whose uppermost boundary is the water table. The pore spaces and cracks in the bedrock and the regolith of this zone are fully saturated.

zoogeographic regions Division of land areas of the world into major realms with characteristic fauna.

Photo and Illustration Credits

TOC.1: NASA/Goddard Space Flight Center Scientific Visualization Studio; TOC.2: Getty Images; TOC.3: Tom L. McKnight Photo; TOC.4: NASA; TOC.5: Homer Sykes/Alamy; TOC.6: Eric Nguyen/Corbis; TOC.7: Sadao Kurita/Minden Pictures; TOC.8 Paul Souders/Getty Images; TOC.9: Craig Tuttle/Corbis; TOC.10, 12: Tom L. McKnight Photo; TOC.13: Douglas Peebles Photography/Alamy; TOC.14: Darrel Hess; TOC.15: NASA Earth Observatory; TOC.17, 18: Darrel Hess; TOC.19: Tom L. McKnight Photo

Chapter 1

Opener: Reto Stöckli, based on data from NASA and NOAA; **Figure 1-2:** John Warden/Getty Images; **Figure 1-3:** Edward J. Tarbuck and Frederick K. Lutgens, *Earth Science,* 11th ed., Upper Saddle River, NJ: Pearson-Prentice Hall, 2006; **Figure 1-4:** From Edward J. Tarbuck, Frederick K. Lutgens, and Dennis Tasa, *Essentials of Geology*, 10th ed.; **Figure 1-13:** Jon Arnold Images Ltd/Alamy; **Figure 1-14:** Tom L. McKnight Photo; **Figure 1-23:** Elena Moiseeva/Shutterstock.

Chapter 2

Opener: Image courtesy of USGS National Center for EROS and NASA Landsat Project Science Office; **Figure 2-1:** NASA Goddard Space Flight Center Image by Reto Stöckli; **Figure 2-2a:** Dennis Tasa; **Figure 2-2b-d:** ESRI; **Figure 2-3:** Courtesy of National Geographic Society; **Figure 2-6:** Aleksandr Kurganov/Shutterstock; **Figure 2-15:** U.S. Geological Survey; **Figure 2-A,b:** U.S. Air Force; **Figure 2-B:** Courtesy of the U.S. Geological Survey; **Figure 2-18:** Courtesy of Garmin; **Figure 2-21:** U.S. Geological Survey, Denver; **Figure 2-23:** Image courtesy NASA/GSFC/MITI/ERSDAC/JAROS, and U.S./Japan; ASTER Science Team; **Figure 2-25:** Earth Observatory/NASA/Robert Simmon, using Landsat data provided by the United States Geological Survey; **Figure 2-26:** Earth Observatory/NASA/Jeff Schmaltz, MODIS Rapid Response, NASA Goddard Space Flight Center; **Figure 2-C,D:** USDA/Google Earth/Richard Green; **Figure 2-27:** Earth Observatory/NASA; **Figure 2-28:** Dan Marrier/Massachusetts Office of Geographic and Environmental Information; **Figure 2-30:** NASA/Goddard Space Flight Center, Scientific Visualization Studio.

Chapter 3

Opener: Image Science & Analysis Laboratory, NASA Johnson Space Center; **Figure 3-1:** NASA; **Figure 3-4:** Tom L. McKnight Photo; **Figure 3-10:** Getty Images; **Figure 3-13 a,b:** NASA; **Figure 3-A:** U.S. National Weather Service/U.S. Environmental Protection Agency; **Figure 3-B:** Naval Research Laboratory, Marine Meteorology Division; **Figure 3-15:** Alexander Chaikin/ Shutterstock; **Figure 3-19:** Tom L. McKnight Photo; **Figure 3-21:** European Space Agency/Photo Researchers.

Chapter 4

Opener: Theo Allofs/Corbis; **Figure 4-1a:** Roc Planck/Photo Researchers, Inc.; **Figure 4-1b:** Robert Semeniuk; **Figure 4-B:** Jupiter Images; **Figure 4-10:** Darrel Hess; **Figure 4-C,D,E:** NOAA; **Figure 4-19:** Getty Images; **Figure 4-26:** Johns Hopkins University Applied Physics Laboratory; **Figure 4-F,G,H:** NASA; **Figure 4-28:** Tom L. McKnight Photo; **Figure 4-33:** National Climate Data Center/NOAA; **Figure 4-34:** Adapted from Michael Collier and Robert Webb, *Floods, Droughts, and Climate Change,* Tucson: University of Arizona Press, 2002; updated with NOAA data.

Chapter 5

Opener: Toby Melville/Reuters; **Figure 5-2:** Darrel Hess; **Figure 5-A:** Stephen Sadler; **Figure 5-19:** Tom L. McKnight Photo; **Figure 5-21:** NASA/Naval Research Laboratory, Marine Meteorology Division; **Figure 5-23:** NASA; **Figure 5-31:** Dieter Telemans/Panos Pictures; **Figure 5-35:** From Edward J. Tarbuck, Frederick K. Lutgens, and Dennis Tasa, *Earth Science,* 12th ed.; **Figure 5-39:** U.S. Geological Survey, Denver; **Figure 5-40:** After Michael Collier and Robert Webb, Floods, Droughts, and Climate Change, Tucson: University of Arizona Press, 2002; **Figure 5-41:** After Edward Aguado and James E. Burt, *Understanding Weather and Climate,* 4th ed., Upper Saddle River, NJ: Pearson-Prentice Hall, 2007; **Figure 5-B:** TAO ProjectOffice, NOAA/PMEL; **Figure 5-C,D,E:** NOAA.

Chapter 6

Opener: John Eastcott and Yva Momatiuk/National Geographic Image Collection; **Figure 6-2:** After Alan P. Trujillo and Harold V. Thurman, *Essentials of Oceanography,* 8th ed., Upper Saddle River, NJ: Pearson-Prentice Hall, 2005; **Figure 6-3:** Fedorov Oleksiy/Shutterstock; **Figure 6-A,B:** Naval Research Laboratory; **Figure 6-C,D,E:** Courtesy of Naval Research Laboratory/Marine Meteorology Division; **Figure 6-8:** After Edward Aguado and James E. Burt, *Understanding Weather and Climate,* 4th ed., Upper Saddle River, NJ: Pearson-Prentice Hall, 2007; **Figures 6-10, 6-22, 6-27:** From Edward J. Tarbuck, Frederick K. Lutgens, and Dennis Tasa, *Earth Science,* 12th ed.; **Figure 6-11:** Fedorov Oleksiy/ Shutterstock; **Figure 6-15a:** Trevor Mein/Getty Images; **Figure 6-15b:** Steven J. Kazlowski/Alamy; **Figure 6-15c:** Alexandru Chiriac/ Shutterstock; **Figure 6-17:** Tom L. McKnight Photo; **Figure 6-F:** Jacques Descloitres, MODIS Rapid Response Team, NASA/GSFC; **Figure 6-20a:** Lawrence Pringle/Photo Researchers; **Figure 6-20b:** Tom L. McKnight Photo; **Figure 6-26:** Homer Sykes/Alamy; **Figure 6-29:** Tom L. McKnight Photo; **Figure 6-30:** Tom L. McKnight Photo; **Figure 6-39:** Data from National Atmospheric Deposition Program.

Chapter 7

Opener: Gene Blevins/LA Daily News/Corbis; **Figures 7-3, 7-4, 7-10, 7-18, 7-22, 7-31, 7-32:** From Edward J. Tarbuck, Frederick K. Lutgens, and Dennis Tasa, *Earth Science,* 12th ed.; **Figure 7-7:** SeaWiFS Project/Orbimage/NASA/Goddard Space Flight Center; **Figure 7-16:** NASA MODIS Response Team; **Figure 7-A:** NOAA; **Figure 7-B:** Vincent Laforet/POOL/epa/Corbis; **Figure 7-C:** National Hurricane Center/NOAA; **Figure 7-D:** NOAA; **Figure 7-19:** After W. M. Gray, "*Tropical Cyclone Genesis,*" Atmospheric Science Paper 234, Fort Collins, CO: Department of Atmospheric Science, Colorado State University, 1975; **Figure 7-20:** © 2005 Ray Sterner and Steve Babin, Johns Hopkins University Applied Physics Laboratory; **Figure 7-23 a,b:** NASA MODIS Response Team; **Figure 7-24:** Associated Press; **Figure 7-E, F:** American Meteorological Society; **Figure 7-28:** A & J Verkaik/Corbis; **Figure 7-30:** Eric Nguyen/Corbis; **Figure 7-G:** Courtesy of Oklahoma Mesonetwork Project.

Chapter 8

Opener: Sadao Kurita/Minden Pictures; **Figure 8-6:** Juan Carlos Munoz/age fotostock; **Figure 8-8:** Renee Lynn/Corbis; **Figure 8-10:** Biswaranjan Rout/Associated Press; **Figure 8-14:** Tom L. McKnight Photo; **Figure 8-18:** Tom L. McKnight Photo; **Figure 8-21:** Tom L. McKnight Photo; **Figure 8-24:** Clint Farlinger/Alamy; **Figure 8-25:** After U.S. Census of Agriculture, 1992; **Figure 8-26:** Tom L. McKnight Photo; **Figure 8-28:** Tom L. McKnight Photo; **Figure 8-31:** Tom L. McKnight Photo; **Figure 8-A:** From Frederick K. Lutgens, Edward J. Tarbuck, and Dennis Tasa, *Atmosphere,* 11th ed.; **Figure 8-B:** NASA; **Figure 8-33:** Natalie Fobes/Corbis; **Figure 8-34:** Maria Stenzel/National Geographic Image Collection; **Figure 8-38:** Tom L. McKnight Photo; **Figure 8-41:** After Edward Aguado and James E. Burt, *Understanding Weather and Climate,* 4th ed., Upper Saddle River, NJ: Pearson-Prentice Hall, 2007, and Frederick K. Lutgens and Edward J. Tarbuck, *The Atmosphere,* 9th ed., Upper Saddle River, NJ: Pearson-Prentice Hall, 2004; **Figure 8-42:** Chris Gilbert/British Antarctic Survey; **Figure 8-D:** Chris Gilbert/British Antarctic Survey; **Figure 8-E:** NOAA; **Figure 8-43:** NASA; **Figure 8-45:** Simon Frasier/Photo Researchers; **Figure 8-F:** Slovegrove/Istockphoto.com.

Chapter 9

Opener: Gavin Hellier/Corbis; **Figure 9-3:** Craig Tuttle/Corbis; **Figure 9-6:** Andrew Syred/Photo Researchers; **Figure 9-8:** *J. L. Davies, Geographical Variation in Coastal Development, 2d ed., Longman: New York, 1980, 179);* **Figure 9-9b:** J.H. Robinson/ Photo Researchers; **9.A:** NOAA Marine Debris Program; **9.B:** © 2009 Scripps Institution of Oceanography, UC San Diego; **Figure 9-10** *Adapted from Alan P. Trujillo and Harold V. Thurman,* Essentials of Oceanography, *9th ed., Upper Saddle River, NJ: Pearson-Prentice Hall, 2008;* **9.12:** Paul Souders/Getty Images;

9.16: Ashley Cooper/Alamy; **9.D**: Courtesy of Benjamin Jones, USGS; **9.C.a, b**: NOAA; **Figure 9-18** Tom L. McKnight Photo; **9.19a1, 2**: NASA Earth Observatory; **9.19b**: Gerd Ludwig/National Geographic Image Collection; **9.20a,b**: NASA; **9.21**: ksteen5/Istockphoto.com; **9.22a, b**: USGS; **9.23**: Mark E. Gibson/Corbis; **Figures 9-26, 9-27, 9-30**: From Edward J. Tarbuck, Frederick K. Lutgens, and Dennis Tasa, *Essentials of Geology*, 10th ed.; **Figure 9-28**: Tom L. McKnight photo; **9-E,F**: U.S. Geological Survey, Denver.

Chapter 10

Opener: Bill Lea/Dembinsky Associates; **Figure 10-1**: (a) Tom L. McKnight photo (b) Bates Littlehales/Animals Animals/Earth Scenes; **Figure 10-2**: Craig Tuttle/Corbis; **Figure 10-4**: MODIS/NASA; **Figure 10-5**: NASA Earth Observatory/Reto Stockli and Jesse Allen; **Figure 10-A**: Scott Sinklier/AgStock Images/Corbis; **Figure 10-b**: Global Warming Images/ Alamy; **Figure 10-11**: Alan Carey/Photo Researchers; **Figure 10-12**: Tom L. McKnight Photo; **Figure 10-13**: Jaime Plaza Van Roon/Auscape; **Figure 10-14**: Tom L. McKnight Photo; **Figure 10-15**: Alan D. Carey/ Photo Researchers; **Figure 10-16**: Robert McGouey/Alamy; **Figure 10-18**: Darrel Hess; **Figure 10-C**: Jack Thomas/Alamy; **Figure 10-D**: Jim West/ Alamy; **Figure 10-20**: vlynder/Istockphoto.com; **Figure 10-21**: Tom L. McKnight Photo; **Figure 10-22**: Tom L. McKnight Photo; **Figure 10-23a,b**: Tom L. McKnight Photo; **Figure 10-23c**: M. Phillip Kahl/Photo Researchers; **Figure 10-25**: Darrel Hess.

Chapter 11

Opener: Mitsuaki Iwago/Minden Pictures; **Figure 11-5**: Darrel Hess; **Figure 11-6a**: Tom L. McKnight Photo; **Figure 11-6b**: John Eastcott/Yva Momatiuk/Photo Researchers; **Figure 11-8**: C. Prescott-Allen; **Figure 11-9**: Niebrugge Images/Alamy; **Figure 11-11**: Darrel Hess; **Figure 11-13**: Kevin Ebi/Alamy; **Figure 11-14**: Tom L. McKnight Photo; **Figure 11-15**: Darrel Hess; **Figure 11-16**: Tom L. McKnight Photo; **Figure 11- 17**: Juniors Bildarchiv/Alamy; **Figure 11-A**: Courtesy of the Audobon Society; **Figure 11-B**: Jim Zipp/Photo Researchers; **Figure 11-18a**: Tom L. McKnight Photo; **Figure 11-18b**: Tom McHugh/Photo Researchers; **Figure 11-19a**: Phillip Gouravitch/Photo Researchers; **Figure 11-19b**: Joe Austin Photography/Alamy; **Figure 11-20a**: John Cancolosi/Peter Arnold; **Figure 11-20b**: John Serafin; **Figure 11-21**: Darrel Hess; **Figure 11-22**: Tom L. McKnight Photo; **Figure 11-26**: Dr. Morley Read/Photo Researchers; **Figure 11-27**: worldswildlifewonders/ Shutterstock; **Figure 11-28**: Tom McHugh/Photo Researchers; **Figure 11-29**: Tom L. McKnight Photo; **Figure 11-30**: Tom L. McKnight Photo; **Figure 11-31**: Darrel Hess; **Figure 11-32**: Darrel Hess; **Figure 11-33a**: Design Pics Inc./ Alamy; **Figure 11-33b, c**: Tom L. McKnight Photo; **Figure 11-34**: Tiziana and Gianni Baldizzone/Corbis; **Figure 11-35**: Darrel Hess; **Figure 11-36**: Tom L. McKnight Photo; **Figure 11-37**: Darrel Hess; **Figure 11-38**: Cornforth Images/Alamy; **Figure 11-39**: Darrel Hess; **Figure 11-C**: NASA/Earth Observatory; **Figure 11-D**: Landsat Image courtesy of the United Nations Environmental Programme; **Figure 11-E**: NASA/Earth Observatory; **Figure 11-40**: Tom L. McKnight Photo; **Figure 11-44**: Nigel Cattlin/Holt Studios International/Photo Researchers; **Figure 11-45**: Tom L. McKnight Photo; **Figure 11-46**: Getty Images/Time Life Pictures; **Figure 11-47**: Bob Gossington/ Photoshot Holdings Ltd.

Chapter 12

Opener: Michael Collier; **Figure 12-1**: Superstock; **Figure 12-6**: Tom McHugh/Photo Researchers, Inc; **Figure 12-7**: David M. Dennis/Animals Animals/Earth Scenes; **Figure 12-8**: Richard R. Hansen/Photo Researchers; **Figure 12-16a, b**: Tom L. McKnight Photo; **Figure 12-19** Adapted from Brady, Nyle C., and Ray R. Weil, *The Nature and Properties of Soils*, Upper Saddle River, NJ: Prentice Hall, 1999; **Figure 12-22**: From Edward J. Tarbuck, Frederick K. Lutgens, and Dennis Tasa, *Essentials of Geology*, 10th ed.; **Figure 12-23**: Tom L. McKnight Photo; **Figure 12-A,B,C**: Randy Schaetzl Photo; **Figure 12-27b**: Randy Schaetzl Photo; **Figure 12-28b**: USDA; **Figure 12-29b**: Loyal A. Quandt, National Soil Survey Center; **Figure 12-30b**: Courtesy of USDA Natural Resources Conservation Service, National Soil Survey Center; **Figure 12-31b**: Randy Schaetzl Photo; **Figure 12-32b**: USDA; **Figure 12-33b**: USDA; **Figure 12-34b**: Loyal A. Quandt, National Soil Survey Center; **Figure 12-35b**: Randy Schaetzl Photo; **Figure 12-36b**: USDA; **Figure 12-37b**: Randy Schaetzl Photo; **Figure 12-38b**: Randy Schaetzl Photo.

Chapter 13

Opener: Kevin Taylor/Aurora; **Figure 13-1**: From Edward J. Tarbuck, Frederick K. Lutgens, and Dennis Tasa, *Essentials of Geology*, 10th ed.; **Figure 13-2**: G. Tompkinson/Photo Researchers; **Figure 13-3**: Tasa Graphic Arts, Inc; **Figure 13-6a**: Breck P. Kent/Animals Animals/Earth Scenes; **Figure 13-6b,c,e**: Breck P. Kent; **Figure 13-6d,f**: Breck P. Kent/Animals Animals/Earth Scenes; **Figure 13-7a**: Tom L. McKnight Photo; **Figure 13-7b**: Michael Szoenyi/Photo Researchers; **Figure 13-8a**: Tom L. McKnight Photo; **Figure 13-8b**: Donna Tucker; **Figure 13-10a**: Tom L. McKnight Photo; **Figure 13-10b**: Andreas Einsiedel/Dorling Kindersly Media Library; **Figure 13-11**: Tom L. McKnight Photo; **Figure 13-12a**: Tom L. McKnight Photo; **Figure 13-12b**: Harry Taylor © Dorling Kindersley, Courtesy of the Royal Museum of Scotland, Edinburgh; **Figure 13-14a**: Tom L. McKnight Photo; **Figure 13-14b**: Richard M. Busch; **Figure 13-15a**: Kennan Ward/Corbis; **Figure 13-15b**: Joel Arem/Photo Researchers; **Figure 13-16a**: Kevin Schafer/Peter Arnold Inc.; **Figure 13-16b**: Richard M. Busch; **Figure 13-21**: Tom L. McKnight Photo; **Figure 13-22**: Tom L. McKnight Photo.

Chapter 14

Opener: Nick Turner/npl/ Minden Pictures; **Figures 14-1-7, 14-12, 14-13, 14-15, 14-17–23, 14-38, 14-45, 14-47-49, 14-55, 14-59, 14-62, 14-64**: From Edward J. Tarbuck, Frederick K. Lutgens, and Dennis Tasa, *Essentials of Geology*, 10th ed.; **Figures 14-9, 14-10**: *Adapted with permission from R. L. Larson and others,* The Bedrock Geology of the World, *New York: W. H. Freeman and Company, 1985;* **Figures 14-16, 14-44**: From Edward J. Tarbuck, Frederick K. Lutgens, and Dennis Tasa, *Earth*, 10th ed.; **Figure 14-26**: Douglas Peebles Photography/Alamy; **Figure 14-28**: Dana Stephenson//Getty Images; **Figure 14-29**: Darrel Hess; **Figure 14-30**: John Gerlach/ Animals Animals/Earth Scenes; **Figure 14-35**: Ulises Ruiz/ Associated Press; **Figure 14-37**: Darrel Hess; **Figure 14-39**: Dennis Tasa; **Figure 14-40**: Greg Vaughn/Alamy; **Figure 14-42**: *From U.S. Geological Survey Fact Sheet 092–02, Mount Mazama and Crater Lake: Growth and Destruction of a Cascade Volcano, 2002;* **Figure 14-43**: Dennis Tasa; **Figure 14-A.a**: Tom L. Mcknight Photo; **Figure 14-A.b**: Tom L. Mcknight Photo; **Figure 14-45**: U.S. Geological Survey/U.S. Department of the Interior; **Figure 14-46**: Michael S. Yamashita/ Corbis; **Figure 14-50**: Tom L. McKnight Photo; **Figure 14-51c**: Tom L. McKnight Photo; **Figure 14-53**: USGS; **Figure 14-56**: Tom L. McKnight Photo; **Figure 14-58**: Darrel Hess; **Figure 14-61**: Michael Collier; **Figure 14-B**: U.S. Geological Survey, Denver; **Figure 14-C**: NASA/JPL; **Figure 14-63**: Roger Hess; **Figure 14-65**: Tom L. McKnight Photo.

Chapter 15

Opener: Carr Clifton; **Figure 15-2**: Douglas Peebles Photography/Alamy; **Figure 15-5**: Tom L. McKnight Photo; **Figure 15-6**: Tom L. McKnight Photo; **Figure 15-8**: Tom L. McKnight Photo; **Figure 15-9**: Darrel Hess; **Figure 15-10**: Michael Runkel/SuperStock; **Figure 15-12**: Tom L. McKnight Photo; **Figure 15-13a,b**: Darrel Hess; **Figures 15-14, 15-18, 15-21, 15-23-25**: From Edward J. Tarbuck, Frederick K. Lutgens, and Dennis Tasa, *Essentials of Geology*, 10th ed.; **Figure 15-15**: Charlie Ott/Photo Researchers; **Figure 15-16**: Martin Bond/Photo Researchers; **Figure 15-20**: Darrel Hess; **Figure 15-22**: U.S. Geological Survey/U.S. Department of the Interior; **Figure 15-26**: L. W. Smith, Waterways Experiment Station USACE/U.S. Geological Survey, Denver; **Figure 15-B**: Photo by Robert L. Shuster/ USGS; **Figure 15-C**: Mark Reid, U.S. Geological Survey; **Figure 15-29**: Tom L. McKnight Photo; **Figure 15-30**: Tom L. McKnight Photo.

Chapter 16

Opener: Patricio Robles Gil/npl/Minden Pictures; **Figure 16-1**: Tom L. McKnight Photo; **Figures 16-3, 16-11, 16-28, 16-38-39, 16-41, 16-43**: From Edward J. Tarbuck, Frederick K. Lutgens, and Dennis Tasa, *Essentials of Geology*, 10th ed.; **Figure 16-5**: USDA; **Figure 16-6**: Tom L. McKnight Photo; **Figure 16-8**: Tom L. McKnight Photo; **Figure 16-10**: Brendan Evans/U.S. Coast Guard/Associated Press; **Figure 16-13**: Darrel Hess; **Figure 16-14**: Michael Collier; **Figure 16-15**: Bill Bachmann/Photo Researchers; **Figure 16-B**: Calvin Larsen/Photo Researchers; **Figure 16-23**: Darrel Hess; **Figure 16-26**: Associated Press; **Figure 16-30**: Michael Collier; **Figure 16-33**: NASA Earth Observatory; **Figure 16-37**: Tom L. McKnight Photo; **Figure 16-42**: Michael Maples/U.S. Army Corps of Engineers Digital Visual Library; **Figure 16-C**: U.S. Geological Survey, Denver; **Figure 16-D**: USGS; **Figure 16-E**: Superstock; **Figure 16-44**: Tom L. McKnight Photo.

Chapter 17

Opener: Jose B. Ruiz/npl/Minden Pictures; **Figure 17-3**: Dennis Tasa; **Figure 17-5**: Alex S. MacLean/Peter Arnold;

Figure 17-8: BIHAIBO/ Istockphoto.com; **Figure 17-9:** Orlando Sentinel; **Figure 17-11:** Darrel Hess; **Figure 17-A:** Tom Brakefield/Getty Images; **Figure 17-12:** Tom L. McKnight Photo; **Figure 17-13:** Darrel Hess; **Figure 17-14:** Cyril Ruoso/Minden; **Figure 17-17:** Tom L. McKnight Photo.

Chapter 18

Opener: Carr Clifton; **Figure 18-2:** Tom L. McKnight Photo; **Figure 18-4:** Darrel Hess; **Figure 18-6:** Darrel Hess; **Figure 18-8:** Superstock; **Figure 18-9:** Tom L. McKnight Photo; **Figure 18-10:** Tom L. McKnight Photo; **Figure 18-B:** NASA Earth Observatory; **Figure 18-14:** Mlenny/ Istockphoto.com; **Figure 18-15:** Darrel Hess; **Figure 18-16:** Tom L. McKnight Photo; **Figure 18-17:** Darrel Hess; **Figure 18-20:** From Edward J. Tarbuck, Frederick K. Lutgens, and Dennis Tasa, *Essentials of Geology*, 10th ed.; **Figure 18-21:** Frantisek Staud/Alamy; **Figure 18-22:** Tom L. McKnight Photo; **Figure 18-23:** Darrel Hess; **Figure 18-24:** Tom L. McKnight Photo; **Figure 18-28b:** Michael Collier; **Figure 18-29:** Dennis Tasa; **Figure 18-30:** Darrel Hess; **Figure 18-31:** Tom L. McKnight Photo; **Figure 18-34:** Tom L. McKnight Photo; **Figure 18-35:** Darrel Hess; **Figure 18-36:** Darrel Hess.

Chapter 19

Opener: Carr Clifton/Minden Pictures; **Figure 19-1:** Tom L. McKnight Photo; **Figures 19-3, 19-6, 19-12, 19-21, 19-23:** From Edward J. Tarbuck, Frederick K. Lutgens, and Dennis Tasa, *Essentials of Geology*, 10th ed.; **Figure 19-8:** Howard Platt/Getty; **Figure 19-B:** NASA; **Figure 19-C.a,b:** Jesse Allen & Robert Simon/NASA; **Figure 19-10:** NASA maps by Robert Simmon, based on ASTER data; **Figure 19-14:** Darrel Hess; **Figure 19-15:** Walter H. Hodge/Peter Arnold; **Figure 19-16:** Tom L. McKnight Photo; **Figure 19-17:** Tom L. McKnight Photo; **Figure 19-18:** Tom L. McKnight Photo; **Figure 19-19:** NASA/ Photri; **Figure 19-20:** Darrel Hess; **Figure 19-25:** Ward's Natural Science Establishment; **Figure 19-D.a:** U.S. Geological Survey, Denver; **Figure 19-D.b:** U.S. Geological Survey, Denver; **Figure 19.A:** Dennis Tasa; **Figure 19-E:** NOAA/Landsat/ASTER; **Figure 19-27:** Tom L. McKnight Photo; **Figure 19-29:** Darrel Hess; **Figure 19-30:** Tom L. McKnight Photo; **Figure 19-32:** Tom L. McKnight Photo; **Figure 19-34:** Tom L. McKnight Photo; **Figure 19-35:** Tom L. McKnight Photo; **Figure 19-36:** Tom L. McKnight Photo; **Figure 19-37:** Tom L. McKnight Photo; **Figure 19-38:** Tom L. McKnight Photo; **Figure 19-39:** *After François Matthes, U.S. Geological Survey, Professional Paper 160;* **Figure 19-40:** Darrel Hess; **Figure 19-42:** Tom L. McKnight Photo; **Figure 19-43:** Darrel Hess; **Figure 19-44:** Tom L. McKnight Photo; **Figure 19-45:** Steven J. Kazlowski/Alamy ; **Figure 19-46:** Michael Collier; **Figure 19-47:** *After Edward Aguado and James E. Burt,* Understanding Weather and Climate, *4th ed., Upper Saddle River, NJ: Pearson-Prentice Hall, 2007.*

Chapter 20

Opener: Tim Fitzharris/Minden Pictures; **Figure 20-1:** Tom L. McKnight Photo; **Figures 20-2, 20-5, 20-12, 20-8, 20-11, 20-15, 20-16, 20-23:** From Edward J. Tarbuck, Frederick K. Lutgens, and Dennis Tasa, *Essentials of Geology*, 10th ed.; **Figure 20-4:** Tom L. McKnight Photo; **Figure 20-6a,b:** Tom L. McKnight Photo; **Figure 20-9:** Tom L. McKnight Photo; **Figure 20-A:** NOAA; **Figure 20-B:** Getty Images; **Figure 20-C:** Getty Images; **Figure 20-D:** Image courtesy of Marine Current Turbines Limited; **Figure 20-E:** JoeFox CountyDown/Alamy; **Figure 20-10:** Darrel Hess; **Figure 20-11:** NASA; **Figure 20-13:** Tom L. McKnight Photo; **Figure 20-14:** Tom L. McKnight Photo; **Figure 20-17:** David Wall/Danita Delimont.com; **Figure 20-18:** David Wall/Alamy; **Figure 20-20:** Padre Island National Seashore; **Figure 20-22:** USGS; **Figure 20-24:** LOOK Die Bildagentur der Fotografen GmbH/Alamy; **Figure 20-25:** NASA/Goddard Space Flight Center; **Figure 20-26:** Peter Hendrie/Getty Images; **Figure 20-2:** Darrel Hess; **Figure 20-F:** Gary Bell/SeaPics.com; **Figure 20-G:** NOAA/; **Figure 20-30:** Darrel Hess; **Figure 20-31:** Tony Roberts/ Sportshot/Photo Researchers; **Figure 20-33:** Jack Fields/Photo Researchers; **Figure 20-34:** Lawrence Ong/NASA Earth Observatory Team.

Index

X

Y

Z